The Physics and Chemistry of Solids

The Physics and Chemistry of Solids

S.R. Elliott

Fellow in Physics and Chemistry,
Trinity College, Cambridge, UK

and

Reader in Solid-State Chemical Physics,
Department of Chemistry,
University of Cambridge, UK

JOHN WILEY & SONS
Chichester · New York · Weinheim · Brisbane · Singapore · Toronto

Baffins Lane, Chichester,
West Sussex PO19 1UD, England

National 01243 779777
International (+44) 1243 779777

e-mail (for orders and customer service enquiries): cs-books @ wiley.co.uk.
Visit our Home Page on http://www.wiley.co.uk.
or http://www.wiley.com

Other Wiley Editorial Offices

John Wiley & Sons, Inc., 605 Third Avenue,
New York, NY 10158–0012, USA

WILEY-VCH Verlag GmbH, Pappelallee 3,
D-69469 Weinheim, Germany

Jacaranda Wiley Ltd, 33 Park Road, Milton,
Queensland 4064, Australia

John Wiley & Sons (Asia) Pte Ltd, 2 Clementi Loop #02–01,
Jin Xing Distripark, Singapore 129809

John Wiley & Sons (Canada) Ltd, 22 Worcester Road,
Rexdale, Ontario M9W 1L1, Canada

British Library Cataloguing in Publication Data

A catalogue record for this book is available from the British Library

ISBN 0 471 98194x; 0471 98195 8 (pbk.)

Typeset in 10/12pt Times by Pure Tech India Ltd, Pondicherry
Printed and bound in Great Britain by Bookcraft (Bath) Ltd
This book is printed on acid-free paper responsibly manufactured from sustainable forestation, for which at least two trees are planted for each one used for paper production.

To Isaac and Hester and, of course, Penelope

Contents

Preface

Why write yet another textbook on the solid state? Most previous books have concentrated on the physics, chemistry, or the materials aspects of solids. However, this historical division is becoming increasingly irrelevant in the interdisciplinary scientific world in which we now find ourselves. So I have tried to write a comprehensive, yet comprehensible, book that deals with solids from all three viewpoints, although the main emphasis is on the physical behaviour of materials. Thus, I hope that the book will be equally accessible for students, or practitioners, in condensed-matter physics, solid-state chemistry, materials science or engineering. It should be suitable for beginning students in solid-state physics, or advanced undergraduates in solid-state chemistry or materials science or engineering, or for graduate students in all such disciplines.

I have deliberately tried to keep the level of the mathematics used in this book to a relatively low level: a knowledge of differential (and vector) calculus is assumed, as is some knowledge of matrix manipulation. There are very many formulae and equations given in this book and I make no apology for this, since a proper understanding of the subject can only be achieved at the quantitative, not the qualitative, level. Nevertheless, I have tried, wherever possible, to introduce each topic in a descriptive manner, so that beginners can get an idea of what is involved, before a more detailed, and mathematical, discussion is given. However, I have tried, again wherever possible, to *derive* the equations used in a description of physical phenomena rather than simply to state them, since I feel that a much greater understanding is achieved thereby. In some cases, however, this has not proved possible, either for reasons of space in the text (although some such derivations have been made the subject of some of the problems), or because the level of mathematics involved is too advanced (in which case references are provided to the derivation given in other sources for the interested reader).

Obviously some prior knowledge on the part of the student has been assumed in the writing of this book. It is assumed that students will already have taken courses in aspects of thermodynamics, statistical mechanics, elementary electricity and magnetism and quantum mechanics. However, I have tried to keep the level of quantum mechanics as simple as possible; for example, second quantization does not feature.

The division of the material in this book is, perhaps, a little unusual. There are just eight chapters. The first four, dealing with Synthesis and Preparation of Materials, Atomic Structure and Bonding, Defects and Atomic Dynamics, consider mainly the properties of *atoms* in the solid state. The remaining four chapters, Electrons in Solids,

Electron Dynamics, Dielectric and Magnetic Properties, and Reduced Dimensionality, concentrate instead on the behaviour of *electrons* in materials. Special features include the following. The book begins with a brief survey of materials- synthesis and preparation techniques. This important topic (the properties of solids cannot be studied if they cannot be produced!) is a central subject of solid-state chemistry, yet it is rarely mentioned in most solid-state physics texts. One theme running through the book is disorder: although the ideal single-crystalline state permits a mathematical description to be made of its physical behaviour, real materials are inevitably defective, and these defects can dominate the physical and chemical behaviour. In view of the importance of defects, this topic is therefore introduced early on (Chapter 3). In the section on electrons in solids (Chapter 5), descriptions of electron states from the viewpoints of both physicists ('bands') and chemists ('bonds') are discussed and compared. The physical behaviour of materials described in Chapters 2–7 is for three-dimensional, bulk solids. In the final chapter, materials systems with dimensionalities *less* than three are discussed. Much of the most interesting solid-state research currently being carried out is on low-dimensional systems and so, to give the reader an impression of the interest and vitality of this field, many topics are discussed that are the subject of current research and which are not usually found in solid-state textbooks: examples include fullerenes, nanotubules and quantum dots, wires and wells.

What has been left out? First, surface physics, that is the study of free surfaces of solids and their interactions with gases (and liquids), has been omitted entirely. This broad field has been the subject of a number of books itself, and a reasonably comprehensive discussion could not be included in this book without making it even longer and more unwieldy than it already is. Second, in retrospect, I would have liked to have included more discussion on polymers, although they do get a mention: perhaps in any future edition, this omission will be rectified.

Throughout the book, I have used SI units, although occasionally I have lapsed into using units convenient in the description of the solid state, particularly the angstrom and the electron volt. I have made a conscious decision *not* to include c.g.s. units at all: students nowadays do not know the c.g.s. system, particularly relating to electrical and magnetic units, and I think that their use, even side-by-side with SI units, can only confuse the reader.

I am grateful to Drs J. P. Attfield, J. Klinowski, M. A. Morris, P. M. S. Monk, A. V. Powell, T. Rayment and A. D. Yoffe and Profs E. A. Davis and T. Hibma for reading and commenting on parts of the book, to Mr J. Portsmouth for helping with the figures and last, but definitely not least, I am particularly grateful to Laura Cordy and Catherine Byfield for transcribing my near-illegible scrawls into a more legible, and thanks to Laura a more systematic, form, a feat second only to the decipherment of the Rosetta Stone. However, all remaining infelicities and outright errors remain, of course, my own responsibility.

Finally, I feel guilty about the neglect of my research group and of my family during the time taken to write this book; I am very grateful for their forbearance and understanding of my physical, and mental, absence.

Notes to students and instructors

This book aims to be a reasonably comprehensive introduction to the physics and chemistry of solid materials. As a consequence, there is too much material in some chapters to be included in a lecture course on these individual subjects. In order to give some guidance as to the relative importance of separate sections in a particular chapter, sections have been categorized according to whether they are essential for a basic understanding of the subject, whether they are optional and suitable for a second reading (*), or whether they are more advanced topics suitable for a higher-level course (**). In addition, portions of a section which could be left for a second reading are marked by a line alongside the text.

The book can be used in very different ways, depending on the subject being studied. Condensed-matter physicists could read, for instance, principally Chapters 2, 4, 5, 6, 7 and parts of Chapter 8. Solid-state chemists could follow Chapters 1, 2, 3 (except § 3.4.3 on mechanical properties), 5 (except perhaps for § 5.8) and parts of Chapter 8. Materials scientists or engineers, on the other hand, could read Chapters 1, 2, 3, 5 and parts of Chapters 7 and 8. Of course, for a fuller appreciation of the field as a whole, it is hoped that sometimes the reader might stray from the chosen path.

A large number of problems and exercises (over 200 in total) appear at the end of each chapter. The purpose of these is two-fold: first to reinforce an understanding of the material presented in the text by performing numerical exercises, and second to supplement the material in the book, either by providing derivations of equations that have been merely stated in the text for reasons of space, or occasionally to go beyond the content of the text by introducing new ideas and concepts in some of the problems. A number of essay questions have also been set: every practising scientist has regularly to write essays, for example, in the form of journal articles or reports describing original research. In order to do this successfully, certain skills in marshalling and presenting information must be mastered, but this aspect of a scientific education is often neglected in university or college courses. Although some of the essay questions can be answered satisfactorily using the information presented in the book, some require outside reading, either of other, more advanced texts or of the primary research literature, and *all* answers to essay questions would benefit from the additional information gained in this way. Worked solutions of all the numerical problems are available, free of charge, from the publishers, to instructors who adopt this book for the course that they are teaching.

S. R. Elliott
Cambridge, July 1997

List of Tables

Glossary of Abbreviations and Acronyms

a.c.	alternating current
AM	air mass
ARPES	angular-resolved photoemission spectroscopy
b.c.c.	body-centred cubic
BCS	Bardeen–Cooper–Schrieffer (theory)
BKBO	barium potassium bismuthate
BPBO	barium lead bismuthate
BSCCO	barium strontium calcium cuprate
CBE	chemical beam epitaxy
c.c.p.	cubic close-packed
CMR	colossal magnetoresistance
c.p.	close-packed
CS	crystallographic shear
CVD	chemical-vapour deposition
c.w.	continuous wave
D	dimension(al)
d.c.	direct current
dHvA	de Haas–van Alphen (effect)
DRP	dense random packing
EELS	electron energy-loss spectroscopy
EL	electroluminescence
EMF	electromotive force
EPM	empirical pseudopotential method
EPR	electron paramagnetic resonance
ESR	electron spin resonance
EXAFS	extended X-ray absorption fine structure (spectroscopy)
f.c.c.	face-centred cubic
FET	field-effect transistor
FF	fill factor

FID	free-induction decay
FT	Fourier transform
GEP	general equivalent point
GMR	giant magnetoresistance
GRIN	graded index
HBCCO	mercury barium calcium cuprate
h.c.p.	hexagonal close-packed
HEMT	high electron mobility transistor
HOMO	highest occupied molecular orbital
IR	infrared
ITO	indium tin oxide
KCP	potassium platinocyanate
KDP	potassium dihydrogen phosphate
LA	longitudinal acoustic
LASER	light amplification by the stimulated emission of radiation
LCAO	linear combination of atomic orbitals
LED	light-emitting diode
LO	longitudinal optic
LPE	liquid-phase epitaxy
LST	Lyddane–Sachs–Teller (equation)
LUMO	lowest unoccupied molecular orbital
MBE	molecular-beam epitaxy
MIGS	metal-induced gap states
MOCVD	metal-organic chemical-vapour deposition
MODFET	modulation-doped field-effect transistor
MOMBE	metal-organic molecular-beam epitaxy
MOS	metal–oxide–semiconductor
MOSFET	metal-oxide-semiconductor field-effect transistor
MQW	multiple quantum well
N	normal (phonon-scattering process)
NEXAFS	near-edge X-ray absorption fine structure (spectroscopy)
NFE	nearly-free electron (model)
NMR	nuclear magnetic resonance
OPW	orthogonal plane wave
PA	*poly*-acetylene
PECVD	plasma-enhanced chemical-vapour deposition
PPP	*poly*-(*p*-phenylene)
PPV	*poly*-(*p*-phenylene vinylene)
PPy	*poly*-pyrrole
PZT	lead zirconate-titanate
RDF	radial distribution function
r.f.	radio frequency
RKKY	Ruderman–Kittel–Kasuya–Yosida (interaction)
s.c.	simple cubic
SCH	separate confinement by heterojunctions
SCL	space-charge-limited (current)
SHG	second-harmonic generation

SQUID	superconducting quantum-interference device
TA	transverse acoustic
TB	tight binding
TBA	tight-binding approximation
TCNQ	tetracyanoquinodimethane
TLS	two-level system
TO	transverse optic
TTF	tetrathiofulvalene
U	umklapp (phonon-scattering process)
UHV	ultra-high vacuum
UPS	ultraviolet photoemission spectroscopy
UV	ultraviolet
XANES	X-ray absorption near-edge structure
XPS	X-ray photoemission spectroscopy
YAG	yttrium aluminium garnet
YBCO	yttrium barium cuprate
YIG	yttrium iron garnet

Glossary of Symbols

It is impossible to avoid the use of the same symbol for more than one quantity in a scientific textbook, and this book is no exception. However, a list of symbols that denote a single quantity is given below: other symbols are defined in the text.

Symbol	*Meaning*
a	cubic unit-cell parameter
a_i	activity of species i
a_{m}	magnetic moment expressed as a number of Bohr magnetons
a_0	Bohr radius
A	Madelung constant
$\boldsymbol{A}$	magnetic vector potential
A_{em}	Einstein coefficient for spontaneous radiative emission
b	neutron scattering length
$\boldsymbol{b}$	Burgers vector
B	bulk modulus
$\boldsymbol{B}$	magnetic flux density
B_{abs}	Einstein coefficient for stimulated radiative absorption
$\boldsymbol{B}_{\mathrm{c}}$	superconducting critical magnetic flux density
$\boldsymbol{B}_{\mathrm{c1}}$	lower superconducting critical magnetic flux density
$\boldsymbol{B}_{\mathrm{c2}}$	upper superconducting critical magnetic flux density
B_{em}	Einstein coefficient for stimulated radiative emission
$\boldsymbol{B}_{\mathrm{ext}}$	external magnetic flux density
$B_J(x)$	Brillouin function
$\boldsymbol{B}_{\mathrm{loc}}$	local magnetic flux density
$\boldsymbol{B}_{\mathrm{mac}}$	macroscopic magnetic flux density
$\boldsymbol{B}_{\mathrm{r}}$	flux remanence
$\boldsymbol{B}_{\mathrm{W}}$	Weiss molecular magnetic flux density
$\bar{c}$	mean gas molecular speed
c_{ij}	component of elastic-stiffness tensor
c_v	heat capacity per unit volume at constant volume
C_p	heat capacity at constant pressure
C_v	heat capacity at constant volume

Symbol	*Meaning*
$\boldsymbol{d}$	piezoelectric tensor
$\tilde{d}$	spectral (fracton) dimension
D	diffusion coefficient (1D)
$\boldsymbol{D}$	electrical displacement
$\mathfrak{D}$	diffusion coefficient tensor (3D)
D_{H}	Hausdorff dimension
$D(\epsilon)$	density of electron states per volume per energy interval
e	electron charge
e_{ij}	component of strain tensor
e_{s}	shear strain
E	electric field
$\boldsymbol{E}_{\mathrm{c}}$	coercive electric field
$\boldsymbol{E}_{\mathrm{loc}}$	local electric field
$\boldsymbol{E}_{\mathrm{mac}}$	macroscopic electric field
E_{y}	Young's modulus
$\mathscr{E}$	energy
$\mathscr{E}_{\mathrm{a}}$	acceptor binding energy
$\mathscr{E}_{\mathrm{B,n}}$	Schottky barrier height on an n-type semiconductor
$\mathscr{E}_{\mathrm{B,p}}$	Schottky barrier height on a p-type semiconductor
$\mathscr{E}_{\mathrm{c}}$	conduction-band-minimum energy
$\mathscr{E}_{\mathrm{cov}}$	covalent energy
$\mathscr{E}_{\mathrm{d}}$	donor binding energy
$\mathscr{E}_{\mathrm{F}}$	Fermi energy
$\mathscr{E}_{\mathrm{F,n}}$	electron quasi-Fermi level
$\mathscr{E}_{\mathrm{F,p}}$	hole quasi-Fermi level
$\mathscr{E}_{\mathrm{g}}$	(direct) bandgap energy
$\mathscr{E}_{\mathrm{g}}^{\mathrm{i}}$	indirect bandgap energy
$\mathscr{E}_{\mathrm{h}}$	hybridization energy
$\mathscr{E}_{\mathrm{i}}$	ionic energy
$\mathscr{E}_{\mathrm{p}}$	Penn gap
$\mathscr{E}_{\mathrm{v}}$	valence-band-maximum energy
f_{c}	diffusion correlation factor
f_{i}	ionicity
f_{j}	X-ray scattering factor of atom j
f_{H}	Helmholtz free energy per volume
F	Faraday
$\boldsymbol{F}$	force
F_{hkl}	scattering amplitude
F_{H}	Helmholtz free energy
g_{e}	electron g-factor
g_J	Landé g-factor
g_{s}	spin degeneracy
g_{v}	valley degeneracy
$g(k)$	density of states per wavevector interval
$g(\mathscr{E})$	density of states per energy interval
$\bar{g}(\mathscr{E})$	density of electron states per energy interval for one spin type

Symbol	*Meaning*
$g_j(\mathscr{E})$	joint density of electron states per energy interval
G	Gibbs free energy
$\mathbf{G}$	reciprocal-lattice vector
G_{ph}	photogeneration rate
G_s	shear modulus
G_0	conductance
$G(r)$	reduced radial distribution function
h	Planck's constant
$\hbar$	Planck's constant divided by 2π
H	enthalpy
$\mathbf{H}$	magnetic field intensity
$\mathbf{H}_c$	flux coercivity
$\mathbf{H}_{ci}$	intrinsic coercivity
$\mathbf{H}_d$	demagnetizing magnetic field intensity
$\mathbf{H}_{mac}$	macroscopic magnetic field intensity
$\mathbf{H}_s$	shape-anisotropy magnetic field intensity
I	nuclear spin quantum number
I_b	base current
I_c	collector current
I_e	emitter current
I_{el}	electron ionization energy
I_{ph}	photocurrent
I_S	Stoner parameter
I_{sc}	short-circuit current
$\mathbf{j}$	current density (charge flux)
J	total angular momentum quantum number
$\mathbf{J}$	total angular momentum operator
$\mathfrak{J}$	exchange constant
J_N	atomic flux of particle N
J_Q	heat flux
$J(r)$	radial distribution function
$\mathbf{k}$	wavevector
k_B	Boltzmann constant
k_D	Debye wavevector
k_F	Fermi wavevector
$\mathbf{K}$	scattering (momentum-transfer) vector
$K(\omega)$	optical-absorption coefficient
l_ϕ	phase-coherence length
L	Lorentz number
$\mathbf{L}$	angular momentum operator
L_{ij}	Onsager coefficient
$\mathfrak{L}(x)$	Langevin function
m^{**}	polaron effective mass
m_c^*	cyclotron effective mass
m_e	electron mass
m_e^*	effective electron mass

Symbol	*Meaning*
$\boldsymbol{m}_e^*$	effective electron mass tensor
m_h^*	effective hole mass
m_l^*	longitudinal component of electron effective mass
m_t^*	transverse component of electron effective mass
$\boldsymbol{M}$	magnetization
$\boldsymbol{M}_r$	remanent magnetization
n	electron concentration
$n^\dagger$	complex refractive index
n_c	critical electron concentration for metal–insulator transition
n_i	intrinsic carrier concentration
n_r	real part of refractive index
N_a	acceptor concentration
N_c	effective concentration of conduction-band states
N_d	donor concentration
N_v	effective concentration of valence-band states
p	pressure
p	hole concentration
$\boldsymbol{p}$	electric dipole moment
$\boldsymbol{p}$	momentum
$\boldsymbol{P}$	electrical polarization
$\boldsymbol{P}_r$	remanent polarization
$\boldsymbol{P}_s$	saturation polarization
$P(\omega)$	mode participation ratio
q	electrical charge
$\boldsymbol{q}$	phonon wavevector
$Q(\boldsymbol{k})$	phonon mode amplitude
$\boldsymbol{r}$	distance vector
r_c	pseudopotential core radius
r_{sp}	small-polaron radius
r_{ws}	Wigner–Seitz radius
R	gas constant
$\boldsymbol{R}$	real-space lattice vector
R^*	exciton binding energy (Rydberg constant)
R_d	donor Rydberg constant
R_H	Hall coefficient
R_0	hydrogen Rydberg constant
s_{ij}	component of elastic-compliance tensor
S	entropy
$\boldsymbol{S}$	spin operator
S_F	cross-sectional area of Fermi surface
S_T	thermopower
$S(K)$	structure factor
t	time
T	temperature
$\boldsymbol{T}$	torque
T_c	superconducting transition temperature

Symbol	*Meaning*
T_{Cf}	ferroelectric Curie temperature
T_F	Fermi temperature
$\boldsymbol{u}$	atomic displacement
U	internal energy
U_H	Hubbard energy
U_K	magnetic anisotropy energy density
U_M	magnetic energy density
$\boldsymbol{v}$	velocity
$\boldsymbol{v}_d$	drift velocity
$\boldsymbol{v}_F$	Fermi velocity
$\boldsymbol{v}_g$	group velocity
V	volume
V_c	unit-cell volume (real space)
V_c^*	unit-cell volume (reciprocal space)
V_g	gate voltage
V_{oc}	open-circuit voltage
$V(r)$	potential energy
W_H	polaron hopping energy
z	atomic coordination number
Z	atomic number
Z	partition function
$\not{Z}$	grand sum
α_i	tight-binding self-energy for orbital i
α_L	inverse localization length
α_p	polarizability
β_i	tight-binding overlap energy for orbital i
β_T	volume coefficient of thermal expansion
γ	Grüneisen constant
γ_i	activity coefficient of species i
γ_m	gyromagnetic ratio
γ_s	surface (interfacial) energy
δ_{ij}	Kronecker delta
$\delta(x)$	Dirac delta function
Δ_{so}	spin–orbit splitting energy
$\Delta(T)$	superconducting energy gap at temperature T
$\Delta\sigma$	photoconductivity
$\varepsilon^\dagger$	complex dielectric constant
ε	electromotive force
ε_0	vacuum permittivity
ε_1	real part of dielectric constant
ε_2	imaginary part of dielectric constant
$\varepsilon(0)$	static dielectric constant
$\varepsilon(\infty)$	high-frequency dielectric constant
ζ_{BCS}	BCS coherence length
ζ_0	superconducting coherence length
η	electrochemical potential

Symbol	*Meaning*
η_q	quantum efficiency
θ	angle
θ_{CW}	Curie–Weiss temperature
θ_{D}	Debye temperature
θ_{N}	Néel temperature
κ	compressibility
κ_{GL}	Ginzberg–Landau parameter
κ_{i}	imaginary part of refractive index
κ_{T}	thermal conductivity
λ	wavelength
λ_{F}	Fermi wavelength
λ_{L}	London penetration depth
λ_{N}	Néel constant
λ_{TF}	Thomas–Fermi screening length
λ_{W}	Weiss constant
Λ	mean-free path
μ	chemical potential
$\boldsymbol{\mu}$	magnetic moment
μ_{B}	Bohr magneton
μ_{e}	electron mobility
μ_{h}	hole mobility
μ_{H}	Hall mobility
μ_{r}	relative permeability
μ_{T}	Thomson heat
μ_0	vacuum permability
ν	frequency
ν_{P}	Poisson ratio
Ξ	spin wavefunction
Π	Peltier coefficient
ρ	electric charge density
$\boldsymbol{\rho}$	electrical resistivity tensor
$\rho(r)$	(atomic/electron) density function
σ	electrical conductivity
σ_{ij}	component of stress tensor
σ_{s}	shear-stress tensor
σ_0	d.c. conductivity
$\sigma(\omega)$	a.c. conductivity
Σ	scattering cross-section
τ	scattering/relaxation time
τ_ϕ	inelastic scattering time
ϕ	packing fraction
ϕ	electrostatic potential
ϕ_{c}	contact potential
ϕ_i	work function of material *i*
Φ	magnetic flux
Φ_{o}	flux quantum (fluxoid)

Symbol	*Meaning*
χ	dielectric susceptibility tensor
χ_i	electron affinity of material i
χ_{m}	magnetic susceptibility
$\chi^{(n)}$	nth-order non-linear electrical susceptibility
$\Psi(r)$	wavefunction
$\psi(r)$	wavefunction
ω	radial frequency
ω_{c}	cyclotron frequency
ω_{D}	Debye frequency
ω_{L}	longitudinal phonon-mode frequency
ω_{p}	plasma frequency
ω_{T}	transverse phonon-mode frequency
Ω	solid angle

CHAPTER

1 Synthesis and Preparation of Materials

Introduction

It is self-evident that, before the physical and chemical characteristics of any material can be investigated, it must first be produced in the required solid-state form. The subject of materials synthesis and preparation forms, not surprisingly, one of the core themes of solid-state chemistry and texts thereon, but is very often missing from books on solid-state physics. This omission is unfortunate, since it is often difficult to gain a proper appreciation of the properties of materials without knowing how they were prepared. For example, structural defects (the subject of Chapter 3) can often control much of the physical and chemical behaviour of solids; the type and concentration of such defects can be determined by the mode of preparation (or subsequent treatment, thermal or otherwise). For this reason, this book starts with a brief survey of some of the main methods employed to produce solid materials.

Synthesis and preparation mean rather different things. Synthesis of a material is the formation of a *new* chemical product from different starting materials in the gas, the liquid or the solid phase. Preparation of a material, on the other hand, refers to the production of a material in a different physical (but the same chemical) state, e.g. a thin film produced from the bulk material, one crystalline form produced from another crystalline (or non-crystalline) form etc. In the following discussion, whilst differentiating between synthesis and preparation techniques as they appear, these procedures will not be discussed separately in different sections. Rather, a more unifying approach is to discuss separately the production (synthesis or preparation) of solid materials from constituents in different physical states of matter, i.e. gas, liquid or solid.

Gas to solid synthesis and preparation 1.1

A number of synthetic and preparative techniques involve the condensation of gas-phase compounds, thereby forming solid-state materials, usually in the form of thin films deposited on a suitable substrate. Some of these techniques are of considerable technological importance, e.g. the production of multilayers of different semiconducting materials (§8.4.4) by molecular-beam epitaxy. In some cases, it is possible to change the composition of a solid by exposing it to a gas. Examples include the insertion of foreign atoms (intercalants) in the vapour phase into a solid material having a rather open structure, e.g. the insertion of fluorine or potassium (vapour) into the layered carbon crystal, graphite (see also §1.2.5), and oxygen into the crystal $YBa_2Cu_3O_6$, forming the high-temperature superconductor, 'YBCO', $YBa_2Cu_3O_{7-\delta}$ (see §6.4.3). Another form of gas–solid reaction, experienced in everyday life, is the tarnishing of metals, in which reaction of a metal (e.g. Ag, Cu, etc.) with a reactive gas (e.g. O_2, Cl_2, H_2S, etc.) forms a solid product at the solid–gas interface. Very often, tarnishing reactions exhibit parabolic growth kinetics for the thickness, $\Delta x \propto t^{1/2}$ (see §1.3.1 for a derivation of this law). 'Nitriding' of steels, the formation of a thin surface nitride layer following exposure to ammonia at high temperature ($\simeq$ 500 °C), is another example; the nitrides, formed by reacting with elements such as Al, Cr or V in the steel, provide a very tough exterior layer, used, for example, in the manufacture of gear wheels.

1.1.1 Vapour deposition

Perhaps the most straightforward form of gas–solid production technique involves the condensation of a vapour onto a (relatively) cooled substrate, the vapour phase of the starting material being achieved by the heating of a solid (sublimation) or a liquid (evaporation). In its simplest guise (see Fig. 1.1), a vapour-deposition apparatus consists of an evacuated chamber (typically pumped by an oil-diffusion pump equipped with a liquid-nitrogen trap (to remove water-vapour and hydrocarbon contaminants) or by a turbomolecular pump) in which the base pressure is $\leq 10^{-4}$ Pa. The starting material, usually in the form of a solid powder, is placed in a 'boat' made from a refractory metal (e.g. Mo or Ta) which is resistively heated in the case of solid charges having relatively low melting points (say $\leq$ 1000 °C). The ensuing vapour, resulting from evaporation of the molten solid contained within the boat, then strikes a substrate positioned over the boat, where it condenses and forms a thin solid film; deposition rates of 0.1–1 μm/s are typical. In order to achieve compositional homogeneity and thickness uniformity, the substrate is often rotated about an axis normal to the plane of the substrate.

This technique as described is one of *physical preparation*: a material in one solid phase (e.g. a bulk crystalline form) is transformed, via an intervening vapour-phase stage, into another physical phase with the *same* (nominal) chemical composition. Very often, metastable solid phases in thin-film form can be prepared in this way, especially if 'cold' substrates (either at room temperature or below) are used. In this case, the vapour-phase atoms/molecules have only a very short time ($\simeq 10^{-12}$s) after collision with the surface to form a low-energy structure; hence, they can be thought of as 'sticking' to the substrate where they strike. In this way, excess free energy is 'frozen' into the resulting films; indeed, vapour deposition is one of the principal

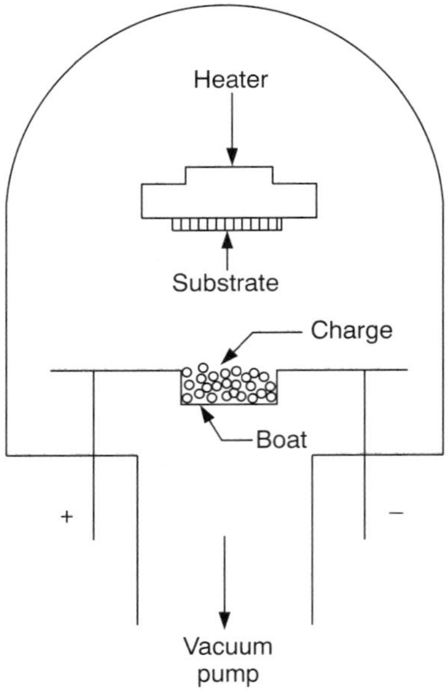

Fig. 1.1 Schematic illustration of a thermal-evaporation vapour-deposition chamber.

techniques for preparing amorphous (non-crystalline) thin films of materials, in which the structural disorder (i.e. configurational entropy) is much higher than that found in the corresponding crystalline phases. Increasing the substrate temperature will concomitantly increase the surface mobility of adsorbed vapour species, thereby allowing (crystalline) structural modifications with lower free energies to be explored.

Heating of the starting material need not be via resistive Joule heating of the metallic container, as in Fig. 1.1. Indeed, it may be desirable to avoid this method to preclude possible chemical interactions between the charge and container. For more highly refractory materials, electron-beam heating can be used, in which a focused, high-current electron beam is directed onto the material to be vaporized; this method can be used to produce thin films of, for example, amorphous silicon or germanium. Alternatively, high-power lasers can be used as the thermal source if the laser light is absorbed efficiently by the material to be vaporized. This technique is known as laser ablation, and is used, for example, to prepare thin films of the cuprate-based 'high-temperature' superconductors (see §6.4.3), e.g. $YBa_2Cu_3O_7$; these films can even be produced in single-crystal form if an appropriate single-crystal substrate is used (e.g. $SrTiO_3$). In addition, laser ablation of a graphitic carbon target (doped with $\simeq 1$ at. % of Co and Ni) held at 1200 °C in flowing Ar produces single-wall carbon nanotubules (§8.3.4) (Thess *et al.* (1996)).

A common problem with the simple vapour-phase deposition techniques outlined above concerns compositional control when the starting material has a complex chemical composition. Fractionation often occurs, in which the compositions of the vapour and the initial solid charge are not the same. This can occur if the composition of the melt is *not* the same as that of the starting solid. In an incongruent melting transformation, at least one of the phases in a solid, containing at least two or more different thermodynamic phases with different compositions, has a different composition from the equilibrium melt. Congruent melting, with *no* change in composition between solid phases and liquid, occurs at the congruent point M in the phase diagram shown in Fig. 1.2. The eutectic point, E, in Fig. 1.2, i.e. the composition and temperature at which the liquidus (solid–liquid equilibrium tie-line) occurs at a minimum temperature, strictly marks an incongruent transformation. The solid obtained by cooling the eutectic melt consists of a lamellar microstructure (dependent on cooling rate) of alternating layers of the two phases, say α and β, each with compositions *different* from that of the eutectic melt; however, the *overall* composition of the eutectic solid and melt is the same, and hence the melting transformation may be regarded as being pseudo-congruent for the present purpose. At compositions other than the eutectic or congruent melting points, the composition of the liquid when the solid first melts, i.e. when the solid is still in thermal equilibrium with the melt (below the liquidus), is *different* from that of the starting solid and hence the vapour under such conditions can exhibit fractionation.

Another cause of compositional variation between evaporated film and starting material is if the vapour consists of molecular species that do not preserve the composition of the melt; an example is the stoichiometric compound As_2S_3 for which the vapour in thermal equilibrium with the melt consists of the molecular species As_4S_4 and S_2. The composition of a film formed by quenching such a vapour can differ from the starting

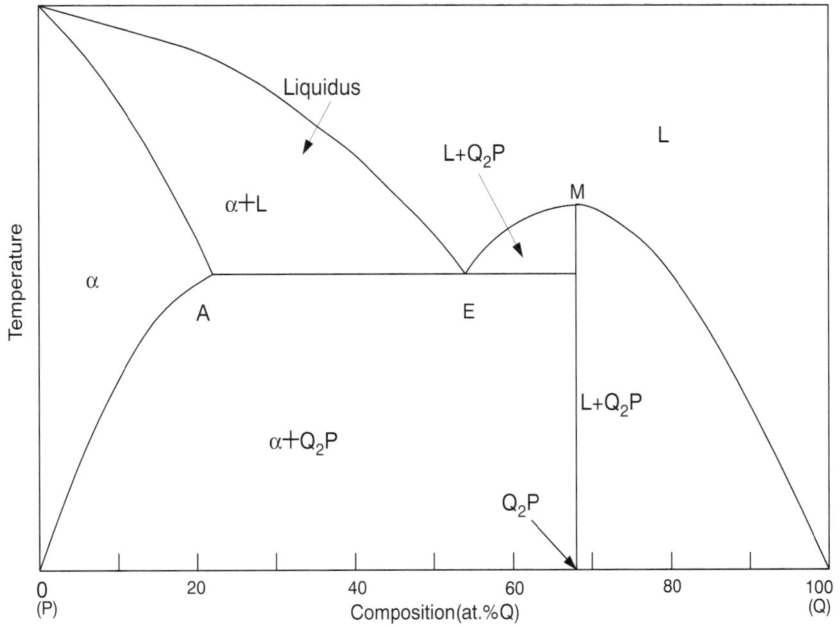

Fig. 1.2 Hypothetical solid–liquid phase diagram showing eutectic (E) and congruent melting (M) points. The two solid phases involved have compositions corresponding to the points A and M.

composition for two reasons: lighter molecular species may be preferentially pumped out of the evaporation chamber, or the different molecules may have different sticking coefficients, i.e. different probabilities for chemisorption or long-lived adsorption, on the substrate. Compositional variations in films prepared by thermal evaporation can often be obviated by using flash evaporation, in which the charge is heated, and vaporized, very rapidly, thereby not allowing time for the above thermal-equilibrium fractionation effects to become important.

A vapour-deposition *synthetic* technique that allows precise compositional control is molecular-beam epitaxy (MBE). In this, separate, well-collimated molecular beams of each of the required constituents (e.g. Ga, Al and As beams for $Ga_{1-x}Al_xAs$ films), each emerging from the very small orifice of an electrically heated Knudsen (effusion) cell containing the component, are directed at a substrate. The evaporation process occurs in an ultra-high vacuum (UHV) chamber (Fig. 1.3), with a base pressure of less than 10^{-9} Pa: such UHV conditions ensure minimal gas-phase contamination of the growing film (see Problem 1.1), and also allow ballistic transport of the molecular beam to the substrate (no scattering of the effusing atoms by gas-phase atoms or molecules). Each Knudsen cell is equipped with a computer-controlled shutter that allows the composition and thickness of the growing film to be controlled precisely.

Typical MBE thin-film growth rates are $1 \mu m\ h^{-1}$ ($\simeq 0.3$ nm s^{-1}), equivalent to about one monolayer per second; if the switching time between sources associated with the

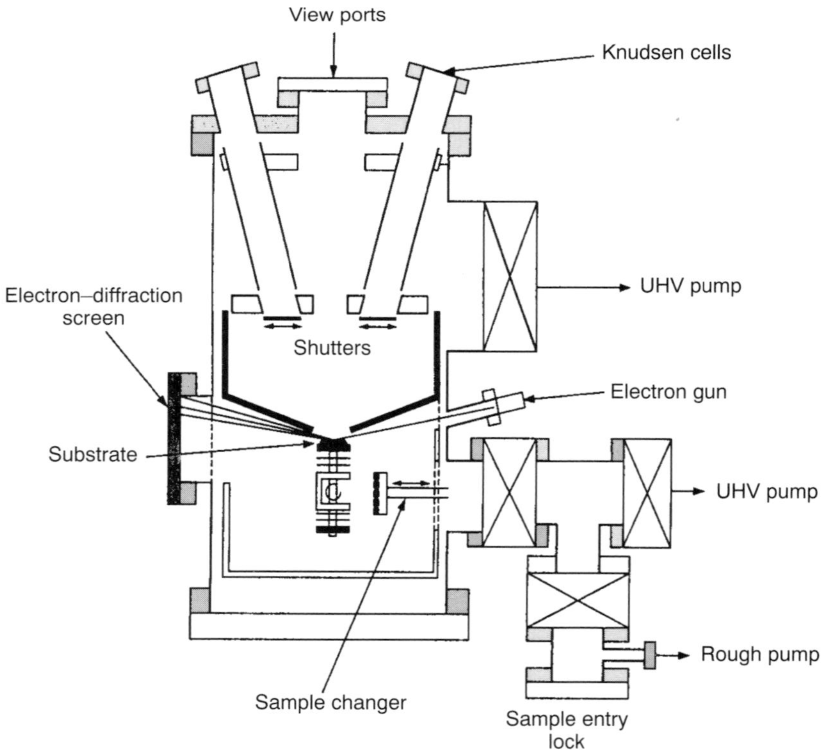

Fig. 1.3 Schematic illustration of an MBE growth chamber.

opening and closing of shutters is appreciably less than 1 s, then sharp changes, of the order of atomic dimensions, in the compositional profiles through the thickness of the film can be achieved. The composition of any particular layer is determined by the flux of atoms in each molecular beam, in turn controlled by the temperatures of the Knudsen cells. In this way, very complicated 'artificial structures' (see §8.4.4) consisting of multilayers comprising different materials, e.g. the semiconductors GaAs and $Ga_{1-x}Al_xAs$, can be fabricated; electrically active atoms ('dopants'—see §6.5.2), e.g. Si or Be for GaAs, can also be differentially incorporated in different layers by using molecular-beam sources of these dopant elements.

If a single-crystal slice, say of GaAs, is used as the substrate for subsequent molecular-beam deposition of a film of, for example, GaAs or $Ga_{1-x}Al_xAs$ (which has very nearly the same size crystalline building block (or 'unit cell'—see §2.1.2) as GaAs), then, under favourable conditions (e.g. at an elevated substrate temperature of 500–600 °C for GaAs to ensure sufficient surface mobility), epitaxial growth of the film occurs; i.e. the film grows in exact structural registry at the interface with the underlying structure of the substrate.

Not only can monolayer, epitaxial films of crystalline GaAs be grown on GaAs crystal substrates, but the chemical stoichiometry of GaAs can be maintained exactly in the vapour-deposited film, even for a *non-stoichiometric* ratio of the molecular-beam fluxes. Ga (and Al) atoms have a sticking coefficient of near unity on GaAs substrates at $\simeq$ 550 °C, whereas arsenic (which vaporizes preferentially as the pyramidal molecule As_4) has a sticking coefficient of nearly zero on a surface deficient in Ga but close to unity on Ga-rich surfaces (at elevated temperatures when the unreactive As_4 molecules can be pyrolytically decomposed to form As atoms that react with the Ga atoms). Thus, the film growth is controlled by the Ga-atom flux.

Although the MBE technique offers precise compositional control over the film growth, its low growth rate and concomitant low sample throughput, coupled with the high cost of the UHV chamber and associated surface-diagnostic facilities, means that it is not suitable for high-volume industrial use.

1.1.2 Chemical vapour deposition

Chemical vapour deposition (CVD) is the process whereby reactive precursor vapour-phase molecular species react, either homogeneously in the gas phase or heterogeneously at the solid–gas interface at the substrate surface, producing a film with a composition *different* from that of the starting materials. (This vapour-phase synthetic technique should be differentiated from MBE (§1.1.1), for example, where the only chemical reaction occurring is that between adsorbate atoms and atoms on the surface of the substrate.) The precursor molecules can be made to decompose by means of heat (pyrolysis), absorption of UV light (photolysis) or in an electrical plasma formed in the gas; thermal-CVD is the most commonly used method. This technique has the advantage that refractory materials can be vapour-deposited at relatively low temperatures. However, process control may be difficult and ultra-pure volatile precursors may not be readily available.

The technique of thermal-CVD is widely used to produce high-purity thin films of semiconducting materials (§5.2.5); such films can grow epitaxially onto single-crystal substrates at elevated substrate temperatures. An important application of thermal-CVD is the production of thin films of polycrystalline silicon by the thermal decomposition of silane:

$$SiH_4(g) \stackrel{\text{heat}}{\rightarrow} Si(s) + H_2(g), \tag{1.1}$$

where the gaseous reaction byproduct (H_2) escapes from the growing film of solid silicon (Si(s)). Electrical dopants (§6.5.2) can also be incorporated substitutionally into the structure of the crystalline Si films prepared by thermal-CVD by using precursor gases such as phosphine (PH_3) and arsine (AsH_3) which provide P and As dopants, respectively.

Another important application of thermal-CVD is in the vapour-phase hydrolysis of silicon tetrachloride, giving 'synthetic-grade', high-purity silica glass:

$$SiCl_4(g) + 2H_2O(g) \stackrel{\text{heat}}{\rightarrow} SiO_2(s) + 4HCl(g). \tag{1.2}$$

Compound semiconductors that are technologically important, such as III-V materials (e.g. GaAs, InP or GaP), can also be grown epitaxially as single-crystal films using thermal-CVD, but in this case the precursor species are gaseous organic molecules; hence, this variant is termed metal-organic chemical-vapour deposition (MOCVD). For the synthesis of GaAs films, for example, trimethyl gallium and arsine can be used as precursors:

$$Ga(CH_3)_3(g) + AsH_3(g) \rightarrow GaAs(s) + 3CH_4(g). \tag{1.3}$$

Phosphorus- and aluminium-containing materials can be made using PH_3 and $Al(CH_3)_3$ as precursors. Gaseous hydride or organic molecules containing dopant atoms for this semiconducting system that can be used in the thermal-CVD process are SiH_4, $Te(C_2H_5)_2$ and $Mg(C_2H_5)_2$.

A schematic illustration of an MOCVD growth apparatus is shown in Fig. 1.4. MOCVD has the advantage compared with MBE that the apparatus is cheap (no ultra-high vacuum equipment is required) and that the various sources of different atoms can easily be controlled by gas-flow regulators. It has the disadvantage, however, that the whole gas volume of the reactor must be exchanged when changing from one gas source to another in order to grow a film of different composition. The associated switching time determines the minimum thickness of one type of film that can be grown by this technique: operation of the reactor at low gas pressures, however, allows this switching time to be reduced sufficiently so that atomically abrupt heterojunctions between two different materials (e.g. GaAs and $Ga_{1-x}Al_xAs$—see §8.4.4) can be produced, similar to those fabricated by MBE.

The growth process in thermal-CVD is much more complicated than for vapour-deposition techniques, such as MBE (§1.1.1); diffusion of vapour species both towards and away from the reacting surface, and in the gas phase, as well as surface-reaction kinetics all determine the overall growth kinetics. The temperature dependence of the MOCVD growth rate for GaAs from the precursors AsH_3 and trimethyl gallium ($Ga(CH_3)_3$, TMG) or triethyl gallium ($Ga(C_2H_5)_3$, TEG) is shown in Fig. 1.5. TEG is more easily dissociated than is TMG, so the growth rate (at lower temperatures) is therefore higher for TEG than TMG. The thermally activated behaviour of the growth rate for both Ga precursors at lower temperatures indicates that, in this régime, thermally activated surface-reaction rates control the film-growth kinetics. At higher

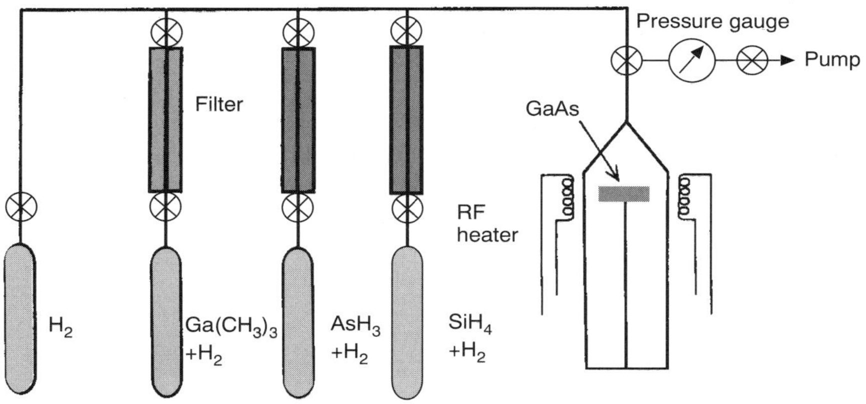

Fig. 1.4 Schematic illustration of an MOCVD growth apparatus.

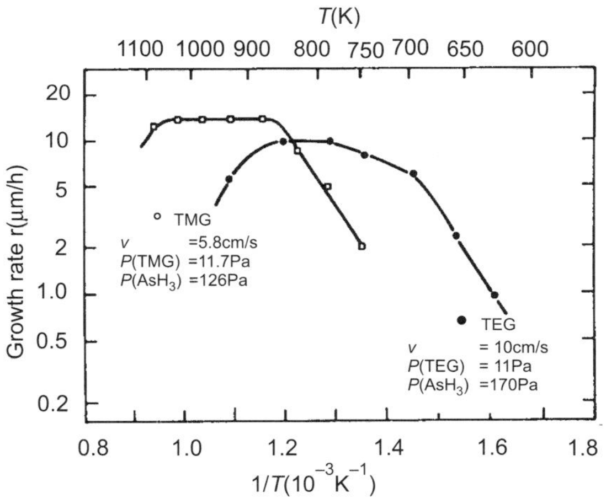

Fig. 1.5 Temperature dependence of the growth rate of GaAs films by MOCVD using the precursor molecules AsH_3 and trimethyl gallium (TMG) or triethyl gallium (TEG). The gas-flow velocities, *v*, and respective partial pressures of the precursor gases are given. (After Plass *et al.* (1988). Reprinted from *J. Crystal Growth*, **88**, 455, Plass *et al.*, © 1988 with kind permission from Elsevier Science – NL, Sara Burgerhartstraat 25, 1055 KV Amsterdam, The Netherlands)

temperatures, the temperature-independent plateau in the growth rate is due to growth being limited by atomic-transport processes in the gas phase, either diffusion of reactants towards, or of gaseous products away from, the substrate; the level of the growth rate in this régime depends on conditions inside the CVD reactor (e.g. gas-flow velocity). The decrease in growth rate observed at the highest temperatures is due to competing reactions occurring also at places in the reactor other than at the substrate (e.g. at the reactor walls).

A variant of MOCVD is called chemical beam epitaxy (CBE) or metal-organic molecular beam epitaxy (MOMBE); as the latter name implies, metal-organic precursor gases are fed into a UHV chamber, each controlled by a gas-flow regulator, and directed at the substrate as collimated molecular beams by injection through capillary inlets. In conventional thermal-CVD, decomposition of precursor molecules such as AsH_3 first takes place via collisions in the gas phase above the heated substrate; in MOMBE, however, because of the very low background gas pressure, such gas-phase reactions do not occur, and instead the molecules are pyrolytically 'cracked' on passing through heated inlet capillaries. Needless to say, carbon is a prevalent contaminant in all forms of deposition techniques using metal-organic precursors.

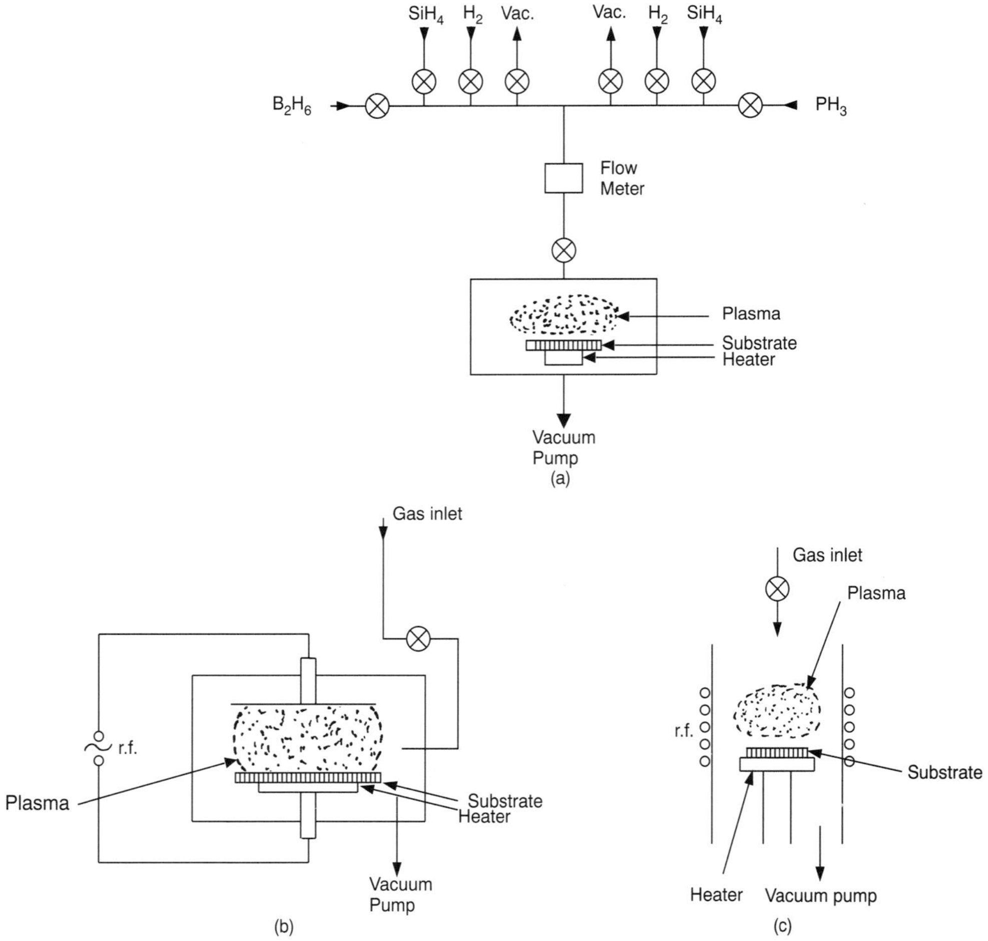

Fig. 1.6 (a) Schematic illustration of a plasma-enhanced chemical-vapour-deposition (PECVD) (or glow-discharge) apparatus. A radio-frequency (r.f.) alternating electric field causes a plasma to be set up in the gas, thereby producing homogeneous gas-phase decomposition of the reactant molecules. (b) Capacitive coupling of the r.f. field to the plasma (shaded). (c) Inductive coupling of the r.f. field.

Energy sources other than heat can be used to initiate molecular decomposition in CVD. One obvious alternative source is *light*: photons (typically in the UV region of the electromagnetic spectrum) with energy sufficient to break (or severely weaken) intramolecular chemical bonds can cause direct dissociation, or facilitate dissociation in association with gas-phase collisions. An example of the application of this photo-CVD technique is the photolysis of disilane (Si_2H_6), thereby forming high-quality films of amorphous hydrogenated silicon, a-Si:H, typically containing a few (1–10) at. % of chemically bonded hydrogen (Yoshida *et al.* (1990)).

Another energy source that can be used to promote CVD is that associated with an electrical plasma (a gas of ionized atoms and electrons); this process is termed plasma-enhanced chemical vapour deposition (PECVD) or equivalently glow-discharge decomposition (see Fig. 1.6). It is commonly used to produce thin films of amorphous or polycrystalline (hydrogenated) silicon, depending on the conditions, from silane as precursor (see reaction (1.1)). The plasma, struck in a mixture of the feedstock gas (e.g. SiH_4) and a buffer gas (e.g. H_2 or He), by the application of an a.c. electrical field (usually radio-frequency, and typically 13.6 MHz) causes the precursor molecules to dissociate into ions and neutral free radicals (e.g. $SiH_3\cdot$) in the gas phase by means of electron collisions. These species then impinge on a moderately heated substrate (at a lower temperature than conventionally used in thermal-CVD, e.g. $\simeq$ 200 °C) to form an amorphous thin film. High dilution of SiH_4 by H_2, as well as higher substrate temperatures, favours the production of microcrystalline Si. Electrical dopants, e.g. P or B, can be incorporated substitutionally into films produced by PECVD by admitting the gas phosphine (PH_3) or diborane (B_2H_6) into the feedstock gas flow.

Plasma-enhanced CVD (involving microwave excitation) is also used to prepare thin films of crystalline diamond from a hydrocarbon-hydrogen gas mixture; usually 0.5–5% methane (CH_4) is used, with substrate temperatures of $\simeq$ 700 °C. Heteroepitaxy, which is the epitaxial growth of one material (diamond) on a substrate of a different material, can take place on lattice-matched substrates having very nearly the same size of crystal unit cell as diamond. Such substrates include boron nitride (BN), nickel and silicon carbide (β-SiC). Single-crystal Si substrates would be ideal, considering their wide availability and use in microelectronics applications. However, the lattice mismatch between diamond and silicon is approximately 34%, precluding direct heteroepitaxial growth. Nevertheless, if an intervening layer of β-SiC is grown on the silicon, then epitaxial diamond growth can take place.

Finally, a variant of CVD is the method of vapour-phase transport, or chemical transport in the vapour phase. In this, a solid compound A is placed at one end of a sealed glass tube (Fig. 1.7) containing a reactive transport gas B and subject to a temperature gradient along the length of the tube. The solid A reacts with gaseous B to form a *gaseous* intermediate. This then transports to the far end of the tube under the influence of the thermal gradient, whereupon it decomposes, regenerating the compound A in purified (and often single-crystal) form:

$$A(s) + B(g) \rightleftharpoons AB(g). \tag{1.4}$$

If the reaction forming AB is *endothermic*, the reactant A is placed at the *hot* end so that the reaction with B is favoured: the gaseous intermediate is transported to the cooler end, whereupon it decomposes to give solid A as the product. For an *exothermic*

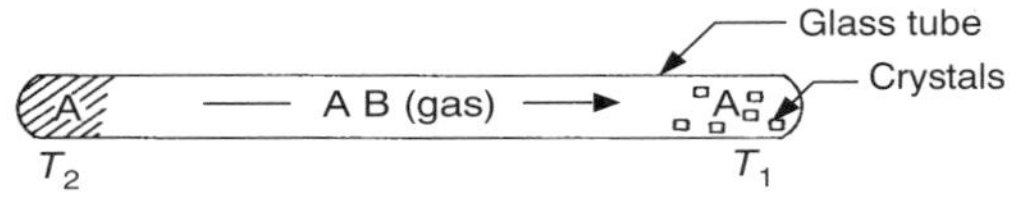

Fig. 1.7 Illustration of the experimental configuration used in vapour-phase transport of a solid substance A via a gaseous intermediate AB, formed by reaction with a gaseous transport agent, B. For an endothermic reaction of formation of AB, the temperatures are such that $T_2 > T_1$, and for an exothermic reaction, $T_2 < T_1$.

reaction of A and B giving AB, the reactant is placed at the cool end, and the product forms at the hot end.

An example of an endothermic reaction of this sort involves the reaction between metallic platinum and gaseous oxygen at temperatures in excess of 1200 °C:

$$Pt(s) + O_2(g) \rightleftharpoons PtO_2(g). \tag{1.5}$$

An exothermic example is the van Arkel method for the purification of certain metals, involving the reaction of the metal with the transporting agent, iodine, for example:

$$Cr(s) + I_2(g) \rightleftharpoons CrI_2(g). \tag{1.6}$$

By this reaction, metals such as Cu, Fe, Hf, Nb, Ta, Ti and V can be extracted from such compounds as oxides, nitrides or carbides. Iodine can also be used as the transporting agent in the purification of the II-VI material, ZnS, via the endothermic (forward) reaction:

$$ZnS(s) + I_2(g) \underset{800^\circ C}{\overset{900^\circ C}{\rightleftharpoons}} ZnI_2(g) + \tfrac{1}{2}S_2(g). \tag{1.7}$$

This reaction can even be used to synthesize ZnS in the first place from the elements, since the direct elemental reaction

$$Zn(l) + S(l) \overset{800^\circ C}{\rightarrow} ZnS(s) \tag{1.8}$$

is very slow because a solid skin of product forms on the liquid metal, thereby inhibiting further reaction. In the presence of iodine, however, reaction (1.8) goes to completion because the ZnS is carried away from the site of reaction by the vapour-transport reaction (1.7).

The III–V semiconductor GaAs can also be transported in a similar fashion, using HCl as the transport agent:

$$GaAs(s) + HCl(g) \rightleftharpoons GaCl(g) + \tfrac{1}{2}H_2(g) + \tfrac{1}{4}As_4(g). \tag{1.9}$$

Synthesis of GaAs, making use of this vapour-transport equilibrium, can be achieved using $AsCl_3$, Ga and H_2 as reactants.

Finally, an intriguing variant is the possibility to transport *two* substances simultaneously in *opposite* directions if one reaction is exothermic and another is endothermic. An example of this is the separation of tungsten and its dioxide, using iodine and water as the vapour-transport species:

$$\mathrm{WO_2(s) + I_2(g) \underset{800^\circ C}{\overset{1000^\circ C}{\rightleftharpoons}} WO_2I_2(g),} \tag{1.10a}$$

$$\mathrm{W(s) + 2H_2O(g) + 3I_2(g) \underset{1000^\circ C}{\overset{800^\circ C}{\rightleftharpoons}} WO_2I_2(g) + 4HI(g).} \tag{1.10b}$$

1.1.3 Sputtering

Sputtering is the process whereby material in a solid target is ablated by bombardment with energetic ions from an electrical, low-pressure plasma struck in a gas. Ejected material from the target, in the form of ionized atoms or clusters of atoms, then subsequently passes to a substrate where a film of the target material is deposited. Most commonly, physical sputtering is used with a chemically inert plasma gas (such as Ar). This is a *preparation* technique where the target material is simply physically transported to the substrate and the resultant film should have (more-or-less) the same chemical composition as the target, since most elements have similar sputtering rates. (However, some of the sputtering gas may be physically entrained in voids in the sputtered film.) If a reactive gas, e.g. O_2 or H_2, is included with the sputtering gas, then chemical sputtering can occur, and a compound of the target material and the gas can be synthesized as a thin film; an example is the formation of SiO_2 films from a silicon target and an oxygen-containing sputtering gas.

Figure 1.8 shows a schematic illustration of a sputtering chamber. A base pressure of $\leqslant 10^{-4}$ Pa is maintained by an oil diffusion pump, and the sputtering gas is introduced into the chamber at a pressure of typically 0.1–1 Pa. The target material is bonded to one electrode and the substrate to another in a parallel-plate (capacitive) configuration. Usually the target is at the bottom, in 'sputter-up' mode, thereby allowing loose powders to be used as target material. The simplest way to induce sputtering is to apply a high negative d.c. voltage to the target, thereby attracting positively charged ions from the plasma struck in the sputtering gas and which sputter away the target surface. However, d.c. sputtering is only feasible for target materials that are metallic, or

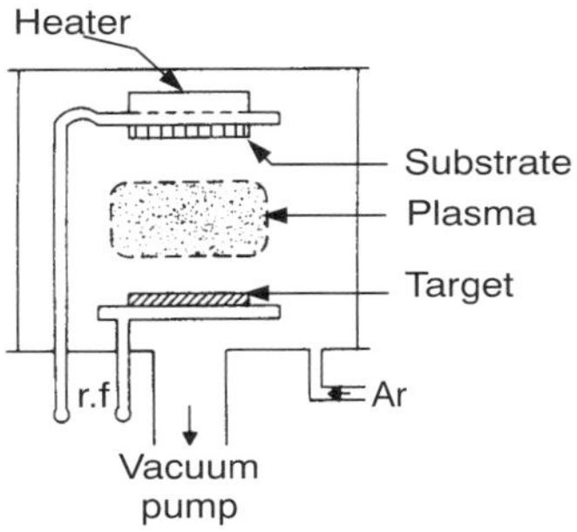

Fig. 1.8 Schematic illustration of an r.f. sputtering chamber in sputter-up configuration. The target is bonded to the lower electrode, and the substrate to the upper electrode (equipped with a heater). A plasma is struck in the sputtering gas (e.g. Ar) fed into the chamber at a relatively high pressure, and pumped away by the vacuum system.

at least sufficiently electrically conducting that the target can act as an electrode. This technique will therefore not work for insulating target materials.

In order to sputter poorly conducting materials, an a.c field (generally r.f., with a typical frequency of 13.6 MHz) is capacitively coupled to the plasma, as in one variant of a PECVD growth chamber (Fig. 1.6b), via capacitive coupling to the target electrode. For metallic targets, this is achieved by connecting a capacitor in series with the target; non-metallic targets are bonded to a metal backing electrode and this arrangement itself forms the capacitive link. In r.f. sputtering, after striking the plasma, positive ions in the plasma are attracted towards the target during each negative half-cycle. However, the mobility of electrons in the plasma is greater than that of the ions, so more electrons than ions are attracted to the top surface of the target during the respective half-cycles, resulting in the build-up of a d.c. negative bias of a few kV on the target. (The value of the bias voltage can also be changed by applying an additional external bias to target and/or substrate.) In steady-state conditions, therefore, the ions are attracted from the plasma to the target by this d.c. bias. Sputtered ions from the target are carried towards the substrate by the following half-cycle of the r.f. field.

Deposition rates for physical r.f. sputtering of insulators are rather low, typically 1–10 $\text{Å}\ s^{-1}$, but are a little higher for metals. Reactive sputtering rates can be appreciably higher still. The deposition rate can be increased significantly by the technique of magnetron sputtering, where a tailored, constant magnetic field around the target electrode concentrates the plasma density in the vicinity of the target surface, thereby increasing the flux of ablating ions from the plasma onto the target. Further details of the sputtering technique can be found in Behrisch (1981, 1983, 1991).

The sputtering technique lends itself to the creation of 'combinatorial libraries', i.e. collections of samples with varying compositions deposited onto the same substrate during a given deposition run. The different compositions are obtained by selectively masking various sputtering targets. This approach has been used to identify promising phosphors (i.e. light-emitting materials) with the general composition $Gd(La,Sr)AlO_x$ (Sun *et al.* 1997).

Liquid to solid synthesis and preparation 1.2

Materials in bulk, rather than thin-film, form are often made starting from the liquid phase, either by solidification of a melt to form single crystals when the cooling rate is very low, or to form non-crystalline materials (glasses) when the cooling rate is sufficiently fast that crystallization is precluded. Alternatively, materials may be crystallized from solution. Often, the materials synthesis takes place in the liquid (particularly molten) phase prior to solidification, in which case the process of forming the solid is one of physical preparation (i.e. change of phase). In other cases, synthesis of new solid compounds can be achieved with the involvement of a liquid. One way of categorizing such preparation and synthesis routes is in terms of whether a melt or a solution is the precursor liquid phase, and this scheme will be used here.

1.2.1 Crystal growth from the melt

Three generic methods of producing (single) crystalline material from a melt may be distinguished, depending on whether the crystal is 'pulled' out of the melt, or crystallization takes place in a crucible container, or is 'container-less', occurring within a solid rod of material. Highly perfect, ultrapure single-crystal specimens can be grown in this way, particularly of semiconductors for use in the electronics industry, where the requirements for sample purity and crystal perfection are extremely stringent: electrically active impurities or structural defects must be precluded, otherwise the intrinsic electronic behaviour of the semiconductor is severely degraded. Crystal-growth technology has now reached the point where electrically active impurity concentrations in crystalline Si and Ge can be reduced below the level of 10^{16}m^{-3} (this corresponds to a purity of better than one part in 10^{12}!), although the concentrations of electrically inactive contaminants (e.g. H, C, O) are typically four orders of magnitude higher. The density of structural defects, e.g. the line defect called a dislocation (§3.1.2), in such crystals can be reduced below the level of 10^{6}m^{-3}. Now, even *isotopically pure* large single crystals can be grown, for example of Ge (which naturally consists of five isotopes). Further details on techniques for crystal growth from the melt are given, for example, in Brice (1973) and Roy (1992).

Perhaps the most important technique for growing perfect single crystals from a melt (that forms congruently—see Fig. 1.2) is the Czochralski method illustrated in Fig. 1.9. In this, a melt of the material contained in a crucible is maintained at a temperature just above its melting point. A single-crystal 'seed' attached to a pulling rod is placed in the surface of the melt and withdrawn slowly (typically with a pull rate of a few millimetres per minute); the melt crystallizes epitaxially on the seed crystal, preserving its crystallographic orientation in the pulled crystal. The pulling rod and crucible are generally rotated, in opposite senses, at a frequency of a few tens of revolutions per minute in order to maintain melt homogeneity (for compound materials) and a constant temperature in the crystal-growth region. As a result, of course, of this rotation about its longitudinal axis, the pulled crystal has a cylindrical geometry; Si crystals can now be grown in this way with diameters greater than 0.2 m. A high-pressure, inert-gas atmosphere (e.g. of Ar) is often used around the growing crystal and the melt in the crucible in order to reduce contamination of the crystal and also, in the case of compound

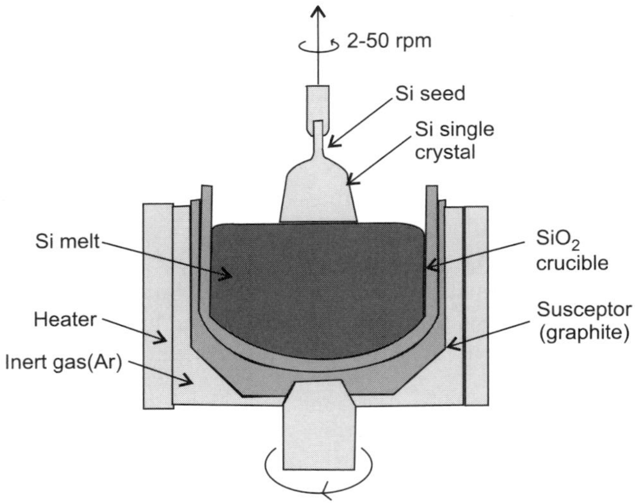

Fig. 1.9 Schematic illustration of the Czochralski method for growing single crystals from a melt, e.g. of Si.

materials, e.g. GaAs and InP, in order to try to reduce the loss of the more volatile anion constituents. The melt can also be encapsulated by means of a layer of different liquid (e.g. molten boron trioxide in the case of the latter compound semiconductor) in order to prevent volatile loss (although obviously boron contamination of the resulting crystals can be a problem); this technique is known as the liquid-encapsulated Czochralski method.

An alternative way of producing single crystals from a congruently melting liquid is to subject the crucible, containing a melt in contact with a seed crystal, to a temperature gradient. In the Bridgman technique, the temperature gradient along the crucible is kept constant while the average temperature of the furnace is reduced with time (Fig. 1.10a). In this way, the position at which the temperature is equal to the melting temperature, T_m, of the material moves progressively along the length of the crucible and, with it, the crystallization front. A closely related technique is the Stockbarger method, in which the crucible is *moved* relative to the furnace in which the temperature gradient *and* average temperature are both kept constant (Fig. 1.10b). The result is the same as for the Bridgman technique; the point at which the melt temperature equals T_m, and hence the crystallization front, moves progressively along the crucible as it is translated relative to the furnace. An example of the use of the Stockbarger method is in the manufacture of directionally solidified (or even single-crystal) turbine blades (see § 3.4.3).

A technique related to the Stockbarger method is the zone-melting technique for single-crystal production. In this, the temperature profile along the crucible is peak-shaped, with the temperature exceeding the melting temperature of the material to be crystallized only in a very small region (Fig. 1.10c); such a temperature profile can be achieved, for example, by means of a ring-shaped heater, the plane of which is perpendicular to the long axis of the crucible. The crucible is translated relative to the heater,

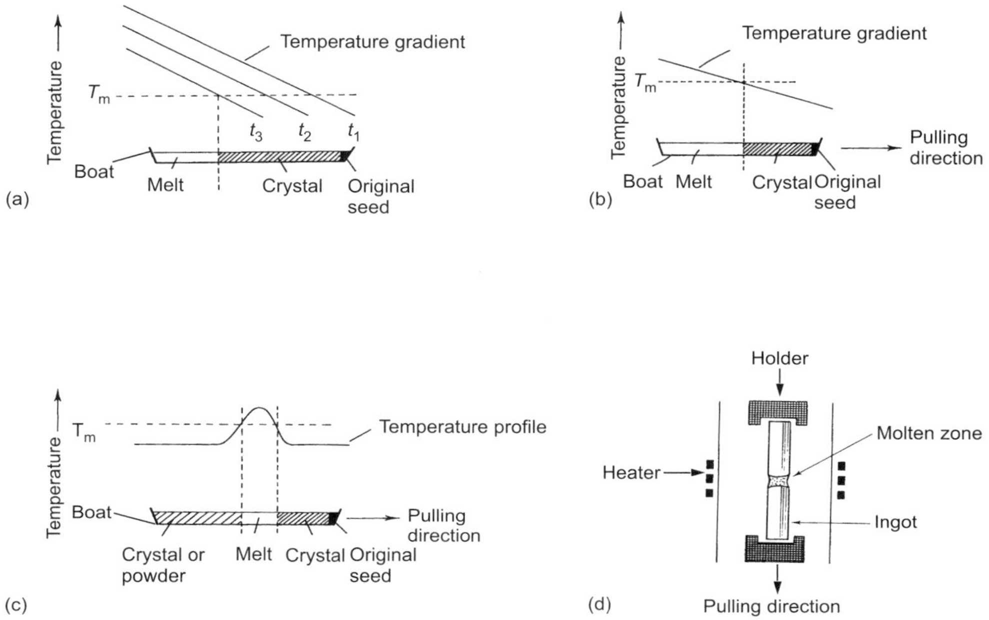

Fig. 1.10 Schematic illustration of crystal-growth techniques employing temperature gradients: (a) Bridgman technique, in which the temperature gradient is maintained, but the average temperature decreases with increasing time, $t_1 > t_2 > t_3$; (b) Stockbarger method, in which a crucible is translated relative to a fixed (linear) temperature gradient; (c) zone-melting technique, in which a crucible is translated relative to a fixed (peaked) temperature profile; (d) floating-zone method in which a molten zone is confined by surface tension between a polycrystalline ingot and a single-crystal seed, both free-standing.

and hence the hot zone, where the material melts, correspondingly moves along the crucible, leaving behind it a solid crystallized from a seed crystal. A modification of this method is used in the floating-zone technique in which a rod of polycrystalline material, clamped at one end and held vertically, is contacted with a seed crystal. This junction region is locally heated, as in zone melting, and the molten layer is held in place by surface tension (Fig. 1.10d). This floating zone is then moved slowly along the polycrystalline rod by moving the rod relative to the heating elements; single-crystalline material is left behind the melt front. This technique can be used to produce low-dislocation-density rods of crystalline Si.

This approach is also used in the zone-refinement method to purify a pre-existing rod-like crystal. The ring-like heater is moved slowly along the rod, and the narrow molten region is therefore moved through the sample. The solubility of impurities will be higher in the hot molten region than in the surrounding cooler solid, and so the impurities preferentially dissolve in the molten zone and hence are swept out of the crystal as the heater is moved along the rod.

Single-crystal films can also be grown epitaxially on suitable substrates from the molten state; this technique is known as liquid-phase epitaxy (LPE). Pockets in a heated slider contain the molten material(s) to be deposited (Fig. 1.11). By moving the slider

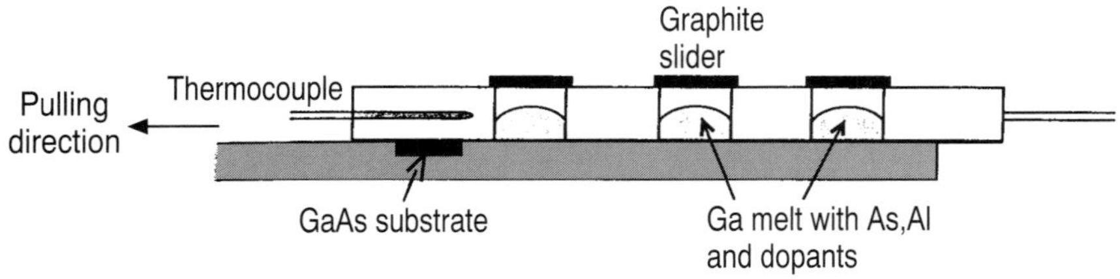

Fig. 1.11 Schematic illustration of the liquid-phase epitaxy technique.

over the surface of a substrate, crystallite nucleation and epitaxial growth of a single-crystalline thin film take place from the melt at the substrate surface. By placing different materials in different pockets of the slider, layers of differing compositions can be grown one on top of another. This technique can be used to prepare thin crystalline layers of GaAs-based semiconductors (for example, for use as laser diodes —see §8.5.2.4). Liquid Ga is used as a solvent for solutes such as As or Al to make the basic semiconductor materials $Ga_xAl_{1-x}As$, and also for solutes which can act as electrically active dopants in the semiconductors. When the liquid, viewed as a solution of say As in Ga (see also §1.2.3), is cooled as it comes into contact with the substrate, it becomes supersaturated with As and nucleation of GaAs occurs at the interface between liquid and substrate. LPE has the advantage that it is simple and inexpensive; however, it is not suitable if precise control over growth conditions, or a very thin film, is required (typical film thicknesses are a few microns).

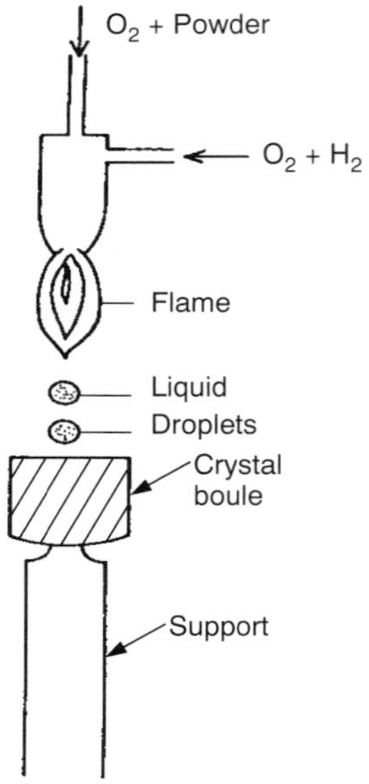

Fig. 1.12 Schematic illustration of the Verneuil method for flame-fusion growth.

An altogether different, container-less method for producing single crystals from the melt is the Verneuil technique illustrated in Fig. 1.12. Powder of the material to be made into a single crystal is dropped through a high-temperature flame (e.g. an oxy-hydrogen torch), wherein it melts and forms liquid droplets. These molten droplets then fall onto a seed crystal, held on a rotating pedestal support, where they solidify and form a single-crystal boule. This simple technique is commonly used to manufacture crystals of high-melting-point oxides, e.g. synthetic corundum (α-Al_2O_3). Although this flame-fusion method is inexpensive, the relatively high crystal growth rate ($\simeq$ 10 mm / h) leads to a somewhat poor quality of the ensuing crystals, having characteristic defects in the form of 'tree-ring'-like striations, resulting from fluctuations in the cooling rate of the molten layer in contact with the top surface of the boule.

1.2.2 Liquid quenching

In the previous section, the goal of the various materials-preparation procedures described therein was to produce as perfect a crystalline solid as possible, starting from a melt. Alternatively, sometimes a material is desired in a solid form that lacks any trace of crystallinity. A liquid (melt) is disordered both structurally (the long-range structural order, characteristic of a crystal—see §2.1.1—is absent in a liquid) and dynamically (the positions of atoms in a liquid fluctuate in time due to both vibrational and diffusive motion). Hence there is a prospect that, if a liquid could be cooled sufficiently quickly (i.e. 'quenched') so that crystallization could be bypassed, then the disordered structure characteristic of a liquid could be frozen-in, and a solid, structurally disordered phase would result; this is known as a glass.

Such a vitrification (or glass-forming) process involves supercooling of a liquid below its normal freezing point (or, equivalently, the melting point of the corresponding *crystalline* phase). The freezing of a crystalline solid is a first-order thermodynamic phase transition: there is a discontinuity in first-order thermodynamic variables, such as entropy, $S = -(\partial G/\partial T)_p$, or volume, $V = (\partial G/\partial p_-)_T$, at the transition (Fig. 1.13). In contrast, the transformation from a melt to the glassy phase is a transition (see Fig. 1.13) in which there is no discontinuity in first-order thermodynamic variables at the glass-transition temperature, T_g, but rather a change in their temperature gradient. Hence, there is a discontinuity in *second*-order thermodynamic variables, such as the calorimetric heat capacity at constant pressure:

$$C_p = \left(\frac{\partial H}{\partial T}\right)_p = T\left(\frac{\partial S}{\partial T}\right)_p = -T\left(\frac{\partial^2 G}{\partial T^2}\right)_p. \quad (1.11)$$

This jump in heat capacity is a clearer marker of the glass transition than is the measurement of the change in slope of the temperature dependence of the density (see Fig. 1.13). Unlike the melt-freezing/crystal-melting first-order phase transition, the glass transition is *not* an equilibrium (second-order) thermodynamic transition because it is *kinetically controlled*; a melt supercooled with a *quicker* quenching rate has a *lower* value of T_g. The nature of the glass transition is still uncertain and the subject of much current research; however, it is clear that the process is mainly one of kinetic arrest. As the temperature of the melt decreases in the supercooled régime, its viscosity correspondingly rapidly increases until the point is reached when, on *the time-scale of*

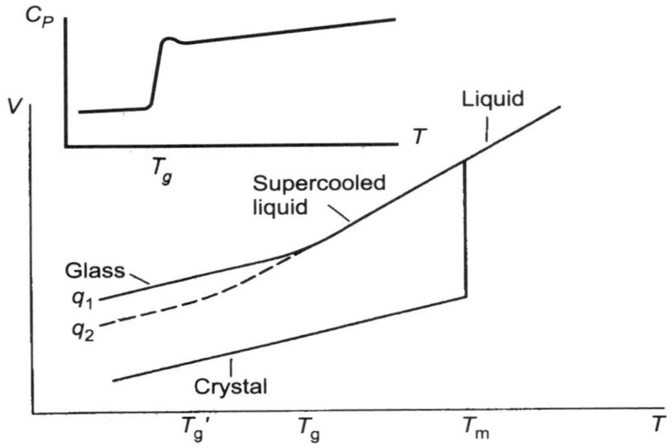

Fig. 1.13 Schematic illustration of the change in volume with temperature as a supercooled liquid is cooled through the glass-transition temperature, T_g, below the first-order crystallization phase transition at the melting/freezing temperature, T_m. A more rapid quenching rate, $q_2 > q_1$, gives a lower glass-transition temperature $T_g' < T_g$. Shown in the inset is the variation of the heat capacity at constant pressure in the vicinity of T_g.

the experiment, no further structural relaxation of the melt appears to take place: the material is then glassy and behaves as a solid.

The above picture considers the glass transition entirely in terms of the behaviour of the supercooled liquid. Another important aspect, of course, is the avoidance of crystallization during quenching. The process of crystallization from a melt can be divided into two stages: nucleation and growth. Crystal nucleation is the formation, by thermally induced structural fluctuations in the melt, of crystal nuclei, i.e. microscopic regions, having the structure of the crystal, that are greater in size than a critical radius so that they can subsequently grow and not disintegrate to re-form the liquid structure. Nucleation may be homogeneous (occurring in a random manner throughout the liquid) or, more commonly, heterogeneous (occurring at surfaces of the container, foreign particles, etc.). Crystal growth is the subsequent process, whereby a crystal nucleus continues to enlarge by the progressive addition of atoms onto its surface from the liquid phase.

The temperature dependences of the (homogeneous) crystallite nucleation rate in unit volume ($I(t)$) and of the crystallite growth rate per unit volume ($u(t)$) in the supercooled melt are shown schematically in Fig. 1.14. The curves are peaked, with the maximum rates I_{max} and u_{max} occurring generally at different temperatures. The rates exhibit such peaked behaviour because at very low (deeply undercooled) temperatures, the viscosity is so high that atomic motion is very difficult and the rates are small; at temperatures very close to the melting temperature, T_m, on the other hand, crystallites preferentially melt rather than form or grow, and again the rates are small (see also Problem 1.4). The critical temperature region for crystal growth is between T_1 and T_2 in Fig. 1.14; here, the nucleation rate is high enough for an appreciable number of crystallite nuclei to form in the first place, and, at the same time, the growth rate is high enough for these nuclei to grow into larger crystallites. Thus, in order to *avoid* crystallization, i.e. in order to prepare a glass, a melt must be quenched to temperatures below T_2, the bottom of the

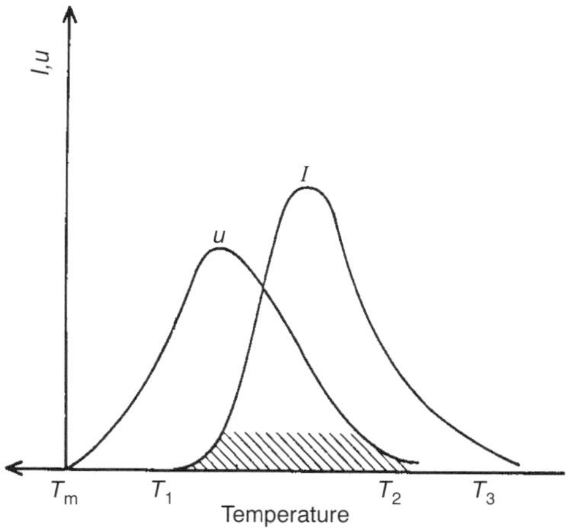

Fig. 1.14 Schematic illustration of the temperature dependence of the crystallite nucleation rate, I, and the crystallite growth rate, u, in the supercooled liquid state below T_m.

growth region, and ideally below T_3, the bottom of the nucleation region, in a sufficiently short time that no crystallite nuclei can form or grow; i.e. the quench rate, q, must be greater than the nucleation or growth rates, $q \geqslant VI$, Vu, where V is the sample volume.

For the case of network-forming glasses, in which a macromolecular, highly cross-linked structure is formed (an everyday example is silica, SiO_2), the critical quench rate for vitrification can be quite small, say a few (tens of) degrees per second. However, for systems in which the atomic bonding is much less directional (e.g. in metals—see §2.2.3.1), the viscosity in the supercooled melt remains relatively much higher and crystallite formation is more prevalent. As a result, ultra-rapid cooling techniques must be employed in order to vitrify metallic alloys, e.g. $Pd_{80}Ge_{20}$. One such technique is melt-spinning, in which a jet of melt is directed at the edge of a rapidly spinning disc of copper; the melt, on striking the surface of the disc, cools extremely rapidly (because of the high thermal conductivity of Cu—see §6.3.2.2) and is transformed into a glassy tape of thickness $\simeq 0.1$ mm (Fig. 1.15). In this way, cooling rates of the order of 10^6 K/s

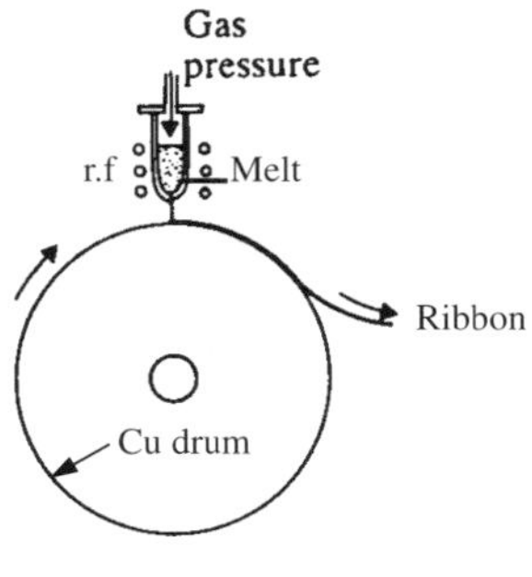

Fig. 1.15 Schematic illustration of a melt-spinning apparatus to produce glassy-metal ribbons.

can be achieved. A modification of this approach, planar-flow casting, allows foils of widths up to $\simeq 0.1$ m to be produced.

1.2.3 Crystallization from solution

One common way of obtaining crystalline material is from solution; everyday examples of this method include the crystallization of salt (NaCl) and sugar from their respective aqueous solutions. This technique can be used both to prepare materials in (single-) crystalline form and also to synthesize crystals. This approach has the advantage that the starting state, a solution, is generally homogeneous and single-phase. As a result, formation of crystalline products is generally much easier than in all-solid-state processes (§1.3.1) because atoms do not need to diffuse so far, and also because atomic diffusion is faster, meaning that solution synthesis mostly occurs at much lower temperatures than in solid-state reactions. Materials synthesis via solution is one example of what is known as *chimie douce*, or 'gentle chemistry'.

Preparation of crystals from solution involves dissolving the material to be crystallized in a suitable non-reactive solvent, and then causing crystallization to occur, either by reducing the solubility of the solute (e.g. simply by reducing the temperature) or by increasing the solute concentration in the solution by removing some of the solvent (e.g. by evaporation), thereby forming a supersaturated solution; in both cases, crystals will then form from the solution, the crystal size being larger, the lower the growth rate, in general. (The rôle played by defects in crystal growth is discussed in §3.4.1.) More controllable crystal growth is obtained, obviously, by immersing a seed crystal into a saturated solution. A problem with this technique, however, is the possibility of inclusion of the solvent in the crystals.

Solvents used for low-temperature solution growth (say, at temperatures below 100 °C) are commonly water for the crystallization of many inorganic compounds (although other solvents, e.g. ammonia or HF may also be used), and various organic solvents (e.g. acetone, ethanol, carbon tetrachloride, etc.) for crystals of organic molecules. A discussion of the low-temperature growth of crystals from solution is given in e.g. Brice (1973).

High-temperature solution growth involves temperatures much higher than the boiling point of water, typically of the order of 1000 °C. The solvents in this case can be liquid metals, e.g. Ga (a solvent for As, as used in the liquid-phase epitaxy (LPE) method — see §1.2.1), Pt (a solvent for B), Pb, Sn and Zn, etc. (solvents for Si, Ge, GaAs, GaP, etc.). Alternatively, they can be inorganic compounds such as oxides or fluorides, e.g. KF (a solvent for $BaTiO_3$), SnF_2 (a solvent for ZnS), $Na_2B_4O_7$ (a solvent for Fe_2O_3 and, with B_2O_3 as a co-solvent, for TiO_2) etc. These inorganic solvents are called fluxes since they effectively reduce the melting point of the pure solute by forming a lower-melting-point solution. This form of high-temperature solution growth is therefore also known as flux growth.

It is somewhat a matter of semantics whether or not crystal growth from liquid metals or inorganic fluxes should be regarded as being high-temperature solution growth or growth from a melt (see §1.2.1). The behaviour of the solvent–solute system can be understood in terms of phase diagrams, introduced in §1.1.1, that display the phase-equilibrium fields as a function of temperature and composition for the binary system

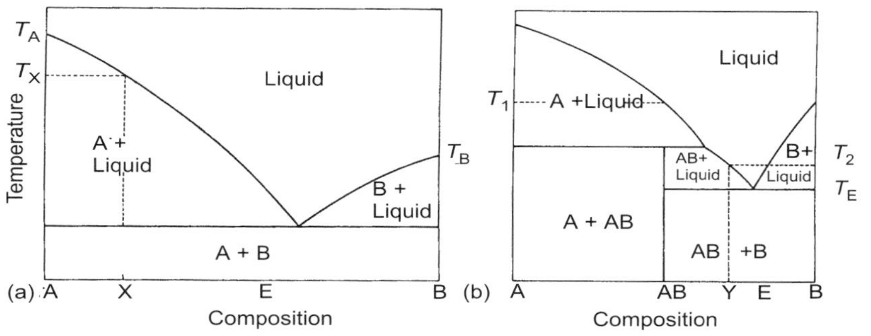

Fig. 1.16 (a) Representative phase diagram demonstrating the use of a flux B to grow crystals of A. The composition E is the eutectic composition. (b) Representative phase diagram for an incongruently melting compound AB. Cooling of the liquid with the composition AB produces only crystals of A; a liquid with composition such as Y is needed to grow crystals of AB.

A + B, where A is the solute, say, and B is the solvent. Perhaps the simplest case to consider is when the solute and solvent form a *eutectic system*: the addition of the flux B causes the melting point of A to decrease as A dissolves in B (Fig. 1.16a). Slow cooling of a composition such as X marked in Fig. 1.16a, between the temperatures T_x and T_E, produces crystals of A together with a liquid (solution of flux and solute) that varies in composition as the temperature changes (see Problem 1.5). Eventually, when the temperature reaches the eutectic temperature, T_E, any remaining liquid will form a fine-grained solid eutectic mixture, consisting of small crystals of both solute A and solvent B, embedded in which will be the much larger crystals of A that formed during the slow-cooling process. These larger crystals can be removed from the eutectic matrix by, for example, mechanical means.

Synthesis of a compound, say AB, may take place when a solute is dissolved in a solvent at high temperatures. The corresponding representative phase diagram for the case when AB is an *incongruently* melting compound is shown in Fig. 1.16b. An attempt to produce crystals of AB simply by cooling a liquid with this composition (for which the liquidus temperature is T_1 (see Fig. 1.16b)) will not succeed, since from the phase diagram it can be seen that what is produced first on cooling is crystalline A, not AB. However, cooling a liquid with the composition Y marked in Fig. 1.16b will give rise to crystals of AB being formed in the temperature range T_2 to T_E; on cooling below the eutectic temperature, T_E, a eutectic crystalline mixture of AB and B will form.

Another, widely used method for synthesizing solid materials from solution is by chemical precipitation, in which a solution-phase reaction occurs, producing an insoluble, often crystalline (albeit finely divided) precipitate. An example is the reaction in aqueous solution producing AgCl:

$$AgNO_3(aq.) + NaCl(aq.) \rightarrow AgCl(s) + NaNO_3(aq.). \tag{1.12}$$

Occasionally, amorphous (non-crystalline) precipitates are formed, for example when H_2S gas is bubbled through a solution of arsenic oxide in dilute hydrochloric acid:

$$As_2O_3(aq.) + 3H_2S(g) \rightarrow As_2S_3(s) + 3H_2O. \tag{1.13}$$

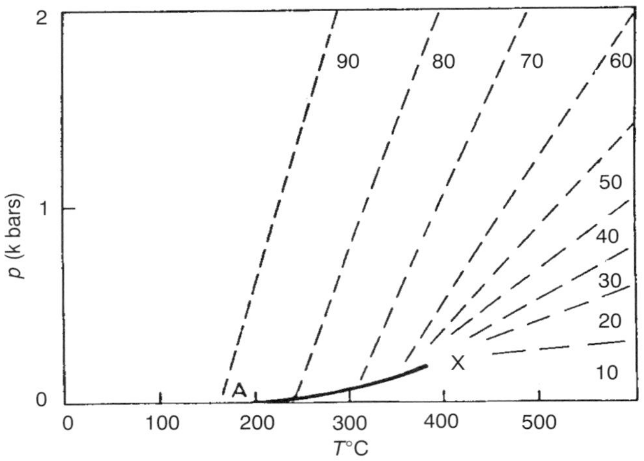

Fig. 1.17 Pressure–temperature phase diagram for water vapour and liquid in the vicinity of the critical point X (T_{cr} = 374.2 °C; p_{cr} = 218.3 bar), showing the vapour–liquid tie-line (A-X) ending at X. The dashed lines indicate the autogeneous pressures exerted in a closed pressure cell at various temperatures for the percentage filling factors of the 'bomb' at ambient temperature and pressure, as marked.

Very often, the solubility of a material in a solvent, e.g. water, is insufficient under normal conditions (atmospheric pressure and temperatures of less than 100 °C for water) to allow solution-phase synthesis and/or crystallization to be carried out. In such cases, solvents may be more effective in the supercritical state, i.e. at temperatures and pressures above the critical point at (T_{cr}, p_{cr}) in the liquid–vapour phase diagram (see Fig. 1.17), where the liquid and gaseous states become indistinguishable and form a common fluid phase. Supercritical CO_2(T_{cr} = 304.2 K; p_{cr} = 7.38 MPa) is increasingly being used as a solvent for organic molecules and polymers.

Supercritical water (T_{cr} = 647.3 K; p_{cr} = 22.12 MPa) is an effective solvent for many inorganic materials, particularly in the presence of a mineralizer, a soluble compound producing ions that otherwise would not be present in significant concentration in solution and which increase the solubility of the solute; this technique involving supercritical water is also known as the hydrothermal-growth method. Crystal growth under hydrothermal conditions takes place in a metal pressure cell, rather worryingly normally called a 'bomb' (Fig. 1.18), in which the material to be dissolved and subsequently crystallized is placed at the bottom, with seed crystals suspended at the top, and the bomb is part-filled with water. (The autogeneous (i.e. self-generated) pressures of supercritical water exerted at temperatures for various degrees of filling are shown in Fig. 1.17.) The bomb is placed in a temperature gradient so that the temperature, T_2, at the bottom is higher than at the top, T_1: material at the bottom of the bomb is therefore dissolved and the resulting solution is conveyed by convection to the top where, because the temperature is lower, the solution becomes supersaturated and crystallization onto the seeds occurs. Hydrothermal growth is commonly used to prepare large single crystals of the α-quartz crystalline modification of silica under conditions of 400 °C

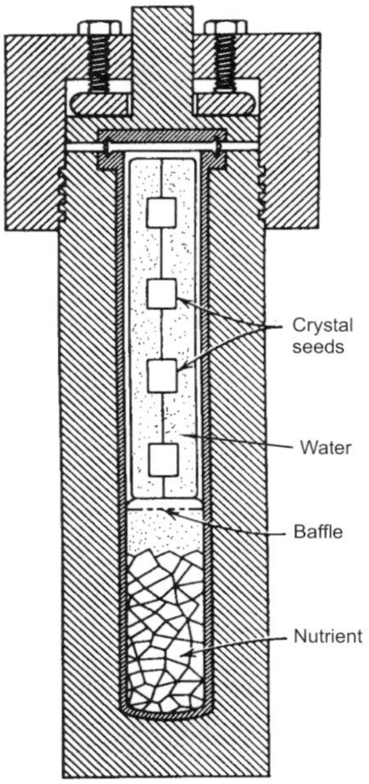

Fig. 1.18 Schematic illustration of a pressure 'bomb' used for hydrothermal crystal growth.

and 170 MPa using NaOH as a mineralizer. The method can also be used to grow single crystals of corundum (or synthetic sapphire, α-Al_2O_3).

Although the hydrothermal-growth method has the disadvantage that very often mineralizer ions (e.g. OH^- in the case of quartz) are incorporated into the growing crystal, it has the advantage, shared by all low-temperature solution techniques, that crystalline phases can be synthesized and prepared that would be metastable or unstable at the much higher temperatures encountered in solid-state synthesis (§1.3.1).

An example of a class of metastable materials that can be made in this way is the zeolites, hydrous aluminosilicates with the general chemical formula

$$M_{x/n}[(AlO_2)_x(SiO_2)_y] \cdot mH_2O,$$

in which the positively charged cations M of valence n (e.g. Na^+ with $n = 1$) electrically compensate the negative charge on the aluminosilicate framework arising from the fact that both Al and Si are four-fold coordinated by oxygen atoms but aluminium is trivalent (Al^{3+}) whereas silicon is tetravalent (Si^{4+}). The name 'zeolite' was coined from the Greek words *zeo* (to boil) and *lithos* (stone), since naturally occurring zeolite minerals readily give up their water of crystallization on heating and appear to 'boil'. This behaviour occurs because the water molecules (and also the charge-compensating cations) lie in the inter-linked cavities and channels (of size $\simeq 5$ Å) that characterize the structure of these highly porous crystalline solids (Fig. 1.19), and hence such species can readily be removed and/or ion-exchanged. The open aluminosilicate framework of zeolites can be regarded as being constructed from ordered aggregates of secondary

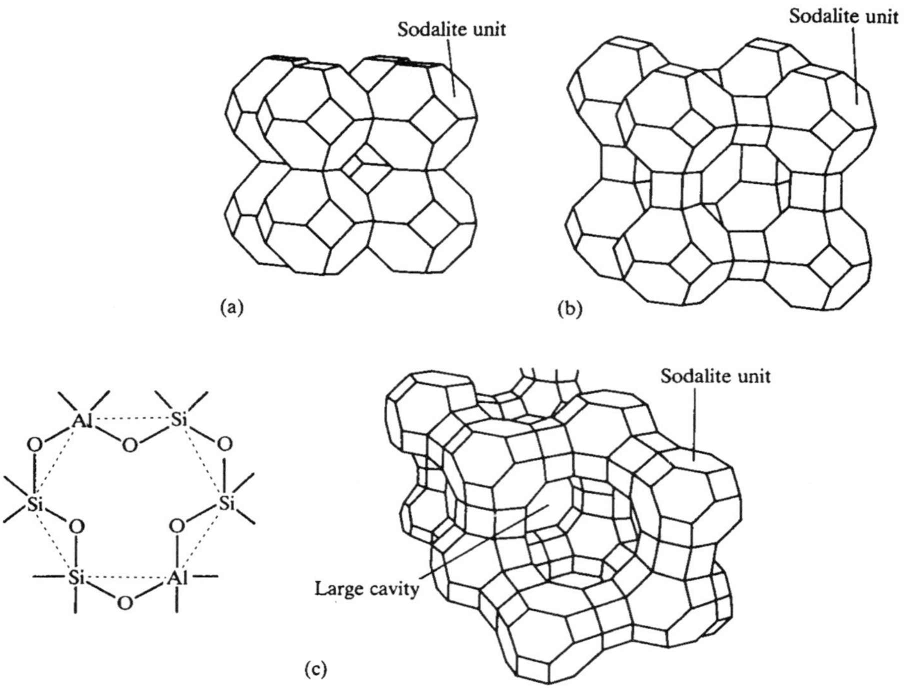

Fig. 1.19 The aluminosilicate frameworks for three zeolitic structures built up from sodalite cages: (a) sodalite; (b) zeolite-A; (c) zeolites –X and –Y (faujasite). Straight lines in these figures denote vectors between the centres of neighbouring tetrahedra (i.e. Si or Al atoms) as shown in the inset; the positions of the oxygen atoms are not marked.

cage-like structural units, in turn comprising a number of corner-sharing SiO_4 and AlO_4 tetrahedra. One common example of such a secondary unit is the sodalite (or β) cage comprising 24 connected silica or alumina tetrahedra, which has the geometry of a truncated octahedron (Fig. 1.19). Different modes of linkage between sodalite units lead to different zeolite structures, e.g. sodium zeolite-A ($x = y = 12$, $n = 1$, $m = 27$ in the general zeolite formula) (Fig. 1.19b), and zeolite-X and -Y synthetic faujasites ($x \simeq 60$, $y \simeq 130$, $m \simeq 250$) (Fig. 1.19c), each having differently sized cavities and channels.

Zeolites are synthesized and prepared in crystalline form via an interesting variation of the hydrothermal-growth method, in which the added mineralizers (or certain organic molecules) act as structural templates around which the aluminosilicate framework grows in generating the porous zeolitic structures. The starting material is a basic aqueous solution of sodium silicate and aluminate (containing $[Al(OH)_4]^-$ ions) at a high pH, the base being either an alkali hydroxide (e.g. NaOH) or an organic base, the cations of the latter acting as efficient templating molecules. A gel (a semi-rigid partly cross-linked polymeric structure) then forms at room temperature by a process of copolymerization (see §1.2.4), for example for a sodium zeolite:

$$NaAl(OH)_4(aq.) + Na_2SiO_3(aq.) + NaOH(aq.) \rightarrow \\ Na_x(AlO_2)_y(SiO_2)_z \cdot NaOH \cdot H_2O(gel). \tag{1.14}$$

Hydrothermal treatment of the gel for several days at temperatures in the range 60–200 °C under autogeneous pressure then yields crystals of the zeolite:

$$Na_x(AlO_2)_y(SiO_2)_z \cdot NaOH \cdot H_2O(gel) \rightarrow Na_x[(AlO_2)_y(SiO_2)_z] \cdot mH_2O. \tag{1.15}$$

Such a synthesis using NaOH as base will tend to produce rather dense zeolites, with relatively small cavities and channels and with Si/Al ratios ($y:x$) of order 1 : 1; an example is Na zeolite-A (Fig. 1.19b). Silicon-rich zeolites can be made by increasing the proportion of silicate in the starting solution, but also by using large-sized (e.g. alkyl ammonium) cations in the base. The large size of cations such as tetramethyl ammonium, $[N(CH_3)_4]^+$, used in the synthesis of zeolite ZK-4 (having the same framework structure as zeolite-A, but with a Si/Al ratio of 2.5 : 1), or tetra-*n*-propyl ammonium, $[N(C_3H_7)_4]^+$, used to synthesize ZSM-5 (with Si/Al ratios from 20 : 1 upwards, and used as a catalyst for the conversion of methanol to gasoline), means that few such cations can be incorporated in the structure and hence the proportion of Al (responsible for the negative charge on the framework) must be reduced. Further details about zeolites are given in, for example, Smart and Moore (1992) and Dyer (1988), and details of their synthesis are given in Barrer (1982).

This templating effect is taken to the extreme in the synthesis of the mesoporous MCM-41 molecular sieves, with large pore sizes of the order of 30–100 Å, where long-chain surfactants, such as $C_nH_{2n+1}(CH_3)_3N^+Br^-$ $(8 < n < 16)$, are used as the template molecules (Kresge *et al.* (1992)). At relatively high reaction temperatures (> 100 °C), the templating molecules remain separate and relatively dense, microporous zeolites (e.g. ZSM-5) are produced. However, at low temperatures ($\simeq$ 100°C), these molecules, consisting of a hydrophobic (i.e. water-repelling) tail (i.e. the aliphatic chain) and a hydrophilic (water-attracting) head (i.e. the bromide ion) spontaneously aggregate to form a micelle, i.e. a cluster containing tens or hundreds of molecules. The hydrophobic

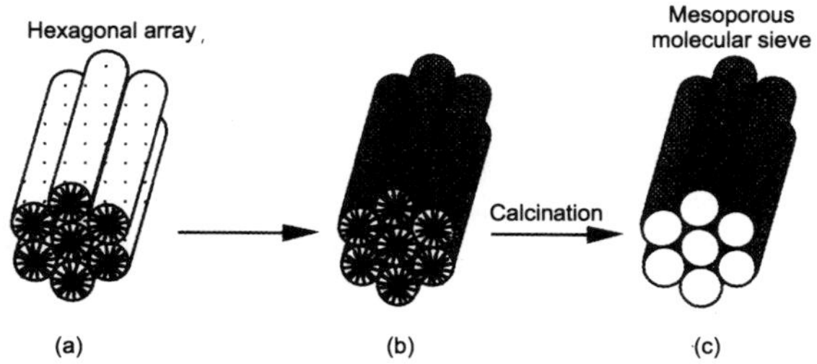

Fig. 1.20 Schematic illustration of the use of micelles (clusters of long-chain molecules having hydrophobic (organic) and hydrophilic heads) as templates for the synthesis of a mesoporous silicate molecular sieve. (a) The hexagonal array formed by cylindrical micelles, with the hydrophobic parts of the molecules being in the centre of the cylinder, and the hydrophilic head-groups on the outside (light grey): (b) Silicate species (dark grey) occupy the water in the space between micelles in the vicinity of the polar head-groups; (c) Calcination leaves a mesoporous silicate framework (MCM- 41) with hexagonal geometry. (After Kresge *et al.* (1992). Reprinted with permission from *Nature* **359**, 710. © 1992 Macmillan Magazines Ltd.)

ends attract each other, leaving the hydrophilic head groups on the outer surface of the cluster. In this case it is believed that cylindrical, rather than spheroidal, micelles are formed, which aggregate to give hexagonal arrays (Fig. 1.20) Hydrated silicate species in solution occupy the continuous region of water surrounding the micelles and can condense to form walls comprising amorphous silicates (see §1.2.4); calcination on heating then 'burns off' the organic templates, leaving a mesoporous silicate framework structure having the same hexagonal geometry as the micellar array (Fig. 1.20).

Hydrothermal synthesis can also be employed to produce '1 2 3' phosphate materials, e.g. $AgTi_2(PO_4)_3$, that have a tunnel structure, somewhat like zeolites. In addition, a variety of transition-metal oxides, including many structures which cannot be prepared otherwise, can also be synthesized by hydrothermal methods (see Whittingham (1996)).

1.2.4 Sol–gel methods

An important technique that can be used to synthesize many materials, and prepare them in a variety of sample shapes and forms, is the sol–gel process; this is another example of *chimie douce*. This method is especially suited for the synthesis and preparation of refractory oxide materials at relatively low temperatures; for example, fully dense SiO_2 glass can be made this way at an operating temperature of 1200 °C, instead of the melting temperature, $T_m \simeq 2000$ °C, although the precursor porous material in this process is formed at much lower temperatures.

The starting point of this method is a sol, which is a colloidal dispersion of small particles suspended in a liquid. A sol can be stabilized by *peptization*, i.e. the addition of peptizing agents (e.g. HNO_3) which form an electrically charged layer around each particle; electrostatic repulsion then prevents particle aggregation. Under suitable chemical and thermal conditions, the particles in some sols can be made to react or interact electrostatically so that they form a continuous, three-dimensional network of connected particles, known as a gel, instead of aggregating to form larger, but discrete, particles, as happens in precipitation or flocculation (Fig. 1.21). The 'wet' gel formed during the sol-gel process consists of a network of connected particles containing the liquid sol in its interstices. An as-prepared gel can behave as a rubbery, easily deformed solid or as an extremely fragile, brittle solid from a mechanical point of view; as a gel ages, more of the particles in the sol in the pores condense out onto the existing framework, thereby progressively stiffening it. The structure of a gel, as shown in Fig. 1.21, is an example of a *fractal* structure (§2.1.1), in which the mass density scales as $\rho(r) \propto r^{D_H}$, where r is the radius of a cluster and the fractal dimension $D_H < 3$: a fully dense material is characterized by $\rho(r) \propto \rho^3$.

The sol–gel process naturally lends itself to the preparation of gel samples in a wide variety of shapes and forms. Bulk gels can be made in the shape of any container that can be filled with a liquid sol. In addition, gel films can be made by coating a substrate with the sol prior to gellification. Finally, even gel fibres can be made by drawing (pulling) a fibre from a gelling, viscous sol.

Removal of the interstitial liquid from a wet gel produces a dry gel, known as a xerogel. However, drying a gel is not a trivial process if the structural integrity of a bulk (monolithic) gel sample is required to be maintained, i.e. if the sample must remain crack-free. Drying a gel simply by heating, to evaporate the interstitial liquid, inevitably

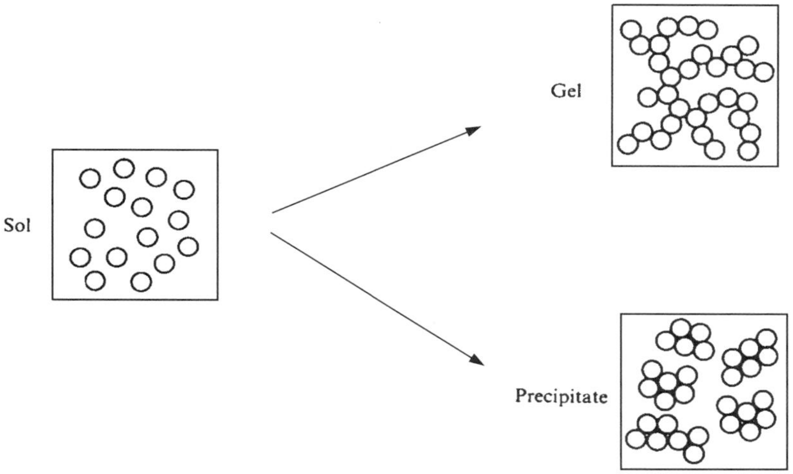

Fig. 1.21 Illustration of gel formation and precipitation (flocculation) from a sol.

is problematic: the gel shrinks and, worse, cracks form as a result of the differential stresses generated in the drying gel due to *capillary forces*. The origin of these internal stresses can be seen by reference to Fig. 1.22 which illustrates two pores of different radii in a gel. When the pores are completely full, in the 'wet' gel (Fig. 1.22a), there are no capillary forces, but as the gel dries, leaving the pores part-filled with liquid (Fig. 1.22b), menisci form at the liquid–vapour interface in each pore. There is a pressure difference Δp across the two sides of a meniscus (responsible for capillary rise of liquids in narrow tubes) that is associated with the surface energy γ_s needed to create the free surface of the liquid, and is given by Laplace's equation:

$$\Delta p = (2\gamma_s \cos\theta)/r. \tag{1.16}$$

Here, r is the radius of the pore and θ is the contact angle between meniscus and pore ($\theta = 0°$ for perfect wetting of the surface of the pore by the liquid); thus, $r/\cos\theta$ is the

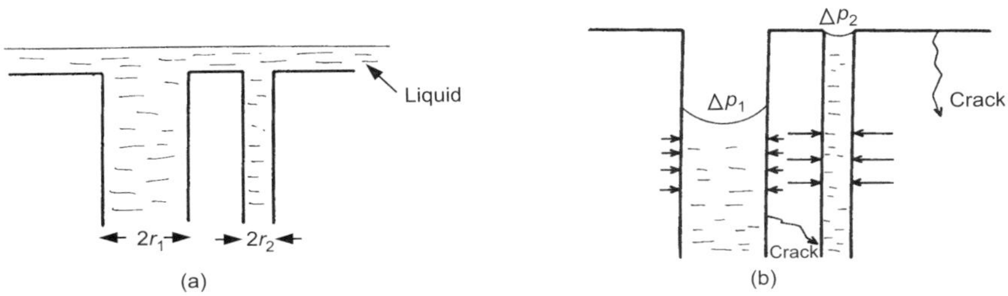

Fig. 1.22 Two pores in a gel in: (a) the 'wet' state, where the pores are completely filled with sol liquid; (b) a partially dried gel, where pressure differences across the menisci are different, $\Delta p_2 > \Delta p_1$, for pores with different radii, $r_2 < r_1$. The associated differential stresses can cause the gel to crack.

radius of curvature of the meniscus. Hence, pores of different sizes will experience different capillary stresses and, if these stress differences exceed the tensile strength of the gel material (§3.4.3), cracking occurs.

One way of avoiding the effect of capillary forces is to use supercritical (or hypercritical) drying. The gel is placed in an autoclave, together with some of the same liquid that is trapped in the pores of the gel, and the temperature is raised until the critical point (cf. X in Fig. 1.17) is exceeded. Under these supercritical conditions, the distinction between liquid and vapour vanishes: there are, therefore, *no* capillary forces in this supercritical fluid régime. Successive flushing of the autoclave by dry Ar then removes all traces of the fluid, leaving a dry, structurally intact gel called an aerogel. In practice, this method only works, in the case of silica, for alcogels, made with alcohol as the liquid comprising the original sol; hydrogels, where water is the liquid, cannot be dried in this way, since silica *dissolves* in water under supercritical conditions (§1.2.3). Aerogels, as prepared, can have extremely high values of porosity ($\geqslant 99\%$). They can subsequently be densified by sintering, i.e. heating so that the pore volume collapses as a result of viscous flow of the gel framework.

Two methods of producing the sol precursor, e.g. of SiO_2, can be distinguished: either acidification of silicate solutions, or hydrolysis of Si-containing compounds with subsequent condensation of silanol groups. In the former, alkali silicate aqueous solutions are converted into solutions of silicic acid on acidification; when its concentration exceeds the solubility limit of 100 p.p.m., monomeric $Si(OH)_4$ species are produced which can then polymerize by the condensation reaction:

$$-\mathrm{Si}-\mathrm{OH}+\mathrm{HO}-\mathrm{Si}- \rightarrow -\mathrm{Si}-\mathrm{O}-\mathrm{Si}- +\mathrm{H_2O}. \qquad (1.17)$$

At low pH, spheroidal silica clusters of diameter 1–2 nm are formed by successive polymerization/condensation reactions. At high pH, larger particles are generated, forming a stable sol with SiO_2 particles of diameter 10–20 nm for a pH of 10 (e.g. the commercial product, Ludox®). Such a high-pH silica sol is destabilized, leading to gel formation, by acidification. Similar methods are used to convert the precursor $TiCl_4$ into particles of TiO_2 used in paint manufacture.

An alternative method is to hydrolyse reactive metal compounds, for example alcoholates (also known as alkoxides), $M(OR)_n$, where M is a 'metal' (e.g. Si, Ge, B, Al, Ti, Y, Zr) and R is an alkyl group (e.g. methyl, CH_3, ethyl, C_2H_5, or propyl, C_3H_7). Since alcoholates and water are immiscible, the compounds are generally dissolved in an alcohol, e.g. methanol or ethanol. (However, intimate mechanical mixing of alcoholates and water can be achieved using ultrasound, the resulting (dense) gels being termed sonogels.) Addition of water to the alcoholate solutions causes hydrolysis to take place:

$$\mathrm{M(OR)}_n + n\mathrm{H_2O} \rightarrow \mathrm{M(OH)}_n + n\mathrm{ROH}. \qquad (1.18)$$

This is followed by a series of condensation reactions as in reaction (1.17), the overall reaction of which is

$$m\mathrm{M(OH)}_n \rightarrow m\mathrm{MO}_{n/2} + \frac{mn}{2}\mathrm{H_2O}; \qquad (1.19)$$

the aggregates of the product form a sol. It can be seen that this method allows mixed oxide gels to be produced readily, since intimate mixing of the alcoholate solutions occurs prior to hydrolysis.

1.2.5 Ion exchange and intercalation

A final class of *chimie douce* involves the exchange/insertion of chemical species, often ions, into materials, thereby changing their composition and, sometimes, also their structure. Ion exchange, as its name implies, is the replacement of one type of ion present in the structure of a material by another type. Intercalation is the insertion of a chemical species into the structure of a material where none existed before. A prerequisite for both processes is an *open structure* so that the chemical species being exchanged or inserted can move relatively freely between the interior and exterior of a sample. In addition, ion exchange will only be feasible, under moderate reaction conditions, if the ions to be exchanged are not too strongly bonded in the structure.

One class of materials exhibiting easy ion exchange is the aluminosilicates known as zeolites (§1.2.3), the structure of which is characterized by having large pores interconnected by channels; the well-defined channel diameter of a given zeolite (see Fig. 1.19) allows such materials to be used as molecular sieves. However, cations, often alkali ions, that compensate for the negative charge on the aluminosilicate framework, also lie in these pores and channels and hence are easy to exchange, simply by passing a solution containing the ion to be inserted over a zeolite containing another ion (as in water softeners, where Ca^{2+} ions in water are exchanged for Na^+ ions in the zeolite). Even the aluminium ions in the zeolitic framework can be removed, albeit by somewhat more vigorous reactions involving acid treatment, in which decationation accompanies dealumination:

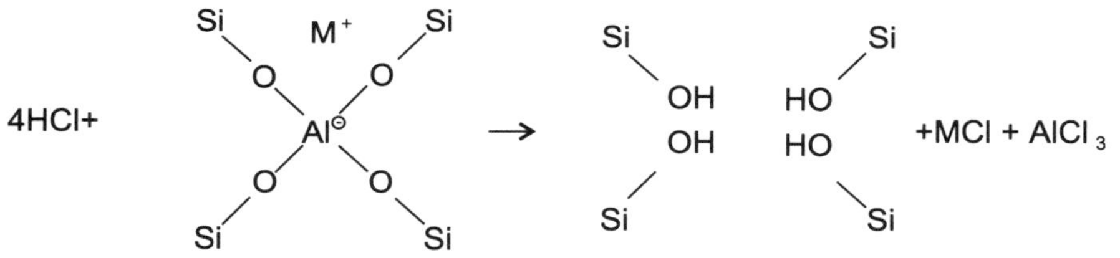

$$4HCl+ \quad \begin{array}{c} Si \diagdown \quad M^+ \quad \diagup Si \\ O \diagdown \quad \diagup O \\ Al^{\ominus} \\ O \diagup \quad \diagdown O \\ Si \diagup \quad \diagdown Si \end{array} \quad \rightarrow \quad \begin{array}{c} Si \diagdown \qquad \qquad \diagup Si \\ OH \quad HO \\ OH \quad HO \\ Si \diagup \qquad \qquad \diagdown Si \end{array} \quad +MCl + AlCl_3 \quad (1.20)$$

followed by a condensation (dehydroxylation) reaction to give two Si-O-Si sites.

In other cases, channels in the structure are not so big and hence ion exchange is somewhat more difficult. The structure of Na-β-alumina (§3.4.2.2) consists of layers of aluminium oxide, between which the Na^+ cations are situated and can move; ion exchange in this case can be achieved by immersion in a molten salt (at, e.g., 300 °C), containing the cation (e.g. Li^+, K^+, Rb^+, Ag^+, Cu^+, etc.) to be inserted and exchanged for Na^+ in the β-alumina. Note that easily exchangeable cations are *monovalent*: insertion of divalent cations (e.g. Ca^{2+}), replacing *two* Na^+ ions, is possible, but exchange only takes place at high temperatures (e.g. $\simeq$ 800 °C) since ionic transport of divalent cations is very slow. This is because the ionic binding energy between the cations and oxygen anions is much larger than for monovalent cations as a result of the increased electrical charge on the cation.

Many crystals are lamellar, i.e. layer-like, but do not have exchangeable cations. In such cases, it is often possible, however, to intercalate chemical species between the layers, which are correspondingly pushed apart. Graphite, the layer-crystal polymorph of carbon (see §8.4.1) can be readily intercalated with either electropositive or electronegative atoms (which subsequently ionize when intercalated), or with molecules from the liquid phase (as well as from the vapour — see §1.1). Examples include the following intercalation reactions:

$$\text{Graphite} + \text{K(melt)} \rightarrow C_8K, \tag{1.21}$$

$$\text{Graphite} + Br_2 \rightarrow C_8Br, \tag{1.22}$$

$$\text{Graphite} + FeCl_3 \rightarrow \text{Graphite} : FeCl_3. \tag{1.23}$$

Note that well-defined stoichiometries of intercalated product can often be achieved, although non-stoichiometric compositions can also be synthesized. Many intercalation reactions are reversible; e.g. by subjecting the intercalated product to a vacuum, the intercalant can be removed. Further details on intercalation reactions in graphite, particularly concerning the electron transfer that occurs between intercalant and graphite host, are given in §8.4.1.

A great number of layered, or tunnel-containing, compounds can be intercalated, including the layer crystals of transition-metal dichalcogenides MX_2 (M = transition metal; X = S or Se), for example TiS_2, MoS_2 etc. (see §8.4.1), as well as crystals of V_2O_5, MoO_3, TiO_2, MnO_2. Intercalation of the tunnel compound crystalline WO_3 by Na gives the compounds Na_xWO_3 which are known as sodium-tungsten bronzes after their metallic appearance (colour and optical reflectivity): the incorporation of metallic sodium in the optically transparent, electrically insulating material WO_3 is accompanied by electron transfer from intercalant to host, resulting in metallic behaviour of the intercalated compound for $x \geqslant 0.2$.

Lithium can readily be intercalated between the layers of the layered crystal TiS_2 (see §8.4.1), either chemically or electrochemically, to form a continuous range of

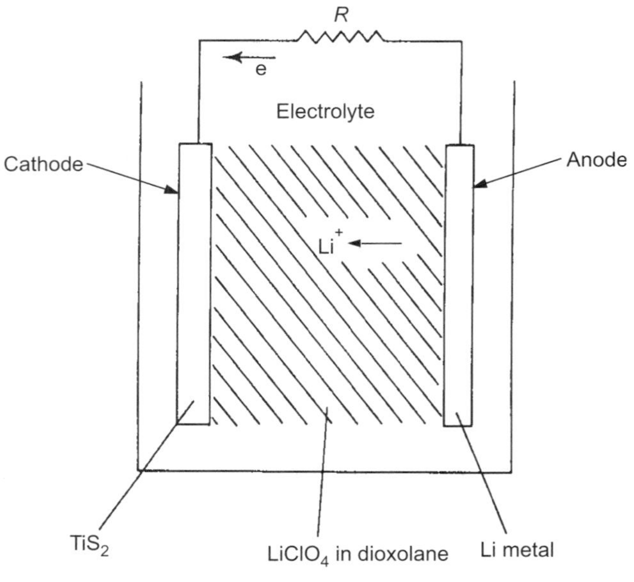

Fig. 1.23 Illustration of the electrochemical intercalation of TiS_2 by Li in an electrochemical cell.

intercalation products $Li_xTiS_2(0 < x < 1)$). Chemical intercalation can be carried out by using *n*-butyl lithium in an organic solvent (e.g. hexane) as the source of lithium:

$$xC_4H_9Li + TiS_2 \rightarrow Li_xTiS_2 + \frac{x}{2}C_8H_{18}. \tag{1.24}$$

More active lithiating agents, e.g. metallic lithium itself, can cause reduction of the dichalcogenide to lower chalcogenides or even to metallic titanium.

Alternatively, the intercalation may be carried out electrochemically, i.e. in an electrochemical cell, in which the powdered TiS_2, mixed with electrically conducting graphite powder to enhance its electrical conductivity, forms the cathode (§3.5.2), with metallic lithium being the anode (Fig. 1.23). An ionically-conducting, Li^+-containing electrolyte (e.g. $LiClO_4$ dissolved in a non-aqueous solvent, e.g. dioxolane) transports Li^+ ions from anode to cathode, where they intercalate the TiS_2. The corresponding half-cell reactions are therefore:

$$Li \rightarrow Li^+ + e^- \qquad \text{(anode)}, \tag{1.25a}$$

$$TiS_2 + Li^+ + e^- \rightarrow LiTiS_2 \qquad \text{(cathode)}. \tag{1.25b}$$

The electrons liberated at the anode in reaction (1.25a) travel to the cathode, where they take part in reaction (1.25b), via an external electrical circuit (Fig. 1.23).

Amorphous (non-crystalline) thin films of WO_3 can be electrolytically intercalated with protons (using an aqueous acid as electrolyte) to give the compounds $H_xWO_3(x \leqslant 0.2)$ which have a deep blue colour instead of the very pale yellow colour of pure WO_3; this phenomenon of electrochromism, i.e. a colour change induced electrochemically, is due to the reduction of some W^{6+} ions to W^{5+} ions.

Solid to solid synthesis and preparation 1.3

The synthesis of solid materials, particularly in polycrystalline powdered form, is frequently carried out using solid precursors. However, solid-state reactions are often very slow and difficult to carry out to completion unless performed at very high temperatures, where reacting atoms can diffuse through solid materials to the reaction front more easily. Transformation of one phase to another (with the same chemical composition) can also occur in the solid state, either at elevated temperatures or elevated pressures (or both).

1.3.1 Solid-state reactions

Solid-state reactions can be divided into two stages: nucleation of the reaction product and its subsequent growth. Consider the simplest case of two single crystals of different reactant solids, in contact along a planar interface (Fig. 1.24a), and for the sake of definiteness consider the reaction between a divalent oxide AO and a trivalent oxide B_2O_3 in a 1 : 1 molar ratio to form a product having the spinel structure (see §7.2.5.6 — the archetype mineral spinel has the composition $MgAl_2O_4$). The reaction is thus:

$$AO + B_2O_3 \rightarrow AB_2O_4. \tag{1.26}$$

Nucleation is, in general, difficult because product and reactants have different structures and hence considerable structural rearrangement of the lattice of the reactants is required to form nuclei of the product. This reconstructive nucleation is obviously energetically costly, and hence will only occur to a significant extent at elevated temperatures. Nucleation is made easier, however, if the product and one of

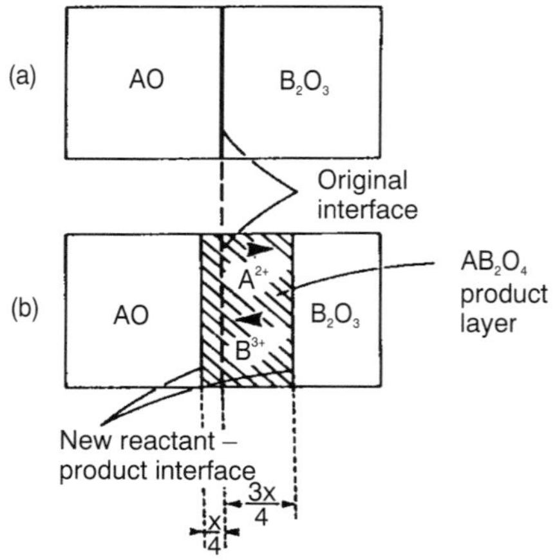

Fig. 1.24 Schematic illustration of the progress of a solid-state reaction occurring by cationic interdiffusion, for example between single crystals of a divalent oxide AO and a trivalent oxide, B_2O_3, forming a spinel material AB_2O_4. (*a*) before reaction occurs; (*b*) after growth of the product layer, of thickness Δx.

the reactants have very similar structures, in terms of both the atomic arrangement and the bond lengths involved. This is the case, for example, for MgO and spinel, $MgAl_2O_4$ (although not for alumina and spinel): both have a cubic array of oxide ions (see §2.2.4). Thus, nuclei formation at the MgO interface is favoured since the oxide-ion arrangement (although obviously not the cation arrangement) can be effectively continuous across the reactant–product interface of a nucleus. Nucleation is particularly favoured when the structural matching of the two materials is so precise (say, bond lengths the same to within 10–15%) that epitaxial (or epitactic) nucleation occurs (§1.1.1), where there is structural registry in the plane of the interface, or particularly when topotactic nucleation occurs, where the structural registry also extends into the reactant and product nucleus in a direction normal to the interface.

Once nuclei of the product have formed, growth of the product layer is generally limited by the diffusion of ions from reactant, through the product layer, to the reaction front. In the case of spinel formation, for example, there are *two* reaction fronts, one at the $AO - AB_2O_4$ interface and the other at the $B_2O_3 - AB_2O_4$ interface (Fig. 1.24b); this is an example of the Wagner reaction mechanism. In these systems, cation (A^{2+}, B^{3+}) diffusion is dominant, being faster than that of oxygen, and hence there is a *counterdiffusion* of the two cation species, A^{2+} from AO to B_2O_3, and B^{3+} from B_2O_3 to AO. Thus, the reactions occurring at the two interfaces may be written as

$$2B^{3+} + 4AO - 3A^{2+} \rightarrow AB_2O_4 \quad (AO - AB_2O_4 \text{ interface}), \tag{1.27a}$$

$$3A^{2+} + 4B_2O_3 - 2B^{3+} \rightarrow 3AB_2O_4 \quad (B_2O_3 - AB_2O_4 \text{ interface}), \tag{1.27b}$$

giving the overall reaction as in (1.26):

$$4AO + 4B_2O_3 \rightarrow 4AB_2O_4,$$

where the positive and negative signs for the cations in reactions (1.27) indicate their arrival or departure, respectively, from the particular interface.

Note that, from reactions (1.27), it is apparent that the $B_2O_3 - AB_2O_4$ interface should move at *three* times the rate of the $AO - AB_2O_4$ interface (Fig. 1.24b); this differential movement of diffusion fronts is known as the Kirkendall effect (§3.4.2.1). However, at the temperature ($\simeq 1500$ °C) at which reaction for (Mg, Al) spinel naturally takes place, a range of $(MgO)_x(Al_2O_3)_{1-x}$ solid solutions exist, with $0.5 < x < 0.6$, i.e. corresponding to formulae ranging from $Mg_{0.73}Al_{2.18}O_4$ to $MgAl_2O_4$. Thus, the product formed at the MgO–spinel interface will be the most magnesium-rich, i.e. the stoichiometric composition with $x = 0.5$, but that formed at the interface with Al_2O_3 will be the most aluminium-rich, i.e. $x \simeq 0.4$. Hence, the two reaction interfaces do not move with rates having the predicted ratio 1: 3; for the reaction between MgO and Fe_2O_3, for example, the growth-rate ratio has been found to be 1: 2.7.

For the growth stage of spinel formation, as for many solid-state reactions (including tarnishing — §1.1), interdiffusion of ions through the product layer is the rate-controlling step. In such a case, the thickness Δx of the product layer increases *parabolically* with time, $(\Delta x)^2 \propto t$. Diffusion of atoms in solids is almost invariably mediated by structural defects (e.g. atomic vacancies, etc. — see §3.4.2). There will be a linear concentration gradient of defects in the product layer under steady-state conditions, i.e. the defect concentration will be inversely proportional to Δx. The defect, and concomitantly ion, flux (the number of diffusing species passing through unit area per unit time) is proportional to the concentration gradient (§3.4.2.1), and since the rate of change of thickness of the product layer, $d(\Delta x)/dt$, is proportional to the ion flux, the growth rate can therefore be written as:

$$\frac{d(\Delta x)}{dt} = \frac{k}{\Delta x}, \tag{1.28}$$

where k is a rate constant. This rate law can be integrated to give parabolic growth kinetics:

$$(\Delta x)^2 = 2kt. \tag{1.29}$$

This parabolic growth law has been confirmed experimentally, for example for the case of the formation of the spinel $NiAl_2O_4$ from NiO and Al_2O_3 (Fig. 1.25).

The rate of solid-state reactions also depends on the area of contact between the solid reactants, i.e. on their surface areas. The surface area of a given mass of solid can be increased enormously simply by making the particle size smaller (Problem 1.8). Very fine powders can be produced most simply, in general, by precipitation from solution (§1.2.3); such a method also has the advantage that the reactant particles are intimately mixed, thereby ensuring homogeneity of the final product of the ensuing solid-state reaction. However, solid-state reactions involving powders invariably produce solid products having pore defects.

In some solid-state processes, nucleation can be homogeneous, i.e. it can take place throughout the volume of the solid. This is often the case for thermal decomposition of solids, for example the reaction:

$$CaCO_3(s) \rightarrow CaO(s) + CO_2(g) \tag{1.30}$$

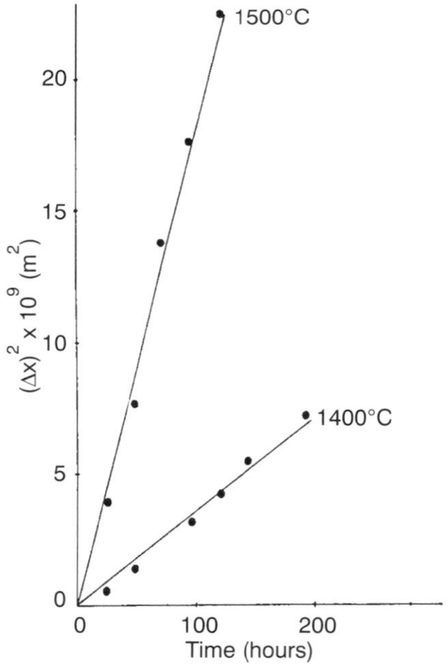

Fig. 1.25 Parabolic-growth kinetics for the formation of the spinel $NiAl_2O_4$ from NiO and Al_2O_3. (After Pettit *et al.* (1966). Reproduced by permission of American Ceramic Society)

Suppose that in a solid there are N potential nucleation sites per volume at which thermal decomposition may occur, each of which has an equal *a priori* probability of becoming an actual nucleus. If n is the concentration of nuclei already present, then

$$\frac{dn}{dt} = k(N - n), \tag{1.31}$$

where k is a rate constant or, on integration:

$$n = N[1 - \exp(-kt)]. \tag{1.32}$$

For small values of kt, the exponential in eqn. (1.32) can be expanded, giving a linear nucleation law:

$$n = kNt. \tag{1.33}$$

Assuming that the nuclei are spherical in shape, and that they grow with a constant linear velocity v, and that no new nuclei can grow in those regions where reaction has already occurred, then it can be shown that the fractional volume extent of the decomposition, α, is given by (see Problem 1.9):

$$\alpha = [1 - \exp(-\beta t^4)], \tag{1.34}$$

where

$$\beta = \frac{\pi}{3} v^3 kN. \tag{1.35}$$

Equation (1.34) is a particular example of the general, empirical Avrami–Erofeev equation:

$$\alpha = [1 - \exp(-\beta t^n)]. \tag{1.36}$$

Further details on solid-state reactions are given in Schmalzried (1981).

1.3.2 High-pressure preparation and synthesis

The application of high pressure (often accompanied also by high temperature) to a solid can cause a transition to a different phase of the material, having a higher density and, often, higher coordination numbers of the atoms involved. Although such high-density phases can be thermodynamically stable only at high pressures (and temperatures), nevertheless they can be kinetically metastable under ambient conditions if pressure-quenched (i.e. decompressed, and cooled, rapidly), since the reconstructive transformation that must take place for the high-pressure phase to revert to the low-pressure form is kinetically hindered at low temperatures where there is insufficient thermal energy available to initiate the necessary bond-breaking.

An example of high-pressure solid-state preparation is the transformation of the graphite modification of carbon (the most stable form at ambient pressure and temperature) to diamond: the coordination of the carbon atoms increases from three to four, and the density increases from 2.25×10^3 to 3.52×10^3 kg m^{-3}. The pressure-temperature phase diagram for these two polymorphic forms of carbon, together with the liquid phase, is shown in Fig. 1.26. Typical transformation conditions for the production of synthetic diamond are 13 GPa and 3000 °C; even at this elevated temperature, however, the transformation is sluggish.

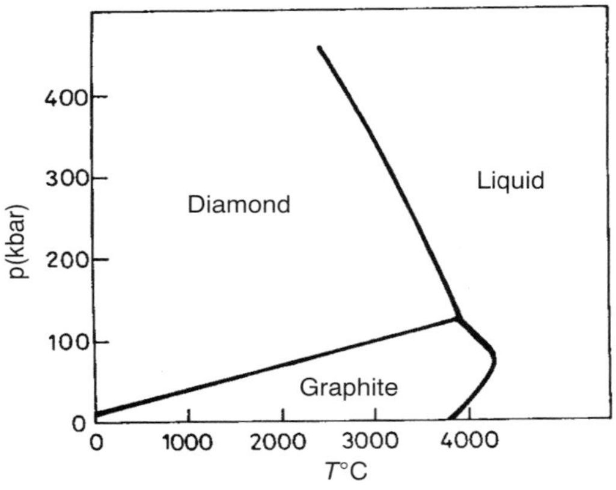

Fig. 1.26 Pressure–temperature phase diagram for carbon. (After Bundy (1963). Reprinted with permission from *J. Chem. Phys.* **38**, 631. © 1963 American Institute of Physics)

Other examples of pressure-induced transformations include that between the quartz and stishovite modifications of silica (SiO_2), typically occurring at 12 GPa and 1200 °C: quartz has 4: 2 coordination of silicon and oxygen, respectively, whereas stishovite (having the rutile structure — §2.2.4.2) is characterized by 6: 3 coordination. Potassium chloride (KCl), having the rocksalt structure (§2.2.4.2) and 6: 6 coordination at ambient conditions, transforms to the caesium chloride structure (§2.2.4.2) with 8 : 8 coordination, at a typical pressure of 2 GPa at ambient temperature. Cadmium sulphide, with the wurtzite structure (§2.2.4.2) and 4 : 4 coordination, is transformed to the rocksalt structure at 3 GPa at ambient temperature.

Occasionally, high-pressure synthesis can be employed to synthesize solids that are not stable at ambient conditions; again, pressure quenching is used to obtain the high-pressure form in a metastable state at ambient conditions. An example is the synthesis of the distorted perovskite structure of $PbSnO_3$:

$$SnO_2(s) + Pb_2SnO_4(s) \underset{400\,^\circ C}{\overset{>7 GPa}{\longrightarrow}} 2PbSnO_3(s). \tag{1.37}$$

1.3.3 Glass-ceramics

A broad definition of a ceramic is a solid compound of metallic elements and non-metallic elements (commonly oxygen) in which *ionic* bonding is predominant (see §2.2.4). In terms of this definition, inorganic (non-metallic) glasses form a sub-set of ceramic materials. However, the term glass-ceramic is normally reserved for those materials in which a glass has been controllably devitrified (crystallized) to give a *fine-grained* polycrystalline microstructure. The size of the crystallites in a glass-ceramic is typically 0.1 μm; for comparison, crystallite sizes in conventional ceramics are much larger, e.g. $\simeq$ 10 μm for alumina and 40 μm for porcelain (heat-treated ('fired') aluminosilicate).

Heating a normal glass to temperatures comparable to the crystallite growth temperature (see Fig. 1.14) in general does not produce a glass-ceramic as the crystallized product: the resulting crystallites are much too large on average and have too wide a range of sizes. This is because crystallization in this case occurs at relatively few (heterogeneous) nucleation centres, generally distributed on the glass interfaces. In order to prepare a glass-ceramic, a large number of extrinsic nucleation centres (nucleants) must be distributed homogeneously through the matrix when preparing the glass precursor. This 'seeding' can be achieved by incorporating, say, thermally unstable metal oxides that decompose to give a colloidal suspension of very small metal particles during the melting, prior to vitrification; alternatively, metals such as Cu or Pt at a concentration of $\simeq 0.05\ \%$, when dissolved in the melt, form colloidal metal dispersions. In addition, certain oxides, notably TiO_2, can induce phase separation in the melt, leading to an extremely fine dispersion. All these forms of structural inhomogeneity, dispersed uniformly throughout a melt-quenched glass, can act as heterogeneous nucleation centres.

In order to form a glass-ceramic, a seeded glass is subjected to two thermal treatments. The first, at a lower temperature, T_{nuc}, nucleates crystallites in the seeded glasses; the second, at a higher temperature, T_{gr}, causes the growth of the crystal nuclei (cf. Fig. 1.14). Most of the volume of the glass is crystallized, but a small proportion of the glassy phase persists in the interstices between crystallites. Since the viscosity of the glass is greatly decreased at elevated temperatures, e.g. T_{gr}, the glass tends to flow into the interstices, resulting in near-zero porosity for glass-ceramics, in contrast to the appreciable porosity of conventional (aluminosilicate) ceramics.

One common family of glassy materials used to produce glass-ceramics is $Li_2O - Al_2O_3 - SiO_2$, which crystallizes to give β-eucryptite ($Li_2O \cdot Al_2O_3 \cdot 2SiO_2$) and β-spodumene ($Li_2O \cdot Al_2O_3 \cdot 4SiO_2$). These glass-ceramics have very low thermal-expansion coefficients and are used for cookware, astronomical-telescope mirror supports, etc. Another family is glassy $MgO - Al_2O_3 - SiO_2$, which crystallizes to cordierite ($2MgO \cdot 2Al_2O_3 \cdot 5SiO_2$) and solid solutions of β-quartz, producing high-strength, electrically insulating glass-ceramics. Further details of glass-ceramics are given in McMillan (1979).

Problems

1.1 Estimate how long it takes to form a monolayer of nitrogen contaminant on a clean surface at 300 K for background nitrogen pressure corresponding to:
(a) high vacuum (10^{-5} Pa);
(b) ultra-high vacuum (10^{-9} Pa).

1.2 What microstructure might be expected to occur in a thin film produced by thermal evaporation at an *oblique* angle of incidence onto a cooled substrate?

1.3 Suggest a reaction scheme that would lead to separation of Cu_2O from Cu or CuO using HCl as a transporting agent.

1.4 Explain why, although the reaction between molten aluminium and sulphur proceeds very slowly at 800 °C, the addition of iodine results in the ready formation of crystals of Al_2S_3 at a cooler place.

1.5 (a) By considering the changes in free energy accompanying the formation from a liquid of crystallite nuclei, assumed to be spherical in shape for simplicity, show that there is a critical radius, r_c, below which 'embryos' are unstable and redissolve and above which nuclei are stable and can grow, given by $r_c = 2\sigma/\Delta g$, where σ is the surface (interfacial) energy per unit area, and Δg is the difference in free energy per unit volume between liquid and solid. Hence, show that the free-energy (thermodynamic) barrier to nucleation, associated with the formation of critical-sized embryos is $\Delta G^* = 16\pi\sigma^3/3(\Delta g)^2$. By making suitable approximations, show further that $\Delta G^* \simeq 16\pi\sigma^3 T_m^2 V_m^2/3(\Delta T)^2(\Delta H_f)^2$, where T_m is the melting temperature, ΔH_f the latent heat of fusion, V_m the molar volume of the liquid and ΔT the degree of undercooling of the melt below T_m.

(b) By considering also the rate at which atoms attempt to cross the liquid–embryo interface, governed by the *kinetic* barrier to nucleation, ΔG_a, show that the rate of nucleation is given by $I = k\ \exp(-\Delta G_a/RT)\exp(-\Delta G^*/RT)$. Hence account for the form of the temperature dependence of I shown in Fig. 1.14.

1.6 For the eutectic binary phase diagram for solute A and flux B shown in Fig. 1.16a, where the eutectic composition, E, is, say, $A_{0.33}B_{0.67}$, what is the fractional amount of liquid at temperatures T_x and the eutectic temperature, T_E, for the composition X, say, $A_{0.8}B_{0.2}$. (Hint: use a 'lever rule' for fractional amounts, derived from the principle of moments.) What happens to the concentration of the liquid, starting at X, as the melt is cooled from T_x to T_E?

1.7 Essay: Compare and contrast the various techniques for preparing single crystals from the liquid phase. What are the advantages and disadvantages in each case?

1.8 What is the surface area in each case of the same volume of a material existing as:
(a) a single cubic crystal of side 10^{-2} m;
(b) a powder of cube edge 10 μm;
(c) an ultra-fine precipitate of cube edge 10 nm?

1.9 Derive the thermal-decomposition rate law, eqn. (1.34), for spherical nuclei, the number of which increases linearly with time (eqn. (1.33)), and which grow at a constant velocity, v.

1.10 Essay: Explain the advantages that '*chimie douce*' has over conventional solid-state reactions in the synthesis of materials, giving examples of its application.

References

Barrer, R. M., 1982, *Hydrothermal Chemistry of Zeolites* (Academic Press: London).
Behrisch, R., ed., 1981, *Sputtering by Particle Bombardment, I*, Topics Appl. Phys. **47** (Springer-Verlag: Berlin).
Behrisch, R., ed., 1983, *Sputtering by Particle Bombardment, II*, Topics Appl. Phys. **52** (Springer-Verlag: Berlin).
Behrisch, R., ed., 1991, *Sputtering by Particle Bombardment, III*, Topics Appl. Phys. **64** (Springer-Verlag: Berlin).
Brice, J. C., 1973, *The Growth of Crystals from Liquids*, in Selected Topics in Solid State Physics, ed. E. P. Wohlfarth (North-Holland: Amsterdam).
Bundy, F. B., 1963, *J. Chem. Phys.* **38**, 631.
Dyer, A., 1988, *An Introduction to Zeolite Molecular Sieves* (Wiley: Chichester).
Kresge, C. T., Leonowicz, M. E., Roth, W. J., Vartuli, J. C. and Beck, J. S., 1992, *Nature* **359**, 710.
McMillan, P. W., 1979, *Glass Ceramics*, 2nd edn (Academic Press: London).
Pettit, E. H., Randklev and Felten, E. J., 1966, *J. Am. Cer. Soc.* **49**, 199.
Plass, C., Heinecke, H., Kayser, O., Lüth, H. and Balk, J., 1988, *J. Cryst. Growth* **88**, 455.
Roy, B. N., 1992, *Crystal Growth from Melts* (Wiley: Chichester).
Schmalzried, H., 1981, *Solid State Reactions* (Verlag Chemie: Weinheim).
Smart, L. and Moore, E., 1992, *Solid State Chemistry* (Chapman and Hall: London).
Sun, X.-D., Gao, C., Wang, J. and Xiang, X.-D., 1997, *Appl. Phys. Lett.* **70**, 3353.
Thess, A., Lee, R., Nikolaev, P., Dai, H., Petit, P., Robert, J., Xu, C., Lee., Y. H., Kim, S. G., Rinzler, A. G., Colbert, D. T., Scuseria, G. E., Tomanek, D., Fischer, J. E. and Smalley, R. E., 1996, *Science* **273**, 483.
Whittingham, M. S., 1996, *Current Opinion in Solid State and Materials Science* **1**, 227.
Yoshida, A., Inoue, K., Ohashi, H. and Saito, Y., 1990, *Appl. Phys. Lett.* **57**, 484.

CHAPTER
2 Atomic Structure and Bonding

Most properties of solids are determined by their atomic arrangement, i.e. structure. Thus, in order properly to understand the physical behaviour exhibited by solid materials, it is essential at the outset to gain an understanding of their atomic structure. This chapter, therefore, consists of a classification scheme for solids, followed by a discussion of the different types of structures adopted by solids, and a brief survey of symmetry aspects. A discussion is then given of the different forms of empirical and exact quantum-mechanical interatomic potentials used in computer simulations of the structure of materials. Finally, a survey is given of various experimental techniques used in structure determination.

Most texts on the solid state commence with a discussion of the atomic structure. This book differs from most in the type of materials included for discussion (e.g. quasicrystals and fractals), the inclusion of a section on computer simulation (an approach that is becoming increasingly more useful with the introduction of ever more powerful computing facilities) and the discussion of experimental techniques for structure determination (such as X-ray absorption spectroscopy and holography) that normally do not feature in solid-state physics textbooks.

Order and disorder 2.1

2.1.1 Classification of solids

In principle, all solid materials may be classified, in structural terms, into one of three distinct categories:

- crystals
- fractals
- amorphous solids

Crystals are the most structurally ordered of solids. This order is associated with the existence of translational periodicity, whereby an atom, or group of atoms, forming the unit cell, when translationally repeated in a periodic fashion in all vector directions defining the dimensional space in which the crystal is defined, thereby generates the crystal (see Fig. 2.1). Normally, of course, the embedding space is three-dimensional, but in the example shown in Fig. 2.1 it is two-dimensional for ease of illustration. The embedding space may also have a dimensionality *greater* than three; although of mathematical interest, such hyper-space crystals are of no physical interest except in the case of *quasicrystals* (see below).

Another distinct structural class of materials comprises fractals. These structures are characterized by being self-similar, that is the structure looks identical at all length scales. An everyday culinary example is the cauliflower: the overall shape (morphology) is superficially similar to that of the major florets forming the whole cauliflower, in turn the same as the individual florets making up the major florets etc. A two-dimensional mathematical example of a fractal structure is the Sierpinski gasket (see Fig. 2.2), but there are many others (see e.g. Mandelbrot (1982), Peitgen *et al.* (1992)). Real materials exhibiting fractal structures are silica aerogels (very low-density materials, 0.1 g cm^{-3})

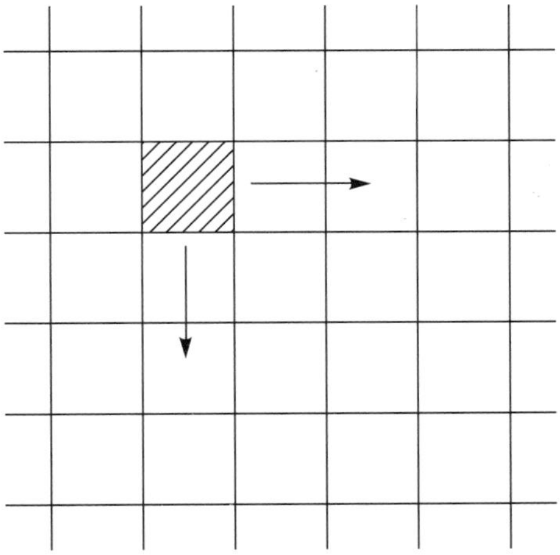

Fig. 2.1 A two-dimensional crystal generated by the translationally periodic displacement of a unit cell (in this case, a square).

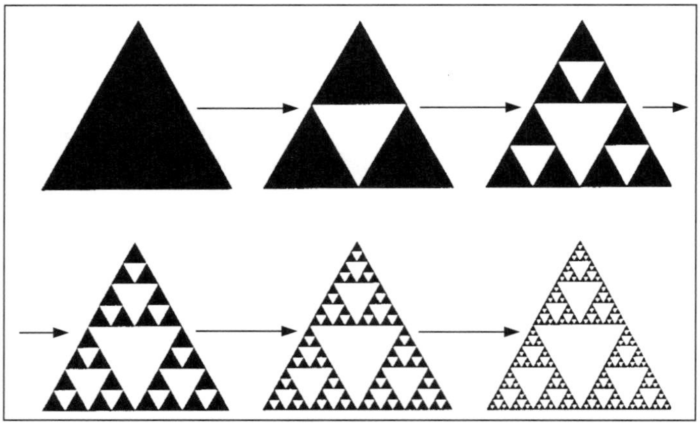

Fig. 2.2 A two-dimensional example of a fractal structure: the Sierpinski gasket. The basic steps to construct the Sierpinski gasket are shown: at a given step, each black triangle has an inverted triangular hole (one quarter the size) inserted, touching at the midpoints of the edges of the larger triangle.

made by gelling a silica sol under the appropriate conditions; the fractal structure in this can be envisaged as an ever-branching tree comprised of silica particles.

Amorphous (or, synonymously, non-crystalline) materials form the third generic class of materials. Such materials are devoid of both the long-range order (translational periodicity) characteristic of crystals and the self-similarity characteristic of fractals. Their structure, as will be seen later, can only be described in a statistical sense. This is not to say, however, that amorphous materials exhibit no structural order whatsoever; i.e. they are not necessarily completely random structurally. In many cases, particularly for those amorphous materials in which directed (i.e. covalent) bonding is prevalent, the short-range order, characterized by the nearest-neighbour bond length and coordination number and bond angle, quantities defining local coordination polyhedra, is relatively well-defined, with only relatively small statistical fluctuations. However, much greater fluctuations characterize the connection of such coordination polyhedra, with the result that structural order is destroyed for distances beyond that corresponding to a few coordination polyhedra (see Fig. 2.3). The resulting structure for an amorphous covalent material is termed a continuous random network.

It should be stressed that the above three general categories of structures of solids represent, in many cases, idealized situations. Thus, perfect single crystals are mathematically describable but physically unrealizable. Real crystals inevitably contain structural defects (see Chapter 3) or extrinsic impurities that destroy the complete translational periodicity characteristic of perfect crystallinity; note, however, that the disorder in this case is much more heterogeneous and localized than the more homogeneous disorder characteristic of amorphous solids (see Fig. 2.3a). Even the surfaces of finite-sized, otherwise structurally perfect crystals cause the destruction of infinite periodic structural order.

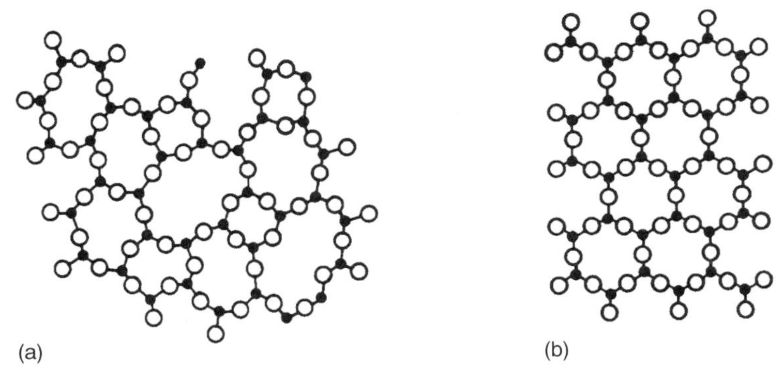

Fig. 2.3 (a) A two-dimensional representation of an amorphous, covalently bonded structure, in which one type of atom (•) is trigonally coordinated and another (o) is two-fold coordinated. A real example might be B_2O_3. (b) The corresponding crystalline modification of the amorphous structure shown in (a).

For the case of fractal structures, self-similarity in real materials only extends over a certain range of distances. The lower limit is determined by the size of the 'bead' from which the fractal structure is generated; obviously, the size of the bead cannot be less than that of an individual atom. The upper limit of the self-similarity length scale is determined by the onset of structural homogeneity and isotropy (often necessary to ensure the mechanical integrity of a fractal structure in which the density would otherwise decrease with increasing distance).

An interesting class of materials are the so-called quasicrystals (see e.g. Jaric (1988), Janot (1995)) which appear, at first sight, to fall into none of the above three generic structural categories. Such materials exhibit 'icosahedral' order, evident in the five-fold symmetry of the sharp diffraction spots comprising the electron diffraction patterns (see §2.6.1), first observed in a rapidly quenched alloy of $MnAl_6$ (Shectman *et al.* (1984)), or the icosahedral morphology of the growth habit exhibited by more stable quasicrystalline alloy materials (e.g. AlCuFe or Al_6CuLi_3) for which slow cooling of the melt can produce single quasicrystals (Fig. 2.4). The significance of the discovery of these materials is that no three-dimensional periodic lattice can have icosahedral symmetry, or any other point-group symmetry containing five-fold (or seven-fold or higher) rotations (see §2.3.2). Thus, these materials are *aperiodic*, yet simultaneously exhibit *orientational order* (like conventional crystals), albeit with five-fold symmetry. A further conundrum is that the diffraction patterns of such materials consist of an arrangement of sharp spots: conventional crystallographic wisdom has it that sharp diffraction spots are characteristic only of single, translationally periodic crystals (see §2.6.1). The structure of quasicrystals can be understood, and the above paradoxes resolved, by noting that an aperiodic quasicrystalline structure in three dimensions (3D) is simply the projection onto 3D of a *periodic* (hypercubic) structure in 6D space (see e.g. Bak and Goldman (1988)). A two-dimensional representation of quasicrystalline order is given by the 'Penrose tiling' (Penrose (1974))—see Fig. 2.5—in which an aperiodic structural arrangement, exhibiting local five-fold rotation symmetry, can be achieved by an

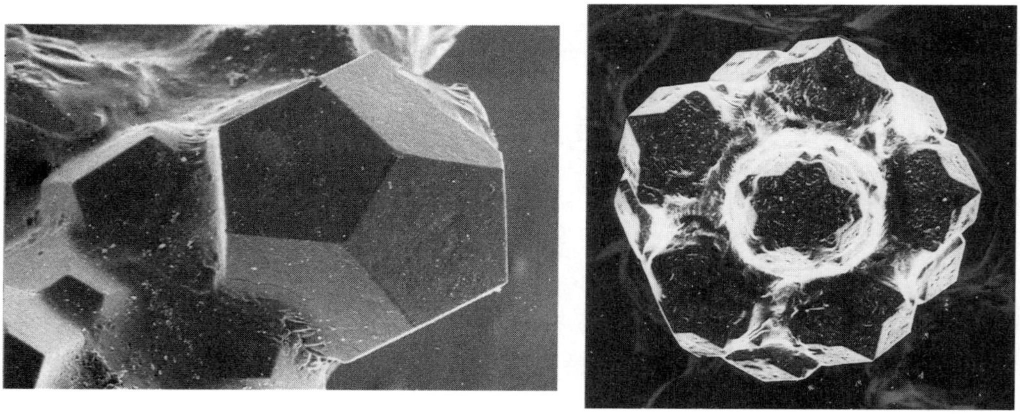

(a) (b)

Fig. 2.4 Photographs of quasicrystals showing their characteristic five-fold symmetry: (a) a dodecahedron of AlCuFe; (b) Al_6CuLi_3, showing 12 triacontahedra centred at the vertices of an icosahedron. (Courtesy of Dr. M. Audier.)

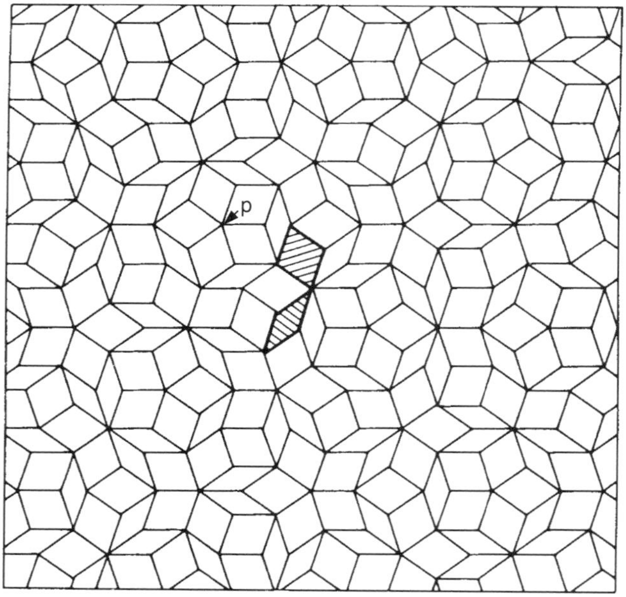

Fig. 2.5 A 2D Penrose tiling illustrating quasicrystalline order: the structure is aperiodic, yet has local five-fold rotation symmetry (see point P). The two types of rhombi (fat and thin) comprising the tiling are shaded.

appropriate arrangement of 'fat' and 'thin' rhombi, having acute angles of 72° and 36°, respectively (note 360°/5 = 72°). The Penrose tiling shown in Fig. 2.5 is, in fact, an appropriate projection of a 5D hypercubic lattice onto 2D (Whittaker and Whittaker (1990)).

2.1.2 Lattices and unit cells

The description of the structure of (ideal) crystals is greatly simplified by the presence of translational periodicity. For further simplification, the discussion will be restricted to the case of three-dimensional objects (since these, of course, make up the vast majority of structures of interest to the solid-state physicist or chemist or materials scientist); however, the extension of crystallography to higher dimensions (of use in describing structures of quasicrystals, for example) is moderately straightforward (Bak and Goldman (1988), Janot (1995)).

A 3D perfect crystal can be generated by decorating each point in an appropriate lattice with the same basis in each case (a basis being an atom or a group of atoms)—see Fig. 2.6. A lattice is an infinite array of mathematical points in 3D space having the translational periodicity characteristic of the crystal. It is defined in terms of three

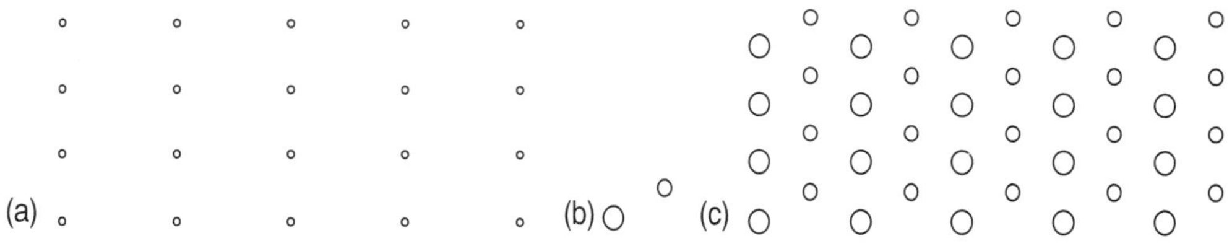

Fig. 2.6 2D illustration of the relationship between (a) a lattice; (b) the atomic basis; (c) and the actual crystal structure.

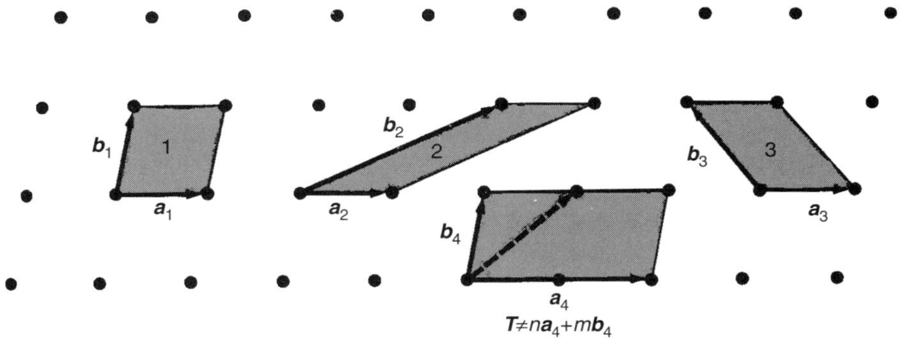

Fig. 2.7 Different ways of choosing translation vectors for a 2D lattice. All except ($\boldsymbol{a}_4$, $\boldsymbol{b}_4$) are primitive vectors.

fundamental translation vectors, $\boldsymbol{a}$, $\boldsymbol{b}$, $\boldsymbol{c}$: two lattice points are thus connected by a translation vector of the form

$$\boldsymbol{R} = u\boldsymbol{a} + v\boldsymbol{b} + w\boldsymbol{c}, \tag{2.1}$$

where u, v and w are integers. There are many ways of choosing the translation vectors for a given lattice (see Fig. 2.7), but those for which eqn. (2.1) is satisfied are said to be primitive. The geometric object defined by the primitive axes (a parallelpiped in 3D) is termed the unit cell, and this has a volume (in 3D) given by:

$$V_c = \boldsymbol{a} \cdot (\boldsymbol{b} \times \boldsymbol{c}). \tag{2.2}$$

Different types of unit cell can be envisaged, depending on the relative disposition of lattice points within the cell. A cell containing lattice points only at the corners (Fig. 2.8a) is termed primitive (P); others include additional lattice points at the body centre (body-centred, I—Fig. 2.8b), at the centre of opposite faces (side-centred, e.g. C—Fig. 2.8c), and at the centre of all faces (face-centred, F—Fig. 2.8d).

There are just fourteen 3D crystal lattices with different symmetries, the so-called Bravais lattices. These may be further subdivided into *seven* crystal systems, each having one or more different types of unit cell, as indicated in Table 2.1. The relationships between the unit vectors $\boldsymbol{a}$, $\boldsymbol{b}$, $\boldsymbol{c}$ and the intervector angles α, β, γ defining the various unit cells are shown in Fig. 2.9.

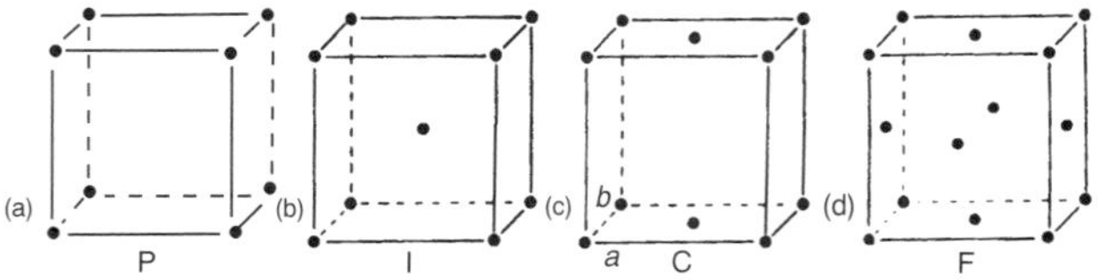

Fig. 2.8 Different types of unit cells: (a) primitive (P); (b) body-centred (I); (c) side-centred, in the $\boldsymbol{c}$ direction (C); (d) face-centred (F).

Table 2.1 The seven 3D crystal systems and their unit cells

Crystal system	*Unit-cell coordinates*	*Symmetry*	*Bravais lattices*
Cubic	$a = b = c; \alpha = \beta = \gamma = 90°$	Four 3-fold axes	P, F, I
Tetragonal	$a = b \neq c; \alpha = \beta = \gamma = 90°$	One 4-fold axis	P, I
Orthorhombic	$a \neq b \neq c; \alpha = \beta = \gamma = 90°$	Three 2-fold axes or mirror planes	P, F, I, C (or A or B)
Hexagonal	$a = b \neq c; \alpha = \beta = 90°, \gamma = 120°$	One 6-fold axis	P
Trigonal*	$a = b \neq c; \alpha = \beta = 90°, \gamma = 120°$	One 3-fold axis	P
Monoclinic†	$a \neq b \neq c; \alpha = \beta = 90°, \gamma \neq 90°$	One 2-fold axis or mirror plane	P, C
Triclinic	$a \neq b \neq c; \alpha \neq \beta \neq \gamma$	None	P

* It should be noted that there is an alternative definition of the trigonal cell, i.e. $a = b = c, \alpha = \beta \neq 90°$, but with a rhombohedral (R) unit cell. Note also that some sources (particularly American) subsume the trigonal system into the hexagonal system, making six crystal systems in total.
† Note that an equivalent definition of the monoclinic cell, favoured by crystallographers, has $\alpha = \gamma = 90°, \beta \neq 90°$.

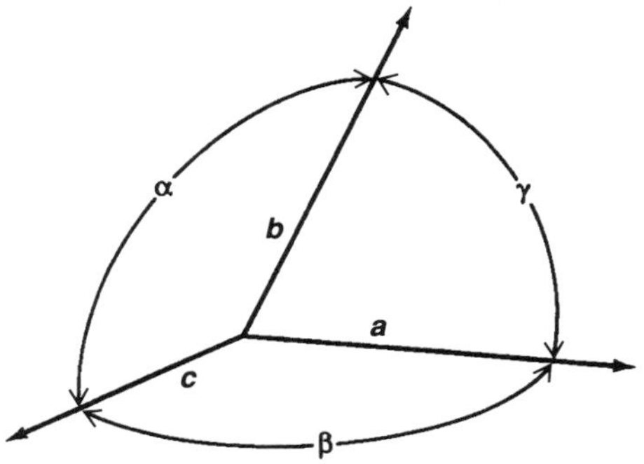

Fig. 2.9 Relationship between unit-cell vectors ***a***, ***b***, ***c*** and the intervector angles α, β, γ defining the coordinates of the seven crystal systems.

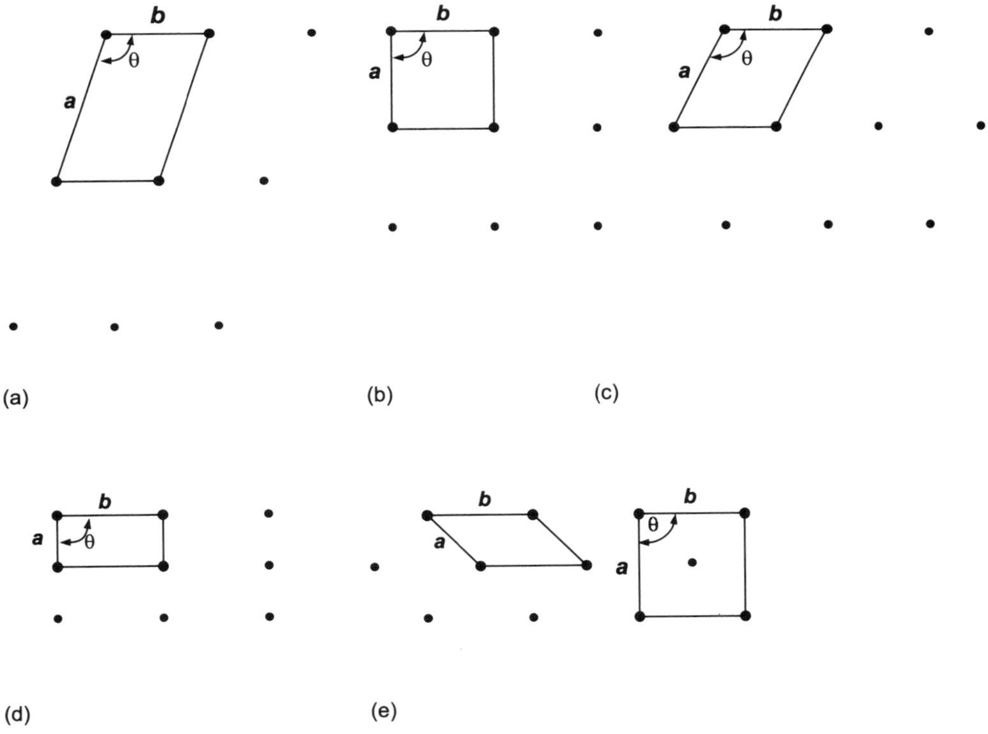

Fig. 2.10 2D Bravais lattices: (a) Oblique lattice; (b) square lattice ($|a| = |b|; \theta = 90°$); (c) hexagonal lattice ($|a| = |b|; \theta = 120°$); (d) rectangular lattice ($|a| \neq |b|; \theta = 90°$); (e) centred rectangular lattice showing the primitive cell and the rectangular cell ($|a| \neq |b|; \theta = 90°$).

In 2D, there are only five Bravais lattices (Fig. 2.10). The one with the least symmetry is the oblique lattice with the intervector angle $\theta \neq 90°$ or $120°$ and sides $a \neq b$

(Fig. 2.10a). Restricting the angle θ to 90° or 120° imposes additional rotation or mirror symmetries, resulting in four other Bravais lattices (Fig. 2.10b–e).

2.1.3 Nomenclature for crystal directions and planes

It is often necessary to refer to a particular vector direction in a crystal, or to a particular plane of atoms, and for this a convention is needed which is general and valid for all crystal types.

For the case of directions, consider a vector z connecting a general point P and the unit-cell origin O (Fig. 2.11). The vector z makes a projection of u' on the **a** axis, v' on the $\boldsymbol{b}$ axis and w' on the $\boldsymbol{c}$ axis (Fig. 2.11). The three numbers u', v', w' are then each divided by their highest common denominator to reduce them to a set of smallest integers u, v, w: the direction of **z** is then denoted $[u\ v\ w]$. If the projection of z along a particular unit-cell axis is in the *negative* direction, then the integer involved is distinguished by placing a bar on top, e.g. $\bar{u}$. A *complete* set of equivalent directions in a crystal is denoted as $\langle uvw \rangle$ (note the use of angular, rather than square brackets). Thus, for the case of the cubic crystal system as an example, the complete set of cube-edge directions is $\langle 100 \rangle$, the face diagonals $\langle 110 \rangle$ and the body diagonals $\langle 111 \rangle$.

Atomic planes in crystals are labelled in a similar way to directions. Consider a general plane intersecting the unit-cell axes at $u'\boldsymbol{a}$, $v'\boldsymbol{b}$ and $w'\boldsymbol{c}$ (see Fig. 2.12). The reciprocal quantities $1/u'$, $1/v'$ and $1/w'$ are then transformed to the smallest three integers, h, k, l, having the same ratios (the Miller indices), and the plane is then denoted as $(h\ k\ l)$. Again, for a negative intersection of a plane with a unit-cell direction, the relevant index is distinguished by having a bar on top. Thus, the six faces of a cubic crystal are represented as (100), (010), (001), $(\bar{1}00)$, $(0\bar{1}0)$, $(00\bar{1})$. The complete set of planes of a particular type is denoted as $\{h\ k\ l\}$: the set of cube faces is thus $\{100\}$. The most important atomic planes in cubic crystals are shown in Fig. 2.13. However, it is important to understand that not all Miller planes coincide with actual atomic planes.

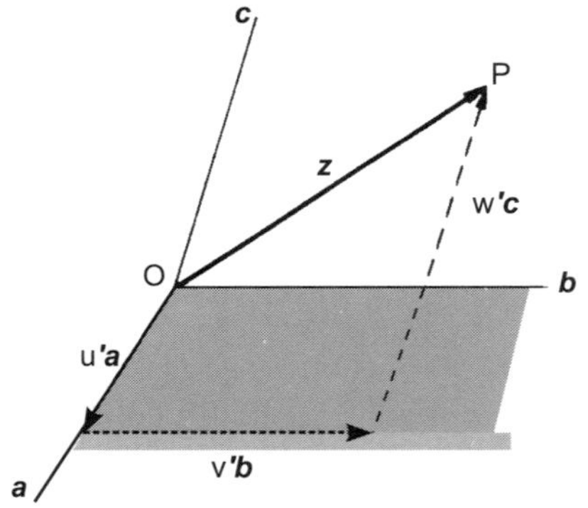

Fig. 2.11 Representation of directions in crystals. A vector z makes projections u', v', w' onto the unit- cell axes $\boldsymbol{a}, \boldsymbol{b}, \boldsymbol{c}$, respectively. Division of the set u', v', w' by the highest common denominator produces the set u, v, w: the direction is then denoted as $[uvw]$.

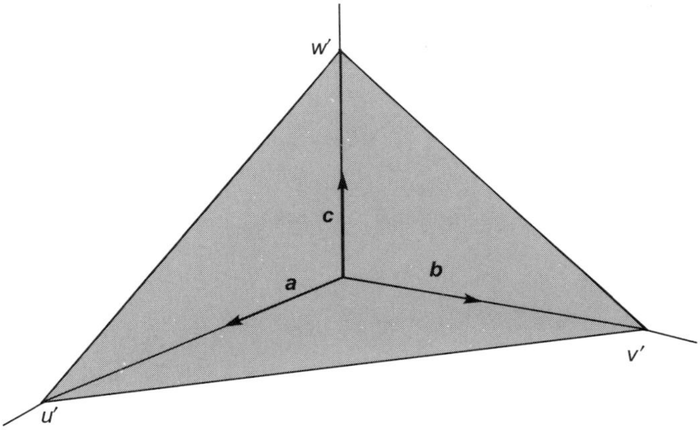

Fig. 2.12 Representation of atomic planes in crystals in terms of Miller indices. The plane intersects the unit-cell axes at u', v', w'. The reciprocal quantities $1/u', 1/v'$ and $1/w'$ are then converted into the smallest commensurate integers, hkl. The plane is denoted (hkl).

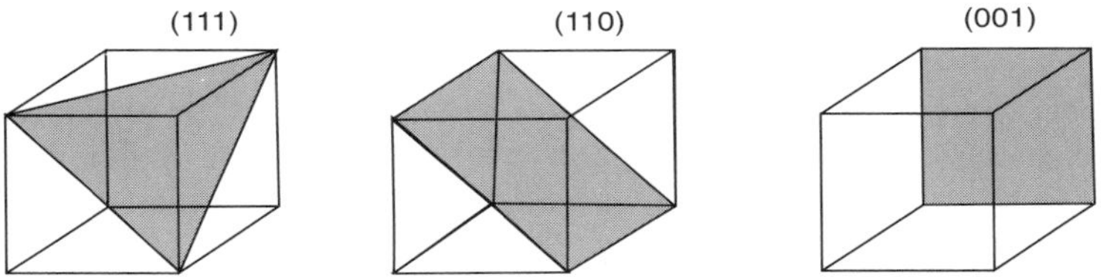

Fig. 2.13 Important atomic planes in the cubic crystal system.

Note that in cubic crystals (but not necessarily for other crystal systems), the direction $[hkl]$ is orthogonal to the plane (hkl). The use of particular types of brackets in different contexts should be noted.

The Miller-index system has the advantage that the *spacings* between adjacent planes may be calculated readily (such lattice-plane spacings determine the condition for diffraction of waves (e.g. X-rays) by crystals—see §2.6.1). Thus, for a cubic crystal with unit-cell dimension a, the spacing d_{hkl} between the planes $\{hkl\}$ is given by (see Problem 2.1b):

$$\mathrm{d}_{hkl} = a/(h^2 + k^2 + l^2)^{1/2}. \tag{2.3}$$

For the case of hexagonal crystals, it is conventional and convenient to use *four* Miller indices $(hkil)$ to characterize atomic planes, even though, strictly, one of the indices must be redundant. As before, the Miller indices are obtained by forming into the smallest integers the reciprocals of the intersections along four crystal axes. Three of the axes, $\boldsymbol{a}$, $\boldsymbol{b}$, $\boldsymbol{c}$, are taken to lie in the basal plane at 120° to each other; the fourth axis, **d**, is normal to the other three (Fig. 2.14). Thus, for example, the basal plane is denoted (0001) and the set of hexagonal faces is represented as $\{10\bar{1}0\}$.

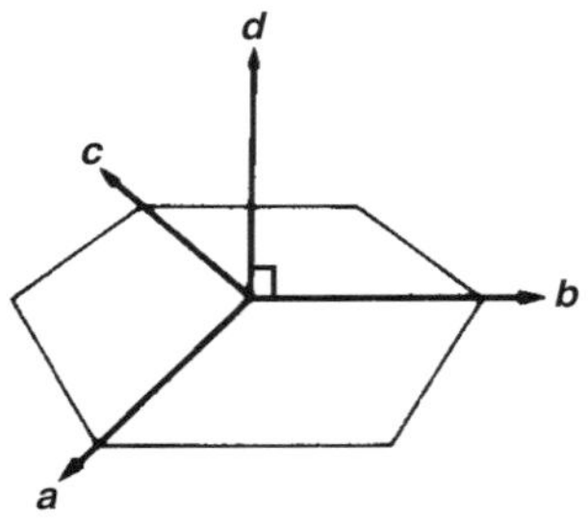

Fig. 2.14 Axes used to obtain the Miller indices (*hkil*) for the hexagonal lattice. The *d* axis is perpendicular to the basal plane.

2.1.4 Non-crystalline structures

Non-crystalline, or amorphous, structures are devoid of long-range translational periodicity, and hence many of the simplifications inherent in the structural description of crystals (e.g. in terms of a unit cell containing only a very few atoms periodically continued in space) are not applicable. Thus, the unit cell of an amorphous solid can be regarded, albeit not very usefully, as being infinite in extent. Because of the degree of

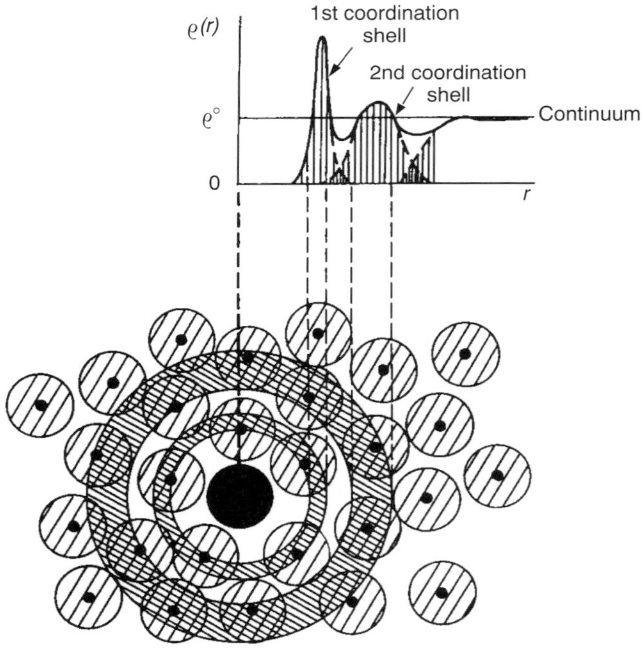

Fig. 2.15 Schematic illustration of the structural origin of the density function, $\rho(r)$, used to characterize the structure of non-crystalline materials.

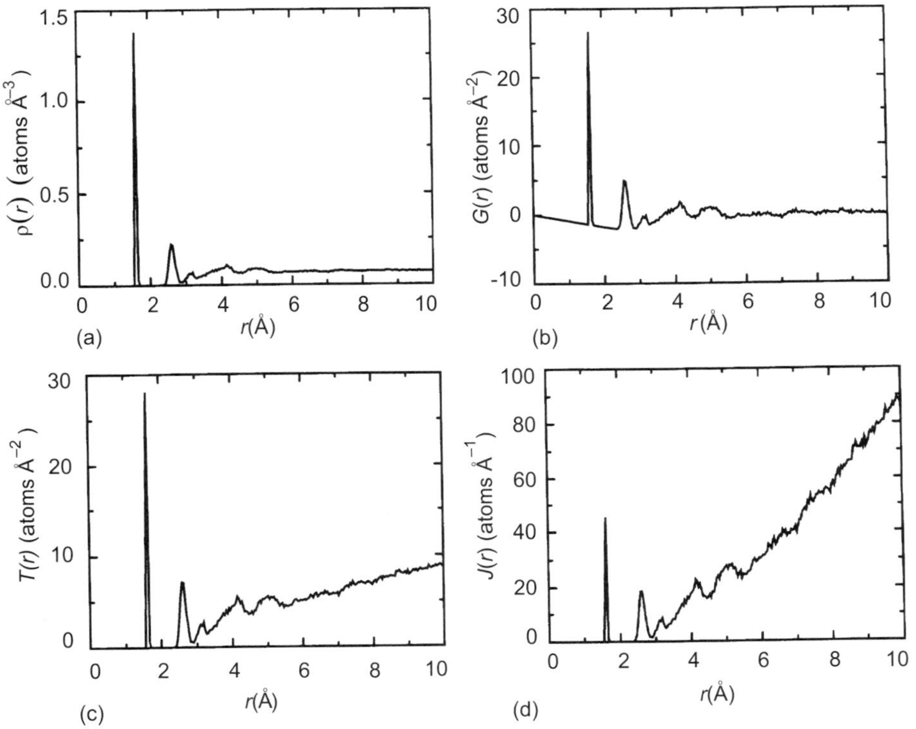

Fig. 2.16 Illustration of various real-space correlation functions used to characterize the structure of amorphous solids, in this case a model of vitreous SiO_2 (figure courtesy of Dr. S. Taraskin): with average atomic density $\rho^0 = 7.15$ Å^{-3}. (a) $\rho(r)$; (b) $G(r) = 4\pi r\rho(r) - 4\pi r\rho^\circ$; (c) $T(r) = 4\pi r\rho(r)$; (d) $J(r) = 4\pi r^2\rho(r)$.

randomness characterizing non-crystalline structures, a *statistical* description (in terms of spatial distributions or probability functions) is more appropriate. However, one simplifying feature which can be invoked in the case of amorphous materials is *orientational isotropy*: averaged over the whole solid, the structural environment at a given distance from a particular (type of) atom is the same in all directions. Thus, if the structure is described in the general case in terms of the density function $\rho(\boldsymbol{r})$, being the atomic density at a vector distance $\boldsymbol{r}$ from a given origin atom, in the case of amorphous materials, $\rho(\boldsymbol{r})$ becomes simply a one-dimensional function, $\rho(r)$, dependent only on the scalar distance r, as a consequence of orientational isotropy (see Fig. 2.15). A number of real-space structural probability (correlation) functions, all related to $\rho(r)$, are used to describe the structure of non-crystalline solids, and these are illustrated in Fig. 2.16. Of these, the most useful is the radial distribution function (RDF), defined as

$$J(r) = 4\pi r^2\rho(r). \tag{2.4}$$

Since the RDF is simply the average probability of finding an atom in the distance interval $\mathrm{d}r$ between distances r and $r + \mathrm{d}r$ from a given origin atom, the integral of the curve of $J(r)$ over a given peak, i.e. the area under the peak (say the first) gives the appropriate average atomic coordination number (e.g. the average nearest-neighbour coordination number).

Atomic packing and geometry 2.2

2.2.1 Types of interatomic bonding

The structure of a solid is determined by a balance of the attractive and repulsive forces acting between the atoms making up the solid. Structural parameters such as equilibrium bond lengths and angles, coordination numbers and atomic densities result from this balance of opposing forces. The cohesion of materials results essentially from the attractive electrostatic interaction between the negative charge of valence electrons and the positive charges of atomic nuclei. However, it is conventional to distinguish between different types of bonding, depending on the electron distribution in the solid. (Note that the gravitational attraction between atoms is much too small to contribute significantly to the bonding in solids—see Problem 2.3.)

Thus, if there is no charge transferred between atoms, and the charge distribution of each atom is spherically symmetric, the only attractive interatomic interaction that can exist is the so-called van der Waals (or London) interaction, resulting from induced electric dipole–dipole effects. Although relatively weak, this attractive interaction is responsible for the cohesion of rare-gas solids, such as crystalline argon, and many molecular solids made up of organic molecules. It is also present in all other types of solids but this interaction is often swamped by the other, stronger interactions operative in those cases. The attractive van der Waals interaction acts pairwise between atoms, and depends on the interatomic separation to the inverse sixth power: this r^{-6} dependence ensures that the interaction is short-ranged, acting essentially only between nearest-neighbour pairs of atoms. It is also centrosymmetric, i.e. there is no orientational dependence to the interaction. Thus, the cohesive energy of van der Waals solids is maximized by maximizing the atomic packing.

In the opposite extreme, where there is complete charge transfer between different atoms commensurate with their valencies, i.e. atoms become positively charged cations and negatively charged anions, the dominant attractive interaction between the ions is ionic in character, with the $1/r$ dependence characteristic of Coulomb electrostatic interactions. This ionic interaction is *long-ranged*; unlike the case of the van der Waals attractive interaction, the ionic force is *not* confined to nearest neighbours only, but extends to very distant neighbours, making the evaluation of Coulombic cohesive energies rather complicated (see §2.2.4.1). The ionic interaction is also centrosymmetric; thus the lowest energy structures are those with large numbers of conjugate ions around a given ion (subject, however, to the avoidance of near-neighbour like-atom pairings, which would greatly destabilize the structure owing to the very high Coulomb *repulsive* interactions involved).

Another scenario is when electropositive atoms become ionized, forming cations, but in the absence of other electronegative atoms (which would otherwise trap the liberated electrons, thereby forming conjugate ions, i.e. anions), the electrons instead form a delocalized electron 'gas' within which the ion cores are dispersed. Such materials are called metals. Contributions to the binding energy of metals come from a number of sources, including the electrostatic attraction between the electron gas and the ion cores (see §2.2.3.1), but some of these interactions can only be understood following a thorough discussion of the behaviour of electrons in solids (Chapter 5). The interaction is both long-ranged and also many-body (involving many particles simultaneously). The metallic bonding interaction involving the delocalized electron gas is again non-directional, and so atomic structures are favoured in which the ion packing density is maximized.

The fourth major type of bonding is the covalent interaction which, like the van der Waals interaction, is short-ranged and occurs between pairs of atoms. The overlap of partially occupied electron orbitals on each of two neighbouring atoms leads to a lowering of the overall electronic energy, and hence a net bonding interaction. However, such covalent bonds, i.e. regions of high electron density between atoms, are very directional in space, and so covalent materials are not close-packed like van der Waals, ionic or metallic solids, but instead adopt structures with low atomic densities.

The final type of bond that is often distinguished is the so-called hydrogen bond. Since the hydrogen atom contains only one electron, it can only form a single covalent bond to another atom. However, fully or partially ionized, a hydrogen atom can form essentially ionic bonds with other (electronegative) atoms, e.g. O, N or F, a process facilitated by the very small size of the proton compared with other atoms or ions. This interaction plays an important rôle in certain inorganic materials, notably water. In the case of crystalline ice, H_2O molecules are arranged so that each O atom is surrounded in a tetrahedral arrangement by four other O atoms with H atoms lying between each pair of O atoms, each O atom being essentially covalently bonded to two of the H atoms (thereby making an H_2O molecule) and hydrogen-bonded to the other two H atoms—the so-called 'ice rule' (see Fig. 2.17). Hydrogen bonding also plays an essential rôle in biology, for example being responsible for the stability of the α-helix structure of DNA.

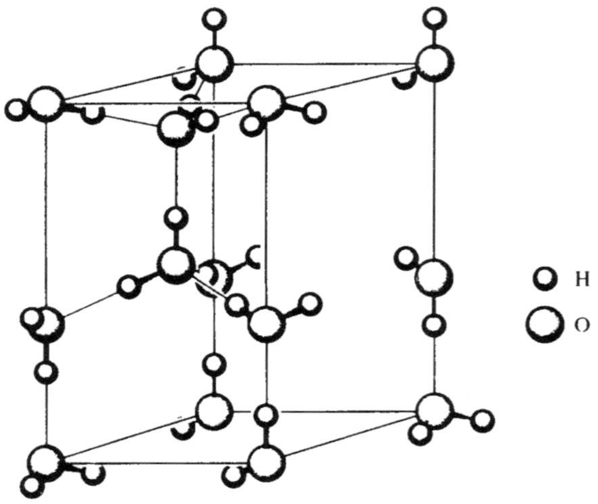

Fig. 2.17 Illustration of the structure of ice, a hydrogen-bonded tetrahedral network of oxygen atoms.

Finally, it should be noted that the bonding types described above represent extreme cases. Very often, for materials containing more than one type of atom and where *partial* charge transfer takes place, distinctions between, say, ionic and covalent bonding become inappropriate and the actual interatomic bonding is intermediate in character between the two extremes. For such cases of iono-covalent bonding, the fractions f_i of ionic (or heteropolar) character and f_c of covalent (or homopolar) character in the bond are important quantities. Obviously, these are related via $f_i + f_c = 1$. For an elemental material like Ge, $f_c = 1$ and $f_i = 0$, whereas for an ionic solid like NaCl, $f_i \simeq 0.9$.

Phillips (1973) has given a scheme for estimating in solids the ionic fractions of bonds, or ionicities, in terms of appropriate energy gaps. For the case of a *homopolar* material, like Ge, only covalent interactions exist. The covalent interaction V_{AA} between two orbitals on two neighbouring atoms, A, results in the formation of bonding and antibonding molecular-orbital levels, lying in energy below and above, respectively, the atomic-orbital energy level, and separated by the covalent energy $\mathscr{E}_{cov} = 2V_{AA}$. In the case of a solid, additional interactions with neighbouring atoms cause a broadening of the bonding and antibonding molecular-orbital levels into bands of electron states, forming the valence band and conduction band, respectively (see Chapter 5 for a fuller discussion). In the solid, therefore, $\mathscr{E}_{cov}$ can be associated with the separation between the *centres* of the two bands.

In the case of *heteropolar* materials, e.g. the compound AB, where partial charge transfer takes place, the additional ionic interactions cause an increase of the energy gap from $\mathscr{E}_{cov}$ to $\mathscr{E}_p$. The origin of this energy can be seen from a molecular-orbital (MO) picture of the bonding interaction between orbitals on neighbouring atoms A and B, having orbital energies $\mathscr{E}_A$ and $\mathscr{E}_B$, respectively (Fig. 2.18). If V_{AB} is the (covalent)

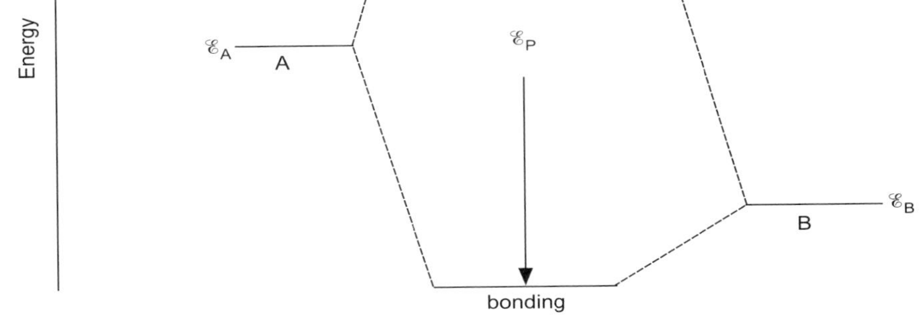

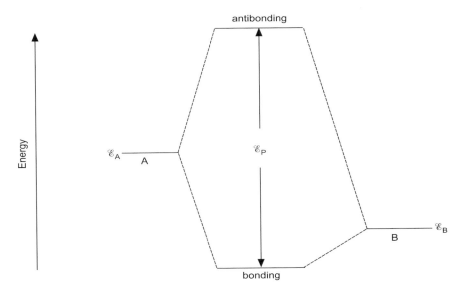

Fig. 2.18 Formation of bonding and antibonding states for a heteropolar diatomic molecule.

interaction energy between the orbitals, the resulting bonding and antibonding orbital combinations have energies given by the solutions of the secular determinant:

$$\begin{vmatrix} \mathscr{E}_A - \mathscr{E} & V_{AB} \\ V_{AB} & \mathscr{E}_B - \mathscr{E} \end{vmatrix} = 0, \tag{2.5}$$

i.e.

$$\mathscr{E} = (\mathscr{E}_A + \mathscr{E}_B)/2 \pm \{V_{AB}^2 + (\mathscr{E}_A - \mathscr{E}_B)^2/4\}^{1/2}. \tag{2.6}$$

The energy difference $\mathscr{E}_p$ (the Penn gap) between the upper antibonding state and the lower bonding state in the molecular-orbital picture, or between band centres for the solid state, is therefore from eqn. (2.6)

$$\mathscr{E}_p = \{4V_{AB}^2 + (\mathscr{E}_A - \mathscr{E}_B)^2\}^{1/2} \tag{2.7a}$$

or

$$\mathscr{E}_p = (\mathscr{E}_{cov}^2 + \mathscr{E}_i^2)^{1/2}, \tag{2.7b}$$

where $\mathscr{E}_i (= \mathscr{E}_A - \mathscr{E}_B)$ is the ionic contribution to the gap. For a covalent system, the Penn gap is $\mathscr{E}_p = \mathscr{E}_{cov}$. In this approach, the ionicity (fraction of ionic character) of a solid is given by (Phillips (1973)):

$$f_i = (\mathscr{E}_i/\mathscr{E}_p)^2. \tag{2.8}$$

Estimates for the parameters in this approach, $\mathscr{E}_{cov}$ and $\mathscr{E}_i$, can be obtained from optical absorption experiments (see §5.8): peaks in optical-absorption spectra can be associated with the energy separation between bands. Thus, for homopolar covalent materials (e.g. C, Si, Ge) $\mathscr{E}_{cov}$ is obtained immediately. In an *isoelectronic* series of compounds, e.g. Ge, GaAs, ZnSe, CuBr, the bond length is almost invariant, and hence it can be assumed that $\mathscr{E}_{cov}$ remains approximately constant. In other cases, it can be assumed that $\mathscr{E}_{cov}$ scales with the interatomic spacing d, e.g. as $\mathscr{E}_{cov} \propto d^{-2.5}$ (valid for the elements). For heteropolar materials, $\mathscr{E}_p$ can be found from optical-absorption spectra, and hence $\mathscr{E}_i$ can be obtained from eqn. (2.7b). The ionic energy is proportional to the *electronegativity difference* between the elements forming the heteropolar solid.

2.2.2 Van der Waals solids

2.2.2.1 Interatomic bonding

Although the van der Waals interaction is ubiquitous, because it is so relatively weak it is only important in those solids where other interactions (ionic, covalent, metallic) are absent. This attractive interaction arises from induced dipole–dipole interactions, between even closed-shell atoms with spherically symmetric charge distributions, as a result of (zero-point) fluctuations in the charge distribution. The characteristic r^{-6} dependence of the attractive van der Waals interaction can be derived from classical electrostatics considerations (although properly the interaction is due to a quantum-mechanical effect and should be treated as such). A spontaneous fluctuation of an otherwise spherically symmetric charge distribution of an atom will produce an instantaneous dipole moment, p_1. This gives rise to an electric field at a distance r from the perturbed atom of the form $E \sim p_1/r^3$, which will polarize a second atom located at r, having an electronic polarizability α, giving rise in it to an induced dipole moment $p_2 \sim \alpha p_1/r^3$. The potential of a dipole $\boldsymbol{p}$ in a field $\boldsymbol{E}$ is proportional to $\boldsymbol{p} \cdot \boldsymbol{E}$, and hence the attractive part of the van der Waals interaction varies as $-Ar^{-6}$. (For a simplified quantum-mechanical derivation of this result, using coupled simple harmonic oscillators with equal and opposite charges at either end, representing an atomic dipole consisting of a negative charge associated with the displaced electron distribution and the positive charge of the nucleus, see Borg and Dienes (1992) and Kittel (1996).)

If the distance r between two atoms is progressively decreased, ultimately the electronic charge distributions of the two atoms will begin to overlap. The Pauli exclusion principle for fermions precludes the occupation of the same region of space by two electrons having the *same* quantum numbers. As the electron clouds overlap, the only way to satisfy the Pauli exclusion principle is for electrons to be promoted to higher-lying states, thereby causing an increase in the (electronic) energy. Obviously, this behaviour leads to a *repulsive* interaction, and one which varies very rapidly with r. An empirical form often used to account for the repulsive term in the potential is a power law, such as B/r^n. Conventionally, the exponent n is chosen to be 12, which is sufficiently large to mimic well the abrupt increase in the actual potential with decreasing distance at short distances, but also, it must be admitted, for reasons of computational expediency (combining well with the exponent 6 in the van der Waals attractive part of the potential in algebraic manipulations). Thus, the overall potential energy for a pair of atoms can be written in the form:

$$U(r) = 4\varepsilon\left[\left(\frac{\sigma}{r}\right)^{12} - \left(\frac{\sigma}{r}\right)^{6}\right] \tag{2.9}$$

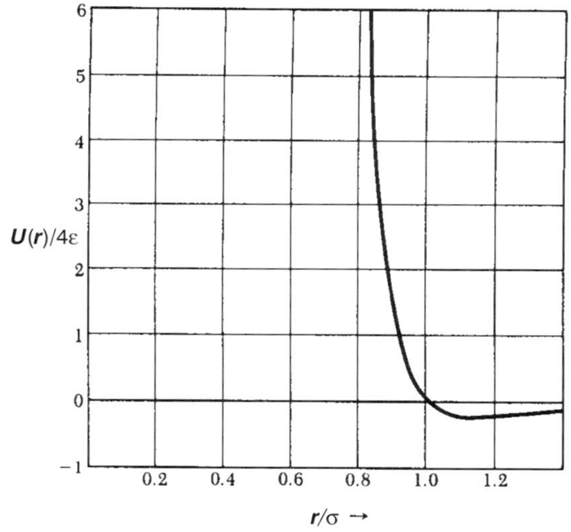

Fig. 2.19 Form of the Lennard-Jones potential modelling the interaction between two rare-gas atoms.

where the parameters ε and σ have the units of energy and length, respectively. This, the Lennard-Jones potential, is illustrated in Fig. 2.19. Note that $U(r) = 0$ when $r = \sigma$ (or infinity). The position of the minimum in the pair potential can be found as usual by setting $\mathrm{d}U(r)/\mathrm{d}r = 0$, giving $r_{\min} = (2)^{1/6}\sigma$ and $U_{\min} = -\varepsilon$.

Although the r^{-12} term gives a reasonably satisfactory representation of the repulsive part of the potential due to Pauli-exclusion effects, nevertheless other empirical functional forms are also used which are equally algebraically manipulable, e.g. the exponential repulsive term

$$U_{\text{rep}}(r) = \lambda \exp(-r/\rho), \tag{2.10}$$

where λ and ρ are parameters representing the strength and range of the interaction, respectively.

The cohesive energy of a rare-gas crystal can be estimated by summing the Lennard-Jones potential (eqn. (2.9)) over all pairs of atoms in the crystal, assuming that the kinetic energy of the atoms may be neglected. Thus, the total potential energy is

$$U_{\text{tot}} = \frac{1}{2} N 4\varepsilon \left[\sum_{i \neq j} \left(\frac{\sigma}{a_{ij} R} \right)^{12} - \sum_{i \neq j} \left(\frac{\sigma}{a_{ij} R} \right)^{6} \right] \tag{2.11}$$

where N is the number of atoms in the crystal, the factor of 1/2 is to correct for overcounting of pair interactions when all atoms in turn are taken as the origin in evaluating the summations, and the interatomic distance is written as $r = a_{ij}R$, R being the nearest-neighbour separation. The quantities $m = \Sigma_{ij} a_{ij}^{-12}$ and $n = \Sigma_{ij} a_{ij}^{-6}$ can be evaluated for any particular crystal structure: for a cubic close-packed (face-centred cubic, f.c.c.) array, $m = 12.131\,88$, $n = 14.453\,92$; for a hexagonal close-packed (h.c.p.) crystal, $m = 12.132\,29$ and $n = 14.454\,89$; while for a body-centred cubic (b.c.c.) structure, $m = 9.114\,18$ and $n = 12.2533$ (Kittel (1996)). A detailed description of these structural types will be given shortly, but suffice it to say now that both f.c.c. and h.c.p. structures have 12 nearest neighbours surrounding a given atom, whereas the coordination number is 8 for a b.c.c. structure. Thus, it can be seen that most of the cohesive-energy contribution in rare-gas crystals (which adopt the f.c.c. structure) comes from the nearest neighbours. The equilibrium interatomic spacing is found, from setting $\mathrm{d}U_{\text{tot}}/\mathrm{d}r = 0$, to be

$$R = \left(\frac{2m}{n} \right)^{1/6} \sigma = 1.09\sigma \tag{2.12}$$

for the f.c.c. structure; substituting this value into eqn. (2.11), together with the values of m and n appropriate for the f.c.c. structure, gives as an estimate for the cohesive energy

$$U_{\text{tot}} = -8.6 N \varepsilon. \tag{2.13}$$

Experimental values of the cohesive energy of rare-gas crystals are in the range 0.02 to 0.2 eV/atom (Ne to Xe); eqn. (2.13) predicts values that are within 10% or so of these values.

2.2.2.2 *Cubic close-packed structure*

The archetypal examples of van der Waals solids are the rare-gas crystals, Ne, Ar, Kr and Xe (He is a liquid at zero pressure due to large zero-point fluctuations). These crystallize in a close-packed (c.p.) structural arrangement, specifically the cubic close-packed (c.c.p.), or, equivalently, the face-centred cubic (f.c.c.), structure. The c.c.p. (f.c.c.) crystal structure is also favoured by other spherically symmetric entities, e.g. the C_{60} molecule 'buckminsterfullerene' (see §8.2.1), mutually interacting via the van der Waals potential. The c.c.p. (f.c.c.) and its close relation, the hexagonal close-packed (h.c.p.) structure (see §2.2.3.2), are the structures for which the nearest-neighbour coordination number is maximized (equal to 12) for a 3D packing of equal-sized spheres. 3D close-packed structures are built up from stackings of close-packed 2D layers, in which every sphere (atom) is in contact with six others in a hexagonal arrangement (Fig. 2.20). Another c.p. layer can be laid on top of the first, such that each atom in the second layer sits in a hollow formed by three atoms in the first (A) layer; there are two such positions, labelled B and C in Fig. 2.20. The c.c.p. (f.c.c.) structure then results from the packing sequence ABCABC... Thus a given atom has a total coordination number of 12, comprising six atoms from the same layer, and three atoms each from the adjacent layers.

The conventional (non-primitive) unit cell of the c.c.p. structure is the f.c.c. cell (Fig. 2.21a) which clearly exhibits the underlying cubic symmetry. This cell contains a total of four lattice points (each of the eight corner sites is shared between eight different cells; each of the six face-centred sites is shared between two cells). The *primitive* unit cell of the c.c.p. structure, on the other hand, is a rhombohedron (Fig. 2.21b) containing only one lattice site, and is therefore only one-quarter the size of the f.c.c. cell (Fig. 2.21a). The close-packed arrangement (Fig. 2.20) is not readily discernible by casual inspection of the f.c.c. unit cell in Fig. 2.21a. In fact, the c.p. layers are the {111} planes of the f.c.c. unit cell (see Fig. 2.22).

The close-packed structures have the highest packing fraction of all crystalline structures comprising spheres (atoms) of one size; the packing fraction ϕ is defined as the ratio of the volume actually occupied by the spheres (atoms) to the volume of the

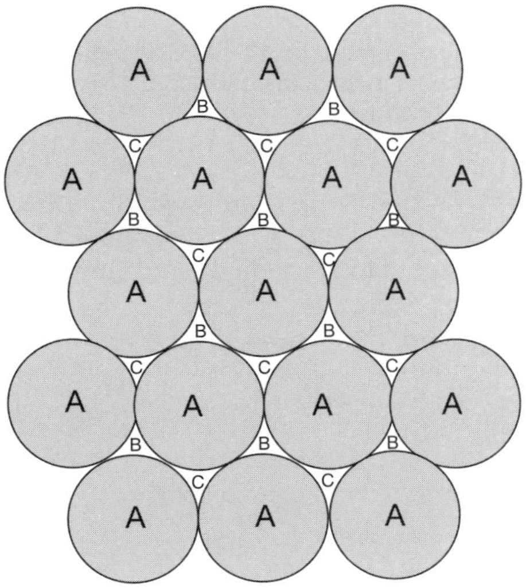

Fig. 2.20 A 2D close-packed layer of spheres (atoms) occupying A sites. Adjacent c.p. layers can occupy either B or C sites.

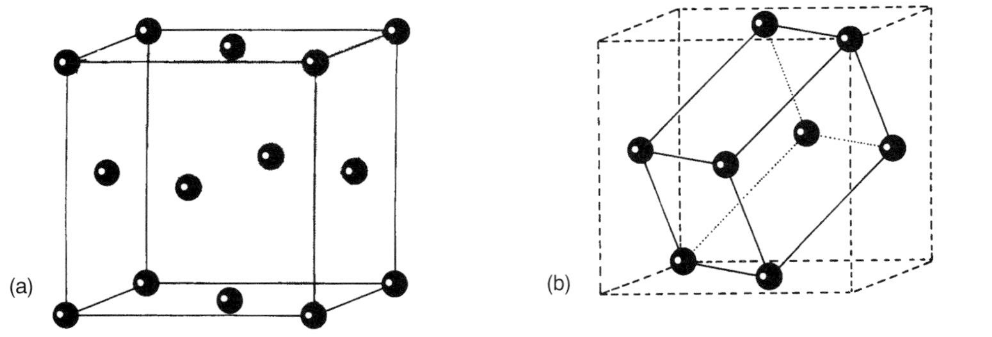

Fig.2.21 Unit cells for the cubic close-packed structure: (a) conventional (non-primitive) f.c.c. unit cell; (b) primitive rhombohedral unit cell.

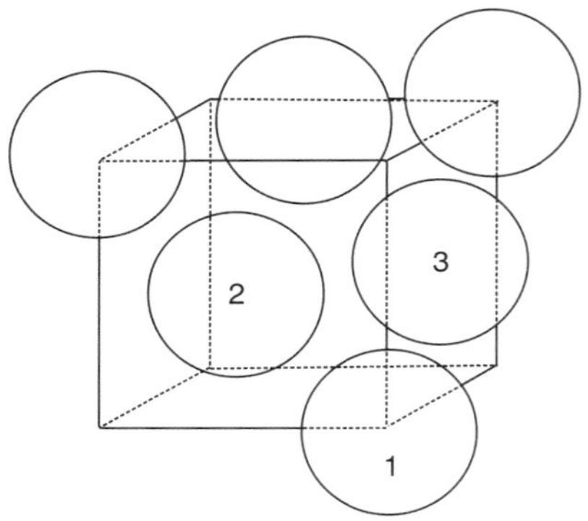

Fig.2.22 A (111) plane of the f.c.c. unit cell of the c.c.p. structure showing the close-packed layer (the atoms completing the hexagonal arrangement around atom 1 characteristic of the c.p. layers derive from other cells and are not shown).

unit cell. For the c.c.p. (f.c.c.) structure composed of spheres of radius r, the face diagonal is given by $4r$ (since the atoms 1, 2, 3 in Fig. 2.22, forming part of the c.p. layer in the (111) plane, are in contact), and hence the face edge length is given by $2\sqrt{2}r$; the cell volume is thus $16\sqrt{2}r^3$. Since the volume of each sphere is $4\pi r^3/3$ and there are four lattice sites (spheres) per cell, the packing fraction is thus

$$\phi = \frac{4 \times 4\pi r^3}{16\sqrt{2}r^3 \times 3} = \frac{\sqrt{2}\pi}{6} = 0.7405. \tag{2.14}$$

The same value of ϕ characterizes the h.c.p. structure as well.

2.2.2.3 *Wigner–Seitz cell*

An equivalent, informative way of regarding crystalline structures is in terms of a different type of unit cell, the Wigner–Seitz unit cell. (This formalism is also particularly

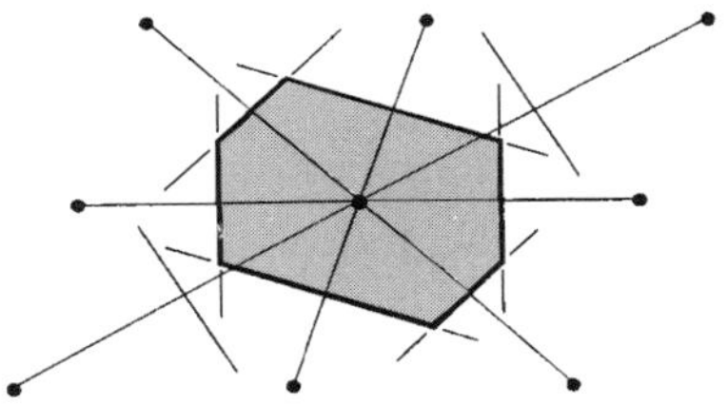

Fig.2.23 Illustration of the generation of the Wigner–Seitz cell in 2D.

useful for the discussion of electronic properties in terms of the so-called reciprocal lattice–see §2.4.1) The rules for generating the Wigner–Seitz cell are illustrated in Fig. 2.23 for the 2D case. Lines are drawn from a given lattice point to connect all other neighbouring lattice points. Lines in 2D (planes in 3D) are then drawn normal to these lines at the midpoints. The smallest area in 2D (volume in 3D) enclosed in this manner is the Wigner–Seitz cell. It is a primitive unit cell, and is atom-centred; all space can be covered by a periodic translation of the cell as for other unit cells. In the case where each lattice point represents an atom, the Wigner–Seitz cell also represents the atomic coordination polyhedron.

For the c.c.p. (f.c.c.) structure, the corresponding Wigner–Seitz cell is a rhombic dodecahedron (see Fig. 2.24). Since this polyhedron has 12 faces, this indicates directly the 12–fold nearest-neighbour coordination of the c.c.p. lattice. The symmetries inherent in the cubic lattice (see §2.3.2) are also clearly evident: four 3–fold rotation axes in the $\langle 111 \rangle$ directions, three 4–fold axes in the $\langle 100 \rangle$ directions and six 2–fold axes in the $\langle 110 \rangle$ directions.

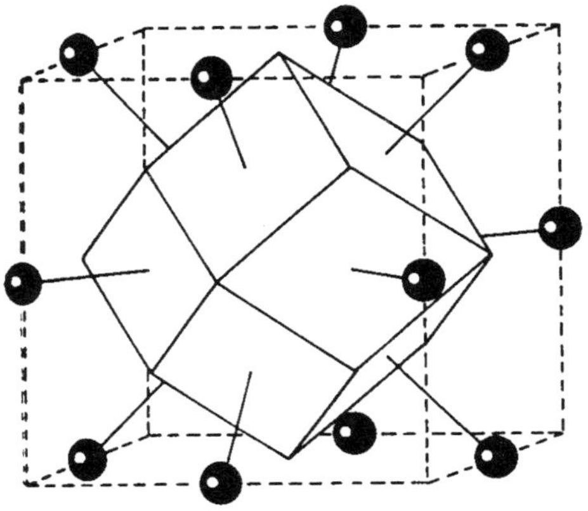

Fig.2.24 The Wigner–Seitz cell of the c.c.p. (f.c.c.) lattice: the rhombic dodecahedron. The origin of the crystallographic axes has been shifted so that an atom lies at the centre of the cell. Bonds between this central atom and its 12 near neighbours are indicated.

2.2.2.4 *Finite clusters*

Although the c.c.p. (f.c.c.) structure appears to be the most stable structure for van der Waals materials, such as rare-gas solids, this is true only for solids effectively *infinite* in extent; for smaller clusters of atoms, the f.c.c. structure may *not*, in fact, have the lowest free energy. The large surface-to-volume ratios of small clusters of atoms means that an additional destabilizing, surface-strain energy exists. It is interesting and significant that neutral (uncharged) rare-gas clusters exist containing 'magic' numbers of atoms (13, 19, 23, 26, 29, etc.). These can all be obtained by the serial addition of atoms to an initial 13–atom icosahedral seed (Fig. 2.25a), thereby creating so-called Mackay icosahedra (Mackay (1982))—see Fig. 2.25b for a 55–atom example. It is significant that the two cluster structures shown in Fig. 2.25 exhibit *five-fold* rotational symmetry, which is incompatible with translational periodicity in 3D (see §2.3.2). See §8.2.2.2 for a discussion of clusters of Na atoms, where small magic numbers (<1500) are determined by the numbers of electrons needed to fill completely the energy levels for electrons confined to a spherical potential well.

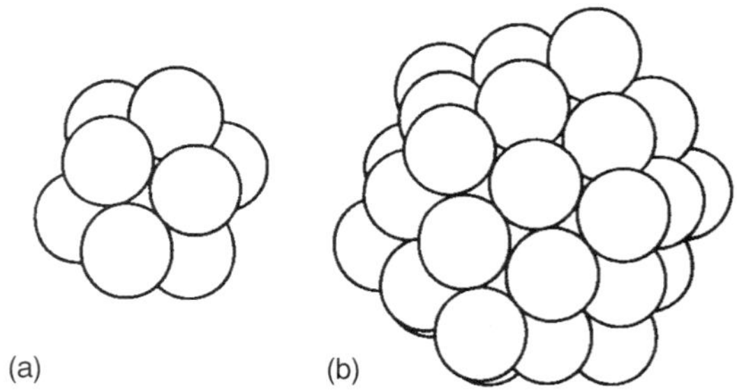

Fig.2.25 Structures for rare-gas clusters: (a) 13-atom icosahedron; (b) 55-atom Mackay icosahedron.

2.2.3 Metallic solids

2.2.3.1 *Interatomic bonding*

Materials in which the valence electrons are spatially delocalized throughout the solid (and not confined to regions of high electron density, i.e. bonds, between atoms), and for which, concomitantly, the electrical conductivity is very high are termed metals. (Another succinct definition of a metal, in terms of the *electronic* structure, will be given in §5.2.5) The metallic binding interaction is impossible to understand properly without a prior discussion of electronic properties, and so a more thorough discussion of metallic cohesion will be deferred to §5.1.3.4. Nevertheless, some insight can be obtained from the following simple model, where it is assumed that the attractive part of the potential, binding atoms together, comes from the attraction between the positively charged ion cores and the negatively charged delocalized electron gas.

Consider the model of an idealized metal, consisting of ion cores located at the lattice sites of a crystal and bathed in a uniform negative charge density, representing the

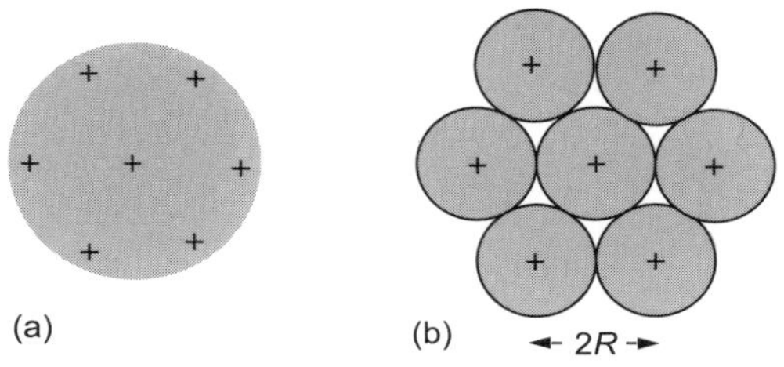

Fig.2.26 (a) Schematic representation of a model for metals, positive ion cores bathed in a uniform sea of negative charge. (b) Model for calculating the cohesive energy of metals, in which the negative charge density, representing the delocalized electron gas, is confined to spherical volumes around each ion core, each sphere containing one electron.

delocalized electron gas (Fig. 2.26a), represented instead by spherical volumes, of radius R, containing a uniform density of negative (electronic) charge, surrounding each ion core (Fig. 2.26b). R can be taken to be the Wigner–Seitz radius. The overall attractive potential energy per atom is evaluated by calculating the potential experienced by a spherical shell of charge (of thickness $\mathrm{d}r$, lying between $r(\leqslant R)$ and $r + \mathrm{d}r$) due to the net charge (ion core plus electronic) lying within the spherical volume of radius r. This charge is given by

$$q(r) = e - e(r^3/R^3) \tag{2.15}$$

and therefore the electrostatic potential at the shell is $\phi(r) = q(r)/4\pi\varepsilon_0 r$, with ε_0 being the permittivity of free space. Thus the attractive potential energy of the charge in the spherical shell, $\mathrm{d}q = 4\pi r^2 \rho \mathrm{d}r$, where ρ is the negative (electronic) charge density given by

$$\rho = -e/(4\pi R^3/3) \tag{2.16}$$

experiencing the potential $\phi(r)$ is given by $\mathrm{d}U_\mathrm{A} = \phi(r)\mathrm{d}q$ or

$$\mathrm{d}U_\mathrm{A} = \frac{4\pi r^2 \rho[e - e(r^3/R^3)]}{4\pi\varepsilon_0 r}\mathrm{d}r.$$

Integrating over the volume of the sphere, of radius R, gives the net attractive potential energy per atom (ion core):

$$\begin{aligned} U_\mathrm{A} &= \frac{-3e^2}{4\pi\varepsilon_0 R^3}\int_0^R [r - r^4/R^3]\mathrm{d}r \\ &= -9e^2/40\pi\varepsilon_0 R. \end{aligned} \tag{2.17}$$

Note that this term includes a contribution from the attractive Coulombic interaction between electrons and ions and a repulsive contribution from electron–electron interactions (see §5.1.3.4).

The repulsive energy associated with the electron gas/ion core system is a little more subtle. At first sight, it might be thought that the electrostatic Coulombic repulsion between the positively charged ion cores (Fig. 2.26) would give rise to a repulsive contribution to the energy. However, the intervening electron gas very effectively screens such repulsive interactions, and so this contribution can be neglected. The real origin of the repulsive interaction lies instead in the *kinetic energy* of the electron gas: due to quantum effects (effectively the Pauli exclusion principle), the kinetic energy of the electron-gas system *increases* when it becomes more compressed, i.e. it exhibits a repulsive behaviour. A proper discussion of this effect is given in §5.1.2; here, we merely quote the result for the repulsive energy (i.e. the mean kinetic energy of the electron gas, cf. eqn. (5.17)), viz.

$$U_R = \frac{3\hbar^2(3\pi^2 N/V)^{2/3}}{10m_e} \tag{2.18}$$

where m_e is the electron mass, N is the total number of electrons and V the volume. The electron density in the model of Fig. 2.26b is given by

$$n = \frac{N}{V} = \frac{1}{4\pi R^3/3} \tag{2.19}$$

and thus

$$U_R = \frac{3\hbar^2(9\pi/4)^{2/3}}{10m_e R^2}. \tag{2.20}$$

The total energy per atom, U_T, in this model is then given by the sum of eqns. (2.17) and (2.20).

The equilibrium separation, $2R_0$, between ion cores in this model is found as usual by setting $\mathrm{d}U_T/\mathrm{d}R = 0$, giving

$$2R_0 = \frac{16\hbar^2\pi\varepsilon_0}{3m_e^2}\left(\frac{9\pi}{4}\right)^{2/3} = 4.9a_0 = 2.6\text{Å}, \tag{2.21}$$

where a_0 is the hydrogenic Bohr radius ($a_0 = \hbar^2 4\pi\varepsilon_0/e^2 m_e$). This estimate is in (surprisingly) good agreement with experimental values for free-electron-like metals (e.g. Au, 2.88 Å; Cu, 2.56 Å).

Substitution of eqn. (2.21) into the expression for U_T gives an estimate for the cohesive energy per atom

$$U_0 = -9e^2/80\pi\varepsilon_0 R_0. \tag{2.22}$$

Evaluation of eqn. (2.22) gives $U_0 \simeq -5\text{eV/atom}$, in reasonable agreement with experimental values (e.g. Au, 3.81 eV/atom; Cu, 3.49 eV/atom).

2.2.3.2 *Close-packed structures*

The non-orientational nature of the metallic bonding interaction in the simplest case ensures that metals usually adopt close-packed structures. The cubic close-packed (face-centred cubic) arrangement has already been discussed (§2.2.2.2); metals such as Al, Cu, Ni, Ag and Au adopt the c.c.p. (f.c.c.) structure.

Another way of stacking the close-packed planes shown in Fig. 2.20 is in the sequence ABAB..., instead of the sequence ABCABC characteristic of the c.c.p. (f.c.c.) structure. The resulting structure is called the hexagonal close-packed (h.c.p.) structure. (Note that the h.c.p. structure is *not* a Bravais lattice.) In this case, the close-packed layers lie parallel to the basal plane of the structure (see Fig. 2.27). The packing density

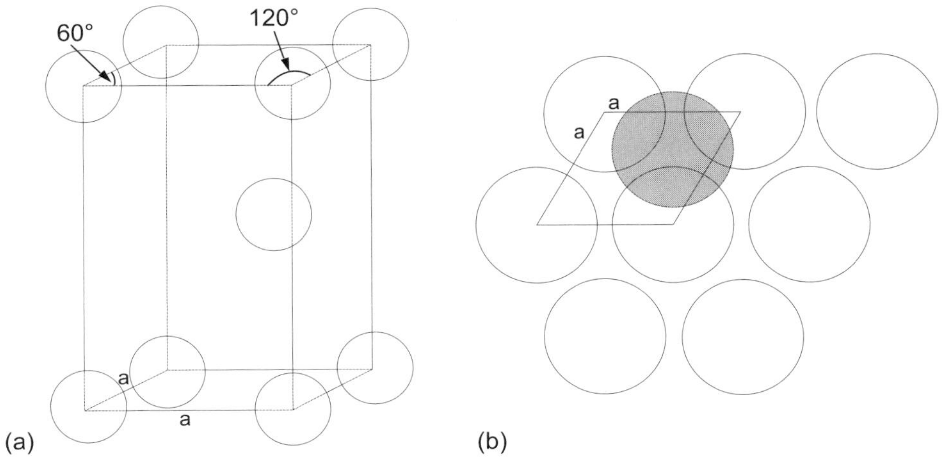

Fig.2.27 Hexagonal unit cell of the h.c.p. structure: (a) elevation view; (b) plan view. The atom shown shaded is the second atom of the basis at (1/3, 2/3, 1/2).

is the same as that for the c.c.p. (f.c.c.) structure (0.7405), and the atomic coordination is also 12. The unit cell contains *two* atoms as the basis: one is at (0, 0, 0) and the other at (1/3, 2/3, 1/2). The ratio of unit-cell parameters for the ideal h.c.p. structure is given by (see Problem 2.8):

$$\frac{c}{a} = \left(\frac{8}{3}\right)^{1/2} = 1.633. \tag{2.23}$$

A number of metals adopt the h.c.p. structure, even though the c/a values often differ somewhat from the ideal case (e.g. Co, $c/a = 1.622$; Mg, 1.623; Zn, 1.861; Cd, 1.886).

Another structure commonly adopted by metals (e.g. the alkali metals, and Cr, Fe, W) is the body-centred cubic (b.c.c.) structure (for the conventional unit cell containing 2 atoms per cell, see Fig. 2.8b). Unlike the c.c.p. and h.c.p. structures, the b.c.c. structure is *not* close-packed; the nearest-neighbour coordination number is 8, and the packing fraction, $\phi = 0.68$, is lower than that of the c.c.p. and h.c.p. structures (see Problem 2.9).

The differences in cohesive energies for metals crystallizing, for example, in the c.c.p., h.c.p. or b.c.c. structures are very small. As a result, metals may adopt different crystal structures depending on external parameters such as temperature or pressure; i.e. they are polymorphic. Understanding why certain metals adopt particular structures is a difficult task (see e.g. Sutton (1993)); the expression for the cohesive energy given by eqn. (2.22) is structure-independent, and such a structure-independent contribution makes up 90% of the total cohesive energy. The residual, structure-deciding contributions derive, for example, from the orientational effects of d-electrons in transition metals.

2.2.3.3 *Amorphous metals*

Thus far, the supposition has been that metals always form *crystalline* structures. However, in the case, say, of very rapid cooling of a melt, there may be insufficient time for the structure with the lowest free energy (that is, the crystal) to form, and a higher-energy, more disordered structure will be quenched in.

The most extreme examples of such disordered structures are amorphous metals (see e.g. Elliott (1990)), which completely lack all translational and orientational order. They are commonly made by 'melt-spinning', in which a jet of the molten metal is directed at the edge of a rapidly spinning Cu disc (see §1.2.2); effective cooling rates of the order of $10^6 Ks^{-1}$ can be achieved in this way, and the metallic glass is produced in the form of a very long ribbon, a few mm (or cm) wide, and a few μm thick. Metals which can be rendered glassy in this way tend to be eutectic alloys of transition metals and 'metalloid' atoms (C, B, P, Si, etc.), e.g. $Pd_{80}Si_{20}$ or $Ni_{80}B_{20}$, although many other glass-forming alloy compositions exist.

The simplest model for the structure of an amorphous metal is the self-explanatory dense-random packing (DRP) of equal-sized hard spheres. Each atom is surrounded by a different number of atoms; the average nearest-neighbour coordination number is $\simeq 13$. The packing fraction of a DRP structure ($\phi \simeq 0.64$) is appreciably lower than for the crystalline c.c.p. or h.c.p. structures ($\phi = 0.74$). However, ultra-pure elemental metals do not readily form glasses, and the DRP model does not adequately describe the structure of, say, metal–metalloid alloys. Structural studies using diffraction (§2.6.1) indicate that, in the case of metal–metalloid alloys, the nearest-neighbour coordination number around metalloid atoms is smaller ($\simeq 9$) than that around metal atoms ($\simeq 13$). Also, metalloid–metalloid nearest neighbours appear to be absent, although they would be expected to occur in a binary-atom DRP structure. Gaskell (1981) has suggested that there persist in the amorphous material local structural elements that are characteristic of the corresponding *crystalline* structure, namely the (capped) trigonal prismatic coordination polyhedron found for example in cementite, crystalline Fe_3C, which has nine-fold coordination of the metalloid and no metalloid–metalloid close contacts (see Fig. 2.28). The occurrence of such stereochemically favoured structural units may be a result of directional-bonding effects associated with the transition-metal d-electrons.

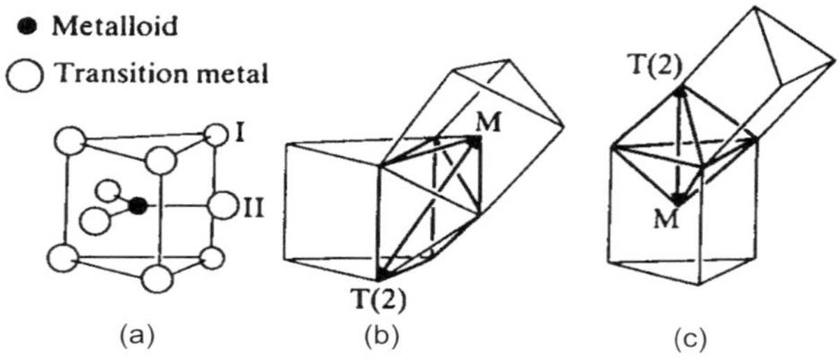

Fig.2.28 Trigonal prismatic packing of transition metals around a metalloid atom occurring in crystalline metal–metalloid alloys and also, possibly, in glasses of the same materials (Gaskell (1981)): (a) 'Capped' trigonal prism, showing the capping atoms (II) and prism atoms (I). (b) Edge-sharing arrangement in the cementite (crystalline Fe_3C) structure. (c) Edge-sharing arrangement in the crystalline Fe_3P structure. The metalloid–second-neighbour-metal distances, M–T(2), are different in each case. (Reprinted with permission from *Nature* **289**, 474. © 1981 Macmillan Magazines Ltd.)

2.2.4 Ionic solids

2.2.4.1 Interatomic bonding

The electrostatic attractive interaction between oppositely charged cations and anions in ionic solids provides the major contribution to the overall binding part of the interaction energy is such materials; the magnitude of the van der Waals attractive interaction (see eqn. (2.9)) is only about 1% of the Coulomb term. The Coulomb interionic interaction U_{ij}^{c} (positive between like ions, negative between unlike ions) can be written as

$$U_{ij}^{\mathrm{c}} = \frac{\pm q^2}{4\pi\varepsilon_0 a_{ij} R}, \tag{2.24}$$

where q is the ionic charge (assumed to be the same for all types of ion, i, j), and the interionic distance r_{ij} is written as $r_{ij} = a_{ij}R$, where R is the nearest-neighbour separation (cf. eqn. (2.11)). Summing over all ions in the solid gives

$$U^{\mathrm{c}} = \sum_{i \neq j} \frac{\pm q^2}{4\pi\varepsilon_0 a_{ij} R} = \frac{-Aq^2}{4\pi\varepsilon_0 R}, \tag{2.25}$$

where the so-called Madelung constant A is given by the sum

$$A = \sum_{i \neq j} \frac{\pm 1}{a_{ij}}, \tag{2.26}$$

and the positive sign refers to pairs of ions with opposite charge, and the negative sign to ion pairs with the same charge. Obviously, the value of the Madelung constant depends on both the particular crystal structure adopted by the ions (see §2.2.4.2), and also whether it is defined in terms of the nearest-neighbour distance (as in eqn. (2.26)) or in terms of other distances such as the unit-cell parameter a, etc.

The repulsive interaction between ions (having closed-shell electronic configurations) is comparable to that between rare-gas atoms, and so a power-law form, β/r^n, is often chosen for the repulsive part of the potential. Thus, the total lattice energy U_{tot} of a crystal comprising $2N$ ions (in the case of equally and oppositely charged ions, as in NaCl) is given by the Born–Landé equation:

$$\begin{aligned} U_{\mathrm{tot}} &= \frac{1}{2} \times 2N \left(\sum_{i \neq j} \frac{\beta}{a_{ij}^n R^n} - \frac{Aq^2}{4\pi\varepsilon_0 R} \right) \\ &= N \left(\frac{\beta b}{R^n} - \frac{Aq^2}{4\pi\varepsilon_0 R} \right), \end{aligned} \tag{2.27}$$

where

$$b = \sum_{i \neq j} \frac{1}{a_{ij}^n}, \tag{2.28}$$

and where the factor 1/2 is used to avoid overcounting of pairs of interactions. Alternatively, the exponential form (eqn. (2.10)) for the repulsive potentials is often used, giving the Born–Mayer potential:

$$U_{\text{tot}} = N\left(z\lambda e^{-R/\rho} - \frac{Aq^2}{4\pi\varepsilon_0 R}\right), \tag{2.29}$$

where it has been assumed that the repulsive interaction is so extremely short-ranged that a summation as in eqn. (2.28) simply gives the nearest-neighbour coordination number, z, of the crystal in question.

The expressions (2.27) and (2.29) can be simplified somewhat by evaluating them at the equilibrium position $R = R_0$, subject to the condition $dU_{\text{tot}}/dR = 0$. Thus, for the power-law repulsive potential case (eqn. (2.27)):

$$U^0_{\text{tot}} = -\frac{NAq^2}{4\pi\varepsilon_0 R_0}\left(1 - \frac{1}{n}\right) \tag{2.30}$$

and for stability ($U^0_{\text{tot}} < 0$), n must be greater than unity. For the exponential repulsive case (eqn. (2.29)), the result is:

$$U^0_{\text{tot}} = -\frac{NAq^2}{4\pi\varepsilon_0 R_0}\left(1 - \frac{\rho}{R_0}\right) \tag{2.31}$$

where, for stability reasons, $\rho < R_0$. The Madelung constant can be calculated for given ionic crystal structures (see below). The repulsive-potential parameters can be evaluated from measured values of the bulk modulus (or compressibility).

Use of eqn. (2.30) or (2.31) enables the lattice (or cohesive) energy of ionic crystals (with respect to free ions) to be calculated: they give very good agreement with experimental values. For the case, for example, of NaCl, the cohesive energy is 7.9 eV per formula unit (NaCl). For calculating the energy of a crystal of NaCl with respect to *neutral* atoms, the ionization energy (5.14 eV per atom) for the process $Na \rightarrow Na^+ + e^-$, and the electron affinity (3.61 eV per atom) for the process $Cl + e^- \rightarrow Cl^-$, must also be taken into account. Thus, relative to *neutral* atoms, the lattice energy of the NaCl crystal is $7.9 + 3.6 - 5.1 = 6.4$ eV per formula unit.

The compressibility, κ, or its inverse, the bulk modulus, B, is defined as (see §3.4.3):

$$\kappa \equiv \frac{1}{B} = -\frac{1}{V}\left(\frac{dV}{dp}\right)_T. \tag{2.32}$$

From the first law of thermodynamics, and at low temperatures ($T \rightarrow 0$K), the application of an external pressure p to a solid causing a volume change $-dV$ does external work on the solid, resulting in an increase in the internal energy $dU = -p\,dV$, or

$$p = -\left(\frac{dU}{dV}\right)_T.$$

Thus, from eqn. (2.32), the bulk modulus at the *equilibrium* volume (denoted by the subscript zero), is given by

$$B = V\left(\frac{\partial^2 U}{\partial V^2}\right)_0. \tag{2.33}$$

Now

$$\frac{\partial U}{\partial V} = \left(\frac{\partial U}{\partial R}\right) \cdot \left(\frac{\partial R}{\partial V}\right)$$

and hence

$$\frac{\partial^2 U}{\partial V^2} = \left(\frac{\partial U}{\partial R}\right)\left(\frac{\partial^2 R}{\partial V^2}\right) + \left(\frac{\partial^2 U}{\partial R^2}\right)\left(\frac{\partial R}{\partial V}\right)^2. \tag{2.34}$$

At the equilibrium distance, $R = R_0$, $(\partial U/\partial R)_0 = 0$, and so the first term in eqn. (2.34) vanishes. Using, as an example $V = 2NR^3$ for the NaCl structure (see below), then $(\partial R/\partial V)_0^2 = (36N^2R_0^4)^{-1}$, and hence the bulk modulus is given by

$$B = \frac{1}{18NR_0}\left(\frac{\partial^2 U_{\text{tot}}}{\partial R^2}\right)_0. \tag{2.35}$$

For the case of the power-law repulsive interaction (eqn. (2.27)), the bulk modulus becomes

$$B = \frac{Aq^2(n-1)}{72\pi\varepsilon_0 R_0^4}. \tag{2.36}$$

Comparison of eqn. (2.36) with experimental values of B and R_0 for the case of say NaCl gives $n \simeq 9$ (i.e. significantly smaller than the value 12 conventionally used in the Lennard-Jones potential for rare-gas solids (eqn. (2.11)). For the case of the exponential repulsive interaction (eqn. (2.29)), the bulk modulus is instead given by

$$B = \frac{Aq^2}{72\pi\varepsilon_0 R_0^4}\left(\frac{R_0}{\rho} - 2\right), \tag{2.37}$$

whence the parameter ρ can also be obtained by comparison with experimental values of the compressibility ($\rho \simeq 0.1R_0$).

2.2.4.2 *Structures of ionic solids*

In discussing the various crystal structures that ionic solids may adopt, the following considerations need to be borne in mind. In order to maximize the net electrostatic attraction between ions in a crystal, nearest-neighbour hetero-ionic coordination numbers should be as high as possible (subject to a central ion maintaining 'contact' with the surrounding ions), and repulsive interactions between like ions should be minimized by ensuring that like ions are situated as far away from each other as possible. Furthermore, local electroneutrality is preserved, i.e. the valence of a given ion is equal to the sum of the electrostatic bond strengths between it and adjacent ions of opposite charge, where the electrostatic bond strength, E_b, of the n bonds between M^{m+} cations and X^{x-} anions is defined as

$$E_b = m/n. \tag{2.38}$$

For each anion, the sum of the electrostatic bond strengths of the surrounding cations must equal the negative charge on the anion, viz.

$$\sum m/n = x. \tag{2.39}$$

This balance of competing requirements for attractive and repulsive components of the electrostatic interaction ensures that highly symmetric structures ensue with *maximized* volumes (to reduce repulsive interactions).

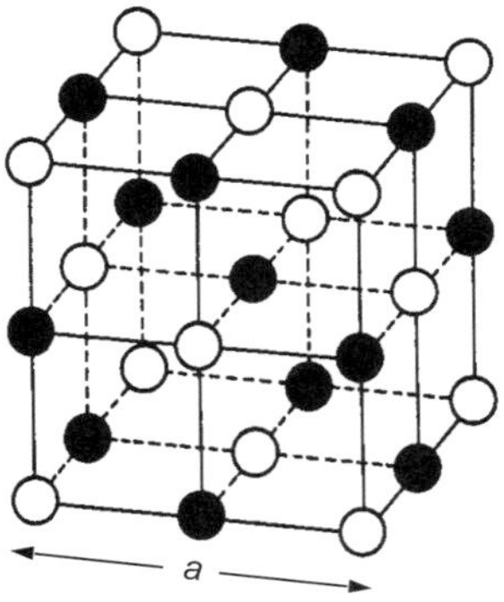

Fig.2.29 Representation of the rocksalt (NaCl) structure (Na = •; Cl = o).

Perhaps the simplest ionic crystal structure of all is the so-called rocksalt structure adopted by NaCl (as well as KBr, AgBr, PbS, MgO, etc.). One way of generating the NaCl structure is to occupy alternately with Na^+ or Cl^-ions the lattice sites of a simple cubic crystal. The result (Fig. 2.29) is an f.c.c. crystal with a basis containing one Na^+ ion (e.g. at the position 1/2 1/2 1/2, referred to the unit cell) and one Cl^- ion (at 0 0 0); there are four formula units (NaCl) in each unit cell—see Fig. 2.29). Each ion is surrounded by *six* nearest-neighbour ions of the opposite type in the ⟨100⟩ directions.

An alternative, and instructive, way of regarding ionic structures is in terms of close-packed arrangements of anions (generally larger in size than cations), with the smaller cations occupying the interstitial voids in the close-packed array. It has already been demonstrated (§2.2.2.2) that the f.c.c. crystal can be constructed by assembling stacks of close-packed planes of atoms (Fig. 2.20) in the sequence ABCABC... Between any two such stacks, interstitial voids are unavoidably created, and these voids are of two types (Fig. 2.30): octahedral sites (O), in which the number of atoms bordering the void is six, three from one layer and three from the adjacent layer, and tetrahedral sites (T), in which each void is bordered by three atoms from one layer and one atom from the adjacent layer. In fact, two types of tetrahedral holes can also be distinguished: T_+ sites are those in which the apex of the tetrahedron of atoms defining the void points *up* with respect to a layer normal, and T_-sites are those in which the apex points *down*. There are equal numbers of O, T_+ and T_-sites, one each per anion.

The NaCl structure is then generated by inserting Na^+ cations in each of the octahedral holes occurring in an f.c.c. (c.c.p.) lattice made up of Cl^- anions (the close-packed planes of Cl^- ions lie along the {111} planes of the resulting f.c.c. rocksalt structure). The octahedral coordination of Cl^- ions about the Na^+ ions is thus self-evident.

This idea of generating ionic crystalline structures by decorating the interstitial voids in (more-or-less) close-packed arrangements of one type of ion with the ions of the opposite type is general and useful. Thus, for example, the NiAs structure is similar to the rocksalt structure in that the octahedral interstices are also occupied by cations, but the anion sub-lattice has a different (h.c.p.) stacking sequence, viz. ABAB... Another structure based on an h.c.p. stacking of anions is the wurtzite (hexagonal ZnS) structure, in which only one type (T_+ or T_-) of tetrahedral interstice is occupied by

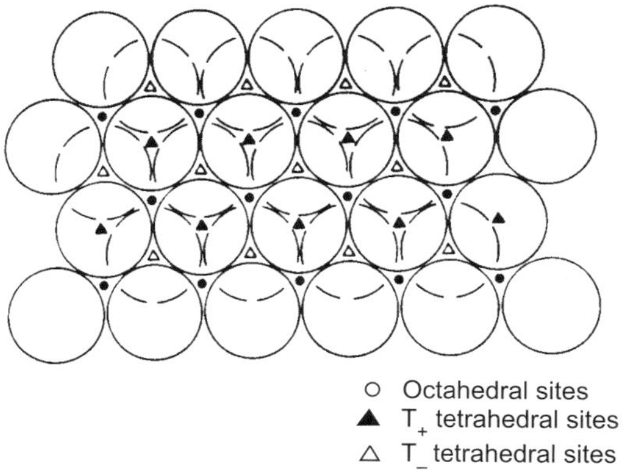

Fig.2.30 Distribution of interstitial voids between two adjacent close-packed layers of atoms. Dashed circles represent atoms lying below the plane of atoms represented by full circles.

cations: in this case, the structure consists of layers of ZnS_4 (or equivalently SZn_4) tetrahedra, connected via the apices, and arranged in an ABAB stacking sequence (alternate layers rotated by 180° about the *c*-axis). It should be remarked that, in many cases, the interstitial voids are too small to accept the cations if the anion sub-lattice is *truly* close-packed. In this case, an expansion of the anion sub-lattice occurs, so that anions no longer contact each other in the {111} planes, but nevertheless the underlying lattice structure is retained; such structures are sometimes called eutactic. Table 2.2 gives a number of examples of common ionic crystal types based on such interstitial decorations of close-packed structures. In some structures with relatively

Table 2.2 Some ionic crystal structures and the site occupancy of interstitial voids in close-packed arrangements of one type of ion

	Interstitial sites			
Anion arrangement	T_+	T_-	O	*Example*
c.c.p. (f.c.c.)	—	—	1	rocksalt (NaCl)
	1 —	—		zincblende, sphalerite (ZnS)
	1/8	1/8	1/2	spinel ($MgAl_2O_4$)
	—	—	1/2	$CdCl_2$
	1	1	—	antifluorite (Na_2O)
h.c.p.	—	—	1	NiAs
	1	—	—	wurtzite (ZnS)
	—	—	1/2	CdI_2
	—	—	1/2	rutile* (TiO_2)
	—	—	2/3	Al_2O_3
	1/8	1/8	1/2	olivine (Mg_2SiO_4)
c.c.p. CaO_3 layers	—	—	1/4	perovskite ($CaTiO_3$)

* The oxide layers are not planar but buckled.

large cations and *small anions*, the cations instead form the close-packed layers and the anions then occupy the interstitial holes. An example is the fluorite (CaF_2) structure (also adopted by the dioxides of large tetravalent cations, e.g. PbO_2, UO_2), the inverse of which is the antifluorite structure where the anions are close-packed (see Table 2.2).

Other ionic structures exist which are *not* based on close-packed structures. For example, the caesium chloride structure is based on the simple cubic structure. There is one formula unit per primitive cell, with one type of ion located at 000 and the other type at the body-centre position (1/2 1/2 1/2) of the simple cubic lattice (see Fig. 2.31). The nearest-neighbour coordination is evidently eight in this case. The structure is *not* b.c.c. since different types of atoms occur at the two lattice sites.

Yet another way of regarding such structures is in terms of (almost) space-filling arrangements of linked polyhedra. Thus, for the rocksalt (NaCl) structure, as an

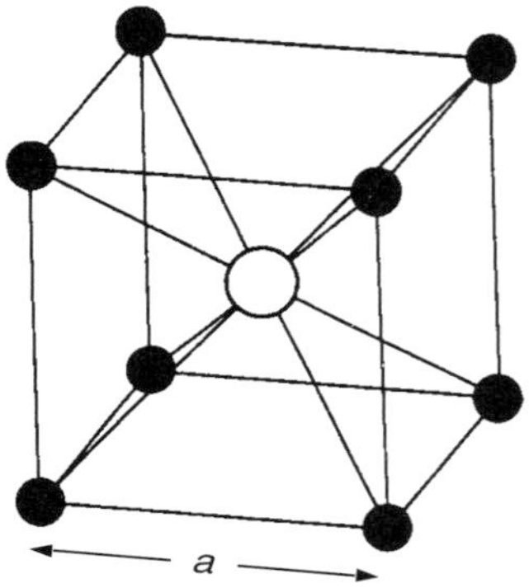

Fig.2.31 Representation of the CsCl structure (Cs = •; Cl = o).

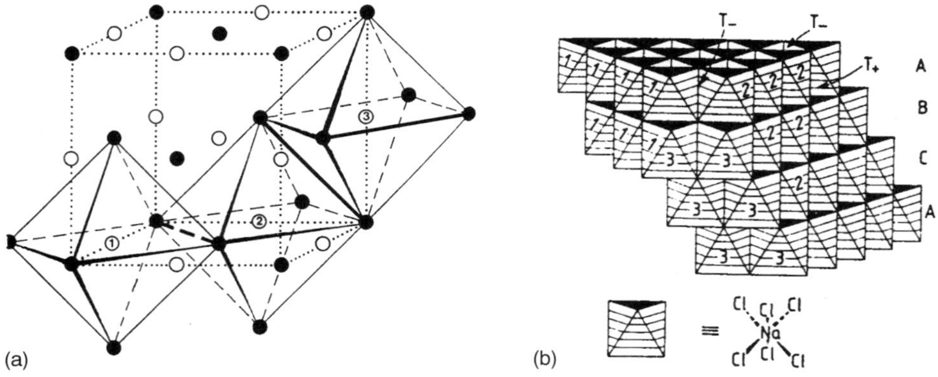

Fig.2.32 (a) Unit cell of the rocksalt structure illustrating the edge-sharing of $NaCl_6$ octahedra. (b) The rocksalt structure characterized as an array of edge-sharing octahedra. The tetrahedral interstices, T_+, T_-, are also marked.

example, the crystal structure can be regarded as being made up from, say, $NaCl_6$ octahedra with all 12 edges of a given octahedron shared with those of neighbouring octahedra (Fig. 2.32). Such a structure, however, is not completely space-filling; empty tetrahedral cavities exist within the framework of linked octahedra (Fig. 2.32b). (These are the same as the tetrahedral interstices T_+, T_- in the close-packed atomic representation.) The rutile (TiO_2) structure, as another example, can be regarded as a framework of octahedra each sharing two edges and six corners.

2.2.4.3 Criteria governing adoption of ionic structures

If ionic structures are regarded as comprising one type of ion (e.g. cations) occupying interstitial voids in a (close-packed) arrangement of another type of ion (e.g. anions), then one factor governing whether a given pair of ions (say M^+X^-) will crystallize in a particular crystal structure is whether the cation will fit into a particular (tetrahedral or octahedral) interstice. Thus, a factor determining the occurrence of certain crystal types is the radius ratio r_M/r_X of cation radius to anion radius. Minimum radii of interstices (and hence of cations) can be determined for a given structure by geometry.

For the case of an f.c.c. lattice (e.g. the rocksalt structure), in which the anions form a close-packed array, anions 1, 2, 3 in Fig. 2.33a form one close-packed layer in the (111) plane, in contact with another close-packed layer (anions 4, 5, 6), and thereby enclose an octahedral interstice, O. The diameter of the interstice is determined by the four atoms 2, 3, 4, 5 arranged in the (100) equatorial plane (Fig. 2.33b). Thus, from Fig. 2.33b it is evident that

$$[2(r_M + r_X)]^2 = (2r_X)^2 + (2r_X)^2$$

and hence

$$\frac{r_M}{r_X} = \sqrt{2} - 1 = 0.414. \tag{2.40}$$

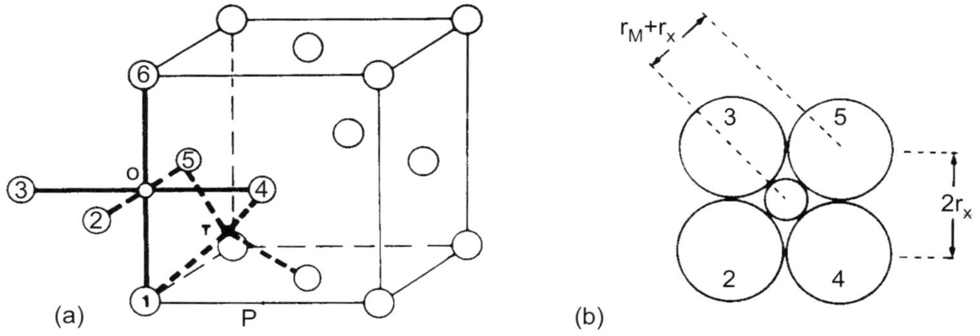

Fig.2.33 (a) Octahedral (O) and tetrahedral (T) interstitial (cation) sites in an f.c.c. anion array. (b) Depiction of the atomic arrangement around an octahedral interstice in the (100) equatorial plane. The numbers of the atoms correspond to those in (a).

For the tetrahedral interstices (T) in an f.c.c. structure, the distance between point P and anion 5 in Fig. 2.33a is equal to $2(r_M + r_X)$ and is equal to the body diagonal of one of the eight small cubes making up the f.c.c. unit cell. Thus,

$$[2(r_M + r_X)]^2 = (2r_X)^2 + (\sqrt{2}r_X)^2$$

or

$$\frac{r_M}{r_X} = \frac{1}{2}(\sqrt{6} - 2) = 0.225. \tag{2.41}$$

Cations having radius ratios $r_M/r_X > 0.414$ fitting into octahedral holes in an f.c.c. anionic lattice would push apart the anions (so that they were no longer close-packed but eutactic) until a critical radius ratio is reached where an increased coordination (eight-fold, characteristic of the CsCl structure) of anions in contact with the cation becomes possible. For the CsCl structure with eight-fold coordination, the sum of the anion and interstice diameters equals the body-diagonal distance of the cubic cell

$$[2(r_M + r_X)] = \sqrt{3}(2r_X)$$

or

$$\frac{r_M}{r_X} = (\sqrt{3} - 1) = 0.732. \tag{2.42}$$

However, it should be admitted that radius-ratio considerations are generally not particularly successful for predicting preferred crystal types, especially if there is any partial covalent contribution to the bonding in the case of heteropolar systems. In such cases, the Phillips approach (§2.2.1) based on bond ionicities is much more successful. A plot of the ionic energy $\mathscr{E}_i$ versus covalent energy $\mathscr{E}_c$ for a number of crystals of the type $A^N B^{8-N}$ (where N is the number of valence electrons), including Group IV solids (C, Si, Ge), III–V compounds (GaAs, etc.), II–VI materials (e.g. ZnS), etc. shows that a critical value of ionicity (eqn. (2.8)), $f_i^c = 0.785$, separates all four-fold (predominantly covalent) structures from six-fold coordinated ionic structures (see Fig. 2.34).

2.2.4.4. The Madelung constant

Having discussed the crystal structures that ionic materials may adopt, we are now in a position to evaluate the Madelung constant, A, introduced in eqn. (2.26). As mentioned earlier, the value of A depends on the particular crystal under consideration. However, evaluation of the summation in eqn. (2.26) cannot be carried out straightforwardly by simply inserting the appropriate interionic distances, since the resulting sum is not obviously convergent. Instead, different ways of grouping the terms must be found in order to make the sum convergent. As an example, for the case of the rocksalt (NaCl) structure (fig. (2.29)), the charges on the ions can be formally assigned to be either inside or outside the cubic unit cell. Thus, for a given cell, ions on faces retain 1/2 their charge, those at edges 1/4 their charge and those at corners just 1/8 their charge (two faces, four edges and eight corners being shared between neighbouring cells). Thus, evaluating eqn. (2.26) for ions within the cell having these formal charges gives $\frac{6/2}{1} - \frac{12/4}{\sqrt{2}} + \frac{8/8}{\sqrt{3}} = 1.45$ as the contribution to A. Including the contribution from the next largest cube gives an additional contribution of 0.3, making the estimate

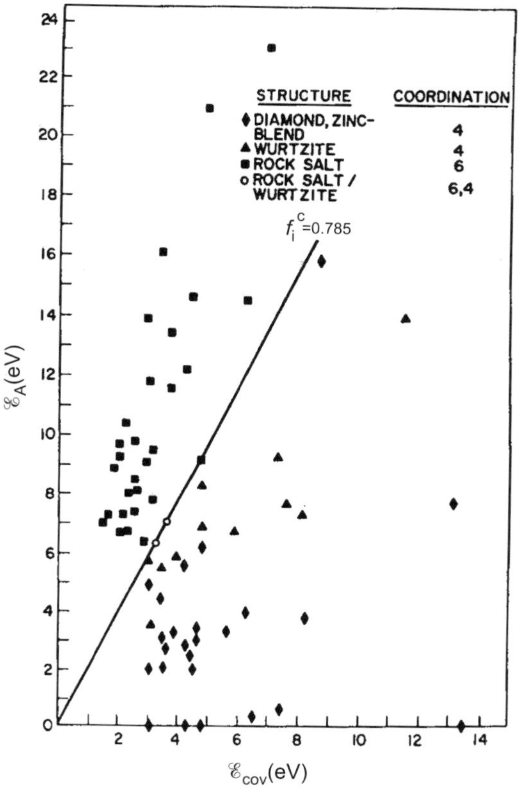

Fig.2.34 Plot of values of the ionic energy $\mathscr{E}_i$ versus $\mathscr{E}_c$ (representing the strength of the covalent interaction) for crystals of the type $A^N B^{8-N}$. The line corresponding to the critical ionicity $f_i^c = 0.785$ divides four-fold from six-fold coordinated crystals (Phillips (1973). Reprinted by permission by Academic Press, Inc.).

Table 2.3 Values of the Madelung constant for various crystal structures

Crystal type	*A*
Zincblende (ZnS)	1.638 06
Wurtzite (ZnS)	1.641 32
Rocksalt (NaCl)	1.747 558
CsCl	1.762 670
CdI_2	4.71
Rutile (TiO_2)	4.816
Fluorite/antifluorite	5.038 78
Corundum (Al_2O_3)	25.031 2

(After Borg and Dienes (1992). Reproduced by permission of Academic Press, Inc.)

$A \simeq 1.75$, which is already close to the precise value. Evaluation of A for even larger cells produces a progressively more accurate estimate. This method only works satisfactorily if the surfaces of the cell are charge neutral. A powerful general method of evaluating the Madelung constant is via the Ewald summation method (§2.5.3.3). Table 2.3 gives accurate values of the Madelung constant for the f.c.c. rocksalt and other ionic structures.

2.2.5 Covalent solids

2.2.5.1 Interatomic bonding

Covalent bonding differs from the previous types of bonding interactions by being strongly *directional.* The electron density, associated with a pair of electrons with opposing spins being characteristic of covalency, is spatially localized between pairs of atoms and, particularly when orbital hybridization occurs, the electron charge density of bonding orbitals is concentrated also in certain orientations: for example, sp^2 hybrids are arranged in a trigonal planar arrangement, with an angle of 120° between orbitals; sp^3 hybrids point towards the four corners of a tetrahedron, with an angle of 109° 28′ between orbitals (see §5.3.2). The formation of such covalent bonds is favoured energetically essentially because of the enhanced electrostatic interaction between the two positively charged nuclei of the atoms forming the bond and the negatively charged electron distribution lying between the nuclei. Covalent bonds are generally strong, and comparable in strength to ionic bonds. A full description of covalent-bond formation in *isolated* molecules is given in every physical chemistry textbook, and so will not be repeated here. However, in the solid state, due to interactions between an orbital on a given atom and orbitals on many other atoms, the discrete energy levels associated with molecular orbitals become broadened into energy *bands*; a full discussion of such solid-state effects will be deferred until Chapter 5.

2.2.5.2 Covalent structures

Due to the strong orientational characteristic of covalent bonding, structures comprising atoms bonded in this way have low densities: the nearest-neighbour coordination number is determined by the type of hybridization present (e.g. four for sp^3 hybridization), and is very much less than the value of 12 characteristic of close-packed structures.

The diamond structure (see Fig. 2.35a) is perhaps the archetypal covalent structure, adopted by the Group IV elements C (in the diamond polymorph), Si, Ge and Sn (in the α polymorph). The primitive basis contains *two* atoms (at 000 and 1/4 1/4 1/4), and hence there are eight atoms in the conventional cubic unit (Fig. 2.35). The packing density of the diamond crystal is only $\phi = 0.34$ (recall that $\phi = 0.74$ for close-packed f.c.c. or h.c.p. crystals). Although the representation of the structure shown in Fig. 2.35a in terms of a network of covalent bonds is the most usual, an equivalent way of regarding the structure is to represent it as a (non-space-filling) framework of CC_4 tetrahedra, each apex of a tetrahedron being connected to four others (Fig. 2.35b). Finally, but not particularly usefully, the diamond structure can also be regarded,

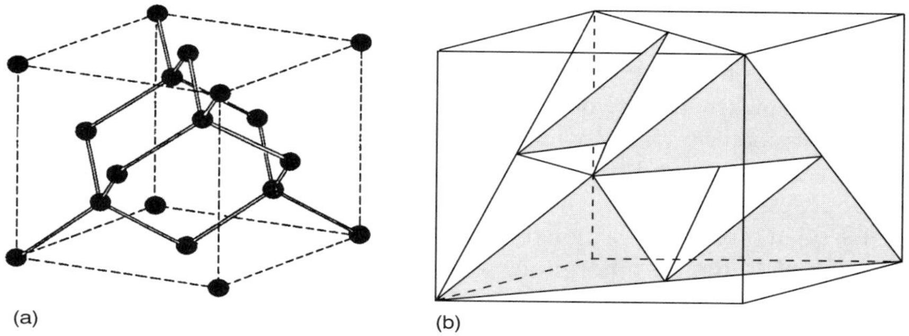

Fig.2.35 (a) The diamond structure, showing the tetrahedral arrangement of covalent bonds between atoms. (b) The diamond structure represented as a framework of corner-sharing CC_4 tetrahedra, i.e. with a C atom at the centroid of each tetrahedron.

following the discussion in §2.2.4.2, as a cubic close-packed array of C atoms, with an equivalent number of (equal-sized) C atoms occupying T_+ interstices.

The sphalerite (cubic ZnS) structure is simply derived from the diamond structure by decorating one of the atoms of the basis with the cation and the other by the anion; heteropolar bonding is thereby assured. This structure is commonly adopted by binary AB compounds having mixed iono-covalent bonding, e.g. the materials SiC, GaAs, AlAs, GaP, AlP and InSb, all of which are semiconductors and exhibit interesting electronic behaviour (see Chapters 6 and 8). The sphalerite structure, too, can be represented either in terms of bonds (cf. Fig. 2.35a) or coordination polyhedra (cf. Fig. 2.35b), the latter being ZnS_4 (or equivalently SZn_4) tetrahedra. The ideal sphalerite structure can also be regarded instead as a decoration by, say, the Zn^{2+} ions of the T_+ interstices in a cubic close-packed array of S^{2-} anions (with an ABCABC stacking repeat). (Compare this structure with the wurtzite (hexagonal ZnS) structure—§2.2.4.2.) However, this simple stacking sequence is often not preserved in real crystals. Both ZnS and SiC, for example, exhibit polytypism, where the stacking sequence is much longer and more complex than the simple ABC repeat of the c.c.p. structure.

Symmetry 2.3

An understanding of the structure, and concomitant physical properties, of crystals can be greatly enhanced by the use of symmetry aspects. However, a proper understanding of the use of symmetry in this way relies on having a thorough prior knowledge of group theory. For this reason, and because this book does not set out to be a text on crystallography, only essential features of symmetry-related aspects of crystalline solids will be discussed here. (Moreover, symmetry is, obviously, of no use in the discussion of aperiodic materials, such as amorphous solids.) For further details, the reader is referred to the books by Burns (1990), Burns and Glazer (1990), Altmann (1991) and Nichols (1995), for example.

2.3.1 Symmetry operations

A symmetry operation applied to an object (e.g. a molecule or a crystal) interchanges the positions of various parts of the object (e.g. the atoms) in such a way that the object appears *exactly* as it did before the operation took place (e.g. it is in an equivalent position). A trivial example, applicable to all objects, is the so-called identity operator, which leaves the object unchanged. Other operations involve rotation, inversion, reflection and translation, and combinations thereof, as will be seen shortly.

It is useful to distinguish, at the outset, two general types of symmetry operations. Point-symmetry operations are those carried out with respect to a *fixed point* in space; this point therefore does not translate during the operation. Such symmetry operations apply obviously to, say, individual molecules, but can also be used to characterize crystal structures. A point group is a self-consistent set of point-symmetry operations satisfying the rules for the existence of a mathematical group.† In contrast, the space group of a crystal is the set of *all* the symmetry operations, including the (infinite number of) unit-cell translations (cf. eqn. (2.1)), and symmetry operations (also involving translations) such as screw operations and glide planes (see later), if they exist, as well as the point-group operations. There are 32 different crystallographic point groups and a total of 230 different space groups.

One particularly irksome and confusing feature is that *two* different schemes for denoting symmetry operations are in use: the Schönflies notation (commonly used by solid-state physicists and chemists and molecular spectroscopists), and the Hermann–Mauguin notation (used by crystallographers), sometimes referred to as the International notation (after the standard reference in which the system is outlined, the *International Tables for X-ray Crystallography* (Henry and Lonsdale (1952)). Since both systems are so widely used, a given symmetry operation described in this book will be denoted as Schönflies (International).

The various point-symmetry operations of interest are listed in Table 2.4, together with their labelling according to the two notations. The identity operator $E(1)$, already

† A **group**, in the strict mathematrical sense, is a collection of elements that satisfy the four conditions: (i) **closure** (the result of multiplying any two elements must also be member of the set); (ii) **identity** (one of the elements must be the identity operator E, such that $EA_i = A_iE = A_i$ for all members A_i of the set); (iii) **inverse** (every element must possess an inverse, also in the set, such that $A_i^{-1}A_j = E$, where $A_j^{-1} = A_i$ and A_j is a member of the set); (iv) **associativity** (such that $(AB)C = A(BC)$).

Table 2.4 Point-symmetry operations

Symmetry element	*Schönflies*	*International (Hermann–Mauguin)*
Identity	E	1
Inversion (centre of symmetry)	i	$\bar{1}$
Rotation	C_n	n
Rotatory inversion	iC_n	$\bar{n}$
Reflection (mirror)	σ	m
Improper rotation (rotoreflection)	S_n	—

mentioned, simply leaves an object unchanged (or equivalently rotated by either 0 or 2π radians about any axis); thus for any coordinate set (x, y, z), $E(1)$ $(x, y, z) \rightarrow (x, y, z)$. The inversion operator inverts an object through an origin (the centre of inversion or symmetry), i.e. i $(\bar{1})$ $(x, y, z) \rightarrow (-x, -y, -z)$. The ('proper') rotation operator $C_n(n)$ rotates an object by $360°/n$ ($2\pi/n$ radians) about an axis (according to the right-hand screw rule by convention), i.e. $C_4(4)$ $(x, y, z) \rightarrow (y, -x, z)$. If this axis is *not* the principal axis (conventionally denoted the c-or z-axis), i.e. that having the highest symmetry of the object, then the operator is denoted C'_n, and a superscript in the Schönflies notation denotes the actual rotation axis; e.g. C_n^x is an n-fold rotation about the x-axis, i.e. C_2^x $(x, y, z) \rightarrow (x, -y, -z)$. Note that a *numerical* superscript means that the simple rotation operator should be repeated that number of times; e.g. C_n^m denotes the m-fold application of the n-fold rotation operator about the principal axis. The reflection (or mirror-plane) operator σ(m) displaces a given point of an object to the other side of the plane in question by an equal distance along the perpendicular to the plane; thus, for reflection across the z-plane (containing the x-y-axes), σ(m) $(x, y, z) \rightarrow (x, y, -z)$. Different planes of reflection can be considered: reflection in the horizontal plane, σ_h, the plane of reflection being perpendicular to the principal axis and containing the origin; reflection in the vertical plane, σ_v, the plane of reflection containing the principal axis; and reflection in a diagonal plane, σ_d, the reflection plane containing the principal axis but also bisecting the angle between the two-fold axes normal to the principal axis. Finally, the improper rotation (or rotoreflection) operator S_n refers to a rotation by $2\pi/n$ followed by a reflection in the horizontal plane, i.e. $S_n = \sigma_h C_n$ (note the convention for writing a subsequent symmetry operation to the *left* of the prior operation). Thus, S_4 $(x, y, z) \rightarrow (y, -x, -z)$. These point-symmetry operations are illustrated in Fig. 2.36.

As an example, there are 12 point-symmetry operations characteristic of the molecule PF_3Cl_2 (see Fig. 2.37), namely E, C_3, C_3^2, $3C_2^1$, σ_h, $3\sigma_v$, S_3 and S_3^5. This complete set of symmetry operations makes up the point group of the molecule. A point group can be represented graphically by means of a stereogram. This is a projection of points lying on the surface of a sphere surrounding the object onto the x–y plane, the intersections with the plane being with lines from the points to the opposite pole. A point in the $+z$ or $-z$ hemisphere is represented by an open circle (o) or dot (•), respectively. (N.B. the difference in notation for points above and below a plane between point groups (o, •) and space groups (+, −).) For the case of PF_3Cl_2, starting with an arbitrary point, a general equivalent position (GEP) (*not* on a symmetry point, line or plane), and denoted as 1, the first symmetry operation $E(1)$ applied to 1 leaves it unchanged. The second operation, $C_3(3)$, changes 1 into the GEP 2, rotated by 120° about the principal c-axis

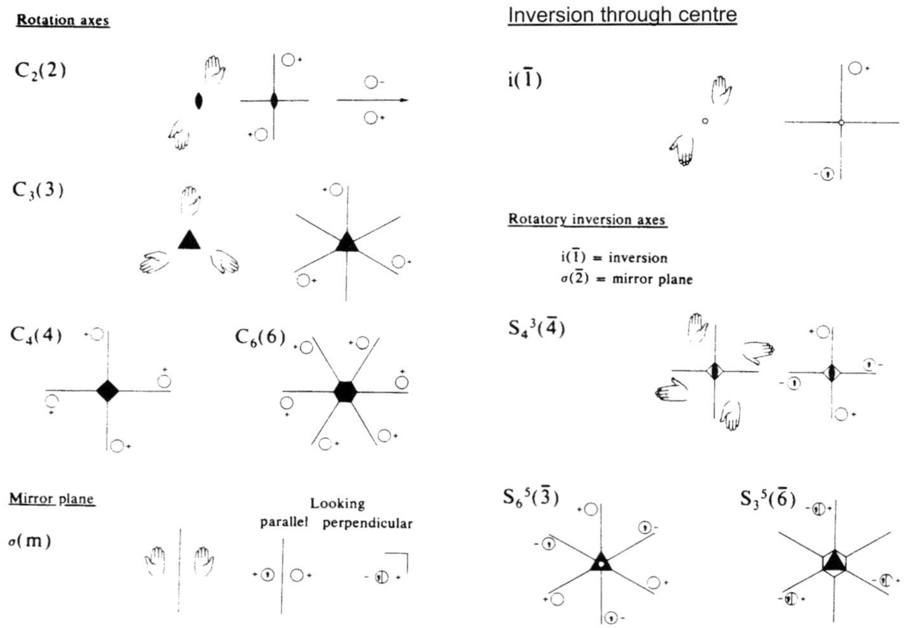

Fig.2.36 Illustration of point-symmetry operations and their notation in the Schönflies (International) system. Conventional diagrammatic symbols, e.g. filled oval, triangle, square, etc., used to represent such operations are also shown. Open circles denote a general object, such as a set of atoms, to which the symmetry operation is applied. The symbols +, − denote objects above, and below, the plane of the paper respectively. A split circle denotes one circle above and another below the plane. A comma inside a circle represents an enantiomorphic (or mirror) image of a circle without a comma (hands are also used to illustrate graphically such enantiomorphic pairs). This (+−,) notation is used for space groups, but not for point groups. (After Burns (1985). Reprinted by permission of Academic Press, Inc.)

(the Cl—P—Cl bond in Fig. 2.36), $C_3^2(3^2)$ applied to 1 generates the GEP 3, rotated by 240°, etc. The stereogrammatic representation of the whole point group is shown in the inset to Fig. 2.37, where the first number of each pair refers to o and the second to •. The particular symmetry elements are also represented on the diagram. Thus, the thick radial lines represent the three σ_v operations and the thick line for the outer circle represents σ_h. The symbol at the centre refers to $S_3(\bar{6}^5)$, and the three C_2' operations in the xy plane are also indicated. Note that, in general, there are as many GEPs as there are symmetry elements in the point group (12 in this case).

2.3.2 Crystallographic point groups

The various point groups are given symbols from which all symmetry elements contained within the group may be inferred. The simplest examples are those containing purely *rotational* symmetry elements. For individual molecules, there are *no* restrictions

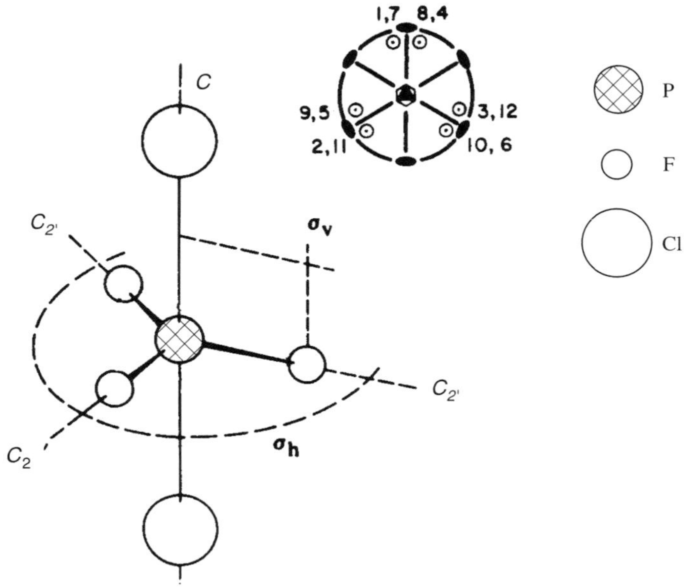

Fig.2.37 Point-symmetry elements for the molecule PF_3Cl_2. Shown inset is the stereogram for the point group of the molecule.

on the rotational symmetry operations C_n or S_n; e.g. five-fold rotational symmetry exists for the pentagonal molecule ferrocene ($Fe(C_5H_5)_2$). However, because we are interested in *crystals*, having translational periodicity, rotational symmetries C_n with $n = 5$ and $n > 6$ are incompatible with such long-range order. This restriction can be demonstrated in two dimensions as follows.

Consider two lattice points A, B, separated by a unit translation vector $\boldsymbol{r}$. Applying a rotational symmetry operation, R, to A generates a new point A′ (rotated by an angle α). Similarly, the inverse operator R^{-1} (also a symmetry operation) applied to B generates B′ (rotated by an angle α in the opposite sense)—see Fig. 2.38. Since R and R^{-1} are both symmetry operations, A′ and B′ must also be lattice points, and so the vector $\boldsymbol{r}'$ connecting them must be an integral multiple m of $\boldsymbol{r}$ (because of periodicity), i.e.

$$r' = mr \tag{2.43a}$$

and from the geometry of Fig. 2.38,

$$r' = -2r\cos\alpha + r. \tag{2.43b}$$

Combining these two equations gives

$$\cos\alpha = (1 - m)/2 = M/2 \tag{2.44}$$

where M is also an integer. For closure of R, $0 < \alpha < 180°$, and so $|\cos\alpha| \leqslant 1$ or $|M| \leqslant 2$. Thus, from eqn. (2.44), α can only have values $0, \pi/3, \pi/2, 2\pi/3, \pi$; the only

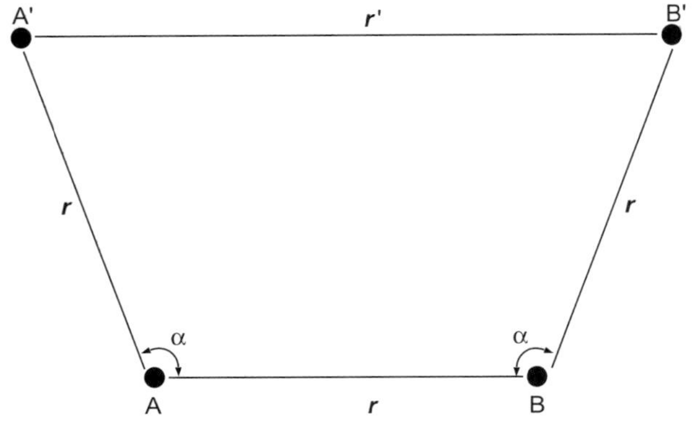

Fig.2.38 Lattice points A′ and B′ obtained by applying rotational symmetry operations to the lattice points A and B, respectively. Translational periodicity is only maintained for $\alpha = 0$, $\pi/3$, $\pi/2$, $2\pi/3$ and π.

allowed rotations consistent with lattice periodicity are hence $2\pi/n$, where $n = 1, 2, 3, 4, 6$, i.e. not 5 and not > 6. Of course, five-fold rotational symmetry is not incompatible with being able to tile a plane in 2D, for example—see the Penrose tiling in Fig. 2.5—but such a structure is translationally aperiodic.

Thus, of the 32 *crystallographic* point groups (consistent with lattice periodicity) there are only 5 purely rotational point groups, given the Schönflies (International) symbols $C_1(1)$, $C_2(2)$, $C_3(3)$, $C_4(4)$ and $C_6(6)$, where the group $C_6(6)$, for example, contains the set of symmetry elements $\{E(1), C_6(6), C_3(3), C_2(2), C_3^2(3^2), C_6^5(6^5)\}$. (Note that confusingly, but only in this case of pure rotational point groups, the symbol for the point group is the *same* as that of one of its elements.)

Further point groups are generated by including symmetry operations other than rotations. Thus, addition of a σ_h mirror plane to each of the above 5 rotational point groups generates another 5 groups, written as $C_{nh} \equiv C_n \times \{E, \sigma_h\}$, which means the set of operations resulting from the multiplication of every operation in C_n by E plus those obtained by multiplying by σ_h. The new point groups thus include twice as many symmetry operations. They are $C_{1h}(m)$, $C_{2h}(2/m)$, $C_{3h}(\bar{6})$, $C_{4h}(4/m)$, C_{6h}('$6/m$'); all, but the last, are crystallographic point groups. (The n/m symbol in the International notation means that the m-mirror plane is *perpendicular* to the n-fold rotation axis.)

Inclusion of a vertical mirror plane equivalently generates another set of point-groups, $C_{nv} \equiv C_n \times \{E, \sigma_v\}$. However, $C_{1v} \equiv C_{1h}$, and so only four new ones are created ($C_{2v}(2mm)$, $C_{3v}(3m)$, $C_{4v}(4mm)$ and $C_{6v}(6mm)$). (The lack of a slash symbol between the n-fold axis and the mirror planes in the International notation means that the mirror plane *contains* the axis.)

Improper rotation (rotoreflection) operations generate another three distinct point groups. S_2 contains operations $\{E,i\}$ and the group is conventionally denoted $C_i(\bar{1})$. $S_3(\bar{6})$ is identical to the already mentioned $C_{3h}(\bar{6})$. Additional point groups are therefore $S_4(\bar{4})$ and $S_6(\bar{3})$.

Adding a two-fold axis perpendicular to the principal axis of the five purely rotational point groups ($C_1(1)$ etc.) generates four new point groups (D_n) with, again, twice as many symmetry operations, viz. $D_n \equiv C_n \times \{E, C_2'\}$, i.e. $D_2(222), D_3(32), D_4(422)$ and $D_6(622)$. (Note that $D_1 \equiv C_2$, and so is not included.)

Four additional point groups are created by adding a C_2' axis to the C_{nh} point groups, viz. $D_{nh} \equiv C_{nh} \times \{E, C_2'\}$, i.e. $D_{2h}(mmm), D_{3h}(\bar{6}m2), D_{4h}(4/mmm)$ and D_{6h} $(6/mmm)$.

Adding a C_2' axis to the S_n point groups generates two new point groups, viz. $D_{nd} \equiv S_n \times \{E, C_2'\}$, i.e. $D_{2d}(\bar{4}2m)$ and $D_{3d}(\bar{3}m)$.

There are, in addition, five cubic point groups that do *not* possess a unique axis like all the previously cited point groups, but instead are characterized by having four three-fold axes (along the $\{111\}$ directions of a cube). The point group with the smallest number of (purely rotational) symmetry elements is $T = \{E, 4C_3, 4C_3^2, 3C_2\}$, where the C_2 operations are along the *a*-, *b*-and *c*-axes. The point groups T_h and T_d are obtained by including a horizontal mirror plane (σ_h) and a diagonal mirror plane (σ_d), respectively, i.e. $T_h \equiv T \times \{E, \sigma_h\}$ and $T_d \equiv T \times \{E, \sigma_d\}$. The point group O has six four-fold axes and 24 symmetry elements, and the group with the highest number (48) of symmetry operations is $O_h \equiv O \times \{E, \sigma_h\}$. In summary, the cubic point groups are thus $T(23)$, $T_h(m3)$, $T_d(\bar{4}3m)$, O (432) and O_h (m3m).

The 32 crystallographic point groups are important because macroscopic physical properties of crystals have at least the symmetry of the point group. They are listed for convenience in Fig. 2.39, with their symmetry operations (and GEPs) represented in stereogrammatic form. They can be assigned to one of the seven different crystal systems (see Table 2.1) by ascertaining which symmetry operations maintain a lattice (*devoid* of a basis of atoms) within a given crystal system. For example, for the case of a monoclinic lattice, the symmetry operations $C_2(2)$ and $\sigma(\bar{2})$ are the only ones that keep the lattice as being monoclinic. The point groups $C_2(2) \equiv \{E, C_2\}$ and $C_{1h} = \{E, \sigma_h\}$ are thus consistent with the monoclinic system, as is $C_{2h}(2/m) = \{E, C_2, i, \sigma_h\}$. Assignment of the 32 crystallographic point groups to the seven crystal systems is given in Table 2.5.

2.3.3 Space groups

As mentioned previously, the *space group* of a crystal is determined both by the *point group* (relating to the lattice) and by *translational* symmetry operations (relating to the decoration of the lattice by a basis of atoms to form the crystal). The latter operations include the (infinite number of) lattice translations (eqn. (2.1)), screw operations and glide planes. Screw operations and glide planes are composite operations: a screw operation is a rotation followed by a translation, τ, which is a fraction of the unit-cell dimension, and a glide plane is a reflection across a plane followed by a translation τ. A screw operation can be written in the so-called Seitz notation (used by physicists) as $\{R/\tau\}$, where R is a rotational symmetry operation. Note that, in this case, the screw operation is applied to a *single origin* in a unit cell and, as a result, τ can have components perpendicular to the axis of rotation. (Crystallographers, conventionally and confusingly, apply screw-symmetry operations to *different* origins in the unit cell,

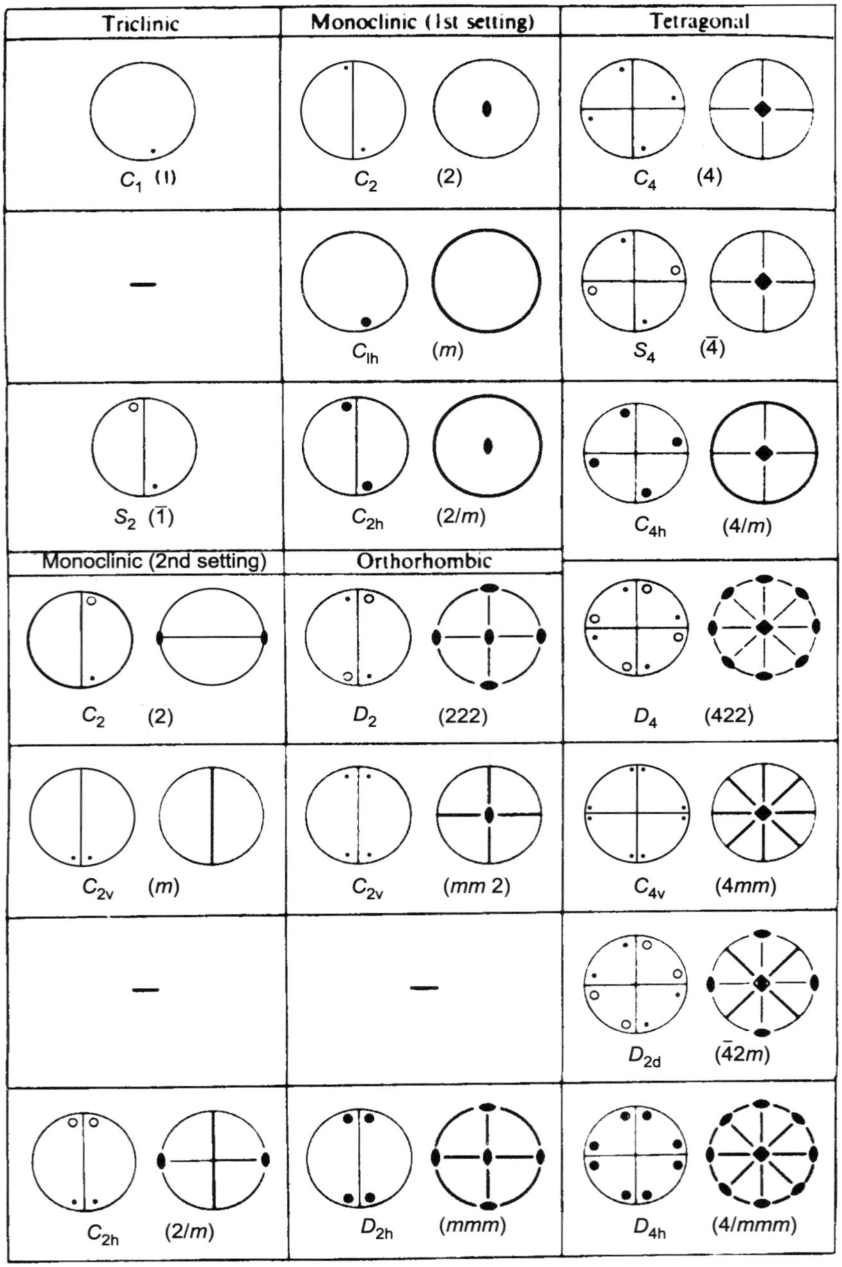

Fig.2.39 (a) Stereograms of the 32 crystallographic point groups, arranged according to the seven different crystal systems. The left-hand stereograms illustrate the GEPs, and the right-hand stereograms the symmetry elements, of each point group. (After Burns (1985). Reprinted by permission of Academic Press, Inc.)

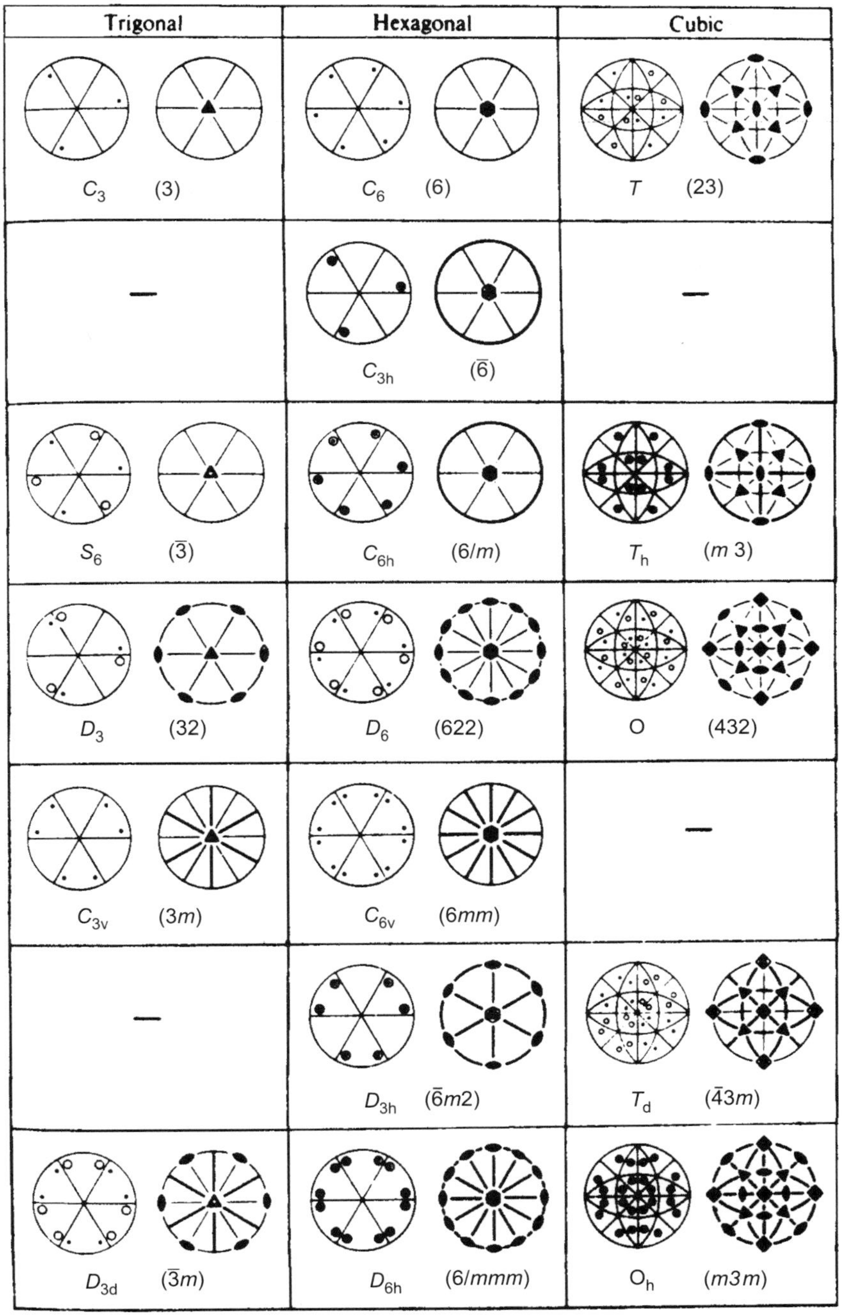

Trigonal	Hexagonal	Cubic
C_3 (3)	C_6 (6)	T (23)
—	C_{3h} ($\bar{6}$)	—
S_6 ($\bar{3}$)	C_{6h} (6/m)	T_h (m 3)
D_3 (32)	D_6 (622)	O (432)
C_{3v} (3m)	C_{6v} (6mm)	—
—	D_{3h} ($\bar{6}$m2)	T_d ($\bar{4}$3m)
D_{3d} ($\bar{3}$m)	D_{6h} (6/mmm)	O_h (m3m)

Fig.2.39 (b)

Table 2.5 The 32 crystallographic point groups assigned to the 7 crystal systems

Schönflies	*International*	*Symmetry Elements*	*Generating Elements*
Triclinic			
C_1	1	E	E
$S_2(C_i)$	$\bar{1}$	$E\ i$	i
Monoclinic			
C_2	2	$E\ C_2$	C_2
$C_{1h}(C_s)$	m	$E\ \sigma_h$	σ_h
C_{2h}	$2/m$	$E\ C_2\ i\ \sigma_h$	iC_2
Orthorhombic			
$D_2(V)$	222	$E\ C_2\ C_2'\ C_2'$	$C_2C_2^y$
C_{2v}	$mm2$	$E\ C_2\ \sigma_v\ \sigma_v$	$C_2\sigma_v^y$
$D_{2h}(V_h)$	mmm	$E\ C_2\ C_2'\ C_2'\ i\ \sigma_h\ \sigma_v\ \sigma_v$	$i\sigma_v^y C_2$
Tetragonal			
C_4	4	$E\ 2C_4\ C_2$	C_4
S_4	$\bar{4}$	$E\ 2S_4\ C_2$	S_4^3
C_{4h}	$4/m$	$E\ 2C_4\ C_2\ i\ 2S_4\ \sigma_h$	iC_4
D_4	422	$E\ 2C_4\ C_2\ 2C_2'\ 2C_2''$	$C_2^y C_4$
C_{4v}	$4mm$	$E\ 2C_4\ C_2\ 2\sigma_v\ 2\sigma_d$	$\sigma_v^y C_4$
$D_{2d}(V_d)$	$\bar{4}2m$	$E\ C_2\ 2C_2'\ 2\sigma_d\ 2S_4$	$C_2^y S_4^3$
D_{4h}	$4/mmm$	$E\ 2C_4\ C_2\ 2C_2'\ 2C_2''$ $i\ 2S_4\ \sigma_h\ 2\sigma_v\ 2\sigma_d$	$iC_2^y C_4$
Trigonal (Rhombohedral)			
C_3	3	$E\ 2C_3$	C_3
$S_6(C_{3i})$	$\bar{3}$	$E\ 2C_3\ i\ 2S_6$	iC_3
D_3	32	$E\ 2C_3\ 3C_2'$	$C_2^y C_3$
C_{3v}	$3m$	$E\ 2C_3\ 3\sigma_v$	$\sigma_v^y C_3$
D_{3d}	$\bar{3}m$	$E\ 2C_3\ 3C_2'\ i\ 2S_6\ 3\sigma_v$	$iC_2^y C_3$
Hexagonal			
C_6	6	$E\ 2C_6\ 2C_3\ C_2$	C_2C_3
C_{3h}	$\bar{6}$	$E\ 2C_3\ \sigma_h\ 2S_3$	$\sigma_h C_3$
C_{6h}	$6/m$	$E\ 2C_6\ 2C_3\ C_2\ i\ 2S_3\ 2S_6\ \sigma_h$	iC_2C_3
D_6	622	$E\ 2C_6\ 2C_3\ C_2\ 3C_2'\ 3C_2''$	$C_2C_2^y C_3$
C_{6v}	$6mm$	$E\ 2C_6\ 2C_3\ C_2\ 3\sigma_v\ 3\sigma_d$	$C_2\sigma_v^y C_3$
D_{3h}	$\bar{6}m2$	$E\ 2C_3\ 3C_2'\ \sigma_h\ 2S_3\ 3\sigma_v$	$C_2^y \sigma_h C_3$
D_{6h}	$6/mmm$	$E\ 2C_6\ 2C_3\ C_2\ 3C_2'\ 3C_2''$ $i\ 2S_3\ 2S_6\ \sigma_h\ 3\sigma_v\ 3\sigma_d$	$iC_2^y C_2C_3$
Cubic			
T	23	$E\ 8C_3\ 3C_2$	$C_2C_3[111]$
T_h	$m3$	$E\ 8C_3\ 3C_2\ i\ 8S_6\ 3\sigma_h$	$iC_2C_3[111]$
O	432	$E\ 8C_3\ 3C_2\ 6C_2\ 6C_4$	$C_4C_3[111]$
T_d	$\bar{4}3m$	$E\ 8C_3\ 3C_2\ 6\sigma_d\ 6S_4$	$S_4^3 C_3[111]$
O_h	$m3m$	$E\ 8C_3\ 3C_2\ 6C_2\ 6C_4$ $i\ 8S_6\ 3\sigma_h\ 6\sigma_d\ 6S_4$	$iC_4C_3[111]$

and thereby ensure that the screw operation always has the translation τ *parallel* to the rotation axis.)

Of the 230 different space groups, two categories can be distinguished. Symmorphic space groups are specified completely by symmetry operations acting at a common point (but not necessarily involving a fractional unit-cell translation t) as well as the unit-cell displacements (eqn. (2.1)) characteristic of translational periodicity. They are obtained by combining the 32 crystallographic point groups with the 14 Bravais lattices (see Table 2.1). Each lattice point of a Bravais lattice belonging to a particular crystal system is decorated with a basis of atoms arranged in such a way that it satisfies the symmetry of a point group belonging to the *same* crystal system. In this way, the 73 symmorphic space groups can be generated. Non-symmorphic space groups, in contrast, are those for which at least one symmetry operation involving a translation τ (i.e. a glide plane or a screw operation) is *required*. Note that the 32 crystallographic point groups can thus be recovered from the 230 space groups by setting all translations equal to *zero*.

As for the point groups, the space groups are denoted using two parallel systems, the Schönflies symbol and the International notation (given in parentheses, as before). The International notation is somewhat more informative. For example, one of the orthorhombic space groups is $C_{2v}^1(Pmm2)$; the International notation denotes that, in this case, a primitive *P*-lattice has been used to derive the space group.

Reciprocal space 2.4

2.4.1 Reciprocal lattice

Thus far, the structures of solids have been discussed in relation to *atomic positions*, usually in three dimensions (although a 6D description is appropriate for quasi-crystals— see §2.1.1). That is to say, (3D) crystal structures, for example, can be characterized in terms of lattice-translation vectors $\boldsymbol{R}_{uvw}$ (cf. eqn. (2.1)) of the Bravais lattice, i.e.

$$\boldsymbol{R}_{uvw} = u\boldsymbol{a} + v\boldsymbol{b} + w\boldsymbol{c} \tag{2.45}$$

where u, v and w are positive or negative integers and a, b and c are, say, vectors defining the real-space unit cell.

However, instead of describing the crystal structure conventionally in terms of atomic positions (associated with lattice points), a completely equivalent structural description is in terms of *crystal planes*. A set of planes, denoted as (hkl) in Miller indices, is completely specified by the direction vector normal to the planes, represented, say, by a unit vector $\hat{\boldsymbol{n}}_{hkl}$, and the interplanar spacing, d_{hkl}. The overall structure of the crystal is then completely specified in terms of the set of values $\{\hat{\boldsymbol{n}}_{hkl}, d_{hkl}\}$.

Although valid, this particular way of describing crystal structures in terms of $\{\hat{\boldsymbol{n}}_{hkl}, d_{hkl}\}$ values is cumbersome and not particularly useful. However, a much more useful representation is to define the vector

$$\boldsymbol{G}_{hkl} = 2\pi\hat{\boldsymbol{n}}_{hkl}/\mathrm{d}_{hkl}. \tag{2.46}$$

(Note that it is conventional in crystallography to omit the factor of 2π.) These vectors, having the dimension of inverse length, then define the so-called reciprocal lattice. This plays a central rôle in enabling a simple understanding to be achieved of, for example, diffraction by crystals of waves (be they X-rays and neutrons incident externally, or electrons present within the crystal), as will be seen later. Thus, every crystal structure has both a real-space (Bravais) lattice and a reciprocal lattice associated with it.

Each point in the reciprocal lattice corresponds to a family of planes in real (lattice) space. However, it should be emphasized that the crystal planes (hkl) corresponding to reciprocal-lattice vectors $\boldsymbol{G}_{hkl}$ do *not* necessarily coincide with real atomic planes (or planes through lattice points) in the real-space crystal structure. The Miller indices of a plane are defined (§2.1.3) as the *smallest* three integers corresponding to the reciprocals of the intersections of the plane with the axes (see Fig. 2.12). Thus, for a cubic real-space lattice, for instance, the (100) plane does indeed coincide with the atomic (lattice) planes of the crystal, but not all the planes denoted as (n00) do so. For example, for the planes (200), having *half* the spacing of the (100) planes, only every other plane coincides with a real crystal atomic plane. Thus, most of the points in the reciprocal lattice do *not* correspond to real atomic planes of the crystal. Nevertheless, the concept of the reciprocal lattice is of great utility, as will become apparent.

A reciprocal-lattice vector $\boldsymbol{G}_{hkl}$ can also be defined in terms of basis vectors $\boldsymbol{a}^*$, $\boldsymbol{b}^*$ and $\boldsymbol{c}^*$ of the reciprocal lattice, viz.

$$\boldsymbol{G}_{hkl} = h\boldsymbol{a}^* + k\boldsymbol{b}^* + l\boldsymbol{c}^*. \tag{2.47}$$

where, for example, the vector $\boldsymbol{a}^*$ is given in terms of the real-space lattice basis vectors as

$$\boldsymbol{a}^* = \frac{2\pi(\boldsymbol{b} \times \boldsymbol{c})}{\boldsymbol{a} \cdot (\boldsymbol{b} \times \boldsymbol{c})}, \tag{2.48}$$

and corresponding expressions for $\boldsymbol{b}^*$ and $\boldsymbol{c}^*$ are obtained by cyclic permutation. The quantity in the denominator is simply the volume of the real-space unit cell (eqn. (2.2)).

Thus, the following relationships between real-space and reciprocal-lattice basis vectors follow immediately:

$$\boldsymbol{a} \cdot \boldsymbol{a}^* = 2\pi \quad \text{etc.} \tag{2.49a}$$

and

$$\boldsymbol{a} \cdot \boldsymbol{b}^* = 0 \quad \text{etc.}, \tag{2.49b}$$

and hence

$$\boldsymbol{G}_{hkl} \cdot \boldsymbol{R}_{uvw} = 2\pi(hu + kv + lw). \tag{2.50}$$

Therefore, an alternative definition that a general vector $\boldsymbol{k}$ in reciprocal space (or 'k-space')

$$\boldsymbol{k} = k_1\boldsymbol{a}^* + k_2\boldsymbol{b}^* + k_3\boldsymbol{c}^* \tag{2.51}$$

(for all values of coefficients k_i) be a *reciprocal-lattice vector* is that the relation

$$\exp(\mathrm{i}\boldsymbol{k} \cdot \boldsymbol{R}_{uvw}) = \exp(\mathrm{i}\boldsymbol{G} \cdot \boldsymbol{R}_{uvw}) = 1 \tag{2.52}$$

be satisfied, which is the case if the coefficients k_i are all integers (cf. eqn. (2.50)).

One example of the use of reciprocal-lattice vectors is in the representation of, say, a periodic physical quantity of a crystal, such as the atomic density $n(\boldsymbol{r})$, as a Fourier series. It is well known that any periodic function $f(x)$ with period a can be expanded as an infinite Fourier series of sine and cosine functions:

$$f(x) = \sum_{p=0}^{\infty}[c_p\cos(2\pi px/a) + s_p\sin(2\pi px/a)], \tag{2.53}$$

or more succinctly as

$$f(x) = \sum_{p=0}^{\infty} n_p\exp(2\pi \mathrm{i}px/a), \tag{2.54}$$

where now p can take both positive and negative integral values.

Generalizing this for the 3D case, the atomic density, say, can then be written in terms of reciprocal-space vectors as;

$$n(\boldsymbol{r}) = \sum_{k=0}^{\infty} n_k\exp(\mathrm{i}\boldsymbol{k} \cdot \boldsymbol{r}). \tag{2.55}$$

For $n(\boldsymbol{r})$ to be translationally periodic with respect to a real-space lattice vector $\boldsymbol{R}$ (eqn. (2.1)), i.e. for the relation

$$n(\boldsymbol{r} + \boldsymbol{R}) = n(\boldsymbol{r}) \tag{2.56}$$

to hold, implies that the general reciprocal-space vector $\boldsymbol{k}$ is restricted to be a reciprocal-*lattice* vector $\boldsymbol{G}$. This can be seen by substituting the expression for the real-space lattice translation vector $\boldsymbol{R}$ (eqn. (2.1)) into eqn. (2.56) and expanding $n(\boldsymbol{r}+\boldsymbol{R})$ as a Fourier series (cf. eqn. (2.55)), giving

$$n(\boldsymbol{r}+\boldsymbol{R})=\sum_{k=0}^{\infty} n_k \exp(\mathrm{i}\boldsymbol{k}\cdot\boldsymbol{r})\exp(\mathrm{i}\boldsymbol{k}\cdot\boldsymbol{R}). \tag{2.57}$$

The right-hand side of eqn. (2.57) becomes equal to $n(\boldsymbol{r})$ (cf. eqn. (2.55)) only if $\exp(\mathrm{i}\boldsymbol{k}\cdot\boldsymbol{R})=1$, which is only satisfied if $\boldsymbol{k}\equiv\boldsymbol{G}$ (cf. eqn. (2.52)). Thus, the Fourier components of a function having the periodicity of the real-space crystal lattice are the corresponding reciprocal-lattice vectors. Note that Fourier inversion of eqn. (2.55) (with $\boldsymbol{k}\equiv\boldsymbol{G}$) gives the Fourier coefficients

$$n_G=\frac{1}{V_\mathrm{c}}\int_{\text{cell}} n(\boldsymbol{r})\exp(-\mathrm{i}\boldsymbol{G}\cdot\boldsymbol{r})\,\mathrm{d}V, \tag{2.58}$$

where V_c is the volume of a (real-space) unit cell of the crystal (eqn. (2.2)).

It is instructive to ascertain the reciprocal lattices of a number of simple real-space lattices. The easiest example is the case of the simple cubic lattice for which the primitive translation vectors in real space are

$$\boldsymbol{a}=a\hat{\boldsymbol{x}};\quad \boldsymbol{b}=a\hat{\boldsymbol{y}};\quad \boldsymbol{c}=a\hat{\boldsymbol{z}}, \tag{2.59a}$$

where $\hat{\boldsymbol{x}}$, $\hat{\boldsymbol{y}}$ and $\hat{\boldsymbol{z}}$ are unit-length orthogonal vectors. The volume of the real-space unit cell is $V_\mathrm{c}=\boldsymbol{a}\cdot(\boldsymbol{b}\times\boldsymbol{c})=a^3$. The reciprocal-lattice primitive vectors are (from eqn. (2.48)) thus

$$\boldsymbol{a}^*=(2\pi/a)\hat{\boldsymbol{x}};\quad \boldsymbol{b}^*=(2\pi/a)\hat{\boldsymbol{y}};\quad \boldsymbol{c}^*=(2\pi/a)\hat{\boldsymbol{z}}. \tag{2.59b}$$

Hence, the reciprocal lattice of the simple cubic lattice is *also* a single cubic lattice, with lattice constant $2\pi/a$ and cell volume $V_\mathrm{c}^*=8\pi^3/a^3$. Note that it is true generally that the volume of the primitive cell in reciprocal space is simply given by (see Problem 2.16)

$$V_\mathrm{c}^*=(2\pi)^3/V_\mathrm{c}. \tag{2.60}$$

The reciprocal lattice of a real-space b.c.c. lattice can also be established in a similar manner. The primitive real-space translation vectors of the b.c.c. lattice are (by inspection of Fig. 2.40a) given by:

$$\boldsymbol{a}=\frac{1}{2}a(-\hat{\boldsymbol{x}}+\hat{\boldsymbol{y}}+\hat{\boldsymbol{z}});\quad \boldsymbol{b}=\frac{1}{2}a(\hat{\boldsymbol{x}}-\hat{\boldsymbol{y}}+\hat{\boldsymbol{z}});\quad \boldsymbol{c}=\frac{1}{2}a(\hat{\boldsymbol{x}}+\hat{\boldsymbol{y}}-\hat{\boldsymbol{z}}) \tag{2.61a}$$

where a is the cube length and $\hat{\boldsymbol{x}}$, $\hat{\boldsymbol{y}}$ and $\hat{\boldsymbol{z}}$ are orthogonal unit-cell vectors as before. The volume of the real-space unit cell is thus $V_\mathrm{c}=a^3/2$. The primitive reciprocal-lattice vectors are thus

$$\boldsymbol{a}^*=(2\pi/a)(\hat{\boldsymbol{y}}+\hat{\boldsymbol{z}});\quad \boldsymbol{b}^*=(2\pi/a)(\hat{\boldsymbol{x}}+\hat{\boldsymbol{z}});\quad \boldsymbol{c}^*=(2\pi/a)(\hat{\boldsymbol{x}}+\hat{\boldsymbol{y}}). \tag{2.61b}$$

Inspection of Fig. 2.40b shows that these are in fact the primitive vectors of an f.c.c. lattice: the reciprocal lattice of a real-space b.c.c. lattice is an f.c.c. lattice. It has a cell volume of $V_\mathrm{c}^*=\boldsymbol{a}^*\cdot(\boldsymbol{b}^*\times\boldsymbol{c}^*)=2(2\pi/a)^3$.

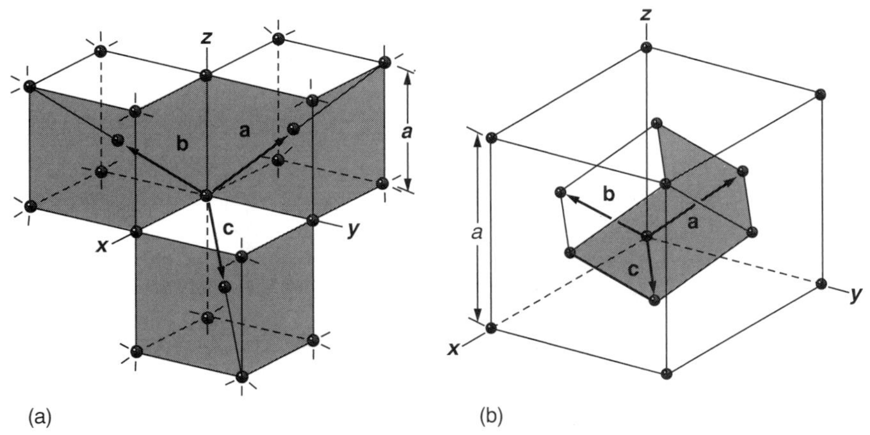

Fig.2.40 Primitive translation vectors of: (a) the b.c.c. lattice; (b) the f.c.c. lattice.

Likewise, the reciprocal lattice of an f.c.c. real-space lattice is a b.c.c. lattice. The primitive translation vectors of the f.c.c. lattice (Fig. 2.40b) are

$$\boldsymbol{a} = \frac{1}{2}a(\hat{\boldsymbol{y}} + \hat{\boldsymbol{z}}); \quad \boldsymbol{b} = \frac{1}{2}a(\hat{\boldsymbol{x}} + \hat{\boldsymbol{z}}); \quad \boldsymbol{c} = \frac{1}{2}a(\hat{\boldsymbol{x}} + \hat{\boldsymbol{y}}) \tag{2.62a}$$

and the volume of the primitive cell is $V_c = a^3/4$. The primitive vectors of the reciprocal lattice are thus

$$\boldsymbol{a}^* = (2\pi/a)(-\hat{\boldsymbol{x}} + \hat{\boldsymbol{y}} + \hat{\boldsymbol{z}}); \quad \boldsymbol{b}^* = (2\pi/a)(\hat{\boldsymbol{x}} - \hat{\boldsymbol{y}} + \hat{\boldsymbol{z}}); \quad \boldsymbol{c}^* = (2\pi/a)(\hat{\boldsymbol{x}} + \hat{\boldsymbol{y}} - \hat{\boldsymbol{z}}), \tag{2.62b}$$

i.e. those of a b.c.c. lattice. The volume of the reciprocal-space primitive cell is $V_c^* = 4(2\pi/a)^3$.

Note that Mermin (1992) has reformulated the derivation of space groups (§2.3.3) in Fourier, rather than real, space. In this way, crystals, quasicrystals and incommensurately modulated crystals can be treated in a unified way (since all such structures produce sharp diffraction peaks in reciprocal space—see §2.6.1).

2.4.2 Brillouin zones

A space-filling representation of a cell of a lattice, equivalent to that of the primitive cell, is the so-called Wigner–Seitz cell (see §2.2.2.3 for a discussion). This is as true for a reciprocal lattice as for a real-space lattice; the Wigner–Seitz cell of the reciprocal lattice is conventionally referred to as the Brillouin zone.

As mentioned previously, the Wigner–Seitz construction (see Fig. 2.23) consists of drawing vectors between a given lattice point and all other lattice points. Lines in 2D (planes in 3D) are then drawn perpendicular to these vectors at the midpoints. The smallest area in 2D (volume in 3D) enclosed in this manner about a lattice point is called, for the case of the reciprocal lattice, the first Brillouin zone. This construction is

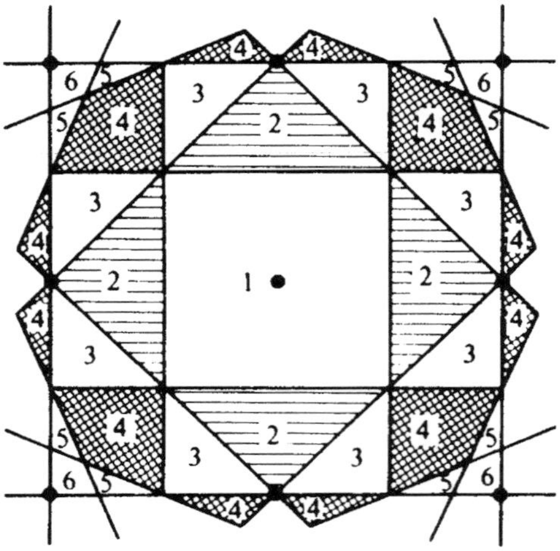

Fig.2.41 Construction of Brillouin zones (Wigner–Seitz cells of the reciprocal lattice) for a 2D square lattice. The different order zones are labelled (only parts of the fifth and sixth zones are shown).

shown in Fig. 2.41 for the case of a 2D square reciprocal lattice (corresponding to a 2D square real-space lattice). The square cell denoted as 1 is the first Brillouin zone, and results from taking vectors between the central lattice point and the four nearest lattice points in the $\langle 10 \rangle$ directions.

Vectors can also be taken between the central lattice point and the next-nearest lattice points, i.e. in the $\langle 11 \rangle$ directions, and the area so enclosed by the bisectors (minus the area of the first Brillouin zone) forms the *second* Brillouin zone, denoted as 2 in Fig. 2.41. Higher-order Brillouin zones are generated in a similar fashion by taking vectors to ever more distant reciprocal-lattice points from the origin. Note that, by displacing a given higher-order Brillouin zone into the region of the first zone by means of reciprocal-lattice-vector translations, the first zone is completely tiled: the area in 2D (volume in 3D) of all higher-order Brillouin zones is the *same* as that of the first zone (in turn, equal to that of the primitive cell of the reciprocal lattice, cf. eqn. (2.60)).

It is instructive to construct the first Brillouin zones of the b.c.c. and f.c.c. reciprocal lattices (these will prove useful later in a discussion of electronic properties—see Chapter 5). A real-space b.c.c. lattice has an f.c.c. reciprocal lattice with primitive vectors given by eqn. (2.61b). The 12 *shortest* vectors for the f.c.c. reciprocal lattice are thus

$$(2\pi/a)(\pm\hat{\boldsymbol{y}} \pm \hat{\boldsymbol{z}}); \quad (2\pi/a)(\pm\hat{\boldsymbol{x}} \pm \hat{\boldsymbol{z}}); \quad (2\pi/a)(\pm\hat{\boldsymbol{x}} \pm \hat{\boldsymbol{y}}). \tag{2.63}$$

Hence, the first Brillouin zone is bounded by planes normal to the midpoints of these vectors, i.e. the vectors from the central lattice point of the zone to the faces of the polyhedron so formed (a regular rhombic dodecahedron—see Fig. 2.42a) are given by

$$(\pi/a)(\pm\hat{\boldsymbol{y}} \pm \hat{\boldsymbol{z}}); \quad (\pi/a)(\pm\hat{\boldsymbol{x}} \pm \hat{\boldsymbol{z}}); \quad (\pi/a)(\pm\hat{\boldsymbol{x}} \pm \hat{\boldsymbol{y}}). \tag{2.64}$$

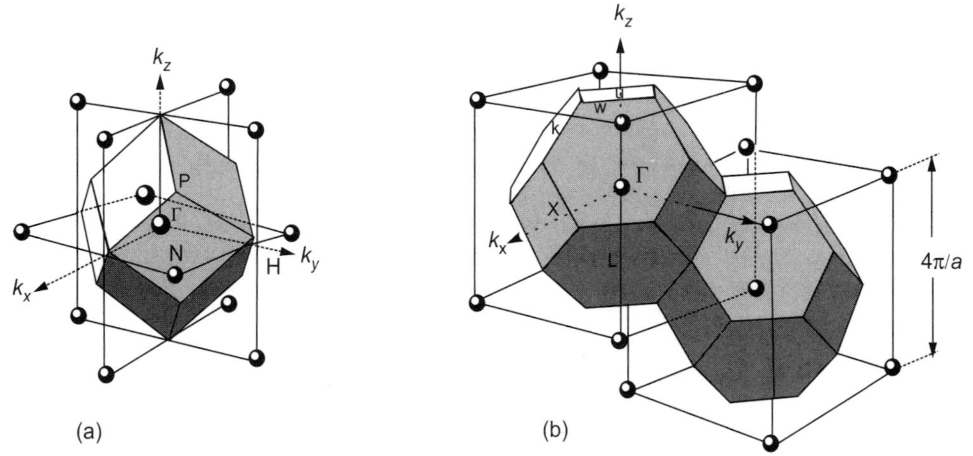

Fig.2.42 First Brillouin zone for: (a) the b.c.c. real-space lattice (f.c.c. reciprocal lattice)—a rhombic dodecahedron. Special high-symmetry points in k-space are $\Gamma\langle 0,0,0\rangle$; $H\langle 1,0,0\rangle$; $N\langle 1,1,0\rangle$; $P\langle 1,1,1\rangle$ (b) the f.c.c. real-space lattice (b.c.c. reciprocal lattice)—a truncated octahedron. Special points are $\Gamma\langle 0,\ 0,\ 0\rangle$; $X\langle 1,\ 0,\ 0\rangle$; $W\langle 1,\ 1/2,\ 0\rangle$; $K\langle 3/4,\ 3/4,\ 0\rangle$; $L\langle 1/2,\ 1/2,\ 1/2\rangle$.

For the case of a real-space f.c.c. lattice, with a b.c.c. reciprocal lattice, the shortest reciprocal-lattice vectors are now the eight vectors

$$(2\pi/a)(\pm\hat{\boldsymbol{x}} \pm \hat{\boldsymbol{y}} \pm \hat{\boldsymbol{z}}), \tag{2.65}$$

but the octahedron formed by the eight planes perpendicular to these vectors at their midpoints is truncated at the apices by six other reciprocal lattice vectors

$$(2\pi/a)(\pm 2\hat{\boldsymbol{x}}); \quad (2\pi/a)(\pm 2\hat{\boldsymbol{y}}); \quad (2\pi/a)(\pm 2\hat{\boldsymbol{z}}) \tag{2.66}$$

where $(2\pi/a)(2\hat{\boldsymbol{x}})$, for example, is a reciprocal lattice vector since it is equal to $\boldsymbol{b}^* + \boldsymbol{c}^*$—see eqn. (2.62b). The first Brillouin zone in this case is therefore a truncated octahedron (see Fig. 2.42b).

Computer simulation of materials 2.5

As will be seen shortly (§2.6.1), the structure of a perfect single crystal may be determined experimentally using diffraction methods. However, for materials containing structural disorder, experimental techniques often do not permit the structure to be determined completely (or sometimes at all). In such cases, computer simulation of the structure, using as input an appropriate interatomic potential, may be the only way of acquiring detailed structural information about a material. In addition, computer-simulation techniques permit physical properties to be investigated that otherwise might be very difficult to study experimentally, e.g. atomic diffusion. Such simulations may also be used to predict new structures of materials and properties thereof. For instance, various crystalline phases of C_3N_4, with tetrahedral coordination of the (sp^3) carbon and trigonal planar coordination of the nitrogen, have been predicted to have hardnesses comparable to or even in excess of that of diamond. Thus, for the hexagonal β-phase, the bulk modulus is predicted to be $B = 427$ GPa (Liu and Cohen (1989)), and for the cubic phase $B = 496$ GPa (Teter and Hemley (1996)), compared with $B = 440$ GPa for diamond. However, reliable synthesis of these materials has not yet proved possible. A summary of computer-simulation techniques and the kinds of interatomic potentials used to model the structures of materials is given in the following: more detail is given, for example, in the book by Allen and Tildesley (1987), albeit for the simulation of liquids.

2.5.1 Models and boundary conditions

The aim of a computer simulation is to produce a structural model that has the same structural arrangement, the same lattice energy and physical properties as the actual material under investigation. Ideally, the model should be as large as possible, but constraints associated with computer memory and, more importantly, execution speed of the computer program implementing the algorithm used in the simulation commonly limit the size of models to be a few thousand particles (atoms). However, developments in computing have recently been so rapid that the massively parallel architectures now available for some computers have allowed the simulation of very large (million-atom) systems to be performed (albeit with the use of empirical, phenomenological interatomic potentials). On the other hand, even with the most powerful computers available, simulations properly evaluating the quantum-mechanical interactions (see §2.5.3.4) between atoms are limited for the foreseeable future to the order of a hundred atoms.

For a typical simulation involving, say, 1000 atoms arranged in a cubic box, approximately half of the atoms lie at the surface of the box. Atoms at a free surface will experience very different interatomic interactions from those deep in the box due to their different local environments. One way to obviate this difficulty is to impose periodic boundary (Born–von Karman) conditions on the simulation box. A 2D representation of the application of periodic boundary conditions is shown in Fig. 2.43. The simulation box containing the particles under study is replicated periodically through space to give an infinite 'lattice'. As a particle moves in the simulation box during the simulation run under the action of the interatomic forces exerted by its neighbours, its

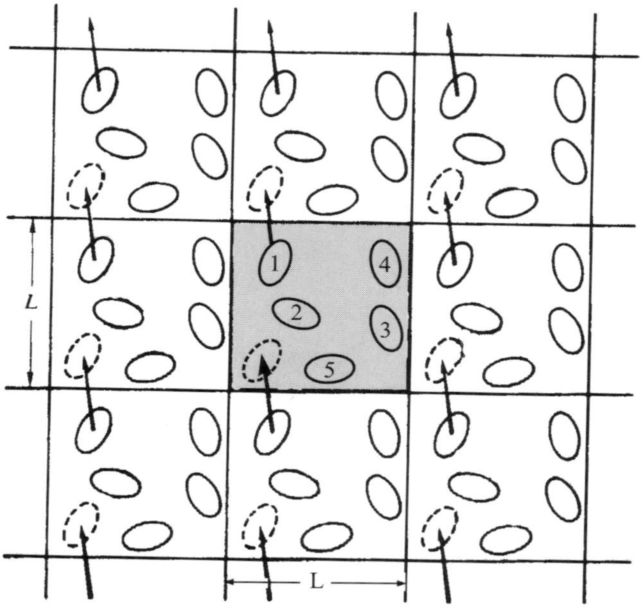

Fig.2.43 Illustration of periodic boundary conditions applied in 2D to a square simulation box containing five particles. As the particle labelled 1 leaves the central box, its image moves into the box from a neighbouring replica box.

image in all replica boxes moves in exactly the same way, even to the extent that should a particle (labelled 1 in Fig. 2.43) actually leave the simulation box, its image enters the box through the opposite face. There are thus no walls at the boundary of the central simulation box but the number of particles in the box is conserved. There are no 'surface' particles either, since a given particle situated near the surface of the central box can interact with other image particles in neighbouring replica boxes.

Although of great utility, periodic boundary conditions do have some drawbacks. If the interatomic potential, U, is sufficiently long-ranged, i.e. if $U(r) \sim r^{-\alpha}$, where α is less than the dimensionality of the system, a given particle will interact appreciably with its *own* images in neighbouring replica boxes. Hence, the periodicity of the Born–von Karman boundary conditions is imposed in such a case on what might otherwise be a spatially isotropic system (e.g. a liquid). Although cubic boxes, and their periodic replicas, are most commonly used in simulations because of the associated geometric simplicity, nevertheless other completely space-filling polyhedra can also be used as simulation boxes. Two such polyhedra are the rhombic dodecahedron and the truncated octahedron illustrated in Fig. 2.42; these have the advantage over the cubic box that they are more nearly spherical, and hence can be used to simulate structurally isotropic materials, such as liquids and glasses, more realistically.

A final consideration concerning boundary conditions concerns the *thermodynamic* conditions under which a simulation is run. Normally, the number, N, of particles in the simulation box is kept constant. It is also easiest to keep constant the volume, V, of the box. Thus, if the total energy $\mathscr{E}$ (kinetic plus potential) of the system is also kept constant, the system corresponds to the microcanonical ensemble in statistical thermodynamics (constant-NVE). Alternatively, the temperature T of the system may be kept constant; this corresponds to the canonical ensemble (constant-NVT). A scenario more closely mimicking physical reality is when the *pressure* of the system is constrained to be constant; correspondingly, the volume of the simulation box must vary. Such constant-NpT conditions correspond to the isothermal–isobaric ensemble.

2.5.2 Simulation methods

There are basically two techniques for the computer simulation of structures that are in common use: the molecular-dynamics method and the Monte Carlo approach. These two methods will be discussed briefly in the following; for more detail, the reader is referred to Allen and Tildesley (1987).

2.5.2.1 Molecular dynamics

As the name implies, this simulation technique follows the time evolution of a system of N particles, mutually interacting via a potential U, by solving the classical equations of motion to which they are subject. The equation of motion for the ith particle in the system is given by the expression relating force and acceleration:

$$m_i \frac{\mathrm{d}^2 \boldsymbol{r}_i}{\mathrm{d}t^2} = \boldsymbol{f}_i, \tag{2.67}$$

where $\boldsymbol{r}_i$ is the coordinate, and m_i the mass, of the particle, and $\boldsymbol{f}_i$ is the total force exerted on it as a result of the interactions with all the other particles, and

$$\boldsymbol{f}_i = -\nabla_{\boldsymbol{r}_i} U(\boldsymbol{r}). \tag{2.68}$$

Differential equations, such as eqn. (2.67), can be solved by a finite-difference approach which is particularly suitable for implementation using a computer. Knowing the dynamical history of the system of particles at time t (particle positions, velocities, accelerations, etc.), the dynamical variables at a later time $t + \delta t$ are calculated using an appropriate approximation method to integrate eqn. (2.67). One of the most commonly used approximation schemes is the Verlet algorithm, which requires knowledge of the particle position $\boldsymbol{r}_i(t)$ and the acceleration $\boldsymbol{a}_i(t)$ at time t, as well as the position at the *previous* time step, $\boldsymbol{r}_i(t - \delta t)$. Performing Taylor expansions for $\boldsymbol{r}_i(t \pm \delta t)$ about $\boldsymbol{r}_i(t)$ gives

$$\boldsymbol{r}_i(t + \delta t) = \boldsymbol{r}_i(t) + \boldsymbol{v}_i(t)\delta t + \frac{1}{2}\boldsymbol{a}_i(t)(\delta t)^2 + \cdots \tag{2.69a}$$

and

$$\boldsymbol{r}_i(t - \delta t) = \boldsymbol{r}_i(t) - \boldsymbol{v}_i(t)\delta t + \frac{1}{2}\boldsymbol{a}_i(t)(\delta t)^2 - \cdots \tag{2.69b}$$

Substituting eqn. (2.69b) into eqn (2.69a) gives the Verlet algorithm:

$$\boldsymbol{r}_i(t+\delta t) = 2\boldsymbol{r}_i(t) - \boldsymbol{r}_i(t-\delta t) + \boldsymbol{a}_i(t)(\delta t)^2 + \cdots \tag{2.70}$$

Note that this expression does not explicitly involve the velocity, and is accurate to $0(\delta t)^4$. The velocity can be computed from

$$\boldsymbol{v}_i(t) = \frac{r_i(t+\delta t) - r_i(t-\delta t)}{2\delta t} \tag{2.71}$$

which is accurate to $0(\delta t)^2$. The Verlet expression (eqn. (2.70)) is also time-reversible.

The time step δt ideally should be as long as possible, so that a simulation run can encompass as large an elapsed time as possible for a given number of time steps (usually set by limitations on computer time). However, although the optimum size of δt is determined somewhat by the particular finite-difference method used, nevertheless it must be much smaller than, say, a characteristic vibrational period of the solid; typically $\delta t \simeq 10^{-15}$s.

Molecular dynamics is a powerful simulation technique because the dynamical history of the interacting system of particles is known for all times, t. Thus, not only can the static structure of the model be found by averaging over a sufficiently long time (once the system is in equilibrium), but dynamical quantities such as space–time correlation functions relating, for example, to atomic diffusion (§3.4.2.1) or vibrational behaviour (see Chapter 4), may be calculated.

2.5.2.2 *Monte Carlo simulations*

Another commonly used computer-simulation technique for modelling the structures of materials is the Monte Carlo approach, so called because it uses random numbers in the evaluation of the particle displacements during the course of a simulation run.

One popular way of implementing a Monte Carlo simulation is to use the Metropolis method. Starting from an initial configuration, say a random distribution of particles (atoms), one atom i is picked randomly and displaced in a random direction by a random amount, from $\boldsymbol{r}_{\rm i}^m$ to $\boldsymbol{r}_{\rm i}^n$, subject to the maximum displacement being the adjustable parameter $\delta r_{\max}$. The change in potential energy of the system, δV_{mn}, resulting from the movement of this atom is then calculated for an assumed form for the interatomic potential. If the atomic displacement is 'downhill' in energy terms ($\delta V_{mn} \leq 0$), then the new position is accepted unconditionally. If, however, the move is 'uphill' in energy ($\delta V_{mn} > 0$), then the move is accepted only conditionally, subject to the Boltzmann probability factor, exp $(-\delta V_{mn}/k_{\rm B}T)$.

2.5.3 Interatomic potentials

The success of any computer simulation of the structure of a material is ultimately governed by the accuracy of the interatomic potential used in the simulation, along the lines of the computing maxim: 'garbage in, garbage out'. Often in the past, empirical, phenomenological potentials have been used, and these are mostly chosen so that the interatomic forces resulting from them (eqn. (2.68)) are easily and rapidly computable. However, there is an increasing trend for interatomic interactions to be calculated by *ab initio* quantum-mechanical methods, although this approach is very costly in terms of computing time and, as yet, only very small models can be studied in this way.

In the following, aspects of the interatomic potentials used in the simulation of various types of materials (ionic, covalent, metallic, etc.) will be discussed, starting with atomic systems.

2.5.3.1 Van der Waals systems

The interatomic potential between atoms may be written, generally, as the sum of terms involving interactions between pairs of atoms, triplets, quartets, etc., viz.

$$U = \sum_i \sum_{j>i} U_2(\mathbf{r}_i, \mathbf{r}_j) + \sum_i \sum_{j>i} \sum_{k>j>i} U_3(\mathbf{r}_i, \mathbf{r}_j, \mathbf{r}_k) + \sum_i \sum_{j>i} \sum_{k>j>i} \sum_{l>k>j>i} U_4(\mathbf{r}_i, \mathbf{r}_j, \mathbf{r}_k, \mathbf{r}_l) + \ldots \tag{2.72}$$

where no pairs, triplets, etc. are counted twice. Of these contributions, the pairwise interaction, U_2, is the most important. Moreover, it depends only on the (scalar) separation between pairs of atoms $r_{ij} = |\mathbf{r}_i - \mathbf{r}_j|$, and not on the *vector* positions of the atoms, i.e. $U_2 = U_2(r_{ij})$, making it particularly easy to compute. The pairwise (non-metallic, non-covalent) interaction between uncharged atoms is well represented by the Lennard-Jones 12–6 potential (eqn. (2.9)).

In general, the triplet contribution, U_3, is not negligible (although the other higher-order terms effectively are): for the case of, say, f.c.c. crystalline Ar, the triplet term makes a contribution of roughly 10% to the overall potential energy. Triplet (and higher-order) terms are very costly in computer time to compute because of the required multiple summations over atoms. However, in certain cases, some averaged form of three-body (triplet) interactions can be incorporated into an *effective* pair potential, i.e.

$$U(r) = \sum_i \sum_{j>i} U_2^{\text{eff}}(r_{ij}), \tag{2.73}$$

where U_2^{eff} depend on external variables, such as density (pressure) or temperature, as a result of the artificiality of incorporating many-body effects in a pairwise representation. (The pairwise or triplet contributions, U_2 and U_3, etc., of course do *not* suffer from this unphysical feature.) However, this disadvantage is more than offset by the computational advantages in dealing only with pairwise interactions for the case of U_2^{eff}.

2.5.3.2 Covalent interactions

The interatomic interaction associated with the build-up of electron charge density between atoms, in other words, the covalent bond, is modelled empirically in the case of diatomic *molecules* reasonably well by the anharmonic Morse potential

$$U = U_0\{1 - \exp[-a(r - r_e)]\}^2, \tag{2.74}$$

where the three adjustable parameters are r_e, the equilibrium bond length, U_0, the depth of the potential energy minimum and a, which effectively determines the radius of curvature near the bottom of the well.

However, for the case of polyatomic systems, notably solids, the dominant characteristic of the covalent interaction is that it is markedly non-centrosymmetric, i.e. covalent bonds are strongly directed in space as a result of orbital hybridization effects. As a result, it is evident that three-body (triplet) interactions, reflecting the restoring forces associated with maintaining *angles* between covalent bonds at their equilibrium values, must be very important. (Higher-order interactions, e.g. the four-body interaction giving rise to restoring forces maintaining torsion (or dihedral) angles for rotation about the common bond between two neighbouring pairs of atoms are weaker but non-negligible.)

A simple form of phenomenological interatomic potential which has been widely used to simulate covalent interactions in, principally, tetrahedrally bonded materials, such as Si or Ge, is the Keating (1966) potential:

$$\begin{aligned} U = {} & \frac{3}{4}\alpha_K \sum_i \sum_{j\neq i} [(\boldsymbol{u}_i - \boldsymbol{u}_j)\cdot\hat{\boldsymbol{r}}_{ij}]^2 \\ & + \frac{3}{16}\beta_K \sum_i \sum_{j\neq i} \sum_{k\neq i,j} [(\boldsymbol{u}_i - \boldsymbol{u}_j)\cdot\hat{\boldsymbol{r}}_{ik} + (\boldsymbol{u}_i - \boldsymbol{u}_k)\cdot\hat{\boldsymbol{r}}_{ij}]^2 \end{aligned} \tag{2.75}$$

where $\boldsymbol{u}_i$ and $\boldsymbol{u}_j$ are displacements of nearest-neighbour atoms i and j, respectively, and $\hat{\boldsymbol{r}}_{ij}$ is the unit vector connecting i and j. The first term in eqn. (2.75) corresponds to bond *stretching* (i.e. a two-body interaction) and the second term to bond *bending* (a three-body interaction), having force constants α_K and β_K, respectively. Note that the Keating potential is harmonic, since it only contains terms that are quadratic in displacements (see §4.2.2 and eqn. (4.30)).

* 2.5.3.3 *Ionic interactions*

The Coulombic potential (eqn. (2.25)) is extremely simple, yet very difficult to cope with computationally because of its long-ranged nature. (In contrast, the very short-ranged nature of the non-Coulombic repulsive term appearing in the total potential energy, whether of power-law form (eqn. (2.27)) or of exponential form (eqn. (2.29)), confines it to acting effectively between nearest-neighbour atoms only.) The problem in calculating the lattice sum for the Coulombic potential, i.e. evaluating the Madelung constant (eqn. (2.26)—see §2.2.4.4), is that the sum is *conditionally* convergent; that is, the result depends on the order in which terms are taken in the summation.

One way of circumventing this problem of conditional convergence in the case of periodic systems (or where periodic boundary conditions are applied to the simulation box) is to use the trick inherent in the Ewald method. Namely, a broad, spherically symmetric charge distribution of equal magnitude but *opposite* sign is added to each point charge (ion) in the actual system (Fig. 2.44a), with the result that interionic interactions between neighbouring ions become screened and *short-ranged*; the total screened potential can then be readily summed over all ions in the unit cell (or simulation box) and all (image) ions in neighbouring, periodically repeated cells. A cancelling charge distribution, of the same shape as that of the first fictitious charge distribution but of opposite sign (i.e. the *same* as that of the actual ions) is also added (Fig. 2.44b). The potential energy of this second charge distribution is evaluated by first summing over charges in *reciprocal* space by expressing the translationally periodic functions, the electrostatic potential and the charge density of Fig. 2.44b, as Fourier series (eqn. (2.55))

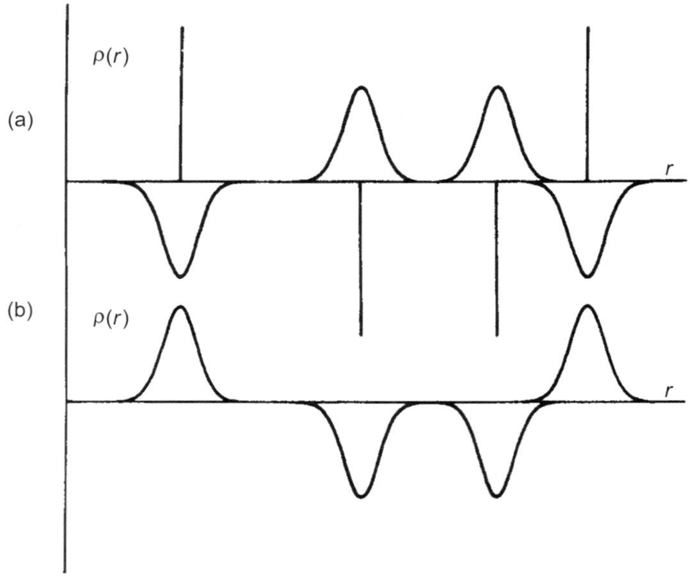

Fig.2.44 Fictitious charge distributions used in the Ewald method for evaluating the lattice sums of electrostatic interactions between ions in a periodic lattice: (a) the actual point charges (represented as delta functions) plus the fictitious screening charge distribution; (b) the fictitious cancelling charge distribution.

in terms of reciprocal-lattice vectors $\boldsymbol{G}$; the result is then Fourier transformed back to real space. (For details of the calculation, see for example Kittel (1996).)

For the case of a *Gaussian* fictitious charge distribution,

$$\rho(r) = q_i \left(\frac{\kappa^2}{\pi}\right)^{3/2} \exp(-\kappa^2 r^2), \tag{2.76}$$

where q_i is the charge of the ith ion and the adjustable parameter κ determines the breadth of the charge distribution (and is used to control the rapidity of convergence of the summation), the result of the Ewald summation for the potential energy is (Allen and Tildesley (1987)):

$$U = \frac{1}{8\pi\varepsilon_0} \sum_{i=1}^{N} \sum_{j=1}^{N} \left[\sum_{|\boldsymbol{n}|=0}^{\infty} q_i q_j \frac{\operatorname{erfc}(\kappa|\boldsymbol{r}_{ij} + \boldsymbol{n}|)}{|\boldsymbol{r}_{ij} + \boldsymbol{n}|} + \frac{1}{\pi L^3} \sum_{\boldsymbol{G} \neq 0} q_i q_j \left(\frac{4\pi^2}{G^2}\right) \exp\left(\frac{-G^2}{4\kappa^2}\right) \cos(\boldsymbol{G} \cdot \boldsymbol{r}_{ij}) \right]$$
$$- \frac{1}{4\pi\varepsilon_0} \left(\frac{\kappa}{\pi^{1/2}}\right) \sum_{i=1}^{N} q_i^2. \tag{2.77}$$

In the first term, the vector $\boldsymbol{n} = (n_x L, n_y L, n_z L)$ defines the periodic array of unit cells (or replica cells) of a cube of side L. The complementary error function that appears in the expression is defined as

$$\mathrm{erfc}(x) = \frac{2}{\sqrt{\pi}} \int_x^{\infty} \exp(-y^2)\mathrm{d}y. \tag{2.78}$$

Since erfc(x) decreases to zero with increasing x, the only term contributing to the real-space summation is that with $\boldsymbol{n} = 0$, i.e. for the central cell, if the parameter κ is large enough.

A large value of κ, however, corresponds to a narrow distribution of fictitious charge (cf. eqn. (2.76)), which needs many reciprocal-lattice vectors $\boldsymbol{G}$ to represent it, as in the second term in eqn. (2.77). This is computationally very expensive, so a compromise is sought: typically, k is set to the value $5/L$, necessitating the use of a hundred or so wavevectors in the reciprocal-space summation.

The final term in eqn. (2.77) is the self-energy due to the electrostatic potential of the cancelling fictitious charge distribution (Fig. 2.44b) at the site of ion i.

**2.5.3.4 *Quantum-mechanical methods*

The interatomic potentials discussed so far have all suffered from the fact that they are, to a greater or lesser extent, empirical and phenomenological; the parameters involved need to be found by fitting to experiment. This is obviously an unsatisfactory state of affairs and, in principle, it is far preferable to evaluate the potential *ab initio* by solving the Schrödinger equation for the electrons in the system. However, for interacting electrons, the wavefunction involved is a function of some 10^{23} variables, and thus the problem appears to be one of intractable difficulty.

Fortunately, the density functional theory (valid for all types of materials) proposes that in fact it is the electronic *charge density*, $n(\boldsymbol{r})$, that completely determines all features of the electronic behaviour of a system of interacting electrons in the *ground state*: the ground-state energy of the interacting electron gas is a unique functional (function of a function) of the charge density (Hohenberg and Kohn (1964)). Although the functional is not known, the Hohenberg–Kohn theorem allows a variational principle to be established to find the ground-state energy: the energy functional takes its minimum value (i.e. the ground-state energy) when the charge density $n(\boldsymbol{r})$ is the *true* ground-state charge density, $\rho(\boldsymbol{r})$.

Kohn and Sham (1965) further showed that the ground-state energy functional could be written as:

$$\mathscr{E}_G[\rho(\boldsymbol{r})] = T[\rho(\boldsymbol{r})] + \int [\rho(\boldsymbol{r})]\phi_{\mathrm{N}}(\boldsymbol{r})\mathrm{d}\boldsymbol{r} + \frac{1}{2}\int \rho(\boldsymbol{r})\phi_{\mathrm{H}}(\boldsymbol{r})\mathrm{d}\boldsymbol{r} + \mathscr{E}_{xc}[\rho(\boldsymbol{r})], \tag{2.79}$$

where the first term represents the electronic kinetic energy. The second term is the electrostatic energy associated with the electronic charge and the potential due to the nuclei (ion cores), with

$$\phi_{\mathrm{N}}(\boldsymbol{r}) = \sum_i \frac{q_i}{4\pi\varepsilon_0|\boldsymbol{r} - \boldsymbol{R}_i|} \tag{2.80}$$

where q_i and $\boldsymbol{R}_i$ are the charge and position, respectively, of the ith nucleus. The third term in eqn. (2.79) is the so-called Hartree term, and represents the electrostatic energy of an electron moving independently in the mean electrostatic field due to all the other electrons; the Hartree potential is given by

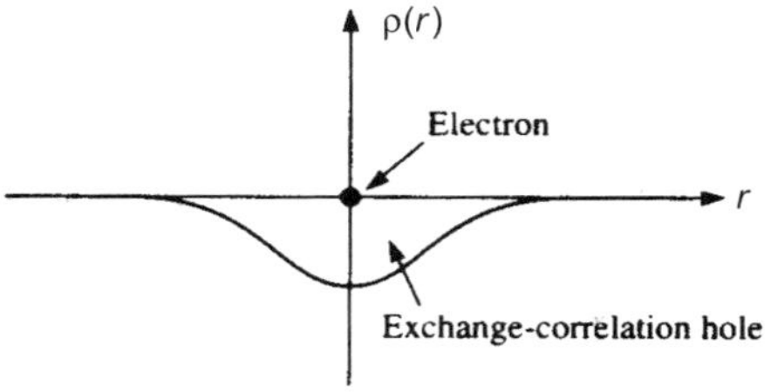

Fig.2.45 Representation of the exchange-correlation hole, the charge-density depletion surrounding each electron due to the Pauli exclusion principle (see eqn. (7.212)).

$$\phi_{\mathrm{H}}(\boldsymbol{r}) = \int \frac{\rho(\boldsymbol{r}')}{4\pi\varepsilon_0|\boldsymbol{r}-\boldsymbol{r}'|}\mathrm{d}\boldsymbol{r}'. \tag{2.81}$$

The final term in eqn. (2.79) is the exchange-correlation energy (see §7.2.5.1) which corrects for the overestimation in the Hartree approximation of the Coulomb repulsion between electrons, both of an electron with itself, and also with all other electrons as a result of the neglect of the exchange-correlation hole. This is the depletion in the probability of finding an electron in the vicinity of another electron (Fig. 2.45) as a result of the exchange interaction (see eqn. (7.212)): electrons with parallel spins repel each other due to the Pauli exclusion principle. The exchange-correlation hole has a charge of exactly minus one electron, and so for distances beyond the radius of the electron and exchange-correlation hole pair given by the distance at which the charge density of the hole screens (§5.6.1) that of the electron (typically of order 1 Å), the quasiparticle appears to be electrically *neutral*. Thus, the electrons behave *as if* they were independent, with no Coulomb interactions between them.

Minimization of the Kohn–Sham ground-state energy functional (eqn. (2.79)) subject to the constraint that the total number, N, of electrons be conserved, i.e.

$$\int \rho(\boldsymbol{r})\mathrm{d}\boldsymbol{r} = N, \tag{2.82}$$

leads to the set of Kohn–Sham equations for all electrons i, in the form of the time-independent Schrödinger equation, viz.

$$\frac{-h^2}{8\pi^2 m_e}\nabla^2\Psi_i(\boldsymbol{r}) + V_{\mathrm{eff}}(\boldsymbol{r})\Psi_i(\boldsymbol{r}) = \mathscr{E}_i^{\mathrm{KS}}\Psi_i(\boldsymbol{r}) \tag{2.83}$$

where

$$V_{\mathrm{eff}}(\boldsymbol{r}) = V_{\mathrm{H}}(\boldsymbol{r}) + V_{\mathrm{N}}(\boldsymbol{r}) + V_{\mathrm{XC}}(\boldsymbol{r}) \tag{2.84}$$

and

$$\rho(\boldsymbol{r}) = \sum_{\text{i occupied}} \Psi_i(\boldsymbol{r})\Psi_i^*(\boldsymbol{r}). \tag{2.85}$$

The exchange-correlation potential, $\phi_{\mathrm{xc}}(\boldsymbol{r})$, is equal to the derivative $\delta\mathscr{E}_{\mathrm{xc}}[\rho(\boldsymbol{r})]/\delta\rho(\boldsymbol{r})$. Thus, the interacting N-electron problem is transformed into N single-electron equations (eqn. (2.83)), where each electron moves in the effective potential V_{eff} due to all other electrons (and ions) (eqn. (2.84)).

Although the Kohn–Sham equations (eqn. (2.83)) are an exact solution of the many-body interacting-electron problem, the exchange-correlation energy functional, $\mathscr{E}_{\text{xc}}[\rho(\boldsymbol{r})]$, is not known generally in the case of a spatially varying charge density, as in a solid. In the local density approximation, $\mathscr{E}_{\text{xc}}$ for a non-uniform charge distribution is approximated at a given position by the value of $\mathscr{E}_{\text{xc}}$ corresponding to a system of *uniform* charge density (whose value of $\mathscr{E}_{\text{xc}}$ *is* known) having the same charge density as at the site in question in the real charge distribution. This approximation is valid, in practice, even for rapidly varying charge-density fluctuations, since effectively all that is being done is approximating the shape of the exchange-correlation hole in the non-uniform electron distribution. Further details of the density-functional approach are given, for example, in the articles by Srivastava and Weaire (1987) and Remler and Madden (1990).

Conventionally, the Kohn–Sham set of equations (eqn. (2.83)) are solved by expressing the wavefunction $\Psi_i(\boldsymbol{r})$ in terms of some basis set, often plane waves, and written as a Fourier series in terms of reciprocal-lattice vectors (cf. eqn. (2.55)), and solved using matrix diagonalization. This procedure is repeated until self-consistency is achieved (since $V_{\text{eff}}(\boldsymbol{r})$ depends on $\rho(\boldsymbol{r})$, and hence on the occupied orbitals).

An example of the use of this *ab initio* simulation approach, in the local density approximation, is given in Fig. 2.46, where the total energies of Si and Ge are given as a function of volume for a variety of possible crystal structures. The diamond cubic structure is correctly predicted to be the most stable structure under ambient conditions, and a phase transition to the β-tin structure under pressure is also predicted, again in

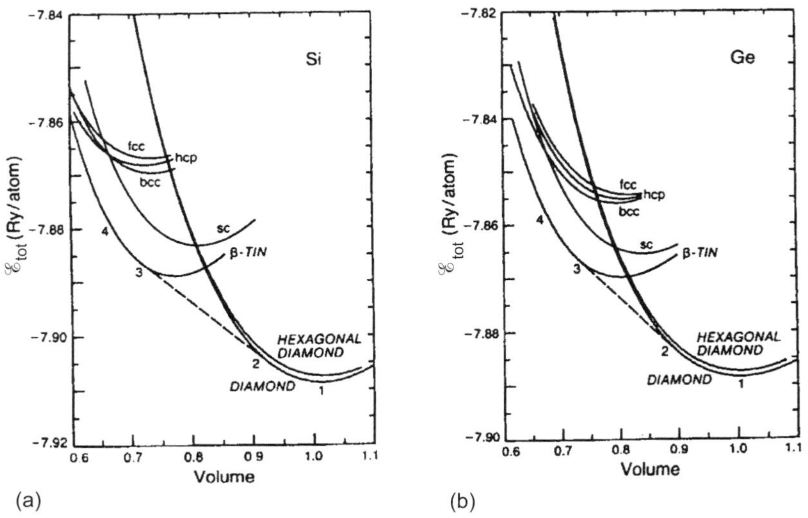

Fig.2.46 Total energy (in units of rydberg per atom) as a function of volume (normalized to the experimental volume) in seven different crystal structures for (a) Si; (b) Ge. (Reproduced with permission from Yin and Cohen (1982), *Phys. Rev.* **B26**, 5668. © 1982. The American Physical Society)

agreement with experimental observation. The charge density between atoms in GaAs calculated in this way is shown in Plate II.

A powerful alternative method to solving the optimization problem involving the Kohn–Sham equations (eqn. (2.83)) is the approach due to Car and Parrinello (1985). In this, the atomic configuration corresponding to the electronic ground state (calculated using density-functional theory), is found by a process of simulated annealing: variables characterizing electronic orbitals and ionic positions are both varied, subject to certain constraints, so that the total energy decreases until the ground-state configuration is found. In the Car–Parrinello approach, this procedure is implemented using molecular dynamics (see §2.5.2.1).

An electronic state can be represented by a set of occupied orbitals $\phi_i (i = 1, \cdots, n)$, each described by an expansion in terms of some basis $\chi_k (k = 1, \cdots, m)$, i.e.

$$\phi_i = \sum_k c_k^i \chi_k. \tag{2.86}$$

The simulated-annealing electronic-structure calculation is just a search among the space $\mathscr{C}$ made up of all coefficients c_k^i, on a hypersurface determined by the constraint that the orbitals are orthonormal, to find the optimum set $\mathscr{C}_{\text{opt}}$ corresponding to the electronic ground state. This is done using constraint dynamics involving a classical Lagrangian

$$L = T - V, \tag{2.87}$$

where T is the kinetic energy and V the potential energy.

The kinetic energy is given by

$$T = \frac{1}{2} \sum_I m_I (\dot{\boldsymbol{r}}_I)^2 + \frac{1}{2} \mu \sum_i \sum_k (\dot{c}_k^i)^2 \tag{2.88}$$

where the first term represents the real ionic kinetic energy ($\dot{\boldsymbol{r}} \equiv d\boldsymbol{r}/dt$) and the second term is a fictitious kinetic energy associated with the electronic orbital coefficient 'velocities' defined as

$$\dot{c}_k^i = \frac{\mathrm{d}c_k^i}{\mathrm{d}t} \tag{2.89}$$

and where μ is a fictitious inertial mass assigned to the 'motion' through coefficient space $\mathscr{C}$. The potential energy includes both electronic-energy and purely ionic contributions and thus is a function of both the ion positions $\boldsymbol{r}_i$ and the electronic orbital coefficient, $V = V\{\boldsymbol{r}_i, c_k^i\}$. The equation of motion associated with the Lagrangian is

$$\frac{\mathrm{d}}{\mathrm{d}t} \frac{\partial L}{\partial \dot{c}_k^i} + \frac{\partial L}{\partial c_k^i} = 0. \tag{2.90}$$

The electronic orbitals must satisfy the constraint of orthonormality, i.e.

$$\begin{aligned} \sigma_{ij} &= \frac{1}{\Omega} \int_\Omega \Psi_i^*(\boldsymbol{r}) \Psi_j(\boldsymbol{r}) \mathrm{d}\boldsymbol{r} - \delta_{ij} = 0 \\ &= \sum_k c_k^{i*} c_k^j - \delta_{ij} = 0 \end{aligned} \tag{2.91}$$

where Ω is the volume and $\delta_{ij} = 1(i = j), = 0$ otherwise. Such constraints lead to additional constraint 'forces' of the form $-\lambda_{ij}\partial\sigma_{ij}/\partial c_k^i = \lambda'_{ij}c_k^i$, where λ_{ij} is the appropriate Lagrange multiplier, and $\lambda'_{ij} = \lambda_{ij}(i \neq j)$ and $\lambda'_{ij} = 2\lambda_{ii}(i = j)$. Thus, two coupled equations of motion can be derived, a fictitious one for the orbital coefficients

$$\mu\ddot{c}_k^i = -\frac{\partial V}{\partial c_k^i} - \sum_j \lambda'_{ij}c_k^i \tag{2.92}$$

and another for the ions

$$m_I\ddot{\boldsymbol{r}}_I = -\nabla V\{\boldsymbol{r}_I, c_k^i\}. \tag{2.93}$$

The Car–Parrinello scheme solves these two equations of motion simultaneously in the same time step, i.e. the ions are moved and the electronic configuration adjusted to keep it on the orthonormal hypersurface until the ground-state energy configuration is found.

The Car–Parinello method has the advantage that not only can (the energies of) static structures be found, but that the *dynamics* of the system can also be explored. For example, the diffusive behaviour of atoms in solids (e.g. H in crystalline Si) can be calculated *ab initio*. A disadvantage, if the electronic structure is treated using density-functional theory, is that ground-state properties only can be investigated: electronically excited states are not accessible. Moreover, only small systems (consisting of a few tens of atoms) can be simulated in this way at present, even using the most powerful supercomputers, essentially because of limitations associated with computer memory and the need to store the many thousands of basis functions necessary to represent the highly curved wavefunctions of light elements, or of d- or f-electrons in metals.

Experimental structure determination 2.6

A knowledge of the atomic structure of a material is a prerequisite to an understanding of its physical behaviour. For the case of crystalline materials, diffraction methods allow the structure to be determined (see §2.6.1). However, for materials containing disorder, e.g. defective crystalline solids and amorphous materials, diffraction can, at best, only provide partial structural information. In such cases, other experimental techniques can provide complementary structural information.

2.6.1 Diffraction

Waves diffract when they meet an obstacle; the interference pattern appearing at a plane remote from the diffracting object, caused by the constructive and destructive interference of wavefronts emanating from different parts of the obstacle, is characteristic of the diffracting obstacle. This principle is used to determine the structural arrangement of atoms in a material, where the diffracting object is now the collection of atoms themselves. Diffraction is most effective when the wavelength of the incident radiation is comparable to the size of the diffracting object. Interatomic distances in condensed phases are of the order of 2 Å (0.2 nm), and three types of radiation having wavelengths of this order of magnitude are commonly used in diffraction experiments on materials, namely X-rays, neutrons and electrons.

2.6.1.1 X-rays, neutrons and electrons as diffraction probes

X-rays are electromagnetic waves having a wavelength of $\simeq 1$ Å, i.e. intermediate between the ultra-violet and γ-ray regions of the electromagnetic spectrum. They are commonly generated in the laboratory by the bombardment of a metal target (the anode) by electrons accelerated from a cathode by a high voltage, V. Deceleration of the electron beam by collisions in the target metal causes bremsstrahlung (or braking) radiation to be emitted in the form of X-rays having a wide range of photon energies. The maximum photon energy (minimum wavelength) corresponds to the case when an electron is completely stopped in *one* collision; normally, however, many collisions take place during the deceleration process, and as a result there is a spectrum of photon energies below (wavelengths greater than) the maximum (minimum) value (see Fig. 2.47), i.e.

$$h\upsilon = \frac{hc}{\lambda} \leqslant eV. \tag{2.94}$$

At sufficiently high acceleration voltages, the impinging electrons have enough energy to eject electrons from the core levels of the target metal atoms, whereupon higher-lying electrons drop down to fill the core holes, thereby emitting X-rays of well-defined energies characteristic of the target metal. If 1s (K-shell) electrons are ionized by the impinging electron beam, the resulting holes can be filled with electrons from the L shell ($2p_{3/2}$, $2p_{1/2}$) or the M shell ($3p_{3/2}$, $3p_{1/2}$), resulting in the K_{α_1}, K_{α_2} and K_{β_1}, K_{β_2} X-ray lines, respectively (see Fig. 2.47). For the common target materials Cu and Mo, the

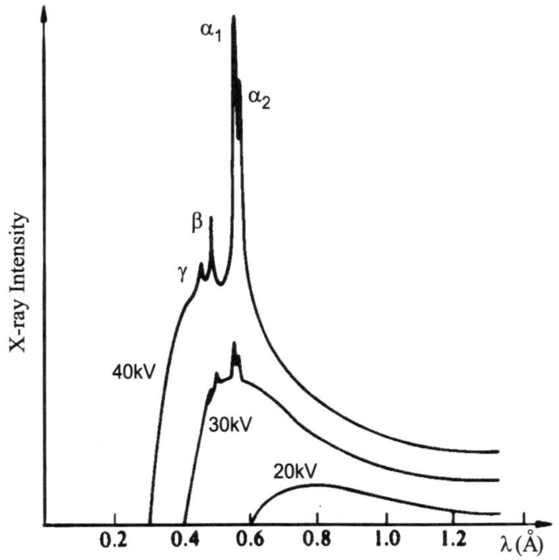

Fig.2.47 Intensity of X-rays emitted from an Ag target versus wavelength for the values of anode potential marked. At sufficiently high electron acceleration voltages, sharp peaks due to X-ray emission from intra-atomic electronic transitions are superimposed on the broad bremsstrahlung background. (After Burns (1985). Reproduced by permission of Academic Press, Inc.)

average values of the X-ray wavelengths for the dominant lines are CuK_α (1.5418 Å) and MoK_α (0.7107 Å).

The intense, almost monochromatic, K_α X-rays can be selected for use in diffraction experiments by using a suitable filter (e.g. Ni for CuK_α radiation), or by using as a monochromator a suitable single crystal (e.g. Si) oriented such that the Bragg diffraction condition is obeyed for the wavelength of interest, i.e.

$$2d\sin\theta = n\lambda \tag{2.95}$$

where d is the interplanar spacing, θ is *half* the angle subtended by the incident and diffracted X-ray beams (see Fig. 2.48) and n is the order of the reflection. This equation can be derived by assuming that X-rays (or other waves) reflect specularly from lattice planes: constructive interference occurs when the path difference $2d\sin\theta$ is an integral number of wavelengths, $n\lambda$.

An alternative source of short-wavelength X-rays is the so-called synchrotron radiation emitted when high-energy (GeV) electrons are forced to move in a circular trajectory under the action of bending magnets. Relativistic effects cause the dipole electromagnetic radiation to be shifted upward in energy from the radio frequency corresponding to the acceleration frequency of the electrons in the synchrotron ring up to the hard X-ray region (see Fig. 2.49). In addition, relativistic effects cause the normal 'dumb-bell' intensity distribution of dipole radiation to be concentrated into the (narrowed) forward lobe. The white synchrotron X-ray beam is monochromatized in the same way as for a conventional X-ray generator.

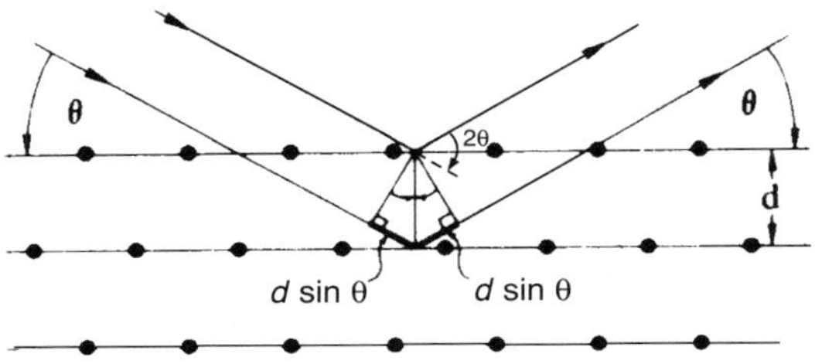

Fig.2.48 Bragg diffraction ('reflection') from a set of lattice planes of spacing d. The Bragg angle is half that subtended by the incident and diffracted beams. The path differences between the two rays are indicated.

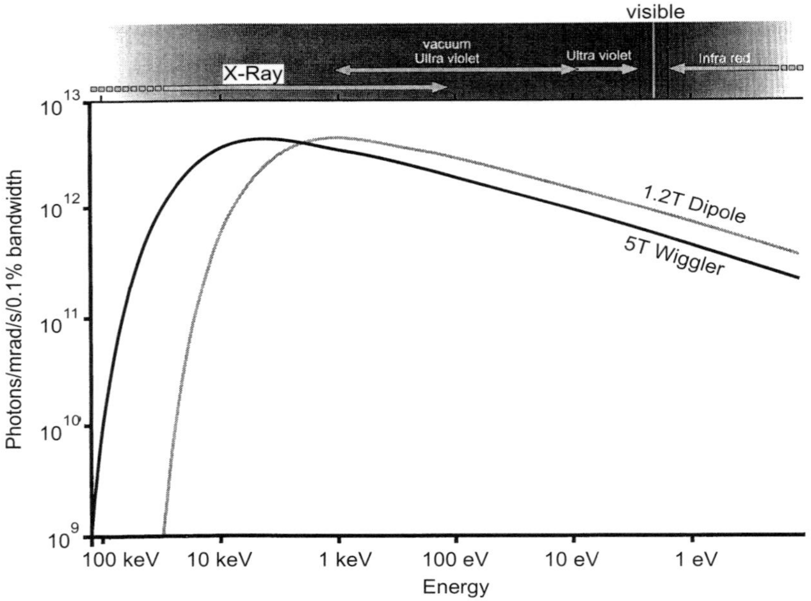

Fig.2.49 Spectral dependence of the photon flux emitted by the Daresbury 2 GeV Synchrotron Radiation Source at a bending-magnet or 'wiggler' output. (Figure courtesy of CLRC)

Neutrons may also be used to perform diffraction experiments on materials. The de Broglie relation relating the momentum p of a free particle (e.g. a neutron) and the wavelength of the corresponding quantum-mechanical wave representation is

$$\lambda = \frac{h}{p} = \frac{h}{mv} = \frac{h}{(2m\mathscr{E})^{1/2}} \tag{2.96}$$

where m and $\mathscr{E}$ are the mass and energy of the particle, respectively. High-energy ('epithermal') neutrons are produced as a byproduct of nuclear reactions, either in a reactor core or from the so-called spallation reaction when a high-energy proton beam from a synchrotron bombards a heavy metal target, typically uranium. Neutrons from a reactor source are produced effectively continuously in time, whereas those from a spallation source are pulsed and follow the bunches of protons impinging on the target. Passage of epithermal neutrons through a moderator causes a thermalization of their energy due to repeated collisions, and effective energy transfer, with the light atoms in the moderator (e.g. graphite). Thus, if the moderator is held at a temperature T, the average energy of the emerging 'thermal' neutrons is $3k_B T/2$. For a moderator operated at room temperature, eqn. (2.96) gives the wavelength of the thermal neutrons as $\lambda \simeq 1.5$ Å, i.e. comparable to interatomic spacings in solids. Neutrons have the advantage of being very penetrating (several cm) in matter.

Thermal neutrons emanating from a steady-state reactor source can be monochromatized using a single-crystal monochromator, as for X-rays. In the case of a pulsed spallation source, neutrons with a specific wavelength can be selected or analyzed by time-of-flight means: neutrons traversing the distance L from source to sample to detector in a time t have a velocity L/t and hence momentum $m_n L/t$. Thus the de Broglie relation (eqn. (2.96)) gives for the corresponding wavelength

$$\lambda = ht/m_n L. \tag{2.97}$$

Finally, electrons can be used for diffraction measurements also. Use of eqn. (2.96) with the electronic mass shows that 150 eV electrons have a wavelength of 1 Å, suitable for diffraction from arrays of atoms. However, such low-energy electrons are scattered very strongly by the electron clouds of atoms, with the result that their penetration depth is only a few ångströms, and so the use of such electrons is limited to surface studies. Electrons with energies of several keV (produced in an electron microscope) have a greater penetrating power (of several hundred Å) but a correspondingly shorter wavelength, confining the diffracted beams to very small angles.

2.6.1.2 The Laue formulation of diffraction

Consider a collimated, monochromatic beam, say of X-rays, incident on a sample from which it diffracts (Fig. 2.50a). Von Laue assumed that each site (set of atoms), say in the Bravais lattice of a crystal, would radiate the incident radiation in all directions at the same frequency. Constructive interference then gives rise to enhanced radiation intensity in certain directions. The incident beam, represented as a plane wave, can be written as $e^{i\boldsymbol{k}\cdot\boldsymbol{r}}$ and the diffracted beam as $e^{i\boldsymbol{k}'\cdot\boldsymbol{r}}$, where $\boldsymbol{k}$ and $\boldsymbol{k}'$ are the respective wavevectors. The scattering vector $\boldsymbol{K}$ is simply the vector difference between $\boldsymbol{k}$ and $\boldsymbol{k}'$ (Fig. 2.50b), i.e.

$$\boldsymbol{K} = \boldsymbol{k}' - \boldsymbol{k}. \tag{2.98}$$

In the case of *elastic* scattering where no energy, and hence momentum $\hbar k$, is exchanged between wave and scatterer, $|\boldsymbol{k}| = |\boldsymbol{k}'| = 2\pi/\lambda$, where λ is the wavelength of the wave. Thus, from geometrical considerations in Fig. 2.50b, the magnitude of the scattering vector is given by

$$|\boldsymbol{K}| = \frac{4\pi \sin\theta}{\lambda}. \tag{2.99}$$

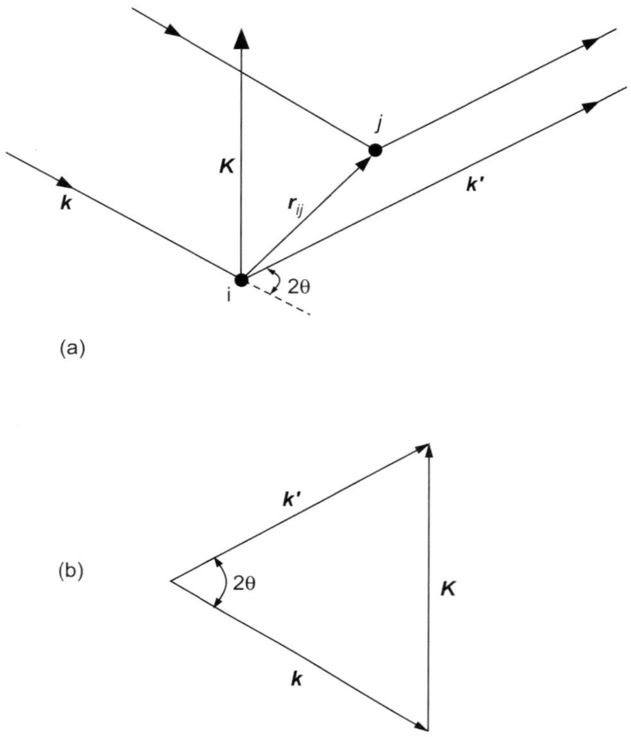

Fig.2.50 (*a*) Path difference for waves diffracted from two arbitrary scattering centres i, j, related by a vector r_{ij}: the incident beam has wavevector $\boldsymbol{k}$ and the diffracted beam has wavevector $\boldsymbol{k}'$. The scattering vector $\boldsymbol{K}$ is also shown. (*b*) Definition of the scattering vector $\boldsymbol{K} = \boldsymbol{k}' - \boldsymbol{k}.\boldsymbol{K}$ has the value $(4\pi/\lambda)\sin\theta$ for elastic scattering ($|\boldsymbol{k}| = |\boldsymbol{k}'| = 2\pi/\lambda$).

The maximum available value of $|K|$ is thus equal to $4\pi/\lambda$ for backscattering ($2\theta = 180°$).

The path difference for the incident wave between points i and j in Fig. 2.50a is equal to $-\hat{\boldsymbol{k}} \cdot \boldsymbol{r}_{ij}$ (where $\hat{\boldsymbol{k}} = \boldsymbol{k}/|\boldsymbol{k}|$), and likewise that for the scattered wave is $\hat{\boldsymbol{k}}' \cdot \boldsymbol{r}_{ij}$. Thus the total path difference is $(\hat{\boldsymbol{k}}' - \hat{\boldsymbol{k}}) \cdot \boldsymbol{r}_{ij}$, or equivalently the total difference in phase angle is $2\pi\boldsymbol{r}_{ij} \cdot (\hat{\boldsymbol{k}}' - \hat{\boldsymbol{k}})/\lambda$ or $(\boldsymbol{k}' - \boldsymbol{k}) \cdot \boldsymbol{r}_{ij} \equiv \boldsymbol{K} \cdot \boldsymbol{r}_{ij}$, giving a phase factor of $\exp[\boldsymbol{K} \cdot \boldsymbol{r}_{ij}]$ for the wave scattered from i relative to that scattered from j.

This result can be generalized for the case of a 3D Bravais lattice, where the vector r_{ij} now becomes the 3D translation vector $\boldsymbol{R}$ (eqn. (2.1)). Thus, for constructive interference to take place for *all* the waves scattered from each of the lattice points, the overall phase factor must equal unity (equivalent to the path difference being an integral number of wavelengths), i.e.

$$\exp(\mathrm{i}\boldsymbol{K} \cdot \boldsymbol{R}) = 1. \tag{2.100}$$

From the previous discussion on the reciprocal lattice (§2.4.1), it is apparent that eqn. (2.100) is *only* satisfied when the scattering vector $\boldsymbol{K}$ is equal to a reciprocal-lattice vector $\boldsymbol{G}$ (cf. eqn. (2.52)). Thus, the Laue condition for diffraction from a 3D translationally periodic object (a crystal) can thus be written succinctly as

$$\boldsymbol{K} = \boldsymbol{G}. \tag{2.101}$$

An alternative expression for the Laue diffraction condition can be derived in terms only of the incident wavevector, $\boldsymbol{k}$, in the case of elastic diffraction. Equation (2.101) can be rewritten (from eqn. (2.98)) as $\boldsymbol{k} + \boldsymbol{G} = \boldsymbol{k}'$ or $(\boldsymbol{k} + \boldsymbol{G})^2 = k'^2 = k^2$, or

$$2\boldsymbol{k} \cdot \boldsymbol{G} + G^2 = 0. \tag{2.102}$$

However, since $\boldsymbol{G}$ is a reciprocal-lattice vector, then so is $-\boldsymbol{G}$, and thus eqn. (2.102) can be rewritten also as

$$\frac{\boldsymbol{k} \cdot \boldsymbol{G}}{G} = \frac{G}{2}. \tag{2.103}$$

Equation (2.103) has a very powerful geometrical representation since it describes the locus of all points, called the Bragg plane, which is the perpendicular bisector of a reciprocal-lattice vector $\boldsymbol{G}$ (Fig. 2.51). The Laue condition is thus only satisfied for those incident wavevectors whose tip lies on the Bragg plane in reciprocal space when the wavevector origin is tied to a reciprocal-lattice point. Note also that the construction shown in Fig. 2.51 based on eqn. (2.103) is exactly the same as that used to construct the *first Brillouin zone* (cf. Fig. 2.41): the Bragg planes are simply the faces of the first Brillouin zone.

The Laue condition (eqn. (2.101)) gives a more satisfactory derivation of the Bragg diffraction law (eqn. (2.95)) than that given originally by Bragg in terms of the *ad hoc* assumption that planes of atoms specularly reflect X-rays, neutrons, etc., which subsequently constructively interfere (see Fig. 2.48). Substituting in eqn. (2.101) for the magnitude of the scattering vector $|\boldsymbol{K}| = (4\pi/\lambda)\sin\,\theta$ using eqn. (2.99) and for the general reciprocal-lattice vector $\boldsymbol{G}$, being an integral multiple (n) of the shortest vector $\boldsymbol{G}_0$ parallel to $\boldsymbol{G}$, where $|\boldsymbol{G}_0| = 2\pi/d$ (eqn. (2.46)) and d is the spacing of the Bravais lattice planes normal to $\boldsymbol{G}$, i.e.

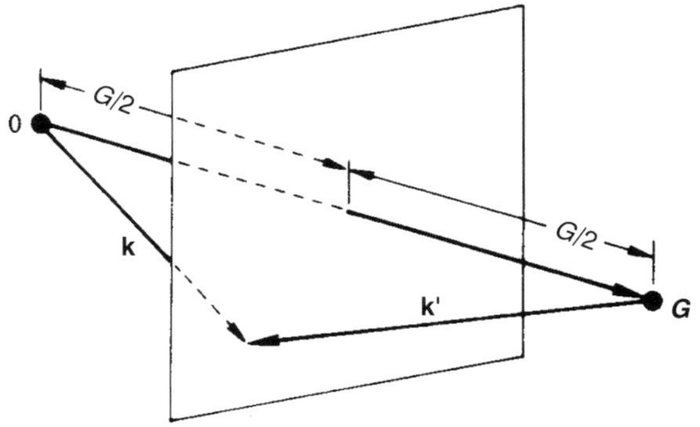

Fig.2.51 Geometrical representation of the Laue condition for diffraction. Diffraction only occurs for an incident wavevector whose tip lies on the Bragg plane (or equivalently the face of the first Brillouin zone) if its origin is tied to a reciprocal-lattice point.

$$|\boldsymbol{G}| = 2\pi n/d, \tag{2.104}$$

which immediately gives the Bragg condition $2d\sin\theta = n\lambda$ (eqn. (2.95)). Thus, the order, n, of the Bragg reflection is simply the length of the reciprocal-lattice vector, $\boldsymbol{G}$, characterizing the diffraction peak, normalized by the smallest parallel reciprocal-lattice vector.

Another useful geometrical representation of the Laue condition for diffraction (eqn. (2.101)) is the so-called Ewald construction (Fig. 2.52). The wavevector $\boldsymbol{k}$ corresponding to the incident beam of radiation with wavelength λ is drawn with its origin placed such that the vector $\boldsymbol{k}$ terminates at a reciprocal-lattice point. A sphere of radius $|\boldsymbol{k}| = 2\pi/\lambda$ is drawn about the origin of $\boldsymbol{k}$. A diffracted beam will exist if the surface of this Ewald sphere intersects any other point in the reciprocal lattice. If this occurs, then $\boldsymbol{k}' = \boldsymbol{k} + \boldsymbol{G}$, i.e. the Laue condition (eqn. (2.101)) is satisfied.

The overall scattering amplitude of waves diffracted from planes (hkl) in a crystal can be written as a sum over all atoms i as

$$F_{hkl} = \sum_i f_i \exp[\mathrm{i}\boldsymbol{G} \cdot \boldsymbol{R}_i], \tag{2.105}$$

where f_i is the atomic scattering (or form) factor (which depends on the type of radiation used—see §2.6.1.3) and the exponential term is the phase factor, eqn. (2.100). The argument in the exponential term can be expanded by rewriting $\boldsymbol{R}_i$ and $\boldsymbol{G}$ in terms of their respective basis vectors (eqns. (2.1) and (2.47), respectively), giving

$$F_{hkl} = \sum_i f_i \exp\{2\pi\mathrm{i}(hu_i + kv_i + lw_i)\}, \tag{2.106}$$

where eqn. (2.49a) has been used.

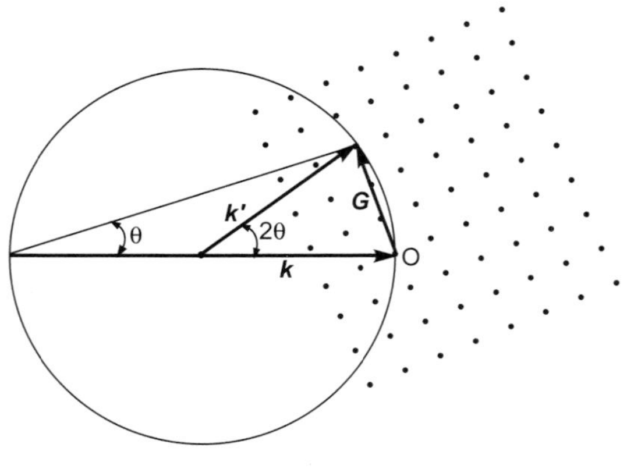

Fig.2.52 The Ewald construction for the Laue condition for elastic diffraction from a crystal. The wavevector $\boldsymbol{k}$ corresponding to the incident beam of radiation of wavelength λ is drawn with its origin such that $\boldsymbol{k}$ terminates at a reciprocal-lattice point, O. A sphere of radius $|\mathbf{k}| = 2\pi/\lambda$ is drawn about the origin of $\boldsymbol{k}$, and a diffracted beam will exist if the surface of this sphere intersects any other point in the reciprocal lattice, i.e. when $\boldsymbol{k}' = \boldsymbol{k} + \boldsymbol{G}$.

Depending on the particular type of Bravais lattice involved, eqn. (2.106) can be used to show that the scattering amplitude is identically zero, i.e. there are systematic absences in the expected number of Bragg diffraction peaks, for particular combinations of values of *hkl* (known as extinction rules). Thus, for example, for a b.c.c. lattice (with identical atoms at the positions 000 and 1/2 1/2 1/2 in the unit cell), use of eqn. (2.106) shows that

$$\boldsymbol{F}_{hkl} = \begin{cases} 2f & (h+k+l = \text{even}), \\ 0 & (h+k+l = \text{odd}). \end{cases}$$

The extinction rules for a number of Bravais lattices are given in Table 2.6. Further systematic absences can arise if additional translational symmetry elements are present. For example, for a 2_1 screw axis parallel to x, reflections for which $h = 2n+1$ are also absent. Systematic absences for various glide planes are also given in Table 2.6.

Thus far, the discussion has been devoted entirely to diffraction from perfectly crystalline materials. What of *amorphous materials*—do they diffract X-rays and neutrons as well? They do, but the diffraction intensity is no longer confined only to those directions determined by the reciprocal-lattice vectors (cf. eqn. (2.101)) since there is *no* reciprocal lattice for an amorphous material because of the lack of translational periodicity. Instead the diffraction intensity is diffusely spread out with respect to *all* scattering vectors $\boldsymbol{K}$, although the scattering intensity at some values of $\boldsymbol{K}$ is still greater than at others.

★ In order to calculate the diffuse scattering intensity from an amorphous solid, the (measurable) scattering intensity I is calculated as the product of the scattering amplitude F_i from one atom i (with respect to some arbitrary origin atom O) and the complex conjugate of the amplitude for atom j, viz.

$$I = F_i F_j^*. \tag{2.107}$$

Generalizing eqn. (2.105) for the scattering amplitude for the case of an arbitrary scattering vector $\boldsymbol{K}$, and general atomic positions $\boldsymbol{r}_i$, $\boldsymbol{r}_j$ not necessarily related by symmetry, gives

$$I = \sum_i f_i \exp[\mathrm{i}\boldsymbol{K}\cdot\boldsymbol{r}_i] \sum_j f_j \exp[-\mathrm{i}\boldsymbol{K}\cdot\boldsymbol{r}_j] \tag{2.108}$$

Table 2.6 Extinction rules for diffraction for some lattice types and symmetry elements

Symmetry element	*Reflection affected*	*Systematic-absence condition*
Centred cells		
Body-centred, I	hkl	$h+k+l = 2n+1$
Face-centred, F	hkl	$h+k, h+l, k+l = 2n+1$
Side-centred, C	hkl	$h+k = 2n+1$
Screw axis		
2_1 along a	$h00$	$h = 2n+1$
Glide planes $\perp b$		
Translation (a/2) (a-glide)	$h0l$	$h = 2n+1$
Translation (a/2 + c/2) (n-glide)	$h0l$	$h+l = 2n+1$
Translation ($a/4 + c/4$) (d-glide)	$h0l$	$h+l = 4n+1, 2, 3$

or

$$I = \sum_i \sum_j f_i f_j \exp[\mathrm{i}\boldsymbol{K} \cdot \boldsymbol{r}_{ij}], \tag{2.109}$$

where the interatomic vector $\boldsymbol{r}_{ij} = \boldsymbol{r}_i - \boldsymbol{r}_j$ has been used. A simplifying assumption generally valid for amorphous materials is that the material is *isotropic*, i.e. the vector $\boldsymbol{r}_{ij}$ may adopt all directions with equal probability (or equivalently the tip of the vector may take all positions on the surface of a sphere whose centre is the origin of the vector). The orientational average of the phase factor in eqn. (2.109) is thus given by

$$\begin{aligned} < \exp[\mathrm{i}\boldsymbol{K} \cdot \boldsymbol{r}_{ij}] > &= \frac{1}{4\pi r_{ij}^2} \int_{\phi=0}^{\pi} \exp[\mathrm{i}K r_{ij} \cos\phi] 2\pi r_{ij}^2 \mathrm{d}(\cos\phi) \\ &= \frac{\sin K r_{ij}}{K r_{ij}}, \end{aligned} \tag{2.110}$$

where ϕ is the angle subtended by the vectors $\boldsymbol{K}$ and $\boldsymbol{r}_{ij}$ (see Fig. 2.50a), and $K \equiv |\boldsymbol{K}|$ and $r_{ij} = |\boldsymbol{r}_{ij}|$. Substituting eqn. (2.110) into eqn. (2.109) yields the so-called Debye equation for the diffuse scattering from a random array of atoms,

$$I(K) = \sum_i \sum_j f_i f_j \frac{\sin K r_{ij}}{K r_{ij}}. \tag{2.111}$$

Note that the diffuse scattering intensity is a function only of the *magnitude* of the scattering vector $\boldsymbol{K}$; the scattering intensity is the same in all directions in $\boldsymbol{K}$-space as a consequence of the real-space amorphous structure being spatially isotropic. ★

2.6.1.3 Atomic form factor

The atomic scattering factor, or form factor, f, introduced in eqn. (2.105) describes the scattering amplitude for a wave scattered by an individual atom (i.e. without the interatomic interference effects associated with the phase factor). The atomic form factor has a different functional dependence on the magnitude of the scattering vector, K, depending on the type of radiation involved.

For the case of X-ray diffraction, the X-rays (being electromagnetic waves) interact with the *electrons* in the atoms: the scattering process can be thought of as one involving absorption of the X-ray photon, accompanied by the excitation of the electronic system, immediately followed by de-excitation and re-radiation of an X-ray photon. Since the size of an atom is comparable to the X-ray wavelength, *intra*-atomic interference effects are important, and the X-ray atomic form factor is thus strongly dependent on K.

It is reasonable to assume that the amplitude of an X-ray wave scattered from a particular volume element $\mathrm{d}V$ is proportional to $n(\boldsymbol{r})dV$ (times a phase-factor term), where $n(\boldsymbol{r})$ is the electronic charge density contained within the volume element at position $\boldsymbol{r}$. For a whole sample, the scattering amplitude is

$$F = \int n(\boldsymbol{r}) \exp[\mathrm{i}\boldsymbol{K} \cdot \boldsymbol{r}] \mathrm{d}V \tag{2.112}$$

for scattering with a general value of reciprocal-lattice vector, $\boldsymbol{K}$. The total electron concentration of the sample can be written as the sum over the electron concentrations n_i of individual atoms i, i.e.

$$n(\boldsymbol{r}) = \sum_i n_i(\boldsymbol{r} - \boldsymbol{r}_i). \tag{2.113}$$

Writing $\boldsymbol{r} - \boldsymbol{r}_i = \boldsymbol{\rho}$, substitution of eqn. (2.113) into eqn. (2.112) gives

$$F = \sum_i \exp[\mathrm{i}\boldsymbol{K} \cdot \boldsymbol{r}_i] \int n_i(\boldsymbol{\rho}) \exp[\mathrm{i}\boldsymbol{K} \cdot \boldsymbol{\rho}]\mathrm{d}V. \tag{2.114}$$

Comparison of eqn. (2.114) with the generalized form of eqn. (2.105), i.e. $F = \sum_i f_i \exp[\mathrm{i}\boldsymbol{K} \cdot \boldsymbol{r}_i]$, shows that the X-ray atomic form factor is given by the relation

$$f_i^{\mathrm{x}} = \int n_i(\boldsymbol{r}) \exp[\mathrm{i}\boldsymbol{K} \cdot \boldsymbol{r}]\mathrm{d}V, \tag{2.115}$$

where the substitution $\boldsymbol{r} = \boldsymbol{\rho}$ has been made. The integral in eqn. (2.115) is over the volume of the charge distribution of the atom; the volume element is given by $\mathrm{d}V = 2\pi r^2 \sin\phi \, \mathrm{d}\phi \, \mathrm{d}r$, where ϕ is the angle that $\boldsymbol{K}$ makes with $\boldsymbol{r}$. In the case of a spherically symmetric charge distribution, $n_i(\boldsymbol{r}) = n_i(r)$, the evaluation of the integral over ϕ is as for eqn. (2.110), and thus

$$f_i^{\mathrm{x}}(K) = 4\pi \int n_i(r) r^2 \frac{\sin Kr}{Kr} \mathrm{d}r. \tag{2.116}$$

For *forward* scattering, $2\theta = 0$, and hence $K = 0$ (cf. eqn. (2.99)). Thus, in this case,

$$f_i^{\mathrm{x}}(K = 0) = 4\pi \int n_i(r) r^2 \mathrm{d}r = Z, \tag{2.117}$$

where Z is the atomic number of the element (i.e. the total number of electrons in the atom). Thus, X-ray scattering is very weak for very light atoms, such as H or Li. The X-ray atomic form factor decreases strongly with increasing K (eqn. (2.116)), as can be seen in Fig. 2.53 for the case of Al ($Z = 13$). Thus, unavoidably, the intensity of an X-ray diffraction pattern decreases markedly with increasing scattering angle, making the measurement of high-order reflections difficult using X-ray diffraction.

In the case of *electron* diffraction, the electrons are scattered both by the electron distribution in the atom *and* by the charged nucleus. The electron form factor is given by the expression

$$f^{\mathrm{e}}(K) = \frac{2m_{\mathrm{e}}e^2}{\hbar^2}[(Z - f^{\mathrm{x}}(K))/K^2], \tag{2.118}$$

where m_e is the electronic mass. The electron scattering factor has an even stronger dependence on K than $f^x(K)$.

Neutrons, in contrast, interact only with atomic nuclei via the nuclear force. The scattering process can be regarded as the momentary capture of the incoming neutron by a nucleus, and then re-emission of the neutron. If a particular nucleus has no nuclear spin, then the scattering from all such nuclei in a solid will be the same (coherent), analogous to that of X-rays, and interatomic interference (diffraction) effects can occur. However, if an elemental constituent, A, of a sample consists of a number of different isotopes, i, each with a *different* neutron scattering factor, or scattering length $b_{\mathrm{A},i}$, then the *coherent* neutron-scattering cross-section, Σ_c, is related to the *mean* scattering length $\bar{b}_{\mathrm{A}} = \sum_i c_i b_{\mathrm{A},i}$, where c_i is the abundance of isotope i, via the equation

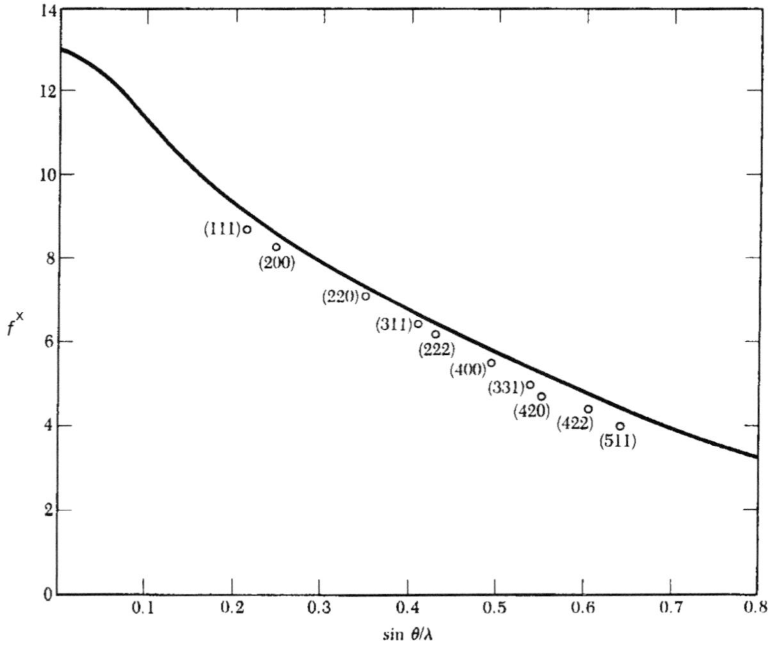

Fig.2.53 Absolute atomic scattering factors for metallic Al measured for the reflections labelled, compared with a theoretical Hartree–Fock calculation (solid line). (After Batterman *et al.* (1961). Reprinted with permission from *Phys. Rev.* **122**, 68. © 1961. APS)

$\Sigma_c = 4\pi\bar{b}^2_{\text{AA}}$. The mean-square deviation of scattering lengths from the mean gives rise to incoherent scattering, with a cross-section $\Sigma_i = 4\pi(\overline{(b^2_{\text{A}}} - \bar{b}^2_{\text{A}}))$ that does not contain a phase factor and hence does not contain structural information through an interference term. In addition to this isotopic incoherence, nuclei with non-zero nuclear spins ($I > 0$) can also give rise to spin incoherence. This is because the $(2I + 1)$ different nuclear-spin states all have different scattering lengths, b_I. A nucleus exhibiting predominantly incoherent scattering is ^{51}V.

The neutron scattering factor does not vary systematically with Z like the X-ray form factor (eqn. (2.117)). Instead, there is considerable variation in b from element to element (see Fig. 2.54). Thus, neutron diffraction can readily be performed for light-element-containing (e.g. hydrogenous) materials. For certain isotopes (e.g. ^{1}H, ^{7}Li, ^{48}Ti, ^{62}Ni) the neutron scattering length is even *negative*, associated with a phase change of π of the neutron on scattering.

A feature of particular advantage is that the neutron-scattering length is *independent* of K; this is due to the fact that the size of the nucleus is very much smaller than the neutron wavelength, thereby precluding intra-atomic interference effects as found for X-ray scattering. This behaviour can be demonstrated by assuming that eqn. (2.116) is valid for neutron scattering as well; if the size of the nucleus is confined to the vicinity of $r = 0$, then the factor $(\sin Kr)/Kr \simeq 1$, and hence $f^n \equiv b$ is independent of K.

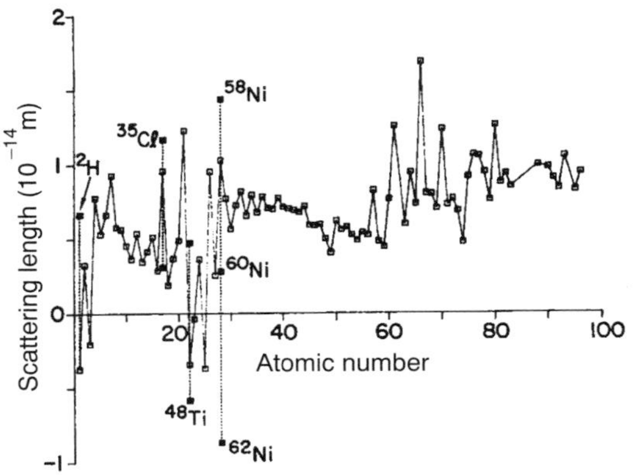

Fig.2.54 Variation of the neutron scattering length b with atomic number.

Finally, since the neutron has an intrinsic spin ($I = 1/2$), and hence a magnetic moment, it can interact with other unpaired spins, such as electrons in magnetic materials, for example. Thus, the *magnetic* structures of such materials may be probed using neutron diffraction (see §7.2.5.6). Spin polarization of neutrons allows nuclear and magnetic scattering to be separated when peaks overlap in the diffraction pattern.

2.6.1.4 Experimental methods

Diffraction experiments require a source of radiation (X-rays, neutrons, or electrons), a monochromator if appropriate, a sample, and a detector for the scattered radiation. Such experiments can essentially be divided into two categories: single-crystal experiments and powder experiments.

Single-crystal X-ray or neutron-diffraction measurements in general provide the most complete set of diffraction data for crystalline materials. Such experiments nowadays are performed using a four-circle diffractometer (see Fig. 2.55), in which each set of *hkl* planes of the crystal is successively brought into the diffraction condition (eqn. (2.95)) by systematically varying the position of the crystal with respect to the incident beam direction using the angles of rotation (ϕ, χ, ω); the detector can then scan across the reflections using a 2θ-scan. Older single-crystal methods used X-rays, in which the detector consisted of a sheet of photographic film wrapped cylindrically around the sample, and again the crystal was moved relative to the beam direction and film in certain ways. In precession photographs, in particular, the diffraction spots form a direct map of the reciprocal lattice. For example, Fig. 2.56 shows a schematic illustration of two precession photographs (i.e. two 'slices' through the 3D reciprocal space) for a b.c.c. (I) crystal, with some of the reflections labelled according to the Miller indices of the real-space planes to which they correspond; the axes $\boldsymbol{a}^*$, $\boldsymbol{b}^*$ of the reciprocal lattice are also indicated. Note that reflections such as 100, 300, 120, 140, etc. are absent in

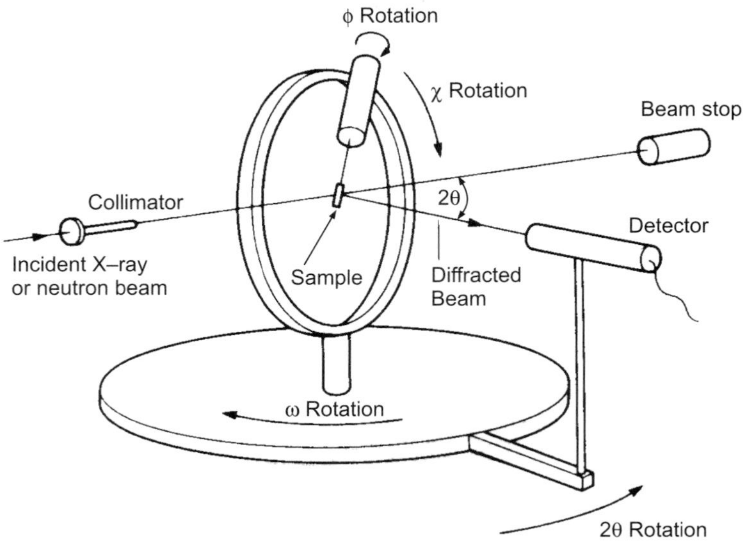

Fig.2.55 Schematic illustration of a four-circle diffractometer used for single-crystal diffraction.

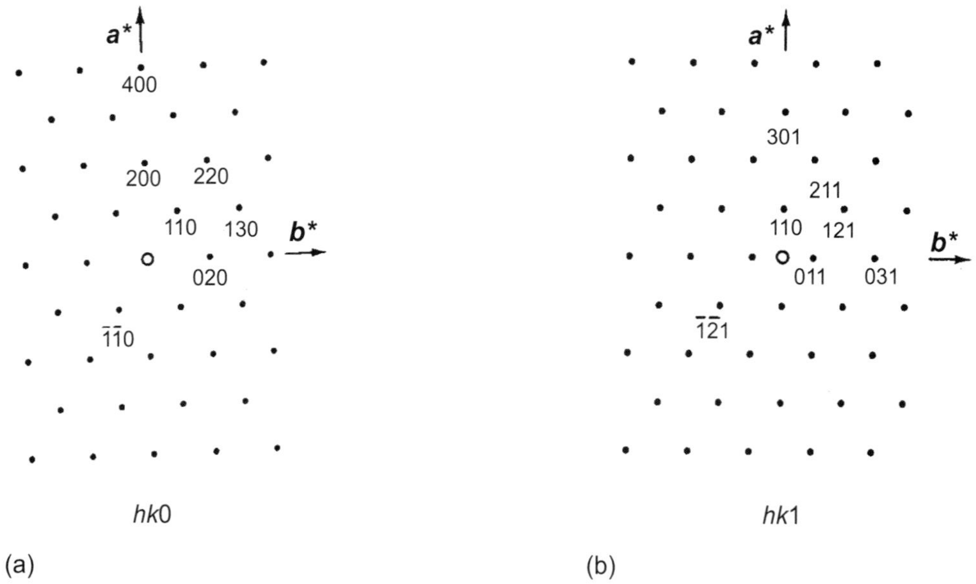

Fig.2.56 Schematic illustration of precession photographs for a b.c.c. single crystal for (a) *hk*0 reflections; (b) *hk*1 reflections.

Some of the diffraction spots are marked according to the *hkl* planes to which they correspond. The reciprocal-lattice axes are labelled a^* and b^*.

Fig. 2.56a, and 201, 401, 021, 041, etc. are missing in Fig. 2.56b. These are examples of the systematic-absence condition $(h + k + l) = 2n + 1$ characteristic of a b.c.c. crystal (Table 2.6). The f.c.c. nature of the reciprocal-space lattice of the real-space b.c.c. lattice (see §2.4.1) is also evident in Fig. 2.56.

A four-circle diffraction measurement of a single crystal yields a set of diffraction intensities $|F_{hkl}|^2$, from which the *magnitude* (modulus) of the scattering amplitude $|F_{hkl}|$ may be found. Unfortunately, the structural information needed to solve the crystal structure, namely the atomic density $n(\boldsymbol{r})$ for neutron diffraction (electron density for X-ray diffraction), is related via the Fourier series

$$F_{hkl} = \sum_h \sum_k \sum_l n(\boldsymbol{r}) \exp[2\pi \mathrm{i}(hx + ky + lz)] \tag{2.119}$$

to the *total* amplitude $F_{hkl} = |F_{hkl}|e^{i\phi_{hkl}}$, where the phase factor ϕ_{hkl} is unknown. (Note that a summation over the discrete values of hkl is taken, rather than the integral form of the Fourier transform of eqn. (2.112).) This phase problem is a major obstacle to the determination of crystal structures.

One way of obviating the phase problem is to use the Patterson method. Although $n(\boldsymbol{r})$ cannot be obtained from eqn. (2.119) directly by Fourier inversion (cf. eqn. (2.58)) (since the F_{hkl} are not known), instead a Fourier inversion is effected using the measured *intensities* $|F_{hkl}|^2$ as coefficients to produce a Patterson map, viz.

$$P_{uvw} = \frac{1}{V} \sum_h \sum_k \sum_1 |F_{hkl}|^2 \exp[-2\pi \mathrm{i}(hu + kv + lw)], \tag{2.120}$$

where again a summation over the discrete set of hkl values has been used. For X-ray diffraction, the resulting Patterson map looks somewhat like a 3D electron-density map of a solid, but whereas in an electron-density map the highest intensity contours correspond to the atomic *positions*, in a Patterson map, the peaks correspond instead to *interatomic vectors* between pairs of atoms. Since, for X-ray diffraction, the atomic scattering factors scale as the atomic number, Z (eqn. (2.117)), the intensity of a peak in the Patterson map is proportional to the product $Z_i Z_j$ for a pair of atoms i and j. Thus, this method provides a way of finding the positions of *heavy* atoms in the unit cell of a crystal since the strongest peaks in the Patterson map correspond to these. An approximate value of the atomic scattering amplitude, F_{hkl}, can then be calculated from eqn. (2.119) using only the atomic positions of the heavy atoms: although the magnitude of F_{hkl} will not necessarily be very accurate, the *phase* should be more-or-less correct. Inverse Fourier transformation of eqn. (2.119) to yield the electron-density map $n(\boldsymbol{r})$, now using *measured* values of $|F_{hkl}|$ with the estimated phases, should lead to the lighter atoms also being located. The atomic positions in the unit cell are then optimized by a least-squares minimization, structure-refinement procedure (taking into account thermal displacements of the atoms) in terms of the quality-of-fit (or 'reliability') R factor, given by

$$R(\%) = \frac{\sum_{hkl} |F_{hkl}^{\mathrm{obs}} - F_{hkl}^{\mathrm{calc}}|}{\sum_{hkl} |F_{hkl}^{\mathrm{obs}}|}, \tag{2.121}$$

where R values of $\lesssim 5\%$ are typical for a refined structure. For further details, see e.g. West (1987).

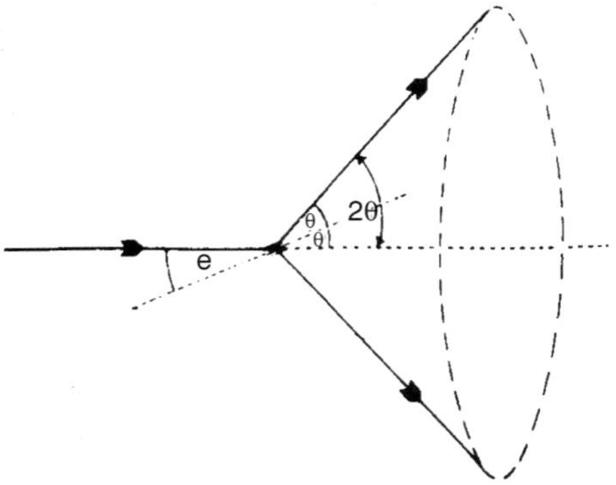

Fig.2.57 Schematic illustration of the formation of a cone of diffracted radiation from a given lattice plane in the powder-diffraction method.

The other common type of X-ray or neutron diffraction measurement is powder diffraction, used for samples which cannot be produced in single-crystal form of sufficient size. Assuming that the crystallites are completely *randomly* oriented, a given real-space lattice plane making an angle of θ to the incident beam of radiation will produce a *cone* of diffracted radiation (with included angle 4θ) when all angular orientations about the incident-beam axis are taken into account (see Fig. 2.57): the individual Bragg diffraction spots characteristic of a single-crystal diffraction pattern, e.g. in a precession photograph (Fig. 2.56), become transformed into a series of concentric arcs of scattered intensity about the incident-beam direction. As a result, the structural information available in 3D from a single-crystal experiment (i.e. the reciprocal lattice) is effectively reduced to a 1D representation in the case of powder diffraction (i.e. the *radii* of the conical arcs).

Interplanar spacings d, for a polycrystalline sample can be calculated using the Bragg law (eqn. (2.95)) from the values of diffraction angle, 2θ, corresponding to peaks in a powder pattern. Often, an unknown structure can be identified by 'finger-print' means by comparing d-spacings, and the corresponding powder-diffraction peak intensities, with values tabulated for some 30 000 known inorganic structures in the JCPDS (1984) powder-diffraction file. This approach is particularly suitable for relatively high-symmetry crystals where the diffraction peaks are well separated in angle, but less so for low-symmetry materials where appreciable peak overlap occurs.

A more modern approach uses Rietveld profile analysis of powder-diffraction patterns measured using neutrons or X-rays to *refine* structures. A least-squares optimization procedure is applied to the fitting of diffraction-peak *profiles* (i.e. intensity and width), taking into account peak overlap for low-symmetry structures (see Fig. 2.58 for an example). Very recently, however, the *ab initio* determination of crystal structures from high-resolution X-ray or neutron powder-diffraction patterns has been achieved, i.e. without the assumption of a trial structure as needed in the Rietveld method (see Cheetham (1987)); such a method has the advantage that single crystals are no longer needed for a complete structure determination.

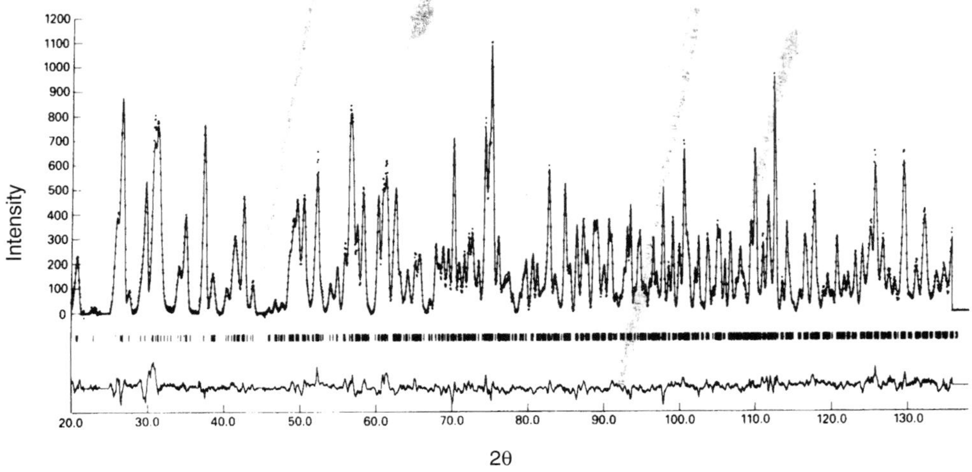

Fig.2.58 Neutron powder-diffraction pattern of polycrystalline $Fe_2(SO_4)_3$ showing the results of a Rietveld refinement (full line) to the measured data (dots). The reflection positions and the difference curve (measured minus calculated) are also shown (Cheetham (1987), in *Solid-State Chemistry: Techniques*, eds A. K. Cheetham and P. Day, 1987. Reproduced by permission of Oxford University Press)

Electron-diffraction measurements are carried out in a transmission electron microscope (Fig. 2.59a), but this instrument has the added advantage that the diffracted beams can also be combined together again using electrostatic lenses to form a magnified image of the sample (Fig. 2.59b). An example of a diffraction pattern and corresponding real-space *image*, for the case of a quasicrystalline (icosahedral) sample of an AlFeCu alloy, is shown in Fig. 2.60. Electron diffraction/imaging has the advantage that it may be carried out on very small areas (since the size of a focused electron beam is of the order of a few μm in diameter), although the sample thickness must also be very small (a few 100 Å) to avoid absorption problems. It should be noted that transmission electron microscope images are *projections* of the atomic structure along the beam direction.

Thus far, the discussion has concentrated on crystalline materials. However, the diffuse scattering exhibited by *amorphous* materials, as exemplified by the Debye equation (eqn. (2.111)), also contains structural information. A diffraction measurement on an amorphous material is equivalent to a powder-diffraction experiment for a polycrystalline material: the diffracted intensity now consists of a pattern of concentric *diffuse* haloes, i.e. the diffracted intensity function is a *one-dimensional* function of the magnitude of the scattering vector, $K = |\boldsymbol{K}|$ (eqn. (2.99)).

★ The Debye expression (eqn. (2.111)) for the diffuse scattering from an amorphous solid can be analysed most simply if the material is monatomic, with atomic scattering factor f, in which case eqn. (2.111) becomes

$$I = \sum_{i=1}^{N} f^2 + \sum_{i=1}^{N} \sum_{j \neq i}^{N} f^2 \frac{\sin K r_{ij}}{K r_{ij}}. \tag{2.122}$$

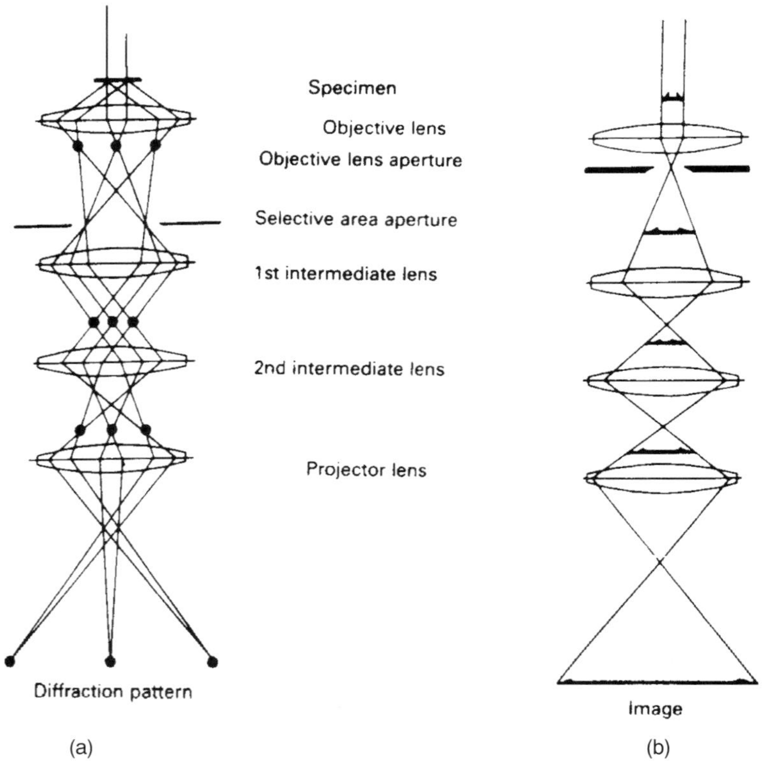

Fig.2.59 Schematic ray diagrams for the formation of: (a) a diffraction pattern, and (b) an image in the transmission electron microscope.

The summation over i simply yields the number of atoms, N, in the sample. The summation over j can be replaced by an integral involving the atomic-density function $\rho(r)$ (averaged over all atoms) for each atom j taken in turn as origin, with $r = r_{ij}$, i.e.

$$I = Nf^2 + Nf^2 \int \rho(r) \frac{\sin Kr}{Kr} \mathrm{d}V. \tag{2.123}$$

Adding and subtracting a term in ρ^0, the (macroscopic) average density, and assuming spherical symmetry gives:

$$I = Nf^2 + Nf^2 \int_0^\infty 4\pi r^2 [\rho(r) - \rho^0] \frac{\sin Kr}{Kr} \mathrm{d}r + Nf^2 \int_0^\infty 4\pi r^2 \rho^0 \frac{\sin Kr}{Kr} \mathrm{d}r. \tag{2.124}$$

The third term in eqn. (2.124) represents small-angle scattering due to the finite sample size and is completely masked by the undiffracted transmitted beam and so may be neglected. Thus, eqn. (2.124) may be rewritten succinctly as

$$K[S(K) - 1] = \int_0^\infty G(r) \sin Kr \, \mathrm{d}r \tag{2.125}$$

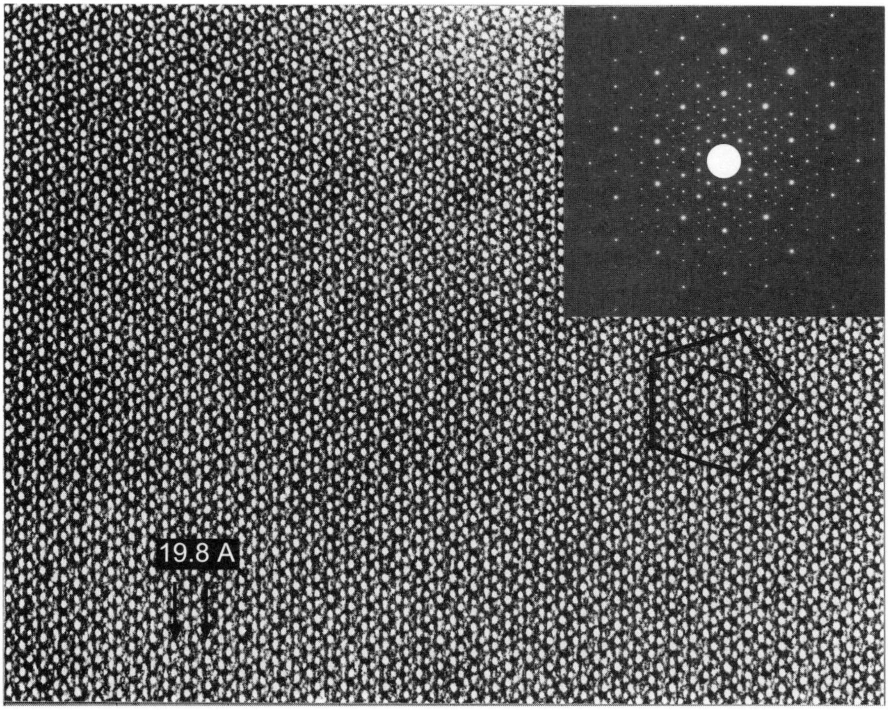

Fig.2.60 High-resolution real-space image, and the corresponding diffraction pattern, of quasi-crystalline AlCuFe obtained using an electron microscope. The diffraction pattern has pentagonal symmetry, and pentagonal orientational arrangements in the real-space image are also indicated. The lattice planes in the image, when viewed from a grazing angle, are not periodic: the spacings are related by the golden mean, $\tau = 2\cos 36^\circ = (1 + \sqrt{5})/2$. (Courtesy of Dr. M. Audier)

where the structure factor, $S(K)$, is defined as

$$S(K) = I(K)/Nf^2, \tag{2.126}$$

which oscillates about unity, and the reduced radial distribution function, $G(r)$, (so-called because it oscillates about zero) is given by

$$G(r) = 4\pi r[\rho(r) - \rho^0] \equiv J(r)/r - 4\pi r\rho^0. \tag{2.127}$$

$J(r)$ is the radial distribution function (RDF) (see §2.1.4, Fig. 2.16 and eqn. (2.4)).

Thus, eqn. (2.125) is suggestive of the *form* of a Fourier transform, and so can be Fourier inverted to yield the reduced RDF in terms of the measured structure factor $S(K)$, viz.

$$G(r) = \frac{2}{\pi}\int_0^\infty K[S(K) - 1]\sin Kr\, dK. \tag{2.128}$$

Note that the inverse Fourier transform (eqn. (2.128)) requires diffraction data for an *infinite* range of scattering vectors. In practice, however, the maximum value is

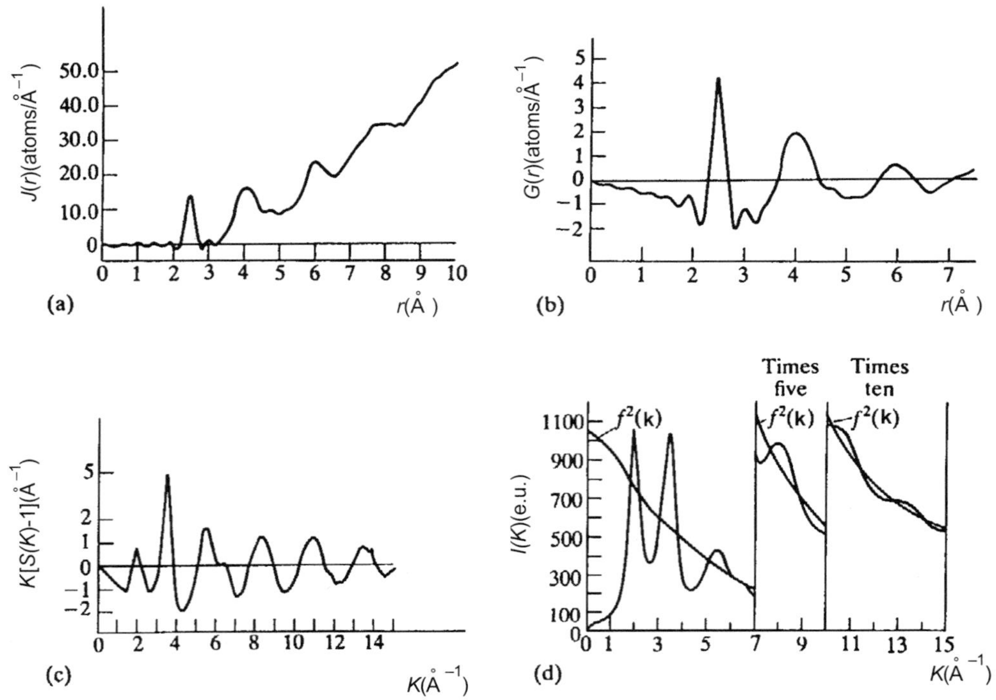

Fig.2.61 Structural correlation functions as a function of distance and scattering vector obtained from X-ray diffraction from amorphous Ge (Temkin *et al.* (1973)): (a) RDF; (b) reduced RDF; (c) reduced structure factor; (d) measured X-ray scattering intensity. (Reprinted and permission of Taylor & Francis Group Ltd.)

$K = 4\pi/\lambda$ (for the case of backscattering, $2\theta = 180°$—cf. eqn. (2.99)). The unavoidable use of truncated diffraction data-sets in the Fourier transform of eqn. (2.128) produces spurious 'termination' oscillations in $G(r)$, particularly at small r.

Examples of various real- and reciprocal-space pair-correlation functions obtained from diffraction measurements on amorphous Ge are shown in Fig. 2.61. Peak positions in the RDF, $J(r)$, correspond to interatomic separations, and areas under peaks correspond to average atomic coordination numbers (although peak overlap is very pronounced and hence a severe problem for larger r). The approach outlined above is also valid for the case of (poly)crystalline materials.

*2.6.2 X-ray absorption fine-structure spectroscopy

The X-ray absorption coefficient of atoms, $\mu_x(\mathscr{E})$, is generally a monotonically decreasing function of X-ray photon energy, $\mathscr{E}$, except when the photon energy becomes equal to an ionization energy for the atom, corresponding to the excitation of a bound electron to an energy above the vacuum level of the atom in the solid, whereupon the

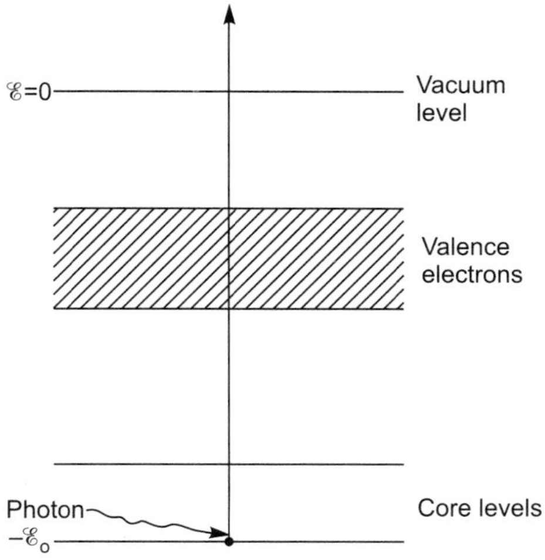

Fig.2.62 Schematic illustration of the photo- excitation by an X-ray photon of an electron from a core level (e.g. a ls, K electron) with binding energy $-\mathscr{E}_0$ to beyond the vacuum level ($\mathscr{E} = 0$) of the atom in the solid, thereby forming a free photoelectron.

electron becomes free of the atom (see Fig. 2.62). The threshold photon energy required to create such a photoelectron is

$$h\nu = \mathscr{E}_0, \tag{2.129}$$

where $\mathscr{E}_0$ is the binding energy of the core electron. At the photon energy $\mathscr{E}_0$, therefore, the X-ray absorption coefficient discontinuously increases in magnitude, producing an absorption edge (see Fig. 2.63). The positions of the X-ray absorption edges are *different* for each element. A photoelectron created by an X-ray photon with energy $h\nu > \mathscr{E}_0$ will carry away the excess energy as translational kinetic energy, and thus the wavevector k of the associated propagating electron wave can be found using the de Broglie relation between wavelength and momentum (eqn. (2.96)), viz.

$$k = \frac{2\pi}{\lambda} = \left(\frac{2m_e(h\nu - \mathscr{E}_0)}{\hbar^2}\right)^{1/2}, \tag{2.130}$$

where m_e is the electron rest mass. (Note that this is *half* the maximum value $4\pi/\lambda$ of the diffraction scattering vector K (cf. eqn. (2.99)).)

An outgoing photoelectron freed from an atom in a solid can then undergo a process of internal diffraction, i.e. the electron wave can diffract from atoms neighbouring the atom that absorbed the X-ray, and interference can then occur between the backscattered electron waves and the outgoing electron wave (see Fig. 2.64). The final-state photoelectron wavefunction thus contains contributions from both outgoing and backscattered wave components, and the interference between these two waves therefore causes a modulation of the photon-energy dependence of the matrix element for the X-ray absorption process and hence of the absorption coefficient, μ_x, itself. These

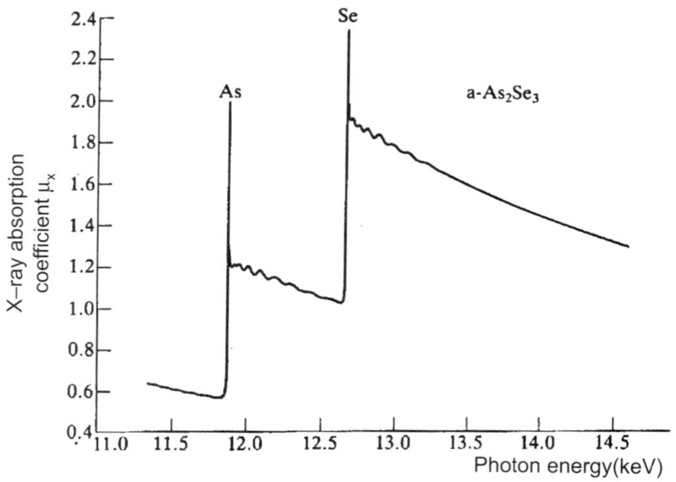

Fig.2.63 X-ray absorption spectra for amorphous As_2Se_3 showing the absorption edges for excitation from the K(ls) core level of As and Se atoms, together with the fine structure at photon energies above the edges. (Data by R. Pettifer, in Elliott (1990).)

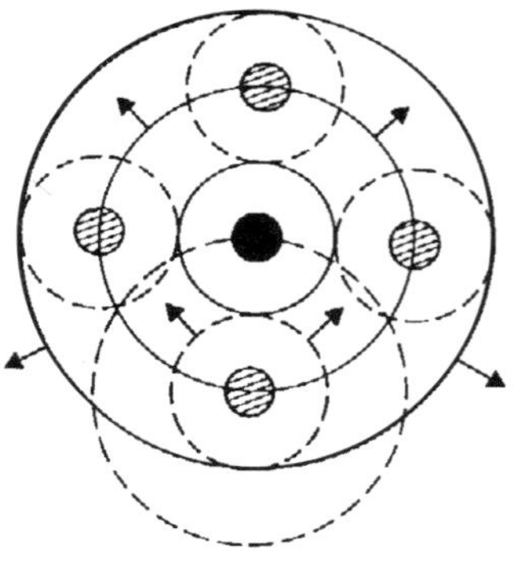

Fig.2.64 Schematic illustration of the internal-diffraction process leading to X-ray absorption fine structure. An atom (filled circle) on absorbing an X-ray photon with energy greater than the electron binding energy, creates an outgoing photoelectron wave (solid circular line) that can backscatter (dashed circular lines) from neighbouring atoms (hatched circles). Interference takes place where outgoing and backscattered waves overlap, and this modulates the X-ray absorption coefficient.

oscillations in μ_x are known as X-ray absorption fine structure (XAFS): see Fig. 2.63 for examples for the case of amorphous As_2S_3.

Two types of internal-diffraction events can be distinguished. In the first, for photon energies much greater than the threshold energy ($h\nu > \mathscr{E}_0 + 50$ eV), the outgoing electron wave scatters only *once* from a neighbouring atom; i.e. *pairs* of atoms are involved in the process (the atoms absorbing the X-rays and the backscattering atoms) and, in this regard, the resulting *extended* X-ray absorption fine structure (EXAFS) is similar in origin to a conventional diffraction pattern (§2.6.1) which is also due to a single-scattering process and hence also involves only pairwise structural correlations. In the second type of internal-diffraction process, operative at much lower photon energies ($\mathscr{E}_0 < h\nu < \mathscr{E}_0 + 50$ eV), electron scattering from *more* than one backscattering atom takes place. In the simplest case of such multiple scattering, three atoms are involved (the atom absorbing the X-ray and two backscattering atoms), and thus the resulting X-ray absorption near-edge structure (XANES), or near-edge X-ray absorption fine structure (NEXAFS), contains structural information about triplet, and higher, correlations. XANES oscillations can only be investigated theoretically by performing multiple-scattering calculations for electrons in an *assumed* structural model of the solid.

On the other hand, the single-scattering process characteristic of EXAFS oscillations allows an analytic expression to be obtained for the normalized X-ray absorption fine-structure amplitude

$$\chi(k) = (\mu_x(k) - \mu_0(k))/\mu_0(k), \tag{2.131}$$

where μ_0 is the background X-ray absorption coefficient, i.e.

$$\chi(k) = -\sum_i \frac{N_i}{R_i^2}\frac{|f_i(\pi)|}{k} \exp(-2R_i/\lambda_e)\exp(-2\sigma_i^2 k^2) \times \sin[2kR_i + 2\delta(k) + \eta_i(k)], \tag{2.132}$$

where the summation is over all shells of backscattering atoms, i, each containing N_i atoms at a distance R_i from the absorbing atom. The EXAFS amplitude depends on N_i and on the backscattering amplitude $|f_i(\pi)|$ of atoms i, and is attenuated by a factor of $1/R_i^2$ (since both outgoing and backscattered electron waves are assumed to be spherical, each decreasing in amplitude as $1/R$), because of the finite mean-free path λ_e of electrons in the material, and by the Debye Waller factor (see §4.2.6) involving static and thermal r.m.s. displacements σ_i about the equilibrium bond length along the vector joining absorbing and backscattering atoms. Finally, the EXAFS amplitude is sinusoidally modulated by an interference term involving the phase shift of the photoelectron; additional phase shifts $\delta(k)$ and $\eta(k)$ arise because the photoelectron is emitted from, and backscattered by, atomic potentials, respectively.

Structural information from EXAFS measurements can be extracted by fitting the experimental amplitude (eqn. (2.127)) to that calculated using eqn. (2.132) using the structural quantities (N_i, R_i and σ_i) as variable parameters, and taking values for phase shifts, backscattering amplitude and electron mean-free path from corresponding fits of EXAFS data for known crystalline structures. Note that the single-scattering expression for $\chi(k)$ given by eqn. (2.132) is a local function, and does *not* depend on translational periodicity; the technique is therefore equally applicable to crystalline and disordered materials. Furthermore, it is *atom-specific* since the X-ray absorption-edge energy is characteristic of a given element (see Fig. 2.63), and so the *local* structure around a particular type of element, acting as X-ray absorber, can be investigated in multi-element materials.

Obviously, X-ray absorption fine-structure measurements require X-rays having photon energies spanning a wide range, both below and above an X-ray absorption edge. A convenient, high-brightness source of such X-rays is synchrotron radiation (Fig. 2.49).

*2.6.3 Holography

The ultimate goal in determining the structure of solids is to use so-called direct methods to produce a faithful, 3D real-space image of the atomic arrangement in the sample, without the use of assumed trial structures in the analysis of the data. Transmission electron microscopic images do not satisfy this criterion since essentially they are 2D projections of the 3D image along the electron-beam direction. Conventional diffraction experiments (§2.6.1) cannot be used for this purpose either, since what is measured is the diffracted *intensity*, and all phase information which encodes the relative positions of the atoms is lost. However, one technique, which can be viewed as being a variant of the diffraction method, and which *does* retain the phase information, is holography, first proposed by Gabor (1948).

Holography involves the interference between *two* waves, one, the reference beam (R), which travels unimpeded from source (S) to detector (D), and the other, the object beam (O), which results from scattering of the wave emitted by the source by an intervening object P: the hologram is the interference pattern observed at D caused by the interference between the R and O waves (see Fig. 2.65a). If the wave amplitude at D is written as

$$\psi = R + O, \tag{2.133}$$

then the intensity there is given by

$$I = \psi^* \psi = R^* R + O^* O + R^* O + O^* R. \tag{2.134}$$

The last two interference terms in eqn. (2.134) contain the holographic information. If, say, a photographic film recording the intensity variations at D were then placed at P, instead of the scattering object, and illuminated with a conjugate beam R^* (e.g. a converging spherical wave, if R were a diverging spherical wave emanating from the point source S), a 3D image of the object can be reconstructed. The transmitted amplitude would be proportional to R^*I, i.e.

$$A = |R|^2 R^* + |O|^2 R^* + |R|^2 O^* + (R^*)^2 O. \tag{2.135}$$

The second term in eqn. (2.135) is negligible if $|R| \gg |O|$, which is true if O is derived from R by scattering. The fourth term produces the holographic image of the object (the third term is associated with a ghost, or twin, image).

The amplitude of the spherical waves R and O can be written as

$$R = F_0(\boldsymbol{k}) \exp(\mathrm{i}\boldsymbol{k} \cdot \boldsymbol{r}) \tag{2.136}$$

and

$$O = \sum_{i \neq 0} F_i(\boldsymbol{k}) \exp[\mathrm{i}\boldsymbol{k} \cdot (\boldsymbol{r} - \boldsymbol{r}_i)], \tag{2.137}$$

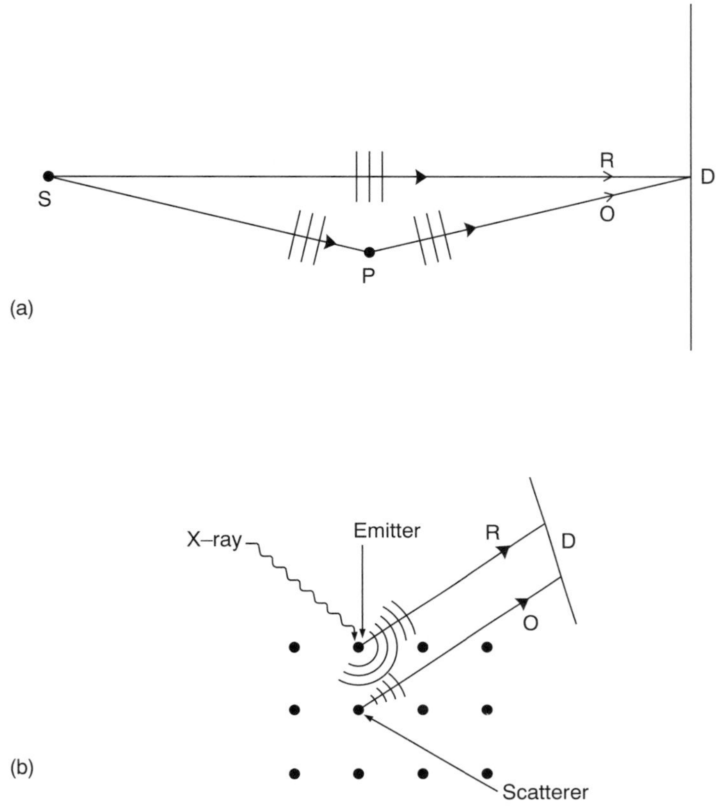

Fig.2.65 (a) Schematic illustration of the experimental arrangement used in Gabor holography with an external source (S) of coherent radiation, providing both a direct reference beam R and an object beam O scattered by an object (P). Interference between beams R and O in the detector plane (D) results in hologram formation. (b) Schematic illustration of the principle of an internal atomic X-ray source used in the holographic imaging of atoms. An incident X-ray creates a core-level hole in an atom; subsequent recombination of a higher-lying electron with the hole creates a fluorescent X-ray, which can travel to the detector directly (R beam), or be scattered by neighbouring atoms (O beam).

where $F_i(\boldsymbol{k})$ represents the angular variation of the spherical waves, $\boldsymbol{k}$ their wavevector and $\boldsymbol{r}_i$ is the position of the ith scatterer. The reconstructed wave amplitude of the holographic interference terms, in analogy to eqn. (2.135), can thus be written mathematically using the Helmholtz–Kirchhoff formula, akin to a Fourier integral, as

$$A(\boldsymbol{r}) = \int (R^* O + O^* R) \exp(-\mathrm{i}\boldsymbol{k} \cdot \boldsymbol{r}) \mathrm{d}\hat{\boldsymbol{k}} \tag{2.138}$$

where $\hat{\boldsymbol{k}} = \boldsymbol{k}/|\boldsymbol{k}|$. Peaks in $A(\boldsymbol{r})$ occur whenever $\boldsymbol{r} = \pm\boldsymbol{r}_i$ (the negative sign corresponding to twin images), i.e. when the phase factor in eqn. (2.137) equals unity.

In conventional optical holography, the wave source (a laser) provides a coherent beam that illuminates the object O, as well as providing the reference beam (Fig. 2.65a). For the purpose of forming holographic images of *atoms* in solids, such an arrangement is not feasible, since no source is yet available of coherent X-rays (required so that the wavelength of the radiation is comparable to the interatomic spacing for optimum diffraction conditions).

Instead, X-rays emitted from the atoms themselves in the solid may be used as *internal* point-like sources for holographic purposes. An X-ray emitted from an atom may proceed unimpeded direct to the detector D (*R* beam), or be scattered by a neighbouring atom, giving an *O* beam (see Fig. 2.65b). The atomic X-ray emission can be stimulated by illuminating the solid with high-energy X-rays, thereby creating holes in core-level states following the excitation of photoelectrons (see Fig. 2.62); the filling of such core holes by downward transitions of other higher-lying electrons results in fluorescent X-ray emission at well-defined energies, or wavelengths.

Such an experiment has been performed by Tegze and Faigel (1996) on a single crystal of $SrTiO_3$, which has a perovskite structure, with lattice constant $a = 3.9$ Å. The Sr atoms were the internal sources of X-rays, following K-shell excitation, and they also acted as scatterers (the X-ray atomic scattering factors of Ti and O being much smaller). The structural situation is thus rather simple, since the Sr atoms are arranged on a simple cubic lattice. The experimental configuration used to record the atomic hologram resembles a two-angle diffraction experiment (see fig 2.66), for which the components of $\hat{\boldsymbol{k}}$ are $\hat{k}_x = \sin\theta\sin\phi, \hat{k}_y = \sin\theta\cos\phi$, and the reconstructed image is then obtained using

$$A(\boldsymbol{r}) = \int\int I(\boldsymbol{k})\exp\left[-\mathrm{i}kz(1 - \hat{k}_x^2 - \hat{k}_y^2)^{1/2}\right]\exp[-\mathrm{i}k(x\hat{k}_x + y\hat{k}_y)]\mathrm{d}\hat{k}_x\mathrm{d}\hat{k}_y. \qquad (2.139)$$

The resulting hologram (projected onto the $k_x - k_y$ plane) is shown in Plate I, together with the reconstructed image showing only the Sr atoms. The cubic array of atoms is clearly evident in the image, although there is some variation in intensity from site to site (a result of the real and twin holographic images coinciding and causing a partial cancellation of intensity). However, this holographic technique is not yet sufficiently developed to be widely applicable.

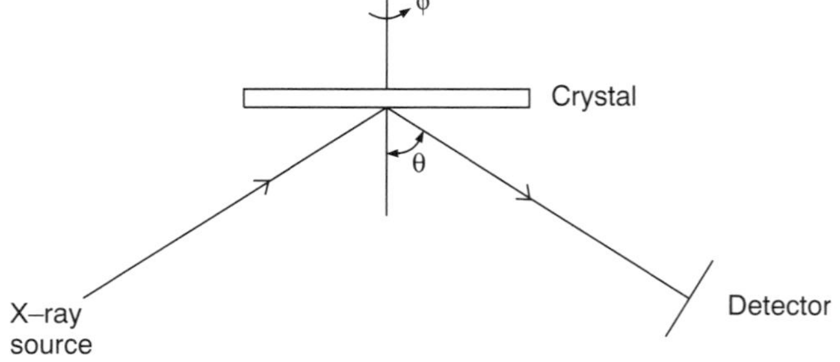

Fig.2.66 Schematic illustration of the two-angle diffraction experiment used to record an X-ray atomic hologram.

Problems

2.1 (a) Show that the angle ϕ between two planes in a cubic system with Miller indices $(h_1k_1l_1)$ and $(h_2k_2l_2)$ is given by

$$\cos\phi = \frac{h_1h_2 + k_1k_2 + l_1l_2}{(h_1^2 + k_1^2 + l_1^2)^{1/2}(h_2^2 + k_2^2 + l_2^2)^{1/2}}.$$

(b) Show that the interplanar spacing d_{hkl} between planes hkl in a cubic crystal with lattice parameter a is given by (eqn. (2.3)):

$$d_{hkl} = a/(h^2 + k^2 + l^2)^{1/2}.$$

2.2 The primitive translation vectors of a hexagonal lattice may be written as

$$\boldsymbol{a} = (\sqrt{3}a/2)\hat{\boldsymbol{x}} + (a/2)\hat{\boldsymbol{y}};\ \boldsymbol{b} = -(\sqrt{3}a/2)\hat{\boldsymbol{x}} + (a/2)\hat{\boldsymbol{y}};\ \boldsymbol{c} = c\hat{\boldsymbol{z}}.$$

Show that the volume of the primitive cell is given by $V_c = (\sqrt{3}/2)a^2c$.

2.3 Show that gravitational attraction makes a negligible contribution to the cohesive energy of solids.

2.4 Show that the residual entropy of ice associated with the disorder of proton positions in the structure is given by $R\ \ln(3/2)$. Use the Boltzmann equation for the entropy per atom, $S = k_B \ln\Omega$, where Ω is the number of possible available configurations. For 1 mole of ice with $2N_A$ hydrogen atoms, show that there are a total of 2^{2N_A} possible configurations but, of these, only a fraction of 3/8 satisfy the ice rule.

2.5 Derive an expression for the bulk modulus of an f.c.c. Lennard-Jones solid at the equilibrium interatomic distance.

2.6 Prove that the Wigner–Seitz cell for any 2D Bravais lattice (Fig. 2.10) is either a hexagon or a rectangle.

2.7 Show that the bulk modulus of a free-electron gas is given by $B = \frac{2n\mathscr{E}_F}{15}$. (Hint: take the potential energy to be the sum of eqns. (2.17) and (2.18).)

2.8 Prove that the ratio of lattice parameters for an ideal h.c.p. structure is given by $c/a = \sqrt{8/3}$.

2.9 Show that the packing fractions of the b.c.c., s.c. and diamond lattices are $\sqrt{3}\pi/8 = 0.68, \pi/6 = 0.52$ and $\sqrt{3}\pi/16 = 0.34$, respectively (cf. the value of $\sqrt{2}\pi/6 = 0.74$ for an f.c.c. / h.c.p. structure).

2.10 Show that the Madelung constant for a 1D ionic crystal of alternate positive and negative charges (of the same magnitude) is $A = 2\ln 2$.

2.11 Generate the lattice for the antifluorite (Na_2O) structure, and hence demonstrate that the cations are four-fold coordinated and the anions are eight-fold coordinated. How can the structure be described in terms of linked polyhedra?

2.12 Show that

(a) the h.c.p. structure can be regarded as two interpenetrating simple hexagonal (triangular) Bravais lattices (with primitive vectors $\boldsymbol{a}, \boldsymbol{b}, \boldsymbol{c}$), displaced by $(1/3)\boldsymbol{a} = (1/3)\boldsymbol{b} + (1/2)\boldsymbol{c}$

(b) the diamond structure can be regarded as two interpenetrating f.c.c. Bravais lattices displaced along the body diagonal of the cubic cell by $(1/4)(\boldsymbol{a} + \boldsymbol{b} + \boldsymbol{c})$.

2.13 The molecule H_2O has the four symmetry operations E, C_2 and $2\sigma_v$; the molecule NH_3 has the six symmetry operations E, C_3, C_3^2 and $3\sigma_v$. Derive in each case the corresponding stereogrammatic representations.

2.14 Show that the basis consisting of the four circles shown in the diagram has a point-group symmetry of $C_{2v}(mm2)$. Combination of this basis with the P-lattice of the orthorhombic crystal system generates the space group $C_{2v}^1(Pmm2)$. Generate the space group $C_{2v}^{11}(Cmm2)$ by using instead a C-centred lattice.

2.15 A plane with Miller indices (hkl) perpendicular to the reciprocal-lattice vector $\boldsymbol{G} = h\boldsymbol{a}* + k\boldsymbol{b}* + l\boldsymbol{c}*$ is contained within the plane $\boldsymbol{G}\cdot\boldsymbol{r} = A$. Show that the intercepts of this plane with the axes determined by the real-space lattice primitive vectors $\boldsymbol{a}$, $\boldsymbol{b}$, $\boldsymbol{c}$ are inversely proportional to the Miller indices.

2.16 Prove that the volume of the primitive cell in reciprocal space V_c^* is $(2\pi)^3$ times the reciprocal of the volume of the real-space primitive cell, V_c, i.e.

$$\boldsymbol{a}^* \cdot (\boldsymbol{b}^* \times \boldsymbol{c}^*) = \frac{(2\pi)^3}{(\boldsymbol{a} \cdot (\boldsymbol{b} \times \boldsymbol{c}))}.$$

(Hint: use the vector identity $(\boldsymbol{A} \times \boldsymbol{B}) \times (\boldsymbol{C} \times \boldsymbol{D}) = \boldsymbol{B}(\boldsymbol{A} \cdot (\boldsymbol{C} \times \boldsymbol{D})) - \boldsymbol{A}(\boldsymbol{B} \cdot (\boldsymbol{C} \times \boldsymbol{D}))$.)

2.17 Prove that the reciprocal lattice of a reciprocal lattice is the real-space lattice, i.e.

$$\frac{2\pi \boldsymbol{b}^* \times \boldsymbol{c}^*}{\boldsymbol{a}^* \cdot (\boldsymbol{b}^* \times \boldsymbol{c}^*)} = \boldsymbol{a}, \text{etc.}$$

2.18 (a) Show that the primitive translations of the reciprocal lattice of the hexagonal real-space lattice described in Problem 2.2 are given by

$$\boldsymbol{a}^* = (2\pi/\sqrt{3}a)\hat{\boldsymbol{x}} + (2\pi/a)\hat{\boldsymbol{y}};\ \boldsymbol{b}^* = -(2\pi/\sqrt{3}a)\hat{\boldsymbol{x}} + (2\pi/a)\hat{\boldsymbol{y}};\ \boldsymbol{c}^* = (2\pi/c)\hat{\boldsymbol{z}}.$$

(b) How are the two lattices related?
(c) Describe and sketch the first Brillouin zone of the hexagonal space lattice.

2.19 (a) Prove that the density of lattice points per unit area in a lattice plane is given by d/V_c, where d is the interplanar spacing and V_c is the primitive cell volume.
(b) Prove that the lattice planes having the greatest area densities of lattice points are the $\{111\}$ planes in an f.c.c. Bravais lattice and the $\{110\}$ planes in a b.c.c. lattice. (Hint: use the relation between families of lattice planes and reciprocal lattice vectors.)

2.20 Use the Ewald construction to find the Bragg reflections allowed in a Laue experiment, where a 'white' incident X-ray beam is used, having a continuous range of wavelengths corresponding to wavevectors between k_0 and k_1.

2.21 Obtain the allowed X-ray reflections for a crystal of KBr (isostructural with NaCl). How would this change in the case of isostructural crystalline KCl? (Hint: the numbers of electrons in the ions K^+ and Cl^- are the same.)

2.22 Show that the extinction rules for diffraction by a crystal of diamond are such that the allowed reflections satisfy the relation $h + k + l = 4n$ for hkl all even, or else hkl are all odd.

2.23 Prove that for the H atom in the ls state, for which the charge density is given by $n(r) = (\pi a_0^3)^{-1}\exp(-2r/a_0)$, a_0 being the Bohr radius, the X-ray form factor is given by $f^x = 16/(4 + K^2a_0^2)^2$.

2.24 Powder diffraction patterns of three different monatomic crystals, A, B, C, one being f.c.c., another b.c.c. and the other having the diamond structure, give the following values for the diffraction angle, $2\theta(\circ)$.

A	B	C
42.2	28.8	42.8
49.2	41.0	73.2
72.0	50.8	89.0
87.3	59.6	115.0

(a) Ascertain the crystal structures of A, B and C.
(b) If the wavelength of the X-rays used is 1.54 Å, what is the unit-cell parameter in each case?
(c) If the diamond structure were replaced by a zincblende structure having a cubic unit cell with the same length, at what angles would the first four diffraction rings now occur?

2.25 Deduce values for the average nearest-neighbour bond length and bond angle for amorphous Ge from the reduced RDF given in Fig. 2.61.

2.26 Essay: Discuss the problems involved in extracting structural information on non-crystalline materials from diffraction data.

References

Allen, M. P. and Tildesley, D. J. 1987, *Computer Simulation of Liquids* (Clarendon Press: Oxford).
Altmann, S. L., 1991, *Band Theory of Solids: An Introduction from the Point of View of Symmetry* (Clarendon Press: Oxford).

Bak, P. and Goldman, A. I., 1988, in *Introduction to Quasicrystals*, ed. M. V. Jaric (Academic Press: London), p. 143.
Batterman, W., Chipman D. R. and de Marco J. J., 1961, *Phys. Rev.* **122**, 68.
Borg, R. J. and Dienes, G. J., 1992, *The Physical Chemistry of Solids* (Academic Press: London).
Burns, G., 1985, *Solid State Physics* (Academic Press: London).
Burns, G. and Glazer, A. M., 1990, *Space Groups for Solid State Scientists*, 2nd edn. (Academic Press: New York).
Car, R. and Parrinello, M., 1985, *Phys. Rev. Lett.* **55**, 2471.
Cheetham, A. K., 1987, in *Solid-State Chemistry: Techniques*, eds A. K. Cheetham and P. Day (Clarendon Press: Oxford), p. 39.
Elliott, S. R., 1990, *Physics of Amorphous Materials*, 2nd edn (Longman: Harlow).
Gabor, D., 1948, *Nature* **161**, 777.
Gaskell, P. H., 1981, *Nature* **289**, 474.
Henry, N. F. M. and Lonsdale, K., eds, 1952, *International Tables for X-ray Crystallography*, vol. 1 (Kynoch Press: Birmingham).
Hohenberg, P. and Kohn, W., 1964, *Phys. Rev.* **136**, 864.
Janot, C., 1995, *Quasicrystals: A Primer* (Oxford University Press: Oxford).
Jaric, M. V., ed., 1988, *Introduction to Quasicrystals* (Academic Press: London).
JCPDS International Centre for Diffraction Data, 1984, *Powder Diffraction File — Inorganic Components* (Philadelphia).
Keating, P. N., 1966, *Phys. Rev.* **145**, 637.
Kittel, C., 1996, *Introduction to Solid State Physics*, 7th edn (Wiley: New York).
Kohn, W. and Sham, L. J., 1965, *Phys. Rev.* **140**, 1133.
Liu, A. Y. and Cohen, M. L., 1989, *Science* **245**, 841.
Mackay, A. L., 1982, *Physica* **A114**, 609.
Mandelbrot, B., 1982, *The Fractal Geometry of Nature* (W. H. Freeman: New York).
Mermin, N. D., 1992, *Rev. Mod. Phys.* **64**, 3.
Nichols, C. S., 1995, *Structure and Bonding in Condensed Matter* (Cambridge University Press: Cambridge).
Peitgen, H-O., Jürgens, H. and Saupe, D., 1992, *Chaos and Fractals: New Frontiers of Science* (Springer-Verlag: New York).
Penrose, R., 1974, *Bull. Inst. Math. Appl.* **10**, 266.
Phillips, J. C., 1973, *Bonds and Bands in Semiconductors* (Academic Press: New York).
Remler, D. K. and Madden, P. A., 1990, *Mol. Phys.* **70**, 921.
Shectman, D., Blech, I., Gratias, D. and Cahn, J. W., 1984, *Phys. Rev. Lett.* **53**, 1951.
Srivastava, G. P. and Weaire, D., 1987, *Adv. Phys.* **36**, 463.
Sutton, A. P., 1993, *Electronic Structure of Materials* (Clarendon Press: Oxford).
Tegze, M. and Faigel, G., 1996, *Nature* **380**, 49.
Temkin, R. J., Paul, W. and Connell, G. A. N., 1973, *Adv. Phys.* **22**, 581.
Teter, D. M. and Hemley, R. J., 1996, *Science* **271**, 53.
West, A. R., 1987, *Solid State Chemistry and its Applications* (Wiley: New York).
Whittaker, E. J. W. and Whittaker, R. M., 1990, in *Quasicrystals, Networks and Molecules of Fivefold Symmetry*, ed. I. Hargittai (VCH: New York), p. 107.
Yin, M. T. and Cohen, M. L., 1982, *Phys. Rev.* **B26**, 5668.

CHAPTER

3 Defects

Introduction

In the previous chapter, it was assumed implicitly that the structures of the solids considered have been *ideal*, i.e. devoid of defects. Real materials, however, invariably contain structural defects which can dominate their physical and chemical behaviour. The concept of a structural defect can only be understood by reference to a standard, structurally perfect state of the material. Thus, for crystals, the reference state is the ideal single crystal, for which translational symmetry (eqn. (2.1)) holds everywhere throughout the structure. (Note that this statement implies that ideal crystals are of *infinite* extent; even the presence of surface boundaries in the case of finite-sized crystals is sufficient to destroy perfect translational order.) For the intermediate case of quasicrystals (§2.1.1), the reference state would be the ideal hypercubic crystal in 6D space from which the 3D quasicrystal structure is obtained by projection.

This concept can even be applied to the case of amorphous solids, even though they might naïvely be thought to be the most defective of all materials. For the case of covalently bonded amorphous materials, the reference state can be the continuous random network (§2.1.1), in which the nearest-neighbour coordination of atoms is in accord with the chemical valence at every site, even though structural disorder, in the form of fluctuations in dihedral (torsion) and bond angles and, to a lesser extent, bond lengths, ensures that translational periodicity is destroyed. For the case of amorphous metals, the reference state can be a dense random packing of spheres (§2.2.3.3); in this case, however, even the nearest-neighbour coordination is not completely defined, but is instead a statistical quantity determined by local constraints on atomic packing. On the other hand, the stereochemical (trigonal prismatic) model for the structure of transition metal–metalloid glasses (§2.2.3.3), like the corresponding continuous random network structure for covalent amorphous solids, makes it a particularly suitable reference structure.

A structural defect is thus a configuration in which an atom, or group of atoms, does not satisfy the structural rules pertaining to the ideal reference state of the material. There are very many different types of structural defects, as will be discussed in the next section (§3.1), but all share the same characteristic, namely that the structural disturbance associated with the defects is spatially localized in some way, i.e. defects are structurally *inhomogeneous*.

One form of structural disturbance that is structurally *homogeneous*, i.e. shared to a greater or lesser extent by *all* the atoms in the solid, but which is not normally regarded as a structural defect, is associated with vibrational excitations. Such small, time-varying displacements, present at all temperatures (even at absolute zero as zero-point motions), can be regarded as averaging essentially to zero, i.e. the reference structure can be recovered, in a time-averaged picture (particularly if the interatomic potential is harmonic) even though in an *instantaneous* picture atoms are displaced from their reference sites. Vibrational excitations in solids will be discussed in the next chapter. Useful surveys of defects and their related properties are given in Catlow (1994), Hayes and Stoneham (1985) and Henderson (1972).

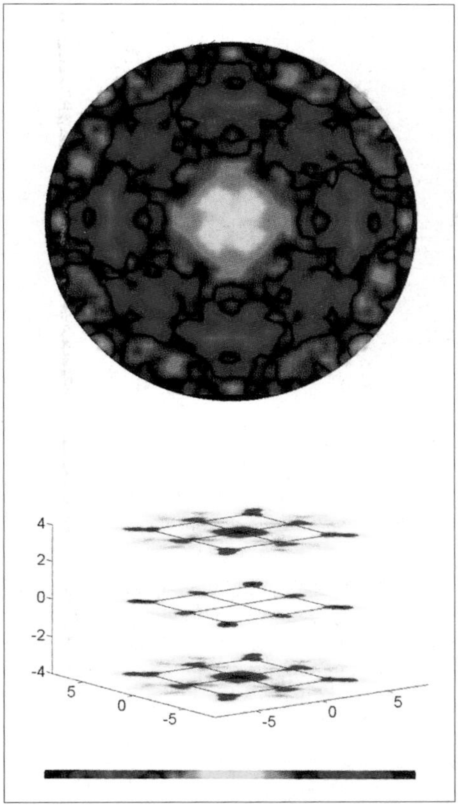

Plate I X-ray hologram of a single crystal of $SrTiO_3$ (top) with, below, the reconstructed 3D image of the crystal structure showing only the Sr atoms. (Figure courtesy of Dr. M. Tegze.)

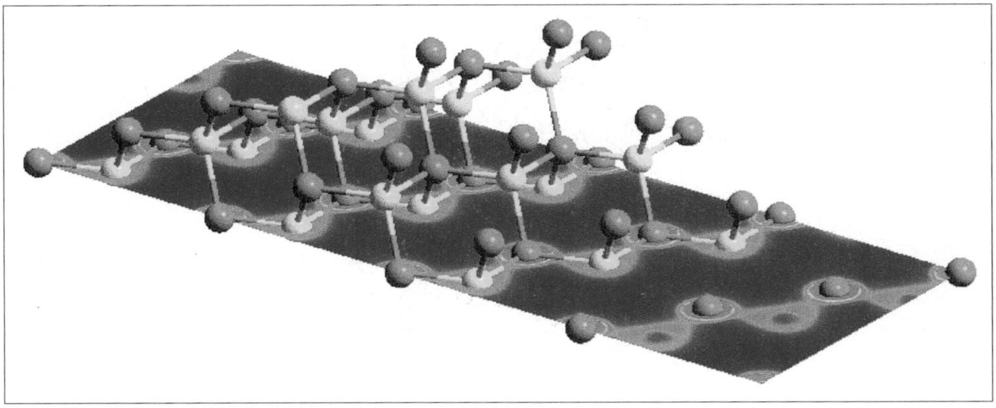

Plate II Electronic charge density in GaAs for a slice through the crystal structure, calculated using the CASTEP code in the Cerius suite of programs from Molecular Simulations Inc. (This is a plane-wave pseudopotential code based on density functional theory and within the local density approximation.) Red indicates a high charge density (on As atoms) and blue indicates a low charge density (on Ga atoms). (Figure courtesy of C. J. Pickard.)

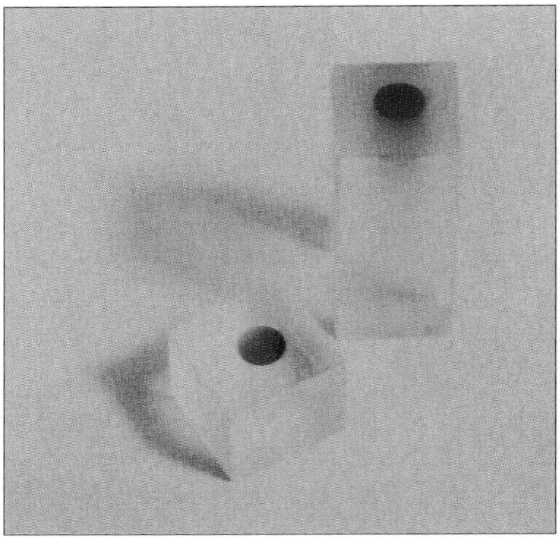

Plate III Colour centres in alkali halide crystals induced by X-irradiation: top, NaCl, below KCl. (Figure courtesy of Dr. A. D. Yoffe.)

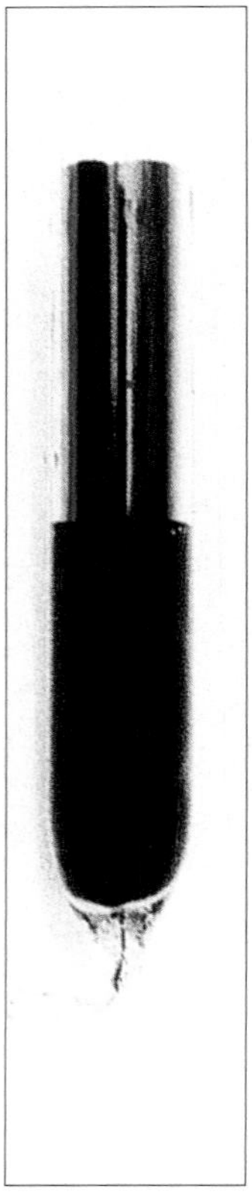

Plate IV Metal–insulator phase transition in an Na–NH_3 solution (2 mol. % Na) at approximately –60°C. The upper (bronze) phase is metallic and the lower (dark-blue) phase is non-metallic. (Figure courtesy of Prof. P. P. Edwards.)

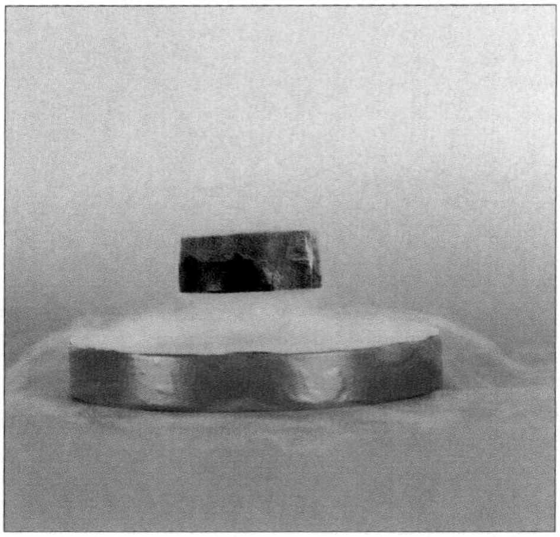

Plate V Magnetic levitation above a permanent magnet of a single-crystal grain of superconducting YBCO ($YBa_2Cu_3O_{7-\delta}$) held at a temperature below its superconducting transition temperature. (Figure courtesy of Dr. D. A. Cardwell.)

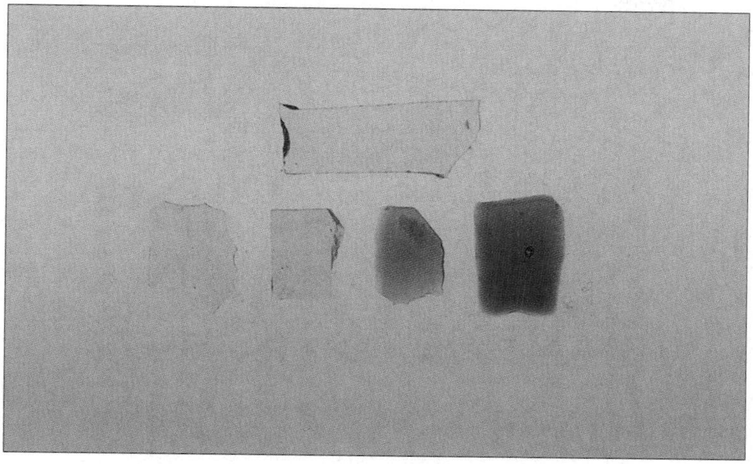

Plate VI Size-quantization effects on the optical properties of a borosilicate glass matrix, containing approx. 2 x 10^{25} CdSe molecules per m^3, with differently sized nanocrystals of CdSe depending on heat treatment. The clear sample at the top is the as-quenched glass, and is optically transparent up to approx. 4 eV. The four samples below are from the same batch and have had different heat treatments. Left: annealed at 625°C for 1 h, producing nanocrystals of approx. 1.5 nm diamenter and a band edge at approx. 2.9 eV; left middle: annealed at 650°C for 1 h, producing nanocrystals of approx. 2.5 nm diameter and a band edge at approx. 2.5 eV; right middle: annealed at 700°C for 1 h, producing nanocrystals of approx. 5 nm diameter and a band edge at 1.9 eV; right: annealed at 800°C for 1 h, producing nanocrystals of approx. 10 nm diameter and a band edge at 1.75 eV. (Figure courtesy of Prof. P. D. Persans.)

(a)

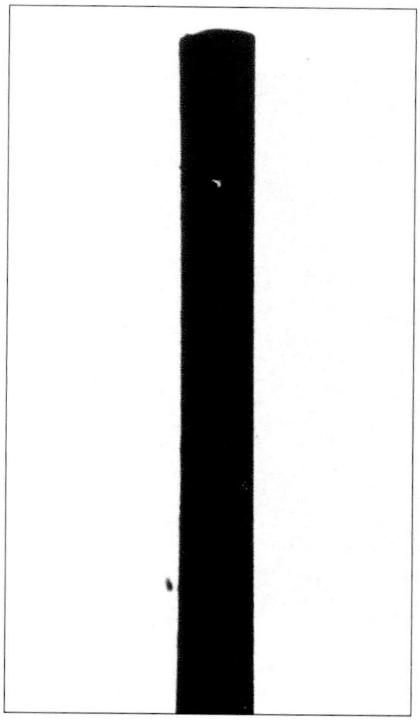

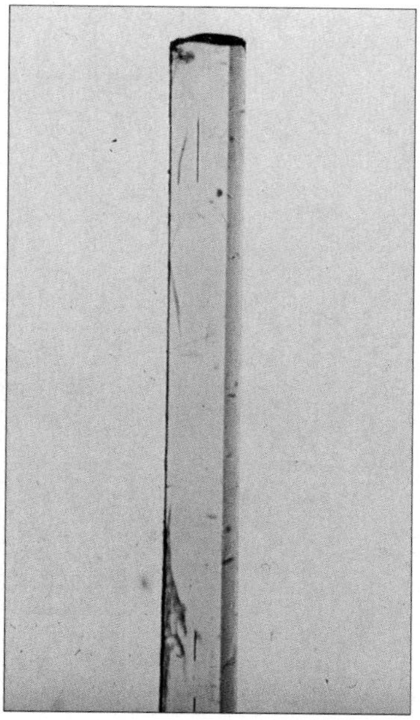

(b)

Plate VII The one-dimensional metal, KCP ($K_2Pt(CN)_4Br_{0.3} \cdot 3H_2O$), photographed in linearly polarized light showing: (a) metallic-like reflectivity for the ***E***-vector of light parallel to the crystal axis: (b) optical transparency, characteristic of an insulator, for ***E*** perpendicular to the crystal axis. (Figure courtesy of Prof. A. E. Underhill.)

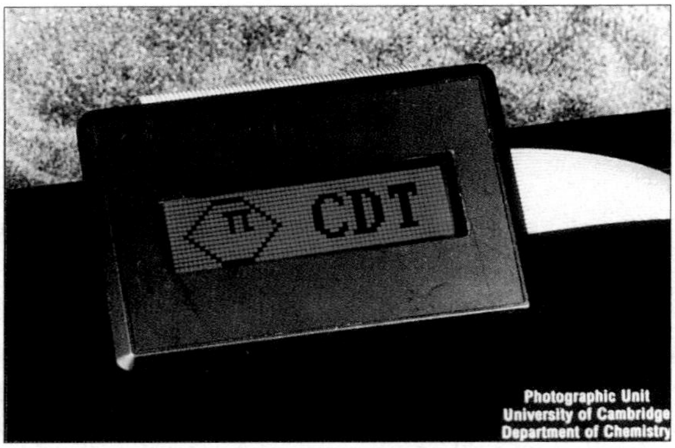

(a)

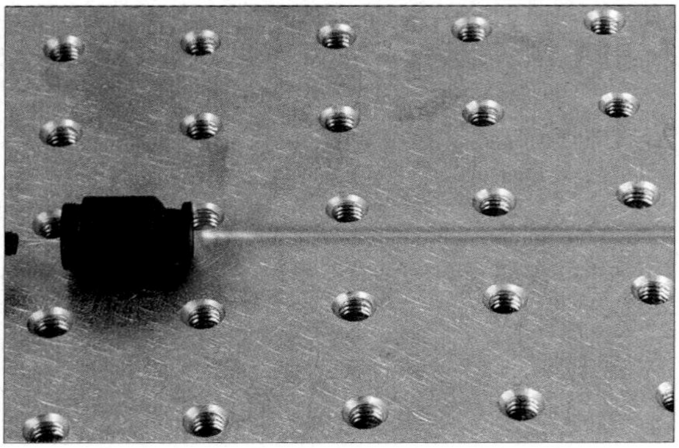

(b)

(c)

Plate VIII (a) Passively addressed dot-matrix display fabricated by Cambridge Display Technology using the light-emitting organic polymer, poly(p-phenylene vinylene). (Figure courtesy of Dr. A. B. Holmes.) (b) Blue light emission from an InGaN multiple-quantum-well laser. (Figure courtesy of Dr. S Nakamura.) (c) Traffic light utilizing inorganic-semiconductor light-emitting diodes (LEDs): red, AlInGaP LED; amber, AlInGaP single-quantum-well LED; green, InGaN single-quantum-well LED. (S. Nakamura, *Solid State Communications*, **102**, 237 (1997).)

Types of defects 3.1

Structural defects, as defined above, can arise from a variety of causes: they may be (unavoidably) thermally generated, or may arise in the course of fabrication of the solid, incorporated either unintentionally or deliberately.

As mentioned above, defects can be defined as being inhomogeneous structural disturbances of a reference structure. For ease of discussion, structural defects may be divided into three categories, depending on the spatial dimensionality associated with the defect. Thus, one can envisage defects of zero dimensionality, often called point defects, that are associated with a single atomic site or a (globular) cluster of sites. In addition, extended defects can occur, being aggregates of point defects, that are either one-dimensional in character (i.e. linear in extent) or else two-dimensional (sometimes called planar defects). In this classification scheme, a dimensionality of three corresponds to the situation in which most, if not all, atoms in the solid do not correspond to the ideal sites of a reference structure. In the case of dynamic disorder, associated say with cooperative vibrational excitations (see Chapter 4), such time-varying deviations from the reference structure are not normally thought of as being defects in the normal sense. Static 3D variations in structure from the reference state are usually discussed in terms of phase transitions, e.g. the inclusion of a second phase (heterogeneous phase separation), or disordering transitions in which atomic sites of a reference structure are occupied statistically by, say, two types of atoms.

Each of the above three categories of structural defect will now be discussed in turn, and examples given of each.

3.1.1 Zero-dimensional (point) defects

Unassociated point defects at the atomic scale in crystals are of three types. An atom may be missing from a lattice site, thereby forming an atomic vacancy. Alternatively, another type of atom may be present at a lattice site, forming a substitutional impurity defect. Finally, an atom may be situated, not at a normal lattice site, but at an interstitial site between lattice sites (see §3.4.2): the extra atom may be of the same type as in the rest of the crystal, in which case the defect is termed a self-interstitial; alternatively, the interstitial site may be occupied by a (small) impurity atom. These three basic types of point defect found in crystals are illustrated schematically in Fig. 3.1. An atom that leaves a lattice site, thereby forming an atomic vacancy, can end up in one of two positions: either on the surface, in which case the isolated vacancy is termed a Schottky defect, or as a self-interstitial. When an interstitial and a vacancy are sufficiently well separated that their opposing lattice strain fields do not overlap, hence stabilizing the defects against reconstructive recombination, the defect pair is known as a Frenkel defect.

Defects occurring in *ionic* crystals are subject to the additional constraint that electroneutrality must be maintained overall. Thus, for example, intrinsic Schottky defects in stoichiometric ionic crystals occur in *pairs*, one vacancy on the cation sub-lattice and the other on the anion sub-lattice (see Fig. 3.2). An anion vacancy in e.g. NaCl has a charge associated with it of + e: since the lattice with a Cl^- anion surrounded by $6Na^+$ cations is originally locally electroneutral, removal of the

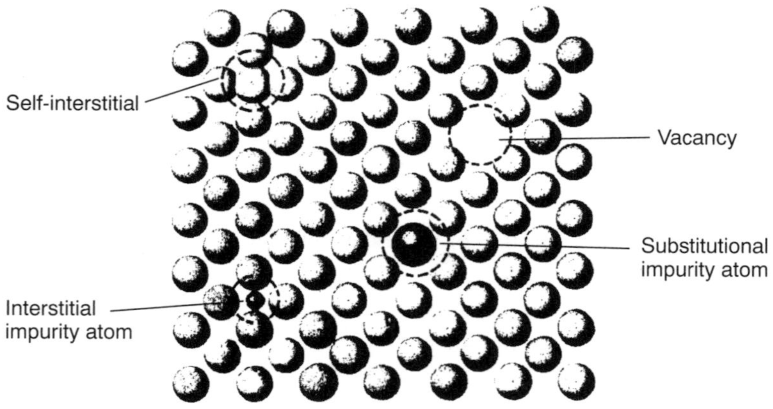

Fig. 3.1 Schematic illustration of point defects occurring in crystalline monatomic solids.

Cl	Na	Cl	Na	Cl	Na	Cl	Na	Cl
Na	Cl	Na	Cl	Na	Cl	Na	Cl	Na
Cl	Na	Cl	Na	Cl	Na	Cl	Na	Cl
Na	Cl	□	Cl	Na	Cl	Na	Cl	Na
Cl	Na	Cl	Na	Cl	Na	□	Na	Cl
Na	Cl	Na	Cl	Na	Cl	Na	Cl	Na
Cl	Na	Cl	Na	Cl	Na	Cl	Na	Cl
Na	Cl	Na	Cl	Na	Cl	Na	Cl	Na

Fig. 3.2 2D schematic illustration of Schottky defects in an ionic crystal, e.g. NaCl, showing one vacancy on the cation sub-lattice and another on the anion sub-lattice.

negatively charged anion to create the vacancy must result in a remanent charge of $+e$; likewise, a cation vacancy has a charge of $-e$ associated with it. Schottky defects are the dominant intrinsic defects in crystalline alkali halides and alkaline-earth oxides.

Frenkel defect pairs can also occur in ionic crystals, and *cation* Frenkel defects are the predominant defects in silver halides, e.g. AgCl or AgBr, which have the rocksalt (NaCl) structure (Fig. 3.3a). The interstitial site in such a lattice has eight-fold coordination, i.e. tetrahedral coordination by both Ag^+ and halide ions (see Fig. 3.3b). Since the Ag^+ ion is so polarizable, it is probable that there is also some covalent interaction between interstitial Ag^+ ions and neighbouring halide ions which acts to stabilize the defect. For the case of ionic crystals having small anions and large cations and having the fluorite structure (e.g. alkaline-earth fluorides, ZrO_2 and ThO_2), *anion* Frenkel defects are favoured instead, with the anions occupying octahedral interstices in the structure.

Thus far, the discussion has concentrated on intrinsic point defects that can occur in thermal equilibrium (see §3.2) in stoichiometric crystalline ionic salts. However,

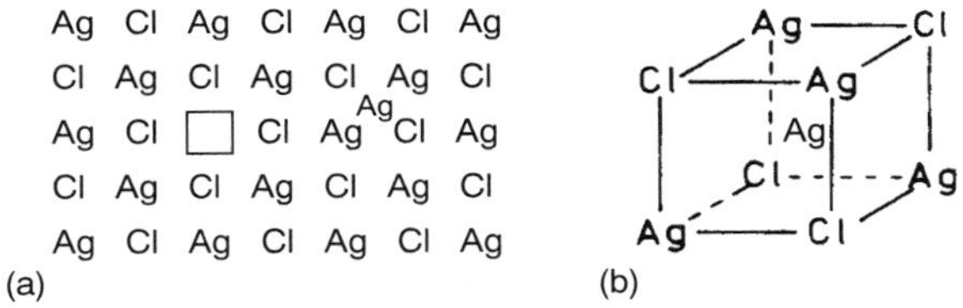

Fig. 3.3 (a) 2D schematic illustration of a Frenkel defect in AgCl. (b) The interstitial site has tetrahedral coordination by both Ag^+ and Cl^- ions.

extrinsic defects can also be created in such materials by external agents. Thus, ionizing radiation (e.g. X-rays, γ-rays, neutrons and high-energy electrons) can create atomic vacancies in solids by knock-on collisions between atoms and the bombarding particle.

Alternatively, extrinsic defects can be created chemically, e.g. by making the material non-stoichiometric. Thus, exposure of, for example, an NaCl crystal to Na vapour leads to in-diffusion and incorporation of Na atoms, resulting in an alkali concentration in excess of that set by stoichiometry. In this case, the extra Na atoms are *not* incorporated interstitially; the mass density of the non-stoichiometric crystal does not increase with increasing Na content as would be expected from interstitial involvement, but instead *decreases*, indicating the creation of an equal number of anion vacancies, with the excess Na atoms occupying normal cation sites. The excess Na atoms are ionized, and the freed electrons are bound to the positively charged anionic vacancy sites (see Fig. 3.4). Such localized electrons can undergo photo-induced transitions between discrete quantized energy levels, leading to optical absorption in the visible part of the spectrum; the otherwise colourless alkali halide crystals become *coloured* on becoming non-stoichiometric (see §3.3.1). The fact that the colour, associated with the position of the absorption band, is not materially affected if *another* type of alkali is incorporated (e.g. K in NaCl) indicates that it is the anion vacancy–electron complex (the so-called F-centre), and not the cation, that is the colour centre. The striking blue–purple colour of the rare 'Blue John' mineral form of fluorite (CaF_2) found only in Derbyshire, England, is due to naturally occurring F-centres.

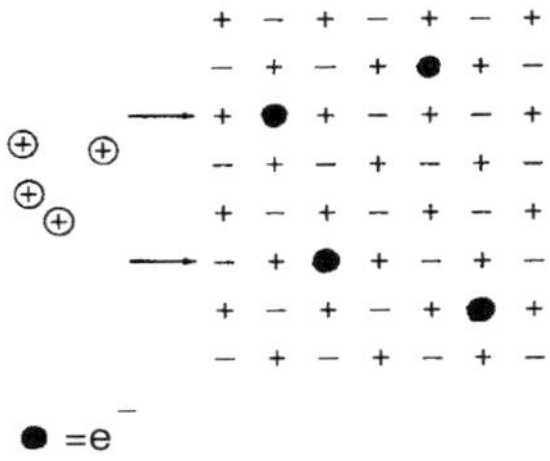

Fig. 3.4 Schematic illustration of the formation of colour centres in alkali-halide crystals by the incorporation of excess cations on exposure of the crystal to an alkali vapour. The ionized excess alkali atoms occupy normal cation positions, thereby creating anion vacancies; the ionized electrons bound to the positively charged anionic vacancy sites are the colour centres (known as F-centres).

Interesting variations arise when a particular ionic crystal is doped (e.g. during crystal growth) with aliovalent impurities (usually cations) having a *different* valency to that of the host ion for which it substitutes. Substitution by a *higher-valence* cation can create either *cation vacancies* (e.g. $CaCl_2$ in NaCl) or *interstitial anions* (e.g. YF_3 in CaF_2) in order to compensate for the increased positive charge at the substituted cation site. In a similar manner, substitution by a *lower-valence* cation can create *anion vacancies* (e.g. CaO in ZrO_2, stabilizing the cubic form of zirconia) or *interstitial cations* (e.g. 'stuffed silicas', i.e. aluminosilicates, in which the substitution of an Si^{4+} ion by an Al^{3+} at a tetrahedral site is electrically compensated by the occupation of an empty interstice in the open silica framework by an alkali-metal cation, e.g. Li^+).

Point defects in crystals are prone to form defect complexes through their mutual interaction via elastic-strain, electrostatic or electronic forces. A simple example is the divacancy, i.e. a pair of atomic vacancies occupying nearest-neighbour sites in the lattice. In *metals*, divacancy formation is energetically favourable essentially because of the associated decrease in elastic strain energy. In *ionic crystals*, pairing of cation and anion vacancies occurs because of the electrostatic Coulomb attraction between the oppositely charged defects (Fig. 3.5a); similarly, pairing occurs between aliovalent impurities and cation vacancies (Fig. 3.5b). In both the latter cases, the overall charge of the defect pair is zero, but nonetheless they form electrical dipoles, and hence such defect pairs can attract other defect pairs via dipolar interactions, thereby forming yet larger defect clusters. This is well known to occur in the anion-deficient material, CeO_{2-x}.

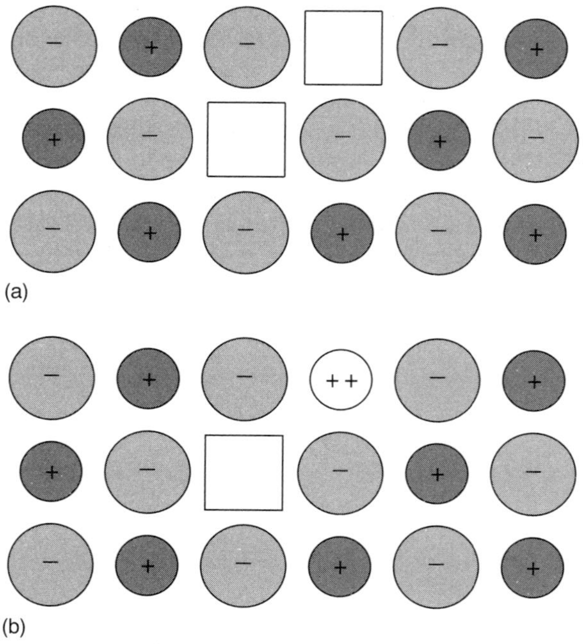

Fig. 3.5 Illustration of defect pairing in ionic crystals: (a) a pair of Schottky defects (cation and anion vacancies); (b) an aliovalent impurity–cation vacancy pair (e.g. $CaCl_2$ in KCl).

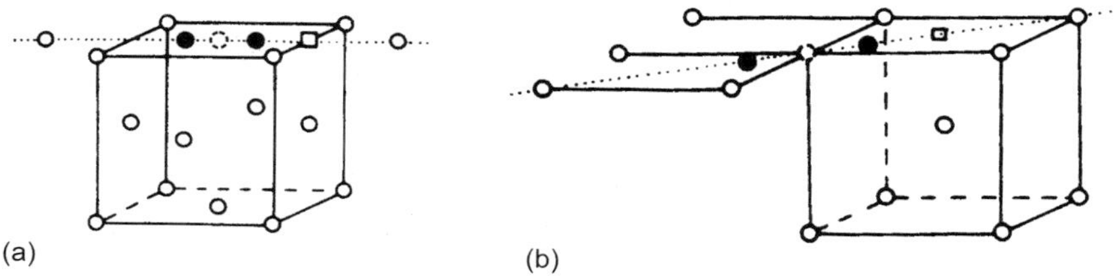

Fig. 3.6 Examples of a split (or dumb-bell-shaped) interstitial in metallic crystals having: (a) an f.c.c. structure; (b) a b.c.c. structure. The normal interstitial sites are denoted as □, a normally occupied lattice site as ◌, and the atoms forming the split interstitial as •.

Defect pairing of a different kind is found for interstitial atoms in metals. An interstitial atom placed at an ideal interstitial site in the structure interacts with a neighbouring atom on a normal lattice site (via elastic strain), displacing it from its site and producing a displacement of the interstitial atom as well. The result is a neighbouring pair of atoms displaced from their theoretical positions, called a split interstitial or a dumb-bell-shaped interstitial. An example is shown in Fig. 3.6a for the case of an f.c.c. crystal (e.g. a Pt interstitial in Pt). Instead of the interstitial atom occupying an octahedral interstice (on the cubic cell edge), both it and a neighbouring face-centred atom are displaced in the $\langle 100 \rangle$ direction to form a split interstitial. Another instance is given in Fig. 3.6b for the case of a b.c.c. metal (interstitial carbon in b.c.c. α-Fe, forming steel, is the most important example). Again, the interstitial atom is displaced from its ideal interstitial position (centre of a cube face) together with a neighbouring atom; split interstitials in this case can form in the $\langle 110 \rangle$ direction, as in Fig. 3.6b, as well as the $\langle 100 \rangle$ and $\langle 111 \rangle$ directions.

In the case of amorphous materials, where translational periodicity does not exist, it is sometimes no longer possible to define some of the point defects illustrated in Fig. 3.1; for example, the dense random-packing amorphous reference structure is sufficiently structurally flexible with regard to coordination numbers that interstitial sites and vacancies in amorphous metals essentially lose their identities. For the case of covalently bonded amorphous solids, where the local coordination is well defined and set by valency requirements, such distinctions are easier to make. However, the presence of structural disorder means that, for covalent amorphous materials, there can exist a new type of defect, the isolated dangling bond, which has no counterpart in the corresponding crystalline solid.

As an example, consider the tetrahedrally bonded material, Si. For the *crystalline* form, removal of an atom to create a single vacancy defect V necessarily involves the simultaneous breaking of the four covalent bonds to the atom; as a result, four dangling-bond orbitals point towards the centre of the void from the atoms bordering it. The electrical charge state of the vacancy so created must be neutral, i.e. V^0 (an uncharged atom is removed from an originally charge-neutral crystalline configuration), and each orbital contains one electron. However, because of an electronic-degeneracy-lowering (Jahn–Teller) instability, a vacancy having this charge state is unstable with respect to a structural distortion. The ideal tetrahedral arrangement of the monovacancy, having T_d point-group symmetry, is adopted, in fact, by the charge state V^{2+},

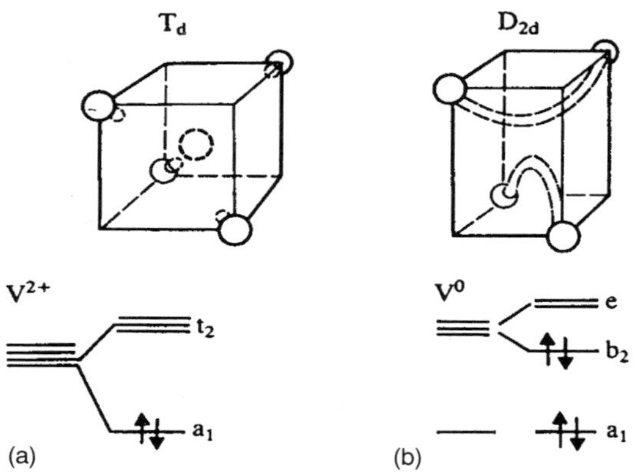

Fig. 3.7 Atomic configurations and electronic level occupancies of the monovacancy in crystalline Si in two charge states: (a) undistorted V^{2+} centre (T_d symmetry); (b) tetragonally distorted V^0 centre (D_{2d} symmetry).

in which two electrons have been withdrawn from the orbitals bordering the vacancy by electrical doping of the crystal (see §6.5.2); the remaining two electrons occupy in a spin-paired fashion the lowest, singly degenerate symmetric a_1 electronic state (see Fig. 3.7a). The two extra electrons in the neutral vacancy, V^0, occupy the higher-lying, triply-degenerate t_2 state, which is susceptible to a Jahn–Teller splitting upon occupation by electrons, associated with a structural distortion, resulting in a distorted vacancy with tetragonal D_{2d} symmetry; the four electrons associated with the original dangling-bond orbitals have reconstructed to form two (bent) bonds (see Fig. 3.7b).

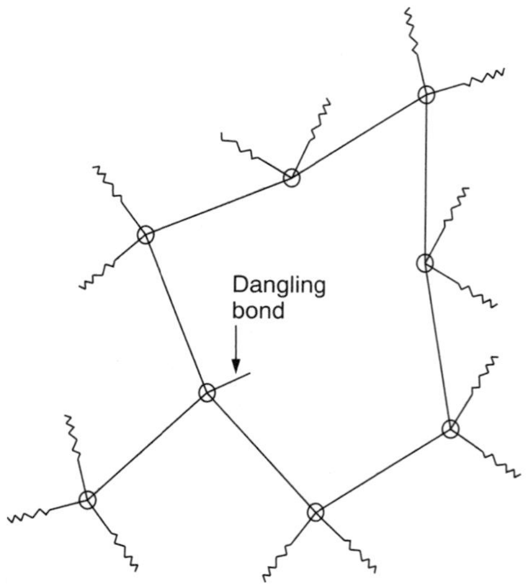

Fig. 3.8 Schematic 2D illustration of an isolated dangling bond in amorphous Si.

For the case of non-crystalline covalent materials, such as amorphous silicon, however, *isolated* dangling bonds can be created during preparation and can survive because of the structural flexibility inherent in the amorphous structure (Fig. 3.8). Thus, electronic rebonding such as occurs in the monovacancy in crystalline Si can no longer easily take place, and the electrons associated with the isolated dangling-bond orbitals can remain unpaired, i.e. in the D^0 configuration (where the superscript denotes the charge state of the defect).

3.1.2 One-dimensional (line) defects

Several extended defects can be distinguished in which the structural disturbance with respect to the reference structure is confined to one dimension. Two types can be distinguished: line defects associated with translational displacements of atoms (dislocations), and those associated with rotational displacements (disclinations). The former category of line defects is the more prevalent, and controls properties as varied as mechanical properties and crystal growth.

Two extreme types of dislocations can be distinguished in turn, depending on the direction of the structural disturbance associated with the defect relative to its length. The easiest to visualize is the so-called edge dislocation (Fig. 3.9), which can be thought of as resulting from the insertion of an extra half-plane of atoms (in the upper half-plane in Fig. 3.9) or, equivalently, the removal of a half-plane, followed by rebonding of the atoms in the vicinity of the defect. The line of the defect is obviously determined by the *edge* of the inserted (removed) half-plane of atoms in the interior of the material; this comprises the core of the dislocation.

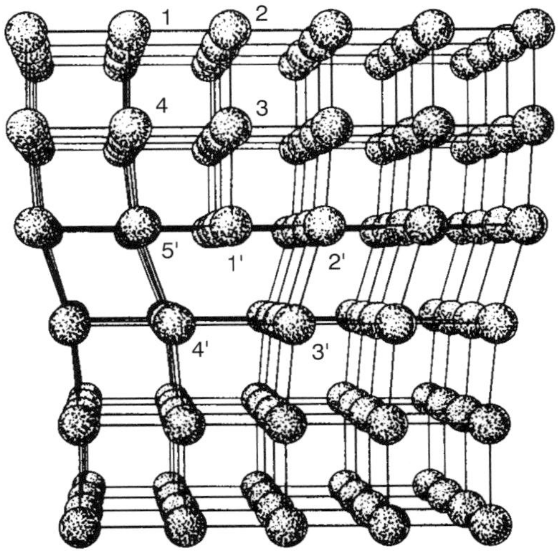

Fig. 3.9 Representation of the structure near the core of an edge dislocation. The circuit of atoms (1′–5′) used to determine the Burgers vector is indicated, as is a circuit (1–4) in the bulk of the material well away from the defect.

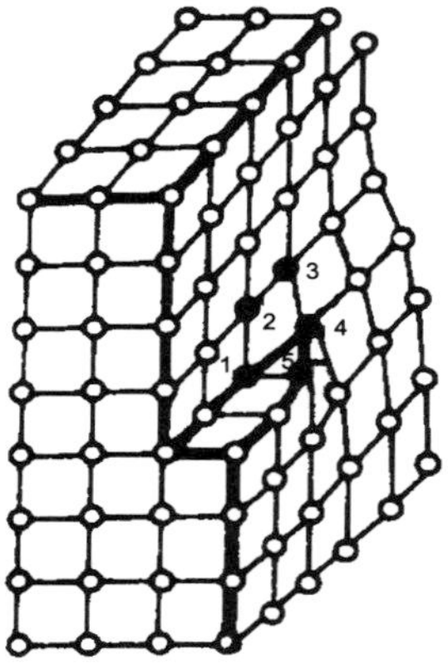

Fig. 3.10 Representation of the structure near the core of a screw dislocation. The circuit of atoms used to determine the Burgers vector is indicated.

The character of the atomic disturbance associated with a dislocation can be quantified, in terms of both magnitude and direction, by means of the so-called Burgers vector, $\boldsymbol{b}$. The Burgers vector is found by performing a circuit from atom to atom, where each step corresponds to an interatomic spacing. If such a circuit does *not* enclose a dislocation, a closed path is achieved by traversing the same number of steps in each direction (e.g. the path 1–2–3–4–1 in Fig. 3.9). On the other hand, if the circuit *does* enclose the core of a dislocation, a closed path *cannot* be achieved (e.g. the path 1′–2′–3′–4′–5′–1′ in Fig. 3.9). The displacement vector (as a multiple of the interatomic spacing) needed to close the path is the Burgers vector, $\boldsymbol{b}$ (i.e. 1′–5′ in Fig. 3.9). Thus, for an *edge* dislocation, $\boldsymbol{b}$ is *perpendicular* to the line of the defect. (Note, however, that a circuit enclosing a line of, say, vacancies will produce a *zero* Burgers vector—a dislocation is unique in being characterized by a finite value of $\boldsymbol{b}$.)

The other form of dislocation is termed a screw dislocation (see Fig. 3.10). The atomic disturbance in this case can be thought of as arising from making a (half-)cut into the solid and displacing the material on either side of the faces of the cut in a direction *parallel* to the line of the cut. Performing an atomic circuit as before to determine the Burgers vector (1–2–3–4–5–1 in Fig. 3.10) shows that $\boldsymbol{b}$ is *parallel* to the core for a *screw dislocation*.

The Burgers vector $\boldsymbol{b}$ is constant throughout the length of a given dislocation, even though the direction of the dislocation may vary along its length. Since a dislocation cannot terminate in the middle of a crystal, if it does not end at a surface it must either form a dislocation loop or be part of a branched network of dislocations. Different parts of a dislocation loop must necessarily have different characters: screw-like when $\boldsymbol{b}$ is parallel to the core, edge-like when $\boldsymbol{b}$ is perpendicular to the core, and a complex mixed character otherwise. For a node of a dislocation network, the following conservation law for the Burgers vectors must be obeyed:

$$\sum \boldsymbol{b} = 0. \tag{3.1}$$

For further information on dislocations, the reader is referred to Hull and Bacon (1984).

One further distinct type of line defect in solids is the disclination that is associated with *rotational* displacements, rather than the translational displacements characteristic of dislocations. A disclination can be viewed as a defect that changes the local *curvature* of a structure. The principle behind its operation may perhaps be seen most readily using a two-dimensional example. A disclination *point* in 2D is generated by making a half-cut in the plane (the limit of the cut marking the core of the disclination), and then either adding or removing material in the form of a wedge, thereby creating negative or positive local curvature, respectively. In a 2D hexagonal lattice, for example, a five-fold ring is created at the site of a disclination causing positive local curvature, and a seven-fold ring is the site of one producing negative local curvature (see Fig. 3.11a,b). In 3D, the disclination is a *line* defect and, as with dislocations, there are two types: wedge and screw disclinations (see Fig. 3.12a,b).

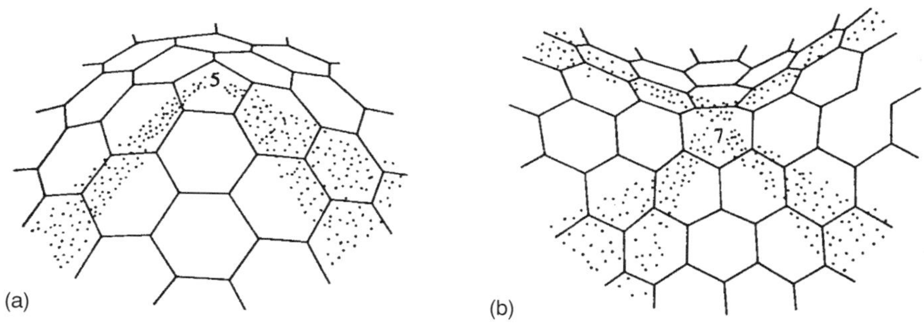

Fig. 3.11 Types of disclination points for a 2D hexagonal lattice creating: (a) a five-fold ring, associated with positive local curvature; (b) a seven-fold ring, associated with negative local curvature.

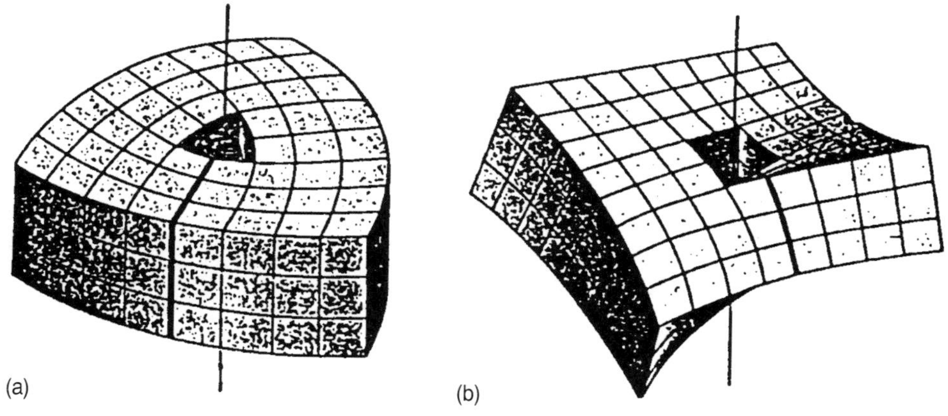

Fig. 3.12 Types of disclination lines in 3D showing: (a) a wedge disclination; (b) a screw disclination.

3.1.3 Two-dimensional (planar) defects

It has been mentioned previously that the external surfaces of a finite crystal can be regarded as defects, in the sense that they destroy the infinite translational periodicity characteristic of an ideal single crystal. However, there may also be *internal* surfaces in what appears to be, ostensibly, a single crystal. These are associated with the presence of a mosaic (or domain) texture to the crystal: the overall crystal in such a case actually consists of an aggregate of domains of perfect crystalline material (typically 10 000 Å in size), there being a structural mismatch at the interface, the grain boundary, between domains. Typically, this mismatch is very small, amounting to an angular misorientation between crystallites of fractions of a degree.

Such low-angle grain boundaries can be regarded as comprising linear arrays of dislocations. An example, a low-angle tilt boundary, is shown in Fig. 3.13; the misorientation in this case can be viewed as an angular rotation by the angle θ about the common axis of one part of the crystal relative to the other (in the plane of the boundary). The tilt boundary can be represented as a line of edge dislocations with separation D; the tilt angle is thus given by

$$\theta = b/D \tag{3.2}$$

where b is the magnitude of the Burgers vector of the dislocations. There exist also twist grain boundaries formed from an array of *screw* dislocations in which the rotation axis is *perpendicular* to the boundary plane. In general, however, grain boundaries are composed of a mixture of both tilt and twist components.

Another type of grain boundary, but one which does not involve angular misorientations, is the so-called antiphase boundary, illustrated in Fig. 3.14 for the case of a 2D

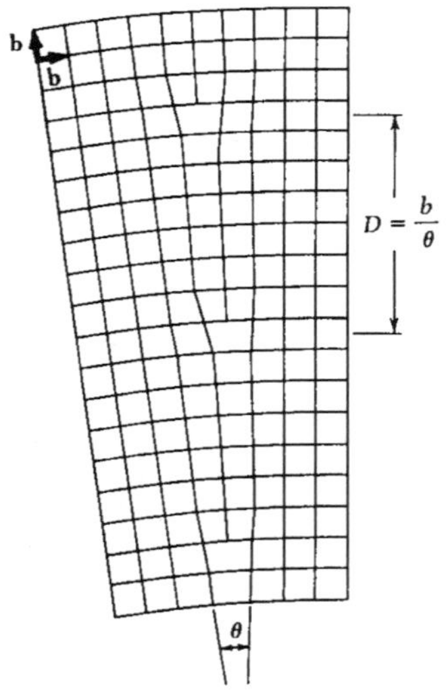

Fig. 3.13 Representation of a low-angle tilt grain boundary as a linear array of edge dislocations.

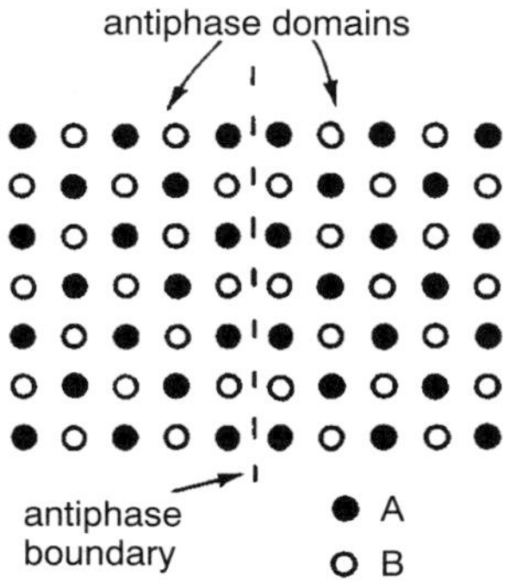

Fig. 3.14 2D representation of an antiphase boundary for a crystal AB.

crystal AB. At the antiphase boundary, like atoms face each other and the normal stacking sequence ABAB is reversed. Thus, the antiphase boundary is an example of a stacking fault, and can be regarded as arising when an entire plane of atoms is removed. Stacking faults are particularly prevalent in materials crystallizing in the c.c.p. structure characterized by the stacking sequence ABCABC.... It is obviously relatively easy for such a sequence to be interrupted.

Finally, a type of planar defect found in crystals can be associated with non-stoichiometry, particularly in transition-metal oxides such as MoO_{3-x}, WO_{3-x} and TiO_{2-x}. In such cases, chemical reduction of the parent stoichiometric compound produces oxygen vacancies that are not distributed at random throughout the structure but, instead, are concentrated on certain planes in the crystal. Crystallographic shear planes arise when such planar arrays of vacancies are eliminated by a structural condensation, resulting in a local change in structure; they have well-defined non-stoichiometric compositions, e.g. Mo_nO_{3n-1} and Ti_nO_{2n-1} (known as Magneli phases). An example of a crystallographic shear plane in MoO_3, having the composition Mo_8O_{23}, is shown in Fig. 3.15 (see Problem 3.3): the defect in this case consists of a plane containing groups of four *edge-sharing* MoO_6 octahedra, which can be thought of as resulting from a shear of a

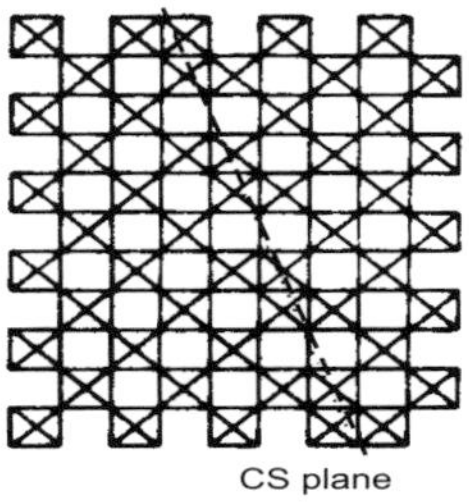

Fig. 3.15 A crystallographic shear (CS) plane in MoO_3 having the composition Mo_8O_{23}. The squares containing crosses represent, in projection, chains of corner-sharing MoO_3 octahedra. The CS plane corresponds to groups of four *edge*-sharing octahedra.

plane containing the corner-sharing linkages characteristic of the parent stoichiometric compound, MoO_3. In TiO_2, crystallographic shear planes are associated with groups of *face-sharing* TiO_6 octahedra, instead of the edge-sharing linkage found in the rutile form of TiO_2. An arbitrary, non-stoichiometric reduced composition of these transition-metal oxides is thus achieved by varying the spacing between crystallographic shear planes separating blocks of stoichiometric material.

Energetics and thermodynamics of defects 3.2

Real crystals are often imperfect, i.e. defective, because the presence of structural defects up to a certain concentration can cause a lowering of the overall free energy: the creation of a defect in an otherwise perfect structure always costs energy, the enthalpy of formation ΔH, which can be offset, however, by an increase in the entropy term $-T\Delta S$, where ΔS is the entropy associated with the large number of positions that the defect may occupy. Thus, the Gibbs free energy, $G = H - TS$, can have a minimum at a *finite* defect concentration (see Fig. 3.16). Thus, in this case, *thermodynamics* determines the defect concentration. If, on the other hand, the energy of defect formation is very high, the minimum value of ΔG corresponds to the ideal, defect-free structure, and in this case (e.g. for the case of extended defects), the defects are present only as metastable configurations, for instance resulting from the method used to prepare the material.

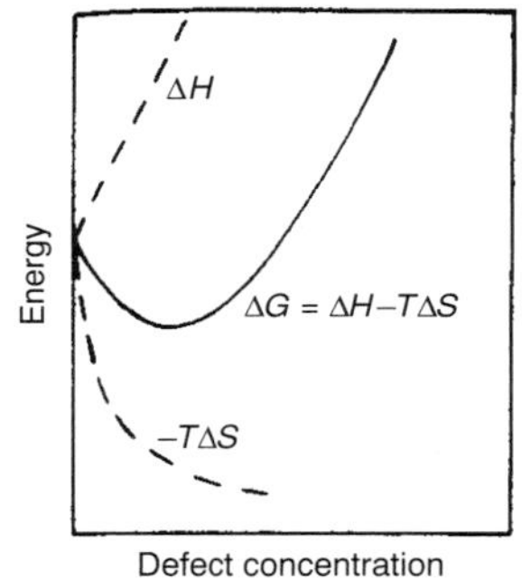

Fig. 3.16 Schematic illustration of the energy changes associated with the incorporation of defects into a perfect crystal. The defect concentration corresponding to the minimum in the Gibbs free energy is thus determined thermodynamically.

3.2.1 Point defects

Point defects are usually *intrinsic*, that is their concentrations are normally determined by thermodynamic considerations, unless a greater concentration of extrinsic defects is introduced by outside means (e.g. irradiation, chemical doping, etc.). Thus, the equilibrium concentrations of intrinsic defects can be found by thermodynamic methods, by evaluating the entropy associated with the formation of the defects.

Perhaps the easiest case to consider first is that of *atomic vacancies*. If N_V vacancies are created among N_L lattice sites, the number of ways of permuting the vacancies over the sites is given by the combinatorial result:

$$\Omega = \frac{N_L!}{N_V!(N_L - N_V)!}. \tag{3.3}$$

Thus, the configurational (mixing) entropy associated with these permutations is given by the Boltzmann expression $\Delta S = k_B \ln\Omega$, and the ensuing expression can be simplified by use of Stirling's approximation (valid for large numbers), i.e. $\ln x! \simeq x\ln x - x$, giving

$$\Delta S_m \simeq k_B[N_L \ln N_L - N_V \ln N_V - (N_L - N_V)\ln(N_L - N_V)]. \tag{3.4}$$

The change in Gibbs free energy for the creation of N_V vacancies can thus be written as

$$\Delta G_V \simeq N_V(\Delta h_V - T\Delta s_{vib}) + k_B T\left[(N_L - N_V)\ln\left(\frac{N_L - N_V}{N_L}\right) + \left(\frac{N_V}{N_L}\right)\ln\left(\frac{N_V}{N_L}\right)\right], \tag{3.5}$$

where the lower case for the enthalpy h and entropy s indicates that the quantities are per atom, and where the last term corresponds to eqn. (3.4) in a rearranged form. (Δs_{vib} is a correction term due to the change in *vibrational* entropy that arises from a lowering of parts of the vibrational frequency spectrum for atoms bordering the vacancy due to structural relaxation.) The free energy may be minimized with respect to the vacancy concentration to give the equilibrium defect concentration (cf. Fig. 3.16), i.e.

$$\frac{\partial \Delta G_V}{\partial N_V} = 0 = \Delta h_V - T\Delta s_{vib} + k_B T \ln\left(\frac{N_V}{N_L - N_V}\right). \tag{3.6}$$

Since $N_L \gg N_V$, eqn. (3.6) can thus be written approximately in terms of the atomic fraction of vacancies, $x_V = N_V/N_L$, as the Boltzmann factor

$$x_V \simeq \exp[-\Delta h_V/k_B T], \tag{3.7}$$

neglecting the small vibrational correction term (see Problem 3.4).

The energetics involved in vacancy formation can be understood with reference to Fig. 3.17 showing the atomic processes involved in the creation of vacancies in the interior of a crystal. A vacancy is first created by the jump of an atom out of a surface layer onto the surface: the surface may be an external or an internal one (or a dislocation). The vacancy then progressively moves into the interior of the crystal via a series of *outward* atomic jumps.

The vacancy-formation enthalpy must obviously be related to the *cohesive energy* of the solid, U_0, i.e. the energy required to separate the atoms in the solid to infinity. However, the final configuration after forming a vacancy (Fig. 3.17) is with the displaced atom on the *surface* of the solid and not at infinity; thus an energy associated with surface adsorption (equivalent to minus the sublimation enthalpy, h_s) is recovered.

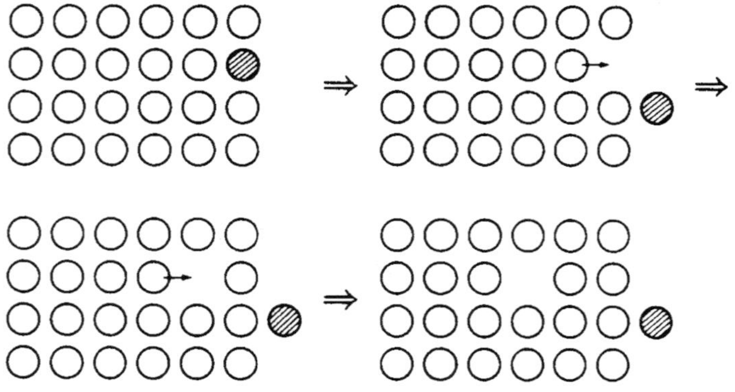

Fig. 3.17 Mechanism of vacancy formation in a crystal. An atom (shaded) moves onto a surface and the resulting surface vacancy progressively moves into the interior of the crystal via a sequence of atomic jumps.

If it is assumed that approximately one-half of the bonds broken in removing an atom from the interior of a solid are broken on removing it from a surface, then $h_s \simeq U_0/2$. Finally, the atoms surrounding a vacancy relax in order to lower the strain energy that they experience. Atoms around a vacancy in a metal or rare-gas solid move *inward* to relieve the compressive strain. Ions around a vacancy in an ionic crystal move *outward* because of the repulsive electrostatic forces between the like charges of the ions surrounding the vacancy and that associated with the vacancy itself (e.g. an Na^+-ion vacancy in NaCl, having a negative charge, is surrounded by negatively charged anions). Thus, the overall vacancy-formation energy can be written as

$$\Delta h_V \simeq \frac{1}{2} U_O - h_r \tag{3.8}$$

where h_r is the strain-relaxation enthalpy.

In the case of ionic crystals, of course, vacancies cannot be generated on only *one* ion sub-lattice; in practice, cation *and* anion vacancies (Schottky defect pairs) are created simultaneously to maintain charge neutrality. In this case, the analysis leading to eqn. (3.7) for the equilibrium concentration of vacancies of a single type must be suitably modified. The number of ways of permuting N_V^c cation vacancies and N_V^a anion vacancies over N_L lattice sites is

$$\Omega = \frac{N_L!}{N_V^c!(N_L - N_V^c)!} \times \frac{N_L!}{N_V^a!(N_L - N_V^a)!}, \tag{3.9}$$

where the number of Schottky pairs, $N_S = N_V^c = N_V^a$. The equilibrium fractional concentration, x_S, of Schottky pairs is then obtained by minimizing the expression for the free energy, as before, leading to the approximate expression

$$x_S \simeq \exp[-\Delta h_S / 2k_B T], \tag{3.10}$$

where Δh_S is the formation enthalpy of a Schottky defect pair, and the vibrational entropy corrections (one each for cation and anion vacancies) are neglected. Thus, the measured activation energy for vacancy formation in ionic crystals consists of a sum of two terms, one for cation vacancies and one for anion vacancies.

Experimental values of Schottky-pair formation enthalpies are given in Table 3.1 for some representative ionic crystals. For comparison, Δh_V for metals is of the order of 1 eV (being higher, the more refractory the metal). Note that the formation energy per *vacancy* is thus rather similar between metals and ionic crystals.

The equilibrium fractional concentration of *isolated interstitial defects*, x_i (i.e. *not* associated with a corresponding vacancy as in a Frenkel pair) can be calculated in exactly the same way as for isolated vacancies (cf. eqn. (3.7)), giving

$$x_i \simeq \exp[-\Delta h_i / k_B T], \tag{3.11}$$

where Δh_i is the formation enthalpy of an isolated interstitial (see Problem 3.4).

Isolated interstitial defects in, say, metals have a very small probability of formation because of a correspondingly large enthalpy of formation, Δh_i, associated with the large dilational strain produced by introducing an atom into an interstitial site. For Cu, Δh_i is estimated to be ~ 3 eV, compared with $\Delta h_V \sim 1$ eV for vacancy formation.

Frenkel defects, i.e. interstitial–vacancy pairs, can be treated formally in a similar manner as for Schottky-vacancy pair defects (eqn. (3.10)). The number of permutations

Table 3.1 Formation enthalpies and entropies of defects

Material	Δh_f *(eV)*	$\Delta s_f/k_B$
Vacancies		
Al	0.75	2.4
Cu	1.18	1.6–3.0
Ag	1.09	$\simeq$ 1.5
Au	0.94	1.0
Schottky defects		
NaCl	2.44	9.8
NaI	2.00	7.6
KCl	2.54	9.0
KBr	2.53	10.3
Frenkel defects		
AgCl	1.45–1.55	5.4–12.2
AgBr	1.13–1.28	6.6–12.2

(After Allnatt and Lidiard (1993). Reproduced by permission of Cambridge University Press)

of placing N_i interstitials on N_I interstitial sites and N_V vacancies on N_L lattice sites is given by

$$\Omega = \frac{N_L!}{N_V!(N_L - N_V!} \times \frac{N_I!}{N_i!(N_I - N_i)!}. \tag{3.12}$$

Minimizing the free energy in the usual way with respect to the number N_F of Frenkel pairs to find the equilibrium concentration of defects, subject to $N_F = N_i = N_V$, gives

$$N_F = [(N_L - N_V)(N_I - N_i)]^{1/2}\exp(\Delta s_F/2k_B)\exp(-\Delta h_F/2k_BT), \tag{3.13}$$

where Δs_F is the vibrational entropy change associated with the Frenkel defect. Neglecting this correction term, and assuming that $N_i \ll N_I$, $N_V \ll N_L$, gives

$$N_F \simeq (N_L N_I)^{1/2}\exp(-\Delta h_F/2k_BT). \tag{3.14}$$

Note that in both eqns. (3.10) and (3.14), applicable to *pairs* of defects, a factor of 2 appears in the denominator of the exponent. Experimental values of Δh_F for crystalline silver halides are given in Table 3.1.

3.2.2 Extended defects

For the case of *point* defects, the formation energy, which is *independent* of N, the number of atoms in the sample, is more than offset by a gain in entropy (of order ln N), as seen in §3.2.1, resulting in the defect concentration being determined by thermal-equilibrium considerations.

This is not so for extended defects. For the case of line defects, e.g. dislocations, the formation energy must be proportional to the linear dimensions of the sample ($\propto N^{1/3}$), and for planar defects it must be proportional to the cross-sectional area ($\propto N^{2/3}$). Thus, entropy gains of order ln N are insufficient in such cases to give rise to minima in the free energy at finite defect concentrations (cf. Fig. 3.16). Hence, line and planar defects are likely to be *metastable* configurations rather than being subject to thermal-equilibrium constraints. As a result, thermal annealing can greatly alter the concentration of such metastable defects. For example, the dislocation density in the most perfect Si or Ge crystals available can be as low as $10^6\ \mathrm{m^{-2}}$, whereas in heavily deformed metals it can be up to $10^{16}\ \mathrm{m^{-2}}$.

The (elastic) energy associated with dislocations may be calculated using simple continuum elasticity theory. The case of a screw dislocation (Fig. 3.10) is the most straightforward to consider. The region (far from the core) around a screw dislocation can be represented, as in Fig. 3.18, by a cylindrical shell of material of circumference $2\pi r$ sheared by an amount equal to the Burgers vector b of the dislocation; i.e. the shear strain is given by $e_s = b/2\pi r$. For an elastic continuum, the associated shear stress is given by

$$\sigma_s = G_s e_s = G_s b/2\pi r, \tag{3.15}$$

where G_s is the shear modulus (eqn. (3.111)). The elastic energy per unit length stored in the shell is thus

$$\mathrm{d}\mathscr{E}_s = \frac{1}{2} G_s e_s^2 \mathrm{d}V = (G_s b^2/4\pi r)\mathrm{d}r. \tag{3.16}$$

Integrating over the range from the radius of the dislocation core r_0 (comparable to b) to R (limited by the size of the crystal for an isolated dislocation, or half the separation between dislocations otherwise) gives for the elastic energy per unit length of a screw dislocation (see Problem 3.5):

$$\mathscr{E}_s = \frac{G_s b^2}{4\pi} \ln\left(\frac{R}{r_0}\right). \tag{3.17}$$

Evaluating the similar quantity for the case of an edge dislocation is rather more complicated because it involves both shear and normal stresses. The result is

$$\mathscr{E}_e = \frac{G_s b^2}{4\pi(1-\upsilon_p)} \ln\left(\frac{R}{r_0}\right), \tag{3.18}$$

where υ_p is the Poisson ratio (the ratio of the compressive strain, in an orthogonal direction to that of an applied uniaxial stress, to the dilational strain in that direction, eqn. (3.107)). Note that both strain energies are proportional to the quantity $G_s b^2$. For representative values of $G_s \simeq 4 \times 10^{10}\ \mathrm{N\,m^{-2}}$ and $b \simeq 2.5$ Å, the typical dislocation energy per length, $G_s b^2$, is about $2.5 \times 10^{-9}\ \mathrm{J\,m^{-1}}$, or about 4 eV for each atomic plane threaded by the dislocation.

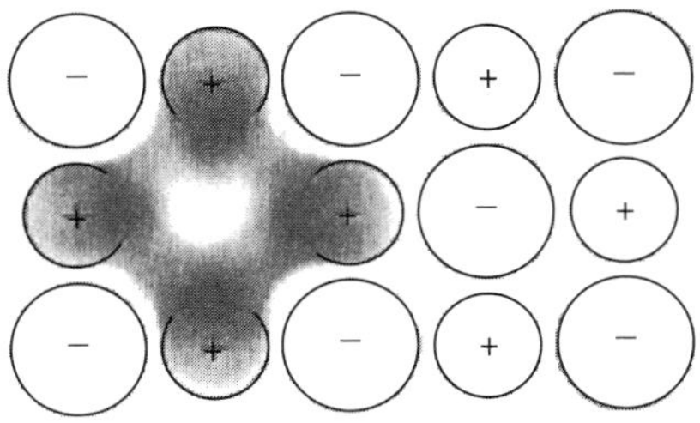

Fig. 3.18 Illustration of the F-centre in alkali-halide crystals: an electron bound to an anion vacancy. The charge distribution of the trapped electron is represented schematically, although the electron wavefunction does extend beyond the nearest-neighbour cation coordination shell.

Spectroscopy and microscopy of defects 3.3

Defects in solids can be detected using a variety of spectroscopic probes. Thus, the electronic states associated with defects can be probed optically in terms of electronic transitions between the electronic energy levels. Unpaired electron spins associated with defects can be detected via their paramagnetism, e.g. using electron spin resonance. Finally, extended defects can often be imaged directly using electron microscopy.

3.3.1 Optical spectroscopy

The spatial localization of electrons trapped at point-defect sites causes a quantization of the related electronic energy levels, and light can be used to probe such quantized levels: absorption of incident photons causes transitions upward in energy and downward electronic transitions result in photon emission.

One important example of the use of optical spectroscopy in probing defects is for the case of 'colour centres' in crystalline alkali halides (see §3.1.1). These materials are extremely ionic, with the result that the valence-electron charge density is strongly localized on the anions, requiring a great deal of energy to remove it (much greater than the photon energy for light in the visible part of the electromagnetic spectrum). These solids are therefore transparent (non-absorbing) for visible light when in a structurally defect-free state. (An alternative description of optical absorption in terms of excitations between an occupied valence band and unoccupied conduction band of electrons separated by a very large ($\simeq$10 eV) forbidden energy gap is given in §5.8.2.)

However, alkali-halide crystals, exposed to alkali-metal vapour or ionizing radiation (e.g. X-rays), can contain anion vacancies which, being effectively positively charged after removal of the negatively charged anion to satisfy charge neutrality, can subsequently trap an electron. The simplest example is the so-called F-centre (after *Farbe*, the German word for colour), which is a single anion vacancy containing the *same* number of electrons as the charge of the displaced anion: the electron density is mostly associated with the cations adjacent to the vacancy (Fig. 3.18). The electronic energy levels in the case of F-centres in alkali-halide crystals (containing a single electron) can be calculated approximately by regarding the system as being equivalent to the 3D particle-in-a-box quantum-mechanical problem, in which an electron is confined to a cubic box, of side length a, by infinitely high potential barriers at the cube faces. For the case where the potential energy within the box is constant (and equal to zero), the quantized electronic energy levels are given (see Problem 3.6) by:

$$\mathscr{E} = \frac{h^2}{8m_e a^2}(n_1^2 + n_2^2 + n_3^2), \tag{3.19}$$

where n_1, n_2 and n_3 are the quantum numbers characterizing the number of nodes $(n_i - 1)$ in the electronic wavefunction in each of the three orthogonal directions of the box and m_e is the electronic mass. The ground state corresponds to the set of quantum numbers $(n_1\, n_2\, n_3) = (111)$, and the first excited state to the (triply degenerate) state with quantum numbers of the form (211). Thus, the transition energy between ground and first excited states, corresponding to the photon absorption energy of this model of the F-centre, is given by

$$\Delta\mathscr{E} = \frac{3h^2}{8m_e a^2}. \tag{3.20}$$

If KCl is used as an example, and the size of the box is taken to be the lattice constant $a = 6.29$ Å (see Fig. (3.18)), eqn. (3.20) predicts that the photon absorption energy should be 2.85 eV, compared with the experimental value of 2.3 eV. Considering the crudeness of the model, in particular that the excess electron is confined precisely to the region bounded by the nearest-neighbour anions (see Fig. 3.18) and that it moves in a constant (zero) potential instead of that (of the order of the Madelung energy) associated with the surrounding positively charged cations, the agreement is reasonable. The importance of eqn. (3.20), however, is that it predicts that the F-centre optical absorption energy should depend on the lattice spacing. This form of relationship (the Mollwo–Ivey law) is exhibited, for example, by alkali halides (Fig. 3.19 and Plate III), where the empirical relationship $\Delta\mathscr{E} \propto a^{-1.77}$ is found, rather than the exact inverse-square relation predicted by the simple square-well model (eqn. (3.20)).

Although the F-centre, say in alkali-halide crystals, is perhaps the simplest form of defect-related colour centre, other defect configurations can also act as colour centres,

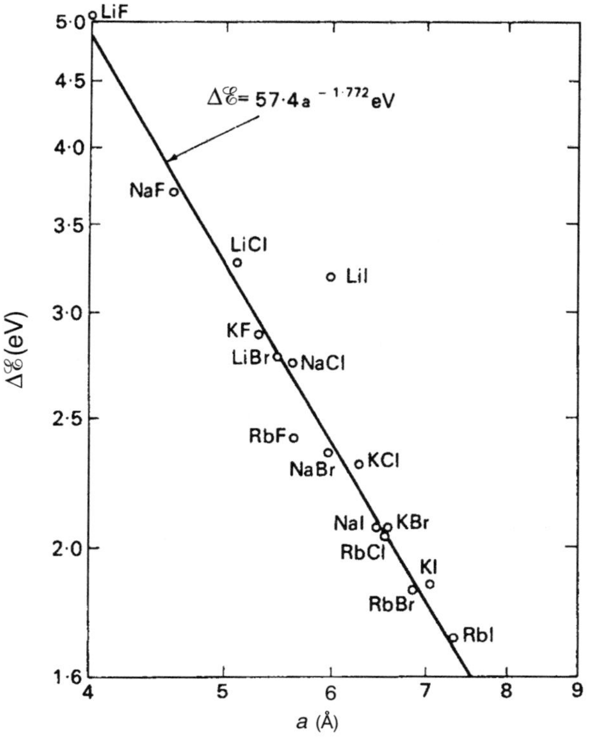

Fig. 3.19 Dependence of the F-centre optical absorption energy on lattice parameter (a) for alkali halide crystals. (After Dawson and Pooley (1969). Reproduced by permission of Akademie Verlag GmbH)

but they have different optical-absorption characteristics and the simple particle-in-a-box model (eqn. (3.19)) can no longer be used to calculate these. Some colour-centre variants are shown schematically in Fig. 3.20. The F_A-centre is the same as the simple F-centre, except that one of the cations surrounding the anion vacancy has been replaced by a monovalent impurity ion. An F′-centre is a simple F-centre, but containing an *extra* electron. Aggregates of F-centres may also occur, and one form of nomenclature is to denote by a subscript the number of anion vacancies comprising the defect complex. An F_2-centre (sometimes called an M-centre) consists of two neighbouring F-centres lying in a ⟨100⟩ direction, and is shown in Fig. 3.20. An F_3-centre (also termed an R-centre) is an aggregate of three F-centres in a triangular arrangement in the (111) plane (not shown).

In principle, the antimorph of an F-centre, i.e. a positive charge (a missing electron, or 'hole'—see §6.2.2) trapped at a cation vacancy and denoted a V-centre, should also exist; although it appears to be unstable in alkali halides, it *is* found in oxides. A more common hole-related centre is the so-called 'molecular ion', $[X_2^-]$, formed when a hole is trapped by a negatively charged anion, X^-, having an originally filled p^6 shell. The

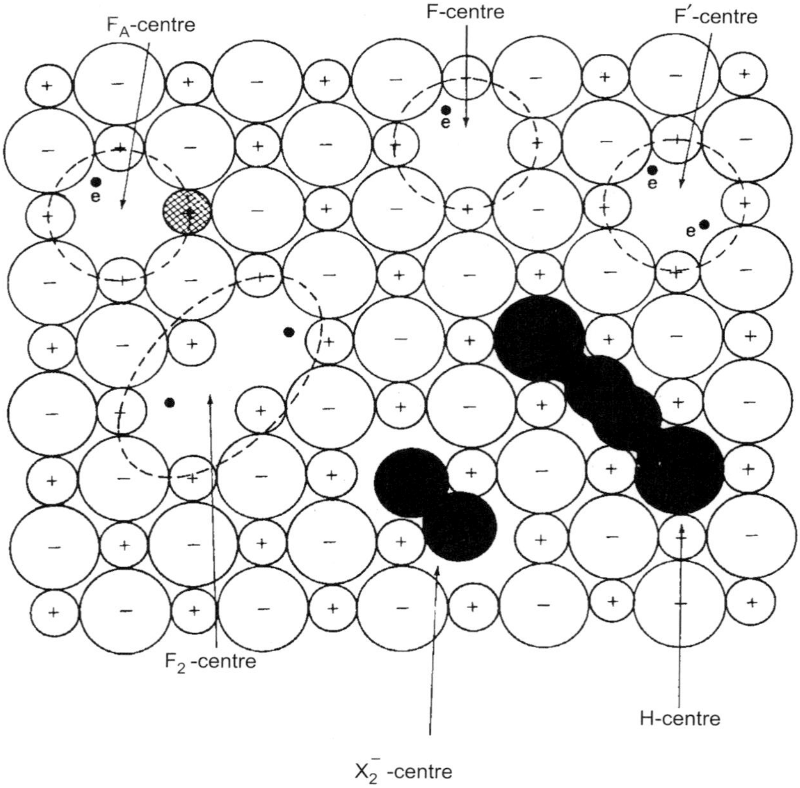

Fig. 3.20 Schematic representations of different types of colour centres in alkali-halide crystals. A monovalent cation impurity is shown as a shaded circle.

resulting p^5 electronic configuration, being spatially asymmetric, is subject to a structural distortion that acts to lower the electronic energy, in this case by the formation of effectively a covalent bond between the anion associated with the trapped hole and the nearest-neighbour anion in the $\langle 100 \rangle$ direction, resulting in the formation of an $[X_2^-]$ entity (Fig. 3.20). (Somewhat confusingly, this defect configuration is also sometimes referred to as a V_K-centre in the literature, even though it does *not* involve a cation vacancy.) The hole associated with the $[X_2^-]$ (or V_K) centre is referred to as being 'self-trapped'; its electronic wavefunction is localized (in the vicinity of the bond between the two anion species involved), and the total energy is lowered as a result of the structural distortion involved in forming the $[X_2^-]$ entity. Consequently, the self-trapped hole is relatively immobile (certainly in comparison with a hole in an otherwise filled (valence) band of electrons in the case where lattice distortion effects are negligible—see §6.5.1.2); the self-trapped hole *can* move from site to site, but only by a thermally activated 'hopping' process, where the activation energy for the hopping rate is related to the self-trapping distortion energy. The molecular-ion $[X_2^-]$ centre in alkali-halide crystals is one example of a small polaron (§6.6), i.e. a charge carrier (in this case, a hole) associated with a structural distortion strong enough to cause the excess carrier to self-trap, the spatial extent of the distortion being comparable to that of the wavefunction of the carrier.

A more complicated centre, somewhat related to the $[X_2^-]$ (V_K) centre, is the so-called H-centre, which comprises a neutral anion occupying an interstitial position between two neighbouring anions, in the $\langle 100 \rangle$ direction in the case of alkali halides (see Fig. 3.20). A molecular-ion-like configuration can form by means of structural distortion as in the $[X_2^-]$ (V_K) centre, but this is now located at a normal anion site, and this central molecular ion then interacts weakly with the two neighbouring anions in the same $\langle 100 \rangle$ direction. Such a defect configuration is an example of a crowdion. Further details about colour centres and related defects can be found, for instance, in Henderson (1972), Townsend and Kelly (1973) and Stoneham (1975).

*3.3.2 Electron spin resonance

Many defect configurations, e.g. the F-centre and the $[X_2^-]$ (V_K) centre in alkali-halide crystals, the electrically neutral dangling bond, D^0, in covalent amorphous materials and the singly charged state (V^+) of the monovacancy in crystalline Si, etc. contain *unpaired* electrons. The magnetic moment associated with such unpaired electron spins can be detected using the technique of electron spin (sometimes also called paramagnetic) resonance, ESR (or EPR). (See §7.2 for a discussion of magnetic behaviour.)

The value of the magnetic moment μ_m of an electron can be expressed as

$$\mu_m = -\gamma_m \hbar S \tag{3.21}$$

where S is the value of the electron spin (1/2), and γ_m is called the gyromagnetic ratio and can be written as

$$\gamma_m = g_e e / 2m_e. \tag{3.22}$$

The quantity $e\hbar/2m_e$ that correspondingly occurs in eqn. (3.21) is termed the Bohr magneton (μ_B). The Landé g-factor in eqn. (3.22) for the case of a *free* electron spin is given by

$$g_e \simeq 2\left(1 + \frac{\mu_0 c e^2}{2\pi h}\right) = 2.0023 \tag{3.23}$$

where μ_0 is the vacuum permittivity. However, the spin of the unpaired electron, when in condensed matter, may couple to angular momentum, either the intrinsic orbital angular momentum of the electron or, more interestingly from the present point of view, through overlap of the electron wavefunction with the electron orbitals of atoms neighbouring the spin. The latter situation is obviously dependent on the local atomic structural environment, and measurement of the g-factor can thus lead to a structural identification of the defect concerned.

In the absence of an applied magnetic field, the energy levels of a spin-1/2 particle (e.g. an electron) are doubly degenerate (i.e. the energies of the two quantum states with magnetic quantum numbers $m_s = \pm 1/2$ are equal), with a value, say, of $\mathscr{E}_0$. The application of an external magnetic field with constant value B_0 lifts the spin degeneracy via the Zeeman interaction, and the energy levels split to become (see Fig. 3.21):

$$\mathscr{E} = \mathscr{E}_0 + m_s g_e \mu_B B_0. \tag{3.24}$$

In a semi-classical sense, these two energy levels can be regarded as corresponding to the electron spin pointing parallel or antiparallel to the magnetic-field direction.

The aim of the ESR technique is to detect the Zeeman–split energy levels by causing magnetic-dipole transitions to occur between them (equivalent to flipping the electron spin) by means of the application of an *alternating* magnetic field B_1, of frequency ν, in a direction *perpendicular* to the steady field B_0. A photon of energy $h\nu$ is resonantly absorbed (i.e. causes a transition) when this energy is equal to the energy separation between Zeeman-split levels (eqn. (3.24)) and when the selection rule $\Delta m_s = \pm 1$ is obeyed, i.e.

$$h\nu = g_e \mu_B B_0. \tag{3.25}$$

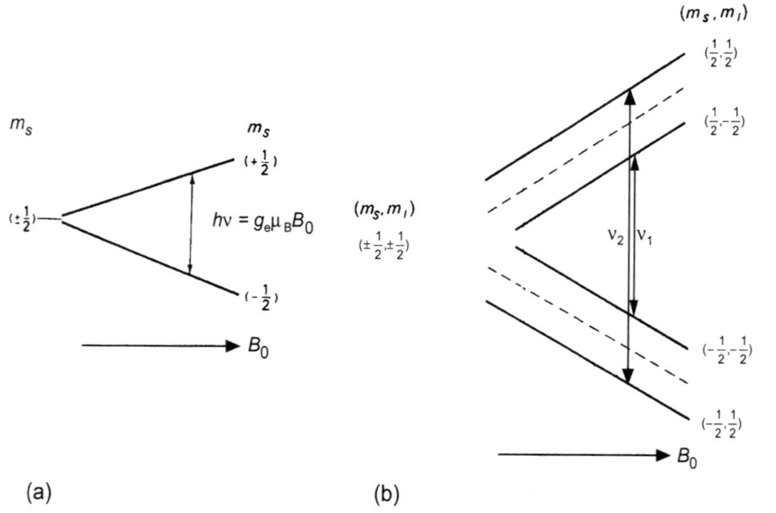

Fig. 3.21 Illustration of the energy levels and ESR transitions for a defect having: (a) an unpaired electron ($S = 1/2$) but no hyperfine interaction; (b) an unpaired electron having a hyperfine interaction with an atom with nuclear spin $I = 1/2$.

For the case of a free spin ($g_e \simeq 2$), and an applied magnetic field of $B_0 \simeq 0.3$T, the resonant frequency is $\nu \simeq 9$ GHz, i.e. in the microwave region of the electromagnetic spectrum

The resonance condition can be detected in one of two ways: either the frequency ν of the alternating magnetic field (associated with microwaves incident on the sample) is kept constant and the Zeeman field B_0 is varied and swept through the resonance condition (eqn. (3.25)), or, conversely, the Zeeman magnetic field B_0 is kept constant and the frequency of the alternating field is varied. The former procedure was used in continuous-wave (c.w.) ESR spectrometers in the past; frequency modulation ($\simeq$ 100 kHz) of the alternating field B_1 in such instruments ensures that the resonance signal is measured as a *derivative* of the resonance peak. The latter way of detecting resonance forms the basis of the latest generation of Fourier-transform (FT) ESR spectrometers in which an appropriately chosen pulse of microwave radiation is used to irradiate the sample. The pulse, in the simplest case having a square-wave shape, can be regarded (cf. eqn. (2.54)) as being comprised of a sum of Fourier components having a wide range of frequencies, including the electron-spin resonant frequency, and so will cause a resonant spin–flip transition. The response of the sample to the microwave pulse is detected as a change of magnetization with time (the 'free-induction decay', FID); Fourier transformation of the FID yields the resonant absorption-peak spectrum as a function of frequency (equivalent to the integrated derivative ESR signal obtained from a c.w. spectrometer). An example of a derivative ESR spectrum obtained using a conventional c.w. spectrometer for the case of a simple defect, the unpaired spin associated with an electrically neutral dangling bond (D^0) in amorphous (hydrogenated) silicon is shown in Fig. 3.22a.

Considerably more information about the local structure around unpaired spins at point defects than that provided simply by the value of the g-factor can be obtained, in principle, from a measurement of the so-called hyperfine interaction between the spin of the unpaired electron and the nuclear spin I of surrounding atoms. The hyperfine interaction between an unpaired electron spin ($S = 1/2$) and a single nucleus with nuclear spin $I > 0$ results in an extra term for the energy (cf. eqn. (3.24)), viz.

$$\mathscr{E} = \mathscr{E}_0 + m_s g_e \mu_B B_0 + m_s m_I \mathscr{E}_{HFS}, \tag{3.26}$$

where m_I is the nuclear-spin quantum number. The hyperfine interaction energy consists of a so-called isotropic term (A), and an anisotropic term (A') which depends on the classical dipole–dipole interaction between electron and nuclear spins (and which is a function of the angle between $\boldsymbol{B}_0$ and the line connecting the nucleus and the electron spin. The isotropic term depends on the probability of the electron wavefunction being at the nucleus of the atom in question via the Fermi contact expression

$$A = \frac{8\pi}{3} g_e \mu_B g_N \mu_N F \mid \chi(0) \mid^2, \tag{3.27}$$

where g_N and μ_N are the nuclear g-factor and magneton, respectively, F is an amplification factor and $\chi(0)$ is the electron wavefunction amplitude at the nucleus.

The hyperfine interaction results in an extra splitting of the energy associated with the electron spin (see Fig. 3.21b) characterized by the additional quantum number m_I, and magnetic-dipole transitions can occur between such levels if the extra selection rule $\Delta m_I = 0$ is also satisfied. Thus, for an electron ($S = 1/2$) interacting with a nuclear spin

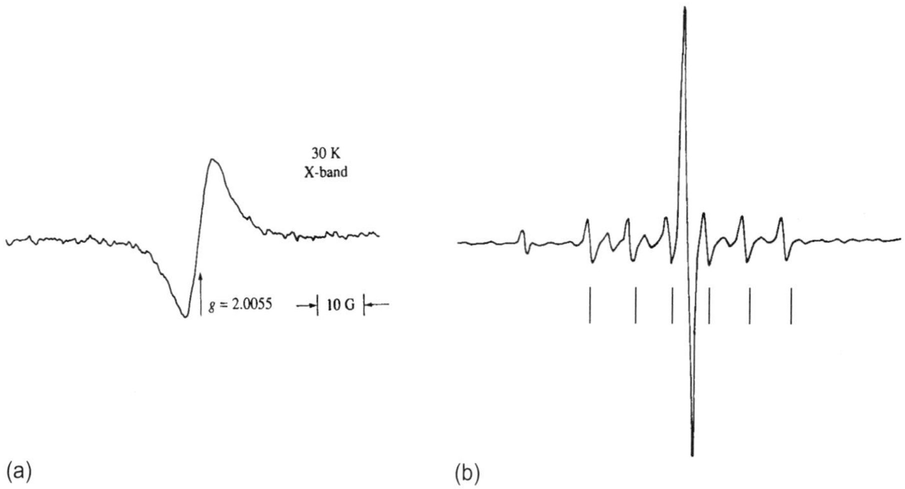

Fig. 3.22 Derivative ESR spectra measured using a c.w. spectrometer for the defects: (a) the electrically neutral dangling bond (D^0) in amorphous (hydrogenated) silicon (Street (1991). Reproduced by permission of Cambridge University Press); (b) the F^+-centre in MgO, measured with B_0 parallel to the $\langle 111 \rangle$ direction. The central intense part of the spectrum corresponds to the free-spin case of electrons in anion vacancies being surrounded only by Mg ions with zero nuclear spin. The six-fold hyperfine-split spectrum is characteristic of the interaction between an electron and a single ^{25}Mg nucleus with spin $I = 5/2$. (Henderson and Wertz (1968). Reproduced by permission of Taylor & Francis Group Ltd.)

$I > 0$, there are $(2I + 1)$ equally spaced lines in the hyperfine-split spectrum for a given magnetic field B_0, i.e. two lines in the simplest case of the $S = I = 1/2$ system (Fig. 3.21b).

A clear example of the ESR spectrum of an F-centre is that of the F^+-centre (i.e. containing a single electron) in crystalline MgO (which has the rocksalt structure). Each anion vacancy has six nearest-neighbour Mg^{2+} ions, 90% of which have zero nuclear spin and hence give rise to a single (derivative) ESR line characterized by the free-spin g-value of 2.0023 (see Fig. 3.22b). The remaining 10% of Mg^{2+} ions consist of the ^{25}Mg isotope ($I = 5/2$), and for the case of a *single* ^{25}Mg nucleus present in the nearest-neighbour shell of an anion vacancy a six-fold hyperfine-split spectrum results, as shown in Fig. 3.22b. A higher spectrometer gain reveals an additional 11–fold hyperfine-split set of lines characteristic of the interaction between the electron and a nearest-neighbour environment of *two* ^{25}Mg nuclei (see Problem 3.7). In the alkali halide NaCl, for example, the situation is greatly complicated by the fact that the magnetic nucleus ^{23}Na($I = 3/2$) is 100% abundant, thereby appreciably increasing the number of hyperfine-split lines. The overlap of the electron wavefunction with the nuclei of more distant shells (e.g. the nearest-neighbour Cl nuclei ($I = 3/2$)) increases the number yet further (see Problem 3.8), often resulting in a broad, featureless ESR line being exhibited by F-centres in alkali halides.

3.3.3 Electron microscopy

Under certain conditions, extended (e.g. linear or planar) defects can be imaged using various microscopy techniques. For example, the structural disruption around dislocations makes this region more susceptible to chemical reaction, e.g. etching by liquid reagents. The resulting etch pits, much larger of course than the size of the dislocation cores, can be revealed using high-magnification optical microscopy. Imaging of defects at atomic-scale resolution, however, can be achieved, in principle, by using electron microscopy.

The principles underlying the operation of the transmission electron microscope have been mentioned in §2.6.1.4 and the two modes of operation, viz. diffraction and imaging, are illustrated in Fig. 2.58. In the present context, however, it is the *imaging* of structural imperfections that is of primary interest. There are essentially three basic ways of performing imaging experiments in a transmission electron microscope, depending on which beams are used for imaging, as indicated in Fig. 3.23. In bright-field imaging, only the main transmitted (undiffracted) electron beam is allowed to pass through the objective aperture: the corresponding image of an ideal, perfectly flat, defect-free crystal would be homogeneously bright. Any defects in a specimen will change the path of some diffracted beams, causing some of them to pass through the aperture and interfere with the transmitted beam, thereby causing variations in the

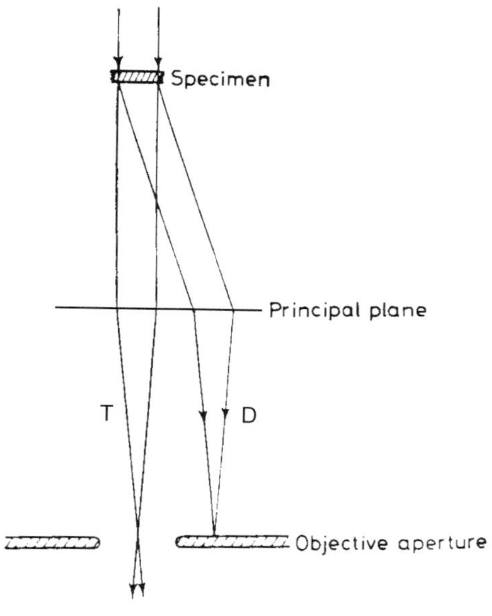

Fig. 3.23 Illustration of the bright-field imaging operation of a transmission electron microscope, in which only the main transmitted electron beam (T) is allowed to pass through the objective aperture. If, instead, a single diffracted beam (D) is allowed to pass through the aperture, a dark-field image is obtained. If the main beam and several diffracted beams are allowed to pass through the aperture, interference between these beams can give a high-resolution image under appropriate conditions.

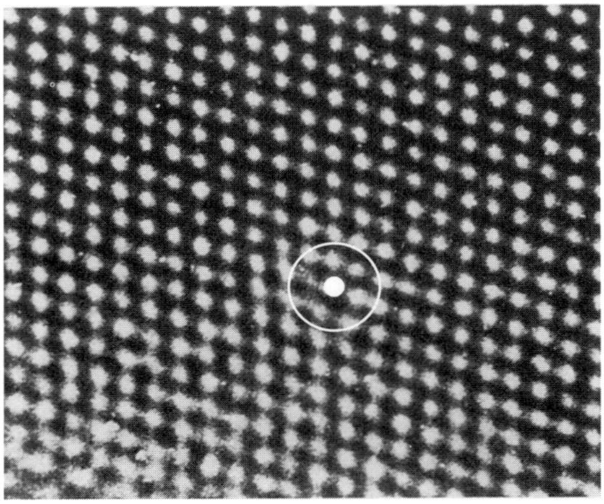

Fig. 3.24 High-resolution lattice image (along $\langle 110 \rangle$) of crystalline Si, obtained by electron microscopy, showing a dislocation core with $\boldsymbol{b} = \frac{1}{2}[110]$. (Reprinted from *Ultramicroscopy* **15**, 51, Hutchison, © 1984 with kind permission from Elsevier Science - NL, Sara Burgerhartstraat 25, 1055 KV Amsterdam, The Netherlands

intensity of the image, termed diffraction contrast. The contrast may be reversed by allowing instead just a single *diffracted* beam to pass through the objective aperture (e.g. by tilting the electron beam); this is known as dark-field imaging. More information on these aspects of electron microscopy is given in the book by Hirsch *et al.* (1965). If the main beam *and* several diffracted beams are allowed to pass through the aperture, interference between these beams can give rise to a high-resolution image of a perfect crystal and, of course, any structural imperfections therein, under appropriate conditions. Atomic resolution (of the order of an ångström) is possible using this technique, the resolution essentially being limited by unavoidable imperfections (aberration) in the objective lens. An example of a high-resolution electron-microscope image of an edge dislocation is shown in Fig. 3.24; the extra half-plane of atoms can be clearly seen.

The mechanism by which structural defects, e.g. an edge dislocation, can produce diffraction contrast, and hence be visible in either a bright- or a dark-field image, is illustrated schematically in Fig. 3.25. Atomic planes near a dislocation core, being displaced from the orientation of planes in an ideal crystal, can satisfy the condition for diffraction. Thus, intensity is removed from the transmitted electron beam (thereby leading to a *dark* image of the dislocation line in bright-field mode) and is transferred to a diffracted beam which, if used for dark-field imaging, would give rise to a *bright* image.

The Burgers vector of a dislocation in a crystal can be determined by ascertaining the conditions under which it can be imaged in a bright-or dark-field electron-microscope image. The scattering amplitude for waves (e.g. electrons) diffracted by a set of atomic planes (*hkl*) in a perfect crystal is given by eqn. (2.105), viz. as the summation over all atoms *i*:

$$F_{hkl} = \sum_i f_i \exp(\mathrm{i}\boldsymbol{G}.\boldsymbol{R}_i),$$

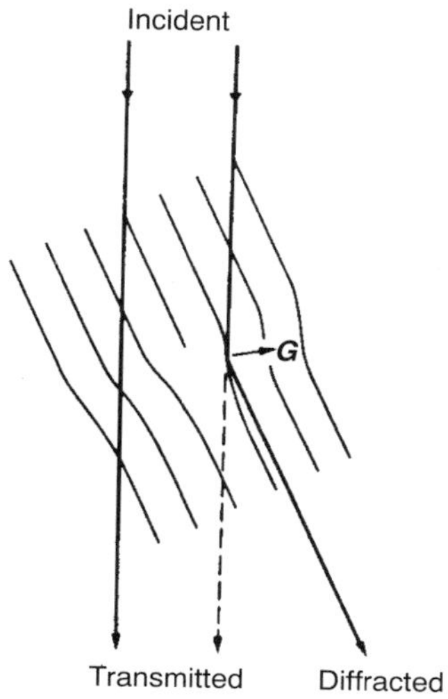

Fig. 3.25 Schematic illustration of the origin of diffraction contrast for an edge dislocation. Crystal-lattice planes near the dislocation core satisfy the diffraction condition, and intensity is thereby removed from the transmitted beam (giving a dark image in bright field) and transferred to a diffracted beam (giving a bright image in dark field).

where $G \equiv G_{hkl}$ is the reciprocal lattice vector corresponding to the lattice planes. For the case of a crystal containing a defect, e.g. a dislocation, the interatomic vector $\boldsymbol{R}_i$ no longer satisfies the condition for perfect translational periodicity (eqn. (2.1)), but instead can be written as

$$\boldsymbol{R}_i' = \boldsymbol{R}_i + \boldsymbol{u}_i \tag{3.28}$$

where $\boldsymbol{u}_i$ is the displacement vector of an atom in the unit cell from its ideal position $\boldsymbol{R}_i$. (In practice, $\boldsymbol{u}_i$ need not be a constant, but could vary with the distance z away from an extended defect in a crystal.) Thus, an extra phase factor, $\exp(\mathrm{i}\, \boldsymbol{G} \cdot \boldsymbol{u}_i)$, is introduced into the expression for the amplitude of the diffracted wave. The condition for *invisibility* of the defect is when this phase factor is unity, i.e. when

$$\boldsymbol{G} \cdot \boldsymbol{u} = 0. \tag{3.29}$$

For the case of screw dislocations, atomic planes *parallel* to the line of the dislocation remain undisturbed, i.e. $\boldsymbol{u} \propto \boldsymbol{b}$ (see Fig. 3.10), and hence the invisibility criterion is thus

$$\boldsymbol{G} \cdot \boldsymbol{b} = 0. \tag{3.30}$$

The case of *edge* dislocations is a little more complicated since the displacement vector $\boldsymbol{u}$ is non-zero in all directions normal to the line of the dislocation. The invisibility criterion $\boldsymbol{G} \cdot \boldsymbol{u} = 0$ (eqn. (3.29)) thus corresponds both to eqn. (3.30), as for screw dislocations, and also to

$$\boldsymbol{G} \cdot (\boldsymbol{b} \times \boldsymbol{t}) = 0 \tag{3.31}$$

where $\boldsymbol{t}$ is a vector along the dislocation line. In other words, only when $\boldsymbol{G}$ is *parallel* to the dislocation line is the edge dislocation invisible in the bright/dark-field image.

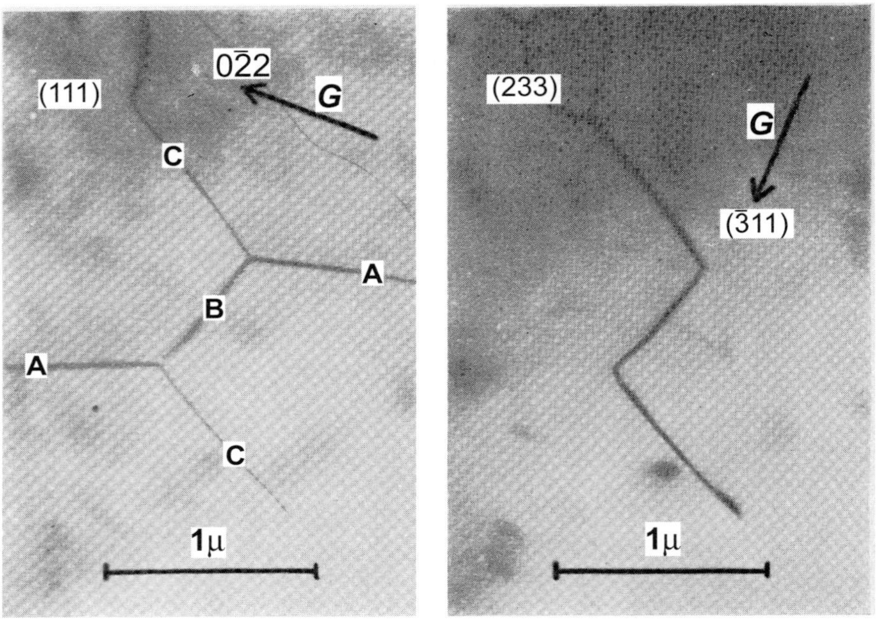

Fig. 3.26 Use of the invisibility criterion, $\boldsymbol{G} \cdot \boldsymbol{b} = 0$, to ascertain the Burgers vector of screw dislocations in crystalline Si. The dislocations in the network lying in the (111) plane, marked A, are (a) visible; (b) absent when $\boldsymbol{G} = \bar{3}11$, consistent with the Burgers vector being $\boldsymbol{b} = \frac{1}{2}[0\bar{1}1]$. (Booker (1964). Reproduced by permission of The Royal Society of Chemistry)

An example of this method of ascertaining the Burgers vector is illustrated in Fig. 3.26 for the case of screw dislocations in crystalline silicon.

Planar defects may also be imaged using an electron microscope under certain circumstances. For example, stacking faults (§3.1.3) can be imaged, subject to the invisibility criterion (eqn. (3.29)), as illustrated in Fig. 3.27. A stacking fault is characterized by a *constant* displacement vector $\boldsymbol{u}$ parallel to the fault plane. Thus, when the diffracting atomic planes are *parallel* to the fault plane, as in Fig. 3.27a, $\boldsymbol{G} \cdot \boldsymbol{u} = 0$ and

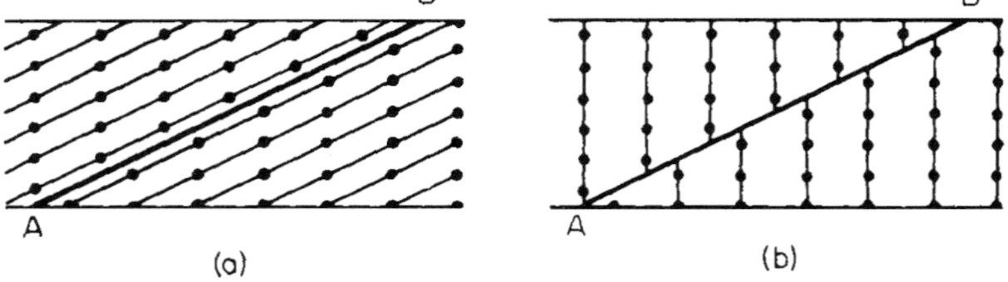

Fig. 3.27 Illustration of the invisibility criterion for the imaging of a stacking fault AB: (a) the diffracting atomic planes are parallel to the fault plane and so $\boldsymbol{G} \cdot \boldsymbol{u} = 0$ and the fault is invisible; (b) diffracting planes that are non-parallel to the fault plane allow it to be imaged.

the fault is invisible in an electron-microscope image. For other atomic planes that are non-parallel to the fault plane (Fig. 3.27b), the quantity $\boldsymbol{G} \cdot \boldsymbol{u}$ is non-zero and hence the fault is visible (as a series of light and dark fringes parallel to the line of intersection of the fault plane with the sample surface for a fault inclined at an angle to the plane of a thin crystal specimen).

3.4 Defect-related behaviour

Defects are a ubiquitous, and often unavoidable, feature of the structure of crystalline and non-crystalline materials. Thus, it is not too surprising that often the physical (and sometimes chemical) behaviour of materials is dominated by that of the structural defects that they contain. Some examples have been mentioned previously, namely the optical and magnetic properties of unpaired electrons associated with 'colour-centre' point defects and dangling bonds (§§3.3.1, 3.3.2). Three other generic types of behaviour generally mediated by defects will be discussed in this section, namely crystal growth, atomic transport and mechanical properties.

3.4.1 Crystal growth

Crystals grow by the addition of atoms, from either the vapour or the liquid phase, onto growing crystal faces. A number of different types of surface sites can exist at which such atomic adsorption events can occur. Figure 3.28 shows some representative surface sites for the case of a simple cubic (s.c.) crystal. Three basic types of site can be distinguished: face (or terrace) sites (denoted 8 in Fig. 3.28), ledge (or edge) sites (labelled 6) and kink sites (denoted '1/2'). In the schematic diagram in Fig. 3.28, each cube represents an atom: a shared face indicates the formation of a bond between two

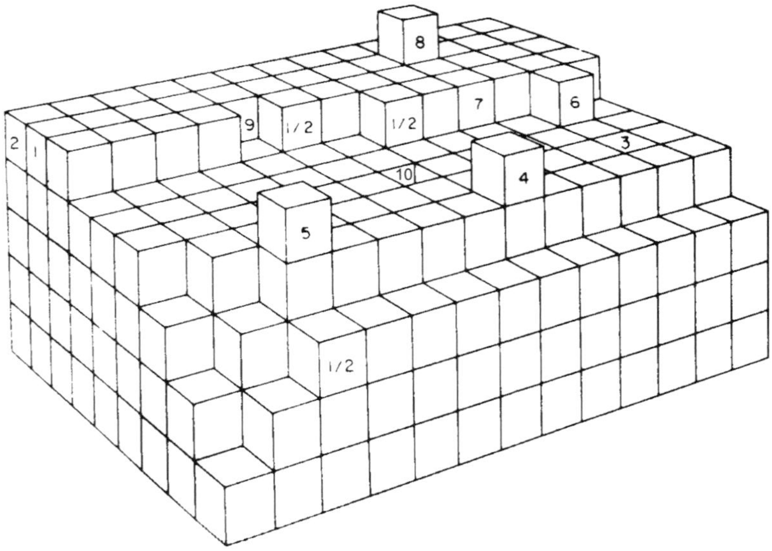

Fig. 3.28 Schematic illustration of atomic and crystal-growth sites for a simple cubic crystal; the cubes represent atoms. The various sites are: (1) completed edge; (2) completed corner; (3) completed surface; (4) incomplete edge; (5) incomplete corner; (6) step edge or ledge site; (7) completed step; (8) face site; (9) edge vacancy; (10) face vacancy. The kink sites are denoted as 1/2 because their atomic binding energy is half that of a bulk atom.

nearest-neighbour atoms (of binding energy, or bond strength, ϕ_1) and a common edge between two cubes represents the formation of a second (next-nearest) neighbour bond (of bond strength $\phi_2 < \phi_1$). Thus, in this scheme and for the $\{100\}$ face of an s.c. crystal, a face site has a binding energy of $(\phi_1 + 4\phi_2)$, a ledge site an energy of $(2\phi_1 + 6\phi_2)$ and a kink site an energy of $(3\phi_1 + 6\phi_2)$ — see Problem 3.9; i.e. atoms at kink sites are the most strongly bound. For comparison, a fully coordinated atom in the bulk of an s.c. crystal has a binding energy of $(6\phi_1 + 12\phi_2)$ or, since each bond in the bulk is shared between two atoms, the mean atomization energy of the crystal is $(6\phi_1 + 12\phi_2)/2 = (3\phi_1 + 6\phi_2)$, i.e. the *same* as the binding energy of a kink site. Thus, an atom adsorbed on a face site is not bound very strongly and can therefore easily subsequently desorb. On the other hand, an adsorbed atom that manages to diffuse over the surface to a ledge, or especially a kink, site will be strongly bound there and can thus contribute to the growth of the crystal face.

Crystal growth only occurs under conditions of supersaturation, i.e. when the actual vapour pressure is greater than the equilibrium vapour pressure (equilibrium being when the chemical potentials of crystal and vapour are the same) in the case of vapour-phase growth, or when a melt is supercooled below the equilibrium melting temperature for crystal and melt, or when a solution has a higher solute concentration than the concentration at which crystal and solution are in equilibrium (Lewis (1980)). In the case of very *low* degrees of supersaturation, crystal growth can be sustained only if an inexhaustible supply of kink or ledge sites is available at growing surfaces, since it is very difficult to nucleate a new monolayer on top of a perfect, complete layer. The types of kink and ledge sites shown in Fig. 3.28 do not satisfy this requirement since they are not regenerated once the crystal face is fully grown.

Fig. 3.29 Scanning tunnelling microscope image of spiral growth (associated with a screw dislocation) of crystalline GaSb (100), grown epitaxially on GaAs (100) at 475°C. The image size is 400 × 400 nm (Brown *et al.* (1996)).

However, the so-called Frank source, a *screw dislocation* (Fig. 3.10) emerging onto a growing crystal face, *does* provide an inexhaustible supply of high-binding-energy ledge sites. Adatoms occupying these ledge sites produce a *spiral* growth pattern, the centre being the dislocation core, with new adsorption sites being created continuously as the growth front rotates. When the growth rate is independent of the edge direction in the plane of the growing surface, the resulting growth pattern is in the form of an Archimedes spiral, with the radius r and rotation angle θ being related via $r = c\theta$, c being a constant. An illustration of the spiral crystal-growth pattern associated with a screw dislocation is given in Fig. 3.29 for the case of GaSb.

3.4.2 Atomic transport

A solid can be distinguished from a liquid in terms of either macroscopic or microscopic criteria. On the macroscopic scale, a material can be said to behave in a *solid-like* manner if it has a shear viscosity greater than say 10^{15} poise (10^{14}N s m^{-2}); this somewhat arbitrary condition is equivalent to the statement that a small shear stress applied to a sample for one day produces no discernible permanent deformation (see Problem 3.10).

On the microscopic (atomic) scale, a material behaves in a liquid-like manner if an atom, on average, diffuses away from the site it occupied at time $t = 0$ at later times t. In terms of the time-dependent density pair-correlation function $\rho(\boldsymbol{r}, t)$, a generalization of the time-independent function related to the RDF (eqn. (2.4)), a material behaves in a *liquid-like* manner if the configuration-and time-averaged auto-correlation function

$$\langle \rho(\boldsymbol{r}, t)\rho(\boldsymbol{r}, 0)\rangle_n = \left\langle \int \int \rho(\boldsymbol{r}, t)\rho(\boldsymbol{r}, 0)\mathrm{d}\boldsymbol{r}\,\mathrm{d}t \right\rangle_n \tag{3.32}$$

tends to zero, where $\langle\,\rangle_n$ denotes a configurational average over all atoms. Otherwise, for non-zero values of this averaged function, atoms generally vibrate about their equilibrium positions and the material is solid-like. If the microscopic flow process in a material can be represented by a single (perhaps average) characteristic relaxation time τ, then a solid may also be distinguished from a liquid in terms of relative time-scales by the Deborah number, D_{f}, (so called after the prophetess who foretold that mountains would flow before the Lord (Judges 5:5)), defined as

$$D_{\mathrm{f}} = \tau/t \tag{3.33}$$

where t is the measurement time. For very small values of $D_{\mathrm{f}}(t \gg \tau)$ a material behaves as a liquid, whereas for very large values of $D_{\mathrm{f}}(t \ll \tau)$ it behaves as a solid.

An interesting situation intermediate between liquids and solids defined in this atomistic way is the case of plastic crystals. These materials consist of nearly spherical molecules (e.g. C_{60}, P_4Se_3, adamantane, etc.) held together by van der Waals interactions (§2.2.2.1). Above the plastic-crystal transition temperature, the thermal energy is sufficiently great partially to overcome the intermolecular bonding and to allow the molecules to rotate freely about one or more axes, but with the molecular centres of mass remaining at the sites of the underlying (usually f.c.c.) crystal lattice. Such materials represent an example of dynamic orientational disorder.

The atomistic criterion based on eqn. (3.32), distinguishing liquid-like and solid-like behaviour, does not imply, however, that *individual* atoms in, say, a crystalline solid do not make diffusive (non-vibrational) or other translational excursions away from their normal positions. Indeed, atoms can and do move in a lattice under the influence of suitable driving forces (e.g. concentration gradients) or external agencies (e.g. electric fields or mechanical stresses). However, almost invariably such atomic motion is mediated by *structural defects*; individual atomic transport generally involves point defects (e.g. vacancies and interstitials) and cooperative atomic motion is mediated by extended defects (e.g. dislocations). Since point defects are generally present in thermal equilibrium (§3.2.1), their concentration increases with increasing temperature and so, at elevated temperatures (say greater than half the melting temperature, $T_m/2$, measured in kelvin) point-defect-mediated atomic diffusion can become appreciable. At low temperatures, when the concentration of point defects is small, atomic diffusion can only take place in association with metastable extended defects, such as along dislocation cores or at grain boundaries, etc.

Perhaps the simplest (defect-free) mechanism of atomic transport is the direct-exchange mechanism, in which two atoms simply exchange places simultaneously in the lattice. If more than two atoms are involved in the exchange, it is known as a ring mechanism. However, in general, this mechanism is seldom found because of the high (strain) energetic cost involved in the exchange.

The processes by which individual atoms may move in a crystalline lattice in association with point defects are represented schematically in Fig. 3.30. The commonest defect-related mechanism is the vacancy mechanism (Fig. 3.30), in which an atom moves into a neighbouring vacant site (or, equivalently, a vacancy moves in the *opposite* direction). There is also an interstitial mechanism (Fig. 3.30), in which an atom at an

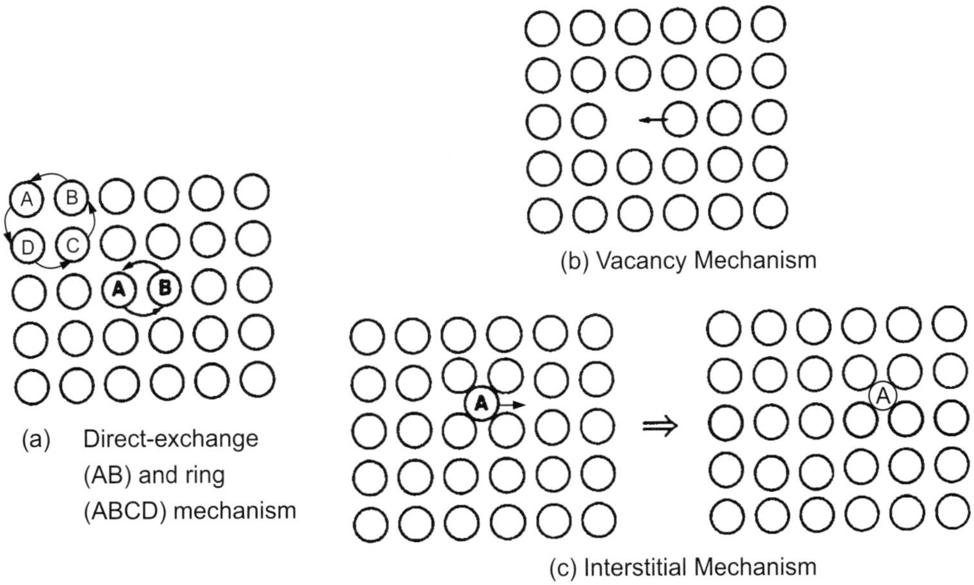

Fig. 3.30 Schematic illustration of three mechanisms of atomic transport in solids: (a) direct-exchange (ring) mechanism; (b) vacancy mechanism; (c) interstitial mechanism.

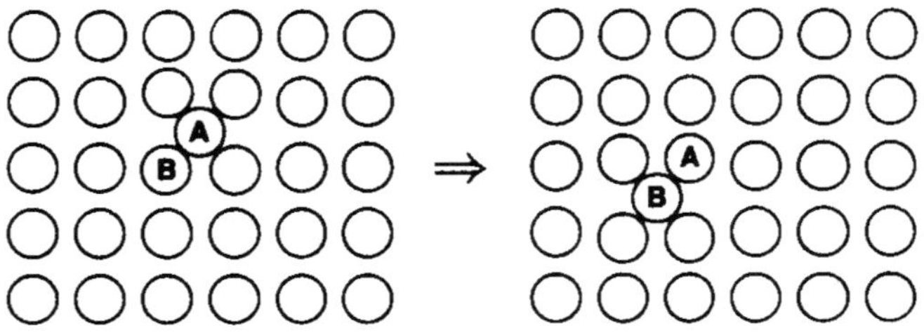

Fig. 3.31 Schematic illustration of the interstitialcy mechanism of atomic transport in solids.

interstitial site moves to a neighbouring interstitial site; the most important example of this is the case of the impurity carbon in α–(b.c.c.) iron. A related mechanism, found in materials containing cation Frenkel defects (e.g. silver halides) or anion Frenkel defects (e.g. some alkaline earth halides), is the interstitialcy mechanism (Fig. 3.31), in which an atom at an interstitial site moves to a lattice site, in turn displacing the atom originally there to an interstitial position. Of these four mechanisms, all but the interstitial mechanism involve substitutional exchange of atoms.

Structural defects, notably dislocations, also greatly facilitate the motion of *planes* of atoms relative to other planes in crystals undergoing plastic deformation when subjected to applied shear stresses. The (ideal) critical yield stress required to cause the *simultaneous* slip of one layer of atoms over another layer is very high (a simple calculation gives $\simeq G_s/6$, where G_s is the shear, or rigidity, modulus—see Problem 3.11). However, the yield stress is reduced by several orders of magnitude with the involvement of line defects, i.e. dislocations, for which atomic motion occurs *consecutively*, a line of atoms at a time. This slip process is illustrated schematically in Fig. 3.32. Note that the dislocation line marks the boundary between the slipped and unslipped material: an *edge* dislocation is *perpendicular* to the direction of slip, whereas a *screw* dislocation is *parallel* to the slip direction (see Fig. 3.32). If a dislocation moves right through the crystal, the entire slip plane in the crystal will have slipped by an amount equal to the Burgers vector $\boldsymbol{b}$, i.e. one lattice spacing. The atomic motion in the region of the dislocation core associated with the motion ('glide') of an edge dislocation along the slip plane is illustrated in Fig. 3.33.

3.4.2.1 *Atomic diffusion*

Two variants of atomic diffusion may be distinguished: self-diffusion refers to the motion of an atom in a solid consisting of atoms of the *same* type; impurity diffusion, on the other hand, refers to the transport of a (dilute) solute atom in a host (solvent) structure consisting of atoms of a *different* type. Often, and particularly for self-diffusion, the diffusive transport of atoms in solids is measured by monitoring the evolution of a known spatial distribution of a radioactive isotope tracer as a function of time (or distance).

A concentration gradient, of impurity atoms, vacancies or interstitials as the case may be, gives rise to atomic diffusion, and this is characterized by the quantity called the diffusion coefficient (or diffusivity), D_i, which is simply the constant of proportionality

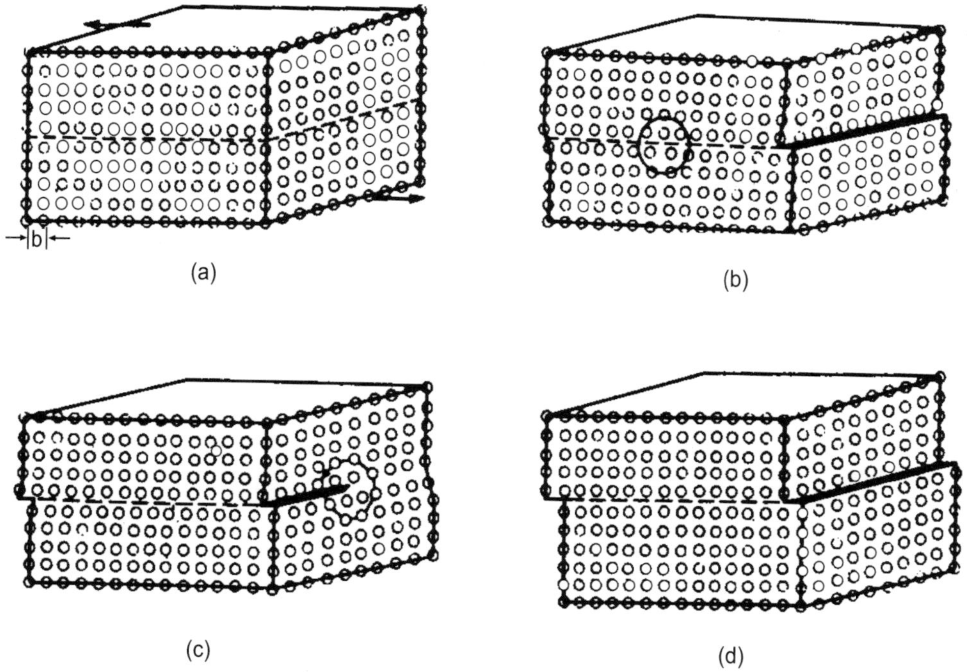

Fig. 3.32 Slip in a crystal mediated by dislocations: (a) the unslipped crystal; (b) mediation by an edge dislocation moving from right to left; (c) mediation by a screw dislocation moving from front to back; (d) the fully slipped crystal.

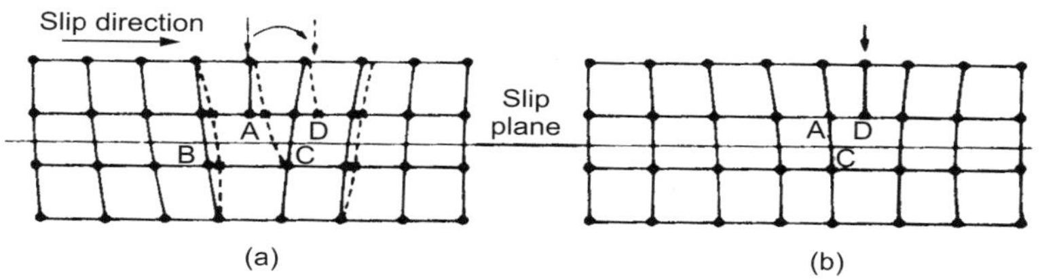

Fig. 3.33 Motion of an edge dislocation in the slip plane caused by the breaking of the line of bonds CD (projected into the paper) and the making of the line of bonds AC.

between the net flux, $\boldsymbol{J}_i$, of atoms of type i (the number crossing a unit area perpendicular to the flux direction in unit time) and the concentration gradient causing the diffusion.

For the case of diffusion in *one dimension* (the x-direction), this relationship can be expressed as Fick's first law:

$$\boldsymbol{J}_i = -D_i \frac{\partial n_i}{\partial x}, \tag{3.34a}$$

generalized for three dimensions (3D) to

$$\boldsymbol{J}_i = -\mathfrak{D}_i \boldsymbol{\nabla} n_i. \tag{3.34b}$$

where the concentrations are conventionally taken to be *negative* so that there is a positive flux of atoms along the x-axis. Note that in eqn. (3.34b) the flux $\boldsymbol{J}_i$ is a vector and the diffusion coefficient $\mathfrak{D}_i$ is a (symmetric) second-rank tensor.

The diffusion process must also satisfy the continuity equation describing the conservation of atoms of type i, namely that any variation of the concentration with time must produce a concomitant spatially varying flux. In one dimension (1D) this can be expressed as

$$\frac{\partial n_i}{\partial t} + \frac{\partial J_i}{\partial x} = 0 \tag{3.35a}$$

and in 3D as

$$\frac{\partial n_i}{\partial t} + \boldsymbol{\nabla} \cdot \boldsymbol{J}_i = 0. \tag{3.35b}$$

Combination of eqns. (3.34) and (3.35) yields the diffusion equation, or Fick's second law, i.e. in 1D

$$\frac{\partial n_i}{\partial t} = \frac{\partial}{\partial x}\left(D_i \frac{\partial n_i}{\partial x}\right) \tag{3.36a}$$

and in 3D

$$\frac{\partial n_i}{\partial t} = \boldsymbol{\nabla} \cdot (\mathfrak{D}_i \boldsymbol{\nabla} n_i). \tag{3.36b}$$

Note that the possibility of a spatial variation of the diffusion coefficient itself has not been excluded in these equations. If the material is sufficiently homogeneous that the diffusion coefficient is everywhere constant, then eqns. (3.36) reduce to

$$\frac{\partial n_i}{\partial t} = D_i \frac{\partial^2 n_i}{\partial x^2} \quad (1\text{D}) \tag{3.37a}$$

and

$$\frac{\partial n_i}{\partial t} = \mathfrak{D}_i \boldsymbol{\nabla}^2 n_i \quad (3\text{D}). \tag{3.37b}$$

In general, coupled flows may also exist, where a concentration gradient in one atomic species j can cause a flux in *another*, i. Such a situation may be represented by a generalization of, say, eqn. (3.34b) as

$$\boldsymbol{J}_i = \sum_{j=1}^{n} \mathfrak{D}_{ij} \boldsymbol{\nabla} n_j. \tag{3.38}$$

The form of the diffusion equation (eqns. (3.36), (3.37)) describes not only atomic transport but also the flow of heat in a temperature gradient. However, a temperature gradient can also give rise to an *atomic* flux, in the *absence* of a concentration gradient; this is known as the Soret effect. By analogy with eqn. (3.34b), the flux may be written as

$$\boldsymbol{J}_i = -\$\boldsymbol{\nabla} T \tag{3.39}$$

where the Soret constant \$ is, like the diffusion constant, a symmetric second-rank tensor but may have either sign (whereas the diffusion coefficient is always positive)—see Problem 3.12.

The second-order differential equation that is the diffusion equation (say the 1D form, eqn. (3.37a)) can be solved to give the spatial and temporal variation of the atomic concentration, $n_i(x,t)$, given suitable boundary conditions, e.g. the starting profile, $n_i(x,0)$ (see Problem 3.13). Thus, for the case of N_i atoms initially located at the point $x = x_0$ at $t = 0$, i.e. in the form of a Dirac delta function, the diffusion profile at a later time, t, is given by a Gaussian distribution (see Problem 3.13(b)), i.e.

$$n_i(x,t) = \frac{N_i}{(4\pi D_i t)^{1/2}} \exp\{-(x-x_0)^2/4D_i t\}. \tag{3.40}$$

For the case of a semi-infinite *constant* (step-function) initial distribution, $n_i(x,0) = n_i^0(-\infty < x < 0)$, the diffusion profile is given instead (see Problem 3.13(c)) by

$$n_i(x,t) = \frac{n_i^0}{2} \operatorname{erfc}(x/(4D_i t)^{1/2}), \tag{3.41}$$

where the complementary error function, erfc(y), is defined as:

$$\operatorname{erfc}(y) = (2/\pi)^{1/2} \int_y^{\infty} \mathrm{e}^{-w^2} \mathrm{d}w. \tag{3.42}$$

The two diffusion profiles embodied in eqns. (3.40) and (3.41) are illustrated in Fig. 3.34. Other solutions of the diffusion equation are detailed in Crank (1975).

★ Diffusion is, by its very nature, an *irreversible* process: it is concerned with the transition of a system from a state of non-equilibrium to one of equilibrium (i.e. in general, to a state of compositional and thermal homogeneity). Thus, diffusion processes can also be discussed in the framework of the theory of non-equilibrium thermodynamics (or thermodynamics of irreversible processes). From this viewpoint, diffusion in general can be described in terms of a set of linear, phenomenological equations relating material fluxes $\boldsymbol{J}_i$ (of atoms, ions, electrons, etc.), and heat flow $\boldsymbol{J}_Q$ if appropriate, and the appropriate thermodynamic 'driving forces' $\boldsymbol{X}_j$, i.e.

$$\boldsymbol{J}_i = \sum_j L_{ij} \boldsymbol{X}_j \tag{3.43}$$

where the Onsager coefficients L_{ij} are each second-rank tensors, independent of the driving forces.

The Onsager theorem states that in the absence of magnetic fields, and as a result of time-reversal symmetry, the off-diagonal elements of the L_{ij}-coefficients are symmetric, i.e.

$$L_{ij}^{\alpha\beta} = L_{ji}^{\beta\alpha} \tag{3.44}$$

for all chemical-species indices i, j and Cartesian components α, β. Further, the diagonal L-coefficients must be semi-positive definite,

$$L_{ii} \geqslant 0, \tag{3.45a}$$

with the off-diagonal coefficients obeying the inequality

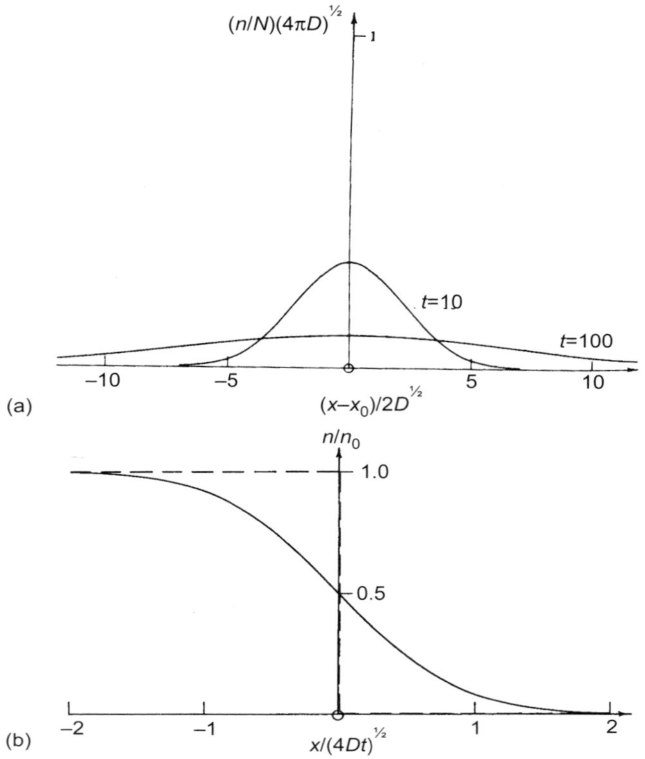

Fig. 3.34 Illustration of the diffusion profile, obtained by solving the 1D diffusion equation, resulting from an initial concentration distribution which is: (a) a delta function at $x = x_0$; (b) a semi-infinite constant concentration ($x \leqslant 0$).

$$L_{ii}L_{jj} - L_{ij}L_{ji} \geqslant 0 \tag{3.45b}$$

as a result of the second law of thermodynamics, which follows by substituting eqn. (3.43) into the expression for the rate of entropy production per unit volume, $\bar{s}$, given by (Allnatt and Lidiard (1993))

$$T\bar{s} = \sum_i \boldsymbol{J}_i \cdot \boldsymbol{X}_i + \boldsymbol{J}_Q \cdot \boldsymbol{X}_Q. \tag{3.46}$$

The driving forces $\boldsymbol{X}_j$ may consist of mechanical forces $\boldsymbol{F}_j$, such as electrostatic forces exerted on ions of charge q_j by an electric field $\boldsymbol{E}, \boldsymbol{F}_j = q_j\boldsymbol{E}$, and/or driving 'forces' due to, say, compositional variations, related to the gradient of the chemical potential, μ_j, of component j; thus, in general:

$$\boldsymbol{X}_j = -\boldsymbol{\nabla}\mu_j + \boldsymbol{F}_j. \tag{3.47}$$

In the presence of a temperature gradient, the corresponding thermal driving force responsible for the heat flux $\boldsymbol{J}_Q$ is the thermal 'force' $\boldsymbol{X}_Q = -(\boldsymbol{\nabla} T)/T$. For a system in mechanical equilibrium, there is a restriction on the driving forces as expressed by Prigogine's theorem (a generalized form of the Gibbs–Duhem relation):

$$\sum_i N_i \boldsymbol{X}_i = 0. \tag{3.48}$$

★ Consider now a diffusion couple, composed of, say, two dissimilar metals A and B (Fig. 3.35), for which there will be two Fickian first-law equations for the atomic fluxes (eqn. (3.34)), characterized by the two intrinsic diffusion coefficients, D_A and D_B. Obviously overall there is only a *single* diffusional process, namely the interdiffusion of A and B, characterized by the chemical interdiffusion coefficient $\tilde{D}_{AB}$. However, if $D_A \neq D_B$ (e.g. as a result of vacancy involvement), the diffusion zone, the region where the diffusive fluxes are the greatest, i.e. the effective interface between the two metals, itself moves with respect to the region of the couple where no diffusion is occurring (Fig. 3.35). This behaviour is known as the Kirkendall effect. Both interdiffusion and the Kirkendall effect are determined by D_A and D_B. The 3D atomic fluxes relative to the *fixed* parts of the lattice are given by

$$\boldsymbol{J}'_A = \boldsymbol{J}_A - c_A(\boldsymbol{J}_A + \boldsymbol{J}_B) \tag{3.49a}$$

and

$$\boldsymbol{J}'_B = \boldsymbol{J}_B - c_B(\boldsymbol{J}_A + \boldsymbol{J}_B) \tag{3.49b}$$

since the velocity of the diffusion zone relative to the fixed lattice is $-V(\boldsymbol{J}_A + \boldsymbol{J}_B)$, where V is the volume per site and c_A and c_B are site mole fractions, $c_A = n_A/n$ and $c_B = n_B/n$

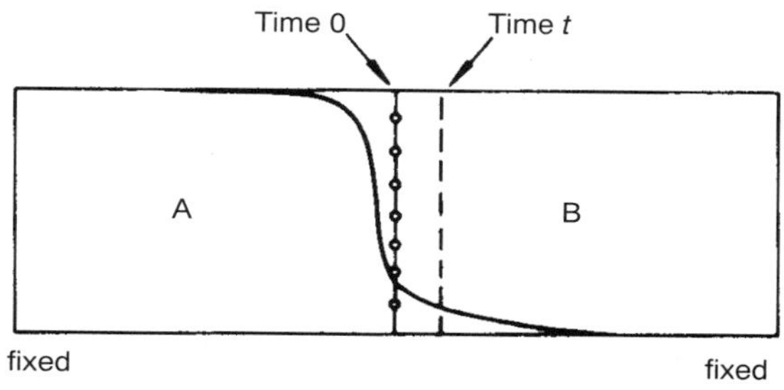

Fig. 3.35 Schematic diagram of a diffusion couple comprising two dissimilar metals A and B. The circles at the original interface denote inert markers used to monitor the movement of the diffusion zone relative to the fixed ends of the couple, i.e. the Kirkendall effect. The superimposed graph represents the concentration of A atoms in the couple at a later time, t. The diffusion zone moves to the right because it has been assumed that $D_B > D_A$.

with $c_{\rm A} + c_{\rm B} = 1$ (if the concentration of vacancies is negligible). Thus, from eqns. (3.34b) and (3.49a)

$$\begin{aligned} \boldsymbol{J}'_{\rm A} &= -(c_{\rm B}\mathfrak{D}_{\rm A} + c_{\rm A}\mathfrak{D}_{\rm B})\boldsymbol{\nabla} n_{\rm A} \\ &\equiv -\tilde{\mathfrak{D}}\,\boldsymbol{\nabla} n_{\rm A} \end{aligned} \tag{3.50}$$

with a similar equation for B. Thus the chemical interdiffusion coefficient is given by the Darken equation:

$$\tilde{\mathfrak{D}}_{\rm AB} = c_{\rm A}\mathfrak{D}_{\rm B} + c_{\rm B}\mathfrak{D}_{\rm A}. \tag{3.51}$$

The Kirkendall velocity of the diffusion zone relative to the fixed parts of the sample is given by

$$\boldsymbol{v}_{\rm K} = (\mathfrak{D}_{\rm A} - \mathfrak{D}_{\rm B})\boldsymbol{\nabla} c_{\rm A}. \tag{3.52}$$

This situation can also be analysed in terms of the Onsager L-coefficients (cf. eqn. (3.43)), viz.

$$\boldsymbol{J}_{\rm A} = \boldsymbol{L}_{\rm AA}\boldsymbol{X}_{A} + \boldsymbol{L}_{\rm AB}\boldsymbol{X}_{\rm B}, \tag{3.53a}$$

$$\boldsymbol{J}_{\rm B} = \boldsymbol{L}_{\rm BA}\boldsymbol{X}_{\rm A} + \boldsymbol{L}_{\rm BB}\boldsymbol{X}_{\rm B} \tag{3.53b}$$

with the Prigogine relation (eqn. (3.48)) becoming in this case

$$c_{\rm A}\boldsymbol{X}_{\rm A} + c_{\rm B}\boldsymbol{X}_{\rm B} = 0. \tag{3.54}$$

Combining eqns. (3.53) and (3.54) gives for the flux of A (and similarly for B):

$$\boldsymbol{J}_{\rm A} = \left(\boldsymbol{L}_{\rm AA} - \frac{c_{\rm A}}{c_{\rm B}}\boldsymbol{L}_{\rm AB}\right)\boldsymbol{X}_{\rm A}. \tag{3.55}$$

The chemical potential of component A can be written as

$$\mu_{\rm A} = \mu^0_{\rm A}(T,p) + k_{\rm B}T\ln(\gamma_{\rm A}c_{\rm A}), \tag{3.56}$$

where $\gamma_{\rm A}$ is the activity coefficient (a measure of non-ideality in terms of interactions between species), and the quantity $a_{\rm A} = \gamma_{\rm A}c_{\rm A}$ is the activity of A. Thus, use of eqn. (3.47), with the assumption that only concentration gradients are important, gives

$$\boldsymbol{J}_{\rm A} = -\left(\frac{L_{\rm AA}}{c_{\rm A}} - \frac{L_{\rm AB}}{c_{\rm B}}\right)k_{\rm B}T\left(1 + \frac{\partial\ln\gamma_{\rm A}}{\partial\ln c_{\rm A}}\right)\nabla c_{\rm A}. \tag{3.57}$$

For the case of a dilute alloy ($c_{\rm A} \ll c_{\rm B}$), the activity coefficient $\gamma_{\rm A}$ tends to unity as $c_{\rm A} \to 0$, and assuming further that the ratio $L_{\rm AB}/L_{\rm AA}$ becomes independent of $c_{\rm A}$ in this limit, a simple relation between the diffusion coefficient and the Onsager coefficient is found, viz.

$$\mathfrak{D}_{\rm A} = \frac{k_{\rm B}TL_{\rm AA}}{nc_{\rm A}} = \frac{k_{\rm B}TL_{\rm AA}}{n_{\rm A}}. \tag{3.58}$$

Instead of the macroscopic approach outlined above, atomic diffusion can also be treated from a microscopic point of view, in terms of a random-walk process of the diffusing atoms (like the Brownian motion of a suspended particle in a fluid). Consider a dilute concentration of a species (A) diffusing in 1D in a host matrix (B) as a result of a concentration gradient $\partial n_A/\partial x$. If an A atom jumps a distance a (of the order of the lattice spacing) between two sites at every diffusion event, the concentrations of A at two adjacent planes, 1 and 2, separated by a distance a, are related by

$$n_{A,2} = n_{A,1} + a\frac{\partial n_A}{\partial x}. \tag{3.59}$$

If the average number of jumps per unit time made by a diffusing A atom is Γ_A, half of which (in 1D) are forwards and half are backwards (in the absence of external forces), then the number of atoms per unit area moving from plane 1 to plane 2 is $\Gamma_A a n_{A,1}/2$, and that from 2 to 1 is $\Gamma_A a n_{A,2}/2$, and hence the overall flux is given by

$$J_A = \frac{\Gamma_A}{2} a\,(n_{A,1} - n_{A,2}) = -\frac{\Gamma_A}{2} a^2 \frac{\partial n_A}{\partial x}. \tag{3.60}$$

Thus, by comparison with eqn. (3.34a), the diffusion coefficient is related to microscopic quantities characterizing individual atomic jumps via the Einstein relation, given (in 1D) by

$$D_A^0 = \frac{\Gamma_A a^2}{2}. \tag{3.61}$$

For *three-dimensional diffusion*, if each of the principal axes along which diffusion takes place is *equivalent* (as in cubic crystals), the principal components of the diffusion-coefficient matrix are

$$\mathfrak{D}^0_{A,ii} = \frac{\Gamma_A a^2}{6}, \qquad i = x, y \text{ or } z. \tag{3.62}$$

The diffusion coefficients appearing in eqns. (3.61) and (3.62) have been given the superscript zero to denote that they refer to the case when there is *no* correlation between consecutive atomic jumps.

The quantity Γ_A, the average number of jumps of A atoms per second, is related to the atomic jump frequency γ_A by

$$\Gamma_A = p\,\gamma_A \tag{3.63}$$

where p is a probability factor that depends on the jump mechanism and the local structure. For interstitial atoms, $p = z$, the number of distinct sites that such atoms can reach in a single jump; for defect-related diffusion (e.g. involving vacancies), p is the probability that there is a neighbouring defect available to mediate the atomic jump.

For the case of pure interstitial diffusion, every jump to a neighbouring site is uncorrelated with (i.e. statistically independent of) the preceding jump. This is *not* true for vacancy-mediated and interstitialcy mechanisms, however. Following a jump of an atom into a neighbouring vacancy (Fig. 3.30), for example, the subsequent jump of the atom is likely simply to be back to its *original* position. Thus, in this case, subsequent atomic jumps are strongly correlated with each other. (Note, however, that the motions of the *vacancies* themselves are *uncorrelated*.) The effect of such correlation effects is to cause the observed diffusion coefficient to be *less* than the microscopic quantity given by the Einstein relation (eqn. (3.61)), i.e.

$$D_A = f_c D_A^0 \tag{3.64}$$

where the correlation factor $f_c \leqslant 1$.

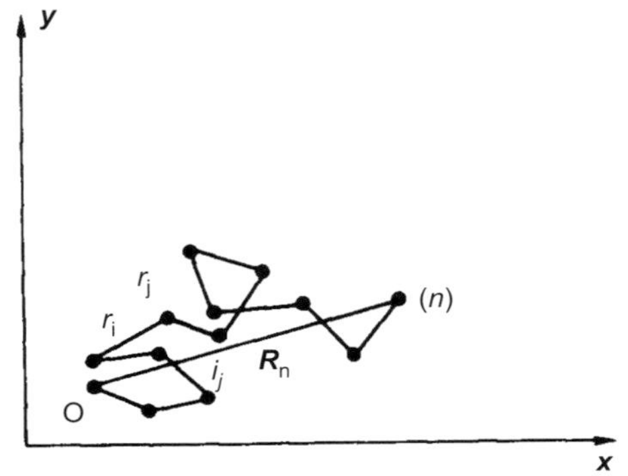

Fig. 3.36 Two-dimensional illustration of the path, after n jumps, of a particle performing a random walk, with equal-length jumps $\boldsymbol{r}$, starting at the origin O.

The correlation factor can be evaluated by considering a random walk of atomic jumps with equal length r (see Fig. 3.36). After n jumps, the vector $\boldsymbol{R}_n$ connecting the initial and final sites is given by

$$\boldsymbol{R}_n = \sum_{i=1}^{n} \boldsymbol{r}_i. \tag{3.65}$$

For a true random walk, the *average* of $\boldsymbol{R}_n$, i.e. $\bar{\boldsymbol{R}}_n$, equals zero as $n \to \infty$, since there will be as many jumps in one direction as in the reverse direction. However, the mean-square displacement is *not* zero. Thus, taking the dot product of both sides of eqn. (3.65) gives

$$\begin{aligned} R_n^2 &= \sum_{i=1}^{n} \boldsymbol{r}_i \cdot \boldsymbol{r}_i + 2\sum_{i=1}^{n-1} \boldsymbol{r}_i \cdot \boldsymbol{r}_{i+1} + 2\sum_{i=1}^{n-2} \boldsymbol{r}_i \cdot \boldsymbol{r}_{i+2} + 2\boldsymbol{r}_{n-1} \cdot \boldsymbol{r}_n \\ &= \sum_{i=1}^{n} r_i^2 + 2\sum_{j=1}^{n-1}\sum_{i=1}^{n-j} \boldsymbol{r}_i \cdot \boldsymbol{r}_{i+j} \\ &= nr^2\left(1 + \frac{2}{n}\sum_{j=1}^{n-1}\sum_{i=1}^{n-j} \cos\theta_{i,i+j}\right), \end{aligned} \tag{3.66}$$

where $\theta_{i,i+j}$ is the angle between the directions of the ith and the $(i+j)$th jump. The mean-square displacement is thus given by

$$\overline{R_n^2} = nr^2\left(1 + \frac{2}{n}\sum_{j=1}^{n-1}\sum_{i=1}^{n-j} \cos\theta_{i,i+j}\right). \tag{3.67}$$

For a true, uncorrelated random-walk process, such as involving the motion of interstitials, the double-summation term in eqn. (3.67) is zero as $n \to \infty$ (there are as many jumps with a positive value of $\cos\theta_{i,i+j}$ as with equal and opposite values), and $\overline{R_n^2} = nr^2$. However, the effect of correlations between subsequent atomic jumps is to give a finite, *negative* value for the double-summation term.

After n jumps or, equivalently, a time t given by

$$t = n/\Gamma_A, \tag{3.68}$$

the diffusion coefficient in 3D can be expressed in terms of *macroscopic* mean-square displacements by analogy with eqns. (3.61) and (3.62), i.e.

$$\mathfrak{D}_{A,ii} = \frac{\overline{R_n^2}}{6t}, \quad i = x, y \text{ or } z \tag{3.69a}$$

or, in 1D,

$$D_A = \overline{x_n^2}/2t. \tag{3.69b}$$

Substituting eqn. (3.67) for $\overline{R_n^2}$ in eqn. (3.69a) gives an expression for the effective diffusion coefficient in the form of eqn. (3.64), with $\mathfrak{D}^0_{A,ii} = nr^2/6t$ and the correlation factor given by

$$f_c = \lim_{n \to \infty} \left(1 + \frac{2}{n} \sum_{j=1}^{n-1} \sum_{i=1}^{n-j} \overline{\cos\theta_{i,i+j}} \right). \tag{3.70}$$

This expression can be simplified (see Problem 3.14) to become

$$f_c = \frac{1 + \overline{\cos\theta_1}}{1 - \overline{\cos\theta_1}}, \tag{3.71}$$

where θ_1 is the angle between the directions of consecutive jumps. Thus, f_c is simply a geometric factor, determined both by the microscopic mechanism of diffusion and the structure of the material within which diffusion takes place. Some representative values of f_c are given in Table 3.2.

★ Finally, we discuss the temperature dependence of the diffusion coefficient. In general, this temperature dependence arises from two distinct thermodynamic factors, a defect-creation term and an atomic-mobility term: it costs thermal energy to create the defects which mediate diffusion, and it also costs thermal energy to transfer an atom from one site to another.

Table 3.2 Calculated tracer correlation factors, f_c, for self-diffusion in the limit of infinitely low defect concentrations for different host structures and diffusion mechanisms

Lattice	*Mechanism*	f_c
Honeycomb (2D)	Vacancy	1/3
Square (2D)	Vacancy	$1/(\pi - 1) = 0.47$
Triangular (2D)	Vacancy	$(\pi + 6\sqrt{3})/(11\pi - 6\sqrt{3}) = 0.56$
Diamond	Vacancy	1/2
s.c.	Vacancy	0.65
b.c.c.	Vacancy	0.73
f.c.c.	Vacancy	0.78
NaCl	Collinear interstitialcy	2/3

(After Allnatt and Lidiard (1993). Reproduced by permission of Cambridge University Press)

The thermal creation of defects is determined (under isobaric conditions) by the Gibbs free energy of formation of a single defect, related to the formation enthalpy and entropy by

$$\Delta g_{\mathrm{f}} = \Delta h_{\mathrm{f}} - T\Delta s_{\mathrm{f}}, \tag{3.72}$$

where the lower case has been used for the extensive thermodynamic quantities to denote that they refer to single defects. The fraction of, say, vacancies, x_{v}, existing in thermal equilibrium is thus given by

$$x_{\mathrm{v}} = \exp(-\Delta g_{\mathrm{v}}/k_{\mathrm{B}}T), \tag{3.73}$$

where $g_{\mathrm{v}} \equiv g_{\mathrm{f}}$ (vacancy). Formation enthalpies and entropies for various defects are given in Table 3.1.

The energetics involved in the motion of an atom between two sites can be understood by reference to Fig. 3.37. There is an energy barrier between initial (Fig. 3.37a) and final (Fig. 3.37c) configurations associated with the transition state or activated configuration (Fig. 3.37b) where the diffusing atom pushes aside the intervening atoms. The free-energy difference (associated with strain energy) between the activated and the initial configurations, $\Delta g^{\ddagger}$, is thus equal to the activation free energy for atomic mobility, $\Delta g_{\mathrm{m}} = \Delta g^{\ddagger}$, since local thermal fluctuations in energy giving an atom sufficient energy to surmount the barrier occur with a frequency proportional to the Boltzmann factor $\exp(-\Delta g_{\mathrm{m}}/k_{\mathrm{B}}T)$. The mobility free energy can be decomposed into enthalpy and entropy terms as for the formation free energy (eqn. (3.72)), i.e.

$$\Delta g_{\mathrm{m}} = \Delta h_{\mathrm{m}} - T\Delta s_{\mathrm{m}}. \tag{3.74}$$

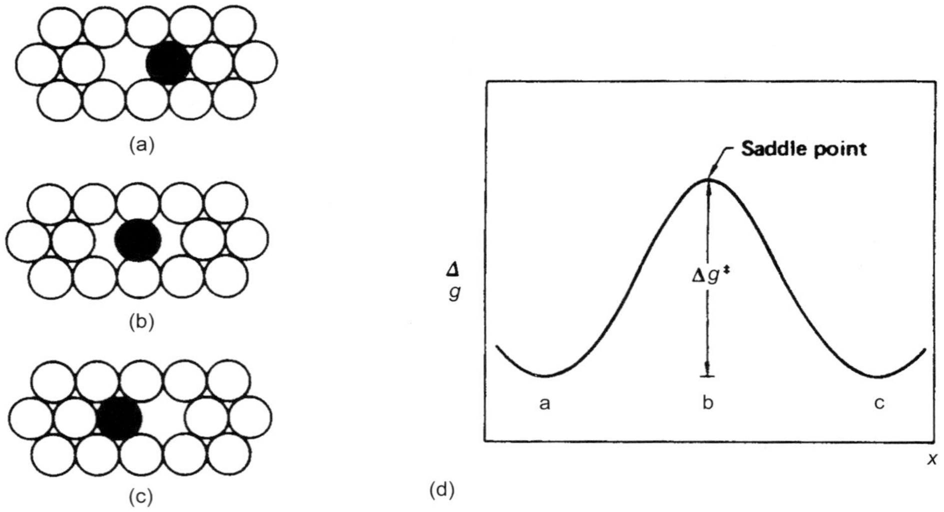

Fig. 3.37 Schematic illustration of the vacancy-mediated motion of an atom, shown shaded, from one site (a) to another site (c) via the activated transition state (b). The free-energy profile corresponding to configurations (a)–(c) is shown in (d). For self-diffusion in a simple crystal, the free energies of the initial (a) and final (c) states are identical, and the activation free energy $\Delta g^{\ddagger}$ for the transport process is equal to the difference in free energy between the saddle-point configuration (b) and the initial state (a).

For vacancy-mediated atomic motion, $\Delta h_m \leqslant \Delta h_f$, but for interstitial motion $\Delta h_m \ll \Delta h_f$ (e.g. $\Delta h_m \sim 0.1$ eV for Cu).

Thermally activated atomic transport thus proceeds via a series of discrete jumps, or hops, over activation barriers separating equilibrium lattice sites, the diffusing atom spending an appreciable time (typically many vibrational periods) at each lattice site before acquiring sufficient energy via a thermal fluctuation to enable it to hop to an adjacent site.

Thus, in the expression for the mean jump rate Γ_A of an atom A (eqn. (3.63)), the probability of finding a defect (e.g. a vacancy) adjacent to an atom is $p = \exp(-\Delta g_f/k_B T)$ (cf. eqn. (3.73)), and the jump frequency γ_A is given by the product of the atomic vibrational frequency υ_A and the probability of executing a jump, i.e.

$$\gamma_A = \upsilon_A \exp(-\Delta g_m/k_B T). \tag{3.75}$$

Combining these relations with the Einstein relation (eqn. (3.62)), taking into account the correlation factor (eqn. (3.64)), yields an expression for the diffusion coefficient in the form of the Arrhenius relation

$$\mathfrak{D}_A = \mathfrak{D}_A^0 \exp(-\mathscr{E}_D/k_B T) \tag{3.76}$$

where the activation energy $\mathscr{E}_D$ is given by a sum of defect-formation and mobility terms

$$\mathscr{E}_D = \Delta h_f + \Delta h_m \tag{3.77}$$

and the prefactor of the exponential is

$$\mathfrak{D}_A^0 = \frac{f_c a^2 \upsilon_A}{6} \exp[(\Delta s_f + \Delta s_m)/k_B]. \tag{3.78}$$

The Arrhenius temperature dependence embodied in eqn. (3.76) is illustrated in Fig. 3.38 for the case of the interstitial diffusion of carbon in b.c.c. iron. Some values of formation and migration enthalpies, and the corresponding diffusion activation energies, for vacancy-mediated diffusion in some f.c.c. metals are given in Table 3.3. Representative values of pre-exponential factors and activation energies for diffusion of various atoms in a selection of crystalline materials are given in Table 3.4. (See also Problem 3.15.)

Further details on atomic transport in solids are to be found in Borg and Dienes (1988) and Allnatt and Lidiard (1993).

Table 3.3 Values of the vacancy formation enthalpy (Δh_v) and atomic migration enthalpy (Δh_m), and their sum, together with the activation energy for diffusion, in some f.c.c. metals

	Cu	Ag	Au	Pt	Al
Δh_v(eV)	1.18	1.09	0.94	1.3	0.75
Δh_m(eV)	0.88	0.83	0.78	1.21	0.56
$\Delta h_v + \Delta h_m$(eV)	2.06	1.92	1.72	2.51	1.31
$\mathscr{E}_D$(eV)	2.04	1.91	1.82	2.51	1.39

(Data from Myers (1990), Allnatt and Lidiard (1993) (Reprinted by permission of Cambridge University Press), and Kittel (1996))

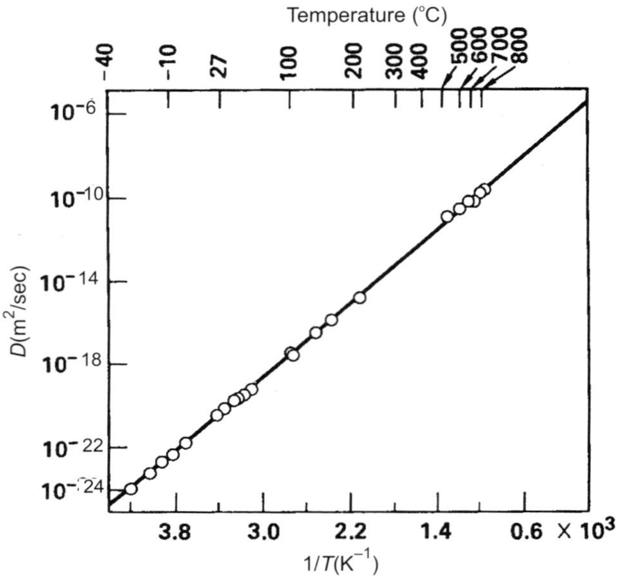

Fig. 3.38 Arrhenius plot of the diffusion coefficient for carbon diffusing interstitially in b.c.c. iron.

Table 3.4 Pre-exponential factors and activation energies of diffusion for various diffusing species in a range of crystals

Host crystal	*Diffusing atom*	$D^0(m^2s^{-1})$	$\mathscr{E}_D(eV)$
Cu	Cu	2×10^{-5}	2.04
Ag	Ag	4×10^{-5}	1.91
Ag	Cu	1.2×10^{-4}	2.00
Na	Na	2.4×10^{-5}	0.45
Si	Al	8.0×10^{-4}	3.47
Si	Ga	3.6×10^{-4}	3.51
Si	As	3.2×10^{-5}	3.56
Si	Li	2×10^{-7}	0.66
Ge	Ge	1.0×10^{-3}	3.1
Fe	C	2.0×10^{-6}	0.87

(After Kittel (1996). Reprinted by permission of John Wiley & Sons Inc.)

3.4.2.2 Ionic conductivity

Mobile atoms subject to an external driving force experience an overall drift in their average position with time. This is the case, for example, for charged mobile ions in a solid subject to an external electric field $\boldsymbol{E}$. Although the ionic conductivity of a material might result from the *same* microscopic atomic-transport mechanism (e.g. a vacancy-mediated or interstitialcy mechanism) that controls normal atomic diffusion, the conductivity is characterized by the mean displacement $\langle x \rangle$ being non-zero, while diffusion is characterized by a non-zero value of the mean-square displacement $\langle x^2 \rangle$.

In general, the overall atomic flux in a solid after a time t can be written, in terms of the atomic concentration, as the Einstein–Smoluchowski equation (see Problem 3.16) for motion in the x-direction

$$J_x = \frac{n\langle x\rangle}{t} - \frac{\langle x^2\rangle}{2t}\frac{\partial n}{\partial x} - n\frac{\partial}{\partial x}\left(\frac{\langle x^2\rangle}{2t}\right), \tag{3.79}$$

where the first term represents the drift of atoms subject to an external force, the second term is the normal diffusion expression (cf. eqns. (3.34a) and (3.69b)) and the third term represents the case when the diffusion coefficient is spatially varying.

The ionic mobility, μ, is then defined as the mean drift velocity per unit electric field, viz.

$$\mu = \frac{\langle x\rangle}{Et} \equiv \frac{J_x}{nE}. \tag{3.80}$$

The electrical conductivity is defined as the constant of proportionality between the flux $\boldsymbol{j}$ of charge, carried in this case by ions, and the electric field

$$\boldsymbol{j} = \sigma \boldsymbol{E}. \tag{3.81}$$

It is evident that, in general, the conductivity will be a *tensor*, although for simplicity we consider it here to be a scalar quantity. Hence, combining eqns. (3.80) and (3.81), with $\boldsymbol{j} = q\boldsymbol{J}$, q being the ionic charge, yields an expression relating conductivity, mobility and ion concentration, viz.

$$\sigma = nq\mu. \tag{3.82}$$

One might ask why the application of a steady electric field to a system containing mobile ions should not result in an *infinite* ionic mobility or conductivity, since naïvely it might be thought that the field would cause the ions to accelerate without limit. However, the ionic motion is not unimpeded as this picture would imply. We have already seen that atoms spend a considerable time 'trapped' at normal lattice sites before jumping to another site in diffusive motion. It is only during the flight between sites that the electric field accelerates the ions, but this acquired velocity is reduced to zero every time that an ion comes to rest at a site. Thus, a finite drift velocity, or equivalently mobility, is established.

A relation, the Einstein equation, between the ionic mobility and the diffusion coefficient can also be demonstrated. Consider a solid containing a uniform distribution of mobile cations A subject to an external electric field $\boldsymbol{E}$. From eqn. (3.43) for the ionic flux written in terms of the Onsager L-coefficients and eqn. (3.80), the ionic mobility can be written as

$$\mu_{\mathrm{A}} = \frac{|q_{\mathrm{A}}|L_{\mathrm{AA}}}{n_{\mathrm{A}}}. \tag{3.83}$$

Taking the infinite-dilution-limit expression for the diffusion coefficient also expressed in terms of the L-coefficients (eqn. (3.58)), one thereby obtains the Einstein relation

$$\frac{\mu_{\mathrm{A}}}{D_{\mathrm{A}}} = \frac{|q_A|}{k_{\mathrm{B}}T}. \tag{3.84a}$$

or in terms of the ionic conductivity, using eqn. (3.82), the Nernst–Einstein relation:

$$\frac{\sigma_A}{D_A} = \frac{n_A q_A^2}{k_B T}. \tag{3.84b}$$

(An alternative derivation of the Einstein relation is given in Problem 3.17.) Thus, by measuring the ionic conductivity, a much simpler procedure than measuring the ionic mobility, the corresponding diffusion coefficient can be evaluated using eqn. (3.84b).

The temperature dependence of the ionic conductivity can be obtained by substituting the temperature dependence of the diffusion coefficient (eqns. (3.76)–(3.78)) into the Nernst–Einstein relation (eqn. (3.84b)), giving

$$\sigma_A T = \frac{f_c n_A q_A^2 a^2 \upsilon_A}{6k_B} \exp[(\Delta s_f + \Delta s_m)/k_B] \exp[-(\Delta h_f + \Delta h_m)/k_B T]. \tag{3.85}$$

Thus, a plot of ln (σT) versus inverse temperature should yield a straight line, the slope of which gives the conductivity activation energy $\mathscr{E}_\sigma = (\Delta h_f + \Delta h_m)$. The temperature dependence of the ionic conductivity of a number of materials, both cationic and anionic conductors, is shown in Fig. 3.39.

It can be seen that, in general, the magnitude of the conductivity is rather low and its activation energy is rather large. However, the ionic conductivity of certain materials, e.g. α-AgI, $RbAg_4I_5$ and Na-β-alumina, is very high and is comparable to that of liquid electrolytes, such as concentrated sulphuric acid. The conductivity activation energies are correspondingly rather small. Such materials are termed superionic, or fast-ion, conductors, or solid electrolytes (since they exhibit very high ionic conductivities but are electronically insulating). Superionic materials can be distinguished from more poorly conducting ionic solids by having a high concentration of mobile ions, e.g. as a result of a defective structure, so that the defect-formation contribution Δh_f to the activation energy (eqn. (3.77)) is not significant. Moreover, superionic conductors are characterized by having 'easy' pathways for ionic transport, so that the ionic-mobility contribution to the activation energy (eqn. (3.77)) is also small.

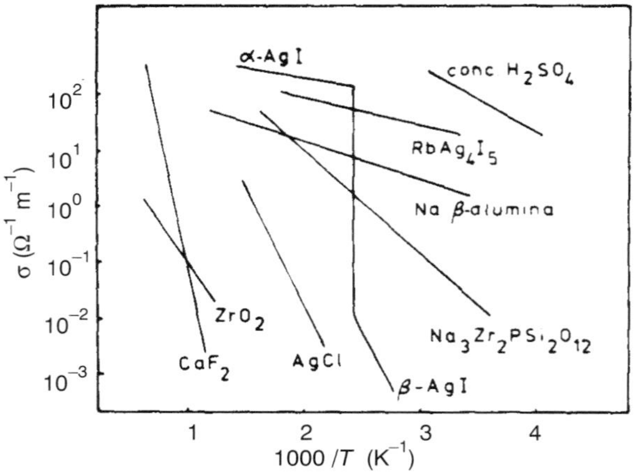

Fig. 3.39 Arrhenius plot of the ionic conductivity of some ionically conducting materials, compared with the liquid electrolyte H_2SO_4 (West (1987). Reproduced by permission of John Wiley & Sons Inc.)

Na-β-alumina is one example of a solid electrolyte: its atomic structure consists of cubic close-packed layers of oxygen ions stacked in 3D (blocks as in spinel, $MgAl_2O_4$), but every fifth layer has 75% of the oxygens missing, and the Na^+ ions sit in these oxygen-deficient layers. Ionic conductivity is very easy *within* such planes (but not in the perpendicular direction, through the close-packed spinel blocks), with the result that Na-β-alumina is a very good *two-dimensional* ionic conductor.

AgI is an interesting example of an ionically conducting material. At temperatures below 146°C, the β-phase (an h.c.p. wurtzite-like structure of I^-ions, with Ag^+ ions in the tetrahedral interstices) is stable, and it is a relatively poor ionic conductor. However, a phase transition to the b.c.c. α-phase occurs at 146°C, and this structure is characterized by I^-ions occupying the corner and body-centre positions, with the two Ag^+ ions per unit cell being distributed statistically over twelve tetrahedrally coordinated sites. The Ag^+ ions can move (in an almost 'liquid-like' way) between tetrahedral sites via 24 intermediate trigonal sites, the intermediate configurations being stabilized (i.e. the mobility enthalpy Δh_m being lowered) by an effective covalent interaction between the polarizing Ag^+ cations and the strongly polarizable I^-anions. As a result, the ionic conductivity of the α-phase of AgI is some four orders of magnitude higher than that of the β-phase. The abrupt decrease in the ionic conductivity at the $\alpha - \beta$ phase transition makes AgI unsuitable for applications. However, replacement of 20% of the Ag^+ ions by Rb^+ in $RbAg_4I_5$ stabilizes a superionic structure (somewhat different from that of α-AgI) with the Ag^+ ions again randomly distributed over a greater number of tetrahedral sites.

Certain glasses also exhibit superionic behaviour, e.g. $AgPO_3$–AgI. However, these materials mostly have somewhat lower ionic conductivities and higher activation energies than the best crystalline fast-ion conductors. Nevertheless, glasses have the advantage of their ionic transport being spatially *isotropic*, rather than being confined to certain channels or planes, as is often the case for crystalline superionic materials. The reader is referred to, for example, West (1987) for more details on superionic materials.

*3.4.3 Mechanical properties

Materials deform when subject to an applied mechanical stress. For small deformations, all solids behave in an elastic manner: that is, the deformation only exists whilst the stress is applied. Furthermore, the strain (the proportional displacement of shape or volume) is proportional to the applied stress (Hooke's law) for small deformations. This linearity in elastic response is a direct consequence of a limiting harmonic form (potential energy quadratic in the displacement) for the interatomic potential (§2.5) for small strains. For higher levels of strains, Hooke's law may break down, either because the potential is no longer harmonic (but the mechanical response is still elastic) or, more likely, structural defects (e.g. dislocations) facilitate plastic (i.e. permanent) deformation under an applied stress. A discussion of the static elastic properties of solids in the harmonic limit, unrelated to defects, will be given first, followed by a description of the influence of defects in causing plastic deformation.

A solid is stressed by applying external forces such that both the net force and the net torque are zero. The first is achieved by exerting forces equal in magnitude, but in opposite directions, *perpendicularly* to opposite faces, tending to compress/elongate the sample; the second is achieved by applying equal and opposite forces in a *parallel* fashion to opposite faces, tending to shear the sample (see Fig. 3.40). The value of the stress depends on both the magnitude and direction of the applied force and also the particular face on which it acts. A force $\boldsymbol{F}_x$ applied in, say, the x-direction to a plane with area A_x whose normal is also in the x-direction produces a stress component

$$\sigma_{xx} = F_x/A_x. \tag{3.86a}$$

Likewise, the same force applied to a plane whose normal is in the y-direction produces the stress component

$$\sigma_{xy} = F_x/A_y. \tag{3.86b}$$

Thus, stress in general is described by a second-rank tensor σ with nine components $\sigma_{ij}(i,j = x, y, z)$, conveniently written as a (3×3) matrix

$$\sigma = \begin{bmatrix} \sigma_{xx} & \sigma_{xy} & \sigma_{xz} \\ \sigma_{yx} & \sigma_{yy} & \sigma_{yz} \\ \sigma_{zx} & \sigma_{zy} & \sigma_{zz} \end{bmatrix}. \tag{3.87}$$

However, only *six* of these stress components are independent because of the constraint of zero applied torque, which implies that

$$\sigma_{xy} = \sigma_{yx}; \sigma_{yz} = \sigma_{zy}; \sigma_{zx} = \sigma_{xz}. \tag{3.88}$$

Special cases of stress have most of the six independent stress components equal to zero. For instance, for uniaxial stress in the x-direction, the only non-zero component is $\sigma_{xx} = \sigma$; for hydrostatic stress (pressure), the non-zero components are the diagonal components $\sigma_{xx} = \sigma_{yy} = \sigma_{zz} = -p$; and pure shear (in which the shape, but not the volume, of the solid is changed) is represented by, say, $\sigma_{xy} = \sigma_{yx} = \sigma$ being the only non-zero components.

A particular orientation of the spatial coordinates can always be found such that the general stress matrix (eqn. (3.87)) has *diagonal* components only, i.e.

$$\sigma' = \begin{bmatrix} \sigma_x & 0 & 0 \\ 0 & \sigma_y & 0 \\ 0 & 0 & \sigma_z \end{bmatrix}, \tag{3.89}$$

the six independent variables defining the stress system now being the three principal stresses $\sigma_i(i = x, y, z)$ acting along the principal axes, and the three variables needed to determine the orientation of the principal axes with respect to the original coordinate

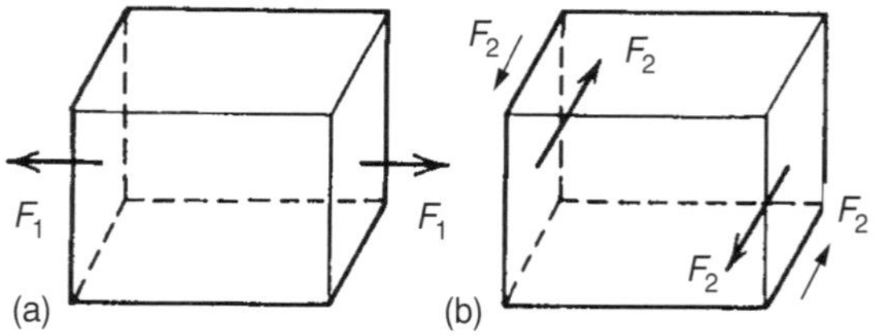

Fig. 3.40 Illustration of the forces required to stress a sample: (a) uniaxially (F_1); (b) in a purely shear-like fashion (F_2).

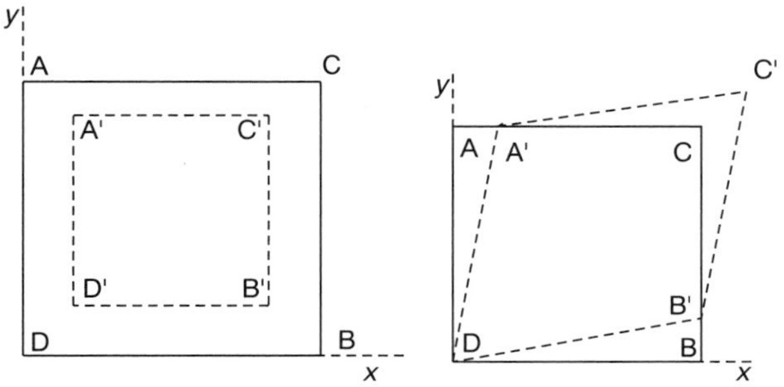

Fig. 3.41 Representation of: (*a*) uniform compression resulting from a purely hydrostatic compressive stress; (*b*) pure shear deformation (at constant volume).

system. A general stress matrix, transformed to the principal axis system (eqn. (3.89)), can always be rewritten as

$$\begin{bmatrix} \sigma_x & 0 & 0 \\ 0 & \sigma_y & 0 \\ 0 & 0 & \sigma_z \end{bmatrix} = \begin{bmatrix} \sigma_0 & 0 & 0 \\ 0 & \sigma_0 & 0 \\ 0 & 0 & \sigma_0 \end{bmatrix} + \begin{bmatrix} (\sigma_x - \sigma_0) & 0 & 0 \\ 0 & (\sigma_y - \sigma_0) & 0 \\ 0 & 0 & (\sigma_z - \sigma_0) \end{bmatrix}, \tag{3.90}$$

where the new stress component is given by

$$\sigma_0 = (\sigma_x + \sigma_y + \sigma_z)/3. \tag{3.91}$$

The first term of eqn. (3.90) thus represents a purely *hydrostatic* term, with

$$\sigma_0 = -p = (\sigma_x + \sigma_y + \sigma_z)/3; \tag{3.92}$$

this causes a change in volume, but not of shape (Fig. 3.41), of an elastically isotropic solid. The second term in eqn. (3.90) represents a *pure shear* or deviatoric stress term, causing a change in shape, but not of volume, of a solid (Fig. 3.41), since the sum of the diagonal components equals zero, i.e. $\sum_{x,y,z} \sigma_{ii} = 0$; this is the same as in the general representation of the pure shear-stress tensor: $\sigma_{ij} = \sigma$ (e.g. $i = x, j = y$), $\sigma_{ij} = 0$ otherwise.

The application of an external force to a solid causes a deformation because different points in the material are displaced by different amounts. Consider a point P initially at $\boldsymbol{r}$ and a neighbouring point Q initially at $\boldsymbol{r} + \Delta\boldsymbol{r}$, displaced under the action of a stress to P' (at $\boldsymbol{r} + \boldsymbol{u}$) and Q' (at $\boldsymbol{r} + \Delta\boldsymbol{r} + \boldsymbol{u} + \Delta\boldsymbol{u}$) respectively—see Fig. 3.42. For small relative displacements ($|\Delta\boldsymbol{u}| \ll |\Delta\boldsymbol{r}|$), the components of the relative displacement $\Delta\boldsymbol{u}$ are given by

$$\Delta u_i = (\partial u_i/\partial x)\Delta x + (\partial u_i/\partial y)\Delta y + (\partial u_i/\partial z)\Delta z \quad (i = x, y, z). \tag{3.93}$$

The strain components e_{ij} are then defined in terms of the dimensionless displacement gradients as

$$e_{ii} = \partial u_i/\partial i \quad (i = x, y, z) \tag{3.94a}$$

and

$$e_{ij} = e_{ji} = (\partial u_i/\partial j + \partial u_j/\partial i) \quad (i = x, y, z). \tag{3.94b}$$

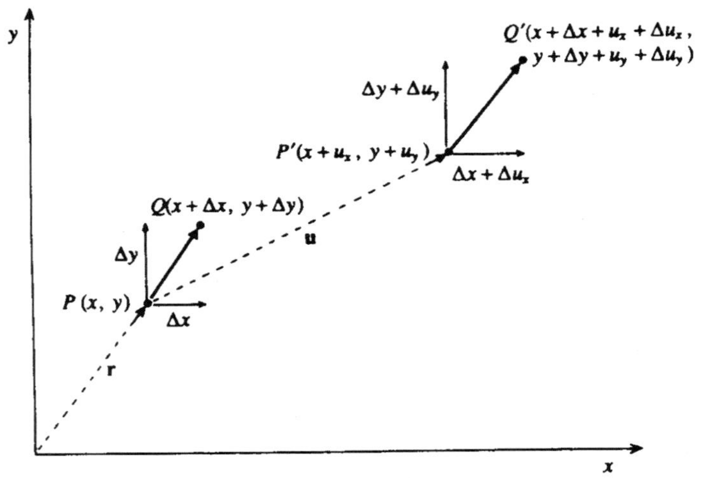

Fig. 3.42 Two-dimensional representation of infinitesimal homogeneous elastic strain. Two points, P and Q, are displaced to points P′ and Q′ under the action of an external stress. The strain components e_{ij} $(i = x, y)$ are defined in terms of the differential displacement gradients, $\partial u_i/\partial j$.

Thus, the strain components can, like the stress components, be written as a (3×3) matrix, with again only six of the nine components being independent because of the requirement that the off-diagonal components obey the condition $e_{ij} = e_{ji}$ in order to exclude rigid rotations.

Special cases of strain include *uniaxial* strain in the x-direction with the only non-zero strain component being $e_{xx} = e$; *uniform dilatation/compression*, resulting from hydrostatic stress, with $e_{ii} = e(i = x, y, z)$; and *pure shear* with $e_{ij} = e_{ji} = e(i = x, j = y$, for example).

As with a general stress (eqn. (3.90)), a general strain can be separated into a *dilatational-strain* component and a *pure shear* or *deviatoric-strain* component; i.e. for a general component

$$e_{ij} = e_0\delta_{ij}/3 + (e_{ij} - e_0\delta_{ij}/3) \tag{3.95}$$

where the Kronecker delta symbol has the properties:

$$\delta_{ij} = 1, \quad i = j, \tag{3.96a}$$

$$\delta_{ij} = 0, \quad i \neq j. \tag{3.96b}$$

Since there are only six independent stress or strain components, a convenient shorthand notation is to relabel the component indices as follows:

$$\begin{array}{cccccc} xx & yy & zz & yz & zx & xy \\ \downarrow & \downarrow & \downarrow & \downarrow & \downarrow & \downarrow \\ 1 & 2 & 3 & 4 & 5 & 6 \end{array}$$

Hooke's law states that the stress and strain are directly proportional to each other. Thus, in terms of elastic-stiffness coefficients, c_{ij}, a stress coefficient can be written as a function of strain as

$$\sigma_i = \sum_{j=1}^{6} c_{ij} e_j. \qquad (3.97)$$

The relationship between stress and strain coefficients can also be written in matrix form as

$$\begin{bmatrix} \sigma_1 \\ \sigma_2 \\ \sigma_3 \\ \sigma_4 \\ \sigma_5 \\ \sigma_6 \end{bmatrix} = \begin{bmatrix} c_{11} & c_{12} & c_{13} & c_{14} & c_{15} & c_{16} \\ c_{21} & c_{22} & c_{23} & c_{24} & c_{25} & c_{26} \\ c_{31} & c_{32} & c_{33} & c_{34} & c_{35} & c_{36} \\ c_{41} & c_{42} & c_{43} & c_{44} & c_{45} & c_{46} \\ c_{51} & c_{52} & c_{53} & c_{54} & c_{55} & c_{56} \\ c_{61} & c_{62} & c_{63} & c_{64} & c_{65} & c_{66} \end{bmatrix} \begin{bmatrix} e_1 \\ e_2 \\ e_3 \\ e_4 \\ e_5 \\ e_6 \end{bmatrix}. \qquad (3.98)$$

Alternatively, a strain coefficient can be expressed in terms of the elastic-compliance coefficients, s_{ij}, as a function of stress:

$$e_i = \sum_{j=1}^{6} s_{ij} \sigma_j. \qquad (3.99)$$

Note that both elastic compliance and stiffness are quantities describing a material as an elastic continuum.

Of the 36 elastic-stiffness (or compliance) coefficients (see eqn. (3.98)), in the most general case (e.g. for triclinic crystals) 21 are independent and non-zero as a result of the general condition

$$c_{ij} = c_{ji}. \qquad (3.100)$$

The presence of higher symmetry reduces the number of non-zero, independent values of elastic-stiffness coefficients even further. Thus, for the case of *cubic* crystals, just *three* components are independent, viz. c_{11}, c_{12} and c_{44}, with

$$c_{11} = c_{22} = c_{33}, \qquad (3.101a)$$
$$c_{12} = c_{13} = c_{23}, \qquad (3.101b)$$
$$c_{44} = c_{55} = c_{66}, \qquad (3.101c)$$

and all other coefficients, not related via eqn. (3.100), being zero. These elastic-stiffness constants are related to the corresponding elastic-compliance coefficients via the relations

$$S_{11} = \frac{c_{11} + c_{12}}{(c_{11} - c_{12})(c_{11} + 2c_{12})} > 0, \qquad (3.102a)$$
$$S_{12} = \frac{-c_{12}}{(c_{11} - c_{12})(c_{11} + 2c_{12})} \leq 0, \qquad (3.102b)$$
$$S_{44} = \frac{1}{c_{44}}. \qquad (3.102c)$$

as obtained from inversion of the matrix in eqn. (3.98).

In the case of *elastically isotropic* solids (e.g. polycrystalline materials with randomly oriented microcrystals, or amorphous solids), the number of independent elastic coefficients decreases to *two* since, in addition to the equalities applicable to cubic materials (eqn. (3.101)), the coefficients are further linearly related to each other by the equation

$$c_{11} = c_{12} + 2c_{44}. \tag{3.103}$$

For such isotropic materials, the two independent elastic-stiffness coefficients are conventionally referred to as the Lamé constants:

$$\lambda \equiv c_{12}, \tag{3.104a}$$

$$\mu \equiv c_{44}, \tag{3.104b}$$

with $c_{11} \equiv (\lambda + 2\mu)$.

Certain special forms of elastic moduli result for particular patterns of the applied stress. In the case of *uniaxial* stress (with the faces of the solid not subject to the stress being unclamped), the appropriate elastic constant is Young's modulus, defined as

$$E_{\mathrm{y}} = \frac{\text{normal stress}}{\text{normal strain}} = \frac{\sigma_{xx}}{e_{xx}} \tag{3.105}$$

and thus (see Problem 3.19):

$$E_{\mathrm{y}} = \frac{1}{s_{11}} = \frac{(c_{11} - c_{12})(c_{11} + 2c_{12})}{(c_{11} + c_{12})} = \mu\frac{(3\lambda + 2\mu)}{(\lambda + \mu)} \tag{3.106}$$

for the case of isotropic media. Under such stress loading, the sample also deforms in directions *perpendicular* to the stress direction. This behaviour is quantified by Poisson's ratio, defined as:

$$\upsilon_{\mathrm{p}} = \frac{|\text{transverse strain}|}{\text{normal strain}} = \frac{|e_{yy}|}{e_{xx}} \tag{3.107}$$

with (see Problem 3.20)

$$\upsilon_{\mathrm{p}} = \frac{-s_{12}}{s_{11}} = \frac{c_{12}}{(c_{11} + c_{12})} = \frac{\lambda}{2(\lambda + \mu)}, \tag{3.108}$$

where $0 \leqslant \upsilon_{\mathrm{p}} < 0.5$.

In the case of *hydrostatic* stress (pressure), the bulk modulus B (equal to the inverse of the compressibility, $\kappa = 1/B$)—see eqn. (2.32)—relates the pressure to the dilatation (fractional change in volume) e_0, viz.

$$B = \frac{p}{|e_0|}, \tag{3.109}$$

with (see Problem 3.19)

$$B = \frac{1}{3(s_{11} + 2s_{12})} = \frac{(c_{11} + 2c_{12})}{3} = \frac{(3\lambda + 2\mu)}{3}. \tag{3.110}$$

(The bulk modulus is connected to *microscopic* quantities relating to the interatomic potential in eqns. (2.36) and (2.37) and Problem 2.5.)

Finally, for the case of *pure shear* stress, the shear (or rigidity) modulus is defined as:

$$G_{\mathrm{s}} = \frac{\text{shear stress}}{\text{shear strain}} = \frac{\sigma_{xy}}{e_{xy}}, \tag{3.111}$$

with (see Problem 3.19) for isotropic solids:

$$G_{\mathrm{s}} = \frac{1}{2(s_{11} - s_{12})} = \frac{(c_{11} - c_{12})}{2} \equiv c_{44} = \mu. \tag{3.112}$$

The various elastic moduli are inter-related, as can be seen by examining eqns. (3.106), (3.108), (3.110) and (3.12); for example:

$$G_s = \frac{E_y}{2(1+\upsilon_p)} \tag{3.113}$$

and

$$B = \frac{E_y}{3(1-2\upsilon_p)}. \tag{3.114}$$

Table 3.5 lists various elastic constants for some representative elastically isotropic and cubic materials. (See Problem 3.21 for a discussion of the elastic properties of isotropic *liquids*.)

Thus far, only the static elastic properties of solids in the linear stress–strain régime have been considered. What happens to a solid if the limit of proportionality (point A in Fig. 3.43) is exceeded? For brittle solids (e.g. silica glass or cast iron), the end of the elastic régime corresponds to *bond breaking*, resulting in catastrophic rupture of the material when subject to tensile stress (Fig. 3.43). Brittle (Griffiths-type) fracture is always associated with the existence or creation of *cracks*, either at the surface or in the bulk of the material. Such cracks may occur at inclusions (e.g. graphite flakes in cast iron) or at grain boundaries between microcrystallites. The stress at a crack tip is much higher than the average value of the stress in the bulk of the material, and this local stress magnification can cause the crack to open up and move rapidly through the sample, causing fracture.

However, (pure) metals more commonly exhibit plastic deformation rather than brittle fracture. Plastic-deformation characteristics are also illustrated in Fig. 3.43. Beyond the elastic limit (point B in Fig. 3.43), permanent strain is produced by the stress, and beyond the yield stress (or strength) point (C), plastic deformation becomes increasingly easy. Removal of the stress at, say, point D in Fig. 3.43 results in a permanent strain e_p remaining at zero stress; reapplication of the stress causes elastic deformation of the solid until the point D is very nearly reached, i.e. the stress to which the material was previously subjected. Thus, the material has become work-hardened by

Table 3.5 Values of Young's modulus (E_y), shear modulus (G_s), Poisson's ratio (υ) and elastic-stiffness coefficients for some elastically isotropic and cubic materials at ambient temperature and pressure

Material	E_y	G_s	υ_p	c_{11}	c_{12}	c_{44}
Isotropic						
Silica glass	73.1	31.2	0.17	78.5	16.1	31.2
Polyethylene	1.2	0.4	0.45	0.1	0.03	0.04
Cubic						
Al	72.5	27.6	0.31	107.5	60.8	28.5
Pb	26	8.5	0.41	47.7	40.3	14.4
Fe	215	77	0.29	228.1	133.5	110.9
C(dia)	900	360	0.24	950	390	430
Si	130	51	0.28	165.7	63.9	79.6
NaCl	53.5	23.1	0.20	48.6	11.9	12.8

(Elastic moduli in GPa)
(After Barber and Loudon (1989). Reproduced by permission of Cambridge University Press)

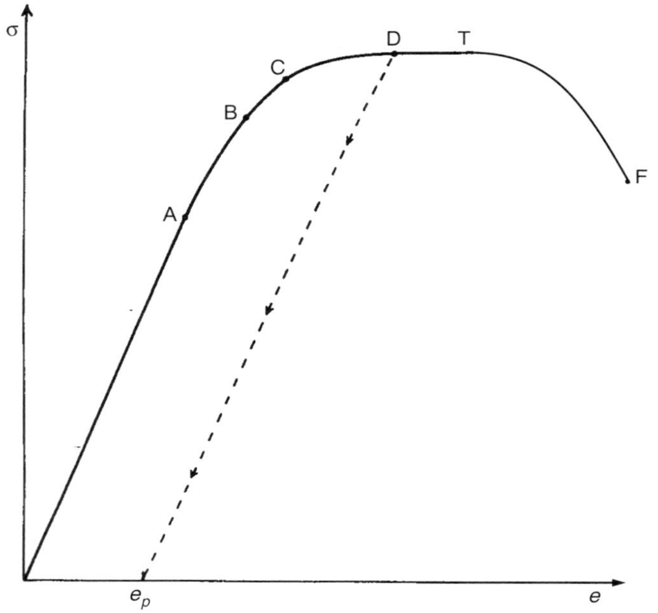

Fig. 3.43 Schematic illustrations of the stress–strain behaviour for materials stressed beyond the elastic limit in the vicinity of the onset of plastic deformation. Point A marks the limit of the Hooke's law régime; point B is the elastic limit, marking the onset of plastic deformation, with point C being the yield point at which plastic deformation occurs readily. Removal of the stress from a point (D) above the yield point results in the production of a permanent plastic deformation e_p at zero applied stress. Reapplication of stress to the plastically strained material results in elastic deformation until the point D is nearly reached. Point T gives the tensile strength (stress), and point F marks the position at which fracture occurs.

the plastic deformation (the new yield stress is greater than the initial value at C). The tensile stress (strength) corresponds to the *maximum* in the stress–strain curve (point T in Fig. 3.43); this is the maximum stress that a sample can withstand. Applied stresses greater than this inevitably cause fracture (at point F in Fig. 3.43). Griffiths-type brittle fracture does not occur because the atomic motion at the tips of cracks associated with the deformation does not allow stress magnification to occur to such an extent.

Plastic deformation must be associated with abnormally easy relative motion of atoms; such facile motion is mediated by *dislocations* causing slip (relative shear motion) of certain planes of atoms (Fig. 3.32). The critical stress needed to cause slip on a particular plane of atoms is given by *Schmid's law*. Application of a tensile force F to a cylindrical single crystal results in a shear stress σ_s on a plane whose normal vector $\hat{\boldsymbol{n}}$ makes an angle θ with the cylinder axis (see Fig. 3.44). If the slip direction within the plane is given by OP, making an angle ψ with the cylinder axis, then the component of the force acting in the slip direction is $F\cos\psi$ and this acts over a plane of area $(A/\cos\theta)$, where A is the cross-sectional area of the cylindrical crystal. Thus, the resolved shear stress is

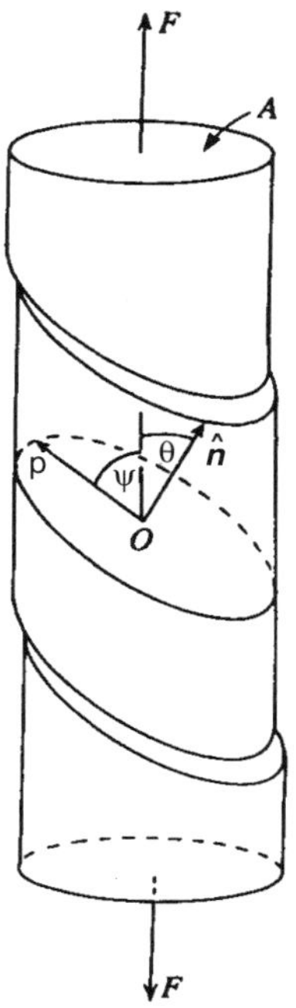

Fig. 3.44 Derivation of the resolved shear stress in the slip direction and plane of a cylindrical single crystal undergoing plastic deformation by dislocation-mediated slip under the action of axial, tensile forces.

$$\sigma_s = (F/A)\cos\theta \cos\psi, \tag{3.115}$$

and Schmid's law states that slip occurs when $\sigma_s = \sigma_s^c$, the critical resolved shear stress of the material. The applied stress, $\sigma_s = F/A$, corresponding to this condition is known as the yield stress; in a stress–strain curve it corresponds to the stress where Hooke's law begins not to be obeyed. The critical resolved shear stress of most materials is considerably less than that predicted by assuming that *simultaneous* relative slip motion of whole planes of atoms occurs (Problem 3.11). The onset of glide of dislocations along the slip plane, caused by the *consecutive* motion of atoms (see Fig. 3.33), greatly reduces the value of σ_s^c.

Plastic deformation of a solid by a slip mechanism is specified in terms of the slip system (*hkl*) [*uvw*], represented as the Miller indices of the slip plane (*hkl*) and the direction indices [*uvw*] of the slip direction within it. In general, the slip planes correspond to the *close-packed planes* and the slip directions to the *close-packed directions* within such planes. Thus, for f.c.c. metals, there are 12 slip systems $\{111\}\ \langle 1\bar{1}0\rangle$ (four $\{111\}$ planes each containing three independent slip directions), for b.c.c. metals there are 12 $\{110\}\ \langle\bar{1}11\rangle$ slip systems, 12 $\{211\}\ \langle\bar{1}11\rangle$ systems and 24 $\{321\}\ \langle\bar{1}11\rangle$ systems, while for h.c.p. crystals there are just three principal $\{0001\}\ \langle 11\bar{2}0\rangle$ slip systems associated with the basal plane, together with three subsidiary $\{10\bar{1}0\}\ \langle 11\bar{2}0\rangle$ systems and six $\{10\bar{1}1\}\ \langle 11\bar{2}0\rangle$ systems. The multiplicity of possible slip systems in f.c.c.

and b.c.c. metals makes them much more ductile (i.e. susceptible to plastic deformation) than h.c.p. crystals which are relatively brittle (see Problem 3.21). In the case of *ionic* crystals, electrostatic considerations also play a rôle in determining the slip system since atomic slip motion leading to close like-ion neighbours is precluded on electrostatic energetic grounds. Thus, for NaCl, the slip system is $\{110\}$ $\langle 110 \rangle$, and the restricted number of slip systems for ionic solids accounts for their brittleness.

The application of an external shear stress σ_s to the crystal exerts a force on a dislocation and causes it to move. If a length l of the dislocation, with magnitude of the Burgers vector b, moves a distance δx along the slip plane, a proportionate amount of slip $(l\delta x/A)b$ results. The external stress σ_s applied to the area A of the crystal thus does work (force × distance) equal to $\sigma_s A(l\delta x/A)b$ and this work is equivalent to that resulting from the movement by a distance δx of the dislocation subject to an effective force F per unit length. Thus:

$$Fl\delta x = \sigma_s lb\delta x$$

or

$$F = \sigma_s b. \tag{3.116}$$

Plastic deformation is accompanied by a great increase in dislocation density. One mechanism for dislocation multiplication is the Frank–Read source (see Fig. 3.45), which consists of a dislocation pinned at each end by impurities or nodes in a dislocation network. The force per unit length F (eqn. (3.116)) acting on this dislocation segment will cause it to bow outward increasingly until the two lobes in Fig. 3.45c just touch. These two parts of the same dislocation, moving in opposite directions, produce the same deformation and must therefore be of *opposite* signs and therefore mutually annihilate to form a region of perfect crystal, leaving behind a dislocation loop and the pinned dislocation segment able to repeat the process.

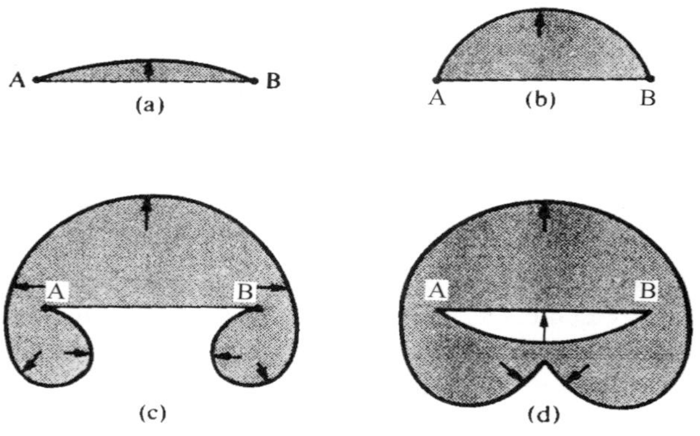

Fig. 3.45 The Frank–Read source for dislocation multiplication. A dislocation segment, pinned at A and B, is bowed out by an applied force (a, b). The lobes of the dislocation loop (c) eventually meet and recombine to give a perfect region of crystal (d) and a dislocation segment still pinned at A and B that is able to repeat the process indefinitely.

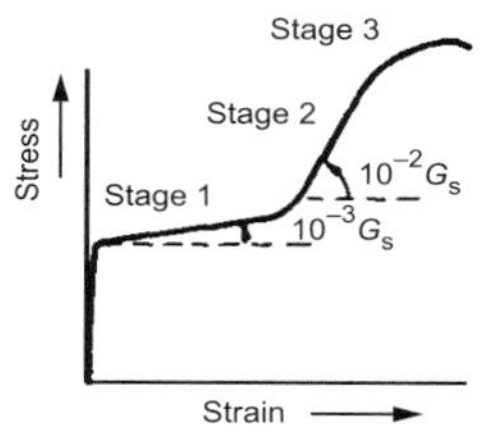

Fig. 3.46 Schematic illustration of the stress–strain curve for f.c.c. metals showing, sequentially, stage 1, 2 and 3 behaviour.

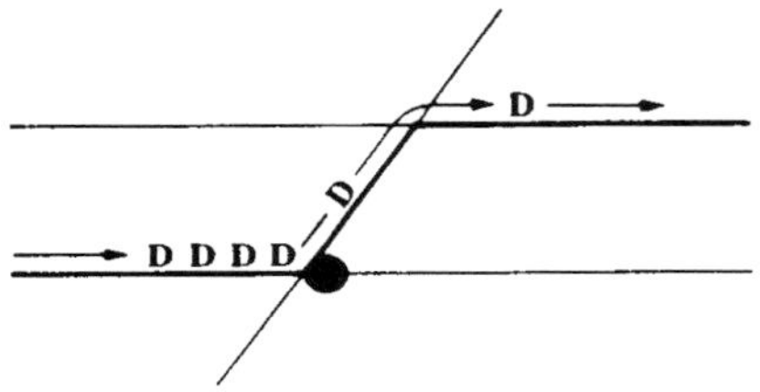

Fig. 3.47 Schematic illustration of the process of cross-slip responsible for stage-3 work-hardening in f.c.c. and b.c.c. metal crystals. A screw dislocation (D) can move along a slip plane intersecting the original slip plane at the point where a dislocation-pinning obstruction exists.

Stress–strain curves for metal single crystals can exhibit several stages of plastic-deformation behaviour (see Fig. 3.46). Stage 1 is called the easy-glide region because the slope of the stress-strain curve is small ($\simeq 10^{-3}\, G_s$) and there is little work-hardening. This is sometimes followed by a linear stage-2 region, characterized by a much greater slope ($\simeq 10^{-2}\, G_s$) characteristic of a greater amount of work-hardening. Finally, a stage-3 region can be reached where the work-hardening becomes smaller again. Stage 1 is associated with the easy glide of dislocations in the slip plane, at first for the slip system having the lowest Schmid factor (eqn. (3.115)), and then at higher stresses, slip can be initiated, in general, on other (non-parallel) slip planes. The intersection of dislocations on different slip planes stops the easy glide and this occurrence marks the end of stage-1 behaviour. H.c.p. metals, with only *one* dominant slip plane (the basal plane), do not suffer as much from this dislocation locking, and hence (unlike f.c.c. or b.c.c. metals) such materials exhibit stage-1 behaviour until fracture occurs. The increased work-hardening associated with stage 2 can arise from a number of reasons, such as the difficulty a dislocation has in moving through a region densely populated with other dislocations threading the slip plane, or the pinning of a dislocation by, for example, a precipitate inclusion. The decrease in rate of work-hardening characteristic of stage 3 results from cross-slip, i.e. the easy motion of a (screw) dislocation at high levels of stress away from a pinning obstruction along another intersecting slip plane onto a plane parallel to the original slip plane (Fig. 3.47). This process only occurs in f.c.c. and b.c.c. metals having a multiplicity of slip planes, but not in h.c.p. crystals where slip is generally restricted to a single (basal) plane.

It has been assumed so far that the equilibrium strain is achieved *instantaneously* on the application of an external stress. However, sometimes (particularly at elevated temperatures), the strain becomes time-dependent, and the establishment of the maximum strain lags behind that of the stress: this behaviour is known as anelasticity. The phenomenon in which plastic deformation occurs with time at constant applied stress at high temperatures ($> 0.4T_{\text{melt}}$) is known as creep (see Fig. 3.48), and can be caused by several microscopic mechanisms, including motion of dislocations, shearing motion of microcrystals at grain boundaries and diffusion of vacancies to grain boundaries. Creep is a serious problem in circumstances where materials are subjected to a tensile stress at high temperatures (e.g. turbine blades in jet engines). Ceramic materials (compounds of metals and non-metallic elements, e.g. oxides, nitrides and carbides), unlike metals, are generally very resistant to creep up to very high temperatures ($\leqslant 2000\,°\text{C}$). Alternatively, a strategy for obviating vacancy-mediated creep is to ensure that no grain boundaries exist that are perpendicular to the tensile-strain axis. This has been achieved in the case of turbine blades by *directional solidification*, in which the metal alloy is poured into a mould and cooled at one end: columnar crystallites then grow from the root to the tip of the blade, and even single-crystal blades containing no grain boundaries at all can be fabricated in this way (see §1.2.1). Creep resistance can also be achieved by the incorporation of structural inhomogeneity, e.g. a dispersed phase as in the 'superalloys' (alloys of Co and Ni with refractory metals, such as Nb, Mo, W or Ta, or Ti) Solid-solution alloying also contributes to the reduction of creep.

Mechanical properties of materials are discussed in many materials science and engineering books; for further details see, for example, Barber and Loudon (1989), Callister (1997), Nabarro and de Villiers (1995) and Haasen (1996).

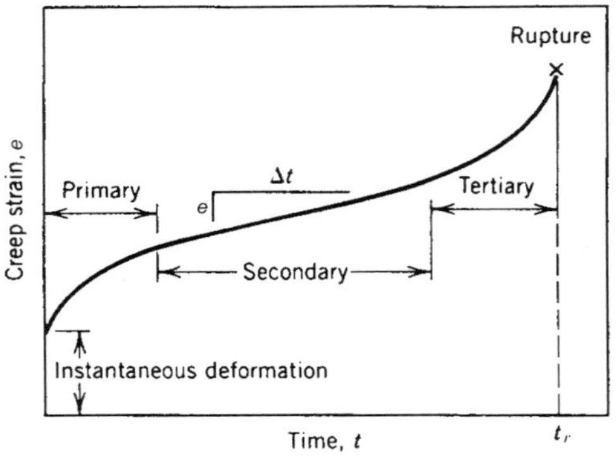

Fig. 3.48 Typical time dependence of the strain of a material undergoing creep at constant stress and at an elevated temperature. The primary creep region is characterized by the Andrade law $e(t) = e_0 + at^{1/3}$. The secondary (or steady-state) creep region results from a balance between the competing processes of strain-hardening and thermally activated recovery (annealing). In the tertiary creep régime, the creep rate accelerates due to the onset of some mechanical instability (e.g. a crack), leading to rupture.

Applications 3.5

Although the presence of structural defects in solids is very often an unwanted circumstance, since their presence is frequently deleterious to the intrinsic behaviour of the materials, nevertheless a number of technological applications of materials do exploit defect-controlled behaviour. Examples discussed here include the photographic process and the use of solid electrolytes.

3.5.1 The photographic process

Photographic film consists of sub-micron-sized grains of iodine-containing crystalline AgBr dispersed in a binder (gelatin) and supported on paper or plastic film. The photographic process depends on the physical properties and behaviour of the point defects present in the AgBr:I grains: these are cation Frenkel defects (Ag^+ interstitials and vacancies) — see §3.1.1.

A latent image (which needs to be developed subsequently to be made visible) is formed by the production of clusters of just three to six silver atoms on the surface of the AgBr grains following the absorption of light photons. Latent-image formation involves two distinct processes: one is opto-electronic and the other is ionic in nature. The primary opto-electronic event is the absorption of a photon by a grain, causing the excitation of an electron (e^-) from the filled electron states in the 'valence band', across the forbidden energy gap (bandgap) where no electron states are allowed, into a state in the empty 'conduction band' of AgBr. (See Chapter 5 for a full discussion of electron states in solids.) The electronic state in the valence band vacated by the optically excited electron is, perforce, *positively* charged and is called a hole (h^+). Thus, the photon-absorption event can be written as

$$h\upsilon \rightarrow e^- + h^+, \tag{3.117}$$

and the photo-created holes are mostly trapped by I^- ions. However, the bandgap of AgBr is rather large (2.7 eV), and so only light with an energy greater than this, or a wavelength shorter than 460 nm, can be absorbed. In order to extend the spectral range, sensitizers (such as sulphur or organic dyes) are adsorbed onto the surface of the AgBr grains, and photo-created electrons in an excited energy level, say of a dye molecule, can subsequently transfer into the conduction band of the AgBr.

Photo-created electrons, trapped at states at the grain surfaces, can then take part in a sequence of reactions involving mobile interstitial silver ions, Ag_i^+, diffusing to the surface, for example:

$$\begin{aligned}
Ag_i^+ + e^- &\rightarrow Ag^0, \\
Ag^0 + e^- &\rightarrow Ag^-, \\
Ag^- + Ag_i^+ &\rightarrow Ag_2^0, \\
Ag_2^0 + e^- &\rightarrow Ag_2^-, \\
Ag_2^- + Ag_i^+ &\rightarrow Ag_3^0 \text{ etc.}
\end{aligned} \tag{3.118}$$

The film containing grains having a latent image (Ag_n^0 clusters) is then developed by first treating it with a reducing agent (e.g. an alkaline solution of hydroquinone) which reduces to metallic silver only those AgBr grains containing a latent image since such metallic Ag_n^0 clusters act as a *catalyst* for the reduction reaction. Finally, all remaining AgBr grains that have not been exposed to light and hence that have not been reduced, are dissolved by a solution of 'hypo' ($Na_2S_2O_3$) which forms a water-soluble complex with Ag^+ ions.

It should be noted that, although only about 10 photons need to be absorbed to create a latent image in a grain of AgBr, the overall gain (number of silver atoms produced per incident absorbed photon) is of the order of $\simeq 10^8$, since a typical grain of AgBr might contain $\simeq 10^9$ Ag^+ ions, *all* of which are converted to Ag^0 neutral atoms on development. This enormous enhancement factor accounts for the high light sensitivity of photographic films. Note also that the AgBr-based photographic film produces a *negative* image, i.e. dark where illuminated.

3.5.2 Solid electrolytes

Materials exhibiting superionic conductivity, but which are electronically insulating, i.e. solid electrolytes (see §3.4.2.2), are used in a number of applications, e.g. in all-solid-state batteries and as ion sensors.

A battery is an electrochemical cell that converts free energy, liberated in a chemical reaction or from a change in concentration of a species, into electrical energy. The reactants taking part in a reaction are either chemically reduced or oxidized at an electrode: oxidation occurs at the negatively charged anode, say made of metal A, where the following reaction occurs:

$$A \rightarrow A^{\nu+} + \upsilon e^-, \tag{3.119a}$$

and the electrons so released travel around an external electrical circuit to the positively charged cathode, say of metal B, where reduction occurs:

$$B^{\upsilon+} + \upsilon e^- \rightarrow B. \tag{3.119b}$$

The overall reaction is thus:

$$A + B^{\upsilon+} + X^{\upsilon-} \rightleftharpoons A + X^{\upsilon-} + B, \tag{3.119c}$$

where $X^{\upsilon-}$ is a common anionic species.

The change in (molar) Gibbs free energy ΔG for the reaction given in eqn. (3.119c) is related to the electromotive force (EMF), ε, of the cell, which is the electrical potential difference between anode and cathode, via

$$\Delta G = -\upsilon F \varepsilon, \tag{3.120}$$

where F is the Faraday (the charge of a mole of electrons). Moreover, the Gibbs free energy, or equivalently the chemical potential, of each reacting species, i, is related to its activity, a_i, by eqn. (3.56), and thus for the overall reaction (eqn. (3.119c)), the free-energy change is also given by

$$\Delta G = \Delta G^{\ominus} + RT \ln\left[\frac{a(A^{\upsilon+})a(B)}{a(A)a(B^{\upsilon+})}\right], \tag{3.121}$$

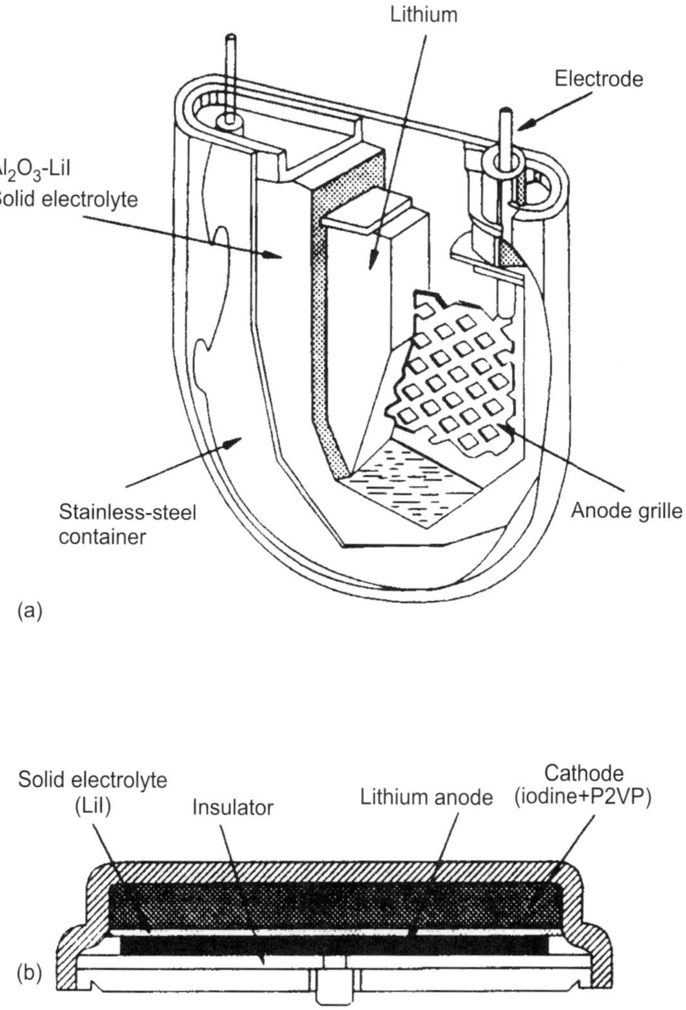

Fig. 3.49 Schematic illustrations of the construction of Li/LiI/I_2 batteries for (a) heart pacemakers; (b) pocket calculators.

where $\Delta G^{\ominus}$ is the *standard* free energy (the value when all species have unit activity). Combining eqns. (3.120) and (3.121) yields the Nernst equation for the EMF

$$\varepsilon = \varepsilon^{\ominus} - \frac{RT}{\upsilon F} \ln\left[\frac{a(\mathrm{A}^{\upsilon+})a(\mathrm{B})}{a(\mathrm{A})a(\mathrm{B}^{\upsilon+})}\right], \tag{3.122}$$

where $\varepsilon^{\ominus}$ is the standard EMF of the cell (for all species at unit activity).

The purpose of the electrolyte is simply to allow free passage between the electrodes of the ions involved in the chemical reaction (eqn. (3.119c)) but, at the same time, to prevent the electrons liberated at the anode from passing directly through the cell, thereby short-circuiting it. Solid electrolytes have a number of advantages over conventional liquid electrolytes: they can be used, in thin-film form, to produce integrated batteries for microelectronics applications and, moreover, they do not leak!

One example of a primary battery (i.e. one that operates just once and cannot be recharged) employing a solid electrolyte is the $Li/LiI/I_2$ cell used in heart-pacemaker (Fig. 3.49a) and pocket-calculator (Fig. 3.49b) applications. The standard EMF for this cell, and hence the open-circuit voltage, is $\varepsilon^{\ominus} = 2.8\,V$, and the anode and cathode reactions are $2Li \rightarrow 2Li^+ + 2e^-$(anode) and $I_2 + 2e^- \rightarrow 2I^-$ (cathode), with the overall reaction being $2Li + I_2 \rightarrow 2LiI$. The solid electrolyte in this case is crystalline LiI, through which Li^+ ions migrate because of the presence of Schottky defects. This material can function as an electrolyte even though its room-temperature ionic conductivity is rather low ($\simeq 10^{-4}\,S\,m^{-1}$). The iodine cathode material by itself is insufficiently electronically conducting to act as an electrode; thus, it is mixed with 5% of a conducting polymer (poly-2–vinylpyridine). This type of battery has a very long discharge lifetime ($\simeq$ 10 years).

The same solid electrolyte is used in another type of cell used as button batteries, viz. $Li/LiI–Al_2O_3/PbI_2$, PbS. The addition of high-surface-area alumina causes a significant increase in the level of the ionic conductivity of the LiI by several orders of magnitude because of enhanced surface conduction at the alumina grains. The overall cell reactions in this case can be written as $2Li + PbI_2 \rightarrow 2LiI + Pb$ and $2Li + PbS \rightarrow Li_2S + Pb$. This cell has the advantage that the very small electronic conductivity of the electrolyte means that the level of self-discharge is very low, and hence its shelf-life before use is very long (> 2 years).

Secondary or storage batteries are *rechargeable*: once the cell is discharged, the overall chemical reaction that has taken place can be reversed, and the reactant

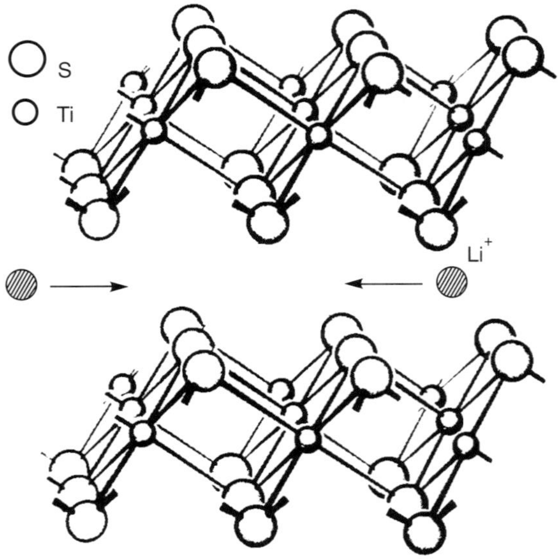

Fig. 3.50 Schematic representation of the intercalation of Li between the layers of crystalline TiS_2 (having the CdI_2–type structure).

concentrations associated with the anode and cathode restored, by the application of a reverse electrical current. One way of making reversible cathodes is to use insertion or intercalation compounds, in which foreign atoms (e.g. Li) can be reversibly inserted into (and removed from) the structure of the host material (see §8.4.1). An example is intercalation of atoms into layer-like crystals (e.g. TiS_2, with the CdI_2-type structure), between those layers that are weakly bonded together by van der Waals interactions (Fig. 3.50), for which the cathode reaction can be written as

$$TiS_2 + \upsilon Li^+ + \upsilon e^- \rightarrow Li_\upsilon TiS_2. \tag{3.123}$$

One interesting type of secondary cell making use of such a reversible cathode material is the lithium battery employing a polymer–salt complex as electrolyte. A highly polarizable salt such as lithium perchlorate, $LiClO_4$, or LiI, is dissolved into an amorphous polymer, e.g. polyethylene oxide (PEO), $(-CH_2-CH_2-O-)_n$, that acts as an aprotic solvent for the salt. The conductivity of the Li+ ions so liberated is rather high, but since the glass-transition ('softening') temperature, T_g, of the amorphous polymer–salt complex is generally lower than room temperature, the temperature dependence of the Li+ ionic conductivity is markedly *non-Arrhenian*, as a result of the motion of the ions being linked cooperatively with that of polymer chain segments. However, a battery made using such a polymer–salt electrolyte (see Fig. 3.51a) has one significant advantage: it is *mechanically flexible*, and hence novel tape-based battery configurations can be fabricated with such materials (see Fig. 3.51b).

Secondary batteries, with high energy-to-mass ratios, are of great interest for use in powering electric automobiles. One that has been extensively studied is the sodium–

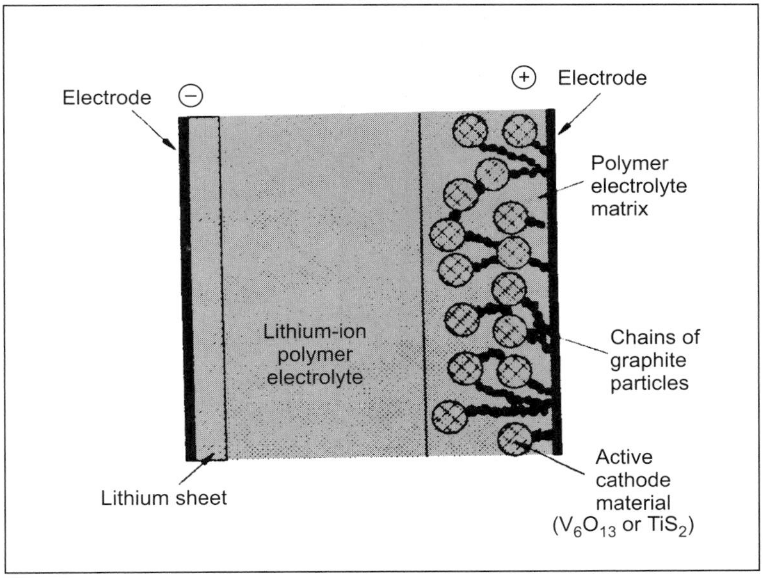

Fig. 3.51 (a) Schematic illustration of a secondary battery employing a polymer–salt electrolyte and a reversible cathode insertion material (e.g. TiS_2). Graphite is mixed with the cathode particles to make electrical contact with the cathode. (b) Different configurations of batteries making use of the mechanical flexibility of polymer–salt electrolytes.

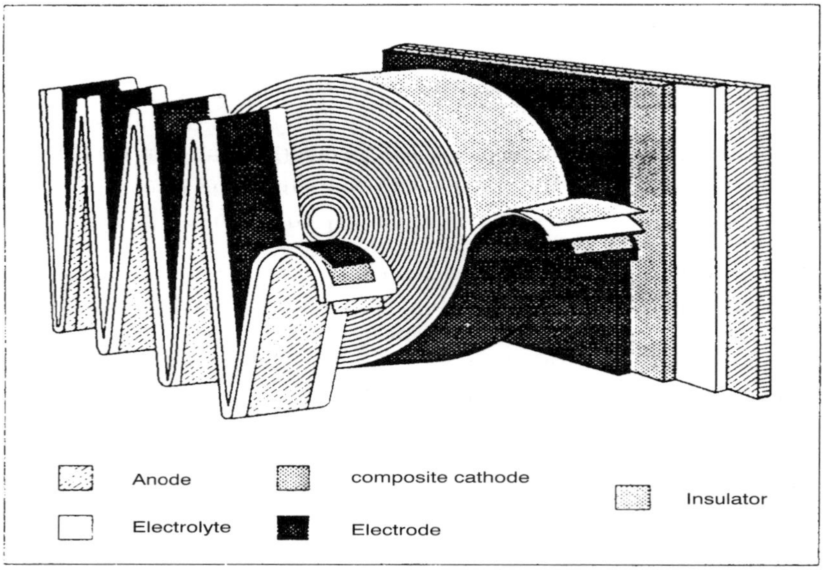

Fig. 3.51 (b)

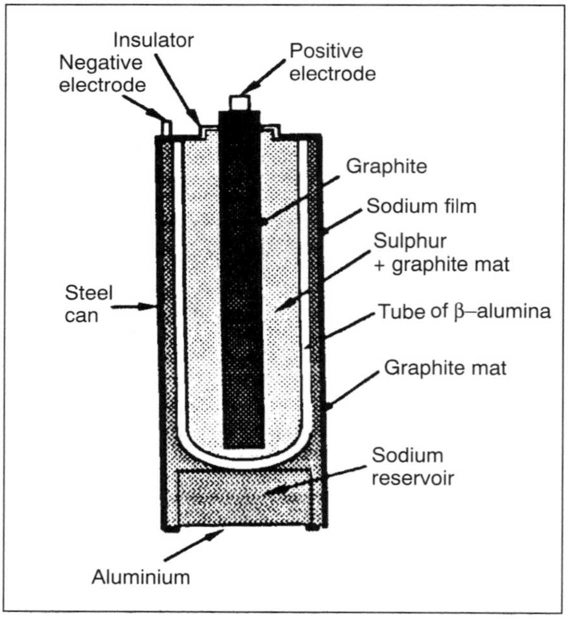

Fig. 3.52 Schematic illustration of the sodium–sulphur battery.

sulphur battery (Fig. 3.52) which uses Na-β-alumina as the solid-electrolyte material. This battery operates at 300°C, and the electrolyte separates molten sodium (the anode material) from molten sulphur (the cathode material) dispersed in a graphite felt matrix to facilitate electrical contact. The overall cell reaction involves the formation of sodium polysulphides:

$$2Na + xS \rightarrow Na_2S_x. \tag{3.124}$$

During the initial stage of discharge, the pentasulphide ($x = 5$) is formed, but as the discharge progresses lower polysulphides are formed; if the discharge is terminated when $x \simeq 2.7$, complete recharging of the battery is possible. The theoretical energy density of this battery is rather high at $\simeq 1000$ W h kg^{-1} (as a result of the very exothermic nature of the reaction (eqn. (3.124)), but actual Na/S batteries only achieve $\simeq 100$ W h kg^{-1}. (For comparison, the energy density of a conventional lead–acid automobile accumulator battery is only $\simeq 30$ W h kg^{-1}.)

Solid electrolytes can also be used in chemical sensors for ions in solution or for trace gases, making use of the Nernst equation (eqn. (3.122)) relating the (half-)cell EMF to the activity (or, approximately, the concentration) of the active species. One of the most widely used chemical sensors is the 'glass electrode' for measuring the pH ($= -\log_{10} a_{H_3O^+}$) of aqueous solutions. The surface of silica glass consists of 'non-bridging' oxygen sites $\equiv Si - O^-$, i.e. oxygen atoms bonded only to a *single* Si atom. These then establish the following equilibrium with the protons in solution when the glass surface is immersed in the solution:

$$SiOH + H_2O \rightleftharpoons SiO^- + H_3O^+. \tag{3.125}$$

The Nernst equation (eqn. (3.122)) shows that the electrode EMF is proportional to the pH.

Problems

3.1 Wustite (ferrous oxide) is always non-stoichiometric. For a particular crystal with an Fe:O ratio of 0.945:1, and a measured density of 5.728×10^3 kg m^{-3}, deduce whether the non-stoichiometry is associated with iron vacancies or oxygen interstitials. (The unit-cell parameter of the f.c.c. crystal is $a = 4.3$ Å.)

3.2 Show, by considering the shortest lattice-translation vectors in the structure, that the Burgers vector $\boldsymbol{b}$ and its magnitude (length) b are respectively:
(a) $(a/2)\langle 111\rangle$ and $a\sqrt{3}/2$ for b.c.c. crystals and
(b) $(a/2)\langle 110\rangle$ and $a/\sqrt{2}$ for f.c.c. crystals.

3.3 Confirm that the composition of the structure of reduced MoO_3 containing crystallographic shear planes shown in Fig. 3.15 corresponds to Mo_8O_{23}.

3.4 Estimate the atomic fraction of atomic vacancies and intersititials in crystalline Cu at
(a) 300 K;
(b) 1000 K.

3.5 For a Cu crystal containing a homogeneous dislocation density of 10^{10} m^{-2}, estimate the elastic strain energy per length of a screw dislocation. ($G_S = 4 \times 10^{10}$ N m^{-2}; $b = 2.6$ Å.)

3.6 Obtain an expression (eqn. (3.19)) for the energy levels of a quantum particle confined to a cubic box, of side length a, by infinitely high potential barriers at the cube faces, and subject to zero potential energy within the box. What is the energy of the ground-state configuration and why?

3.7 (a) For the case of the F^+-centre in MgO (§3.3.2), show that 53% of such anion-vacancy sites have no magnetic nuclei as nearest neighbours, and 35% have a single magnetic nucleus (^{25}Mg ($I = 5/2$; 10% abundance)) as nearest neighbour. (Hint: use the binomial distribution.)

(b) Show that, for the case of *two* ^{25}Mg isotopes in the nearest-neighbour cation shell of an F^+-centre, an 11-fold hyperfine-split ESR spectrum results, with the intensities of the lines being in the ratio 1:2:3:4:5:6:5, etc.

3.8 Show that for the F-centre in NaCl, the hyperfine interaction between the unpaired electron and the nearest-neighbour ^{23}Na nuclei ($I = 3/2$; 100% abundance) results in a set of 19 lines, each of which is further split into 37 lines due to the hyperfine interaction between the electron and the nearest-neighbour ^{35}Cl and ^{37}Cl nuclei ($I = 3/2$, 100% total abundance).

3.9 (a) Show that the atomic binding energies of the surface sites on a simple cubic crystal shown in Fig. 3.28 are, in terms of the nearest-neighbour (shared-face) bond strength ϕ_1, and next-nearest-neighbour (shared-edge) bond strength ϕ_2: (1)$4\phi_1 + 4\phi_2$; (2)$3\phi_1 + 3\phi_2$; (3)$5\phi_1 + 8\phi_2$; (4)$\phi_1 + 3\phi_2$; (5)$\phi_1 + 2\phi_2$; (6)$2\phi_1 + 6\phi_2$; (7)$4\phi_1 + 8\phi_2$; (8)$\phi_1 + 4\phi_2$; (9) $-4\phi_1 - 6\phi_2$; (10) $-5\phi_1 - 8\phi_2$, with the kink ('1/2') site having the value $3\phi_1 + 6\phi_2$.
(b) Show that the surface energies per atom of the crystal planes $\{hkl\}$ for the simple cubic structure are $\sigma_{100} = \phi_1/2 + 2\phi_2, \sigma_{110} = \phi_1 + 4\phi_2$ and $\sigma_{111} = 3\phi_1/2 + 6\phi_2$.
(c) Hence show that the surface energies per unit area, $(\sigma n)_{hkl}$, where n_{hkl} is the atomic areal density (see Problem 2.19) are in the ratio $(\sigma n)_{100} : (\sigma n)_{110} : (\sigma n)_{111} = 1 : \sqrt{2} : \sqrt{3}$. (N.B. in general, closest-packed planes have the lowest surface energies.)

3.10 Show that, if the application of a force of 100 N for a day to a cubic sample of material with a volume of 1 cm^3 produces no discernible permanent deformation ($\leqslant$0.01 mm), this behaviour is equivalent to the criterion that solid-like behaviour is characteristic of materials having a shear viscosity $\eta = \sigma_x/(\mathrm{d}v_x/\mathrm{d}z)$ greater than 10^{14} N s m^{-2} (σ_x is the shear stress in the x-direction causing a velocity gradient $\mathrm{d}v_x/\mathrm{d}z$, where $\mathrm{d}z$ is the thickness of an element perpendicular to the applied stress.

3.11 Derive an estimate for the critical shear (yield) stress, σ_s^c, to displace simultaneously one layer of atoms in a crystal over another.
(a) First show that, for small elastic displacements, x, the shear stress σ_s is related to the shear modulus G_s via $\sigma_\mathrm{s} = G_\mathrm{s}x/d$, where d is the interplanar spacing normal to x.
(b) By assuming that the shear stress σ_s can also be written as a sinusoidal function of x, with the periodicity a of atoms lying in the slip plane, show that $\sigma_\mathrm{s}^\mathrm{c} = G_\mathrm{s}/2\pi$. (More realistic calculations give $\sigma_\mathrm{s}^\mathrm{c} \simeq G_\mathrm{s}/30$.)

3.12 Show that the Soret effect (eqn. (3.39)) gives rise to a temperature-gradient-induced concentration gradient (in 1D) given by $-(\$_i/D_i)(\partial T/\partial x)$ in a closed system.

3.13 (a) Show, for example by substitution into eqn. (3.37a), that

$$n_i(x, t) = n_i^0 \exp(-k^2 D_i t)\exp(\mathrm{i}kx)$$

is a general solution to the 1D diffusion equation, where k is a constant. Since the diffusion equation is a *linear* equation, a sum of terms each of which has the form of the general solution above, but with different values of k, is also a solution.
(b) Hence show that the diffusion profile resulting from an atomic distribution initially in the form of a sheet, with N_i atoms positioned as a Dirac delta function at position $x = x_0$, i.e. $n_i(x, 0) = N_i\delta(x - x_0)$, is given by eqn. (3.40). (Hints: (i)) the mathematical definition of the delta function is

$$\delta(x - x_0) = (1/2\pi) \int_{-\infty}^{\infty} \exp[\mathrm{i}k(x - x_0)]\mathrm{d}k;$$

(ii) evaluation of the integral $\int_{-\infty}^{\infty} \exp(-k^2 Dt + \mathrm{i}k(x - x_0))\mathrm{d}k$ can be achieved by using the answer for the integral of a gaussian function

$$\int_{-\infty}^{\infty} \exp(-\lambda y^2)\mathrm{d}y = (\pi/\lambda)^{1/2},$$

together with the changes of variable $\lambda \equiv D_i t$ and $y \equiv (k - i(x - x_0)/2Dt)$.)

(c) Show that the diffusion profile resulting from a semi-infinite, constant-composition initial profile, $n_i(x,0) = n_i^0(-\infty < x < 0)$ is given by eqn. (3.41). (Hint: represent the constant-composition profile as a sum of delta functions, or an integral in the limit, with $N_i = n_i^0 \mathrm{d}x$.)

3.14 The expression for the diffusion-coefficient correlation factor, f_c, given by eqn. (3.70), can be rewritten as $f_c = 1 + 2\sum_{j=1}^{n-1} (\overline{\cos\theta_1})^j$ (see Borg and Dienes (1988)). Show that this can be simplified, in turn, to give eqn. (3.71).

3.15 The vacancy-mediated self-diffusion of Cu has an experimental activation energy $\mathscr{E}_D = 2.04$ eV and pre-factor $D_0 = 2 \times 10^{-5} \mathrm{m^2 s^{-1}}$. Obtain theoretical estimates for these two quantities, taking the vibrational frequency of Cu atoms to be $\upsilon_{Cu} = 10^{13}$ Hz, and given that the unit-cell parameter of f.c.c. Cu is 3.61 Å.

3.16 Derive the Einstein–Smoluchowski relation (eqn. (3.79)) using the conditional probability $p(X, t : x)$ that, given an A atom on the plane at x at zero time, A is on the plane at $x + X$ at time t.

(a) Show that the flux of A atoms in the positive x-direction is given by

$$J_x = (1/t)\left[\int_{-\infty}^{x_0} n(x)\mathrm{d}x \int_{x_0-x}^{\infty} p(X,t:x)\mathrm{d}X - \int_{x_0}^{\infty} n(x)\mathrm{d}x \int_{-\infty}^{x_0-x} p(X,t:x)\mathrm{d}X\right],$$

where the first term represents the number of atoms which, at $t = 0$ were at $x < x_0$ and which at t are at $x > x_0$, the second term represents the number of atoms which at $t = 0$ were at $x > x_0$ and which at t are at $x < x_0$, with $n(x)$ being the concentration of A atoms at x at $t = 0$.

(b) By assuming that the spatial variation of $n(x)$ is smaller than that of $p(X, t : x)$ and expanding $n(x)$ in a Taylor series about x_0 and truncating the expansion at the first term, show, by integration by parts, that eqn. (3.79) results, with $< x^n > = \int X^n p(X, t : x)\mathrm{d}X$.

3.17 Derive the Einstein relation (eqn. (3.84a)) from the Einstein–Smoluchowski equation (eqn. (3.79)).

Assume that an initially homogeneous ion-conducting material is subject to an electric field E_x in the positive x-direction, and that no ions are added or removed from the system. By integrating the expression obtained from eqn. (3.79) describing the equilibrium condition, and comparing the result for $n(x)$ with that resulting from Boltzmann statistics, derive the Einstein relation.

3.18 Essay: Discuss, by reference to α-AgI and Na-β-alumina, the factors that make some materials exhibit superionic behaviour. How does the behaviour of superionic crystals differ from that of glasses?

3.19 Write down the elastic-stiffness matrix for a cubic crystal.

3.20 Derive expressions, valid for an elastically isotropic solid material for:

(a) Young's modulus (eqn. (3.106));

(b) Poisson's ratio (eqn. (3.108));

(c) the bulk modulus (eqn. (3.110)); and

(d) the shear modulus (eqn. (3.112)).

(Hint: express the strains in terms of the stresses using the elastic-compliance coefficients and use eqn. (3.102) to express the answers in terms of elastic-stiffness coefficients.)

3.21 Discuss the elastic properties of isotropic *liquids*.

(a) What are the values of elastic-stiffness and compliance coefficients for a liquid?

(b) What are the values of Young's modulus, Poisson's ratio, shear modulus and bulk modulus?

3.22 Calculate the magnitude of the tensile stress, applied in the [010] direction to a single crystal of b.c.c. Fe, necessary to initiate slip on the slip system (110) [$\bar{1}$11], given that the critical resolved shear stress is $\sigma_s^c = 30$ MPa.

3.23 Essay: What factors influence the performance of solid-state batteries, e.g. the open-circuit voltage, energy density and self-discharge lifetime?

References

Allnatt, A. R. and Lidiard, A. B., 1993, *Atomic Transport in Solids* (Cambridge University Press: Cambridge).

Barber, D. J. and Loudon, R., 1989, *An Introduction to the Properties of Condensed Matter* (Cambridge University Press: Cambridge).

Booker, G.R., 1964, *Disc. Farad. Soc.* **38**, 298.

Borg, R. J. and Dienes, G. J., 1988, *Solid State Diffusion* (Academic Press: New York).

Brown, S.J., Grimshaw, M.P., Ritchie, D.A. and Jones, G.A.C., 1996, *Appl. Phys. Lett.* **69**, 1468.

Callister, W. D., 1997, *Materials Science and Engineering: An Introduction,* 4th edn (Wiley: New York).

Catlow, C. R. A., ed., 1994, *Defects and Disorder in Crystalline and Amorphous Solids,* NATO ASI C418 (Kluwer: Dordrecht).

Crank, J., 1975, *The Mathematics of Diffusion*, 2nd edn (Oxford University Press: Oxford).

Dawson, R. K. and Pooley, D., 1969, *Phys. Stat. Sol.* **35**, 95.

Haasen, P., 1996, *Physical Metallurgy,* 3rd edn (Cambridge University Press: Cambridge).

Hayes, W. and Stoneham, A. M., 1985, *Defects and Defect Processes in Nonmetallic Solids* (Wiley: New York).

Henderson, B., 1972, *Defects in Crystalline Solids* (Edward Arnold: London).

Henderson, B. and Wertz, J. E., 1968, *Adv. Phys.* **17**, 749.

Hirsch, P. B., Howie, A., Nicholson, R. B., Pashley, D. W. and Whelan, M. J., 1965, *Electron Microscopy of Thin Crystals* (Butterworth: London).

Hull, D. and Bacon, D. J., 1984, *Introduction to Dislocations*, 3rd edn (Pergamon: Oxford).

Hutchison, J. L., 1984, *Ultramicroscopy* **15**, 51.

Kittel, C., 1996, *Introduction to Solid State Physics*, 7th edn (Wiley: New York).

Lewis, B., 1980, in *Crystal Growth*, 2nd edn, ed. B. R. Pamplin (Pergamon: Oxford).

Myers, H. P., 1990, *Introductory Solid State Physics* (Taylor and Francis: London).

Nabarro, N. R. N. and de Villiers, H. L., 1995, *The Physics of Creep* (Taylor and Francis: London).

Stoneham, A. M., 1975, *Theory of Defects in Solids* (Clarendon Press: Oxford).

Street, R. A., 1991, *Hydrogenated Amorphous Silicon* (Cambridge University Press: Cambridge).

Townsend, P. D. and Kelly, J. C., 1973, *Colour Centres and Imperfections in Insulators and Semiconductors* (Sussex University Press: Brighton).

West, A. R., 1987, *Solid State Chemistry and its Applications* (Wiley: New York).

CHAPTER 4 Atomic Dynamics

Introduction

Thus far, we have discussed mainly *static* atomic properties of materials, e.g. structure and elastic properties (although atomic diffusion and ionic conductivity in crystals were covered in §3.4.2 because of their intimate connection to structural defects in such solids). In this chapter, the emphasis will be on aspects of atomic dynamics, mainly related to the *vibrational* behaviour of atoms in solid materials. Because vibrations are the principal type of excitation in many types of solids that can be activated thermally, for which the characteristic energy is $k_B T$, many *thermal* properties of materials (e.g. the heat capacity and thermal conductivity in electrical insulators) are controlled by their atomic vibrational behaviour.

In certain special cases, the vibrational amplitude can be concentrated on just a few atoms, e.g. an impurity mode centred on a light impurity atom. However, in general, vibrational excitations in solids are collective modes: essentially *all* atoms in the material take part in the vibrational mode. The influence of translational periodicity characteristic of the structure of crystals has a profound influence on the vibrational behaviour when the wavelength of the vibrations becomes comparable to the size of the unit cell. On the other hand, when the vibrational wavelength is much *larger* than any structural variation in the material, the solid essentially behaves as an elastic continuum, and this chapter begins with a discussion of the dynamical properties of solids in this limit, following on naturally from the discussion of static elastic properties already given in §3.4.3.

Dynamics of continuous media 4.1

Although real materials are obviously composed of atoms, nevertheless, at length scales much greater than the interatomic spacing or the unit-cell parameter, solids behave essentially as *continuous media*. In such a limit, a material might still be *anisotropic*, as a result of its particular crystal symmetry, but the 'graininess' associated with variations in atomic density at shorter length scales is unimportant. We will first consider the propagation of sound waves in materials in this continuum approximation, and this will be followed by a discussion of the problem of counting the number of modes in such solids.

4.1.1 Propagation of sound

A sound (or acoustic) wave is simply an elastic wave travelling in a medium. For a material regarded as an elastic continuum, the sound velocity is then directly related to the elastic modulus of the material. This relationship can be demonstrated straightforwardly for the simple case of an isotropic elastic medium characterized by a single elastic constant, c, in other words a *fluid* for which $c = B$, the bulk modulus (see Problem 3.21).

Consider a cube of material, of mass density ρ, subject to a spatially varying stress σ_x in the x-direction (see Fig. 4.1), in turn causing an instantaneous displacement u_x. The net force acting on the volume element is thus

$$\mathrm{d}F_x = [\sigma_x(x + \Delta x) - \sigma_x(x)]\Delta y \Delta z = \frac{\partial \sigma_x}{\partial x} \Delta x \Delta y \Delta z. \tag{4.1}$$

Hence, Newton's equation of motion, equating force to rate of change of linear momentum (or mass times acceleration), gives

$$\rho \frac{\partial^2 u_x}{\partial t^2} = \frac{\partial \sigma_x}{\partial x}, \tag{4.2}$$

which determines the time dependence of the response of the material in returning to the position of static equilibrium (when stresses on opposing faces of the sample are equal

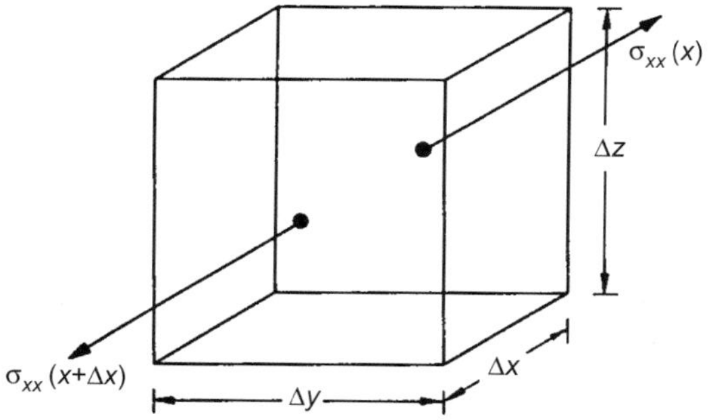

Fig. 4.1 Volume element of a uniform fluid subject to a differential strain in the x-direction.

and opposite). Since the stress σ_x is linearly proportional to the strain e_x (cf. eqn. (3.97)), viz. $\sigma_x = ce_x$, and the strain is related to the displacement via $e_x = \partial u_x/\partial x$ (cf. eqn. (3.94a)), eqn. (4.2) becomes

$$\rho\frac{\partial^2 u_x}{\partial t^2} = c\frac{\partial^2 u_x}{\partial x^2}. \tag{4.3}$$

Equation. (4.3) is in the form of the standard wave equation, for which (for this one-dimensional example) the solution is the plane wave

$$u_x(x,t) = u_x^0 \exp\{\mathrm{i}(k_x x - \omega t)\}, \tag{4.4}$$

where k_x is the x-component of the wavevector $\boldsymbol{k}$, related to the wavelength λ of the wave by

$$|\boldsymbol{k}| = \frac{2\pi}{\lambda}, \tag{4.5}$$

and where $\boldsymbol{k}$ is *parallel* to the propagation direction of the wave (perpendicular to the wavefront) and ω is the (radial) frequency of the wave.

The wave (or phase) velocity, defined as

$$v = \omega/k, \tag{4.6}$$

is thus

$$v = \left(\frac{c}{\rho}\right)^{1/2}, \tag{4.7}$$

as found by substituting the trial solution (eqn. (4.4)) into the wave equation (eqn. (4.3)). In general, another velocity associated with a travelling wave can also be defined, and this is important for waves travelling in dispersive media where the linearity between ω and $\boldsymbol{k}$ breaks down. The group velocity, defined as

$$v_{\mathrm{g}} = \partial\omega/\partial k, \tag{4.8a}$$

or, in general,

$$\boldsymbol{v}_{\mathrm{g}} = \nabla_k \omega(\boldsymbol{k}), \tag{4.8b}$$

is a measure of the velocity of a wave packet, composed of a group of plane waves (cf. eqn. (4.4)), and having a narrow spread of frequencies about some mean value, ω. For acoustic waves with long wavelengths, i.e. in the elastic-continuum limit, the phase and group velocities are obviously *equal*.

Note that the solution given by eqn. (4.4) represents a *longitudinal* vibrational wave, consisting of successive compressional and rarefactional displacements, with the displacement amplitude *parallel* to the propagation direction of the wave, and this is the only sort of acoustic wave that an isotropic fluid can support. The other type of vibrational wave, namely a transverse wave (with the displacement amplitude *perpendicular* to the propagation direction) cannot propagate in a fluid because it involves *shear* displacements and the shear modulus is identically zero in a fluid (see Problem 3.21).

The situation is more complicated in solids, where more than one elastic modulus is non-zero: for example, even for an *isotropic* solid, there are *two* non-zero values (λ, μ), and for *cubic* materials, there are three (c_{11}, c_{12} and c_{44}) (see §3.4.3). As a consequence, both longitudinal and transverse acoustic modes exist in isotropic solids, and have *different* sound velocities (see Problem 4.1). In fact, there are *two* independent transverse (shear) modes, with the polarization vectors $\boldsymbol{u}^0$ perpendicular to each other and both perpendicular to the propagation direction vector $\boldsymbol{k}$. However, longitudinal and transverse waves cannot be so easily distinguished in anisotropic crystals: for a general $\boldsymbol{k}$ direction, $\boldsymbol{u}^0$ is not necessarily parallel to $\boldsymbol{k}$ for a longitudinal-type mode.

★ The wave equation for an elastic wave travelling in a solid characterized by an elastic-stiffness tensor $\boldsymbol{c}$ can be obtained from the expression for the elastic-energy density (energy per unit volume):

$$\tilde{U} = \frac{1}{2}\sum_{i,j,k,l} c_{ijkl} e_{ij} e_{kl}^{*}, \tag{4.9}$$

where the full fourth-rank tensor nature of $\boldsymbol{c}$ (rather than the abbreviated form used in eqn. (3.97)) has been employed, and the complex conjugate of the strain is included to ensure that $\tilde{U}$ is a real quantity, because the displacement amplitude of the wave is complex (cf. eqn. (4.4)), viz.

$$\boldsymbol{u}(\boldsymbol{r}, t) = \boldsymbol{u}^0 \exp[\mathrm{i}(\boldsymbol{k} - \omega t)], \tag{4.10}$$

where $\boldsymbol{k} = (k_1, k_2, k_3)$. Thus, from the definition of strain (eqn. (3.94b)) and eqn. (4.10), the time-dependent strain is:

$$e_{ij}(t) = \mathrm{i}[k_j u_i(t) + k_i u_j(t)] \tag{4.11a}$$

$$= \mathrm{i}(k_j u_i^0 + k_i u_j^0) \exp[\mathrm{i}(\boldsymbol{k}\cdot\boldsymbol{r} - \omega t)]. \tag{4.11b}$$

For a volume element $\mathrm{d}V$ of the material, Newton's equation of motion becomes

$$\rho \frac{\partial^2 u_i}{\partial t^2} \mathrm{d}V = -\frac{\partial}{\partial u_i} \mathrm{d}\mathscr{E} \tag{4.12a}$$

where the right-hand side represents the force and $\mathrm{d}\mathscr{E} = \tilde{U}\mathrm{d}V$. The left-hand side of eqn. (4.12a) becomes

$$\rho \frac{\partial^2 u_i}{\partial t^2} \mathrm{d}V = -\rho\omega^2 u_i^0 \mathrm{d}V \tag{4.12b}$$

using eqn. (4.10), and the right-hand side can be written as

$$\begin{aligned} -\frac{\partial}{\partial u_i}(\mathrm{d}\mathscr{E}) &= -\frac{\mathrm{d}V}{2}\sum_{j,k,l} c_{ijkl}\left(\frac{\partial e_{ij}}{\partial u_i}\right) e_{kl}^{*} \\ &= -\frac{\mathrm{d}V}{2}\sum_{j,k,l} c_{ijkl} k_j (k_k u_l + k_l u_k) \\ &= -\mathrm{d}V \sum_{j,k,l} c_{ijkl} k_j k_k u_l. \end{aligned} \tag{4.13}$$

Thus, eqn. (4.12) becomes

$$\rho\omega^2 u_i^0 = \sum_{j,k,l} c_{ijkl} k_j k_k u_l^0, \tag{4.14}$$

which may be written in matrix notation as

$$\rho\omega^2 \boldsymbol{u}^0 = \boldsymbol{M} \cdot \boldsymbol{u}^0 \tag{4.15}$$

where the components of the dynamical matrix $\boldsymbol{M}$ are given by

$$M_{il} = \sum c_{ijkl} k_j k_k. \tag{4.16}$$

For the case of a cubic crystal, the symmetric dynamical matrix has the form (for the upper right-hand corner only—cf. Problem 3.19):

$$\boldsymbol{M} = \begin{bmatrix} c_{11}k_1^2 + c_{44}(k_2^2 + k_3^2) & (c_{12} + c_{44})k_1k_2 & (c_{12} + c_{44})k_1k_3 \\ & c_{11}k_2^2 + c_{44}(k_1^2 + k_3^2) & (c_{12} + c_{44})k_2k_3 \\ & & c_{11}k_3^2 + c_{44}(k_1^2 + k_2^2) \end{bmatrix}. \tag{4.17}$$

The sound velocities for elastic waves propagating along different high-symmetry directions in a cubic material can be obtained by solving the matrix eqn. (4.15). The results are given in Table 4.1. Note that longitudinal-acoustic (LA) sound velocities are always greater than transverse-acoustic (TA) sound velocities. Fig 4.2 shows the displacements associated with representative sound waves propagating along certain symmetry directions and with particular displacement vectors $\boldsymbol{u}^0$.

Finally, note that if $c_{11} \leqslant c_{12}$, a cubic crystal is *unstable* with respect to the atomic motions associated with the TA mode along [110] with the polarization vector along

Table 4.1 Sound velocities of acoustic modes in cubic crystals propagating in high-symmetry directions

Polarization	$\boldsymbol{k} = [100]$	$\boldsymbol{k} = [110]$		$\boldsymbol{k} = [111]$	
L	c_{11} ([100])	$\frac{1}{2}(c_{11} + c_{12} + 2c_{44})$	([110])	$\frac{1}{3}(c_{11} + 2c_{12} + 4c_{44})$	([111])
T	c_{44} ([010])	$\frac{1}{2}(c_{11} - c_{12})$	([$1\bar{1}0$])	$\frac{1}{3}(c_{11} - c_{12} + c_{44})$	([$1\bar{1}0$])
T	c_{44} ([001])	c_{44}	([001])	$\frac{1}{3}(c_{11} - c_{12} + c_{44})$	([$11\bar{2}$])

The values given in the table correspond to ρv^2, where ρ is the density, with the polarizations of the modes given in parentheses.

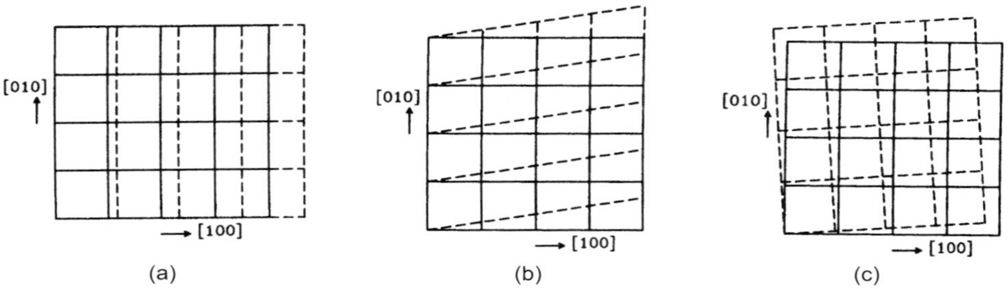

Fig. 4.2 Displacements of unit cells in cubic crystals associated with long-wavelength acoustic modes: (a) a longitudinal mode propagating along [100]; (b) a transverse mode (polarized along [010]) propagating along [100]; (c) a transverse mode (polarized along [$1\bar{1}0$]) propagating along [110]. (After Dove (1993). Reproduced by permission of Cambridge University Press)

$[1\bar{1}0]$ (see Table 4.1): a ferroelastic phase transition (§7.1.5.4) then occurs as a result of this *softening* of the TA mode (see Dove (1993) and Salje (1990)).

4.1.2 Counting vibrational states

The plane-wave-like vibrational wave represented by eqn. (4.10) is characterized by the wavevector $\boldsymbol{k}$ (the frequency ω being linearly related to $|\boldsymbol{k}|$ through eqn. (4.6)). As a general solution of the wave equation, eqn. (4.10) can take *all* values of $\boldsymbol{k}$: restrictions on the allowed values of $\boldsymbol{k}$ appear through the imposition of boundary conditions. Two types of boundary conditions can be envisaged, depending on whether standing waves or propagating waves (§4.2.1) are involved.

For the case of standing (stationary) waves, resulting from the interference between waves propagating in a particular direction and waves reflected back in the opposite direction from the surface of a sample of material, assumed for simplicity to be a cube of side L, the appropriate boundary condition for vibrational waves reflecting from mechanically *free* surfaces is that an *antinode* of the vibrational amplitude should exist at each surface (see Fig. 4.3). This then corresponds to there being an integral number of half-wavelengths of the standing wave along the length of the cube, i.e. the allowed values of the standing-wave wavevectors are given by

$$k_i^{\mathrm{s}} = n_i(\pi/L) \qquad (i = x, y, z), \tag{4.18}$$

where the n_i are non-zero *positive* integers (negative values do not give different standing waves). Thus, each allowed standing-wave solution of the wave equation (4.12) consistent with the boundary conditions is represented by a *point* in the reciprocal space containing the $\boldsymbol{k}$-vectors (see Fig. 4.4 for a 2D illustration). The spacing between

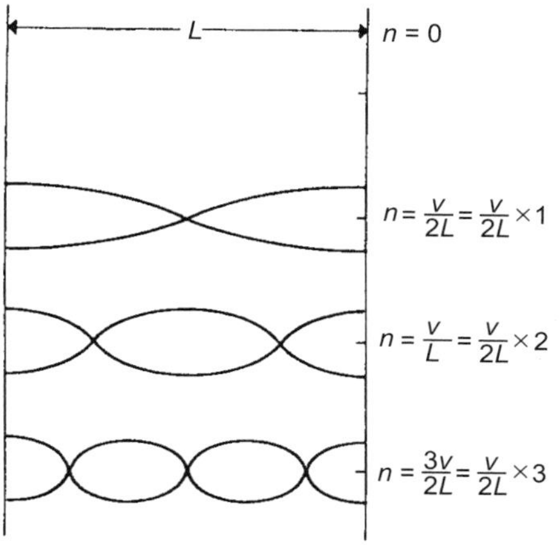

Fig. 4.3 Schematic 1D illustration of standing transverse elastic waves set up between the free surfaces of a cube of an elastic continuum with antinodes at the free surfaces. The boundary condition corresponds to there being an integral number of half-wavelengths along the length L of the cube.

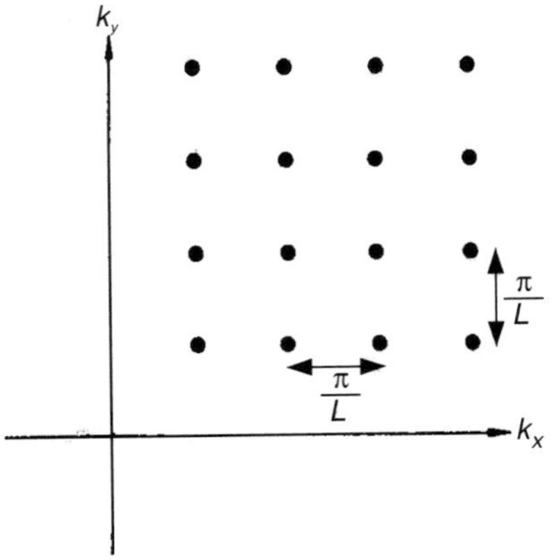

Fig. 4.4 2D illustration of the allowed values in reciprocal (***k***−) space of the standing-wave solutions of the vibrational wave equation for a square of side L of a continuous elastic medium.

allowed ***k***-values is $\Delta k^s = \pi/L$ from eqn. (4.18), and hence the volume of ***k***-space corresponding to one ***k***-value (standing-wave state) is

$$V_k^s = \left(\frac{\pi}{L}\right)^3 \tag{4.19}$$

or, equivalently, the number of ***k***-values (standing-wave states) contained in unit volume of ***k***-space is

$$\rho_k^s = \frac{V}{\pi^3}, \tag{4.20}$$

where V is the sample volume ($= L^3$). For large samples, the spacing between the points in ***k***-space is effectively infinitesimal, and k can be taken as a continuous variable rather than as a discrete quantity.

Thus, for an isotropic solid, the number of distinct standing-wave states, for given polarization type i, having wavevectors between k and $k + dk$ is equal to the volume of the *positive octant* ($k_i \geqslant 0, i = x, y, z$) of a spherical shell of radius k and thickness dk, multiplied by the ***k***-space density ρ_k^s, i.e.

$$g_i(k)\mathrm{d}k = \frac{1}{8}\frac{V}{\pi^3}4\pi k^2 \mathrm{d}k = \frac{Vk^2\mathrm{d}k}{2\pi^2}. \tag{4.21}$$

The density of states for a mode i in terms of frequency, $g_i(\omega)$ (the number of allowed standing-wave states with frequencies between ω and $\omega + d\omega$) is obtained from eqn. (4.21), since $g_i(\omega)d\omega = g_i(k)\mathrm{d}k$, by making use of the linear dispersion relation (eqn. (4.6)) valid for long-wavelength acoustic modes, giving:

$$g_i(\omega)\mathrm{d}\omega = \frac{V\omega^2}{2\pi^2 v_i^3}\mathrm{d}\omega, \tag{4.22}$$

where v_i is an appropriate sound velocity for mode i. However, *three* acoustic modes, one LA mode and two degenerate TA modes, can propagate in continuous elastic media (see Problem 4.1), and so the total vibrational density of states is thus given by

$$g(\omega)\mathrm{d}\omega = \frac{V\omega^2}{2\pi^2}\left(\frac{1}{v_1^3} + \frac{2}{v_t^3}\right)\mathrm{d}\omega \equiv \frac{3V\omega^2}{2\pi^2 v_0^3}\mathrm{d}\omega, \tag{4.23}$$

where $v_0 = (v_1^{-3} + 2v_t^{-3})^{-1/3}$ is an appropriate average of the LA-and TA-mode velocities. This quadratic frequency dependence is the so-called Debye density of vibrational states.

Although the analysis so far has treated a solid as being an elastic continuum, nevertheless real solids are comprised of atoms. Each atom has *three* degrees of dynamical freedom, and so for a sample containing N atoms, there is a total of $3N$ degrees of freedom which are vibrational in character (except for the $k = 0$ state which corresponds to uniform translational motion of the sample). This constraint on the number of degrees of freedom imposes a limit on the maximum frequency $\omega_{max} = \omega_D$ (or equivalently, the maximum wavevector, k_D) that can exist; this can be evaluated by integrating over the vibrational density of states up to the limiting frequency and setting the total number of states equal to $3N$, viz.

$$3N = \frac{3V}{2\pi^2 v_0^3}\int_0^{\omega_\mathrm{D}} \omega^2 \mathrm{d}\omega, \tag{4.24}$$

with the Debye frequency given by

$$\omega_\mathrm{D} = \left(\frac{6\pi^2 N}{V}\right)^{1/3} v_0. \tag{4.25}$$

The Debye density of states is illustrated in Fig. 4.5.

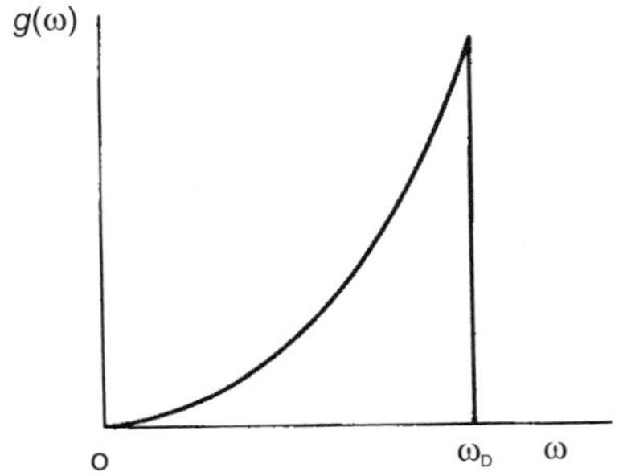

Fig. 4.5 The Debye density of vibrational states.

Vibrations of periodic systems 4.2

The presence of translational periodicity has a profound effect on the vibrational behaviour when the wavelength of the vibrational excitations becomes comparable to the periodic repeat distance, a. For $\lambda \gg a$, however, the behaviour characteristic of an elastic continuum is recovered.

4.2.1 Counting running-wave states

The presence of structural periodicity in a solid imposes translational periodicity on the excitations (e.g. vibrational, electronic) that exist within it. Thus, for a *running wave* of the form of eqn. (4.10) that is the solution of the vibrational equation of motion for a periodic array of atoms, with length L, periodic (or Born–von Karman) boundary conditions are appropriate, i.e.

$$u(r) = u(r + L). \tag{4.26}$$

Such a boundary condition can be envisaged as follows. In the case of a linear chain of N particles, where nearest neighbours are connected by springs (representing bonds between atoms), with equilibrium spacing a, periodic boundary conditions are achieved by connecting one end of the chain to the other to form a ring of length $L = Na$ (Fig. 4.6a). Hence, an integral number of wavelengths must fit into the length L, resulting in the allowed $\boldsymbol{k}$-values for running-wave states:

$$k_i^{\mathrm{r}} = 0, \pm n_i\left(\frac{2\pi}{L}\right) \ldots, \frac{N\pi}{L} \quad (i = x, y, z). \tag{4.27}$$

An equivalent way of understanding periodic boundary conditions, which is more realistic for other than 1D systems, involves the imposition of a mechanical constraint forcing atom N to interact with atom 1 via a massless, rigid rod and a spring (see Fig. 4.6b).

Note that, in contrast to the case for fixed boundary conditions leading to stationary waves (eqn. (4.18)), both positive and negative integers are allowed for running-wave solutions, and moreover the spacing between allowed $\boldsymbol{k}$-values is $\Delta k^{\mathrm{r}} = 2\pi/L$, *twice* that for standing-wave states. Thus, the number of $\boldsymbol{k}$-values, corresponding to running-wave states, contained in unit volume of $\boldsymbol{k}$-space is now

$$\rho_k^{\mathrm{r}} = \frac{V}{8\pi^3}. \tag{4.28}$$

Hence, the number of distinct states, for a given polarization type i, having wavevectors between k and $k + \mathrm{d}k$ is ρ_k^{r} multiplied by the volume of an entire spherical shell in $\boldsymbol{k}$-space (since both positive and negative $\boldsymbol{k}$-values are allowed), i.e.

$$g_i(k)\mathrm{d}k = \frac{V}{8\pi^3} 4\pi k^2 \mathrm{d}k = \frac{Vk^2 \mathrm{d}k}{2\pi^2}.$$

This is identical to eqn. (4.21).

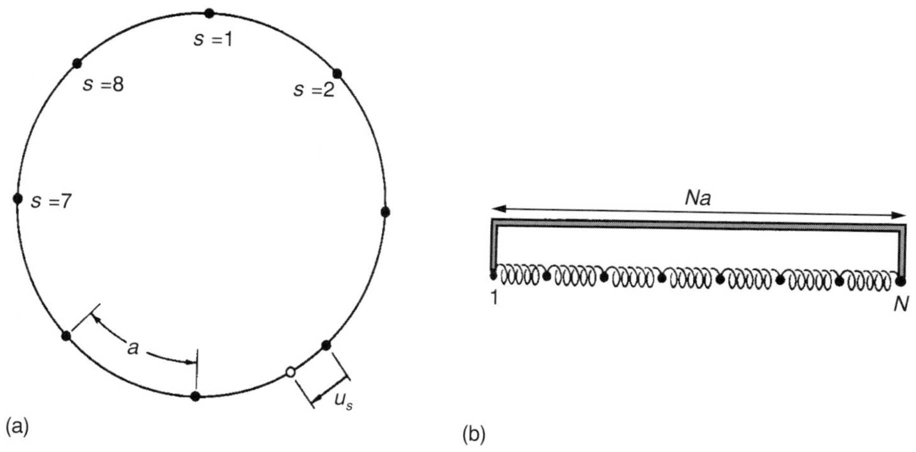

Fig. 4.6 Representation of periodic boundary conditions for eight particles interacting via springs. (a) A linear chain connected to form a ring of length $L = 8a$. For modes of the form $u_s \propto \exp(iska)$, periodic boundary conditions lead to eight modes (one per atom) with $k = 0$, $\pm 2\pi/L$, $\pm 4\pi/L$, $\pm 6\pi/L$, $8\pi/L$. (b) A mechanical constraint (a massless, rigid rod and a spring) connecting particle $N = 8$ with particle 1.

4.2.2 One-dimensional monatomic chain

The analysis of the vibrational behaviour of real 3D solids, even those with the simplifying feature of translational periodicity, is extremely complex and can only be achieved by numerical means using a computer. Nevertheless, certain simple models that can be treated analytically serve to illustrate many of the vibrational features exhibited by real materials. The simplest such model is the 1D monatomic chain.

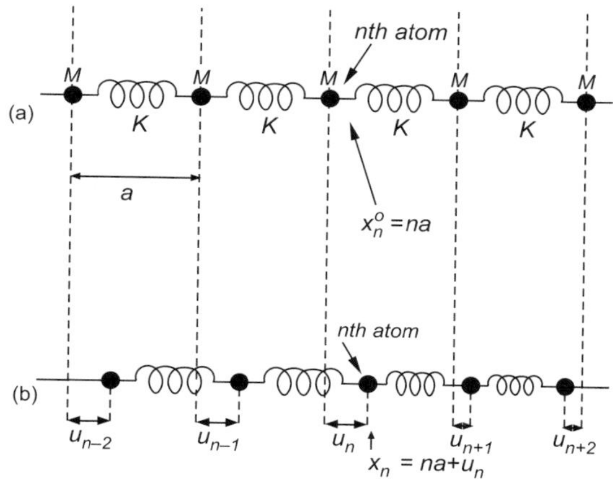

Fig. 4.7 A periodic one-dimensional chain of identical masses M connected by springs: (a) at the equilibrium positions $x_n^0 = na$; (b) at displaced positions $x_n = na + u_n$.

At equilibrium, a chain of N atoms, each of mass M, has an interatomic spacing equal to the unit-cell length a (Fig. 4.7a). If nearest-neighbour interactions are dominant, and the energy between two neighbours at a distance a is $\phi(a)$, then the total potential energy of the chain at rest (subject to periodic boundary conditions—§4.2.1) is:

$$U = N\phi(a). \tag{4.29}$$

If now the nth atom is displaced by a distance u_n, with all atoms being displaced by concomitant amounts (see Fig. 4.7b), the total potential energy of the chain can be calculated using a Taylor series expansion about a, and summing over all atoms, viz.:

$$U = N\phi(a) + \sum_{m\geqslant 1} \frac{1}{m!}\frac{\partial^m \phi}{\partial u^m} \sum_{n}^{N} (u_n - u_{n+1})^m. \tag{4.30}$$

Since a is the equilibrium separation, the first derivative of $\phi(m = 1)$ is identically zero (there is no net force on an atom at equilibrium) and the next, quadratic term $(m = 2)$ will be the dominant term. Neglecting the anharmonic terms $(m > 2)$ gives the harmonic approximation. Thus, the harmonic vibrational energy of the 1D chain is, from eqn. (4.30):

$$U^{\text{harm}} = \frac{1}{2}K\sum_{n}(u_n - u_{n+1})^2 \tag{4.31}$$

where the spring constant is given by

$$K = \left(\frac{\partial^2 U}{\partial u^2}\right)_{u=0}. \tag{4.32}$$

The equation of motion for the nth atom with mass M can be written, using Newton's equation, as the harmonic expression involving coupled atomic displacements:

$$M\frac{\partial^2 u_n}{\partial t^2} = -\frac{\partial U^{\text{harm}}}{\partial u_n} = -K(2u_n - u_{n+1} - u_{n-1}), \tag{4.33}$$

where the force, represented by the right-hand side, is obtained by differentiation of the two terms involving u_n in the series expansion of eqn. (4.31), namely $(u_{n-1} - u_n)^2$ and $(u_n - u_{n+1})^2$. A trial solution is a linear superposition of travelling waves having the form of eqn. (4.4), with the amplitude of displacement of the nth atom being:

$$u_n(t) = \sum_k u_k^0 \exp\{\mathrm{i}(kx_n^0 - \omega_k t)\}, \tag{4.34}$$

where the discrete values of k arising from the periodic boundary conditions (eqn. (4.27)) are used as labels for the waves, and where $x_n^0 = na$. Substitution of eqn. (4.34) into eqn. (4.33) for *one* particular k-value gives

$$\begin{aligned} &- M\omega_k^2 u_k^0 \exp\{\mathrm{i}(kna - \omega_k t)\} = \\ &- K u_k^0 \exp(-\mathrm{i}\omega_k t)[2\exp(\mathrm{i}kna) - \exp[\mathrm{i}k(n-1)a] - \exp[\mathrm{i}k(n+1)a]] \end{aligned} \tag{4.35}$$

or, on cancelling the terms u_k^0 and $\exp\{\mathrm{i}(kna - \omega_k t)\}$,

$$\begin{aligned} \omega_k^2 &= \frac{2K}{M}(1 - \cos ka) \\ &= \frac{4K}{M}\sin^2(ka/2) \end{aligned}$$

so that

$$\omega_k = 2\left(\frac{K}{M}\right)^{1/2} |\sin(ka/2)|. \tag{4.36}$$

The maximum ('cut-off') value of the vibrational frequency of the 1D periodic chain is thus when $\sin(ka/2) = 1$, i.e.

$$\omega_k^{\max} = 2\sqrt{\frac{K}{M}} \tag{4.37}$$

for **k**-values such that

$$k' = \frac{n\pi}{a} \quad (n = \pm 1, \pm 2, \text{etc.}). \tag{4.38}$$

The vibrational frequency ω_k is no longer proportional to the wavevector k for all k-values as is the case for continuous elastic media (eqn. (4.6)), but has a periodic (sinusoidal) behaviour (eqn. (4.36)) with the periodicity $(2\pi/a)$ of the *reciprocal lattice* (§2.4.1). The dispersion curve (variation of ω_k with k) for the linear monatomic chain is shown in Fig. 4.8.

Note that the dispersion relation (eqn. (4.36)) does *not* depend on the index n for an individual atom in the chain: *all* atoms in the chain contribute equally to a *collective* vibrational mode, and each allowed k-state corresponds to a normal mode of the system that oscillates independently of all other normal modes (in the harmonic approximation).

Inspection of the allowed k-states consistent with periodic boundary conditions (eqn. (4.27)) shows that there are a total of N k-states (i.e. one per atom) in the range $\Delta k = 2\pi/a$, (more explicitly in the range $-\pi/a < k \leqslant \pi/a$) corresponding to the region between two maxima either side of the origin in the dispersion curve (eqn. (4.38)). This region of k-space marks the extent of the first Brillouin zone (see §2.4.2) which is the primitive, lattice-point-centred cell of the reciprocal lattice. Although allowed solutions of the wave equation do exist for k-values lying in higher Brillouin zones in the repeated

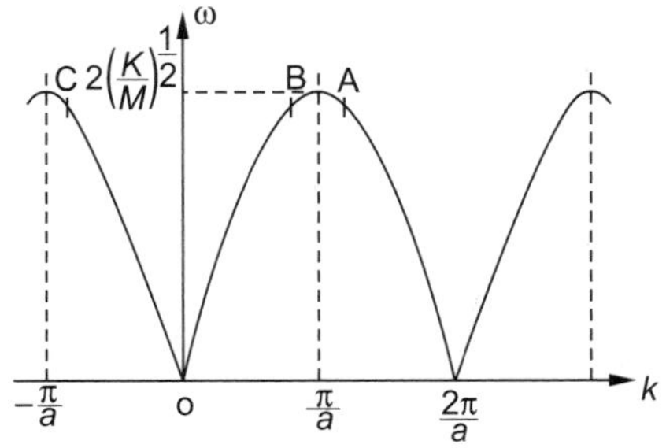

Fig. 4.8 Dispersion relation for the normal-mode frequencies of a monatomic chain with periodicity a drawn in the repeated-zone scheme. It has been assumed that the allowed k-values are sufficiently closely spaced that the dispersion relation can be drawn as a continuous curve. Points A, B, C, having the same frequency, correspond to the same instantaneous displacements (see Fig. 4.9): point B represents a wave travelling to the right, whereas points A and C refer to a wave moving to the left. (After Hook and Hall (1991). Reproduced by permission of John Wiley & Sons Inc.)

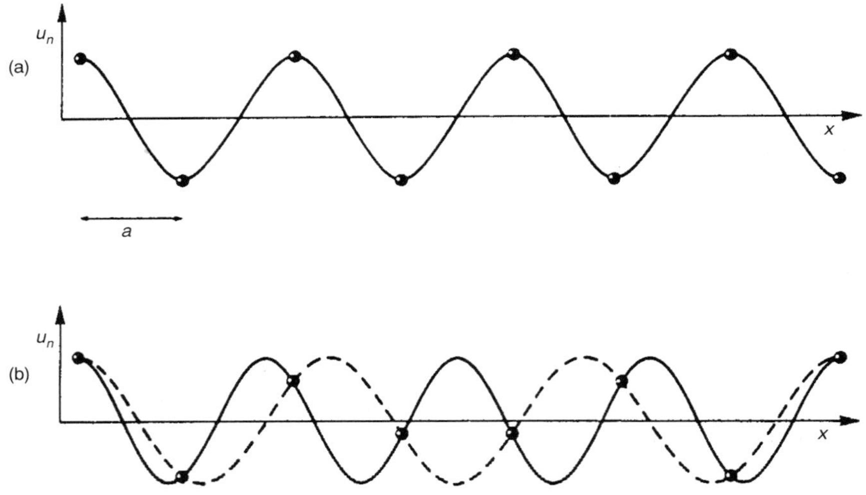

Fig. 4.9 Atomic displacements of the linear monatomic chain (shown as transverse, rather than longitudinal, for clarity) for the wavevectors: (a) $k = \pi/a$ (zone boundary); (b) $k = 8\pi/7a$ (second Brillouin zone), full curve; and $k = 6\pi/7a$ (first Brillouin zone), dashed curve. The *atomic* displacements in (b) are identical, and the two k-values correspond to points A and B in Fig. 4.8. (After Hook and Hall (1991). Reproduced by permission of John Wiley & Sons Inc.)

zone scheme (see Fig. 4.8), such states do *not* represent physically distinct solutions: these are contained entirely within the first Brillouin zone. States separated by a *reciprocal-lattice vector* $\boldsymbol{G} = 2\pi/a$ (e.g. A and C in Fig. 4.8) are *identical* because $u_{k+G} = A\exp\{i(k+G)na\} = A\exp(ikna)\exp(iGna) = u_k$, since $\exp(iGna) = 1$ (cf. eqn. (2.52)). This behaviour is illustrated in Fig. 4.9.

For very small values of wavevector, $ka \ll 1$, i.e. in the long-wavelength limit, the dispersion relation (eqn. (4.36)) becomes

$$\omega_k = a\left(\frac{K}{M}\right)^{1/2} k. \tag{4.39}$$

Thus, the phase velocity (eqn. (4.6)) and group velocity (eqn. (4.8)) are equal, i.e. the (sound) waves are dispersionless, with

$$v = v_g = a\left(\frac{K}{M}\right)^{1/2}. \tag{4.40}$$

Since the mass per unit length of the 1D chain is $\rho = M/a$, and the elastic modulus c, defined as force = elastic modulus times strain for a 1D system (cf. eqn. (3.97)), is $c = Ka$, then eqn. (4.40) can be rewritten as

$$v = \left(\frac{c}{\rho}\right)^{1/2},$$

in agreement with the result obtained from elastic-continuum theory (eqn. (4.7)).

The group velocity (eqn. (4.8)) calculated for the dispersion relation of the linear monatomic chain (eqn. (4.36)) is, in general:

$$v_g = a\left(\frac{K}{M}\right)^{1/2} \cos(ka/2). \tag{4.41}$$

Thus, the group velocity goes to *zero* at the Brillouin zone boundaries, $k = \pm\pi/a$. Hence, the states corresponding to these k-values are *not* travelling waves, but instead are *standing* waves. Such behaviour can be understood as the result of interference between travelling waves moving in *opposite* directions as a consequence of *Bragg reflection* (§2.6.1.1) from the periodic array of atoms. The Bragg condition (eqn. (2.95)) can be written as

$$n\lambda = n2\pi/k' = 2a \sin\theta = 2a$$

or

$$k' = n\pi/a, \tag{4.42}$$

where back-reflection along the chain corresponds to a Bragg scattering angle of $2\theta = 180°$. Thus, eqn. (4.42) corresponds to the Brillouin-zone boundaries (eqn. (4.38)).

4.2.3 One-dimensional chain with a basis

Most crystalline solids contain more than one atom per unit cell, often as a result of the composition but sometimes as a consequence of the structure even for monatomic systems (e.g. the diamond-cubic structure, §2.2.5.2). Thus, it is of interest to examine the vibrational behaviour of a periodic, one-dimensional chain consisting of *two* types of atoms, as a model for real materials. Qualitatively different results are obtained from those characteristic of the monatomic chain discussed in the previous section.

The model is illustrated in Fig. 4.10, and is a generalization of that in Fig. 4.7, with alternating unequal masses m and $M > m$ connected by identical springs and separated by an equilibrium distance $a/2$; the unit-cell spacing is thus a. Two equations of motion can be formulated (cf. eqn. (4.33)), one for the masses M, i.e.

$$M\frac{\partial^2 u_n^M}{\partial t^2} = -K(2u_n^M - u_{n+1}^m - u_{n-1}^m) \tag{4.43a}$$

and the other for the masses m, i.e.

$$m\frac{\partial^2 u_{n-1}^m}{\partial t^2} = -K(2u_{n-1}^m - u_n^M - u_{n-2}^M). \tag{4.43b}$$

Assume that the displacements of the M atoms have a running-wave solution of the form of eqn. (4.34) for *one* k-vector, i.e.

$$u_n^M = A \exp\{i(kx_n^0 - \omega_k t)\} \tag{4.44a}$$

where $x_n^0 = na/2$, and a similar expression exists for the m masses, but multiplied by a complex factor α representing the relative amplitude and phase, i.e.

$$u_n^m = \alpha A \exp\{i(kx_n^0 - \omega_k t)\}. \tag{4.44b}$$

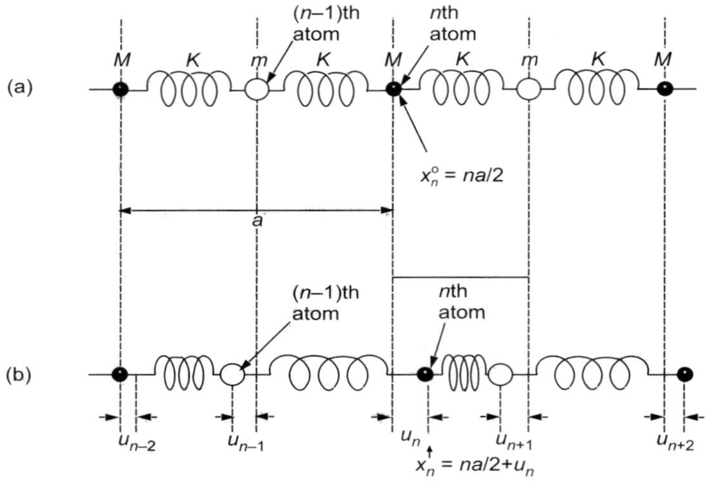

Fig. 4.10 A periodic one-dimensional chain, with unit-cell dimension a, consisting of alternating unequal masses connected by identical springs: (a) at the equilibrium positions $x_n^0 = na/2$; (b) at displaced positions $x_n = na/2 + u_n$.

Substituting eqns. (4.44) into (4.43) gives

$$\begin{aligned} -\omega_k^2 M \exp\{\mathrm{i}(kna/2 - \omega_k t)\} = &- K[2\exp\{\mathrm{i}(kna/2 - \omega_k t)\} \\ &- \alpha \exp\{\mathrm{i}[k(n+1)a/2 - \omega_k t]\} \\ &- \alpha \exp\{\mathrm{i}[k(n-1)a/2 - \omega_k t]\}] \end{aligned} \tag{4.45a}$$

and

$$\begin{aligned} -\alpha\omega_k^2 m \exp\{\mathrm{i}[k(n-1)a/2 - \omega_k t]\} = &- K[2\alpha \exp\{\mathrm{i}[k(n-1)a/2 - \omega_k t]\} \\ &- \exp\{\mathrm{i}(kna/2 - \omega_k t)\} \\ &- \exp\{\mathrm{i}[k(n-2)a/2 - \omega_k t]\}] \end{aligned} \tag{4.45b}$$

or, simplifying:

$$\omega_k^2 M = 2K[1 - \alpha\cos(ka/2)\,] \tag{4.46a}$$

and

$$\alpha\omega_k^2 m = 2K[\alpha - \cos(ka/2)]. \tag{4.46b}$$

Solving for α, eqn. (4.46) may be rewritten as

$$\alpha = \frac{2K\cos(ka/2)}{2K - \omega_k^2 m} = \frac{2K - \omega_k^2 M}{2K\cos(ka/2)}, \tag{4.47}$$

which may be further rearranged in the form of a quadratic equation in ω_k^2, i.e.

$$mM\omega_k^4 - 2K(M+m)\omega_k^2 + 4K^2\sin^2(ka/2) = 0. \tag{4.48}$$

This has the solutions:

$$\omega_k^2 = \frac{K(M+m)}{Mm} \pm K\left[\left(\frac{M+m}{Mm}\right)^2 - \frac{4}{Mm}\sin^2(ka/2)\right]^{1/2}. \tag{4.49}$$

Thus, for any wavevector k, there are *two* frequencies corresponding to two branches (see Fig. 4.11), the upper branch corresponding to the positive sign in eqn. (4.49) and the lower branch to the negative sign. Each branch is periodic in reciprocal space, with a period equal to the reciprocal-lattice vector $2\pi/a$. The lower branch is very similar to the dispersion curve characteristic of the monatomic chain. Figure 4.8 shows the dispersion curves in the repeated-zone scheme.

It is instructive to investigate the behaviour of the dispersion curves of the two branches both near the zone centre ($k \simeq 0$) and also near the zone boundary ($k \simeq \pm\pi/a$) (see Problem 4.5). In the long-wavelength limit ($k \ll 1/a \simeq 0$), the *lower branch* has the behaviour:

$$\omega_k^2 \simeq \frac{Ka^2k^2}{2(M+m)} \tag{4.50a}$$

with

$$\alpha = 1. \tag{4.50b}$$

Thus, the two types of atom oscillate with the *same* amplitude and phase, and there is a linear dispersion of the frequency characterized by the sound velocity

$$v_l = a\left(\frac{K}{2(M+m)}\right)^{1/2}, \tag{4.51}$$

consistent with the expression eqn. (4.7) for an elastic continuum, since $\rho = (M+m)/a$ and $c = Ka/2$. Thus, this branch corresponds to a longitudinal acoustic (LA) mode.

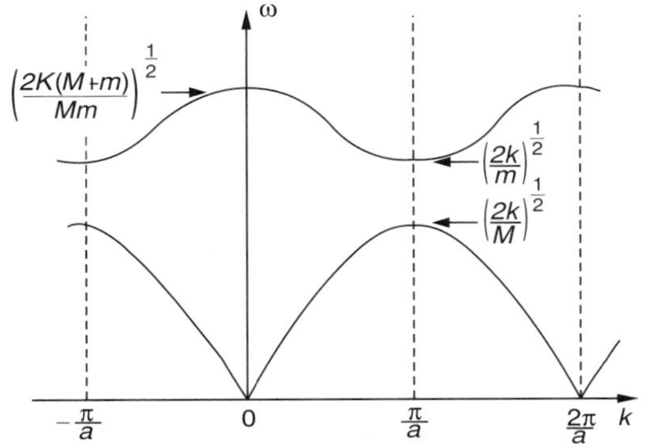

Fig. 4.11 Dispersion curves for the normal-mode frequencies of a chain, with periodicity a, consisting of two types of atom with masses m and M, drawn in the repeated-zone scheme. The lower branch corresponds to the longitudinal acoustic (LA) modes, and the upper branch to the longitudinal optic (LO) modes at $k = 0$.

At $k \simeq 0$, the *upper branch* has the behaviour

$$\omega_k^2 \simeq \frac{2K(M+m)}{Mm} \tag{4.52a}$$

with

$$\alpha \simeq 1 - M/m. \tag{4.52b}$$

Thus, this mode is independent of k in the vicinity of the zone centre. The value of α characterizing the mode at this k-value (eqn. (4.52b)) indicates that the two types of atom move in *antiphase* with their centres of mass at rest. If the two types of atom carry opposite electrical charges, as say in an alkali halide, this type of motion may be excited by the electric field of a light wave, and so this branch is termed the longitudinal optic (LO) mode.

At the zone boundary ($k = \pm\pi/a$), the maximum frequency of the *lower* (acoustical) branch is given by

$$\omega_k = \left(\frac{2K}{M}\right)^{1/2} \tag{4.53a}$$

with

$$\alpha = 0. \tag{4.53b}$$

Thus, the masses m are at rest and only the masses M oscillate. The corresponding frequency of the *upper* branch is

$$\omega_k = \left(\frac{2K}{m}\right)^{1/2} \tag{4.54a}$$

with

$$\alpha = \infty. \tag{4.54b}$$

Thus, in contrast, the M masses are at rest, and only the m masses vibrate. Figure 4.12 gives a representation of the atomic displacements corresponding to the four cases. Although acoustic- and optic-mode behaviour can be distinguished at the *zone centre*,

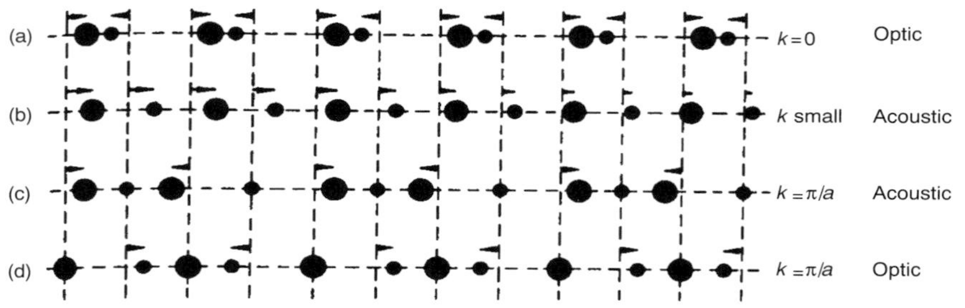

Fig. 4.12 Illustration of longitudinal optic (LO) and longitudinal acoustic (LA) modes for a linear diatomic chain, with equal force constants, at the zone centre and zone boundary. (After Burns (1985). Reproduced by permission of Academic Press, Inc.)

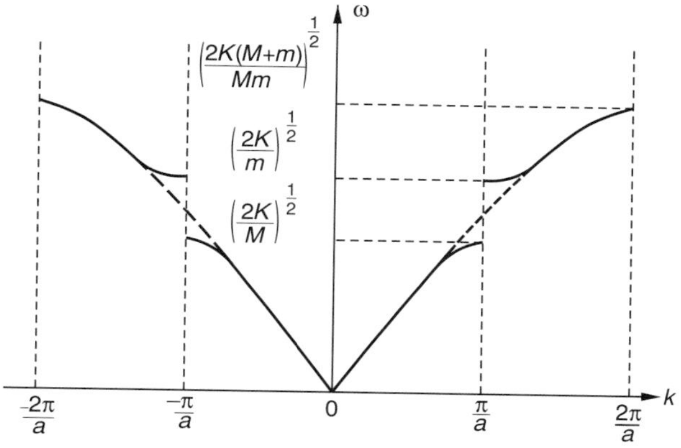

Fig. 4.13 Dispersion curve for the normal-mode frequencies of a diatomic chain represented in the extended-zone scheme. Note the forbidden gap in allowed vibrational frequencies at the zone boundaries, $k = \pm\pi/a$.

this is *not* generally the case at other values of $\boldsymbol{k}$. Nevertheless, the convention is to refer to the *whole* of the branch as being acoustic or optic, as the case may be.

For the dispersion curve plotted in Fig. 4.11, in the first Brillouin zone, $-\pi/a < k \leqslant \pi/a$, there are N modes associated with the lower branch and N modes for the upper branch if there are N masses in the chain.

An alternative way of representing the dispersion of the vibrational modes of a diatomic chain is in terms of the extended-zone scheme (Fig. 4.13), where only *one frequency* ω_k is associated with *each k-value* (optic modes in the first Brillouin zone (Fig. 4.11) are translated by a reciprocal-lattice vector to the second zone) and the $2N$ modes are now distributed in the range $-2\pi/a < k \leqslant 2\pi/a$. It can be seen that a forbidden gap has been opened up at $k = \pm\pi/a$. Wave-like solutions to the equation of motion do not exist in the gap region (B–C in Fig. 4.13); solutions with real values of frequency correspond to *complex* values of k in this region and so the wave is spatially damped. If the masses m and M are allowed to become equal, the gaps in Fig. 4.13 disappear, and the dispersion curve reverts to that for the monatomic chain (see Fig. 4.8, but recall that the value of a in Fig. 4.8 is *half* that in Fig. 4.13). In general, doubling the unit cell halves the Brillouin zone, and this causes folding of the dispersion relation into the reduced zone.

4.2.4 Three-dimensional crystals

The one-dimensional models discussed in the previous two sections provide a reasonable understanding for the behaviour of real, three-dimensional crystals. The principal difference is that in 3D solids transverse (shear) modes are allowed, as well as longitudinal modes. Except for high-symmetry directions in the Brillouin zone, for say cubic crystals, the transverse acoustic (TA) modes are non-degenerate and have different frequencies from the LA mode as a result of their differing sound velocities (see e.g. Table 4.1).

If there are N primitive cells in the crystal, each containing p atoms, the crystal has a total of $3pN$ dynamical degrees of freedom (since each atom has 3 degrees of freedom). Thus, since each individual branch of the vibrational dispersion curve has N allowed $\boldsymbol{k}$-values (in the first Brillouin zone), this implies that there must be $3p$ different branches. Of these, three are acoustic (LA and TA) modes, and the remaining $3(p-1)$ are optic modes (of which $(p-1)$ are longitudinal optic (LO) modes and $2(p-1)$ are transverse optic (TO) modes).

For 3D solids, the dispersion relation $\omega(\boldsymbol{k})$ is a surface in *four*-dimensional space and hence is impossible to display in its entirety. Instead, particular directions (usually associated with high symmetry) are chosen within the first Brillouin zone, and the dispersion relation can be plotted as a 2D graph for such one-dimensional trajectories in $\boldsymbol{k}$-space.

The calculation of the vibrational behaviour of 3D solids, starting say from an assumed interatomic potential, is in principle a straightforward procedure, along the lines used to analyse 1D models (§4.2.2 and 4.2.3), but in practice is complicated by the need to use matrix notation and the necessity of solving such matrix equations numerically using a computer, although symmetry relations can simplify the problem. Details are given, for example, in Ashcroft and Mermin (1976), Srivastava (1990) and Dove (1993).

An example of a vibrational dispersion curve for a crystalline solid having *one* atom as the basis of the unit cell, and hence which is a real 3D analogue of the linear monatomic chain model (§4.2.2), is for Ar, which crystallizes in the f.c.c. structure (§2.2.2.2); its dispersion curves are shown in Fig. 4.14. Qualitatively the same behaviour is exhibited as for the monatomic linear chain (Fig. 4.8) except that two additional TA modes are evident (non-degenerate along the $(0, \xi, \xi)$ direction with respect to the Brillouin zone). Another example is for b.c.c. K (Fig. 4.15): see Problem 4.6 for a discussion of the behaviour of these dispersion curves.

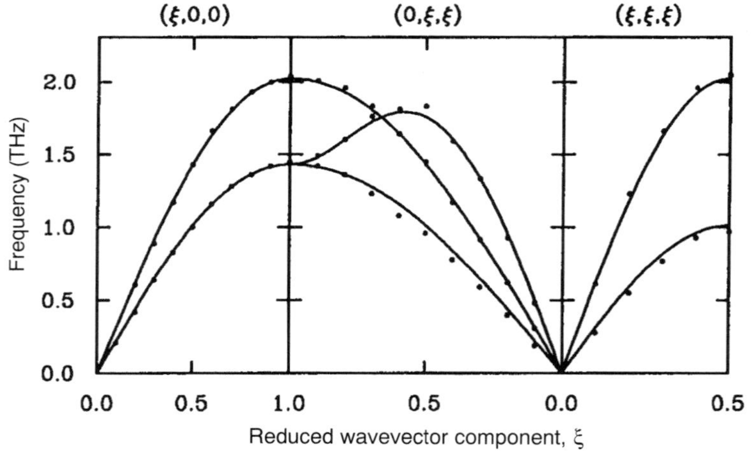

Fig. 4.14 Measured (dots) and calculated (lines) vibrational dispersion curves for crystalline Ar for three high-symmetry directions in reciprocal space with respect to the Brillouin zone. The reduced wavevector is normalized by the particular reciprocal-lattice vector (Reprinted with permission from Fujii *et al.* (1974), *Phys. Rev.* **B10**, 3647. © 1974. The American Physical Society)

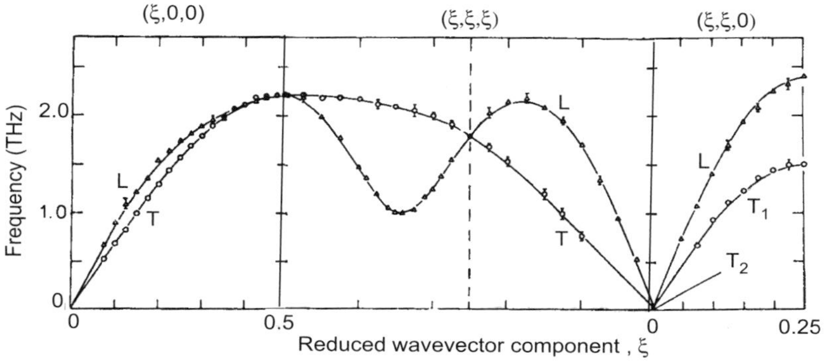

Fig. 4.15 Vibrational dispersion curves for b.c.c. potassium for different directions in reciprocal space. The dashed line in (b) corresponds to the position of the first Brillouin zone boundary in the [111] direction in reciprocal space. See Problem 4.6 for a discussion of aspects relating to this figure. Reprinted with permission from Cowley *et al.* (1966), *Phys. Rev.* **150**, 487. © 1966. The American Physical Society.

The alkali halides are examples of crystals having a basis of *two* (different) atoms in the unit cell. Since $p = 2$, three acoustic branches ($\omega \rightarrow 0$ as $\boldsymbol{k} \rightarrow 0$) and three optic ($\omega \neq 0$ at $\boldsymbol{k} = 0$) are expected. This behaviour is shown for NaCl in Fig. 4.16. Note that the TO and LO modes at the zone centre are *non*-degenerate: this so-called LO–TO splitting is characteristic of most ionic crystals (where the atomic-vibrational mode is associated with an oscillating electrical dipole), and is due to the fact that for LO modes at long wavelengths there is an additional restoring force resulting from the electric polarization field set up by the oscillating dipoles that does not affect the TO modes (see §4.4).

Another type of material that is analogous to the linear diatomic chain (§4.2.3) consists of monatomic crystals having structures with two atoms in the unit cell (e.g. the

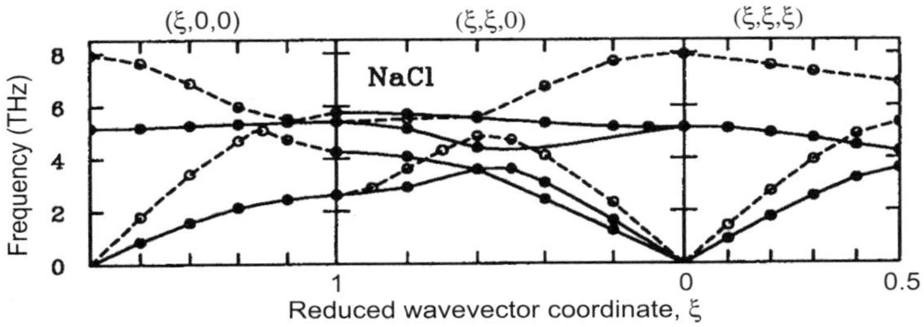

Fig. 4.16 Measured vibrational dispersion curves for NaCl. Transverse modes are shown as filled circles connected by continuous curves and longitudinal modes are denoted by open circles and dashed curves. Note the LO–TO splitting at $k = 0$, and the degeneracy of the transverse branches in high-symmetry directions. (Reprinted with permission from Raunio *et al.* (1969), *Phys. Rev.* **178**, 1496. © 1969. The American Physical Society)

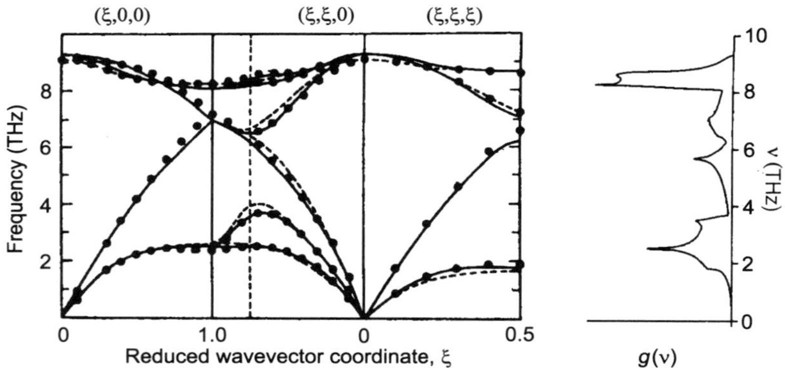

Fig. 4.17 Vibrational dispersion curves for Ge: the points are experimental data and the lines are calculated curves for the bond–charge model, where a covalent bond is modelled by positioning charges at the mid-bond sites. Note the degeneracy of the transverse modes in high-symmetry directions (($\xi, 0, 0$) and (ξ, ξ, ξ)) and the absence of an LO–TO splitting at $k = 0$ (Reprinted with permission from Weber (1977), *Phys. Rev.* **B15**, 4789. © 1977. The American Physical Society). Also shown is the corresponding density of states.

diamond-cubic structure (§2.2.5.2) and the h.c.p. structure (§2.2.3.2)). An example is germanium with the former structure, and its vibrational dispersion curves are shown in Fig. 4.17. Note that, because there is no electrical dipole associated with the two crystallographically distinct atoms, there is, concomitantly, no LO–TO splitting. Both transverse branches are doubly degenerate in the high-symmetry $k = (\xi, \xi, \xi)$ direction: the total number of branches is six since $p = 2$.

Compilations of measured vibrational dispersion curves have been given for metals by Willis and Pryor (1975), and for insulators by Bilz and Kress (1979) with an update by Dove (1993).

It was seen for the simple 1D models (§§4.2.2 and 4.2.3) that, at the zone boundary, the dispersion curve becomes *flat*, i.e. $\partial\omega/\partial k$ (and hence the group velocity) tends to zero. For the case of 3D crystals, which have 3D Brillouin zones, the component of $\nabla_k \omega$ *perpendicular* to the surface of the Brillouin zone boundary must vanish for there to be a flattening of the dispersion curve. This is exemplified by the dispersion curves of potassium shown in Fig. 4.15: the [111] direction in reciprocal space does *not* intersect at right angles the face of the rhombic dodecahedral Brillouin zone corresponding to the b.c.c. real-space lattice (see Fig. 2.41a), and so there is no flattening of the dispersion curve there.

Any flattening of the dispersion curve in $\boldsymbol{k}$-space can produce marked features in the corresponding vibrational density of states, $g(\omega)$, since then very many modes can exist in a small range of frequencies corresponding to many different $\boldsymbol{k}$-values. A general expression is required for the density of states in order to analyse this behaviour, and this can be obtained by generalizing the treatment given in §4.1.2. Since $g(\omega)$ is the number of modes (states) with frequencies between ω and $\omega + d\omega$, consider two constant-frequency surfaces in reciprocal space, one corresponding to the frequency ω and the other to the frequency $\omega + d\omega$ (see Fig. 4.18). The generalization of the expression

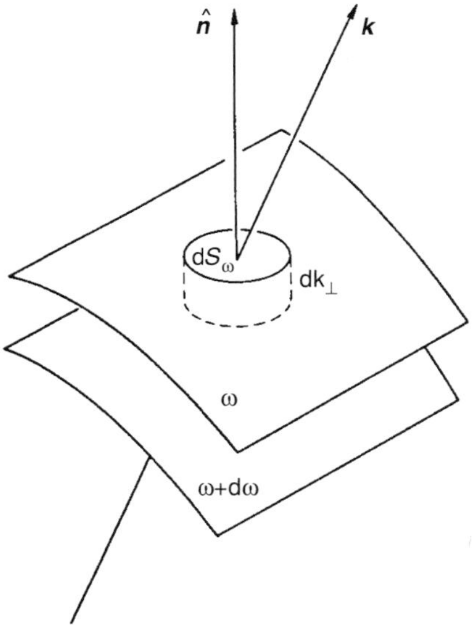

Fig. 4.18 Illustration of two constant-frequency surfaces in $\boldsymbol{k}$-space showing a surface element dS_ω and the wavevector interval $dk_\perp$ perpendicular to both surfaces at this point.

for $g(k)dk$ (eqn. (4.21)) valid in 1D (or effectively in one dimension for an isotropic solid) becomes for a general 3D solid:

$$g_{\rm i}(\omega)d\omega = \rho_k^r \int_{\rm shell} d\boldsymbol{k} = \frac{V}{(2\pi)^3} \int_{\rm shell} d^3k, \tag{4.55}$$

where V is the volume of the crystal, and where the integral is taken over the volume of the shell in $\boldsymbol{k}$-space between the two constant-frequency surfaces (Fig. 4.18). If dS_ω is an element of area on the surface ω = constant corresponding to a particular $\boldsymbol{k}$-vector and $dk_\perp$ is the distance in $\boldsymbol{k}$-space between the two surfaces at this point, normal to both surfaces (Fig. 4.18), then

$$\int_{\rm shell} d^3k = \int dS_\omega dk_\perp. \tag{4.56}$$

The vector gradient of the frequency, $\nabla_k\omega$, is also normal to the surface ω = constant, by definition, and so the change in frequency from one surface to the other is simply

$$d\omega = |\nabla_k\omega| dk_\perp. \tag{4.57}$$

Thus, the density of states (eqn. (4.55)) is given by an expression involving a *surface* integral in $\boldsymbol{k}$-space:

$$g_i(\omega) = \frac{V}{(2\pi)^3} \int_s \frac{dS_\omega}{|\nabla_k\omega|}, \tag{4.58}$$

where $|\nabla_k \omega| = v_g$, the magnitude of the group velocity, and where the index i indicates that the expression is for a single branch i only. The total density of states is then given by a sum over all branches of expressions of the form of eqn. (4.58). The Debye expression (eqn. (4.22)), valid for an elastic continuum, is recovered by noting that $|\nabla_k \omega| = v_g = v$ (the phase velocity) and the surface integral in eqn. (4.58) just gives the quantity $4\pi k^2 = 4\pi\omega^2/v^2$.

For the case of a *two-dimensional system*, by analogy with eqn. (4.58) the density of states is given by

$$g(\omega) = \frac{A}{(2\pi)^2} \int_1 \frac{dl_\omega}{|\nabla_k \omega|} \tag{4.59}$$

where the line integral is taken over a constant-frequency curve. In the case of a *one-dimensional system* (e.g. the monatomic chain, §4.2.2), the density of states is simply (see Problem 4.4):

$$g(\omega) = \frac{L}{\pi} \frac{1}{|\nabla_k \omega|}. \tag{4.60}$$

Thus, the vanishing of the group velocity ($\nabla_k \omega = 0$) at the zone boundary of a 1D crystal causes a van Hove singularity in the density of states itself (cf. eqn. (4.60)). For the case of 3D crystals, the infinities occur in the *curvature* of the dispersion curve, but not in the density of states. In the vicinity of a critical point at $\boldsymbol{k} = \boldsymbol{k}_c$, since $\partial\omega/\partial\boldsymbol{k} = 0$, the behaviour of the dispersion relation can be approximated using a Taylor expansion, i.e.

$$\omega(\boldsymbol{k}) \simeq \omega(\boldsymbol{k}_c) + \sum_{j=1}^{3} a_j (k_{cj} - k_j)^2, \tag{4.61}$$

where k_j is a component of $\boldsymbol{k}$. This implies that the density of states for 3D solids behaves as $(\omega_c - \omega)^{-1/2}$ in the vicinity of the van Hove singularities. Figure 4.17 shows the vibrational density of states for crystalline Ge: the origins of the van Hove critical points can be identified in the corresponding dispersion curves.

4.2.5 Quantization of vibrational modes: phonons

Thus far, the treatment of the vibrational behaviour of materials has been entirely *classical*. For a harmonic solid, the vibrational excitations are the collective, independent normal modes, having frequencies ω_k determined by the dispersion relation $\omega_k(\boldsymbol{k})$ with the allowed values of $\boldsymbol{k}$ set by the boundary conditions. In a proper, quantum-mechanical treatment, the eigenvalues of the Hamiltonian, involving the kinetic and (harmonic) potential energies of the atoms in a solid, need to be obtained. This is a standard problem in quantum mechanics (see e.g. Ashcroft and Mermin (1976), Dove (1993) and Kittel (1996)) but, because the level of analysis required is somewhat beyond the level assumed in this book, we will simply quote the result here.

In the classical limit, the energy of a given normal mode with frequency ω_k, determined by the wave amplitude, can take *any* value. By contrast, the quantum-mechanical result, treating each normal mode as an independent harmonic oscillator with frequency

ω_k, is that the energy is *quantized* and can only take the values characterized by the *integers* $n(\boldsymbol{k}, p)$:

$$\mathscr{E}(\boldsymbol{k}, p) = (n(\boldsymbol{k}, p) + 1/2)\hbar\omega_k(p) \tag{4.62}$$

for a particular branch, p. A vibrational state of the whole crystal is thus specified by giving the excitation numbers $n(\boldsymbol{k}, p)$ for each of the $3N$ normal modes, and the total vibrational energy is hence the sum of the energies of each of the $3N$ normal modes

$$\mathscr{E} = \sum_{\boldsymbol{k},p} \mathscr{E}(\boldsymbol{k}, p). \tag{4.63}$$

Instead of describing the vibrational state of a crystal in terms of the excitation number $n(\boldsymbol{k}, p)$ of the normal mode with wavevector $\boldsymbol{k}$ in branch p, it is more convenient and conventional to say, equivalently, that there are $n(\boldsymbol{k}, p)$ phonons (i.e. particle-like entities representing the quantized elastic waves). This is by direct analogy with the case of the quantized *electromagnetic field*, where the allowed energies of a normal mode of the radiation field in a cavity are given by $(n + 1/2)\hbar\omega$ (cf. eqn. (4.62)), where ω is the mode frequency, and n is taken to be the number of *photons* present with frequency ω, rather than, equivalently, as the excitation quantum number of the cavity mode. Note that the quantum-mechanical expression for the energy (eqn. (4.62)) implies that the vibrational energy of a solid is *non-zero* even when there are no phonons present: the residual energy of a given mode, $\mathscr{E}_0(\boldsymbol{k}, p) = (1/2)\hbar\omega_k(p)$, is the zero-point energy.

The number of phonons, with a particular wavevector or frequency, is determined by the *temperature* when a solid is in thermal equilibrium. The exact functional form for the dependence of $n(\boldsymbol{k}, p)$ on temperature can be obtained using the methods of statistical mechanics, namely by use of the partition function, Z, defined generally as

$$Z = \sum_{i=1}^{\infty} \exp(-\mathscr{E}_i/k_B T), \tag{4.64}$$

where $\mathscr{E}_i$ is the energy of the ith excited state. The partition function for the phonons associated with the normal modes of a crystal is thus, using eqn. (4.62) (but dropping the branch label for clarity)

$$\begin{aligned} Z &= \sum_{n=0}^{\infty} \exp[-(n + 1/2)\hbar\omega_k/k_B T] \\ &= \exp(-\hbar\omega_k/2k_B T)\{1 + \exp(-\hbar\omega_k/k_B T) + \exp(-2\hbar\omega_k/k_B T) + \ldots\} \\ &= \frac{\exp(-\hbar\omega_k/2k_B T)}{1 - \exp(-\hbar\omega_k/k_B T)}, \end{aligned} \tag{4.65}$$

which has been evaluated by noting that the summation is simply a geometric progression. The mean (thermal-equilibrium) energy is given by the standard statistical-mechanical result:

$$\langle \mathscr{E} \rangle = k_B T^2 \frac{\partial(\ln Z)}{\partial T}, \tag{4.66}$$

whence, from eqn. (4.65),

$$\langle \mathscr{E} \rangle = \frac{1}{2}\hbar\omega_k + \frac{\hbar\omega_k}{\exp(\hbar\omega_k/k_B T) - 1} \tag{4.67}$$

where the angular bracket here denotes a *thermal* average:

$$\langle A\rangle = \sum_n A \exp(-\mathscr{E}_n/k_B T)/\sum_n \exp(-\mathscr{E}_n/k_B T).$$

Comparison of eqns. (4.62) and (4.67) shows that the mean, thermal-equilibrium phonon-occupation number is thus

$$n(\boldsymbol{k},p) \equiv \langle n(\omega_k(p),T)\rangle = \frac{1}{\exp(\hbar\omega_k(p)/k_B T) - 1}. \tag{4.68}$$

The Planck distribution law (eqn. (4.68)) (also valid for *photons*) is a special form of the Bose–Einstein distribution function, valid for bosons (particles with zero or integral spin) for which there is no restriction on the occupancy of the various energy levels, but with the chemical potential, μ, taken to be *zero* (the total number of phonons present in thermal equilibrium is not an independent variable, as it is, say, for a gas of ^{4}He atoms, but is determined entirely by the temperature). Note that for bosons, e.g. phonons, satisfying Bose–Einstein statistics (eqn. (4.68)), the number of such particle-like entities in a given state is unlimited; this is in marked contrast to the case of fermions (particles with non-integral spin, e.g. electrons) obeying Fermi–Dirac statistics (see §5.1.2) where only *one* particle is allowed to occupy each state.

* 4.2.6 Normal-mode amplitudes

The quantum-mechanical result that the energies of the normal modes of a crystal are quantized (eqn. (4.62)) implies that, concomitantly, the mode *amplitudes* are also quantized in general.

The solution of the wave equation for a 1D monatomic crystal (eqn. (4.34)) can be generalized for the case of a 3D crystal, namely the displacement amplitude of the ith atom in the lth unit cell is:

$$\boldsymbol{u}_{il}(t) = \sum_{k,p} \boldsymbol{u}_i^0(\boldsymbol{k},p) \exp\{\mathrm{i}[\boldsymbol{k}\cdot\boldsymbol{r}_{il} - \omega_k(p)t]\}, \tag{4.69}$$

where, as before, p is the branch label. This expression can be rewritten as:

$$\boldsymbol{u}_{il}(t) = \frac{1}{(Nm_i)^{1/2}} \sum_{k,p} \boldsymbol{e}_i(\boldsymbol{k},p) \exp(\mathrm{i}\boldsymbol{k}\cdot\boldsymbol{r}_{il}) Q(\boldsymbol{k},p), \tag{4.70}$$

where N is the number of unit cells in the crystal and the complex time-dependent scalar quantity $Q(\boldsymbol{k},p)$ represents the wave amplitude which, because an atomic displacement must be real, satisfies the relation

$$Q(-\boldsymbol{k},p) = Q^*(\boldsymbol{k},p). \tag{4.71}$$

The vector quantity $\boldsymbol{e}_i(\boldsymbol{k},p)$, on the other hand, is the mode eigenvector (or displacement or polarization vector) and gives the *direction* of the atomic displacement. It is normalized such that

$$\sum_i |\boldsymbol{e}_i(\boldsymbol{k},p)|^2 = 1, \tag{4.72}$$

and because the normal modes are mutually orthogonal (in the mathematical sense), for two modes labelled p and p',

$$\sum_i \boldsymbol{e}_i(\boldsymbol{k},p) \cdot \boldsymbol{e}_i(-\boldsymbol{k},p') = \delta_{p,p'}, \tag{4.73}$$

where $\delta_{p,p'}$, is the Kronecker delta function (eqn. (3.96)).

The average vibrational kinetic energy T of a crystal is

$$\langle T\rangle = \frac{1}{2}\sum_{i,l} m_i \langle|\dot{\boldsymbol{u}}_{il}(t)|^2\rangle. \tag{4.74}$$

Thus, from eqn. (4.70), it can be shown (see Problem 4.7(a)) that eqn. (4.74) becomes

$$\langle T\rangle = \frac{1}{2}\sum_{k,p} \omega_k^2(p)\langle|Q(\boldsymbol{k},p)|^2\rangle. \tag{4.75}$$

Since, for a harmonic oscillator, the average kinetic energy equals the average potential energy, the *total* harmonic vibrational energy of a crystal is thus twice that given in eqn. (4.75), i.e.

$$\langle\mathscr{E}\rangle = \sum_{k,p} \omega_k^2(p)\langle|(Q(\boldsymbol{k},p)|^2\rangle. \tag{4.76}$$

Hence for a *single* mode, the average energy is

$$\langle\mathscr{E}(\boldsymbol{k},p)\rangle = \omega_k^2(p)\langle|(Q(\boldsymbol{k},p)|^2\rangle, \tag{4.77}$$

and by comparison with eqns. (4.67) and (4.68), the mean normal-mode amplitude can therefore be written as:

$$\langle|Q(\boldsymbol{k},p)|^2\rangle = \frac{\hbar}{\omega_k(p)}(n(\omega_k, T) + 1/2). \tag{4.78}$$

In the *high-temperature limit*, $T \gg \hbar\omega_k/k_{\mathrm{B}}$, the Bose–Einstein relation for the thermal-equilibrium phonon-occupation number (eqn. (4.68)) becomes

$$\langle n(\omega_k, T)\rangle \simeq \frac{k_{\mathrm{B}}T}{\hbar\omega_k} \gg 1, \tag{4.79}$$

i.e. the number of phonons with a given frequency increases linearly with temperature. Thus, in this limit

$$\langle|Q(\boldsymbol{k},p)|^2\rangle \simeq \frac{k_{\mathrm{B}}T}{\omega_k^2(p)}, \tag{4.80}$$

i.e. all modes have the *same* total energy, $\langle\mathscr{E}\rangle = k_{\mathrm{B}}T$ (from eqn. (4.77)), in agreement with the classical theory of equipartition of energy.

The actual mean atomic displacement can be obtained (see Problem 4.7(b)) from eqn. (4.70) as

$$\langle|\boldsymbol{u}_i|^2\rangle = \frac{1}{Nm_i}\sum_{k,p}\langle|(Q(\boldsymbol{k},p)|^2\rangle. \tag{4.81}$$

In the *high-temperature limit*, eqn. (4.80) can be used to approximate the terms in the summation, and if, moreover, the Einstein model is assumed for the vibrational frequencies, namely that they are all *identical*, $\omega_k = \omega_0$ (more valid for optic modes than for acoustic modes), eqn. (4.81) reduces to:

$$\langle |\boldsymbol{u}_i|^2 \rangle = \frac{3k_B T}{m_i \omega_0^2}, \tag{4.82}$$

where account has been taken of the fact that there are a total of $3N$ modes. For example, for Ge (mass 72.6 amu), for which the optic-mode frequency is $v_0 = 8$ THz (Fig. 4.17), eqn. (4.82) predicts that the root-mean-square displacement in this case is $\simeq$ 0.06 Å at 300 K.

The above treatment of the vibrational displacements of atoms in crystalline solids has a bearing on the temperature dependence of the intensity of X-rays or neutrons scattered by such materials (§2.6.1). In the analysis of diffraction given in §2.6.1.2, it was implicitly assumed that the positions of the atoms were *fixed*. Naïvely, it might be thought that thermal fluctuations in atomic positions, associated with normal-mode vibrations, would lead to the Laue diffraction condition (eqn. (2.101) et seq.) not being satisfied and hence to a complete destruction of coherent Bragg-diffracted beams. In fact, this is not the case: thermal vibrations of atoms in a crystal cause a partial, but not complete, decrease in the scattering amplitude, but no change in the width of a diffraction peak.

The scattering amplitude (eqn. (2.105)) of waves (e.g. X-rays or neutrons) elastically diffracted from planes (*hkl*) in a crystal containing oscillating atoms can be written as:

$$F_i = f_i \exp[\mathrm{i}(\boldsymbol{G} \cdot \boldsymbol{R}_i^0)] \langle \exp[\mathrm{i}(\boldsymbol{G} \cdot \boldsymbol{u}_i)] \rangle, \tag{4.83}$$

where the instantaneous position of atom i is displaced by a distance $\boldsymbol{u}_i$ from the equilibrium position $\boldsymbol{R}_i^0$, i.e. $\boldsymbol{R}_i(t) = \boldsymbol{R}_i^0 + \boldsymbol{u}_i(t)$, and where the $\langle\ \rangle$ brackets denote a thermal average as before. The last term in eqn. (4.83) can be expanded as a series, i.e.

$$\langle \exp[\mathrm{i}(\boldsymbol{G} . \boldsymbol{u}_i)] \rangle \simeq 1 + \mathrm{i}\langle \boldsymbol{G} \cdot \boldsymbol{u}_i \rangle - \tfrac{1}{2}\langle (\boldsymbol{G} \cdot \boldsymbol{u}_i)^2 \rangle + \ldots . \tag{4.84}$$

However, the term $\langle \boldsymbol{G} \cdot \boldsymbol{u}_i \rangle$ is zero, since $\boldsymbol{u}_i(t)$ represents a random thermal displacement presumably uncorrelated with $\boldsymbol{G}$. Moreover, the last term in eqn. (4.84) can be expressed as

$$\tfrac{1}{2}\langle (\boldsymbol{G} \cdot \boldsymbol{u}_i)^2 \rangle = \tfrac{1}{2} G^2 \langle u_i^2 \rangle \langle \cos^2\theta \rangle = \tfrac{1}{6} G^2 \langle u_i^2 \rangle, \tag{4.85}$$

where the spherical average of $\cos^2\theta$ gives $\langle \cos^2\theta \rangle = 1/3$. Now the factor

$$\exp\left\{ -\frac{1}{6} G^2 \langle u_i^2 \rangle \right\} \simeq 1 - \tfrac{1}{6} G^2 \langle u_i^2 \rangle + \ldots \tag{4.86}$$

is the same as eqn. (4.84) (in fact to all orders, for the case of a harmonic oscillator), and hence the scattered *intensity* ($\propto |F_i|^2$) is thus

$$I = I_0 \exp\{ -\tfrac{1}{3} G^2 \langle u_i^2 \rangle \}, \tag{4.87}$$

where I_0 is the scattering intensity for a *fixed* array of atoms (eqn. (2.109)). The exponential factor in eqn. (4.87) is known as the Debye–Waller factor. In the high-temperature limit, and assuming that all atoms vibrate independently with the *same* frequency ω_0 (the Einstein approximation), eqn. (4.82) can be used as an approximation for the mean-square atomic displacement, with the result that (see also Problem 4.8):

$$I \simeq I_0 \exp\{-k_B T G^2 / m_i \omega_0^2\}. \tag{4.88}$$

The scattering intensity that eqn. (4.87) or (4.88) predicts is lost from the elastically (Bragg-) diffracted beams is *inelastically* scattered and appears as a diffuse background: an incident X-ray or neutron can cause the excitation, or de-excitation, of a phonon, in a scattering event, causing a change in both the direction *and* energy of the scattered wave-like entity.

4.2.7 Crystal momentum

The physical momentum of a *free particle* is given by the expression

$$\boldsymbol{p} = \hbar \boldsymbol{k} \tag{4.89}$$

where $\boldsymbol{k}$ is the wavevector of the wave-like representation of the particle. This can be demonstrated by applying the momentum operator, $-i\hbar\nabla$, to the plane-wave state (cf. eqn. (4.10)) that represents a free particle, viz.

$$-i\hbar\nabla(A \exp(i\boldsymbol{k}\cdot\boldsymbol{r})) = \hbar\boldsymbol{k}(A \exp(i\boldsymbol{k}\cdot\boldsymbol{r})) \tag{4.90}$$

An eigenstate of the momentum operator therefore has a momentum given by eqn. (4.89).

Hence, it is natural to ascribe a momentum $\hbar\boldsymbol{k}$ to a phonon with an allowed wavevector $\boldsymbol{k}$. Although this is formally possible, such a quantity does *not* represent a true, physical momentum of the phonon quasiparticle, but instead is termed crystal momentum. The reason why crystal momentum is not a true momentum can be seen from two viewpoints. First, unlike the case of a free particle, where the wavevector of the wave-like representation (eqn. (4.90)) can take *any* value, the physically distinct values of the wavevector of a phonon excited in a crystalline lattice are restricted by the periodic boundary conditions to take only discrete values (cf. eqn. (4.27)) lying within the first Brillouin zone. This has the result that the phonon wavevector, and hence its associated crystal momentum, is not single-valued, but is only known modulo a reciprocal-lattice vector (see §4.2.2), i.e.

$$\boldsymbol{k}_i' = \boldsymbol{k}_i + \boldsymbol{G}. \tag{4.91}$$

The second aspect is that a phonon involves the *relative* vibrational motion of atoms (relative to the centre of mass), and for all $\boldsymbol{k}$-values (except $\boldsymbol{k} = 0$) the centre of mass of a crystal does not move when a phonon is excited within it. A zone-centre, acoustic-branch phonon *does* correspond to a *uniform* motion of all atoms and hence, in principle, could carry true linear momentum but, since $\boldsymbol{k} = 0$, the magnitude of this true momentum is zero.

However, for most purposes, crystal momentum does behave like a real momentum, e.g. in interactions of phonons with other particles, e.g. mobile electrons present in the crystal, or X-rays or neutrons incident externally upon it. The total wavevector of waves interacting in a crystal is conserved, modulo a reciprocal-lattice vector. Thus, for example, for the case of an *inelastic* scattering event involving an external particle (e.g. an X-ray photon), with initial wavevector $\boldsymbol{K}$, in which energy is lost from the particle (creating a phonon) or gained by the particle (associated with the destruction of a phonon), the wavevector selection rule becomes

$$\boldsymbol{K}' = \boldsymbol{K} \pm \boldsymbol{k} + \boldsymbol{G}, \tag{4.92}$$

where $\boldsymbol{K}'$ is the final wavevector of the scattered particle, and the plus and minus signs of the phonon wavevector refer to phonon absorption and creation, respectively. A more detailed discussion of crystal momentum is given in Ashcroft and Mermin (1976).

Vibrations of disordered materials 4.3

Thus far, the vibrational behaviour of two generic classes of materials has been analysed, namely continuous elastic media (§4.1) and systems with perfect translational periodicity (§4.2). In the latter case, the presence of translational periodicity greatly simplified the solution of the relevant equations of motion by the use of appropriate symmetry relations, and the allowed solutions correspond to discrete values of the wavevector, $\boldsymbol{k}$, set by the boundary conditions and hence the $\boldsymbol{k}$-values can be used as labels for such normal-mode solutions.

However, most real materials are *not* perfect single crystals, nor do they behave as elastic continua. As we have seen in Chapter 3, structural defects and disorder are ubiquitous in real materials, and the presence of disorder can have profound effects on the atomic-vibrational behaviour of solids (as also for other types of excitations).

**4.3.1 Impurity modes

The simplest form of disorder associated with defects is perhaps the substitutional defect, an atom having a different mass and/or force constant from the other atoms in an otherwise perfect crystal. This system exhibits behaviour, e.g. spatial localization of the wave amplitude, that is a characteristic feature of the behaviour of vibrations (and electrons) in completely disordered 3D materials, e.g. amorphous solids.

A system with substitutional disorder that can be treated analytically is the monatomic linear chain (§4.2.2) containing a single substitutional isotopic impurity (i.e. an atom with a different mass from, but subject to the same force constant as, the other atoms in the chain. Although this system containing a single defect is the simplest possible, nevertheless its solution requires a new mathematical approach, involving the Green's function. It is found that, if the mass μ of the defect is *less* than that of all other atoms in the chain (M), then a single vibrational mode is split off from the band of allowed states for the monatomic chain and lies above that maximum frequency (eqn. (4.37)). Moreover, the mode amplitude of this split-off state is strongly spatially localized in the vicinity of the defect, the more so the lighter the impurity atom. This problem will be analysed in some detail (Donovan and Angress (1971)) because it illustrates the usefulness of the Green's function approach.

We begin the analysis by writing the trial solution of the ideal 1D chain containing N atoms (eqn. (4.34)) in the form of eqn. (4.70), i.e. as

$$u_l = \sum_k \chi_l(k) Q(k) \tag{4.93a}$$

with

$$\chi_l(k) = \frac{1}{(NM)^{1/2}} e(k) \mathrm{e}^{\mathrm{i}kR_l} \tag{4.93b}$$

where $Q(k)$ contains the time dependence of the wave amplitude, $e(k)$ is the mode eigenvector and l is a label for the unit cells (one atom per cell).

The equation of motion for the chain containing the isotopic defect of mass μ_l at position l can be written as:

$$M\omega^2\chi_l - K\sum_{l'}\chi_{l'} = \sum_{l'} C_{ll'}\chi_{l'} \tag{4.94}$$

where

$$C_{ll'} = \omega^2 \Delta M_l \delta_{l0}\delta_{l'0} \tag{4.95}$$

with the mass defect given by

$$\Delta M_l = M - \mu_l \tag{4.96}$$

and it has been assumed that the defect is situated at $l = 0$ for the sake of definiteness. The frequency ω of the defect-containing chain appears in eqn. (4.94) as a result of the operation of the acceleration operator $(\partial^2/\partial t^2)$ on $Q(k) \propto \exp(\mathrm{i}\omega t)$. The constant K in eqn. (4.94) is the force constant (eqn. (4.32)), assumed to be constant for all bonds between atoms in the chain.

Equation (4.94) can be written in matrix notation as

$$\boldsymbol{L}\chi = \boldsymbol{C}\chi \tag{4.97}$$

where the inverse matrix

$$\boldsymbol{L}^{-1} = \boldsymbol{G} \tag{4.98}$$

is the Green's function. Multiplying both sides of eqn. (4.97) by the Green's function yields

$$\chi = \boldsymbol{GC}\chi \tag{4.99}$$

and the eigenvalues (i.e. the mode frequencies) are found by solving the determinantal equation

$$|\boldsymbol{GC} - \boldsymbol{E}| = 0, \tag{4.100}$$

where $\boldsymbol{E}$ is the unit matrix. From eqn. (4.94), the involvement of the Green's function can be written as

$$M\omega^2 G_{ll''} - K\sum_{l'} G_{l'l''} = \delta_{ll''}, \tag{4.101}$$

and multiplying through by χ_1 and summing over l gives:

$$M\omega^2\sum_l \chi_l G_{ll''} - M\sum_{l'}\omega_k^2\chi_{l'}G_{l'l''} = \chi_{l''} \tag{4.102}$$

where the equation for the *defect-free* chain has been substituted, viz.

$$M\omega_k^2\chi_{l'} = K\sum_l \chi_l. \tag{4.103}$$

However, since l' in eqn. (4.102) is simply a dummy index, this equation can be rewritten more compactly as:

$$M\sum_l \chi_l G_{ll'''} = \frac{\chi_{l''}}{\omega^2 - \omega_k^2}. \tag{4.104}$$

Multiplying through by $\chi_{l'}^*$ and summing over k then yields (cf. eqn. (4.73)):

$$G_{ll''}(\omega^2) = \sum_k \frac{\chi_{l'}^* \chi_{l''}}{\omega^2 - \omega_k^2}$$
$$= \frac{1}{NM} \sum_k \frac{\exp[\mathrm{i}k(R_{l''} - R_{l'})]}{(\omega^2 - \omega_k^2)}. \tag{4.105}$$

From eqn. (4.99),

$$\chi_l = \sum_{ll'} G_{ll''}(\omega^2) C_{l''l'} \chi_{l'}$$
$$= G_l \omega^2 \Delta M \chi_0, \tag{4.106}$$

where use has been made of eqn. (4.95), and $G_l = G_{ll'}$ since it depends on l and l' only through the difference in lattice vectors $(R_l - R_{l'})$.

The summation involved in the Green's function (eqn. (4.105)) can be evaluated (see e.g. Maradudin (1964)) as

$$G_l = \frac{1}{2K\sin\theta}\left[\cot\left(\frac{N\theta}{2}\right)\cos l\theta + \sin|l|\theta\right], \tag{4.107}$$

where $\theta = ka, a$ being the interatomic spacing in the chain, and with

$$\sin^2(\theta/2) = \frac{M\omega^2}{4K} = \frac{\omega^2}{\omega_{\max}^2}, \tag{4.108}$$

where $\omega_{\max}$ is the maximum cut-off frequency for the 1D monatomic, defect-free chain (eqn. (4.37)). For the case of the isotopic defect situated at $l = 0$, eqn. (4.106) gives

$$\varepsilon M \omega^2 G_0 = 1, \tag{4.109}$$

where the mass-defect parameter is defined as

$$\varepsilon = \frac{\Delta M_0}{M} = \frac{M - \mu_0}{M}. \tag{4.110}$$

Substituting the expression for the Green's function (eqn. (4.107)), with $l = 0$, into eqn. (4.109) gives:

$$\varepsilon \tan(\theta/2) = \tan(N\theta/2), \tag{4.111}$$

the roots of which give the perturbed frequencies when the defect is present. The frequencies of a defect-containing chain containing a heavier impurity are lower, and those for a chain containing a lighter impurity are higher, than the frequencies characteristic of the monatomic chain.

However, the mode, that in the perfect 1D chain corresponds to $k = \pi/a$ (i.e. $\theta = \pi$) and $\omega_{\max}$, is, for the case of a light-defect-containing chain (with $\varepsilon\rangle 0$, i.e. $\mu\langle M$), displaced to a frequency lying *above* $\omega_{\max}$. Since this lies in a frequency range beyond that for which propagating waves can exist, it is a *localized* non-propagating wave, and hence the argument θ is complex, viz.

$$\theta = \pi + \mathrm{i}\phi. \tag{4.112}$$

Substitution of eqn. (4.112) as the argument in eqn. (4.111) gives

$$-\varepsilon \cot(\mathrm{i}\phi/2) = \tan\left[\frac{N}{2}[\pi + \mathrm{i}\phi]\right] \simeq \mathrm{i} \tag{4.113}$$

for large N, which can be rearranged in the form

$$\exp(\phi) = \frac{1+\varepsilon}{1-\varepsilon}. \tag{4.114}$$

The frequency of this localized impurity mode is then, from eqn. (4.108) (see also Problem 4.10):

$$\omega_{\mathrm{L}}^2 = \frac{\omega_{\max}^2}{2}(1 + \cos \mathrm{i}\phi) = \frac{\omega_{\max}^2}{4}(2 + \mathrm{e}^{\phi} + \mathrm{e}^{-\phi})$$

or

$$\omega_{\mathrm{L}}^2 = \frac{\omega_{\max}^2}{1-\varepsilon^2}. \tag{4.115}$$

The spatial variation of the wave amplitude for the localized impurity can be obtained from eqn. (4.106), using eqn. (4.107) with $l = 0$, i.e.

$$\begin{aligned}\chi_l &= (\cos(l\theta) + \varepsilon \tan(\theta/2) \sin(|l|\theta))\chi_0 \\ &= [(-1)^l \cos(\mathrm{i}l\phi) + \mathrm{i}(-1)^l \sin(\mathrm{i}|l|\phi)]\chi_0\end{aligned} \tag{4.116}$$

using eqn. (4.112) and hence

$$\chi_l = (-1)^l \left(\frac{1-\varepsilon}{1+\varepsilon}\right)^{|l|} \chi_0 = \chi_{-l}. \tag{4.117}$$

Since $0 < \varepsilon \leq 1$ for a light impurity atom, the amplitudes χ_l decrease according to a geometric progression as l increases, i.e. with increasing distance from the defect. This localized behaviour of the impurity mode is illustrated in Fig. 4.19.

In a similar manner, isotopic substitutional defects in a linear diatomic chain (§4.2.3) give rise to localized impurity vibrational modes in the forbidden frequency gap

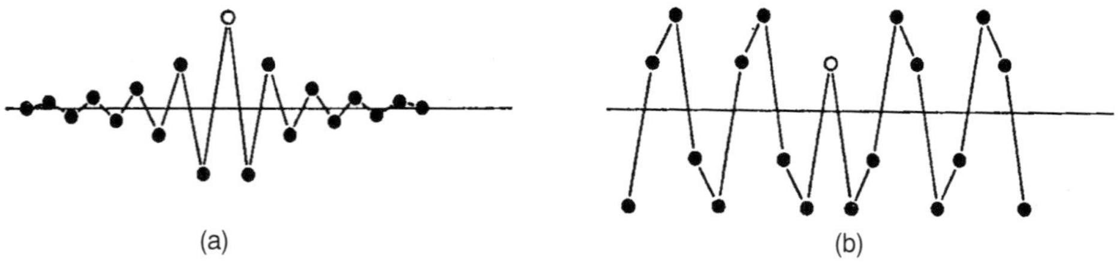

Fig. 4.19 (a) A localized vibrational mode on a chain of atoms associated with a single substitutional isotopic defect with a lighter mass than the other atoms ($\epsilon = 0.2$—see eqn. (4.110)); (b) Atomic displacements for a chain with a heavier impurity atom ($\varepsilon = -2.0$) and $k = \pi/a$ (corresponding to the maximum frequency for the ideal monatomic periodic chain).

$(2K)^{1/2}(m^{-1/2} - M^{-1/2})$ between the bands of allowed vibrational states (the acoustic and optic branches). This is a model for the vibrational behaviour of substitutional impurities in, for example, alkali-halide crystals.

The 1D models give a reasonable picture of the behaviour associated with impurities in 3D crystalline solids. Localized gap modes are found for light impurities ($\varepsilon \rangle 0$), but for density-of-states distributions more realistic than the simple Debye form (eqn. (4.23)), local modes appear only for ε greater than some critical value, ε_c (Dawber and Elliott (1963)). (Localized *electron* states also occur as a consequence of disorder, and the existence of a critical degree of disorder is necessary for these as well (see §6.7).)

Finally, for *heavy* impurities ($\varepsilon < 0$), a qualitatively different behaviour is exhibited by 3D solids compared with the 1D-chain models. Instead of the downward shift in frequencies caused by heavy impurities in 1D chains (cf. eqns. (4.108) and (4.111)), an enhancement of the mode amplitude occurs at a particular frequency (depending on the value of ε) in the band of allowed frequencies of the monatomic chain, $0 < \omega < \omega_{\max}$ (eqn. (4.37)). Such impurity-related modes are known as in-band resonances; the vibrational amplitude of such modes is *not* spatially localized in the vicinity of the impurity, as it is for local modes for light impurities (eqn. (4.117)). Further details of impurity vibrational modes are given in Donovan and Angress (1971) and Stoneham (1975).

4.3.2 Amorphous solids

Amorphous solids have no real-space periodic lattice, by definition, and consequently no reciprocal lattice either (Elliott (1990)). As a consequence, the wavevector $\boldsymbol{k}$ of vibrational modes excited in non-crystalline materials is no longer restricted to the discrete values (eqn. (4.27)) related to reciprocal-lattice vectors (as imposed by periodic boundary conditions). Instead, the wavevector becomes ill-defined, in the sense that several vibrational modes with the same frequency but *different* values of $\boldsymbol{k}$ can coexist in amorphous solids (see Fig. 4.20). Although $\boldsymbol{k}$ is relatively well defined at small values of wavevector (in the long-wavelength limit, materials behave like elastic continua (§4.1) and so the disordered nature of the structure is of no consequence), at larger values of $\boldsymbol{k}$ the uncertainty in $\boldsymbol{k}$ becomes very great indeed and can become comparable to the magnitude of $\boldsymbol{k}$ itself, $\Delta k \simeq k$ (the so-called Ioffe–Regel limit—see eqn. (4.235)). Thus dispersion relations, $\omega(\boldsymbol{k})$, are of little use in describing vibrational states in disordered solids, and $\boldsymbol{k}$ cannot be used as an unambiguous label for the modes.

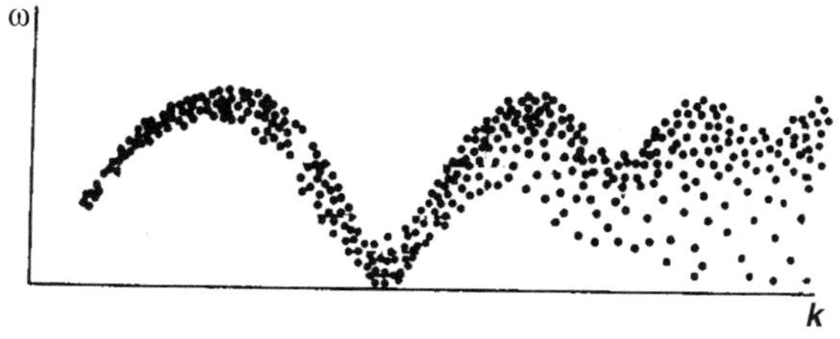

Fig. 4.20 Schematic illustration of the dispersion relation $\omega(k)$ of an acoustic branch for vibrational excitations in an amorphous solid.

However, one quantity which is equally valid in describing vibrational excitations in crystalline and non-crystalline solids is the *density of states*, the number of vibrational states having frequencies between ω and $\omega + d\omega$. For a crystal, this is defined by eqn. (4.58). For an amorphous solid, an alternative definition is simply in terms of a sum over delta functions corresponding to the allowed frequencies of modes, i.e.

$$g(\omega) = \sum_k \delta(\omega - \omega_k), \tag{4.118}$$

where k is simply a label for the modes and has no other physical significance.

Since dispersion relations for vibrational modes are so ill-defined (Fig. 4.20), obviously features in the density of states that are characteristic of crystals, such as van Hove singularities (where $\nabla_k \omega = 0$) do not appear in the density of states for the corresponding amorphous materials. Experimental and theoretical vibrational densities of states for amorphous Si are shown in Fig. 4.21, together with the corresponding calculated density of states for crystalline Si. It can be seen that, overall, the densities of states for the crystalline and amorphous phases are rather similar, which is not too surprising since the atomic masses are identical, and the force constants associated with nearest-neighbour covalent interactions are very similar in the two cases. The prominent peaks at the low-and high-frequency extremities in the crystalline density of states, associated with the TA and TO modes (see Fig. 4.17) also appear in the density of states for amorphous Si, although the van Hove singularities are absent.

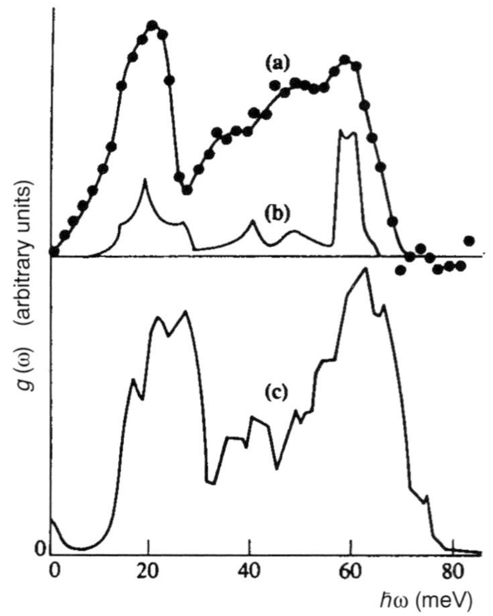

Fig. 4.21 (a) Experimental density of vibrational states for amorphous Si; (b) calculated density of states for crystalline Si (on a reduced vertical scale); (c) the same for amorphous Si. (Reprinted with permission from Kamitakahara *et al.* (1984), *Phys. Rev. Lett.* **52**, 644. © 1984. The American Physical Society)

One aspect of the behaviour of the excitations exhibited by amorphous solids is qualitatively different from that displayed by the corresponding crystals: vibrational states at the edges of bands of allowed states are *spatially localized.* This is true both for vibrational states *and* electronic states (§6.7). (Note that this behaviour is different from the localized nature of impurity *gap* states (see §4.3.1).) The existence of localized, non-propagating excitations in a range of frequencies, which in a crystalline solid would correspond to extended, propagating states (for the cases both of vibrations *and* electrons) is a special consequence of structural disorder. This behaviour is illustrated in Fig. 4.22 for the case of vitreous silica, showing the total vibrational density of states compared with the participation ratio, $P(\omega)$, which is a measure of the number of atoms effectively contributing to a given mode, and defined as:

$$P(\omega) = \frac{M_1^2}{M_0 M_2}, \tag{4.119}$$

where M_r is the rth moment of the kinetic energy of a mode with frequency ω_j, i.e.

$$M_r(\omega_j) = \sum_i |u_i^{(j)}|^{2r}. \tag{4.120}$$

An *extended* mode, in which all atoms participate equally, gives $P = 1$, and a *localized* mode, involving only a *single* atom, gives a value $P = 1/N$, where N is the total number of atoms. It can be seen from Fig. 4.22b that the vibrational modes in silica glass are indeed localized at the band edges, particularly at high frequencies.

*4.3.3 Fractons

As mentioned in §2.1.1, fractals are materials (e.g. aerogels) exhibiting self-similarity in their structure over a certain length scale $a < r < \zeta$, where the lower limit cannot be less

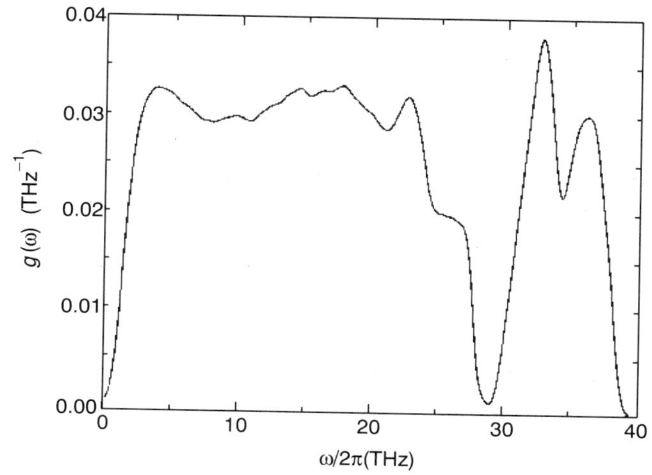

Fig. 4.22 (a) Density of vibrational states calculated for a model of glassy silica;

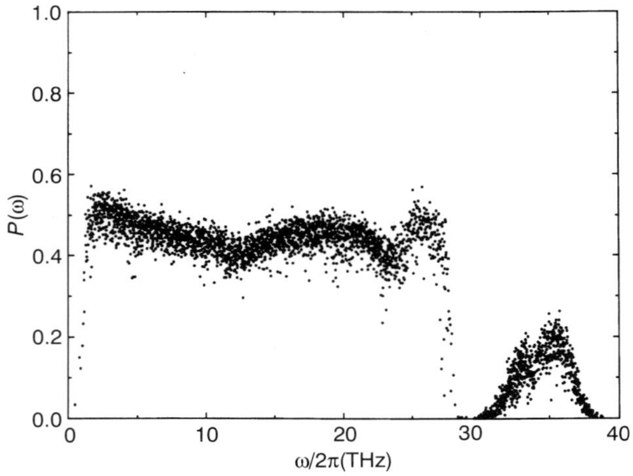

Fig. 4.22 (b) the corresponding participation ratio, $P(\omega)$, for the modes, showing evidence for vibrational localization at the band edges ($P \rightarrow 0$). (Courtesy of Dr S. Taraskin)

than the interatomic spacing, and the upper limit marks the onset of structural homogeneity. Fractal-like structures exhibit a scaling of their mass with increasing distance as

$$M(r) \propto r^{D_{\mathrm{H}}}, \tag{4.121}$$

where D_{H}, the Hausdorff dimension, has the (not necessarily integer) value $D_{\mathrm{H}} \leq d$, where d is the normal Euclidean dimension ($d = 3$ for 3D structures). The behaviour of vibrational excitations of fractal structures ($D_{\mathrm{H}} < d$) is qualitatively different from that exhibited by homogeneous crystalline solids ($d = 3$), but it bears some resemblance to that characteristic of homogeneous *amorphous* solids (albeit with $d = 3$), namely that, above a certain frequency, ω_{c}, vibrational modes are *spatially localized*; these excitations are termed fractons in the case of fractal structures.

In the long-wavelength ($k < \zeta^{-1}$), low-frequency ($\omega < \omega_{\mathrm{c}}$) limit, even a fractal structure will seem like an elastic continuum, and hence the Debye density of vibrational states $g_{\mathrm{D}}(\omega) \propto \omega^{d-1}$ (eqn. (4.23)) will describe the propagating (acoustic) phonon modes in this régime. However, when the vibrational wavelength is sufficiently short that it is comparable to the length scale where self-similarity holds ($k\rangle\zeta^{-1}$), i.e. above a critical frequency ($\omega\rangle\omega_{\mathrm{c}}$), there is a cross-over from extended phonon-like behaviour to localized fracton behaviour, for which the fracton density of states can be written as

$$g_{\mathrm{f}}(\omega) \propto \omega^{\tilde{d}-1}, \tag{4.122}$$

where $\tilde{d}$ is the so-called fracton (or spectral) dimension, and $\tilde{d} \leq D_{\mathrm{H}} \leq d$. The cross-over frequency scales with distance as

$$\omega_{\mathrm{c}} \propto \zeta^{-D_{\mathrm{H}}/\tilde{d}}. \tag{4.123}$$

* LO–TO splitting in ionic solids: polaritons 4.4

In §4.2.4 it was noted that the LO and TO modes at the zone centre are *non-degenerate* for *ionic* crystals (see e.g. Fig. 4.16), whereas they are *degenerate* for *non*-ionic (homopolar) crystals (see e.g. Fig. 4.17). This difference in behaviour must obviously derive from the electrically charged nature of the ions in ionic solids, resulting in an extra contribution to the force constants as a consequence of electrostatic interactions that do not occur in homopolar covalent solids.

In the long-wavelength ($\boldsymbol{k} = 0$) limit of the optic branch, the basis atoms in individual Wigner–Seitz cells vibrate relative to each other: the two ion sub-lattices of an ionic crystal with two atoms as a basis vibrate in antiphase, as in the 1D chain analogue (§ 4.2.3—see Fig. 4.12).

Let $u_\pm$ and $\pm q$ be the displacements and charges of the ions, where the signs refer to the type of ion (cation or anion). The electrical polarization of a cell is thus proportional to the induced dipole moment $p = q(u_+ - u_-) \equiv qu$. However, the charge displacements can produce internal electric fields that induce further dipole moments on the ions of the lattice, thereby contributing an extra polarization contribution $\alpha_\mathrm{p}\boldsymbol{E}_\mathrm{loc}$, where $\alpha_\mathrm{p}(= \alpha_\mathrm{p}^+ + \alpha_\mathrm{p}^-)$ is the polarizability of the ions and $\boldsymbol{E}_\mathrm{loc}$ is the effective (local) electric field at the ions (see §7.1 for a discussion of dielectric properties). This local field is related to the macroscopic field $\boldsymbol{E}$ within the sample by the Lorentz relation (see §7.1.2 and eqn. (7.24)):

$$\boldsymbol{E}_\mathrm{loc} = \boldsymbol{E} + \frac{\boldsymbol{P}}{3\varepsilon_0} \tag{4.124}$$

where $\boldsymbol{P}$ is the electric polarization (the dipole moment per unit volume) and ε_0 is the permittivity of free space; eqn. (4.124) is valid for *cubic* Bravais lattices. If it is assumed that these two sources of polarization can simply be added, the total polarization for N Wigner–Seitz cells in a volume V is therefore:

$$\boldsymbol{P} = \frac{N}{V}(q\boldsymbol{u} + \alpha_\mathrm{p}\boldsymbol{E}_\mathrm{loc}) = \frac{N}{V}\frac{(q\boldsymbol{u} + \alpha_\mathrm{p}\boldsymbol{E})}{(1 - N\alpha_\mathrm{p}/3\varepsilon_0 V)}. \tag{4.125}$$

The equations of motion for the ions are given by

$$M_+\ddot{\boldsymbol{u}}_+ = -k(\boldsymbol{u}_+ - \boldsymbol{u}_-) + q\boldsymbol{E}_\mathrm{loc} \tag{4.126a}$$

and

$$M_-\ddot{\boldsymbol{u}}_- = k(\boldsymbol{u}_+ - \boldsymbol{u}_-) - q\boldsymbol{E}_\mathrm{loc} \tag{4.126b}$$

where k is the restoring force constant and $M_\pm$ are the ionic masses. Equations (4.126) can be reduced to a single equation by making use of the reduced mass $\mu_M = M_+M_-/(M_+ + M_-)$ with $\boldsymbol{u} = \boldsymbol{u}_+ - \boldsymbol{u}_-$, i.e.

$$\mu_M\ddot{\boldsymbol{u}} = -k\boldsymbol{u} + q\boldsymbol{E}_\mathrm{loc}. \tag{4.127}$$

By writing the ionic displacement in the renormalized form, $\boldsymbol{w} \equiv (N\mu_M/V)^{1/2}\boldsymbol{u}$, eqns. (4.125) and (4.127) become:

$$\ddot{\boldsymbol{w}} = b_{11}\boldsymbol{w} + b_{12}\boldsymbol{E} \tag{4.128a}$$

$$\boldsymbol{P} = b_{21}\boldsymbol{w} + b_{22}\boldsymbol{E}, \tag{4.128b}$$

where the off-diagonal coefficients are symmetric, $b_{12} = b_{21}$.

The coefficients b_{ij} can be related to macroscopic, measurable quantities, namely the dielectric constant $\varepsilon(\omega)$ at low and high frequencies (see §7.1.1). In the static case, $\ddot{\boldsymbol{w}} = 0$ and so from eqn. (4.128)

$$\boldsymbol{P}(0) = \left(b_{22} - \frac{b_{12}^2}{b_{11}}\right)\boldsymbol{E} \equiv [\varepsilon(0) - 1]\varepsilon_0\boldsymbol{E}, \tag{4.129}$$

since $\boldsymbol{E} + \boldsymbol{P}/\varepsilon_0 = \varepsilon\boldsymbol{E}$. The quantity $\varepsilon(0)$ is the *static* dielectric constant. For very-high-frequency fields $\boldsymbol{E}(\omega)$, the ions ultimately cannot respond to the rapidly changing forces due to their inertia and hence cannot contribute to the polarization. In this case, $\boldsymbol{w} = 0$, and hence

$$\boldsymbol{P}(\infty) = b_{22}\boldsymbol{E} \equiv [\varepsilon(\infty) - 1]\varepsilon_0\boldsymbol{E}, \tag{4.130}$$

where $\varepsilon(\infty)$ is the *high-frequency* dielectric constant associated with the polarization of *electrons* in the atoms which, being much lighter than the ions, *can* still respond to electric fields having frequencies of the order of atomic vibrational frequencies. $\varepsilon(\infty)$ is evidently only constant, i.e. independent of frequency, for frequencies (in the IR region) much less than that characterizing electronic motion.

In principle, solving the two coupled equations (eqn. (4.128)), derived from simple electrostatics, produces longitudinal and transverse (optic phonon) wave solutions. In practice, the assumption, implicit in the electrostatic approximation, that the Coulombic forces act *instantaneously* is incorrect: account must be taken of the fact that there is a temporal retardation effect due to the finite velocity of light, and this affects the transverse phonons since light is a transverse wave. Hence Maxwell's equations must also be included in the problem: it will be seen later that the actual modes (called polaritons) at very low $\boldsymbol{k}$-values ($\boldsymbol{k} \to 0$) consist partly of mechanical waves (i.e. phonons) and partly of electromagnetic radiative waves.

Maxwell's equations for this situation (no free charges) are:

$$\nabla \cdot \boldsymbol{D} = 0, \tag{4.131}$$

$$\nabla \cdot \boldsymbol{B} = 0, \tag{4.132}$$

$$\nabla \times \boldsymbol{E} = -\dot{\boldsymbol{B}}, \tag{4.133}$$

$$\frac{1}{\mu_0}\nabla \times \boldsymbol{B} = \dot{\boldsymbol{D}} = \varepsilon_0\dot{\boldsymbol{E}} + \dot{\boldsymbol{P}}, \tag{4.134}$$

where $\nabla\cdot$ and $\nabla\times$ are the div and curl operators, respectively, μ_0 is the vacuum permeability, and the electrical displacement $\boldsymbol{D} = \varepsilon_0\boldsymbol{E} + \boldsymbol{P} = \varepsilon\varepsilon_0\boldsymbol{E}$ (eqn. (7.4)) It has been assumed that the ionic material is non-magnetic and so the magnetic permeability $\mu = 1$.

If it is assumed that $X = X^0\exp[\mathrm{i}(\boldsymbol{k}\cdot\boldsymbol{r} - \omega t)]$, where $\boldsymbol{X} = \boldsymbol{w}, \boldsymbol{E}, \boldsymbol{B}$ or $\boldsymbol{P}$, is a trial solution, eqns. (4.128a) and (4.131–4) then yield:

$$-\omega^2\boldsymbol{w} = b_{11}\boldsymbol{w} + b_{12}\boldsymbol{E}, \tag{4.135}$$

$$\boldsymbol{k}\cdot(\varepsilon_0\boldsymbol{E} + \boldsymbol{P}) = 0, \tag{4.136}$$

$$\boldsymbol{k} \cdot \boldsymbol{B} = 0, \tag{4.137}$$

$$\boldsymbol{k} \times \boldsymbol{E} = -\omega \boldsymbol{B}, \tag{4.138}$$

$$\boldsymbol{k} \times \boldsymbol{B} = -\omega\mu_0(\varepsilon_0 \boldsymbol{E} + \boldsymbol{P}). \tag{4.139}$$

Substituting eqn. (4.135) into eqn. (4.128b) and eliminating $\boldsymbol{w}$ gives

$$\boldsymbol{P} = \left\{ b_{22} - \frac{b_{12}^2}{b_{11} + \omega^2} \right\} \boldsymbol{E} \equiv [\varepsilon(\omega) - 1]\varepsilon_0 \boldsymbol{E}. \tag{4.140}$$

Combining eqns. (4.136) and (4.140) then yields

$$\varepsilon(\omega)(\boldsymbol{k} \cdot \boldsymbol{E}) = 0. \tag{4.141}$$

One solution of eqn. (4.141) is when $\varepsilon(\omega) = 0$, which corresponds to a *longitudinal* wave. Thus $\boldsymbol{P} = -\varepsilon_0 \boldsymbol{E}$ (from eqn. (4.140)) and hence $(\varepsilon_0 \boldsymbol{E} + \boldsymbol{P}) = 0$. Consequently, from eqn. (4.139), $\boldsymbol{k} \times \boldsymbol{B} = 0$ and $\boldsymbol{B}$ either vanishes or is parallel to $\boldsymbol{k}$. On the other hand, eqn. (4.137) implies that $\boldsymbol{B}$ either vanishes or is *perpendicular* to $\boldsymbol{k}$. Thus, $\boldsymbol{B} = 0$, and hence eqn. (4.138) indicates that $\boldsymbol{E}$ is parallel to $\boldsymbol{k}$, i.e. constituting a longitudinal wave, since now $\boldsymbol{k} \times \boldsymbol{E} = 0$. Thus, the frequency of the longitudinal mode, ω_L, is found by solving for $\varepsilon(\omega) = 0$, where

$$\varepsilon(\omega) = \varepsilon(\infty) + \frac{(\varepsilon(0) - \varepsilon(\infty))\omega_0^2}{\omega_0^2 - \omega^2} \tag{4.142}$$

from eqns. (4.129) and (4.130), and where the substitution

$$\omega_0^2 = -b_{11} = \frac{1}{\mu_M}\left[k - \frac{Nq^2}{3\varepsilon_0 V(1 - N\alpha_p/3\varepsilon_0 V)}\right] \tag{4.143}$$

has been made. Hence, from eqn. (4.142), $\varepsilon(\omega) = 0$ when

$$\omega_L = \left(\frac{\varepsilon(0)}{\varepsilon(\infty)}\right)^{1/2} \omega_0. \tag{4.144}$$

Note that this mode is dispersionless in this model.

The other way of satisfying eqn. (4.141) is for $\boldsymbol{k} \cdot \boldsymbol{E} = 0$, i.e. $\boldsymbol{E}$ and $\boldsymbol{k}$ are mutually perpendicular, thereby constituting a *transverse* wave. From eqn. (4.138), these are both also perpendicular to $\boldsymbol{B}$, with their scalar magnitudes satisfying in this case the relation

$$k = \omega B/E. \tag{4.145}$$

Equation (4.139), the last equation of the set, reduces to the scalar equation

$$kB = \omega\mu_0(\varepsilon_0 E + P). \tag{4.146}$$

Eliminating B and P using eqns. (4.140) and (4.145), together with eqn. (4.142), gives

$$\frac{k^2 c^2 E^2}{\omega^2} = \varepsilon(\omega)E, \tag{4.147}$$

where the speed of light is given by $c = (\varepsilon_0\mu_0)^{-1/2}$. Since $\boldsymbol{E}$ is not zero, therefore (from eqn. (4.142))

$$\frac{k^2c^2}{\omega^2} = \varepsilon(\infty) + \frac{(\varepsilon(0) - \varepsilon(\infty))}{(\omega_0^2 - \omega^2)}\omega_0^2 \tag{4.148a}$$

or, in rewritten form:

$$\omega^4 - \omega^2\left[\omega_L^2 + \frac{k^2c^2}{\varepsilon(\infty)}\right] + \frac{c^2k^2\omega_0^2}{\varepsilon(\infty)} = 0. \tag{4.148b}$$

In the limit $k \to \infty$, eqn. (4.148a) predicts that $\omega \to \omega_0$, which we identify with the TO-phonon frequency, ω_T. Thus, eqn. (4.144) relates the longitudinal and transverse optic-phonon frequencies via the so-called Lyddane–Sachs–Teller (LST) relation (valid for cubic crystals)

$$\omega_L = \left(\frac{\varepsilon(0)}{\varepsilon(\infty)}\right)^{1/2}\omega_T. \tag{4.149}$$

(See also Problem 4.11.)

However, near the zone centre (specifically in the vicinity of the intersection of the dispersion curve for the photon, $\omega = ck$, and the TO-phonon dispersion curve), the transverse-mode dispersion curve becomes very complicated, as indicated by eqn. (4.148). In general, for a given $\boldsymbol{k}$-value, there are *two* branches (each doubly degenerate), separated by a forbidden gap (see Fig. 4.23a). Note that this gap does *not* arise because of periodicity (as in the gap between acoustic and optic modes for ionic crystals—see Fig. 4.16), but because the dielectric constant $\varepsilon(\omega)$ is *negative* between the frequencies ω_L

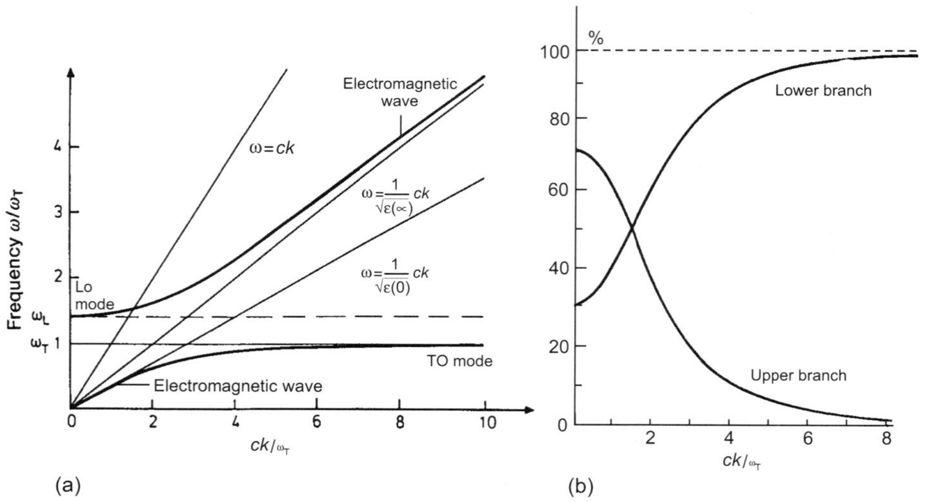

Fig. 4.23 (a) Polariton (coupled transverse phonon-photon) dispersion curves for an ionic crystal. (b) Percentage mechanical (phonon-like) energy in the transverse modes.

and ω_T. From eqn. (4.147), this implies an imaginary value of wavevector $\boldsymbol{k}$, and hence a non-propagating (evanescent) wave solution. Note also that as $k \to 0$, eqn. (4.148b) predicts that

$$\omega^4 = \omega^2 \omega_L^2, \tag{4.150}$$

having solutions $\omega = 0$ (lower branch) or $\omega = \omega_L$ (upper branch). Thus, the *transverse* mode at the zone centre has a finite frequency equal to that of the *longitudinal phonon*.

The transverse mode with the dispersion curve given by eqn. (4.148) is called a polariton: it is a *coupled mode* with both mechanical (phonon-like) character and electromagnetic-radiation (photon-like) character associated with the electric fields, the relative percentage of which changes with wavevector (Fig. 4.23b). Equation (4.148a) predicts for the photon-like parts that, for $\omega \ll \omega_T$,

$$\omega = \frac{ck}{(\varepsilon(0))^{1/2}} \tag{4.151}$$

and, for $\omega \gg \omega_T$,

$$\omega = \frac{ck}{(\varepsilon(\infty))^{1/2}}, \tag{4.152}$$

characteristic of light propagating in media with refractive indices of $\varepsilon(0)^{1/2}$ and $\varepsilon(\infty)^{1/2}$, respectively. Figure 4.24 shows experimental dispersion curves for the polaritons and LO phonons in crystalline GaP.

Finally, for a crystal composed of atoms *without* net electrical charges, e.g. homopolar systems (such as Si), there is obviously no ionic polarization and so $\varepsilon(0) = \varepsilon(\infty)$, and hence $\omega_L = \omega_T$ at $k = 0$ from the LST relation: there is no LO–TO splitting (see Fig. 4.17).

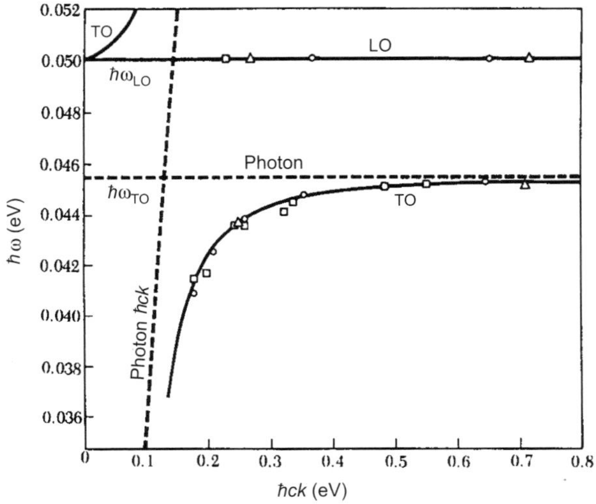

Fig. 4.24 Dispersion curves for transverse polariton modes and LO phonons measured for crystalline GaP (symbols, experimental data; lines, theoretical curves). (After Kittel (1996). Reproduced by permission of John Wiley & Sons Inc.)

Experimental probes of vibrational states 4.5

A number of experimental techniques can be used to investigate vibrational excitations in materials. These can be essentially divided into two categories, depending on whether the probe particle (a photon) is wholly *absorbed* (transferring all of its energy to the phonon system) or whether the particle (photon or neutron) is *inelastically scattered* (losing, or gaining, only part of its energy to, or from, the phonon system). The photon energy at which absorption takes place, or the change in energy of the scattered particle, therefore gives the frequency ω of the vibration involved, since $\mathscr{E} = \hbar\omega$. However, methods involving scattering processes have the advantage that, in principle, information about the (crystal) *momentum* of the vibrational modes involved can also be obtained via the momentum-conservation law.

4.5.1 Infrared absorption spectroscopy

Optic modes near the zone centre typically have frequencies of the order of a few THz (see Fig. 4.16), corresponding to wavelengths of light in the region $10^3 - 10^4$ nm, i.e. in the infrared (IR) part of the electromagnetic spectrum. Thus, if a TO mode gives rise to a change in electrical dipole moment of the unit cell of a crystal (a prerequisite for a dipole-allowed optical transition), an IR photon of the appropriate frequency will be absorbed, and a phonon is correspondingly created. The magnitude of the wavevector of IR photons, related to the radial frequency by the speed of light, $k = \omega/c$, has a typical value of $k = 10^5 \mathrm{m}^{-1}$, minuscule compared with the extent of the first Brillouin zone ($k = \pi/a \simeq 10^9 \mathrm{m}^{-1}$) — see Problem 4.11 (d). Hence, only TO modes very near the zone centre of crystals can be probed by direct photon absorption if only a single phonon is involved, since, by the conservation law for crystal momentum

$$\boldsymbol{k} = \boldsymbol{k}' + \boldsymbol{G}, \tag{4.153}$$

where $\boldsymbol{k}$ is the wavevector of the absorbed photon and $\boldsymbol{k}'$ that of the phonon created in the crystal. However, phonons away from the zone centre can be probed if, say, *two* phonons are created simultaneously on the absorption of a single photon. For such two-phonon absorption, the conservation laws are for frequency (or energy)

$$\omega = \Omega' + \Omega'' \tag{4.154a}$$

and for wavevector (or momentum)

$$\boldsymbol{k} = \boldsymbol{k}' + \boldsymbol{k}'' + \boldsymbol{G}. \tag{4.154b}$$

where the single- and double-primed quantities refer to the two created phonons. Since the wavevector for light, $\boldsymbol{k}$, is so very small relatively, $\boldsymbol{k}'$ and $\boldsymbol{k}''$ must be nearly equal and opposite, but may have an appreciable magnitude. In fact, two-phonon absorption is greatest for vibrational modes near the zone boundary where the density of modes is highest.

IR absorption experiments traditionally have used a source of IR radiation (e.g. a resistively heated element), an IR detector (e.g. a thermocouple) and a dispersing element (e.g. an alkali halide, IR-transmitting, prism, or a diffraction grating) to vary

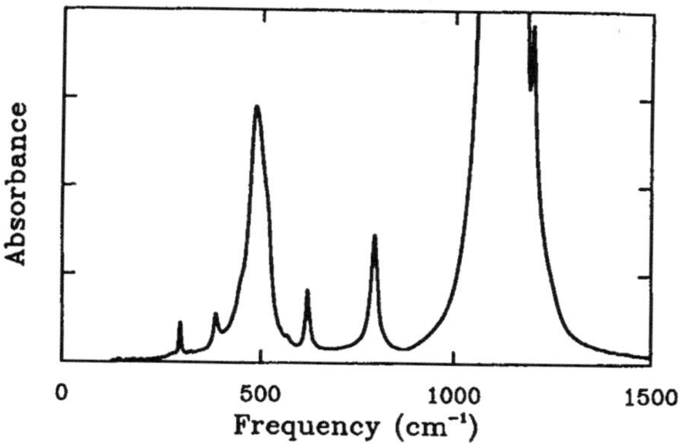

Fig. 4.25 Infrared absorption spectrum for the cristobalite polymorph of crystalline SiO_2. (After Dove (1993) Reproduced by permission of Cambridge University Press)

the wavelength. More modern spectrometers utilize a Michelson interferometer instead of the dispersing element, by which an absorption spectrum is obtained as a function of time (for the motion of one of the mirrors in the interferometer), and Fourier transformation converts the time-dependent data into a conventional spectrum as a function of frequency, as obtained directly from a grating spectrometer. Peaks in absorption spectra correspond to allowed vibrational transitions (near-zone-centre TO-mode frequencies) — see Fig. 4.25 for an example.

The requirement that a photon-induced vibrational transition in a solid be associated with a change in electrical dipole moment means that either permanent dipoles need to be present (as in ionic solids) with particular vibrational (optic) modes causing a change in dipole moment, or else, e.g. in the case of covalent materials, a particular vibrational mode leads to a *dynamic* dipole moment associated with an instantaneous compression of some bonds and extension of others with corresponding changes in local electronic charge densities, even though there is no static dipole moment associated with the equilibrium configuration. Thus, for example, optic modes in crystalline Si or Ge are IR-inactive (the perfect tetrahedral local symmetry ensures that the overall dynamic dipole moment is zero). However, for the case of *amorphous* Si or Ge, where, although the local coordination is still four, the perfect tetrahedral symmetry characteristic of the crystal is lifted, and the vibrational modes become IR-active because dynamic dipole moments are created.

There is another important difference in the behaviour of crystalline and amorphous materials in this regard. Because of the lack of translational periodicity in amorphous solids, reciprocal space has no meaning and k is no longer a good quantum number for vibrational modes (§4.3.2). Consequently, the k-selection rule (eqn. (4.153)), restricting photon-induced one-phonon transitions to zone-centre modes in crystals, is relaxed in amorphous materials and in principle *all* modes can contribute to an IR absorption spectrum (although not necessarily with equal weight). This behaviour is illustrated in Fig. 4.26 for the case of a-Ge: comparison with the form of the density of vibrational states for the isostructural material a-Si (Fig. 4.21a) shows

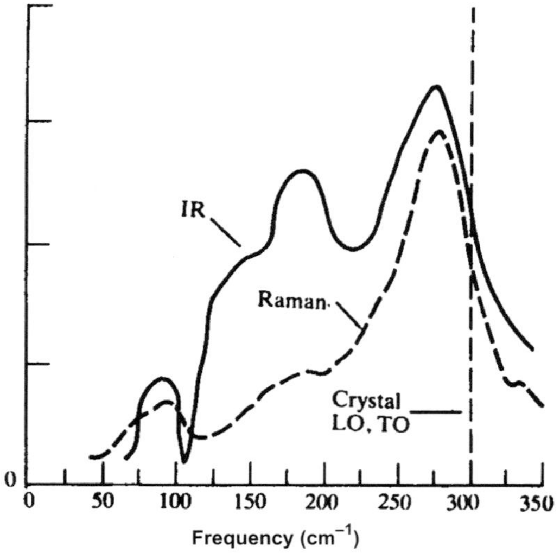

Fig. 4.26 Comparison of experimental IR absorption and Raman scattering spectra of amorphous Ge (Reprinted with permission from Alben *et al.* (1975), *Phys. Rev.* **B11**, 2271. © 1975. The American Physical Society). The frequency of the degenerate LO, TO zone-centre phonon in crystalline Ge is also shown, even though these modes are IR-inactive.

that modes throughout the band of states contribute to the measured IR absorption spectrum.

In addition to the *position* of a peak in an IR absorption spectrum giving the frequency of an IR-active vibrational mode, the *intensity* of an absorption peak is determined by a macroscopic quantity, the absorption coefficient, related to the dielectric constant of a non-metallic material. A monochromatic, plane electromagnetic wave propagating in a medium with refractive index n_r has an electric vector given by

$$\boldsymbol{E} = \boldsymbol{E}_0 \exp[\mathrm{i}(\boldsymbol{k} \cdot \boldsymbol{r} - \omega \mathrm{t})], \tag{4.155}$$

with the magnitude of the wavevector given by (cf. eqn. (4.6)):

$$k = \frac{n_r \omega}{c}. \tag{4.156}$$

For the case of a *lossy medium*, within which *absorption* of the electromagnetic wave takes place (e.g. by excitation of phonons), the amplitude of the wave (i.e. the $\boldsymbol{E}$-vector) propagating, say, in the z-direction, decays with z, implying that the refractive index is *complex*, i.e.

$$n^\dagger = n_r + \mathrm{i}\kappa_i, \tag{4.157}$$

and

$$E = E_0 \exp(-\omega\kappa_i z/c) \exp[\mathrm{i}(\omega n_r z/c - \omega t)]. \tag{4.158}$$

Since the (complex) refractive index and dielectric constant are related via the expression

$$n^\dagger = \sqrt{\varepsilon^\dagger} \tag{4.159}$$

where, in general, the complex dielectric constant is given by

$$\varepsilon^\dagger = \varepsilon_1 + \mathrm{i}\varepsilon_2 \tag{4.160}$$

(sometimes also written as $\varepsilon^\dagger = \varepsilon' + \mathrm{i}\varepsilon''$), the optical constants n_r and κ_i are related to the real and imaginary components of the dielectric constant by

$$n_\mathrm{r}^2 - \kappa_\mathrm{i}^2 = \varepsilon_1 \tag{4.161}$$

and

$$2n_\mathrm{r}\kappa_\mathrm{i} = \varepsilon_2. \tag{4.162}$$

Thus, for an optically thin material, in which the real part of the refractive index n_r is close to unity (the vacuum value), the intensity of transmitted light ($\propto EE^*$) is governed (cf. eqns. (4.158) and (4.162)) by the quantity known as the absorption coefficient (see Problem 4.14), viz.

$$K(\omega) = \frac{2\omega\kappa_\mathrm{i}}{\mathrm{c}} \tag{4.163a}$$

or, from eqn. (4.162),

$$K(\omega) = \frac{\omega\varepsilon_2(\omega)}{n_\mathrm{r}c}. \tag{4.163b}$$

The treatment given in §4.4, based on the *harmonic approximation* (cf. eqn. (4.127)), gave rise to an expression for the dielectric constant $\varepsilon(\omega)$ that was entirely *real* (eqn. (4.142)), and obviously this cannot describe optically absorbing materials which are characterized by *complex* values of the dielectric constant (eqn. (4.160)) or refractive index (eqn. (4.157)). This behaviour can be modelled, however, by assuming that instead a *damped* harmonic equation describes the motion of ions in a solid in the presence of a local electric field, viz.

$$\mu_M\ddot{\boldsymbol{u}} + \mu_M\gamma\dot{\boldsymbol{u}} = -k\boldsymbol{u} + q\boldsymbol{E}_\mathrm{loc}, \tag{4.164}$$

where the damping constant, $\gamma(\rangle 0)$, accounts for the optical absorption (i.e. loss of electromagnetic energy) in the medium. This equation of motion, together with the expression for the polarization, $\boldsymbol{P}$, (eqn. (4.125)), can be solved (see Problem 4.12) to give for the complex dielectric constant (cf. eqn. (4.142)):

$$\varepsilon^\dagger(\omega) = \varepsilon(\infty) + \frac{(\varepsilon(0) - \varepsilon(\infty))\omega_0^2}{(\omega_0^2 - \omega^2 - \mathrm{i}\gamma\omega)}. \tag{4.165}$$

The real and imaginary parts of $\varepsilon^\dagger(\omega)$ — see Problem 4.12 — are plotted against frequency in Fig. 4.27; the imaginary part, $\varepsilon_2(\omega)$, has the (Lorentzian) form of a damped resonance curve, peaking near ω_0 and with γ being the full width at half maximum.

In fact, the real and imaginary parts of the complex dielectric constant are inter-related via the Kramers–Kronig relations (see e.g. Kittel (1996) for a derivation):

$$\varepsilon_1(\omega) - 1 = \frac{2}{\pi}\mathfrak{P}\int_0^\infty \frac{\omega'\varepsilon_2(\omega')}{\omega'^2 - \omega^2}\,\mathrm{d}\omega', \tag{4.166a}$$

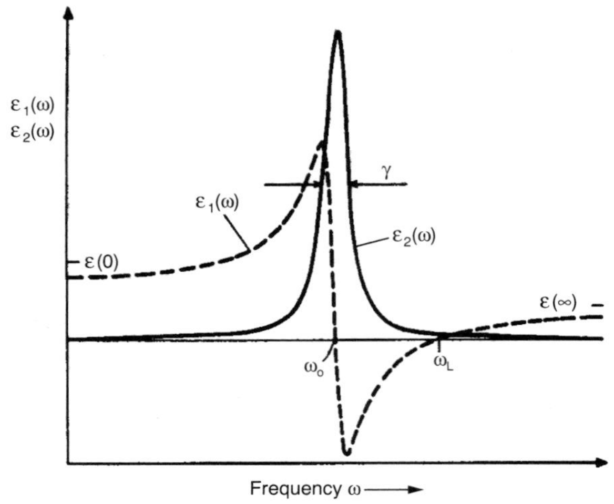

Fig. 4.27 Frequency dependence of the real and imaginary parts of the complex dielectric constant $\epsilon^{\dagger}(\omega)$ for the model of a damped dipolar oscillator.

$$\varepsilon_2(\omega) = \frac{2\omega}{\pi} \mathfrak{P} \int_0^{\infty} \frac{\varepsilon_1(\omega')}{\omega^2 - \omega'^2} \mathrm{d}\omega', \qquad (4.166b)$$

where $\mathfrak{P}$ represents the Cauchy principal value of the integral (omitting the contribution to the integral of the singularity at $\omega = \omega'$). Thus, if one function is known for all frequencies (in practice, it is sufficient to measure it only over a frequency range somewhat larger than that in which it varies with frequency) then the conjugate function can be calculated at any frequency. If ε_2 is only measured to frequencies *less* than those characteristic of electronic excitations ($\omega \simeq 10^{16}\mathrm{s}^{-1}$), then the factor of unity on the left-hand side of eqn. (4.166a) should be replaced by the high-frequency dielectric constant, $\varepsilon(\infty)$.

The polariton (coupled phonon–electromagnetic wave) dispersion curve, characteristic of ionic crystals (Fig. 4.23a), has a profound effect on the optical behaviour of such materials in the IR region. In the frequency range between ω_{T} and $\omega_{\mathrm{L}}(= (\varepsilon(0)/\varepsilon(\infty))^{1/2}\omega_{\mathrm{T}})$, eqn. (4.142) predicts that the real part of the dielectric constant is *negative* (see also Fig. 4.27) implying, from eqn. (4.159), that the refractive index is imaginary, and hence that no electromagnetic wave can propagate in the solid in this frequency band. Consequently, in the harmonic approximation, the IR reflectivity of ionic solids should be 100% in this restrahlen (residual wave) band. In practice, ionic damping causes the reflectivity to be somewhat less than total, since some of the light is absorbed in the solid rather than being reflected. The behaviour of the IR reflectivity of NaCl in the polariton region is shown in Fig. 4.28.

4.5.2 Inelastic photon scattering

Particles, such as photons (or neutrons, see §4.5.3), incident externally on a solid, can be scattered by the atoms within it and, if exchanging energy and momentum with

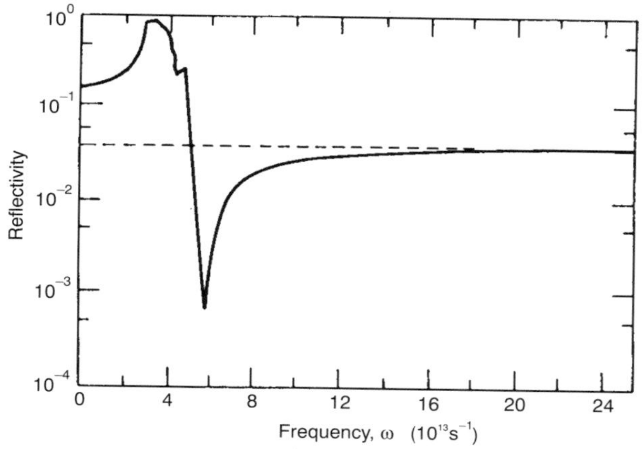

Fig. 4.28 Reflectivity of crystalline NaCl in the IR region. The reflectivity is very high in the frequency range between $\omega_T = 3.1 \times 10^{13}\ s^{-1}$ and $\omega_L = 5.0 \times 10^{13}\ s^{-1}$, although not unity because of finite absorption. The dashed line shows the reflectivity for visible light. (Smith and Manogue (1981), *J. Opt. Soc. Am.*, **71**, 935. Reproduced by permission of Optical Society of America)

excitations associated with the atoms (e.g. phonons), the scattered probe particles emerge with a different energy and direction from that of the incident particles. Thus, in principle, by analysing inelastically scattered particles for energy and momentum, information on the energy and momentum of the excitations can be gleaned.

In the case of scattering of *photons*, the scattering process can be envisaged as the momentary absorption of an incident photon accompanied by an excitation of the electronic distribution of an atom, followed by an electronic de-excitation and the emission of another photon. Inelastic photon scattering occurs only if there is a change in *electronic polarizability* associated with the excitations (e.g. phonons) with which energy is being exchanged.

Normally, photons from the visible part of the electromagnetic spectrum are used in scattering experiments since lasers are readily available; high-intensity sources are necessary, since the probability of inelastic photon scattering is very low. Two types of phonon-related inelastic light-scattering experiments can be distinguished: light scattering from *acoustic* phonons is termed Brillouin scattering, and light scattering from *optic* phonons is referred to as Raman scattering. As was seen for the case of IR photon absorption (§4.5.1), the wavevector of light is very much smaller than phonon wavevectors corresponding to the Brillouin zone boundary, and hence, as will be seen shortly, one-phonon inelastic light scattering also only probes vibrational states near the zone centre. However, this is not the case for inelastic *X-ray* scattering, where the X-ray photons can have appreciable values of wavevector, comparable to zone-boundary phonon wavevectors, as seen already in the case of diffraction (see §2.6.1.2). However, X-rays (e.g. from a synchrotron-radiation source) need to be very highly monochromated so that the very small inelastic energy transfers corresponding to vibrational (de-) excitations can be detected.

The selection rules associated with inelastic scattering can be found by generalizing the treatment given in §2.6.1.2 for the case of elastic scattering to that in which the atoms can undergo time-dependent displacements, as done in the discussion of the Debye–Waller factor given in §4.2.6. The amplitude of a wave scattered by an assembly of atoms i, each undergoing time-dependent displacements $\boldsymbol{u}_i(t)$ about an equilibrium position R_i^0, i.e. where

$$\boldsymbol{R}_i(t) = \boldsymbol{R}_i^0 + \boldsymbol{u}_i(t), \tag{4.167}$$

can be written as (cf. eqn. (4.83)):

$$F \propto \exp(-\mathrm{i}\omega_0 t) \sum_i \exp[\mathrm{i}\boldsymbol{K} \cdot \boldsymbol{R}_i(t)], \tag{4.168}$$

where the time dependence of the incident wave, with frequency ω_0, has now been explicitly included. Substitution of eqn. (4.167) into eqn. (4.168) yields

$$F \propto \exp(-\mathrm{i}\omega_0 t) \sum_i \exp[\mathrm{i}\boldsymbol{K} \cdot \boldsymbol{R}_i^0] \exp[\mathrm{i}\boldsymbol{K} \cdot \boldsymbol{u}_i(t)] \tag{4.169}$$

and for small displacements, the second exponential factor can be expanded to give

$$F \propto \exp(-\mathrm{i}\omega_0 t) \sum_i [1 + \mathrm{i}\boldsymbol{K} \cdot \boldsymbol{u}_i(t)] \exp[\mathrm{i}\boldsymbol{K} \cdot \boldsymbol{R}_i^0]. \tag{4.170}$$

The phonon displacement $\boldsymbol{u}_i$ can be written as an expansion in plane waves (denoting the phonon wavevector as $\boldsymbol{q}$ to distinguish it from $\boldsymbol{k}$ used for light-waves):

$$\boldsymbol{u}_i(t) = \boldsymbol{u}_i^0 \exp[\pm(\boldsymbol{q} \cdot \boldsymbol{R}_i^0 - \omega_q t)]. \tag{4.171}$$

The scattered wave amplitude then can be separated into two terms, one being for *elastic* scattering, i.e. having the *same* frequency as the incident wave, viz.

$$F_{\mathrm{el}} \propto \exp(-\mathrm{i}\omega_0 t) \sum_i \exp[\mathrm{i}\boldsymbol{K} \cdot \boldsymbol{R}_i^0], \tag{4.172}$$

and the other representing *inelastic* scattering, viz.

$$F_{\mathrm{inel}} \propto \sum_i (\boldsymbol{K} \cdot \boldsymbol{u}_i^0 \exp[\mathrm{i}(\boldsymbol{K} \pm \boldsymbol{q}) \cdot \boldsymbol{R}_i^0] \exp[-\mathrm{i}(\omega_0 \pm \omega_q)t]. \tag{4.173}$$

Thus, the conservation law for frequency (or energy) of the scattered wave is

$$\omega = \omega_0 \pm \omega_q \tag{4.174}$$

and since the summation over i in eqn. (4.173) only yields non-zero contributions for crystals when $\boldsymbol{K} + \boldsymbol{q}$ is equal to a reciprocal-lattice vector, the scattered wavevector is given by:

$$\boldsymbol{k} = \boldsymbol{k}_0 \mp \boldsymbol{q} + \boldsymbol{G}, \tag{4.175}$$

where the scattering vector $\boldsymbol{K}$ is the difference between scattered and incident light wavevectors (eqn. (2.98)), i.e. $\boldsymbol{K} = \boldsymbol{k} - \boldsymbol{k}_0$. For *one-phonon* light scattering, as above, since $|\ \boldsymbol{k}\ |, |\ \boldsymbol{k}_0\ | \ll |\ \boldsymbol{q}\ |$, only zone-centre vibrational modes are involved.

Photons emitted with a *lower* frequency than that of the incident light, $\omega = \omega_0 - \omega_q$, corresponding to the *creation* of a phonon, form the Stokes line, and scattered photons with a *higher* frequency, $\omega = \omega_0 + \omega_q$, corresponding to the *destruction* of a phonon, form the anti-Stokes line; the *elastically* scattered light ($\omega = \omega_0$) is called the Rayleigh line. The formation of Stokes and anti-Stokes photons by one-phonon scattering processes is represented schematically in Fig. 4.29 and the actual Stokes Raman spectrum of the cristobalite polymorph of SiO_2 is shown in Fig. 4.30. The intensity of the Stokes line depends on the matrix element for the creation of one phonon, which can be shown quantum-mechanically to be just that for the harmonic oscillator (see e.g. Dove (1993), Kittel (1996)), i.e.

$$I(\omega_0 - \omega_q) \propto (n(\omega_q, T) + 1), \tag{4.176}$$

where $n(\omega_q, T)$ is the Planck distribution function (eqn. (4.68)). For the case of anti-Stokes scattering, the scattered intensity is proportional to the number of phonons with frequency ω_q present in thermal equilibrium,

$$I(\omega_0 + \omega_q) \propto n(\omega_q, T). \tag{4.177}$$

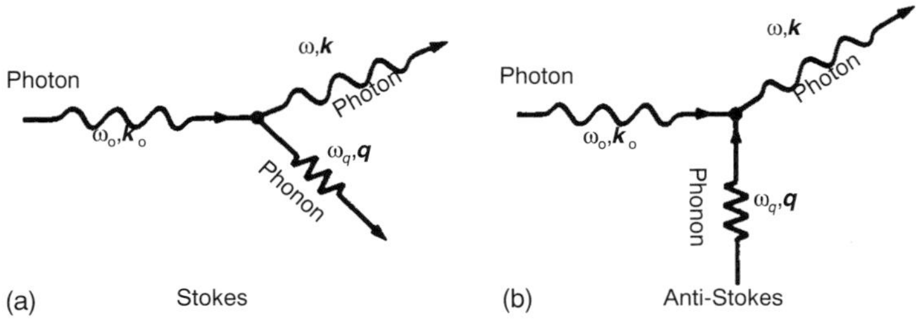

Fig. 4.29 Schematic illustration of the inelastic scattering of photons by a single phonon for: (a) Stokes emission; (b) anti-Stokes emission.

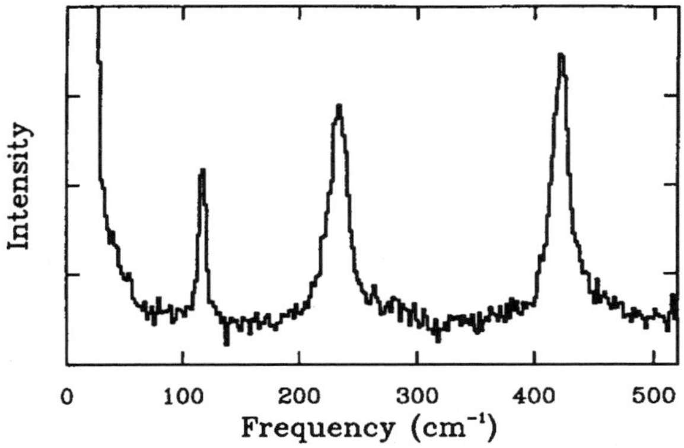

Fig. 4.30 Stokes Raman scattering spectrum of the cristobalite polymorph of crystalline SiO_2. (Compare with the corresponding IR absorption spectrum, Fig. 4.25.) (After Dove (1993). Reproduced by permission of Cambridge University Press)

Thus, in thermal equilibrium, the ratio of anti-Stokes to Stokes light scattering is simply

$$\frac{I(\omega_0 + \omega_q)}{I(\omega_0 - \omega_q)} = \frac{n(\omega_q, T)}{(n(\omega_q, T) + 1)} = \exp(-\hbar\omega_q/k_B T). \tag{4.178}$$

The anti-Stokes intensity vanishes as $T \to 0$ K.

As mentioned above, light scattering is associated with a change in polarizability of the valence electrons. The electric field $\boldsymbol{E}$ of an incident light wave produces a polarization $\boldsymbol{P}$ given by (see also eqn. (7.1)):

$$\boldsymbol{P} = \varepsilon_0 \chi \boldsymbol{E}, \tag{4.179}$$

where χ is the dielectric susceptibility tensor. Any periodic modulation of $\boldsymbol{P}$ leads to the emission of an electromagnetic wave — the scattered wave. The electronic susceptibility is a function of the nuclear coordinates, and hence any modulation of these (e.g. via phonon excitation) will perturb χ as well. The susceptibility may therefore be expanded as a Taylor series in terms of atomic displacements u, viz.

$$\chi = \chi^0 + \left(\frac{\partial \chi}{\partial u}\right) u + \frac{1}{2}\left(\frac{\partial^2 \chi}{\partial u^2}\right) u^2 + \ldots. \tag{4.180}$$

The first term in eqn. (4.180) corresponds to Rayleigh scattering: the scattered wave frequency is the same as that of the incident light wave, $\boldsymbol{E} = \boldsymbol{E}_0 \cos\omega_0 t$, since χ^0 is unperturbed. The second and third terms correspond to first-order and second-order light-scattering processes, involving *one* and *two* phonons, respectively.

For first-order light scattering, where the atomic displacement can be written as $u = u_0 \cos\omega_q t$, the polarization (eqn. (4.179)) becomes

$$\begin{aligned} \boldsymbol{P}(t) &= \varepsilon_0 \frac{\partial \chi}{\partial u} u_0 \boldsymbol{E}_0 \cos\omega_q t \cdot \cos\omega_0 t \\ &= \frac{\varepsilon_0}{2} \frac{\partial \chi}{\partial u} u_0 \boldsymbol{E}_0 \{\cos(\omega_0 + \omega_q)t + \cos(\omega_0 - \omega_q)t\}. \end{aligned} \tag{4.181}$$

Hence, the scattered radiation contains components with frequencies $\omega_0 + \omega_q$ (anti-Stokes) and $\omega_0 - \omega_q$ (Stokes). The scattered intensity is proportional to $|P(t)|^2$. It can be seen from eqn. (4.181) that, for first-order light scattering to be observed, the quantity $\partial\chi/\partial u$ must be non-zero (for the particular polarization of incident and scattered light used). This is the case for zone-centre TO phonons in crystalline Si or Ge, which can therefore be probed by first-order Raman scattering (see Fig. 4.31) even though such modes are unobservable by one-phonon IR absorption spectroscopy (§4.5.1). In fact, it is generally true that for crystals with *centres of inversion* (i (1) symmetry operations, e.g. alkali halides, the diamond cubic structure, but not ZnS), IR-active TO phonons are *not* Raman-active, and vice versa: this is the rule of mutual exclusion.

Second-order Raman scattering involves *two* phonons, either the simultaneous creation or destruction of two phonons, or the creation of one and the destruction of another. The frequencies of the phonons need not be the same; as for the case of two-phonon IR absorption (§4.5.1), phonons having wavevectors near that corresponding to the zone boundary can be involved, and the probability of this otherwise low-probability process is enhanced because of the high density of vibrational modes associated generally with zone-boundary states.

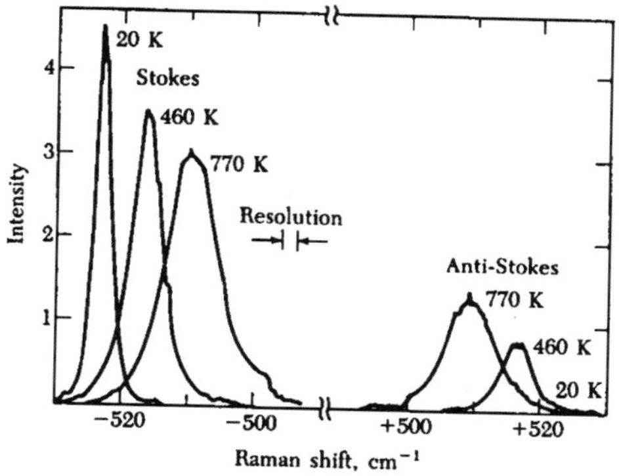

Fig. 4.31 First-order Raman spectrum of the zone-centre TO mode in crystalline Si at three temperatures; both Stokes and anti-Stokes spectra are shown. (After Kittel (1996). Reproduced by permission of John Wiley & Sons Inc.)

Brillouin scattering involves the scattering of light by *acoustic* phonons, for which the frequency ω_q^i of branch i is related to the magnitude of the phonon wavevector q by the appropriate sound velocity v_i via

$$\omega_q^i = qv_i. \tag{4.182}$$

Since the incident and scattered light frequencies are much greater than the phonon frequency, $(\omega_0, \omega) \gg \omega_q$, the magnitudes of incident and scattered light wavevectors are nearly equal ($|\boldsymbol{k}| \simeq |\boldsymbol{k}_0|$). From the conservation law for crystal momentum (eqn. (4.175) with $\boldsymbol{G} = 0$), the magnitude of the phonon wavevector is given by the relation (cf. eqn. (2.99)):

$$q \simeq 2k_0 \sin(\phi/2), \tag{4.183}$$

where the arrangement of vectors is as for elastic scattering (diffraction) with $\boldsymbol{q}$ taking the place of the scattering vector $\boldsymbol{K}$ (see Fig. 2.50b) and ϕ is the angle between $\boldsymbol{k}$ and $\boldsymbol{k}_0$. Since light propagating in a medium with refractive index given by $n_r = \sqrt{\varepsilon(\infty)}$ (cf. eqn. (4.159) has the frequency and wavevector of light related by the phase velocity of light in the solid, viz.

$$\omega_0 = \frac{c}{\sqrt{\varepsilon(\infty)}} k_0, \tag{4.184}$$

the frequency of Brillouin-scattered photons is thus given by

$$\omega_q^i = 2v_i \frac{\omega_0 \sqrt{\varepsilon(\infty)}}{c} \sin(\phi/2). \tag{4.185}$$

Hence, the positions of the Brillouin lines are determined by sound velocities, or equivalently the elastic constants (see e.g. Table 4.1), of the solid. A Brillouin spectrum of the quartz polymorph of crystalline SiO_2 is shown in Fig. 4.32: the single longitudinal and two transverse acoustic modes can be distinguished. Further details on light scattering are given, for example, in Cardona (1983).

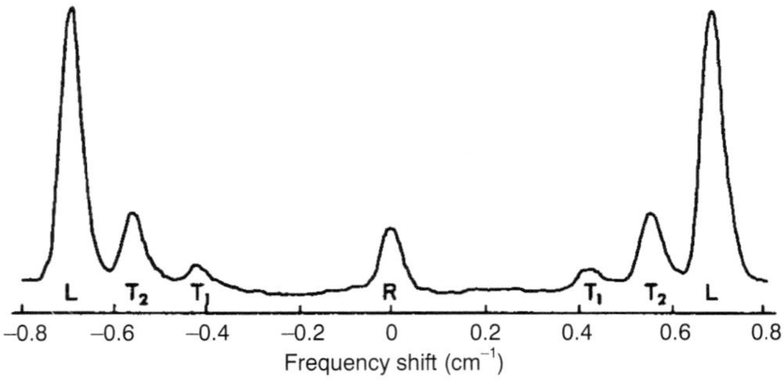

Fig. 4.32 Brillouin spectrum of the quartz polymorph of crystalline SiO_2 showing the two transverse and longitudinal acoustic modes. (Shapiro *et al.* (1966))

4.5.3 Inelastic neutron scattering

The advantages of using neutrons for the structural study of materials has been summarized in §§2.6.1.1 and 2.6.1.3. However, thermal neutrons are also particularly suited for the investigation of vibrational modes in solids because their energy ($\simeq k_B T$) is comparable to that of the excitations, thereby ensuring that inelastic scattering is easily measurable. Moreover, the magnitude of the neutron wavevector, given by the expression (cf. eqn. (2.96)):

$$k = \frac{(2m_n \mathscr{E})^{1/2}}{\hbar} = \frac{(3m_n k_B T)^{1/2}}{\hbar}, \tag{4.186}$$

where m_n is the neutron mass and T is the temperature of the moderator producing the thermal neutrons, has a value ($k \simeq 4 \times 10^{10} \text{m}^{-1}$ for $T = 300$ K) that is much greater than typical phonon wavevectors corresponding to the Brillouin zone boundary. Hence, unlike one-phonon light absorption or scattering, one-phonon inelastic neutron scattering can probe vibrational excitations throughout the zone as a function of phonon wavevector. The phonon dispersion curves shown in Figs. 4.14–17 were obtained in this way.

The neutron scattering function (or dynamical structure factor) $S(\boldsymbol{K}, \omega)$ is given by the time Fourier transform of the intermediate scattering function, $F(\boldsymbol{K}, t)$:

$$S(\boldsymbol{K}, \omega) = \int F(\boldsymbol{K}, t) \exp(-i\omega t) dt, \tag{4.187}$$

with

$$F(\boldsymbol{K}, t) = \langle \rho(\boldsymbol{K}, t) \rho^*(\boldsymbol{K}, 0) \rangle, \tag{4.188}$$

where the angular brackets here denote a time average over all starting times. The density function $\rho(\boldsymbol{K}, t)$ is the sum of the scattering amplitudes (eqns. (2.105), (4.83)) over all atoms, i, viz.

$$\rho(\boldsymbol{K}, t) = \sum_i b_i \exp(\mathrm{i}\boldsymbol{K} \cdot \boldsymbol{R}_i). \tag{4.189}$$

Thus, the neutron-scattering function from a time-varying structure with $\boldsymbol{R}_i(t) = \boldsymbol{R}_i^0 + \boldsymbol{u}_i(t)$ is

$$\begin{aligned} S(\boldsymbol{K}, \omega) = \sum_{i,j} \{ & b_i b_j \exp\{(\mathrm{i}\boldsymbol{K} \cdot (\boldsymbol{R}_i^0 - \boldsymbol{R}_j^0)\} \times \\ & \int \langle \exp\{\mathrm{i}\boldsymbol{K} \cdot [\boldsymbol{u}_i(t) - \boldsymbol{u}_j(0)]\}\rangle \exp(-\mathrm{i}\omega t)\mathrm{d}t\}. \end{aligned} \tag{4.190}$$

This expression describes diffraction of neutrons (with no phonon involvement), as well as inelastic neutron scattering involving the creation or destruction of one or more phonons.

The selection rules for one-phonon inelastic neutron scattering are the *same* as those for inelastic light scattering (i.e. eqns. (4.174), (4.175)). Further details on inelastic neutron scattering can be found in Willis (1973), Squires (1978) and Sköld and Price (1986).

Vibration-related properties 4.6

Many thermal properties of solids are determined by their vibrational behaviour, e.g. heat capacity (specific heat), thermal conductivity and thermal expansion. Although other types of excitations may also contribute to some of these quantities if present (e.g. electronic excitations in metals), *all* materials have atomic vibrational excitations and for electrical insulators (non-metals) these are dominant.

4.6.1 Heat capacity

The heat capacity, or specific heat, at constant volume (C_v) is defined as the change in internal energy U with temperature:

$$C_v = \left(\frac{\partial U}{\partial T}\right)_v. \tag{4.191}$$

It will be assumed here that the only excitations that can be thermally excited and contribute to an increase in the internal energy are vibrational modes, e.g. phonons. Note also that, because the thermal expansion of solids is very small (negligible compared with that of gases, although non-zero — see §4.6.2.1), the heat capacity at constant volume is practically identical to that at constant pressure, $C_v \simeq C_p$ $(= (\partial H/\partial T)_p)$ (see Problem 4.17); the equality is exact for a harmonic crystal.

The vibrational contribution to the internal energy is given by the product of the energy of a given phonon, $\hbar\omega_k(p)$ for the branch p, the phonon occupation number, $n(\boldsymbol{k},p)$ (eqn. (4.68)) and the density of modes in $\boldsymbol{k}$-space (eqn. (4.28)), i.e.

$$C_v = \frac{\partial}{\partial T}\sum_p \int \frac{V}{(2\pi)^3}\frac{\hbar\omega_k(p)\mathrm{d}k}{[\exp(\hbar\omega_k(p)/k_\mathrm{B}T) - 1]} \tag{4.192}$$

where the integral is taken over allowed $\boldsymbol{k}$-vectors in the first Brillouin zone or, in terms of the density of states as a function of frequency:

$$C_v = \frac{\partial}{\partial T}\sum_p \int \frac{V g(\omega_k(p))\hbar\omega_k(p)\mathrm{d}\omega_k(p)}{[\exp(\hbar\omega_k(p)/k_\mathrm{B}T) - 1]}. \tag{4.193}$$

Although, in general, these expressions must be evaluated numerically, for actual density-of-states distributions, nevertheless certain simplifications can be made at high temperatures (irrespective of the precise form of the density of states) and at very low temperatures in the case of the Debye-like density of states (eqn. (4.23)).

Using the general form for the vibrational density of states (eqn. (4.118)), eqn. (4.193) can be rewritten as

$$C_v = \frac{\partial}{\partial T}\sum_{k,p} \frac{\hbar\omega_k(p)}{[\exp(\hbar\omega_k(p)/k_\mathrm{B}T) - 1]}. \tag{4.194}$$

At *high* temperatures, $\hbar\omega_k(p) \ll k_\mathrm{B}T$, the summand in eqn. (4.194), expressed in terms of the variable $x \equiv \hbar\omega_k(p)/k_\mathrm{B}T \ll 1$, can be expanded as:

$$\frac{\hbar\omega_k(p)}{e^x - 1} = \frac{\hbar\omega_k(p)}{x + x^2/2 + x^3/6 + \cdots} \simeq \frac{\hbar\omega_k(p)}{x}\left[1 - \frac{x}{2} + \frac{x^2}{12}\right], \tag{4.195}$$

and if only the first term in the expansion of eqn. (4.195) is kept, the summation of the term $\hbar\omega_k(p)/x = k_B T$ over all $\boldsymbol{k}$ and p gives for the specific heat the constant value:

$$C_v = 3Nk_B = 3R, \tag{4.196}$$

for one mole, which is the Dulong–Petit value in the classical limit. It should be noted that this result is valid only in the *harmonic* approximation.

The other limit for which an analytic result can be obtained for the heat capacity is at *low* temperatures. At low temperatures, the phonon occupation number (eqn. (4.68)) has the form $n(\omega_k(p), T) \simeq \exp(-\hbar\omega_k(p)/k_B T)$ for *optic* modes which have a non-zero lower limit to their frequency (see Fig. 4.11), and thus vanishes as $T \to 0K$. *Acoustic* modes, on the other hand, continue to be excited even at very low temperatures, since for them $\omega \to 0$ as $k \to 0$. Moreover, if it is assumed that the acoustic modes obey a *linear* dispersion law, i.e. $\omega_k(p) = v(p)k$ (valid in reality at very low temperatures, where the frequencies of excited vibrational modes are correspondingly low), the Debye approximation (§4.1.2), valid for elastic continua, is recovered.

Hence, using the Debye density of states (eqn. (4.23)), the heat capacity (eqn. (4.193)) can be written approximately as

$$C_v = \frac{3V\hbar}{2\pi^2 v_0^3}\frac{\partial}{\partial T}\int_0^{\omega_D}\frac{\omega^3 d\omega}{\exp(\hbar\omega/k_B T) - 1} \tag{4.197a}$$

$$= \frac{3V\hbar^2}{2\pi^2 v_0^3 k_B T^2}\int_0^{\omega_D}\frac{\omega^4 \exp(\hbar\omega/k_B T)}{[\exp(\hbar\omega/k_B T) - 1]^2} d\omega, \tag{4.197b}$$

where ω_D is the Debye frequency (eqn. (4.25)). Changing the integration variable to $x = \hbar\omega/k_B T$, and denoting the Debye temperature θ_D as

$$\theta_D = \frac{\hbar\omega_D}{k_B} = \frac{\hbar v_0}{k_B}\left(\frac{6\pi^2 N}{V}\right)^{1/3}, \tag{4.198}$$

eqn. (4.197) can be rewritten as:

$$C_v = 9Nk_B\left(\frac{T}{\theta_D}\right)^3\int_0^{x_D}\frac{x^4 e^x dx}{(e^x - 1)^2}, \tag{4.199}$$

where $x_D = \hbar\omega_D/k_B T = \theta_D/T$. This Debye form for the heat capacity is plotted in Fig. 4.33: it approaches zero as $T \to 0$ K, and tends to the Dulong–Petit limiting value (eqn. (4.196)) at high temperatures. Values of θ_D are typically in the range 50–1000 K (see Table 4.2).

An analytic expression for the heat capacity, valid at low temperatures, can be obtained readily by changing the variable in the integral in eqn. (4.197a) to $x = \hbar\omega/k_B T$ and extending the upper integration limit from x_D to infinity. The resulting integral can be evaluated as

$$\int_0^\infty \frac{x^3 dx}{(e^x - 1)} = \int_0^\infty\left(\sum_{n=1}^\infty e^{-nx}\right)x^3 dx = 6\sum_{n=1}^\infty \frac{1}{n^4} = \pi^4/15, \tag{4.200}$$

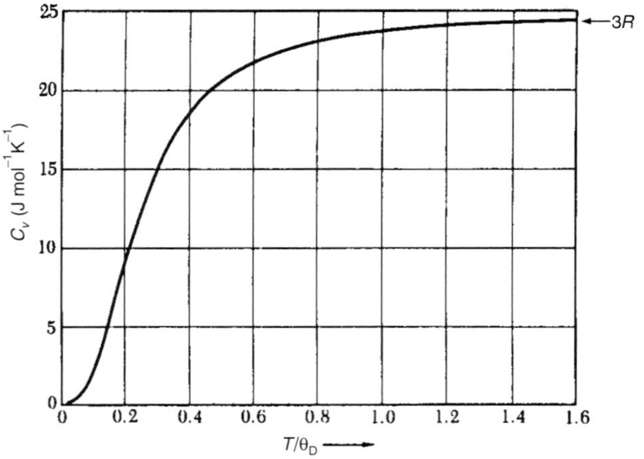

Fig. 4.33 The form of the heat capacity of a solid calculated in the Debye approximation plotted against temperature normalized by the Debye temperature θ_D. The temperature dependence follows the T^3 law for $T \leq 0.1\theta_D$.

Table 4.2 Debye temperatures for some elements and compounds

Element	θ_D (K)	*Element*	θ_D (K)	*Element*	θ_D (K)
Ag	225	Fe	470	Rb	56
Al	428	Ge	374	Se	90
Au	165	Hg	71.9	Si	645
B	1220	K	91	Sn (grey)	260
Be	1440	Li	344	Sn (white)	170
C (dia)	2230	Mo	450	Ti	420
C (graphite)	760	Na	158	U	207
Co	445	Ni	450	V	380
Cs	38	Pb	105	W	400
Cu	343	Pt	240	Zn	327
Compound	θ_D (K)	*Compound*	θ_D (K)	*Compound*	θ_D (K)
LiF	670	KF	335	CsF	245
LiCl	420	KCl	240	CsCl	175
LiBr	340	KBr	192	CsBr	125
LiI	280	KI	173	CsI	102
NaF	445	RbF	267	AgCl	180
NaCl	297	RbCl	194	AgBr	140
NaBr	238	RbBr	149	BN	600
NaI	197	RbI	122	SiO_2 (quartz)	255

(After Kittel (1996) (Reproduced by permission of John Wiley & Sons Inc.) and Burns (1985) (Reproduced by permission of Academic Press, Inc.))

where the exponential factor $(e^x - 1)$ has been replaced by a geometric progression and the ensuing integral evaluated by parts. Thus, performing the differentiation with respect to temperature in eqn. (4.197) gives:

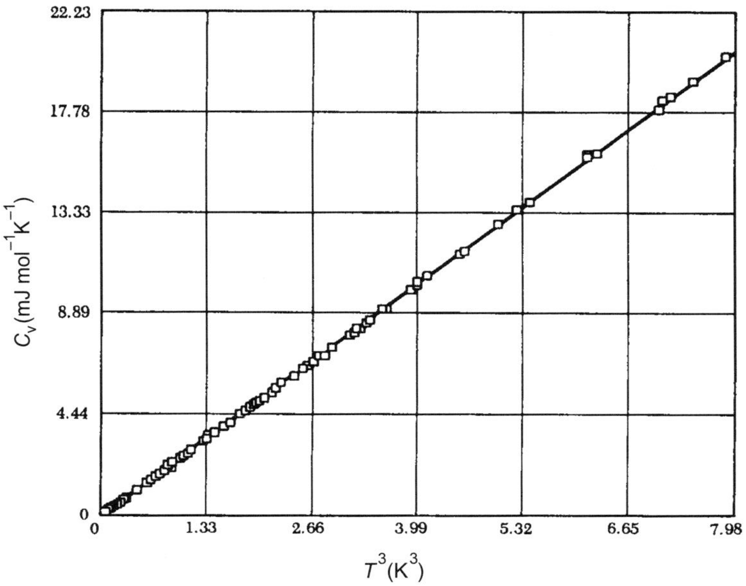

Fig. 4.34 Low-temperature heat capacity of solid Ar plotted versus T^3 to reveal the agreement with the Debye law ($\theta_D = 92$ K). (Kittel (1996). Reproduced by permission of John Wiley & Sons Inc.)

$$C_v = \frac{12\pi^4}{5} Nk_B \left(\frac{T}{\theta_D}\right)^3. \qquad (4.201)$$

The Debye model predicts that the limiting low-temperature behaviour of the vibrational contribution to the heat capacity has a T^n dependence for n-dimensional solids (see Problem 4.20). Figure 4.34 shows an example of the Debye T^3 law obeyed by solid Ar. This low-temperature behaviour is masked in the case of metals by an *electronic* contribution to the heat capacity which varies *linearly* with temperature (see §5.1.3.1).

A qualitative illustration of the origin of the Debye T^3 law can be gained by a geometric argument, assuming that only vibrational modes for which $\hbar\omega \leqslant k_B T$ will be appreciably excited at a low temperature T, each of which may be taken to have an energy of approximately $k_B T$. The maximum allowed magnitude of the (Debye) phonon wavevector is k_D, related to the Debye frequency by

$$k_D = \omega_D / v_0, \qquad (4.202)$$

and thus only the fraction of modes in $\boldsymbol{k}$-space given approximately by $(k_T/k_D)^3 = (\omega_T/\omega_D)^3$ (see Fig. 4.35) are thermally excited (where $\hbar\omega_T = \hbar v_0 k_T = k_B T$). Thus, there are of order $3N(T/\theta_D)^3$ modes that are thermally excited, and hence the corresponding energy is $U \simeq 3Nk_B T(T/\theta_D)^3$. Hence, the heat capacity should be $C_v \simeq 12Nk_B(T/\theta_D)^3$, the form of which agrees with eqn. (4.201), although the numerical factor is incorrect because of the assumption that each excited vibrational mode carries an energy of $k_B T$ (strictly valid only at *high* temperatures), and the uncertainty in defining k_T.

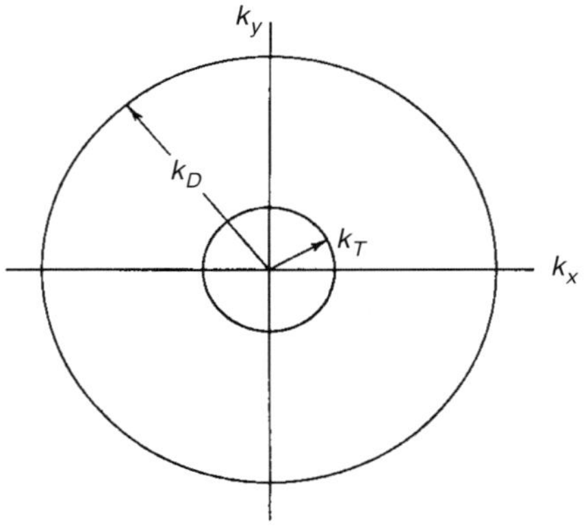

Fig. 4.35 Geometric illustration of the origin of the Debye T^3 law for the low-temperature heat capacity. Of the allowed volume in $\boldsymbol{k}$-space given by the radius $k_{\rm D}$, only modes in the volume with radius $k_T \simeq k_{\rm B}T/\hbar v_0$ are thermally excited. Thus, of the $3N$ possible modes, only a fraction $(k_T/k_{\rm D})^3 = (T/\theta_{\rm D})^3$ are excited. The corresponding internal energy is $U \propto T^4$ and hence $C_v \propto T^3$.

Another model which can be used to obtain an analytic expression for the heat capacity of solids uses the *Einstein approximation*, in which *all* atoms are assumed to vibrate independently with the *same* frequency, ω_0. Obviously this approximation is invalid for acoustic modes, but is reasonable for optic modes. The heat capacity calculated from this model (see Problem 4.21) reduces to the Dulong–Petit limit at high temperatures, in agreement with experiment, but has a much more rapid (approximately exponential) temperature dependence at low temperatures than is observed experimentally or predicted by the Debye model. The two different ways of approximating the phonon dispersion curves of crystals are illustrated in Fig. 4.36.

It might be thought that the Debye model for the heat capacity, based on the assumption that vibrational modes excited at low temperatures are of sufficiently long wavelength (small wavevector) that the material behaves as an elastic continuum, would be valid for *all* types of solids. In fact, *disordered* materials (e.g. glasses) exhibit an extra contribution to the vibrational heat capacity at very low temperatures ($T < 1K$) in addition to the Debye term (eqn. (4.201)), as can be seen in Fig. 4.37. This additional contribution varies approximately *linearly* with temperature, and so the total heat capacity at low temperatures can be written as

$$C_v = aT + bT^3, \tag{4.203}$$

where a and b are temperature-independent constants. Thus, a plot of C_v/T versus T^2 reveals the extra contribution as a non-zero intercept (a) on the abscissa (see Fig. 4.38). The linear term in the heat capacity of disordered materials is *not* due to an *electronic* contribution (§5.1.3.1), since glassy silica, for example, is an extremely good electrical insulator with a negligible concentration of mobile electrons. Instead, the extra contribution to the heat capacity must come from additional low-frequency vibrational modes that are not included in the Debye model.

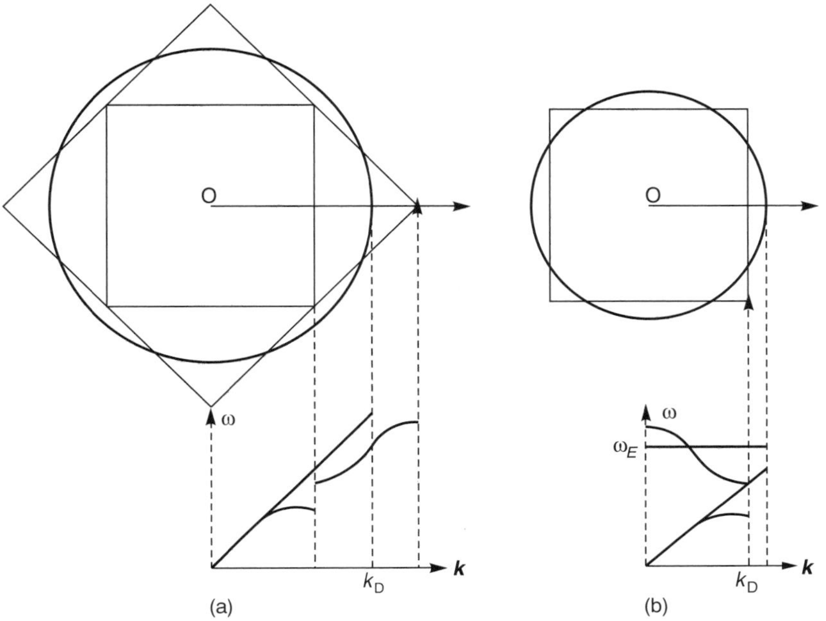

Fig. 4.36 Two ways of approximating the acoustic and optic branches of a diatomic crystal (for the case of two dimensions and along a high-symmetry direction for simplicity) for calculating the vibrational heat capacity. (a) The Debye approximation. The first two Brillouin zones of the square lattice are replaced by a circle with the same total area and radius k_{D}, and the actual phonon dispersion curve in the extended-zone scheme is approximated by a single linear curve for $k \leq k_{\mathrm{D}}$. (b) The Debye approximation for the acoustic branch and the Einstein approximation for the optic branch. The first zone is replaced by a circle of radius k_D with the same area, and the acoustic branch replaced by a linear curve for $k \leq k_{\mathrm{D}}$, and the optic branch is replaced by a constant-frequency line in the same region.

★ A phenomenological model for the anomalous linear specific-heat term in disordered solids ascribes it to the excitations of two-level systems, i.e. configurations characterized by just two energy levels at $\pm\mathscr{E}$. The heat capacity of a *single* two-level system is (see Problem 4.21):

$$C_v = k_{\mathrm{B}}(\mathscr{E}/k_{\mathrm{B}}T)^2 \operatorname{sech}^2(\mathscr{E}/k_{\mathrm{B}}T). \tag{4.204}$$

In the case of materials containing structural disorder, it is reasonable to assume a variety of atomic configurations giving two-level systems, characterized by a *distribution* of energy levels, $n(\mathscr{E})$. Thus, the net heat capacity can be written in this case as

$$C_v = k_{\mathrm{B}} \int n(\mathscr{E})(\mathscr{E}/k_{\mathrm{B}}T)^2 \operatorname{sech}^2(\mathscr{E}/k_{\mathrm{B}}T)\mathrm{d}\mathscr{E}. \tag{4.205}$$

If it is assumed that in an amorphous solid the density of two-level systems is *constant*, $n(\mathscr{E}) = n_0$, and changing the integration variable to $x = \mathscr{E}/k_BT$, eqn. (4.205) becomes

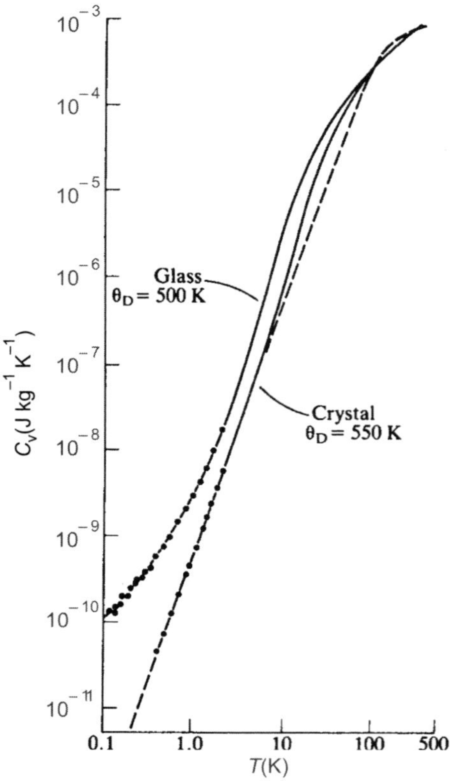

Fig. 4.37 Heat capacity of glassy SiO_2 compared with that of the crystalline polymorph quartz. (The dashed line for the crystal corresponds to the Debye T^3 law.) Note the excess (linear) contribution heat capacity in excess of the Debye T^3 behaviour exhibited by the glass for temperatures less than 1 K. (Afer Zeller and Pohl (1971). Reprinted with permission from *Phys. Rev.* **B4**, 2029. © 1971. The American Physical Society)

$$C_v = k_B^2 n_0 T \int_0^\infty x^2 \operatorname{sech}^2 x \mathrm{d}x, \tag{4.206}$$

where for very low temperatures ($x \gg 1$), the upper limit of the integral can be extended to infinity. In this case, the integral yields a constant ($\pi^2/6$) and so the heat capacity due to a broad distribution of two-level systems is

$$C_v = \frac{\pi^2}{6} n_0 k_B^2 T. \tag{4.207}$$

The microscopic origin of the two-level systems in glasses is still uncertain. One possibility is that they are associated with highly *anharmonic* atomic configurations, characterized by double-well potentials (Fig. 4.39): atomic tunnelling at very low temperatures through the potential-energy barrier separating two configurations, corresponding to the two wells, leads to pairs of tunnel-split energy levels, which could act as the two-level systems.

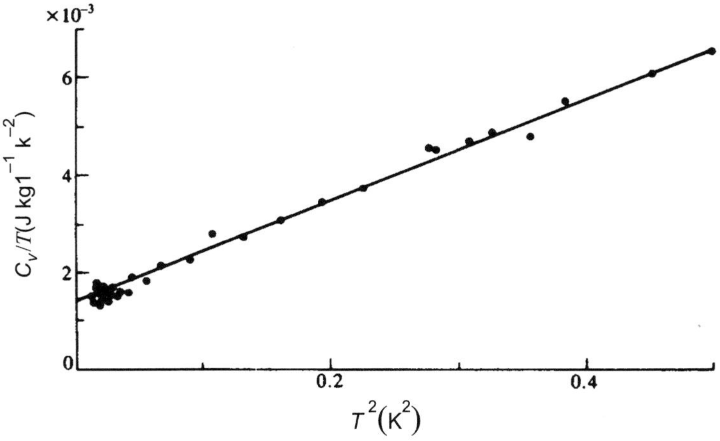

Fig. 4.38 Heat capacity of glassy B_2O_3 plotted as C_v/T versus T^2 to demonstrate the existence, via the non-zero intercept, of a linear term. (Reprinted with permission from Stephens (1976), *Phys. Rev.* **B13**, 852. © 1976. The American Physical Society)

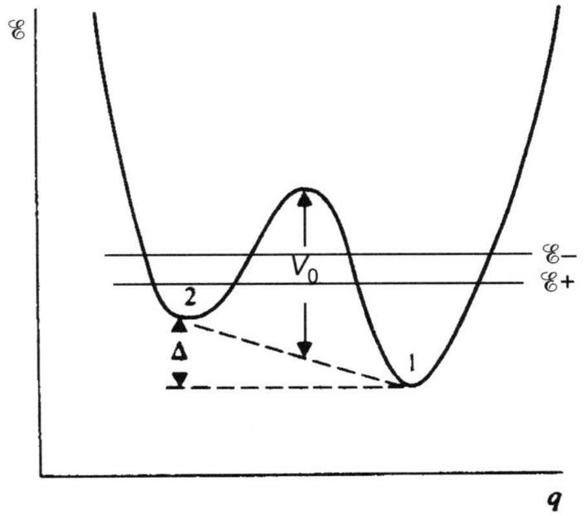

Fig. 4.39 Double-well vibrational potential for which the tunnel-split allowed energy levels are a possible origin of two-level systems responsible for the linearly temperature-dependent term in the heat capacity of amorphous solids.

4.6.2 Anharmonicity

The theory of atomic dynamics developed so far has assumed that the *harmonic approximation* is appropriate; that is, the potential energy (eqn. (4.30)) has been expanded only to terms quadratic in the atomic displacement so that the potential-energy curve is parabolic. This approximation has a number of significant consequences which are summarized below.

1. For the case of crystalline materials, the allowed vibrational excitations are normal modes, i.e. independent vibrational waves (phonons) that do not mutually interact, or evolve with time.
2. The harmonic approximation predicts that there is no thermal expansion of the lattice: the average interatomic spacing does not increase with increasing atomic displacement (temperature) for a harmonic oscillator. As a consequence, the heat capacities at constant volume and pressure are predicted to be identical, as are the adiabatic and isothermal elastic constants (which are, moreover, predicted to be independent of temperature and pressure).
3. Finally, the harmonic approximation implies that the heat capacity reaches a constant (Dulong–Petit) limit, $3R$, at high temperatures ($T \rangle \theta_D$).

In practice, real materials are characterized by non-parabolic dependences of potential energy on displacement (see Figs. 2.19 and 5.9) and they exhibit clear signs of anharmonic behaviour. Two collinear phonons injected into a crystal can combine to give a third beam of phonons with a frequency equal to the sum of the other two. Solids have non-zero values of the thermal-expansion coefficient (§4.6.2.1). The heat capacity increases steadily above the Dulong–Petit limit with increasing temperature for $T \rangle \theta_D$ (Fig. 4.40). Peaks in one-phonon inelastic neutron-scattering experiments are not infinitesimally narrow, but have finite widths as a result of the phonons having a finite, rather than infinite, lifetime and due to the Heisenberg uncertainty principle ($\Delta E \Delta t \simeq \hbar/2$).

The *cubic* term in the Taylor expansion of the potential energy (eqn. (4.30)) involves the mixing of *three* phonons, i.e. the coalescence of two phonons to give a third, or the

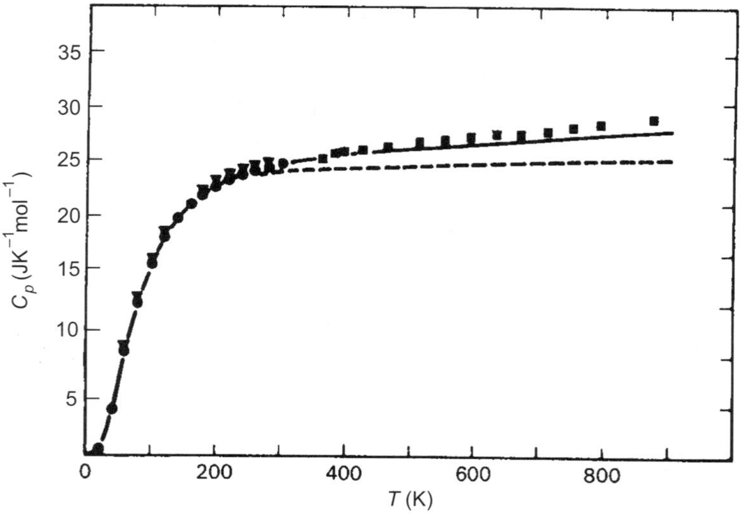

Fig. 4.40 Lattice heat capacity of Cu: points, experimental data; dashed line, theoretical calculation based on the harmonic approximation; solid line, theoretical calculation taking account of anharmonicity. (Miiller and Brockhouse (1971). Reproduced by permission of NRC Research Press)

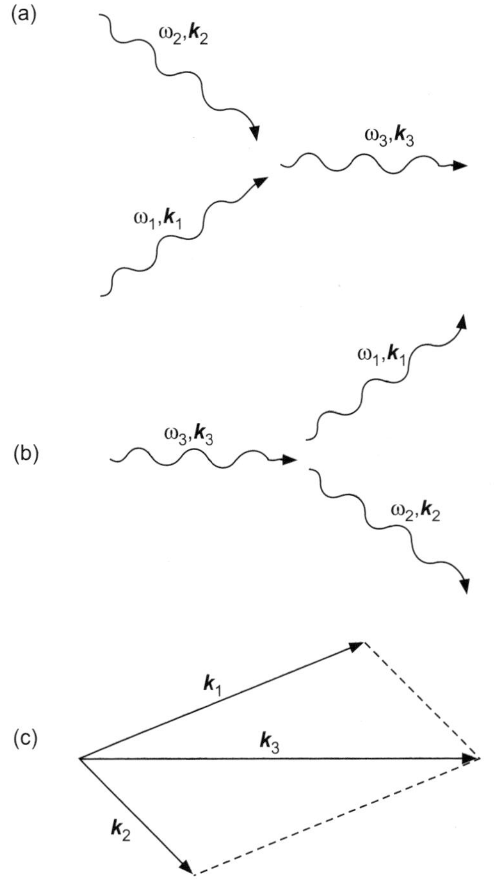

Fig. 4.41 Representation of the interactions between three phonons as a result of the cubic anharmonic term in the vibrational potential energy: (a) coalescence of two phonons to give a third; (b) decay of a phonon into two others; (c) 2D illustration of the conservation of crystal momentum in the first zone ($G = 0$) (normal, 'N', process).

decay of a single phonon into two others (Fig. 4.41). The conservation laws for energy and crystal momentum dictate that in this case (cf. Fig. 4.41):

$$\omega_3 = \omega_1 + \omega_2 \tag{4.208a}$$

and

$$\boldsymbol{k}_3 = \boldsymbol{k}_1 + \boldsymbol{k}_2 + \boldsymbol{G}. \tag{4.208b}$$

In the case where $\boldsymbol{k}_1$ and $\boldsymbol{k}_2$, lying in the first Brillouin zone, are such that $\boldsymbol{k}_3$ also lies in the first zone, $\boldsymbol{G} = 0$, and the process is called a normal (or N) process (Fig. 4.41c). The allowed solutions of the conservation equations (4.208) can be found by a geometric construction (Fig. 4.42): it can be seen that, in this case, a TA phonon $(\omega_1, \boldsymbol{k}_1)$ coalesces

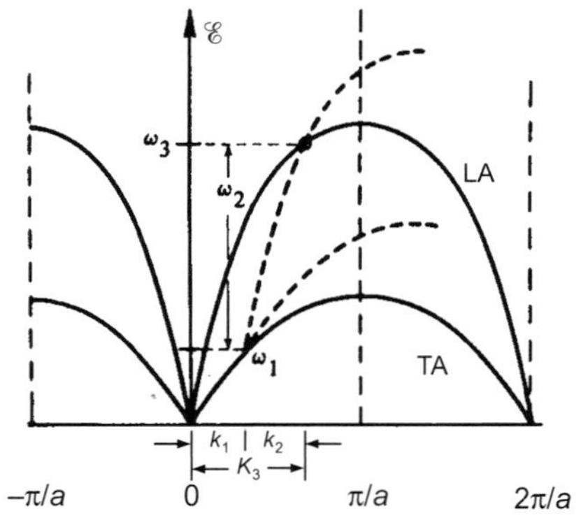

Fig. 4.42 Geometric construction to find the allowed values of frequency ω_3 and wavevector $\boldsymbol{k}_3$ of a phonon created, in an N-process, by the coalescence of two others with frequencies ω_1 and ω_2 and wavevectors $\boldsymbol{k}_1$ and $\boldsymbol{k}_2$, satisfying the conservation laws of energy and crystal momentum. For simplicity, it is assumed that all three wavevectors are collinear. A replica of the dispersion curves (dashed curves) is constructed with its origin at the position $(\omega_1, \boldsymbol{k}_1)$ of one of the phonons (in the TA branch). The intersections of the dispersion curves and their displaced replica give the allowed values of the other two phonons, in this case phonons in the LA branch with ω_2, $\boldsymbol{k}_2$ and ω_3, $\boldsymbol{k}_3$.

with an LA phonon $(\omega_2, \boldsymbol{k}_2)$ to give another LA phonon $(\omega_3, \boldsymbol{k}_3)$ (see Problem 4.22). Likewise, the *quartic* term in the potential energy is associated with the mixing of *four* phonons. Thus, the finite lifetime of phonons is due to their decay into two or more different phonons.

*4.6.2.1 *Thermal expansion*

Real materials, having anharmonic vibrational potentials, generally exhibit thermal expansion (see Fig. 4.43) because, unlike the case for harmonic potentials, the average interatomic spacing *increases* with increasing vibrational excitation, i.e. with increasing temperature. The volume coefficient of thermal expansion is defined as

$$\beta_T = \frac{1}{V}\left(\frac{\partial V}{\partial T}\right)_p, \tag{4.209}$$

and the linear coefficient is simply one-third this value, $\alpha_T = \beta_T/3$. Experimental values of β_T are typically of the order of 10^{-5} K^{-1} at room temperature. Equation (4.209) can be rewritten as

$$\beta_T = -\frac{1}{V}\left(\frac{\partial V}{\partial p}\right)_T\left(\frac{\partial p}{\partial T}\right)_V = \frac{1}{B}\left(\frac{\partial p}{\partial T}\right)_V, \tag{4.210}$$

where B is the bulk modulus (eqn. (2.32)). The pressure is given by the volume derivative of the Helmholtz free energy:

$$p = -\left(\frac{\partial F_{\mathrm{H}}}{\partial V}\right)_T, \tag{4.211}$$

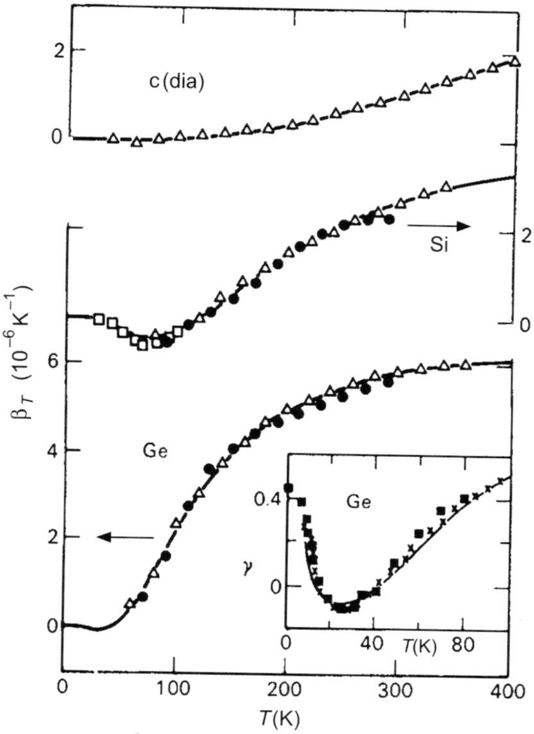

Fig. 4.43 Experimental (points) and calculated (curves) values of the volume coefficients of thermal expansion for crystalline diamond, Si and Ge. Both Si and Ge exhibit *negative* thermal-expansion coefficients at low temperatures arising from negative values of the average Grüneisen parameter shown in the inset for Ge. (Dolling and Cowley (1966). Reproduced by permission of IOP Publishing Ltd.)

with

$$F_{\mathrm{H}} = U - TS. \tag{4.212}$$

Since the vibrational internal energy U of a solid depends on the frequencies of the modes, anharmonicity, leading to thermal-expansion effects, is thus associated with a dependence of the mode frequencies on the *volume* of the crystal (the frequencies are volume-independent in the harmonic approximation).

The pressure may be expressed entirely in terms of the internal energy by making use of the second law of thermodynamics, written in the form

$$T\left(\frac{\partial S}{\partial T}\right)_V = \left(\frac{\partial U}{\partial T}\right)_V, \tag{4.213}$$

viz.

$$p = -\frac{\partial}{\partial V}\left[U - T\int_0^T \frac{\mathrm{d}T'}{T'}\frac{\partial}{\partial T'}U(T', V)\right]. \tag{4.214}$$

The internal energy is given by eqn. (4.67), plus a term U^{eq} representing the energy of the solid with its atoms at their equilibrium positions; substituting this into eqn. (4.214) gives

$$\begin{aligned} p = & -\frac{\partial}{\partial V}\left[U^{\mathrm{eq}} + \sum \frac{1}{2}\hbar\omega_k(p)\right] \\ & + \sum_{k,p}\left[-\frac{\partial}{\partial V}(\hbar\omega_k(p))\right]\frac{1}{\exp(\hbar\omega_k(p)/k_{\mathrm{B}}T) - 1}\end{aligned} \tag{4.215}$$

The first term in eqn. (4.215) gives the pressure of the solid at $T = 0\mathrm{K}$ in general, and at any temperature in the harmonic approximation, while the second term gives rise to a temperature-dependent contribution to the pressure (and hence a finite value of the thermal-expansion coefficient—see eqn. (4.210)) if the mode frequencies are volume-dependent.

Substitution of eqn. (4.215) into eqn. (4.210) gives

$$\beta_T = \frac{1}{B}\sum_{k,p}\left[-\frac{\partial}{\partial V}\hbar\omega_k(p)\right]\frac{\partial}{\partial T}n(\boldsymbol{k},p) \tag{4.216}$$

where $n(\boldsymbol{k},p)$ is the phonon-occupation number (eqn. (4.68)). This may be compared with eqn. (4.194) for the heat capacity which can be written in similar form, i.e.

$$C_V = \sum_{k,p}\hbar\omega_k(p)\frac{\partial}{\partial T}n(\boldsymbol{k},p) \equiv \sum_{\boldsymbol{k},p} C_V(\boldsymbol{k},p), \tag{4.217}$$

where $C_V(\boldsymbol{k},p)$ is the contribution of the phonon $\boldsymbol{k}$ in branch p to the overall heat capacity.

The so-called Grüneisen parameter for the normal mode characterized by $(\boldsymbol{k},p)$ is defined as

$$\gamma_{k,p} = -\frac{\partial(\ln\omega_k(p))}{\partial(\ln V)}, \tag{4.218}$$

and a mean Grüneisen constant can be defined as the average of the contributions of the modes, weighted in terms of the mode specific heat, viz.

$$\gamma = \frac{\sum_{k,p}\gamma_{k,p}C_V(\boldsymbol{k},p)}{\sum_{k,p}C_V(\boldsymbol{k},p)}. \tag{4.219}$$

Combining eqns. (4.216–219) gives the Grüneisen law:

$$\beta_T = \frac{\gamma C_V}{BV}. \tag{4.220}$$

Values of the mean Grüneisen constant γ typically fall in the range 1–3 (and are in fact weakly temperature-dependent). Neglecting the weak temperature dependences of γ and of the bulk modulus in eqn. (4.220), it can be seen from eqn. (4.220) that the temperature dependence of the volume thermal-expansion coefficient should follow that of C_V (see §4.6.1), viz. $\beta_T \propto T^3$ at very low temperatures, tending to a constant value at high temperatures ($T \gg \theta_D$). This general behaviour is evident in Fig. 4.43; the negative

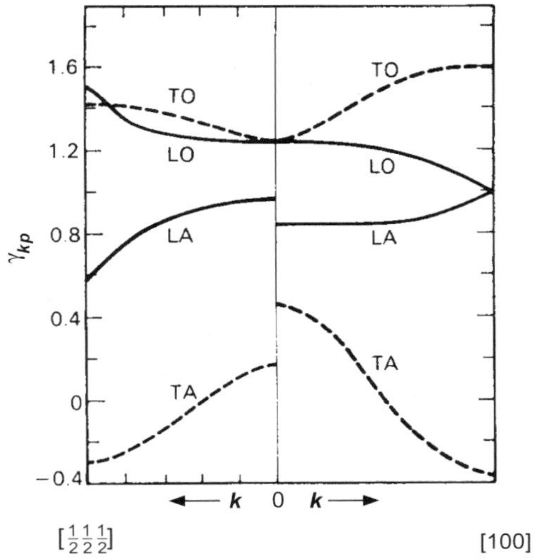

Fig. 4.44 Calculated values of the mode Grüneisen parameter for Ge for ***k***-vectors parallel to [100] and [111] directions. (Dolling and Cowley (1966). Reproduced by permission of IOP Publishing Ltd.)

values of β_T for Si and Ge at low temperatures are due to negative values of the mode Grüneisen parameters for the TA mode (see Fig. 4.44) which contributes significantly to the heat capacity at low temperatures. Negative thermal expansion is also found in certain crystalline oxide materials, e.g. eucryptite ($Li_2Al_2Si_2O_8$), $NaTi_2P_3O_{12}$ and ZrW_2O_8, the last material exhibiting a negative value of β_T in the very wide temperature range 0.3–1050 K (Mary *et al.* (1996)). Negative thermal expansion in these materials is also believed to be due to peculiarities in the behaviour of the TA modes, in these cases associated with coupled rotational motion of coordination polyhedra (e.g. SiO_4, PO_4 or WO_4).

It should be noted that the above discussion is for the case of *microscopic* thermal expansion, associated with changes in the bond length with temperature, as might be monitored experimentally by measuring the temperature dependence of the crystal lattice parameter $a(T)$ using, for example, diffraction techniques (§2.6.1). However, measurement of the *macroscopic* thermal-expansion coefficient (using a dilatometer), e.g. the linear expansion coefficient $\Delta l/l$, may yield a *larger* value than that $(\Delta a/a)$ deduced from lattice-parameter measurements if vacancy formation (§3.1.2) is important. In such a case, the macroscopic expansion coefficient is given by

$$\frac{\Delta l}{l} = \frac{\Delta a}{a} + \frac{x_V}{3}, \tag{4.221}$$

where x_v is the fractional vacancy concentration (eqn. (3.7)).

*4.6.2.2 *Thermal conductivity*

In addition to having an effect on equilibrium properties of materials, e.g. heat capacity, thermal expansion, etc., anharmonicity in interatomic potentials can have a pronounced influence on *non-equilibrium* (transport) processes, e.g. heat conduction. As mentioned previously in §3.4.2.1, the heat flux or thermal current density, $\boldsymbol{J}_Q$, between two points (i.e. the energy transported per unit time through unit area normal to the vector connecting the two points) is proportional to the temperature gradient, i.e.

$$\boldsymbol{J}_Q = -\kappa_T \nabla T \tag{4.222}$$

where the constant of proportionality, κ_T, is the thermal conductivity. (For the case of non-cubic crystals, κ_T is a *tensor* quantity.)

In the absence of free, itinerant electrons (i.e. for the case of electrical insulators or non-metals), heat is carried through a solid by *phonons*. A temperature variation at one end of a crystal, localized spatially within a region Δx, also produces a non-equilibrium distribution of phonons, and this wave packet consists of phonons having a distribution $|\Delta \boldsymbol{k}| \simeq 1/\Delta x$ of wavevectors about the value $\boldsymbol{k}$. This phonon wave packet can propagate through the solid with the group velocity $\boldsymbol{v}_g(\boldsymbol{k})$ (eqn. (4.8)). For the case of an infinitely large, perfect single crystal for which the interatomic potential is *harmonic*, such a wave packet will propagate in a ballistic fashion, without any resistance or impediment to its motion, since the phonons are the independent normal modes of the system. In such an idealized case, the thermal conductivity is *infinite*. In reality, structural defects (including surfaces of the crystal) can cause *scattering* of the phonons, and if such scattering events result in a net change in momentum of the phonon wave packet, there is a corresponding resistance to the thermal current, i.e. the thermal conductivity becomes *finite*. Anharmonicity-induced phonon–phonon scattering, in certain cases, can also contribute to the resistance to heat flow.

A general expression for the thermal conductivity can be obtained as follows. The heat flux $\boldsymbol{J}_Q$ is simply the sum of the energies carried by all the phonon modes, multiplied by the velocity at which the heat is transported (the phonon group velocity) divided by the volume:

$$\boldsymbol{J}_Q = \frac{1}{V}\sum_{k,p} n(\boldsymbol{k},p)\hbar\omega_k \boldsymbol{v}_g(\boldsymbol{k},p), \tag{4.223}$$

where $n(\boldsymbol{k},p)$ is the phonon occupation number (eqn. (4.68)) and $\boldsymbol{v}_g(\boldsymbol{k},p)$ is the group velocity of the mode $(\boldsymbol{k},p)$ (eqn. (4.8b)). Under equilibrium conditions, for a uniform temperature ($\nabla T = 0$), $\boldsymbol{J}_Q = 0$ since the group velocities are distributed isotropically ($v_g^x(\boldsymbol{k}) = -v_g^x(-\boldsymbol{k})$) and the equilibrium phonon occupation numbers $n(\boldsymbol{k},p) \equiv n_{k,p} = n_{k,p}^0$ are equal for positive and negative $\boldsymbol{k}$-values.

Hence, a thermal current can occur only if the phonon occupation number differs from the equilibrium value, $n_{k,p} = n_{k,p}^0$; therefore the heat flux can also be written in terms of this difference in phonon numbers:

$$\boldsymbol{J}_Q = \frac{1}{V}\sum_{k,p} \hbar\omega_k(p)[n - n^0]_{k,p}\boldsymbol{v}_g(\boldsymbol{k},p). \tag{4.224}$$

The phonon number may change with time in a particular region of the crystal for two reasons: either because of a difference in the net *diffusion* of phonon wave packets into or out of the region, and/or because of anharmonic phonon *decay* into other phonons (see §4.6.2). Thus, the overall change in phonon number is given by the Boltzmann equation:

$$\frac{\mathrm{d}n_{k,p}}{\mathrm{d}t} = \left.\frac{\partial n_{k,p}}{\partial t}\right|_{\mathrm{diff}} + \left.\frac{\partial n_{k,p}}{\partial t}\right|_{\mathrm{decay}}. \tag{4.225}$$

In the case when the temperature gradient does not change in time, the phonon number is also time-independent, i.e. $\mathrm{d}n_{k,p}/\mathrm{d}t = 0$.

For phonon-decay processes, the time dependence can be expressed approximately in terms of a constant relaxation time $\tau_{k,p}$ for a non-equilibrium phonon distribution to relax back to its equilibrium value, i.e.

$$\left.\frac{\partial n_{k,p}}{\partial t}\right|_{\mathrm{decay}} = -(n_{k,p} - n^0_{k,p})/\tau_{k,p}. \tag{4.226}$$

The time dependence of the phonon diffusion term is related to the phonon (not heat) flux $\boldsymbol{J}_{k,p} = n_{k,p}\boldsymbol{v}_{\mathrm{g}}(\boldsymbol{k},p)$ via the continuity equation (cf. eqn. (3.34b)), i.e.

$$\left.\frac{\partial n_{k,p}}{\partial t}\right|_{\mathrm{diff}} = -\nabla \cdot \boldsymbol{J}_{k,p} = -v_{\mathrm{g}}(\boldsymbol{k},p) \cdot \nabla n_{k,p}. \tag{4.227}$$

This expression can be rewritten as follows:

$$\left.\frac{\partial n_{k,p}}{\partial t}\right|_{\mathrm{diff}} = -\left(\frac{\partial n^0_{k,p}}{\partial T}\right)\boldsymbol{v}_{\mathrm{g}}(\boldsymbol{k},p) \cdot \nabla T, \tag{4.228}$$

where $n_{k,p}$ has been replaced by $n^0_{k,p}$ because we are considering steady-state, *local* thermal-equilibrium conditions. From the Boltzmann equation (eqn. (4.225)) with $\mathrm{d}n_{k,p}/\mathrm{d}t = 0$, we have therefore from eqns. (4.227) and (4.228)

$$[n - n^0]_{k,p} = -\tau_{k,p}\left(\frac{\partial n^0_{k,p}}{\partial T}\right)\boldsymbol{v}_{\mathrm{g}}(\boldsymbol{k},p) \cdot \nabla T, \tag{4.229}$$

and substituting eqn. (4.229) into eqn. (4.224), and comparing with eqn. (4.222), gives for the thermal conductivity (assumed to be scalar):

$$\kappa_T = \frac{1}{3V}\sum_{k,p} \hbar\omega_k(p) v^2_{\mathrm{g}}(\boldsymbol{k},p)\tau(\boldsymbol{k},p)\frac{\partial n^0_{k,p}}{\partial T}, \tag{4.230}$$

where the spherical average of $\boldsymbol{v}^2_{\mathrm{g}}$, with $\boldsymbol{v}_{\mathrm{g}}$ projected along a direction parallel to the temperature gradient ∇T, involves the factor $\overline{\cos^2\theta} = 1/3$ (see e.g. §4.2.6). Equation (4.230) can be simplified further by noting that the term $\hbar\omega_{k,p}\partial n^0_{k,p}/\partial T$ is simply the mode heat capacity $C_V(\boldsymbol{k},p)$ (eqn. (4.217)), and defining the mode mean-free path $\Lambda_{k,p}$ as

$$\Lambda_{k,p} = v_{\mathrm{g}}(\boldsymbol{k},p)\tau_{k,p}, \tag{4.231}$$

which is the mean distance travelled by a phonon between scattering events which change the net momentum, gives finally

$$\kappa_T = \frac{1}{3V}\sum_{k,p} C_V(\boldsymbol{k},p)\Lambda_{k,p} v_{\mathrm{g}}(\boldsymbol{k},p). \tag{4.232}$$

Note that eqn. (4.232) has the *same* form as the expression for the thermal conductivity of a *gas*, obtained from kinetic theory, i.e. $\kappa_T = C_V\Lambda\bar{c}/3V$, where $\bar{c}$ is the mean molecular speed.

A number of phonon-scattering mechanisms exist which can serve to limit the phonon mean-free path; these include phonon–phonon scattering, scattering by defects (point defects, different isotopes of atoms and dislocations) and also boundary (surface) scattering from the sample. Each will be associated with a characteristic mean-free path Λ_i and, if more than one scattering mechanism is present, the mean-free paths add *reciprocally*

$$1/\Lambda = 1/\Lambda_1 + 1/\Lambda_2 + \ldots, \tag{4.233}$$

because the corresponding relaxation times τ_i add reciprocally. Thus, as for all types of conductivity which add in *parallel*, the thermal conductivity is determined by the sum of the reciprocals of individual constituents. The dominant contribution is that which has the *shortest* relaxation time or mean-free path.

Normal (N)-type anharmonicity-induced three-phonon scattering events (Fig. 4.41c), in which all three wavevectors lie within the *same* zone, conserve the total momentum of the phonons; the momentum after the event ($\hbar \boldsymbol{k}_3$) is equal to that before it ($\hbar \boldsymbol{k}_1 + \hbar \boldsymbol{k}_2$). Thus, such three-phonon collisions *cannot* act as a mechanism for providing resistance to the heat flow.

Phonon–phonon collisions which *do* cause a change in the phonon momentum, and hence which can limit the thermal conductivity, are those in which, say, the two wavevectors of the interacting phonons lie within the first Brillouin zone but that of the resultant phonon lies *outside* it (see Fig. 4.45a). This wavevector can be brought back into the first zone by the addition of an appropriate reciprocal-lattice vector $\boldsymbol{G}$. It can be seen that, in this manner, the direction of $\boldsymbol{k}_3$ is *opposite* to that of $\boldsymbol{k}_1$ and $\boldsymbol{k}_2$ and hence phonon momentum is *not* conserved and the energy flow is reversed. Such an event is called an umklapp (U) process (folding over). The geometric construction shown in Fig. 4.45b to find allowed solutions of the conservation laws of energy and

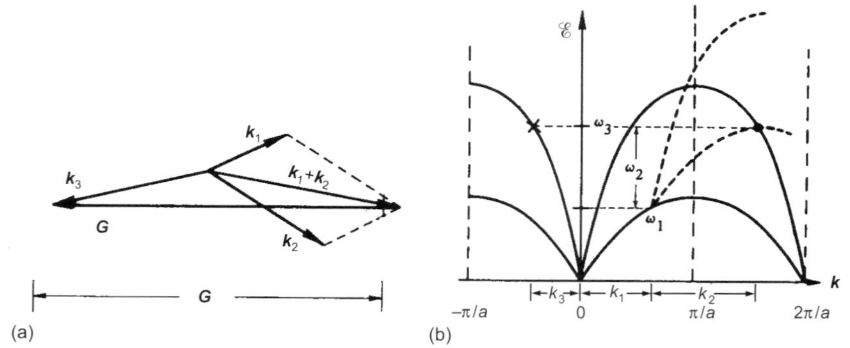

Fig. 4.45 Umklapp (U) phonon–phonon scattering process: (a) The wavevector $\boldsymbol{k}_3$ of the phonon resulting from the combination of two others ($\boldsymbol{k}_1$, $\boldsymbol{k}_2$) lies outside the first Brillouin zone, and is brought back into it by the addition of a suitable reciprocal-lattice vector, $\boldsymbol{G}$ (b) Geometric construction to find the allowed values of frequency (ω_3) and wavevector ($\boldsymbol{k}_3$) of a phonon created in a U-process from the coalescence of two other phonons ($\omega_1 \boldsymbol{k}_1$, $\omega_2 \boldsymbol{k}_2$). It is assumed for simplicity that all three wavevectors are collinear. See Fig. 4.42 for an explanation of the construction.

momentum is for an umklapp process involving two TA phonons (ω_1, $\boldsymbol{k}_1$; ω_2, $\boldsymbol{k}_2$) transforming into an LA phonon (ω_3, $\boldsymbol{k}_3$). (An additional N-process involving TA + LA $\rightarrow$ LA is also evidently a solution in this case.) Obviously U-processes can only occur when $\boldsymbol{k}_1$ and $\boldsymbol{k}_2$ are comparable to or greater than half the extent of the first Brillouin zone.

At very high temperatures ($T \gg \theta_D$), U-processes will be highly probable, and since the total number of phonons in a crystal is proportional to T (the high-temperature limit for the phonon-occupation number, eqn. (4.79)), it is expected therefore that the phonon mean-free path Λ (limited by phonon–phonon collisions) will be inversely proportional to the number of phonons present; hence $\kappa_T \propto T^{-1}$ in this region, since C_v is constant.

In the intermediate temperature régime ($T \leq \theta_D$), the number of umklapp events freezes out as the temperature is lowered. Umklapp processes can only occur for phonons with wavevectors $k \geq \pi/a$; such phonons therefore must have energies of order $\simeq k_B\theta_D/2$. The probability of such phonons existing is thus (from eqn. (4.68)) proportional to $\exp(-\theta_D/2T)$, and the temperature dependence of the phonon mean-free path will be the inverse of this. Therefore, in this régime, the thermal conductivity should increase exponentially with decreasing temperature according to $\kappa_T \propto \exp(\theta_D/2T)$ (neglecting the relatively weak temperature dependence of the heat capacity at such temperatures).

At yet lower temperatures, the mean-free path increases exponentially with inverse temperature until it becomes comparable to either the average separation between structural defects in the material (which can scatter phonons and hence limit Λ) or, for a sufficiently perfect crystal, until it becomes comparable to the dimensions of the specimen, and where the mean-free path is limited by boundary scattering of phonons

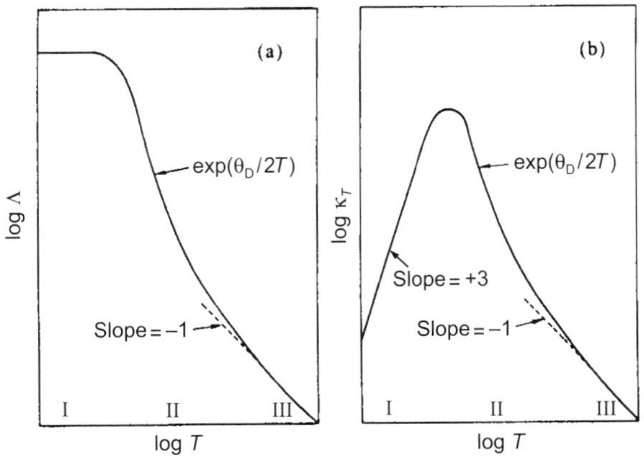

Fig. 4.46 (a) Schematic illustration of the temperature dependence of the phonon mean-free path in non-metallic crystals showing three régimes: I, temperature-independent region limited by impurity or boundary scattering; II, exponential dependence due to freeze-out of umklapp processes; III, T^{-1} dependence reflecting the temperature dependence of the number of phonons. (b) Schematic illustration of the temperature dependence of the thermal conductivity of crystals corresponding to the behaviour of the mean-free path shown in (a).

from the external surfaces of the crystal (the Casimir limit). At low temperatures, the dominant phonon wavelength, related to the unit-cell parameter via

$$\lambda \simeq \frac{\theta_D}{T} a, \tag{4.234}$$

is much larger than the size of point defects; hence the scattering of phonons by defects is analogous to the Rayleigh scattering of long-wavelength light by small particles, for which the scattering probability is proportional to k^4 (or ω^4) or, in the present case, to T^4 using eqn. (4.234), implying that the mean-free path behaves as $\Lambda \propto T^{-4}$. Thus, at low temperatures, where $C_V \propto T^3, \kappa_T \propto T^{-1}$. In the case of boundary scattering, the mean-free path will become *independent* of temperature and hence the temperature dependence of the heat capacity will determine that of the thermal conductivity: at low temperatures, therefore, $\kappa_T \propto T^3$ (cf. eqn. (4.201)). The temperature dependence of the phonon mean-free path for crystals is illustrated schematically in Fig. 4.46a, and that for the corresponding thermal conductivity in Fig. 4.46b; note that a pronounced peak in $\kappa_T(T)$ is evident.

Experimental data for κ_T of highly perfect crystals of sapphire (crystalline Al_2O_3) of different sizes are shown in Fig. 4.47: in the low-temperature Casimir régime, where

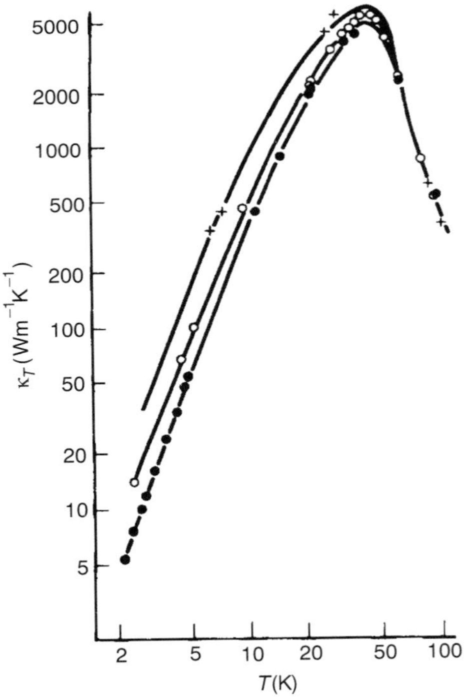

Fig. 4.47 Thermal conductivities of cylindrical specimens of sapphire (crystalline alumina) having diameters of 1.02 mm (•), 1.55 mm (o) and 2.8 mm (+). (Berman *et al.* (1955). Reproduced by permission of The Royal Society)

$\kappa_T \propto T^3$, the phonon mean-free paths are determined by the diameters of the samples. Note that this size effect vanishes at high temperatures, where the mean-free path is determined by U-processes.

Sapphire is a good conductor of heat, and diamond is even better; for the case of good quality diamonds $\kappa_T(300\ \mathrm{K}) \simeq 2000\ \mathrm{W\ m^{-1}\ K^{-1}}$, appreciably larger than the thermal conductivity of metallic Cu at the same temperature ($\simeq 400\ \mathrm{W\ m^{-1}\ K^{-1}}$) for which the heat is carried predominantly by electrons. The high phonon thermal conductivity for diamond is the result of the very high values of sound velocity and Debye temperature characteristic of this material.

Even a structurally perfect crystal containing a negligible concentration of structural defects may not exhibit boundary scattering at low temperatures if the solid contains a mixture of *isotopes* of the atoms: the associated differences in isotopic mass ensure that appreciable phonon scattering takes place since normal vibrational modes are independent even in the harmonic limit only when all the masses of the lattice bases are identical. This isotope effect is shown in Fig. 4.48: it can be seen that the isotopically purer sample exhibits a factor of three enhancement in κ_T at the peak maximum and has the T^3 temperature dependence at low temperatures characteristic of boundary scattering, i.e. where the mean-free path is comparable to the size of the specimen and not limited by defects, isotopic or otherwise.

Non-metallic glasses exhibit a behaviour of the thermal conductivity which is quantitatively and qualitatively different from that exhibited by crystals (Fig. 4.46b): the

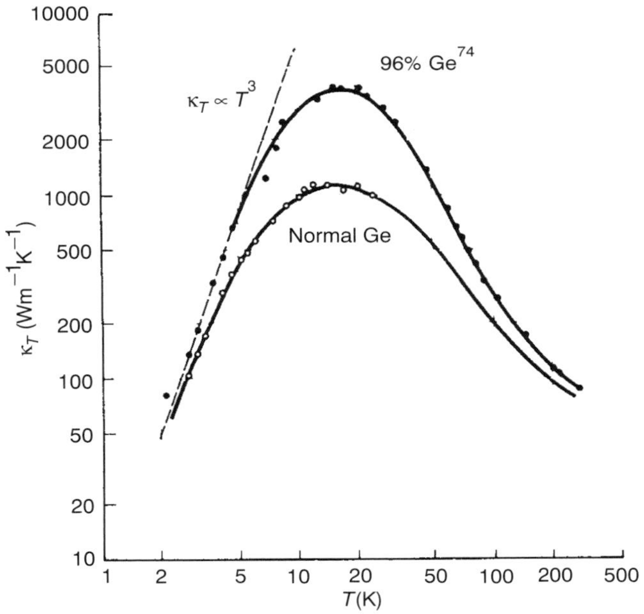

Fig. 4.48 Thermal conductivity of crystalline Ge showing the reduction in κ_T caused by phonon scattering from different isotopes in the natural abundance sample (20% Ge^{70}, 27% Ge^{72}, 8% Ge^{73}, 37% Ge^{74} and 8% Ge^{76}). (Reprinted with permission from Geballe and Hull (1958), *Phys. Rev.* **110**, 773. © 1958. The American Physical Society)

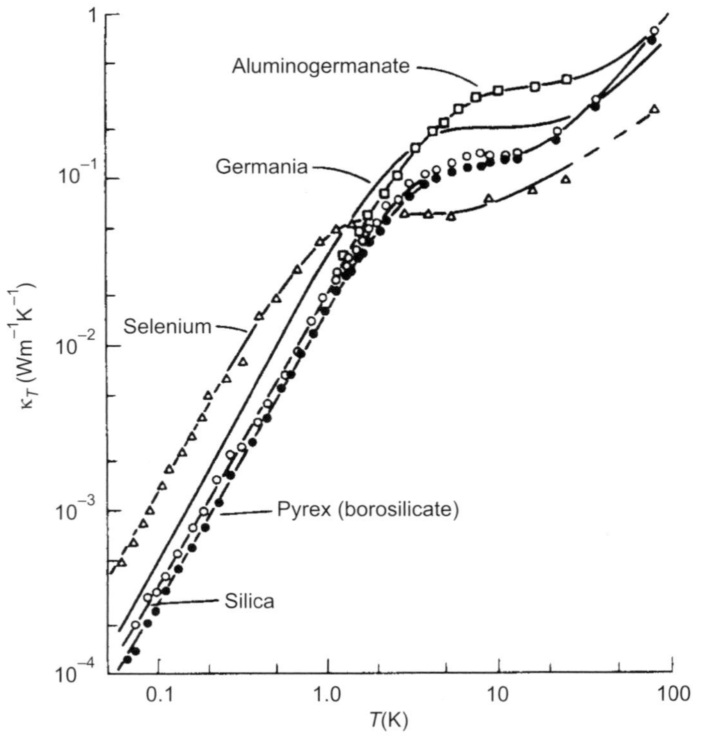

Fig. 4.49 Thermal conductivity as a function of temperature for a number of non-metallic glasses. (Reprinted with permission from Zeller and Pohl (1971), *Phys. Rev.* **B4**, 2029. © 1971. The American Physical Society)

overall level of κ_T for glasses is smaller by several orders of magnitude than that of crystals (see Fig. 4.49) due to the strong scattering of phonons by the disordered atomic structure. Furthermore, no peak in $\kappa_T(T)$ like that found for crystals is exhibited: instead, a plateau in κ_T is found at $\simeq 10K$, with a T^2 dependence at lower temperatures (Fig. 4.49).

The low-temperature T^2 behaviour of κ_T can be explained by assuming that propagating phonons are scattered by an additional mechanism involving the *two-level systems* (TLS) that dominate the heat capacity at such temperatures (§4.6.1). A resonant scattering mechanism can take place, whereby a phonon with energy $\hbar\omega$ equal to the energy-level spacing of the TLS can be absorbed, resulting in an excitation of the TLS from its ground state; the excited TLS can subsequently decay, thereby emitting an incoherent phonon, but with the same frequency. The probability of such resonant scattering processes will be proportional to both the density of TLS states $n(\mathscr{E})$ (taken to be constant, n_0—see §4.6.1) and to the square of the phonon field, i.e. to their energy $\hbar\omega$. Such phonons are dominant at a temperature $T \simeq \hbar\omega/k_B$ (see eqn. (4.234)), and hence the phonon mean-free path should vary as $\Lambda \propto T^{-1}$ for this mechanism. Since the heat capacity due to *phonons* in this temperature régime is $C_v \propto T^3$ (although the localized

TLS states contribute to the heat capacity, they do *not* help to carry the heat since they are non-propagating), eqn. (4.232) predicts that $\kappa_T \propto T^2$, as observed. As the temperature increases, the dominant phonon wavelength decreases, and the phonon mean-free path also decreases (due to elastic scattering of phonons by spatial fluctuations in density or force constants in the disordered structure), until eventually the mean-free path becomes comparable to the dominant phonon wavelength λ. This condition

$$\Lambda \simeq \lambda, \tag{4.235}$$

is called the Ioffe–Regel limit and is indicative of the onset of diffusive, rather than propagating, excitations, in this case phonons. The minimum value of $\Lambda \simeq a$ corresponds to spatial localization of the excitations. Although the picture is still not entirely clear, it appears that phonons are very strongly scattered in the vicinity of the plateau in the temperature dependence of the thermal conductivity. The upturn in κ_T beyond the plateau may be due to the anharmonic mixing of diffusive and propagating phonons to give other propagating phonons able to carry the heat.

Problems

4.1 Show that for the case of an isotropic solid, the longitudinal sound velocity is $v_l = (c_{11}/\rho)^{1/2}$ and the transverse sound velocity is $v_t = [(c_{11} - c_{12})/2\rho]^{1/2}$.

4.2 Derive an expression for the density of vibrational states in *two* dimensions for an isotropic material (see Fig. 4.4).

4.3 Extend the harmonic vibrational analysis of the monatomic linear chain to include interactions between neighbouring atoms more distant than the nearest neighbours. Show that the dispersion relation obtained is of the same form as eqn. (4.36) with an appropriate sum over neighbours.

4.4 Calculate the vibrational density of states, $g(\omega)$, for the periodic one-dimensional, monatomic chain and compare, by means of a sketch, its behaviour with that of the Debye density of states characteristic of an elastic continuum. How does the behaviour change for the linear diatomic chain model?

4.5 (a) Investigate the zone-centre and zone-boundary behaviour of the dispersion curve for the vibrations of a diatomic linear chain with equal force constants (eqn. (4.49)) and prove eqns. (4.50–4.54).
(b) What will the behaviour be at the zone boundary for a chain with equal masses but unequal force constants K', K''?

4.6 (a) Calculate the reciprocal-lattice vectors $\boldsymbol{G}_{100}$, $\boldsymbol{G}_{110}$ and $\boldsymbol{G}_{111}$, and hence deduce the positions of the first Brillouin zone boundaries in these directions for the case of b.c.c. potassium with $a = 5.23$ Å.
(b) Why are the T and L vibrational modes degenerate for $\boldsymbol{k} = \boldsymbol{G}_{100}/2$ (see Fig. 4.15)?
(c) Why do the dispersion curves match at $\boldsymbol{k} = \boldsymbol{G}_{100}/2$ and $\boldsymbol{G}_{111}/2$?
(d) Why does the slope of the dispersion curves not vanish at the first Brillouin zone boundary in the [111] direction?

4.7 Derive
(a) eqn. (4.75) for the average vibrational kinetic energy of a crystal;
(b) eqn. (4.81) for the mean atomic displacement.

4.8 Obtain an expression for the Debye–Waller factor (eqn. (4.87)) in the Einstein approximation at 0 K. Estimate how much of an incident (X-ray or neutron) beam is scattered inelastically from a crystal of Ge at 0 K for the first allowed Bragg peak (unit-cell parameter $a = 5.66$ Å).

4.9 Essay: Discuss the validity of applying cyclic (Born–von Karman) boundary conditions to the discussion of vibrational excitations in crystalline solids. (See, for example, Appendix IV in Born and Huang (1954).)

4.10 Show that the frequency of the localized vibrational mode associated with a light substitutional impurity atom (mass μ) in an otherwise periodic, linear monatomic chain (with masses M) is given by $\omega_{\rm loc}^2 = 2K/\mu$ when $\mu \ll M$. Rationalize this result in terms of the degree of localization of the mode. Comment on the fact that the ratios of the frequencies of the localized modes of ^{6}Li and ^{7}Li in AgBr:Li and AgCl:Li are 1.08 and 1.04, respectively.

4.11 (a) Derive the Lyddane–Sachs–Teller relation (eqn. (4.149)) in the *electrostatic* approximation, i.e. using only eqns. (4.131) and (4.133) of the Maxwell equations. (Hint: make use of eqn. (4.142)) for the dielectric constant $\varepsilon(\omega)$.)
(b) In this picture, account for the fact that $\omega_{\rm L} > \omega_{\rm T}$.
(c) Compare the condition that gives the TO mode in this model with that which appears in a proper treatment based on the full set of Maxwell's equations (see §4.4).
(d) Estimate the wavevector where polariton effects are important, and compare this value with a typical phonon zone-boundary wavevector.

4.12 Derive the expression (eqn. (4.165)) for the complex dielectric constant $\varepsilon^\dagger(\omega)$ for the model of a damped dipolar oscillator having the equation of motion given by eqn. (4.164)). Obtain expressions for the real (ε_1) and imaginary (ε_2) parts of $\varepsilon^\dagger$, and show that ε_2 peaks at a frequency $\omega_0 = \omega_{\rm T}$, and ε_1 has zeros at frequencies ω_0 and $\omega_{\rm L}$ (in the limit of small damping).

4.13 (a) Derive the Fresnel relations for light waves passing from vacuum ($n_{\rm r} = 1$) through a medium with complex refractive index $n^\dagger$. (Hint: make use of the fact that the values of E and H of the electromagnetic wave *parallel* to the interface are continuous across the boundary.) Hence, show that for normal incidence, the amplitude transmission factor from vacuum into the medium is $t_1 = 2/(n^\dagger + 1)$, and from the medium back into the vacuum is $t_2 = 2n^\dagger/(n^\dagger + 1)$, and the amplitude reflectivity factor for reflection within the medium at the medium–vacuum interface is $r = (n^\dagger - 1)/(n^\dagger + 1)$. Show that the reflectance for intensities is correspondingly $R = (n_r - 1)^2 + \kappa_i^2/((n_r + 1)^2 + \kappa_i^2)$.
(b) Show that the amplitude of a beam transmitted through the medium, of thickness d, is given by

$$E = E_0 t_1 t_2 \exp(in^\dagger \omega d/c)/[1 - r^2 \exp(2in^\dagger \omega d/c)].$$

Hence show that for an optically thin sample ($n^\dagger \sim 1$), the transmitted intensity obeys the Beer–Lambert law $I = I_0 \exp(-Kd)$, where the absorption coefficient is given by eqn. (4.163).

4.14 The normal experimental arrangement for Raman scattering has right-angle geometry in which the incident and scattered beams are at 90°. What is the value of the phonon wavevector in such a case, if laser light of wavelength 6000 Å is used? The polariton dispersion curves for GaP shown in Fig. 4.29 were measured using Raman scattering. What scattering geometry do you think was used in this case?

4.15 (a) Show that, for example by considering the conservation laws for energy and momentum, multiphonon (e.g. two-phonon) scattering produces a broad, featureless background inelastic scattering spectrum, with the only peaks being due to one-phonon scattering processes.
(b) Show that for an incident neutron with zero energy, one-phonon Stokes scattering is not possible but one-phonon anti-Stokes scattering is allowed. Use a graphical method to find the allowed solutions.

4.16 Show that multiphonon inelastic scattering, where n phonons of a single vibrational mode are created or destroyed, can be described equivalently as nth-order Bragg scattering from a grating moving in the solid with the phonon phase velocity, and where the grating itself comprises density fluctuations in an otherwise continuous elastic medium. (Hint: consider scattering from a frame in which the grating is at rest, and consider the Doppler shift in frequencies between the laboratory and moving frames.)

4.17 Estimate how good is the approximation for a real solid at room temperature of equating the heat capacities at constant volume and pressure. (Hint: derive the equation $C_p - C_v = VT\beta_T^2/\kappa_T$.)

4.18 Show that the zero-point energy in the Debye model is given by $\mathscr{E}_0 = \frac{9}{8} N\hbar\omega_{\rm D}$ for a solid containing N atoms.

4.19 Obtain expressions for the heat capacity of two-and one-dimensional solids in the Debye approximation and show that the limiting low-temperature behaviour is $C_v \propto T^n$, where n is the dimensionality. (For the appropriate vibrational densities of states, see Problems 4.2 and 4.4.)

4.20 Calculate the vibrational heat capacity of a solid in the Einstein approximation, i.e. assume that all vibrational modes have the *same* frequency ω_0. What is the limiting behaviour of C_v at very high and very low temperatures in this model?

4.21 Obtain an expression for the heat capacity C_v associated with a *single* two-level system, having energy levels at $\pm\mathscr{E}$. Sketch the temperature dependence of C_v and account for the behaviour at high temperatures. The peaked nature of $C_v(T)$ is termed a Schottky anomaly.

4.22 Show that, for the case of three-phonon mixing events (with the wavevectors all collinear), there is no allowed process involving three phonons from the *same* branch.

4.23 Estimate the volume thermal expansion coefficient of Ar at 77 K by making use of the Grüneisen law (eqn. (4.220)). The coefficients in the Lennard-Jones potential for Ar are $\epsilon = 1.67 \times 10^{-21}$ J and $\sigma = 3.4$ Å, and $\theta_D = 92K$ (see Problem 2.5 for the isothermal bulk modulus). (Hint: show, by expanding the interatomic potential V in a Taylor's series, that the Grüneisen constant can be written as

$$\gamma = -\frac{a}{6}\left(\frac{\partial^3 V}{\partial r^3}\right)_{r=a} \Big/ \left(\frac{\partial^2 V}{\partial r^2}\right)_{r=a}.$$

4.24 Essay: Compare and contrast the processes underlying heat conduction in a solid (mediated by phonons) and in a gas.

4.25 Calculate the thermal conductivity at 1 K of a rod of crystalline synthetic sapphire with a diameter of 3 mm (the molecular weight of Al_2O_3 is 102). The velocity of sound may be taken to be 5×10^3 m s^{-1}, the density $= 4 \times 10^3$ kg m^{-3} and $\theta_D = 10^3$K.

References

Alben, R., Weaire, D., Smith, J.E. and Brodsky, M.H., 1975, *Phys. Rev.* **B11**, 2271.

Ashcroft, N. W. and Mermin, N. D., 1976, *Solid State Physics* (Holt, Rinehart and Winston: New York).

Berman, R., Foster, E. L. and Ziman, J. M., 1955, *Proc. Roy. Soc.* A**231**, 130.

Bilz, H. and Kress, W., 1979, *Phonon Dispersion Relations in Insulators* (Springer-Verlag: Berlin).

Born, M. and Huang, K., 1954, *Dynamical Theory of Crystal Lattices* (Clarendon Press: Oxford).

Burns, G., 1985, *Solid State Physics* (Academic Press: London).

Cardona, M., ed., 1983, *Light Scattering in Solids I*, 2nd edn, Topics in Applied Physics, vol. 8 (Springer-Verlag: Berlin).

Cowley, R. A., Woods, A. D. B. and Dolling, G., 1966, *Phys. Rev.* **150**, 487.

Dawber, P. G. and Elliott, R.J., 1963, *Proc. Roy. Soc. (Lond.)* A**273**, 222.

Dolling, G. and Cowley, R.A., 1966, *Proc. Phys. Soc.* **88**, 463.

Donovan, D. and Angress, J. F., 1971, *Lattice Vibrations* (Chapman and Hall: London).

Dove, M. T., 1993, *Introduction to Lattice Dynamics* (Cambridge University Press: Cambridge).

Elliott, S.R., 1990, *Physics of Amorphous Materials*, 2nd edn (Longman: Harlow).

Fujii, Y. Lurie, N. A., Pynn, R. and Shirane, G., 1974, *Phys. Rev.* **B10**, 3647.

Geballe, T.H. and Hull, G.W., 1958, *Phys. Rev.* **110**, 773.

Hook, J.R. and Hall, H.E., 1991, *Solid State Physics* (Wiley: Chichester).

Kamitakahara, W. A., Shanks, H. R., McClelland, J. F., Buchenau, U., Gompf, U. and Pintschovius, L., 1984, *Phys. Rev. Lett.* **52**, 644.

Kittel, C., 1996, *Introduction to Solid State Physics* (Wiley: New York).

Maradudin, A. A., 1964, in *Phonons and Phonon Interactions*, ed. T.A. Bak (W. A. Benjamin: New York).

Mary, T.A., Evans, J.S.O., Vogt, T. and Sleight, A.W., 1996, *Science* **272**, 90.

Miiller, A. P. and Brockhouse, B. N., 1971, *Can. J. Phys.* **49**, 704.

Raunio G., Almqvist, L. and Stedman, R., 1969, *Phys. Rev.* **178**, 1496.

Salje, E. K. H., 1990, *Phase Transitions in Ferroelastic and Co-elastic Crystals* (Cambridge University Press: Cambridge).
Shapiro, S. M., Gammon, R. W. and Cummins, H. Z., 1966, *Appl. Phys. Lett.* **9**, 157.
Sköld, K. and Price, D. L., eds, 1986, *Neutron Scattering* (Academic Press: New York).
Smith, D.Y. and Manogue, C. A., 1981, *J. Opt. Soc. Am.* **71**, 935.
Squires, G. L. 1978, *Introduction to the Theory of Thermal Neutron Scattering* (Cambridge University Press: Cambridge).
Srivastava, G.P., 1990, *The Physics of Phonons* (Adam Hilger, Bristol).
Stephens, R. B., 1976, *Phys. Rev.* B**13**, 852.
Stoneham, A. M., 1975, *Theory of Defects in Solids* (Clarendon Press: Oxford).
Weber, W., 1977, *Phys. Rev.* **B15**, 4789.
Willis, B. T. M., ed., 1973, *Chemical Applications of Thermal Neutron Scattering* (Oxford University Press: Oxford).
Willis, B. T. M. and Pryor, A. W., 1975, *Thermal Vibrations in Crystallography* (Cambridge University Press: Cambridge).
Zeller, R. C. and Pohl, R. O., 1971, *Phys. Rev.* B**4**, 2029.

CHAPTER

5 Electrons in Solids

Introduction

Solids are composed of a mixture of atomic nuclei and electrons. Thus far, we have tended to concentrate more on the properties associated with *atoms* in materials, and the behaviour of the electrons has in general only been alluded to in passing, for example in connection with the force constants between atoms determining their vibrational behaviour. In the remaining part of this book, however, we will be concerned with the properties of the *electrons* in solids, specifically here with the electronic structure, i.e. the distribution of electrons (either spatially or as a function of energy) and the behaviour of the electron sub-system itself.

Two, complementary, ways of treating electrons in solids have evolved over the years, and it is unfortunate that different scientific disciplines have, until recently, generally adopted one of these approaches and neglected the other. Thus, it has been traditional for *physicists* to treat electrons (in ideal crystals) as *delocalized waves* extending throughout the solid. The allowed wave-like solutions of the quantum-mechanical Schrödinger equation for electrons moving in an appropriate potential can be labelled according to the wavevectors of the waves, resulting in a concise description of the allowed electron states as bands of allowed energies in reciprocal space $\mathscr{E}(\boldsymbol{k})$ (in exact analogy with the dispersion curves $\omega(\boldsymbol{q})$ of phonons in crystals — see §4.2.2). In contrast, *chemists* have traditionally regarded electrons in molecules and, by extension, in solids in terms of atomic orbitals, i.e. as spatially *localized bonds* (e.g. pairs of electrons) between pairs of atoms; thus this is a local, real-space picture.

These two approaches can be equivalent, but they stress different aspects of the problem. Moreover, they each have particular advantages and disadvantages in their description of electron states. It is obviously very helpful to be able to describe, and hence to visualize, electron states in a real-space description, and this is why the chemist's approach is so appealing. This local picture is especially useful for describing electrons in *insulating* (non-metallic) materials in general and in amorphous (non-crystalline) solids in particular. The reciprocal-space description conventionally used by physicists is a very powerful approach which allows electronic properties to be calculated relatively straightforwardly, employing the simplifying features of symmetry. However, it has the disadvantage that thinking in reciprocal-space terms is somewhat unnatural; moreover, this approach is only valid for perfectly crystalline systems and cannot be used if structural disorder is present. Furthermore, the neglect of electron–electron interactions inherent in the wave-like electron band theory means that many materials are wrongly predicted to be metallic conductors by this model and are, in fact, insulators.

In this chapter we will discuss the electronic structure of materials, i.e. the spatial and energetic distribution of electrons in solids, together with their equilibrium and optical properties. In discussing electron states in solids, we commence with the physicist's view involving the description of free electrons in a wave-like way (band theory), followed by a real-space description more in accord with the chemist's bonding viewpoint, and an attempt will be made to reconcile these two apparently opposing points of view. Finally, a discussion of optical properties of materials, i.e. the interaction between photons and electrons in solids, will be given. Non-equilibrium (transport) electronic properties of solids form the subject of Chapter 6, while Chapter 7 is devoted to a discussion of the behaviour of spatially localized electrons (e.g. magnetic and dielectric properties).

Electrons as a gas 5.1

5.1.1 The free-electron gas

The simplest model for describing electrons in solids is to assume that the valence electrons of an atom (that is, electrons from the outermost unfilled electron shells) are, in solids, free to move anywhere throughout the volume of the material; thus they behave as a delocalized 'gas' of free electrons. This is an appropriate model for *metallic* solids. For the case of alkali metals, for example Na, the neutral atom has the electronic configuration $1s^2 2s^2 2p^6 3s^1$, and the valence electron in this case is the single 3s electron. Ionization of the valence electron leaves behind the core electrons, having an inert-gas-like configuration, and they remain tightly bound to the nucleus (in the case of Na^+, the core electrons are the set $1s^2 2s^2 2p^6$). In the case of transition-metal and rare-earth elements, the d- and f-electrons, even in part-filled shells, are sufficiently tightly bound to the nucleus that they do not contribute to the free-electron gas. The model assumes that the ion cores (the atomic nuclei plus core electrons) have a negligible size (i.e. the valence-electron gas is free to explore *all* points in the volume). This is not a bad approximation in practice, since the amount of free volume outside the ion cores is appreciable in such materials (e.g. the internuclear separation in solid Na is 3.66 Å, considerably larger than the distance of 1.94 Å equal to twice the ionic radius of Na^+).

Another assumption underlying the free-electron-gas model is that the valence electrons experience everywhere in the solid a *constant* electrostatic potential. This is obviously untrue in practice, since the ion cores will actually be the sites of deep minima in the potential, but nevertheless it is often a reasonably good approximation that the potential associated with the ion cores is constant *between* ionic sites (see §5.2.3). In the simplest version of the free-electron-gas model, any 'graininess' associated with individual atoms is smoothed out. Hence it is assumed that the net charge associated with the ion cores (as well as the rest of the valence electrons) is uniformly distributed throughout the volume of the solid; this is known as the jellium model. A valence electron experiences a *constant potential* everywhere, except at the surfaces of the solid where there is a potential barrier preventing escape of electrons out of the solid. The potential model is therefore that of the 3D square-well, particle-in-a box problem (Fig. 5.1): an electron having a total energy $\mathscr{E}$ will therefore have a work function (the extra energy needed for an electron escape from the solid) given by $\phi = W - \mathscr{E}$, where W is the well depth (see Fig. 5.1).

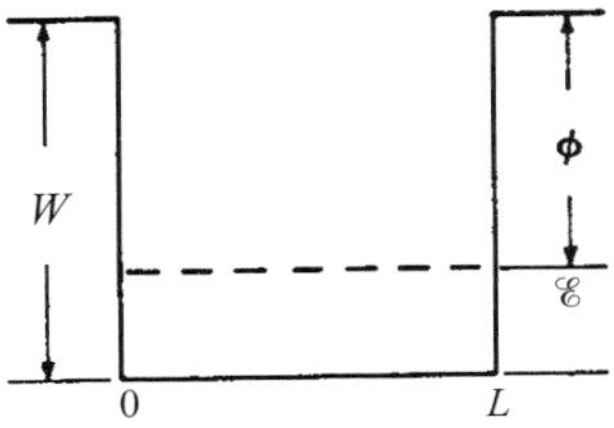

Fig. 5.1 Illustration of the square-well potential experienced by an electron in the free-electron-gas model. The potential due to the charged ion cores is assumed to be spatially uniform. A free valence electron, having a total energy $\mathscr{E}$, is therefore bound within a solid as long as $\mathscr{E} < W$ (the well depth); the work-function energy is $\phi = W - \mathscr{E}$.

The scenario whereby all free electrons have the *same* average kinetic energy $\mathscr{E}$ and the electrons are treated as *classical distinguishable particles*, as in a normal gas, is known as the Drude model.

5.1.2 The Fermi electron gas

Electrons are quantum, not classical, particles and so the allowed electronic states must be solutions of the Schrödinger equation which, in the general (non-relativistic) case where the potential energy V is a function of both space and time, is

$$-\frac{\hbar^2}{2m_e}\nabla^2\Psi(\boldsymbol{r},t) + V(\boldsymbol{r},t)\Psi(\boldsymbol{r},t) = \mathrm{i}\hbar\frac{\partial}{\partial t}[\Psi(\boldsymbol{r},t)], \tag{5.1}$$

where m_e is the electron mass and $\Psi(\boldsymbol{r},t)$ is the electron wavefunction, i.e. the wave-like solution of eqn. (5.1). For the case of a potential energy that is time-independent, $V(r,t) = V(r)$, the wavefunction can be separated into the product of a spatially varying and a time-dependent part, viz.

$$\Psi(\boldsymbol{r},t) = \psi(\boldsymbol{r})\phi(t), \tag{5.2}$$

and the spatially varying wavefunction $\psi(\boldsymbol{r})$ is the stationary-state solution of the time-independent Schrödinger equation

$$-\frac{\hbar^2}{2m_e}\nabla^2\psi(\boldsymbol{r}) + V\psi(\boldsymbol{r}) = \mathscr{E}\psi(\boldsymbol{r}), \tag{5.3a}$$

or equivalently as the eigenvalue equation

$$\mathscr{H}\psi(\boldsymbol{r}) = \mathscr{E}\psi(\boldsymbol{r}), \tag{5.3ab}$$

where $\mathscr{H}$ is the Hamiltonian operator $(= -(\hbar^2/2m_e)\nabla^2 + V)$.

For the case where the potential energy is time-independent, the time-dependent Schrödinger equation (eqn. (5.1)) becomes

$$\mathrm{i}\hbar\frac{\partial}{\partial t}\Psi(\boldsymbol{r},t) = \mathscr{E}\Psi(\boldsymbol{r},t) \tag{5.4a}$$

or, using eqn. (5.2),

$$\mathrm{i}\hbar\psi(\boldsymbol{r})\frac{\partial\phi(t)}{\partial t} = \mathscr{E}\psi(\boldsymbol{r})\phi(t). \tag{5.4b}$$

The solution of this is

$$\phi(t) = \exp(-\mathrm{i}\mathscr{E}t/\hbar) \equiv \exp(-\mathrm{i}\omega t) \tag{5.5}$$

where ω is the radial frequency $(= \mathscr{E}/\hbar)$ and so

$$\Psi(\boldsymbol{r},t) = \psi(\boldsymbol{r})\exp(-\mathrm{i}\omega t). \tag{5.6}$$

Allowed solutions of eqn. (5.3) depend on the appropriate boundary conditions to which the electrons are subject, and are quantized; that is, the energies of the allowed states are functions of *discrete* quantum numbers.

There is obviously a formal equivalence between the quantum electron-gas model and the elastic-continuum model for vibrational excitations in solids described in §4.1: in both, the 'graininess' associated with the presence of atoms in real solids is averaged out. However, there is an important difference between electrons and phonons (or indeed the classical distinguishable particles of the Drude model, §5.1.1), namely that electrons are indistinguishable particles having a *non-integral spin* ($s = 1/2$) and, as such, are fermions and obey the Pauli exclusion principle. The occupancy of allowed electron states is determined by the Fermi–Dirac distribution function (very different from the Planck distribution function characteristic of phonons (eqn. (4.68)) or the Maxwell–Boltzmann distribution applicable to classical distinguishable particles, e.g. eqn. (3.7)). The model of a quantum free-electron gas subject to Fermi–Dirac statistics is called the Sommerfeld model; this will be analysed in the rest of this section.

In general, the solution of the Schrödinger equation (eqn. (5.3)) is an almost impossible task because of the complications caused by electron–electron interactions. As a result, the Hamiltonian (through the potential-energy term) is a function of the positions of *all* the electrons, $\mathscr{H} = \mathscr{H}(\boldsymbol{r}_1, \boldsymbol{r}_2, \ldots)$, and hence the electron wavefunction is a many-body function, $\psi(\boldsymbol{r}_1, \boldsymbol{r}_2, \ldots)$. The problem is greatly simplified, however, by invoking the independent-electron approximation, in which the electrons mutually interact only via an averaged potential energy (the Hartree approximation, but see eqn. (2.84)). In this way, the Hamiltonian can be written as the *sum* of individual one-electron terms:

$$\mathscr{H}(\boldsymbol{r}_1, \boldsymbol{r}_2, \ldots) = \mathscr{H}(\boldsymbol{r}_1) + \mathscr{H}(\boldsymbol{r}_2) + \ldots, \tag{5.7}$$

and, as a result, the many-body wavefunction becomes the *product* of individual one-electron wavefunctions:

$$\psi(\boldsymbol{r}_1, \boldsymbol{r}_2, \ldots) = \psi(\boldsymbol{r}_1) \times \psi(\boldsymbol{r}_2) \times \ldots \tag{5.8}$$

The independent-electron approximation works essentially because of the *screening* (i.e. decrease) by all other electrons of the interaction between any two electrons (see §§2.5.3.4 and 5.6.1 for further details). The free particles are in reality quasiparticles, comprising an electron and its associated 'orthogonality hole' (§2.5.3.4).

For simplicity, assume that a solid can be represented as a cubical box, of side L, with a *constant* potential energy V (comprising the ion-core and averaged electron–electron potential energies) in its interior (and taken to be $V = 0$ without any loss of generality), and with a potential barrier W at the boundaries (Fig. 5.1). The general, one-electron wavefunction that is a solution to the Schrödinger equation (eqn. (5.3a)) is

$$\psi(\boldsymbol{r}) = A \exp(\mathrm{i}\,\boldsymbol{k} \cdot \boldsymbol{r}) = A \exp[\mathrm{i}(k_x x + k_y y + k_z z)] \tag{5.9}$$

where A is a normalization constant set by the condition that there is one electron in the box, viz.

$$1 = \int \psi^* \psi \, \mathrm{d}\boldsymbol{r}. \tag{5.10}$$

The appropriate boundary conditions to take in the case of an electron trapped in a well and subject to a uniform potential, as in Fig. 5.1, is that the wavefunction $\psi(\boldsymbol{r})$ goes to *zero*, i.e. has a node, at the boundaries of the box (strictly valid only for infinitely high barriers at the boundaries). Note that this is different in detail to the case of vibrations in an elastic continuum (§4.12), where an *antinode* of the vibrational amplitude at the mechanically free surfaces was taken as the boundary condition. However, the final result concerning the density of modes is the same. The use of these boundary conditions leads to the *standing-wave solutions* (see Problem 3.6):

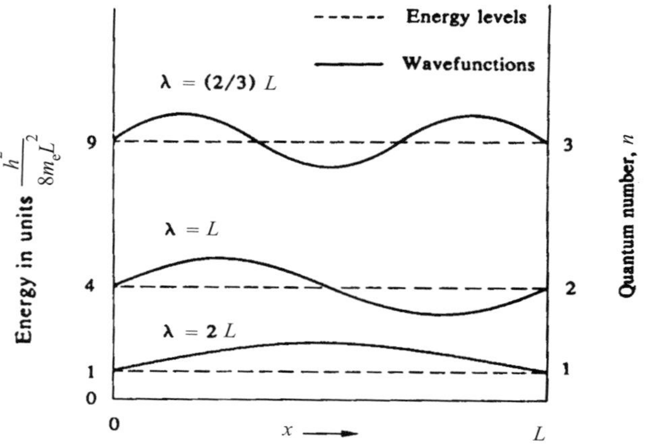

Fig. 5.2 The first three allowed wavefunctions for an electron trapped in a box of length L (taken to be one-dimensional for ease of visualization), displaced according to the corresponding energy levels.

$$\psi(x, y, z) = \left(\frac{8}{L^3}\right)^{1/2} \sin\left(n_1 \frac{\pi x}{L}\right) \sin\left(n_2 \frac{\pi y}{L}\right) \sin\left(n_3 \frac{\pi z}{L}\right), \tag{5.11}$$

where (n_1, n_2, n_3) are quantum numbers (which are *non-zero positive* integers, and the normalization constant is $A = (8/L^3)^{1/2}$. The first three wavefunctions are illustrated (for the case of one dimension, $\psi(x) \propto \sin(n\pi x/L)$, for ease of visualization) in Fig. 5.2. As in §4.1.2, the allowed standing-wave solutions can be represented instead as points in reciprocal space, having the values $k_i^s = n_i(\pi/L)(i = x, y, z)$ (eqn. (4.18)), corresponding to the wavevectors of the standing waves. However, because electrons are spin-1/2 particles, each allowed $\boldsymbol{k}$-state can contain *two* electrons (the spin degeneracy is two in the absence of a magnetic field).

The energy levels of the 3D particle-in-a-box problem are given by eqn. (3.18) (see also Problem 3.6), obtained by substituting eqn. (5.11) into eqn. (5.3a) which, rewritten in terms of wavevectors, are given by

$$\mathscr{E}(\boldsymbol{k}) = \frac{\hbar^2}{2m_e}\left(k_x^2 + k_y^2 + k_z^2\right) \equiv \frac{\hbar^2 k^2}{2m_e}, \tag{5.12}$$

where the superscript s has been omitted for clarity. $\mathscr{E}(\boldsymbol{k})$ is a simple parabolic function of k.

From the previous discussion of vibrational modes of continuous media (§4.1.2), the allowed $\boldsymbol{k}$-states corresponding to physically distinct standing-wave solutions of the Schrödinger equation occupy only the *positive octant* in $\boldsymbol{k}$-space (since n_i, and hence k_i, > 0). Moreover, for macroscopic samples with large size L, the separation in $\boldsymbol{k}$-space between points corresponding to allowed states becomes infinitesimal, and hence the number of allowed electron states with k-values between k and $k + \mathrm{d}k$ (eqn. (4.21)) is simply the volume of the spherical shell with radius k and thickness $\mathrm{d}k$ in the positive octant (Fig. 5.3), multiplied by the $\boldsymbol{k}$-space density of standing-wave points, $(L/\pi)^3$ (eqn. (4.20)), multiplied by the spin degeneracy $g_s = 2$, i.e.

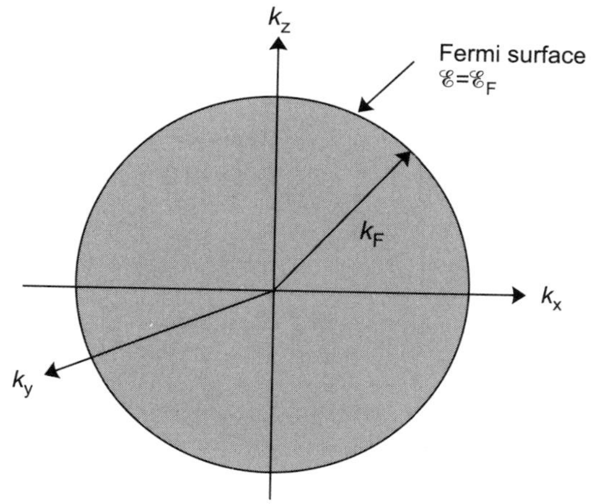

Fig. 5.3 Representation of the Fermi sphere for a 3D free-electron gas in ***k***-space. The Fermi surface, corresponding to electron states with the Fermi energy $\mathscr{E}_F$, is the surface of the sphere, and marks the demarcation at zero kelvin between occupied (shaded) and unoccupied ***k***-states. For the case of boundary conditions leading to standing-wave electron states in a box, only the *positive octant* of this sphere contains physically distinct states. For the case of Born–von Karman periodic boundary conditions, leading to running-wave solutions, *all* states in the spherical volume correspond to physically distinct states. (The spacings between allowed ***k***-states differ by a factor of two for these two situations, although the value of the Fermi wavevector, the radius of the Fermi sphere, is identical in both cases.)

$$g(k)\mathrm{d}k = \frac{Vk^2}{\pi^2}\mathrm{d}k, \tag{5.13}$$

where $V = L^3$ is the volume of the box. (In fact, this result is independent of the type of boundary conditions chosen, whether leading to standing waves, as here, or running waves — see §4.2.1).

The density of free-electron states as a function of energy, $g(\mathscr{E})$, is given by

$$g(\mathscr{E}) = g(k)\frac{\mathrm{d}k}{\mathrm{d}\mathscr{E}}, \tag{5.14}$$

and hence, using eqns. (5.12) and (5.13):

$$g(\mathscr{E}) = \frac{V}{2\pi^2}\left(\frac{2m_e}{\hbar^2}\right)^{3/2}\mathscr{E}^{1/2}. \tag{5.15}$$

Each state can accommodate only *one* electron by the Pauli exclusion principle (recall that the spin degeneracy of two has already been included), and the ground-state configuration (at $T = 0$ K) of the free-electron gas of N electrons corresponds to filling the N lowest states, up to an energy $\mathscr{E} = \mathscr{E}_F$ (the Fermi energy), viz.

$$N = \int_0^{\mathscr{E}_F} g(\mathscr{E})\mathrm{d}\mathscr{E} = \frac{V}{3\pi^2}\left(\frac{2m_e}{\hbar^2}\mathscr{E}_F\right)^{3/2} \equiv \frac{2}{3}\mathscr{E}_F g(\mathscr{E}_F), \tag{5.16}$$

with the Fermi energy given by

$$\mathscr{E}_{\mathrm{F}} = \frac{\hbar^2}{2m_{\mathrm{e}}}\left(\frac{3\pi^2 N}{V}\right)^{2/3}. \tag{5.17}$$

The Fermi energy is determined by the electron density, N/V, or the atomic density if there is a contribution of just one valence electron per atom to the free-electron gas, as for alkali metals. A corresponding reference temperature, the Fermi temperature of the electron gas (which is not a true temperature), can be defined as

$$T_{\mathrm{F}} = \mathscr{E}_{\mathrm{F}}/k_{\mathrm{B}}, \tag{5.18}$$

and the Fermi wavevector (the radius of the Fermi sphere — see Fig. 5.3) is, from eqn. (5.12),

$$k_{\mathrm{F}} = \left(\frac{2m_{\mathrm{e}}}{\hbar^2}\mathscr{E}_{\mathrm{F}}\right)^{1/2} = \left(\frac{3\pi^2 N}{V}\right)^{1/3}. \tag{5.19}$$

Values of $\mathscr{E}_F$ are *very* high, of the order of several electron volts, corresponding to temperatures of $\simeq 20\,000$ K (see Table 5.1 and Problem 5.1). The fact that there is a spread in electron energies from $\mathscr{E} = 0$ to $\mathscr{E}_F$, even for the ground-state configuration at $T = 0$ K, is a direct consequence of the quantum statistics (i.e. the Pauli exclusion principle) that govern the state occupancy of such particles. This behaviour is in stark contrast with the behaviour of a gas of classical, distinguishable particles (as assumed for electrons in the Drude theory — §5.1.1), where the average kinetic energy of every particle is $\mathscr{E} = 3k_{\mathrm{B}}T/2$, and hence is zero at the absolute zero of temperature. For the case of a 3D gas of N electrons, the average kinetic energy at $T = 0$ K can be calculated simply from the density of states (eqn. (5.15)), viz.

Table 5.1 Values of Fermi energy, temperature and wavevector for some metallic elements

Element	$\mathscr{E}_{\mathrm{F}}$ (eV)	$10^{-4}T_{\mathrm{F}}$(K)	k_{F}(Å^{-1})
Ag	5.49	6.38	1.20
Al	11.7	13.6	1.75
Au	5.53	6.42	1.21
Ba	3.64	4.23	0.98
Be	14.3	16.6	1.94
Cs	1.59	1.84	0.65
Cu	7.0	8.16	1.36
Fe	11.1	13.0	1.71
K	2.12	2.46	0.75
Li	4.74	5.51	1.12
Na	3.24	3.77	0.92
Pb	9.47	11.0	1.58
Rb	1.85	2.15	0.70
Sn	10.2	11.8	1.64
Zn	9.47	11.0	1.58

These values are calculated by assuming that the number of free electrons per atom contributing to the electron gas is equal to the conventional valence (i.e. including s- and p-electrons, but neglecting d-electrons).

$$\langle \mathscr{E} \rangle = \int_0^{\mathscr{E}_F} \mathscr{E} g(\mathscr{E}) d\mathscr{E} = \frac{V}{2\pi^2} \left(\frac{2m_e}{\hbar^2} \right)^{3/2} \int_0^{\mathscr{E}_F} \mathscr{E}^{3/2} d\mathscr{E}$$
$$= \frac{3}{5} N \mathscr{E}_F. \tag{5.20}$$

This value is not simply $N\mathscr{E}_F/2$ because the density of states is not constant, but increases with energy.

At temperatures above absolute zero, electrons can be thermally excited from filled states below $\mathscr{E}_F$ to empty states above it. The occupancy of electron states is governed by the Fermi–Dirac distribution function, $f(\mathscr{E})$. This function can be derived by making use of the grand partition function (or grand sum) which is the normalizing factor for the occupation probabilities of energy levels when the number of particles N is *not* held constant (see e.g. Kittel and Kroemer (1980)), viz.

$$\mathcal{Z} = \sum_{N=0}^{\infty} \sum_{1} \exp[N\mu - \mathscr{E}_1(N)]/k_B T. \tag{5.21}$$

This is a sum of Gibbs functions over all states l of the system and for all N; the quantity μ is the chemical potential. (Note that the *partition function*, used for discussing the statistics obeyed by phonons (eqn. (4.64)), is obtained from (eqn. (5.21) by simply setting $\mu = 0$; in the case of vibrational excitations, only the energy levels of a *single* oscillator are considered.) For the case of an electron orbital (a solution of the Schrödinger equation for a single electron), taken to be the system, in thermal contact and in equilibrium with all other orbitals, forming the reservoir and having an energy equal to the chemical potential μ, evaluation of the grand sum (eqn. (5.21)) is simple, since an orbital can only be unoccupied or occupied by an single electron in accordance with Pauli's exclusion principle. Thus:

$$\mathcal{Z} = 1 + \exp[(\mu - \mathscr{E})/k_B T], \tag{5.22}$$

where the first term corresponds to the case of zero occupancy ($N = 0$; $\mathscr{E} = 0$). The thermal average value of the occupancy of the orbital is therefore given by the Gibbs factor for the occupied orbital divided by the grand sum, viz.

$$f(\mathscr{E}) \equiv \langle n(\mathscr{E}) \rangle = \frac{\exp[(\mu - \mathscr{E})/k_B T]}{1 + \exp[(\mu - \mathscr{E})/k_B T]} = \frac{1}{\exp[(\mathscr{E} - \mu)/k_B T] + 1}. \tag{5.23}$$

It is conventional to denote the average occupancy $\langle n(\mathscr{E}) \rangle$ by $f(\mathscr{E})$. At $T = 0$ K, the Fermi–Dirac function is a step function, $f(\mathscr{E}) = 1$ for $\mathscr{E} < \mu$, and $f(\mathscr{E}) = 0$ for $\mathscr{E} > \mu$, and the chemical potential at this temperature is called the Fermi energy, i.e.

$$\mu(T = 0\,\text{K}) \equiv \mathscr{E}_F. \tag{5.24}$$

In general, the chemical potential corresponds to the energy at which $f(\mathscr{E}) = 0.5$, as can be seen from eqn. (5.23).

The Fermi–Dirac distribution function varies with temperature as shown in Fig. 5.4: at finite temperatures, electrons are thermally excited from filled states lying below the chemical potential to empty states lying above it. For the case where the total number of electrons in a (3D) solid is constant, the chemical potential must *decrease* with increasing temperature in order to compensate for the broadening of the distribution to

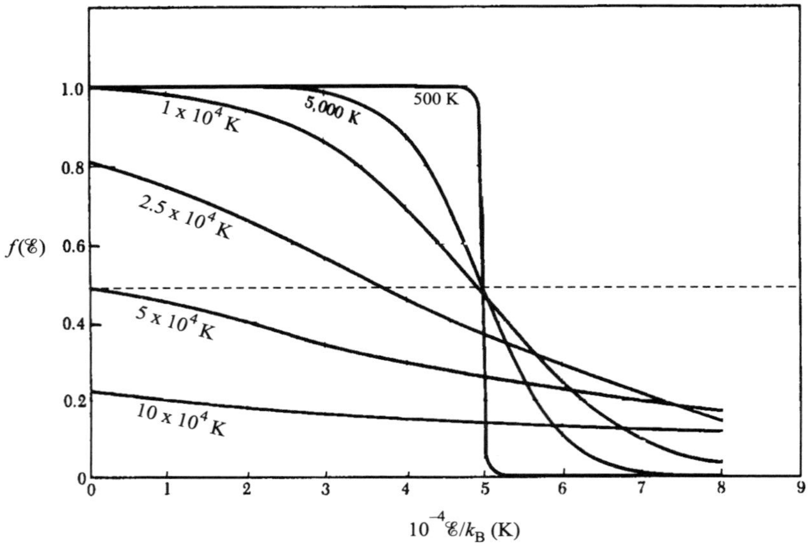

Fig. 5.4 The Fermi–Dirac distribution function at various temperatures for the value of Fermi temperature $T_F = \mathscr{E}_F/k_B = 50\,000$K. The total number of particles is constant. The chemical potential μ at each temperature is the energy at which $f(\mathscr{E}) = 0.5$.

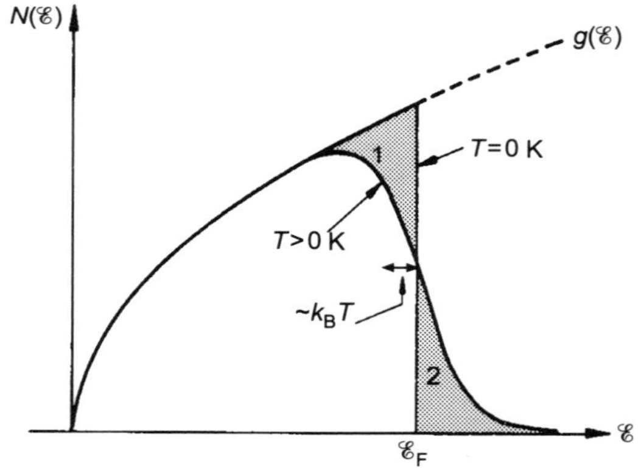

Fig. 5.5 Density of occupied electron states as a function of energy $N(\mathscr{E}) = f(\mathscr{E}, T)g(\mathscr{E})$ for a 3D free-electron gas at a finite temperature $T \ll T_F$ (solid curve). The shaded area represents the filled states at $T = 0$ K. The dashed line is the density of states for a 3D free-electron gas. Electrons that are in states in region 1 at 0 K, of width of order k_BT, are thermally excited to region 2 at a finite temperature, T. As a result, the average energy of the electron gas increases with temperature.

maintain this constancy. The number of electrons is given by an integral over the free-electron density of states (eqn. (5.15)) multiplied by the Fermi–Dirac function (eqn. (5.23)), i.e. an integral over occupied orbitals (see Fig. 5.5), with the (energy) density of occupied orbitals at energy $\mathscr{E}$ being

$$N(\mathscr{E}) = f(\mathscr{E}, T)g(\mathscr{E}); \tag{5.25}$$

Hence, at a finite temperature (cf. eqn. (5.16)):

$$\begin{aligned} N &= \int_0^\infty f(\mathscr{E}, T)g(\mathscr{E})\mathrm{d}\mathscr{E} \\ &= A\int_0^\infty \frac{\mathscr{E}^{1/2}\mathrm{d}\mathscr{E}}{\exp[(\mathscr{E}-\mu)/k_\mathrm{B}T+1]} = BT^{3/2}\int_0^\infty \frac{x^{1/2}\mathrm{d}x}{(\mathrm{e}^{x-\eta}+1)}, \end{aligned} \tag{5.26}$$

where A and B are constants, and the changes of variables $x = \mathscr{E}/k_\mathrm{B}T$ and $\eta = \mu/k_\mathrm{B}T$ have been made. The full temperature dependence of $\mu(T)$ may be obtained from eqn. (5.26) by numerical integration (see e.g. Kittel and Kroemer (1980)); the result is shown graphically in Fig. 5.6.

★ Alternatively, by considering the general integral

$$I = \int_0^\infty f(\mathscr{E})\left\{\frac{\mathrm{d}\Gamma(\mathscr{E})}{\mathrm{d}\mathscr{E}}\right\}\mathrm{d}\mathscr{E}, \tag{5.27a}$$

where $\Gamma(\mathscr{E})$ is *any* function that is zero at $\mathscr{E} = 0$, which can be evaluated approximately as (see Problem 5.3)

$$I = \Gamma(\mu) + \frac{(\pi k_\mathrm{B}T)^2}{6}\frac{\mathrm{d}^2\Gamma(\mu)}{\mathrm{d}\mathscr{E}^2}, \tag{5.27b}$$

it can be demonstrated (see Problem 5.3) that the two leading terms in the expansion of the temperature of μ for the 3D Fermi gas are given by:

$$\mu(T) \simeq \mathscr{E}_\mathrm{F}\left[1 - \frac{\pi^2}{12}\left(\frac{k_\mathrm{B}T}{\mathscr{E}_\mathrm{F}}\right)^2\right]. \tag{5.28}$$

★

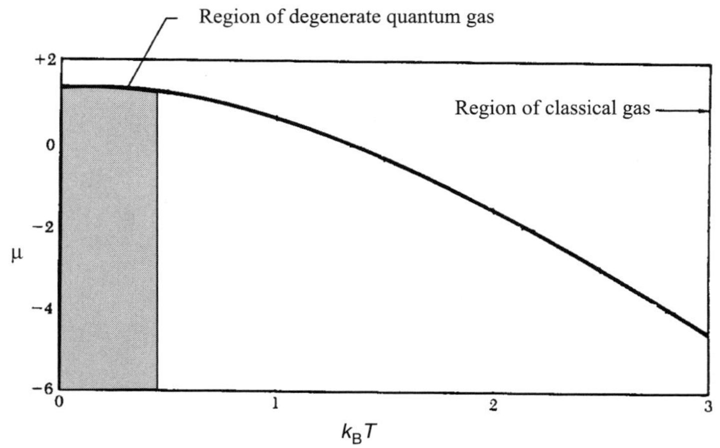

Fig. 5.6 Temperature dependence of the chemical potential of a non-interacting 3D free-electron gas. The units of μ and $k_\mathrm{B}T$ have been chosen to be $0.763\mathscr{E}_\mathrm{F}$, corresponding to the value $\mathscr{E}_\mathrm{F} = (3/2)^{2/3}$. The shaded portion marks the region where the electrons behave as a degenerate quantum gas. (After Kittel (1996). Reproduced by permission of John Wiley & Sons Inc.)

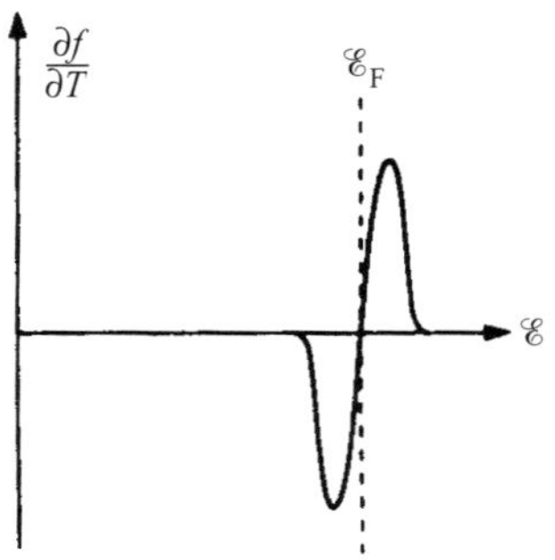

Fig. 5.7 Plot of the temperature derivative of the Fermi–Dirac distribution function, $\partial f(\mathscr{E}, T)/\partial T$. Electrons that occupy states just below $\mathscr{E}_F$ can be excited into states just above $\mathscr{E}_F$.

Non-interacting fermions, for which $T \ll T_F$, form what is called a degenerate Fermi gas. (Since T_F is so high for metallic solids (see Table 5.1), this limit always applies for the case of conduction electrons in metals at temperatures below their melting points.) Note from Fig. 5.6 and eqn. (5.28) that, in the degenerate régime, the chemical potential is almost temperature-independent and close in value to that characteristic of zero kelvin, i.e. $\mu(T) \simeq \mathscr{E}_F$. Therefore, the temperature derivative of the Fermi–Dirac function in the degenerate limit (taking $\mu = \mathscr{E}_F$) can be written approximately as:

$$\frac{\mathrm{d}f(\mathscr{E}, T)}{\mathrm{d}T} \simeq \frac{(\mathscr{E} - \mathscr{E}_F)}{k_B T^2} \frac{\exp[(\mathscr{E} - \mathscr{E}_F)/k_B T]}{\{\exp[(\mathscr{E} - \mathscr{E}_F)/k_B T] + 1\}^2}; \tag{5.29}$$

it can be readily seen from Fig. 5.7 that this quantity is only significant for energies in the vicinity of $\mathscr{E}_F$. As a result, the Fermi surface is quite sharp, even at relatively high temperatures (which are, of course, still much smaller than T_F).

5.1.3 Properties of the Fermi electron gas

A number of equilibrium physical properties of metallic solids can be understood in terms of the free-electron Fermi-gas model. These include the electronic contributions to the bulk modulus, the heat capacity and thermal expansion. Non-equilibrium transport properties (e.g. electrical and thermal conductivities) associated with the free-electron gas are discussed in Chapter 6 and magnetic properties in Chapter 7.

5.1.3.1 Electronic heat capacity

The heat capacity at constant volume is the change in internal energy with temperature (eqn. (4.191)). There will be an electronic contribution to the heat capacity, in addition to the vibrational term (eqn. (4.192)), if free electrons are present in a solid since they too can be thermally excited to higher energies. If the free electrons are regarded as a *classical* gas of N particles (the Drude model), the thermal energy is entirely translationally kinetic in nature, and in the Dulong–Petit limit, i.e. when $k_B T \gg \Delta\mathscr{E}$, where

$\Delta\mathscr{E}$ is the energy spacing between translational energy levels, it is expected that the heat capacity would be temperature-independent, i.e. $C_v = 3Nk_B/2$. In fact, the electronic contribution to C_v is found experimentally to be *linearly dependent on temperature*, and moreover much smaller in magnitude than the Dulong–Petit value. This behaviour can only be understood in terms of the Fermi-gas model.

For a Fermi gas, unlike a classical gas of distinguishable particles, the Pauli exclusion principle restricts thermal excitations *only* to those electrons occupying states within an energy range of approximately $k_B T$ below the Fermi level which can be excited into *empty* states above $\mathscr{E}_F$ (see Fig. 5.5). Thus only a very small fraction of the conduction electrons can contribute to the heat capacity, namely those lying at the top of the energy distribution: lower-lying electrons cannot be thermally excited because this would lead to the double occupancy of states by electrons with the same spin, which is forbidden by the Pauli exclusion principle.

An approximate expression for the electronic heat capacity can be obtained from an examination of Fig. 5.5. If the shaded areas, representing electron states involved in thermal excitation at a temperature T compared with the zero-kelvin electron distribution, are approximated as triangles of height $(1/2)g(\mathscr{E}_F)$ and base $2k_B T$, then Fig. 5.5 implies that an approximate number of electrons given by $(1/2)g(\mathscr{E}_F)k_B T$ have their energies increased on average by $k_B T$. The internal energy then increases by the amount

$$U(T) - U(0) \simeq \frac{1}{2} g(\mathscr{E}_F) k_B^2 T^2,$$

whence differentiation with respect to temperature gives for the heat capacity

$$C_v \simeq g(\mathscr{E}_F) k_B^2 T = \frac{3}{2} \frac{N}{\mathscr{E}_F} k_B^2 T = \frac{3}{2} N k_B \frac{T}{T_F}, \tag{5.30}$$

using eqn. (5.16).

Although this approximate treatment produces a numerical factor that is too small by a factor of about three (see below), nevertheless the functional form is correct. It can be seen that eqn. (5.30) predicts that the electronic heat capacity is linearly dependent on temperature. Hence the *total* heat capacity (electronic and vibrational) at low temperatures can be written as (cf. eqn. (4.201)):

$$C_v = \gamma T + \alpha T^3. \tag{5.31}$$

(Note that the linear term in this expression, due to the Fermi gas, should *not* be confused with that associated with two-level (vibrational) systems in insulating glasses — see eqn. (4.207).) A plot of C_v/T versus T^2 therefore allows the electronic and lattice contributions to be separated, as the zero-temperature intercept (i.e. the Sommerfeld parameter γ) and slope (α), respectively. This behaviour is shown in Fig. 5.8.

★ A more exact calculation of the electronic heat capacity in the Fermi-gas approximation should take account of the fact that the chemical potential is temperature-dependent (albeit weakly — cf. eqn. (5.28)) and should properly treat the thermal occupancy of the electron states. The internal energy of the electron gas at a temperature T is given by

$$U(T) = \int_0^\infty \mathscr{E} f(\mathscr{E}, T) g(\mathscr{E}) \mathrm{d}\mathscr{E}. \tag{5.32}$$

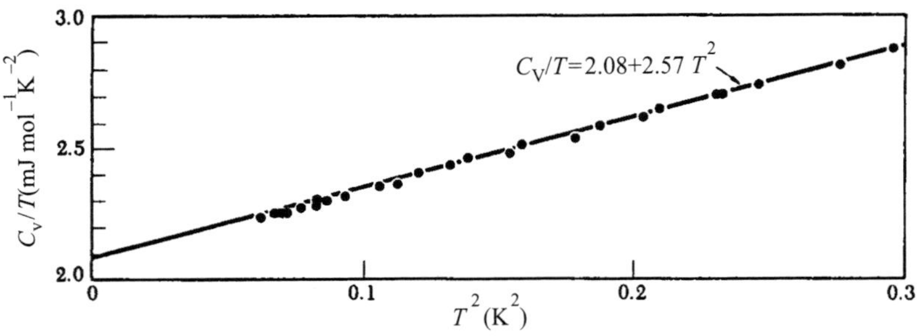

Fig. 5.8 Plot of C_v/T versus T^2 for potassium, revealing a linearly temperature-dependent term (as a finite intercept on the zero-temperature axis) due to the electronic contribution to the heat capacity. (Reprinted with permission from Lien and Phillips (1964), *Phys. Rev.* **133**, A1370. © 1964. The American Physical Society)

★

Using the same approach as employed to evaluate the chemical potential using the general integral given by eqn. (5.27a), the function $\Gamma(\mathscr{E})$ is thus given by $\Gamma = \int_0^\infty \mathscr{E}' g(\mathscr{E}')\mathrm{d}\mathscr{E}'$. Hence, using eqn. (5.27b), the internal energy can be rewritten as:

$$U(T) = \int_0^\mu \mathscr{E} g(\mathscr{E})\mathrm{d}\mathscr{E} + \frac{(\pi k_\mathrm{B} T)^2}{6}\left\{\frac{\mathrm{d}}{\mathrm{d}\mathscr{E}}[\mathscr{E} g(\mathscr{E})]\right\}_\mu . \tag{5.33}$$

Making use of eqn. (5.15), i.e. $g(\mathscr{E}) = A(\mathscr{E})^{1/2}$, the quantity $\{\mathrm{d}[\mathscr{E} g(\mathscr{E})]/\mathrm{d}\mathscr{E}\}_\mu$ equals $(3/2)g(\mu) \simeq (3/2)g(\mathscr{E}_\mathrm{F})$. The integral in eqn. (5.33) can be evaluated as:

$$I = A\int_0^\mu \mathscr{E}^{3/2}\mathrm{d}\mathscr{E} = \frac{2A}{5}\mu^{5/2} \simeq \frac{2A}{5}\mathscr{E}_\mathrm{F}^{5/2}\left[1 - \frac{\pi^2}{12}\left(\frac{k_\mathrm{B}T}{\mathscr{E}_\mathrm{F}}\right)^2\right]^{5/2},$$

by making use of eqn. (5.28). Expanding the term in square brackets gives

$$I \simeq \frac{2A}{5}\mathscr{E}_\mathrm{F}^{5/2} - \frac{\pi^2}{12}A\mathscr{E}_\mathrm{F}^{5/2}\left\{\frac{k_\mathrm{B}T}{\mathscr{E}_\mathrm{F}}\right\}^2 = \frac{2}{5}\mathscr{E}_\mathrm{F}^2 g(\mathscr{E}_\mathrm{F}) - \frac{\pi^2}{12}(k_\mathrm{B}T)^2 g(\mathscr{E}_\mathrm{F}). \tag{5.34}$$

Thus, from eqns. (5.33) and (5.34), the internal energy of the Fermi gas is given by:

$$U(T) \simeq \frac{2}{5}\mathscr{E}_\mathrm{F}^2 g(\mathscr{E}_\mathrm{F}) + \frac{\pi^2}{6}(k_\mathrm{B}T)^2 g(\mathscr{E}_\mathrm{F}) \tag{5.35a}$$

or

$$U(T) \simeq \frac{3}{5}N\mathscr{E}_\mathrm{F} + \frac{\pi^2}{4}Nk_\mathrm{B}\frac{T^2}{T_\mathrm{F}} \tag{5.35b}$$

from eqn. (5.16). Thus, differentiating eqn. (5.35) with respect to temperature gives the heat capacity (cf. eqn. (5.30)): ★

$$C_v \simeq \frac{\pi^2}{3}k_\mathrm{B}^2 g(\mathscr{E}_\mathrm{F})T. \tag{5.36a}$$

This is a general expression valid for any density of states. For the specific case of a free-electron gas:

$$C_v \simeq \frac{\pi^2}{2} N k_B \frac{T}{T_F}. \quad (5.36b)$$

Although the functional form of eqn. (5.36) is well obeyed by simple metals (see Fig. 5.8), there is usually a discrepancy in quantitative agreement (by a factor of two or three) between experimental values of C_v and those calculated on the basis of the Fermi-gas model using the *free-electron mass*, m_e, in the expression for $\mathscr{E}_F$ (eqn. (5.17)). This discrepancy can be rationalized by invoking an effective mass, m_e^*, for the electron that is different from m_e. Such an effective mass arises because of the breakdown for real materials of two assumptions implicit in the Fermi-gas model: (i) a constant electrostatic potential experienced by the electrons (continuum approximation), and (ii) neglect of electron interactions. In reality, electrons move in the spatially varying potential associated with the ion cores, and this can have pronounced effects on the allowed energy states that an electron may have. Correspondingly the parabolic dispersion relation, $\mathscr{E}(k) \propto k^2$ (eqn. (5.12)), characteristic of free particles is no longer valid, and the associated effective mass is no longer that of a free particle (see §6.2.1). Electron–electron interactions also contribute to departures of the effective electron mass from the free-electron value, due to an inertial reaction between electrons, as do electron–phonon interactions (the motion of an electron causes a concomitant change in the local atomic coordinates, i.e. it 'drags' the lattice with it). In certain cases, particularly for f-electron-containing metallic alloys, e.g. UBe_{13}, $CeAl_3$, the effective electron mass inferred from heat-capacity measurements is some three orders of magnitude higher than m_e; such materials are termed, not surprisingly, heavy-fermion compounds.

*5.1.3.2 *Cohesive energy*

The electronic energy considered so far, namely the kinetic energy of the Fermi gas (eqn. (5.20)), is entirely *repulsive* in character: it provides no attractive term to bind atoms together in a solid. A simple model for the cohesive energy was given in §2.2.3.1 in terms of the Coulomb attraction between the electrons and the ion cores. In reality, the picture is a little more complicated, since even for the Fermi gas in the jellium model, where the positive ionic charges are assumed to be uniformly distributed, there is an effective attractive interaction between the free electrons and the positive background charge. This results from the exchange interaction which is a manifestation of the Pauli exclusion principle: the electron motion is *correlated*, with two parallel-spin electrons tending to avoid each other, thereby creating an *exchange-correlation hole* (Fig. 2.45). Since the hole is associated with the exclusion of exactly one electron, this therefore exposes one positive unit of background jellium charge, resulting in an attractive interaction between an electron and this positive charge.

This exchange interaction is somewhat complicated to evaluate (see e.g. Ashcroft and Mermin (1976) or Madelung (1978)); the result is that the free-electron eigenenergy of the Schrödinger equation (eqn. (5.12)) is modified to:

$$\mathscr{E} = \frac{\hbar^2 k^2}{2m_e} - \frac{e^2 k_F}{8\pi^2 \varepsilon_0}\left(2 + \frac{k_F^2 - k^2}{k k_F} \ln\left|\frac{k + k_F}{k - k_F}\right|\right), \quad (5.37)$$

where the Fermi wavevector is given by eqn. (5.19). Integrating this energy over the Fermi sphere gives for the average electron energy at zero kelvin:

$$\langle \mathscr{E} \rangle = \frac{3}{5}\mathscr{E}_F - \frac{3e^2 k_F}{16\pi^2 \varepsilon_0}. \tag{5.38}$$

The exchange energy, the second term in eqn. (5.38), results from the solution of the Hartree–Fock equations, obtained by insisting that the one-electron wavefunctions characteristic of the Hartree equation (the Schrödinger equation with the Hartree potential, eqn. (2.81), and no electron correlations) be *antisymmetric* (i.e. obeying Fermi–Dirac statistics). However, even this is an approximation, and the correction terms to the Hartree–Fock exchange energy (eqn. (5.38)) are often referred to as the correlation energy, interpreted as resulting from the Coulombic repulsion between antiparallel-spin electrons, although strictly they are just a higher-order approximation to the exchange term. The electron energy is often written in terms of atomic units, i.e. with energies measured in Rydbergs (the ionization energy of the H atom, 13.6 eV) and distances measured in Bohr radii ($a_0 = 0.529$ Å), with $e^2 = 2$ and $4\pi\varepsilon_0 = 1$. The total energy of the Fermi gas in these units is (Pettifor (1995)):

$$U_{Fg} = \frac{2.21}{r_s^2} - \frac{0.916}{r_s} - (0.115 - 0.0313 \ln r_s). \tag{5.39}$$

The first two terms are just the Hartree–Fock terms (eqn. (5.38)), and the third term in parentheses is the 'correlation' energy, rewritten in terms of the quantity r_s, the radius of a sphere containing on average a *single* electron. For a gas of N electrons in a volume V, this is given by:

$$r_s = \left(\frac{3V}{4\pi N}\right)^{1/3} \equiv Z^{-1/3} r_{ws}, \tag{5.40}$$

where r_{ws} is the Wigner–Seitz radius (§2.2.2.3) and the charge on an ion is taken to be Ze. The energy of the Fermi gas (eqn. (5.39)) is plotted in Fig. 5.9: it has a minimum at $r_s = 2.23$ Å (4.2 atomic units (au)), with a well depth of 2.2 eV (0.16 Ryd).

The assumption in the jellium model of a *uniform* positive charge density, representing that of the ion cores, is obviously unrealistic. The interaction energy between Z electrons in a Wigner–Seitz cell (assumed for simplicity to be spherical) and a point ion of charge $+Ze$, taking into account also the inter-electron repulsion within the cell, is given by (see Problem 5.8 and eqn. (2.17)):

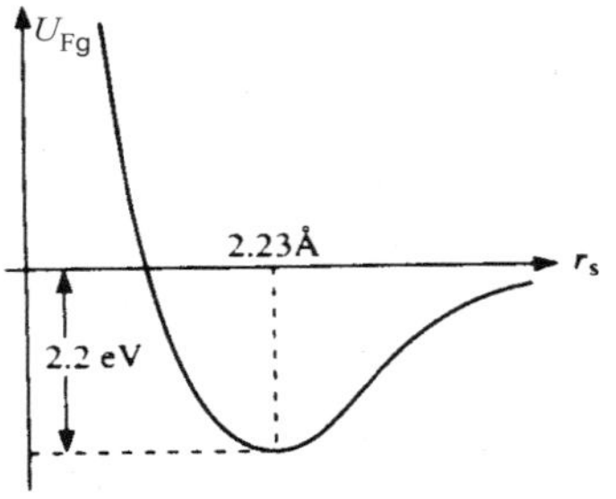

Fig. 5.9 The energy of the Fermi gas (eqn. (5.39)) plotted in electron volts and ångströms. (After Sutton (1993), *Electronic Structure of Materials*, by permission of Oxford University Press)

$$U_{\text{ws}} = U_{\text{ei}} + U_{\text{ee}} = -\frac{3Z^2e^2}{8\pi\varepsilon_0 r_{\text{ws}}} + \frac{3Z^2e^2}{20\pi\varepsilon_0 r_{\text{ws}}} = -\frac{9Z^2e^2}{40\pi\varepsilon_0 r_{\text{ws}}}. \tag{5.41}$$

However, this is an overestimate, since the repulsion between core and valence electrons is neglected.

This effect can be taken into account by assuming that instead of experiencing the true attractive Coulomb potential of the ion, $-Ze/4\pi\varepsilon_0 r$, a valence electron experiences a much *weaker* potential in the vicinity $r < r_c$ of the ion core; this is called a pseudopotential. This behaviour is another consequence of the Pauli exclusion principle. Valence electrons are excluded from the core region if the core-electron orbitals are occupied. Since the valence-electron wavefunction is orthogonal to the core states, the resulting depletion in valence-electron charge density is termed the orthogonality hole (Fig. 5.10). Thus, the valence electrons experience an apparent *repulsive* component of the potential in the core region, which can almost cancel that of the attractive Coulombic term. When this cancellation is complete, the potential is called the Ashcroft empty-core pseudopotential (Fig. 5.11), i.e.

$$\begin{aligned} V_{\text{ps}}(r) &= 0, & r < r_{\text{c}}, \\ &= -\frac{Ze^2}{4\pi\varepsilon_0 r} & r > r_{\text{c}}. \end{aligned} \tag{5.42}$$

This type of pseudopotential, which does not differentiate between electron states with different angular momenta, is said to be local. The form of the Ashcroft pseudopotential

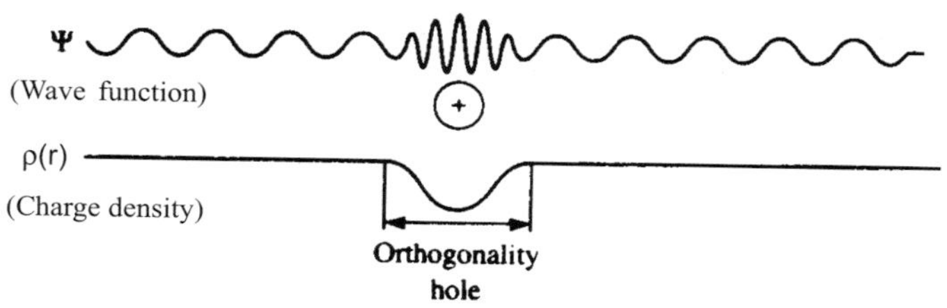

Fig. 5.10 The wavefunction ψ and charge density ρ of a valence electron in the vicinity of an ion core. The rapid oscillations in ψ in the core region, required so that ψ is orthogonal to the tightly bound core electrons, are associated with a depletion in the valence-electron charge density (the orthogonality hole).

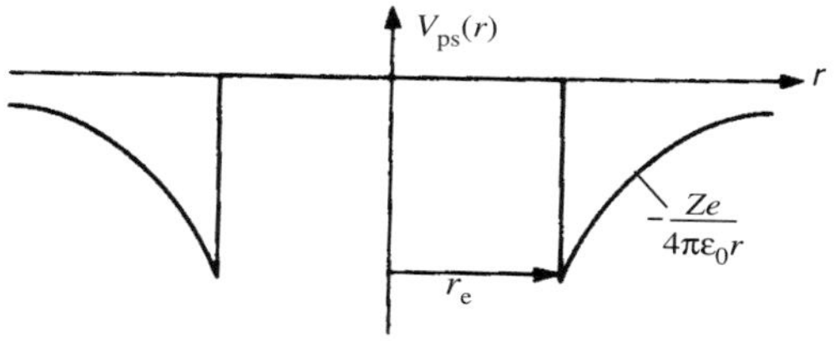

Fig. 5.11 The Ashcroft empty-core pseudopotential: $V_{\text{ps}} = 0$ for $r < r_{\text{c}}$.

causes a modification to the electron–ion attractive interaction (the first term in eqn. (5.41)), namely:

$$U_{\mathrm{ei}} = -\frac{3Z^2e^2}{8\pi\varepsilon_0 r_{\mathrm{ws}}}\left[1-\left(\frac{r_{\mathrm{c}}}{r_{\mathrm{ws}}}\right)^2\right]. \tag{5.43}$$

The total expression for the cohesive energy is thus given by

$$U_{\mathrm{coh}} = ZU_{\mathrm{Fg}} + U_{\mathrm{ws}}. \tag{5.44}$$

In atomic units, this is

$$U_{\mathrm{coh}} = Z\left\{\frac{2.21}{r_{\mathrm{s}}^2} - \frac{0.916}{r_{\mathrm{s}}} - (0.115 - 0.0313\ln r_{\mathrm{s}})\right\} - \frac{3Z^2}{r_{\mathrm{ws}}}\left[1-\left(\frac{r_{\mathrm{c}}}{r_{\mathrm{ws}}}\right)^2\right] + \frac{1.2Z^2}{r_{\mathrm{ws}}}. \tag{5.45}$$

*5.1.3.3 *Bulk modulus*

Since the conduction electrons in a solid are being treated as a gas of particles, albeit a quantum gas, they will exert a pressure given, at zero kelvin, by

$$p = -\left(\frac{\partial U_0}{\partial V}\right)_N, \tag{5.46}$$

where U_0 is the internal energy at $T = 0\,\mathrm{K}$ of the electron gas, i.e. the average kinetic energy $(3/5)N\mathscr{E}_{\mathrm{F}}$ (eqn. (5.20)), since it has been assumed that the potential energy is zero. The origin of the pressure can be regarded as being due to the repulsion experienced by electrons, caused by the Pauli exclusion principle, when they are compressed and tend to occupy the same region of space. Since $\mathscr{E}_{\mathrm{F}} \propto V^{-2/3}$ (eqn. (5.17)), it is straightforward to show from eqn. (5.46) that

$$p = \frac{2U_0}{3V}. \tag{5.47}$$

(In fact, an expression of this form is valid for *all* temperatures — see Problem 5.9.) The bulk modulus is defined as (see eqn. (2.32)):

$$B = -V\left(\frac{\partial p}{\partial V}\right)_{T,N}, \tag{5.48}$$

or, from eqn. (5.46):

$$B = V\left(\frac{\partial^2 U_0}{\partial V^2}\right). \tag{5.49}$$

Thus, considering only the kinetic energy of the free-electron gas, for which $p \propto V^{-5/3}$ from eqns. (5.17) and (5.47), the zero-kelvin electronic contribution to the bulk modulus is

$$B_{\mathrm{KE}} = \frac{10U_0}{9V} = \frac{2}{3}\frac{N}{V}\mathscr{E}_{\mathrm{F}}, \tag{5.50a}$$

or, in atomic units,

$$B_{\mathrm{KE}} = 0.586/r_{\mathrm{s}}^5, \tag{5.50b}$$

Table 5.2 Bulk moduli and cohesive energies of simple metals

Metal	Z	U_{coh}/Z (eV/electron)	B/B_{KE} (calc)	B/B_{KE} (expt)	r_s (au)	r_c (au)
Li	1	1.7	0.63	0.50	3.27	1.32
Na	1	1.1	0.83	0.80	3.99	1.75
K	1	0.9	1.03	1.10	4.86	2.22
Be	2	1.7	0.45	0.27	1.87	0.76
Mg	2	0.8	0.73	0.54	2.66	1.31
Ca	2	0.9	0.95	0.66	3.27	1.73
Zn	2	0.7	0.60	0.45	2.31	1.07
Cd	2	0.6	0.71	0.63	2.59	1.27
Al	3	1.1	0.69	0.32	2.07	1.11
Ga	3	0.9	0.74	0.33	2.19	1.20
Cu	1	3.5	0.45	2.16	2.67	0.91
Ag	1	3.0	0.71	2.94	3.02	1.37
Au	1	3.8	0.69	4.96	3.01	1.35

Z is the valence, U_{coh} is the cohesive energy, B is the bulk modulus and B_{KE} the bulk modulus resulting solely from the kinetic energy contribution of the free-electron gas, r_s is the radius of a sphere containing a single valence electron and r_c is the radius of the Ashcroft empty-core pseudopotential.
(After Pettifor (1995), *Bonding and Structure of Molecules and Solids*, by permission of Oxford University Press)

where r_s is the radius of a sphere containing a single valence electron (eqn. (5.40)).

However, the *attractive* part of the electron energy associated with the ion cores (see §5.1.3.2) makes the solid more compressible (i.e. the bulk modulus is smaller) by a factor of five when the Wigner–Seitz attractive energy (eqn. (5.41)) is included (see Problem 2.7). Neglecting the small contribution from correlation effects, but including the effect of the empty-core pseudopotential of radius r_c, the bulk modulus can be written in atomic units as (Pettifor (1995)):

$$B/B_{KE} = 0.2 + 0.815 r_c^2/r_s. \qquad (5.51)$$

Values of the bulk modulus calculated using eqn. (5.51) are compared with experimental values for some simple metals in Table 5.2, together with values of the cohesive energy and r_c and r_s. It is seen that there is good agreement for B values in general, except for the noble metals where the discrepancy is due to the presence of core d-electrons.

*5.1.3.4 *Thermal expansion*

The delocalized free-electron gas of a metal also contributes to the thermal expansion of the material, in addition to the anharmonic atomic vibrational contribution (§4.6.2.1). The pressure of the Fermi gas is given by (Problem 5.9):

$$p = \frac{2U(T, V)}{3V}, \qquad (5.52)$$

where $U(T, V)$ is given by eqn. (5.35). Since the volume thermal-expansion coefficient is given by

$$\beta_T = \frac{1}{B}\left(\frac{\partial p}{\partial T}\right)_V, \quad (4.210)$$

where B is the bulk modulus, there is a positive electronic contribution to the thermal expansion because $U(T,V)$ is an increasing function of T. The thermal-expansion coefficient is therefore given by

$$\beta_T = \frac{2}{3}\frac{C_v}{BV}, \quad (5.53)$$

since $C_v = (\partial U/\partial T)_v$. (See eqn. (4.220) for the corresponding expression for the anharmonic atomic vibrational term.)

Equation (5.53) can be rewritten making use of the expressions derived previously for the heat capacity (eqn. (5.36b)) and the bulk modulus (eqns. (5.50a) and (5.51)), giving

$$\beta_T = \frac{\pi^2 k_B^2 T}{2A\mathscr{E}_F^2}, \quad (5.54)$$

where the ratio B/B_{KE} (eqn. (5.51)) has been replaced by the temperature-independent constant A. Thus, at low temperatures ($T \lesssim 10$ K), where the electronic heat capacity of metals is greater than that of the lattice-vibrational term, the thermal-expansion coefficient of metals is predicted to be *linearly* dependent on temperature, in contrast to the T^3 dependence exhibited by insulators (§4.6.3).

Electrons in periodic solids 5.2

Although the Fermi-gas model for valence electrons in solids has the virtues of being simple and being capable of accounting for some physical properties of metals, nevertheless it has a number of severe limitations. First and foremost, it predicts that *every* solid containing elements with part-filled electron shells should exhibit *metallic* behaviour associated with the free-electron gas (for a discussion of electronic transport properties, see Chapter 6). This prediction fails spectacularly in the case of elements such as, say, Si or Ge which have an incomplete p-shell (s^2p^2) but which are, nevertheless, *electrical insulators* at the absolute zero of temperature (see Chapter 6): there are no mobile free electrons in this case. Moreover, certain electrical transport measurements, e.g. of the Hall effect (see §6.3.3.2), indicate that the electron mass is apparently *negative*: this behaviour is completely incomprehensible in the free-electron picture. We shall see that these inconsistencies with the free-electron model can be explained by taking into account the fact that the motion of the electrons is *not* free, but is constrained by the spatial arrangement of the potentials associated with the ion cores. This effect is particularly marked in crystals having a periodic array of ion-core potentials.

5.2.1 Bloch wavefunctions

The spatial periodicity of the electrostatic potentials associated with the ion cores in a crystal (Fig. 5.12) imposes a concomitant constraint on the wavefunctions that are the solutions of the Schrödinger equation describing the motion of an electron in the periodic potential. The wavefunction for a periodic potential is the product of a plane wave and a function which has the periodicity of the crystal lattice: this is known as a Bloch function.

Consider for simplicity a 3D lattice having translational periodicity, such that the position of atom n from the origin is given by $\boldsymbol{R}_n = u\boldsymbol{a} + v\boldsymbol{b} + w\boldsymbol{c}$ (eqn. (2.1)), where u, v, w are integers. If it is assumed for the present that the wavefunctions are non-degenerate,

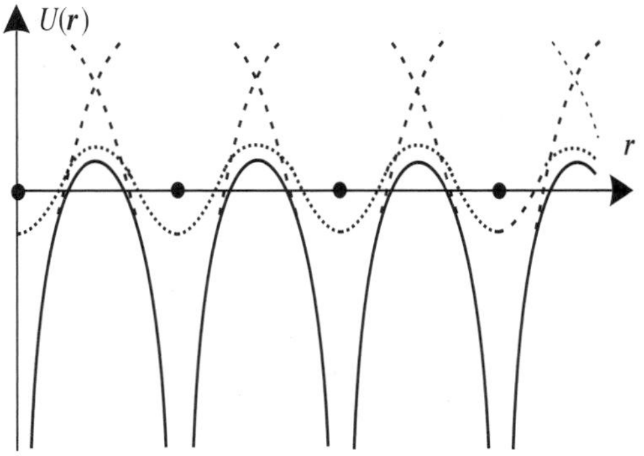

Fig. 5.12 Representation of a periodic electrostatic potential associated with the ion cores in a crystal. The filled circles represent the ion positions. The solid curves represent the potential along a line of atoms, and the dashed curves are the potentials for isolated ions. The dotted curve is the potential along a line midway between planes of ions.

and that the wavefunctions for neighbouring sites differ by a constant phase factor, A, B and C in the $\boldsymbol{a}$, $\boldsymbol{b}$ and $\boldsymbol{c}$ directions, respectively, then imposition of periodic (Born–von Karman) boundary conditions for a parallelopiped of material of sides $N_1\boldsymbol{a}$, $N_2\boldsymbol{b}$ and $N_3\boldsymbol{c}$ gives

$$\begin{aligned}\psi(\boldsymbol{r}+N_1\boldsymbol{a}) &= \psi(\boldsymbol{r}) = A^{N_1}\psi(\boldsymbol{r}),\\ \psi(\boldsymbol{r}+N_2\boldsymbol{b}) &= \psi(\boldsymbol{r}) = B^{N_2}\psi(\boldsymbol{r}),\\ \psi(\boldsymbol{r}+N_3\boldsymbol{c}) &= \psi(\boldsymbol{r}) = C^{N_3}\psi(\boldsymbol{r}).\end{aligned} \tag{5.55}$$

Thus, for example,

$$A^{N_1} = 1, \tag{5.56a}$$

with A being one of the N_1 roots of unity, i.e.

$$A = \exp(2\pi \mathrm{i}\, n_1/N_1), \qquad n_1 = 0, \pm 1, \pm 2, \ldots \tag{5.56b}$$

and there are similar equations for B and C in terms of n_2 and N_2 and n_3 and N_3, respectively. Equation (5.56a) can be rewritten in the form:

$$1 = \mathrm{e}^{\mathrm{i}\boldsymbol{k}.N_1\boldsymbol{a}} = \mathrm{e}^{\mathrm{i}\boldsymbol{k}.N_2\boldsymbol{b}} = \mathrm{e}^{\mathrm{i}\boldsymbol{k}.N_3\boldsymbol{c}}, \tag{5.57}$$

where $\boldsymbol{k}$ is given in terms of reciprocal-lattice vectors (eqn. (2.48)) by $\boldsymbol{k} = q_1\boldsymbol{a}^* + q_2\boldsymbol{b}^* + q_3\boldsymbol{c}^*$ (eqn. (2.51)), with $q_1 = n_1/N_1$, etc.

Thus, by extension, one form of the Bloch function is

$$\psi_k(\boldsymbol{r}+\boldsymbol{R}_n) = \mathrm{e}^{\mathrm{i}\boldsymbol{k}.\boldsymbol{R}_n}\psi_k(\boldsymbol{r}). \tag{5.58}$$

An equivalent form for the Bloch function is that the wavefunction is the modulated plane wave

$$\psi_k(\boldsymbol{r}) = \mathrm{e}^{\mathrm{i}\boldsymbol{k}.\boldsymbol{r}}u_k(\boldsymbol{r}), \tag{5.59}$$

where $u_k(\boldsymbol{r})$ is a function with the *same periodicity* as that of the Bravais lattice:

$$u_k(\boldsymbol{r}+\boldsymbol{R}_n) = u_k(\boldsymbol{r}). \tag{5.60}$$

The validity of eqn. (5.59) can be demonstrated by the substitution into it of eqn. (5.58). A one-dimensional example of a Bloch wave is illustrated in Fig. 5.13.

A more general derivation of Bloch functions, free from the assumption about degeneracy, can be obtained from a consideration of the Schrödinger equation for a translationally periodic potential energy $V(\boldsymbol{r}+\boldsymbol{R}_n) = V(\boldsymbol{r})$, expressed in terms of a Fourier expansion (eqn. (2.55)) in terms of reciprocal-lattice vectors $\boldsymbol{G}$, viz.

$$V(\boldsymbol{r}) = \sum_G V_G \exp(\mathrm{i}\,\boldsymbol{G}.\boldsymbol{r}). \tag{5.61}$$

If the wavefunction is also expressed as a general Fourier series in terms of reciprocal-space vectors $\boldsymbol{k}$ (compatible with the boundary conditions), i.e.

$$\Psi_k(\boldsymbol{r}) = \sum_k C_k \exp(\mathrm{i}\,\boldsymbol{k}.\boldsymbol{r}), \tag{5.62}$$

the time-independent Schrödinger equation (eqn. (5.3a)) can be rewritten as

$$\sum_k \frac{\hbar^2k^2}{2m_\mathrm{e}} C_k\,\mathrm{e}^{\mathrm{i}\boldsymbol{k}.\boldsymbol{r}} + \sum_{k',G} C_{k'}V_G\,\mathrm{e}^{\mathrm{i}(\boldsymbol{k}'+\boldsymbol{G}).\boldsymbol{r}} = \mathscr{E}\sum_k C_k\,\mathrm{e}^{\mathrm{i}\boldsymbol{k}.\boldsymbol{r}}, \tag{5.63}$$

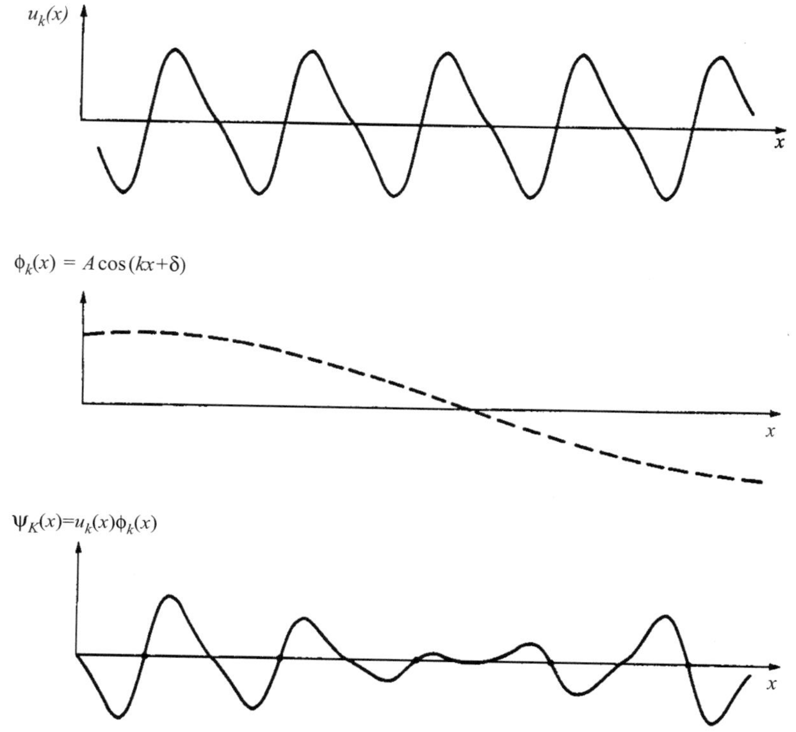

Fig. 5.13 One-dimensional example of the construction of a Bloch wave from a periodic function $u_k(x)$ with p-type bonding character and a plane wave $\cos(kx + \delta)$. Note that the Bloch wavefunction itself is *not* periodic in real space. (After Ibach and Lüth (1995), *Solid State Physics*, p. 132, Fig. 7.1, © Springer-Verlag Gmb & Co. KG)

or, in rearranged form, and calling $\boldsymbol{k}' = \boldsymbol{k} - \boldsymbol{G}$,

$$\sum_k e^{i\boldsymbol{k}.\boldsymbol{r}} \left[\left\{ \frac{\hbar^2 k^2}{2m_e} - \mathscr{E} \right\} C_k + \sum_G V_G C_{k-G} \right] = 0. \tag{5.64}$$

For eqn. (5.64) to be valid at every point $\boldsymbol{r}$, the expression in square brackets must be identically zero for all k, i.e.

$$\left\{ \frac{\hbar^2 k^2}{2m_e} - \mathscr{E} \right\} C_k + \sum_G V_G C_{k-G} = 0. \tag{5.65}$$

This set of algebraic equations is just a re-expression of the wave equation for a periodic lattice. Note that the $\boldsymbol{k}$-values of the wavefunction expansion coefficients C_k differ only by reciprocal-lattice vectors $\boldsymbol{G}$: thus, only a small sub-set of the many $\boldsymbol{k}$-values allowed by the periodic boundary conditions are involved.

Writing the expansion of the wavefunction as

$$\psi_k(\boldsymbol{r}) = \sum_{k'} C_{k'} e^{i\boldsymbol{k}'.\boldsymbol{r}} = \sum_G C_{k-G} e^{i(\boldsymbol{k}-\boldsymbol{G}).\boldsymbol{r}}, \tag{5.66}$$

this may be rearranged as

$$\Psi_k(\boldsymbol{r}) = \left(\sum_G C_{k-G}\,\mathrm{e}^{-\mathrm{i}\boldsymbol{G}.\boldsymbol{r}}\right)\mathrm{e}^{\mathrm{i}\boldsymbol{k}.\boldsymbol{r}} = u_k(\boldsymbol{r})\mathrm{e}^{\mathrm{i}\boldsymbol{k}.\boldsymbol{r}}, \tag{5.67}$$

which has the form of the Bloch function (eqn. (5.59)), since

$$u_k(\boldsymbol{r}) = \sum_G C_{k-G}\,\mathrm{e}^{-\mathrm{i}\boldsymbol{G}.\boldsymbol{r}} \tag{5.68}$$

is a periodic function (eqn. (5.60)). This may be demonstrated by substituting $\boldsymbol{r} = \boldsymbol{r} + \boldsymbol{R}_n$ in eqn. (5.68) and recalling that $\exp(\mathrm{i}\boldsymbol{G}.\boldsymbol{R}_n) = 1$ (eqn. (2.52)).

The translational real-space periodicity of the lattice potential imposes periodicity on the electronic wavefunction and eigenenergy in *reciprocal space*. Thus, from eqn. (5.66), the wavefunction for a wavevector differing from $\boldsymbol{k}$ by a reciprocal-lattice vector $\boldsymbol{G}$ is

$$\begin{aligned}\psi_{k+G}(\boldsymbol{r}) &= \sum_{G'} C_{k+G-G'}\,\mathrm{e}^{\mathrm{i}(\boldsymbol{k}+\boldsymbol{G}-\boldsymbol{G}').\boldsymbol{r}} \\ &= \left(\sum_{G''} C_{k-G''}\,\mathrm{e}^{-\mathrm{i}\boldsymbol{G}''.\boldsymbol{r}}\right)\mathrm{e}^{\mathrm{i}\boldsymbol{k}.\boldsymbol{r}} = \psi_k(\boldsymbol{r})\end{aligned} \tag{5.69}$$

where $\boldsymbol{G}'' = \boldsymbol{G}' - \boldsymbol{G}$. Similar considerations hold for the electron energy obtained from the Schrödinger equation (eqn. (5.3b))

$$\mathscr{H}\psi_k = \mathscr{E}(\boldsymbol{k})\psi_k. \tag{5.70}$$

For a wavevector differing by a reciprocal-lattice vector

$$\mathscr{H}\psi_{k+G} = \mathscr{E}(\boldsymbol{k}+\boldsymbol{G})\psi_{k+G} \tag{5.71}$$

and, using eqn. (5.69), this becomes

$$\mathscr{H}\psi_k = \mathscr{E}(\boldsymbol{k}+\boldsymbol{G})\psi_k. \tag{5.72}$$

Comparison with eqn. (5.70) shows that

$$\mathscr{E}(\boldsymbol{k}+\boldsymbol{G}) = \mathscr{E}(\boldsymbol{k}). \tag{5.73}$$

It should be stressed that, as for phonons, the wavevector $\boldsymbol{k}$ for (Bloch) electron wavefunctions in a periodic potential is *not* a measure of the true momentum: instead $\hbar\boldsymbol{k}$ is termed the *crystal momentum* (§4.2.7).

5.2.2 Energy bands

The function describing the dependence of the electron energy on wavevector, $\mathscr{E}(\boldsymbol{k})$, is called the electronic band structure, and for a 3D solid it is a 4D quantity. In order to visualize it, the energy is plotted versus k for particular trajectories in $\boldsymbol{k}$-space (usually between high-symmetry points in the Brillouin zone), thereby generating lines representing the allowed energy states, called energy bands. The spacing of points in $\boldsymbol{k}$-space, corresponding to allowed states, is so high that such lines of points appear like continuous curves.

All allowed $\boldsymbol{k}$-values fall within the *first Brillouin zone* (see §2.4.2). In the case of a 1D crystal with N atoms, for which periodic boundary conditions correspond to one of the equations of eqn. (5.55), the allowed k-values are given by $k_i = n_i 2\pi/Na = 0, \ldots, \pm n_i(\frac{2\pi}{Na}), \ldots, \pi/a$ (where a is the periodic repeat distance); i.e. there are N allowed values. These correspond to the points in the first Brillouin zone, centred about $\boldsymbol{k} = 0$. Although this in fact contains $(N + 1)$ points, two of these are degenerate since they are connected by a reciprocal-lattice vector; they lie on opposite zone boundaries (cf. eqn. (5.73)). The same considerations also hold in *three* dimensions; the density ρ_k of allowed points in $\boldsymbol{k}$-space is $V/(2\pi)^3$ (eqn. (4.28)), where V is the volume of the real-space primitive unit cell.

An idea of how electron energy bands arise can be gained by considering first the empty-lattice approximation, the hypothetical case where the ion-core potentials are infinitesimally small, i.e. essentially zero. For the case where all Fourier coefficients V_G of the potential are identically zero, the algebraic form of the Schrödinger equation (eqn. (5.65)) reduces simply to $(\hbar^2 k^2/2m_e - \mathscr{E})C_k = 0$; i.e. all coefficients C_{k-G} are zero except for C_k. Thus $u_k(\boldsymbol{r})$ is unity from eqn. (5.68) and $\psi_k(\boldsymbol{r}) = \exp(i\boldsymbol{k}.\boldsymbol{r})$. Hence, the free-electron solution is recovered, and $\mathscr{E}(\boldsymbol{k})$ consists just of a single paraboloid (eqn. (5.12)).

For the case of a vanishingly small periodic potential, the periodicity manifests itself through eqn. (5.73), and for an electron moving in a 1D array of periodic potentials, the parabolic free-electron band is periodically continued in reciprocal space as in Fig. 5.14.

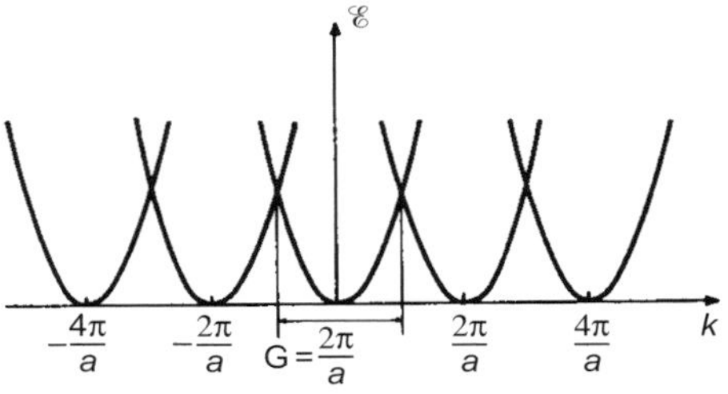

Fig. 5.14 Energy bands for an electron moving in a 1D periodic array of potentials, with periodicity a, in the empty-lattice approximation, plotted in the repeated-zone scheme. The first Brillouin zone is marked.

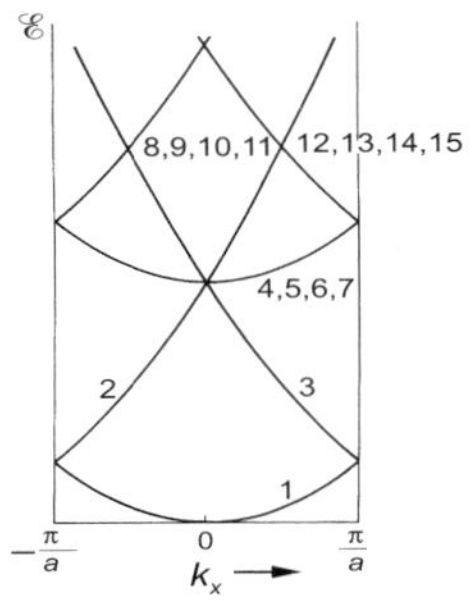

Fig. 5.15 Band structure in the reduced-zone scheme of a free-electron gas (in the empty-lattice approximation) for a simple cubic Bravais lattice plotted along a section in the k_x direction. The labels of the bands are the band indices.

Such a representation is termed the repeated-zone scheme. In three dimensions, a choice must be made as to which section through reciprocal space is to be used, and it is also conventional to represent the electronic band structure in the reduced-zone scheme, i.e. by translating sections of bands lying in higher zones into the first Brillouin zone by the use of appropriate reciprocal-lattice vectors (cf. eqn. (5.73)). The results of this procedure for a free-electron gas in the empty-lattice approximation, for the case of a simple cubic Bravais lattice, are shown in Fig. 5.15 for a section along the k_x direction in the reciprocal space. The various bands are denoted by a band index n according to a particular reciprocal-lattice vector that is needed to bring the band into the first Brillouin zone (see Problem 5.10).

5.2.3 Nearly-free-electron model

Consider now the case of *small* potentials in the core region, with an approximately constant potential in the volume between the cores, and the effects that this has on the band structure of the free-electron gas. Note that, in reality, the Coulombic potentials associated with the ion cores are anything but weak in the vicinity of the cores (see Fig. 5.12), and it might therefore be thought that the nearly-free-electron (NFE) approximation could never be valid. However, as seen in §5.1.3.2, the requirements of orthogonality between the valence (nearly-free) electrons and the tightly bound core electrons ensures, in fact, that the effective potential, the pseudopotential, is very nearly zero in the core region (Fig. 5.11). In addition, the valence electrons very effectively screen the residual Coulombic potential of the pseudopotential in the region between the ion cores (see §5.6.1). Thus, in practice, the requirements of the NFE model are approximately met.

A qualitative picture of the consequences of the NFE model can be gained from a consideration of a 1D crystal with lattice repeat a. If the valence electrons are regarded as waves with wavevector $\boldsymbol{k}$, they will diffract from the periodic ion array when the Laue condition (eqn. (2.101)) is satisfied. For the case of backscattering along the chain of atoms, the scattering vector is $\boldsymbol{K} = 2\boldsymbol{k}$ (see eqn. (2.98)), and hence the diffraction condition $\boldsymbol{K} = \boldsymbol{G}$ is given by

$$k = \pm\frac{G}{2} = \pm\frac{n\pi}{a}, \tag{5.74}$$

(where the negative sign arises because $-\boldsymbol{G}$ is also a reciprocal-lattice vector). Note that these values just correspond to the boundaries of the Brillouin zones (see eqn. (2.103)). At the zone boundaries, in the *empty-lattice approximation*, the electron energy is doubly degenerate (where two parabolas intersect — see Fig. 5.14), corresponding to the two free-electron wavefunctions $\exp(\mathrm{i}Gx/2)$ and $\exp(-\mathrm{i}Gx/2)$. These will have equal weights, and so two combinations of the wavefunctions may be written as:

$$\psi^+ \propto (\mathrm{e}^{\mathrm{i}Gx/2} + \mathrm{e}^{-\mathrm{i}Gx/2}) \propto \cos(Gx/2), \tag{5.75a}$$

$$\psi^- \propto (\mathrm{e}^{\mathrm{i}Gx/2} - \mathrm{e}^{-\mathrm{i}Gx/2}) \propto \sin(Gx/2). \tag{5.75b}$$

These are *not* travelling (running) waves, but instead represent *standing-wave solutions* resulting from the combination of propagating and counterpropagating backscattered (Bragg-diffracted) electron waves.

The corresponding electron charge densities are proportional to $\psi^*\psi$, which are thus (writing $G = 2\pi/a$):

$$\rho^+ \propto \cos^2(\pi x/a), \tag{5.76a}$$

$$\rho^- \propto \sin^2(\pi x/a), \tag{5.76b}$$

and these functions are illustrated schematically in Fig. 5.16. It can be seen that for the solution ψ^+, the charge density is piled up preferentially in the regions of the ion cores, and for ψ^-, the charge density is instead concentrated in the region between the cores. In the empty-lattice approximation, where the potential is everywhere zero, these two solutions obviously have the *same* energy, i.e. they are degenerate (Fig. 5.14).

However, when the ion-core potentials are non-zero (as in Fig. 5.16a), the solution given by ψ^+ will have a *lower energy* $\mathscr{E}^+$ than that for $\psi^-(\mathscr{E}^-)$. Thus, an energy gap, or bandgap, is opened up at the zone boundaries in the NFE approximation. Such a gap is also termed a forbidden gap because there are *no* allowed wave-like solutions in the energy interval between $\mathscr{E}^+$ and $\mathscr{E}^-$. Note, however, that in general, bandgaps need not open up at *every* zone boundary. For some crystal structures, particularly those with a basis, there are some reciprocal-lattice vectors G for which the *extinction rules* of diffraction (§2.6.1.2) are satisfied (in this case for the electron waves travelling inside the crystal). As a result, the corresponding Fourier components of the potential energy V_G are identically zero, and consequently there is *no* bandgap at that particular zone boundary (see Problem 5.13).

An alternative view of the bandgap is that it is the energy necessary to create an electron and hole (i.e the absence of an electron in an otherwise filled band of states — see §6.2.2) at rest with respect to the lattice and sufficiently far apart that their mutual Coulomb attraction is negligible (i.e. exciton formation is precluded — see §5.8.3).

The energy bands for a 1D crystal in the NFE picture are illustrated in Fig. 5.17 in three different representations, namely the extended-, reduced- and repeated-zone schemes. The NFE band structure in the repeated-zone scheme (Fig. 5.17c) should be compared

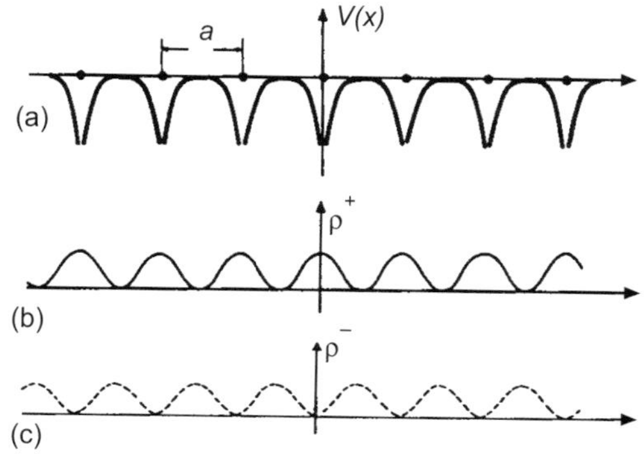

Fig. 5.16 Schematic illustration of the relationship between electron charge density in the NFE model and the positions of the ion cores in a 1D crystal. (a) A periodic array of potentials, with lattice constant a. (b) Charge density (proportional to $\psi^*\psi$) given by ρ^+. (c) Charge density given by ρ^-.

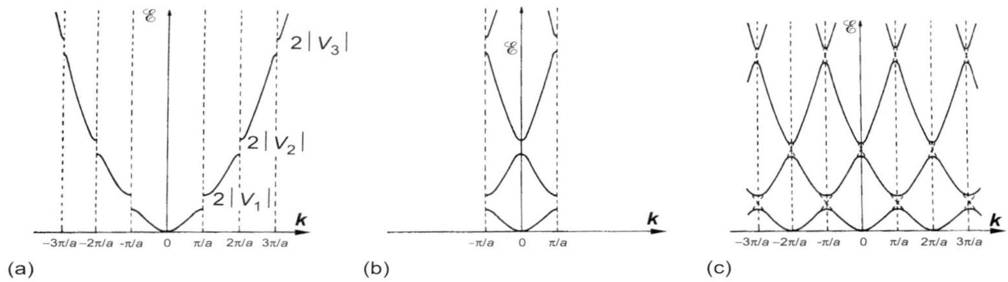

Fig. 5.17 Electronic band structure for a 1D crystal with periodicity a, in the nearly-free-electron approximation, represented in (a) the extended-zone scheme; (b) the reduced zone scheme; (c) the repeated-zone scheme. The magnitudes of the bandgaps at the zone boundaries are indicated in (a) in terms of the Fourier components of the periodic potential. The free-electron parabolas are shown in (c) by the dashed lines.

with Fig. 5.14, corresponding to the free-electron, empty-lattice model. The extended-zone representation has only a single band at any k-value (excepting the zone boundaries), and thus consists of the free-electron parabola punctuated by gaps at the zone boundaries. (The generation of the bands in the reduced-zone scheme (Fig. 5.17b) is particularly easy to understand in terms of translations by reciprocal-lattice vectors of the bands in the extended-zone scheme (Fig. 5.17a).) Of course, exactly the same information is contained in all three representations, but since the physically distinct solutions are confined to the first Brillouin zone (i.e. in the reduced-zone scheme), the repeated-zone representation therefore contains redundant information.

Analytic expressions for the behaviour of the bands in the vicinity of the bandgaps can be obtained by considering the algebraic form of the Schrödinger equation (eqn. (5.65)). At the boundaries of the first Brillouin zone, $\boldsymbol{k} = \pm\boldsymbol{G}/2$, eqn. (5.65) becomes for $\boldsymbol{k} = \boldsymbol{G}/2$

$$(\lambda - \mathscr{E})C_{G/2} + V_G C_{-G/2} = 0, \tag{5.77a}$$

and for $\boldsymbol{k} = -\boldsymbol{G}/2$

$$(\lambda - \mathscr{E})C_{-G/2} + V_{-G} C_{G/2} = 0, \tag{5.77b}$$

where $\lambda = \hbar^2(G/2)^2/2m_e$. Non-zero values for the coefficients C occur if the determinant is zero, i.e.

$$\begin{vmatrix} (\lambda - \mathscr{E}) & V_G \\ V_{-G} & (\lambda - \mathscr{E}) \end{vmatrix} = 0, \tag{5.78}$$

whence

$$(\lambda - \mathscr{E})^2 - |V_G|^2 = 0$$

or

$$\mathscr{E}^{\mp} = \lambda \pm |V_G| \equiv \frac{\hbar^2}{2m_e}(G/2)^2 \pm |V_G|. \tag{5.79}$$

Thus, the energy of one standing-wave solution, ψ^+, is lower by $|V_G|$, and that of the other solution, ψ^-, is higher by $|V_G|$, than the free-electron value: the magnitude of the bandgap is therefore

$$\mathscr{E}_g = \mathscr{E}^+ - \mathscr{E}^+ = 2|V_G|. \tag{5.80}$$

The ratio of the wavefunction coefficients is obtained from eqns. (5.77), i.e.

$$\begin{aligned}\frac{C_{G/2}}{C_{-G/2}} &= \frac{V_G}{(\mathscr{E} - \lambda)} \\ &= \pm 1\end{aligned} \tag{5.81}$$

from eqn. (5.79). This justifies the previous assumption in taking equal weightings of the propagating and backscattered electron waves (eqns. (5.75)).

An approximate solution for electron wavevectors close to the zone boundaries can be found by assuming that just *two* wavefunction coefficients, C_k and C_{k-G}, are significant; this is approximately true if the potential energy V_G is very small and much less than the kinetic energy $\mathscr{E}$. Then, from eqn. (5.62)

$$\psi_k(\boldsymbol{r}) = C_k \, e^{i\boldsymbol{k}.\boldsymbol{r}} + C_{k-G} \, e^{i(\boldsymbol{k}-\boldsymbol{G}).\boldsymbol{r}} \tag{5.82}$$

and the algebraic form of the Schrödinger equation (eqn. (5.65)) becomes

$$(\lambda_k - \mathscr{E})C_k + V_G C_{k-G} = 0 \tag{5.83a}$$

and

$$(\lambda_{k-G} - \mathscr{E})C_{k-G} + V_{-G} C_k = 0 \tag{5.83b}$$

with $\lambda_k = \hbar^2 k^2/2m_e$. Thus the secular determinant

$$\begin{vmatrix} (\lambda_k - \mathscr{E}) & V_G \\ V_{-G} & (\lambda_{k-G} - \mathscr{E}) \end{vmatrix} = 0 \tag{5.84}$$

gives the quadratic equation

$$\mathscr{E}^2 - \mathscr{E}(\lambda_k + \lambda_{k-G}) + \lambda_k \lambda_{k-G} - |V_G|^2 = 0. \tag{5.85}$$

This has the roots

$$\mathscr{E}_k^{\mp} = (\lambda_k + \lambda_{k-G})/2 \pm \left[(\lambda_k + \lambda_{k-G})^2/4 - \lambda_k \lambda_{k-G} + |V_G|^2\right]^{1/2}. \tag{5.86}$$

Note that it is only close to the Brillouin-zone boundaries that the NFE solution differs appreciably from that for the simple free-electron model.

5.2.4 Brillouin zones and energy bands

It is instructive to examine the electronic structure generated by the NFE approach, not just in terms of the band structure, $\mathscr{E}(\boldsymbol{k})$, but also in terms of constant-energy contours in $\boldsymbol{k}$-space with respect to the Brillouin zone. This is of particular importance in discussing the differences between metals and insulators (§5.2.5).

It is simplest at the outset to consider 2D lattices, and, again for simplicity, we consider the 2D square real-space lattice, the Brillouin-zone construction for which has already been given in Fig. 2.41. For the case of the free-electron model in the empty-lattice approximation, the contours of constant energy are just circles in $\boldsymbol{k}$-space (see Fig. 5.18). In the NFE approximation, the electron energy is lowered, below the free-electron value, for $\boldsymbol{k}$-values just below the zone boundary (see Fig. 5.17) and, as a result,

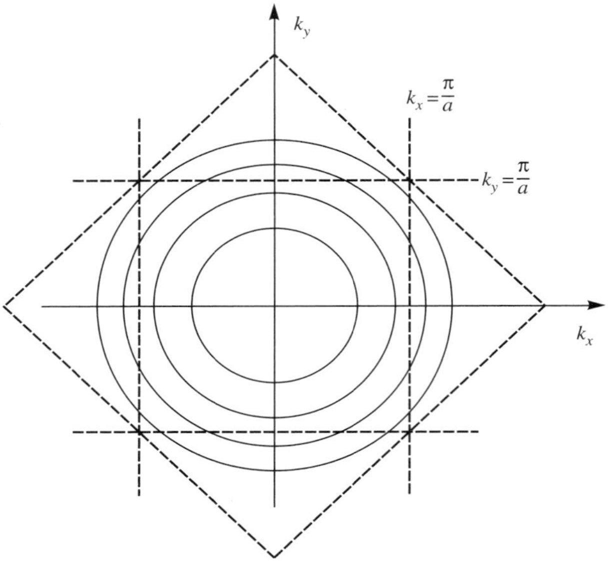

Fig. 5.18 Circular constant-energy contours (at equal energy intervals) for a free-electron gas in the empty-lattice approximation superimposed on the boundaries of the first two Brillouin zones of the 2D square real-space lattice.

the energy contours bend outwards from the free-electron circles towards the zone boundary. Likewise, the increase in energy for $\boldsymbol{k}$-values just above the zone boundary causes the constant-energy contours to fall below the free-electron circular contours towards the zone boundary. The perturbed NFE contours meet the zone boundaries at *right angles*. Since the solutions of the Schrödinger equation at the zone boundaries are *standing waves* (eqn. (5.75)), the electron group velocity, $\partial\omega/\partial k = (1/\hbar)\nabla_k \mathscr{E}$, must vanish there. The gradient of $\mathscr{E}$ in $\boldsymbol{k}$-space must therefore be *parallel* to the zone boundary, and consequently the constant-energy contour is normal to the boundary.

The NFE constant-energy contours for the 2D square lattice, superimposed on the boundaries of the first three, and part of the fourth, Brillouin zones (cf. Fig. 2.41) are shown in Fig. 5.19. The discontinuities in the energy contours at the zone boundaries correspond to the bandgaps in the band-structure ($\mathscr{E}(\boldsymbol{k})$) representation (Fig. 5.17). Figure 5.19 corresponds to the *extended-zone scheme* (cf. Fig. 5.17a). The energy contours can also be represented in the *reduced-zone scheme* (cf. Fig. 5.17b) by translating contours from zones higher than the first back into the first zone by means of appropriate reciprocal-lattice vectors. This is illustrated for the second zone in Fig. 5.20. Periodic continuation in $\boldsymbol{k}$-space of these first and second zones (the contours already lying in the first zone in Fig. 5.19 and those translated into it from the second zone in Fig. 5.20, respectively) generates the *repeated-zone representation* (Fig. 5.21).

In general, of course, the most important energy contour to consider is that corresponding to the *Fermi energy*, $\mathscr{E}_F$, since it is electrons having this energy that control most of the electronic behaviour. A 2D example, for a square lattice containing four

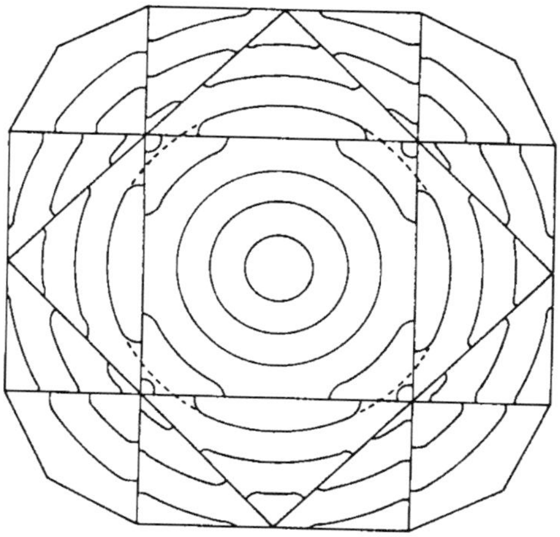

Fig. 5.19 Constant-energy contours for the NFE model applied to a 2D square lattice, superimposed on the boundaries of the first three, and part of the fourth, Brillouin zones in the extended-zone scheme. The dashed curves show one of the undistorted circular contours of the free-electron case.

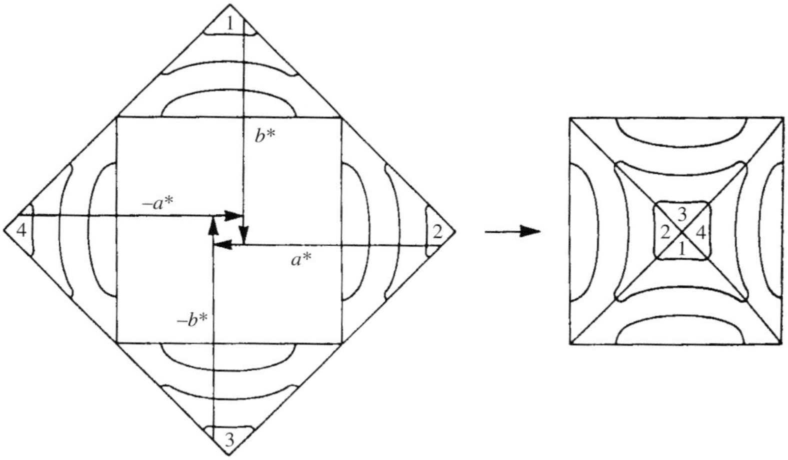

Fig. 5.20 Constant-energy contours from the second Brillouin zone in the extended-zone scheme (Fig. 5.19) represented in the reduced-zone scheme. The reciprocal-lattice vectors necessary to translate the segments from the second zone are indicated.

electrons per primitive unit cell in the free-electron approximation, is shown in Fig. 5.22a; the Fermi circle intersects four zones in this case. Those parts of higher zones that are occupied by electrons can be folded back into the first zone and periodically continued to generate repeated-zone representations, as in Figs. 5.22b–d. Note that there are two topologically distinct contours represented in Figs. 5.22b–d. In one,

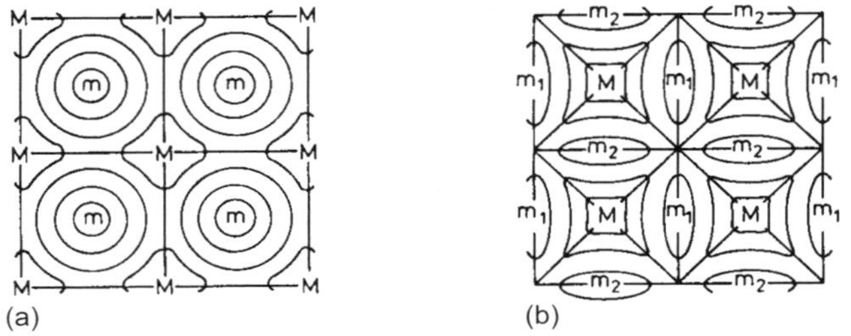

Fig. 5.21 Repeated-zone representation of the constant-energy contours generated from the reduced-zone plots. The labels M and m refer to energy maxima and minima, respectively. (a) The lowest energy band (central part of Fig. 5.19). (b) The second band (Fig. 5.20).

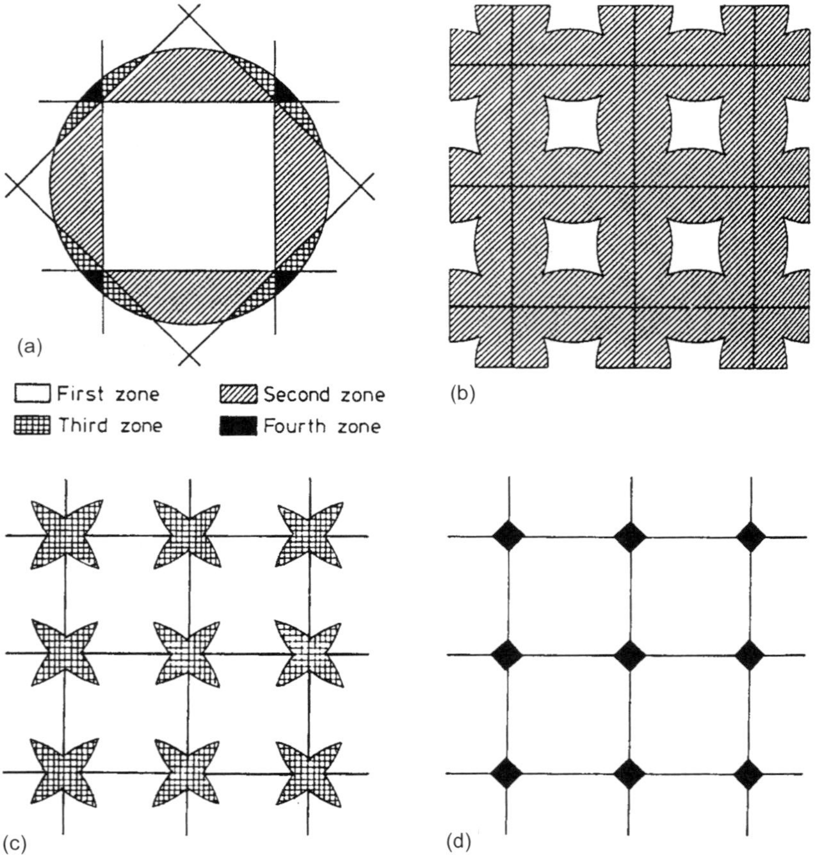

Fig. 5.22 Free-electron energy contour for the Fermi energy for a 2D square lattice with four electrons per primitive cell. (a) Fermi circle superimposed on the boundaries of the first four Brillouin zones. The parts of the various zones occupied by electrons (within the Fermi circle) are indicated. (b) The occupied part of the second zone in the repeated-zone scheme. (c) The occupied part of the third zone. (d) The occupied part of the fourth zone.

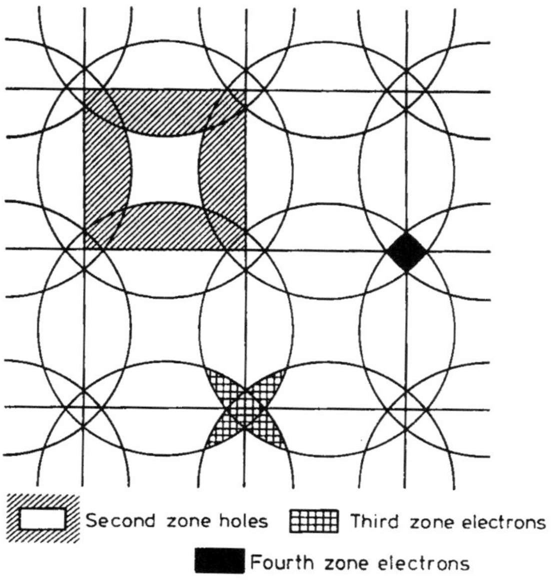

Fig. 5.23 The Harrison construction for generating the repeated-zone representation of the Fermi surface for a 2D square lattice containing four electrons per cell (see Fig. 5.22).

regions of ***k***-space devoid of electrons are entirely surrounded by regions filled with electrons (Fig. 5.22b): such empty regions are termed hole-like, for reasons that will become apparent later (see §6.2.2). In contrast, Figs. 5.22c and 5.22d exhibit regions filled with electrons entirely surrounded by regions devoid of electrons: these filled regions are called electron-like.

A simple geometric construction can be used to generate the repeated-zone patterns for the higher zones given in Figs. 5.22b–d. In the Harrison construction, circles in 2D (spheres in 3D) with radii equal to k_F are drawn centred on each point in the reciprocal lattice (i.e. at the centres of the first Brillouin zones in the repeated-zone scheme) as in Fig. 5.23. The various intersections of these circles (or spheres) generate the repeated-zone patterns according to the following rule: the Fermi surface in the nth Brillouin zone is the boundary separating regions covered by n circles (spheres) from those covered by $n-1$ circles (spheres). If the region covered by the large (n) number of circles (spheres) is outside this surface, it generates a hole-like surface, and if inside, an electron-like surface.

The very sharp geometric shapes of the Fermi surface in different zones shown in Fig. 5.22 result from the free-electron (empty-lattice) approximation. In the NFE case, the Fermi surface is distorted near the zone boundaries, with the effect that the cusps disappear, and for a sufficiently strong ion-core potential (opening up large gaps at the zone boundaries), the Fermi surface can completely disappear from particular zones.

A 3D analogue of the 2D case already considered is shown in Fig. 5.24 for the case of an f.c.c. crystal containing four electrons per primitive cell (e.g. Pb), for which the first Brillouin zone has the form of a truncated octahedron (§2.4.2). The Fermi surfaces in the

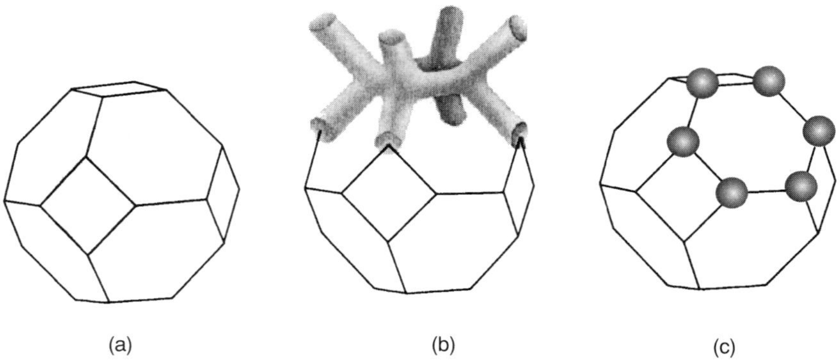

Fig. 5.24 The Fermi surface for an f.c.c. metal with four electrons per cell in the NFE approximation. (a) Second-zone hole-like surface. (b) Third-zone monster. (c) Four-zone electron-like surface.

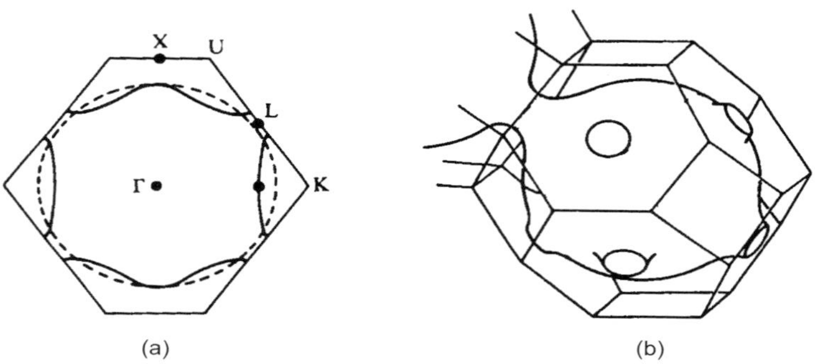

Fig. 5.25 The Fermi surface of Cu. (a) A (110) section through the first Brillouin zone. The dashed line represents the free-electron Fermi sphere. (b) The 3D Fermi surface inserted in the truncated octahedral Brillouin zone.

second and fourth zones are isolated and are hole-like and electron-like, respectively. However, the Fermi surface for the third zone is multiply connected in the repeated-zone scheme and cannot therefore be described as simply electron- or hole-like. Such topological forms for the Fermi surface are called, rather picturesquely, *monsters*.

For the monovalent alkali metals (Na, etc.), the Fermi wavevector is appreciably *less* than the shortest distance in reciprocal space to the first Brillouin-zone boundary (see Problem 5.1). Thus, NFE effects are negligible, and consequently the Fermi surface is simply a sphere lying within the first zone. Thus, such metals essentially behave like quantum free-electron systems. Cu also has one conduction electron per atom, and so the first Brillouin zone is also half-filled. However, k_F is rather close to the L-point in the first Brillouin zone (a truncated octahedron, since Cu has an f.c.c. structure), and so the electron energies in the $\langle 111 \rangle$ directions in $\boldsymbol{k}$-space are strongly perturbed, and 'necks' occur near the L-points (Fig. 5.25).

5.2.5 Metals versus insulators

We have noted previously that the simple free-electron model predicts that all materials with part-filled electron shells should be *metals*, since the resulting conduction-electron gas should be free to respond to an applied electric field at all temperatures. However, this picture is contrary to experience, since many materials are known to be electrical insulators, i.e. they have a zero electrical conductivity at zero kelvin. Although electronic transport properties will be discussed thoroughly in Chapter 6, it is useful, nonetheless, to mention briefly here how the NFE picture allows metals and insulators to be distinguished.

It is instructive to consider at the outset a 1D crystal, for which the band structure consists of a single band, with bandgaps at the zone boundaries, $\pm\pi/a$. Since, in general, the first Brillouin zone contains N states (where N is the number of atoms in the chain of length L), each of which can contain *two* electrons (because of the spin degeneracy), for the case of a monovalent chain (one conduction electron per atom) the zone is only half-filled with electrons, and $\mathscr{E}_F$ lies in the middle of the band (Fig. 5.26a). The intersection

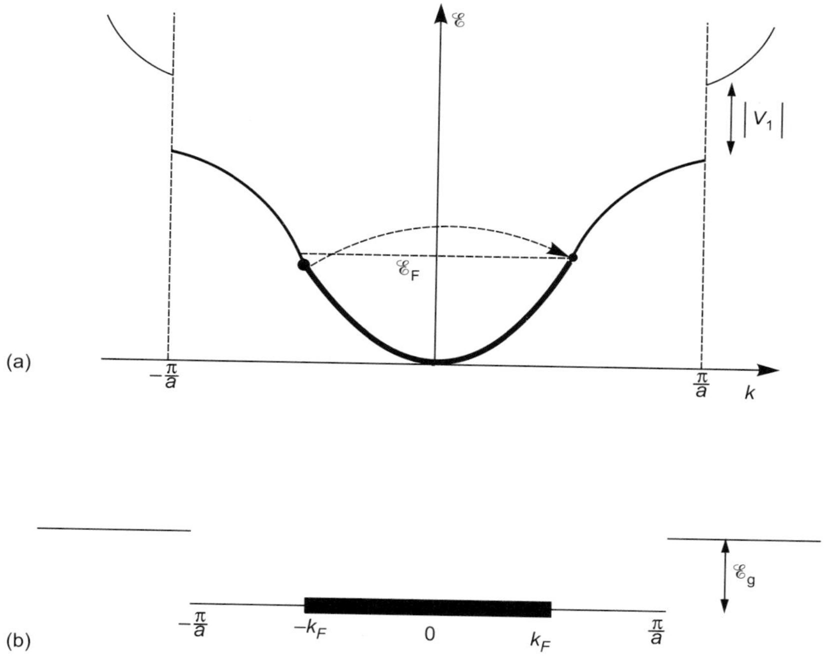

Fig. 5.26 (a) Band structure for a monovalent 1D crystal with periodicity a. Occupied states are shown by the bold curve. The system is metallic because the Fermi level lies in the middle of the band. A possible electronic transition, involving a change in momentum, and hence contributing to the finite electrical resistance, is indicated. (b) Schematic representation of the allowed $\boldsymbol{k}$-states for the 1D monatomic chain in the NFE approximation. There is an energy gap at the zone boundaries, $\pm\pi/a$. If $|k_F| < \pi/a$, the electron distribution (shown by the bold line) can be moved bodily in $\boldsymbol{k}$-space under an applied electric field and so gives a finite conductivity at $T = 0$ K; i.e. metallic behaviour is exhibited.

of the Fermi energy with the energy band is at a *point* in 1D, i.e. $\pm k_F$. The application of an external electric field accelerates free electrons, causing them to increase their momentum (and hence k) in the direction of the field. This process is the basis of metallic conduction (§6.3.2.1). Thus, in principle, the entire electron distribution (the line between $+k_F$ and $-k_F$ in Fig. 5.26b) is moved bodily in $\boldsymbol{k}$-space. However, this can only readily happen if there are *vacant* $\boldsymbol{k}$-states immediately above k_F available to be occupied. In the case of the half-filled zone (Fig. 5.26b), characteristic of the monatomic 1D crystal, evidently there are such states, and so the system should be a metal.

The situation is markedly different for a *diatomic* linear chain, however, since now the first zone is entirely filled with electrons ($k_F = \pm\pi/a$) and the higher zones remain empty at $T = 0$ K. Because there is a bandgap at the zone boundaries, there are now *no* available $\boldsymbol{k}$-states immediately above k_F at the same energy that can be occupied, and so (at $T = 0$ K) the application of an electric field can cause no shift in the $\boldsymbol{k}$-distribution of the electrons: the system is an insulator. At finite temperatures, thermal excitation of electrons across the bandgap can occur, leading to a finite conductivity: this is the basis of semiconduction.

This demarcation is clearest in the monovalent 1D case, since then there is only a *single* band at any general $\boldsymbol{k}$-value (excepting the zone boundaries): all states in one band are lower in energy than those in all other bands (Fig. 5.26a), and hence the lowest-energy band is filled preferentially at $T = 0$ K if, in a gedanken (thought) experiment, more electrons are progressively introduced into the system. However, this is *not* necessarily true for 2D or 3D crystals: a higher band may lie at a *lower* energy than a lower band, but at *different* places in $\boldsymbol{k}$-space (see Fig. 5.27). Thus, even if there are sufficient valence electrons available completely to fill the first zone, hence giving an

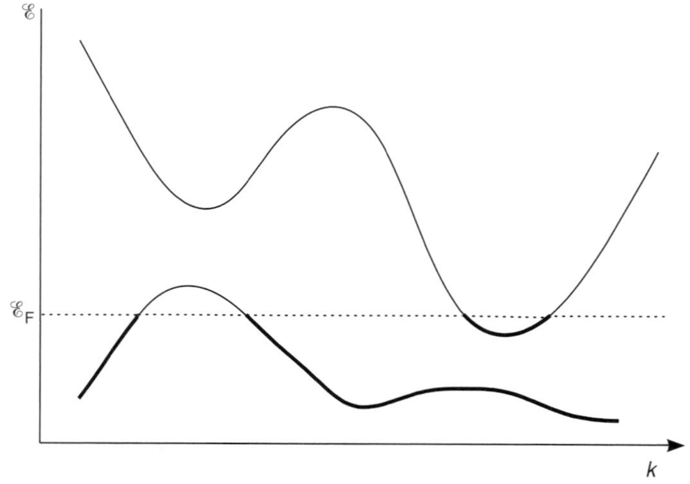

Fig. 5.27 Schematic illustration of the band overlap (at different points in $\boldsymbol{k}$-space) responsible for divalent elements being metallic. Although, in principle, enough electrons are present in order completely to fill the lower band, in practice electrons that would otherwise occupy the topmost states in the lower band instead occupy the lowest-energy states in the upper band. Thus, even at $T = 0$ K, the Fermi level lies in part-filled bands, and hence the material is a metal. Occupied states are shown as bold curves.

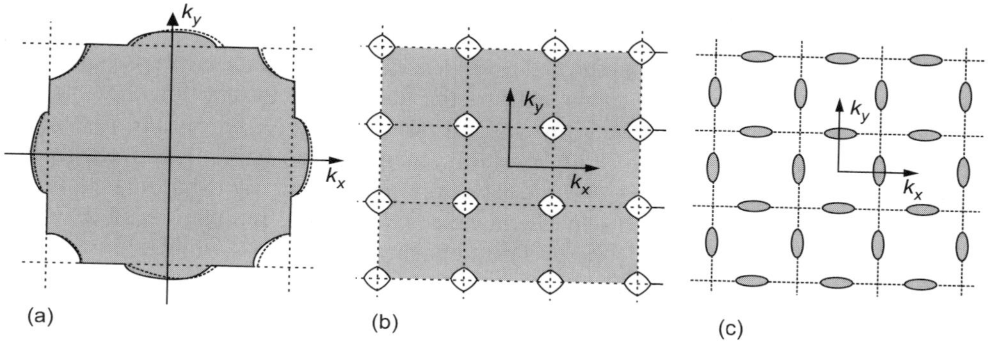

Fig. 5.28 (a) Band overlap at $T = 0$ K for a divalent element with a 2D square-lattice structure. The occupied states are shaded: it can be seen that there is a spill-over of electrons into the second zone. The displacement of that part of the electron distribution able to move under the action of an applied electric field is indicated by the dashed lines. Those parts having an interface at the zone boundaries are characterized by having an energy gap and so cannot move. (b) Filled states in the first zone represented in the repeated- zone scheme, revealing hole-like pockets. (c) Filled states in the second zone in the repeated-zone scheme forming electron-like pockets.

insulator in principle, nevertheless such materials, e.g. the divalent Group IIA elements, Ca, etc.) are *metallic*. Even at $T = 0$ K, the lowest band is not completely filled, but its top most states remain empty with the electrons populating the lowest-energy states in the upper band (Fig. 5.27). Such a material is termed a semi-metal. This behaviour is also indicated, for a 2D square-lattice example, in terms of the filling of Brillouin zones, in Fig. 5.28: it can be seen that electrons 'spill-over' from the first zone into the second (Fig. 5.28a), resulting in hole-like pockets in the first zone (Fig. 5.28b) and electron-like pockets in the second zone (Fig. 5.28c), when represented in the repeated-zone scheme. Similar considerations also hold for 3D crystals.

Thus, a metal can be defined succinctly as a material that has a Fermi surface (provided that the density of delocalized electron states is non-zero at $\mathscr{E}_F$). In other words, if there are delocalized electron states at the Fermi energy (i.e. $\mathscr{E}_F$ lies in a part-filled band of delocalized states), the material will exhibit metallic behaviour.

An insulator can be defined as a material for which an energy gap occurs between (at $T = 0$ K) a completely filled lower band, the valence band, and a higher, empty band, the conduction band, and where the lower band is lower in energy than the higher band for *all* $\boldsymbol{k}$-values.

A semiconductor is an insulator with a relatively small bandgap (say less than 3 eV), so that thermal excitation of electrons across the gap from the filled valence band to the empty conduction band can occur with a reasonable probability at finite temperatures.

It should be noted that, in this picture, insulators are non-conducting, not because the electrons are spatially localized in bonds (the wavefunctions are still delocalized throughout the crystal), but because when a band is full there are no further states available at the same energy to which an electron can be accelerated by an electric field.

5.2.6 Pseudopotentials

Although the nearly-free-electron model provides a correct qualitative explanation for the difference between metals and insulators, the underlying assumption that the ion-core potentials be very weak is rarely valid. It is certainly inappropriate for tightly bound electrons, e.g. d-electrons in transition metals. Two general approaches for calculating electronic structures are the pseudopotential and tight-binding methods. The pseudopotential approach, which can be regarded as a refinement of the NFE model, will be described briefly here. The tight-binding model, applicable to tightly bound electrons for example, will be discussed in §5.3.1.

The pseudopotential is the *effective* potential associated with the ion cores that is experienced by the conduction electrons. It is different from the true Coulombic potentials of the ion cores for two reasons (see §5.1.3.2): (i) cancellation of the attractive part of the potential in the vicinity of the ion cores by an effectively repulsive component, resulting from the requirement that the wavefunction of the conduction electrons be orthogonal to those of the core electrons, the basis of the orthogonal-plane-wave (OPW) approximation (the conduction electrons are essentially excluded from the core region by the Pauli exclusion principle); (ii) screening by the conduction-electron gas of

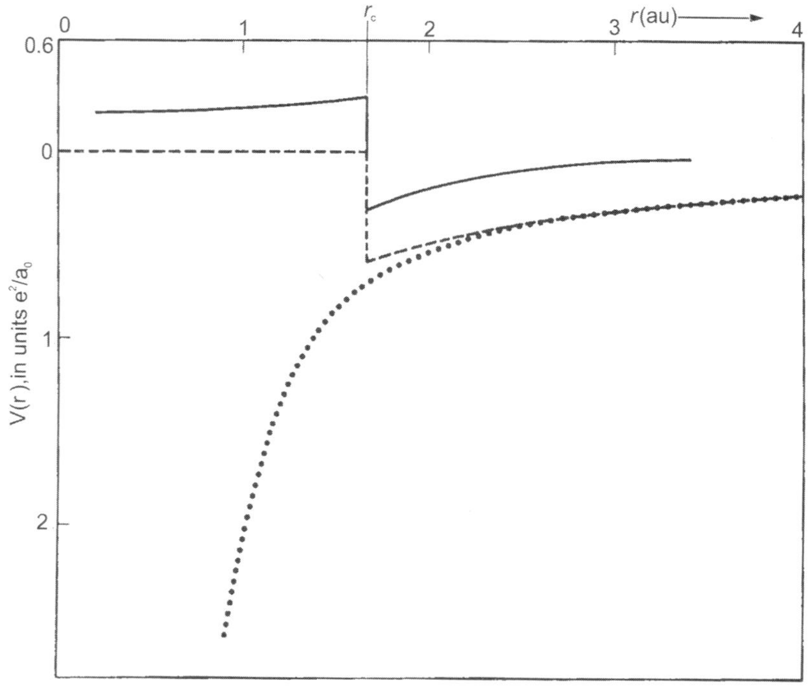

Fig. 5.29 Pseudopotentials for Na in real space. The dashed curve is the unscreened Ashcroft empty-core pseudopotential, with $r_c = 1.66$au (1au = a_0, the Bohr radius), and the solid curve is the (Thomas–Fermi) screened pseudopotential with screening parameter a_0/λ_{TF}. The dotted curve is the true Coulombic potential of the ion core. (Kittel (1996). Reproduced by permission of John Wiley & Sons Inc.)

the ionic charge in the region far from the ion cores. The Ashcroft empty-core pseudopotential assumes *complete* cancellation of the potential in the core region ($r < r_c$) — see eqn. (5.42); this unscreened pseudopotential is shown is Fig. 5.29, together with the screened version. (For a fuller discussion of screening, see §5.6.1.) It can be seen that both pseudopotentials are appreciably smaller in magnitude than the bare Coulombic potential in the region of the ion core. Note also that the screened pseudopotential is *repulsive* in the core region ($r < r_c$).

The general Fourier components of the Ashcroft pseudopotential (eqn. (5.42)) can be found by using eqn. (2.58). The result is

$$V_q = \frac{-Ze^2 \cos qr_c}{\varepsilon_0 \Omega q^2}, \tag{5.87}$$

where Ω is the volume per atom. This function has a first node given by $\cos qr_c = 0$, i.e. at

$$q_0 = \frac{\pi}{2r_c}. \tag{5.88}$$

The reciprocal-space form of the pseudopotential is shown in Fig. 5.30 for the case of f.c.c. Al (Heine and Abarenkov (1964)). In practice, values of the pseudopotential are needed only at reciprocal-*lattice* positions, i.e. $\boldsymbol{q} = \boldsymbol{G}$ (cf. eqn. (5.61)). Note that the pseudopotential is small but positive for such values for Al (Fig. 5.30). The $q = 0$ value of the pseudopotential tends to the *screened-ion limit*, $V_0 = -2\mathscr{E}_F/3$ (see Problem 5.24).

In the empirical pseudopotential method (EPM), values of V_G are estimated by fitting the results of calculated electronic band structures to experimental data, for example optical data (§5.8.2). It is often reasonable to assume that such EPM values are simply *additive* for different types of atom making up a solid, and so the band structures for new materials can be predicted. A more reliable approach, however, is to use *ab initio* methods, e.g. based on the local-density approximation for dealing with correlation effects (§2.5.3.4), to obtain the pseudopotential.

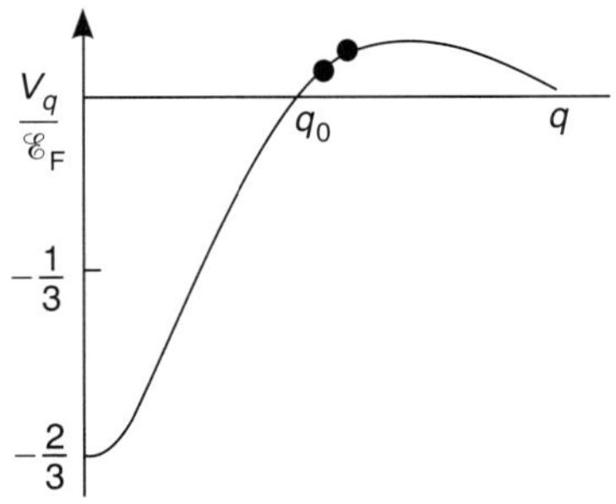

Fig. 5.30 Pseudopotential for f.c.c. Al, (Heine and Abarenkov (1964)), plotted in reciprocal space and normalized by the Fermi energy. The first node occurs at q_0. The values of the pseudopotential at the two reciprocal-lattice values used to calculate the electronic structure are marked with dots: they correspond to $\boldsymbol{G} = (111)$ with magnitude $\sqrt{3}(2\pi/a)$, and (200) with magnitude $2(2\pi/a)$, being the L- and X-points, respectively, in the truncated octahedral Brillouin zone. (After Pettifor (1995), *Bonding and Structure of Molecules and Solids*. Reproduced by permission of Oxford University Press)

The simple picture of a pseudopotential being very weak in the core region in real space (Fig. 5.29) or at reciprocal-lattice values in reciprocal space (Fig. 5.30) is only valid if there are electron states in the ion core available to be orthogonal to equivalent valence-electron states. Thus, for example Na, Mg, Al, Ga and In satisfy this condition (i.e. the NFE approximation is valid), and the electronic density of states is closely free-electron-like ($g(\mathscr{E}) \propto \mathscr{E}^{1/2}$, cf. eqn. (5.15)) — see Fig. 5.31. On the other hand, it can be seen that significant deviations from the NFE picture occur for Li and Be. Since they are first-row elements, they have *no* p-electrons in the core, and so valence-electron states that have p-like symmetry (i.e. with nodes at lattice sites, as for the charge distribution shown in Fig. 5.16c) will suffer large shifts in energy from free-electron values. Indeed, a gap is nearly opened up at $\mathscr{E}_F$ in the case of Be (Fig. 5.31). Third- and fourth-row elements in Groups IA, IIA and IIB also exhibit marked departures from free-electron-like behaviour arising, in this case, from the presence of d-electrons in the core.

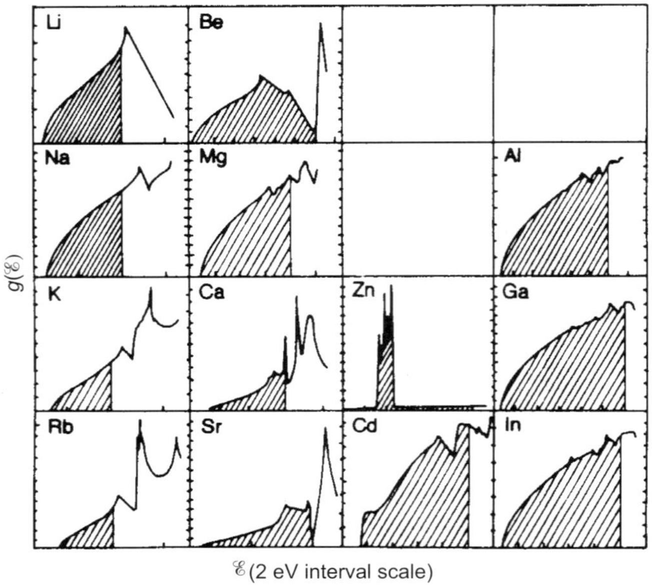

Fig. 5.31 Electronic densities of states of metals with s- and p-electron valence states. The shaded area corresponds to filled states at and below $\mathscr{E}_F$ at $T = 0$ K. (After Pettifor (1995) *Bonding and Structure of Molecules and Solids*, by permission of Oxford University Press)

Electrons in bonds

5.3

In many cases, it is not helpful to regard electrons as forming a (nearly) free-electron gas. It is obviously inappropriate as a model to describe the energetically deep-lying and spatially localized core electrons that have filled electron shells, and it is also inapplicable to electrons that might lie in energy close to the Fermi level but which, nevertheless, are tightly bound to the ion cores (e.g. d-electrons and f-electrons in transition metals and rare earths, respectively). The NFE model is also not appropriate for insulating materials where the terms in the pseudopotential are strong enough to give rise to sizeable bandgaps and for which, therefore, the use of only a very few plane waves (e.g. just two, as in eqn. (5.82)), forming the basis of the NFE approximation, is invalid.

It is also not appropriate for those materials where *hybridization* of atomic orbitals takes place, to give spatially directed hybrid orbitals which can combine to form chemical (covalent) bonds (e.g. as in the Group IV elements). In these latter cases, an approach more akin to that used by chemists in describing bonding in molecules can be more fruitful (§5.3.2). However, this purely local picture of chemical bonds must be modified somewhat when considering the solid state, since for the case of crystals, for example, the presence of translational periodicity means that the electronic wavefunctions involved must satisfy Bloch's theorem (eqns. (5.58) and (5.59)), and a particular electron has an equal probability of being found in any cell anywhere in the crystal. The chemical-bond picture is perhaps most closely realized in the case of amorphous insulators, where the structural disorder can cause the electron states to be spatially localized in hybrid orbitals (see §6.6).

We begin this section by discussing the tight-binding approximation, which can be regarded as representing the extreme opposite case to the free-electron picture.

5.3.1 Tight-binding (LCAO) approximation

The tight-binding approximation (TBA) takes as its starting point the *isolated atom*, for which the electron eigenstates correspond to a series of discrete energy levels. If now a monatomic solid is considered to be progressively formed by the bringing together of a large number of identical atoms, the atomic-like energy levels will be split due to the interactions between atoms, as in molecular orbitals in molecules, corresponding to bonding and antibonding combinations of wavefunctions. However, because of the very large number of atoms present in a typical solid sample, the electron states will therefore form a quasi-continuous band of states with respect to energy (Fig. 5.32). The closer together the atoms are forced, the greater will be the strength of the interactions and hence the broader the bands will be. Thus, electron bands in solids can also result from the overlap with, and interactions between, atomic orbitals (the tight-binding (TB) limit) as well as from (nearly) free-electron-like behaviour.

It is assumed at the outset that the electron wavefunction $\phi_i(\boldsymbol{r})$ corresponding to the ith discrete energy level, $\mathscr{E}_i$, is known from a solution of the Schrödinger equation for the isolated atom, viz.

$$\mathscr{H}_{\mathrm{A}}(\boldsymbol{r}-\boldsymbol{R}_n)\phi_i(\boldsymbol{r}-\boldsymbol{R}_n) = \mathscr{E}_i\phi_i(\boldsymbol{r}-\boldsymbol{R}_n). \tag{5.89}$$

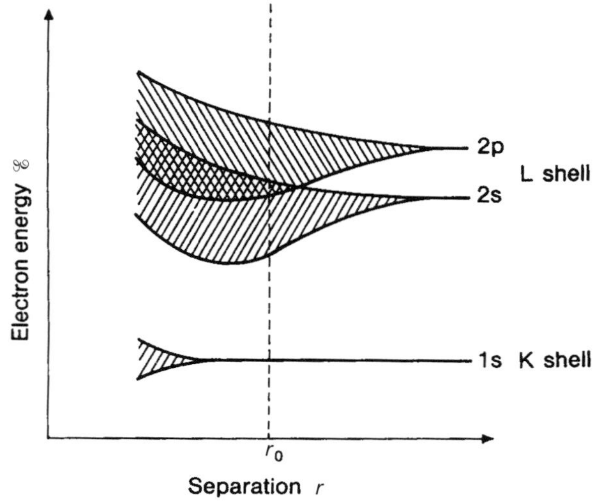

Fig. 5.32 Schematic illustration of band formation in solids from discrete atomic electron energy levels as atoms are progressively forced closer together.

where, for example, $\boldsymbol{R}_n$ is the translation vector for the nth site in a lattice (eqn. (2.1)). The basic assumption in the TBA is that the overlap between neighbouring atomic wavefunctions is very small, so that the extra potential experienced by an electron in a solid is only a small perturbation to the potential characteristic of isolated atoms. Thus, the Hamiltonian in a solid can be written approximately as

$$\mathscr{H} \simeq \mathscr{H}_{\mathrm{A}} + v = \frac{-\hbar^2}{2m_{\mathrm{e}}}\nabla^2 + V_{\mathrm{A}}(\boldsymbol{r} - \boldsymbol{R}_n) + v(\boldsymbol{r} - \boldsymbol{R}_n), \tag{5.90}$$

where the perturbation $v(\boldsymbol{r} - \boldsymbol{R}_n)$ can be approximated as a sum over the atomic potentials for all sites i *apart from* that (n) at which the electron is localized, i.e.

$$v(\boldsymbol{r} - \boldsymbol{R}_n) \simeq \sum_{j \neq n} V_{\mathrm{A}}(\boldsymbol{r} - \boldsymbol{R}_j). \tag{5.91}$$

This is illustrated schematically in Fig. 5.33.

The eigenstates ψ_{ki} of the Hamiltonian (eqn. (5.90)) for eigenvalues $\mathscr{E}_i$ must be Bloch states if the solid is translationally periodic. A good approximation to the true wavefunction is to invoke the linear combination of atomic orbitals (LCAO) approximation, viz.

$$\psi_{ki} \simeq \Phi_{ki} = \sum_n a_{kn}\phi_i(\boldsymbol{r} - \boldsymbol{R}_n). \tag{5.92}$$

The expansion coefficients must comply with translational periodicity (i.e. generate Bloch functions) and contain the wavefunction normalization constant. In order to satisfy both these conditions, $a_{kn} = \exp(\mathrm{i}\boldsymbol{k}\cdot\boldsymbol{R}_n)/N^{1/2}$, where N is the number of atoms in the crystal (see Problem 5.14). Thus, the wavefunction has the form:

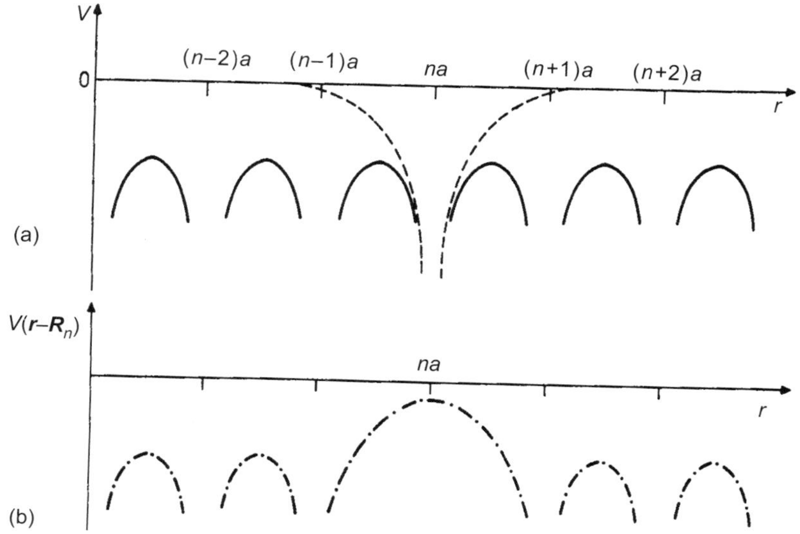

Fig. 5.33 Schematic illustration of potentials experienced by an electron in a 1D solid. (a) The lattice potential (solid line) obtained by summing the potentials V_A of free atoms (dashed line). (b) The perturbation potential $v(\boldsymbol{r}-\boldsymbol{R}_n)=\sum_{i\neq n} V_A(\boldsymbol{r}-\boldsymbol{R}_i)$ used in the tight-binding model.

$$\Phi_{ki} = N^{-1/2}\sum_n \mathrm{e}^{\mathrm{i}\boldsymbol{k}.\boldsymbol{R}_n}\phi_i(\boldsymbol{r}-\boldsymbol{R}_n). \tag{5.93}$$

In general, Bloch functions for any band can always be written in the form of eqn. (5.93), where the functions $\phi(\boldsymbol{r}-\boldsymbol{R}_n)$ are known as Wannier functions. These have the advantage, *not* shared by the atomic orbitals of the TBA, that they are mutually orthogonal.

It is instructive to examine the form of the tight-binding wavefunction (eqn. (5.93)), for example for a particular atomic orbital, say $i = 3\mathrm{s}$, for several $\boldsymbol{k}$-values. This is shown schematically in Fig. 5.34 for the case of a 1D crystal for simplicity. For the state at the centre of the Brillouin zone, $\boldsymbol{k}=0$, the tight-binding wavefunction $\Phi_{\boldsymbol{k}=0,\mathrm{s}}$ corresponds to 3s atomic wavefunctions centred at each lattice site, all with the *same* amplitude and phase. (In other words, the wave corresponding to the state $\boldsymbol{k}=0$ has an infinite wavelength.) At the zone-boundary value, $\boldsymbol{k}=\pi/a$, on the other hand, the wavefunction amplitudes are still equal at each lattice point but the phases alternate since, from eqn. (5.93), $\exp(\mathrm{i}\pi)=-1$. At an arbitrary $\boldsymbol{k}$-value, the amplitudes and phases of the 3s-like wavefunctions at each site are such as to generate an electron wave with a wavelength lying between $2a$ and ∞.

An estimate for the electron energy (higher than the true energy $\mathscr{E}_i$) can be found by using the trial wavefunction Φ_{ki} (eqn. (5.93)) to form the expectation value of the Hamiltonian, i.e.

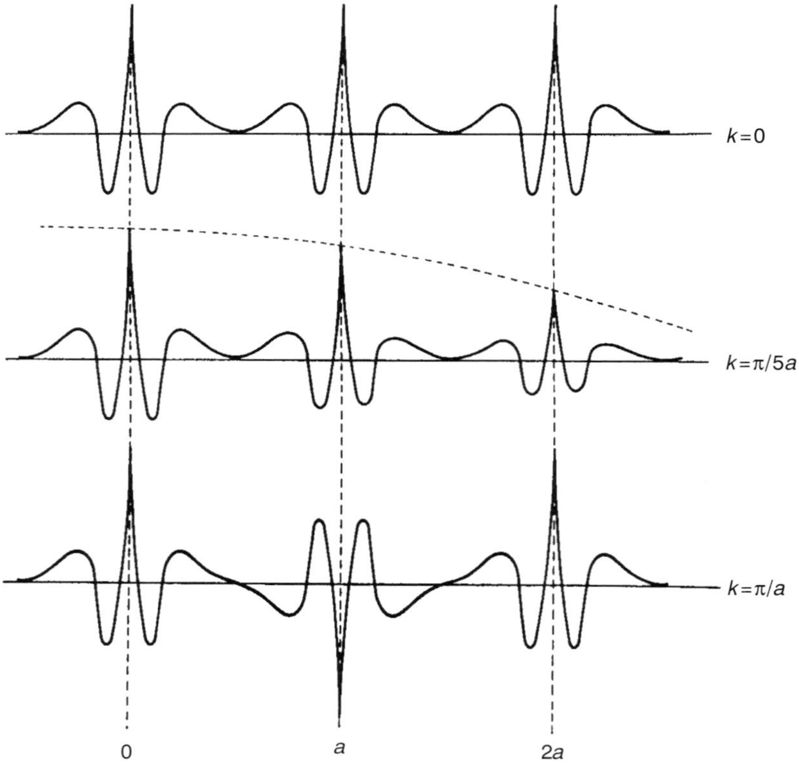

Fig. 5.34 Tight-binding wavefunction for a 1D crystal for values of k including the zone-centre and Brillouin-zone-boundary values. The atomic orbital chosen is a 3s wavefunction.

$$\begin{aligned}\mathscr{E}(\boldsymbol{k}) &= \int \Phi^*_{k,i}\, \mathscr{H}\, \Phi_{k,i}\,\mathrm{d}\boldsymbol{r} \\ &\simeq \frac{1}{N}\sum_{n,m} \mathrm{e}^{\mathrm{i}\boldsymbol{k}.(\boldsymbol{R}_n-\boldsymbol{R}_m)} \int \phi_i^*(\boldsymbol{r}-\boldsymbol{R}_m)[\mathscr{E}_i + v(\boldsymbol{r}-\boldsymbol{R}_n)]\phi_i(\boldsymbol{r}-\boldsymbol{R}_n)\mathrm{d}\boldsymbol{r}.\end{aligned} \tag{5.94}$$

Here, it is assumed that, for the term involving $\mathscr{E}_i$, the overlap between nearest neighbours may be neglected (i.e. $n = m$), and for the term involving the perturbation $v(\boldsymbol{r} - \boldsymbol{R}_n)$, only nearest-neighbour interactions are included. Thus, the tight-binding expression for the electron energy may be written as

$$\mathscr{E}(\boldsymbol{k}) = \mathscr{E}_i - \alpha_i - \beta_i \sum_m \mathrm{e}^{\mathrm{i}\boldsymbol{k}.(\boldsymbol{R}_n-\boldsymbol{R}_m)}, \tag{5.95}$$

where m and n are nearest neighbours, and where the coefficients α_i and β_i are given by

$$\alpha_i = -\int \phi_i^*(\boldsymbol{r}-\boldsymbol{R}_n)v(\boldsymbol{r}-\boldsymbol{R}_n)\phi_i(\boldsymbol{r}-\boldsymbol{R}_n)\mathrm{d}\boldsymbol{r}, \tag{5.96a}$$

$$\beta_i = -\int \phi_i^*(\boldsymbol{r}-\boldsymbol{R}_m)v(\boldsymbol{r}-\boldsymbol{R}_n)\phi_i(\boldsymbol{r}-\boldsymbol{R}_n)\mathrm{d}\boldsymbol{r}, \tag{5.96b}$$

for the simple case where the atomic wavefunction ϕ_i has spherical symmetry, i.e. is an s-state. The quantity α_i is positive since v is negative. In this case, the overlap term is sometimes denoted as $\beta_i = -\text{ss}\sigma(\boldsymbol{R}_n) > 0$, indicating that the overlap is in the form of a σ-bond between s-orbitals on neighbouring sites. It is also termed the hopping integral.

For the example of a simple cubic lattice, for which the six nearest neighbours of a given atom are at distances

$$\boldsymbol{R}_n - \boldsymbol{R}_m = (\pm a, 0, 0); (0, \pm a, 0); (0, 0, \pm a), \tag{5.97}$$

eqn. (5.95) reduces to the expression:

$$\mathscr{E}(k) = \mathscr{E}_i - \alpha_i - 2\beta_i(\cos k_x a + \cos k_y a + \cos k_z a). \tag{5.98}$$

These energy bands are illustrated in Fig. (5.35). Note that the bandwidth in this case is $12\beta_i$: the minimum energy, $-6\beta_i$, occurs at the zone-centre point $\Gamma(k = 0)$ and the maximum energy, $+6\beta_i$, at the Brillouin-zone-boundary point, M ($\boldsymbol{k} = \pi/a, \pi/a, \pi/a$). The states at the bottom of the band (Γ) are pure bonding states and those at the top (M) are pure antibonding states, as can be seen by examination of eqn. (5.93). This can also be seen pictorially in the case of the 2D analogue, the square lattice (Fig. 5.36). In general, the tight-binding band width $\Delta\mathscr{E}_i$ depends on the nearest-neighbour coordination number z and the wavefunction-overlap term β_i (see also Problem 5.15) via

$$\Delta\mathscr{E}_i = 2z\beta_i. \tag{5.99}$$

Note that such widths are much smaller than those of NFE-like bands.

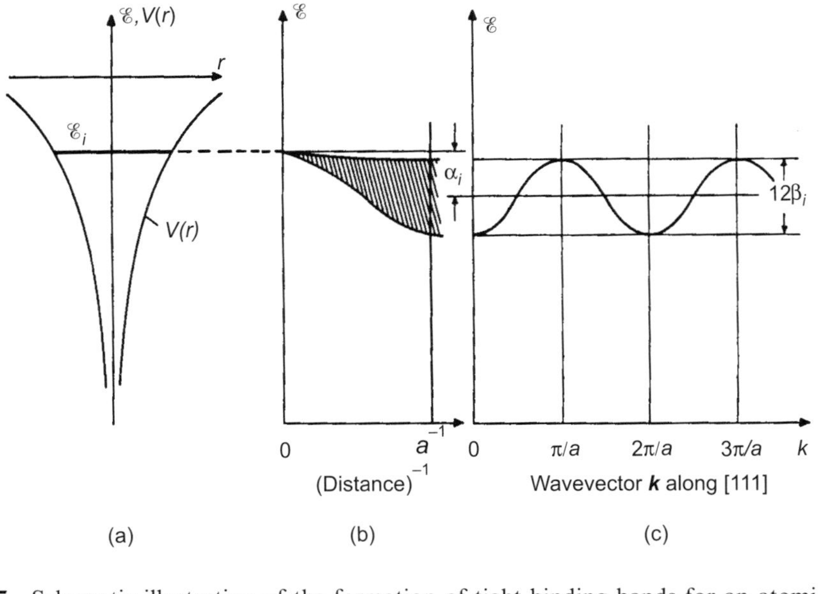

Fig. 5.35 Schematic illustration of the formation of tight-binding bands for an atomic energy level for a simple cubic lattice with lattice parameter a. (a) Free-atom potential energy with, superimposed, an electron energy level $\mathscr{E}_i$. (b) Representation of the dependence of the tight-binding bandwidth with inverse distance (or, equivalently, orbital overlap). (c) Tight-binding band structure for the $\boldsymbol{k}$-direction [111].

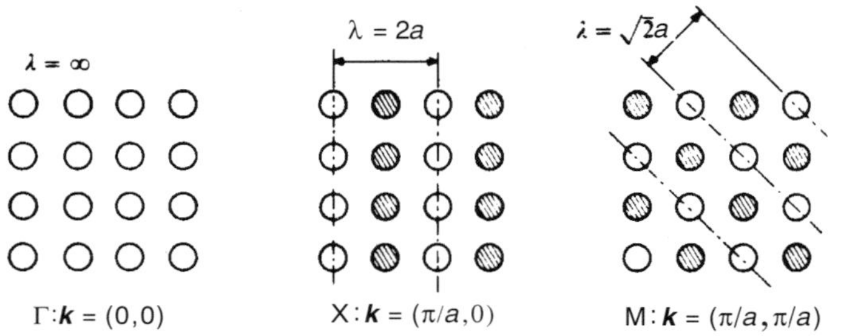

Fig. 5.36 Relative phases of s-orbital coefficients a_{kn} of Bloch functions in a tight-binding treatment for a 2D square lattice for different $\boldsymbol{k}$- values. Open circles represent positive values and shaded circles negative values (of the same magnitude) of orbital coefficients. The wavefronts of the corresponding electron waves are indicated. Note that for the Γ-point, nearest-neighbour combinations are all bonding, whereas they are all antibonding for M. (After Cox (1987), *Electronic Structure and Chemistry of Solids*, by permission of Oxford University Press)

For small k-values, the cosine terms in eqn. (5.98) may be expanded to second order, giving

$$\mathscr{E}(k) \simeq \mathscr{E}_i - \alpha_i - 6\beta_i + \beta k^2 a^2, \tag{5.100}$$

where $k^2 = k_x^2 + k_y^2 + k_z^2$. Therefore, the $\mathscr{E}(k)$ surface is spherically symmetric near $k = 0$, and the parabolic dependence on k is the same as that for the NFE model (cf. eqn. (5.86)). The TB s-band structure for a simple cubic lattice is shown in Fig. 5.37a for two directions in $\boldsymbol{k}$-space. The corresponding electronic density of states is shown in Fig. 5.37b: note the van Hove singularities, where $\nabla_k \mathscr{E} = 0$, resulting from the flat bands at the Brillouin-zone boundaries in the $(k, 0, 0)$ and $(k, k, 0)$ directions.

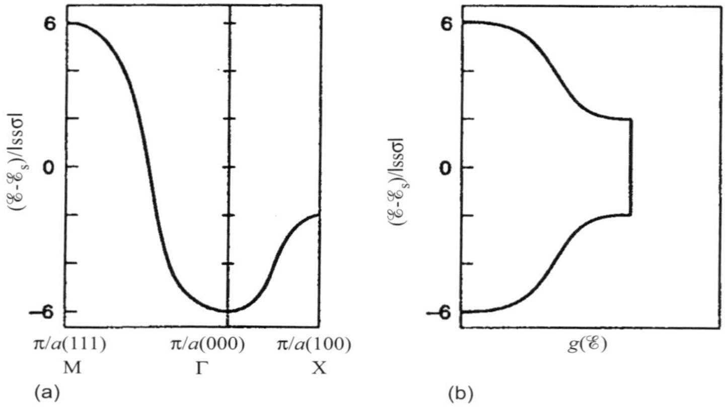

Fig. 5.37 Tight-binding results for s-states on a simple cubic lattice: (a) Band structure in the [100] and [111] directions in $\boldsymbol{k}$-space. (b) Density of states.

A TB treatment of other types of atomic orbitals, e.g. p- or d-states, is similar to that described above for s-states, except that the orbital degeneracy inherent in such atomic states results in a 3×3 secular problem for p-states or a 5×5 secular problem for d-states. A qualitative understanding of the effect of including, say, p-orbitals in a TB analysis can be obtained from a generalization of Fig. 5.36, where now p_x or p_y orbitals are placed on sites in a 2D square lattice and the combinations of orbital coefficients can be examined for various $\boldsymbol{k}$-values (Fig. 5.38). Now, both σ and π interactions between nearest neighbours need to be considered. Thus, at Γ, the two Bloch-wave states are degenerate, with all σ-antibonding and π-bonding interactions occurring between nearest neighbours. Likewise, at the M-point, the two states are degenerate, but now the interactions are all σ-bonding and π-antibonding between nearest neighbours. However, for the X-point, the two states split in energy, the lowest-energy configuration consisting of all σ-bonding and π-bonding nearest-neighbour interactions, and the upper level corresponds to all σ-antibonding and π-antibonding interactions. The corresponding band structure is shown in Fig. 5.39. ★

The three degenerate atomic p-orbitals may be written as $\phi_i = x\zeta(r), y\zeta(r)$ and $z\zeta(r)$ where $\zeta(r)$ is a spherically symmetric function. Thus, the tight-binding wavefunction (eqn. (5.93)) now takes the form:

$$\Phi_{k,\mathrm{p}} = N^{-1/2} \sum_{\alpha=x,y,z} C_\alpha \sum_n \mathrm{e}^{\mathrm{i}\boldsymbol{k}.\boldsymbol{R}_n} \phi_\alpha(\boldsymbol{r} - \boldsymbol{R}_n). \tag{5.101}$$

Substitution of this into the Schrödinger equation leads to the 3×3 secular determinant for the TB energies, viz.

$$\left|(\mathscr{E}_\mathrm{p} - \mathscr{E}(\boldsymbol{k}))\delta_{\alpha\alpha'} + T_{\alpha\alpha'}\right| = 0, \tag{5.102}$$

where $\delta_{\alpha\alpha'}$ is the Kronecker delta and the matrix elements are given by

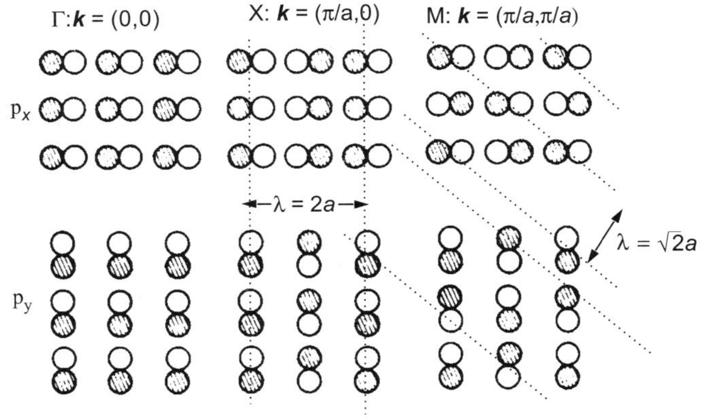

Fig. 5.38 Relative phases of p-orbital coefficients of Bloch functions in a tight-binding treatment for a 2D square lattice for different $\boldsymbol{k}$-values. Open circles of lobes indicate positive values, and hatched circles negative values, of orbital coefficients.

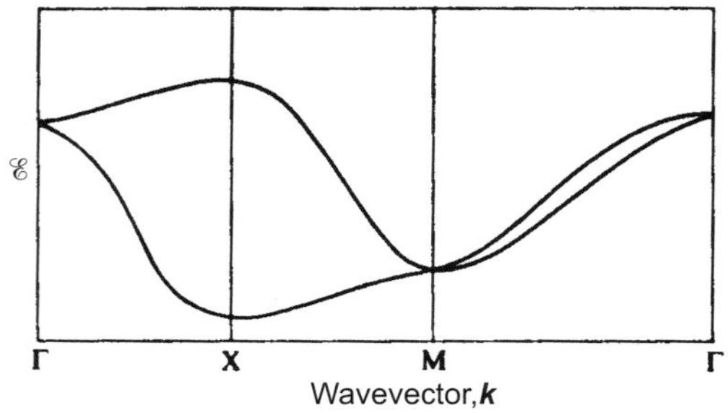

Fig. 5.39 Band structure for a 2D square lattice with p_x and p_y orbitals at each site for different directions in $\boldsymbol{k}$-space.

$$T_{\alpha\alpha'} = \sum_n e^{i\boldsymbol{k}.\boldsymbol{R}_n} \int \phi_\alpha^*(\boldsymbol{r}) v(\boldsymbol{r}) \phi'_\alpha(\boldsymbol{r} - \boldsymbol{R}_n) d\boldsymbol{r}. \tag{5.103}$$

The overlap integrals now depend on the *direction* $\hat{\boldsymbol{R}}_n = (l, m, n)$ (where l, m, n are direction cosines) as well as the distance R_n , and are given by (Slater and Koster (1954)):

$$\int \phi_x^* v \phi_x d\boldsymbol{r} = l^2 \, \mathrm{pp}\,\sigma + (1 - l^2) \, \mathrm{pp}\,\pi, \tag{5.104a}$$

$$\int \phi_x^* v \phi_y d\boldsymbol{r} = lm \, \mathrm{pp}\,\sigma - lm \, \mathrm{pp}\,\pi, \tag{5.104b}$$

$$\int \phi_x^* v \phi_z d\boldsymbol{r} = ln \, \mathrm{pp}\,\sigma - ln \, \mathrm{pp}\,\pi, \tag{5.104c}$$

where pp σ and pp π are the overlap (hopping) integrals for σ and π bonding, respectively, between p-orbitals on nearest-neighbour sites (Fig. 5.40). The ratios of the various overlap integrals for the case of materials (e.g. Si, Ge) with s and p valence electrons was found by Harrison (1980) to be

$$\mathrm{pp}\,\sigma : \ \mathrm{pp}\,\pi : \ \mathrm{sp}\,\sigma : \ \mathrm{ss}\,\sigma = 2.31 : \ -0.58 : \ 1.31 : \ -1.00. \tag{5.105}$$

For a simple-cubic lattice, the diagonal matrix elements are given by

$$T_{xx} = 2\mathrm{pp}\sigma \cos k_x a + 2\mathrm{pp}\pi \,(\cos k_y a + \cos k_z a), \tag{5.106}$$

with T_{yy} and T_{zz} found by cyclic permutation; the off-diagonal matrix elements are identically zero. Hence, say in the [100] direction in k-space:

$$\mathscr{E}_k = \mathscr{E}_\mathrm{p} + \begin{cases} 2\mathrm{pp}\pi \cos ka + 2(\mathrm{pp}\sigma + \mathrm{pp}\pi) \\ 2\mathrm{pp}\sigma \cos ka + 4\mathrm{pp}\pi, \end{cases} \tag{5.107}$$

where the upper term of higher energy corresponds to a doubly degenerate state and therefore there is triple degeneracy at the Γ-point ($\boldsymbol{k} = 0$). The behaviour of eqn. (5.107) is shown in Fig. 5.39 for the case of a 2D square lattice.

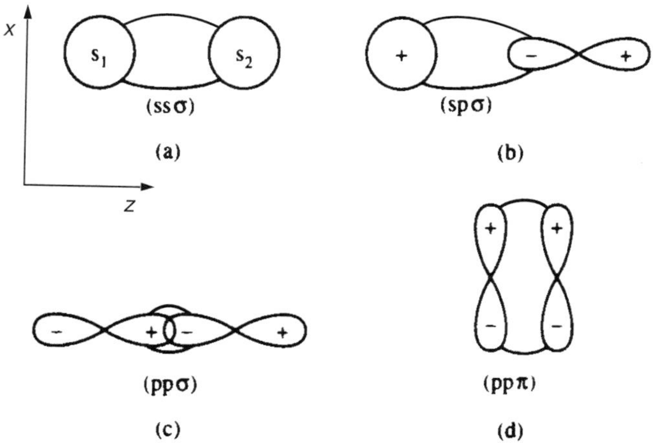

Fig. 5.40 Schematic illustration of the geometric origins of the four non-zero overlap (hopping) integrals between s- and p-orbitals. (a) ssσ: this is *negative* if both s-orbitals have the same sign. (b) spσ: this is *positive* if the negative lobe of the p_z-orbital is closest to a positive s-orbital. (c) ppσ: this is *positive*, since it is dominated by the contributions between the positive orbital on one p_z-orbital overlapping with the closest negative orbital on a neighbouring p_z-orbital. (d) ppπ: this is *negative* because it is dominated by the overlap between two positive lobes and between two negative lobes on neighbouring p_x- or p_y-orbitals.

The Bloch-like functional form of the TB wavefunction (eqn. (5.93)) indicates that an electron in a TB band has an equal probability of being in any cell in the crystal. Thus, electrons in a part-filled TB band may contribute to the electrical conductivity and can be regarded as moving from site to site by means of quantum-mechanical tunnelling. It might be thought that, by progressively decreasing the separation between isolated (i.e. electrically insulating) atoms, the electrical conductivity would *continuously* increase from zero as the tight-binding bandwidth increases (Fig. 5.35b). However, this presupposes that the *independent-electron approximation*, implicit in the TBA, is valid at all interatomic separations. This is not the case, and a discontinuous (Mott) transition from the conducting to the insulating state occurs on increasing the separation beyond a critical value (§5.6.3). This breakdown of the independent-electron approximation occurs because electron–electron repulsions at a particular site are not properly treated.

Finally, it should be noted that, in principle, the tight-binding method does *not* depend on there being periodicity of the lattice: the method can equally well be used to describe electron states in disordered (e.g. amorphous) materials, although of course in such cases the TB wavefunction cannot be written in Bloch form (cf. eqn. (5.93)).

5.3.2 Hybridization

Thus far, we have described valence electrons in solids in terms of two limiting situations: the NFE and TB models. In both cases, since the relevant Bloch wavefunctions extend over all cells in a crystal, part-filled bands (originating from unfilled atomic shells of electrons) are predicted to give *metallic* behaviour. While this picture is successful in many cases, it breaks down for the case, for example, of the Group IV elements that have the tetrahedrally coordinated, diamond-cubic structure (§2.2.5), i.e. the diamond form of carbon, silicon, germanium and the α-phase of tin. For these, the atomic electronic configuration of the outer shells is s^2p^2, and if the p-states were to form a separate TB band, as in Fig. 5.41, then such materials would be expected to be metallic since such a p-band could contain six states per atom but actually would contain only two electrons per atom. In fact, they are *semiconductors*, with bandgaps decreasing in size as the group is descended.

The reason for the apparent breakdown in this picture is that, in such materials, the s-states and p-states do not form *separate* bands, as on the left-hand side of Fig. 5.41, but instead the states hybridize to form new, spatially directed orbital combinations: in the case of the Group IV elements, and related materials like the binary III–V compounds (GaAs, etc.); the hybrid orbitals are termed sp^3-hybrids for reasons that will become apparent shortly. Such sp^3-hybrids can then form *bonding* and *antibonding* combinations, separated by a hybridization gap (as on the right-hand side of Fig. 5.41). The sp^3-bonding (valence) band contains four states per atom, as does the sp^3-antibonding (conduction) band. Thus, for the case of the Group IV elements, the sp^3-bonding

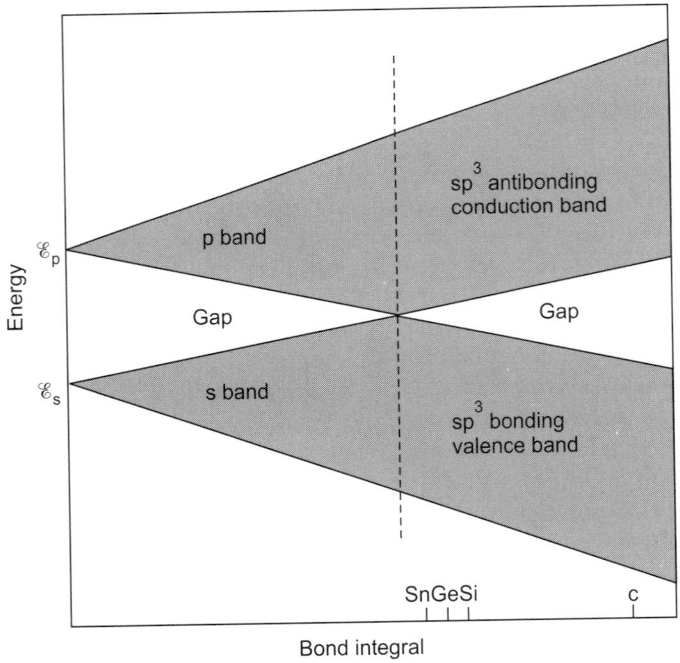

Fig. 5.41 Schematic illustration of the origin of the hybridization gap between bonding and antibonding bands formed from sp^3-hybrids for tetrahedrally coordinated Group IV elements. The dashed line marks the boundary between, on the left, the formation of tight-binding bands from atomic states and, on the right, sp^3-hybrid bands. The hybridization bandgap increases with increasing bond integral (scaling inversely with interatomic separation). The positions of Sn, Ge, Si and C are indicated according to the relative values of the experimental bandgaps. For all of these elements, the atomic s–p splitting energy is approximately constant $\Delta\mathscr{E}_{ps} = \mathscr{E}_p - \mathscr{E}_s = 7.5 \pm 1\text{eV}$.

valence band is completely full and the conduction band is completely empty at $T = 0\,\text{K}$: the materials are therefore predicted to be insulators/semiconductors, as observed experimentally. Band formation from hybridized states is favoured over that from atomic states if the cost in energy to hybridize atomic orbitals (the promotion energy, $\mathscr{E}_p - \mathscr{E}_s$ in Fig. 5.41) is more than compensated by the extra bonding energy gained via more efficient overlap between spatially directed hybrid orbitals than between atomic orbitals.

Hybrid orbitals are states constructed from linear combinations of atomic states on a given atom. For the example of Group IV elements, III–V compounds, etc., the atomic states involved are s- and p-states, although for different systems other orbitals (e.g. d-states) can also be involved. Two equivalent s–p hybrid orbitals on the same atom can thus be written as

$$\psi_1 = f(\psi_s + \lambda\psi_{p1}), \tag{5.108a}$$

$$\psi_2 = f(\psi_s + \lambda\psi_{p2}), \tag{5.108b}$$

where f is a normalization constant, ψ_{p1} and ψ_{p2} are normalized p-orbitals having arbitrary orientations and λ^2 is termed the mixing ratio of p-states to s-states in the hybrids. The normalization constant is found from the relation

$$1 = \int \psi_1^* \psi_1 \mathrm{d}\mathbf{r} = f^2(1 + \lambda^2), \tag{5.109a}$$

$$f = (1 + \lambda^2)^{-1/2}. \tag{5.109b}$$

We also require the two hybrid states ψ_1 and ψ_2 on the same atom to be orthogonal, i.e.

$$0 = \int \psi_1^* \psi_2 \mathrm{d}\mathbf{r} = f^2\left[1 + \lambda \int \psi_{p1}^* \psi_s \mathrm{d}\mathbf{r} + \lambda \int \psi_s^* \psi_{p2} \mathrm{d}\mathbf{r} + \lambda^2 \int \psi_{p1}^* \psi_{p2} \mathrm{d}\mathbf{r}\right]. \tag{5.110}$$

The two overlap integrals between s- and p-states are zero by symmetry. In order to evaluate the overlap integral involving two p-states, it may be assumed without loss of generality that ψ_{p1} points along the x-axis, and the other orbital ψ_{p2} can be resolved into contributions that are parallel ($\psi_{p2}\cos\theta$) and perpendicular ($\psi_{p2}\sin\theta$) to the x-axis, where θ is the angle between the two orbitals (Fig. 5.42). Thus

$$\int \psi_{p1}^* \psi_{p2} \mathrm{d}\mathbf{r} = \cos\theta, \tag{5.111}$$

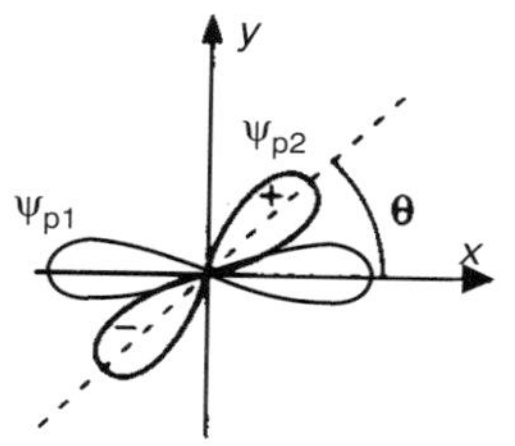

Fig. 5.42 Two p-orbitals on an atom. One is placed arbitrarily along the x-axis, and the other lies at an angle θ in the same plane and has a component $\psi_{p2}\cos\theta$ parallel to the x-axis.

and hence the orthogonality condition (eqn. (5.110)) reduces to the requirement that

$$\lambda^2 = -1/\cos\theta. \tag{5.112}$$

Different hybrid-orbital combinations may be formed between s-orbitals and varying numbers of p-orbitals on the same atom. Mixing of an s-orbital with a single p-orbital creates the sp-hybrids:

$$\psi_1 = \frac{1}{\sqrt{2}}(\psi_s + \psi_x), \tag{5.113a}$$

$$\psi_2 = \frac{1}{\sqrt{2}}(\psi_s - \psi_x), \tag{5.113b}$$

where ψ_x stands for the p_x-orbital. These combinations are illustrated in Fig. 5.43a.

Mixing of an s-orbital with two p-orbitals creates three sp^2-hybrids (Problem 5.18) that lie in the same (say x–y) plane (see Fig. 5.43b). The p_z-orbital remains unchanged and points in the z-direction, perpendicular to the sp^2-hybrids.

If an s-orbital is mixed with all three p-orbitals, four sp^3-hybrids are produced. This is the situation of interest for the diamond-cubic materials. The s-like content of the hybrids is $1/(1+\lambda^2) = 1/4$, i.e. $\lambda^2 = 3$. Thus, from eqn. (5.112), the sp^3-hybrids are directed in a tetrahedral arrangement to the four corners of a cube (Fig. 5.43c), with an included angle $\theta = \cos^{-1}(-1/3) = 109°28'$ subtended between them. The four sp^3-hybrids can be written as:

$$\psi_1 = \frac{1}{2}(\psi_s + \psi_x + \psi_y + \psi_z), \tag{5.114a}$$

$$\psi_2 = \frac{1}{2}(\psi_s + \psi_x - \psi_y - \psi_z), \tag{5.114b}$$

$$\psi_3 = \frac{1}{2}(\psi_s - \psi_x + \psi_y - \psi_z), \tag{5.114c}$$

$$\psi_4 = \frac{1}{2}(\psi_s - \psi_x - \psi_y + \psi_z). \tag{5.114d}$$

The energy of a hybrid orbital (eqn. (5.108)) is given by

$$\mathscr{E}_h = \int \psi_i^* \, \mathscr{H} \, \psi_i \mathrm{d}\mathbf{r} = \frac{\mathscr{E}_s + \lambda^2 \mathscr{E}_p}{(1+\lambda^2)}. \tag{5.115}$$

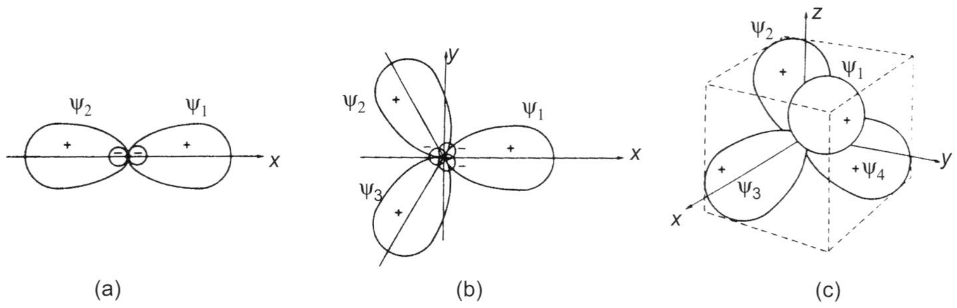

Fig. 5.43 Illustration of the geometries of various s–p hybrid orbitals: (a) sp-hybrids, pointing linearly in opposite directions. (b) sp^2-hybrids, lying in the same plane with a subtended angle of 120°. (c) sp^3-hybrids, in a tetrahedral arrangement with a subtended angle of 109° 28′.

For the case of sp^3-hybrids, the hybrid energy is therefore

$$\mathscr{E}_h = (\mathscr{E}_s + 3\mathscr{E}_p)/4, \quad (5.116)$$

since $\lambda^2 = 3$. Thus, if the four sp^3-hybrids on an atom contain one electron each, the on-site energy is $(\mathscr{E}_s + 3\mathscr{E}_p)$, compared with the corresponding energy $(2\mathscr{E}_s + 2\mathscr{E}_p)$ for the unhybridized atom with an s^{2}p^2 electronic configuration. The difference between these is the promotion energy

$$\Delta\mathscr{E}_{ps} = \mathscr{E}_p - \mathscr{E}_s, \quad (5.117)$$

required to promote an s-electron to a p-state energy so that it can mix.

The origin of the bandgap between valence and conduction bands in hybridized systems, e.g. Si or Ge, can be seen clearly by reference to Fig. 5.44. Atomic s- and p-states are hybridized to form sp^3-hybrids with an energy $\mathscr{E}_h$, which can then form bonding and antibonding combinations between orbitals on neighbouring atoms pointing along the common bond, resulting in bonding and antibonding energy levels separated by $2|\beta_1|$, where $|\beta_1|$ is the magnitude of the hybrid bond integral. Such molecular orbitals become broadened into bands, the valence band being formed

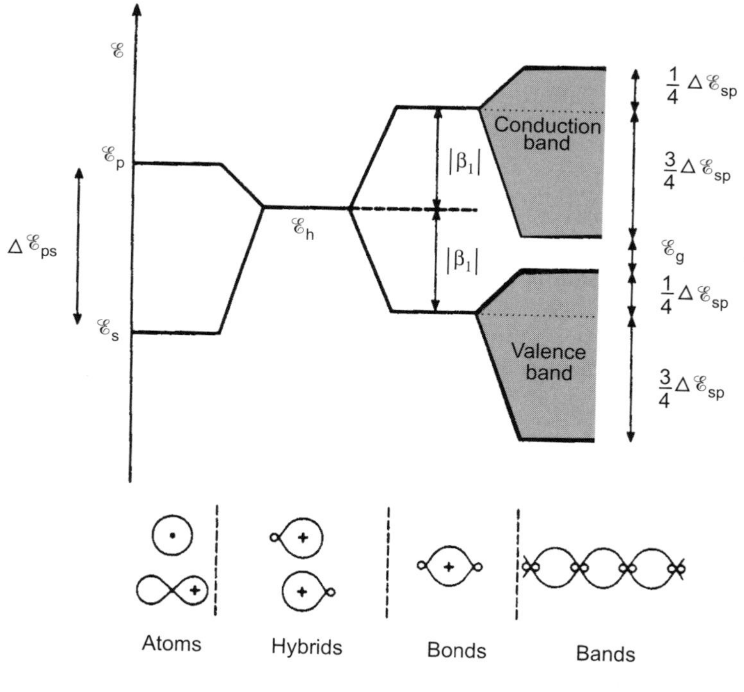

Fig. 5.44 Illustration of bandgap formation in hybridized systems, e.g. for sp^3-hybrids in Si. Atomic s- and p-states, with energy levels separated by $\Delta\mathscr{E}_{ps}$, are hybridized to form sp^3-hybrids with energy $\mathscr{E}_h$. These form bonding and antibonding combinations, giving molecular-orbital levels separated by $2|\beta_1|$. Valence and conduction bands form from these levels as a result of an intrasite hopping interaction between sp^3-hybrids on the same atom; the bandwidth is $\Delta\mathscr{E}_{ps}$. A gap is formed if $2|\beta_1| > \Delta\mathscr{E}_{ps}$. (From: *Electronic Structure and the Properties of Solids* by Harrison © 1980 by W.H. Freeman and Company. Used with permission.)

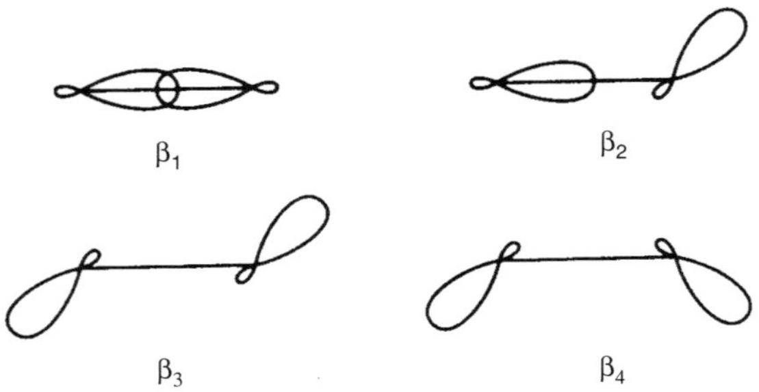

Fig. 5.45 Four types of hopping (overlap) integral between sp^3-hybrids on neighbouring atoms. β_1 is the largest and the only one retained in the Weaire–Thorpe model.

from bonding orbitals and the conduction band from antibonding orbitals, by, in the simplest case (the Weaire–Thorpe (1971) model), interactions between hybrid orbitals on the *same* atom. Such on-site interactions thereby allow an electron to hop from bond to bond between different atoms in the structure: this delocalization is essentially the cause of the band formation.

We shall show shortly that the width of each of the two bands is simply $\Delta\mathscr{E}_{ps}$, the atomic s–p splitting, and so from Fig. 5.44 a simple criterion can be gleaned for the existence of a bandgap, namely that the magnitude of the bonding–antibonding separation of the molecular orbitals should be greater than the bandwidth, i.e.

$$2|\beta_1| \geqslant \Delta\mathscr{E}_{ps}. \tag{5.118}$$

Note that this result is based solely on *local* interactions, both intersite and intrasite, and therefore does *not* invoke translational periodicity. Hence, this approach explains why, for example, both amorphous and crystalline forms of Si have gaps of comparable size between filled valence and empty conduction band states, since the average short-range order is identical (i.e. tetrahedral) in both phases.

The simple Weaire–Thorpe model considers only two interactions: an intersite and an intrasite hopping integral. Of the possible intersite hopping integrals between hybrid orbitals on neighbouring sites (Fig. 5.45), only the largest (β_1) is retained, corresponding to overlap between orbitals pointing at each other along the line of the bond. This has the value

$$\beta_1 = (\mathrm{ss}\sigma - 2\sqrt{3}\mathrm{sp}\sigma - 3\mathrm{pp}\sigma)/4, \tag{5.119}$$

and, from eqn. (5.105), is negative.

The intrasite hopping integral is

$$\int \psi_\alpha^* \mathscr{H} \psi_\beta \, d\mathbf{r} = -\frac{1}{4}(\mathscr{E}_p - \mathscr{E}_s) = -\frac{1}{4}\Delta\mathscr{E}_{ps} \tag{5.120}$$

for different hybrid orbitals on the same atom $\alpha \neq \beta$, from eqns. (5.108) and (5.111). If now bonding and antibonding orbitals, ϕ_{ij}^+ and ϕ_{ij}^-, between atoms i and j (see Fig. 5.46a) are written as

$$\phi_{ij}^{\pm} = \frac{1}{\sqrt{2}}\left(\psi_{ij}^{(i)} \pm \psi_{ij}^{(j)}\right), \tag{5.121}$$

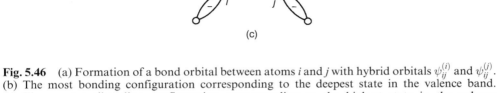

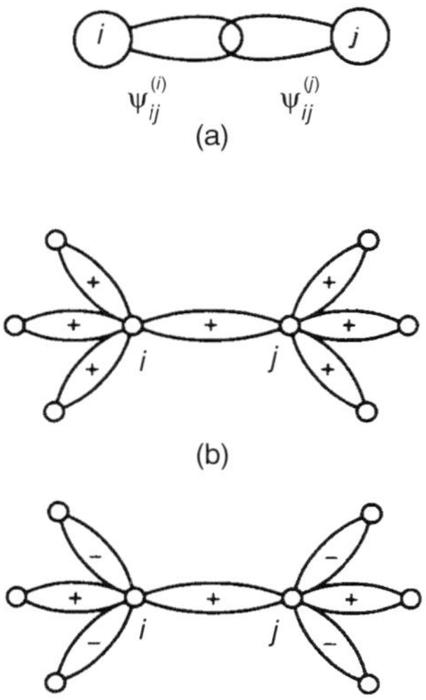

Fig. 5.46 (a) Formation of a bond orbital between atoms *i* and *j* with hybrid orbitals $\psi_{ij}^{(i)}$ and $\psi_{ij}^{(j)}$. (b) The most bonding configuration corresponding to the deepest state in the valence band. (c) The most antibonding configuration corresponding to the highest state in the valence band.

where $\psi_{ij}^{(i)}$ is an sp^3-hybrid on atom *i* pointing along the bond between atoms *i* and *j*, then the interaction between bond orbitals on neighbouring atoms is (from eqn. (5.120)):

$$\int \phi_{ij}^{+*}\,\mathscr{H}\,\phi_{ik}^{+}\mathrm{d}\mathbf{r} = \frac{1}{2}\int \left(\psi_{ij}^{*(i)} + \psi_{ij}^{*(j)}\right)\mathscr{H}\left(\psi_{ik}^{(i)} + \psi_{ik}^{(k)}\right)\mathrm{d}\mathbf{r}. \tag{5.122}$$

Since the model only considers intersite overlap between hybrids along common bonds, the only term remaining from eqn. (5.122) is

$$\begin{aligned}\int \phi_{ij}^{+*}\,\mathscr{H}\,\phi_{ik}^{+}\mathrm{d}\mathbf{r} &= \frac{1}{2}\int \psi_{ij}^{*(i)}\,\mathscr{H}\,\psi_{ik}^{(i)}\mathrm{d}\mathbf{r}\\ &= -\frac{1}{8}\Delta\mathscr{E}_{\mathrm{ps.}}\end{aligned} \tag{5.123}$$

from eqn. (5.120).

The extremities of the valence and conduction bands can now be found. The bottom of, say, the valence band corresponds to the most bonding configuration of the hybrids (Fig. 5.46b): its energy is

$$\begin{aligned}\mathscr{E}_{\mathrm{v}}^{\min} &= \mathscr{E}_{\mathrm{h}} - |\beta_1| + 6(-\frac{1}{8}\Delta\mathscr{E}_{\mathrm{ps}})\\ &= \mathscr{E}_{\mathrm{h}} - |\beta_1| - \frac{3}{4}\Delta\mathscr{E}_{\mathrm{ps,}}\end{aligned} \tag{5.124}$$

since the hopping integrals associated with the bond orbital and the six orbitals at either end of the *ij* bond have the same sign. The top of the valence band, on the other hand, corresponds to the most antibonding situation, shown in Fig. 5.46c: its energy is correspondingly

$$\begin{aligned}\mathscr{E}_{\mathrm{v}}^{\max} &= \mathscr{E}_{\mathrm{h}} - |\beta_1| + (2-4)(-\tfrac{1}{8}\Delta\mathscr{E}_{\mathrm{ps}}) \\ &= \mathscr{E}_{\mathrm{h}} - |\beta_1| + \tfrac{1}{4}\Delta\mathscr{E}_{\mathrm{ps}}\end{aligned} \quad (5.125)$$

Thus, the width of the valence band is just $\Delta\mathscr{E}_{\mathrm{ps}} = \mathscr{E}_{\mathrm{p}} - \mathscr{E}_{\mathrm{s}}$. Similar considerations for the conduction band give the same result (see Problem 5.19).

Although the Weaire–Thorpe model provides a clear physical picture of the origin of the bandgap in solids like Si, the neglect of the other intersite interactions $\beta_i (i = 2-4)$, see Fig. 5.45, means that it gives a poor quantitative description of the band structure. However, rather than include ever higher terms β_i, it is simpler just to consider hopping interactions between s- or p-states on one atom with s- or p-states on another atom without invoking complete hybridization; such a situation is referred to as a minimal atomic basis set. In fact, it is found that, for crystalline Si, the s–p mixing ratio is actually 1.7, rather than the value of 3 expected for ideal sp^3-hybridization (see Sutton (1993)).

Understanding band structures 5.4

We are now in a position to interpret features in band structures ($\mathscr{E}(\boldsymbol{k})$ diagrams) of crystalline materials from both physical and chemical points of view. It will be seen that the band structures for crystalline materials characterized by different forms of interatomic bonding are qualitatively different. Insulators can be distinguished from metals, if we know the extent of band filling for a particular material. Insulators (large bandgap) and semiconductors (small bandgap), defined as those materials for which the highest occupied (valence) bands are everywhere in $\boldsymbol{k}$-space lower in energy than the lowest unoccupied (conduction) bands, can be subdivided into two categories for the case of crystals. Those in which the highest occupied valence state (HOMO in chemists' parlance) is at the *same* point in $\boldsymbol{k}$-space (usually the Γ-point) as the lowest unoccupied conduction state (LUMO) are termed direct-gap materials; those crystalline solids for which the minimum bandgap occurs at a *different* $\boldsymbol{k}$-value from the conduction-band minimum are called indirect-gap materials (see Fig. 5.47).

At points of very high symmetry in the Brillouin zone, bands originating from a given set of atomic orbitals are very often degenerate. Some limited degree of degeneracy between such bands can also hold for trajectories in $\boldsymbol{k}$-space between high-symmetry points. At a general $\boldsymbol{k}$-point in the band structure, however, the full number of bands is revealed, equal to the number of atoms in the basis of the unit cell multiplied by the number of atomic orbitals involved. Each band can then accept two electrons per atom as a result of electron-spin degeneracy. A detailed discussion of symmetry aspects relating to band structures is given, for example, in Burns (1985), Altmann (1991) and Nichols (1995).

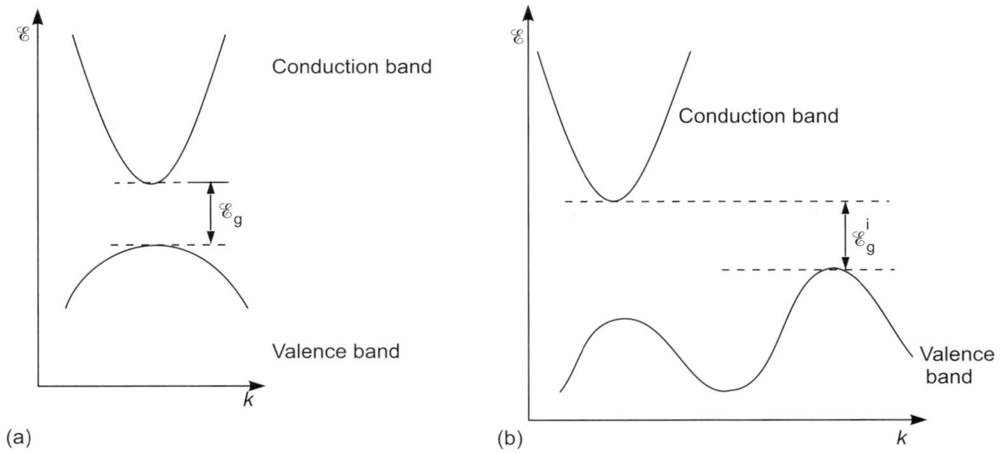

Fig. 5.47 Types of bandgap in insulators/semiconductors: (a) direct gap; (b) indirect gap.

5.4.1 Ionic crystals

The calculated band structures of two representative ionic crystals, KCl and AgCl, are given in Fig. 5.48. KCl is an archetypal highly ionic solid with near-complete charge

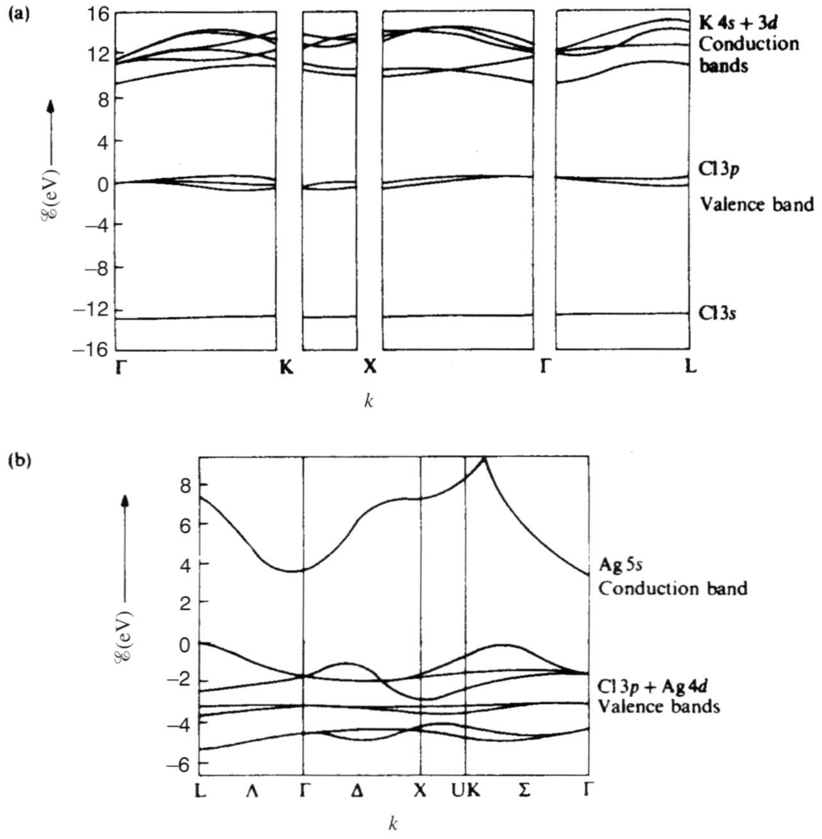

Fig. 5.48 Calculated band structures of two ionic crystals: (a) KCl (Cohen and Heine (1970). Reproduced by permission of Academic Press, Inc.); (b) AgCl (Wong *et al.* (1976). Reproduced by permission of Akademie Verlag GmbH).

transfer from the more electropositive atom (K) to the more electronegative (Cl). Thus, the valence electrons are entirely associated with the tightly bound 3p-states on the anion (Cl); correspondingly, the 4s- and 3d-states of the cation (K) are empty. It can be seen from Fig. 5.48a that the Cl 3p-bands exhibit very little variation in energy with $\boldsymbol{k}$: the width of the valence band is therefore extremely narrow, as expected from a tight-binding treatment (§5.3.1) of such tightly bound electrons. (The 3s-states of Cl, being core states, are yet more tightly bound, and form an even narrower band, essentially a discrete atomic level). In the case of complete charge transfer from K to Cl, all three Cl 3p-bands (the valence band) will be completely filled with electrons and the K 4s- and 3d-states will be completely empty. Since there is a gap of ~ 8 eV between these two bands, KCl is an extreme electrical insulator, with a direct gap (at Γ). Values of the bandgaps for alkali-halide crystals are given in Table 5.3. To a first approximation, the bandgaps are approximately equal to the difference in energies of cation s- and anion

Table 5.3 Experimental bandgaps for alkali halides

	Li	Na	K	Rb	Cs
F	13.6	11.6	10.7	10.3	9.9
	(17.1)	(12.9)	(9.7)	(8.7)	—
	(11.5)	(11.9)	(12.8)	(13.1)	(13.4)
Cl	9.4	8.5	8.4	8.2	8.3
	(10.5)	(8.7)	(7.0)	(6.4)	(5.4)
	(6.8)	(7.2)	(8.1)	(8.4)	(8.8)
Br	7.6	7.5	7.4	7.4	7.3
	(9.2)	(7.8)	(6.4)	(5.8)	(5.0)
	(5.7)	(6.1)	(7.0)	(7.3)	(7.6)

Bandgap values are in eV. Values in parentheses are values from the bond-length scaling formula (eqn. (5.126)), above, and below, values of ($\mathscr{E}_s^c - \mathscr{E}_p^a$), where $\mathscr{E}_s^c$ and $\mathscr{E}_p^a$ are the cation and anion s- and p-state atomic energy levels, respectively. (From: *Electronic Structure and the Properties of Solids* by Harrison © 1980 by W.H. Freeman and Company. Used with permission.)

p-states, $\mathscr{E}_s^c - \mathscr{E}_p^a$, although the variation of $\mathscr{E}_g$ with cation atomic number for a given anion is not reproduced. This behaviour is captured by use of the formula

$$\mathscr{E}_g = A\hbar^2/m_e d^2 \tag{5.126}$$

linking the bandgap with the nearest-neighbour spacing, d; the value of the constant is $A = 18.2$. The same formula also accounts for the variation of $\mathscr{E}_g$ for rare-gas solids and covalent solids, but with different values of the constant, A (Harrison (1980)).

In the case of AgCl (Fig. 5.48b), the Ag 4d-orbitals (with a filled shell) are too far apart from each other to produce any significant width in a TB band. However, they lie in the same energy region as the Cl 3p-states, and consequently there are appreciable covalent-like interactions between d- and p-states, leading to an increase in the width of the valence band. Nevertheless, the degree of mixing is *not* the same for all $\boldsymbol{k}$-values. At $\boldsymbol{k} = 0\,(\Gamma)$, there is *no* interaction between p- and d-states because of their different symmetries. Fig 5.49 shows that there is a net cancellation in overlaps between lobes on neighbouring orbitals, whereas the interaction (mixing) is maximal (i.e. the bandwidth is greater) at a non-zero value of $\boldsymbol{k}$. As a result, AgCl is an indirect-gap semiconductor (the gap between valence and conduction bands is much less than for KCl).

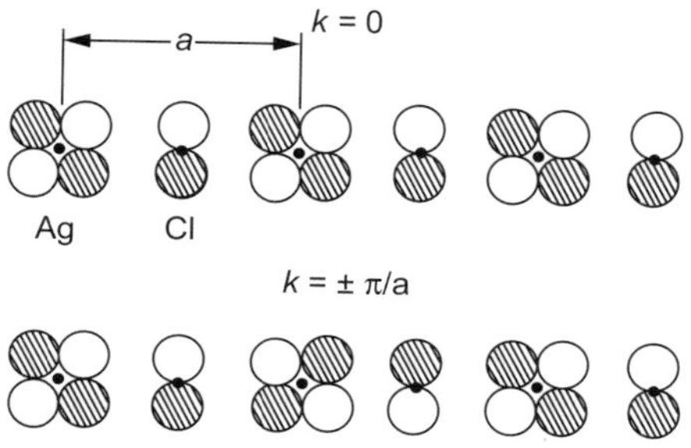

Fig. 5.49 Schematic illustration showing the interactions between Ag 4d and Cl 3p orbitals for $\boldsymbol{k} = 0$ (no mixing) and $\pm\pi/a$ (maximal mixing). (After Cox (1987), *Electronic Structure and Chemistry of Solids*, by permission of Oxford University Press)

5.4.2 Covalent crystals

The very strong orbital mixing characteristic of covalent bonding results in electron states being delocalized through a crystalline solid and not spatially localized in the vicinity of ion cores as in completely ionic materials. As a result, the widths of electron bands for crystalline covalent materials are much greater than those of ionic crystals (Fig. 5.48). Stoichiometric covalent materials, like ionic solids, are insulators (or semiconductors, depending on the magnitude of the bandgap): at $T = 0\,\mathrm{K}$, fully occupied (valence) bands are separated in energy from completely empty (conduction) bands. (Non-stoichiometry, however, may shift the Fermi level into one of the bands, making the material metallic.)

Energy bands for crystalline Ge are shown in Fig. 5.50, calculated in three ways: (a) in the LCAO approximation, (b) using a full pseudopotential calculation, and (c) in the empty-lattice approximation. It can be seen that the free-electron-like bands in Fig. 5.50c are remarkably similar to those at the bottom of the valence band and the upper part of the conduction band of the 'true' band structure (Fig. 5.50b), except of course that bandgaps at certain points in the Brillouin zone (notably the Γ- and L-points) do not occur in the free-electron band structure. There are only eight LCAO bands in Fig. 5.50a (as can be seen at the low symmetry points K, U) because only one s- and three p-states per atom in the unit cell were considered: the conduction band in the pseudopotential band structure is much more complicated because other states (e.g. 4d-states) were included. Nevertheless, for the valence band, the agreement between the LCAO and pseudopotential band structures is rather good.

As for the case of p–d mixing in AgCl (§5.4.1), mixing between s- and p-states in tetrahedrally coordinated crystals, like Ge, is *absent* on symmetry grounds at the Γ-point. Thus, at Γ, the states are purely s-like or p-like. The bottom of the valence band is therefore s-like, and the top is p-like and triply degenerate (in the absence of spin-orbit coupling — see later). Likewise, at Γ, the bottom of the conduction band is purely s-like.

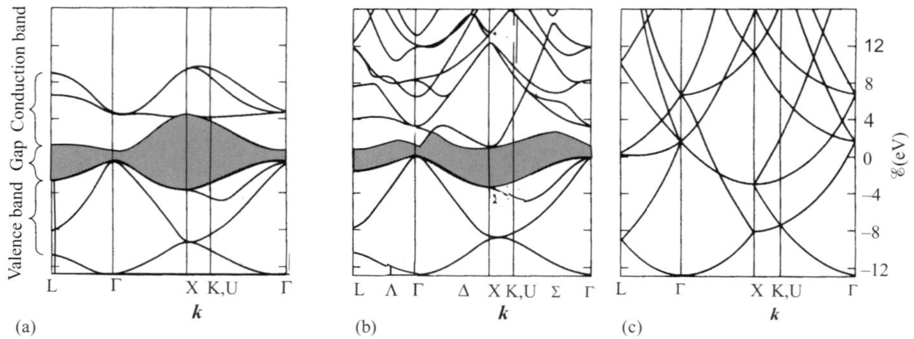

Fig. 5.50 Calculated band structures for crystalline Ge (neglecting spin–orbit coupling): (a) LCAO approximation involving only s- and p-like valence states; (b) pseudopotential calculation; (c) free-electron-like bands in the empty-lattice approximation. (From: *Electronic Structure and the Properties of Solids* by Harrison © 1980 by W.H. Freeman and Company. Used with permission.)

On moving along a given band away from the Γ-point, the degree of s–p mixing increases and the bands move up (for the s-state) and down (for the p-states), as seen already in Figs. 5.37a and 5.39. Thus, the width of the valence band is given by the atomic s–p splitting energy, $\Delta\mathscr{E}_{ps}$, as deduced previously from a bond-orbital picture (Fig. 5.44).

As can be seen from Fig. 5.50b, Ge is an indirect-gap semiconductor, the maximum of the valence band being at Γ and the minima in the conduction band being in the $\langle 111 \rangle$ directions at the L-point. Crystalline Si is also an indirect-gap material, but for it (and diamond and GaP) the conduction-band minima are in the $\langle 100 \rangle$ directions near the X-point. Constant-energy surfaces ('electron pockets') in k-space, referred to the first Brillouin zone, are shown in Fig. 5.51 for regions near the conduction-band minima for these two materials. In a principal-axis representation for say Si (the principal axes in $\boldsymbol{k}$-space are $\langle 100 \rangle$), the energy relative to the conduction-band minimum at $(0, 0, k_0)$ for an ellipsoid aligned along the z-direction in $\boldsymbol{k}$-space, can be written approximately as

$$\mathscr{E}(\boldsymbol{k}) = \frac{\hbar^2}{2}\left[\frac{(k_x^2 + k_y^2)}{m_t^*} + \frac{(k_z - k_0)^2}{m_l^*}\right], \tag{5.127}$$

where m_t^* and m_l^* are the so-called transverse and longitudinal components of the effective mass (§6.5.1.4). The constant-energy surfaces are therefore ellipsoids around the $\langle 100 \rangle$ directions in Si, and the $\langle 111 \rangle$ directions for Ge.

It is instructive to examine the trends in the band structures of tetrahedrally bonded materials (with either the diamond-cubic or wurtzite structures) that are *isoelectronic* with the Group IV elements, i.e. the III–V and II–VI compounds exhibiting mixed iono-covalent bonding. Some of these band structures are shown in Fig. 5.52. On going from left to right in Fig. 5.52, there is an increase in ionicity (i.e. an increase of charge transfer from the more electropositive atom to the more electronegative); on progressing from top to bottom, there is an increase in metallicity (i.e. metallic tendency). These

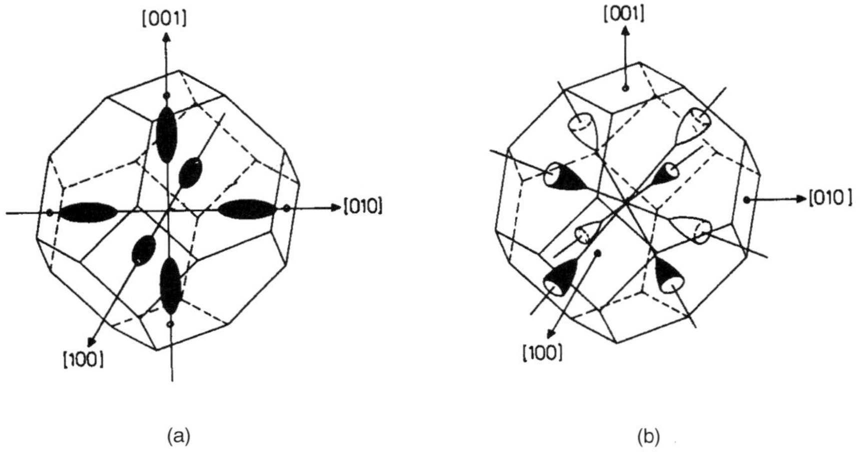

Fig. 5.51 Constant-energy surfaces near the conduction-band minima for (a) Si; (b) Ge.

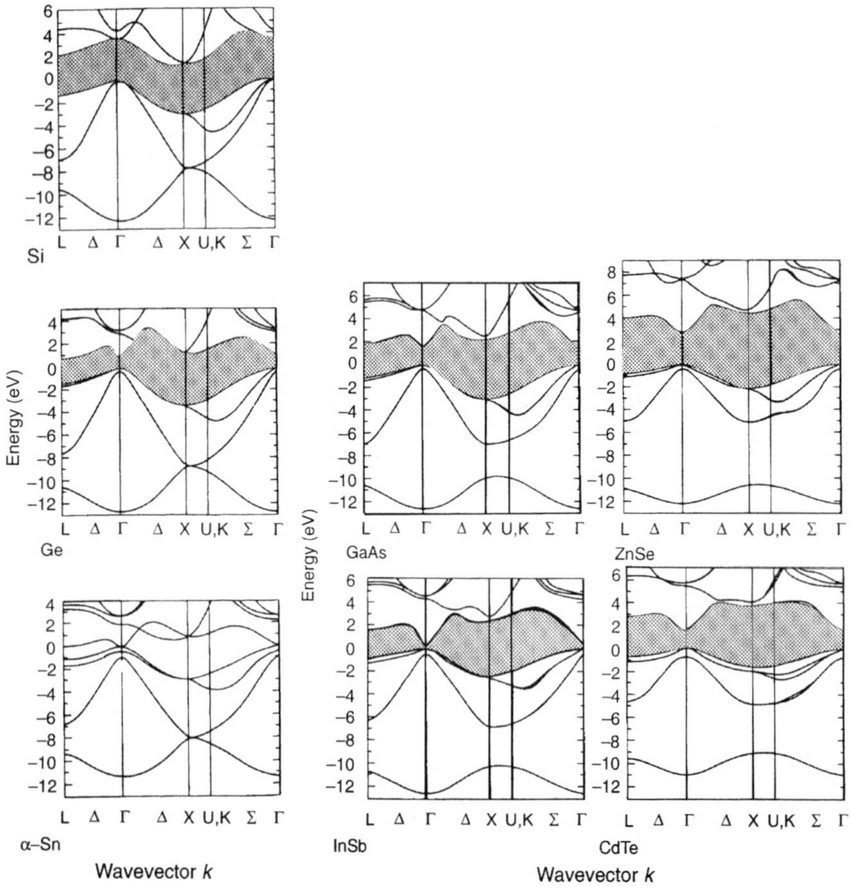

Fig. 5.52 Calculated band structures of Si, Ge and α-Sn, together with III–V and II–VI compounds isoelectronic with Ge and Sn. (After Chelikowsky and Cohen (1976). Reprinted with permission from *Phys. Rev.* **B14**, 556. © 1976. The American Physical Society)

behaviours can be quantified in terms of three energies (Harrison (1980)): $V_1 = (\mathscr{E}_p - \mathscr{E}_s)/4$, the intrasite hopping integral (eqn. (5.120)) between hybrid orbitals on the same atom and a measure of metallicity (the greater the bandwidth, the more likely is the material to be metallic); $V_2 = |\beta_1|$, the intersite overlap integral (eqn. (5.119)) and a measure of covalency; and $V_3 = (\mathscr{E}_s^c - \mathscr{E}_s^a)/2$, half the difference between atomic s-levels for cation and anion, and a measure of the polarity. Thus, one definition for the *ionicity* of a bond between dissimilar atoms is (Harrison (1980)):

$$\alpha_i = V_3/(V_2^2 + V_3^2)^{1/2}. \tag{5.128}$$

(Another definition, according to Phillips, is given by eqn. (2.8).) In a similar fashion, the *metallicity*, representing the relative dominance of the 'banding' interaction (V_1) to the 'bonding' interaction (V_2), can be defined as (Harrison (1980)):

$$\alpha_m = V_1/(V_2^2 + V_3^2)^{1/2}. \tag{5.129}$$

Examination of Fig. 5.52 shows that, with increasing metallicity, the gap at Γ decreases, as expected. Indeed, the lowest conduction band drops more quickly at Γ than at say X or L with increasing metallicity, so that the compound materials become *direct-gap* semiconductors (except GaP and AlSb). Representative values of bandgaps for tetrahedral materials are given in Table 5.4. With increasing ionicity, the gap at Γ increases; moreover, the degree of s–p mixing decreases so that the bands become narrower, as in ionic crystals (Fig. 5.48). The symmetry-required degeneracy at the X-point between s- and p-bands for Si and Ge is lifted for the compound materials containing dissimilar atoms where the symmetry is broken.

Finally, it should be noted that the degeneracy at the Γ point of the p-states forming the top of the valence band is increasingly lifted with increasing metallicity. This effect is due to spin-orbit coupling of the electron spin $\boldsymbol{s}$ to the orbital angular momentum $\boldsymbol{l}$ to give the total angular momentum $\boldsymbol{j}$:

$$\boldsymbol{j} = \boldsymbol{l} + \boldsymbol{s}. \tag{5.130}$$

For the p-states ($l = 1$), two spin-orbit-coupled levels result, with $j = 3/2$ and $1/2$. The $p_{3/2}$-derived band lies above the 'split-off' band derived from $p_{1/2}$ states by the spin–orbit splitting energy Δ_{so}. This quantity increases with increasing atomic number (or, equivalently, metallicity), essentially because the orbiting electron experiences a greater

Table 5.4 Bandgap for tetrahedrally coordinated semiconducting crystals

Crystal	*Type of gap*	$\mathscr{E}_g$ (eV)	*Crystal*	*Type of gap*	$\mathscr{E}_g$ (eV)
C	i	5.4	InP	d	1.42
Si	i	1.17	InAs	d	0.43
Ge	i	0.74	InSb	d	0.23
α-Sn	d	0.00	AlSb	i	1.65
GaP	i	2.32	CdS	d	2.58
GaAs	d	1.52	CdSe	d	1.84
GaSb	d	0.81	CdTe	d	1.61
			ZnSe	d	2.82

(Bandgap values refer to $T = 0\,\text{K}$; i = indirect; d = direct.)
(After Kittel (1996). Reproduced by permission of John Wiley & Sons Inc.)

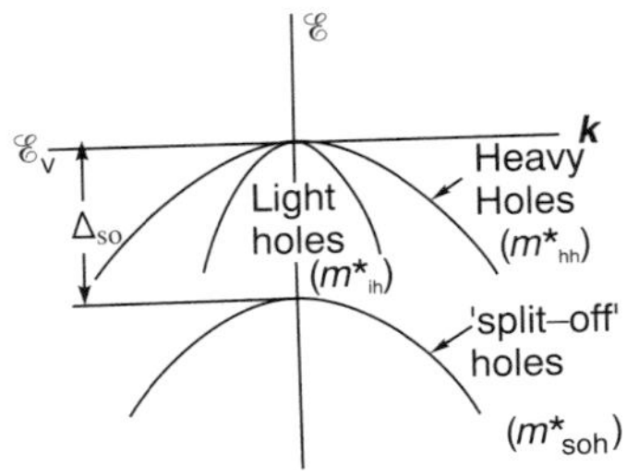

Figure 5.53 Schematic illustration of the effects of spin–orbit coupling on the p-like bands at the top of the valence band (near Γ) for tetrahedral semiconductors. A split-off $p_{1/2}$ band is separated by Δ_{so}, the spin-orbit coupling energy, from a $p_{3/2}$ band which is doubly degenerate (excluding spin degeneracy) at Γ. At other $\boldsymbol{k}$-values, two bands with different curvatures result: the heavy-hole and light-hole bands.

internal atomic magnetic field the greater is the nuclear charge. Δ_{so} is very small for Si (0.04 eV), fairly large for Ge (0.29 eV) and GaAs (0.34 eV), and quite sizeable for InSb (0.82 eV).

The local band structure in the vicinity of Γ is illustrated in Fig. 5.53 for the case of appreciable spin–orbit coupling. The higher-lying $p_{3/2}$ band is quadrupally degenerate (including spin degeneracy) at Γ ($m_j = \pm 3/2, \pm 1/2$), whereas the split-off $p_{1/2}$ band is doubly degenerate ($m_j = \pm 1/2$) there. However, away from the Γ-point, the degeneracy of the $p_{3/2}$ band is partially lifted, and two bands with different radii of curvature in $\boldsymbol{k}$-space result (in the absence of a magnetic field). Since the radius of curvature of a band is inversely proportional to the electron effective mass (see §6.2.1), the two higher bands are called the heavy-hole and light-hole bands. ('Holes', or an absence of electrons in an otherwise completely filled band, are the effective charge carriers in the valence band of semiconductors — see §6.2.2.)

5.4.3 Metallic crystals

Metallic crystals are characterized by the Fermi level lying within a band or bands of delocalized states. Band structures for two representative elemental metallic crystals, Al and Cu, are shown in Fig. 5.54. The energy bands of the s-p ($3s^2 3p^1$) metal, Al, are very nearly free-electron-like (the energy dependence of the corresponding density of states is also very close to the $\mathscr{E}^{1/2}$ behaviour expected for a free-electron gas — see Fig. 5.31). However, it can be seen from Fig. 5.54a that the Fermi level for Al intersects several bands, and so the Fermi surface is not a continuous spherical surface contained entirely within the first Brillouin zone, as for alkali metals, but, because the number of conduction electrons (3) is greater than that (2) required exactly to fill the first zone, the free-electron Fermi surface intersects parts of the next three zones (somewhat as in Fig. 5.24).

In contrast, the Fermi level for Cu ($3d^{10} 4s^1$) intersects only a single s-band in the Γ–X and Γ–K directions (see Fig. 5.54b), and so the Fermi surface is approximately spherical within the first Brillouin zone. However, it can be seen that it fails to intersect the band in the Γ–L direction: there is therefore a gap at L, as seen from the 'necks' in the Fermi

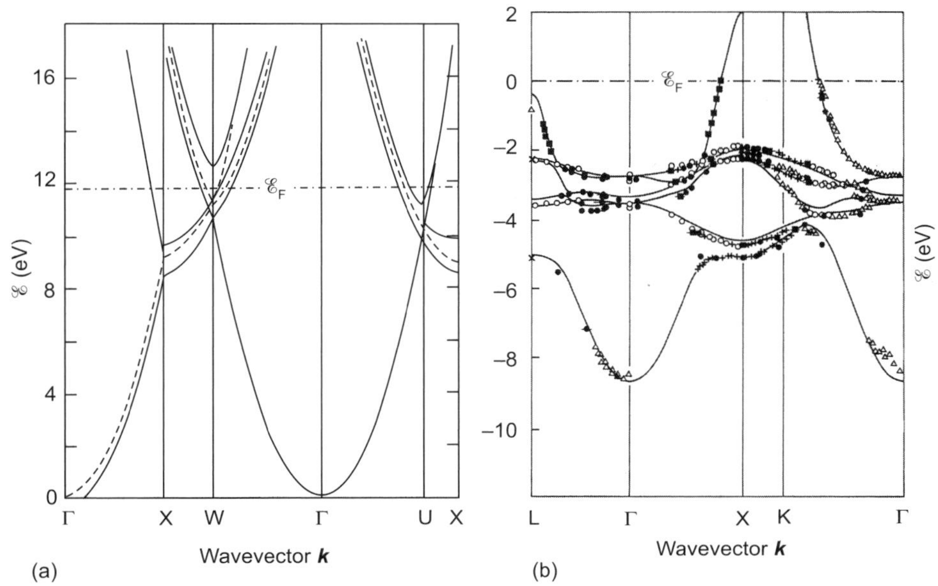

Fig. 5.54 Calculated band structures for elemental metallic crystals: (a) Al (the dashed curves are free-electron-like bands). (After Segal (1961)); (b) Cu (the points are experimental data). (Courths and Hüfner (1984), after Eckhardt *et al.* (1984) Reprinted from *Phys. Rep.* **112**, 53, Courths and Hüfner, © 1984 with kind permission from Elsevier Science - NL, Sara Burgerhartstraat 25, 1055 KV Amsterdam, The Netherlands)

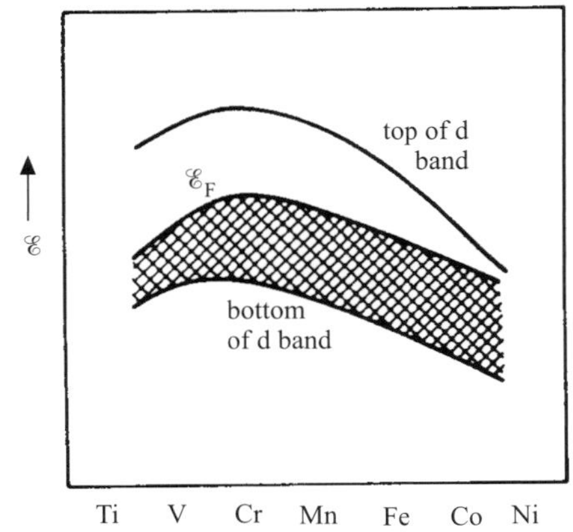

Fig. 5.55 Schematic illustration of the variation in energy of the d-band, and the position of $\mathscr{E}_F$ within it, for transition metals.

surface at such positions in the Brillouin zone (Fig. 5.25b). As for Al, the lowest bands are also free-electron-like, but in the middle of the energy range of the band structure are five rather flat bands corresponding to tightly-bound filled d-states (the Fermi level lies higher in energy than the d-bands in this case). It can be seen that s–d mixing causes the free-electron-like parabolic form of the lowest part of the s-band to be severely distorted in the region of the d-band: the two types of bands obey the no-crossing rule for bands of different symmetry (see Burns (1985) or Altmann (1991) for further details).

For transition metals, with partly filled d-bands, the Fermi level lies in the d-band, but its absolute position and relative position (say with respect to the bottom of the band) do not vary smoothly with band-filling (see Fig. 5.55). The centre of the d-band falls in energy with increasing band-filling (in excess of the half-filled state) because of increasingly ineffective shielding of the nuclear charge for one d-electron by the other d-electrons. At the same time, the states become more tightly bound and hence the bandwidth decreases.

Density of states 5.5

The density of electron states (the number of states in unit energy interval) is obtained from the band structure $\mathscr{E}(\boldsymbol{k})$ via an integration in $\boldsymbol{k}$-space over the Brillouin zone of a constant-energy surface of the band structure, as for vibrational modes (eqn. (4.58)), according to

$$D(\mathscr{E}) = \frac{2}{(2\pi)^3} \int_{\mathscr{E}(\boldsymbol{k})=\text{const}} \frac{dS_{\mathscr{E}}}{\nabla_k \mathscr{E}}, \tag{5.131}$$

where the electron-spin degeneracy has been explicitly included and $D(\mathscr{E})$ has units per energy per volume. Thus, flat bands in $\boldsymbol{k}$-space give rise to sharp features in the density

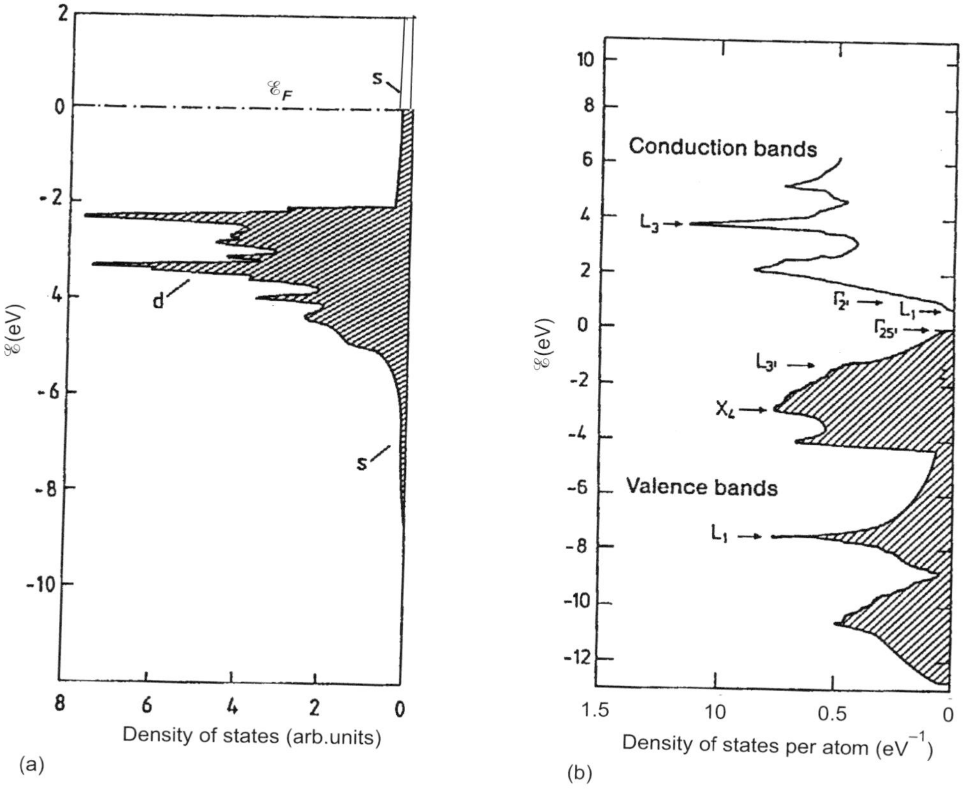

Fig. 5.56 Density of electron states for: (a) Cu, derived from the band structure given in Fig 5.54b. The parts of the density of states originating from s- and d-states are marked. Occupied states at $T = 0$ K are shaded. (After Ibach and Lüth (1995), *Solid State Physics*, p. 145, Fig. 7.12(a), © Springer-Verlag GmbH & Co. KG); (b) Ge, derived from a band structure as in Fig . 5.50b. The labelling of features corresponds to that of the band structure where $\nabla_k \mathscr{E} = 0$. Occupied states at $T = 0$ K are shaded. (After Herman *et al.* (1967). Reproduced by permission of Addison-Wesley-Longman)

of states (called van Hove singularities when $\nabla_k \mathscr{E} = 0$). The density of states of Cu (Fig. 5.56a) shows this behaviour. Superimposed on a very broad ($\simeq$ 12 eV wide) density-of-states curve corresponding to the s-bands is a highly structured, intense and narrow series of peaks corresponding to the narrow d-bands in the band structure. It can be seen that the density of (s-)states at the Fermi level is finite, and so the material is a metal.

It is instructive to compare Fig. 5.56a with Fig. 5.56b for the semiconductor crystalline Ge. The density of states for Ge has a lower band of states (the valence band) completely separated by a forbidden energy gap from an upper band of states (the conduction band). At $T = 0$ K, the valence band is completely filled with electrons and the conduction band is completely empty: the Fermi level lies in the gap where there are no states (see §6.5.1.1), and so the material is insulating at zero kelvin. Note that the gap in the density of states of an insulator/semiconductor corresponds to the *minimum* energy gap in the corresponding band structure (Fig. 5.50b); in the case of Ge, this is the indirect gap of $\mathscr{E}_g = 0.74$ eV between the valence band at Γ and the conduction band at L.

For the case of disordered materials, e.g. amorphous solids, where there is no translational periodicity, electron states cannot be described in terms of a band structure $\mathscr{E}(\boldsymbol{k})$, since the electron wavevector is ill-defined. However, the density of states is still valid as a description of electron states in disordered materials, although evidently it cannot be evaluated using eqn. (5.131). Instead, as for vibrational states (§4.3.2), it can be calculated, for example, as a sum of delta functions at allowed electron energies (cf. eqn. (4.118)). For the case of amorphous Si or Ge, for example, since the local atomic structure (short-range order) around a given atom is, on average, very similar to that in

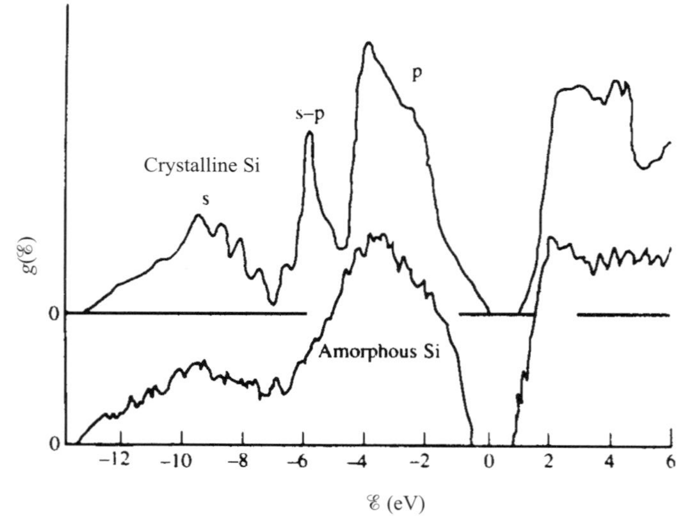

Fig. 5.57 Density of states for amorphous and crystalline Si calculated using the tight-binding approximation. (After Street (1991). Reproduced by permission of Cambridge University Press)

the crystal (i.e. tetrahedral coordination), it would be expected from the Weaire–Thorpe picture of the origin of the bands in terms of hybridized bond orbitals, presented in §5.3.2, that the overall density of states would be very similar for amorphous and crystalline modifications of the same material. This is indeed observed (Fig. 5.57), although several qualitative differences exist. Obviously, van Hove singularities are smeared out in the amorphous case due to the absence of translational periodicity. Furthermore, band edges in the density of states are sharp in the case of crystalline materials, varying as $(\mathscr{E} - \mathscr{E}_c)^{1/2}$ near the bottom of the conduction band at $\mathscr{E}_c$, or as $(\mathscr{E}_v - \mathscr{E})^{1/2}$ near the top of the valence band at $\mathscr{E}_v$ (since $\mathscr{E}(\boldsymbol{k})$ can be expanded quadratically in the vicinity of a critical point — see §4.2.4). However, the structural and, concomitantly, energetic disorder characteristic of amorphous solids results in a broadening of the bands at the extremities, thereby forming band tails. The origin of band tails can be understood by reference to Fig. 5.44. In the Weaire–Thorpe model, the bandwidth is governed by the on-site hopping integral between hybrid orbitals on the *same* atom; fluctuations in local structural order (e.g. bond-angle variations) in the amorphous phase result in fluctuations in this hopping integral, and hence lead to a broadening of the bands and the creation of band tails. The tail states deepest in the gap, furthest from the band edges, correspond to the most distorted sites.

**5.5.1 Local density of states: the moments theorem

The density of states defined by eqn. (5.131) is the *total* density of states for all (types of) atoms in a crystal. In certain cases, however, for example crystals with different types of atoms in the basis, or disordered materials such as those containing point defects, it is often informative to know also the *local* density of electron states, i.e. that associated with a particular single atom. As well as providing insight thereby into the origin of specific features in the total density of states, the local density of states is significant because information about its overall shape as a function of energy can be obtained very simply, as will be seen shortly, from a knowledge of the local topology of the atomic structure around the atom in question. Thus, much information about the total density of states (which is just a sum of the local densities of states) can be obtained from *real-space* considerations (See Sutton (1993), Pettifor (1995)), rather than via detailed calculations in reciprocal space using eqn. (5.131).

A general expression for the total density of states, valid for all types of materials, is as a sum of delta functions (cf. eqn. (4.118)):

$$g(\mathscr{E}) = \sum_{\mathscr{E}_i} \delta(\mathscr{E} - \mathscr{E}_i). \tag{5.132}$$

The *local* density of states, associated with a particular atom n, can be obtained by 'projecting' the contributions to the total quantity onto the atom in question. This is achieved by weighting each contribution, at $\mathscr{E}_i$, by a coefficient, P_n^i, that is the probability of finding an electron in an atomic state ϕ_n, localized on atom n. The eigenstate Ψ_i associated with energy $\mathscr{E}_i$ is expanded in an orthonormal basis set of such atomic states, i.e.

$$\Psi_i = \sum c_a^i \phi_a, \tag{5.133}$$

with the expansion coefficients being written as

$$\begin{aligned} c_a^i &= \int \phi_i^* \Psi_i \mathrm{d}\mathbf{r} \\ &\equiv \langle \phi_\mathrm{a} | \Psi_i \rangle \end{aligned} \tag{5.134}$$

in the so-called bra-ket notation. The probability coefficient α_n^i is thus given by

$$P_n^i = |\langle n | \Psi_i \rangle|^2, \tag{5.135}$$

where $\langle n| \equiv \phi_n^*$, and hence the local density of states on atom n is given by the expression

$$d_n(\mathscr{E}) = \sum_{\mathscr{E}_i} P_n^i \delta(\mathscr{E} - \mathscr{E}_i). \tag{5.136}$$

The total density of states is recovered by summing over all local contributions:

$$\begin{aligned} g(\mathscr{E}) = \sum_n d_n(\mathscr{E}) &= \sum_n |\langle n | \Psi_i \rangle|^2 \sum_{\mathscr{E}_i} \delta(\mathscr{E} - \mathscr{E}_i) \\ &= \sum_{\mathscr{E}_i} \delta(\mathscr{E} - \mathscr{E}_i), \end{aligned} \tag{5.137}$$

since the normalization condition for the orthonormal atomic basis set is $\sum_n |\langle n | \Psi_i \rangle|^2 = 1$.

The moments theorem of Cyrot-Lackmann (1968) states that, in the tight-binding approximation, the mth moment of the local density of states on atom n is just related to the sum of all closed paths, consisting of m hops of an electron between nearest-neighbour atoms in the structure, starting and finishing at atom n.

The mth moment of any distribution function $f(x)$ about the origin is defined as

$$\mu^{(m)}(0) = \int_{-\infty}^{\infty} x^m f(x) \mathrm{d}x \tag{5.138a}$$

and the moment about, say, the mean $\bar{x}$, is

$$\mu^{(m)} = \int_{-\infty}^{\infty} (x - \bar{x})^m f(x) \mathrm{d}x. \tag{5.138b}$$

For the case of the local density of states on atom n, the appropriate expression for the mth moment is

$$\mu_n^{(m)} = \int_{\text{band}} (\mathscr{E} - \mathscr{H}_{nn})^m d_n(\mathscr{E}) \mathrm{d}\mathscr{E}, \tag{5.139}$$

where the Hamiltonian matrix element $\mathscr{H}_{nn}$ is given by

$$\mathscr{H}_{nn} = \int \phi_n^* \mathscr{H} \phi_n \mathrm{d}\mathbf{r} \equiv \langle n | \mathscr{H} | n \rangle; \tag{5.140}$$

this is equal to $\mathscr{E}_n - \alpha_n$, the tight-binding on-site energy (cf. eqn. (5.96a)). Substitution of eqn. (5.136) into eqn. (5.139) gives for the mth moment:

$$\mu_n^{(m)} = \int_{\text{band}} \sum_{\mathscr{E}_i} (\mathscr{E} - \mathscr{H}_{nn})^m \langle n | \Psi_i \rangle \langle \Psi_i | n \rangle \delta(\mathscr{E} - \mathscr{E}_i) \mathrm{d}\mathscr{E}. \tag{5.141}$$

The operation of the delta function in eqn. (5.141) ensures that all terms in the integration are zero except for $\mathscr{E} = \mathscr{E}_i$, and hence

$$\mu_n^{(m)} = \sum_{\mathscr{E}_i} \langle n|\Psi_i\rangle(\mathscr{E}_i - \mathscr{H}_{nn})^m\langle\Psi_i|n\rangle \tag{5.142}$$

Equation (5.142) can be regarded as simply the matrix element O_{nn} of an operator $\hat{O}$, where

$$\hat{O} = \sum_{\mathscr{E}_i} |\Psi_i(\mathscr{E}_i - \mathscr{H}_{nn})^m\langle\Psi_i| \tag{5.143}$$

or, alternatively,

$$\hat{O} = (\mathscr{H} - \mathscr{H}_{nn})^m, \tag{5.144}$$

since $\mathscr{H} = \sum_{\mathscr{E}_i} |\Psi_i\rangle\mathscr{E}_i\langle\Psi_i|$. Thus, finally, the mth moment can be written as

$$\mu_n^{(m)} = \langle n|(\mathscr{H} - \mathscr{H}_{nn})^m|n\rangle \tag{5.145}$$

The zeroth moment, $\mu_n^{(0)}$, is unity, since $\langle n|n\rangle = 1$. The first moment, $\mu_n^{(1)}$ is zero, since $\langle n|(\mathscr{H} - \mathscr{H}_{nn})|n\rangle = \mathscr{H}_{nn} - \mathscr{H}_{nn}\langle n|n\rangle = 0$. Therefore, the first moment, which gives the relative position of the mean of a distribution, predicts, for the case of the local density of states, that the mean coincides with the matrix element $\mathscr{H}_{nn}$.

The second moment gives the *variance* of a distribution: the square root of $\mu^{(2)}$ therefore gives the standard deviation with respect to the mean, and hence is a measure of the width of the distribution. For the local density of states, the second moment can be written as

$$\begin{aligned}\mu_n^{(2)} &= \langle n|(\mathscr{H} - \mathscr{H}_{nn})(\mathscr{H} - \mathscr{H}_{nn})|n\rangle \\ &= \sum_{n'} \langle n|(\mathscr{H} - \mathscr{H}_{nn})|n'\rangle\ \langle n'|(\mathscr{H} - \mathscr{H}_{nn})|n\rangle.\end{aligned} \tag{5.146}$$

However, in the tight-binding approximation, where only interactions between nearest neighbours are significant, the only non-zero matrix elements are therefore $\mathscr{H}_{nn'} = \langle n|\mathscr{H}|n'\rangle$ for n and n' being nearest neighbours. Thus, eqn. (5.146) reduces to

$$\mu_n^{(2)} = \sum_{n' \neq n} \mathscr{H}_{nn'}\mathscr{H}_{n'n}, \tag{5.147}$$

since $\langle n|\mathscr{H} - \mathscr{H}_{nn}|n\rangle = 0$ and $\langle n|\mathscr{H} - \mathscr{H}_{nn}|n'\rangle = \mathscr{H}_{nn'}$ for $n \neq n'$. The interpretation of eqn. (5.147) is therefore that the second moment is the sum over all paths in which an electron starts at atom n, hops to neighbouring atom n' (with associated matrix element $\mathscr{H}_{nn'} = \beta$, the nearest-neighbour hopping integral, eqn. (5.96b)) and back again to n. For a structure with z nearest neighbours, the second moment is therefore given by

$$\mu_n^{(2)} = z\beta^2; \tag{5.148}$$

the r.m.s. width (standard deviation) of the local density of states thus scales as $z^{1/2}$.

The expression for the third moment, $\mu_n^{(3)}$, can be inferred from eqn. (5.147), viz.

$$\mu_n^{(3)} = \sum_{n' \neq n} \sum_{n'' \neq n,n'} \mathscr{H}_{nn'}\mathscr{H}_{n'n''}\mathscr{H}_{n''n}, \tag{5.149}$$

and involves closed paths of three hops. It determines the *skewness* of the distribution about the mean $\mathscr{H}_{nn}$: a large negative value of $\mu_n^{(3)}$, for example, corresponds to a distribution skewed to values *less* than the mean (as in Fig. 5.58a).

The value of the fourth moment, $\mu_n^{(4)}$, defined by analogy with eqns. (5.148) and (5.149), determines whether the distribution is unimodal or bimodal via the shape parameter, defined as

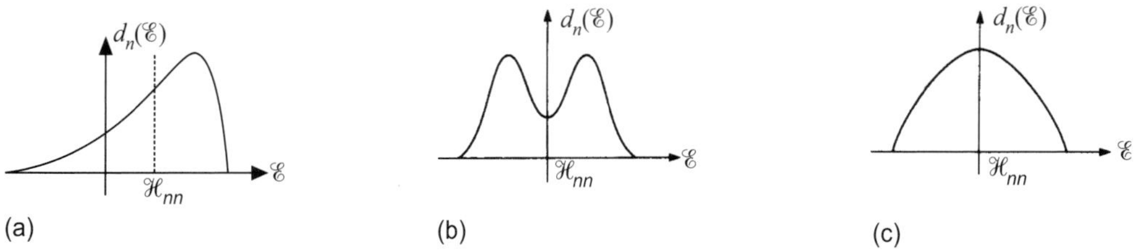

Fig. 5.58 (a) Illustration of a local density of states distribution for which the third moment is negative, $\mu_n^{(3)} < 0$. (b) Bimodal local density of states, for which the shape parameter s involving the fourth moment is less than unity. (c) Unimodal local density of states, for which $s \geqslant 1$.

$$s = (\mu^{(4)}\mu^{(2)} - [\mu^{(2)}]^3 - [\mu^{(3)}]^2)/[\mu^{(2)}]^3 \tag{5.150a}$$

$$= \left(\mu^{(4)}/[\mu^{(2)}]^2\right) - 1 \quad : \mu^{(3)} = 0. \tag{5.150b}$$

The condition $s < 1$ corresponds to bimodal behaviour (Fig. 5.58b) and $s \geqslant 1$ to unimodal behaviour (Fig. 5.58c).

An illustration of the use of the moments approach is for the case of a 3D simple-cubic lattice with identical s-states at each site. The nearest-neighbour coordination number is $z = 6$ so, from eqn. (5.148), $\mu_n^{(2)} = 6\beta^2$. The third moment is zero because there are *no* closed three-membered paths between nearest neighbours for this lattice. The fourth moment may be computed for the paths shown in Fig. 5.59: $\mu_n^{(4)} = 90\beta^4$. Thus, the shape parameter from eqn. (5.150b) is $s = 1.5$, and so the local density of states is predicted to be unskewed (symmetric about the mean) and unimodal. This is seen in Fig. 5.37b. Problem 5.22 uses the moments method to discuss the local density of states for a disordered system, consisting of a single substitutional impurity (akin to the impurity vibrational mode discussed using a Green function method in §4.3.1).

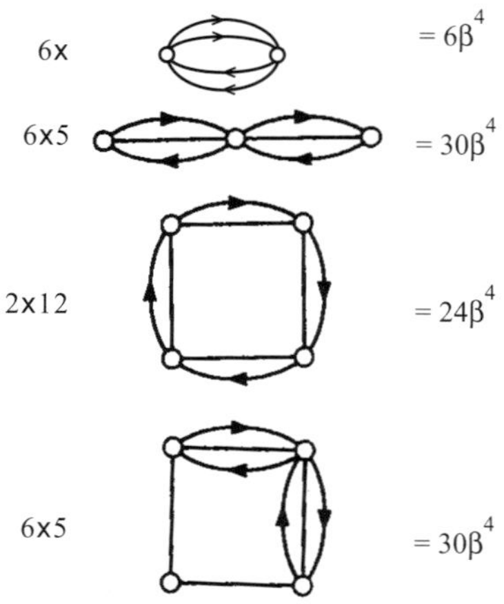

Fig. 5.59 Illustration of four-fold paths contributing to $\mu_n^{(4)}$ in the simple cubic lattice.

Breakdown of the independent-electron approximation 5.6

Much of the discussion so far of electron states in solids has assumed that the electrons behave effectively independently: e.g. electron–electron interactions have been neglected. At first sight, this behaviour appears very surprising. In Cu, for instance, the conduction-electron concentration is $8.45 \times 10^{28}\ \mathrm{m}^{-3}$, corresponding to an average inter-electron separation of $r = 2.56$ Å. Thus, the Coulombic repulsive potential energy between two such electrons is $e^2/4\pi\varepsilon_0 r \simeq 5.6$ eV, compared with the average kinetic energy per electron of the free-electron gas (eqn. (5.20)) $3\mathscr{E}_F/5 \simeq 4.2$ eV. It seems remarkable that such large electron–electron interactions can be neglected and that, for example, band theory built on the concept of independent-electron Bloch wavefunctions (§5.2.1) should work so well. The resolution of this apparent paradox depends on a number of features. One aspect is that, as will be seen in the next section (§5.6.1), the electrostatic interaction between two electrons is, to a considerable extent, screened (i.e. decreased) by other mobile conduction electrons. Secondly, in fact, it is not the electrons themselves that behave independently, but *quasiparticles* comprising electrons and their conjugate 'exchange-correlation holes' (see §§2.5.3.4 and 5.1.3.2). Such an electron system, where interactions are explicitly considered, is termed a Landau Fermi liquid to distinguish it from the *Fermi gas* where such interactions are absent. Thirdly, the probability of electron–electron collisions is very greatly decreased by the influence of the Pauli exclusion principle (see §6.3.1). Finally, on close examination, the band theory of electrons is not so universally successful as it appears at first sight. For example, although band theory predicts that any material for which the Fermi level lies within a band of delocalized electron states should be metallic, in fact perhaps half of all such binary materials for which this prediction might expect to be satisfied are insulators! This discrepancy is a rather clear indication of the breakdown of the independent-electron approximation. Various aspects of this problem will be explored in the following sections. Superconductivity, which is a spectacular instance of the breakdown of the independent-electron approximation, is discussed in § 6.4.

5.6.1 Screening

If a positive charge, say *Ze*, is immersed in a distribution of electrons that are free to move (e.g. the free-electron gas in a metal), electrons are attracted towards the extra charge, thereby leading to a surplus of negative charge density in the vicinity of the positive charge; this space charge therefore shields, or screens, the electrostatic potential $\phi(r)$ associated with the extraneous charge at distances remote from the charge, the reduction factor being the dielectric constant ε:

$$\phi(r) = \frac{Ze}{4\pi\varepsilon\varepsilon_0 r}. \tag{5.151}$$

For the case of *dielectric materials* (see §7.1), where electrical charges are *bound* to ion cores, the screening is only partial, and ε has a relatively small value. For metals with mobile electrons, the screening is *total* at large distances from the extraneous ion, and so $\varepsilon \to \infty$. However, the screening is effectively complete at much shorter distances, of the order of 1 Å, and this is one reason why electrons appear to behave independently at distances greater than this screening distance.

An approximate treatment of electron screening can be made in the Thomas–Fermi approximation, where the spatial variation of the perturbation in electrostatic potential is assumed to be much more long-ranged than that associated with the electrons, e.g. the wavelength at the Fermi level $\lambda_F(= 2\pi/k_F)$. A local perturbation in the potential, $\delta\phi(r)$, causing the free-electron density-of-states curve, $D(\mathscr{E}) = g(\mathscr{E})/V \propto \mathscr{E}^{1/2}$, (where V is the volume) to be lowered locally by an energy $|e|\delta\phi$ is equivalent to the placing of a positive charge in the electron gas. Since the chemical potential must everywhere be constant in the solid in equilibrium, electrons must therefore flow towards the local perturbation. As long as the perturbation potential is not too large ($|e|\delta\phi \ll \mathscr{E}_F$), the change in electron concentration, $\delta n(r)$, is given by

$$\delta n(r) = |e|D(\mathscr{E}_F)\delta\phi(r). \tag{5.152}$$

Away from the immediate vicinity of the inserted point charge, Poisson's equation also relates δn and $\delta\phi$ via

$$\nabla^2(\delta\phi) = \frac{e}{\varepsilon_0}\delta n = \frac{e^2}{\varepsilon_0}D(\mathscr{E}_F)\delta\phi, \tag{5.153}$$

where ε_0 is the vacuum permittivity. This equation can be solved straightforwardly in real space if spherical symmetry of the perturbation potential is assumed, e.g. if a point charge is introduced into the electron gas. In this case, spherical polar coordinates can be used, for which

$$\nabla^2 \equiv \frac{1}{r^2}\frac{\partial}{\partial r}\left(r^2\frac{\partial}{\partial r}\right),$$

whence it can be shown that the solution is of the form (see also Problem 5.23):

$$\delta\phi(r) = \frac{\alpha e^{-r/\lambda_{TF}}}{r}, \tag{5.154}$$

where λ_{TF} is the Thomas–Fermi screening length. The boundary condition is such that as $\lambda_{TF} \to \infty$, the screening effect must vanish and the normal Coulomb electrostatic potential must be recovered, i.e. $\alpha = e/4\pi\varepsilon_0$. The Thomas–Fermi screening length λ_{TF} is obtained by solving the Poisson equation, substituting eqn. (5.154) into eqn. (5.153), giving

$$\lambda_{TF} = \left[\frac{e^2D(\mathscr{E}_F)}{\varepsilon_0}\right]^{-1/2} \tag{5.155a}$$

for the case of the free-electron gas, where $D(\mathscr{E}_F)$ is given by combining eqns. (5.16) and (5.17), and so

$$\lambda_{TF} = \frac{\pi\hbar}{e}\left(\frac{\varepsilon_0}{m_e}\right)^{1/2}(3\pi^2 n_0)^{-1/6} \equiv \frac{(\pi a_0)^{1/2}}{2}(3\pi^2 n_0)^{-1/6} \tag{5.155b}$$

$$\simeq 0.5\left(\frac{a_0^3}{n_0}\right)^{1/6}, \tag{5.155c}$$

where a_0 is the Bohr radius ($= 4\pi\hbar^2\varepsilon_0/m_e e^2$) and n_0 is the average electron density. Thus, for Cu, for example, the Thomas–Fermi screening length is $\lambda_{TF} = 0.55$ Å.

The screened potential given by eqn. (5.154) is shown in Fig. 5.60, compared with the unscreened Coulomb potential. It can be seen that for distances greater than about $3\lambda_{TF}$

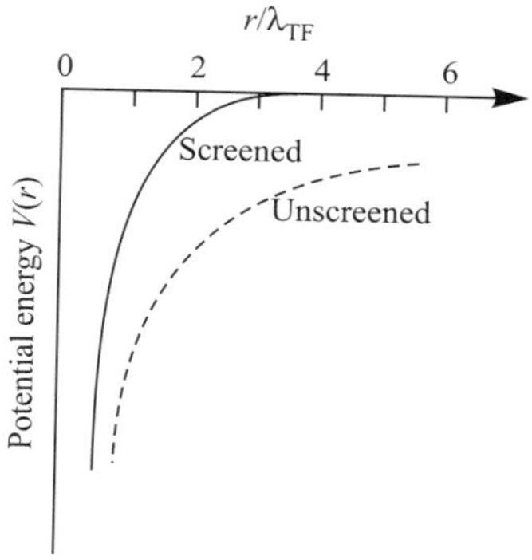

Fig. 5.60 Thomas–Fermi screened electrostatic potential (solid curve) and unscreened Coulomb potential (dashed curve) plotted as a function of distance normalized by the Thomas–Fermi screening length λ_{TF}.

from the origin of the perturbation, the screening is practically complete. Note also that the Thomas–Fermi form of the screened potential (eqn. (5.154)) implies that the (spatially varying) dielectric function is given by $\varepsilon(r) = \exp(r/\lambda_{TF})$. As r becomes greater than λ_{TF}, ε increases strongly and tends to infinity as $r \to \infty$.

Unfortunately, the Thomas–Fermi model is a poor approximation for real metals since the length scale of the perturbing ionic potential is set by the cut-off distance r_c in the pseudopotential (eqn. (5.42)) which is comparable to λ_{TF}, and so the slowly varying condition necessary for the Thomas–Fermi approximation to be valid is not met.

★ A more exact treatment, taking into account the decrease of the screening efficiency with decreasing distance for $r \lesssim \lambda_{TF}$, finds that the screening electron density δn and the potential $\delta\phi$ are related via the Lindhard screening response function, $\chi(\boldsymbol{q})$,

$$\delta n(\boldsymbol{q}) = \chi(\boldsymbol{q})\delta\phi(\boldsymbol{q}), \tag{5.156}$$

where (see Ashcroft and Mermin (1976) and Pettifor (1995)):

$$\chi(\boldsymbol{q}) = -eD(\mathscr{E}_F)\left[\frac{1}{2} + \frac{1-x^2}{4x}\ln\left|\frac{1+x}{1-x}\right|\right], \tag{5.157}$$

with $x = q/2k_F$. Note that this expression exhibits a logarithmic singularity at $x = 1$, manifest as a singularity (Kohn anomaly) in the phonon dispersion curves (§8.3.2), since the Lindhard term describes the electronic screening of ion–ion Coulombic interactions. (Compare this with eqn. (5.152) for the Thomas–Fermi approximation where $\chi = eD(\mathscr{E}_F)$ is *independent* of $\boldsymbol{q}$.) Instead of the screening electron density falling off exponentially, as in the Thomas–Fermi model according to eqn. (5.152), rewritten making use of $D(\mathscr{E}_F) = 3n_0/2\mathscr{E}_F$ (eqn. 5.16)), where n_0 is the average electron density of the electron liquid, i.e.

$$\begin{aligned}\delta n(\mathbf{r}) &= \frac{3n_0e^2}{8\pi\varepsilon_0\mathscr{E}_F}\frac{e^{-r/\lambda_{TF}}}{r} \\ &\equiv \frac{e^{-r/\lambda_{TF}}}{4\pi\lambda_{TF}^2 r},\end{aligned} \tag{5.158}$$

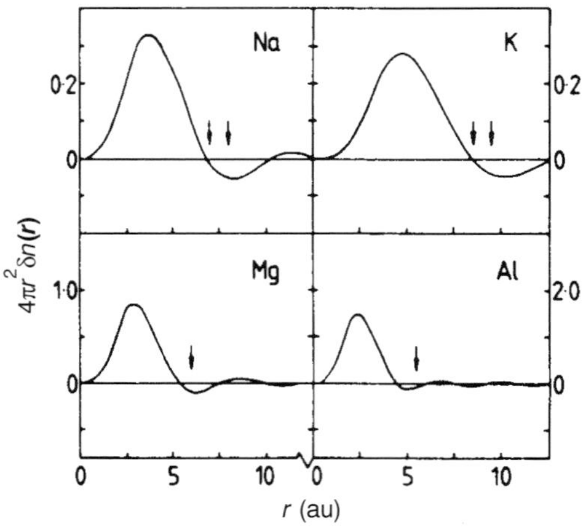

Fig. 5.61 Radial screening charge density of simple metals showing Friedel oscillations. The arrows mark the positions of nearest neighbours in h.c.p. Mg and f.c.c. Al, and nearest and next-nearest neighbours in b.c.c. Na and K. (After Pettifor (1995), *Bonding and Structure of Molecules and Solids*, by permission of Oxford University Press)

an oscillatory algebraic spatial decay is predicted by the Lindhard model, viz.

$$\delta n(\boldsymbol{r}) \simeq \frac{A_0 \cos 2k_F r}{r^3}, \tag{5.159}$$

where k_F is the Fermi wavevector. Such Friedel oscillations in charge density are shown in Fig. 5.61 and give rise to oscillatory interatomic pair potentials that determine the detailed atomic structure of simple metals (see Pettifor (1995)).

★

5.6.2 Plasmons

In metals, the electron concentration is so high that the charge redistribution associated with screening occurs on such very short distance and time scales that the equations of motion for the electron liquid do not involve electron scattering or collisions. Thus, the length scale of charge redistribution is shorter than the electron mean free path, and the characteristic time scale shorter than the times between collisions. At first sight, this circumstance seems unlikely: the electron–electron interactions in a dense electron distribution, producing Fermi-liquid rather than Fermi-gas behaviour, might be expected to result in very short times between collisions of quasiparticles. However, as will be seen in Chapter 6, the Pauli exclusion principle means that not all possible collisions between electrons are allowed, and the average time between collisions is correspondingly much longer. Hence, instead of having to use *diffusion* equations of motion based on scattering (see §3.4.2), as for electrons in semiconductors (§6.5), in metals Newton's laws of motion can be used directly for the electron motion. A collective vibrational mode, involving *all* electrons, is found to be possible — another example of the breakdown of the independent-electron approximation.

In the case where electrons are subject to an electric field $\boldsymbol{E}$ and a concentration gradient ∇n, or the associated driving 'force', the pressure gradient ∇p, the equation of motion is given by

$$m_e n \frac{\mathrm{d}v}{\mathrm{d}t} = -ne\mathscr{E} - \nabla p, \tag{5.160}$$

where v is the electron velocity. The jellium model (uniform positive-ion charge density) is assumed for simplicity. Fluctuations of the electron density from the average value n_0 generate an electric field that is given by Gauss's law:

$$\operatorname{div} \mathscr{E} = \frac{-e(n - n_0)}{\varepsilon_0}. \tag{5.161}$$

Taking the pressure to be that of a Fermi gas, given by $p = 2U/3V$ (see Problem 5.9), where the internal energy is $U \simeq 3n\mathscr{E}_F/5$ (neglecting the small temperature-dependent term in eqn. (5.35)), therefore gives

$$\nabla p \simeq \frac{2}{3}\mathscr{E}_F \nabla n. \tag{5.162}$$

Hence, making use of the continuity equation

$$\frac{\partial n}{\partial t} + \operatorname{div}(n\boldsymbol{v}) = 0, \tag{5.163}$$

the equation of motion of the electron system (eqn. (5.160)) becomes

$$\frac{\partial^2 n}{\partial t^2} = \frac{2}{3}\frac{\mathscr{E}_F}{m_e}\nabla^2 n - \frac{n_0 e^2}{m_e \varepsilon_0}(n - n_0). \tag{5.164}$$

The time-independent case ($\partial^2 n/\partial t^2 = 0$) simply gives the static screening described in §5.6.1.

In the other limit of a slow spatial variation of n (i.e. $\nabla^2 n = 0$), eqn. (5.164) reduces to

$$\frac{\partial^2 n}{\partial t^2} = -\frac{n_0 e^2}{m_e \varepsilon_0}(n - n_0). \tag{5.165}$$

This can be recognized as the simple harmonic oscillator equation, with the angular frequency given by

$$\omega_p = \left(\frac{n_0 e^2}{m_e \varepsilon_0}\right)^{1/2}. \tag{5.166}$$

Note that damping effects, related to the electrical conductivity and electron–phonon interactions, are neglected.

These long-wavelength collective oscillations are termed plasma oscillations, the quantized versions of which are called plasmons: ω_p is the plasma frequency. (A plasma is normally a gas of an equal number of negatively and positively charged particles that are free to move: in a metal, however, only the electrons are mobile.) Note that, in contrast to the vibrations of the ion cores, i.e. acoustic phonons (§4.1.1) where the dispersion curve is such that $\omega \to 0$ as $k \to 0$ (or the wavelength $\lambda \to \infty$), plasma oscillations have a *finite* frequency in the long-wavelength limit. This difference in behaviour is due to the interatomic interactions being short-ranged, in contrast to the long-ranged Coulomb interaction between electrons.

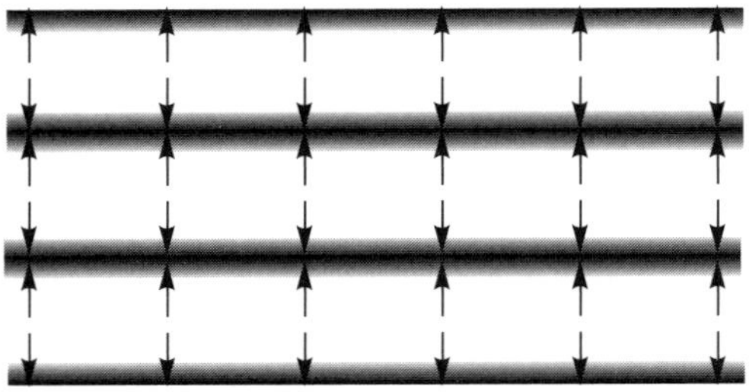

Fig. 5.62 Illustration of a longitudinal plasma oscillation (plasmon). The arrows indicate the displacement directions of the electrons.

Plasma oscillations are collective vibrational modes of the entire electron liquid: these modes are *longitudinal* in polarization (see Fig. 5.62) since fluids cannot support shear, or transverse, vibrational modes (see Problem 3.21 and §4.1.1). By analogy with the treatment of polaritons given in §4.4, the longitudinal plasma-oscillation frequency, ω_L, is given by the condition when the dielectric constant of the electron liquid is zero, $\varepsilon(\omega = \omega_L) = 0$. The dielectric constant of a metal (in the absence of damping) is given by (see Problem 5.25):

$$\varepsilon(\omega) = 1 - \left(\frac{\omega_p}{\omega}\right)^2, \tag{5.167}$$

where ω_p is given by eqn. (5.166). Hence, the plasma frequency is $\omega_L = \omega_p$. A more complete expression for $\varepsilon(\omega)$ is represented by eqn. (5.196).

For the case of Cu, for example, for which $n_0 = 8.45 \times 10^{28}\ \mathrm{m}^{-3}$, eqn. (5.166) gives for the plasma frequency $\omega_p = 1.6 \times 10^{16}\ \mathrm{s}^{-1}$, or a plasmon energy of $\hbar\omega_p = 10.8$ eV. Since this energy is so large, collective-mode plasmons are not thermally excited at normal temperatures: the electron liquid therefore remains in its ground state as far as such long-wavelength excitations are concerned, and this possible instance of the breakdown of the independent-electron approximation is usually averted.

Since plasmons are *longitudinal* waves, they do not couple to transverse electromagnetic waves, and hence cannot be probed directly (but see §5.8.1). However, external electrons can couple to the longitudinal electric field associated with the plasma oscillations. This property forms the basis of the use of electron energy-loss spectroscopy (EELS) as a probe for plasmons: inelastic scattering of electrons incident on a film of a solid can occur by the creation of plasmons in the conduction band of metals, or the filled valence band of semiconductors. The plasma frequency corresponds to peaks in the energy-loss function, $-\mathrm{Im}\{1/\varepsilon(\omega, \boldsymbol{k})\}$ (see Problem 5.26), in other words, when $\varepsilon(\omega) = 0$.

5.6.3 Metal–insulator transitions

One of the most spectacular manifestations of the breakdown of the independent-electron picture is the occurrence of discontinuous transitions between metallic and insulating states as the electron concentration in the system is changed in some way, e.g.

as a function of composition, and the corresponding occurrence of solids that are electrically insulating when band theory predicts metallic properties. The reason for this behaviour can be regarded as being due to a competition between kinetic and potential energies of the electrons: the kinetic energy of electrons is lowered by their *delocalization* (because of Heisenberg's uncertainty principle), i.e. favouring *metallic* behaviour; on the other hand, the potential energy of electrons (which can involve electron–electron interactions) is lowered by their *localization*, i.e. favouring *insulating* behaviour.

The discontinuous nature of metal–insulator transitions in certain cases is also associated with electron–electron interactions, specifically screening (§5.6.1), as in the Mott model for the occurrence of metal–insulator transitions. An isolated charged ion core in a solid (e.g. a phosphorus 'donor' impurity atom (see § 6.5.2) in crystalline silicon) associated with a simple Coulombic potential has an infinite number of bound hydrogenic-like electron energy levels associated with it. An electron with a total energy less than the ionization energy will always be bound by such a potential. However, in the presence of a sufficiently high concentration of mobile conduction electrons, the ion potential is screened to give a much weaker potential away from the ion core (see Fig. 5.60). This screened potential only supports a bound state if

$$\lambda_{TF} \gtrsim a_0^*, \tag{5.168}$$

where a_0^* is the effective Bohr radius of an electron in the solid, modified from the hydrogenic value of the Bohr radius valid *in vacuo* by the inclusion of the dielectric constant (to account for screening) and the band-structure effective mass $m_e^* \neq m_e$ (§6.2.1) arising from the fact that the electrons are not completely free to move in crystals but must move according to allowed energy bands $\mathscr{E}(\boldsymbol{k})$; thus

$$a_0^* = \frac{\varepsilon\varepsilon_0 h^2}{\pi m_e^* e^2}. \tag{5.169}$$

Hence delocalized, metallic behaviour arises if the screening length is less than the effective Bohr radius, i.e.

$$\lambda_{TF} \lesssim a_0^*. \tag{5.170}$$

or, in other words, when the screening is so strong that no electrons can be bound to ion cores. The Mott criterion can also be rewritten as

$$n_c^{1/3} a_0^* = 0.25, \tag{5.171}$$

using eqn. (5.156) for λ_{TF}, where n_c is the critical electron density for metallic behaviour.

It is remarkable that several, apparently very different, models for the metal–insulator transition give very similar conditions for metallic behaviour to eqn. (5.171), and a comprehensive plot of n_c versus a_0^* for many different systems showing metal–insulator transitions (Fig. 5.63) gives an empirical value for the constant $C = n_c^{1/3} a_0^* = 0.26$.

The breakdown of the conventional band picture can be illustrated by the following gedanken experiment involving alkali-metal atoms, e.g. Li. In the condensed crystalline phase, the electron states form nearly-free-electron-like bands which, because they are half-filled, means that the material is metallic. If, now, the crystal is considered to be progressively uniformly dilated, band theory dictates that the system should remain metallic for *all* interatomic separations. Even in going from the NFE limit to the

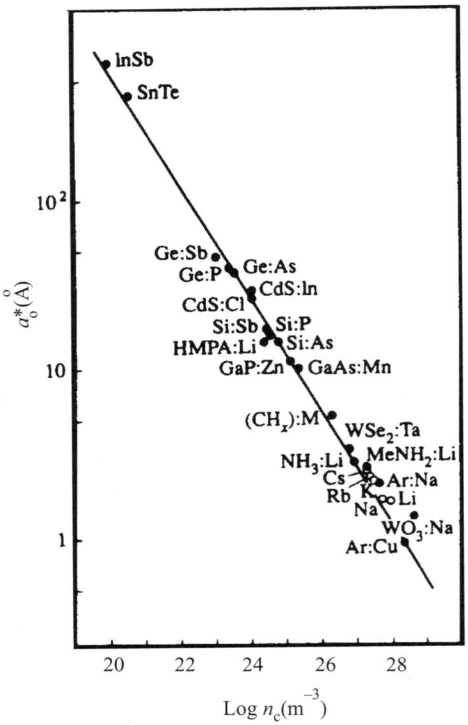

Fig. 5.63 Logarithmic plot of the effective Bohr radius, a_0^*, versus critical electron concentration, n_c, for the metal–insulator transition in a variety of doped (impurity-containing) semiconductors (•) and supercritical alkali metal vapours (○). The straight line corresponds to the criterion $n_c^{1/3} a_0^* = 0.26$. (After Edwards and Sienko (1981). Reprinted with permission from *J. Am. Chem. Soc.* **103**, 2967. © 1981 American Chemical Society)

tight-binding limit with increasing separation, bands (albeit narrow) will still form in this picture and, because such bands are still half-filled with the Fermi level lying within the band, the metallic character will persist. However, it is obvious that, at some critical interatomic separation, this picture must break down for, in the other extreme limit, isolated alkali atoms are electrically insulating.

In fact, the metal–insulator transition occurs near the thermodynamic critical point (i.e. at the extremity of the vapour–liquid tie-line in the phase diagram, where vapour and liquid become an indistinguishable fluid): for Li, for example, the critical liquid density where the metal–insulator transition occurs is (from Fig. 5.63) about one-fifth of the normal crystal density. A somewhat related example is for solutions of alkali metals in liquid ammonia. At low alkali concentrations, the free electrons liberated by the alkali atoms are solvated by surrounding NH_3 molecules and such metal–ammonia solutions are electrically insulating. However, above a critical metal concentration (e.g. $\simeq 15$ mole % for Na) such materials become *metallic* in appearance (e.g. high optical reflectivity — see Plate IV) and behaviour (e.g. high electrical conductivity).

It is instructive to examine the situation regarding electronic excitations for the extreme limit where isolated alkali atoms occur. Instead of there being, as in the band case, vacant electron energy levels immediately above $\mathscr{E}_F$, so that electronic excitation is not thermally activated (i.e. metallic behaviour), electron excitation between neutral isolated alkali atoms requires a large energy. In order to remove an electron from an atom, the ionization energy I_{el} must be supplied and, although some energy is recovered as the electron affinity χ when this electron is added to another neutral atom to give a negatively charged ion, the difference between these two energies, the Hubbard (on-site) energy U_H, the net energy to place two electrons in a given orbital, i.e.

$$U_H = I_{el} - \chi \tag{5.172}$$

is non-zero and large (typically several eV). In the condensed state, on-site electron–electron repulsion still occurs, and can be expressed as (Mott (1974)):

$$U_H = \iint \frac{e^2}{4\pi\varepsilon_0 r_{12}} |\phi(r_1)|^2 |\phi(r_2)|^2 \mathrm{d}\boldsymbol{r}_1 \mathrm{d}\boldsymbol{r}_2, \tag{5.173}$$

where r_{12} is the inter-electron separation at a site, and $\phi(r)$ is the electron wavefunction at the site. In other words, in this model, inter-electron repulsion is neglected except for two electrons at the same site. Conventional band theory, which neglects electron interactions, therefore does not take into account the effects of this Hubbard energy.

Consideration of the Hubbard energy is particularly important for materials such as transition-metal oxides (see e.g. Cox (1992)) that have narrow (d-)bands where, when the electron concentration is not very high and hence screening is ineffective, the correlation between electron motion is much more pronounced than is predicted within the conventional band picture (see §5.1.3.2) and the Hubbard model comes into play. For the case of NiO, for example, which conventional band theory predicts to be metallic, motion of an electron between Ni d-states ($\mathscr{E}_F$ lies within the d-manifold) can be represented as the disproportionation reaction:

$$Ni^{2+} + Ni^{2+} \rightarrow Ni^{+} + Ni^{3+} \tag{5.174a}$$

or equivalently as the d-electron 'reaction'

$$d^8 + d^8 \rightarrow d^9 + d^7. \tag{5.174b}$$

In the Hubbard picture, it costs an energy U_H to place an extra electron on an Ni^{2+} site to make an Ni^{+} ion; hence a gap must open up in the otherwise continuous band of d-states, making the material electrically insulating. In fact the gap is not generally equal to the Hubbard energy U_H itself because the gap is decreased by bandwidth effects. In the limit of infinite separation between Ni^{+} and Ni^{3+} ions, the gap is equal to U_H, given by eqn. (5.172), and the two states on the right-hand side of eqns. (5.174) are atomic levels. As the interionic separation decreases, Hubbard bands (*not* the same as conventional bands) form, the lower Hubbard band corresponding to the motion of 'holes' (i.e. an absence of electrons) among the ion sites:

$$Ni^{2+} + h^{+} \rightarrow Ni^{3+}, \tag{5.175a}$$

and the upper band to the motion of the extra electrons:

$$Ni^{2+} + e^{-} \rightarrow Ni^{+}, \tag{5.175b}$$

The bandwidth, W, of these bands will increase with decreasing interatomic separation (Fig. 5.64) as it becomes easier for electrons and holes to hop between sites. The Hubbard gap, $\mathscr{E}_g^H = U_H - W$, vanishes when the upper and lower Hubbard bands begin to overlap and this situation then marks the Mott–Hubbard transition from the insulating to the metallic state.

The condition for metallic behaviour, i.e. overlap of the Hubbard bands, occurs when

$$W \geqslant U_H. \tag{5.176}$$

An estimate for the critical interatomic spacing R corresponding to this criterion can be obtained if, for simplicity, it is assumed that the electronic wavefunctions at each site in a simple cubic lattice are hydrogenic ls-like. In this case, the Hubbard correlation energy U_H (eqn. (5.173)) is given by the expression (Schiff (1968)):

$$U_H = \frac{5e^2}{32\pi\varepsilon_0 a_0^*}. \tag{5.177}$$

If it is assumed that the Hubbard bands can be described in tight-binding terms, the width is given by eqn. (5.99), i.e.

$$W = 12|\beta_0|e^{-R/a_0^*}, \tag{5.178}$$

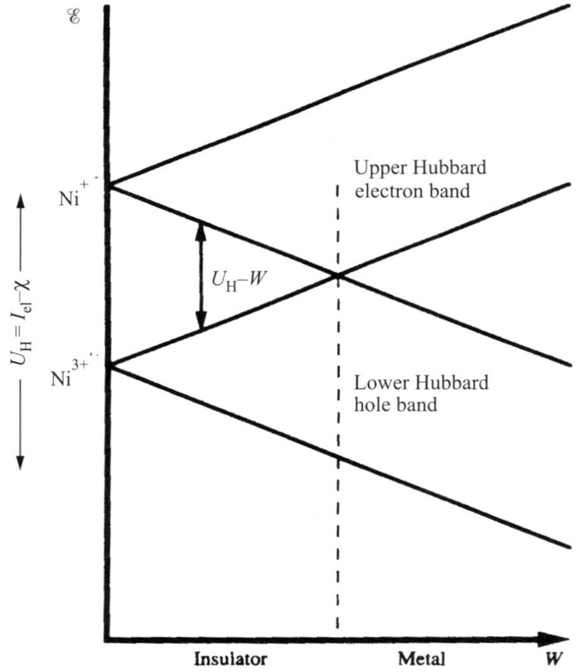

Fig. 5.64 Schematic illustration of the Hubbard bands in NiO as a function of band width W (inversely dependent on the interatomic separation). The lower band corresponds to motion of holes (electrons missing from Ni d-states) and the upper band to electron motion. The insulator–metal transition occurs when the two bands overlap.

since the nearest-neighbour coordination number is $z = 6$ for the simple cubic lattice. The pre-exponential part of the hopping integral for ls-orbitals is given by (Mott (1974)):

$$|\beta_0| = \left(\frac{3}{2}\left(1 + (R/a_0^*) + \frac{(R/a_0^*)^2}{6}\right)\right)\frac{e^2}{4\pi\varepsilon_0 a_0^*}. \tag{5.179}$$

Use of eqns. (5.176) – (5.179) then gives for the criterion for the Mott–Hubbard metal–insulator transition

$$R_c/a_0^* = 5.7, \tag{5.180a}$$

or, since the electron concentration is given by $n_c R_c^3 = 1$ for the s.c. lattice,

$$n_c^{1/3} a_0^* = 0.18. \tag{5.180b}$$

Note that this crude estimate is rather close to that obtained from the very different physical model involving screening (eqn. (5.171)) and that obtained empirically (Fig. 5.63).

Finally, note that a special type of metal–insulator transition, called the Peierls transition, occurs for *one-dimensional metals* (see §8.3.1), whose mechanism does *not* involve electron correlation effects (indeed electron–electron interactions can actually destroy the Peierls transition), but instead is associated with an opening up of a bandgap at the Fermi level in the metallic band, caused by a symmetry-lowering structural distortion (similar to a Jahn–Teller distortion). This phenomenon will be discussed in more detail in §8.3.2.

Experimental probes of electronic structure 5.7

A number of different experimental techniques can be used to obtain information about the electronic structure of solid materials, including the distribution of states as a function of energy (the density of states) and in reciprocal space (band structure). Many of these techniques are optical in nature, that is they rely on the interaction between photons and the electron system. Optical behaviour in general merits a separate section because of its importance and is discussed below (§5.8). Here, we discuss high-energy optical processes (involving UV and X-ray radiation) that can be used to explore the entire energy range of the density of states, rather than just the region in the vicinity of the forbidden gap. In addition, by measuring the electron current generated by the photoelectric effect (photoemission) as a function of emission angle, the electronic band structure, $\mathscr{E}(\mathbf{k})$, can be traced out experimentally.

5.7.1 X-ray spectroscopy

Two types of X-ray spectroscopy can be distinguished, depending on whether the X-ray spectrum corresponds to emission or to absorption. These two possibilities are illustrated in Fig. 5.65. In X-ray emission (or fluorescence) spectroscopy, a high-energy X-ray or electron, incident on the sample, excites an electron from a low-lying state. The resultant empty state can then be subsequently filled by an electron from a filled higher-lying (e.g. valence-band) state in a radiative transition: a soft X-ray is therefore emitted, and a measurement of the intensity of such emitted photons as a function of energy gives the spectrum. The converse process is when the X-ray spectrum corresponds to the measurement of the energies of photons that excite electrons from a low-lying filled state to a higher-lying empty state (e.g. in the conduction band). A variant of the latter absorption technique has already been discussed in §2.6.2, where the final electron state is free of the atom from which it was excited, and interference between such outgoing electron waves and waves backscattered from surrounding atoms results in a modulation of the X-ray absorption coefficient (extended X-ray absorption fine structure — EXAFS) which contains *atomic* structural information. However, X-ray absorption spectroscopy can also be used to probe the *electronic* structure of unoccupied states.

Note that, in both the processes indicated in Fig. 5.65, the electronic transitions are shown involving a deep *core* state. This is because such states have very little overlap with similar states on other atoms in the solid and, as a result, form bands having essentially an infinitesimal width (see Fig. 5.48a). Thus, transitions from filled valence-band states to, or to empty conduction-band states from, such energetically well-defined states allow the energy profiles of the higher-lying bands to be probed. Furthermore, because of the atomic-like nature of the core states involved in the transitions, the selection rule governing such electronic transitions is simply that there must be a change in orbital angular momentum $\Delta l = \pm 1$ for dipole-allowed transitions (i.e. $s \leftrightarrow p$, $p \leftrightarrow d$, etc.).

An example of the use of X-ray emission spectroscopy is shown in Fig. 5.66 for the case of crystalline Si, where two different core levels are utilized. In the K_β spectrum, vacancies (holes) are created in the ls core level (K-shell) and the recombining electrons come from the valence-band electrons that are 3p-like (so as to satisfy the selection rule).

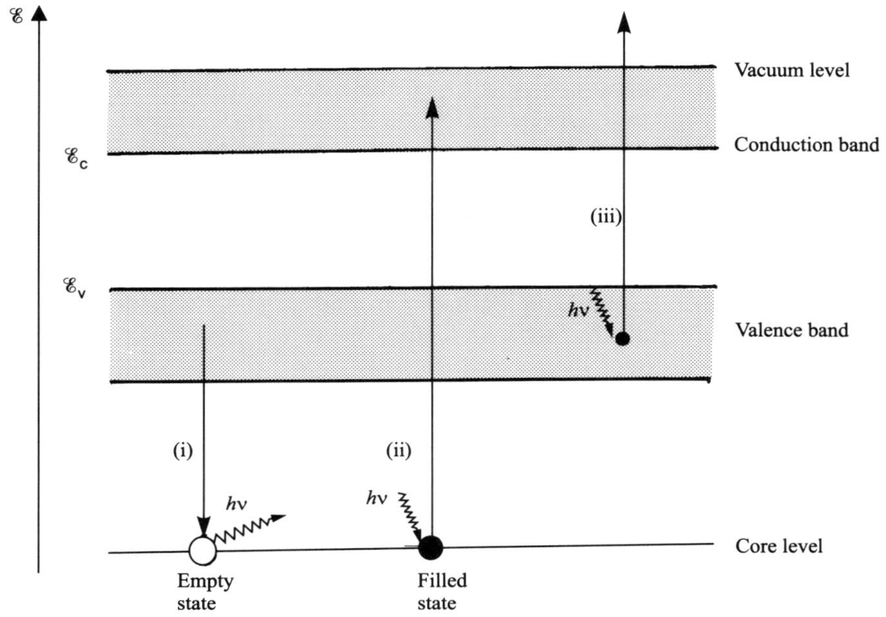

Fig. 5.65 Schematic illustration of the processes underlying (i) X-ray emission (fluorescence) and (ii) X-ray absorption spectroscopies, and (iii) electron photoemission.

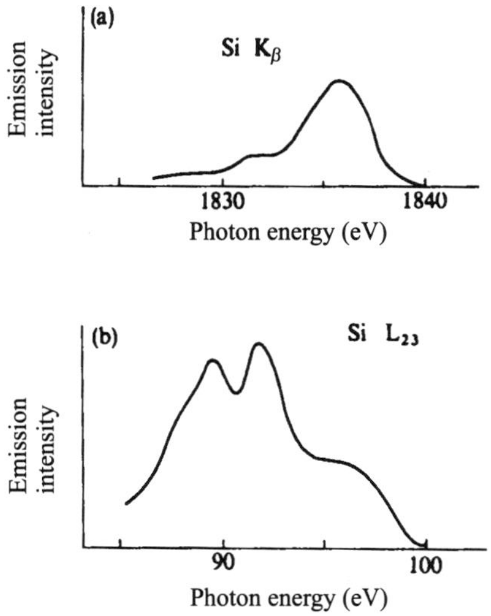

Fig. 5.66 X-ray emission (fluorescence) spectra resulting from electronic transitions from the valence band to core levels for crystalline Si: (a) K_β spectrum (3p contributions in the valence band → ls core level); (b) $L_{2,3}$ spectrum (3s contributions in the valence band → 2p core level). (After Wiech (1981) in *Emission and Scattering Techniques*, ed. P. Day, NATO ASI C**73**, p. 103, Fig. 15(c, d) with kind permission from Kluwer Academic Publishers)

For the $L_{2,3}$ spectrum, the holes are created in the 2p core states of the L-shell, and hence only valence-band 3s-like states can recombine with them so as to emit X-radiation. Note the very different X-ray photon energy scales in Figs. 5.66a and 5.66b corresponding to the different energy levels of the ls and 2p core states. More importantly, the *shapes* of the two X-ray emission spectra are very different. If the 3s- and 3p-states were hybridized *uniformly* through the valence band, as is implicit in the hybridized bond-orbital picture of Fig. 5.44, then it would be expected that the valence-band profiles of the K_β and $L_{2,3}$ spectra would be identical. Instead, the spectra show that 3p-states occur preferentially at the *top* of the valence band (Fig. 5.66a) and 3s-states contribute more at the *bottom* of the band (Fig. 5.66b). These findings are in accord with the predictions made on symmetry grounds in §5.4.2 that, at the Γ-point in the band structure of tetrahedral semiconductors, the top of the valence band is purely p-like and the bottom is purely s-like.

5.7.2 Photoemission spectroscopy

A technique, allied to X-ray absorption spectroscopy, is electron photoemission. This makes use of the photoelectric effect: an incident photon excites an electron from a filled state to beyond the vacuum level, and causes the electron to escape from the binding interaction of the solid. In this case, the distribution of kinetic energies of the emitted electrons is measured under monoenergetic photon excitation. The *maximum* kinetic energy, $\mathscr{E}_{max}$, of the photoelectrons is given by the difference between the incident photon energy $\hbar\omega_L$ and the electron binding energy (ionization energy) I_{el}, i.e.

$$\mathscr{E}_{max} = \hbar\omega_L - I_{el}. \tag{5.181}$$

(Lower values of kinetic energy will also occur due to inelastic scattering of the electrons before emission.) The binding energy, I_{el}, is the energy below the vacuum level from which electron photoemission occurs. The *minimum* value of binding energy, I_{el}^{min}, thus corresponds to excitation from the *highest-lying occupied states*. In semiconductors, these are at the top of the valence band at an energy $\mathscr{E}_v$; in metals, the equivalent position is the Fermi energy, $\mathscr{E}_F$, in the conduction band, and the binding energy in this case is also called the work function ϕ (see Fig. 5.67). The electron affinity χ of a semiconductor, defined as the energy difference between conduction-band minimum and the vacuum level, is thus the difference between I_{el} and the bandgap:

$$\chi = I_{el} - \mathscr{E}_g. \tag{5.182}$$

Different types of photon sources may be used in photoemission experiments. Monochromatic ultraviolet light can be obtained from the discrete lines emitted by rare-gas discharges, e.g. the He I line at 21.22 eV and the He II line at 40.82 eV: the use of such photons forms the basis of ultraviolet photoemission spectroscopy (UPS). Alternatively, soft X-ray lines emitted by metal anodes, e.g. the K_α line (2p → ls transition) of Al at 1486.6 eV, may be used in X-ray photoemission spectroscopy (XPS). However, the X-ray lines of such light elements are much broader ($\gtrsim$ 1eV) than the gas-discharge UV lines because the initial electron states form a broad band and not a discrete core level (see Fig. 5.54a). The best source of electromagnetic radiation is the broad-band intense synchrotron radiation emitted by a high-energy electron synchrotron (see Fig. 2.48).

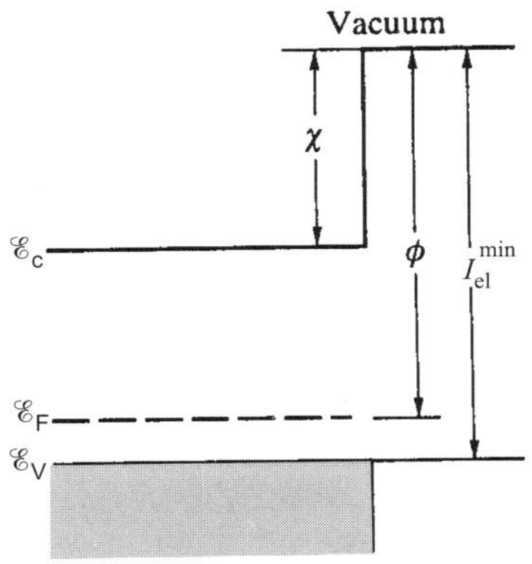

Fig. 5.67 Illustration of the definitions of electron affinity χ, work function ϕ and threshold photoelectron energy I_{el}^{min} for a semiconductor, referenced to the bottom of the conduction band $\mathscr{E}_c$, top of the valence band $\mathscr{E}_v$ and the Fermi energy $\mathscr{E}_F$. The bandgap is $\mathscr{E}_g = \mathscr{E}_c - \mathscr{E}_v$.

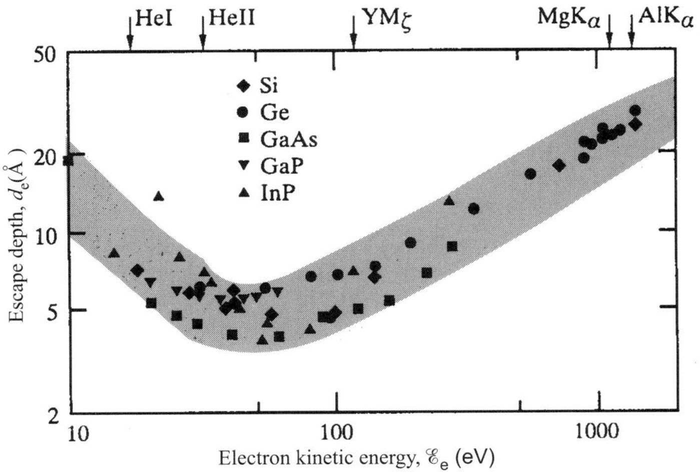

Fig. 5.68 The universal dependence of electron escape depth, d_e, on electron kinetic energy $\mathscr{E}_e$ in various solids. The variation in values of d_e for a given value of $\mathscr{E}_e$ is shown by the shaded area: specific values for some representative crystalline semiconductors are shown. (After Yu and Cardona (1996), *Fundamentals of Semiconductors*, p. 420, Fig. 8.5, © Springer-Verlag GmbH & Co. KG)

Monochromatized photons, produced by single-crystal monochromators, and covering a wide range of energies, can be produced in this way.

Although evidently the photon energy used must be greater than the maximum electronic binding energy of interest in the solid (cf. eqn. (5.181)), another important consideration regarding the choice of photon energy concerns the escape depth d_e of the photoemitted electrons. There is a universal dependence of d_e on electron kinetic energy (related to photon energy via eqn. (5.181)), and this is shown in Fig. 5.68. It should be noted that values of d_e are typically in the range 5–50 Å, and consequently photoelectrons (particularly in UPS) emanate only from the surface region. This is obviously of paramount importance in surface science, but in the study of the properties of *bulk* materials care has to be taken that the surface region is not different from the bulk in terms of structure or composition. As a result, photoemission experiments must be performed in an ultra-high-vacuum (UHV) environment to reduce unintentional surface contamination. The use of X-ray excitation, and the concomitantly larger values of escape depth (Fig. 5.68), helps to ensure that bulk characteristics of the electronic structure are probed in XPS.

The geometry of a photoelectron experiment is shown in Fig. 5.69, where the possible experimental variables (e.g. angular, polarization, energy) are indicated. Often the photoelectron current is measured over a range of angles θ_e and ϕ_e, giving angle-integrated spectra, in which case essentially the electronic density of (occupied) states is probed. In the so-called three-step model, the photoemission process can be regarded as being separable into the following three events: (i) photo-induced excitation of an electron from a filled state to an empty conduction-band state; (ii) ballistic transport of the electron to the surface (without scattering); (iii) transmission of the electron through the surface, resulting in its emission from the solid. In this picture, the photocurrent, I_{ph}, can be written approximately as

$$I_{ph}(\mathscr{E}_e - \hbar\omega_L) = g(\mathscr{E}_e - \hbar\omega_L)P(\mathscr{E}_e), \tag{5.183}$$

where $P(\mathscr{E})$ is a transition probability relating to processes (ii) and (iii), and the photoelectron energy is measured with respect to some reference level (e.g. $\mathscr{E}_F$, $\mathscr{E}_v$ or the vacuum level). Thus, if $P(\mathscr{E})$ is only a slowly varying function of energy, the photocurrent signal should be a measure of the occupied (e.g. valence-band) density of states, $g(\mathscr{E})$.

An example of the application of this technique is shown in Fig. 5.70 for the case of Al metal. The spectrum is a measure of the density of occupied states (the free-electron-like density of states multiplied by the Fermi–Dirac function), with the zero of the binding

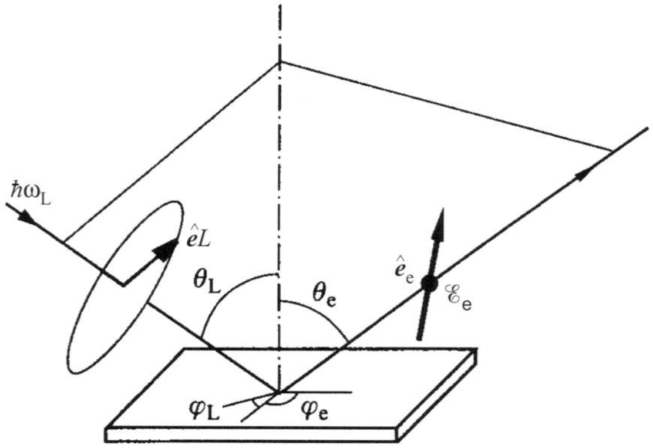

Fig. 5.69 Schematic diagram of a photoelectron experiment indicating possible experimental variables relating to the incident light beam (L) or the emitted electron current (e). The unit vector $\hat{e}$ is a polarization vector.

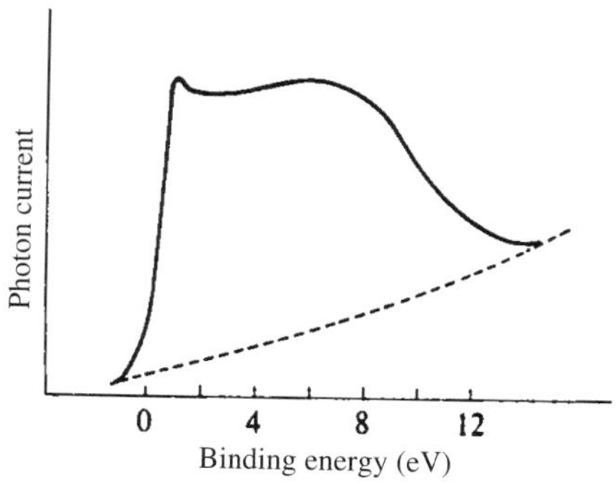

Fig. 5.70 Experimental photoelectron spectrum of Al. The zero of the binding energy is taken at $\mathscr{E}_F$. The dashed line represents a background signal. (After Steiner *et al.* (1979), in *Photoemission in Solids, II*, eds L. Ley and M. Cardona, p. 369, Fig. 7.9(a), © Springer-Verlag GmbH & Co. KG)

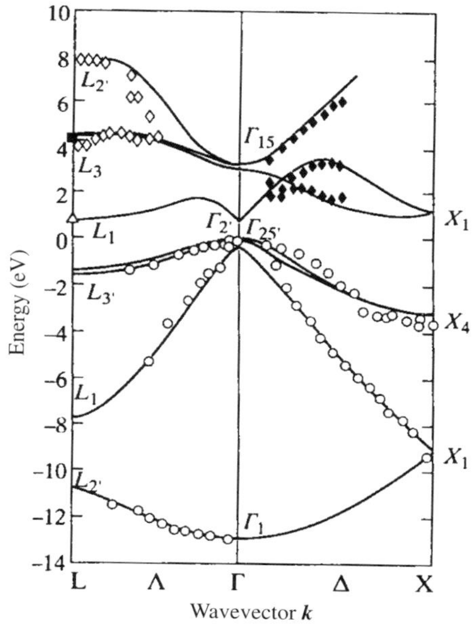

Fig. 5.71 Band structure of crystalline Ge obtained by angle-resolved UPS (valence band) and inverse photoemission (conduction band), compared with theoretical results (solid lines). (After Yu and Cardona (1996), *Fundamentals of Semiconductors*, p. 445, Fig. 8.23, © Springer-Verlag GmbH & Co. KG)

energy being taken at $\mathscr{E}_F$. Figure 5.70 should be compared with the theoretical curve given in Fig. 5.31.

An alternative variant of the technique is not to integrate over the photoelectron angles θ_e and ϕ_e but to measure the photocurrent as a function of them, i.e. $I_{\mathrm{ph}}(\hbar\omega_{\mathrm{L}}, \mathscr{E}_{\mathrm{e}}, \theta_{\mathrm{e}}, \phi_{\mathrm{e}})$, resulting in angle-resolved photoelectron spectroscopy (ARPES). The significance of this approach is that information on the *wavevector* $\boldsymbol{k}$ of the initial electron state can be obtained thereby, and hence the band structure, $\mathscr{E}(\boldsymbol{k})$, can be determined experimentally. It is assumed that $\boldsymbol{k}$ is conserved for processes (i) and (ii) of the three-step model, but, in the final transmission step by an electron through the surface, only the component of wavevector *parallel* to the surface, $\boldsymbol{k}_{\|}$, is conserved (modulo a surface reciprocal-lattice vector, $\boldsymbol{G}_{\|}$); the perpendicular component, $\boldsymbol{k}_{\perp}$, is not conserved because of the loss of translational periodicity at the surface in the direction normal to the surface. As a result, the band structure can be explored experimentally (see e.g. Yu and Cardona (1996) for details). An example of an experimental determination of $\mathscr{E}(\boldsymbol{k})$ obtained in this way for the valence band of crystalline Ge is shown in Fig. 5.71.

A normal photoemission experiment (photon in, electron out) probes *filled* electron states: however, empty conduction-band states cannot be accessed in this way. This restriction can be rectified by performing the experiment *in reverse*. An incident electron, with well-defined energy $\mathscr{E}_{\mathrm{e}}$, is directed at the surface of a solid at an angle $(\theta_{\mathrm{e}}, \phi_{\mathrm{e}})$, passes through the surface and occupies a high-lying conduction-band state (assumed, for simplicity, to be free-electron-like). The electron can then make a transition to a lower empty conduction-band state, thereby emitting a photon whose energy is measured. This process is referred to as inverse photoemission. Angle-resolved $(\theta_{\mathrm{e}}, \phi_{\mathrm{e}})$ measurements can be performed, as for normal photoemission, resulting in the determination of $\mathscr{E}(\boldsymbol{k})$ for the unoccupied conduction-band states. An example of this approach for the case of crystalline Ge is shown in Fig. 5.71.

Optical properties of electrons in solids 5.8

Electromagnetic radiation can interact with materials in a number of ways: e.g. in principle, it can be scattered, reflected or absorbed by a solid medium. The mechanism by which the radiation interacts with solids can also vary. In §§. 4.4, 4.5.1 and 4.5.2, the interactions of photons with phonons (lattice vibrations) has already been discussed. Here, the interest is in the interaction of photons with the *electrons* in materials. In fact, two instances of such an interaction have already been mentioned, namely the (elastic) scattering of X-rays by the electron density associated with atoms, in other words X-ray diffraction (§2.6.1.3), and X-ray spectroscopy (§5.7.1). In this section, other interactions will be explored between electrons in metals and insulators (or semiconductors) and light (in the IR, visible and UV regions of the electromagnetic spectrum).

The optical properties of solids are often expressed in terms either of the complex refractive index $n^\dagger = n_r + i\kappa_i$ (eqn. (4.157)) or the complex dielectric constant $\varepsilon^\dagger = \varepsilon_1 + i\varepsilon_2$ (eqn. (4.160)), where $n^\dagger = \sqrt{\varepsilon^\dagger}$ (eqn. (4.159)). The measurable optical quantities, the reflectivity, transmissivity or absorption of a medium are related via the Fresnel relations to the (components of the) complex refractive index or dielectric constant (see Problem 4.14). The Kramers–Kronig relations (eqn. (4.166)) relate the real and imaginary parts, for example, of the dielectric constant.

In general, the most pronounced optical processes are absorption and reflection, since these involve the lowest order of interaction between the electric vector $\mathscr{E}$ of the light and the excitations within the material (e.g. linear with $\mathscr{E}$ for absorption). Inelastic light scattering (Raman scattering) is much weaker since it involves higher-order interactions (e.g. the intensity of first-order Raman scattering, involving the creation/destruction of say a single phonon, depends on E^2—see eqn. (4.181)). Equally, non-linear optical processes (e.g. sum-frequency generation) involving second- and third-order interactions are also very weak (see §5.8.4).

In the following, the optical behaviour of materials has, for simplicity, been divided into three sections, dealing respectively with intraband optical properties, interband electronic transitions and non-linear optical behaviour.

5.8.1 Intraband optical properties

In this section, we will be concerned with the interactions of photons with electrons resulting in electronic excitations within a *single* band. (Electronic transitions between two bands forms the subject of the next section.) Two types of electronic intraband excitations can be envisaged. The first are transitions of electrons within part-filled bands, from filled states below $\mathscr{E}_F$ to empty states in the same conduction band above $\mathscr{E}_F$, e.g. in metals (or highly-doped 'degenerate' semiconductors—see §6.5.2): such processes give rise to free-carrier absorption of photons (see Fig. 5.72). Another related electronic excitation is the longitudinal vibrational motion of the entire free-electron gas of a metal (or equivalently of the completely filled valence-band states of a semiconductor), termed *plasma oscillations* or *plasmons* (see §5.6.2).

The optical properties are related to the complex, frequency-dependent dielectric constant, $\varepsilon^\dagger(\omega)$, and this can be calculated from the equation of motion of free electrons in the Drude approximation. Consider a gas of electrons with concentration n subject to

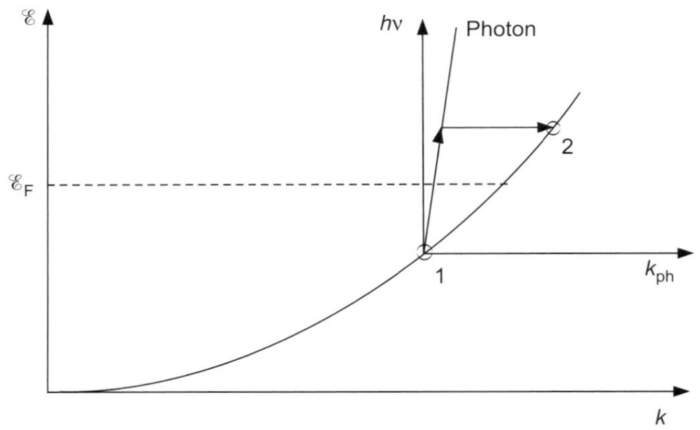

Fig. 5.72 Illustration of free-carrier optical absorption in a metal or heavily doped semiconductor. A photon is absorbed and causes a transition of an electron from a state (1) below $\mathscr{E}_F$ to an empty state (2) in the *same* band above $\mathscr{E}_F$. Since the photon dispersion curve does not intersect the free-electron-like band at 2, a 'direct' transition is not possible by which the momenta of photon and electron alone are conserved in the transition. Instead, an 'indirect' transition involving a scattering event, e.g. with a phonon, is necessary to conserve momentum.

an electric field $\mathscr{E}$ which causes a homogeneous displacement $\boldsymbol{u}$, with velocity $\boldsymbol{v}$, of the electron gas with respect to the ion cores. The equation of motion is thus:

$$nm_e^* \frac{d\boldsymbol{v}}{dt} + \gamma \boldsymbol{v} = -ne\mathscr{E} \tag{5.184}$$

where γ is a damping factor and m_e^* is the effective electron mass. Note that, unlike the corresponding case for ionic motion (eqn. (4.164)), there is *no* mechanical restoring-force term ($\propto \omega_0^2 \boldsymbol{u}$) for the electron gas. The damping term is related to energy losses sustained by the moving electron gas, i.e. due to inelastic electron–phonon scattering events. Such processes are the origin of the electrical resistivity ρ (see §6.1.1). For a constant electron flux (i.e. when $d\boldsymbol{v}/dt = 0$, the electron current density $\boldsymbol{j}$ is given by

$$\boldsymbol{j} = -e\, n\, \boldsymbol{v} = \frac{n^2e^2\mathscr{E}}{\gamma} \tag{5.185}$$

from eqn. (5.184), and hence the conductivity, defined as the constant of proportionality between current density and field

$$\sigma = \boldsymbol{j}/\mathscr{E}, \tag{5.186}$$

in the zero-frequency (d.c.) case is given by

$$\sigma_0 = \frac{n^2e^2}{\gamma} = \frac{ne^2\tau}{m_e^*}, \tag{5.187}$$

where $\tau = nm_e^*/\gamma$ is the scattering relaxation time.

If the electric field is taken to be a transverse plane wave, as in light, i.e. $\mathscr{E}(t) = \mathscr{E}_0 \exp[i(\boldsymbol{k} \cdot \boldsymbol{r} - \omega t)]$, then assuming that the electron velocity has the same time dependence, the equation of motion (eqn. (5.184)) becomes

$$-\mathrm{i}\omega n m_e^* \boldsymbol{v} + \gamma \boldsymbol{v} = -ne\boldsymbol{E} \tag{5.188a}$$

or

$$\boldsymbol{v} = -ne\boldsymbol{E}/(\gamma - \mathrm{i}\omega n m_e^*). \tag{5.188b}$$

Using the definition (eqn. (5.186)) for σ, the frequency-dependent conductivity of the free-electron gas is thus given by

$$\sigma(\omega) = \frac{\sigma_0}{1 - \mathrm{i}\omega\tau} = \frac{\sigma_0(1 + \mathrm{i}\omega\tau)}{(1 + \omega^2\tau^2)}. \tag{5.189}$$

An expression for the complex refractive index $n^\dagger(\omega)$ can be obtained from Maxwell's equations for electromagnetic fields in a medium. These are:

$$\nabla \cdot \boldsymbol{D} = \rho, \qquad \nabla \cdot \boldsymbol{B} = 0, \tag{5.190a, b}$$

$$\nabla \times \boldsymbol{E} = -\frac{\partial \boldsymbol{B}}{\partial t}, \qquad \nabla \times \boldsymbol{B} = \frac{\varepsilon(\infty)}{c^2}\frac{\partial \boldsymbol{E}}{\partial t} + \frac{\sigma \boldsymbol{E}}{\varepsilon_0 c^2}, \tag{5.190c, d}$$

where $\varepsilon(\infty)$ is the effective frequency-independent dielectric constant resulting from processes occurring at higher frequencies than those of interest (in the present case, these would be electronic *interband* transitions from one conduction band to another—see §5.8.2) and ε_0 is the vacuum permittivity. Taking the curl of eqn. (5.190c) and making use of eqn. (5.190d) gives

$$-\nabla \times (\nabla \times \boldsymbol{E}) = \frac{\varepsilon(\infty)}{c^2}\frac{\partial^2 \boldsymbol{E}}{\partial t^2} + \frac{\sigma}{\varepsilon_0 c^2}\frac{\partial \boldsymbol{E}}{\partial t}, \tag{5.191}$$

which, for the case of a plane wave ($\mathscr{E} = \mathscr{E}_0 \exp[\mathrm{i}(\boldsymbol{k} \cdot \boldsymbol{r} - \omega t)]$), gives

$$\left\{\frac{\omega^2\varepsilon(\infty)}{c^2} + \frac{\mathrm{i}\omega\sigma}{\varepsilon_0 c^2}\right\}\boldsymbol{E} = k^2\boldsymbol{E} - \boldsymbol{k}(\boldsymbol{k} \cdot \boldsymbol{E}), \tag{5.192}$$

where use has been made of the vector equality $\boldsymbol{k} \times \boldsymbol{k} \times \boldsymbol{E} = \boldsymbol{k}(\boldsymbol{k} \cdot \boldsymbol{E}) - (\boldsymbol{k} \cdot \boldsymbol{k})\boldsymbol{E}$. Electromagnetic waves are transverse ($\boldsymbol{k}$ perpendicular to $\boldsymbol{E}$) and so from eqn. (5.192)

$$k^2 = \frac{\omega^2\varepsilon(\infty)}{c^2} + \frac{\mathrm{i}\omega\sigma}{\varepsilon_0 c^2}. \tag{5.193}$$

Since the complex refractive index $n^\dagger = n_r + \mathrm{i}\kappa_i$ is related to k and ω by the dispersion relation

$$k = (n_r + \mathrm{i}\kappa_i)\frac{\omega}{c} \tag{5.194}$$

then it follows that

$$(n_r + \mathrm{i}\kappa_i)^2 = \varepsilon(\infty) + \frac{\mathrm{i}\sigma}{\omega\varepsilon_0}. \tag{5.195}$$

Thus, the real and imaginary parts of the dielectric constant, ε_1 and ε_2, related to n_r and κ_i by eqn. (4.161) and (4.162), are given by:

$$\begin{aligned}\varepsilon_1 &= \varepsilon(\infty) - \frac{\mathrm{Im}(\sigma)}{\varepsilon_0\omega} = \varepsilon(\infty) - \frac{\tau\sigma_0}{\varepsilon_0(1 + \omega^2\tau^2)} \\ &= \varepsilon(\infty)\left(1 - \frac{\omega_p^2}{(\omega_0^2 + \omega^2)}\right)\end{aligned} \tag{5.196a}$$

and

$$\begin{aligned}\varepsilon_2 &= \frac{\mathrm{Re}(\sigma)}{\omega\varepsilon_0} = \frac{\sigma_0}{\omega\varepsilon_0(1+\omega^2\tau^2)} \\ &= \varepsilon(\infty)\frac{\omega_0}{\omega}\frac{\omega_\mathrm{p}^2}{(\omega_0^2+\omega^2)},\end{aligned} \tag{5.196b}$$

where $\omega_0 = 1/\tau$, and use has been made of eqns (5.187) and (5.189). The *screened* plasma frequency is given by (compare with eqn. (5.166)):

$$\omega_\mathrm{p}^2 = \frac{ne^2}{m_\mathrm{e}^*\varepsilon(\infty)\varepsilon_0}. \tag{5.197}$$

It is instructive to examine the optical response of a free-electron gas in three frequency régimes, i.e. below, near and above the plasma frequency, ω_p. One optical property of particular interest is the reflectivity, *R*, which, at normal incidence for a weakly absorbing material, is given by (see Problem 4.14):

$$R = \frac{(n_\mathrm{r}-1)^2+\kappa_\mathrm{i}^2}{(n_\mathrm{r}+1)^2+\kappa_\mathrm{i}^2}. \tag{5.198}$$

At very low frequencies $\omega \ll \omega_0 = 1/\tau \ll \omega_\mathrm{p}$ (see Problem 5.29(b)), the optical constants given by eqns. (5.196) and (5.198) reduce to:

$$\begin{gathered}\varepsilon_1 \simeq -\frac{\varepsilon(\infty)\omega_\mathrm{p}^2}{\omega_0^2}, \quad \varepsilon_2 \simeq \frac{\varepsilon(\infty)\omega_\mathrm{p}^2}{\omega\omega_0} \equiv \frac{\sigma_0}{\varepsilon_0\omega}, \\ n_\mathrm{r} \simeq \kappa_\mathrm{i} \simeq (\varepsilon_2/2)^{1/2}, \quad R \simeq 1 - \frac{2}{n_\mathrm{r}} = 1 - \left(\frac{2\varepsilon_0\omega}{\sigma_0}\right)^{1/2},\end{gathered} \tag{5.199}$$

with $|\varepsilon_1| \ll \varepsilon_2$. This approximate expression for the reflectivity is called the Hagen–Rubens relation, and indicates that in this frequency régime, since ε_2, and hence n_r, is very large, the reflectivity is nearly 100%. This is in accord with one's everyday experience of the optical reflectivity of metals.

At intermediate frequencies, such that $\omega < \omega_p$ but $\omega > \omega_0$, the components of the dielectric constant are approximately:

$$\varepsilon_1 \simeq -\frac{\varepsilon(\infty)\omega_\mathrm{p}^2}{\omega_0^2}, \quad \varepsilon_2 \simeq \frac{\varepsilon(\infty)\omega_\mathrm{p}^2\omega_0}{\omega^3}, \tag{5.200}$$

with $\kappa_\mathrm{i} \gg \mathrm{n_r}$ (since $|\varepsilon_1| \gg \varepsilon_2$) and $R \simeq 1$ as before. The optical absorption coefficient $K(\omega)$ is given by eqn. (4.16):

$$K(\omega) = \frac{\omega\varepsilon_2(\omega)}{n_\mathrm{r}c}.$$

Thus, in this frequency régime, the free-carrier absorption should behave as $K(\omega) \propto \omega^{-2}$ (or $\propto \lambda^2$); experimental data for the *n*-doped semiconductor InAs (where the *n*-type impurities introduce extra free carriers into the conduction band — see §6.5.2) shown in Fig. 5.73 do exhibit a power-law behaviour of $K(\omega) \propto \lambda^\alpha$, but $\alpha \simeq 3$. This discrepancy with the simple prediction of the free-carrier absorption model outlined above can be explained by invoking a frequency dependence of the relaxation time, τ.

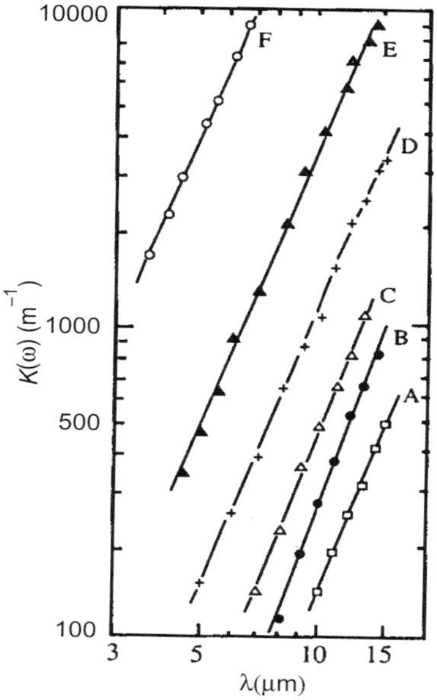

Fig. 5.73 Free-carrier absorption in n-doped InAs at 300 K for six different free-carrier concentrations (in units of 10^{23} m^{-3}): A, 0.28; B, 0.85; C, 1.4; D, 2.5; E, 7.8 and F, 39. The straight lines are $K(\omega) \propto \lambda^3$. (After Dixon (1961))

The most interesting behaviour occurs, however, for frequencies in the vicinity of the plasma frequency, ω_p. At $\omega_1 = \omega_p, \varepsilon_1$ changes in sign from negative to positive, being zero if $\omega_0 \simeq 0(\ll \omega_p)$. Thus, $n_r = 0$, and if $\varepsilon_2 \simeq 0$, from eqn. (5.198) the reflectivity is unity. However, when $\varepsilon_1 = 1$, at a frequency $\omega_2 = [\varepsilon(\infty)/(\varepsilon(\infty) - 1)]^{1/2}\omega_p$ from eqn. (5.196a) (if $\omega_0 \simeq 0$), then $n_r = 1$ (for $\varepsilon_2 \simeq 0$) and from eqn. (5.198) the reflectivity becomes (nearly) *zero*, i.e. the material becomes highly transmitting. For large values of the high-frequency dielectric constant $\varepsilon(\infty)$, such as occur in doped semiconductors, the frequencies ω_1 and ω_2 lie very close together and an abrupt plasma reflection edge is observed. The plasma edge for the free-electron metal Al is in the UV as shown in Fig. 5.74. There is an interband transition that is responsible for the dip in the reflectivity at 1.5 eV and also for the appreciable damping that limits the reflectivity to $\simeq$ 90% for frequencies less than the plasma edge. The plasma edges in the reflectivity exhibited by an n-type doped semiconductor, InSb, with varying free-carrier (i.e. dopant) concentrations, are shown in Fig. 5.75. Note the sharpness of the plasma edges (due to large values of $\varepsilon(\infty)$) and the fact that the edges appear in the *infrared*, not in the UV region as for metals, because of the much smaller free-carrier concentrations. This behaviour is put to good use in another doped semiconductor system, indium oxide doped with tin (ITO), In_2O_3:Sn. The indium oxide becomes metallic when a few percent of the impurity is introduced, but the low concentration of free carriers means that the plasma edge is in the infrared and so the material is optically transparent in the visible part of the spectrum. Thus, ITO can be used as an optically transparent electrode material, and also as an IR-reflective coating.

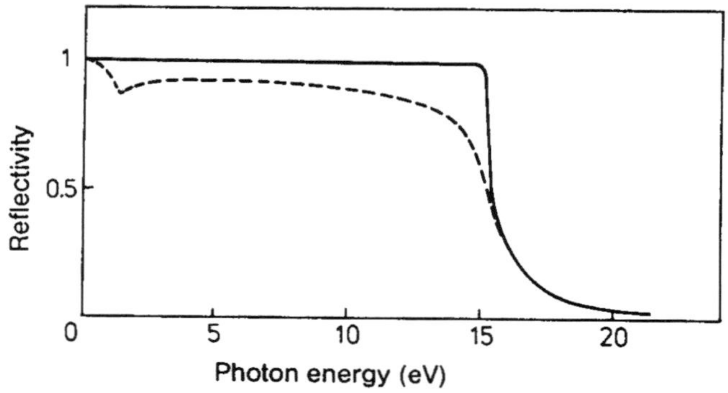

Fig. 5.74 Reflectivity of Al metal (dashed line), together with the theoretical curve for the free-electron gas model (with parameters $\omega_p = 2.3 \times 10^{16}$ Hz, $\sigma = 3.6 \times 10^7\ \Omega^{-1}\ m^{-1}$). (After Ibach and Lüth (1995), *Solid State Physics*, p. 306, Fig. 11.11, © Springer-Verlag GmbH & Co. KG)

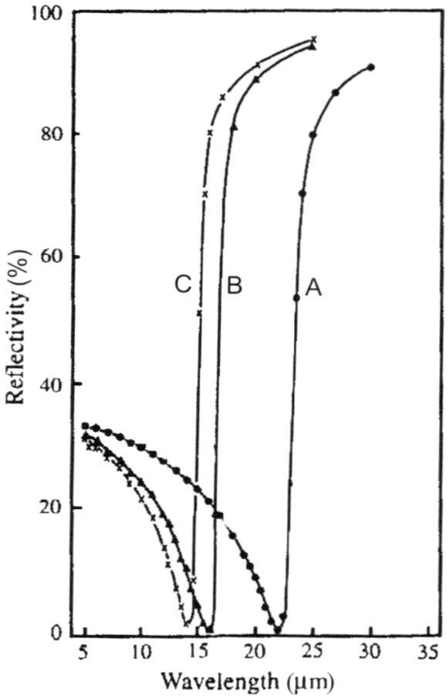

Fig. 5.75 Plasma edges in the room-temperature reflectivity curves of n-type InSb with carrier concentrations (in units of $10^{24}\ m^{-3}$): A, 1.2; B, 2.8; C, 4.0. The solid curves are fits to the data using eqns. (5.196) and (5.198) with m_e^* as the variable parameter. (After Spitzer and Fan (1957). Reprinted with permission from *Phys. Rev.* **106**, 882. © 1957. The American Physical Society)

We have remarked previously that, because the collective electron excitations known as plasmons (§5.6.2) are longitudinal in character, they do not couple directly to transverse electromagnetic waves. The plasmon frequency is revealed as peaks in the energy-loss function, $-\mathrm{Im}(\varepsilon^{-1})$ (Problem 5.26) and, although this cannot be measured directly by optical means, nevertheless the function can be constructed from measured values of ε_1 and ε_2 obtained from optical reflectivity data and by making use of the Kramers–Kronig relations (eqn. (4.166)) between real and imaginary parts of the dielectric constant. An example is shown in Fig. 5.79 for the case of crystalline Ge.

5.8.2 Interband optical behaviour

In this section we consider optically induced electronic transitions *between* two energy bands. For the most part, the two bands involved will be the valence and conduction bands, separated by the forbidden bandgap, in semiconductors or insulators, although it should be remembered that interband transitions can also occur in metals between one conduction band and another higher-lying empty one. It is simplest to consider first electronic transitions caused by the absorption of photons having energies comparable to the bandgap, $\mathscr{E}_g$. This is because, for larger photon energies, a multiplicity of transitions between deep-lying valence-band states and high-lying conduction-band states are possible and these can confuse the picture.

Optical processes in materials must satisfy a number of selection rules. The first is that, for electrical-dipole-allowed transitions, there must be a change in orbital angular momentum of the initial and final electron states of $\Delta l = \pm 1$. Second, energy must be conserved in the transition, i.e.

$$\mathscr{E}_f = \mathscr{E}_i + \hbar\omega, \tag{5.201}$$

where $\mathscr{E}_f$ and $\mathscr{E}_i$ are the final and initial electron energies respectively, and $\hbar\omega$ is the photon energy. Finally, for ideal *crystalline* materials, crystal momentum must also be conserved in the transition:

$$\boldsymbol{k}_f = \boldsymbol{k}_i + \boldsymbol{k}_{ph}, \tag{5.202}$$

where $\boldsymbol{k}_f$ and $\boldsymbol{k}_i$ are the final and initial (crystal) wavevectors of an electron in the final and initial states respectively, and $\boldsymbol{k}_{ph}$ is the photon wavevector. Since $\boldsymbol{k}_{ph}$ is negligible compared with typical electron wavevectors (see Problem 5.31), eqn. (5.202) implies that photo-induced electronic transitions can be regarded as being essentially vertical on the scale of electronic band structures as normally plotted. A diagrammatic representation of the origin of the vertical-transition selection rule is given in Fig. 5.76. For electron states (satisfying the $\Delta l = \pm 1$ atomic selection rule) in different bands, but having the *same* value of electron (crystal) momentum (neglecting the photon contribution), the transition dipoles μ_i are in phase at each atomic site i and hence give a non-zero value when summed over the entire crystal. On the other hand, electron states having *different* values of crystal momentum in the two bands give rise to different, and out-of-phase, values of μ_i at each site, producing no net transition dipole moment when summed over the crystal: i.e. the transition is formally forbidden.

The consequences of the $\Delta\boldsymbol{k} = 0$ (vertical) selection rule on the optical-absorption spectra of crystalline semiconductors and insulators are appreciable; the spectra are significantly different, depending on whether the band structure is of the direct-gap or indirect-gap type (§5.4).

In *direct-gap* materials (e.g. crystalline GaAs), where the minimum of the conduction band and the maximum of the valence band occur at the *same* value of $\boldsymbol{k}$ (usually at $\boldsymbol{k} = 0$, the Γ-point), the $\Delta\boldsymbol{k} = 0$ selection rule ensures that electronic transitions occur

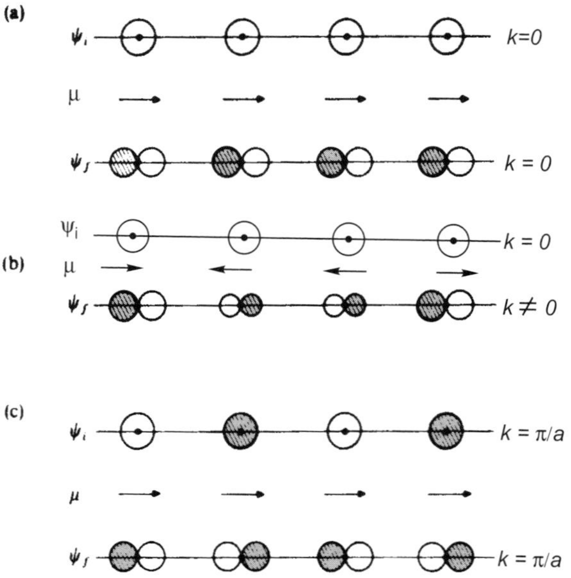

Fig. 5.76 Diagrammatic illustration of the vertical-transition selection rule for optically induced electronic transitions in crystalline materials: $\Delta \boldsymbol{k} = \boldsymbol{k}_f - \boldsymbol{k}_i \simeq 0$, where $\boldsymbol{k}_f$ and $\boldsymbol{k}_i$ are the final and initial (crystal) momenta of the electron states, respectively. Electronic transitions at atoms forming a 1D crystalline array are considered for simplicity. (a) Transition between s- and p-states in two bands, each with $\boldsymbol{k} = 0$, with in-phase transition dipoles (μ). (b) Transition from a $\boldsymbol{k} = 0$ s-state in a band to a p-state at a general $\boldsymbol{k}$-value in a higher band, for which the transition dipoles are out of phase and sum to zero. (c) Transition from a zone-boundary s-state in one band to a zone-boundary p-state in another band, where the transition dipoles are in phase. (After Cox (1987), *Electronic Structure and Chemistry of Solids* by permission of Oxford University Press)

for $\hbar\omega \geqslant \mathscr{E}_g$ (neglecting the effect of the photon wavevector); hence, the optical absorption spectrum of such materials increases discontinuously at $\hbar\omega = \mathscr{E}_g$ (Fig. 5.77), and an extrapolation of the absorption curve to zero absorption yields an estimate for $\mathscr{E}_g$.

In indirect-gap materials (e.g. crystalline Si and Ge), on the other hand, the minimum gap occurs for electron states at *different* values of $\boldsymbol{k}$ (Fig. 5.78a). Thus a vertical electronic transition, involving only a photon, cannot directly connect the two electronic states defining the indirect gap, $\mathscr{E}_g^i$; a photon does not possess enough momentum to ensure conservation of momentum for such a transition. A number of possible additional processes, however, can provide the necessary extra momentum to effect an indirect transition. The most general of these involves *phonons*. A photon with energy equal approximately to $\mathscr{E}_g^i$ causes a vertical transition of an electron in the valence band to a virtual state, which is allowed as long as the lifetime of this state is short enough to satisfy the Heisenberg uncertainty principle. A phonon can then be either emitted or absorbed with enough (but small) energy $\hbar\omega_p$ to cause the electron transition to the conduction-band minimum, i.e. the conservation of energy now reads:

$$\mathscr{E}_f = \mathscr{E}_i + \hbar\omega \pm \hbar\omega_p, \tag{5.203}$$

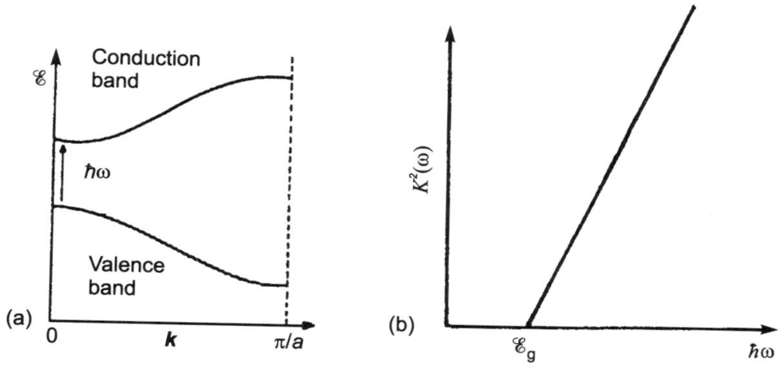

Fig. 5.77 Optical absorption in materials with a direct gap, $\mathscr{E}_g$: (a) vertical, crystal-momentum-conserving direct transition occurring at the minimum bandgap, $\mathscr{E}_g$; (b) the corresponding optical absorption spectrum, $K(\omega)$.

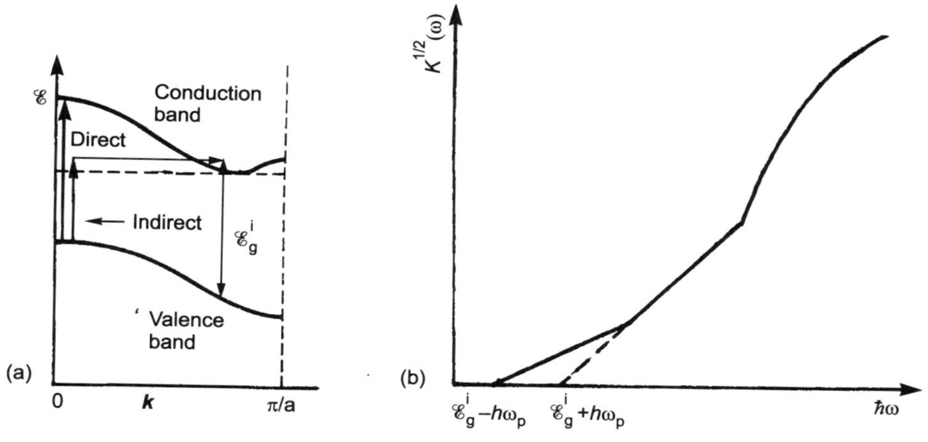

Fig. 5.78 Optical absorption in materials with an indirect gap, $\mathscr{E}_g^i$: (a) Indirect transition between electron states at different points in $\boldsymbol{k}$-space. Photon absorption causes a vertical transition to a virtual state; subsequent phonon emission or absorption provides the necessary momentum to carry the electron to the final state. A higher-energy *direct* transition between valence and conduction bands at $\boldsymbol{k} = 0$ is also shown. (b) The corresponding optical absorption spectrum, $K(\omega)$. The lower-energy intercept at $\mathscr{E}_g^i - \hbar\omega_p$ corresponds to absorption of a phonon with energy $\hbar\omega_p$, and the higher-energy intercept at $\mathscr{E}_g^i + \hbar\omega_p$ corresponds to phonon emission.

where the plus and minus signs refer to phonon absorption and emission, respectively, and the indirect gap is $\mathscr{E}_g^i = \mathscr{E}_f - \mathscr{E}_i$. More importantly, the absorbed or emitted phonon has a sufficiently large momentum that the total (electron plus phonon) momentum can be conserved in the indirect transition between valence-band maximum and conduction-band minimum (Fig. 5.78a), i.e.

$$\boldsymbol{k}_{\rm f} = \boldsymbol{k}_{\rm i} + \boldsymbol{k}_{\rm ph} \pm \boldsymbol{q}_{\rm p}, \tag{5.204}$$

where $\boldsymbol{k}_{\rm ph}$ is the (negligible) wavevector of the photon and $\boldsymbol{q}_{\rm p}$ that of the phonon. Phonon emission is always possible, and in such a case the zero-absorption intercept is at $\mathscr{E}_{\rm g}^{\rm i} + \hbar\omega_{\rm p}$. However, at high temperatures, the number of phonons increases (eqn. (4.68)) and hence the probability of phonon-*absorption* events becomes appreciable, for which the intercept in the optical-absorption spectrum is at $\mathscr{E}_{\rm g}^{\rm i} - \hbar\omega_{\rm p}$ (Fig. 5.78b).

Processes other than those involving phonons can also contribute to momentum conservation in optical transitions, e.g. electron–electron scattering or electron-impurity scattering in heavily doped semiconductors. Impurities are associated with a destruction of perfect translational symmetry and hence with a breakdown of the $\boldsymbol{k}$-selection rule. Indeed for amorphous semiconductors, which have no real-space periodicity and consequently no reciprocal lattice either, the distinction between direct and indirect gaps becomes meaningless and essentially all optical transitions are allowed.

At photon energies much higher than $\mathscr{E}_{\rm g}^{\rm i}$, *direct* vertical transitions between valence- and conduction-band states become possible (see Fig. 5.78a), and these two-particle (electron + photon) interactions then dominate the optical-absorption profile since they have much stronger matrix elements than indirect transitions that involve three-particle (electron + photon + phonon) interactions.

The functional form for the energy dependence of the optical absorption coefficient can be found by calculating the imaginary part of the dielectric constant, $\varepsilon_2(\omega)$ (cf. eqn. (4.163b)). This is done by starting from the expression given by the Fermi Golden Rule, obtained from time-dependent perturbation theory, for the electric-dipole transition probability R for the rate of photon absorption (first assumed to be associated with *direct* processes):

$$R = \frac{2\pi}{\hbar} \sum_{k_{\rm c},k_{\rm v}} \left[\int \psi_{{\rm c},k}^{*} \,\mathscr{H} \psi_{{\rm v},k} {\rm d}\boldsymbol{r} \right]^2 \delta(\mathscr{E}_{\rm c}(\boldsymbol{k}_{\rm c}) - \mathscr{E}_{\rm v}(\boldsymbol{k}_{\rm v}) - \hbar\omega), \tag{5.205}$$

where the subscripts c and v refer to conduction-band and valence-band states, respectively. The Hamiltonian $\mathscr{H}$ describing the interaction between the electric field $\mathscr{E}$ of the electromagnetic radiation and the electron wavefunction (e.g. a Bloch state) is given by

$$\mathscr{H} = -e\boldsymbol{r} \cdot \boldsymbol{E}, \tag{5.206}$$

which is valid for small wavevectors of the electromagnetic wave (the electric-dipole approximation). (A more general expression is in terms of the electron momentum operator $\boldsymbol{p}$ and the vector potential $\boldsymbol{A}$, i.e. $\mathscr{H} = e\boldsymbol{A} \cdot \boldsymbol{p}/(m_{\rm e}c)$.) The integral in eqn. (5.205) can be rewritten as

$$\begin{aligned} \left(\int \psi_{{\rm c},k}^{*} \mathscr{H} \psi_{{\rm v},k} {\rm d}\boldsymbol{r} \right)^2 \equiv |\langle {\rm c}|\mathscr{H}|{\rm v}\rangle|^2 &= \left(\frac{e}{m_{\rm e}c} \right)^2 |\boldsymbol{A}|^2 |P_{\rm cv}|^2 \\ &= \left(\frac{e}{m_{\rm e}\omega} \right)^2 \left| \frac{E(\omega)}{2} \right|^2 |P_{\rm cv}|^2, \end{aligned} \tag{5.207}$$

where the momentum matrix element $|P_{\rm cv}|^2$ is assumed not to be a function of $\boldsymbol{k}$.

The rate of energy lost per unit volume, W, by the electromagnetic field due to absorption is, in terms of the transition probability,

$$W = R\hbar\omega, \tag{5.208}$$

and this is also related to the absorption coefficient K or imaginary part of the dielectric constant ε_2 (cf. eqn. (4.163)) by

$$\begin{aligned} W = -\frac{\mathrm{d}I}{\mathrm{d}t} = -\left(\frac{\mathrm{d}I}{\mathrm{d}x}\right)\left(\frac{\mathrm{d}x}{\mathrm{d}t}\right) &= \frac{c}{n_\mathrm{r}}KI \\ &= \frac{\varepsilon_2 \omega I}{n_\mathrm{r}^2}, \end{aligned} \tag{5.209}$$

where $I = I_0 \exp(-Kx)$ is the transmitted intensity per unit volume of the incident light. Since I is equal to the energy density of the field

$$I = \frac{n_\mathrm{r}^2 \varepsilon_0}{2} |E(\omega)|^2, \tag{5.210}$$

the final expression for the imaginary part of the dielectric constant is

$$\varepsilon_2(\omega) = \frac{\pi e^2}{m_\mathrm{e}^2 \omega^2 \varepsilon_0} \sum_k |P_\mathrm{cv}|^2 \delta(\mathscr{E}_\mathrm{c}(\boldsymbol{k}) - \mathscr{E}_\mathrm{v}(\boldsymbol{k}) - \hbar\omega). \tag{5.211}$$

The real part of the dielectric constant is related to eqn. (5.211) by the Kramers–Kronig transformation (eqn. (4.166)) — see eqn. (7.34).

The summation of the delta functions over $\boldsymbol{k}$ can be replaced by an integration over energy (cf. eqn. (5.132)) of the joint density of states $g_\mathrm{j}(\mathscr{E})$, the density of *pairs* of states in two bands with energy separation $\mathscr{E}_\mathrm{cv}(\boldsymbol{k}) = \mathscr{E}_\mathrm{c}(\boldsymbol{k}) - \mathscr{E}_\mathrm{v}(\boldsymbol{k})$ at the same $\boldsymbol{k}$-value. This is defined as (cf. eqn. (5.131)):

$$g_\mathrm{j}(\mathscr{E}) = \frac{2}{(2\pi)^3} \int \frac{\mathrm{d}S_\mathscr{E}}{|\nabla_k(\mathscr{E}_\mathrm{cv})|}\,. \tag{5.212}$$

The difference in energy between pairs of conduction- and valence-band states with the *same* $\boldsymbol{k}$-vector (i.e. for *direct* transitions) can be written in the vicinity of the direct gap $\mathscr{E}_\mathrm{g}$ as

$$\mathscr{E}_\mathrm{cv}(k) = \mathscr{E}_\mathrm{g} + \frac{\hbar^2 k^2}{2\mu}, \tag{5.213}$$

where it has been assumed that both bands are spherically symmetric for simplicity, and the reduced mass is given by $\mu^{-1} = (m_\mathrm{e}^*)^{-1} + (m_\mathrm{h}^*)^{-1}$, where m_e^* and m_h^* are the effective masses of electrons and holes in the conduction and valence bands, respectively. Thus, from eqn. (5.212), the joint density of states becomes

$$g_\mathrm{j}(\mathscr{E}) = \frac{(2\mu)^{1/2}\mu}{\pi^2 \hbar^3} (\mathscr{E}_\mathrm{cv} - \mathscr{E}_\mathrm{g})^{1/2} \qquad (\mathscr{E}_\mathrm{cv} > \mathscr{E}_\mathrm{g}), \tag{5.214}$$

and zero otherwise. Hence, the final expression for $\varepsilon_2(\omega)$ for *direct* interband transitions is

$$\varepsilon_2(\omega) = \frac{e^2 (2\mu)^{3/2}}{2\pi\varepsilon_0 m_\mathrm{e}^2 \omega^2 \hbar^3} |P_\mathrm{cv}|^2 (\hbar\omega - \mathscr{E}_\mathrm{g})^{1/2}. \tag{5.215}$$

Thus, for direct transitions, a plot of $(\omega^2\varepsilon_2)^2$, or $(\omega K)^2$, versus photon energy should be linear for photon energies somewhat greater than the gap, with an intercept of $\mathscr{E}_g$ on the abscissa.

For higher photon energies, electronic transitions from states deeper in the valence band occur to states higher in the conduction band. For direct transitions in crystals, ε_2, given by eqn. (5.211) with eqn. (5.212), is not featureless but exhibits sharp features where there are van Hove singularities in the *joint* density of states, i.e. where $\nabla_k(\mathscr{E}_{cv}) = 0$. This condition is satisfied for transitions between band extrema, and also for transitions between two states for which the local gradients are non-zero but equal, i.e. $\nabla_k\mathscr{E}_c(\boldsymbol{k}) = \nabla_k\mathscr{E}_v(\boldsymbol{k})$. Figure 5.79a shows experimental reflectance data for crystalline Ge and the real and imaginary parts of the dielectric constant obtained from it by making use of the Kramers–Kronig relations (eqn. (4.166)). The low-energy peak in ε_2 at about $\hbar\omega \simeq 2$ eV is due to transitions between upper-valence-band and lower-conduction-band states having equal, non-zero $\boldsymbol{k}$-space gradients at a $\boldsymbol{k}$-value (Λ) part-way to the L-point $((\pi/a)(111))$ in the band structure (Fig. 5.71). The high-energy peak in ε_2 at about $\hbar\omega \simeq 4.5$ eV is due to transitions between upper-valence-band and lower-conduction-band states at the X-point $((2\pi/a)\ (100))$.

For *indirect* optical transitions, involving an intermediate virtual state and subsequent emission/absorption of a phonon of energy $\hbar\omega_p$ (Fig. 5.78a), the Fermi Golden Rule expression (eqn. (5.205)) must be modified to take account of the intermediate state $|i>$ by making use of second-order perturbation theory, i.e.

$$R_{\text{ind}} = \frac{2\pi}{\hbar}\sum_{k_c,k_v}\left|\sum_i \frac{\langle c|\mathscr{H}'|i\rangle\langle i|\mathscr{H}|v\rangle}{(\mathscr{E}_{iv} - \hbar\omega)}\right|^2 \delta(\mathscr{E}_c(\boldsymbol{k}_c) - (\mathscr{E}_v(\boldsymbol{k}_v) - \hbar\omega \pm \hbar\omega_p), \tag{5.216}$$

where $\mathscr{H}'$ is the Hamiltonian representing the electron-phonon interaction in taking the electron from the intermediate state to the final conduction-band state $|c\rangle$. Assuming that the matrix elements are constant for energies in the region of the indirect gap, $\mathscr{E}_g^i$, the summations over $\boldsymbol{k}_c$ and $\boldsymbol{k}_v$ in eqn. (5.216) reduce to integrations of the separate (*not* joint) valence-band and conduction-band densities of states. Hence, the transition probability becomes

$$R \propto \frac{1}{\omega^2}\int\!\!\int g_v(\mathscr{E}_v)g_c(\mathscr{E}_c)\delta(\mathscr{E}_c - \mathscr{E}_v - \hbar\omega \pm \hbar\omega_p)\mathrm{d}\mathscr{E}_c\mathrm{d}\mathscr{E}_v. \tag{5.217}$$

Assuming, as for the direct-transition case, that the bands in the vicinity of $\mathscr{E}_g^i$ are parabolic and spherically symmetric, then

$$g_v \propto (-\mathscr{E}_v)^{1/2}, \quad \mathscr{E}_v < 0, \tag{5.218a}$$

and

$$g_c \propto (\mathscr{E}_c - \mathscr{E}_g^i)^{1/2}, \quad \mathscr{E}_c > \mathscr{E}_g^i, \tag{5.218b}$$

where the zero of energy has been taken to be at the top of the valence band. The imaginary part of the dielectric constant thus has the form for *indirect* transitions:

$$\varepsilon_2(\omega) \propto \frac{1}{\omega^2}(\hbar\omega \mp \hbar\omega_p - \mathscr{E}_g^i)^2, \quad \hbar\omega \geqslant \mathscr{E}_g^i \pm \hbar\omega_p, \tag{5.219}$$

which should be contrasted with the corresponding expression for direct transitions (eqn. (5.215)).

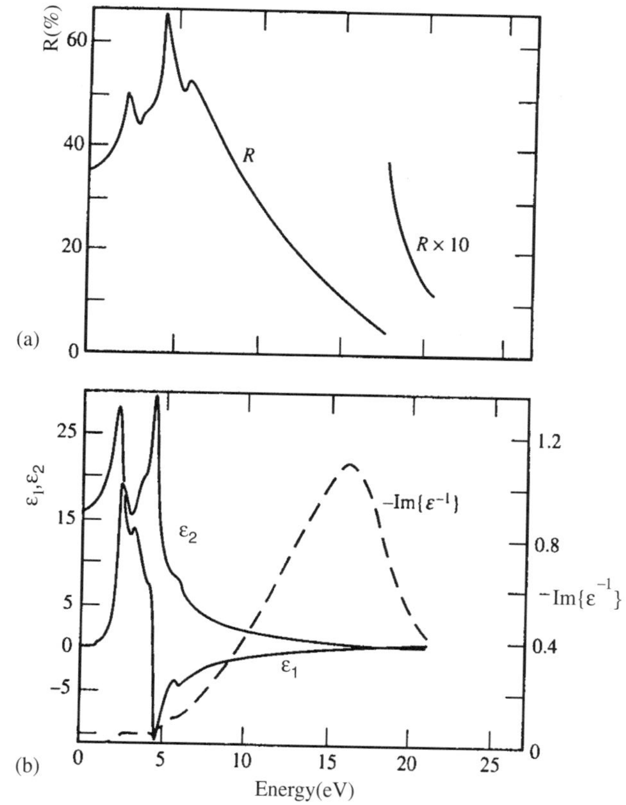

Fig. 5.79 (a) Experimental reflectance curve for crystalline Ge at 300 K. (b) Real (ε_1) and imaginary (ε_2) parts of the complex dielectric function, together with the energy-loss function, -Im $(1/\varepsilon)$, obtained from (a) using the Kramers–Kronig relation. The peaks in ε_1 and ε_2 are associated with van Hove singularities in the joint density of states. The peak in $-\text{Im}(1/\varepsilon)$ is the plasmon peak at $\omega = \omega_p$. (After Phillip and Ehrenreich (1967). Reproduced by permission of Academic Press, Inc.)

Figure 5.80 shows the absorption coefficient of crystalline Ge measured at various temperatures in the vicinity of the indirect gap, which occurs between the Γ-point in the valence band and the L-point in the conduction band (Fig. 5.71). At the lowest temperature (4.2 K), two threshold energies are evident at $\mathscr{E}_g^i + \hbar\omega_p' = 0.75\,\text{eV}$ and $\mathscr{E}_g^i + \hbar\omega_p'' = 0.77$ eV, corresponding to emission of *two* types of phonons. The two lowest-energy phonons are the TA phonon with $\hbar\omega_p' = 8$ meV (at the L-point) and the LA phonon with $\hbar\omega_p'' = 27$ meV at L (see the phonon dispersion curves in Fig. 4.17); they are associated with $\boldsymbol{k}$-values corresponding to the L-point because the conduction-band minimum in the electronic band structure of crystalline Ge occurs at that point. At higher temperatures, phonon absorption becomes important (e.g. at 77 K for the TA and LA phonons), and at the highest temperatures, higher-energy phonons (i.e. TO(L) and LO(L) — see Fig. 4.17) are also involved. Further details on interband transitions in crystalline semiconductors are given in Madelung (1978) and Yu and Cardona (1996).

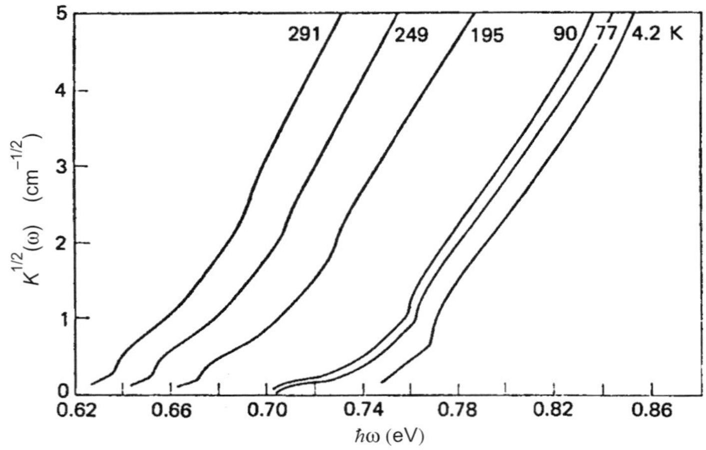

Fig. 5.80 The absorption edge of crystalline Ge, in the vicinity of indirect transitions, measured at various temperatures as indicated. (After MacFarlane *et al.* (1957). Reprinted with permission from *Phys. Rev.* **108**, 1377. © 1957. The American Physical Society)

For the case of non-crystalline materials, where the absence of translational periodicity means that the $\boldsymbol{k}$-selection rule for optical transitions is relaxed, the interband optical spectra are much simpler than for crystals: for example, there are no sharp features in $\varepsilon_2(\omega)$ corresponding to van Hove singularities, and any peaks in such spectra arise from peaks in the electronic density of states. For interband transitions in amorphous solids caused by photons with energy $\hbar\omega \gtrsim \mathscr{E}_g$, it makes no sense to discuss the optical properties in terms of a *joint* density of states (eqn. (5.212)), as in direct transitions in crystalline materials, because $\boldsymbol{k}$ is no longer a good quantum number. Instead, a modification of the approach used to calculate $\varepsilon_2(\omega)$ for indirect transitions may be adopted in the amorphous case; that is, ε_2 is taken to be proportional to an integral over the *product* of the valence- and conduction-band densities of states, i.e.

$$\varepsilon_2(\omega) \propto \frac{|P_{\mathrm{cv}}|^2}{\omega^2} \int_0^{\hbar\omega} g_{\mathrm{v}}(-\mathscr{E}) g_{\mathrm{c}}(\hbar\omega - \mathscr{E}) \mathrm{d}\mathscr{E}, \tag{5.220}$$

where again the zero of energy is taken to be at the top of the valence band. If it is assumed that, in the vicinity of the gap, the valence- and conduction-band profiles have a parabolic form (eqn. (5.218)), then the photon-energy dependence of the optical absorption of amorphous semiconductors becomes:

$$\omega K(\omega) \propto \omega^2 \varepsilon_2(\omega) \propto (\hbar\omega - \mathscr{E}_g)^2. \tag{5.221}$$

This behaviour is shown in Fig. 5.81 for a number of amorphous semiconductors.

Amorphous semiconductors differ from their crystalline counterparts also in the fact that structural disorder causes a 'tailing' of states at the band edges into the otherwise forbidden energy gap. Such tail states give a contribution to the optical absorption at photon energies below the bandgap energy, $\mathscr{E}_g$; this is seen in Fig. 5.81 as the absorption tails below the extrapolated values of the bandgap. It is a remarkable fact that all

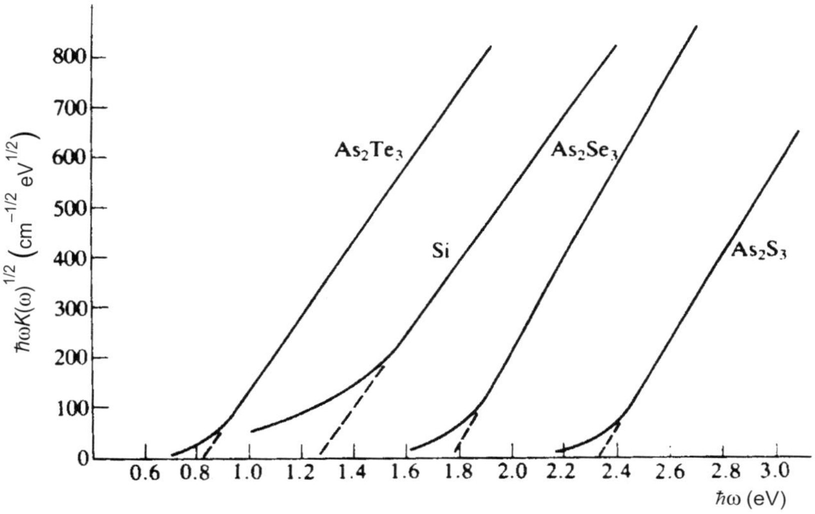

Fig. 5.81 Optical absorption edges of amorphous semiconductors, showing the behaviour $\omega K(\omega) \propto (\hbar\omega - \mathscr{E}_g)^2$, and the extrapolations made to obtain values of $\mathscr{E}_g$. (After Elliott (1990))

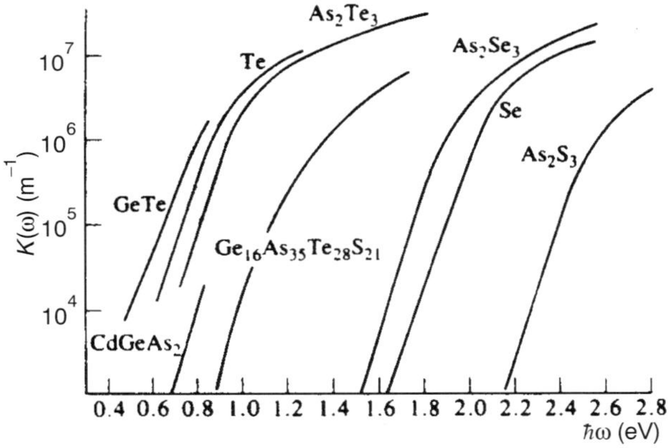

Fig. 5.82 Urbach edges in the subgap region of the optical absorption profiles of a number of amorphous semiconductors. (After Elliott (1990))

amorphous semiconductors and insulators appear to exhibit the same characteristic absorption behaviour in this subgap region, viz. the Urbach edge (also found in crystalline alkali halides):

$$K(\omega) = K_0 \exp[-\Gamma(\mathscr{E}_0 - \hbar\omega)], \tag{5.222}$$

where $\mathscr{E}_0$ is an energy comparable to the optical gap $\mathscr{E}_g$, and Γ is a temperature-dependent constant (above $\simeq$ 77 K), typically having values in the range 10–25 eV^{-1}. Representative Urbach edges for some amorphous semiconductors are shown in Fig. 5.82. It is still not entirely clear what is the origin of the Urbach behaviour, but structural disorder clearly plays a significant rôle: the greater the disorder, the less steep is the Urbach edge (smaller Γ). It appears that the exponential energy dependence of the Urbach edge (eqn. (5.222)) is due to exponential energy profiles of the band tails.

5.8.3 Excitons

Thus far, it has been assumed that photo-induced interband transitions lead to the excitation of an electron to the conduction band, leaving behind an empty electron state (i.e. a 'hole' — see §6.2.2) in the otherwise filled valence band, and that such photo-generated electrons and holes behave *independently* in their separate bands. However, such electron–hole pairs can mutually interact via the Coulombic attraction between negatively charged electron and positively charged hole, and if the binding energy is sufficiently large to bind pairs of electrons and holes, such pairs form new boson-like quasiparticles termed excitons. Exciton formation occurs preferentially at critical points in the joint density of states (eqn. (5.212)), i.e. where $\nabla_k(\mathscr{E}_c) = \nabla_k(\mathscr{E}_v)$, since at such $\boldsymbol{k}$-values the group velocities of the electron and hole are *equal*, facilitating mutual binding to form an exciton. The existence of excitons is another example of the breakdown of the independent-electron approximation (for other examples, see §5.6).

Exciton formation can have dramatic effects on the optical behaviour of semiconductors and insulators, particularly in the subgap region ($\hbar\omega \lesssim \mathscr{E}_g$) where, for a perfect crystal in the absence of exciton formation, no optical absorption would otherwise be expected, since a bound exciton must have a lower energy than a free electron and hole. Excitons can move through a material via exciton bands but, since these quasiparticles are electrically neutral overall, they are not charge carriers and hence do not contribute to the electrical response of solids.

Two types of excitons can be distinguished, depending on their relative binding energies and, as a consequence, on their spatial extent. Very tightly bound excitons are termed Frenkel excitons, and their spatial extent is typically of the order of a single atom (Fig. 5.83a). Weakly–bound excitons, called Mott–Wannier excitons, on the other hand, are characterized by electron–hole separations large compared with interatomic spacings (Fig. 5.83b) and, as such, bear some resemblance to an 'atom' of positronium, consisting of a bound electron–positron pair.

Frenkel excitons are found in materials such as rare-gas crystals and ionic solids, where the dielectric constant is sufficiently small that the Coulombic attraction between electron–hole pairs is not strongly screened (§5.6.1), and hence the exciton is strongly bound and of small extent. Alternatively, they are also commonly found in molecular (organic) crystals (e.g. anthracene, $C_{14}H_{10}$) where the intermolecular interactions are weak (van der Waals-like), and hence excitons tend to be localized on *single* molecules. Frenkel excitons can hop from site to site within a narrow exciton band, whose origin can be understood in terms of the tight-binding approximation (§5.3.1). An example of the effect of Frenkel exciton formation on the optical absorption behaviour of materials is given in Fig. 5.84 for the case of solid Kr. It can be seen that the lowest exciton absorption bands correspond rather closely to the electronic transition energies found in the *atomic* state, implying that the excitons are highly localized in the vicinity of single atoms.

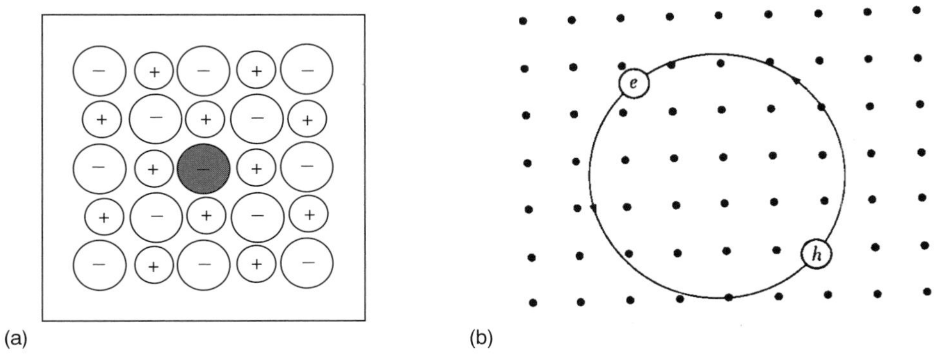

Fig. 5.83 (a) Schematic illustration of a Frenkel exciton in an ionic crystal, e.g. an alkali halide, localized on one anion. (b) Schematic representation of a Mott–Wannier exciton in a crystal.

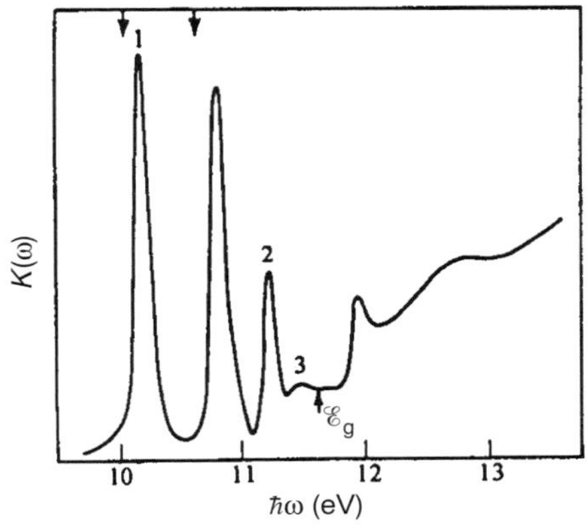

Fig. 5.84 Optical absorption in solid Kr at 20 K showing exciton peaks below the bandgap energy, $\mathscr{E}_g$. The two arrows correspond to electronic transitions in *atomic* Kr, i.e. $4p^6 \rightarrow 4p^6 5s$ (split due to spin–orbit coupling). (After Baldini (1962). Reprinted with permission from *Phys. Rev.* **128**, 1562. © 1962. The American Physical Society)

In the case of Mott–Wannier excitons, the Coulombic interaction is screened by the dielectric constant ε of the material as long as the electron–hole separation is very much larger than the interatomic spacing. For very large separations of electron and hole, the relative (rotational) motion of the two particles is sufficiently slow that the ions can respond, thereby producing a dominant lattice contribution to the polarization and thus $\varepsilon \simeq \varepsilon(0)$, the 'static' dielectric constant. For higher frequencies of electron and hole motion (corresponding to smaller electron–hole separations) that are greater than the LO-phonon frequency, the lattice can no longer respond, and the exciton Coulomb interaction is then screened by the valence electrons, i.e. the effective dielectric constant is then the high-frequency dielectric constant, $\varepsilon \simeq \varepsilon(\infty)$. The general motion of an exciton can be separated into *two* parts: (i) a translational motion of the centre of mass, and (ii) a relative motion with respect to the centre of mass.

The relative motion of the electron–hole pair, interacting via a Coulomb potential, can be analysed in the same way as in the hydrogen-atom problem (electron plus proton). The quantized energy levels for the relative (rotational) motion therefore depend, in general, on three quantum numbers, the principal quantum number n, the orbital-angular momentum quantum number l and the magnetic quantum number m. For spherical (isotropic) electron bands, only n is important and hence the energy levels form the infinite sequence

$$\mathscr{E}_r(n) = \mathscr{E}_r(\infty) - \frac{R^*}{n^2}, \tag{5.223}$$

where R*, the Rydberg constant for the exciton, is given by

$$R^* = \frac{\mu e^4}{32\pi^2\varepsilon^2\varepsilon_0^2\hbar^2} = \frac{\mu R_0}{m_e\varepsilon^2}, \tag{5.224}$$

and where the reduced mass of the exciton is $\mu = ((m_e^*)^{-1} + (m_h^*)^{-1})^{-1}$ and the Rydberg constant for the hydrogen atom is $R_0 = 13.6$ eV. The energy corresponding to the continuum limit is simply the conduction-band edge, i.e. $\mathscr{E}_r(\infty) = \mathscr{E}_g$. Exciton binding energies for Mott–Wannier and Frenkel-like excitons are given in Table 5.5.

The centre-of-mass motion of the exciton is described by motion in an exciton band characterized by an exciton wavevector given by the sum of electron and hole (§6.2.2) wavevectors, $K = \boldsymbol{k}_e + \boldsymbol{k}_h$. The effective mass (see §6.2.1) of the exciton quasiparticle is given by $M = m_e^* + m_h^*$, and so the free-exciton kinetic energy associated with centre-of-mass motion is given by

$$\mathscr{E}_{CM} = \frac{\hbar^2 K^2}{2M}. \tag{5.225}$$

The centre-of-mass exciton motion is translationally invariant in a crystal, and so $\boldsymbol{K}$ (but *not* $\boldsymbol{k}_e$ or $\boldsymbol{k}_h$ separately) is conserved in interactions, e.g. with a photon.

Since an exciton is a two-particle entity, it is not correct simply to superimpose the hydrogenic energy levels (eqn. (5.223)) on the *one-electron* bands representing the translational motion of single electrons or holes; instead, reference should be

Table 5.5 Exciton binding energies R^*

Mott–Wannier type		*Frenkel type*	
Material	R^* (meV)	*Material*	R^* (meV)
Si	14.7	BaO	56
Ge	4.2	KI	480
GaAs	4.9	KCl	400
GaP	3.5	KBr	400
InP	5.1	RbCl	440
CdS	29		
CdSe	15		
CdTe	11		

(After Burns (1985). Reproduced by permission of Academic Press, Inc.)

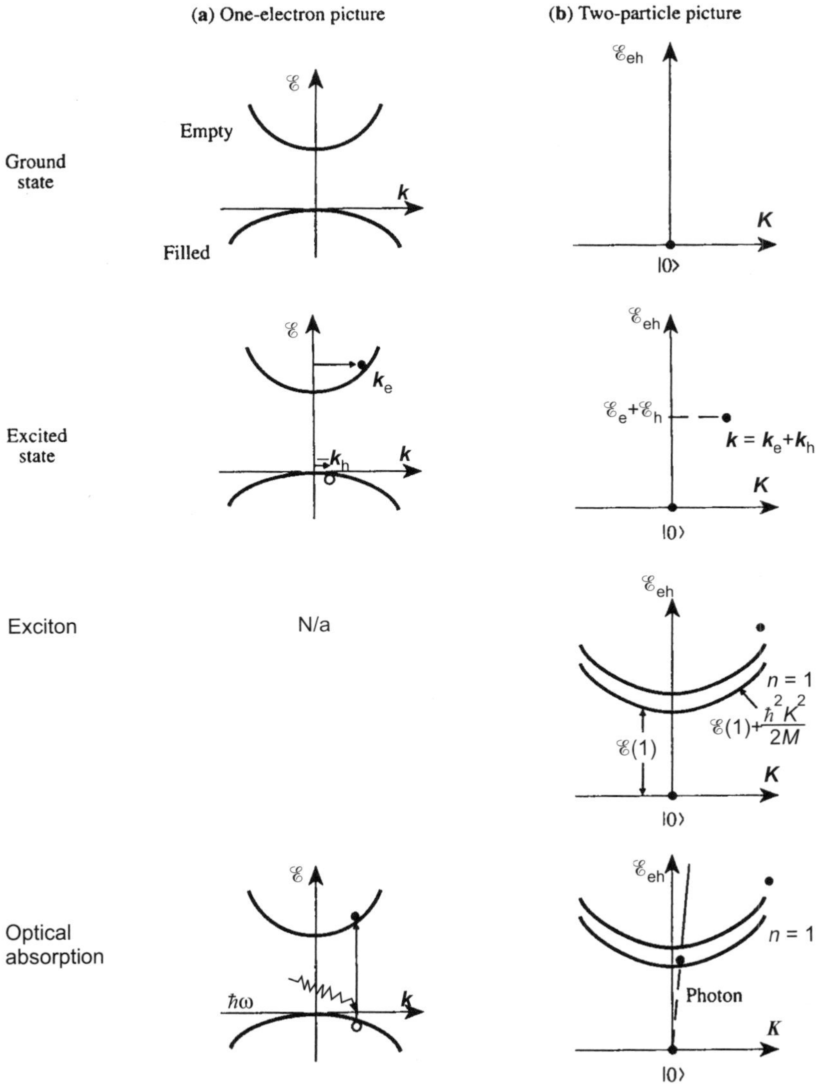

Fig. 5.85 Energy levels of the ground state (filled valence band, empty conduction band) and excited state of a semiconductor, together with photon-absorption processes, in (a) a one-electron band picture (the excitation being an electron in the conduction band and a hole in the valence band); (b) a two-particle picture (the excitation being an exciton). (After Yu and Cardona (1996), *Fundamentals of Semiconductors*, p. 268, Fig. 6.20, © Springer-Verlag GmbH & Co. KG)

made to the *two-particle* picture, for which the appropriate variable is the exciton wavevector, $\boldsymbol{K}$, and this is done in Fig. 5.85. The series of parabolic energy bands of the exciton (a combination of eqns. (5.223) and (5.225)) can only be represented in two-particle $\boldsymbol{K}$-space, not in one-electron $\boldsymbol{k}$-space. In the exciton picture, interaction with a photon occurs when the photon dispersion curve intersects the exciton bands. In fact, the exciton rotational motion and the electromagnetic wave couple together to form a coupled exciton-polariton state (see Problem 5.33(b)), similar to the coupling between ionic vibrations and light involved in polaritons (§4.4). Clear evidence for a Rydberg-like series of absorption peaks associated with exciton formation is shown in Fig. 5.86 for the case of crystalline Cu_2O. In this material, direct transitions at the Γ-point are symmetry-forbidden (see Problem 5.32), and only exciton p-like bound states (with $l = 1$) are weakly electrical-dipole-active. The series of exciton absorption peaks satisfy eqn. (5.223) with $\mathscr{E}_r(\infty) = 2.166$ eV and $R^* = 97$ meV for $n \geqslant 2$ (the $n = 1$ peak is missing, of course, since there is no 1p state).

*5.8.4 Non-linear optical behaviour

Thus far, the discussion of optical properties has assumed that a *single* photon is involved in the electronic excitations. In other words, it has been assumed that the optical response of a solid is *linear* in the electric field of the electromagnetic wave. This is a valid approximation at low light intensities, but for the case of high-power lasers strongly focused on a material (say with a power density of $\simeq 10^{18}$ W m^{-2}), the field strengths of the light can reach levels of order $E \simeq 10^{10}$ V m^{-1}, i.e. comparable to internal fields in solids, and multiphoton optical processes can occur as a result of the ensuing *non-linear* optical processes. One example of such behaviour which is relevant to the discussion given in the immediately preceding sections is two-photon absorption,

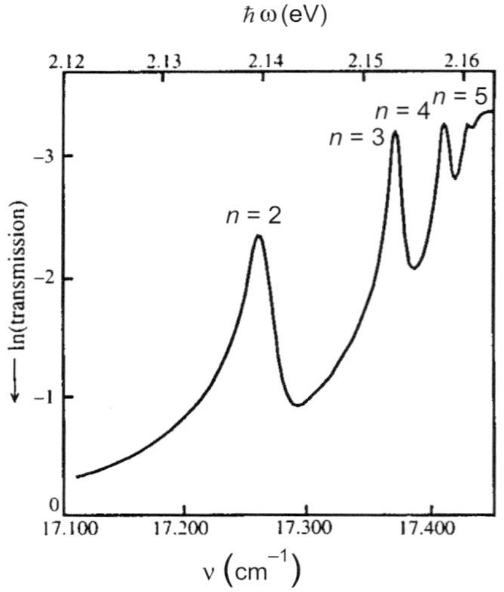

Fig. 5.86 Low-temperature optical absorption spectrum of Cu_2O showing the p-like Rydberg series of exciton peaks for photon energies less than the bandgap (2.166 eV). (After Baumeister (1961). Reprinted with permission from *Phys. Rev.* **121**, 359. © 1961. The American Physical Society)

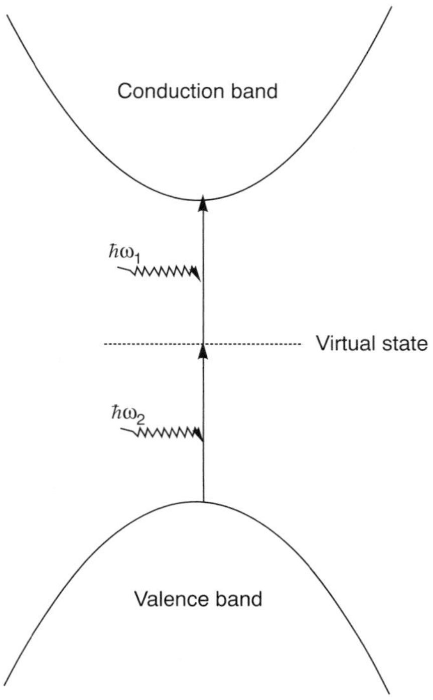

Fig. 5.87 Schematic illustration of two-photon absorption via an intermediate, virtual state.

whereby the absorption of one photon causes an electronic transition to an intermediate virtual state; the subsequent absorption of another photon then causes a further excitation of the electron to the final state (Fig. 5.87). In such a process, energy is conserved only overall, but (crystal) momentum is conserved in each one-photon transition.

Non-linear optical behaviour can be characterized in terms of non-linear susceptibilities linking the induced polarization $\boldsymbol{P}_i(\omega)$ (where the suffix refers to the polarization direction in Cartesian coordinates) and powers of the applied field associated with the incident high-power light, viz.:

$$\boldsymbol{P}_i(\omega) = \varepsilon_0\chi_{ij}^{(1)}\boldsymbol{E}_j(\omega) + \varepsilon_0\chi_{ijk}^{(2)}\boldsymbol{E}_j(\omega_1)\boldsymbol{E}_k(\omega_2) + \varepsilon_0\chi_{ijkl}^{(3)}\boldsymbol{E}_j(\omega_1)\boldsymbol{E}_k(\omega_2)\boldsymbol{E}_l(\omega_3) + \dots \tag{5.226}$$

The first term in eqn. (5.226) is simply the normal linear term, and the first-order dielectric susceptibility $\chi^{(1)}$ is related to the dielectric constant (see eqn. (4.140) and §7.1.1) via

$$\chi^{(1)}(\omega) = \varepsilon(\omega) - 1. \tag{5.227}$$

Note that all dielectric susceptibilities (and the dielectric constant) are tensor quantities, the rank of the tensor increasing with the order of the non-linear susceptibility. The non-linear terms that are the most important involve the second- and third-order susceptibilities, $\chi^{(2)}$ and $\chi^{(3)}$. The second-order susceptibility (and all even-order susceptibilities) is *zero* for crystals having a point group with a centre of symmetry, or for glasses which are structurally isotropic: since in these cases opposing directions are equivalent, the polarization $\boldsymbol{P}$ must change sign when the optical field $\mathscr{E}$ is reversed, and hence the quadratic term $\chi^{(2)}$ must be zero. Of the 32 crystallographic point groups (§2.3.2), 21 do not have a centre of symmetry and hence have non-zero second-order susceptibilities (see Table 2.5 and §7.1.5.1). Symmetry relations further reduce the number of independent non-zero elements in $\chi^{(2)}$ and $\chi^{(3)}$ (see Butcher and Cotter (1990)).

The linearity of the first term involving $\chi^{(1)}$ in eqn. (5.226) implies that the response of the material to an external stimulus of frequency ω is at the *same* frequency. However, this is not necessarily the case for the non-linear terms and therein lies the usefulness of the effect that has made the field of non-linear optics so active (see e.g. Shen (1984), Butcher and Cotter (1990) and Saleh and Teich (1991) for further details). Examples of the types of non-linear effects that can be observed are summarized in the following. Essentially, they can be regarded as wave 'mixing' phenomena, in which various input frequencies ($E(\omega_i)$) are mixed to give several different output frequencies ($P(\omega_j)$).

For the case of the second-order susceptibility, for example, if the incident light consists of two frequencies, ω_1 and ω_2, the electric field in the sample is thus

$$E = E_1 \sin\omega_1 t + E_2 \sin\omega_2 t, \tag{5.228}$$

and the second-order term in eqn. (5.226) involving $\chi^{(2)}$ produces three-wave mixing, giving terms in $P(\omega)$ proportional to

$$E_1^2 \cos 2\omega_1 t, \quad E_2^2 \cos 2\omega_2 t, \quad E_1 E_2 \cos(\omega_1 + \omega_2)t, \quad E_1 E_2 \cos(\omega_1 - \omega_2)t. \tag{5.229}$$

For the case when the optical field contains just one frequency ($\omega_1 = \omega_2 = \omega$), the result is an electrical polarization $\boldsymbol{P}(\omega)$ with a frequency of *twice* the input frequency; this oscillating polarization then leads to light emission with frequency 2ω, i.e. second-harmonic generation (SHG). At the same time, a *static* polarization ($\omega = 0$) is produced, giving rise to a d.c. electrical field in the sample, i.e. optical rectification (by analogy with the electrical case, where an oscillating current is transformed into a d.c. one).

In the general case where the two frequencies are different, $\omega_1 \neq \omega_2$, sum-difference frequency generation takes place. Not all the five possible waves, with frequencies $0, 2\omega_1, 2\omega_2, \omega_1 \pm \omega_2$, are produced by sum-difference frequency generation in any given experimental arrangement. In addition to the frequency-matching condition, say

$$\omega_3 = \omega_1 + \omega_2. \tag{5.230a}$$

the phase-matching condition (i.e. conservation of photon momentum) must also be satisfied, i.e.

$$\boldsymbol{k}_3 = \boldsymbol{k}_1 + \boldsymbol{k}_2. \tag{5.230b}$$

In general, for a *dispersive* medium, where the refractive index is frequency-dependent, $n_r = n(\omega)$, the phase-matching condition, eqn. (5.230b), becomes

$$n_3\omega_3 = n_1\omega_1 + n_2\omega_2. \tag{5.230c}$$

Efficient sum-frequency generation, when eqns. (5.230a) and (5.230c) are simultaneously satisfied, can be achieved, for example by using an optically anisotropic (e.g.

uniaxial) crystal and an appropriate choice of angle of the incident light beam with respect to the optic axis of the crystal (see Problem 5.35).

If one of the frequencies is zero, $\omega_1 = 0$, i.e. light with frequency $\omega_2 = \omega$ is incident on a sample subject to a d.c. electric field $\boldsymbol{E}_0$, the net polarization is proportional to $E_0 E(\omega)$. Hence, from eqn. (5.226), this behaviour is equivalent to the *first-order* dielectric susceptibility $\chi^{(1)}$, and thus the refractive index of the medium, becoming dependent on the applied d.c. electrical field: this is known as the linear electro-optic (or Pockels) effect. The magnitude of second-order non-linear effects is obviously dependent on the magnitude of $\chi^{(2)}$; experimental values of elements of $\chi^{(2)}$ for representative crystalline materials used in SHG applications are given in Table 5.6.

In a similar manner, the third-order susceptibility, $\chi^{(3)}$, gives rise to four-wave mixing, third-harmonic generation and related phenomena. The simplest case is for a light wave incident on a sample subject to a d.c. ($\omega = 0$) electric field $\boldsymbol{E}_0$; this leads to a change in the refractive index of the medium proportional to E_0^2 and is known as the quadratic electro-optic (d.c. Kerr) effect. A related effect caused by third-order non-linearity produces a term in the refractive index, n_r, of the medium proportional to the light intensity I, since the linear susceptibility term $\chi^{(1)}$ can be regarded as being replaced by the factor $\chi^{(1)} + \chi^{(3)} E^2$, or equivalently:

$$n_r = n_0 + n_2 I. \tag{5.231}$$

For silica glass, as used in optic fibres, the Kerr parameter is $n_2 \simeq 6 \times 10^{-23}\ \mathrm{m^2/V^2}$. This effect is very important and is involved in a wide variety of non-linear optical processes, such as self-focusing of laser beams, 'soliton' pulse propagation, etc.

Two other effects involving third-order non-linearities occur when a pump light beam with frequency ω_p and a signal beam with frequency ω_s, chosen so that

$$\omega_p \pm \omega_s = \omega_0, \tag{5.232}$$

where ω_0 is the frequency of a transition in the medium, are mixed by the $\chi^{(3)}$ term. If $\omega_p + \omega_s = \omega_0$, two-photon absorption producing signal attenuation occurs; the effect is

Table 5.6 Values of elements of second-harmonic non-linear susceptibilites

Material	*Point group*	$\chi^{(2)}_{ijk}$ ($\times 10^{-12}\mathrm{mV^{-1}}$)	(*ijk*)
α-SiO_2 (quartz)	$32 - D_3$	0.8	*xxx*
		0.02	*xyz*
$LiNbO_3$	$3m$–C_{3v}	6.14	*yyy*
		−11.6	*zxx*
		81.4	*zzz*
$BaTiO_3$	$4mm$–C_{4v}	−34.4	*xzx*
		−36	*zxx*
		−13.2	*zzz*
KH_2PO_4 (KDP)	$\bar{4}2m$–D_{2d}	0.98	*xyz*
		0.94	*zxy*
CdSe	$6mm$–C_{6v}	62	*zxz*
		57	*zxx*
		109	*zzz*
GaAs	$\bar{4}3m$–T_d	377	*xyz*
GaP	$\bar{4}3m$–T_d	70	*xyz*

(After Shen (1984). Reproduced by permission of John Wiley & Sons Inc.)

proportional to the pump-beam intensity and signal field. The absorption coefficient is proportional to $\mathrm{Im}(\chi^{(3)})$, analogous to the linear case where $K \propto \mathrm{Im}(\chi^{(1)})$. If $\omega_p - \omega_s = \omega_0$, then *emission* of a signal photon occurs on absorption of a pump-beam photon: this is called a stimulated Raman process, resulting in amplification of the input signal beam, the energy coming from the medium, i.e. from the reservoir of energy contained in the transitions of frequency ω_0 that are coupled in. This process should be contrasted with the *spontaneous* Raman process previously discussed in §4.5.2. In general, third-order non-linear optical processes are very weak since $\chi^{(3)}$ is many orders of magnitude (typically ten) smaller than $\chi^{(2)}$. However, the magnitude of $\chi^{(3)}$ may be resonantly enhanced if one or more of the frequencies involved is equal to that of a transition in the medium in which wave mixing is taking place.

Any substance, gas, liquid or solid, exhibits optical non-linearities at sufficiently high optical fields. In the case of solids, two sources of non-linearity are the polarization response of the bound electrons giving a dielectric contribution and of the free conduction-band electrons in doped semiconductors. The latter are of interest because their electronic properties can be controllably varied by changing the dopant concentration (see §6.5.2).

Applications 5.9

There are very many applications of materials that exploit their electronic properties. Many of these, however, involve electrical transport phenomena (the subject of Chapter 6) or dielectric or magnetic properties (the subject of Chapter 7). Yet others rely on devices in which the electronic behaviour at interfaces (or heterojunctions) is exploited, and these form part of the subject of Chapter 8. Here, we consider applications that make use of the *bulk* optical properties of solids.

5.9.1 Optical communication

The most widely used property of inorganic glasses, such as SiO_2 or As_2S_3, is their optical transparency for photon energies below the bandgap ($\mathcal{E}_g \simeq 11$ eV for vitreous silica and 2.4 eV for As_2S_3). The fact that the physical behaviour of glasses is isotropic, that they can be readily processed into required shapes (e.g. plates, fibres) by mechanical extrusion or pulling at temperatures just above the glass-transition (or 'softening') temperature, T_g, and that they are cheap to produce, makes them very attractive for commercial exploitation. The most widely used application of silica glass (containing a few tens of mole percent of network modifiers, e.g. Na_2O and CaO to lower the value of T_g and extend the region of workability in its vicinity) is, of course, as a window material in a wide variety of situations.

A rather more 'hi-tech' application that exploits the optical transparency of silica glass is in optical fibres used for telecommunications applications. A light wave launched into the end of a fibre propagates down the core of the fibre by a process of total internal reflection at the surface of the fibre where, in the simplest step-index case, there is a discontinuity in the refractive index, n_r, between the glass core of the fibre (higher n_r) and a protective polymer cladding (lower n_r)—see Fig. 5.88a. Alternatively, a graded-index fibre can be produced in which n_r decreases steadily (typically parabolically) from the centre of the fibre to the edge; this gradation is achieved by mixing a high-n_r material (GeO_2) to SiO_2 at the centre, and a low-index material (B_2O_3) at the edge, of the glass preform from which the fibre is pulled at temperatures above T_g. In this case, a light ray launched into the fibre is confined to the core region by a process of refraction (Fig. 5.88b). A single transverse, propagating cylindrical waveguide mode can be supported by very thin step-index fibres (typically with a core diameter of $\simeq 10\mu$m), while thicker step-index or graded-index fibres ($\simeq$ 50–200μm diameter) can support many propagating modes simultaneously (multimode operation).

Obviously, for long-distance telecommunication applications, optical-fibre waveguides must be as transparent as possible: light-intensity losses due to optical absorption in, or scattering out of, the fibre must be minimized. Four loss or scattering mechanisms can be identified: absorption due to interband electronic transitions at high (UV) photon energies (the Urbach edge—see §5.8.2); multiphonon absorption at low (IR) frequencies; absorption of light associated with electronic transitions between electronic states in the gap (due to impurities and defects) and the bands of the silica host material; and Rayleigh scattering of the light by density and compositional fluctuations in the glass. Impurities can be removed to a great extent by careful processing, but the three remaining processes are intrinsic and cannot be eliminated; together, they give

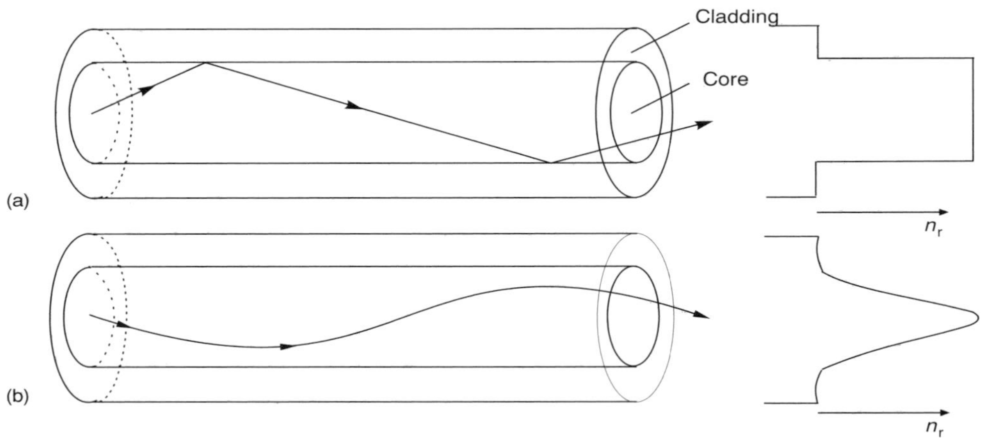

Fig. 5.88 (a) Guiding of a light wave in a step- index glass fibre by total internal reflection at the interface between core and cladding. The step profile of the refractive index is indicated. (b) Guiding of a light wave in a graded-profile glass fibre by refraction. The profile of the refractive index is indicated.

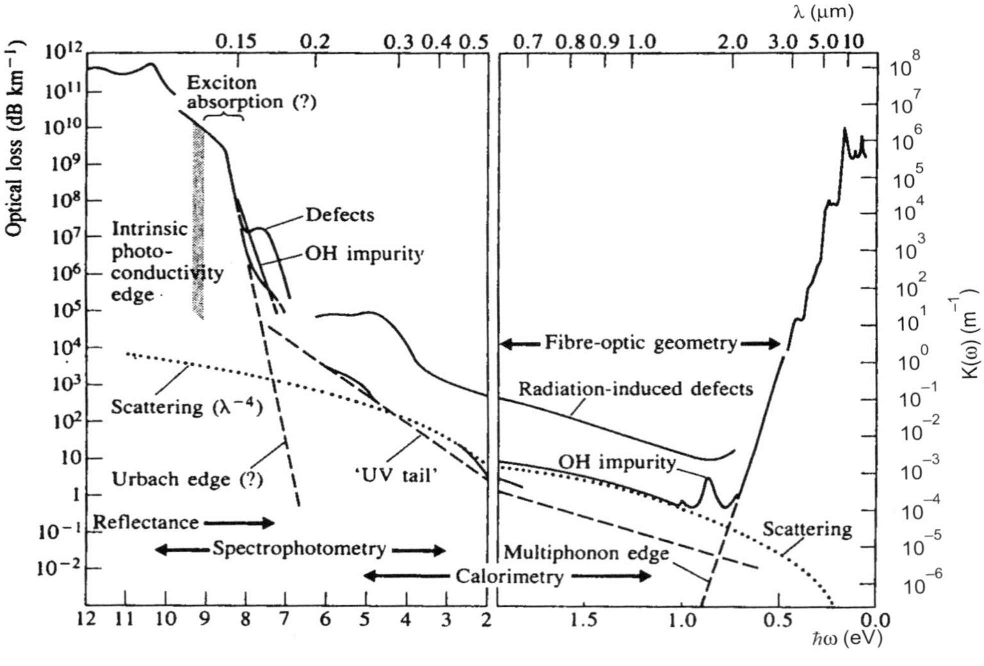

Fig. 5.89 Optical absorption profile of vitreous SiO_2, showing the V-shaped profile formed by the cross-over between Urbach and multiphonon edges. The contribution from Rayleigh scattering is also shown. (Griscom (1985) in *Glass: Current Issues*, ed. A. F. Wright and J. Dupuy, (Martinus Nijhoff), p. 362, Fig. 2 with kind permission from Kluwer Academic Publishers)

a V-shaped effective absorption profile as a function of wavelength (see Fig. 5.89) since $K(\lambda)$ can be written as

$$K(\lambda) = A_1 \exp(A_1/\lambda) + B_1 \exp(-B_2/\lambda) + C/\lambda^4, \tag{5.233}$$

where the first term refers to the Urbach absorption edge (eqn. (5.222)), the second to the multiphonon edge and the last to Rayleigh scattering. Losses of 0.2 dB/km (at $\lambda = 1.6\ \mu$m) have been achieved in real silica-glass fibres, close to the theoretical minimum loss of $\simeq 0.1$ dB/km in SiO_2 (see Fig. 5.89). Rayleigh scattering can be minimized by operating with longer-wavelength light, in which case the multiphonon edge must be pushed to longer wavelengths. This can be achieved by using glasses containing elements heavier than Si and O, for example chalcogenides (e.g. As_2S_3) or fluoride glasses (e.g. ZBLA compositions $(ZrF_4)_{57}(BaF_2)_{36}(LaF_3)_3(AlF_3)_4$), although these materials are harder to process and, in the latter case, more prone to crystallization (thereby causing additional light scattering).

Silica-glass optical fibres are already very widely used in cables carrying telecommunications traffic both locally (in cities) and globally (between countries and continents). They have many advantages over conventional copper-wire cabling for such uses, including extremely large bandwidth ($>$ 1 THz for semiconductor-laser sources), negligible noise, cross-talk and susceptibility to electrical interference and high signal transmission rates ($>$ 1 Gbit/s over 200 km). Further details of fibre-optic applications can be found in Midwinter (1979) and Syms and Cozens (1992).

*5.9.2 Non-linear optical devices

Optical non-linearity of materials forms the basis of the technological revolution, based on the manipulation of light, that will follow the electronics era; by analogy, this new technology has been termed photonics. This is a very active field, and full coverage of it cannot be given here (see e.g. Saleh and Teich (1991) for more details). Instead, mention will be made of a number of optical applications that make use of the novel behaviour imparted by optical non-linearity in materials.

Second-harmonic generation (SHG) is widely used to up-convert the limited frequencies of light emitted by conventional laser sources. Thus, crystals like KDP (see Table 5.6) can be used to convert red ruby-laser light (694 nm) to UV radiation (347 nm). Even doped silica-glass optic fibres (see §5.9.1) can be used to convert, for example, the IR light emitted by a Nd^{3+}: yttrium aluminium garnet (YAG) laser (1.06 mm) to green visible light (530 nm). (Although pure silica glass itself exhibits no second-order optical non-linearities ($\chi^{(2)} = 0$) because of its structural isotropy, doping the central core region of a fibre with GeO_2 produces locally anisotropic defect complexes that result in a finite $\chi^{(2)}$ for the material.)

Three-wave mixing, originating from second-order optical non-linearity, can be used to produce parametric optical amplifiers. In an optical amplifier, two beams are passed through a non-linear medium: one is a high-power (pump) beam, with frequency ω_p, and provides the power for the amplification; the other is a small-intensity signal beam, with frequency ω_s, that is to be amplified. Difference–frequency mixing produces an auxiliary light field (the 'idler' beam), with frequency $\omega_I = \omega_p - \omega_s$, that is proportional to $\chi^{(2)} E_p E_s$. The idler beam then beats (mixes) with the pump beam to produce a term in the electrical polarization, at the signal frequency $\omega_s = \omega_p - \omega_I$, that is proportional to $E_s E_p^2$. Thus, *amplification* of the signal beam is achieved if the phase-matching condition (eqn. (5.230b)) is satisfied.

Cubic optical non-linearity (due to $\chi^{(3)}$) produces a wealth of interesting effects that may be exploited. One such application is the construction of an optical switch using a Fabry–Pérot étalon containing an optical medium with refractive index n_r (Fig. 5.90a). A Fabry–Pérot étalon is non-transmitting for an incident monochromatic light beam of wavelength λ *unless* the expression

$$2n_r d = m\lambda, \tag{5.234}$$

is satisfied, where d is the separation between the mirrors and m is an integer. This relation can be obtained as the condition for constructive interference: note the similarity between this expression and the Bragg equation for diffraction from lattice planes, eqn. (2.95). If the material between the mirrors exhibits a sizeable light-intensity-dependent refractive index (the Kerr effect (eqn. (5.231)) and that, at low light intensities, the refractive index $n_r \simeq n_0$ is such that eqn. (5.234) is *not* satisfied (i.e. the étalon is non-transmitting, i.e. optically 'off '), then at a critical light intensity I_0, the constructive interference condition, eqn. (5.234), will become satisfied and the étalon will suddenly become optically transparent, i.e. switch to being 'on' (Fig. 5.90b).

Four-wave mixing in a third-order non-linear material can produce a very novel effect, viz. optical phase conjugation, that can be used, for example, in the restoration of a distorted wavefront to its original undistorted form. Consider the mixing of three superposed optical waves, with frequencies ω_1, ω_2 and ω_3, with net field given by

$$E(t) = \mathrm{Re}\{E_1 \exp(-\mathrm{i}\omega_1 t)\} + \mathrm{Re}\{E_2 \exp(-\mathrm{i}\omega_2 t)\} + \mathrm{Re}\{E_3 \exp(-\mathrm{i}\omega_3 t)\} \tag{5.235a}$$

or, equivalently, rewritten as a sum of six terms:

$$E(t) = \frac{1}{2} \sum_{i=\pm1,\pm2,\pm3} E(\omega_i) \exp(-\mathrm{i}\omega_i t), \tag{5.235b}$$

where $\omega_{-i} = -\omega_i$ and $E(-\omega_i) = E^*(\omega_i)$. Substitution of eqn. (5.235b) into the expressions for the corresponding induced third-order polarizability (cf. eqn. (5.226)) gives $6^3 = 216$ terms, i.e.

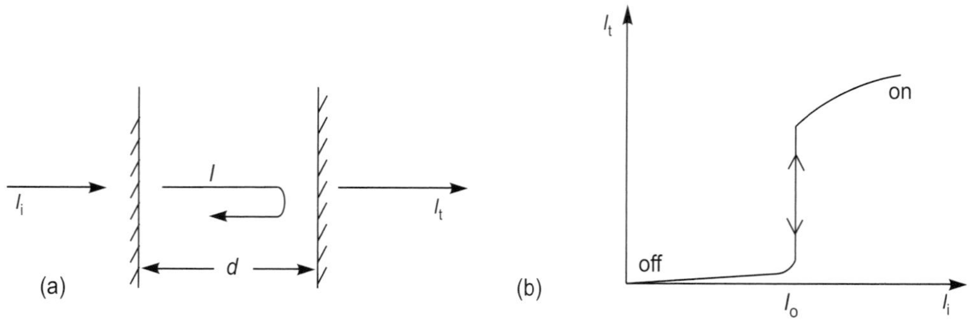

Fig. 5.90 (a) An optical switch based on a Fabry–Pérot étalon comprising an optical medium with an intensity-dependent refractive index between two semi-transparent mirrors. (b) Switching behaviour of the transmitted light intensity I_t.

$$\boldsymbol{P}^{(3)}(t) = \varepsilon_0 \chi^{(3)}_{\alpha\beta\gamma\delta} \sum_{i,j,k=\pm1,\pm2,\pm3} E_\beta(\omega_i) E_\gamma(\omega_j) E_\delta(\omega_k) \exp[-\mathrm{i}(\omega_i + \omega_j + \omega_k)t]. \quad (5.236)$$

Thus, for example, the term $\boldsymbol{P}^{(3)}(\omega_2 = \omega_3 + \omega_4 - \omega_1)$ involves six permutations in the sum of eqn. (5.236), i.e.

$$\boldsymbol{P}^{(3)}(\omega_2) = 6\varepsilon_0 \chi^{(3)}_{\alpha\beta\gamma\delta} E_\beta(\omega_3) E_\gamma(\omega_4) E^*_\delta(\omega_1). \quad (5.237)$$

Thus, the frequency-matching (or photon-energy-conservation) condition is

$$\omega_3 + \omega_4 = \omega_1 + \omega_2, \quad (5.238a)$$

and the phase-matching (or momentum-conservation) condition is

$$\boldsymbol{k}_3 + \boldsymbol{k}_4 = \boldsymbol{k}_1 + \boldsymbol{k}_2. \quad (5.238b)$$

Degenerate four-wave mixing ($\omega_1 = \omega_2 = \omega_3 = \omega_4 = \omega$) satisfies eqn. (5.238a), and if two of the beams, 3 and 4, taken as pump beams, propagate collinearly and in opposite directions (e.g. as achieved by use of a mirror placed behind the optically active medium to reflect a single incident pump beam—see Fig. 5.91), then $\boldsymbol{k}_3 = -\boldsymbol{k}_4$. Hence from eqn. (5.238b), $\boldsymbol{k}_2 = -\boldsymbol{k}_1$ and the resultant beam 2 must propagate in the *opposite* direction to the signal beam 1 *whatever* its angle of incidence (Fig. 5.91). The field amplitude of the resultant beam is, from eqn. (5.237), thus given by

$$E(\boldsymbol{r}, \omega_2) \propto \boldsymbol{A}_\beta(\omega_3) \boldsymbol{A}_\gamma(\omega_4) E^*_\delta(\boldsymbol{r}, \omega_1), \quad (5.239)$$

where $E(\boldsymbol{r}, \omega_i) = \boldsymbol{A}(\omega_i) \exp(\mathrm{i}\boldsymbol{k}_i \cdot \boldsymbol{r})$ and $\boldsymbol{A}(\omega_i) = E^0_i \exp(-\mathrm{i}\omega_i t)$. Therefore, the resulting wave 2 is the (complex) conjugate wave of beam 1. The great peculiarity of the resulting phase-conjugate mirror is that the reflected ray always coincides with the incident ray, unlike in normal mirrors.

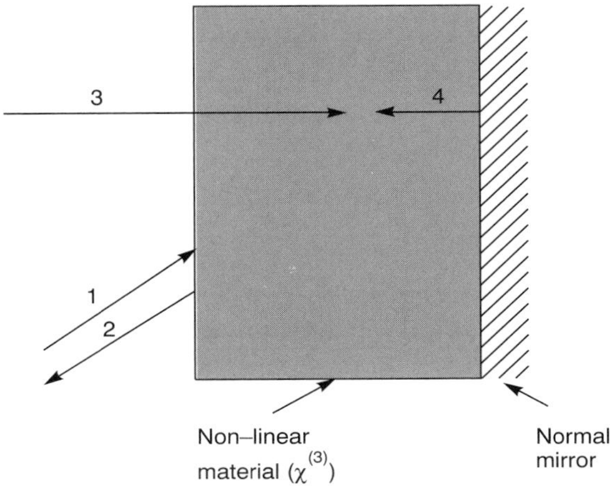

Fig. 5.91 Operation of a phase-conjugate mirror based on degenerate four-wave mixing. The counter-propagating beam 4 is produced by reflection of pump beam 3 by reflection from a normal mirror at the rear surface. An incident signal beam 1 then produces a counter-propagating phase-conjugate reflected beam 2.

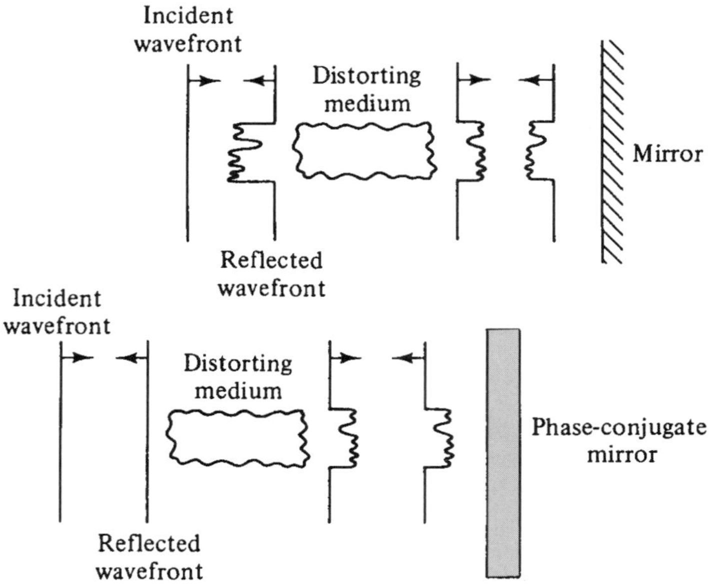

Fig. 5.92 Schematic illustration of the comparison of reflection behaviour of a normal and a phase-conjugate mirror. Phase aberrations induced in a wavefront by passage through a distorting medium are magnified by the second passage through the medium following reflection from an ordinary mirror, but are removed following reflection from a phase-conjugate mirror since the reflected wave is simply the time-reversed form of the incident wave.

The conjugate wave 2 (eqn. (5.239)), i.e. $E_2 \propto \exp[-\mathrm{i}(\boldsymbol{k}_2 \cdot \boldsymbol{r} - \omega_1 t)]$, is simply the *time-reversed* form $(t \rightarrow -t)$ of the incident signal beam $E_1 \propto \exp[\mathrm{i}(\boldsymbol{k}_1 \cdot \boldsymbol{r} - \omega_1 t)] \equiv \exp[\mathrm{i}(-\boldsymbol{k}_2 \cdot \boldsymbol{r} - \omega_1(-t)]$, since $\boldsymbol{k}_2 = -\boldsymbol{k}_1$ from the phase-matching condition. This behaviour enables a phase-conjugate mirror to remove phase aberrations induced in a wavefront by a distorting medium (see Fig. 5.92) since the beam reflected by a phase-conjugate mirror must be a time-reversed replica of the original, undistorted wavefront.

Problems

5.1 For the case of metallic potassium, for which the atomic density is 1.4×10^{28} m^{-3} and the unit-cell parameter is $a = 5.23$ Å, calculate values for the Fermi energy, temperature and wavevector. Comment on the value of the Fermi wavevector compared with the value of k corresponding to the Brillouin-zone boundary. What is the value of the density of states at the Fermi level?

5.2 Obtain expressions for the density of states of a free-electron gas in one and two dimensions. What are the Fermi energy and the average energy of the electron gas in each case?

5.3 Obtain an expression for the leading two terms in an expansion of the temperature dependence of the chemical potential $\mu(T)$ of a Fermi gas (eqn. (5.28)).
(Hint: First demonstrate that eqn. (5.27a) can be rewritten as eqn. (5.27b) by integrating by parts and expressing $\Gamma(\mathscr{E})$ as a Taylor expansion about μ; make use of the standard integral

$$\int_{-\infty}^{\infty} \frac{x^2 e^x}{(1+e^x)^2} dx = \pi^2/3.$$

Then, take $\Gamma(\mathscr{E}) \equiv \int_0^{\mathscr{E}} g(\mathscr{E}')d\mathscr{E}'$ and use eqn. (5.27a) together with eqn. (5.27b) to obtain eqn. (5.28), eliminating N by subtracting $N = \int^{\mathscr{E}_F} g(\mathscr{E}')d\mathscr{E}'$ from both sides of the equation derived from eqn. (5.27a).)

5.4 Estimate the temperature below which the electronic contribution to the heat capacity of potassium becomes greater than the lattice-vibrational contribution. ($\theta_D = 91$ K; atomic density $= 1.4 \times 10^{28}$ m^{-3}.)

5.5 Show that for a 2D Fermi gas, the chemical potential is independent of temperature. Obtain an expression for the corresponding heat capacity at constant volume.

5.6 Obtain an expression for the heat capacity at constant *pressure* for the 3D Fermi gas. Is the difference between it and C_v significant at room temperature for the case, say, of potassium?

5.7 Show, using the general expression for the density of states,

$$g(\omega) = \frac{V}{(2\pi)^3} \int_s \frac{dS_\omega}{|\nabla_k \omega|}$$

(eqn. (4.58)), that eqn. (5.15) is recovered for the Fermi gas.

5.8 Show that the electrostatic energy of Ze valence electrons uniformly distributed in a sphere with a radius given by the Wigner–Seitz radius r_{WS} is given by eqn. (5.41). (Hint: obtain an expression for the electrostatic potential ϕ due to the electrons at a distance $r < r_{WS}$ and use it to calculate separately the attractive interaction between ion and electrons and the repulsive interaction between electrons in the sphere. See §2.2.3.1 for an alternative derivation.)

5.9 Show that, for *all* temperatures, the pressure of the Fermi gas is given by $p = 2U/3V$, where the internal energy U is given by eqn. (5.35). (Hint: make use of eqn. (4.214).)

5.10 Ascertain the reciprocal-lattice vectors needed to translate portions of the free-electron parabolic bands in higher zones to the reduced zone for the case of a cubic Bravais lattice in the empty-lattice approximation (Fig. 5.15). Give expressions for the energies at arbitrary values of k_x for each of the bands.

5.11 The Kronig–Penney potential consists of a 1D array of potential barriers, of height V_0 and thickness t, spaced periodically with a separation a: the potential is zero between the barriers.

(a) Find, by solving the Schrödinger equation for an electron moving back and forth both in the region between the barriers ($t < x < a$) and within a barrier ($0 < x < t$), the energy of the electron in terms of the quantities K and κ, involved in the trial wavefunctions $e^{\pm iKx}$, $e^{\pm \kappa x}$, respectively.

(b) Obtain four equations linking the wavefunctions by matching the wavefunctions, and their first spatial derivative at the boundary $x = t$ and at $x = a$. (Hint: make use of Bloch's theorem in the latter case.) There is a solution to these equations only if the determinant involving the wavefunction coefficients vanishes, resulting in

$$[(\kappa^2 - K^2)/2\kappa K] \sinh\kappa t \sin K(a-t) + \cosh\kappa t \cos K(a-t) = \cos ka \tag{1}$$

where k is the Bloch wavevector.

(c) Consider the case when the potential-barrier height becomes increasingly high ($V_0 \rightarrow \infty$) but at the same time its width becomes infinitesimally narrow ($t \rightarrow 0$) in such a way that

$$V_0 t = \text{constant} = \mu(\hbar^2/m_e a) \tag{2}$$

where the parameter μ is a measure of the strength of the barrier between neighbouring wells. Show that, in this limit, $\kappa \rightarrow \infty$ but $\kappa t \rightarrow 0$, and hence from eqn. (1)

$$\cos Ka + (\mu/Ka) \sin Ka = \cos ka. \tag{3}$$

Indicate on a plot of the left-hand side of eqn. (3) versus Ka the regions where travelling, Bloch-like solutions are forbidden. At what values of Bloch wavevector k do the corresponding energy gaps occur? Make a sketch of the resulting band structure in the reduced-zone scheme.

(d) Comment on the behaviour exhibited by the Kronig–Penney system in the two limits $\mu \to 0$ and $\mu \to \infty$.

5.12 (a) Show that a Bloch wavefunction (eqn. (5.59)) satisfies the following form of the Schrödinger equation, where the operator $\boldsymbol{p} = -\mathrm{i}\hbar\nabla$ has been used:

$$[(\boldsymbol{p} + \hbar \boldsymbol{k})^2/2m_e + V(\boldsymbol{r})]u_k(\boldsymbol{r}) = \mathscr{E}(\boldsymbol{k})u_k(\boldsymbol{r}). \quad (1)$$

In the $\boldsymbol{k} \cdot \boldsymbol{p}$ approximation, the cross term of $(\boldsymbol{p} + \hbar\boldsymbol{k})^2$ in eqn. (1) can be neglected and $u_{k=0}(\boldsymbol{r})$ is an approximate solution: near the ion cores, it is similar to the electron wavefunction for the free atom, and between the cores is almost constant. For finite $\boldsymbol{k}$, the Wigner–Seitz wavefunction is approximately $\psi_k = \exp(\mathrm{i}\boldsymbol{k} \cdot \boldsymbol{r})u_0(\boldsymbol{r})$.

(b) The Schrödinger boundary condition for the wavefunction in a free atom is $\psi(\boldsymbol{r}) \to 0$ as $r \to \infty$. By considering the symmetry of the wavefunction $u_0(\boldsymbol{r})$, deduce the corresponding Wigner–Seitz boundary condition in a crystal.

(c) For Na, the eigenenergy of the 3s conduction electrons is -5.15 eV, and the corresponding energy for $u_0(\boldsymbol{r})$ is -8.2 eV. By considering the average energy per electron of the conduction electrons, treated as making up a Fermi gas, obtain an estimate for the cohesive energy of Na metal.

5.13 Consider the effect on the nearly-free-electron model of a crystal with more than one atom in its basis.

(a) Show, by starting with an expression for the total periodic lattice potential energy $V(\boldsymbol{r})$ written as a sum over atomic potentials $\phi(\boldsymbol{r})$ centred at the ion positions d_j, that the Fourier components of the potential in terms of reciprocal-lattice vectors $\boldsymbol{G}$, can be written as $V_G = e\phi(\boldsymbol{G})S^*_{\mathrm{G}}/\mathfrak{V}_c$, where $\phi(\boldsymbol{G})$ is the Fourier transform of the atomic potential, $\mathfrak{V}_c$ is the volume of the primitive unit cell and S^*_{G} is the complex conjugate of the geometrical scattering amplitude (cf. eqn. (2.102)), i.e. $S_{\mathrm{G}} = \sum_j \exp(\mathrm{i}\boldsymbol{G} \cdot \boldsymbol{d}_j)$.

(b) Hence, show that for an h.c.p. structure there is no energy gap at the top and bottom basal faces of the hexagonal first Brillouin zone.

(c) Show that for the diamond structure (e.g. Ge or Si), there is no energy gap at the X-point in the first Brillouin zone.

5.14 Show that the tight-binding (LCAO) wavefunction, eqn. (5.93), satisfies Bloch's theorem, and that the normalization constant is $N^{-1/2}$, where N is the number of atoms in the crystal.

5.15 Apply the tight-binding approximation to the following lattices containing s-states, and show in each case that the energies of the bands are as given:

(a) 2D square lattice: $\mathscr{E}(k) = \mathscr{E}_{\mathrm{i}} - \alpha_{\mathrm{i}} - 2\beta_{\mathrm{i}}[\cos(k_x a) + \cos(k_y a)]$. Compare the density of states in this case with that for a 3D simple cubic lattice (Fig. 5.37b).

(b) f.c.c. lattice:

$$\begin{aligned}\mathscr{E}(k) = &\mathscr{E}_i - \alpha_i - 4\beta_i[(\cos(k_y a/2)\cos(k_z a/2)) + (\cos(k_z a/2)\cos(k_x a/2)) \\ &+ (\cos(k_x a/2)\cos(k_y a/2))].\end{aligned}$$

(c) b.c.c. lattice:

$$\mathscr{E}(k) = \mathscr{E}_i - \alpha_i - 8\beta_i[\cos(k_x a/2)\cos(k_y a/2)(\cos(k_z a/2)].$$

Find the bandwidths in each case, and show that they are in accord with eqn. (5.99).

5.16 What is the value of the overlap (hopping) integral spπ? Deduce a rule for ascertaining which particular combinations of atomic orbitals forming putative bonds give non-zero overlap integrals.

5.17 Essay: Compare and contrast the nearly-free-electron and tight-binding approaches to describing electron states in solids, making particular reference to situations in which such models break down or are inappropriate.

5.18 Obtain expressions for the three sp^2-hybrids resulting from the combination of an s-orbital with a p$_x$ and a p$_y$ orbital. Show that the hybrids lie in the x–y plane, with a subtended angle of 120°.

5.19 Show that the width of the conduction band in the Weaire–Thorpe hybrid-orbital model (Fig. 5.44) is $\Delta\mathscr{E}_{ps} = \mathscr{E}_p - \mathscr{E}_s$. What will be the effect of inclusion of interactions between bonding and antibonding bond-orbital combinations?

5.20 Why are Cu and Au, in contrast to many other metals, coloured?

5.21 How would you expect the bandgap of semiconductors to vary with temperature?

5.22 Use the moments approach to investigate the effect on the local density of states of incorporating a substitutional impurity in a simple cubic lattice, containing s-states, at site i. Call the on-site matrix element for the impurity $\varepsilon = \mathscr{H}_{ii}$, which is different from that for all other atoms $\alpha = \mathscr{H}_{nn}$. Assume that the hopping integral between the impurity and nearest-neighbour host atoms is the same as that between host atoms, $\mathscr{H}_{in} = \mathscr{H}_{nn'} = \beta$. By considering closed paths for electrons hopping between the impurity (i) and its neighbours, show that:
(a) The zeroth, first and second moments, $\mu_i^{(0)}$ $\mu_i^{(1)}$ and $\mu_i^{(2)}$, are the same as for the perfect crystal.
(b) Show that the third moment is now non-zero and given by $\mu_i^{(3)} = 6\beta^2(\alpha - \varepsilon)$.
(c) Show that, in addition to the four types of four-hop paths shown in Fig. 5.59 for the perfect crystal, there is now an additional term contributing to $\mu_i^{(4)}$, viz. $6\beta^2(\alpha - \varepsilon)^2$.
(d) Sketch the distribution for the two cases $(\alpha - \varepsilon) > 0$ and $(\alpha - \varepsilon) < 0$. (These circumstances are equivalent to the in-band resonance vibrational states of heavy impurities in crystals (§4.3.1).) What will happen for $\alpha - \varepsilon \ll 0$?

5.23 Obtain the Thomas–Fermi screening function $\phi(q) = Ze/[4\pi\varepsilon_0(q^2 + 1/\lambda_{TF}^2)]$ as a function of wavevector q for a point charge Ze immersed in a free-electron gas by solving Poisson's equation (including the effect of the point charge) in reciprocal space. Prove that this result is equivalent to eqn. (5.154). Show that the dependence of the dielectric function on wavevector q in the Thomas–Fermi approximation is given by $\varepsilon(q) = 1 + (\lambda_{TF}q)^{-2}$, where λ_{TF} is the Thomas–Fermi screening length. Prove that the total charge of the electron liquid excluded by the screening interaction from a region of radius of order λ_{TF} around a point charge Ze at the origin is equal to that of the point charge.

5.24 Use the expression for the screened potential $\phi(q)$ derived in Problem 5.23 to show that the $q \to 0$ 'screened-ion' limit of the pseudopotential is $V(0) = -2/3\mathscr{E}_F$ (see Fig. 5.30).

5.25 Derive eqn. (5.167) for the dielectric constant $\varepsilon(\omega)$ of an electron gas in the absence of damping, given that $\varepsilon(\omega) = \boldsymbol{D}(\omega)/\varepsilon_0\boldsymbol{E}(\omega) \equiv 1 + \boldsymbol{P}(\omega)/\varepsilon_0\boldsymbol{E}(\omega)$, where $\boldsymbol{D}$, $\boldsymbol{E}$ and $\boldsymbol{P}$ are the electric displacement, field and polarization, respectively. (Hint: from the equation of motion of a free electron subject to an electric field, obtain an expression for $\boldsymbol{P}$, the dipole moment per unit volume.)

5.26 Show that the time-averaged power dissipation for a charged particle (e.g. a high-energy electron) traversing a solid and undergoing inelastic losses (e.g. associated with plasmon creation) is determined by the energy-loss function, $-\mathrm{Im}\{1/\varepsilon(\omega, \boldsymbol{k})\}$. Show also that the corresponding quantity involved in inelastic losses of *photons* is, instead, $\mathrm{Im}\{\varepsilon(\omega, \boldsymbol{k})\}$. (Hint: the power dissipation per unit volume is given by $W = E \cdot (\partial \boldsymbol{D}/\partial t)$, with $\boldsymbol{D} = \varepsilon\varepsilon_0\boldsymbol{E}$. For the case of electromagnetic waves, the electric-field vector is the dominant quantity, whereas for the case of a charged particle entering a solid, the important quantity is the electric displacement, since div $\boldsymbol{D} = \rho$, where ρ is the charge density.)

5.27 Essay: Discuss various scenarios for a metal–insulator transition in solids.

5.28 Show how, for a *two*-dimensional crystal, the band structure $\mathscr{E}(\boldsymbol{k}_{\|})$ may be obtained from photoemission experiments. ($\boldsymbol{k}_{\|}$ is the electron wavevector parallel to the basal plane.)

5.29 (a) Calculate the plasma frequency ω_p for (i) Al metal (atomic density $= 6.02 \times 10^{28}$ m^{-3}) and (ii) n-type InSb (with carrier concentration $= 1.2 \times 10^{24}$ m^{-3}, effective mass $m_e^* = 0.02m_e$ and high-frequency dielectric constant $\varepsilon(\infty) = 15.68$).
(b) Show that for very low frequencies, $\omega \ll \omega_0, \omega_p$, the free-carrier reflectivity is given by the Hagen–Rubens relation $R \simeq 1 - (2\varepsilon_0\omega/\sigma_0)^{1/2}$.

5.30 Obtain an expression for the skin depth, the depth at which an electromagnetic wave is attenuated in a metal to $1/e$ of its value. What is the skin depth for microwaves ($\upsilon \simeq 2$ GHz) in Cu ($\sigma_0 = 5.9 \times 10^7\ \Omega^{-1}\mathrm{m}^{-1}$)? To what use can this effect be put?

5.31 How much of an approximation is it to assume that photon-induced electronic transitions are vertical? (Hint: compare the magnitudes of a typical photon wavevector and an electron wavevector corresponding to the Brillouin-zone boundary.)

5.32 Calculate the energy dependence of a direct optical transition that is *forbidden* by a selection rule, and show that $\varepsilon_2(\omega) \propto \omega^{-2}(\hbar\omega - \mathscr{E}_g)^{3/2}$. (Hint: expand the momentum matrix element in a Taylor series about a critical point in the joint density of states.)

5.33 (a) Calculate the exciton binding energy for crystalline GaAs, given that $\varepsilon \simeq 13.13$, $m_e^* = 0.067m_e$ and $m_h^* = 0.53m_e$.

(b) Sketch the likely appearance of the exciton–polariton dispersion curve by analogy with that for ordinary polaritons (coupled phonon–photon excitations—see Fig. 4.23a).

5.34 Show that the damped anharmonic oscillator is a model for understanding optical non-linear behaviour. For simplicity, consider only the first (quadratic) anharmonic term ax^2, i.e. the expression $\ddot{x} + \Gamma\dot{x} + \omega_0^2 x + ax^2 = -eE(t)/m_e$ may be taken to describe the motion of an electron bound to an ion and subject to a driving force due to the electric field $E(t)$ of an incident light wave. Obtain expressions for the first- and second-order dielectric susceptibilities, $\chi^{(1)}$ and $\chi^{(2)}$ respectively. (Hint: look for solutions $x = x_1 + x_2$, where x_1 is the solution for the harmonic case, and for small anharmonicity, take $ax^2 \simeq ax_1^2$.)

5.35 Find, by a suitable geometric construction, the angle of propagation, θ, of light with the optic axis of a uniaxial crystal that satisfies the phase-matching condition (eqn. (5.230c)) for second-harmonic generation. The normal modes for a light wave in a uniaxial crystal, with refractive indices n_0 and n_e, are an 'ordinary' wave characterized by an angle-independent refractive index n_0, and an 'extraordinary' wave for which the refractive index is $n(\theta)$, with $1/n^2(\theta) = \cos^2\theta/n_0^2 + \sin^2\theta/n_e^2$.

References

Altmann, S. L., 1991, *Band Theory of Solids: An Introduction from the Point of View of Symmetry* (Clarendon Press: Oxford).

Ashcroft, N. W. and Mermin, N. D., 1976, *Solid State Physics* (Holt, Rinehart and Winston: New York).

Baldini, G., 1962, *Phys. Rev.* **128**, 1562.

Baumeister, P. W., 1961, *Phys. Rev.* **121**, 359.

Burns, G., 1985, *Solid State Physics* (Academic Press: London).

Butcher, P. N. and Cotter, D., 1990, *The Elements of Non-linear Optics* (Cambridge University Press: Cambridge).

Chelikowsky, J. R. and Cohen, M. L., 1976, *Phys. Rev.* B**14**, 556.

Cohen, M. L. and Heine, V., 1970, *Solid State Physics*, **24**, 169.

Courths, R. and Hüfner, S., 1984, *Phys. Rep.* **112**, 53.

Cox, P. A., 1987, *The Electronic Structure and Chemistry of Solids* (Oxford University Press: Oxford).

Cox, P. A., 1992, *Transition Metal Oxides* (Clarendon Press: Oxford).

Cyrot-Lackmann, F., 1968, *J. Phys. Chem. Sol.* **29**, 1235.

Dixon, J. R., 1961, *Proc. Int. Conf. on Physics of Semiconductors* (Academic Press: New York), p. 366.

Eckhardt, H., Fritsche, L. and Noffke, J., 1984, *J. Phys.* F**14**, 97.

Edwards, P. P. and Sienko, M. J., 1981, *J. Am. Chem. Soc.* **103**, 2967.

Elliott, S. R., 1990, *Physics of Amorphous Materials*, 2nd edn (Longman: Harlow).

Griscom, D. L., 1985, in *Glass: Current Issues*, eds A. F. Wright and J. Dupuy, NATO Series E: Applied Sciences, no. 92 (Martinus Nijhoff: Dordrecht), p. 362.

Harrison, W. A., 1980, *Electronic Structure and the Properties of Solids* (Freeman: San Francisco).

Heine, V. and Abarenkov, I., 1964, Phil. Mag. **9**, 451.

Herman, F., Kortum, R. L., Kuglin, C. D. and Shay, J. L., 1967, In *II–VI Semiconducting Compounds*, ed. D. G. Thomas (Benjamin: New York).

Ibach, H. and Lüth, H., 1995, *Solid State Physics*, 2nd edn (Springer: Berlin).

Kittel, C., 1996, *Introduction to Solid State Physics*, 7th edn (Wiley: New York).

Kittel, C. and Kroemer, H., 1980, *Thermal Physics*, 2nd edn (Freeman: New York).

Lien, W. H. and Phillips, N. E., 1964, *Phys. Rev.* **133**, A1370.

MacFarlane, G. G., McLean, T. P. Quarrington, J. E. and Roberts, V., 1957, *Phys. Rev.* **108**, 1377.

Madelung, O., 1978, *Introduction to Solid-State Theory* (Springer-Verlag: Berlin).
Midwinter, J. E., 1979, *Optical Fibers for Transmission* (Wiley: New York).
Mott, N. F., 1974, *Metal–Insulator Transitions* (Taylor and Francis: London).
Nichols, C. S., 1995, *Structure and Bonding in Condensed Matter* (Cambridge Unviersity Press: Cambridge).
Pettifor, D. G., 1995, *Bonding and Structure of Molecules and Solids* (Oxford University Press: Oxford).
Phillip, H. R. and Ehrenreich, H., 1967, *Semiconductors and Semimetals* vol. **3**, eds R. K. Willardson and A. C. Beer, (Academic Press: New York), p. 93.
Saleh, E. A. and Teich, M. C., 1991, *Fundamentals of Photonics* (Wiley: New York).
Schiff, L. I., 1968, *Quantum Mechanics*, 3rd edn (McGraw-Hill: New York), p. 258.
Segal, B., 1961, *Phys. Rev.* **124**, 1797.
Shen, Y. R., 1984, *The Principles of Non-linear Optics* (Wiley: New York).
Slater, J. C. and Koster, G. F., 1954, *Phys. Rev.* **94**, 1498.
Spitzer, W. G. and Fan, H. Y., 1957, *Phys. Rev.* **106**, 882.
Steiner, P., Hoechst, H. and Huefner, S., 1979, in *Photoemission in Solids*, vol. II, eds L. Ley and M. Cardona (Springer Verlag: Berlin), p. 349.
Street, R. A., 1991, *Hydrogenated Amorphous Silicon* (Cambridge University Press: Cambridge).
Sutton, A. P., 1993, *Electronic Structure of Materials* (Clarendon Press: Oxford).
Syms, R. and Cozens, J., 1990, *Optical Guided Waves and Devices* (McGraw-Hill: London).
Weaire, D. and Thorpe, M. F., 1971, *Phys. Rev.* B**4**, 2508.
Wiech, G., 1981, *Emission and Scattering Techniques*, NATO ASI **C73**, ed. P. Day, p. 103.
Wong, J. S-Y., Schluter, M. and Cohen, M. L., 1976, *Phys. Stat. Sol.* **b77**, 295.
Yu, P. Y. and Cardona, M., 1996, *Fundamentals of Semiconductors* (Springer Verlag: Berlin).

CHAPTER 6 Electron Dynamics

Introduction

In the previous chapter, the nature of electron states in solids, i.e. the electronic structure, together with some equilibrium electronic and optical properties, was discussed. In this chapter, the emphasis is on the *dynamical* properties of electrons, i.e. electronic-transport properties. This subject is obviously of great importance, not just from a scientific point of view, but because the entire electronics industry exploits the electronic-transport behaviour exhibited by semiconductors and metals.

After a general introduction to the dynamical behaviour of the free-electron gas and of electrons in periodic solids, the electron-transport behaviour of normal metals, superconductors and semiconductors will be discussed in some detail. These three types of materials will be considered separately because each exhibits very different electronic behaviour.

Dynamics of the free-electron gas 6.1

We begin with a discussion of the transport properties associated with the free-electron gas, since this model serves as a basis for understanding the electronic-transport behaviour of normal metals. Two transport properties will be discussed: electrical conductivity and electronic thermal conductivity.

6.1.1 Electrical conductivity

The electrical-transport characteristics of solids can be represented by the quantities the electrical conductivity, σ (or its inverse, the electrical resistivity, ρ), defined as the constant of proportionality between the electrical current density $\boldsymbol{j}$ and electric field $\boldsymbol{E}$:

$$\boldsymbol{j} = \sigma \boldsymbol{E}. \tag{6.1}$$

Note that, for real anisotropic materials, the direction of the current flow need *not* be in the same direction as the applied field. In this case, the conductivity is a second-rank tensor and not a scalar as is assumed in eqn. (6.1). Ohm's law states that the electrical current I is proportional to the applied potential difference V, i.e.

$$V = IR, \tag{6.2}$$

where R is the electrical resistance ($= 1/G_0$, where G_0 is the conductance) of the sample. Equations (6.1) and (6.2) are equivalent formulations. Since the current density flowing through a sample of cross-sectional area A is $j = I/A$, and the potential difference dropped along the length l of the sample is $V = El$, a comparison of eqns. (6.1) and (6.2) shows that the conductivity and resistivity (assumed to be scalars) are related to the sample resistance via

$$\sigma \equiv 1/\rho = l/RA. \tag{6.3}$$

Hence the units of conductivity are (ohms metre)$^{-1}$ or (Ω m)$^{-1}$, or equivalently Siemens per metre (S m^{-1}).

The equation of motion experienced by a free electron subject to applied electric and magnetic fields is

$$\boldsymbol{F} = m_e \frac{\mathrm{d}\boldsymbol{v}}{\mathrm{d}t} = -e(\boldsymbol{E} + \boldsymbol{v} \times \boldsymbol{B}). \tag{6.4}$$

The above form of Newton's second law, applicable to *particles*, can also be used to describe the motion of electrons described in terms of *plane-wave states* (cf. eqn. (5.9)) if the particle is regarded as being equivalent to a superposition of plane-wave states to give a wave packet (Fig. 6.1), where the velocity in eqn. (6.4) is identified with the *group velocity* (eqn. (4.8)) of the wave packet, given in 1D by

$$\boldsymbol{v}_g = \frac{\partial \omega(\boldsymbol{k})}{\partial \boldsymbol{k}} = \frac{1}{\hbar}\frac{\partial \mathscr{E}(\boldsymbol{k})}{\partial \boldsymbol{k}} \tag{6.5a}$$

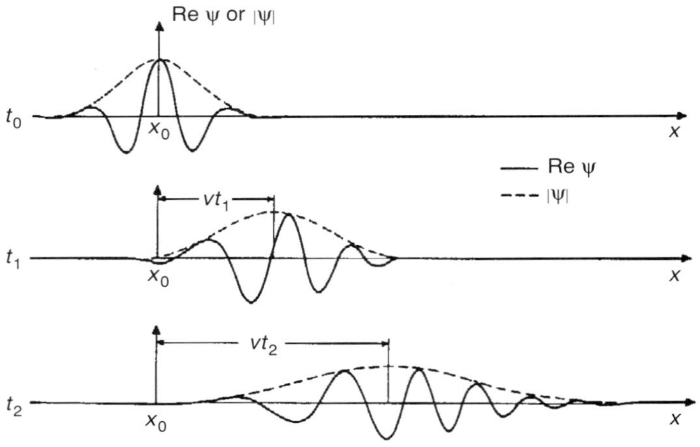

Fig. 6.1 Illustration in real space of a wave packet representing an electron, and its behaviour at three different times: t_0, t_1 and t_2, with $t_2 > t_1 > t_0$. Note the spreading of the wave packet with increasing time due to dispersion in the phase velocity (i.e. $\mathscr{E}(k) \not\propto k$, even for a free electron): the wavelength of oscillations of Re(ψ) becomes larger at the rear, and smaller at the front, of the wave packet. The motion of the centre of the wave packet represents that of the electron. (After Ibach and Lüth (1995), *Solid State Physics*, p. 192, Fig. 9.1, © Springer-Verlag GmbH & Co. KG).

where $\mathscr{E}(\boldsymbol{k})$ is the electron energy, or from eqn. (5.12),

$$\boldsymbol{v}_{\mathrm{g}} = \frac{\hbar \boldsymbol{k}}{m_{\mathrm{e}}} = \frac{\boldsymbol{p}}{m_{\mathrm{e}}}, \tag{6.5b}$$

where $\boldsymbol{p}$ is the particle momentum. In 3D,

$$\boldsymbol{v}_{\mathrm{g}} = \nabla_k \omega(\boldsymbol{k}) = (1/\hbar)\nabla_k \mathscr{E}(\boldsymbol{k}). \tag{6.6}$$

In order for the wave packet to mimic a particle, and for eqn. (6.4) still to be valid, its spatial extent should be smaller than the spatial variation of $\boldsymbol{E}$ and $\boldsymbol{B}$ (i.e. the wavelength of the electromagnetic field, if the fields are not uniform) and also smaller than the mean free path associated with electron collisions (see §6.3.1).

In considering the electrical conductivity, the magnetic field is assumed to be zero (the effects of a magnetic field on electron dynamics will be considered in §6.3.3). In such a case, eqn. (6.4) implies that an applied electric field should accelerate free electrons without limit. Evidently, this is unphysical, and this is a failing of the simple free-electron picture. In *real* metals, the electrons scatter from imperfections in a lattice, e.g. thermal lattice vibrations, defects or impurities and, to a much lesser extent, from other electrons. (Scattering processes will be discussed in more detail in §6.3.1.) These scattering events cause a reversal in the electron momentum and hence act like a damping force in the equation of motion (eqn. (6.4)). If it is assumed that, at every scattering event, the extra drift velocity $\boldsymbol{v}_{\mathrm{d}}(= (\boldsymbol{v} - \boldsymbol{v}_{\mathrm{th}})$, where $\boldsymbol{v}_{\mathrm{th}}$ is the equilibrium thermal velocity) imparted by the electric field, is removed on average, with τ being the average time between electron collisions, then eqn. (6.4) can be modified by the addition of a damping term, $m_e \boldsymbol{v}_d / \tau$, i.e.

$$m_e\left(\frac{d\boldsymbol{v}}{dt}+\frac{\boldsymbol{v}_d}{\tau}\right)=-e(\boldsymbol{E}+\boldsymbol{v}\times\boldsymbol{B}). \tag{6.7}$$

The quantity τ is also known as the electron relaxation time, since it is the time constant characterizing the exponential decay of the drift velocity on removal of the applied fields. In zero field, the mean thermal electron velocity $\bar{\boldsymbol{v}}$ associated with the equilibrium Fermi–Dirac distribution must be *zero* (as many electrons move in one direction as move in the opposite direction). However, when an electric field is applied, there is a finite drift velocity $\boldsymbol{v}_d$ given by eqn. (6.7) (with $\boldsymbol{B}=0$) corresponding to the *steady-state solution* (when $dv/dt=0$, and $\boldsymbol{E}\neq\boldsymbol{E}(t)$):

$$\boldsymbol{v}_d=-\frac{e\tau}{m_e}\boldsymbol{E}. \tag{6.8}$$

The electrical current density is $\boldsymbol{j}=-en\boldsymbol{v}$, where n is the electron density, and hence from eqns. (6.1) and (6.8), the d.c. conductivity σ_0 can be written as the Drude formula:

$$\sigma_0=\frac{ne^2\tau}{m_e}. \tag{6.9}$$

Note that this derivation assumes that *all* conduction electrons contribute to the current, and hence the conductivity. However, although this is incompatible with the Pauli exclusion principle, we shall see (§6.3.2.1) that a proper calculation gives the same result.

The constant of proportionality between $|\boldsymbol{v}|$ and $|\boldsymbol{E}|$ is termed the mobility. Hence the electron mobility is given from eqn. (6.8) by

$$\mu_e=\frac{e\tau}{m_e}. \tag{6.10}$$

It can be seen from a comparison of eqns. (6.9) and (6.10) that the electrical conductivity can also be written as

$$\sigma_0=ne\mu_e. \tag{6.11}$$

Note that the form of eqn. (6.11) is universal, and is the same for *ionic* conductivity (eqn. (3.81)). The a.c. (frequency- dependent) conductivity of the free-electron gas has already been discussed (§5.8.1) and is related to the d.c. value by

$$\sigma(\omega)=\frac{\sigma_0}{1-i\omega\tau}. \tag{5.189}$$

The mean free path Λ of a conduction electron between collisions is the distance that an electron with the Fermi speed v_F travels in time τ, i.e.

$$\Lambda=v_F\tau. \tag{6.12}$$

Here, the Fermi speed

$$v_F=(2\mathscr{E}_F/m_e)^{1/2} \tag{6.13}$$

is taken as a measure of the mean electron speed, since it is those electrons at the Fermi surface, i.e. at the top of the Fermi–Dirac electron distribution, that are able to take part in the transport process.

6.1.2 Thermal conductivity

We have already considered the thermal conductivity κ_T of *non-metals*, in which the heat flux is carried by *phonons* (§4.6.2.2). For the case of metals containing free electrons, the electrons can also transport heat. In fact, this is the dominant contribution to the thermal conductivity of metals, since the phonons are now scattered by the conduction electrons as well as by umklapp phonon-scattering (§4.6.4), with the result that the phonon mean free path is much smaller than in non-metals, and hence, from eqn. (4.232), the phonon contribution to κ_T is reduced.

The electronic contribution to the thermal conductivity can be calculated within the Fermi-gas (Sommerfeld) model. In a material subject to a temperature gradient ∇T, electrons whose last collision was in a hotter region will, on average, carry a greater thermal energy than those electrons emanating from a cooler region. The net heat flux or thermal current density, $\boldsymbol{J}_Q$, can then be evaluated in a similar manner to that used for phonons (§4.6.2.2) (see also Ashcroft and Mermin (1976)), with the result that

$$\boldsymbol{J}_{\mathrm{Q}} = -\kappa_T \nabla T \tag{6.14a}$$

$$= -c_{\mathrm{v}} \tau v_{\mathrm{F}}^2 \nabla T/3, \tag{6.14b}$$

where c_{v} is the electronic heat capacity per *unit volume*.

Hence, the free-electron contribution to the thermal conductivity is given by the gas-kinetic formula

$$\kappa_T = \frac{1}{3} \Lambda v_{\mathrm{F}} c_{\mathrm{v}} \tag{6.15}$$

or

$$\kappa_T = \frac{1}{3} v_{\mathrm{F}}^2 \tau c_{\mathrm{v}}. \tag{6.16}$$

Making use of eqn. (6.13) for v_{F} and eqn. (5.36b) for c_{v} leads to the expression (see also Problem 6.1):

$$\kappa_T = \frac{\pi^2 n k_{\mathrm{B}}^2 T \tau}{3 m_{\mathrm{e}}}, \tag{6.17}$$

where n is the electron density.

The factor $n\tau/m_{\mathrm{e}}$ appears both in the expression for σ_0 (eqn. (6.9)) and in that for κ_T (eqn. (6.17)). Thus, dividing these two quantities eliminates the parameters relating to a particular electron gas, thereby producing a *universal* value for the ratio L of thermal and electrical conductivities,

$$L = \frac{\kappa_T}{\sigma_0 T} = \frac{\pi^2}{3} \left(\frac{k_{\mathrm{B}}}{e} \right)^2, \tag{6.18}$$

known as the Wiedemann–Franz law. The Lorentz number L has the theoretical value $2.45 \times 10^{-8}\,\mathrm{W\,\Omega\,K^{-2}}$.

Dynamics of electrons in periodic solids 6.2

The motion of electrons in crystalline materials is *not* free, as assumed in the previous section, but is strongly constrained by the effects of translational periodicity of the lattice. Thus, as seen in §5.2, electrons are forced to occupy states that have energies given by the allowed band structure and which are often markedly non-free-electron-like. This can lead to dynamical behaviour that is often very different from that predicted by the free-electron model and even, sometimes, is physically counter-intuitive. This is particularly so for the case of the effective (dynamical) mass of electrons in periodic solids.

6.2.1 Effective mass

The inertial mass of a particle is defined as the constant of proportionality between a force $\boldsymbol{F}$ applied to it and the acceleration $\boldsymbol{a}$ imparted thereby: $\boldsymbol{F} = m\boldsymbol{a}$. Normally this mass is identical to the 'true' mass of the particle. However, it is possible to envisage situations in which the *effective* inertial mass is not the same as the true mass. One example is the case of a ball immersed in a viscous fluid. The value of inertial mass of the ball deduced from a knowledge of an applied force and the resulting acceleration of the ball would differ from the true mass because, in its motion, the ball must push the fluid in front of it out of the way, and the ensuing acceleration of the fluid is ignored if attention is paid solely to the motion of the ball. So it is with electrons in periodic structures: because an electron in a crystal cannot strictly be treated in isolation (it forms a system jointly with the lattice), the momentum of such an electron is not a true momentum, but a *crystal momentum* (§4.2.7) and, as such, momentum may be transferred freely between electron and lattice. As a result, it is not expected that the effective inertial mass m_e^* of an electron in a translationally periodic solid should be the same as the 'bare' electron mass, m_e: indeed, in general it is not, and often startlingly not (e.g. sometimes even being *negative*). We shall see, in fact, that the effective mass depends on the geometry of the electronic band structure, being related to the *curvature* of the bands in reciprocal space.

Consider an external d.c. electric field $\boldsymbol{E}$ applied for a time δt to a solid (with $\boldsymbol{B} = 0$). This field will do work on an electron, causing its energy to increase by the amount

$$\delta\mathscr{E} = -e\boldsymbol{E} \cdot \boldsymbol{v}_g \delta t, \tag{6.19}$$

where $\boldsymbol{v}_g$ is the group velocity of the wave packet. This expression can be rewritten as

$$\delta\mathscr{E} = \nabla_k \mathscr{E}(\boldsymbol{k}) \cdot \delta\boldsymbol{k} = \hbar\boldsymbol{v}_g \cdot \delta\boldsymbol{k} \tag{6.20}$$

using eqn. (6.6), or

$$\hbar\delta\boldsymbol{k} = -e\boldsymbol{E}\delta t \tag{6.21}$$

from eqn. (6.19). Hence, the equation of motion in terms of the wave packet can be written as

$$\hbar\dot{\boldsymbol{k}} = -e\boldsymbol{E}. \tag{6.22}$$

The acceleration of the wave packet is given in terms of the temporal rate of change of the component v_i of the group velocity:

$$\dot{v}_i = \frac{1}{\hbar}\frac{\mathrm{d}}{\mathrm{d}t}(\nabla_k \mathscr{E}(\boldsymbol{k}))_i = \frac{1}{\hbar}\sum_j \frac{\partial^2 \mathscr{E}}{\partial k_i \partial k_j}\dot{k}_j \qquad \{i,j\} = \{x, y, z\} \tag{6.23a}$$

or, from eqn. (6.21),

$$\dot{v}_i = \frac{1}{\hbar^2}\sum_j \frac{\partial^2 \mathscr{E}}{\partial k_i \partial k_j}(-eE_j). \tag{6.23b}$$

Comparing this equation with the classical equation of motion of an electron in an electric field $\boldsymbol{E}$,

$$\dot{\boldsymbol{v}} = \frac{1}{m_\mathrm{e}}(-e\boldsymbol{E}), \tag{6.24}$$

shows that the scalar mass in eqn. (6.24) can be formally replaced by an effective-mass tensor $\boldsymbol{m}_\mathrm{e}^*$, with component

$$(\boldsymbol{m}_\mathrm{e}^*)_{ij} = \frac{\hbar^2}{\partial^2 \mathscr{E}(\boldsymbol{k})/\partial k_i \partial k_j} = \frac{\hbar}{\partial v_i/\partial k_j}. \tag{6.25}$$

For a solid having an effective-mass tensor, in general the directions of electron acceleration and applied electric field do not coincide: the electron accelerates along the direction corresponding to least inertia unless the field is directed along that direction (or the direction of greatest inertia) in which case $\boldsymbol{a}$ and $\boldsymbol{E}$ are parallel. Note that the effective mass is inversely proportional to the *curvature* of the band at the point $\boldsymbol{k}$; $\boldsymbol{m}_\mathrm{e}^*$ is correspondingly a function of $\boldsymbol{k}$ in general (but see Problem 6.2). In the case of a 1D band, or spherical symmetry (parabolic bands) in 3D, eqn. (6.25) simplifies to give a *scalar* value of the effective mass:

$$m_\mathrm{e}^* = \frac{\hbar^2}{\partial^2 \mathscr{E}/\partial k^2}. \tag{6.26}$$

Therefore, charge carriers in *flat* bands (in $\boldsymbol{k}$-space) (or narrow band distributions in the density of states) have *high* effective masses.

In general, second-order perturbation theory predicts that the electron effective mass, $\boldsymbol{m}_\mathrm{e}^*$, at the conduction-band edge in *semiconductors* is proportional to the width of the (direct) bandgap (see, e.g., Kittel (1996); Yu and Cardona (1996)). Values of effective masses for electrons and holes (for light-hole, heavy-hole and split-off bands split by spin–orbit coupling — see §5.4.2) are given in Table 6.1 for some tetrahedrally coordinated direct-gap semiconductors, from which it can be seen that this prediction is approximately satisfied.

Figure 6.2 shows the $\boldsymbol{k}$-dependence of the effective mass for a one-dimensional NFE band for simplicity. Note that near the bottom of the band, where the curvature is parabolic, the effective mass has a constant positive value (see Problem 6.2), but at the point of inflection in the band, approximately half-way to the zone boundary, the effective mass in this case becomes infinite (since at that point the curvature is zero). Most intriguing of all is the fact that near the zone boundaries, the effective mass tends

Table 6.1 Effective masses of electrons and holes in some direct-gap semiconductors

Material	$\mathscr{E}_g$(eV)	*Electron* m_e^*/m_e	*Electron* $m_e^*/m_e\mathscr{E}_g(\text{eV}^{-1})$	*Heavy hole* m_{hh}^*/m_e	*Light hole* m_{1h}^*/m_e	*Split-off hole* m_{so}^*/m_e
GaAs	1.52	0.067	0.043	0.53	0.08	0.15
GaSb	0.81	0.047	0.058	0.8	0.05	(0.15)
InP	1.42	0.073	0.055	0.58	0.12	0.12
InAs	0.43	0.026	0.058	0.4	0.026	0.14
InSb	0.23	0.015	0.063	0.42	0.016	(0.12)

Bandgap values correspond to zero kelvin. Values in parentheses are theoretical estimates. Data from Kittel (1996) (Reproduced by permission of John Wiley & Sons Inc. and Yu and Cardona (1996), Fundamentals of Semiconductors, p. 70, Table 2.24, © Springer–Verlag GmbH & Co. KG.

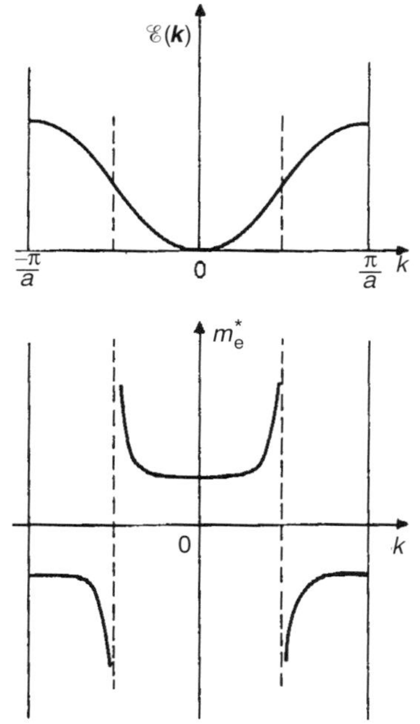

Fig. 6.2 Schematic behaviour of the effective electron mass m_e^* for a 1D band. The effective mass diverges where the curvature of the energy band is zero (point of inflection), at which point the group velocity of the electron wave packet is maximal. Note that a *negative* value of effective mass occurs near the zone boundary due to Bragg reflection of electrons from lattice planes and concomitant transfer of momentum from electrons to the lattice.

to a constant *negative* value. This latter behaviour seems, at first sight, to be counter-intuitive: a negative inertial mass implies that an applied force causes a particle to accelerate in the direction *opposite* to the force. This behaviour would indeed be unphysical for an *isolated* particle, but an electron in a solid is not isolated; its motion is intimately linked with the lattice.

The origin of the negative effective electron mass can be understood as follows. Consider an electron moving (for simplicity) in a 1D band in the $+x$-direction with a $\boldsymbol{k}$-vector close to that corresponding to the zone-boundary value. An increase in the wavevector of the electron wave packet caused by the application of an external electric field (eqn. (6.21)) therefore takes the electron state closer to the condition for *Bragg reflection* (i.e. backscattering) by the lattice planes. (Recall from §5.2.3 that the zone boundaries correspond to perfect Bragg reflection of electron waves.) Thus, instead of the force resulting from the applied electric field steadily increasing the momentum (or $\boldsymbol{k}$) of the electron, as happens for isolated particles, instead the momentum transfer from the electron to the lattice following elastic backscattering of the electron is greater than the transfer of momentum from the field to the electron. Thus the group velocity of the wave packet will increase with increasing $\boldsymbol{k}$ from zero at $\boldsymbol{k} = 0$, reaching a maximum where the energy band has a point of inflection, and then decrease to zero at the zone boundary where Bragg reflection is complete and only standing-wave solutions are allowed (§5.2.3).

It is apparent from Fig. 6.2 that negative effective-mass effects will only be significant for nearly full bands. Hence, for most metals with part (half-filled) bands, the effect is not present. However, for *semi-metals* (§5.2.5), where band overlap at different points in $\boldsymbol{k}$-space causes the lower band to be slightly depleted of electrons (Figs. 5.27, 5.28), or *doped semiconductors*, where certain dopant impurities can also cause a depletion of electrons from the otherwise full valence band (see §6.5.2), negative effective masses are very important in determining the electron dynamical behaviour. Although the negative-effective-mass characteristics of electrons in a nearly full band are an inescapable consequence of lattice periodicity, nevertheless the concept of negative mass is so alien that normally the dynamical behaviour in such situations is discussed instead in terms of an equivalent fictitious quasiparticle having a *positive* effective mass. Such quasiparticles are associated with the unoccupied states and are termed holes; they are discussed in the next section (§6.2.2).

Finally, mention should be made of the effective electron mass that appears in the expression for the electronic heat capacity (§5.1.3.1), e.g. the equation valid for a free-electron-like gas:

$$\begin{aligned} C_v &= \frac{\pi^2 N k_B^2}{\hbar^2} m^*_{C_v} \left(\frac{V}{3\pi^2 N}\right)^{2/3} T \\ &\equiv \frac{\pi^2}{3} k_B^2 g(\mathscr{E}_F) T, \end{aligned} \tag{5.36a}$$

where $g(\mathscr{E})$ is given by eqn. (5.131). Since the density of states at the Fermi level must be a positive quantity, so must the heat-capacity effective mass $m^*_{C_v}$: thus, it is a different quantity from the inertial effective mass discussed above. In fact, $m^*_{C_v} = |\boldsymbol{m}^*|^{1/3}$, where $|\boldsymbol{m}^*|$ is the determinant of the band-structure effective mass (Ashcroft and Mermin (1976)).

6.2.2 Holes

In analysing the dynamical behaviour of electrons in a nearly filled band, such as occurs in semi-metals resulting from band overlap in different parts of $\boldsymbol{k}$-space (see Figs. 5.27

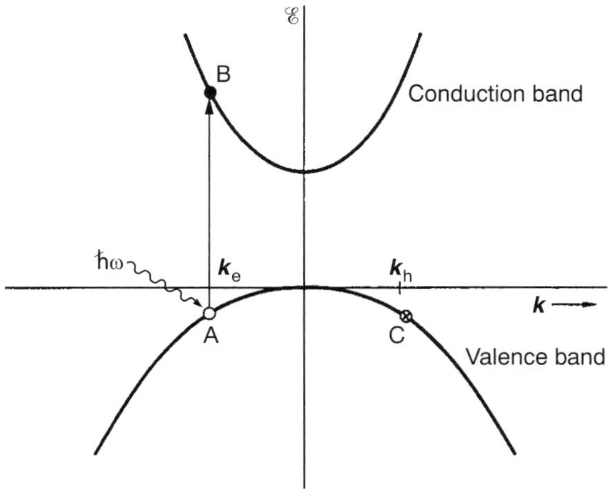

Fig. 6.3 Generation of a hole in a one-dimensional band, otherwise filled with electrons, by a vertical optical transition involving the excitation of an electron with wavevector $\boldsymbol{k}_e$ (at A) to an empty state B in the conduction band. The wavevector of the hole state is $\boldsymbol{k}_h = -\boldsymbol{k}_e$, i.e. at C.

and 5.28 and §5.2.5), or in semiconductors doped with electron 'acceptors' (see §6.5.2), it is often much simpler to consider the effective dynamical behaviour associated with the few empty states, in terms of fictitious quasiparticles called holes, than consider the behaviour of the many electrons in the band. This approach also has the advantage, as we shall see, that the effective mass of the hole quasiparticles is *positive* and hence in accord with common experience, whereas that for electrons near the top of a band is *negative* (§6.2.1). The charge of the hole is also positive.

Consider the simplest case of a hole created in an otherwise full band of electrons by optical excitation (Fig. 6.3). An electron with wavevector $\boldsymbol{k}_e$ is excited from state A in the valence band to state B in the conduction band in a vertical transition (§5.8.2). However, the wavevector of the hole quasiparticle state is *not* $\boldsymbol{k}_e$, as might be thought at first, but the *negative* of this:

$$\boldsymbol{k}_h = -\boldsymbol{k}_e. \tag{6.27}$$

In a completely filled band, the total electron wavevector is zero, i.e. $\sum_i \boldsymbol{k}_i = 0$. This results from the fact that if a lattice has inversion symmetry in real space, the Brillouin zone must also have inversion symmetry, i.e. pairs of electron states with $\pm\boldsymbol{k}_i$ are occupied in the filled state. If one electron with wavevector $\boldsymbol{k}_j$ is removed from the band, as in Fig. 6.3, the net resulting wavevector of the band of electrons, $\sum_{i\neq j} \boldsymbol{k}_i$, must be $-\boldsymbol{k}_j$ and this is therefore the value of wavevector ascribed to the hole (i.e. the absence of an electron in a filled band). Thus, the wavevector of the hole, $\boldsymbol{k}_h$, corresponding to point C in Fig. 6.3, is the same as the wavevector of the electron that remains at C.

The energy of the hole is *opposite in sign* to that of the missing electron, i.e.

$$\mathscr{E}_h(\boldsymbol{k}_h) = -\mathscr{E}_e(\boldsymbol{k}_e). \tag{6.28}$$

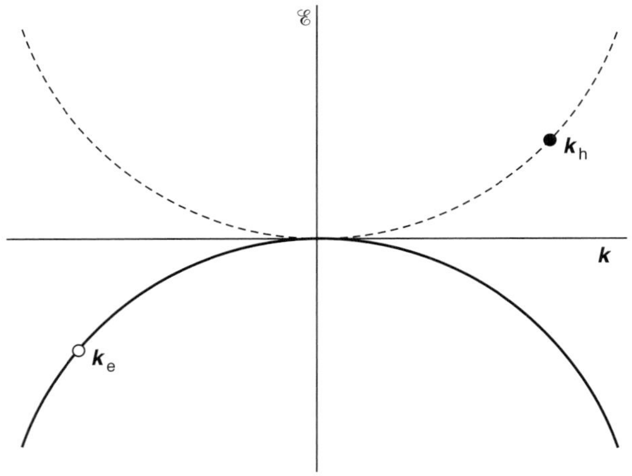

Fig. 6.4 Construction of a one-dimensional hole band (dashed curve) from the corresponding electron band (solid line). The case of a single electron, with wavevector $\boldsymbol{k}_e$, missing from the electron band, and the corresponding case of the hole band occupied by a single hole with wavevector $\boldsymbol{k}_h$, is shown.

In conventional representations of band structures, $\mathscr{E}(\boldsymbol{k})$ is plotted such that *electron* energies increase in an upward vertical direction. Thus, it costs more energy to remove electrons that lie deeper (lower) in a band than those that are less deep (higher): consequently, the corresponding hole energies must increase with increasing depth into the valence band. For symmetric bands resulting from inversion symmetry,† $\mathscr{E}(\boldsymbol{k}) = \mathscr{E}(-\boldsymbol{k})$, and hence, from eqn. (6.27), $\mathscr{E}_e(\boldsymbol{k}_e) = \mathscr{E}_e(-\boldsymbol{k}_e) = -\mathscr{E}_h(-\boldsymbol{k}_e) = -\mathscr{E}_h(\boldsymbol{k}_h)$, demonstrating eqn. (6.28). Consequently, a hole band can be constructed to describe the dynamics of holes (Fig. 6.4): it is, by eqn. (6.28), simply an inversion of the corresponding electron (valence) band, where now hole energies $\mathscr{E}_h$ are also measured positively upwards.

The velocity of the hole (i.e. the group velocity of the wave packet comprising Bloch waves for the whole band in which the missing one electron is in a different $\boldsymbol{k}$-state) is the same as that of the missing electron, i.e.

$$\boldsymbol{v}_h(\boldsymbol{k}_h) = \boldsymbol{v}_e(\boldsymbol{k}_e). \tag{6.29}$$

The group velocity of a wave packet is given by $\boldsymbol{v}_g = (1/\hbar)\nabla_k \mathscr{E}(\boldsymbol{k})$ (eqn. (6.6)), and so inspection of Fig. 6.4 shows that $\nabla\mathscr{E}_h(\boldsymbol{k}_h) = \nabla\mathscr{E}_e(\boldsymbol{k}_e)$, thereby giving eqn. (6.29).

The hole effective mass is the *negative* of the corresponding missing-electron mass,

$$m_h^* = -m_e^* : \tag{6.30}$$

the curvatures in reciprocal space of the hole-like and electron- like bands, giving the respective effective masses (eqn. (6.26)), are obviously equal and opposite (cf. Fig.6.4).

† In general, from the time invariance of the Schrödinger equation when electron spin is taken into account, $\mathscr{E}(\boldsymbol{k}_\downarrow) = \mathscr{E}(-\boldsymbol{k}_\uparrow)$, where the two spin states of an electron are denoted by the arrows.

Finally, the hole quasiparticle has a *positive* charge, $+e$. This can be demonstrated in two ways. The equation of motion of an electron subject to electric and magnetic fields is, from eqns. (6.4) and (6.5b):

$$\hbar \frac{d\boldsymbol{k}_e}{dt} = -e(\boldsymbol{E} + \boldsymbol{v}_e \times \boldsymbol{B}). \quad (6.31)$$

Making the substitutions $\boldsymbol{k}_h = -\boldsymbol{k}_e$ (eqn. (6.27)) and $\boldsymbol{v}_h = \boldsymbol{v}_e$ (eqn. (6.29)) leads to

$$\hbar \frac{d\boldsymbol{k}_h}{dt} = +e(\boldsymbol{E} + \boldsymbol{v}_h \times \boldsymbol{B}). \quad (6.32)$$

Thus, eqn. (6.32) is applicable for a quasiparticle with a positive charge, $+e$.

Alternatively, the electric charge characterizing holes can be deduced from a consideration of the electrical current carried by a band. A band completely full of electrons makes a *zero* contribution to the electrical current. An applied electric field causes the $\boldsymbol{k}$-vectors of each electron in a band to be uniformly displaced in $\boldsymbol{k}$-space with time according to eqn. (6.22). If the $\boldsymbol{k}$-vector of an electron increases so that it crosses a Brillouin-zone boundary at $\boldsymbol{k}_0$, this is equivalent to a $\boldsymbol{k}$-vector at $\boldsymbol{k} = \boldsymbol{k}_0 - \boldsymbol{G}$ (cf. eqn. (5.73)) on the other side of the zone with a negative $\boldsymbol{k}$-vector; i.e. an electron state appears to re-enter the reduced zone on the other side. Thus, the zone always remains full of electrons and does not contribute to the electrical current in an electric field. This conclusion can be demonstrated by calculating the current carried by electrons in a band, with different $\boldsymbol{k}$-vectors, and the approach can be used to show that the effective electrical charge carried by holes is positive.

★ The element of the particle flux of electrons, $\boldsymbol{J}(\boldsymbol{k})$, contributed by a volume element $d\boldsymbol{k}$ at a point $\boldsymbol{k}$ in $\boldsymbol{k}$-space is given by

$$d\boldsymbol{J}(\boldsymbol{k}) = \boldsymbol{v}(\boldsymbol{k}) \frac{d\boldsymbol{k}}{4\pi^3} \quad (6.33a)$$

$$= \frac{\nabla_k(\mathscr{E}(\boldsymbol{k}))d\boldsymbol{k}}{4\pi^3 \hbar}, \quad (6.33b)$$

where the density of electron states in $\boldsymbol{k}$-space is $2V/(2\pi)^3$ (cf. eqn. (4.28)), and the electron-spin degeneracy of two has been taken into account, V is the volume of the crystal, and $\boldsymbol{v}(\boldsymbol{k})$ is the (group) velocity of the electron state at $\boldsymbol{k}$. The electrical current density $\boldsymbol{j}$ due to a completely full band is then given by an integral of eqn. (6.33) over the first Brillouin zone:

$$\boldsymbol{j} = -\frac{e}{4\pi^3 \hbar} \int_{1\text{stB.Z}} \nabla_k(\mathscr{E}(\boldsymbol{k}))d\boldsymbol{k}. \quad (6.34)$$

For each contribution associated with velocity $\boldsymbol{v}(\boldsymbol{k})$ in eqn. (6.34), there is a contribution from $\boldsymbol{v}(-\boldsymbol{k})$ given by $\boldsymbol{v}(-\boldsymbol{k}) = (1/\hbar)\nabla_{-\boldsymbol{k}}\mathscr{E}(-\boldsymbol{k}) = (1/\hbar)\nabla_{-\boldsymbol{k}}\mathscr{E}(\boldsymbol{k}) = (-1/\hbar)\nabla_{\boldsymbol{k}}\mathscr{E}(\boldsymbol{k}) = -\boldsymbol{v}(\boldsymbol{k})$, making use of the symmetry relationship of $\mathscr{E}(k)$ mentioned before. Hence, for a filled band, $\boldsymbol{j} = 0$ from eqn. (6.34).

For the case of a *part-filled* band, $\boldsymbol{j} \neq 0$, since an applied electric field leads to an asymmetric distribution of $\boldsymbol{k}$-states about $\boldsymbol{k} = 0$ resulting from the dynamical evolution $\dot{\boldsymbol{k}}$ given by eqn. (6.22). The equivalent integral to eqn. (6.34) now extends only over *occupied* states, and not the entire Brillouin zone, and this can be written equivalently as an integral over the Brillouin zone *less* the contribution from the unoccupied (i.e. hole) states:

$$j = \frac{-e}{4\pi^3} \int_{k_{occ}} v(k) dk \qquad (6.35a)$$

$$= \frac{-e}{4\pi^3} \int_{1stB.Z.} v(k) dk - \frac{(-e)}{4\pi^3} \int_{k_{empty}} v(k) dk \qquad (6.35b)$$

$$= 0 + \frac{e}{4\pi^3} \int_{k_{empty}} v(k) dk. \qquad (6.35c)$$

Thus, holes behave as *positively* charged carriers. ★

The motion of a single hole in an otherwise filled electron band under the action of an electric field is represented schematically in Fig. 6.5; because of the translational symmetry mentioned above, an electron state passing out of the first Brillouin zone (at A) is exactly equivalent to an electron state entering the zone from the *opposite* side (B), and only the vacant electron state effectively moves. Its motion can be regarded as due to the successive filling of the empty electron state in k-space by electrons moving uniformly in response to the electrical field (eqn. (6.22)). Thus, the empty electron state is dragged along with the motion of the filled states: the motion of the hole in the hole band is in the *opposite* direction (Fig. 6.5), in accord with its opposite charge.

Finally, it should be stressed that the hole and electron descriptions of a particular part-filled band *cannot* be mixed. If the current is regarded as being carried, for example, by positive holes, the electrons make *no* contribution; the filled electron states merely act as potential unfilled hole states. Alternatively, if electrons are considered to be the charge carriers, then the unoccupied states in turn make no contribution to the current.

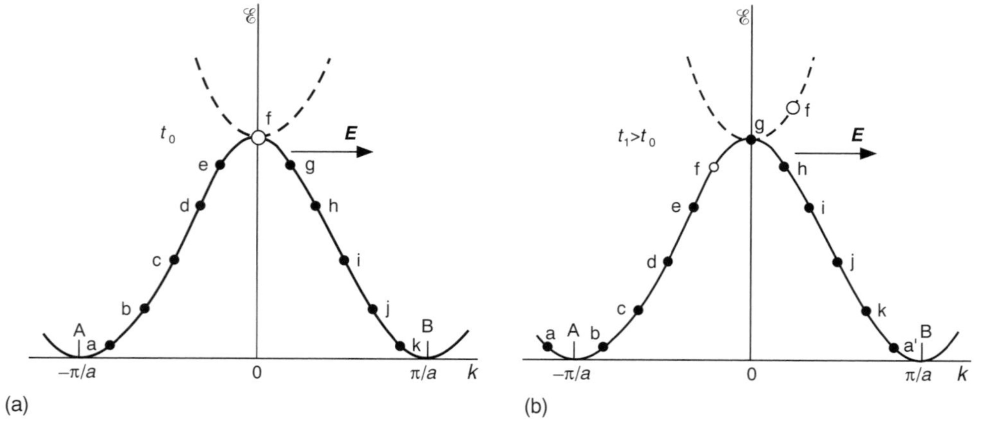

Fig. 6.5 Illustration of the motion of a single hole at two times t_0 and $t_1 > t_0$ in a band, otherwise filled with electrons, subject to a uniform d.c. electrical field, E. The field causes the electron states to move uniformly in k-space according to $\dot{k} = -eE/\hbar$. Whenever a k-state (a) close to a Brillouin-zone boundary (A) passes out of the first zone, it is replaced by an equivalent state (a′) entering through the opposite boundary. The motion of the hole in the hole band (dashed curve) is also shown.

Normal metals 6.3

In this section, we will discuss the electronic transport properties of 'real' metals in the absence of external magnetic fields. The effects of magnetic fields on the transport behaviour of electrons in metals is the subject of §6.3.3. In fact, it is convenient to separate the discussion of electron dynamics in metals into two parts. This section deals with 'normal' metals, as distinguished from 'superconductors' (metals which can exhibit zero resistance below a certain critical temperature) which are the subject of §6.4. Such a division makes sense since the transport behaviour of these two types of metals is qualitatively different, as a result of the quasiparticles responsible for the electron transport being different in each case.

6.3.1 Electron scattering

It was seen in §6.1.1 that, even for a free-electron gas, some sort of electron-scattering mechanism had to be invoked that gave a relaxation time τ, resulting in a finite drift velocity in an electric field and hence a finite value of the d.c. electrical conductivity. The type of scattering processes that are effective in limiting the mean-free path, and hence the conductivity, are those that act to restore a non-equilibrium electron distribution (e.g. resulting from the application of an electric field) to its equilibrium condition given by the Fermi–Dirac distribution function (see Fig. 6.6). As can be seen from the figure, the most effective electron-scattering events are those in which an electron is scattered from one side of the Fermi sphere (e.g. in the free-electron gas approximation) to the

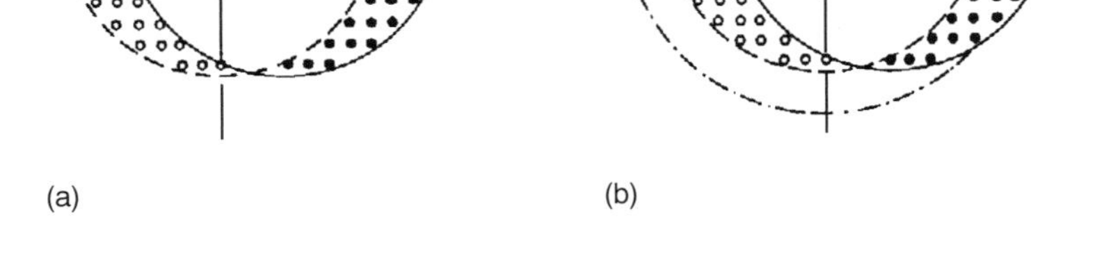

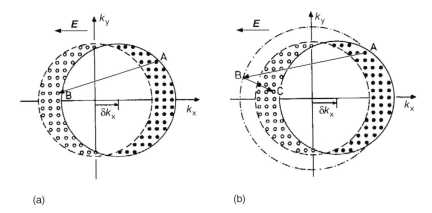

Fig. 6.6 (a) Electron-scattering event in $\boldsymbol{k}$-space that acts to restore a non-equilibrium electron distribution, displaced by $\delta\boldsymbol{k}$ due to the application of a d.c. electric field $\boldsymbol{E}$, to the equilibrium situation (in this case a Fermi sphere at $T = 0$ K (dashed line)) when the field is removed. Occupied electron states are denoted as • and empty states as o. The electron-scattering event A → B is inelastic. (b) Elastic electron-scattering event (A → B) causes an expansion of the electron distribution (dashed–dotted line), not a relaxation to the equilibrium state (dashed line). The equilibrium state can only be achieved by a subsequent inelastic scattering event (B → C).

other side, i.e. the electron undergoes a change in wavevector of order $2\boldsymbol{k}_F$. Note that the relaxational scattering events must be *inelastic* (since the electron states A and B in Fig.6.6a are at different distances from the origin in $\boldsymbol{k}$-space, and hence have different energies). Elastic scattering events (such as A $\rightarrow$ B in fig. 6.6b) merely lead to an *expansion* of the electron distribution in $\boldsymbol{k}$-space and do not by themselves restore the distribution to its equilibrium state.

6.3.1.1 Electron–electron scattering

An obvious potential source of scattering is between electrons themselves since the electron density in metals is very high. Note, however, that in the independent-electron approximation, such electron–electron interactions do *not* exist. The occurrence of electron–electron collisions is another instance of the breakdown of the one-electron approximation (see also §5.6). However, it is found that electron–electron scattering is in general negligible because of the influence of the Pauli exclusion principle, as can be seen by a phase-space argument.

Consider the case of two electrons scattering off each other, going from states 1 and 2 to 3 and 4 (see Fig. 6.7). Energy conservation dictates that

$$\mathscr{E}_1 + \mathscr{E}_2 = \mathscr{E}_3 + \mathscr{E}_4, \tag{6.36}$$

and conservation of crystal momentum stipulates that

$$\boldsymbol{k}_1 + \boldsymbol{k}_2 = \boldsymbol{k}_3 + \boldsymbol{k}_4 + \boldsymbol{G}. \tag{6.37}$$

If it is assumed for the moment that the temperature is just above 0 K, the Fermi sphere, of radius k_F in $\boldsymbol{k}$-space, is nearly entirely filled with electrons. If one electron involved in the scattering has an energy $\mathscr{E}_1 > \mu = \mathscr{E}_\mathrm{F}$, i.e. just above the Fermi level, and

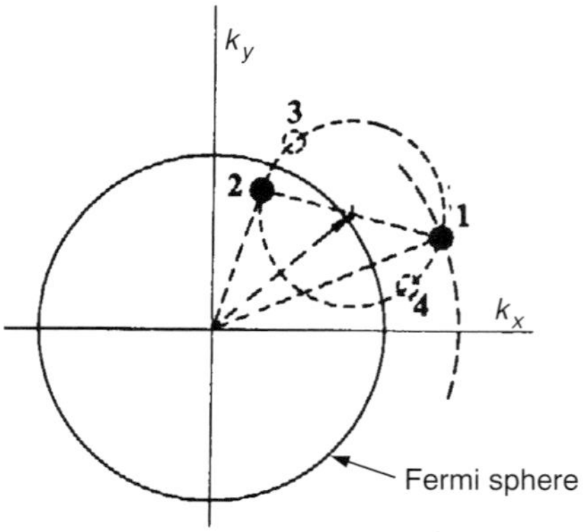

Fig. 6.7 Illustration of electron–electron scattering events in metals satisfying the Pauli exclusion principle. An electron (1), at an energy ε_1 above the Fermi level and outside the Fermi sphere (solid line), can only scatter from another electron (2) lying at an energy $< \varepsilon_1$ below the Fermi level, resulting in the electrons occupying states 3 and 4 *outside* the Fermi surface. Momentum conservation is assured if states 3 and 4 lie on the surface and on the diameter of a sphere in $\boldsymbol{k}$-space (dashed line), constructed so that the centre is at the 'centre of momentum' of states 1 and 2.

the other electron is initially in a state with an energy $\mathscr{E}_2$ such that $\mathscr{E}_2 < \mathscr{E}_F$, then the Pauli exclusion principle dictates that states 3 and 4 must both be *unoccupied*. Thus, quantum statistics require that $\mathscr{E}_3 > \mu$ and $\mathscr{E}_4 > \mu$. From eqn. (6.36), this implies that

$$\mathscr{E}_1 + \mathscr{E}_2 = \mathscr{E}_3 + \mathscr{E}_4 > 2\mu \tag{6.38a}$$

and

$$(\mathscr{E}_1 - \mu) + (\mathscr{E}_2 - \mu) > 0. \tag{6.38b}$$

If state 1 is at an energy $\varepsilon_1 = (\mathscr{E}_1 - \mu) \ll \mu$ above the Fermi level, then eqn. (6.38b) implies that $|\mathscr{E}_1 - \mu| = |\varepsilon_2| < \varepsilon_1$, i.e. state 2 must lie in a shell of thickness ε_1 below the Fermi surface (Fig. 6.7a), from energy considerations. As a consequence, only a fraction $f \simeq \varepsilon_1/\mu$ of all electrons (2) may scatter with the electron 1.

However, this is not the only factor decreasing the probability of electron–electron collisions that is associated with the Pauli exclusion principle: there is an equivalent factor resulting from momentum conservation. For convenience, assume that $\boldsymbol{G} = 0$ in eqn. (6.37). Momentum conservation in the scattering process holds if the scattered electron wavevectors, $\boldsymbol{k}_3$ and $\boldsymbol{k}_4$, lie on the surface and at either end of a diameter of a sphere, whose centre is at the wavevector corresponding to the 'centre of momentum' of electron states 1 and 2 (Fig. 6.7b). The Pauli exclusion principle ensures that states 3 and 4 must lie outside the Fermi sphere, and hence only a fraction $f \simeq \varepsilon_1/\mu$ of pairs of electron states 3 and 4 can satisfy this and also the momentum condition that the $\boldsymbol{k}$-vectors lie on the surface of the constructed sphere. Thus, the overall reduction factor for the probability of electron–electron scattering is $f^2 \simeq (\varepsilon_1/\mu)^2$. For a thermal distribution of electrons such that $T \ll T_F, \varepsilon_1 \simeq k_B T$, and thus the electron–electron scattering cross-section Σ is therefore

$$\Sigma \simeq (k_B T/\mu)^2 \Sigma_0, \tag{6.39}$$

where Σ_0 is the scattering cross-section for a classical gas (i.e. without consideration of quantum statistics). The reduction factor f^2 is very large: since $\mu/k_B \simeq \mathscr{E}_F/k_B \simeq 5 \times 10^4$ K, at a temperature of 1 K, $f^2 \simeq (k_B T/\mu)^2 \simeq 4 \times 10^{-10}$, and even at room temperature the factor is $\simeq 4 \times 10^{-5}$.

The quantity Σ_0 is itself reduced from the value it would have for classical scattering between two charged particles interacting via unscreened Coulomb potentials, i.e. the Rutherford law for the differential cross-section:

$$\frac{d\Sigma}{d\Omega} = \left(\frac{e^2}{8\pi\varepsilon_0 m v^2}\right)^2 \frac{1}{\sin^4\theta}, \tag{6.40}$$

where 2θ is the scattering angle (cf. Fig. 6.8), and $d\Omega = 2\pi \sin(2\theta)\, d(2\theta)$ is the element of solid angle corresponding to scattering angles between 2θ and $2\theta + d(2\theta)$, and v is the velocity of the scattering particle. This decrease arises because of electron screening (§5.6.1); the Coulomb potential is screened (eqn. (5.154)) so that it is negligible beyond distances comparable to the Thomas–Fermi screening length, λ_{TF} (eqn. (5.155)). Typically, in metals, the screened cross-section has a value $\Sigma_0 \simeq 10$ Å$^2 = 10^{-19}$ m^2, so that the actual scattering cross-section (eqn. (6.39)) has a value at 1 K of $\Sigma \simeq 4 \times 10^{-29}$ m^2, and at 300 K $\Sigma \simeq 4 \times 10^{-24}$ m^2.

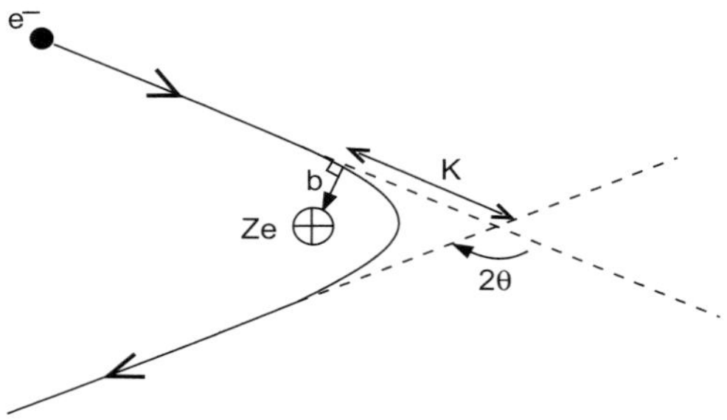

Fig. 6.8 Coulomb scattering of an electron by a positively charged defect with charge Ze. The impact parameter $b = K\cot\theta$, where the characteristic length $K = Ze^2/4\pi\varepsilon_0 m_e v_F^2$. The scattering angle is 2θ.

The mean-free path is related to the collision cross-section by the kinetic formula:

$$\Lambda \simeq \frac{1}{\sqrt{2}n\sum}, \tag{6.41}$$

where n is the electron concentration and the $\sqrt{2}$ factor takes into account that all electrons are moving. For $n \simeq 5 \times 10^{28}$ m^{-3}, the mean-free path due to electron–electron scattering is predicted to be 4×10^{-6} m at 300 K and 0.35 m at 1 K! Mean-free paths of this order of magnitude are indeed observed in very pure metals at very low temperatures. From eqns (6.12), (6.39) and (6.41), the temperature dependence of the electron–electron relaxation time is

$$\tau_e \propto T^{-2}. \tag{6.42}$$

Finally, it should be noted that the elastic scattering processes shown in Fig. 6.7 with $\boldsymbol{G} = 0$, i.e. N-type processes (see §4.6.2.2), even if they were to occur, are not efficient at restoring a non-equilibrium electron distribution to its equilibrium state. For this to occur, inelastic scattering events must occur. This can happen for umklapp (U) processes (see §4.6.2.2), where the wavevector of a scattered electron lying outside the first Brillouin zone is brought back into the reduced zone by a reciprocal-lattice vector, $\boldsymbol{G}$, i.e. the lattice provides the required momentum. However, such umklapp electron–electron collisions are very improbable, for the reasons outlined above.

Nevertheless, electron–electron scattering *is* important in controlling the electrical-transport behaviour of one category of metals, namely transition metals containing a part-filled d-band (§5.4.3). Although it might be thought that the electrical conductivity of transition metals should be high, due to the large effective density of states when the chemical potential lies in the band of d-states (Fig. 5.55), in fact this is not so: the electrical resistivities of the transition metals Ni, Pd, Pt, for example, are about a factor of five larger than those of the noble metals Cu, Ag, Au that immediately follow them in the periodic table. The reason for this behaviour is two-fold: d-electrons have higher effective masses (§6.2.1) associated with the narrow width of the d-band and hence correspondingly a smaller conductivity (eqn. (6.9)); moreover, the strong scattering of s-electrons by the d-electrons in a part-filled d-band significantly increases the contribution to the resistivity of the s-electron channel.

6.3.1.2 Electron–defect scattering

Other electron-scattering processes in crystals involve *imperfections* in the lattice, i.e. fluctuations from translational periodicity. Such fluctuations may be *static* in time (e.g. structural defects, impurities in the lattice) or *time-varying* (e.g. phonons). From Problem 4.17, it can be inferred that electron scattering from moving fluctuations can give rise to inelastic scattering, whereas scattering from fixed fluctuations can only result in elastic scattering. It is reasonable to assume that the scattering behaviour of, say, defects and phonons is independent, so that the respective collision rates are simply additive, i.e.

$$\frac{1}{\tau} = \frac{1}{\tau_{\mathrm{d}}} + \frac{1}{\tau_{\mathrm{ph}}}. \tag{6.43}$$

The scattering of electrons by ionized impurities can be analysed using the Rutherford law, eqn. (6.40). The scattering rate $R_{\mathrm{d}} = 1/\tau_{\mathrm{d}}$ of electrons by a concentration N_{d} of defects is related to the cross-section Σ_{d} by

$$R_{\mathrm{d}} = N_{\mathrm{d}}\Sigma_{\mathrm{d}}v_{\mathrm{F}}, \tag{6.44}$$

from eqns (6.12) and (6.41) (the factor of $\sqrt{2}$ is omitted since the impurities do not move), and where the electron velocity is taken to be the Fermi velocity v_{F}. Thus:

$$\frac{1}{\tau_{\mathrm{d}}} = N_{\mathrm{d}}v_{\mathrm{F}} \int_0^{\pi} \frac{\mathrm{d}\Sigma_{\mathrm{d}}}{\mathrm{d}\Omega}(1 - \cos(2\theta))2\pi\sin(2\theta)\mathrm{d}(2\theta), \tag{6.45}$$

where $\mathrm{d}\Sigma_{\mathrm{d}}/\mathrm{d}\Omega$ is given by eqn. (6.40) (multiplied by a factor of Z^2 if the charge of the impurity is Ze), and the quantity $(1 - \cos(2\theta))$ is a weighting factor to take account of the angular efficiency of electron momentum transfer in the scattering process (the factor is zero for forward scattering, $2\theta = 0$—see Ashcroft and Mermin (1976)). Hence, from eqn. (6.44), it can be seen that τ_{d} is temperature-independent and proportional to v_{F}^3.

Since τ_{d} is *independent* of temperature for metals, this process will be the dominant scattering mechanism in metals at very low temperatures since, as will be seen below, the electron–phonon scattering rate decreases strongly with temperature at low temperatures. For very pure and small samples, electron scattering by the boundaries of the sample may become dominant if electron–electron scattering is negligible; this gives rise to a size effect in electronic-transport properties, as in phonon-transport behaviour (§ 4.6.2.2).

6.3.1.3 Electron–phonon scattering

At temperatures well above absolute zero, the dominant cause of electron scattering is due to *phonons*: a phonon can always be emitted by an electron, thereby scattering the electron to another state (Fig. 6.9a) or at elevated temperatures, when the number of excited phonons is large, a phonon can be absorbed (Fig. 6.9b). Conservation of crystal momentum dictates that

$$\boldsymbol{k}_{\mathrm{f}} = \boldsymbol{k}_{\mathrm{i}} \pm \boldsymbol{q} + \boldsymbol{G} \tag{6.46}$$

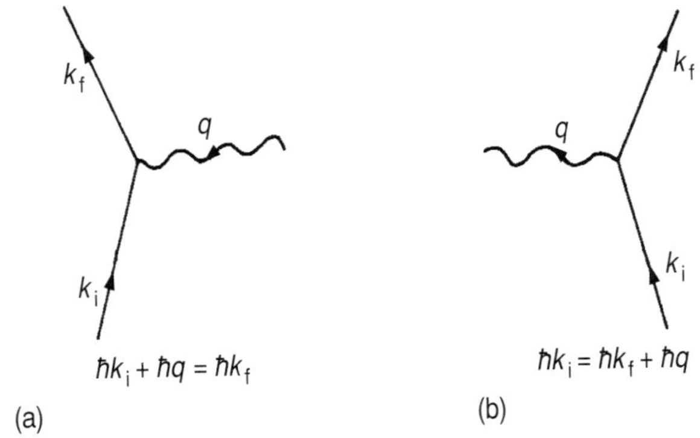

Fig. 6.9 Scattering of an electron by: (a) phonon emission; (b) phonon absorption. The straight lines indicate electrons and the wavy lines denote phonons. The relevant wavevectors are shown.

where $\boldsymbol{k}_{\rm f}$ and $\boldsymbol{k}_{\rm i}$ are the final and initial electron wavevectors, respectively, $\boldsymbol{q}$ is that for the phonon, and the plus and minus signs refer to phonon absorption and emission processes, respectively. Conservation of energy imposes the condition

$$\mathscr{E}(\boldsymbol{k}_{\rm f}) = \mathscr{E}(\boldsymbol{k}_{\rm i}) \pm \hbar\omega(\boldsymbol{q}). \tag{6.47}$$

Since phonons have large values of momentum, these processes are effective in relaxing a non-equilibrium electron distribution (Fig. 6.6), and are usually the dominant cause of electrical resistivity in normal metals at not-too-low temperatures.

The electron–ion Hamiltonian can be written as:

$$\mathscr{H}_{\rm e-ion} = \sum_{l,i} V_{\rm e-ion}(\boldsymbol{r}_l - \boldsymbol{R}_i), \tag{6.48}$$

where $\boldsymbol{r}_l$ is the position of an electron and $\boldsymbol{R}_i$ that of an ion. The *static* part of this Hamiltonian gives the electronic band structures discussed in the previous chapter; the *time-dependent* part, relating to vibrations of the lattice, gives the electron–phonon interaction. If the ion position is written in terms of the time-dependent displacement from an equilibrium position, $\boldsymbol{R}_i(t) = \boldsymbol{R}_i^0 + \boldsymbol{u}_i(t)$, the potential in eqn. (6.48) can then be expanded in powers of the displacement which, including the first-order term, is

$$V_{\rm e-ion} = V(\boldsymbol{r}_l - \boldsymbol{R}_i^0) - \boldsymbol{u}_i \cdot \nabla V(\boldsymbol{r}_l - \boldsymbol{R}_i^0). \tag{6.49}$$

Thus the second term in this equation refers to the electron–phonon interaction. If the potential V is expanded as a Fourier series $V = \sum_\kappa V_\kappa \exp(\mathrm{i}\kappa \cdot (\boldsymbol{r} - \boldsymbol{R}_i))$, where $\kappa = \boldsymbol{q} + \boldsymbol{G}$, then the operation of the grad operator in eqn. (6.49) produces a term linear in $\boldsymbol{q}$, i.e. $V_{\rm e-ph} \propto \boldsymbol{u}_i \cdot \boldsymbol{q}$. Thus, only *longitudinal* phonons (for which $\boldsymbol{u}_i$ is parallel to $\boldsymbol{q}$) couple to electrons. In addition, only LA phonons tend to be important for several reasons. For lattices without a basis, there are no optic phonons (§5.2.2); moreover, because LA phonons have lower energies than LO phonons, they are more easily excited at a given temperature and hence dominate phonon-absorption processes.

The temperature dependence of the electron–phonon scattering relaxation time τ_{ph} appearing in eqn. (6.42) is different in different temperature régimes. At *high temperatures*, greater than the Debye temperature, $T \gg \theta_{\text{D}} = \hbar\omega_{\text{D}}/k_{\text{B}}$, the number of phonons in a normal mode is proportional to the temperature (eqn. (4.79)), and hence:

$$\tau_{\text{ph}} \propto T^{-1}, \qquad T \gg \theta_{\text{D}}. \tag{6.50}$$

At low temperatures ($T \ll \theta_{\text{D}}$), the situation is a little more complex. Only phonons for which $\hbar\omega(\boldsymbol{q}) \leq k_{\text{B}}T$ can be absorbed or emitted by electrons. At low temperatures such phonons are only involved in absorption processes. For phonon emission, an electron must initially be at a level above $\mu \simeq \mathscr{E}_{\text{F}}$ such that the final state is *unoccupied* (i.e. within $\simeq k_{\text{B}}T$ of $\mathscr{E}_{\text{F}}$), and so can only emit a phonon with this energy. Furthermore, at low temperatures, $q \ll k_{\text{D}}$, the Debye wavevector (eqn. (4.202)), and so for LA phonons with the dispersion relation $\omega = v_{\text{L}}q$, the wavevectors are such that $q \leq k_{\text{B}}T/\hbar v_{\text{L}}$. The energy-conservation condition, eqn. (6.47), can be written, together with eqn. (6.46), as

$$\omega(\boldsymbol{q}) = \pm(\mathscr{E}(\boldsymbol{k}+\boldsymbol{q}) - \mathscr{E}(\boldsymbol{k}))/\hbar, \tag{6.51}$$

for $\boldsymbol{G} = 0$, i.e. 'normal' scattering processes. This describes a 2D surface of allowed wavevectors in the 3D phonon- wavevector space, of area proportional to q^2 and hence to T^2 . In addition, the scattering rate, being proportional to the electron–phonon Hamiltonian (eqn. (6.49)), is also proportional to q (see above), and this brings in another factor of $\boldsymbol{T}$, and so

$$\tau_{\text{ph}} \propto T^{-3} \qquad T \ll \theta_{\text{D}}. \tag{6.52}$$

The above considerations are for electron–phonon scattering processes in which $\boldsymbol{G} = 0$, i.e. 'normal' processes (cf. §4.6.2.2): for these, the change in electron momentum is rather small, particularly at the temperatures $T \ll \theta_{\text{D}}$ (see Fig. 6.10). However, 'umklapp' (U) electron–phonon processes (cf. §4.6.2.2) can provide large changes in the electron wavevector with the assistance of a reciprocal-lattice vector, $\boldsymbol{G}$. If an initial electron state ($\boldsymbol{k}_i$) at point A on the Fermi surface (Fig. 6.10) is scattered by a phonon with wavevector $\boldsymbol{q}$ to point B lying on the Fermi surface in a *neighbouring zone*, this is equivalent to the state being scattered to point B' with wavevector $\boldsymbol{k}_{\text{f}}$ on the *other side* of

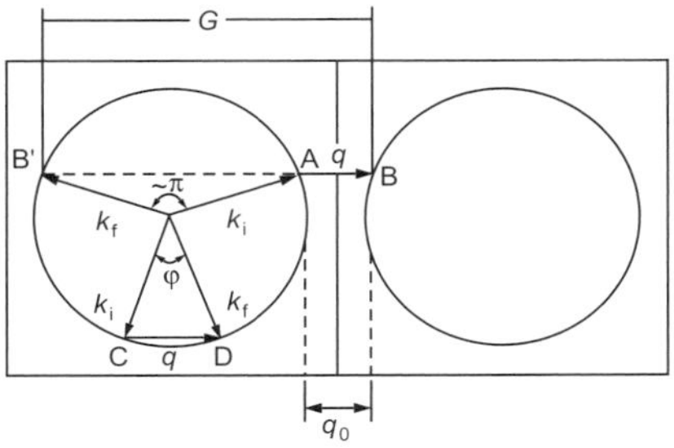

Fig. 6.10 Normal (C → D) and umklapp (A → B → B') electron–phonon scattering processes. The two circles represent Fermi spheres in adjacent zones in the extended-zone scheme.

the *same* Fermi surface. The minimum phonon wavevector capable of effecting umklapp scattering is $\boldsymbol{q}_0$ (see Fig. 6.10). The number of phonons available for umklapp scattering is proportional to $\exp(-\theta_u/T)$ (cf. §4.6.2.2), where θ_{u} is a characteristic temperature determined by the geometry of the Fermi surface; hence umklapp electron–phonon scattering is frozen out at very low temperatures, leaving only inefficient small-angle electron–phonon scattering as the equilibrium-restoring mechanism for the electron distribution. Note that TA phonons can take part in umklapp processes, unlike the case for normal processes.

6.3.2 Electron dynamics in the presence of electric fields and temperature gradients

In this section, we will discuss the behaviour of electrons in metals subject to an applied electric field (giving rise to electrical conductivity or resistivity), or to a temperature gradient (giving rise to electronic thermal conductivity), or to a combination of both (resulting in thermoelectric effects). The effects of magnetic fields on electronic transport in metals will be discussed in §6.3.3.

6.3.2.1 *Electrical conductivity*

The treatment of the electrical conductivity in terms of the Drude model of a free-electron gas given in §6.1.1 is deficient in many respects for real metals. The Fermi–Dirac statistics governing the electron distribution are not incorporated; no physical mechanism for the electron relaxation time is considered; and details of the precise geometry of the Fermi surface associated with the particular crystal structure cannot be included. These deficiencies will now be addressed.

The electrical current density, $\boldsymbol{j}$, was written in eqn. (6.35a) as an integral of electron velocities over occupied $\boldsymbol{k}$-states, viz.:

$$\boldsymbol{j} = -\frac{e}{4\pi^3}\int_{\boldsymbol{k}_{\mathrm{occ}}} \boldsymbol{v}(\boldsymbol{k})\mathrm{d}\boldsymbol{k}.$$

This can be rewritten in terms of the occupation probability for states labelled by $\boldsymbol{k}$ as an integral over the whole of the Brillouin zone, i.e.

$$\boldsymbol{j} = -\frac{e}{4\pi^3}\int_{\mathrm{1stB.Z.}} \boldsymbol{v}(\boldsymbol{k})f(\boldsymbol{k})\mathrm{d}\boldsymbol{k}. \tag{6.53}$$

The distribution function $f(\boldsymbol{k})$ is, in general, *not* equal to the Fermi–Dirac function given by eqn. (5.23), i.e.

$$f_0\{\mathscr{E}(k)\} = \frac{1}{\exp[\mathscr{E}(\boldsymbol{k}) - \mu] + 1} \tag{6.54}$$

since this distribution function is valid only for the *equilibrium case*, i.e. where the electron distribution is spatially homogeneous, in the absence of temperature gradients and with no external fields. However, as will be seen below, $f(\boldsymbol{k})$ can be shown to be a function of $f_0(\boldsymbol{k})$.

The distribution function f is a function of real-space, as well as reciprocal-space (i.e. crystal-momentum) vectors, and $f(\boldsymbol{r}, \boldsymbol{k})\mathrm{d}\boldsymbol{r}\,\mathrm{d}\boldsymbol{k}$ is simply the number of electrons in the 6D volume element $\mathrm{d}\boldsymbol{r}\mathrm{d}\boldsymbol{k}$. Liouville's theorem in classical mechanics states that the distribution is conserved as the system evolves in time (the number of electrons and the 6D volume is preserved — see Ashcroft and Mermin (1976)): after a time interval, $\mathrm{d}t$, therefore,

$$f(t+\mathrm{d}t, \boldsymbol{r}+\mathrm{d}\boldsymbol{r}, \boldsymbol{k}+\mathrm{d}\boldsymbol{k}) = f(t, \boldsymbol{r}, \boldsymbol{k}), \tag{6.55}$$

in the absence of collisions. However, scattering events can also cause electrons to be transferred to states $\boldsymbol{k}+\mathrm{d}\boldsymbol{k}$ and positions $\boldsymbol{r}+\mathrm{d}\boldsymbol{r}$, and so in general

$$f(t+\mathrm{d}t, \boldsymbol{r}+\mathrm{d}\boldsymbol{r}, \boldsymbol{k}+\mathrm{d}\boldsymbol{k}) - f(t, \boldsymbol{r}, \boldsymbol{k}) = \left(\frac{\partial f}{\partial t}\right)_{\mathrm{s}} \mathrm{d}t, \tag{6.56}$$

where the term $(\partial f/\partial t)_{\mathrm{s}}$ refers to the temporal change in f due to scattering. Since $\dot{\boldsymbol{r}} = \boldsymbol{v}(\boldsymbol{k})$, and from eqn. (6.4), $\hbar\dot{\boldsymbol{k}} = -e(\boldsymbol{E} + \boldsymbol{v} \times \boldsymbol{B}) = \boldsymbol{F}(\boldsymbol{r}, \boldsymbol{k})$, where $\boldsymbol{F}$ is the force on the electron, eqn. (6.56) can be rewritten in the form of the Boltzmann transport equation:

$$\frac{\partial f}{\partial t} + \boldsymbol{v} \cdot \nabla_r f + \boldsymbol{F} \cdot \frac{1}{\hbar} \nabla_k f = \left(\frac{\partial f}{\partial t}\right)_{\mathrm{s}}. \tag{6.57}$$

In the present case, where we are considering the electrical conductivity in the absence of magnetic fields, $\boldsymbol{F} = -e\boldsymbol{E}$, and eqn. (6.57) becomes

$$\frac{\partial f}{\partial t} + \boldsymbol{v} \cdot \nabla_r f - \frac{e}{\hbar} \boldsymbol{E} \cdot \nabla_k f = \left(\frac{\partial f}{\partial t}\right)_{\mathrm{s}}. \tag{6.58}$$

In general, the Boltzmann equation is a non-linear integrodifferential equation (because the scattering term is an integral over $\boldsymbol{k}$ of the transition rates between $\boldsymbol{k}$-states—see Ashcroft and Mermin (1976)).

However, a very great simplification can be made by invoking the relaxation-time approximation, i.e. by assuming that the collision term can be represented by

$$\left(\frac{\partial f}{\partial t}\right)_{\mathrm{s}} \simeq -\frac{[f(\boldsymbol{k}) - f_0(\boldsymbol{k})]}{\tau(\boldsymbol{k})}, \tag{6.59}$$

where $f_0(\boldsymbol{k})$ is given by eqn. (6.54). This approximation underlies much of the discussion on scattering mechanisms given in §6.3.1. In steady state, when $\partial f/\partial t = 0$, and if f is not a function of position (i.e. $\nabla_r f = 0$), then eqn. (6.58) reduces to

$$-\frac{e}{\hbar} \boldsymbol{E} \cdot \nabla_k f(\boldsymbol{k}) = -\frac{[f(\boldsymbol{k}) - f_0(\boldsymbol{k})]}{\tau(\boldsymbol{k})}, \tag{6.60}$$

or

$$f(\boldsymbol{k}) = f_0(\boldsymbol{k}) + \frac{e}{\hbar} \tau(\boldsymbol{k}) \boldsymbol{E} \cdot \nabla_k f(\boldsymbol{k}). \tag{6.61}$$

The *linearized* version of this form of the Boltzmann equation is obtained by approximating $f(\boldsymbol{k})$ by $f_0(\boldsymbol{k})$ in the second term on the right-hand side of eqn. (6.61), giving

$$f(\boldsymbol{k}) \simeq f_0(\boldsymbol{k}) + \frac{e}{\hbar}\tau(\boldsymbol{k})\boldsymbol{E}\cdot\nabla_k f_0(\boldsymbol{k}). \tag{6.62}$$

This approximate form of the distribution function may now be used to calculate the current density (eqn. (6.53)), and thereby obtain the conductivity. Thus,

$$\boldsymbol{j} \simeq -\frac{e}{4\pi^3}\int \boldsymbol{v}(\boldsymbol{k})\left[f_0(\boldsymbol{k}) + \frac{e}{\hbar}\tau(\boldsymbol{k})\boldsymbol{E}\cdot\nabla_k f_0(\boldsymbol{k})\right]\mathrm{d}\boldsymbol{k}. \tag{6.63}$$

For spatially isotropic materials and cubic lattices, the conductivity σ is no longer a tensor quantity, but becomes a scalar. Thus, for an electric field, $\boldsymbol{E} = (E_x, 0, 0)$ applied in the x-direction, $\boldsymbol{j} = (j_x, 0, 0)$ and so eqn. (6.63) reduces to

$$j_x = -\frac{e^2}{4\pi^3}\int v_x(\boldsymbol{k})\left[f_0(\boldsymbol{k}) + \frac{e}{\hbar}\tau(\boldsymbol{k})E_x\frac{\partial f_0}{\partial x}\right]\mathrm{d}\boldsymbol{k}. \tag{6.64}$$

The integral over $v_x(\boldsymbol{k})f_0(\boldsymbol{k})$ vanishes because of inversion symmetry about $\boldsymbol{k} = 0$ for the Brillouin zone. Moreover, since

$$\frac{\partial f_0}{\partial k_x} = \hbar v_x\frac{\partial f_0}{\partial \mathscr{E}}, \tag{6.65}$$

eqn. (6.64) becomes

$$j_x = -\frac{e^2}{4\pi^3}E_x\int v_x^2(\boldsymbol{k})\tau(\boldsymbol{k})\frac{\partial f_0}{\partial \mathscr{E}}\mathrm{d}\boldsymbol{k}. \tag{6.66}$$

Thus, the conductivity, $\sigma = j_x/E_x$, is given by

$$\sigma = -\frac{e^2}{4\pi^3}\int v_x^2(\boldsymbol{k})\tau(\boldsymbol{k})\frac{\partial f_0}{\partial \mathscr{E}}\mathrm{d}\boldsymbol{k}. \tag{6.67}$$

The energy derivative of the Fermi–Dirac function can be approximated as a Dirac delta function:

$$\frac{\partial f_0}{\partial \mathscr{E}} \simeq -\delta(\mathscr{E} - \mathscr{E}_{\mathrm{F}}). \tag{6.68}$$

Furthermore, the relation obtained using the construction given in Fig. 4.18 (cf. eqn. (4.57)), viz.

$$\mathrm{d}\boldsymbol{k} = \mathrm{d}S_{\mathscr{E}}\mathrm{d}k_\perp = \mathrm{d}S_{\mathscr{E}}\frac{\mathrm{d}\mathscr{E}}{|\nabla_k\mathscr{E}|} = \mathrm{d}S_{\mathscr{E}}\frac{\mathrm{d}\mathscr{E}}{\hbar v(\boldsymbol{k})}, \tag{6.69}$$

where $\mathrm{d}S_{\mathscr{E}}$ is a surface element in $\boldsymbol{k}$-space at constant energy $\mathscr{E}$, transforms eqn. (6.67) into

$$\sigma \simeq \frac{e^2}{4\pi^3\hbar}\int\frac{v_x^2(\boldsymbol{k})}{v(\boldsymbol{k})}\tau(\boldsymbol{k})\delta(\mathscr{E} - \mathscr{E}_{\mathrm{F}})\mathrm{d}S_{\mathscr{E}}\mathrm{d}\mathscr{E}, \tag{6.70}$$

or, with operation of the delta function $\delta(\mathscr{E} - \mathscr{E}_{\mathrm{F}})$,

$$\sigma \simeq \frac{e^2}{4\pi^3\hbar}\int_{\mathscr{E}=\mathscr{E}_{\mathrm{F}}}\frac{v_x^2(\boldsymbol{k})}{v(\boldsymbol{k})}\tau(\boldsymbol{k})\mathrm{d}S_{\mathscr{E}}. \tag{6.71}$$

Hence, the electrical conductivity of a metal can be expressed as a surface integral over the Fermi surface in $\boldsymbol{k}$-space. Equation (6.71) implies that the d.c. conductivity of a metal is essentially proportional to the area of the Fermi surface available to conduction electrons (see Problem (6.5b)). Thus, materials (for example, alkali metals) for which the Fermi surface is maximal (i.e. a complete spherical surface) have higher conductivities than do metals in which the Fermi surface is greatly reduced. Examples of the latter are semi-metals, such as Mg or Ca, in which the first Brillouin zone is very nearly filled with electrons but band-overlap effects cause a spill-over of electrons into higher zones (see Fig. 5.28); the area of the Fermi surface is rather small in such cases. It can be shown (see Problem 6.5a) that eqn. (6.71) reduces to the Drude expression (eqn. (6.9)) in the case of a free-electron gas, with $\tau = \tau(\mathscr{E}_F)$.

In the relaxation-time approximation, the *resistivity* (inverse of the conductivity) of a metal can be written as the *sum*† of contributions for various scattering mechanisms, e.g. defect (d) and phonon (ph), as follows from eqn. (6.43), i.e. Matthiesen's rule:

$$\rho = \rho_{\mathrm{d}} + \rho_{\mathrm{ph}}. \tag{6.72}$$

The temperature dependence of the resistivity can be obtained by examination of the temperature dependence of the corresponding relaxation times: that for electron–defect scattering is temperature-independent (§6.3.1.2) and for electron-phonon scattering at *high* temperatures ($T \gg \theta_{\mathrm{D}}$), $1/\tau_{\mathrm{ph}} \propto T$, i.e. $\rho_{\mathrm{ph}} \propto T$ (§6.3.1.3). Hence, in this temperature domain,

$$\rho(T) = \rho_{\mathrm{d}} + aT, \qquad T \gg \theta_{\mathrm{D}}, \tag{6.73}$$

where ρ_{d} and a are constants. At very low temperatures, where the electron–phonon scattering is frozen out, ρ becomes temperature-independent, $\rho(T) = \rho_{\mathrm{d}}$, the so-called residual resistivity due to defects.

Between these two limits, the low-temperature behaviour of the electron–phonon scattering (§6.3.1.3) takes over. Although the temperature dependence of the relaxation time in this régime is $1/\tau_{\mathrm{ph}} \propto T^3$ (eqn. (6.52)), this does *not* give the temperature dependence of the resistivity directly in this case. At such low temperatures, the phonon wavevector has a magnitude $|\boldsymbol{q}| \leqslant k_{\mathrm{B}}T/\hbar v_{\mathrm{L}}$, which is very small. The scattering geometry for an N-type process (see Fig. 6.10), which is very nearly elastic in this case since $\boldsymbol{k}_{\mathrm{f}} = \boldsymbol{k}_{\mathrm{i}} + \boldsymbol{q} \simeq \boldsymbol{k}_{\mathrm{i}}$ (the value of the Fermi wavevector), gives

$$|\boldsymbol{q}| \simeq 2k_{\mathrm{F}}\sin(\phi/2) \tag{6.74}$$

where ϕ is the angle between $\boldsymbol{k}_f$ and $\boldsymbol{k}_i$ (see Fig. 6.10). Thus, the process is one of *small-angle scattering*, and in such a situation there is an additional weighting factor of (1–cosϕ) (cf. eqn. (6.45)) to take account of the efficiency of such scattering processes in causing electrical resistivity. Since $(1 - \cos\phi) = 2\sin^2(\phi/2) \simeq 1/2(q/k_{\mathrm{F}})^2$ from eqn. (6.74), this introduces an additional factor of $T^2 (q \propto T)$ into the expression for the resistivity, giving finally the Bloch–Grüneisen T^5 law:

$$\rho_{\mathrm{ph}}(T) \propto T^5, \qquad T \ll \theta_{\mathrm{D}}. \tag{6.75}$$

† In fact, Matthiessen's, rule, that the scattering rates, and hence the inverse relaxation times in the relaxation-time approximation, are simply additive for different processes, is only true if the various τ are *not* dependent on $\boldsymbol{k}$, i.e. the scattering is *isotropic*. For a deeper discussion on the Boltzmann transport equation and the relaxation-time approximation, see, e.g. Ashcroft and Mermin (1976) and Madelung (1978).

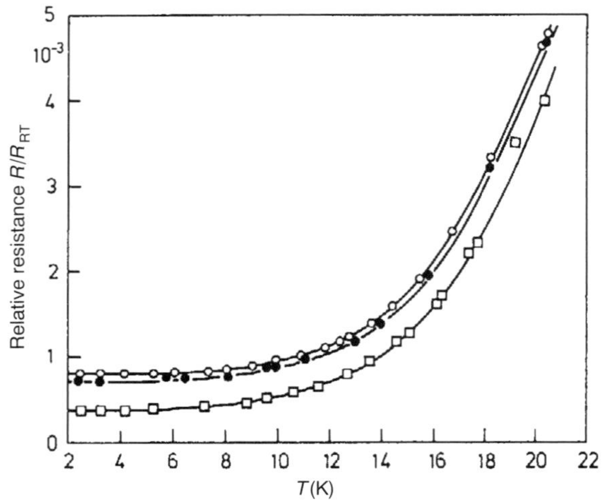

Fig. 6.11 Temperature dependence of the electrical resistance of Na, normalized to the room-temperature value for three samples with different defect concentrations (after McDonald and Mendelssohn (1950); Reproduced by permission of The Royal Society).

The experimental temperature dependence of the electrical resistance of Na is shown in Fig. 6.11: the three régimes of behaviour can easily be discerned. Note the differences in residual (defect-related) resistivity for the three samples with different impurity contents.

*6.3.2.2 *Thermal conductivity*

The electronic thermal conductivity, i.e. the heat carried by electrons, can be calculated using the same Boltzmann transport formulation as was used to discuss the electrical conductivity, except that now the quantity of interest is the heat flux, $\boldsymbol{J}_Q$. A change in heat $đQ$ can be associated with a change in entropy dS of the electrons in a region in a material via

$$đQ = T\mathrm{d}S. \tag{6.76}$$

Since a change in entropy is related to changes in internal energy d$\boldsymbol{U}$ and in the number of electrons d$\boldsymbol{N}$ via the thermodynamic relation

$$T\mathrm{d}S = \mathrm{d}U - \mu\mathrm{d}N, \tag{6.77}$$

where μ is the chemical potential of the electrons ($\mathscr{E}_{\mathrm{F}}$ at absolute zero — see eqn. (5.28)), the heat flux can therefore be written as

$$\boldsymbol{J}_Q = T\boldsymbol{J}_S = \boldsymbol{J}_U - \mu\boldsymbol{J}_N \tag{6.78}$$

where the subscripts S, U and N refer to fluxes in entropy, energy and number of electrons, respectively. By analogy with the expression (eqn. (6.53)) for the electrical current flux $\boldsymbol{j}(= -e\boldsymbol{J}_N)$, the energy and number fluxes can be written as:

$$\boldsymbol{J}_U = \frac{1}{4\pi^3}\int \mathscr{E}(\boldsymbol{k})\boldsymbol{v}(\boldsymbol{k})f(\boldsymbol{k})\mathrm{d}\boldsymbol{k}, \tag{6.79a}$$

$$\boldsymbol{J}_N = \frac{1}{4\pi^3}\int 1\cdot \boldsymbol{v}(\boldsymbol{k})f(\boldsymbol{k})\mathrm{d}\boldsymbol{k} \tag{6.79b}$$

for a single band (a summation over contributions from different bands is necessary if more than one band is involved in the transport). Thus, the heat flux is given by

$$\boldsymbol{J}_Q = \frac{1}{4\pi^3}\int (\mathscr{E}(k)-\mu)\boldsymbol{v}(\boldsymbol{k})f(\boldsymbol{k})\mathrm{d}\boldsymbol{k}. \tag{6.80}$$

The distribution function $f(\boldsymbol{k})$, given by eqn. (6.58), can be approximated by a linearized form, in the relaxation-time approximation, in a similar manner to eqn. (6.61):

$$f(\boldsymbol{k}) \simeq f_0(\boldsymbol{k}) - \tau(\boldsymbol{k})\boldsymbol{v}\cdot\nabla_r f_0(\boldsymbol{k}) \tag{6.81}$$

in the absence of an electric field. Here, the spatial variation is associated with the temperature gradient, i.e. $\nabla_r f_0 = \nabla_r T \partial f_0/\partial T$, and so the heat flux in, say, the x-direction can be written as

$$\boldsymbol{J}_{Q,x} \simeq \frac{1}{4\pi^3}\int (\mathscr{E}(\boldsymbol{k})-\mu)v_x^2(\boldsymbol{k})\tau(\boldsymbol{k})\frac{\partial f_0}{\partial T}\left(-\frac{\partial T}{\partial x}\right)\mathrm{d}\boldsymbol{k}. \tag{6.82}$$

Since the thermal conductivity is given by $\boldsymbol{J}_Q = -\kappa_T\nabla_r T$ (eqn. (6.14)), κ_T can be obtained in general from the above equation. The integral in eqn. (6.82) can be rewritten in terms of the density of electron states per unit volume, $D(\mathscr{E}) = (1/4\pi^3)\int \mathrm{d}S_{\mathscr{E}}/\nabla_k\mathscr{E}$ (eqn. (5.131)). Thus, for example

$$\begin{aligned}\frac{1}{4\pi^3}\int v_x^2(\boldsymbol{k})\tau(\boldsymbol{k})\frac{\partial f_0}{\partial T}\left(-\frac{\partial T}{\partial x}\right)\mathrm{d}\boldsymbol{k} &= \frac{1}{4\pi^3}\int v_x^2(\boldsymbol{k})\tau(\boldsymbol{k})\frac{\partial f_0}{\partial T}\left(-\frac{\partial T}{\partial x}\right)\frac{dS_{\mathscr{E}}}{\nabla_k\mathscr{E}}\mathrm{d}\mathscr{E}\\ &= \frac{2}{3m_\mathrm{e}}\int \mathscr{E}D(\mathscr{E})\tau(\mathscr{E})\frac{\partial f_0}{\partial T}\left(-\frac{\partial T}{\partial x}\right)\mathrm{d}\mathscr{E}\end{aligned} \tag{6.83}$$

where an average of $v_x^2(\langle v_x^2\rangle = \frac{1}{3}v^2)$ has been taken over the Fermi surface (assumed to be spherical). Since the derivative $(\partial f_0/\partial T)$ is only appreciable in the vicinity of $\mathscr{E}_\mathrm{F}$ (cf. eqn. (5.29) and Fig. 5.7), $\tau(\mathscr{E}) = \tau(\mathscr{E}_\mathrm{F})$ can be taken out of the integral if it is assumed that it is only a function of energy. Thus, the energy-dependent terms in eqn. (6.83) can be written as

$$\int \mathscr{E}D(\mathscr{E})\frac{\partial f_0}{\partial T}\mathrm{d}\mathscr{E} = \frac{\partial}{\partial T}\int \mathscr{E}D(\mathscr{E})f_0\mathrm{d}\mathscr{E} = c_v \tag{6.84}$$

where c_v is the electronic heat capacity per unit volume ($= C_v/V$— cf. eqn. (5.36)).

Therefore, eqn. (6.82) for the heat flux can be evaluated (by use of eqn. (5.27)) to give finally the expression for the electronic thermal conductivity for a free-electron gas:

$$\kappa_T = \frac{1}{3}v_\mathrm{F}^2\tau(\mathscr{E}_\mathrm{F})c_v. \tag{6.85}$$

The temperature dependence of κ_T can be understood from eqn. (6.85) in terms of the temperature dependence of $c_v(\propto T$, cf. eqn. (5.36)) and of τ (which is different for different electron-scattering mechanisms — see §6.3.1), i.e. $\kappa_T \propto T\tau(T)$ (eqn. (6.17)). At very low temperatures where impurity or defect scattering is dominant (§6.3.1.2), $\tau_{\rm d}$ is independent of T and hence $\kappa_T(T) \propto T$.

At higher temperatures, phonon scattering (§6.3.1) becomes dominant. For $T < \theta_D, \tau_{\rm ph}(T) \propto T^{-3}$ (eqn. (6.52)) and hence $\kappa_T \propto T^{-2}$; for $T > \theta_{\rm D}, \tau_{\rm ph} \propto T^{-1}$ (eqn. (6.50)) and hence κ_T is temperature-independent. The temperature-dependent behaviour of the *electronic* contribution to the thermal conductivity of metals is shown schematically in Fig. 6.12.

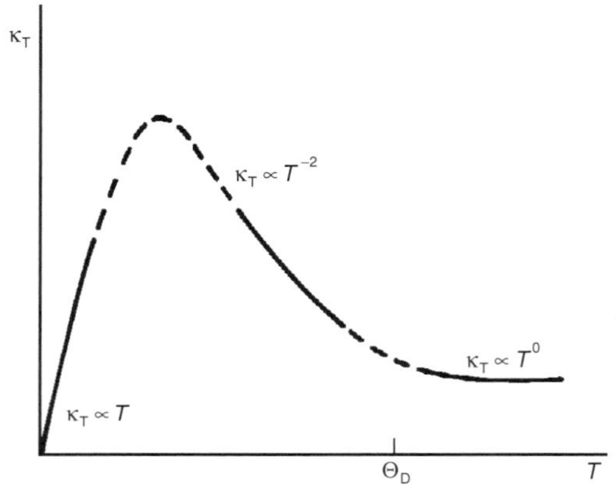

Fig. 6.12 Schematic illustration of the temperature dependence of the electronic contribution to the thermal conductivity of metals, κ_T, showing the limiting forms at various temperatures. Impurity (defect) scattering is dominant at temperatures below the peak and phonon scattering above it.

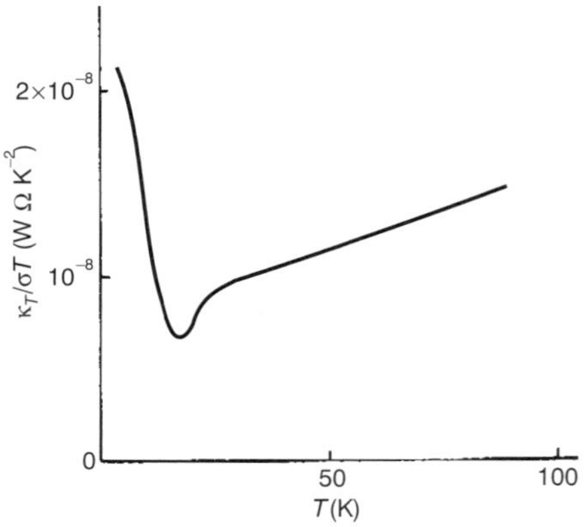

Fig. 6.13 The ratio $\kappa_T/\sigma T (= L$, the Lorentz number) for Na as a function of temperature (after Berman and McDonald (1951); Reproduced by permission of The Royal Society).

The Wiedemann–Franz law (eqn. (6.18)) states that the ratio of thermal and electrical conductivities is simply equal to a universal constant multiplied by the temperature. This relation is also found using the Boltzmann-transport-equation method subject to the approximations used to obtain expressions for σ and κ_T. In fact, the ratio $\kappa_T/\sigma T = L$, the Lorentz number, is *not* generally constant at all temperatures — see Fig. 6.13. This is because the Wiedemann–Franz law is only valid if, in the relaxation-time approximation, a particular electron-scattering mechanism causes a degradation of an electrical current in exactly the same way and at the same rate as for a thermal current. This is the case for *elastic* scattering (or more generally for inelastic scattering processes where the change in electron energy is much smaller than $k_B T$). Thus the Wiedemann–Franz law should be obeyed at very low temperatures where impurity scattering is dominant, and at high temperatures ($T \gg \theta_D$) in the phonon-scattering régime. However, in the intermediate temperature range ($T < \theta_D$), inelastic phonon-scattering events can cause a greater degradation of thermal currents than of electrical currents because the electron *energy* as well as the velocity can be changed (recall that $\boldsymbol{J}_Q$ is proportional to $(\mathscr{E} - \mu)$ — see eqn. (6.82)). Thus, the ratio $L = \kappa_T/\sigma T$ is less than the universal value 2.45×10^{-8} W Ω K^{-2} in this temperature range (see Fig. 6.13).

*6.3.2.3 *Thermoelectric effects*

The analysis of the thermal conductivity given in the previous section considered only the heat flux and ignored the possible effects of any electric charge flux resulting from electric fields: thermal-conductivity measurements are, in fact, performed under open-circuit conditions to ensure that no electrical current flows at the same time as the heat transport takes place. However, in general, heat and current flows can occur simultaneously and this gives rise to a variety of thermoelectric effects. In fact, in the above-mentioned thermal-conductivity experiment, in the presence of a temperature gradient ∇T, an electrical current *will* flow until sufficient charge has built up at the sample surface to give a retarding electric field that exactly cancels the driving force of the temperature gradient on the current flow: under steady-state conditions, therefore, there is no net current. Thus, in general, a temperature gradient in an electrically conducting sample generates an electric field in the opposite direction; this is known as the Seebeck effect.

Thermoelectric behaviour can be discussed in terms of Boltzmann transport theory in the same way as for electrical conductivity (electric current and no thermal flux) and thermal conductivity (thermal flux and no electrical current). Thus, if the electrochemical potential η is defined as

$$\eta = \mu - e\phi, \tag{6.86}$$

where μ and ϕ are the chemical and electrostatic potentials, respectively, then the electrical current density $\boldsymbol{j}$ and heat flux $\boldsymbol{J}_Q$ can be written in terms of the appropriate 'driving forces', and the Onsager coefficients, $\boldsymbol{L}_{ij}$ (see §3.4.2.1), as

$$\boldsymbol{j} = \boldsymbol{L}_{11}\nabla(\eta/e) + \boldsymbol{L}_{12}\frac{(-\nabla T)}{T}, \tag{6.87a}$$

$$\boldsymbol{J}_Q = \boldsymbol{L}_{21}\nabla(\eta/e) + \boldsymbol{L}_{22}\frac{(-\nabla T)}{T}. \tag{6.87b}$$

The Onsager coefficients, which are tensor quantities in general, should not be confused with the Lorentz number (eqn. (6.18)). (Note also that in eqn. (3.42), $\boldsymbol{L}_{ij}$ refers to *particle* fluxes and not electrical current density.) They can be evaluated by expressing the charge and heat fluxes via the Boltzmann formulation, as above. One then finds (see Ashcroft and Mermin (1976), Madelung (1978)):

$$\boldsymbol{L}_{11} = \sigma(\mathscr{E}_{\mathrm{F}}) \equiv \sigma_0, \tag{6.88a}$$

$$\boldsymbol{L}_{12} = \boldsymbol{L}_{21} = \frac{-\pi^2}{3e} k_{\mathrm{B}}^2 T^2 \sigma', \tag{6.88b}$$

$$\boldsymbol{L}_{22} = \frac{\pi^2}{3e^2} k_{\mathrm{B}}^2 T^2 \sigma, \tag{6.88c}$$

where σ' is the energy derivative of the conductivity evaluated at the Fermi level:

$$\sigma' = \frac{\partial}{\partial \mathscr{E}} \sigma(\mathscr{E})_{\mathscr{E}=\mathscr{E}_{\mathrm{F}}}. \tag{6.89}$$

A general expression for the thermal conductivity, without the assumption of degenerate Fermi statistics restricting the validity to metals as in §6.3.3, can be obtained from the set of equations (6.87) by stipulating that the heat flux $\boldsymbol{J}_Q$ arises from the temperature gradient ∇T in the absence of an electrical current. Thus, from eqn. (6.87a), for $\boldsymbol{j} = 0$,

$$\nabla(\eta/e) = -(\boldsymbol{L}_{11})^{-1} \boldsymbol{L}_{12} \left(\frac{-\nabla T}{T} \right), \tag{6.90}$$

and hence, on substituting this into eqn. (6.87b),

$$\boldsymbol{J}_Q = (-\boldsymbol{L}_{21} (\boldsymbol{L}_{11})^{-1} \boldsymbol{L}_{12} + \boldsymbol{L}_{22}) \left(\frac{-\nabla T}{T} \right). \tag{6.91}$$

Therefore, the thermal-conductivity tensor is given in general by

$$\kappa_T = (\boldsymbol{L}_{22} - (\boldsymbol{L}_{21} (\boldsymbol{L}_{11})^{-1} \boldsymbol{L}_{12})/T. \tag{6.92}$$

This expression is valid for both semiconductors (non-degenerate statistics) and metals (degenerate statistics), although in the latter case, since $\sigma' \simeq \sigma/\mathscr{E}_{\mathrm{F}}$, the first term in parentheses in eqn. (6.92) is much larger than the other, and $\kappa_T \simeq \boldsymbol{L}_{22}/T = \pi^2 k_{\mathrm{B}}^2 T\sigma/3e^2$, i.e. the Wiedemann–Franz law is recovered.

The flux equations (6.87) can be rewritten, in terms of quantities related to the Onsager coefficients, and assumed for simplicity to be scalars, so that they can be compared with quantities measured experimentally under either isothermal or open-circuit conditions, i.e.

$$\nabla(\eta/e) = \frac{1}{\sigma} \boldsymbol{j} + S_T \nabla T, \tag{6.93a}$$

$$\boldsymbol{J}_Q = \Pi \boldsymbol{j} - \kappa_T \nabla T, \tag{6.93b}$$

where the thermopower (or thermoelectric power) is given by

$$S_T = L_{12}/TL_{11} \tag{6.94}$$

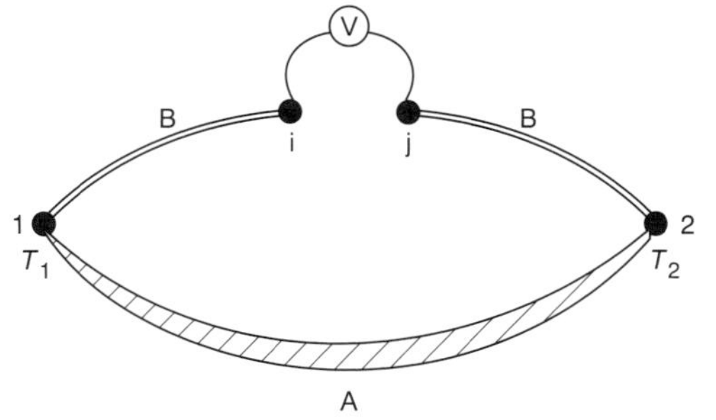

Fig. 6.14 Illustration of the generation of a thermoelectric voltage, measurable by voltmeter V, across a break i–j in a wire (of a metal of type B) when two junctions, 1, 2, between it and another metal A are held at different temperatures, $T_2 \neq T_1$.

and the Peltier coefficient is given by

$$\Pi = L_{21}/L_{11}. \tag{6.95}$$

Thus, from the Onsager relation (eqn. (3.43)) between the Onsager coefficients ($L_{12} = L_{21}$), the Peltier coefficient and the thermopower are related by the Kelvin relation:

$$\Pi = TS_T. \tag{6.96}$$

What are the physical manifestations of the thermoelectrical quantities Π and S_T? Consider first the Seebeck effect involving the thermopower S_T, which is the constant of proportionality between the temperature gradient and the corresponding electrochemical potential gradient in the absence of electrical current flow (cf. eqn. (6.93a)). For the arrangement shown in Fig. 6.14, wires of two dissimilar metals, A and B, are joined together at two junctions 1 and 2, and one of the wires, say B, is cut between the two junctions. A thermoelectric voltage will be developed across this break in the circuit if the two heterojunctions, 1 and 2, are held at different temperatures, T_1 and T_2. The electrochemical potential difference can be obtained by integrating around the circuit shown in Fig. 6.14, i.e. under open-circuit conditions:

$$\begin{aligned}\oint \nabla(\eta/e)\cdot d\boldsymbol{l} &= \delta\eta/e = \oint S_T\nabla T\cdot d\boldsymbol{l} \equiv \oint S_T dT\\ &= \int_{T_1}^{T_2} S_T^{A} dT + \int_{T_1}^{T_2} S_T^{B} dT = \int_{T_1}^{T_2}(S_T^{A} - S_T^{B})dT.\end{aligned} \tag{6.97}$$

Now, $\delta\eta = \delta\mu - e\delta\phi = -e\delta\phi$, since the chemical potential at points i and j in the circuit of Fig. 6.14 is the same (identical material, B) and hence

$$-\delta\phi = \int_{T_1}^{T_2}(S_T^{A} - S_T^{B})dT. \tag{6.98}$$

The electrostatic potential difference generated between points i and j is thus proportional to the difference of the absolute thermopowers of materials A and B, S_T^{A} and S_T^{B}, respectively, and the temperature difference $\delta T = T_2 - T_1$. The Seebeck effect can also occur as a *volume* effect in inhomogeneous conducting materials.

The circuit shown in Fig. 6.14 can be used as a way of measuring temperatures (or, more exactly, the temperature difference of one junction with respect to another at a reference temperature). Absolute values of thermopowers for metals are typically a few microvolts per kelvin (μVK^{-1}). They can be measured by making the wire A in Fig. 6.14 out of a superconducting material (§6.4); superconductors are characterized by having *zero* values of absolute thermopower (§6.4.2.2).

From eqns. (6.88) and (6.94), the thermopower can be expressed in terms of the electrical conductivity as:

$$S_T = -\frac{\pi^2}{3e} k_B^2 T \frac{\sigma'}{\sigma} \tag{6.99a}$$

$$\simeq -\frac{\pi^2 k_B^2 T}{3e\mathscr{E}_F} \tag{6.99b}$$

since $\sigma' \simeq \sigma/\mathscr{E}_F$. The exact derivative of the electrical conductivity, σ', can be obtained by differentiating eqn. (6.67) assuming that $\tau(\boldsymbol{k}) \equiv \tau(\mathscr{E})$, i.e.

$$\sigma'(\mathscr{E}) = \frac{\tau'(\mathscr{E})}{\tau(\mathscr{E})}\sigma(\mathscr{E}) + \frac{e^2\tau(\mathscr{E})}{4\pi^3}\int \delta'(\mathscr{E} - \mathscr{E}(\boldsymbol{k}))\boldsymbol{v}(\boldsymbol{k})\boldsymbol{v}(\boldsymbol{k})\mathrm{d}\boldsymbol{k}. \tag{6.100}$$

The product of electron velocity and energy derivative of the Dirac delta function can be recast in the form (Ashcroft and Mermin (1976)):

$$\boldsymbol{v}(\boldsymbol{k})\delta'(\mathscr{E} - \mathscr{E}(\boldsymbol{k})) = -\frac{1}{\hbar}\frac{\partial}{\partial \boldsymbol{k}}\delta(\mathscr{E} - \mathscr{E}(\boldsymbol{k})), \tag{6.101}$$

and hence an integration by parts gives:

$$\sigma' = \frac{\tau'}{\tau}\sigma + \frac{e^2\tau}{4\pi^3}\int \delta(\mathscr{E}_F - \mathscr{E}(\boldsymbol{k}))(\boldsymbol{m}^*)^{-1}\mathrm{d}\boldsymbol{k}, \tag{6.102}$$

where $\boldsymbol{m}^*$ is the effective-mass tensor (eqn. (6.25)). Thus, the sign of the thermopower is opposite to that of the effective mass, i.e. negative for electrons and positive for holes.

Another thermoelectric phenomenon of interest is the Peltier effect: an electrical current driven around a circuit consisting of two dissimilar metals, kept at constant temperature, causes heat to be given out at one junction and absorbed at the other (Fig. 6.15). From eqn. (6.93b), under isothermal conditions, the heat current is proportional to the electrical current:

$$\boldsymbol{J}_Q = \Pi \boldsymbol{j}. \tag{6.103}$$

Since changes in the heat $đQ$ and entropy $\mathrm{d}S$ are related via $đQ = T\mathrm{d}S$, the corresponding fluxes of these quantities are related by $J_Q = T\boldsymbol{J}_S$. Hence from eqn. (6.103), the quantity $\Pi/T = S_T$ (cf. eqn. (6.96)) is simply the constant of proportionality between an electric current flux and a concomitant entropy flux in a conductor under isothermal conditions. If Π_A and Π_B are the Peltier coefficients of metals A and B respectively, a net heat of $(\Pi_A - \Pi_B)j$ is absorbed at junction 1 and the same heat is given out at junction 2 for the current flow shown in Fig. 6.15 and if $\Pi_A > \Pi_B$. Thus, the Peltier effect can be utilized in solid-state cooling devices.

Finally, another thermoelectric phenomenon, the Thomson effect, concerns the heat generated in a material when *both* electrical currents and thermal gradients are present.

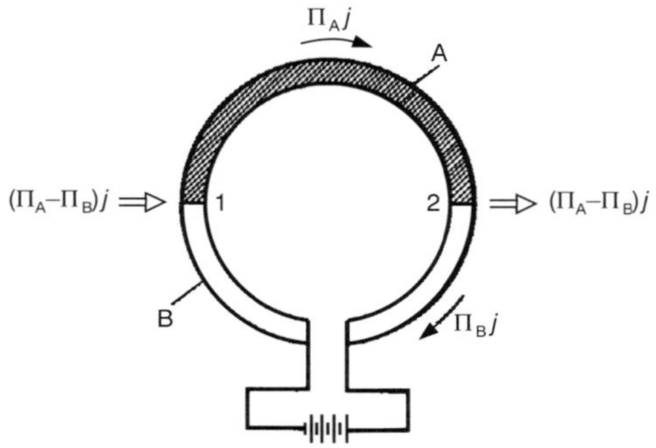

Fig. 6.15 Illustration of the Peltier effect. An electrical current of density j, passing round a bimetallic circuit, held at a uniform temperature, will cause heat equal to $(\Pi_{\mathrm{A}} - \Pi_{\mathrm{B}})j$ to be absorbed at junction 1 and given out at junction 2 if $\Pi_{\mathrm{A}} > \Pi_{\mathrm{B}}$.

The Thomson energy coefficient, μ_{E}, must clearly therefore be related to both the thermopower and the Peltier coefficient. The conservation of energy with time can be written as

$$\frac{\partial u}{\partial t} + \nabla \cdot \boldsymbol{J}_U = \boldsymbol{j} \cdot \boldsymbol{E}, \tag{6.104}$$

where the two terms on the left-hand side refer to the rate of change of internal energy per unit volume, u, associated with the electrons and the spatial variation of the energy flux, $\boldsymbol{J}_U$, respectively, and these equal the rate of work done on the electron current $\boldsymbol{j}$ by an external electric field $\boldsymbol{E}$. Equation (6.104) can be rewritten using $\boldsymbol{J}_U = \boldsymbol{J}_Q + \mu \boldsymbol{J}_N$ (eqn. (6.78)), together with the expression for the electrochemical potential $\eta = \mu - e\phi$ (eqn. (6.86)), as

$$\begin{aligned} \frac{\partial u}{\partial t} &= -\nabla \cdot \boldsymbol{J}_Q - \nabla \cdot (\mu \boldsymbol{J}_N) + \boldsymbol{j} \cdot \boldsymbol{E} \\ &= \boldsymbol{j} \cdot \nabla(\eta/e) - \nabla \cdot \boldsymbol{J}_Q. \end{aligned} \tag{6.105}$$

From eqns. (6.93), this expression becomes:

$$\begin{aligned} \frac{\partial u}{\partial t} &= \frac{j^2}{\sigma} + S_T \boldsymbol{j} \cdot \nabla T - \nabla(\Pi \boldsymbol{j}) + \nabla \cdot (\kappa_T \nabla T) \\ &= \frac{j^2}{\sigma} + \nabla \cdot (\kappa_T \nabla T) - \mu_T \boldsymbol{j} \cdot \nabla T, \end{aligned} \tag{6.106}$$

where it has been assumed that the Peltier coefficient is only a function of temperature (i.e. the material is homogeneous) and the Thomson heat μ_{T} is defined as

$$\mu_T = \frac{\partial \Pi}{\partial T} - S_T. \tag{6.107}$$

The first term in eqn. (6.106) is the Joule heating term, the second refers to the heat entering a region by heat conduction, and the last term describes heat production when both electric currents and temperature gradients are present *simultaneously* (the Thomson effect).

From eqn. (6.96), the Thomson heat can also be rewritten as

$$\mu_T = \frac{\partial S_T}{\partial T}. \tag{6.108}$$

Thus, the Thomson effect enables values of the thermopower to be obtained at temperatures other than that at which it is known. By measuring the change in heat of a sample as the electrical current direction is reversed for a constant temperature gradient, the Thomson heat is obtained from eqn. (6.106); hence, eqn. (6.108) can be used to calculate S_T at any temperature, knowing its value at some reference temperature.

6.3.3 Electron dynamics in the presence of magnetic fields

In the discussion of electron dynamics in metals given in §6.3.2, the effect of an applied magnetic field $\boldsymbol{B}$ was neglected; only electric fields and temperature gradients were considered. However, in general, a magnetic field strongly perturbs the electron motion in solids. As a result, many new phenomena can be distinguished, which can be divided into galvanomagnetic effects (where the primary field is $\boldsymbol{E}$, in addition to $\boldsymbol{B}$) and thermomagnetic effects (where the primary 'field' is ∇T, in addition to $\boldsymbol{B}$). Moreover, the action of a magnetic field by itself causes electron motion due to the Lorentz force (cf. eqn. (6.4)). We will discuss the latter topic first: this will be followed by a discussion of galvanomagnetic and thermomagnetic effects.

6.3.3.1 Cyclotron resonance

The application of a constant magnetic field $\boldsymbol{B}$ to an electrical conductor causes the mobile electrons to be deflected from their original direction of motion by the Lorentz force, producing a rate of change of momentum given by

$$m_{\mathrm{e}} \frac{\mathrm{d}\boldsymbol{v}}{\mathrm{d}t} = -e(\boldsymbol{v} \times \boldsymbol{B}). \tag{6.109}$$

Writing the electron momentum as $m_{\mathrm{e}}\boldsymbol{v} = \hbar\boldsymbol{k}$, and noting that the group velocity of a Bloch wave packet is given by $\boldsymbol{v} = \hbar^{-1}\nabla_{\boldsymbol{k}}\mathscr{E}(\boldsymbol{k})$, the semi-classical equation of motion of an electron in a magnetic field (i.e. neglecting quantization effects) becomes

$$\frac{\mathrm{d}\boldsymbol{k}}{\mathrm{d}t} = \frac{-e}{\hbar^2}(\nabla_k \mathscr{E}(\boldsymbol{k}) \times \boldsymbol{B}). \tag{6.110}$$

In this semi-classical picture, the electron motion can be represented as the motion, in reciprocal space, of allowed $\boldsymbol{k}$-states in a plane perpendicular to the magnetic field and tangential to constant-energy surfaces (Fig. 6.16): the component of $\boldsymbol{k}$ parallel to the magnetic field, and the electron energy $\mathscr{E}(\boldsymbol{k})$, remain constants of the motion. The time taken for an electron to move from state $\boldsymbol{k}_1$ to state $\boldsymbol{k}_2$ is given by

$$t_2 - t_1 = \int_{t_1}^{t_2} \mathrm{d}t = \int_{k_1}^{k_2} \frac{\mathrm{d}k}{|\dot{\boldsymbol{k}}|}, \tag{6.111}$$

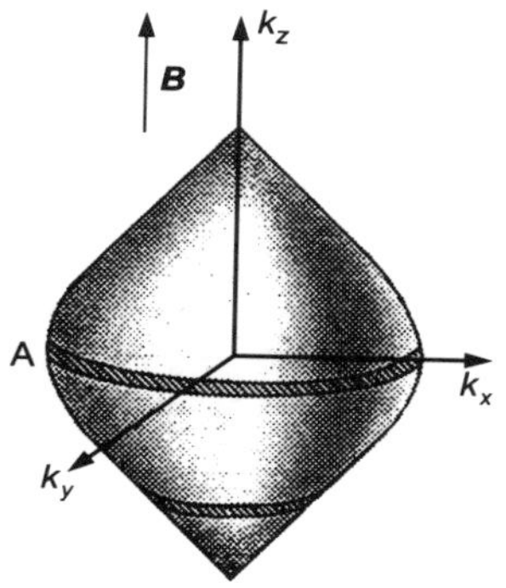

Fig. 6.16 Electron motion in a metal in a constant magnetic field $\boldsymbol{B}$ represented as orbits in $\boldsymbol{k}$-space in planes perpendicular to $\boldsymbol{B}$ cutting a constant-energy surface (i.e. the Fermi surface). Orbit A is known as an extremal orbit.

where $\dot{\boldsymbol{k}}$ is given by eqn. (6.110). Thus, eqn. (6.111) can be rewritten as

$$t_2 - t_1 = \frac{\hbar^2}{eB} \int \frac{\mathrm{d}k}{|(\nabla_k \mathscr{E}(\boldsymbol{k}))_\perp|}, \tag{6.112}$$

where $|(\nabla_k \mathscr{E}(\boldsymbol{k}))_\perp|$ is the component of the energy gradient perpendicular to $\boldsymbol{B}$, i.e. in the plane of the orbit.

This quantity can be related to the area S enclosed by the orbit in $\boldsymbol{k}$-space as follows. Consider two adjacent constant-energy surfaces $\mathscr{E}(\boldsymbol{k}) = \mathscr{E}$ and $\mathscr{E}(\boldsymbol{k}) = \mathscr{E} + \delta\mathscr{E}$, with electron orbits on these surfaces lying in the same plane perpendicular to $\boldsymbol{B}$ (Fig. 6.17). The energy difference is given by

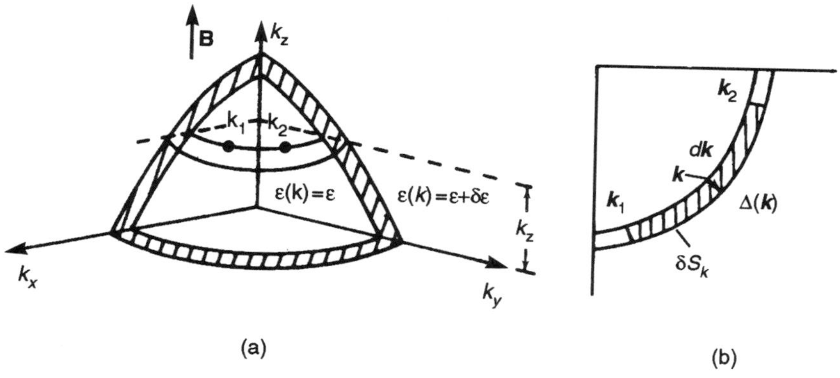

Fig. 6.17 Two constant-energy surfaces in $\boldsymbol{k}$-space, differing in energy by $\delta\mathscr{E}$, with two cyclotron orbits in the same plane normal to the applied magnetic field, $\boldsymbol{B}$. The inset shows a detail of the orbits between states $\boldsymbol{k}_1$ and $\boldsymbol{k}_2$: $\Delta(\boldsymbol{k})$ is the vector joining one orbit at $\boldsymbol{k}$ to the other, and δS_k is the area in $\boldsymbol{k}$-space between the two orbits and $\boldsymbol{k}_1$ and $\boldsymbol{k}_2$.

$$\delta\mathscr{E} = |(\nabla_{\boldsymbol{k}}\mathscr{E}(\boldsymbol{k}))_\perp|\Delta(\boldsymbol{k}) \tag{6.113}$$

where $\Delta(\boldsymbol{k})$ is the magnitude of the vector in $\boldsymbol{k}$-space connecting a point on one orbit at $\boldsymbol{k}$ with the other orbit (Fig. 6.17). Hence, eqn. (6.112) now becomes

$$t_2 - t_1 = \frac{\hbar^2}{eB\delta\mathscr{E}}\int_{k_1}^{k_2} \Delta(\boldsymbol{k})\mathrm{d}k \tag{6.114}$$

and the integral in eqn. (6.114) is simply the area δS_k in $\boldsymbol{k}$-space of the plane between the two orbits and between $\boldsymbol{k}_1$ and $\boldsymbol{k}_2$ (Fig. 6.17). Therefore, the time interval becomes, in the limit as $\delta\mathscr{E} \to 0$,

$$t_2 - t_1 = \frac{\hbar^2}{eB}\frac{\partial S_k}{\partial\mathscr{E}}. \tag{6.115}$$

For complicated Fermi surfaces, such as occur for real metals, e.g. Cu (Fig. 6.18), many different orbits are possible, some of which are 'open' and some of which are closed. The period T of a closed orbit is given by the line integral in eqn. (6.114) having limits $\boldsymbol{k}_1 = \boldsymbol{k}_2$. For a *free-electron gas*, the orbital period can be calculated (see Problem 6.8) to be

$$T_0 = \frac{2\pi m_\mathrm{e}}{eB}, \tag{6.116}$$

from which the cyclotron frequency, $\omega_\mathrm{c} = 2\pi/T_0$, is

$$\omega_\mathrm{c} = \frac{eB}{m_\mathrm{e}}. \tag{6.117}$$

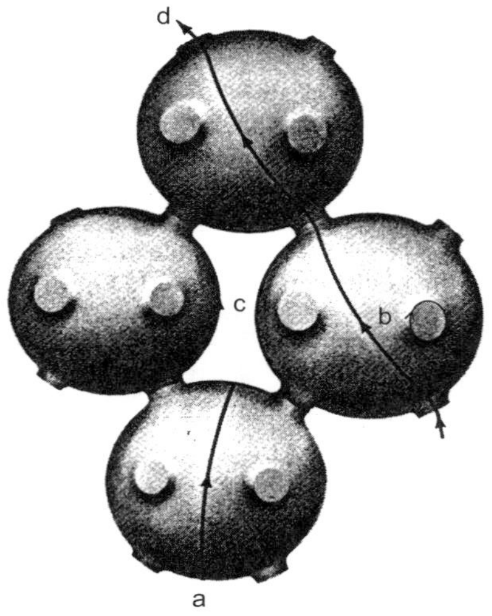

Fig. 6.18 Various cyclotron orbits for electron motion in a magnetic field in Cu. The Fermi surface is shown in the repeated-zone representation. Three types of closed orbit are indicated: electron-like 'belly' (a) and 'neck' (b) orbits and hole-like 'dog-bone' orbit (c). In addition, an open orbit is also indicated (d).

For a real metal, with non-free-electron bands, the cyclotron effective mass m_c^* is defined as

$$m_c^* = \frac{\hbar^2}{2\pi}\frac{\partial S_k}{\partial \mathscr{E}} \tag{6.118}$$

by comparison of eqns. (6.115) and (6.116). Orbits which enclose states for which the energy is *lower* than for states outside the orbit are electron-like (as in Fig. 6.16); orbits for which the converse is true (e.g. the 'dog-bone' orbit in Cu — see Fig. 6.18) are hole-like (i.e. with effective masses of the opposite sign (§6.2.1).

In *real* space, as opposed to reciprocal space, the electron motion in the presence of a magnetic field is a little more complicated. If the magnetic field $\boldsymbol{B}$ is directed along the z-axis, electron motion along this axis is unaffected by the field (cf. eqn. (6.109)). In the plane normal to $\boldsymbol{B}$, an orbit in this plane in $\boldsymbol{k}$-space is equivalent to an orbit in the equivalent plane in real space, but rotated by 90° (Problem 6.9). Thus, in general an electron follows a *helical* path in real space (Fig. 6.19).

The above semi-classical picture neglects orbital quantization, however, and represents a gross simplification of the behaviour. In reality, the whole band picture of crystalline solids developed so far, which is based on discrete $\boldsymbol{k}$-values, corresponding to allowed electron states that can be used as good quantum numbers, breaks down in the presence of a magnetic field. This is because the usefulness of $\boldsymbol{k}$- vectors in describing electron states (in the absence of a magnetic field) stems from the 3D translational invariance (periodicity) characterizing a crystal lattice: in a magnetic field, the Schrödinger equation is no longer translationally invariant in directions normal to $\boldsymbol{B}$. The

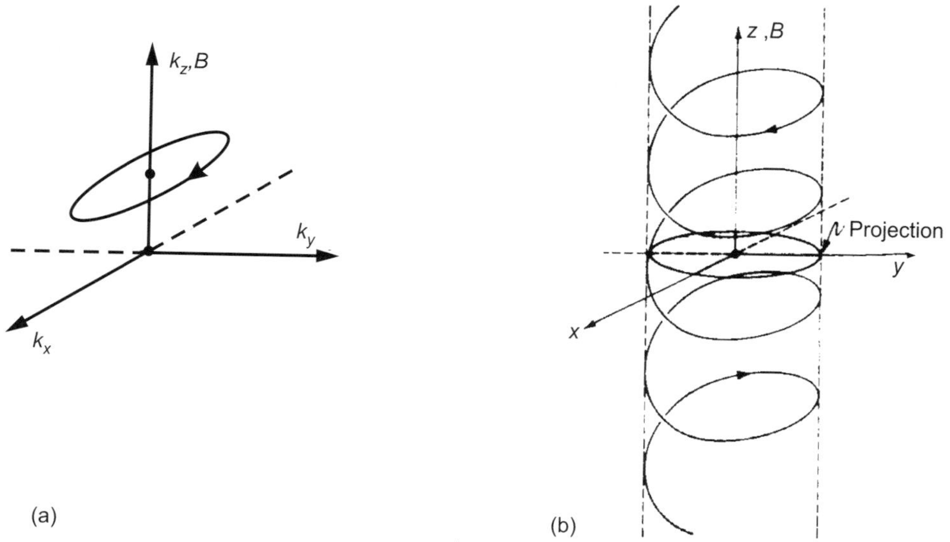

Fig. 6.19 (a) Cyclotron orbit in reciprocal space in a plane perpendicular to the magnetic field $\boldsymbol{B}$. (b) Electron motion in real space with a helical orbit, the projection of which onto the plane normal to $\boldsymbol{B}$ is obtained from the orbit shown in (a), but is rotated by 90° and multiplied by the factor $\hbar/eB$.

momentum $\boldsymbol{p}$ in the Schrödinger equation must be replaced by the term $\boldsymbol{p} + e\boldsymbol{A}$, where $\boldsymbol{A}$ is the vector potential and $\boldsymbol{B} = \text{curl}\boldsymbol{A}$ (see §7.2).

As a result, the imposition of a magnetic field in the z-direction causes the energy of a free electron, confined to say a cubical box of side L oriented parallel to the x, y and z-axes, to be quantized for orbital motion in the x–y plane according to the simple-harmonic oscillator expression (see Problem 6.10), i.e.

$$\hbar^2\left(k_x^2 + k_y^2\right)/2m_e \rightarrow (n + \frac{1}{2})\hbar\omega_c$$

or

$$\mathscr{E}_n = \frac{\hbar^2 k_z^2}{2m_e} + (n + \frac{1}{2})\hbar\omega_c, \quad (n = 0, 1, 2, \ldots), \tag{6.119}$$

where ω_c is the cyclotron frequency (eqn. (6.117)). Motion parallel to $\boldsymbol{B}$ is unaffected by the magnetic field, i.e. it is still free-electron-like. Thus, the dense 3D array of allowed points in $\boldsymbol{k}$-space collapses onto a set of concentric Landau tubes lying parallel to the z-direction (see Fig. 6.20). From eqn. (6.115), writing the derivative $\partial S/\partial \mathscr{E}$ as the difference quotient $(S_{n+1} - S_n)/(\mathscr{E}_{n+1} - \mathscr{E}_n)$ implies (from eqn. (6.119)) that the *area* in $\boldsymbol{k}$-space between successive Landau tubes is quantized, i.e.

$$S_{n+1} - S_n = \frac{2\pi eB}{\hbar}, \tag{6.120}$$

where $t_2 - t_1 = T = 2\pi/\omega_c$. Since the number of 2D $\boldsymbol{k}$-states per unit area of $\boldsymbol{k}$-space is $N(k) = A/2\pi^2$ (see Problem 5.2), where A is the real-space cross-sectional area of the sample normal to $\boldsymbol{B}$, the degeneracy of each Landau tube per unit area of sample for a given value of k_z is given by

$$g_n = \frac{eB}{\pi\hbar}. \tag{6.121}$$

A field of 1 T corresponds to a degeneracy of $4.8 \times 10^{14}/\text{m}^2$. This quantity is so large because, in the semi-classical picture, the motion of an electron in a magnetic field is a helical orbit directed along the z-direction and parallel to $\boldsymbol{B}$ (Fig. 6.19b), with arbitrary x- and y-coordinates.

The occupied electron levels in the presence of a magnetic field lie on the Landau tubes falling inside the Fermi surface in $\boldsymbol{k}$-space (see Fig. 6.20b for an illustration of the free-electron case). In the case of real metals, the Landau tubes are not uniform circular cylinders, as in the free-electron case, but concentric cylinders having shapes determined by the shape of the constant-energy contours in $\boldsymbol{k}$-space of the metal in a plane normal to $\boldsymbol{B}$. However, the quantization of the area in $\boldsymbol{k}$-space between adjacent Landau tubes (eqn. (6.120)) is maintained.

The density of states for a (free-electron) metal in the presence of a magnetic field is very different from the behaviour characteristic of a 3D free-electron gas, viz. $g(\mathscr{E}) \propto \mathscr{E}^{1/2}$ (eqn. (5.15)), because of the formation of the Landau levels. Since each Landau level is associated with *one-dimensional* free-electron-like behaviour along the z-direction, parallel to $\boldsymbol{B}$ (Fig. 6.20a), the corresponding behaviour of the density of states is now $g(\mathscr{E}) \propto (\mathscr{E})^{-1/2}$ (see Problem 5.2); that is, a singularity in the density of states is exhibited at the bottom of each sub-band corresponding to a given Landau level

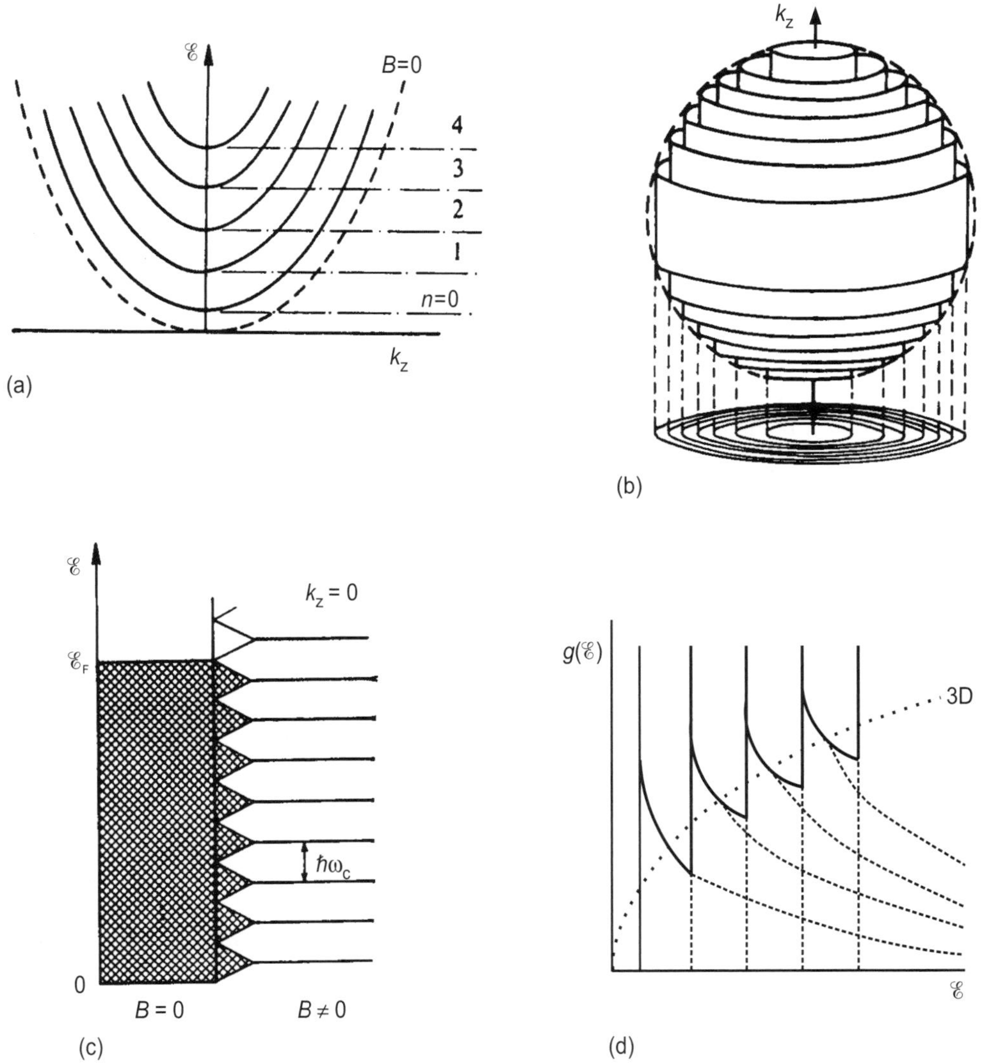

Fig. 6.20 Landau levels for the quantized orbital motion of free electrons in a conductor in a magnetic field $\boldsymbol{B}$ along the z-direction. (a) Free-electron behaviour along the z-direction for different Landau levels with quantum numbers, n. (b) Landau tubes for a free-electron gas. Occupied states are those on the tubes lying within the Fermi sphere in $\boldsymbol{k}$-space. (c) Landau levels in the presence of a magnetic field, $\boldsymbol{B}$, showing the relationship to the band of occupied states below the Fermi level for a metal in the absence of a magnetic field. The transition corresponding to cyclotron resonance is indicated. (d) Density of states for the Landau levels (solid lines), compared with that for a free-electron gas in the absence of a magnetic field (dotted curve).

(Fig. 6.20d). Of course, the total *number* of states must be the same with and without the magnetic field; they are just redistributed.

Orbital motion of electrons in a magnetic field can be probed by a cyclotron resonance experiment, in which an oscillating electromagnetic field with frequency ω, incident on a conductor that is subject to a static magnetic field, is resonantly absorbed when $\omega = \omega_c$. In terms of the quantized Landau levels shown in Fig. 6.20c, the perturbing electromagnetic field causes a transition between adjacent Landau levels at the Fermi level, electrons being excited from a filled level to an empty level with change in quantum number $\Delta n = +1$. Cyclotron resonance is a probe of the shape of the Fermi surface in metals (or constant-energy surfaces in $\boldsymbol{k}$-space near the bottom of the conduction band/top of the valence band for semiconductors — see e.g. Fig. 5.51), in particular permitting values of the (cyclotron) effective mass (eqn. (6.118)) to be measured. The technique is particularly useful for semiconductors where the electromagnetic waves (typically microwaves) are not unduly absorbed by the material.

However, absorption is a problem in the case of metals, where the electromagnetic field will only penetrate to a distance corresponding to the skin depth (Problem 5.30). In this case, the Azbel'–Kaner geometry can be used, in which the static magnetic field is oriented *parallel* to the surface of the metal: the electromagnetic field that does penetrate the metal to within the skin depth can resonantly interact with the electron motion whenever the helical electron orbit passes within the skin depth (Fig. 6.21). Thus, resonance can occur whenever the frequency of the electromagnetic field ω satisfies the condition

$$\omega = p\omega_c, \tag{6.122}$$

where p is an integer. Usually, the cyclotron resonance condition is found by varying the frequency of the electromagnetic radiation (microwaves) whilst the B-field is varied, in which case peaks in the resonant absorption response of the metal occur with a uniform spacing of the inverse field since, from eqns. (6.116), (6.118) and (6.122),

$$\frac{1}{B_{res}} = \frac{ep}{\omega m_c^*} \tag{6.123}$$

An example of a cyclotron resonance curve for Al is shown in Fig. 6.22.

Note that cyclotron resonance can only be detected if the electrons complete many cyclotron orbits before a scattering event occurs, i.e. the condition $\omega_c \tau \gg 1$ must be satisfied, where τ is a characteristic scattering relaxation time (§6.3.1). Moreover, although in principle there are many possible electron cyclotron orbits for a particular constant-energy surface in $\boldsymbol{k}$-space and a given magnetic-field direction (see Fig. 6.16),

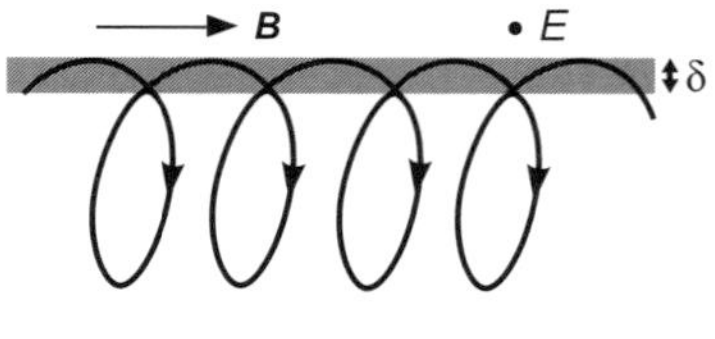

Fig. 6.21 Azbel'–Kaner geometry for observing cyclotron resonance in metals where the penetration of the probing electromagnetic field is confined to the skin depth δ. The application of a static field parallel to the crystal surface causes the helical electron orbit to pass through the skin depth region once per period. Resonance is observed whenever a probe electromagnetic field, applied normal to $\boldsymbol{B}$, has a frequency such that $\omega = p\omega_c$.

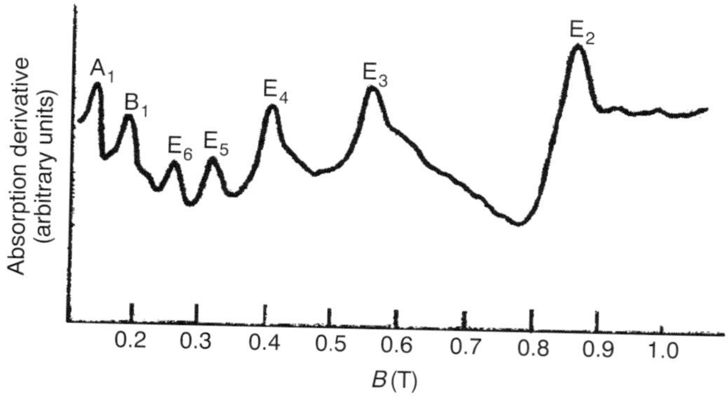

Fig. 6.22 Experimental cyclotron resonance curve for Al. The letters A, B and E refer to different extremal orbits, while the suffixes denote resonant absorption peaks associated with the same extremal orbit; such peaks are evenly spaced when plotted as $1/\boldsymbol{B}$. (After Moore and Spong (1962). Reprinted with permission from *Phys. Rev.* **125**, 846. © 1962. The American Physical Society).

the only important ones in practice are the extremal orbits enclosing the maximum (or minimum) area S of $\boldsymbol{k}$-space and for which $\partial S/\partial k_z = 0$. For these, there are many different neighbouring orbits with approximately the same cyclotron frequency due to the relative constancy of the energy surface with k_z in the vicinity of the plane of extremal orbits.

6.3.3.2 *Galvanomagnetic and thermomagnetic effects*

Electron transport in the presence of a magnetic field, as well as an electric field or a temperature gradient, can be described in terms of galvanomagnetic or thermomagnetic behaviour, respectively, in a similar manner to thermoelectric effects (§6.3.2.3). However, a difference is that, because the magnetic field imposes a preferred direction on a sample even if the material itself is isotropic, the Onsager coefficients in the transport equations (cf. eqns. (6.87)) are always tensors.

Perhaps one of the simplest galvanomagnetic phenomena to understand is the Hall effect in which orthogonal electric and magnetic fields generate in a conductor a secondary electric field, the Hall field, that is normal to both primary fields. Consider the experimental geometry illustrated in Fig. 6.23, in which a constant magnetic field $\boldsymbol{B}$ is applied in the z-direction and an electric field $\boldsymbol{E}$ is applied along the x-direction so that a current density $\boldsymbol{j}$ flows in the same direction. The Lorentz force (eqn. (6.109)) causes a deflection of the trajectory of the electrons (moving in the $-x$-direction in the absence of a magnetic field) in the y-direction, generating negative and positive space charges on the $-y$ and $+y$ faces of the sample, respectively. This charge build-up causes an electric field, the Hall field $\boldsymbol{E}_{\mathrm{H}}$, to be established in the $-y$ direction, the associated force $-e\boldsymbol{E}_{\mathrm{H}}$

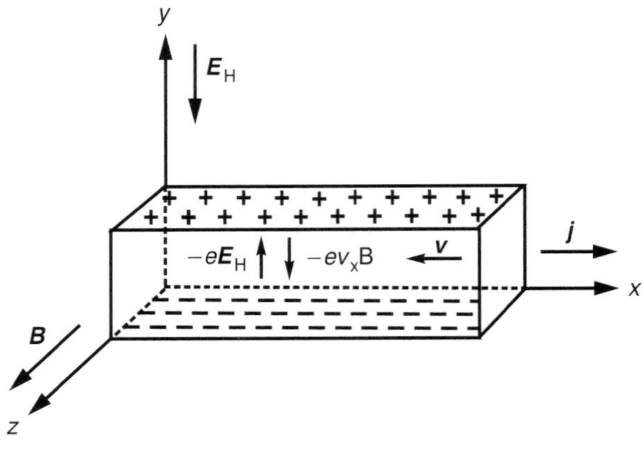

Fig. 6.23 Experimental geometry of the Hall effect. The Hall field $\boldsymbol{E}_\text{H}$ is established due to charge build-up on opposing faces in the direction normal to the applied magnetic and electric fields, resulting from the Lorentz force acting on the moving electrons. In steady state, the force $-e\boldsymbol{E}_\text{H}$ balances the Lorentz force.

opposing the Lorentz force. In steady state, there is therefore no net force on the electrons in the y-direction, and hence no drift-velocity component v_y in that direction; the current flow is thus only in the x-direction.

In steady state, the y-component of the equation of motion given by eqn. (6.7) reduces to

$$0 = -e(E_y - v_x B) \tag{6.124}$$

since $v_y = 0$. Thus,

$$E_y = v_x B = -\frac{j_x B}{ne}. \tag{6.125}$$

The transverse component of the resistivity tensor ρ in $\boldsymbol{E} = \rho \cdot \boldsymbol{j}$, given by

$$\rho_{yx} = -\rho_{xy} \equiv -\frac{E_y}{j_x} = \frac{B}{ne}, \tag{6.126}$$

is called the Hall resistivity.

The Hall coefficient, R_H is defined by the relation

$$\boldsymbol{E}_\text{H} = R_\text{H} \boldsymbol{B} \times \boldsymbol{j}, \tag{6.127}$$

and so, from eqn. (6.125),

$$R_\text{H} = \frac{E_y}{j_x B_z} = \frac{\rho_{xy}}{B_z} = -\frac{1}{ne}. \tag{6.128}$$

Therefore, if *electrons* moving in a single band make up the current, the Hall coefficient is *negative*, reflecting the sign of the charge carried by the electrons, and R_H is inversely proportional to the electron concentration n. If, on the other hand, *holes* carry the current, the velocity direction of the holes is in the opposite direction to that of the electrons, i.e. the same as that of the current, and consequently the Lorentz force is unchanged. Consequently, the Hall field for holes is in the *opposite* direction to that for electrons, and hence the Hall coefficient is correspondingly *positive*, i.e. the same sign as the charge carried by the holes (§6.2.2). Hence, Hall-effect measurements can be used to determine the sign of the charge carriers in conductors, as well as the carrier concentration. Note that the Hall coefficient will be larger for semiconductors than for metals because the carrier concentration in semiconductors (§6.5.1.1) is much smaller than the conduction-electron density in metals.

By analogy with the expression for the d.c. electrical conductivity written (eqn. (6.11)) as $\sigma_0 = ne\mu_e$ for electrons (or a corresponding relation for holes), where μ is the mobility, the Hall mobility can be defined in terms of σ_0 and R_H via the expression (cf. eqn. (6.128)):

$$\mu_H = |R_H|\sigma_0. \tag{6.129}$$

Unless the electron-scattering relaxation time τ is constant for all electrons (i.e. independent of $\boldsymbol{k}$ or energy), the Hall and conductivity mobilities are not the same numerically, but are in a ratio $r_H = <\tau^2> / <\tau>^2$ of order unity: $\mu_H = r_H\mu_e$.

The behaviour of the Hall effect in real metals is often more complex than the previous simple analysis would predict. For instance, although eqn. (6.128) implies that the Hall coefficient should be independent of the applied magnetic field, in reality R_H can depend markedly on $\boldsymbol{B}$ (see Fig. 6.24), e.g. if more than one band contributes to the current (see Problem 6.12). However, at sufficiently high values of magnetic field, in practice such that $\omega_c\tau \gg 1$ (where ω_c is the cyclotron frequency, proportional to B — see eqn. (6.117)), the Hall coefficient reaches a saturation limit; high-field values of the Hall coefficient for a number of metals are given in Table 6.2. It can be seen from the table that the alkali metals behave much as expected: the charge carriers are electrons, with one conduction electron per atom. This is because such metals have a single, free-electron-like conduction band, with a spherical Fermi surface. The Group I_B noble metals also behave more-or-less as expected, although the carrier concentration per atom, n/N, is not exactly unity as a result of the distortions of the Fermi surface from sphericity (see Fig. 5.25b).

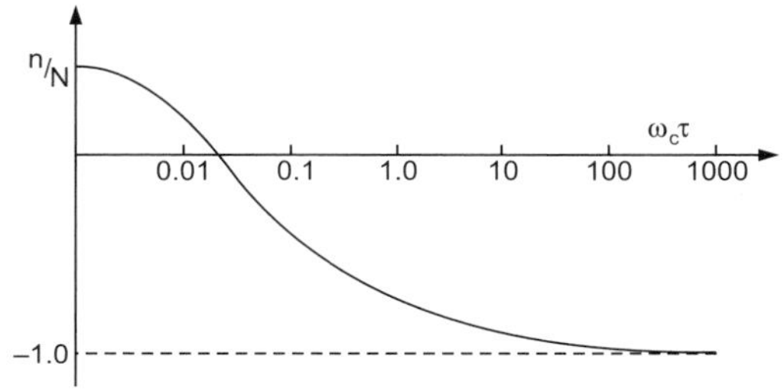

Fig. 6.24 Magnetic-field dependence of the Hall coefficient for Al. The quantity $n/N = -1/R_H Ne$, where N is the atomic concentration, gives the effective number of conduction electrons per atom. The quantity $\omega_c\tau = eB\tau/m_e$ is a measure of the magnetic field. (After Lück (1966). Reproduced by permission of Akademie Verlag GmbH.)

Table 6.2 Hall coefficient, R_H, for various metals

Metal	*Group*	$1/(R_H Ne)$
Li	IA	−0.78
Na		−1.00
K		−1.00
Cu	IB	−1.37
Ag		−1.19
Au		−1.47
Be	IIA	+0.21
Al	IIIB	+1.01
In		+1.02

The quantity $1/(R_H Ne)$, where N is the atomic density, gives the carrier concentration per atom, n/N. A negative value of $1/(R_H Ne)$ indicates that the charge carriers are electrons, and a positive value indicates that holes are the dominant carriers. Data after Kittel (1996). Reproduced by permission of John Wiley & Sons Inc.

The *positive* values of $1/(R_H Ne)$ for the Group II_A and III_B metals cannot be understood at all within the context of the simple Drude free-electron picture (§6.1); instead, they are indicative that *holes* (§6.2.2) are the dominant carriers. In the case of the Group II_A metals, e.g. Be and Mg, band overlap results in semi-metallic behaviour, in which the first Brillouin zone is not completely filled by the two electrons per atom available, but instead some spill-over of electrons into a higher band (second zone) occurs, leaving some holes in the first zone (§5.2.5). A related effect occurs in Al: two of the three available valence electrons per atom completely fill the first zone, but the third electron can occupy hole-like states in the second zone and electron-like states in the third zone (see Fig. 5.24). Thus, if a concentration n of electrons, equivalent to one per atom, is available to fill states in the second (II) and third (III) zones, then

$$n_e^{II} + n_e^{III} = n. \tag{6.130}$$

However, the second zone can hold two electrons per atom, i.e. the total density of electron and hole states must satisfy the relation

$$n_e^{II} + n_h^{II} = 2n, \tag{6.131}$$

and hence from eqns. (6.130) and (6.131):

$$n_e^{III} - n_h^{II} = -n. \tag{6.132}$$

The negative sign on the right-hand side indicates that the net effect of the band filling is equivalent to one hole per atom.

For a material in which *two* bands together contribute to the current, say one electron-like and the other hole-like, the high-field Hall coefficient can be written as (see Problem 6.12):

$$R_H = -\frac{1}{(n_e - n_h)e}. \tag{6.133}$$

Thus, from eqn. (6.132), the high-field Hall coefficient of Al is predicted to behave as if one free hole, rather than three free valence electrons, per atom, is the current charge carrier, in agreement with experiment (Fig. 6.24).

In 2D conducting systems, the behaviour of the Hall effect is very different from that described above for the case of 3D systems; the 2D Hall conductivity exhibits a series of steps with increasing magnetic field. This quantum Hall effect is discussed in §8.4.5.

The transverse magnetoresistivity with $\boldsymbol{B}$ perpendicular to the current flow (as distinct from the longitudinal magnetoresistivity for which $\boldsymbol{B}$ is parallel to the current) is given by the expression

$$\rho_{xx}(B) = \frac{E_x}{j_x}. \tag{6.134}$$

For the simple case of a free-electron gas, the steady-state equation of motion given by eqn. (6.8) predicts a resistivity (inverse conductivity, cf. eqn. (6.9)) that is *independent* of magnetic field. For *real metals*, on the other hand, a field-dependent magnetoresistivity is often found because of an anisotropic Fermi surface (the cause of a finite *longitudinal* magnetoresistance) or a non-constancy (energy dependence) of the relaxation time τ. A field dependence is also exhibited if more than a single band contributes to the current (see Problem 6.12); the (positive) magnetoresistivity increases quadratically with magnetic field, and reaches a saturation value at sufficiently high magnetic fields, unless the material is compensated (equal number of electrons and holes in two bands), in which case the magnetoresistivity increases without limit with increasing field. In the case of *open* electron orbits on the Fermi surface (see Fig. 6.18), the magnetoresistivity also increases without limit (see Ashcroft and Mermin (1976) for further details).

Extremely large, *negative* magnetoresistance behaviour, called colossal magnetoresistance (CMR) has been discovered in certain mixed-valence manganite compounds, $Ln_{1-x}A_xMnO_3$($Ln = La^{3+}, Pr^{3+}$, etc.; $A = Ca^{2+}, Sr^{2+}$, etc.) for $0.15 < x < 0.45$; these materials contain both $Mn^{3+}(d^4)$ and $Mn^{4+}(d^3)$ ions, with the average number of d-electrons per Mn atom equal to $4 - x$. At elevated temperatures, these materials are metallic and magnetically ordered, i.e. ferromagnetic (§7.2.5); below the Curie–Weiss temperature, θ_{CW} (§7.2.5.2), the materials are electrically insulating and magnetically disordered. Therefore, they undergo a metal–insulator transition (§5.6.3) at θ_{CW} with a peak in the temperature dependence of the electrical resistivity (Fig. 6.25a). The mixed-valence state of the Mn ions is responsible both for the ferromagnetism (via 'double exchange' — see §7.2.5.1) and also for the metallic behaviour above θ_{CW}. Three of the four d-electrons in the Mn^{3+} ions occupy the tightly bound set of orbitals d_{xy}, d_{xz}, d_{yz} (t_{2g} states — see §7.2.5.1) and the remaining, electronically active electron resides in the set of orbitals $d_{x^2-y^2}, d_{3z^2-r^2}$ (e_g states) which are doubly degenerate (neglecting spin). This degeneracy leads to a Jahn–Teller structural distortion, resulting in unequal Mn–O nearest-neighbour distances, driven by a lowering of the electronic energy associated with the lifting of this degeneracy; strong polaronic effects (a coupling of electronic and ionic motions—see §6.6) causes localization (§6.7) of the electrons and hence the electrically insulating behaviour below θ_{CW}.

Application of an external magnetic field to the mixed-valence manganites at a temperature close to, but slightly below, the peak in the $\rho(T)$ curve causes a very large drop in electrical resistance, i.e. a negative magnetoresistance ratio $\Delta R/R_B = (R_B - R_{B=0})/R_B$, where R_B is the resistance at a magnetic flux density, B. Values of $\Delta R/R_B \simeq -127\,000\%$ (a factor of more than a thousand) can be attained for

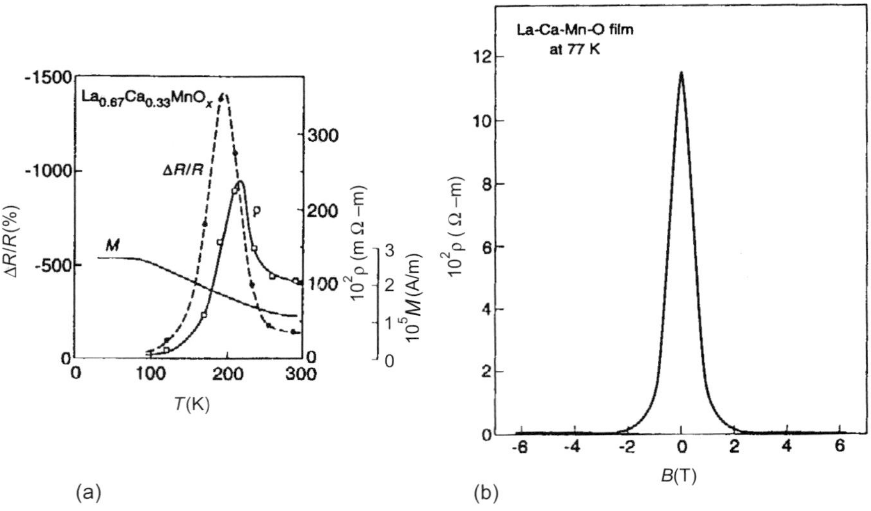

Fig. 6.25 (a) Temperature dependence of the electrical resistivity $\rho(T)$, and the magnetization, $M(T)$, for $La_{0.67}Ca_{0.33}MnO_x$. The negative magnetoresistance ratio, $\Delta R/R_B = (R_B - R_{B=0})/R_B$ is also shown. (b) The magnetoresistance, $\rho(B)$, for an La-Ca-Mn-O film at 77 K. The magnetoresistance ratio in this case is − 127 000%. (a), (b) Reprinted with permission from Jin *et al.*, *Science* **264**, 413. © 1994 American Association for the Advancement of Science.

heat-treated films of $La_{0.67}Ca_{0.33}MnO_x$ (Fig. 6.25b), grown epitaxially on $LaAlO_3$ substrates by laser ablation (§1.1.1) (Jin *et al.* (1994)). This colossal magnetoresistance is *isotropic*, unlike conventional magnetoresistance; that is, it does not depend on the relative orientations of the applied magnetic field and electrical current flow.

In general, galvano- and thermomagnetic properties can be described in terms of transport equations for the electrical current and heat fluxes written in an analogous form to eqns. (6.87) for the case of thermoelectric effects, but expressed as functions of electric and magnetic fields, $\boldsymbol{E}$, $\boldsymbol{B}$, and temperature gradient ∇T, viz.

$$\boldsymbol{j}=\alpha_{11}\boldsymbol{E}+\beta_{11}\boldsymbol{B}\times\boldsymbol{E}+\gamma_{11}\boldsymbol{B}(\boldsymbol{B}\cdot\boldsymbol{E})+\alpha_{12}\nabla T+\beta_{12}\boldsymbol{B}\times\nabla T+\gamma_{12}\boldsymbol{B}(\boldsymbol{B}\cdot\nabla T), \quad (6.135a)$$

$$\boldsymbol{J}_Q=\alpha_{21}\boldsymbol{E}+\beta_{21}\boldsymbol{B}\times\boldsymbol{E}+\gamma_{21}\boldsymbol{B}(\boldsymbol{B}\cdot\boldsymbol{E})+\alpha_{22}\nabla T+\beta_{22}\boldsymbol{B}\times\nabla T+\gamma_{22}\boldsymbol{B}(\boldsymbol{B}\cdot\nabla T). \quad (6.135b)$$

In the simple case where $\boldsymbol{B} = B_z$ is perpendicular to an applied electrical field $\boldsymbol{E} = E_x$ and a temperature gradient $\nabla T = \partial T/\partial x$, eqn. (6.135a) for the electrical current flux reduces to

$$j_x = \alpha_{11}E_x - \beta_{11}B_zE_y + \alpha_{12}\frac{\partial T}{\partial x} - \beta_{12}B_z\frac{\partial T}{\partial y}, \quad (6.136a)$$

$$j_y = \alpha_{11}E_y + \beta_{11}B_zE_x + \alpha_{12}\frac{\partial T}{\partial y} + \beta_{12}B_z\frac{\partial T}{\partial x}, \quad (6.136b)$$

$$j_z = 0, \quad (6.136c)$$

with corresponding equations for the heat flux, $\boldsymbol{J}_Q$. Secondary electric fields and thermal gradients in the y-direction appear as a result of galvano-and thermomagnetic effects. The coefficients α_{11} and β_{11}, for example, are given approximately (Madelung (1978)) by

$$\alpha_{11} = ne\mu_e/(1 + (\omega_c\tau)^2), \quad \beta_{11} = \mu_e\alpha_{11}, \tag{6.137}$$

where n and μ_e are the electron concentration and mobility, respectively.

A number of different magnetic-field-induced phenomena can be distinguished. The appearance of an electric field normal to both temperature gradient and magnetic field is termed the Nernst effect, with coefficient

$$Q_N = \frac{E_y}{B_z(\partial T/\partial x)}. \tag{6.138}$$

The creation of a temperature gradient perpendicular to a magnetic field and a primary temperature gradient is known as the Righi–Leduc effect, with coefficient

$$S_{RL} = \frac{(\partial T/\partial y)}{B_z(\partial T/\partial x)}. \tag{6.139}$$

The Ettingshausen effect is the appearance of a temperature gradient normal to a magnetic field and an electric field, for which the coefficient is

$$P_E = \frac{(\partial T/\partial y)}{j_x B_z}. \tag{6.140}$$

★

Superconductors 6.4

One of the most startling of all phenomena exhibited by solids is superconductivity, which is the sudden loss of all electrical resistance by certain metals when cooled below a certain superconducting critical temperature, T_c. This behaviour is illustrated schematically in Fig. 6.26, together with the temperature dependence of the resistivity of *normal* (i.e. non-superconducting) metals (§6.3.2.1) for comparison. In addition to this extraordinary behaviour, superconducting metals also exhibit a number of remarkable *magnetic* properties in the superconducting state, notably the complete exclusion of magnetic flux (the Meissner effect), at least for certain sample geometries: these properties will be discussed in §7.2.3.3. Superconductors merit a separate section from normal metals, not just because of the spectacular changes in properties that are exhibited at T_c, but because of the very different state adopted by the electrons in the superconducting phase, which is essentially a macroscopic quantum state.

Nearly 30 metallic elements become superconducting at atmospheric pressure (Table 6.3); yet others become superconductors at high pressures when transformed to a crystal structure that is not stable under ambient pressures, or sometimes when prepared in thin-film form (e.g. As, Ba, Bi, Cs, Ge, Se, Si, Te). Very many metallic alloys are superconductors, too (Table 6.3); in some cases (e.g. $CeCu_2Si_2$ or UPt_3), alloys can be superconducting even if the constituent elements are not. Moreover, superconducting behaviour is *not* restricted to crystalline materials exhibiting translational periodicity in their structure; *amorphous metals*, such as $Zr_{70}Pd_{30}$, can also exhibit superconductivity. Finally, superconductors need not be composed of metallic elements at all. Certain inorganic solids, such as the polymeric material $(SN)_x$ are superconducting. Even some low-dimensional (e.g. 1D) materials (see §8.3.1) formed by a stacking of *organic molecules*, such as tetramethyltetraselenofulvalene (TMTSF — see Fig. 8.17 for the structure of a related sulphur-analogue molecule, TTF (tetrathiafulvalene)), in which one-dimensional bands form by overlap of π-orbitals between adjacent molecules, and for which

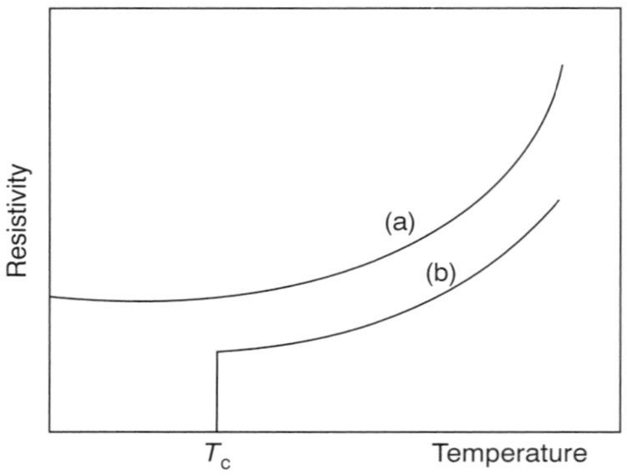

Fig. 6.26 Schematic temperature dependence, $\rho(T)$, of the electrical resistivity at low temperatures for: (a) a normal metal, with $\rho(T) = \rho_d + cT^5$; (b) a superconductor, with a superconducting-transition temperature at T_c (in the absence of a magnetic field).

Table 6.3 Superconducting elements and non-oxide compounds, and their superconducting transition temperatures

Element	T_c (K)	*Element*	T_c (K)	*Compound*	T_c (K)
Al	1.14	Pa	1.4	AuBe	2.64
Am	0.85	Pb	7.193	$CeCu_2Si_2^*$	0.65
Be	0.026	Re	1.4	CuS	1.62
Cd	0.56	Rh	0.0003	$ErRh_4B_4^*$	8.7
Ga	1.091	Ru	0.51	$HoMo_6S_8^*$	1.8
Hf	0.12	Sn	3.722	La_3In	10
α–Hg	4.153	Ta	4.483	Nb_3Al	18.6
β–Hg	3.95	Tc	7.77	Nb_3Ga	20.3
In	3.404	Th	1.368	Nb_3Ge	23.2
Ir	0.14	Ti	0.39	Nb_3Sn	18
α-La	4.88	Tl	2.39	NbTi	10
β-La	6.00	V	5.38	$(SN)_x$	0.26
Mo	0.915	W	0.012	UPt_3^*	0.54
Nb	9.50	Zn	0.875	V_3Ga	16.5
Os	0.655	Zr	0.546	V_3Si	17.1

All materials are 'conventional' (BCS-type) superconductors, except those marked with * which are 'heavy-fermion' superconductors. Data after Burns (1992) (Reproduced by permission of Academic Press, Inc.) and Kittel (1996) (Reproduced by permission of John Wiley & Son Inc.)

metallic behaviour is induced by the inclusion of strongly electron-withdrawing 'acceptor' groups such as the perchlorate ion (ClO_4^-), are also superconductors, albeit generally with rather low values of $T_c(\simeq 1K)$. However, the charge-transfer salt $(ET)_2Cu[N(CN)_2]Br$ (where ET refers to the molecule bis(ethylenedithia)tetrathiafulvalene, related to the molecule TTF) has a T_c of 11.6 K.

It is to be noticed from Table 6.3 that the values of T_c for these materials range up to a maximum temperature of 23.2 K for Nb_3Ge. This value appeared to be the maximum attainable until the discovery in 1986 of the 'high-T_c' oxide (cuprate) superconductors with values of $T_c \geqslant 100$ K. These materials are discussed separately in §6.4.3, since it appears that the mechanism underlying their superconducting behaviour may be very different from that responsible for the behaviour of (most of) the superconductors listed in Table 6.3.

It is instructive also to examine the possible reasons for the non-appearance of superconductivity in elements other than those given in Table 6.3. Superconductivity is a phenomenon in which the conduction electrons in the vicinity of the Fermi level in a metal can attain a lower energy state by forming paired-electron states that give rise to zero electrical resistivity (infinite conductivity) below T_c. Thus, a prerequisite is that the material above T_c be a metal and not an insulator/semiconductor (although elements such as Si or Ge can be transformed into high-density crystalline phases that are metallic by the application of pressure). Paradoxically, it is observed that the normal (non-superconducting) state should not be too *good* a metal: metallic elements, such as the alkali metals, Cu, Ag and Au, which have very high electronic conductivities, are *not* observed to become superconducting, although they may possibly do so at temperatures much below those that are presently easily attainable (in the mK range).

It appears that magnetism and superconductivity are generally mutually incompatible. It is noteworthy that those transition-metal and rare-earth elements that possess

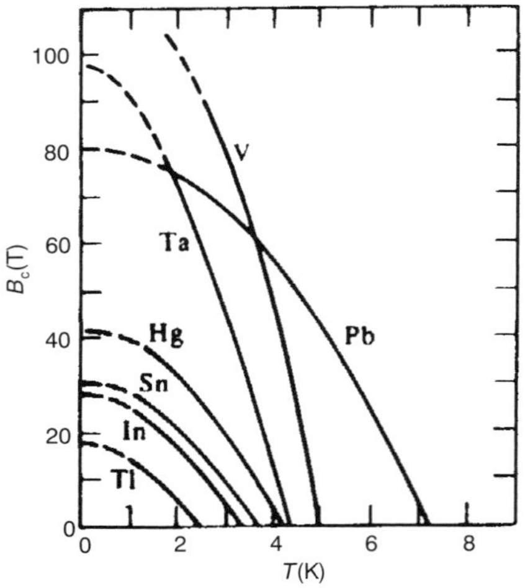

Fig. 6.27 Experimental curves of the critical magnetic flux density, B_c, as a function of temperature for a number of type-I superconducting elements. Materials are superconducting for $B < B_c(T)$, i.e. below the curves, and normal metals above the curves. Note that B_c goes to zero at the (zero-field) superconducting temperature, T_c. (After Burns (1992). Reproduced by permission of Academic Press, Inc.)

magnetic moments by virtue of their electron-spin configurations (see §7.2.4.5) are *not* superconductors. Moreover, the application of an external magnetic field can destroy the superconducting state. There is a critical value of magnetic flux density†, B_c, above which the superconducting state reverts to the normal metallic state; there is a corresponding critical current density, j_c, that generates B_c and which destroys the superconductivity. B_c itself is a function of temperature, vanishing at T_c (Fig. 6.27). The temperature dependence of $B_c(T)$ is given approximately by the empirical expression

$$B_c(T) \simeq B_c(0)[1 - (T/T_c)^2]. \tag{6.141}$$

It should be remarked that the behaviour represented by eqn. (6.141) and illustrated in Fig. 6.27 is the simplest possible; materials exhibiting such behaviour are termed type-I superconductors. There are also materials called type-II, notably Nb, intermetallic alloys and the oxide high-T_c superconductors, that exhibit *two* critical magnetic flux densities $B_c(T)$. Above the upper value, $B_{c2}(T)$, the material is in the normal state, and below the lower value, $B_{c1}(T)$, the material is completely superconducting; in between these two values, the material is in a mixed (or vortex) state, in which normal and superconducting domains coexist. Type-I and type-II materials are differentiated by their *magnetic* behaviour; the Meissner effect is complete in type-I superconductors and type-II superconductors at $B < B_{c1}$ (all magnetic flux is expelled),

† It should be noted that very often in the literature the critical magnetic flux density B_c is re-expressed in terms of the magnetic field strength $\boldsymbol{H}_c = \boldsymbol{B}_c/\mu_o$, where μ_o is the vacuum permeability. This relationship is valid for the geometry of a long cylinder parallel to the field.

whereas it can be incomplete in type-II superconductors in the mixed or vortex state, where domains of normal material, through which vortices of magnetic flux can pass, from within the flux-free superconducting matrix. Further details about type-II behaviour are to be found in §7.2.3.3.

Another manifestation of the destructive influence of a magnetic field on the superconducting state is exhibited by re-entrant superconductors in which, at a temperature T_{c2} below that at which the material transforms from the normal to the superconducting state (T_{c1}), the material reverts to the normal state as a result of the formation of a state that is magnetically ordered (e.g. ferromagnetic — see §7.2.5). This behaviour is exhibited by the Chevrel phases $REMo_6X_8$ (where RE = rare earth; X = S or Se), e.g. $HoMo_6S_8$ for which the superconducting transition is at $T_{c1} \simeq 1.8$ K and the re-entrant transition is at $T_{c2} \simeq 0.7$ K, and by compounds with the formula MRh_4B_4 (M = Y, Th or RE), e.g. $ErRh_4B_4$, with $T_{c1} \simeq 8.7$ K and $T_{c2} \simeq 1$ K.

6.4.1 The superconducting state

The phenomenon of superconductivity is perhaps the most remarkable instance of the breakdown of the independent-electron approximation (§5.6), for it is believed that the superconducting phase consists of a macroscopic coherent quantum state consisting of mutually interacting *pairs* of electron states. This viewpoint will be explored more thoroughly in §6.4.1.2.

*6.4.1.1 Thermodynamic aspects

The transition from the normal metallic state to the superconducting state at T_c, in zero magnetic field, is a *second-order* thermodynamic phase transition involving the electron system: there is no change in the atomic structure, for example, at the transition. This thermodynamic change of state is evident, for instance, in the temperature dependence of the heat capacity C_p (Fig. 6.28): there is a discontinuity in C_p at T_c between the normal and superconducting states, but no change in the sample volume at T_c that would be indicative of a first-order phase transition. This behaviour can be understood from the following thermodynamic argument which shows that, in the superconducting state, the entropy of the (electron) system is *lower* than in the normal state, i.e. the electrons are more *ordered* in the superconducting phase.

The Gibbs free energy per unit *volume*, g, is related to the internal energy per unit volume, u, and entropy per unit volume, s, via

$$g = u - Ts, \tag{6.142}$$

where the pV term has been neglected. If, in the presence of an applied magnetic field of flux density $\boldsymbol{B}_a$, the associated magnetic energy is taken to be $-\boldsymbol{M} \cdot d\boldsymbol{B}_a$, where $\boldsymbol{M}$, the magnetization of a material, is the magnetic dipole moment per unit volume, a change in the free energy is given by

$$dg = -\boldsymbol{M} \cdot d\boldsymbol{B}_a - s dT. \tag{6.143}$$

Integration of this relation gives the magnetic-field dependence of the free energy:

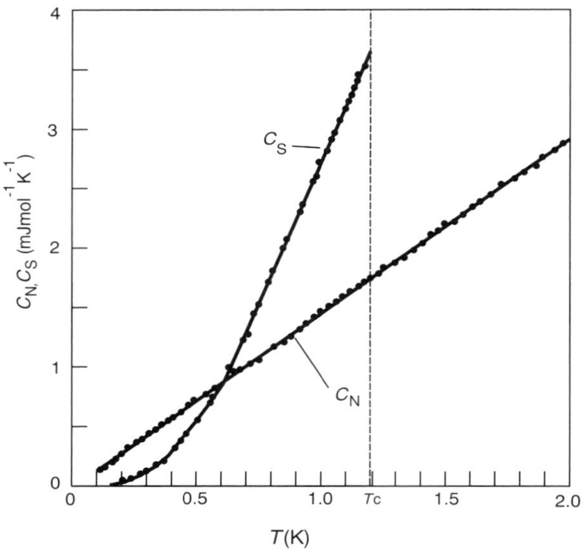

Fig. 6.28 Heat capacity of normal (N) and superconducting (S) Al. The measurements for the N-state below T_c were made by applying an external magnetic field greater than the critical field, B_c. The electronic contribution to the heat capacity is dominant in this temperature régime, because the phonon contribution is small due to the large value of the Debye temperature, θ_D. Note the discontinuity in heat capacity at T_c, indicative of a second-order phase transition and the very different temperature dependence of c_S (exponential) and c_N (linear). (After Phillips (1959). Reprinted with permission from *Phys. Rev.* **114**, 676. © 1959. The American Physical Society)

$$g(B_a, T) = g(0, T) - \int_0^{B_a} \boldsymbol{M} \cdot d\boldsymbol{B}_a. \tag{6.144}$$

The *macroscopic* magnetic flux density $\boldsymbol{B}$ and field $\boldsymbol{H}$ in, and magnetization $\boldsymbol{M}$ of, a sample are related via

$$\boldsymbol{B} = \mu_0(\boldsymbol{H} + \boldsymbol{M}); \tag{6.145}$$

see §7.2.2 for a discussion of internal magnetic fields in solids. For the sample geometry of a long cylinder, $\boldsymbol{B}_a = \mu_0 \boldsymbol{H}$, where $\boldsymbol{H}$ is the macroscopic field inside the sample (see § 7.2.2). For a *type-I superconductor*, due to the Meissner effect (see §7.2.3.3), the flux density inside the material is *zero*, $\boldsymbol{B} = 0$, in the superconducting state, and hence, from eqn. (6.145), $\boldsymbol{M} = -\boldsymbol{H}$. Thus, eqn. (6.144) can be rewritten for the superconducting state as

$$\begin{aligned} g_S(B_a, T) &= g_S(0, T) + \int_0^{B_a} \frac{B_a \mathrm{d}B_a}{\mu_0} \\ &= g_S(0, T) + \frac{B_a^2}{2\mu_0}. \end{aligned} \tag{6.146}$$

The quantity $B_a^2/2\mu_0$ is the extra magnetic energy stored in the field as a result of the exclusion of the flux from the superconductor because of the Meissner effect.

At the transition between superconducting and normal states, occurring at a temperature T and critical magnetic field $B_c(T)$, the Gibbs free energies of the two phases must be identical, i.e.

$$\begin{aligned} g_S(B_c, T) &= g_N(B_c, T) \\ &= g_N(0, T), \end{aligned} \tag{6.147}$$

where the last relation holds if any magnetic behaviour of the normal state is neglected. Thus, from eqn. (6.146),

$$g_S(0, T) - g_N(0, T) = -\frac{B_c^2}{2\mu_0}. \tag{6.148}$$

The superconducting state is therefore thermodynamically more stable than the normal state in zero field at temperatures below T_c.

The difference in entropy per unit volume, Δs, between superconducting and normal states can be obtained from eqn. (6.148) using the relation $s = -(\partial g/\partial T)$ valid at constant (e.g. zero) applied field (cf. eqn. (6.143)):

$$\Delta s = s_S - s_N = \frac{1}{2\mu_0}\frac{\mathrm{d}(B_c^2)}{\mathrm{d}T} = \frac{B_c}{\mu_0}\frac{\mathrm{d}B_c}{\mathrm{d}T}. \tag{6.149}$$

At the superconducting transition temperature, T_c, Δs is *zero* since $B_c(T_c) = 0$ (see Fig. 6.27). Since an extensive thermodynamical variable (entropy) is continuous at the superconducting transition, it is a second-order phase transition. For $T < T_c$, the gradient $\mathrm{d}B_c/\mathrm{d}T$ is negative (Fig. 6.27) and hence the superconducting state has a *lower* entropy than the normal state, i.e. $\Delta s < 0$—see Fig. 6.29.

The heat capacity per unit volume is given by $c = T\partial s/\partial T$, and so from eqn. (6.149):

$$\begin{aligned} \Delta c = c_S - c_N &= \frac{T}{2\mu_0}\frac{\mathrm{d}^2(B_c^2)}{\mathrm{d}T^2} \\ &= \frac{T}{\mu_0}\left[B_c\frac{\mathrm{d}^2 B_c}{\mathrm{d}T^2} + \left(\frac{\mathrm{d}B_c}{\mathrm{d}T}\right)^2\right]. \end{aligned} \tag{6.150}$$

At T_c, even for $B_c = 0$, there is therefore a discontinuous jump in Δc because $(\mathrm{d}B_c/\mathrm{d}T)^2 > 0$. These conclusions also hold for type-II superconductors but, because they show an incomplete Meissner effect, the above analysis is no longer valid.

6.4.1.2 *Electron pairing: the BCS model*

It is evident from the preceding thermodynamic discussion that the superconducting state of the electron sub-system is more ordered (i.e. has a lower entropy) than the normal metallic state at the same temperature. In other words, the free-electron gas is *not* the state with the lowest free energy below T_c (cf. eqn. (6.148)). This implies that there must exist an additional *attractive* interaction between electrons that can overcome the residual screened inter-electron repulsive interaction, in order for the superconducting state of the electrons to become more ordered than a free-electron gas. One candidate for such an attractive inter-electron interaction is the electron–phonon (Fröhlich) interaction.

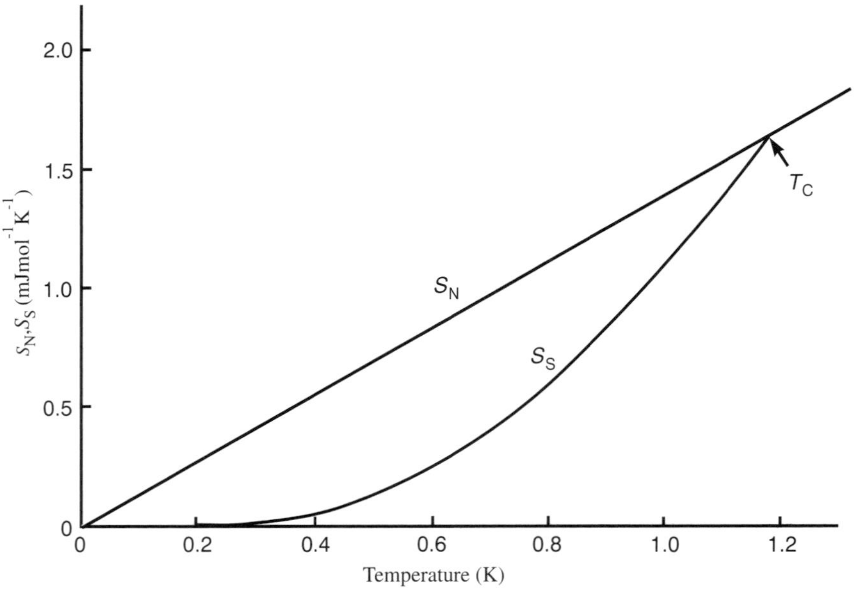

Fig. 6.29 Temperature dependence of the entropy of Al in the normal (N) and superconducting (S) states. Note that the entropies of both states are equal at T_c, and the energy of the superconducting state is the lower for $T < T_c$. (After Kittel (1996). Reproduced by permission of John Wiley & Sons Inc.)

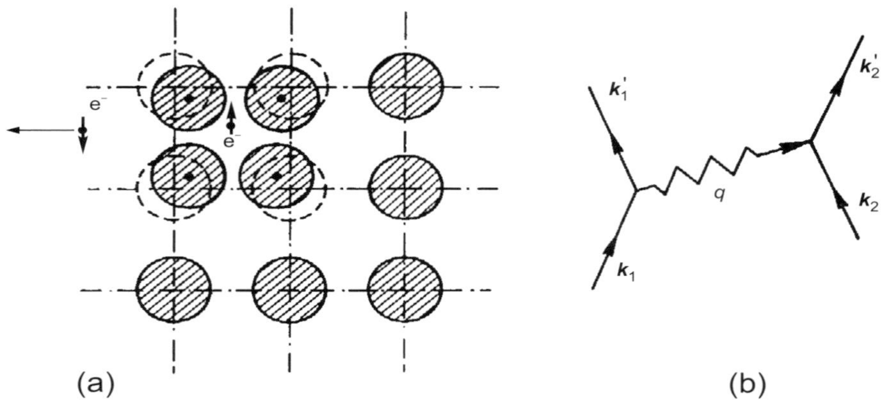

Fig. 6.30 The origin of the attractive electron–phonon interaction in solids. (a) Real-space picture, in which an electron causes a local polarization of the structure by attracting positive ions in its vicinity. A second electron is attracted to this region of higher ionic charge density which persists after the first electron has moved away. (b) Scattering picture, in which a virtual phonon is exchanged between two electrons, causing them to be scattered from states $\boldsymbol{k}_1$ to $\boldsymbol{k}_1'$, and $\boldsymbol{k}_2$ to $\boldsymbol{k}_2'$, respectively.

The physical origin of the attractive electron–phonon interaction can be understood by reference to Fig. 6.30. An electron will cause a local polarization of a material by attracting positive ions in its vicinity, and consequently will leave a trail of such a lattice distortion behind it as the electron moves through the material. The associated local increase in positive-ion density will have an attractive influence for another electron moving in the vicinity, resulting in an effective weak attractive interaction between the two electrons. This interaction is *retarded* because of the very different time-scales associated with electronic and ionic motion: the lattice deformation correspondingly reaches a maximum at a distance *behind* the first electron of

$$d \simeq v_F \frac{2\pi}{\omega_D}, \tag{6.151}$$

where v_F is the velocity of electrons at the Fermi energy $\mathscr{E}_F$ and the Debye frequency ω_D (eqn. (4.25)) is a measure of the time response of the ionic sub-system. Since $v_F \simeq 2 \times 10^6$ m s^{-1} (for $\mathscr{E}_F \simeq 10$ eV — see Table 5.1) and $\omega_D \simeq 5 \times 10^{13}$ s^{-1} (values appropriate for Al), eqn. (6.151) implies that the average separation of electron pairs mutually attracted by the electron–phonon interaction is $d \simeq 2.5 \times 10^{-7}$m $= 2500$ Å. At such large separations, the Coulomb repulsion between the electrons is completely screened (§5.6.1).

This electron–phonon interaction can also be viewed from the standpoint of electron scattering. An electron in a crystal scatters from a state with wavevector $\boldsymbol{k}_1$ to another with wavevector $\boldsymbol{k}_1'$ by emitting a phonon with wavevector $\boldsymbol{q}$ (§6.3.1.3). A second electron can absorb this phonon and correspondingly be scattered from state $\boldsymbol{k}_2$ to $\boldsymbol{k}_2'$. Thus, by conservation of crystal momentum:

$$\boldsymbol{k}_1 = \boldsymbol{k}_1' + \boldsymbol{q}, \tag{6.152a}$$

$$\boldsymbol{k}_2 + \boldsymbol{q} = \boldsymbol{k}_2', \tag{6.152b}$$

and hence, by eliminating $\boldsymbol{q}$ between these two equations,

$$\boldsymbol{k}_1 + \boldsymbol{k}_2 = \boldsymbol{k}_1' + \boldsymbol{k}_2' \equiv \boldsymbol{k}_0. \tag{6.153}$$

Note that, although crystal momentum is conserved for the individual phonon emission/absorption processes (eqns. (6.152)), energy need *not* be conserved (although it must be conserved for the overall scattering process, eqn. (6.153)): a virtual phonon can be emitted and absorbed within a very short time that complies with the Heisenberg energy–time uncertainty relation.

The restriction set by eqn. (6.153) on possible scattering events between two electrons added to a Fermi gas of electrons and which, therefore, occupy states with energy $\mathscr{E} > \mathscr{E}_F$ at $T = 0$ K, is illustrated graphically in Fig. 6.31a. States in $\boldsymbol{k}$-space that can interact are restricted by the Pauli exclusion principle to a shell of width δk corresponding to electron energies between $\mathscr{E}_F$ and $\mathscr{E}_F + \hbar\omega_D$, and eqn. (6.153) is satisfied for two overlapping Fermi spheres whose centres are separated by $\boldsymbol{k}_0$, with the intersections between the spherical shells giving the allowed $\boldsymbol{k}$-states involved in the electron-scattering events. The number of such allowed states is maximized, as is correspondingly the strength of the virtual-phonon-mediated attractive electron–electron interaction, when $\boldsymbol{k}_0 = 0$ or $\boldsymbol{k}_1 = -\boldsymbol{k}_2$ (Fig. 6.31b). Such pairs of electron states with $\boldsymbol{k}_1 = -\boldsymbol{k}_2$ are called Cooper pairs. Cooper first showed that two electrons, interacting via any *attractive* potential,

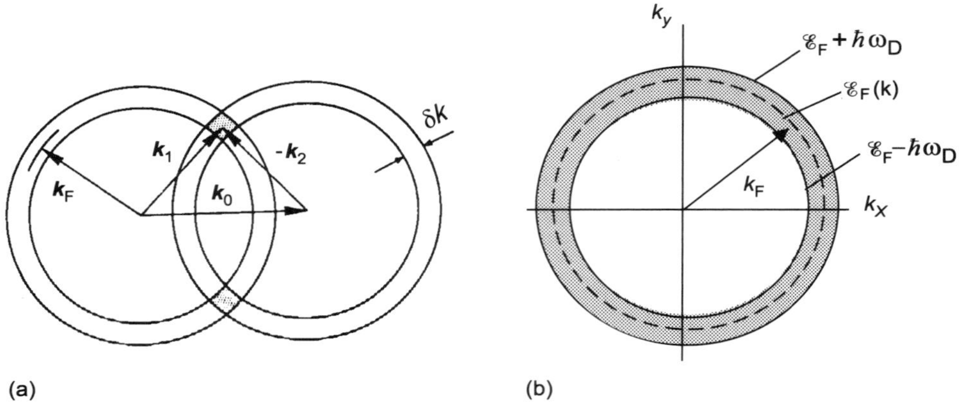

Fig. 6.31 (a) Geometric illustration in reciprocal space of the condition for virtual-phonon exchange between two electrons satisfying the condition $\boldsymbol{k}_1 + \boldsymbol{k}_2 = \boldsymbol{k}_1' + \boldsymbol{k}_2' = \boldsymbol{k}_0$. Two spherical shells, of radius $\boldsymbol{k}_F$ and thickness δk, and whose centres are separated by $\boldsymbol{k}_0$, intersect where the crystal-momentum conservation law is satisfied, The number of pairs of electrons with $\boldsymbol{k}_1$, $\boldsymbol{k}_2$ is proportional to the volume of intersection between the two spheres. (b) The condition for the maximum number of pairs of electrons, $\boldsymbol{k}_1$, $\boldsymbol{k}_2$, and hence the maximum phonon-mediated attractive electron–electron interaction, is when the two spheres in (a) coincide, i.e. when $\boldsymbol{k}_0 = 0$, or $\boldsymbol{k}_1 = \boldsymbol{k}_2$. The interaction potential is assumed, in the BCS model, to be constant $(= -\mathscr{V}_0)$ in the shaded region of reciprocal space, i.e. between shells with energies $\mathscr{E}_F \pm \hbar\omega_D$. For the case of only a single electron pair added to an electron distribution (Cooper pairing), only the states with $\mathscr{E} > \mathscr{E}_F$ are considered.

when added to a sea of electrons have a binding energy that is *less* than $2\mathscr{E}_F$; i.e. the ground state of the non-interacting Fermi gas is unstable to attractive electron–electron interactions (see Problem 6.15). This situation is markedly different from that for two isolated particles (e.g. *in vacuo*), where a minimum attractive potential is necessary to bind the pair.

The two-electron wavefunction corresponding to a stationary Cooper pair can be written as the product of two plane-wave states:

$$\phi(\boldsymbol{r}_1, \boldsymbol{r}_2) = \left\{\frac{1}{V^{1/2}} e^{i\boldsymbol{k}_1\cdot\boldsymbol{r}_1}\right\}\left\{\frac{1}{V^{1/2}} e^{i\boldsymbol{k}_2\cdot\boldsymbol{r}_2}\right\} = \frac{1}{V} e^{i\boldsymbol{k}\cdot(\boldsymbol{r}_1-\boldsymbol{r}_2)}, \tag{6.154}$$

where V is the normalization volume for the wavefunction and $\boldsymbol{k} = \boldsymbol{k}_1 = -\boldsymbol{k}_2$. A general expression for the two-particle wavefunction can be written as

$$\phi(\boldsymbol{r}_1 - \boldsymbol{r}_2) = \frac{1}{V}\sum_k p(\boldsymbol{k}) e^{i\boldsymbol{k}\cdot(\boldsymbol{r}_1-\boldsymbol{r}_2)}, \tag{6.155}$$

where $|p(\boldsymbol{k})|^2$ is the probability of there being one electron in state $\boldsymbol{k}$ and another in state $-\boldsymbol{k}$. If the wavefunction is *spherically symmetric*, i.e. $\phi(\boldsymbol{r}_1, \boldsymbol{r}_2) = \phi(|\boldsymbol{r}_1 - \boldsymbol{r}_2|)$, so that the Cooper pair has no orbital angular momentum (or equivalently $p(\boldsymbol{k}) = p(k)$), as is the case for 'conventional' superconductors (but probably not for high-T_c and heavy-fermion (4f- or 5f-containing) superconductors), it is correspondingly symmetric under interchange of $\boldsymbol{r}_1$ and $\boldsymbol{r}_2$. A total wavefunction that is antisymmetric under interchange of the two electrons (as required for fermions) can be generated by combining the symmetric spatial part with a spin wavefunction $\chi(s_1, s_2)$ having an antisymmetric combination of the electron spins s_1 and s_2, i.e. $\chi(s_1, s_2) = \chi_{\text{odd}}^{S=0}$, or

$$\Phi(1,2) = \phi(|\boldsymbol{r}_1 - \boldsymbol{r}_2|)(\alpha(1)\beta(2) - \beta(1)\alpha(2))/\sqrt{2}. \tag{6.156}$$

Thus, the Cooper pair in this model is a spin-singlet state ($S = 0$), viz. $(\boldsymbol{k}_\uparrow, -\boldsymbol{k}_\downarrow)$, with spherically symmetric (i.e. s-state) orbital angular momentum; it exhibits *s-state pairing*.

Other combinations are also possible in non-conventional superconductors. For the spin-singlet case, d-state pairing is also compatible, for which $\phi_{\text{even}}(\boldsymbol{r}_1, \boldsymbol{r}_2) = \phi_{\text{even}}(\boldsymbol{r}_2, \boldsymbol{r}_1)$ too; in this case, the spatial part of the wavefunction exhibits nodes in certain directions in real or reciprocal space. In principle, *spin-triplet* states ($S = 1$) are also permissible, for which the spin part of the wave function $\chi(s_1, s_2)$ now has the form $\alpha(1)\alpha(2), \beta(1)\beta(2)$, or $[\alpha(1)\beta(2) + \alpha(2)\beta(1)]/\sqrt{2}$, and $\phi(\boldsymbol{r}_1, \boldsymbol{r}_2)$ must be antisymmetric, i.e. $\phi_{\text{odd}}(\boldsymbol{r}_1, \boldsymbol{r}_2) = -\phi_{\text{odd}}(\boldsymbol{r}_2, \boldsymbol{r}_1)$; thus, one has p-state (or f-state) pairing. It appears that d-state pairing might be prevalent in high-T_c materials (§6.4.3).

Although the virtual-phonon-mediated attractive electron–electron interaction, and the consequent formation of isolated Cooper pairs, is central to an understanding of the behaviour of conventional (s-state) superconductors, this does not represent the complete picture. A quantitative theory for the mechanism of superconductivity was provided in 1957 by Bardeen, Cooper and Schrieffer (BCS). The mathematical development of this theory is beyond the level of this book (see e.g. Tinkham (1996) for more detail), but the qualitative results of the theory are relatively easy to understand. The BCS theory asserts that, at $T < T_c$, many electrons form Cooper pairs, thereby lowering the electron energy below that of the non-interacting Fermi-gas ground state: at $T = 0$ K, *all* electrons form Cooper pairs. The superconducting state is a many- body state, in which cooperativity plays an essential rôle. The wavefunction of this BCS state for n electrons can be written in analogy to the single-pair Cooper state (eqn. (6.156)) as

$$\Phi_{\text{TOT}} = P\{\Phi(1,2)\Phi(3,4)\ldots\Phi(n-1,n)\}, \tag{6.157}$$

where P is an antisymmetrization operator for interchange of any two electrons.

It might appear that the BCS superconducting state could equally well be regarded as a Bose–Einstein condensation of Cooper pairs (each of which, because it consists of two spin$-\frac{1}{2}$ fermions, behaves as an integral-spin boson). However, this picture is not strictly accurate since the phenomenon of a Bose–Einstein condensation of bosons into a single ground state is only valid for non-interacting particles. As we have seen, since the size of the Cooper pairs is so very large ($\simeq 1000$ Å — see eqn. (6.151)), and consequently the centres of about 10^6 other pairs are to be found in the volume of a single Cooper pair, there is a very strong overlap between different Cooper pairs in the superconducting condensate. Even though all Cooper pairs reside in the same quantum state, with the same energy, at $T = 0$ K, the situation is a dynamical one: individual electrons in the pairs are continuously scattered between single-electron states with crystal momenta in the range (see Fig. 6.31b and Problem 6.16):

$$\hbar\Delta k \simeq m_e\omega_D/k_F. \tag{6.158}$$

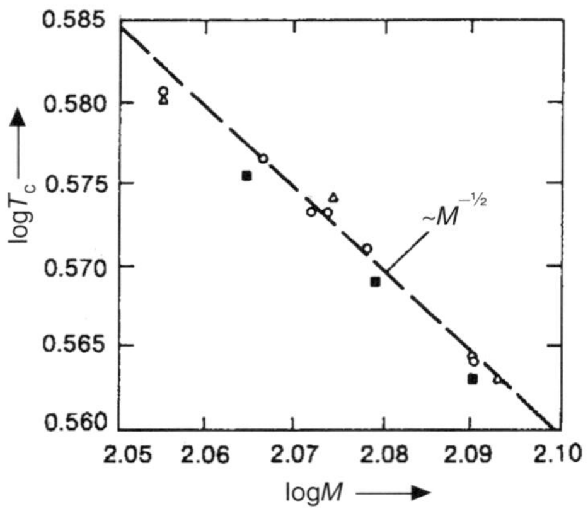

Fig. 6.32 Isotope effect for the superconducting transition temperature in Sn. (After Ibach and Lüth (1995), *Solid State Physics*, p. 245, Fig. 10.15, © Springer-Verlag GmbH & Co. KG)

The BCS model predicts (see e.g. Ibach and Lüth (1995) for a relatively simple derivation) that the superconducting transition temperature, T_c, should have the following form:

$$k_B T_c = 1.14\hbar\omega_D e^{-1/\mathcal{V}_0 g(\mathscr{E}_F)}, \tag{6.159}$$

where $\mathcal{V}_0$ is the electron–phonon interaction strength (Problem 6.15). The linear proportionality between T_c and ω_D in this expression, resulting from the proposed phonon-mediated mechanism for superconductivity, implies that an *isotope effect* for T_c should be observed: the transition temperature should scale with isotopic mass M of the superconductor as $T_c \propto M^{-1/2}$. This predicted behaviour is borne out in the results for Hg and for Sn (Fig. 6.32), although the exact inverse square-root dependence is not found for other materials (e.g. Mg or Nb_3Sn), where the exponent is 0.1–0.3, rather than 0.5. These discrepancies are not too surprising in view of the gross simplifications made in the BCS theory (e.g. constant matrix element $\mathcal{V}_{kk'}$, spherical Fermi surface etc.).

6.4.1.3 *The superconducting gap*

The BCS ground state at $T = 0$ K, consisting of Cooper pairs in the same quantum state, marks a complete breakdown of the independent, one-electron picture. In order to access the first excited state above the ground state, involving 'normal' unpaired electrons, a Cooper pair must be broken up by some external means (e.g. thermal or optical). Since the binding energy must be provided, a minimum energy of 2Δ must be supplied to split up a Cooper pair: there is, therefore, an *energy gap* of Δ between the BCS ground state at $T = 0$ K (all Cooper pairs in the same quantum level) and the first allowed one-electron state (Fig. 6.33a). Alternatively, there is an excitation gap of 2Δ in the electronic density of states, centred at $\mathscr{E}_F$: electron states that, in the normal metallic state, would reside in the energy range $\mathscr{E}_F \pm \Delta$ are pushed out in the superconducting state, as shown in Fig. 6.33b. The superconducting density of states, in the vicinity of the gap, is given by

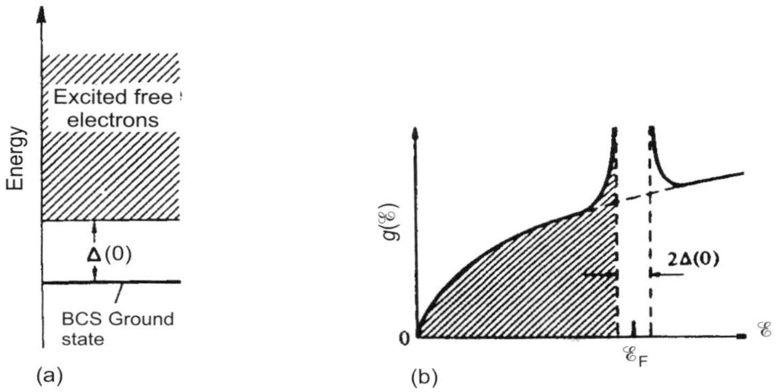

Fig. 6.33 (a) Illustration of the energy gap $\Delta(0)$ in the excitation spectrum at $T = 0$ K between the BCS ground state and the one-electron states of excited free electrons. (b) The density of states in the superconducting state showing the energy gap $2\Delta(0)$ necessary to break up a Cooper at $T = 0$ K. (Note the exaggerated energy scale: Δ is of the order of a few meV, but $\mathscr{E}_F$ is of the order of several eV.)

$$g_s(\mathscr{E}) = g(\mathscr{E}_F)\frac{\mathscr{E}}{(\mathscr{E}^2 - \Delta^2)^{1/2}}. \tag{6.160}$$

This function is singular at $\mathscr{E} = \pm\Delta$, but reverts to the normal one-electron form at energies well away from the gap.

The BCS expression for the gap energy at $T = 0$ K is

$$\begin{aligned}\Delta(0) &= \frac{\hbar\omega_D}{\sinh[1/\mathcal{V}_0 g(\mathscr{E}_F)]} \\ &\simeq 2\hbar\omega_D e^{-1/\mathcal{V}_0 g(\mathscr{E}_F)},\end{aligned} \tag{6.161}$$

where the approximate expression is valid for weak coupling, i.e. $\mathcal{V}_0 g(\mathscr{E}_F) \ll 1$. For *weakly coupled* superconductors, typically $\mathcal{V}_0 g(\mathscr{E}_F) \simeq 0.2$. Strongly coupled superconductors (e.g. Pb, Hg, Nb), on the other hand, are those materials in which the ratio of the superconducting pairing energy to the phonon energy, $2\Delta/\hbar\omega_D$, is larger (see Table 6.4). Values of $\Delta(0)$ for conventional superconductors are in the range 0.2–3 meV, very much smaller than $\mathscr{E}_F$ (typically several eV), i.e. $\Delta(0) \simeq 10^{-4}\mathscr{E}_F$. (Note the similarity between eqn. (6.161) and the expression for the binding energy of a single Cooper pair — eqn. (6) of Problem 6.15.) The functional form of the expression for the superconducting gap $\Delta(0)$ (eqn. (6.161)) is identical to that for the superconducting transition temperature, T_c (eqn. (6.159)): thus, in the BCS theory $\Delta(0)$ and T_c are related in a parameter-free way:

$$2\Delta(0)/k_B T_c = 3.52. \tag{6.162}$$

This relation is satisfied for many elemental and alloy conventional superconductors (see Table 6.4).

As the temperature is increased above zero, an increasing number of Cooper pairs are broken up until, at T_c, no Cooper pairs remain and all electrons are in the

Table 6.4 Parameters for some elemental superconductors

Metals	$\Delta(0)/k_B T_c$	$\mathcal{V}_0 g(\mathscr{E}_F)$	θ_D(K)
Al	1.7	0.18	375
Cd	1.6	0.18	164
Hg	2.3	0.35	70
In	1.8	0.29	109
Pb	2.15	0.39	96
Sn	1.75	0.25	195
Tl	1.8	0.27	100
Zn	1.6	0.18	235

$\Delta(0)$ is the superconducting energy gap at $T = 0$ K; T_c is the superconducting transition temperature; $\mathcal{V}_0$ is the strength of the attractive electron–phonon interaction; $g(\mathscr{E}_f)$ is the density of states at the Fermi level; θ_D is the Debye temperature. (After Ibach and Lüth (1995), *Solid State Physics*, p. 245, Table 10.1, © Springer–Verlag GmbH & Co. KG)

normal, unpaired state. Correspondingly, the superconducting energy gap $\Delta(T)$ decreases to zero at $T = T_c$: just below T_c, the temperature dependence can be approximated as

$$\Delta(T) \simeq AT_c(1 - T/T_c)^{1/2} \tag{6.163}$$

where A is a constant. The experimental results for $\Delta(T)$ for three elemental superconductors (obtained from tunnelling experiments — see §6.4.2.3) are shown in Fig. 6.34; the agreement with the BCS prediction is very impressive.

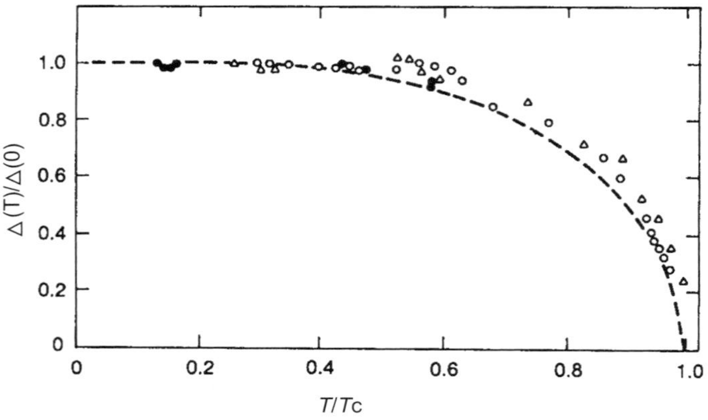

Fig. 6.34 Experimentally measured data for the temperature dependence of the superconducting gap $\Delta(T)$ (from tunnelling experiments) for three elemental superconductors (○, In; △, Sn; •, Pb), normalized to the zero-kelvin value, $\Delta(0)$, plotted against reduced temperature, T/T_c. The dashed curve is the BCS prediction. (After Giaever and Megerle (1961). Reprinted with permission from *Phys. Rev.* **122**, 1101. © 1961. The American Physical Society)

It should be noted that the energy gap in superconductors is of a completely different nature and origin than that in semiconductors or insulators (§5.2.5). In the latter case, the gap is between allowed one-electron states and is due to the electron–lattice interaction. In the former case, the gap is between paired (superconducting) states and one-electron (normal) states, and the interaction is between the electrons themselves. Moreover, the gap in superconductors is very strongly temperature-dependent, decreasing to zero as the critical temperature is approached from below, whereas the gap for an insulator never vanishes although it does decrease slightly with increasing temperature.

The superconducting gap can be probed optically. Photons with energies *less* than the gap, i.e. $\hbar\omega < 2\Delta$, are not absorbed by a superconductor and are reflected from its surface, whereas photons with energy $\hbar\omega \geq 2\Delta$ can break up Cooper pairs and are consequently absorbed. Since the superconducting gaps are of the order of a few meV, far-IR or microwave radiation can be used as a probe. The differences in IR reflectivity of three elements between superconducting and normal states (the latter achieved by applying a magnetic field greater than B_c) are shown in Fig. 6.35: a sharp change in reflectivity occurs at $\hbar\omega = 2\Delta$ when the photons are absorbed by the materials in the superconducting state.

The superconducting gap is also responsible for the non-linear temperature dependence of the heat capacity, c_S, of the superconducting state evident in Fig. 6.28: the fact that the excited levels are separated by an energy gap Δ (Fig. 6.33a) means that c_S depends effectively exponentially on temperature via

$$c_S = 1.34\gamma T_c(\Delta(0)/k_B T)^{3/2}\exp(-\Delta(0)/k_B T), \tag{6.164}$$

where γ is the coefficient of the linearly temperature-dependent electronic heat capacity per unit volume in the normal state (eqn. (5.36)). (The quantity Δ, rather than the gap energy 2Δ, appears in eqn. (6.164) because the heat capacity is related to *single-particle* excitations.) At the critical temperature T_c, the discontinuity in heat capacities between

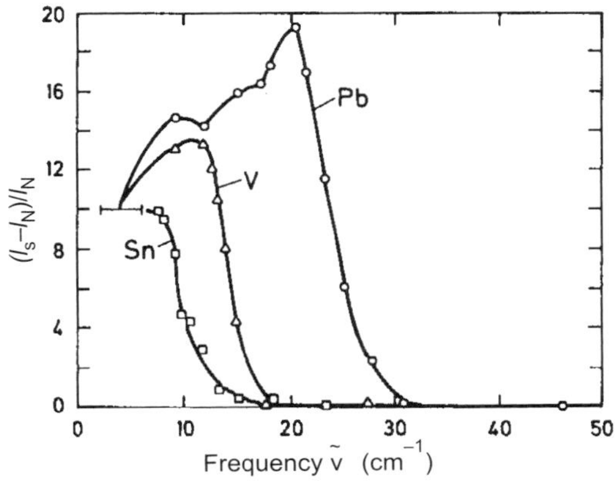

Fig. 6.35 Reflectivity difference between superconducting (S) and normal (N) states at 1.4 K of three elemental superconductors, where $I_{S,N}$ is the intensity of radiation multiply reflected in a cavity of the material in S and N states. (After Richards and Tinkham (1960). Reprinted with permission from *Phys. Rev.* **119**, 575. © 1960. The American Physical Society)

normal and superconducting states (Fig. 6.28) can be expressed in parameter-free form in the BCS model as

$$[c_S(T_c) - c_N(T_C)]/c_N(T_c) = 1.43. \qquad (6.165)$$

6.4.2 Electron dynamics

One of the main distinguishing features of a superconductor is the zero d.c. resistivity (infinite conductivity) at temperatures below T_c. (The other distinct characteristic is the Meissner effect, i.e. the expulsion of magnetic flux from a material in the superconducting state — see §7.2.3.3.) The existence of Cooper pairs of electrons, participating in the many-body BCS ground state, naturally explains this highly unusual electron-transport behaviour. In addition, the quantum- mechanical tunnelling of electrons, particularly as Cooper pairs, between two superconductors is qualitatively completely different in behaviour from that found for normal metals because of the macroscopic extent of the coherence of the superconducting (BCS)-state wavefunction (§6.4.2.3).

6.4.2.1 Supercurrents

An electrical current induced in a ring of superconducting material, e.g. by magnetic induction, decays immeasurably slowly; estimates for the decay time made using nuclear magnetic resonance experiments are of the order of at least 10^5 years, corresponding to the resistivity of the superconducting state being less than $\rho \sim 10^{-25}\Omega$m (cf. the room-temperature resistivity of pure metallic Cu of $\rho \sim 10^{-8}\Omega$m). Such currents are termed persistent currents or supercurrents.

The reason for the existence of supercurrents can be understood as follows. One approach is in terms of the Cooper pairs of electrons that carry the supercurrent and the macroscopic coherent quantum state that such pairs occupy. Since the BCS ground state consists of Cooper pairs all having the same wavefunction (eqn. (6.157)), each pair of which requires an energy equal to 2Δ to unpair the two electrons and hence cause a change in electron state, it is evident that *elastic* scattering of Cooper pairs by, for example, defects (§6.3.1.2) will not perturb the condensate, and hence not contribute to the electrical resistivity. In principle, *inelastic* scattering events could give rise to a finite resistivity. Examples of such events involve phonon emission/absorption and the concomitant creation/destruction of Cooper pairs from/into normal electrons, respectively (Fig. 6.36); such creation and destruction of Cooper pairs takes place continuously at finite temperatures ($< T_c$) since a dynamic equilibrium exists between the Cooper pairs and normal electrons. However, Cooper pairs created by processes such as in Fig. 6.36b have the *same* wavefunction as the pre-existing pairs: unless they have the same centre-of-mass momentum, their binding energy is *zero*. Thus, under normal circumstances, even inelastic scattering processes such as in Fig. 6.36b do not provide a finite resistance to the Cooper-pair supercurrent. Only processes that affect *all* pairs equally (such as the application of an electric field), i.e. that change the coherent many-body wavefunction of the condensate, can alter the current. In this sense the overall wavefunction of the condensate behaves as if it were 'rigid'; that is, it cannot vary spatially and is unaffected by a (weak) magnetic field.

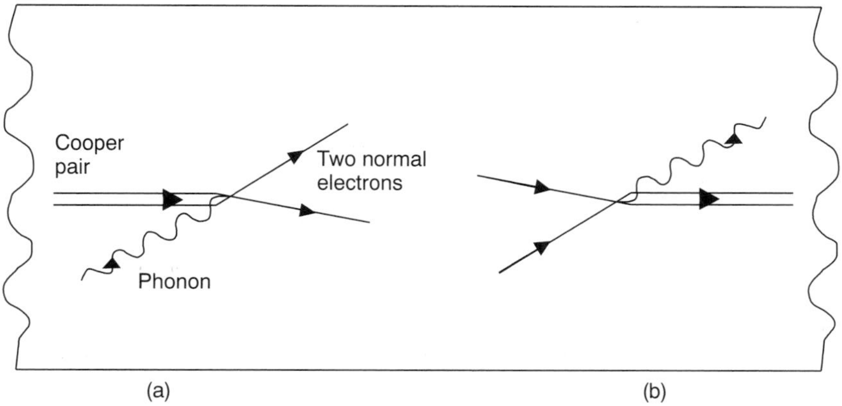

Fig. 6.36 Illustration of inelastic scattering processes involving Cooper pairs in a supercurrent: (a) phonon absorption, leading to the destruction of a Cooper pair and the creation of two normal electrons; (b) phonon emission, leading to the creation of a Cooper pair from two normal electrons.

An alternative viewpoint on the reason for persistent currents is in terms of the magnetic flux generated by the supercurrent carried by say a superconducting ring. Such flux is *quantized* in units of the quantity (the flux quantum) $h/2e = 2.07 \times 10^{-15}$T m^2 (see §7.2.3.3), and hence the generating supercurrent can only decay by discontinuous amounts corresponding to a change of magnetic flux of at least one flux quantum. The energy barrier to a thermal fluctuation that causes the superconducting material to become momentarily normal, and hence able to allow the escape of a flux quantum, is prohibitively high since it involves the destruction of the Cooper pairs involving the pair-breaking energy 2Δ.

At temperatures in the range $0 < T < T_c$, where thermally generated normal electrons coexist with superconducting electrons (Cooper pairs), although the normal-electron current is resisted by the electron-scattering mechanisms operative in metals (§6.3.1), the supercurrent associated with the motion of the Cooper pairs shorts out the normal current flowing in parallel, and hence the superconductor as a whole retains its zero-resistance state until T_c.

Cooper pairs in the BCS condensate all have the *same* wavefunction (eqn. (6.154)), and hence the entire superconducting condensate is described by this wavefunction. In a current-carrying state, the wavefunction is a function of only two spatial variables involving the individual electrons of a pair, $\boldsymbol{r}_1$ and $\boldsymbol{r}_2$, namely the separation, $\boldsymbol{r} = \boldsymbol{r}_1 - \boldsymbol{r}_2$ (as in eqn. (6.155) for the stationary Cooper pair) and the centre-of-mass position $\boldsymbol{R} = (\boldsymbol{r}_1 + \boldsymbol{r}_2)/2$. Neglecting details of the relative motion of the Cooper pairs, the condensate can therefore be characterized by the Ginzberg–Landau order parameter $\psi(\boldsymbol{R})$, which is related to the wavefunction of the Cooper pairs at rest, $\phi(\boldsymbol{r})$ (eqn. (6.155)), via

$$\psi(\boldsymbol{R}) = e^{i\boldsymbol{K}\cdot\boldsymbol{R}}\phi(\boldsymbol{r}), \tag{6.166}$$

where $\boldsymbol{K}$ is the net wavevector of a Cooper pair having charge $-2e$ and mass $2m_e$ (see Problem 6.17).

The current density in the presence of a magnetic field $\boldsymbol{B}$ associated with this order-parameter wavefunction for Cooper pairs is given by the expression (see Problem 6.18):

$$\boldsymbol{j}(\boldsymbol{R}) = \frac{\mathrm{i}\hbar e}{2m_\mathrm{e}}(\psi^*\nabla\psi - \psi\nabla\psi^*) - \frac{2e^2}{m_\mathrm{e}}\psi^*\psi\boldsymbol{A}, \tag{6.167}$$

where $\boldsymbol{A}$ is the magnetic vector potential and $\boldsymbol{B} = \text{curl}\,\boldsymbol{A}$. Insertion of the wavefunction of eqn. (6.166), generalized to

$$\psi(\boldsymbol{R}) = \left(\frac{n_\mathrm{s}}{2}\right)^{1/2}\mathrm{e}^{\mathrm{i}\theta}, \tag{6.168}$$

where $|\phi(r)|^2 = n_\mathrm{s}/2$, the concentration of Cooper pairs, into eqn. (6.167) gives for the current density

$$\boldsymbol{j} = -\left[\frac{\hbar e}{2m_\mathrm{e}}\nabla\theta + \frac{e^2}{m_\mathrm{e}}\boldsymbol{A}\right]n_\mathrm{s}. \tag{6.169}$$

In the gauge $\boldsymbol{A} = 0$ (valid deep inside a type-I superconductor where the Meissner effect is complete, i.e. $\boldsymbol{B} = 0$), and with the phase factor $\theta = \boldsymbol{K}\cdot\boldsymbol{R}$ as in eqn. (6.166), the current density is given by

$$\boldsymbol{j} = -\frac{n_\mathrm{s}e\hbar}{2m_\mathrm{e}}\boldsymbol{K}. \tag{6.170}$$

Taking the curl of both sides of eqn. (6.167), with eqn. (6.168) substituted, gives the London equation (see also Problem 6.19):

$$\text{curl}\boldsymbol{j} = -\frac{n_\mathrm{s}e^2}{m_\mathrm{e}}\boldsymbol{B} \tag{6.171}$$

assuming that the Cooper-pair concentration is spatially homogeneous (i.e. that the wavefunction is *rigid*) and because of the identity $\nabla\times\nabla\theta = 0$. Taking the curl again of both sides of eqn. (6.171), and utilizing the fact that curl curl $\boldsymbol{j} \equiv \nabla\times\nabla\times\boldsymbol{j} = \nabla(\text{div}\,\boldsymbol{j}) - \nabla^2\boldsymbol{j}$ and $\text{div}\,\boldsymbol{j} = 0$, together with the Maxwell relation

$$\boldsymbol{j} = \text{curl}\ \boldsymbol{B}/\mu_0 \tag{6.172}$$

gives an expression for determining the spatial variation of the supercurrent:

$$\nabla^2\boldsymbol{j} = \mu_0\frac{n_\mathrm{s}e^2}{m_\mathrm{e}}\boldsymbol{j}. \tag{6.173}$$

In the case of a semi-infinite geometry for the superconductor (Fig. 6.37) where $\boldsymbol{j} = j(x)\hat{z}$, the one-dimensional form of eqn. (6.173) becomes

$$\frac{\mathrm{d}^2 j}{\mathrm{d}x^2} = \mu_0\frac{n_\mathrm{s}e^2}{m_\mathrm{e}}j, \tag{6.174}$$

with solution

$$j(x) = j_0\mathrm{e}^{-x/\lambda_\mathrm{L}}, \tag{6.175}$$

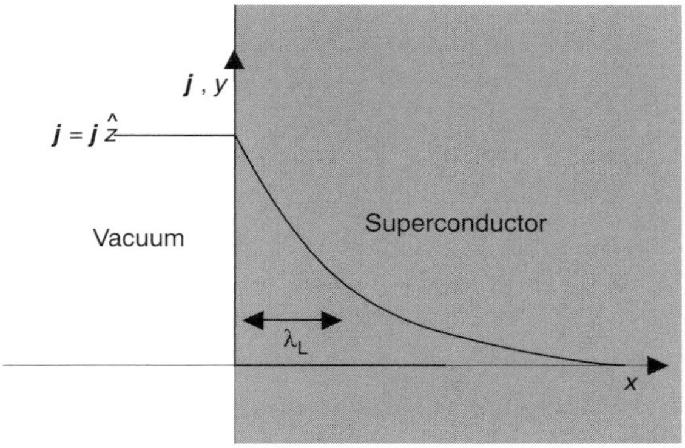

Fig. 6.37 The exponential decrease of the supercurrent into the bulk of a superconductor away from the surface predicted by the London equation. The direction of the current density is in the direction normal to the page ($\hat{z}$).

where the London penetration depth is given by

$$\lambda_L = \left(\frac{m_e}{\mu_0 n_s e^2}\right)^{1/2}. \tag{6.176}$$

Thus, the supercurrent is confined to the free surface and decays exponentially into the bulk of the superconductor. Note that the magnetic flux density also decays into a superconductor from the surface in exactly the same manner (§7.2.3.3); thus, the Meissner effect is complete in a (type-I) superconductor only at depths much greater than λ_L. At $T = 0$ K, when all electrons are superconducting, i.e. $n_s = n$, the conduction-electron density of the metal, eqn. (6.176) gives a value for the London penetration depth of $\lambda_L \simeq 170$ Å for $n_s = 10^{29}$ m^{-3}. Note that eqn. (6.176) implies that the London penetration depth should diverge as T approaches T_c from below, since $n_s \to 0$. This is in accord with experience: at T_c, the Meissner effect vanishes (i.e. an applied magnetic field uniformly penetrates the material) and the electrical current also becomes uniform throughout the material in the normal state.

The London penetration depth is one spatial quantity that is characteristic of a superconductor. Another such quantity is the coherence length ζ_0, which can be regarded as a measure of the extent of the Cooper-pair wavefunction, or alternatively as the distance over which the wavefunction can vary without incurring an appreciable increase in energy. This can be estimated by an argument based on Heisenberg's uncertainty principle. The one-particle electron wavefunctions that constitute the Cooper pairs originate only from a region in energy of approximately $\pm\Delta$ around the Fermi level $\mathscr{E}_F$, where the one-particle occupancy in the superconducting state is appreciably different from that of a metal in the normal state (Fig. 6.38). Thus, from this energy uncertainty, one can calculate the corresponding uncertainty in momentum:

$$\delta\mathscr{E} \simeq 2\Delta = \delta\left(\frac{p^2}{2m_e}\right) \simeq \frac{p_F}{m_e}\delta p, \tag{6.177}$$

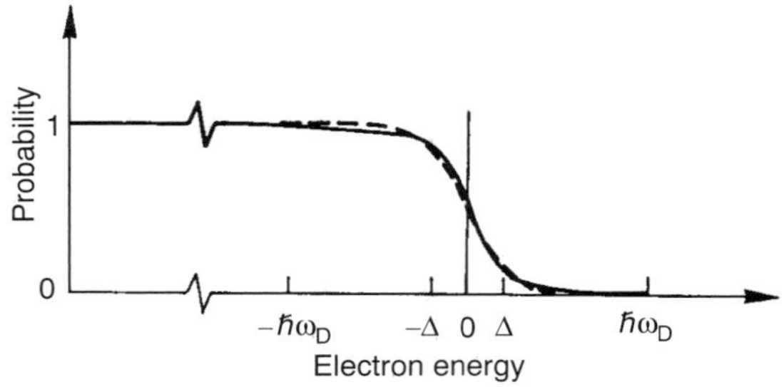

Fig. 6.38 The BCS occupation probability at $T = 0$ K for Cooper pairs in the region of the Fermi energy $\mathscr{E}_F$, taken to be zero on the energy scale (solid line). Note that this is very similar to the Fermi–Dirac distribution function for normal electrons, but at a temperature $T = T_c$ (dashed line).

where p_F is the value of the electron momentum at the Fermi level. Using the Heisenberg uncertainty relation, $\delta x \simeq \hbar/\delta p$, gives

$$\zeta_0 = \delta x \simeq \frac{\hbar v_F}{2\Delta} \equiv \frac{\mathscr{E}_F}{k_F \Delta}. \tag{6.178}$$

A similar result emerges from the BCS theory:

$$\zeta_{BCS} = \frac{\hbar v_F}{\pi \Delta}. \tag{6.179}$$

Typical values of the coherence depth are of the order of $10^3 - 10^4$ Å, consistent with the previous estimation of the size of a Cooper pair using eqn. (6.151).

In practice, particularly for alloys, the measured (magnetic) penetration depth λ_m is often much larger than λ_L (eqn. (6.176)) and dependent on impurity content. This behaviour can be understood in terms of a non-local picture of the electrodynamics where, instead of the electric current density $\boldsymbol{j}$ at a point $\boldsymbol{r}$ depending on the vector potential $\boldsymbol{A}$ at that point (cf. eqn. (6.169)), instead it depends on the vector potential averaged over a *volume* determined by the effective coherence length ζ_{eff}. For a *pure* material, this is given by ζ_{BCS} (eqn. (6.179)), but for impure materials, with a correspondingly high degree of impurity scattering of electrons leading to a short normal-electron mean-free path Λ, it is Λ that determines the length scale for the variation of $\boldsymbol{A}$, and in such a case the effective coherence length is given by a relation of the form

$$1/\zeta_{eff} = 1/\zeta_{BCS} + 1/\Lambda. \tag{6.180}$$

In the so-called dirty limit, where the behaviour is controlled by impurities, $\zeta_{eff} \simeq \Lambda \ll \zeta_{BCS}$. Only for the case where $\zeta_{BCS} \ll \lambda_L$ is the magnetic penetration depth λ_m equal to the London penetration depth λ_L; otherwise, e.g. for dirty superconductors, $\lambda_m \gg \lambda_L$.

The Ginzberg–Landau parameter, which is the ratio of the penetration depth to the coherence length, i.e.

$$\kappa_{GL} = \lambda_m/\zeta_{eff}, \tag{6.181}$$

is important since it allows a numerical distinction to be drawn between type-I and type-II superconductors (see §§. 6.4 and 7.2.3.3). Type-I superconductors are characterized by $\kappa_{GL} \ll 1$ (e.g. for elemental superconductors, $\lambda \simeq 500$ Å and $\zeta \simeq 5000$ Å), whereas type-II superconductors (e.g. alloys and high-T_c (oxide) superconductors) have $\kappa_{GL} \gg 1$.

6.4.2.2 *Thermal behaviour*

In normal metals, heat transport is carried predominantly by the electrons (§6.3.2.2). However, at $T = 0$K, in a superconductor all electrons form Cooper pairs which are in the *same* quantum state. As a consequence, the entropy associated with such a condensate is *zero* (Fig. 6.29). Since the Cooper pairs can carry no entropy, they cannot contribute to the electronic thermal conductivity κ_T which therefore goes to zero as $T \to 0$ K (Fig. 6.39). Superconductors can therefore be used as a thermal switch: operated at a temperature well below T_c, a superconductor acts as a thermal insulator, but conducts heat well if driven into the normal state by the application of a magnetic field in excess of the critical field, B_c.

Since electrons in Cooper pairs carry no entropy (or heat), the Peltier coefficient Π (eqn. (6.103)) associated with a supercurrent must be zero, as must the thermopower S_T (eqn. (6.96)), At temperatures below T_c, any thermal effects in superconductors are therefore only associated with the normal electrons present or the phonons.

6.4.2.3 *Electron tunnelling*

Quantum particles can pass from one classically allowed region (a potential well) to another through a classically forbidden region (a potential barrier), as long as the barrier is low and narrow enough. This process is known as quantum-mechanical tunnelling, and depends on there being an appreciable overlap of the particle wavefunction between the two wells in the classically forbidden region.

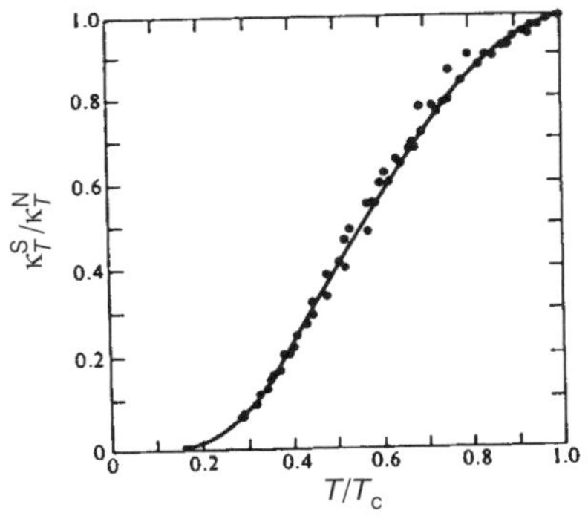

Fig. 6.39 Ratio of the electronic contribution to the thermal conductivity of Al in the superconducting state (κ_T^S) to the normal state (κ_T^N) as a function of reduced temperature, T/T_c. The prediction from the BCS theory is given by the solid curve. (After Satterthwaite (1962). Reprinted with permission from *Phys. Rev.* **125**, 873. © 1962. The American Physical Society)

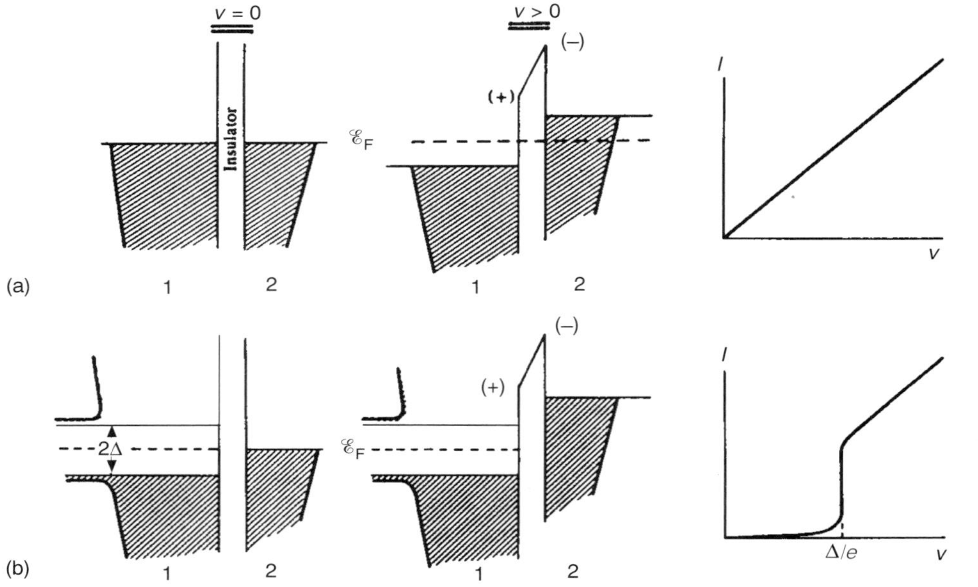

Fig. 6.40 Schematic illustrations of the density of states (occupied states being shown hatched) for two metals separated by a thin electrically insulating layer with zero and a finite applied voltage, V. Also shown is the corresponding tunnelling-current I–V characteristic. (a) Two normal metals, exhibiting an Ohmic tunnelling current-voltage characteristic. (b) A normal metal (2) and a superconducting metal (1) in a tunnel junction. The I–V characteristic is non-Ohmic at low voltages, exhibiting a threshold at $V = \Delta/e$.

An example of this phenomenon is the tunnelling of electrons between two metals separated by a thin electrically insulating layer, made by oxidizing a thin (10–20 Å) layer on one metal before depositing another on top in thin-film form. In the case of dissimilar metals, a tunnelling current of electrons will flow from the metal with the lower work function (i.e. higher initial absolute chemical potential or Fermi energy relative to the vacuum level) to the other until the Fermi levels in the two materials equalize. For the case of two identical normal metals, there is no net current in the absence of an applied electric field (Fig. 6.40a). If an electrostatic potential difference V is applied across the metal–insulator–metal system (where the metals are identical), then the Fermi energy of one metal (2) is raised above that of the other (1); in such a case, it is the electrochemical potential η (eqn. (6.86)) that is constant throughout the system. As a result, a tunnelling current will flow from metal 2 to metal 1, and the magnitude of this current will increase linearly with the relative displacement of the two Fermi levels (since this is proportional to the number of available empty states in metal 1 to which electrons from metal 2 may tunnel); hence Ohm's law is satisfied, $I \propto V$ (Fig. 6.40a).

However, the situation for single-electron tunnelling is very different if one of the metals involved in the junction is a superconductor at a temperature below T_c (Fig. 6.40b). As seen before (Fig. 6.33b), in the superconducting state there is a gap of 2Δ, centred at the Fermi energy, separating occupied, paired electron states and single-particle excitations. Thus, with zero applied potential difference, there is *no* single-electron tunnelling: electrons in the normal metal at $\mathscr{E}_F$ have no states in the superconducting gap of the superconductor to which they can tunnel and, at $T = 0$ K, there are no unpaired electrons in the superconductor able to tunnel to the normal metal. If a voltage is applied which raises the Fermi level of the normal metal relative to that of the superconductor, Giaever tunnelling of single electrons can occur when the Fermi level of the normal metal lies *above* the gap in the superconductor (Fig. 6.40b). As a result, there is a sharp threshold in the current–voltage characteristic at a voltage $V = \Delta/e$ (Fig. 6.40b), and this experiment provides an accurate method of obtaining values for superconducting energy gaps. At temperatures between zero and T_c, an increasing number of unpaired normal electrons occupy states above the gap in the superconducting density of states; these can tunnel into empty states above $\mathscr{E}_F$ in the normal metal and this causes the tail in the tunnelling current below the threshold voltage at finite temperatures below T_c (see also Problem 6.21).

Very different behaviour is observed for the tunnelling of superconducting electron (Cooper) *pairs* between two superconductors, separated by a very thin insulating layer (a Josephson junction), as a result of the macroscopic coherence of the condensate quantum state. This behaviour is referred to as Josephson tunnelling. Two types of phenomena can be distinguished. The d.c. Josephson effect is the occurrence of a d.c. tunnelling current between two superconductors across a tunnel junction in the absence of an external electric or magnetic field. The a.c. Josephson effect is the production of high-frequency (radio-frequency, r.f.) oscillations in the pair tunnelling current when a d.c. voltage is applied to the Josephson junction. Alternatively, an r.f. voltage applied together with the d.c. voltage causes a d.c. current to flow across the junction.

Consider two isolated identical superconductors 1, 2 at the same temperature. The Ginzberg–Landau order parameters (eqn. (6.168)) are $\psi_1 = (n_s/2)^{1/2}e^{i\theta_1}$ and $\psi_2 = (n_s/2)^{1/2}e^{i\theta_2}$, and these are assumed to be spatially uniform in each material. Although the magnitude of the wavefunction is the same $(n_s/2)$ in each case because the temperatures are identical, the phases need *not* be the same. If the two superconductors are *weakly* coupled, e.g. via tunnelling through a thin intervening barrier (or simply a constriction between the two superconductors), it is possible to sustain a phase difference between the superconductors by the passage of a current through the junction or the application of an external voltage — such a situation is termed a weak link. (In the case of *strong* coupling, the phases of the two condensates become irretrievably locked, $\theta_1 = \theta_2$, and cannot be perturbed.)

The behaviour of the superconducting order parameter in the vicinity of a tunnel junction is shown schematically in Fig. 6.41: each order parameter decays exponentially, with decay constant α, into the insulating layer. For a barrier of thickness l with edges at $x = \pm l/2$, the order parameter at a position x within the barrier can be simply written as the sum of the two tunnelling contributions at that point, i.e.

$$\psi = \left(\frac{n_s}{2}\right)^{1/2}\left[e^{i\theta_1-\alpha(x+l/2)} + e^{i\theta_2+\alpha(x-l/2)}\right]. \tag{6.182}$$

The supercurrent density across the junction can be evaluated using eqn. (6.167) in the absence of a magnetic field ($\boldsymbol{A} = 0$), i.e.

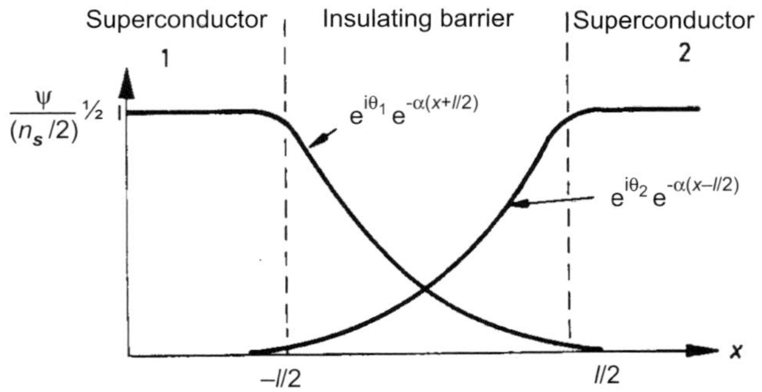

Fig. 6.41 Schematic illustration of the variation of the superconducting order parameter ψ in the vicinity of a Josephson junction, formed from two identical superconductors at the same temperature separated by a thin insulating barrier layer. The exponential decay of ψ in the barrier is due to tunnelling of Cooper pairs.

$$j = \frac{i\hbar e n_s}{2m_e} \alpha e^{-\alpha l} \left(e^{i(\theta_2 - \theta_1)} - e^{-i(\theta_2 - \theta_1)} \right) = j_0 \sin\delta, \tag{6.183}$$

where the maximum electron-pair current density is given by

$$j_0 = \frac{\hbar e n_s}{m_e} \alpha e^{-\alpha l} \tag{6.184}$$

and $\delta = \theta_2 - \theta_1$ is the phase difference between the two components of the Josephson junction. The d.c. Josephson current is maximized, $j = j_0$, when the phase difference is $\delta = \pi/2$.

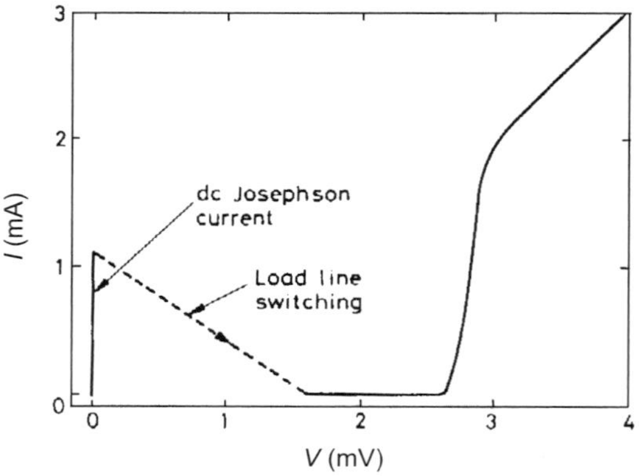

Fig. 6.42 Current-voltage curve for a Pb–PbO–Pb Josephson junction at 1.2 K showing the d.c. Josephson effect (the current spike at zero voltage). The threshold in the current at a finite voltage V_c is due to the onset of tunnelling of normal electrons when $V_c = 2\Delta/e$. (After Langenburg *et al.* (1966), *Proc. IEEE* **54**, 560. © 1966 IEEE)

This behaviour is shown in Fig. 6.42 for a Josephson junction made of superconducting Pb: the current spike at zero voltage is the d.c. Josephson current. The threshold in the d.c. current occurring at a finite voltage V_c in Fig. 6.41 is due to the tunnelling of *normal* electrons, and occurs when the electrons tunnelling through the barrier acquire enough energy from the potential drop to overcome the pair-binding energy of 2Δ, i.e. when $V_c = 2\Delta/e$.

What is not evident in the d.c. response shown in Fig. 6.42 is the *a.c.* Josephson effect observed when a d.c. voltage is applied to the Josephson junction. The time evolution of a quantum state, e.g. as given by the superconducting order parameter ψ, has the form $\psi \propto \exp(-i\mathscr{E}t/\hbar)$ (eqn. (5.55)) where the energy $\mathscr{E}$ here is the chemical potential μ of a Cooper pair. If $\mu = \mu(t)$, the phase of the order parameter is correspondingly time-dependent, $\theta = \theta(t)$, and hence

$$\hbar \frac{\partial \theta}{\partial t} = -\mu. \tag{6.185}$$

The chemical potential in two superconductors separated by a weak link can be made different by the application of a d.c. voltage V across a Josephson junction, and hence from eqn. (6.185)

$$\hbar \left(\frac{\partial \theta_1}{\partial t} - \frac{\partial \theta_2}{\partial t} \right) = -\mu_1 + \mu_2 = 2eV$$

or

$$\hbar \frac{\partial \delta}{\partial t} = -2eV. \tag{6.186}$$

Integrating this equation for constant V gives

$$\delta = -\frac{2eVt}{\hbar} + \delta_0, \tag{6.187}$$

where $\delta_0 \equiv \delta(t = 0)$. Thus, the phase difference $\delta = \theta_2 - \theta_1$ varies linearly with time, and hence the current density (eqn. (6.183)) becomes *alternating* in time:

$$j = j_0 \sin \left[\delta_0 - \frac{2eVt}{\hbar} \right] \tag{6.188}$$

with a frequency $\nu = 2eV/h$ (e.g. 483.6 MHz for $V = 10\mu V$). Thus, the ratio of the applied d.c. voltage and the measured frequency is the flux quantum $h/2e$, and the a.c. Josephson effect can be used to measure this quantity very accurately.

6.4.3 High-T_c superconductors

Perhaps almost as surprising as the initial discovery of superconductivity (in Hg) by Kammerlingh Onnes in 1911 was the finding in 1986 by Bednorz and Muller that certain metallic *oxide* compounds became superconducting at critical temperatures well above the maximum value of T_c then believed to be possible ($T_c = 23.2$ K for Nb_3Ge found in 1972). The first high-T_c material discovered by Bednorz and Muller, namely $(La_{2-x}Ba_x)CuO_4$ had $T_c \simeq 35$ K for $x \simeq 0.15$, but in the intervening period different families of oxide compounds have been found that have even higher values of T_c: e.g. $YBa_2Cu_3O_{7-\delta}$ (YBCO) with $T_c = 92$ K, $Bi_2Sr_2Ca_2Cu_3O_{10}$ (BSCCO) with $T_c = 110$ K and $Tl_2Ba_2Ca_2Cu_3O_{10}$ (TBCCO) with $T_c = 125$ K. The record currently stands at $T_c = 135$ K for a member of yet another family, $HgBa_2Ca_2Cu_3O_{8+\delta}$ (HBCCO). Note that superconducting transition temperatures of these materials are considerably higher than the boiling temperature of liquid nitrogen (77 K), and hence do not require expensive liquid He as a refrigerant to transform them into the superconducting state. It is noteworthy also that these high-T_c materials are mostly all *cuprates*, whose structures comprise sheets with the chemical formula CuO_2; the exceptions are the bismuthate compounds $BaPb_{0.75}Bi_{0.25}O_3$ ($T_c = 12$ K) and $(Ba_{0.6}K_{0.4})BiO_3$(BKBO) with $T_c = 30$ K. Reviews of the properties of high-T_c superconductors can be found for instance in Burns (1992) and Tinkham (1996).

Table 6.5 Superconducting oxides and their superconducting transition temperatures

Formula	T_c(K)	n	*Notations*	
$(La_{2-x}Ba_x)CuO_4$	35	1	La ($n = 1$)	214
$(La_{2-x}Sr_x)CuO_4$	38	1	La ($n = 1$)	214
$(La_{2-x}Sr_x)CaCu_2O_6$	60	2	La ($n = 2$)	—
$YBa_2Cu_3O_7$	92	2	Y123	YBCO
$YBa_2Cu_4O_8$	80	2	Y124	—
$YBa_4Cu_7O_{14}$	40	2	Y247	—
$Bi_2Sr_2CuO_6$	0–20	1	2–Bi ($n = 1$)	Bi 2201 (BSCO)
$Bi_2Sr_2CaCu_2O_8$	85	2	2–Bi ($n = 2$)	Bi 2212
$Bi_2Sr_2Ca_2Cu_3O_{10}$	110	3	2–Bi ($n = 3$)	Bi 2223 (BSCCO)
$TlBa_2CuO_5$	0–50	1	1–Tl ($n = 1$)	Tl 1201
$TlBa_2CaCu_2O_7$	80	2	1–Tl ($n = 2$)	Tl 1212
$TlBa_2Ca_2Cu_3O_9$	110	3	1–Tl ($n = 3$)	Tl 1223
$TlBa_2Ca_3Cu_4O_{11}$	122	4	1–Tl ($n = 4$)	Tl 1234
$Tl_2Ba_2CuO_6$	0–80	1	2–Tl ($n = 1$)	Tl 2201
$Tl_2Ba_2CaCu_2O_8$	108	2	2–Tl ($n = 2$)	Tl 2212
$Tl_2Ba_2Ca_2Cu_3O_{10}$	125	3	2–Tl ($n = 3$)	Tl 2223 (TBCCO)
$HgBa_2CuO_4$	94	1	1–Hg ($n = 1$)	Hg (1201)
$HgBa_2Ca_2Cu_3O_8$	135	3	1–Hg ($n = 3$)	Hg (1223) (HBCCO)
$BaPb_{0.75}Bi_{0.25}O_3$	12	—	—	BPBO
$(Ba_{0.6}K_{0.4})BiO_3$	30	—	—	BKBO

(n is the number of immediately adjacent CuO_2 planes in the unit cell or the number of CuO_2 planes in the unit cell.) Partly after Burns (1992). Reproduced by permission of Academic Press, Inc.

A list of some high-T_c oxide materials with their corresponding transition temperatures is given in Table 6.5. It can be seen immediately that most of these materials are chemically (and hence structurally) very complex; matters are complicated even more by the fact that, in many cases, optimum superconducting behaviour is achieved for slightly non-stoichiometric oxygen compositions (e.g. as for YBCO and HBCCO).

In certain cases, the atomic structure and, concomitantly, the electronic properties, depend crucially on the oxygen stoichiometry. An example is the much-studied YBCO system, $YBa_2Cu_3O_{7-\delta}$, which is metallic and superconducting for $0 < \delta \leq 0.7$, and an insulator with antiferromagnetic ordering of the Cu electron spins (alternatively arranged in an antiparallel fashion — see §7.2.5.6) for $\delta > 0.7$ (Fig. 6.43).

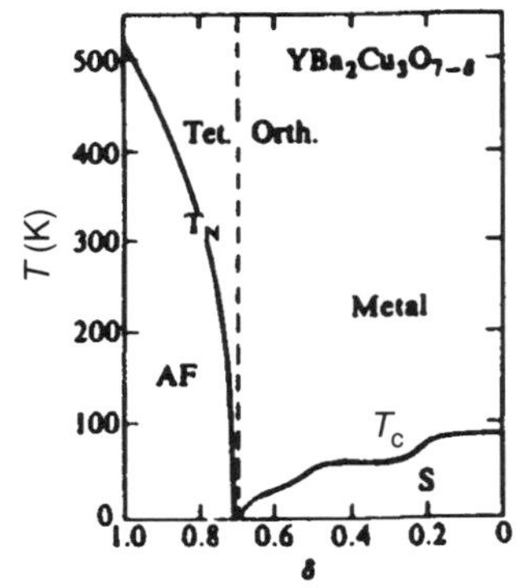

Fig. 6.43 Electronic 'phase diagram' for $YBa_2Cu_3O_{7-\delta}$ (YBCO). For values of oxygen non-stoichiometry $\delta \leqslant 0.7$, the material has an orthorhombic structure and is a superconducting (S) metal; the variation of T_c with δ is shown. For values of $\delta \geqslant 0.7$, the material transforms into an electrical insulator with a tetragonal structure and antiferromagnetic (AF) ordering of the Cu spins; the variation of the Néel temperature, T_N, below which the antiferromagnetic ordering takes place, as a function of δ is also shown. (After Burns (1992). Reproduced by permission of Academic Press, Inc.)

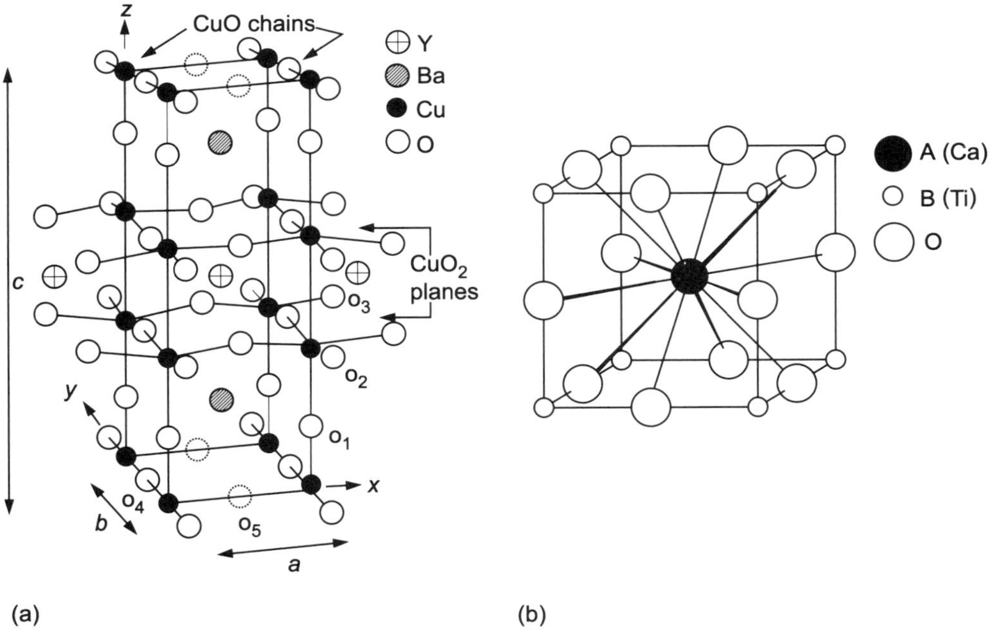

Fig. 6.44 (a) Unit cell of $YBa_2Cu_3O_{7-\delta}$. For the orthorhombic superconducting material ($\delta \simeq 0$), O_4 sites are occupied and O_5 sites are unoccupied. For the tetragonal insulating compound ($\delta = 1$) the O_4 sites are also unoccupied. The CuO_2-containing layers in the $\boldsymbol{a} - \boldsymbol{b}$ plane are marked, as are the chains of linked CuO_2 square-planar units for $\delta = 0$. (*b*) Unit cell of the cubic perovskite structure for a compound ABO_3 (e.g. $CaTiO_3$) for comparison.

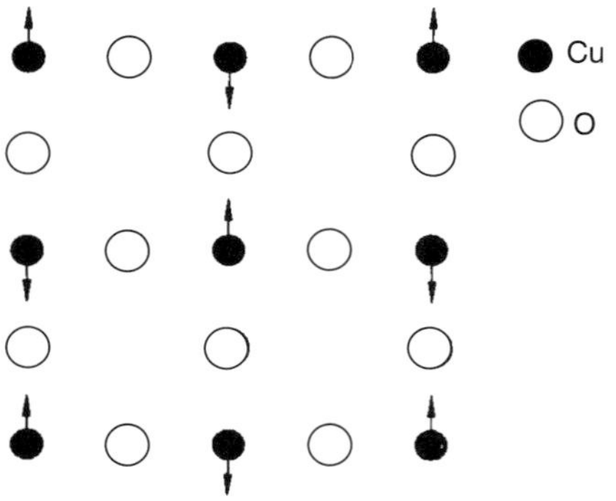

Fig. 6.45 Representation of the antiferromagnetic ordering of the electron spins on the Cu^{2+} ions in the CuO_2 layers in the insulating state of $YBa_2Cu_3O_{7-\delta}(\delta \gtrsim 0.7)$

The structure of $YBa_2Cu_3O_{7-\delta}$ can be understood as an oxygen-deficient modification of the cubic perovskite ($CaTiO_3$) structure (see Fig. 6.44b), with about one-third of the oxygen sites missing and the unit cell tripled in the ***c***-direction. The structural changes that occur for $YBa_2Cu_3O_{7-\delta}$ as δ is varied between 0 and 1 can be understood by reference to Fig. 6.44a. The composition $YBa_2Cu_3O_6(\delta = 1)$ has the oxygen-atom sites O_4 and O_5 unfilled, and hence the structure is *tetragonal* ($a = b \neq c; \alpha = \beta = \gamma = 90°$—cf. Table 2.1). The square-planar CuO_2 layers are clearly evident in Fig. 6.44. In YBCO, the two CuO_2 layers per unit cell ($n = 2$) lying in the ***a***–***b*** plane are separated by Y^{3+} ions. The formula $YBa_2Cu_3O_6$ can be written in terms of formal charges on the ions as $Y^{3+}(Ba^{2+})_2(Cu^{2+})_2Cu^+(O^{2-})_6$: the two Cu^{2+} ions occupy the CuO_2 layers in the $a-b$ plane and the Cu^+ ion is two-fold coordinated along the ***c***-direction in accordance with the usual coordination for Cu^+. Since Cu has the electronic configuration $[Ar]3d^{10}4s^1$, Cu^{2+} ions (but not Cu^+) have an unpaired electron spin ($S = \frac{1}{2}$); these spins order antiferromagnetically in the insulating state ($\delta \geq 0.7$)—see Fig. 6.45.

As excess oxygen is added to $YBa_2Cu_3O_6$ to form $YBa_2Cu_3O_{7-\delta}$, the extra O atoms at first randomly occupy the sites O_4 and O_5 in Fig. 6.44, and hence the material retains its tetragonal symmetry. The additional electronegative O atoms act somewhat like acceptors in covalent semiconductors (see §6.5.2) and 'dope' the crystals by injecting excess holes (§6.2.2). These extra holes reside in states associated with the CuO_2 layers, but for $\delta > 0.7$ the material remains an electrical insulator: both electron–electron interactions (§5.6.3) and electron-localization effects (§6.7) associated with the structural disorder of the oxygen sub-lattice may contribute to the insulating behaviour. At $\delta \simeq 0.7$, there is an insulator–metal transition associated with a change in crystal symmetry to an orthorhombic structure ($a \neq b \neq c; \alpha = \beta = \gamma = 90°$): the excess O atoms preferentially occupy the O_4 sites, leading to Cu–O chains (or linked square-planar CuO_2 units) in the ***b***-direction (see Fig. 6.44a). At the composition $\delta = 0$, the formula $YBa_2Cu_3O_7$ can be written in terms of formal ionic charges as either $Y^{3+}(Ba^{2+})_2(Cu^{2+})_2Cu^{3+}(O^{2-})_7$ or $Y^{3+}(Ba^{2+})_2(Cu^{2+})_3(O^{2-})_6O^-$, depending on whether the extra holes are taken to be associated with the Cu atoms (in the chains) or, more likely, the oxygen atoms in the CuO_2 sheets. The metallic, orthorhombic structure is superconducting for $\delta \leq 0.4$: the excess *holes* form the Cooper pairs responsible for the superconductivity. Holes are the charge carriers in all the cuprate high-T_c superconductors listed in Table 6.5. Electron doping, giving electrons as the majority supercurrent charge carriers, is much less prevalent; it occurs e.g. in $(Ln_{2-x}Ce_x)(CuO_4)$, where the lanthanide ion $Ln = Nd^{3+}$ (or Sm^{3+} or Pr^{3+}) is replaced by Ce^{4+} ions ($T_c = 24$ K for $x \simeq 0.15$).

The CuO_2 layers are common to all the cuprate superconductors, and as the electrical conductivity is higher in the plane of these layers than normal to them, the electrical properties (in both the superconducting and normal states) of these materials are highly *anisotropic*. It is an empirical observation, and one that has been used systematically and successfully to create the highest-T_c superconductors, that T_c increases with an increasing number n of adjacent CuO_2 planes in the unit cell: the present record-holder, HBCCO or Hg(1223) with $T_c = 135$ K has $n = 3$ (Table 6.5).

The field of high-T_c superconductivity based on cuprate materials is under intense research and, at the time of writing, the pairing mechanism underlying superconductivity in these materials is still not fully understood. It appears that a conventional BCS mechanism based on electron–phonon interactions (§6.4.1.2), even with strong coupling, is insufficient to explain many features of these cuprate materials (although it *is* probably operative for the bismuthate materials (BKBO and BPBO—see Table 6.5), having a simple perovskite-like structure (Fig. 6.44b), which appear to be characterized by isotropic *s-state pairing*). Cuprate superconductors are differentiated from conventional superconductors by being characterized by apparent *d-state pairing*, as indicated, for example, by the temperature dependence of the penetration depth, $\lambda(T)$. An s-state pairing mechanism, which gives rise to a superconducting energy gap which is everywhere finite (no nodes) over the Fermi surface, is predicted to have an exponential temperature dependence, $\lambda(T) \propto \exp(-\Delta/k_B T)$, whereas if there are nodes (i.e. states in the overall gap) as would result from d-state pairing, the temperature dependence is expected instead to be a power law, $\lambda(T) \propto (k_B T/\Delta_{max})^n$, with $n = 1$ for line nodes; data for cuprate superconductors appear to fit better to the latter behaviour. Although an electron–phonon interaction, as in conventional superconductors, is probably not operative in the cuprate materials, electron–electron interactions may be important, as is the case in the antiferromagnetic ordering in the normal state.

In fact, high-T_c cuprate materials are anomalous also in many respects in their *normal-state* behaviour: for example, the low-T temperature dependence of the electrical resistivity in the easy-conduction plane of the CuO_2 layers is $\rho \simeq C + DT$, compared with the usual Bloch–Grüneisen behaviour $\rho \simeq A + BT^5$ (eqn. (6.75) and Fig. 6.25a). The electron mean-free path in the $\boldsymbol{a}-\boldsymbol{b}$ plane in YBCO is $\Lambda \simeq 100 - 200$ Å at 100 K.

High-T_c cuprate superconductors are also extreme type-II superconductors, with values of the Ginzburg–Landau parameter (eqn. (6.181)) $\kappa_{GL} \simeq 100$, corresponding to extremely small values of the superconducting coherence length, $\zeta \simeq 10 - 30$ Å in the $\boldsymbol{a}-\boldsymbol{b}$ (CuO_2-layer) plane and $\simeq$ 2–5 Å in the $\boldsymbol{c}$-direction (cf. $\zeta \simeq 10^3 - 10^4$ Å for conventional superconductors; since $\zeta \ll \Lambda$, the cuprate materials lie in the 'clean limit'. High-T_c superconductors have correspondingly extremely high upper critical magnetic fields, i.e. $B_{c2} \simeq 100$ T.

Semiconductors 6.5

Semiconductors can be defined as being those electrical insulators having a forbidden gap of less than, say, 3 eV between the filled valence and empty conduction bands of one-electron states (§5.2.5). (The essential difference in character between the energy gap in a semiconductor/insulator and that in a superconductor (§6.4.1.3) should be noted.) Semiconductors are of interest since their electrical properties can be altered in a controllable way by doping, i.e. by the incorporation of electrically active impurities, thereby producing an extrinsic semiconductor, with concomitant changes in the charge-carrier concentration with respect to the impurity-free intrinsic material. This behaviour is at the heart of the use of semiconductors, particularly crystalline silicon, in electronic devices, and is therefore ultimately responsible for the electronics revolution that society is currently experiencing. In the following, intrinsic semiconductors will be examined first, and this will be followed by a discussion of extrinsic (doped) semiconductors. Excellent reviews of the properties of semiconductors are given by Smith (1978) and Yu and Cardona (1996).

6.5.1 Intrinsic semiconductors

In this section, intrinsic is taken to mean undoped, i.e. free of electrically active impurities. We shall see in §6.5.2.3 that under certain circumstances (i.e. at high temperatures or when 'compensated' with impurities that generate the conjugate type of charge carrier) even extrinsic semiconductors can behave in an effectively intrinsic manner.

6.5.1.1 Intrinsic carrier statistics

In semiconductors, the quantity that essentially controls the electronic behaviour, in the sense that it can be readily varied in extrinsic materials, is the *charge-carrier concentration*; this is conventionally denoted $n \equiv n_e$ for electrons and $p \equiv n_h$ for holes (§6.2.2). At $T = 0$ K, the valence band is completely filled with electrons (or, equivalently, completely empty of holes) and the conduction band is completely empty of electrons. At finite temperatures, thermal excitation (a multiphonon process) of charge carriers across the forbidden gap can take place, increasingly so with increasing temperature. Every electron excited from the valence band into an empty state in the conduction band leaves behind an empty electron state in the valence band that can be regarded as being equivalent to a single hole state (Fig. 6.46). Thus, intrinsic behaviour is characterised by there being an *equal* concentration of thermally generated electrons and holes:

$$n = p \equiv n_i. \tag{6.189}$$

The concentration of electrons in the conduction band, and of holes in the valence band, is determined by the position with respect to the band edges of the *chemical potential*, μ, or the Fermi level $\mathscr{E}_F$, since electrons are fermions and obey Fermi–Dirac

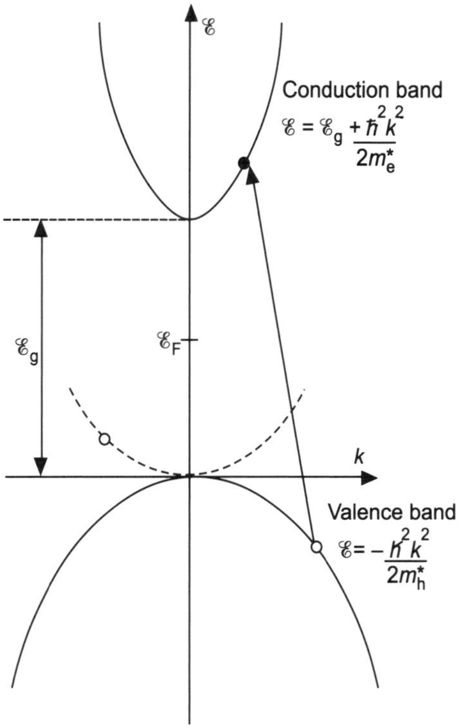

Fig. 6.46 Representation of the thermal generation of an electron–hole pair in a semiconductor with an energy gap $\mathscr{E}_g$ between the valence- and conduction-band minima. The dispersion curves for electrons are the solid curves; the dashed line is the corresponding hole band for the valence band. Note the difference in curvatures of the bands due to the differences in the electron and hole effective masses, m_e^* and m_h^* respectively. The position of the Fermi level is at mid-gap at zero kelvin.

statistics; i.e. the occupation of states is governed by the Fermi–Dirac distribution function (eqn. (5.23)). (Strictly speaking, the Fermi *energy* $\mathscr{E}_F = \mathscr{E}_F^0$ corresponds to the chemical potential at zero kelvin, but the common usage in semiconductor science is to use the expression Fermi level for the chemical potential at finite temperature, $\mu(T)$.) In the simplest case of symmetric valence and conduction bands, i.e. having equal (and opposite) curvatures of the band structure $\mathscr{E}(\boldsymbol{k})$ in the vicinity of the gap, or generally at zero kelvin for any shape of the bands, the chemical potential lies at *midgap* for a semiconductor. Note that it lies, therefore, in an energy region devoid of electron states, in contrast to the case of a *metal*, where μ lies in a band of delocalized states and hence marks the demarcation between filled and vacant states (HOMO in chemical parlance). The situation in semiconductors is different because an excitation *always* involves the creation of an electron–hole pair whose states are separated in energy by the bandgap: for symmetry reasons, therefore, μ must lie part-way between filled and empty electron states (Fig. 6.47).

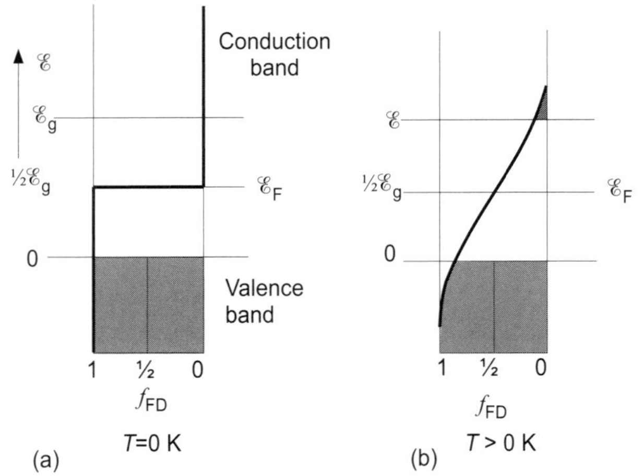

Fig. 6.47 The Fermi–Dirac distribution function for an intrinsic semiconductor at: (a) $T = 0$ K; (b) $T > 0$ K.

A quantitative discussion of carrier concentrations and the position of the chemical potential can be achieved by evaluating the *occupancy* of states in a particular band, which is the product of the volume density of states $D(\mathscr{E})$ (eqn. (5.131)) and the Fermi–Dirac function $f(\mathscr{E})$ (eqn. (5.25)). For the case of say, the conduction band, the electron concentration is given as an integral of the occupancy over the band:

$$n = \int_{\mathscr{E}_c}^{\infty} D_e(\mathscr{E}) f(\mathscr{E}) \mathrm{d}\mathscr{E}, \tag{6.190}$$

where $\mathscr{E}_c$ is the energy corresponding to the bottom of the conduction band. Two approximations may now be invoked in order to simplify the evaluation of the integral in eqn. (6.190). The first is to assume that the chemical potential (Fermi level) for an intrinsic material is far removed in energy from the band edge, i.e. $(\mathscr{E} - \mu) \gg k_B T$, where $\mathscr{E}$ is the energy of an electron in the conduction band; the material is said to be non-degenerate. This is a very good approximation in general for intrinsic semiconductors (see Fig. 6.47). In this case, the Fermi–Dirac distribution function for electrons (eqn. (5.23)) reduces to the Boltzmann function:

$$f_e(\mathscr{E}) = \frac{1}{\exp[(\mathscr{E} - \mu)/k_B T] + 1} \simeq \exp[-(\mathscr{E} - \mu)/k_B T]. \tag{6.191}$$

The other assumption is that the band (at least in the vicinity of the band edge) can be treated as being free-electron-like. Thus, the dispersion relation for electrons in the conduction band can be written as

$$\mathscr{E} = \mathscr{E}_e + \frac{\hbar^2 k^2}{2m_e^*}, \tag{6.192}$$

where m_e^* is the electron effective mass (assumed, for simplicity, to be a scalar quantity). The corresponding density of electron states can be written as (cf. eqn. (5.15)):

$$D_e(\mathscr{E}) = \frac{1}{2\pi^2}\left(\frac{2m_e^*}{\hbar^2}\right)^{3/2}(\mathscr{E} - \mathscr{E}_c)^{1/2}. \tag{6.193}$$

Hence, eqn. (6.190) for the electron concentration in the conduction band becomes:

$$n = \frac{1}{2\pi^2}\left(\frac{2m_e^*}{\hbar^2}\right)^{3/2} e^{\mu/k_B T}\int_{\mathscr{E}_c}^{\infty}(\mathscr{E} - \mathscr{E}_c)^{1/2}e^{-\mathscr{E}/k_B T}d\mathscr{E}$$

or

$$n = 2\left(\frac{2\pi m_e^* k_B T}{h^2}\right)^{3/2} e^{[-(\mathscr{E}_c-\mu)/k_B T]}. \tag{6.194}$$

The pre-exponential term in eqn. (6.194) can be regarded as being the effective concentration N_c of all electron levels in the conduction band if located at the band edge, $\mathscr{E}_c$:

$$N_c = 2\left(\frac{2\pi m_e^* k_B T}{h^2}\right)^{3/2}. \tag{6.195}$$

N_c has the value of $\simeq 2.5 \times 10^{25} m^{-3}$ at 300 K for $m_e^* = m_e$.

The concentration of holes (i.e. missing electrons) in the valence band can be obtained in a similar manner. The Fermi–Dirac function for holes is related to that for electrons (eqn. (6.191)) by

$$\begin{aligned} f_h = 1 - f_e &= 1 - \frac{1}{\exp[(\mathscr{E} - \mu)/k_B T] + 1} \\ &= \frac{1}{\exp[-(\mathscr{E} - \mu)/k_B T] + 1}. \end{aligned} \tag{6.196}$$

For the chemical potential near midgap for an intrinsic semiconductor, $(\mu - \mathscr{E}) \gg k_B T$, and hence

$$\begin{aligned} f_h &\simeq \exp[(\mathscr{E} - \mu)k_B T] \\ &= \exp[-(\mathscr{E}_h + \mu)/k_B T] \end{aligned} \tag{6.197}$$

since $\mathscr{E}_h = -\mathscr{E}$ (eqn. (6.28)). The dispersion relation for the free-hole-like valence band can be written as

$$\mathscr{E}_h = \mathscr{E}_v + \frac{\hbar^2 k^2}{2m_h^*}, \tag{6.198}$$

where $\mathscr{E}_v$ is the energy of the valence-band maximum and $m_h^* = -m_e^*$ is the (scalar) hole effective mass. The corresponding density of states is

$$D_h(\mathscr{E}_h) = \frac{1}{2\pi^2}\left(\frac{2m_h^*}{\hbar^2}\right)^{3/2}(\mathscr{E}_v + \mathscr{E}_h)^{1/2}, \tag{6.199}$$

and hence the hole concentration is given by

$$p = \frac{1}{2\pi^2}\left(\frac{2m_h^*}{\hbar^2}\right)^{3/2} e^{-\mu/k_BT}\int_{\mathscr{E}_v}^{\infty}(\mathscr{E}_v + \mathscr{E}_h)^{1/2} e^{-\mathscr{E}_h/k_BT} d\mathscr{E}_h$$

or

$$p = N_v e^{[(\mathscr{E}_v - \mu)/k_BT]}, \tag{6.200}$$

where $\boldsymbol{N}_v$ is the concentration of levels at the valence-band edge,

$$N_v = 2\left(\frac{2\pi m_h^* k_B T}{h^2}\right)^{3/2}. \tag{6.201}$$

Note that the concentrations of electrons and holes in conduction and valence bands, respectively, given by the general equations (6.194) and (6.200), are determined by the separation in energy between the chemical potential and the respective band edges and depend exponentially on this quantity. An expression *independent* of μ, and therefore valid for *any* doping condition (not just intrinsic materials), can be obtained from eqns. (6.194) and (6.200) by multiplying the expressions together, i.e.

$$np = N_c N_v e^{-\mathscr{E}_g/k_BT} \tag{6.202a}$$

$$= 4\left(\frac{k_BT}{2\pi\hbar^2}\right)^3 (m_e^* m_h^*)^{3/2} e^{-\mathscr{E}_g/k_BT}, \tag{6.202b}$$

where the gap energy $\mathscr{E}_g = \mathscr{E}_c - \mathscr{E}_v$. Since $np = n_i^2$ for an *intrinsic* semiconductor, eqn. (6.202a) implies that

$$n_i = (N_c N_v)^{1/2} e^{-\mathscr{E}_g/2k_BT}. \tag{6.203}$$

For crystalline Si and Ge with bandgaps of 1.1 eV and 0.67 eV respectively, n_i is $1.5 \times 10^{16} m^{-3}$ and $2.4 \times 10^{19} m^{-3}$ respectively at 300 K.

Equation (6.202) has the form of the law of mass action applicable to chemical equilibria. Indeed, this expression can also be obtained by statistical-mechanical means by considering the thermal generation of electrons and holes as an equilibrium process

$$\bullet \rightleftharpoons e^- + h^+ \tag{6.204}$$

where the symbol $\bullet$ denotes the ground-state configuration (filled valence band, empty conduction band). Treating the electrons and holes as free particles, for which the volume-independent partition function $q = Z/V$ (see eqn. (4.64) for a definition of Z) can be written as the product of an electron-spin degeneracy term ($g_s = 2$) and a translational partition function:

$$q^{e,h} = g_s q_t^{e,h} = \left(\frac{2\pi m_{e,h}^* k_B T}{h^2}\right)^{3/2}, \tag{6.205}$$

the equilibrium constant for the excitation 'reaction' of eqn. (6.204) is

$$K = np \tag{6.206}$$

or, in terms of the partition functions,

$$K = (q^e q^h / q^\bullet) e^{-\Delta\mathscr{E}/k_B T}, \tag{6.207}$$

where $\Delta\mathscr{E} = \mathscr{E}_g$ is the activation energy for the excitation, and the ground-state configuration is taken to be immobile (i.e. $q^\bullet = 1$). It can be seen that eqns. (6.205–6.207) reduce to eqn. (6.202).

An explicit expression for the position of the chemical potential can be obtained from eqns. (6.194) and (6.200) for an *intrinsic* semiconductor by invoking eqn. (6.189), i.e.

$$n = p = N_c e^{[-(\mathscr{E}_c - \mu)/k_B T]} = N_v e^{[(\mathscr{E}_v - \mu)/k_B T]}$$

giving

$$\mu = \frac{(\mathscr{E}_c + \mathscr{E}_v)}{2} + \frac{3}{4} k_B T \ln\left(\frac{m_h^*}{m_e^*}\right),$$

or

$$\mu = \frac{\mathscr{E}_g}{2} + \frac{3}{4} k_B T \ln\left(\frac{m_h^*}{m_e^*}\right), \tag{6.208}$$

taking the energy scale to be zero at the valence-band maximum, $\mathscr{E}_v \equiv 0$, $\mathscr{E}_c \equiv \mathscr{E}_g$. Thus, it is seen from eqn. (6.208) that the chemical potential lies *exactly* at midgap for intrinsic semiconductors in two circumstances: in the case of symmetric bands (i.e. $m_e^* = m_h^*$) at any temperature, and at zero kelvin for any shape of the bands (see Problem 6.23).

6.5.1.2 Electrical conductivity

In semiconductors, electrons *and* holes can contribute simultaneously to the electrical current, and hence the expression (eqn. (6.11)) relating the d.c. conductivity, σ_0, the concentration and the mobility μ_α of carriers of type $\alpha(= \mathrm{e, h})$ must be generalized to:

$$\sigma_0 = ne\mu_e + pe\mu_h. \tag{6.209}$$

The mobility of charge carriers of type α is related to the scattering relaxation time τ_α and the effective mass m_α^* by eqn. (6.10):

$$\mu_\alpha = \frac{e\tau_\alpha}{m_\alpha^*}, \quad \alpha = \mathrm{e, h}. \tag{6.210}$$

Since the carrier mobility is only a weak (power-law) function of temperature, as will be seen below, the temperature dependence of the d.c. electrical conductivity of intrinsic semiconductors is governed by that of the carrier concentration, n_i, i.e. the conductivity is *thermally activated*, with an activation energy equal to half the bandgap, $\mathscr{E}_\sigma = \mathscr{E}_g/2$ (neglecting the weak power-law temperature dependence of N_c and N_v—cf. eqns. (6.195), (6.201)):

$$\sigma_0 = \sigma_0(0) e^{-\mathscr{E}_\sigma/k_B T}. \tag{6.211}$$

The conductivity of a semiconductor *increases* with increasing temperature (unlike the case of metals).

For pure intrinsic semiconductors, in the absence of charged impurities, *electron–phonon scattering* (§6.3.1.3) will be the dominant mechanism responsible for limiting the carrier mobility. The temperature dependence of the mobility in this case can be evaluated as follows. The mobility, eqn. (6.210), can be rewritten, in terms of the mean-free path Λ_α and the velocity v_α of a carrier of type α, as

$$\mu_\alpha = \frac{e\Lambda_\alpha}{m_\alpha^* v_\alpha}. \tag{6.212}$$

The velocity of electrons or holes in a semiconductor, treated as free particles, is temperature-dependent, $v_\alpha \propto (k_B T/m_\alpha^*)^{1/2}$. (Compare with the case for *metals* (§6.3.1), where the appropriate velocity, the Fermi velocity, is *temperature-independent*.) The mean-free path is inversely proportional to the number of scattering events, i.e. to the number of phonons, n_{ph}. At relatively high temperatures $(T > \theta_D), n_{ph} \propto T$ (eqn. (4.79)), and hence $\Lambda_\alpha \propto 1/T$ in this classical limit. Thus, the temperature dependence of the electron or hole mobility due to electron–phonon scattering is

$$\mu_\alpha^{e-ph}(T) \propto T^{-3/2}. \tag{6.213}$$

Hence, in this limit, the temperature dependences of $\mu_\alpha^{e-ph}(T)$ and of N_c or N_v cancel when combined according to eqn. (6.209).

Two types of electron–phonon scattering processes can be distinguished for both acoustic and optic phonons: in all cases, it is *longitudinal* phonons that tend to couple to the electrons (§6.3.1.3). For the *acoustic* case, LA phonons modulate the crystal unit cell and hence the lattice potential, resulting in deformation-potential scattering. For crystals with no inversion symmetry, the time-varying strain associated with the LA phonons produces time-varying internal electric fields due to the piezoelectric effect (§7.1.5.1), resulting in additional piezoelectric scattering. LO phonons can give rise to deformation-potential scattering, as for LA phonons. However, in addition, LO phonons produce a time-varying electrical polarization in the unit cell, and this gives rise to polar-mode scattering.

In §6.2.1 it was seen that the electron effective mass scales with the (direct) bandgap and hence, from eqn. (6.210), the highest electron mobilities should be found in those semiconductors having the smallest gaps; in general, this behaviour is observed (Table 6.6). The situation for hole mobilities is complicated by the presence of light-and heavy-hole bands at the valence-band maximum for tetrahedral semiconductors (§5.4.2) and, as a consequence of the possibility of holes being scattered between different bands, the mobility of holes is usually smaller than for electrons (see Table 6.6).

One way of increasing the electrical conductivity of a semiconductor is to shine light on it: photoconductivity arises principally because of the increase in the free-carrier concentration in the valence and/or conduction bands due to optical excitation, either of electron–hole pairs if the light has a photon energy greater than or equal to the bandgap (or of holes or of electrons separately if longer-wavelength light is used directly to excite filled acceptor or donor states, respectively, at low temperatures in extrinsic semiconductors — see §6.5.2). Considering the case of one carrier (say, electron) transport for simplicity, the *dark* d.c. conductivity is given by $\sigma_0 = ne\mu_e$, and the conductivity under light excitation, σ_L, is the sum of σ_0 and the photoconductivity, $\Delta\sigma$, viz.

$$\begin{aligned}\sigma_L &= \sigma_0 + \Delta\sigma \\ &= (n + \Delta n)e(\mu_e + \Delta\mu_e),\end{aligned} \tag{6.214}$$

Table 6.6 Carrier mobilities at 300 K for some semiconductors

Material	$\mu_e(m^2\ V^{-1}\ s^{-1})$	$\mu_h(m^2\ V^{-1}\ s^{-1})$
C (diamond)	0.18	0.12
Si	0.135	0.048
Ge	0.36	0.18
InSb	0.08	0.045
InAs	3.0	0.045
InP	0.45	0.01
GaAs	0.8	0.03
GaSb	0.5	0.1

Data after Kittel (1996). Reproduced by permission of John Wiley & Sons Inc.

where, in general, the light can cause a change in both the carrier concentration, Δn, *and* in the mobility, $\Delta\mu_e$. The increase in carrier concentration is usually proportional to the photogeneration rate G_{ph} (the concentration of carrier (pairs) generated per unit time and per unit volume) and the carrier lifetime, τ_n:

$$\Delta n = G_{ph}\tau_n. \tag{6.215}$$

The lifetime, itself, may also be a function of the photogeneration rate, $\tau_n = \tau_n(G_{ph})$: factors determining the free-carrier lifetime are recombination of electron–hole pairs, e.g. radiative or non-radiative (multiphonon emission or Auger excitation of another electron or hole) or trapping of excess carriers by traps (e.g. defect states in the gap). A photo-induced change in carrier mobility, $\Delta\mu_e$, could arise for a number of reasons: excitation of a carrier between bands characterized by different mobilities, photo-induced change in the concentration or scattering cross-section of charged impurities, or a light-induced lowering of interfacial barriers between grain boundaries. For the case of one-carrier transport at high photogeneration rates such that $\Delta n \gg n$ (but $\Delta\mu \simeq 0$), a figure of merit for the photoconduction can be defined from eqn. (6.214) in terms of the '$\mu\tau$-product', viz.

$$\Delta\sigma/G_{ph}e = \mu_e\tau_n. \tag{6.216}$$

The actual functional form of the photo-carrier concentration, or photoconductivity, on the carrier photogeneration rate, G_{ph}, depends on whether recombination or trapping effects are dominant. Consider an intrinsic semiconductor or insulator in which carriers are created by thermal or optical excitation across the gap, with generation rates of g_{th} and G_{ph}, respectively, and this is balanced by recombination of electron–hole pairs also across the gap. In the *dark*, the thermal generation rate is equal to the recombination rate R in dynamic equilibrium:

$$g_{th} = R = nv\sum p = n^2 v\sum, \tag{6.217}$$

since $n = p$ and where Σ is the recombination cross-section and v the average electron (hole) velocity. The dark carrier lifetime is therefore given by

$$\tau_d = 1/nv\sum. \tag{6.218}$$

Under optical excitation, the total carrier generation rate is

$$g_{\text{th}} + G_{\text{ph}} = (n + \Delta n)v \sum (p + \Delta p) = (n + \Delta n)^2 v \sum . \tag{6.219}$$

For the case of an *insulator*, $g_{\text{th}} \ll G_{\text{ph}}$ and $n \ll \Delta n$, and hence the photocarriers can only recombine with each other to give bimolecular recombination in which $G_{\text{ph}} \propto \Delta n^2$. From eqns. (6.217) and (6.219), under these conditions:

$$G = (2n\Delta n + \Delta n)^2 v \sum \simeq \Delta n^2 v \sum . \tag{6.220}$$

Hence, the photoconductivity behaves as $\Delta\sigma \propto G^{1/2}$.

For the case of a *semiconductor*, $\Delta n \ll n$, and hence eqns. (6.217) and (6.219) give

$$G_{\text{ph}} \simeq 2n\Delta n v \sum . \tag{6.221}$$

This situation is termed monomolecular recombination: a photo-carrier recombines with a conjugate thermally generated carrier, and $\Delta\sigma \propto G_{\text{ph}}$. The lifetime in this case is controlled by the dark behaviour of the semiconductor, with $\tau_{\text{n}} = \tau_{\text{d}}/2$.

Very often, *trapping* of excess carriers by electron or hole traps (typically defect states lying deep in the bandgap) competes with recombination in balancing the carrier generation rate. Consider an insulator containing a total concentration N_{t} of traps (for, say, electrons), n_{t} being the density of occupied traps, with the trapping cross-section given by $\sum_{\text{t}}$. In dynamic equilibrium (neglecting thermal generation)

$$G_{\text{ph}} = (n + \Delta n)v \sum\nolimits_{\text{t}} (N_{\text{t}} - n_{\text{t}}) + (n + \Delta n)v \sum (p + \Delta p),$$

$$\simeq \Delta n v \sum\nolimits_{\text{t}} (N_{\text{t}} - n_{\text{t}}) + \Delta n v \sum \Delta p, \tag{6.222}$$

where now Δn and Δp are no longer equal. If all traps are filled, $n_{\text{t}} = N_{\text{t}}$, after a period of illumination, the first term in eqn. (6.222) vanishes, and since $\Delta p = \Delta n + N_{\text{t}} \simeq N_{\text{t}}$, therefore

$$G_{\text{ph}} \simeq \Delta n v \sum N_{\text{t}}. \tag{6.223}$$

The behaviour is now monomolecular, in contrast to the trap-free case (eqn. (6.220)). Further details on photoconduction can be found in Bube (1992).

*6.5.1.3 *Thermoelectric effects*

The sign of the thermopower, S_T, is the *same* as that of the charge carrier involved in the thermal and electrical current (§6.3.2.3). For intrinsic semiconductors, where electrons and holes both contribute to the current, the overall thermopower is given by the conductivity-weighted sum of the individual thermopowers:

$$S_T = \frac{\sigma_0^{\text{e}} S_T^{\text{e}} + \sigma_0^{\text{h}} S_T^{\text{h}}}{\sigma_0^{\text{e}} + \sigma_0^{\text{h}}}. \tag{6.224}$$

The thermopower of a semiconductor (non-degenerate conductor) associated with a carrier of a *single* type can be obtained as follows (Mott (1993)). The element of electrical current density dj due to, say, electrons with energies between $\mathscr{E}$ and $\mathscr{E} + \mathrm{d}\mathscr{E}$ in the conduction band subject to an electric field $\boldsymbol{E}$ is given by (cf. eqn. (6.67)):

$$\mathrm{d}j = -\sigma(\mathscr{E})\frac{\partial f}{\partial \mathscr{E}} E\mathrm{d}\mathscr{E}, \tag{6.225}$$

where $\sigma(\mathscr{E})$ is the contribution to the d.c. electrical conductivity at energy $\mathscr{E}$ (§6.3.2.3) and $\partial f/\partial\mathscr{E}$ is the energy derivative of the Fermi–Dirac distribution function. The corresponding heat flux is (eqn. (6.78)):

$$\begin{aligned}\mathrm{d}J_Q &= -\frac{(\mathscr{E}-\mu)}{e}\mathrm{d}j \\ &= \frac{\sigma(\mathscr{E})}{e}\frac{\partial f}{\partial \mathscr{E}}(\mathscr{E}-\mu)E\mathrm{d}\mathscr{E}.\end{aligned} \tag{6.226}$$

Integration of this expression gives the Peltier heat Π multiplied by the electrical current flux under isothermal conditions (eqn. (6.93b)), i.e.

$$\Pi j = \frac{E}{e}\int_{\mathscr{E}_c}^{\infty}\sigma(\mathscr{E})\frac{\partial f}{\partial \mathscr{E}}(\mathscr{E}-\mu)\mathrm{d}\mathscr{E}, \tag{6.227}$$

or, since $\Pi/T = S_T$ (eqn. (6.96)),

$$S_T\sigma = \frac{k_B}{e}\int_{\mathscr{E}_c}^{\infty}\frac{\sigma(\mathscr{E})}{k_BT}(\mathscr{E}-\mu)\frac{\partial f}{\partial \mathscr{E}}\mathrm{d}\mathscr{E}. \tag{6.228}$$

Integration of eqn. (6.228) (see e.g. Smith (1978)) gives finally

$$S_T = -\frac{k_B}{e}\left[\frac{(\mathscr{E}_c-\mu)}{k_BT} + \frac{5}{2} + s\right], \tag{6.229a}$$

or, from eqns. (6.194) and (6.195)

$$S_T = -\frac{k_B}{e}\left[\ln\left(\frac{N_c}{n}\right) + \frac{5}{2} + s\right], \tag{6.229b}$$

and corresponding equations (with opposite sign) involving $\mathscr{E}_v$, N_v and p for holes. The factor $s = \partial \ln\tau(\mathscr{E})/\partial\mathscr{E}$ is of order unity, and τ is the relaxation time. Thus, from eqn. (6.219), the thermopower for semiconductors is found to be typically of the order of a few millivolts per kelvin (mV K^{-1}), i.e. approximately a *thousand* times larger than the thermopower of metals at the same temperature. Hence, since the Peltier heat has the relationship $\Pi \propto S_T$, semiconductors are used in solid-state Peltier heating or cooling devices (see §6.3.2.3).

6.5.1.4 *Cyclotron resonance*

As seen previously for the case of metals in §6.3.3.1, cyclotron resonance probes constant-energy surfaces of the electron band structure in reciprocal space: such experiments can provide estimates for the (cyclotron) effective mass and hence some information on the shape of the band structure in $\boldsymbol{k}$-space. The condition for cyclotron resonance to be observed is that many cyclotron orbits be completed before an electron-scattering event takes place, i.e. $\omega_c\tau \gg 1$, where τ is the scattering relaxation time and ω_c is the cyclotron-resonance frequency (cf. eqn. (6.115))

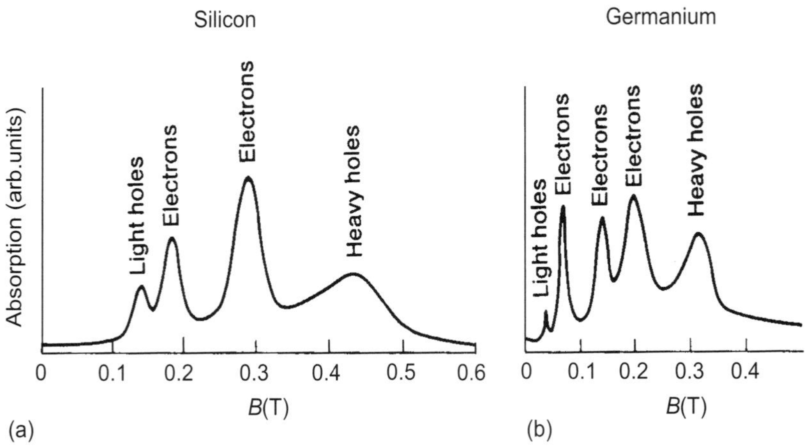

Fig. 6.48 Cyclotron-resonance absorption curves for the crystalline semiconductors: (a) Si; (b) Ge. In both cases, the applied magnetic field is in the (110) plane, at angles of $\theta = 30^\circ$ (a) and 60° (b) to the [001] direction (After Dresselhaus *et al.* (1955). Reprinted with permission from *Phys. Rev.* **98**, 368. © 1955. The American Physical Society)

$$\omega_c = \frac{2\pi eB}{\hbar^2}\left(\frac{\partial S_k}{\partial \mathscr{E}}\right)^{-1} \tag{6.230}$$

with δS_k being the element of area in k-space in the plane of the cyclotron orbit between two constant-energy surfaces separated by an energy $\delta\mathscr{E}$ (Fig. 6.17). In semiconductors, such experiments need to be performed on very pure and perfect crystals (to obviate electron–defect scattering) and at very low temperatures (to reduce electron–phonon scattering). Furthermore, since the carrier densities in the conduction and valence bands of intrinsic semiconductors are very small at low temperatures, excess electron–hole pairs are created by optical excitation (§5.8.2) in order that there are sufficient carriers for cyclotron resonance to be detected.

Electron-like and hole-like constant-energy surfaces (§6.3.3.1) can be distinguished experimentally by using *circularly polarized* electromagnetic radiation as the probe field because of the associated different directions of orbital motion of electrons and holes in a given magnetic field: electrons resonantly absorb right-circularly polarized, and holes left-circularly polarized, radiation. If *plane-polarized* radiation (consisting of equal contributions from both polarizations) is used, both electron-like and hole-like resonances can be detected.

Examples of experimental cyclotron-resonance curves for two semiconductors, crystalline Si and Ge, are shown in Fig. 6.48. The electron resonances correspond to cyclotron orbits on the ellipsoidal constant-energy 'pockets' near the minima of the conduction band, lying in the $\langle 100 \rangle$ directions in $\boldsymbol{k}$-space for Si and the $\langle 111 \rangle$ directions for Ge (Fig. 5.51). These ellipsoidal energy surfaces are characterized by two different values of the effective electron mass, the longitudinal and transverse components m_l^* and $m_t^* (m_l^* > m_t^*)$ with respect to the main axis of the ellipsoids (cf. eqn. (5.127)); cyclotron-resonance data can be used to provide estimates for m_l^* and m_t^* (see Problem

6.25). For a general direction of the applied static magnetic field $\boldsymbol{B}$ with respect to the Brillouin-zone axis, there will be three electron cyclotron resonances in Si (corresponding to the three $\langle 100 \rangle$ axes) and four resonances in Ge (corresponding to the four $\langle 111 \rangle$ axes). For the particular magnetic-field direction used to generate Fig. 6.48, symmetry causes the number of resonances to be reduced to two (Problem 6.25) and three, respectively.

Two hole-like cyclotron-resonance peaks are also observed in the curves of Fig. 6.48. The reason for this is the presence of the light-and heavy-hole bands in the vicinity of the valence-band maximum for Si and Ge (Fig. 5.53).

6.5.2 Extrinsic semiconductors

Extrinsic semiconductors are those in which the carrier concentration is controlled by the presence of electrically active impurities, or dopants, in the host material. This ability to vary the carrier concentration, and hence the electrical conductivity, of semiconductors by simply changing the impurity content is the basis for the use of such materials in technological applications and underpins the entire electronics industry. However, in most cases, electronic devices consist of semiconductors doped *heterogeneously* with impurities. The behaviour in such circumstances is discussed in §8.4.2.2; here the emphasis is on *homogeneous* doping.

Impurities that can *donate* extra electrons to the empty conduction band of the semiconductors are termed donors; impurities that can *accept* electrons from the filled valence band of the semiconductor (or equivalently inject extra holes into the valence band) are called acceptors (Fig. 6.49). Extrinsic semiconductors in which *holes* are the majority carriers are said to be p-type, and those in which *electrons* are the majority carriers are termed n-type. Electrical doping of semiconductors, i.e. the release of extra charge carriers into the bands of the host material, occurs because of the thermal excitation of electrons or holes from shallow dopants, i.e. those for which the electron energy levels associated with the impurity are very close to one or other of the bands, at a level $\mathscr{E}_d$ below the conduction-band edge for donors or at $\mathscr{E}_a$ above the valence-band edge for acceptors (Fig. 6.49). The presence of such energy levels near one or other of the band edges means that the chemical potential is very close to the conduction-band edge for an n-type material, and very near the valence-band edge for a p-type material, as will be seen in §6.5.2.2.

The dopants used in 'conventional' semiconductors such as Si, Ge or GaAs (as compared to, say, polymeric conductors — see §8.3.3) are usually substitutional impurities, in which the impurity atom has a different valence from that of atoms in the host matrix. Thus for the case of tetrahedrally coordinated Group IV monovalent semiconductors Si and Ge, for example, Gp. V elements (e.g. P, As, Sb) having one *extra* electron compared with the host atoms, behave as *donors* when substituted for the Gp. IV atoms (Fig. 6.50a). Similarly Gp. III elements (e.g. B, Al, Ga, In), having one *less* electron than Gp. IV atoms, are *acceptors* when incorporated substitutionally (Fig. 6.50b). Gp. VI elements (e.g. S, Se) have the potential to be double donors, and Gp. III elements (e.g. Be and Zn) can act as double acceptors when incorporated substitutionally into Si or Ge.

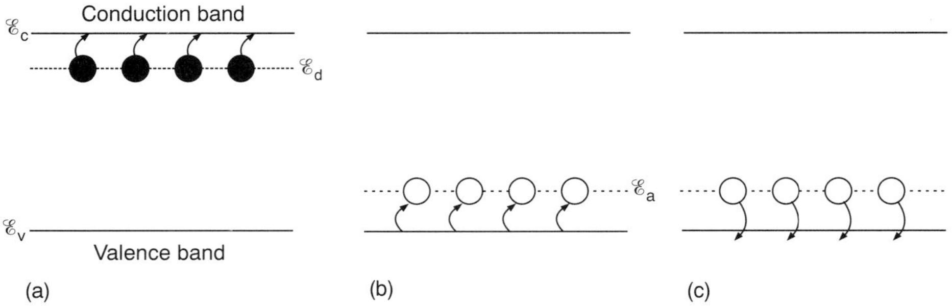

Fig. 6.49 Schematic illustration of the origin of electrical doping in an extrinsic semiconductor: (a) n-type, containing donors that donate extra electrons into the conduction band; (b) p-type, containing acceptors that accept electrons from the valence band; (c) another representation of the action of acceptors, i.e. injecting holes into the valence band.

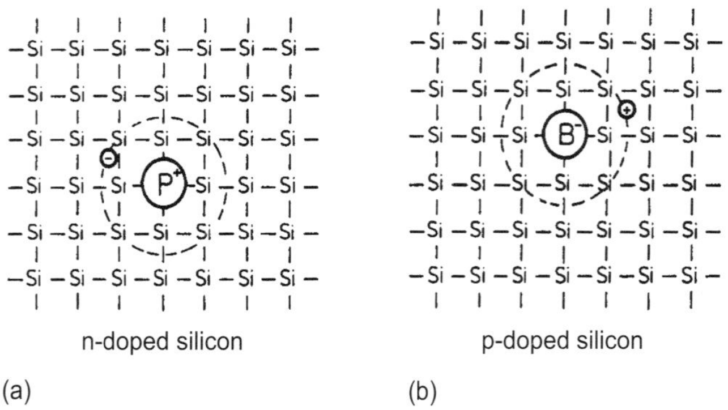

Fig. 6.50 Illustration of substitutional doping in tetrahedrally coordinated Group IV semiconductors: (a) P in Si (donor); (b) B in Si (acceptor). The orbit of the extra electron/hole is shown bound to the positively charged/negatively charged P^+/B^-ion core. The extra electron/hole becomes free of the dopant if thermally excited into a conduction/ valence band of the Si. Note that the scale of the orbit in the figure is reduced by a factor of approximately 10 compared with reality.

The case of doping of *compound* semiconductors, e.g. AB, is a little more complicated since a given impurity atom C could, in principle, substitute for either the cations A or the anions B: such substitutional defects are termed C_A, and C_B, respectively. Depending on the site of substitution, a Gp. IV impurity in say a III–V semiconductor can therefore behave as a donor *or* an acceptor. Thus Si_{Ga} and Ge_{Ga} are donors, whereas Si_{As} and Ge_{As} are acceptors. Gp. VI elements substituted for As, e.g. S_{As}, Se_{As}, act as donors, whereas Zn_{Ga} and Cd_{Ga} are acceptors.

Isovalent (or isoelectronic) defects are those in which the substitutional impurity atom has the *same* valence as the atom for which it substitutes. However, electronegativity differences between the impurity atom and the other atoms mean that it can become electrically active. For example, the N_p centre in GaP can attract an electron to it, thereby becoming negatively charged, because of the large electronegativity difference between N and P. The resultant N_p^- centre can subsequently attract a hole to itself, behaving as an isovalent acceptor. The electron and hole associated with the neutral N_p centre can also be viewed as a *bound exciton* (§5.8.3).

Doping of semiconductors can also be achieved by the incorporation of electrically active impurities in the host material in a non-substitutional manner. For example, impurity atoms that are smaller than the host atoms can occupy *interstitial* sites and, if electropositive, can donate electrons to the conduction band of the semiconductor: Li in Si is an example.

6.5.2.1 Hydrogenic dopants

Shallow substitutional dopants can be reasonably described by a hydrogenic model, in which the extra electron bound (at low temperatures) to a positively charged donor ion core can be treated in the same way as an electron bound to a proton in the hydrogen atom. Holes bound to negatively charged acceptors are correspondingly analogous to a positron bound to a negatively charged muon ('muonium').

The dynamics of an electron bound to a donor (a hole bound to an acceptor can be treated in the same way) can be described by the Schrödinger equation for a particle moving in a Coulombic potential, as long as the radius of the electron orbit is large compared with the interatomic spacing of the host semiconductor so that anisotropy associated with the local atomic configuration is essentially averaged out. There are two important differences between a hydrogen atom and, say, a hydrogenic donor: in the former, the mass of the electron is the free mass m_e and the orbit is *in vacuo*; in the latter; the *effective* mass of the electron, m_e^*, is determined by the shape in $\boldsymbol{k}$-space of the conduction-band states used to construct the defect-related wavefunction, and the electron orbit, if large enough, takes place essentially in the bulk semiconductor, and hence the electrical charge on the ionized donor is *screened* from the electron by the bulk dielectric constant, ε.

As a result, the energy levels of the bound states of a shallow donor (or acceptor) can be written as a Rydberg series in analogy to the solution of the H-atom problem, i.e.

$$\mathscr{E} = \mathscr{E}_c - \frac{R_d}{n^2}, \quad n = 1, 2, 3, \ldots \tag{6.231}$$

where the dopant Rydberg constant, R_d, is related to that for the hydrogen atom, R_0, by

$$R_d = \frac{m_e^*}{m_e \varepsilon^2} R_0, \tag{6.232}$$

or

$$R_d = \frac{e^4 m_e^*}{32\pi^2 \varepsilon^2 \varepsilon_0^2 \hbar^2}. \tag{6.233}$$

The donor-electron energy levels given by the Rydberg series (eqn. (6.231)) tend to a continuum limit—the ionization limit — when $n = \infty$ and the electron is no longer bound to the positively charged core. For the case of a donor in a semiconductor, this ionization level corresponds to the conduction-band edge, $\mathscr{E}_c$, of the semiconductor for, once the donor electron is in the conduction band of the material, it is free to travel anywhere in the solid. The lowest (deepest) bound energy level below $\mathscr{E}_c$, corresponding to $n = 1$, is conventionally termed the donor level, i.e.

$$\mathscr{E}_d = R_d. \tag{6.234}$$

Although $R_0 = 13.6$ eV for the H atom, the corresponding value for a donor electron is greatly reduced from this figure by the combined effects of the inclusion of m_e^* and ε (eqn. (6.232)): typical values of $m_e^* \simeq 0.1 m_e$ and $\varepsilon \simeq 10$ give a maximum binding energy of $\mathscr{E}_d \simeq 13.6$ meV. Note that $\mathscr{E}_d$ depends on quantities pertaining only to the host semiconductor and *not* the dopant itself. The dopant-related Rydberg series of levels can be probed by optical absorption (Fig. 6.51).

The radius of the Bohr orbit of the donor electron, r_d (i.e. the spatial extent of the corresponding 1s-like wavefunction for $n = 1$) is related to the Bohr radius of the H atom, $a_0 (= 0.63$ Å), *viz.*

$$r_d = \left(\frac{m_e}{m_e^*}\right) \varepsilon a_0 \tag{6.235}$$

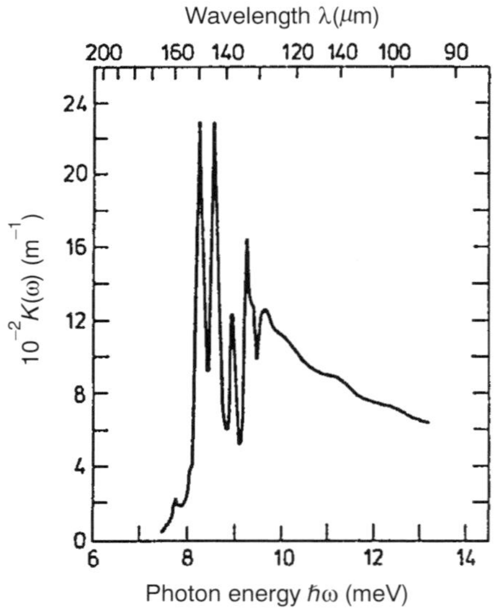

Fig. 6.51 Optical absorption spectrum of bound-electron Rydberg-like levels associated with Sb donors in Ge, measured at 9 K. (After Reuszer and Fischer (1964). Reprinted with permission from *Phys. Rev.* **135**, A1125. © 1964. The American Physical Society)

Table 6.7 Theoretical (hydrogenic) and experimental donor binding energies $\mathscr{E}_d$ for shallow donors in III–V and II–VI semiconductors

Semiconductor	$\mathscr{E}_d^{th}$ (meV)	$\mathscr{E}_d^{exp}$ (meV)
GaAs	5.72	5.84 (Si_{Ga}); 5.88 (Ge_{Ga}); 5.87 (S_{As}); 5.79 (Se_{As})
InSb	0.6	0.6 (Te_{Sb})
CdTe	11.6	14 (In_{Cd}); 14 (Al_{Cd})
ZnSe	25.7	26.3 (Al_{Zn}); 27.9 (Ga_{Zn}); 29.3 (F_{Se}); 26.9 (Cl_{Se})

After Yu and Cardona (1996), *Fundamentals of Semiconductors*, p. 159, Table 4.1, © Springer-Verlag GmbH & Co. KG.

Table 6.8 Experimental donor and acceptor binding energies, $\mathscr{E}_d$ and $\mathscr{E}_a$, respectively, for substitutional dopants in Si and Ge

$\mathscr{E}_d$ (meV)	P	As	Sb	B	Al	Ga	In / $\mathscr{E}_a$ (meV)
Si	45	49	39	45	57	65	16
Ge	12	12.7	9.6	10.4	10.2	10.8	11.2

or

$$r_d = \frac{4\pi\varepsilon\varepsilon_0\hbar^2}{m_e^* e^2}. \tag{6.236}$$

Inserting the same values of m_e^* and ε as used above into eqn. (6.236) gives $r_d = 53$ Å, thereby justifying the approximation used to derive eqns. (6.231) and (6.236) that the orbital radius of an electron bound to a donor be much larger than the interatomic spacing.

This approach gives predicted values in remarkable agreement with experimental values of donor binding energies in *direct-gap* compound semiconductors, where the conduction-band minimum is at the Γ-point and the effective conduction-band effective mass, m_e^*, is approximately isotropic (see Table 6.7). It is not as accurate for dopants in Si or Ge (see Table 6.8) which have conduction-band minima at points in $\boldsymbol{k}$-space far removed from the Γ-point, and hence which have very anisotropic conduction-band electron masses. The problem is also complicated for acceptor binding energies because of the complex valence-band structure near the Γ-point (§5.4.2). Hydrogenic estimates for $\mathscr{E}_d$ in Si ($\varepsilon = 11.7$, $m_e^* = 0.3m_e$) and Ge ($\varepsilon = 15.8$, $m_e^* = 0.12m_e$) are 29.8 meV and 6.5 meV, respectively.

Because the donor wavefunction for shallow, hydrogenic donors is spread over a large distance, corresponding to very many unit cells in real space (cf. eqn. (6.236)), consequently the conduction-band states of the semiconductor used to construct the shallow-donor wavefunctions originate from a very *small* region in $\boldsymbol{k}$-space in the vicinity of the conduction-band minimum at the Γ-point. The donor-electron wavefunction can be written, analogously to a Bloch wavefunction (eqn. (5.59)), approximately as the product of a function with the periodicity of the lattice, but independent of $\boldsymbol{k}$ and equal to the function at $\boldsymbol{k} = 0$, i.e. $u_{k=0}(\boldsymbol{r})$, and an envelope function $\phi(\boldsymbol{r})$ (rather than a plane wave as for a Bloch function) which is localized around the donor site and which is the solution of the Schrödinger equation for the orbital motion of the electron bound to the donor:

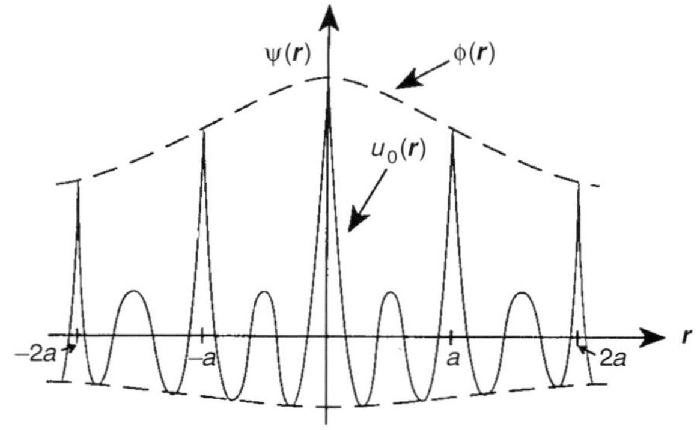

Fig. 6.52 Schematic illustration of a shallow hydrogenic donor-electron wavefunction in real space, being the product of a periodic function $u_0(\boldsymbol{r})$ and an envelope function $\phi(\boldsymbol{r})$ that is the solution of the Schrödinger equation for the orbital motion of an electron, bound to a donor. The lattice separation is a.

$$\psi(\boldsymbol{r}) \simeq u_0(\boldsymbol{r})\phi(\boldsymbol{r}). \tag{6.237}$$

Such a donor wavefunction is shown schematically in Fig. 6.52.

In certain circumstances, the wavefunction $\psi(\boldsymbol{r})$ can be spatially localized, in which case it can be regarded as being made up from Bloch functions from a number of bands and with a wide range of $\boldsymbol{k}$-values, in contrast to the shallow levels. Such electronic configurations are termed deep centres (since the corresponding energy levels usually lie deep within the gap), and they may arise for dopants having large differences in electronegativity or core potential compared with the host atoms, or for structural defects in semiconductors such as dangling bonds (§3.1.1). The occupation of deep levels by electrons is also often accompanied by a lattice distortion.

6.5.2.2 *Dopant carrier statistics*

The carrier concentration in either the valence or conduction band of a semiconductor is determined by the position of the chemical potential in the gap (eqns. (6.194), (6.200)). In general, both acceptor-like *and* donor-like impurities can be present in an extrinsic semiconductor (then said to be compensated) and these centres may both be partially ionized at a given temperature so that the total donor (d) or acceptor (a) concentration is the sum of the un-ionized (neutral) and ionized dopant concentrations:

$$N_{\mathrm{d}} = N_{\mathrm{d}}^{0} + N_{\mathrm{d}}^{+}, \tag{6.238a}$$

$$N_{\mathrm{a}} = N_{\mathrm{a}}^{0} + N_{\mathrm{a}}^{-}. \tag{6.238b}$$

For the case of homogeneous doping, the position of the chemical potential is determined by the constraint of overall charge neutrality:

$$n + N_{\mathrm{a}}^{-} = p + N_{\mathrm{d}}^{+} \tag{6.239}$$

where n and p are the total (extrinsic plus intrinsic) carrier concentrations in the conduction and valence bands, respectively. Unfortunately, the carrier statistics for this general case can only be treated numerically.

For simplicity, and to obtain an analytic expression, it will be assumed henceforth that only a *single* type of dopant, e.g. donors, is present in the semiconductor. It is found that, in such a case, at very low temperatures, the chemical potential is very close to the conduction band edge $\mathscr{E}_c$ (between it and $\mathscr{E}_d$), whereas at very high temperatures, μ falls to midgap (if the conduction and valence bands are symmetric in shape). The same behaviour for acceptors is found relative to the valence-band edge. Three temperature régimes can be distinguished for the carrier statistics of extrinsic semiconductors: the freeze-out régime at the very lowest temperatures ($k_B T \ll \mathscr{E}_d$) where most of the donors are un-ionized; the saturation or exhaustion régime at higher temperatures ($k_B T > \mathscr{E}_d$) where all donors are ionized; and, at the highest temperatures, *intrinsic-like* behaviour is recovered when the concentration of intrinsic carriers thermally excited across the gap becomes greater than the concentration of donor electrons ($n_i \gg N_d$).

The *freeze-out* régime at very low temperatures ($k_B T \ll \mathscr{E}_d$) can be analysed by assuming that $N_d \gg n_i$, which will be satisfied at such temperatures. The equilibrium concentration of conduction electrons (in this approximation, exclusively from donors) can be calculated via a statistical-mechanics approach similar to that used to analyse the ionization of say alkali atoms; here the equilibrium involving the ionization of donors, d, can be written as

$$\mathrm{d} \rightleftharpoons \mathrm{d}^+ + \mathrm{e}^-. \tag{6.240}$$

Thus, the conduction-electron concentration is equal to that of the ionized donors, $n = [\mathrm{d}^+]$, but at very low temperatures most of the donors, are in fact, *un-ionized*, i.e. $[\mathrm{d}] \simeq N_d$. The equilibrium constant for the reaction given by eqn. (2.40) is

$$\begin{aligned} K &= [\mathrm{d}^+][\mathrm{e}^-]/[\mathrm{d}] \\ &\simeq n^2/N_d. \end{aligned} \tag{6.241}$$

Using the statistical-mechanical relation for the equilibrium constant expressed in terms of partition functions (cf. eqn. (6.207)), and noting that since the dopants are immobile, $q^{d+} = q_t^{d+} = 1$, whereas $q^d = g_s q_t^d = 2$ taking into account the fact that a donor state may be occupied by a spin-up or spin-down electron, gives

$$\begin{aligned} K &= \left(\frac{2\pi m_e^* k_B T}{h^2}\right)^{3/2} \mathrm{e}^{-\mathscr{E}_d/k_B T}, \\ &= \frac{N_c}{2} \mathrm{e}^{-\mathscr{E}_d/k_B T}, \end{aligned} \tag{6.242}$$

where the activation energy for the equilibrium (eqn. (6.240)) is taken to be the donor binding energy. Thus, combining eqns. (6.241) and (6.242) gives (see also Problem 6.26):

$$n = \left(\frac{N_d N_c}{2}\right)^{1/2} \mathrm{e}^{-\mathscr{E}_d/2k_B T}. \tag{6.243}$$

Hence, the carrier concentration is thermally activated, with activation energy $\mathscr{E}_d/2$.

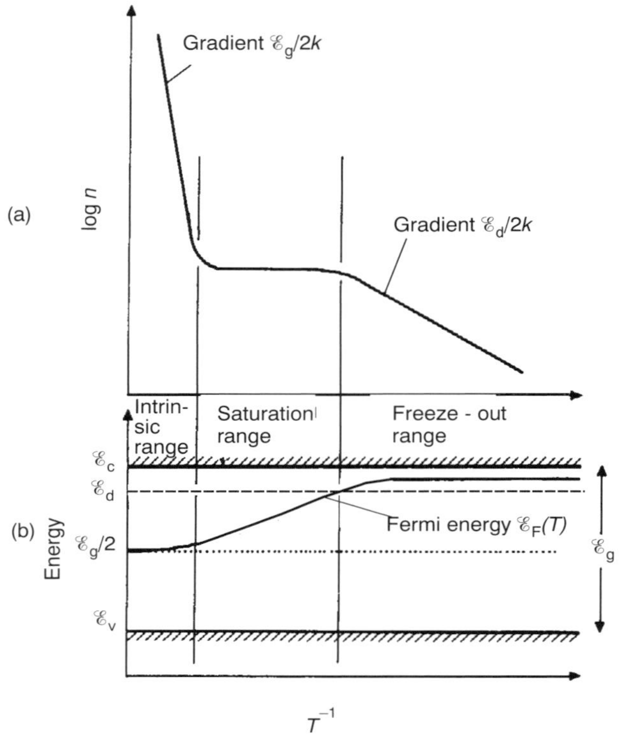

Fig. 6.53 Schematic illustration of the temperature dependence, showing the freeze-out, saturation and intrinsic régimes, in an extrinsic n-type semiconductor, of: (a) the carrier concentration; (b) the chemical potential.

Use of the general expression for n in terms of μ (eqn. (6.194)) with eqn. (6.243) gives for the position of the chemical potential in the freeze-out limit:

$$\mu = \mathscr{E}_\mathrm{c} - \frac{\mathscr{E}_\mathrm{d}}{2} - \frac{k_\mathrm{B}T}{2}\ln\left(\frac{2N_\mathrm{c}}{N_\mathrm{d}}\right). \tag{6.244}$$

At zero kelvin, the chemical potential lies midway between the completely filled donor levels and the conduction-band edge (Fig. 6.53b). As the temperature increases, the chemical potential moves downwards towards midgap (since $N_\mathrm{c} > N_\mathrm{d}$ in general).

In the *saturation* régime ($k_\mathrm{B}T > \mathscr{E}_\mathrm{d}$), all the donors are ionized, and so

$$n = N_\mathrm{d}; \tag{6.245}$$

the carrier concentration is therefore *independent* of temperature (Fig. 6.53a). Equating N_d to eqn. (6.194) for n gives for the temperature dependence of the chemical potential (see also Problem 6.26):

$$\mu = \mathscr{E}_\mathrm{c} - k_\mathrm{B}T\ln\left(\frac{N_\mathrm{c}}{N_\mathrm{d}}\right); \tag{6.246}$$

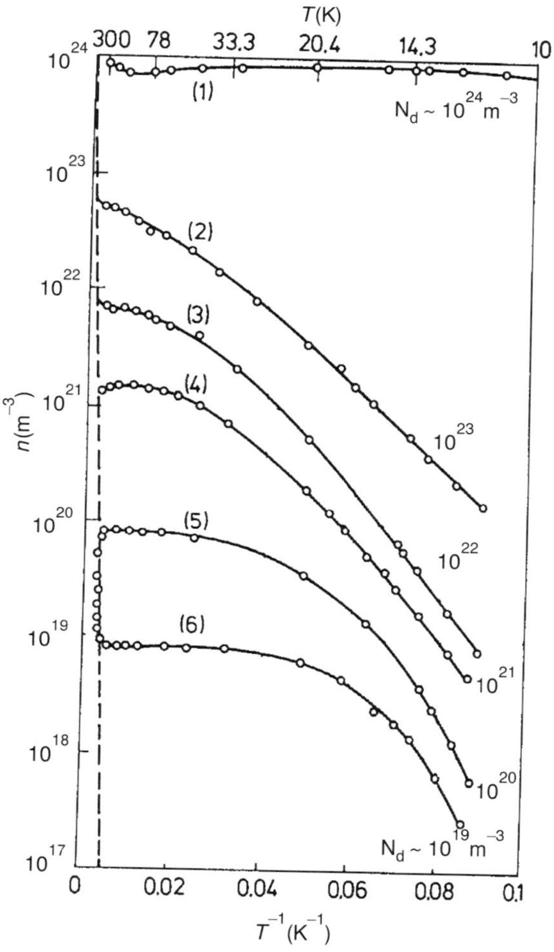

Fig. 6.54 The concentration of electrons in the conduction band measured by the Hall effect, for samples of n-type Ge with donor concentrations as marked. The dashed line at high temperatures represents the activated temperature dependence of the carrier concentration in the intrinsic régime. (After Conwell (1952), *Proc. IRE* **40**, 1327. © 1952 IEEE)

(see Fig. 6.53b). If N_d is comparable to N_c, Fermi–Dirac statistics must be used instead of the approximate Boltzmann expression.

At yet higher temperatures, the thermally generated intrinsic carrier concentration becomes greater than the donor-electron concentration, $n_i \gg N_d$, and the behaviour of the extrinsic semiconductor reverts to being intrinsic-like (Fig. 6.53): the temperature dependence of the carrier concentration is thermally activated, with an activation energy of half the gap (eqn. (6.203)) and the chemical potential lies at midgap if the conduction and valence bands are symmetric in shape (eqn. (6.208)).

For n-type Si, with a phosphorus donor concentration of $N_d = 3 \times 10^{20} m^{-3}$, the saturation régime extends between 50 and 500 K since the intrinsic carrier concentration is so low at these temperatures (§6.5.1.1). For doped Ge, on the other hand, the intrinsic régime starts at $T \simeq 300K$ (Fig. 6.54).

In *compensated* semiconductors, in which both donors *and* acceptors are present simultaneously, the electronic energy of the system is lowered by electrons from the filled donor levels falling down in energy and occupying vacant acceptor centres. In a perfectly compensated extrinsic material, in which $N_d = N_a$, the chemical potential will therefore remain at the position in the gap characteristic of the *intrinsic* material as long as $\mathscr{E}_d = \mathscr{E}_a$.

6.5.2.3 Electrical conductivity

Since $\sigma_0 = ne\mu_e$ (eqn. (6.209)) for an n-type semiconductor, in the freeze-out and intrinsic temperature régimes for an extrinsic semiconductor where the temperature dependence of the carrier concentration, being thermally activated (Fig. 6.54), is greater than that of the mobility, $\mu_e(T)$, the temperature dependence of the conductivity is also thermally activated with the same activation energies as for $n(T)$. In the saturation régime, where $n(T) = N_d$ is constant, the temperature dependence of the conductivity will be controlled by that of the mobility, μ_e.

At higher temperatures in the saturation régime, phonon scattering (§6.3.1.3) will be the dominant mechanism limiting the mobility, for which $\mu_e(T) \propto T^{-3/2}$ in semiconductors (eqn. (6.213)). At lower temperatures, where the number of phonons decreases, electron scattering will be predominantly by the ionized dopants (§6.3.1.2), for which the scattering relaxation time depends on electron velocity v as $\tau_d \propto v^3$ (eqns (6.40) and (6.45), where v in a semiconductor, being controlled effectively by (classical) Boltzmann statistics, has the temperature dependence $v \propto T^{1/2}$. Thus, since $\mu_e \propto \tau_d$ (eqn. (6.210)), the temperature dependence of the electron mobility due to ionized-impurity scattering in semiconductors is

$$\mu_e^{imp}(T) \propto T^{3/2}. \tag{6.247}$$

Hence the overall temperature dependence of the total mobility μ_e, where

$$\frac{1}{\mu_e} = \frac{1}{\mu_e^{e-ph}} + \frac{1}{\mu_e^{imp}}, \tag{6.248}$$

should exhibit a maximum (Fig. 6.55). This behaviour is evident in the experimental curves for $\mu(T)$ for n-type Ge shown in Fig. 6.56.

In the saturation régime for n-Ge, extending between $T \simeq 30K$ and 300 K for the sample with the lowest donor concentration ($N_d \simeq 10^{19} m^{-3}$) in Fig. 6.54, the *decrease* in conductivity with increasing temperature, due to the decrease in $\mu(T)$ because of electron–phonon scattering, is clearly evident (Fig. 6.57).

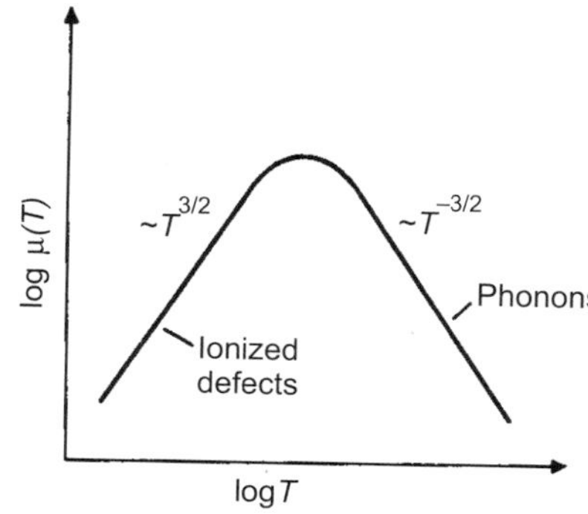

Fig. 6.55 Schematic illustration of the temperature dependence of the carrier mobility, $\mu(T)$, in an extrinsic semiconductor associated with charged-dopant scattering at low temperatures and electron–phonon scattering at high temperatures.

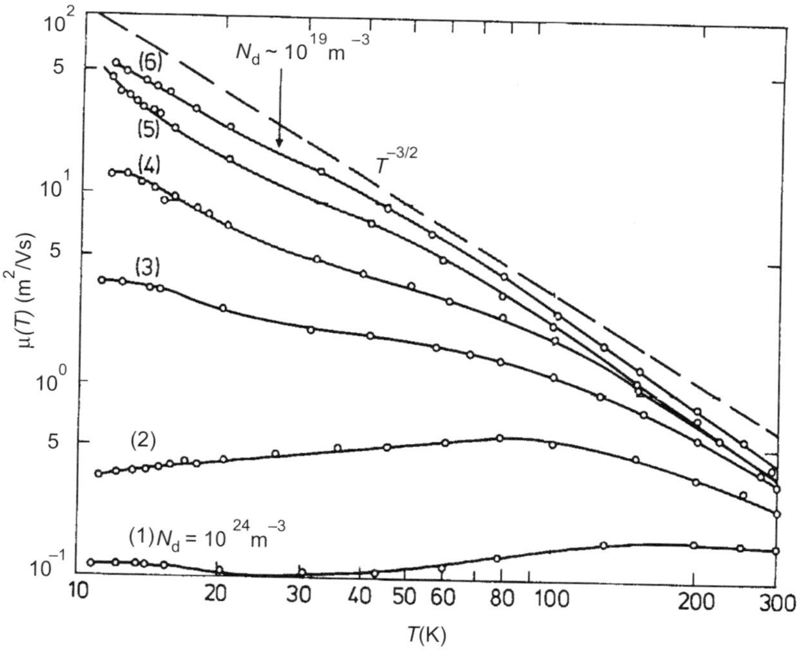

Fig. 6.56 Experimental temperature dependence of the electron mobility in n-type Ge (same samples as in Fig. 6.54). The dashed line shows the $T^{-3/2}$ temperature dependence characteristic of electron–phonon scattering. (After Conwell (1952), *Proc. IRE* **40**, 1327. © 1952 IEEE)

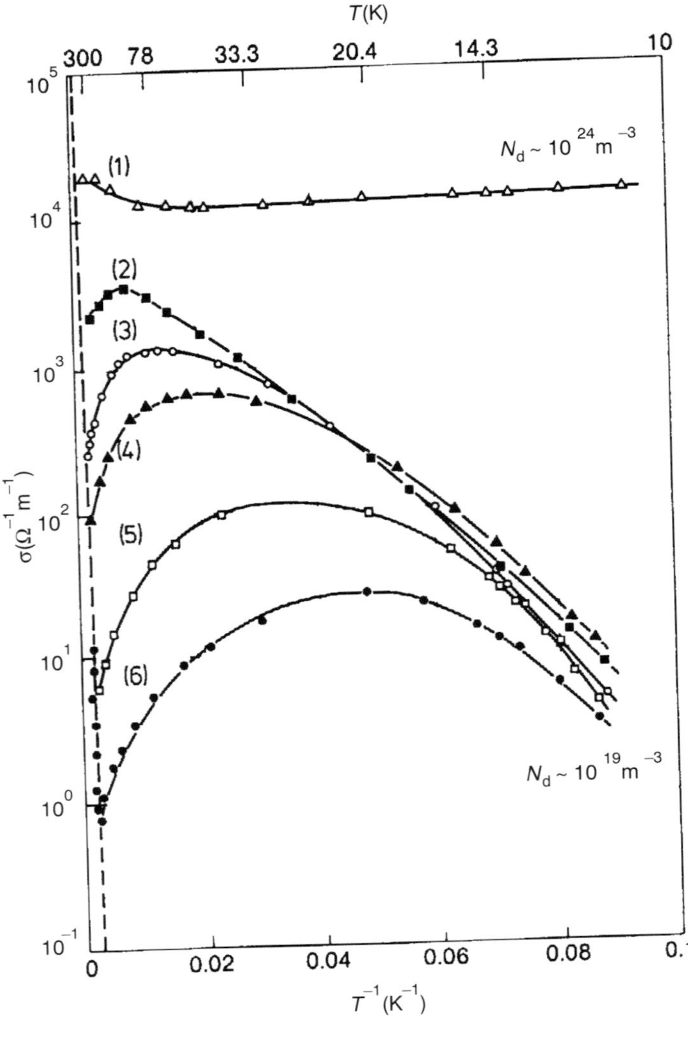

Fig. 6.57 Experimental temperature dependence of the electronic conductivity of n-type Ge (same samples as in Fig. 6.54). The dashed line shows the thermally activated behaviour in the intrinsic régime. (After Conwell (1952), *Proc. IRE* **40**, 1327. © 1952 IEEE).

*Polarons 6.6

In the discussion of electron states in solids given in the previous chapter, and of the dynamical behaviour associated with such states presented in this chapter, it has been assumed generally that the rigid-band approximation is valid; that is, the occupation of a state by an electron does not affect the energy of that state. However, there are a number of exceptions to this simple picture when the coupling between an electron and the lattice, i.e. the *electron–phonon interaction* (§6.3.1.3), is appreciable.

An important example of this behaviour is the case of an electron in a polar lattice, i.e. a lattice having a basis of say two types of atoms with differing charges, e.g. an ionic crystal. An electron in such a material will polarize its immediate surroundings (see Fig. 6.58), and the electron together with the polarization cloud forms a quasiparticle termed a polaron. As the electron moves through the crystal, it will drag the polarization cloud with it and, as a consequence, will behave as if it has a higher inertial mass m^{**}, and hence a lower mobility (eqn. (6.210)), than a free electron. Although in the case of non-polar solids, the dominant interaction of electrons is with LA phonons (§6.3.1.3), in the case of polar materials it is the LO-phonons that are strongly coupled to the electrons because of the large changes in dipole moment associated with the atomic displacements of LO modes (§4.4). Thus, the distortion of the lattice associated with the polarization cloud shown in Fig. 6.58 can be regarded equivalently as being a cloud of excited *virtual* optic phonons associated with the electron. Virtual phonons are emitted and absorbed with a time interval δt that is in accord with Heisenberg's energy–time uncertainty principle so that energy, on average, is conserved even though an electron might not have enough energy to emit a long-lived LO phonon (cf. eqn. (6.51)).

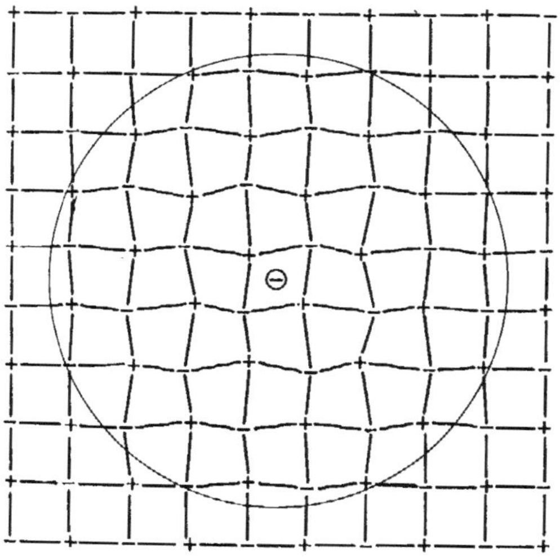

Fig. 6.58 Schematic illustration of a large polaron in a crystal. The electron polarizes its surroundings, and the carrier plus the associated lattice distortion make up the polaron.

Two types of polaron may be distinguished, depending on the spatial extent of the lattice distortion associated with the quasiparticle: small polarons are those for which the distortion is restricted to the immediate vicinity of the electron, and large polarons are those in which the distortion extends over a distance corresponding to many lattice constants (see Fig. 6.58). A review of polarons is given by Alexandrov and Mott (1995).

The strength of the electron–phonon interaction can be described by a dimensionless polaron coupling constant α_p defined as

$$\frac{\alpha_p}{2} = \frac{\mathscr{E}_D}{\hbar\omega_L}, \tag{6.249}$$

where $\mathscr{E}_D$ is the distortion energy associated with the lattice polarization (Fig. 6.58) and ω_L is the $\boldsymbol{k} = 0$ value of the LO-phonon frequency. The quantity $\alpha_p/2$ can also be regarded as the average number of excited virtual phonons around an electron in a polar crystal. Values of α_p range from 0.015 and 0.08 for the III–V materials InSb and InP, through 1.69 and 2.0 for the silver halides AgBr and AgCl, to values of 2.4, 3.97 and 6.6 for the alkali halides LiI, KCl and RbBr, respectively.

A large polaron is characterized by a small value of the coupling constant, $\alpha_p < 1$, and a large spatial extent, $r_{lp} = (\hbar/2m^*\omega_L)^{1/2} \gg a$ where a is the lattice constant. Its energy is reduced by $\alpha_p\hbar\omega_L$ with respect to a bare Bloch state. In such a case, the effective mass enhancement is given by (see e.g. Alexandrov and Mott (1995)):

$$\frac{m^{**}}{m^*} \simeq \left(1 + \frac{\alpha_p}{6}\right). \tag{6.250}$$

Large polarons move in a band, as for Bloch electrons, but with a correspondingly smaller mobility limited by LO-phonon scattering.

With increased polaron coupling constant α_p, the lowering of the polaron energy increases and, particularly for the case of *narrow* bands (such as occur for the d-bands in semiconducting transition-metal oxides), eventually the electron-delocalization tendency associated with band formation is no longer able to resist the localizing tendency associated with the polaron formation, and the carrier becomes self-trapped and forms a *small polaron* whose spatial extent is comparable to the lattice spacing. Different types of small polaron can be distinguished: dielectric small polarons are those in which electrical polarization of a polar lattice (as in Fig. 6.58) is responsible for the carrier localization; molecular small polarons are those in which *covalent* interactions are primarily involved in the lattice distortion, as in the self-trapped-hole V_K-centre (molecular ion) in alkali halides (§3.3.1 and Fig. 3.20).

The binding energy of a dielectric small polaron can be estimated as follows. For a *rigid* polar lattice, the electrostatic potential of an electron in it is given by an expression involving the *static* dielectric constant, $\phi = -e/4\pi\varepsilon(0)\varepsilon_0 r$, whereas if the lattice is allowed to distort because of polarization (Fig. 6.58), the corresponding potential is determined by the high-frequency dielectric constant (i.e. $\phi = -e/4\pi\varepsilon(\infty)\varepsilon_0 r$. The difference between these two quantities is the self-induced potential well for $r > r_{sp}$ (the small-polaron radius). If the potential for $r < r_{sp}$ is assumed to be constant, the potential well looks as in Fig. 6.59, and the depth of the well is given by $\phi_{sp} = -e/4\pi\varepsilon_0 r_{sp}[\varepsilon(\infty)^{-1} - \varepsilon(0)^{-1}] \equiv (-e/4\pi\varepsilon_0\varepsilon_p r_{sp})$, where $\varepsilon_p^{-1} = [\varepsilon(\infty)^{-1} - \varepsilon(0)^{-1}]$. Taking also into account the potential energy associated with the polarization of the surroundings of the polaron, together with the kinetic energy of the electron arising

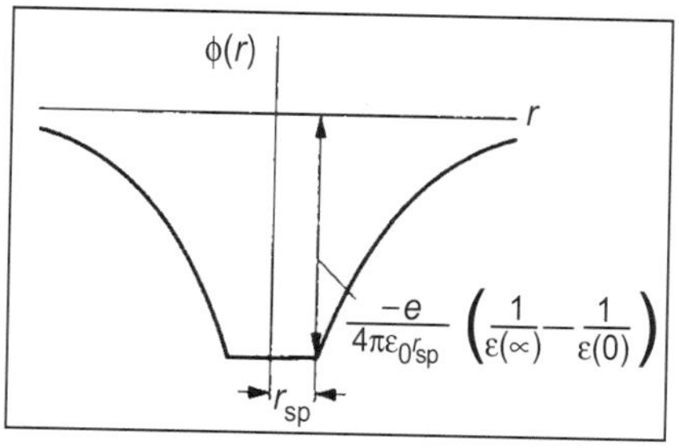

Fig. 6.59 Electrostatic potential well for a dielectric small polaron of radius r_{sp}.

from its confinement to a volume of radius r_{sp}, gives as the dielectric small-polaron binding energy (see Problem 6.28):

$$\mathscr{E}_{sp} = -\frac{e^2}{16\pi\varepsilon_0\varepsilon_p r_{sp}} \tag{6.251a}$$

$$= -\frac{\alpha_p^2}{4}\hbar\omega_L. \tag{6.251b}$$

A small polaron in an ionic lattice can also be regarded in chemical terms as a valence alternation of a cation, caused by a trapped electron or hole, together with the associated lattice distortion surrounding the trapped carrier. An example is afforded by the non-stoichiometric semiconducting oxide $Mn_{1-x}O$, in which a hole trapped on an Mn^{2+} ion, thereby becoming an Mn^{3+} centre, is the site of a small polaron. The hole can then move between Mn^{2+} sites. In certain cases, adjacent small polarons can become stabilized energetically, and form bipolarons (Alexandrov and Mott (1995)). Reduced (non-stoichiometric) WO_3 is believed to contain such bipolarons, in this case consisting of pairs of W^{5+} ions, perhaps stabilized via some degree of metal–metal bonding between edge-shared octahedra, and in Ti_4O_7, neighbouring $Ti^{3+}-Ti^{3+}$ ions can be regarded as comprising an electron bipolaron trapped on Ti^{4+} ions. (See Cox (1992) and Mott (1993) for further details on polarons in semiconducting transition-metal oxides.)

The self-trapping of electrons (or holes) in small-polaron states can be described in a general fashion in terms of a configurational-coordinate model, in which the generalized configurational coordinate q is a measure of the lattice distortion associated with the polaron: q can represent the displacement of an atom neighbouring the site of a trapped hole, thereby forming a *bond* (as in the V_K-centre in alkali halides, or in rare-gas solids); in polar materials, q will be a measure of the polarization caused by the trapped carrier (e.g. the polarization potential energy can be written as $qe^2/4\pi\varepsilon_0\varepsilon_p r$ for $r > r_{sp}$ (see also Problem 6.28)). In these cases, the lattice deformation depends quadratically on q, i.e. $\mathscr{E}_{def} = Aq^2$, and the energy of the electron (or hole) in the trap is taken to be linearly proportional to q, i.e. $\mathscr{E}_{el} = -Bq$. Thus, the total energy is given by

$$\mathscr{E} = Aq^2 - Bq \tag{6.252}$$

which has a minimum at

$$q_0 = B/2A, \tag{6.253}$$

so that eqn. (6.252) can be rewritten as

$$\mathscr{E} = -Aq_0^2 + A(q - q_0)^2. \tag{6.254}$$

Figure 6.60 shows that for this choice of q-dependence of $\mathscr{E}_{\text{el}}$, a localized self-trapped small polaron is always formed since $\mathscr{E}^{\text{min}} < 0$ (see Problem 6.29). The binding energy at $q = q_0$ is given by

$$W_{\text{sp}} = -\mathscr{E}_{\text{sp}} = Aq_0^2 = \frac{1}{2}Bq_0 = \frac{B^2}{4A}. \tag{6.255}$$

Values of W_{sp} are typically several tenths of an electron-volt.

The transport of electrons or holes in localized small-polaron states can be very different from the band-like motion characteristic of *large* polarons. At low temperatures, the small polarons can tunnel between sites in the lattice, resulting in a very narrow small-polaron band (and a correspondingly high effective mass). At higher temperatures ($k_B T \geqslant \frac{1}{2}\hbar\omega_D$), multiphonon-assisted hopping of small polarons occurs

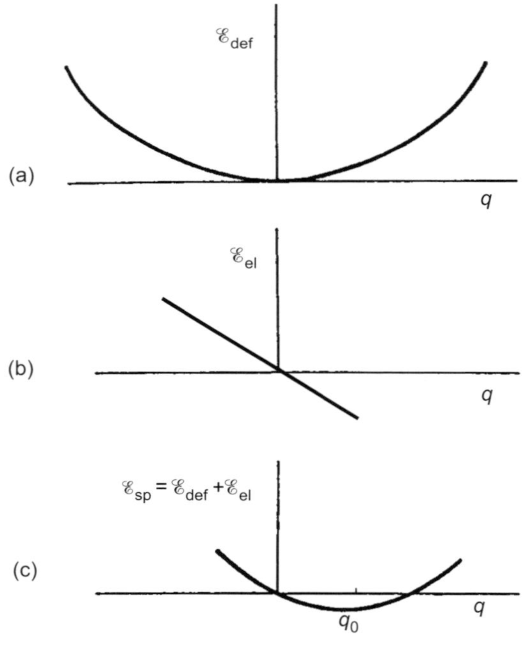

Fig. 6.60 Illustration of small-polaron self-trapping of a carrier in terms of a configurational-coordinate diagram, with configuration coordinate q: (a) the lattice-deformation energy; (b) the electron energy in the trap; (c) the total energy of the small polaron, showing the small-polaron binding energy $\mathscr{E}_{\text{sp}}$ and the configuration coordinate q_0 associated with the minimum-energy configuration.

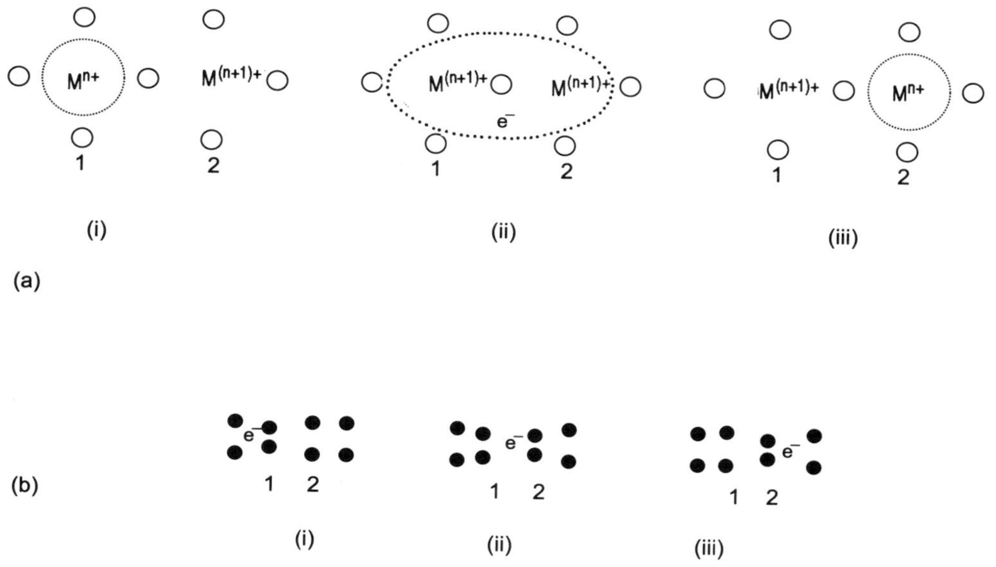

Fig. 6.61 Representation of small-polaron hopping, showing (i) a polaron originally at site 1; (ii) the activated configuration, and (iii) the polaron finally at site 2, for: (a) a valence-alternation ionic system; (b) a molecular-ion system.

between sites; this is a thermally activated process, with a hopping activation energy $W_H = W_{sp}/2$ (i.e. semiconducting-like rather than metallic-like).

This can be seen as follows. Consider a small polaron self-trapped and localized at site 1. In order for the charge carrier to move to site 2, and form a small-polaron centre there, *both* sites must distort by means of thermal fluctuations (a process involving many phonons, since the strain energies involved are so high) so that the electron energy at each site is *equal* (or in other words, the bottoms of the polaron wells must be at the same energy—see Fig. 6.61). The requirement that the electron energies be equal, i.e.

$$Bq_1 = Bq_2, \tag{6.256}$$

where q_1 and q_2 are the configuration coordinates describing the lattice deformation at sites 1 and 2, respectively, means that an electron can readily move (by tunnelling) between the two sites. If the electron moves back and forth between the two sites a number of times during the time that the sites are in the activated state, the polaron motion is said to be adiabatic, and the chance of making the hop between the sites is high. The deformation energy required simultaneously to distort the equilibrium configuration of well 1 to q_1 and to create a distortion q_2 at site 2 (which has no electron at it), where the distortions are equal, $q_1 = q_2$ (see eqn. (6.256)), is given by

$$\mathscr{E}_{\text{def}} = A(q_1 - q_0)^2 + Aq_2^2 = A(q_1 - q_0)^2 + Aq_1^2. \tag{6.257}$$

Minimization of this expression with respect to q_1 shows that the least strain energy is achieved when $\bar{q}_1 = \bar{q}_2 = q_0/2$ (see Fig. 6.62). Substitution of this value into eqn. (6.257)

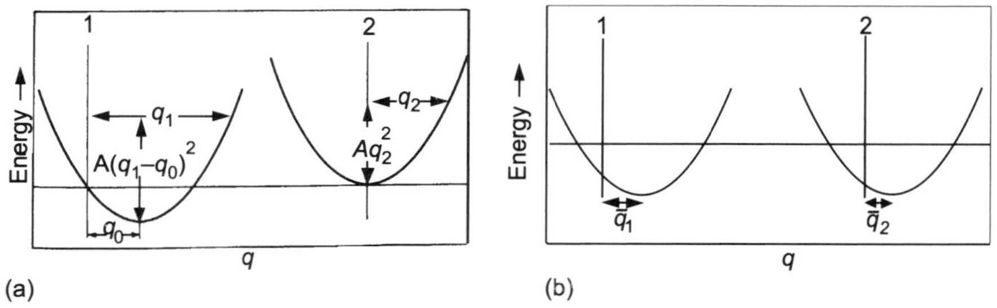

Fig. 6.62 Illustration of the origin of the polaron-hopping energy W_H, involved in the hopping motion of small polarons, in terms of a configurational-coordinate model. (a) Two potential wells representing a lattice distortion at two sites, 1 and 2, site 1 containing an electron. (b) The activated configuration, in which the two wells have minima at the same level, and $\bar{q}_1 = \bar{q}_2 = q_0/2$. The electron energy is then the same for each, and the electron on site 1 can readily tunnel to site 2.

gives as the minimum strain energy required to produce traps with equal well depths on each site, $\mathscr{E}_{\text{def}} = Aq_0^2$; since this quantity is simply the hopping activation energy W_H for polaron mobility, $W_H = W_{sp}/2$ (cf. eqn. (6.255)).

The d.c. electrical conductivity associated with small-polaron hopping motion in valence-alternation systems can be written, from eqn. (3.84b) linking the conductivity and the diffusion coefficient, and eqn. (3.62) for the diffusion coefficient in terms of the jump rate, as

$$\sigma_0 = c(1-c)N\frac{e^2a^2\gamma}{6k_BT}e^{-W_H/k_BT}, \tag{6.258}$$

where c and $(1-c)$ are the relative proportions of the two types of ions, M^{n+} and $M^{(n+1)+}$, of total concentration N, between which electrons (holes) move. The quantity $c(1-c)$ is simply the probability p of finding M^{n+} and $M^{(n+1)+}$ ions on adjacent sites if they are distributed randomly, and hence the average jump rate $\Gamma = p\gamma$, where γ is the atomic jump frequency (eqn. (3.63)). It should be noted that the conductivity activation energy, the polaron hopping energy, appearing in eqn. (6.258) is for the *mobility*; to this should be added any extra activation energy, $\mathscr{E}_0$, required to create charge carriers ($\mathscr{E}_0$ is zero for valence-alternation systems).

The thermopower for small-polaron motion in valence-alternation systems can be written as the temperature-independent Heikes equation:

$$S_T = \frac{k_B}{e}\left[\ln\left(\frac{c}{1-c}\right) + A\right], \tag{6.259}$$

where A is a constant of order unity. For those systems where the conductivity activation energy contains a carrier-creation term, i.e. $\mathscr{E}_\sigma = \mathscr{E}_0 + W_H$, the energy $\mathscr{E}_s$ for the temperature dependence of the thermopower (cf. eqn. (6.229a)) does *not* contain the polaron-hopping energy W_H, and so $\mathscr{E}_s = \mathscr{E}_0$.

**Localization 6.7

In many cases, electrons in solids are *not* characterized by delocalized wavefunctions, as in Bloch functions in ideal crystals (§5.2.1), but instead are spatially *localized*. Several instances of such behaviour have already been touched upon: electrons and holes bound to donors or acceptors, respectively, in semiconductors, and self-trapped small polarons. In most cases, however, as for localization of *vibrational* excitations (§4.3.1), localization of electrons is associated with the presence of *disorder*: dopant impurities are point defects in an otherwise perfect crystal; valence-alternation small polarons are associated with ions having a different charge state from the majority due to non-stoichiometry; and the gross structural disorder characteristic of amorphous semiconductors can cause localization of electron states in a range of energies, as will be seen.

It is a peculiarity of disordered systems that the existence of structural disorder, giving rise to energetic disorder of electrons, can completely localize *all* electron states under certain circumstances. For one-and two-dimensional systems, any amount of disorder, no matter how small, is sufficient to localize all the states, whereas a critical amount of disorder is necessary to localize completely all electron states in 3D; there is the possibility, therefore, of effecting an 'Anderson transition' between delocalized (extended) and localized states in 3D materials as the degree of disorder is increased. Since localized electron states, by definition, cannot contribute to metallic-like conduction, the Anderson transition is another example of a metal–insulator transition in solids (see §5.6.3).

How can a localized state be differentiated from an extended state? Several criteria have been proposed in order to achieve this differentiation. One criterion concerns electron diffusion (or equivalently electrical conductivity): an electron is localized at a site if the electron wavefunction has the form $\psi \propto \exp(-\alpha_{\mathrm{L}} r)$, where α_{L}^{-1} is the localization length and if, at $T = 0$ K, the electron does not diffuse away from the site as $t \to \infty$ or, equivalently, states of energy $\mathscr{E}$ are localized if the ensemble average of the d.c. conductivity is zero at $T = 0$ K. (Recall that delocalized states give a *finite* metallic conductivity at $T = 0$ K—see §6.3.2.1.) Another criterion asserts that localized states in a particular region of a volume of material should be insensitive to the boundary conditions operative at the surface of the box, and hence the linear size L of the sample becomes important: if $\alpha_{\mathrm{L}}^{-1} \gg L$, it is not possible to establish whether an electron state is truly delocalized or whether it would appear to be localized in a yet larger box. A final criterion is the participation ratio

$$P = \sum_i |\psi_i|^2 / \sum_i |\psi_i|^4 \tag{6.260}$$

which is a measure of the number of sites over which a wavefunction has significant amplitude: for a system containing N sites, a wavefunction that is strongly localized on just one site has $P = 1/N \simeq 0$, whereas an extended state is characterized by $P = 1$.

Disorder-induced localization of electron states was first studied by Anderson (1958) in terms of a tight-binding Hamiltonian (§5.3.1) applied to a *periodic* lattice but with disorder present in the form of random site energies $\mathscr{E}_i$ (or diagonal elements of the Hamiltonian)—see Fig. 6.63; such a situation is characterized by what is termed diagonal disorder. The Hamiltonian can be written in bra-ket notation as

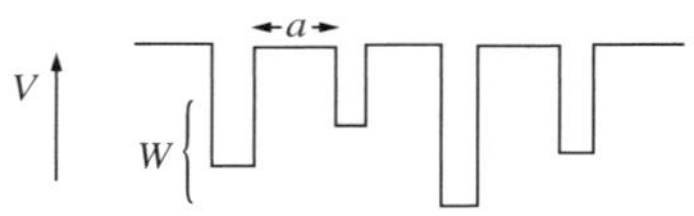

Fig. 6.63 The Anderson model for electron localization: potential wells with random depths on a crystalline lattice.

$$\mathfrak{H} = \sum_i \mathscr{E}_i |i\rangle\langle i| - V \sum_{i,j} |i\rangle\langle j|, \tag{6.261}$$

where the overlap energy (cf. eqn. (5.96b)) is taken to be constant, $-v$, between nearest-neighbour sites i,j (and zero otherwise), and the site energies $\mathscr{E}_i$ (cf. eqn. (5.96a)) are taken to be randomly distributed with a uniform distribution function:

$$\begin{aligned} P(\mathscr{E}) &= 1/W, \quad -W/2 \leqslant \mathscr{E} \leqslant W/2, \\ &= 0, \quad \text{otherwise.} \end{aligned} \tag{6.262}$$

Hence, the only parameter of the model is the dimensionless degree of disorder W/v and the lattice geometry. Consequently there are two competing influences in the Hamiltonian: the overlap term between nearest-neighbour sites tends to cause electrons to be *delocalized*, whereas deep potential wells in the random distribution of well depths tend to cause electrons to be *localized*. For sufficiently large values of $W/v > (W/v)_c$, therefore, *all* the states in a 3D lattice are localized: for a simple-cubic lattice, $(W/v)_c = 16.5$ (see Fig. 6.64). For values of $W/v < (W/v)_c$, only states in the tails of the band near the band edge are localized, and states in the middle of the band are extended: the energy $\mathscr{E}_0$ marking the transition from extended to localized states is called

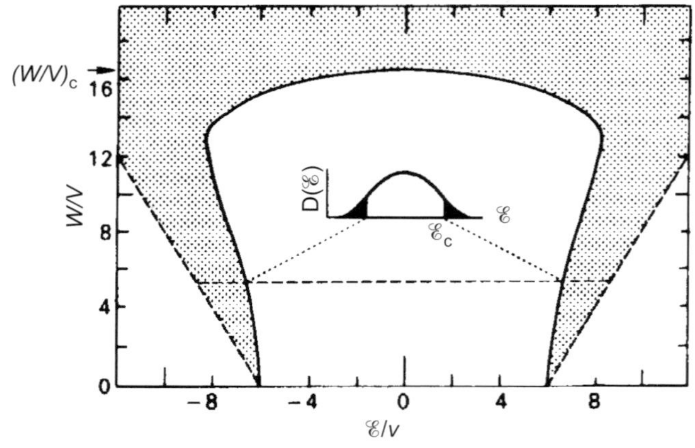

Fig. 6.64 Locus of the position of the mobility edge for a 3D simple-cubic lattice as a function of the normalized diagonal (energetic) disorder, W/v, and the electron energy $\mathscr{E}/v$, where v is the tight-binding overlap energy. The shaded area represents the region in which localized states exist (Kamimura and Aoki (1989), after Zdetsis *et al.* (1985)).

the mobility edge (because the electron conductivity decreases rapidly at this energy). *Off-diagonal* disorder, associated with fluctuations in the overlap energy v in eqn. (6.261), in turn arising from fluctuations in the atomic structure (i.e. nearest-neighbour distances), is found to be much less effective in inducing electron localization than is diagonal disorder. As a result, off-diagonal disorder (as is predominant in real amorphous solids) only causes localization in the band tails.

Delocalized electrons will scatter from the fluctuations in site potentials indicated in Fig. 6.63 and this will limit the mean-free path, Λ; the mean-free path will decrease with increasing disorder, until a minimum value is reached, $\Lambda \simeq a$, where a is the interatomic spacing. Under such conditions of strong scattering, where an electron essentially scatters off every atom, the electron wavevector $\boldsymbol{k}$ is no longer well-defined and $\Delta k/k \simeq 1$; this corresponds to the Ioffe–Regel limit (eqn. (4.235)), $\Lambda \simeq \lambda$, and marks the onset of localization. Light can also be localized in a strongly scattering medium, a fine powder of a high-refractive-index material, e.g. GaAs (Wiersma *et al.* 1997).

The wavefunctions corresponding to extended and strongly localized states are represented schematically in Fig. 6.65: a wavefunction localized at a site $\boldsymbol{r}_0$ is characterized by having an exponentially decaying envelope:

$$\Psi = \exp[-\alpha_{\mathrm{L}}(|\boldsymbol{r} - \boldsymbol{r}_0|)] \sum_n c_n \exp(\mathrm{i}\phi_n)\psi(|\boldsymbol{r} - \boldsymbol{r}_0|), \tag{6.263}$$

where ψ is an atomic wavefunction and the phase ϕ_n varies randomly between sites, n. The localization length, α_{L}^{-1}, is a function of the eigenenergy of the state: at the mobility edge, $\mathscr{E}_0, \alpha_{\mathrm{L}}^{-1} = \infty$, and it decreases with increasing energies deeper into the localized régime, i.e. the degree of localization is stronger for such deep states. The energy dependence of the localization length can be written as

$$\alpha_{\mathrm{L}}^{-1} = a[\mathscr{E}/(\mathscr{E}_0 - \mathscr{E})]^{\nu}, \tag{6.264}$$

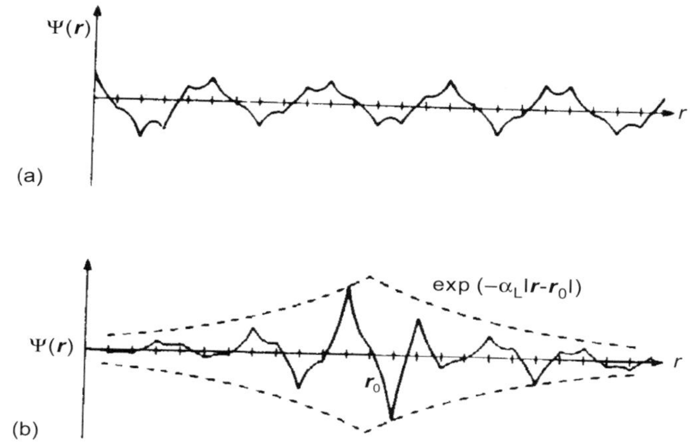

Fig. 6.65 Schematic illustration of the wavefunctions of an electron for: (a) an extended state; (b) a strongly localized state.

where $\mathscr{E}_0$ is the electron energy corresponding to the mobility edge, and the exponent $\nu \simeq 1.5$ (Mott (1993)).

As mentioned previously, electrons in localized states do not diffuse (or contribute to metallic conductivity) at $T = 0$ K. At finite temperatures, a localized electron can only move from one localized site to another by phonon-assisted hopping, a combined thermally activated quantum-tunnelling process. Hence, if the chemical potential lies in a band of *localized* states, the conduction is not metallic-like (as would be the case if extended states were involved), but instead the material is an electrical insulator, and is termed a Fermi glass. The electronic transport between localized states (e.g. between localized dangling-bond defect states (§3.1.1) deep within the bandgap of amorphous semiconductors) can be viewed as an optimization process: an electron will tunnel to a more distant centre if the thermal activation energy needed for the hop is thereby reduced. This Mott variable-range hopping mechanism produces a characteristic temperature dependence for the d.c. conductivity (Problem 6.31(a)):

$$\sigma_0 = \sigma(0)\exp[-A/T^{1/4}], \tag{6.265}$$

where the factor A depends on the localization length α_L^{-1} and the density of states at the Fermi level, $D(\mathscr{E}_F)$, as

$$A = 2.1[\alpha_L^3/k_B D(\mathscr{E}_F))]^{1/4}. \tag{6.266}$$

For very thin films, where the conduction is constrained to be in *two* dimensions, the exponent of the temperature dependence for the variable-range hopping conductivity changes from 1/4 to 1/3 (Problem 6.31(b,c)), and this behaviour is shown for films of a-Ge in Fig. 6.66.

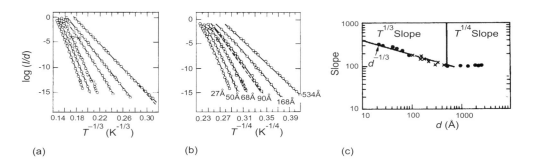

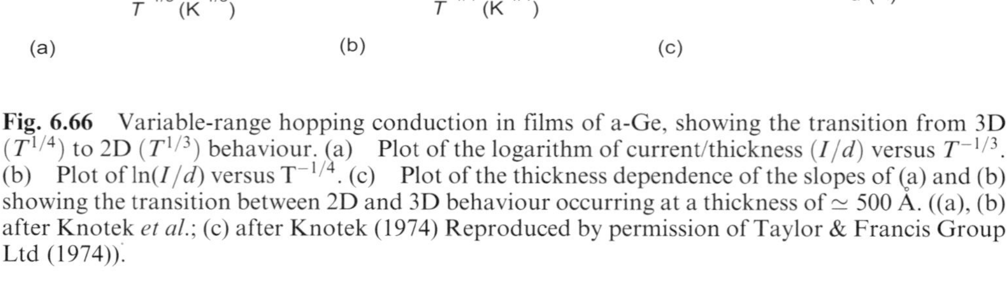

Fig. 6.66 Variable-range hopping conduction in films of a-Ge, showing the transition from 3D ($T^{1/4}$) to 2D ($T^{1/3}$) behaviour. (a) Plot of the logarithm of current/thickness (I/d) versus $T^{-1/3}$. (b) Plot of $\ln(I/d)$ versus $T^{-1/4}$. (c) Plot of the thickness dependence of the slopes of (a) and (b) showing the transition between 2D and 3D behaviour occurring at a thickness of $\simeq$ 500 Å. ((a), (b) after Knotek *et al.*; (c) after Knotek (1974) Reproduced by permission of Taylor & Francis Group Ltd (1974)).

Applications 6.8

Metals are widely used to conduct electricity, or heat; such everyday applications utilize the electron-dynamical properties discussed in §6.3. Semiconductors are also very widely used in electronic (low-power) (in contrast to electrical, i.e. high- power) applications. The most commonly used material for microelectronic applications is crystalline Si, although to a lesser extent III–V compound semiconductors (e.g. GaAs) are also used in applications that exploit their optical characteristics (direct gap) or electrical properties (higher mobility). However, almost invariably, electronic devices comprise extrinsic semiconductors in which the electrical dopants are incorporated in a very inhomogeneous spatial arrangement typically with a step-like profile. In such configurations, essentially all the useful electronic activity takes place in the vicinity of the interfaces ('junctions') between regions of differently doped semiconductor, and hence a discussion of semiconductor-based electronic devices is deferred until §§8.4.2.2 and 8.5 where the electronic behaviour of 2D systems is discussed. In this section, therefore, we discuss some applications of *superconductors* that utilize their bulk electron-dynamical properties.

6.8.1 Electrical applications of superconductors

Superconductors are potentially of considerable significance as loss-less carriers of very large electrical currents. An obvious possible application exploiting the lack of electrical resistance exhibited by materials in the superconducting state is in electrical power transmission systems: large currents can be carried in a loss-free manner without Joule heating as long as the temperature of the superconductor cable is kept below T_c, and also if the current density in the cable is less than the critical value j_c that drives the material into the normal state (see Problem 6.20). The critical current for type-I superconductors is reached when the energy of moving Cooper pairs, due to the supercurrent, becomes comparable to the pair-breaking energy, 2Δ (or equivalently when the magnetic field at the surface of the wire, generated by the current, becomes equal to the critical field, B_c).

In the mixed, vortex state of type-II superconductors (§6.4), such that $B_{c1} < B < B_{c2}$, an electrical current causes a Lorentz force (cf. eqn. (6.7)) to act on the magnetic-flux-containing vortices. If the vortices move as a result, this creates a potential gradient parallel to the current, i.e. it is equivalent to an electrical resistance of the material. Only if the vortices are pinned, by structural defects (e.g. radiation-induced vacancies), or by impurities, is the current loss-less in a type-II superconductor in the vortex régime.

For possible electrical-transmission applications, the savings in costs associated with loss-less transmission with no Joule heating have got to be greater than the costs of cooling the superconductor to below T_c over its entire length. For conventional (non-oxide) superconductors, for which $T_c < 25$ K (Table 6.3), liquid He must be used as the refrigerant, and superconducting operation is not cost-effective. For the high-T_c materials, on the other hand, with values of T_c in excess of the boiling temperature of liquid N_2 (Table 6.5), the costs of refrigeration are correspondingly greatly reduced, but there are considerable technical problems in fabricating these structurally anisotropic, brittle ceramic oxide materials in wire form. Nevertheless, progress in wire fabrication has been made with the micaceous high-T_c material BSCCO (see Table 6.5), in which a silver tube filled with the oxide is drawn and rolled (to a spatial dimension of $\simeq 0.1 \times 2$ mm). The rolling process causes grain alignment in the superconducting material, allowing the current to flow mainly within the a- b planes.

One of the most widespread uses of superconducting materials is as the windings in high-field (1–20 T) magnets, for example used in nuclear magnetic resonance spectrometers and whole-body scanners for magnetic-resonance imaging. An evident requirement for such applications is the capability for the wire in the windings to carry very large currents in order to generate high magnetic fields. Superconducting windings have the advantage that no dynamic cooling is necessary if the current does not produce Joule heating, and hence the current densities can be very high (e.g. the upper limit of $j_c \simeq 10^{11}$ A m^{-2} for Nb–Ti). Existing superconducting magnets are wound using wires fabricated from conventional Nb-alloy superconductors, commonly NbTi, and are operated at the boiling temperature of liquid He, 4.2 K. The superconducting wires are, in fact, extended multifilament wires of about 0.5 mm diameter, comprising filaments of the superconductor embedded in a matrix of a normal metal, e.g. Cu (Fig. 6.67): most of the current is carried by the superconducting filaments. Wires made of high-T_c materials (e.g. silver-sheathed BSCCO), have the potentially significant advantage that, because the upper critical magnetic fields, B_{c2}, are so high for these materials (typically ten to one hundred times that of conventional superconductors), the critical current, j_c, is not affected by the magnetic field as it is for conventional superconductors like Nb alloys (see Fig. 6.68). As a result, very high magnetic fields should be achievable using high-T_c materials.

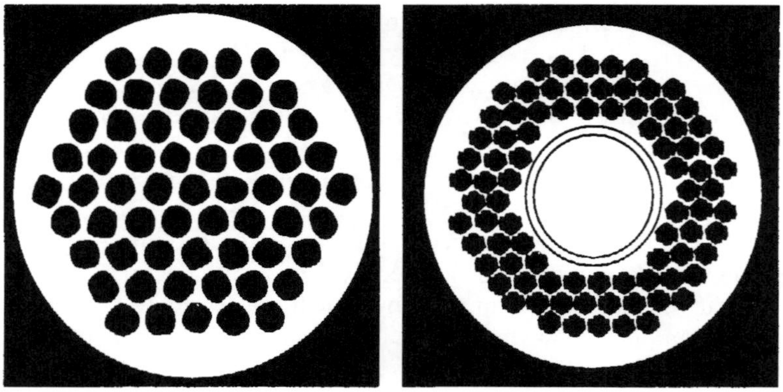

(a) (b)

Fig. 6.67 Two cross-sectional configurations of filamentary superconducting wires. The shaded regions correspond to superconducting filaments, and the light regions correspond to a normal-metal matrix (e.g. Cu). (a) Homogeneous distribution of superconducting filaments. (b) Superconducting filaments arranged around a central core of normal metal.

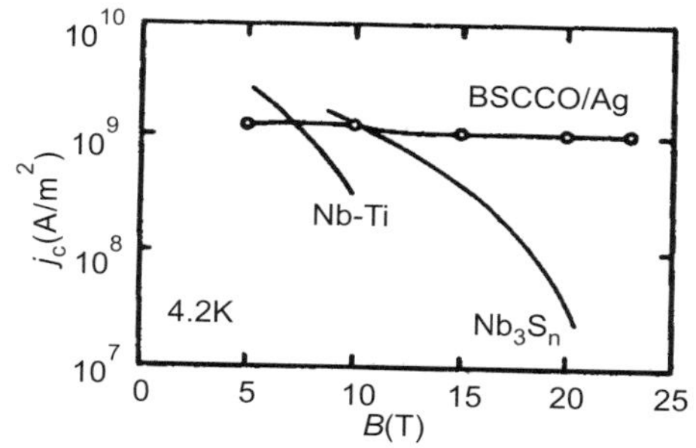

Fig. 6.68 The dependence of the critical current, j_c, on magnetic field for conventional Nb-alloy superconductors and a high-T_c superconductor, BSCCO. (After Sato *et al.* (1991), *IEEE Trans. Mag.* **27**, 1231. © 1991 IEEE)

Other potential applications of superconducting wires in magnetic applications are as windings in electrical motors or electrical transformers. Here, the attraction of the superconducting material is the dissipationless operation, thereby saving the few percent of energy lost as Joule heating when normal-metal windings are used. Again, high-T_c-based wires would be a very attractive proposition in this regard because of the low cost and ease of liquid-nitrogen refrigeration.

One remarkable aspect of the behaviour of superconductors is that, due to the exclusion of magnetic flux from the material in the superconducting state (the Meissner effect), a superconductor will *levitate* in a spatially inhomogeneous magnetic field above a magnet (see Plate V). A simple analysis of this problem can be made by assuming that the magnetic-flux density varies inversely with distance z above the surface of a magnet, i.e. $B(z) = B(a)a/z$, where a is the thickness of the magnet in the z-direction with the origin taken to be at the bottom surface (Fig. 6.69). Since the magnetic-energy density of a field is $u = B^2/2\mu_0$ (cf. eqn. (6.146)), the increase in magnetic energy of the field associated with the complete exclusion of flux from the material of volume V (i.e. a type-I superconductor or a type-II material with $B < B_{c1}$) is given by

$$U_{\text{mag}} = \frac{B^2(a)a^2}{4\mu_0^2 z^2} V. \tag{6.267}$$

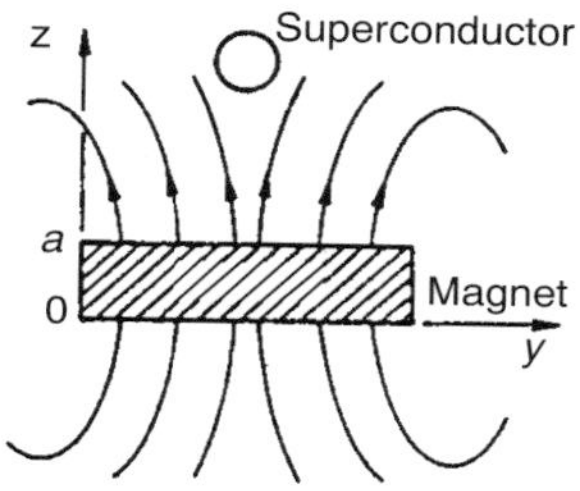

Fig. 6.69 Illustration of the type of magnetic-flux pattern necessary to levitate a superconductor.

The gravitational potential energy of the superconducting object, of density ρ, at a height z above the reference plane is given by

$$U_{\text{grav}} = \rho V g z, \tag{6.268}$$

and hence the total potential energy of the system is $U = U_{\text{mag}} + U_{\text{grav}}$. The net force acting on the particle is $F = -\partial U/\partial z$: in dynamic equilibrium, the net force must be zero, and hence the equilibrium levitation height is given by

$$z_0 = [B^2(a)a^2/2\mu_0^2 \rho g]^{1/3}. \tag{6.269}$$

Lateral stability is achieved (for a type-II material in the vortex régime) by pinning of flux lines (vortices).

Magnetic levitation involving superconducting materials has been considered for use in ground-transportation systems, so-called 'maglev' trains. The levitation configuration involving superconductor and magnet shown in Fig. 6.69 is not feasible for large-scale applications. Instead, it is more likely that levitation could be achieved by utilizing the repulsion between a superconducting magnet in a train and either conventional solenoids on the track or an electrically conducting metallic guide on which it rides: the repulsive force in the latter case is due to the interaction between the magnetic field of the magnet and the eddy currents induced in the guide by the field. Obviously, frictional dissipation of energy between train and rail is greatly reduced in a levitating system, and high speeds of transport are possible (500 km h^{-1}); nevertheless, there is some residual frictional effect, e.g. due to eddy-current damping in the guide rail.

6.8.2 Electronic applications of superconductors

Apart from their use in high-current-carrying superconducting windings in high-field magnets, present applications of superconductors tend to be in low-power devices, e.g. for electronic applications. Perhaps the most widely used such device is the SQUID (superconducting quantum interference device) which makes use of the coherent quantum tunnelling of Cooper pairs of electrons that occurs in Josephson junctions (§6.4.2.3) to measure very small magnetic fields.

The general configuration of a SQUID is shown in Fig. 6.70: a supercurrent is diverted along two paths of a superconductor, each branch containing a weak link (Josephson junction), e.g. a tunnelling barrier. The supercurrents in region I of the device, before division, and in region II, after recombination, are the same except for phase differences δ_a and δ_b which occur at the two junctions a and b, respectively (cf. eqn. (6.183)). The presence of a magnetic field threading the area enclosed by the device causes further shifts in phase, resulting in a periodic modulation of the supercurrent with varying magnetic field. The fact that the operation of a SQUID is unaffected even if the spatial separation of the two tunnel junctions is of the order of a few centimetres means that the coherence of the superconducting order parameter of the condensate is maintained over such distances: this is an example of *macroscopic* quantum coherence. The SQUID has some similarity to an optical interferometer, wherein changes in the path-length difference introduced into one arm of the interferometer cause interference to take place between the two combining light beams, resulting in a periodic modulation

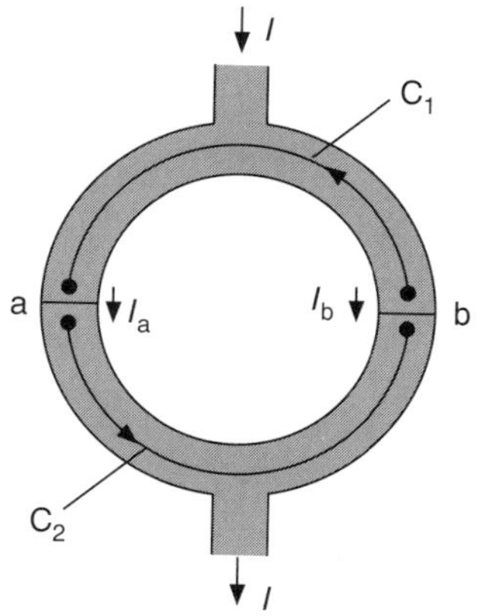

Fig. 6.70 Schematic illustration of a SQUID, consisting of two Josephson junctions, a and b, in two arms of a loop of superconducting material. The current I through the device is modulated by the (quantized) magnetic flux threading the area of the loop. Paths C_1 and C_2 are those along which line integrations are performed to relate phase shifts of the superconducting order parameter across the junctions to the magnetic flux (see text).

of the light intensity. An analogy for a SQUID with very small junction areas is with the Young's (two-slit) arrangement in optics.

The phase shift $\Delta\theta$ of the superconducting order parameter associated with a magnetic field can be obtained from the expression eqn. (6.169) relating electrical-current density to θ and the vector potential $\boldsymbol{A}$ by performing a line integration along a path deep inside the superconductor where, in the Meissner state, the current is *zero*, and for which (from eqn. (6.169))

$$\hbar\nabla\theta = -2e\boldsymbol{A}. \tag{6.270}$$

Performing such a line integration between any two points i and j on the path gives:

$$\Delta\theta\Big|_{\mathrm{i}}^{\mathrm{j}} = \int_{\mathrm{i}}^{\mathrm{j}} \nabla\theta\cdot \mathrm{d}\boldsymbol{l} = -\frac{2e}{\hbar}\int_{\mathrm{i}}^{\mathrm{j}} \boldsymbol{A}\cdot \mathrm{d}\boldsymbol{l}. \tag{6.271}$$

If the supercurrents across the tunnel junctions a and b shown in Fig. 6.70 are written as (cf. eqn. (6.183)):

$$I_{\mathrm{a,b}} = Aj_0 \sin\delta_{\mathrm{a,b}} \tag{6.272}$$

where A is the junction area (assumed to be the same in both cases), then the phase shifts from region I to II via paths a or b in Fig. 6.70 are, from eqn. (6.271), given by

$$\Delta\theta\Big|_{\mathrm{I}}^{\mathrm{II}} = \delta_{\mathrm{a}} - \frac{2e}{\hbar}\int_{\mathrm{a}} \boldsymbol{A}\cdot \mathrm{d}\boldsymbol{l}, \tag{6.273a}$$

$$\Delta\theta\Big|_{\mathrm{I}}^{\mathrm{II}} = \delta_{\mathrm{b}} + \frac{2e}{\hbar}\int_{\mathrm{b}} \boldsymbol{A}\cdot \mathrm{d}\boldsymbol{l}. \tag{6.273b}$$

However, the phase shift between regions I and II must be invariant (modulo 2π) of the path chosen between them, and so by subtraction of eqns. (6.273)

$$\delta_a - \delta_b = \frac{2e}{\hbar} \oint \boldsymbol{A} \cdot d\boldsymbol{l}, \tag{6.274}$$

where the two line integrals of eqns. (2.273), being in opposite directions, combine to give an integral over a closed loop. Stokes's theorem relates a closed-path line integral to an integral over a surface S bounded by the line:

$$\begin{aligned} \oint \boldsymbol{A} \cdot d\boldsymbol{l} &= \int_s (\text{curl}\boldsymbol{A}) \cdot d\boldsymbol{S} \\ &= \int_s \boldsymbol{B} \cdot d\boldsymbol{S} \end{aligned} \tag{6.275}$$

The quantity $\int_s \boldsymbol{B} \cdot d\boldsymbol{S}$ is simply the magnetic flux Φ threading the superconducting loop, i.e.

$$\delta_a - \delta_b = \frac{2e}{\hbar}\Phi. \tag{6.276}$$

Introducing a constant, arbitrary phase factor δ_0 (dependent on the nature of the tunnel junctions, assumed to be identical), the individual junction phase shifts can be written as

$$\delta_a = \delta_0 + \frac{e}{\hbar}\Phi, \tag{6.277a}$$

$$\delta_b = \delta_0 - \frac{e}{\hbar}\Phi. \tag{6.277b}$$

The total supercurrent flowing in regions I and II is given by the sum of the currents flowing through the junctions (eqns. (2.272)):

$$\begin{aligned} I &= I_a + I_b \\ &= 2Aj_0 \cos[(\delta_a - \delta_b)/2] \sin[(\delta_a + \delta_b)/2]. \end{aligned} \tag{6.278}$$

Substituting the phase shifts δ_a and δ_b into this expression gives

$$I = 2Aj_0 \cos[e\Phi/\hbar] \sin\delta_0. \tag{6.279}$$

Hence, the supercurrent flowing through a SQUID is determined by the extent of the magnetic flux threading the device: the maximum supercurrent is given by (Problem 6.32):

$$I_{max} = 2Aj_0|\cos(\pi\Phi/\Phi_0)|, \tag{6.280}$$

where $\Phi_0 = h/2e$ is the flux quantum. Equation (6.280) shows that maxima in the SQUID current occur whenever an *integral* number of flux quanta $\Phi = n\Phi_0$ thread the device. The maximum current passed by a SQUID as a function of applied magnetic field is shown in Fig. 6.71: the effect of the quantization is clearly evident.

SQUIDs can be used as extremely sensitive magnetometers, capable of detecting very small magnetic fields, such as those generated within the living body. They can also be used as very sensitive voltmeters: the voltage source to be measured is connected to the

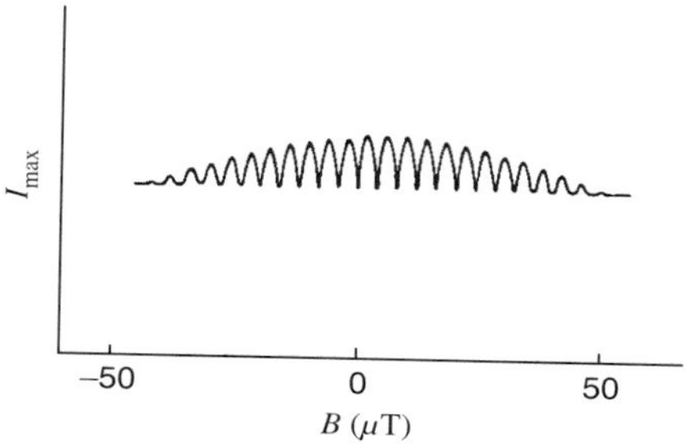

Fig. 6.71 Maximum current passed by a SQUID as a function of applied magnetic field. The short-period oscillations correspond to solutions of eqn. (6.270); the longer-period variation is due to a 'diffraction' effect associated with magnetic flux threading extended Josephson junctions (After Jaklevic *et al.* (1965) Reprinted with permission from *Phys. Rev.* **140**, A1628. © 1965. The American Physical Society)

SQUID through a known resistor R, and the current $I = V/R$ produces a magnetic flux in the SQUID which is measured. High-T_c materials are being developed for SQUID applications operating at $T > 77$ K, although the technical problems involved are challenging. For Josephson junctions to work, the superconducting order parameter must retain its bulk value to within a distance equal to the coherence length of the junction. The very short coherence lengths characteristic of high-T_c cuprate materials (§6.4.3), i.e. $\zeta \simeq 10$–30 Å in the $\boldsymbol{a} - \boldsymbol{b}$ plane, means that control over material processing must be very good. However, grain boundaries in films of say YBCO act as weak-link Josephson junctions and so suitable configurations of single-crystal grains could be used to fabricate SQUID devices.

A number of other applications have been proposed that utilize Josephson junctions, including radiation detectors and very fast switching elements (cf. Fig. 6.42) for computer elements. For further details see e.g. Ruggiero and Rudman (1990), Tinkham (1996).

Problems

6.1 Calculate an expression for the thermal conductivity κ_T of a *classical* electron gas (Drude model). Comment on the relative order of magnitude of this estimate with that calculated for the Fermi gas (eqn. (6.16)). (Hint: start from the gas-kinetic expression $\kappa_T = \Lambda v c_v/3$, where v is an appropriate mean velocity.)

6.2 (a) Show that the expression for the effective electron mass (eqn. (6.26)) reduces to the bare mass m_e for the free-electron gas.

(b) Calculate a general expression for the effective mass, $m_e^*(k)$, for a 1D tight-binding band (cf. eqn. (5.98)), and evaluate it at the zone boundary ($k = \pi/a$). How does m_e^* depend on the width of the band? Account for the behaviour of the electron velocity near $k = \pi/2a$. Why can a d.c. electric field not induce an alternating current (Bloch oscillations) in a metal?

6.3 Essay: Compare and contrast the behaviour of electrons and holes in part-filled bands (e.g. in semi-metals or p-type semiconductors).

6.4 Show, for the case of metallic potassium, that the magnitude of the minimum phonon wavevector capable of effecting electron phonon umklapp scattering is $q_0 = 0.278k_F$. (Hint: see Problem 5.1.)

6.5 Show that eqn. (6.71), for the electrical conductivity of a metal, reduces to (a) eqn. (6.9) for the case of a free-electron gas; (b) $\sigma = e^2 \Lambda S_f / 12\pi^3 \hbar$, where Λ is a mean-free path averaged over the Fermi surface and S_F is the total free area of Fermi surface.

6.6 Show that the change in momentum of an electron, along its original direction of motion, when scattered from one point on the Fermi surface to another via a collision such as in Fig. 6.9, is given by $\delta k \simeq \hbar q^2 / 2k_F$.

6.7 Obtain an expression for the local entropy production in a metal resulting from electrical current flow in the presence of a temperature gradient.

6.8 Show that the cyclotron frequency of a free-electron gas is given by eqn. (6.117). (Hint: use eqn. (6.115).)

6.9 (a) Show that the real-space cyclotron orbit of an electron in a magnetic field $\boldsymbol{B}$ is such that the projection of the orbit on a plane perpendicular to $\boldsymbol{B}$ is the orbit in $\boldsymbol{k}$-space rotated by 90° and multiplied by the scaling factor $\hbar/eB$ (see Fig. 6.19). (Hint: form the vector product of both sides of eqn. (6.109), written as $\hbar\dot{\boldsymbol{k}} = -e(\boldsymbol{v} \times \boldsymbol{B})$, with a unit vector $\hat{\boldsymbol{B}}$ parallel to the $\boldsymbol{B}$-field and integrate the ensuing equation.)

(b) Obtain the expression for the cyclotron frequency of a free-electron gas by considering the forces acting on an electron in a real-space orbit.

6.10 Derive the expression for the quantized Landau levels for electron motion in a magnetic field (eqn. (6.119)).

(a) Show that the Schrödinger equation for a free electron in a constant magnetic field $\boldsymbol{B}$ aligned along the z-direction (represented by the vector potential $\boldsymbol{B} = \text{curl}\,\boldsymbol{A}$, where $\boldsymbol{A} = (0, Bx, 0)$) can be written as

$$-\frac{\hbar^2}{2m_e}\left[\frac{\partial^2\psi}{\partial x^2} + \left(\frac{\partial}{\partial y} + \frac{ieBx}{\hbar}\right)^2 \psi + \frac{\partial^2\psi}{\partial z^2}\right] = \mathscr{E}\psi. \tag{1}$$

(b) Demonstrate that this equation has a solution of the form $\psi(x, y, z) = f(x)\exp[i(\lambda y + k_z z)]$, where the function $f(x)$ satisfies the equation

$$-\frac{\hbar^2}{2m_e}\frac{\partial^2 f(x)}{\partial x^2} + \frac{e^2B^2}{2m_e}\left(x + \frac{\hbar\lambda}{eB}\right) f(x) = \mathscr{E}_0 f(x) \tag{2}$$

with $\mathscr{E} = \mathscr{E}_0 + \hbar^2 k_z^2 / 2m_e$.

(c) Show that eqn. (2) is the Schrödinger equation for a simple harmonic oscillator, centred at the point $x = -\hbar\lambda/eB$, and hence obtain an expression for the oscillator frequency and the energy eigenvalue $\mathscr{E}$.

6.11 Consider the effect of electron spin on the Landau levels, and obtain an expression for the energy levels (cf. eqn. (6.119)) taking spin into account. What is the level spacing for: (a) free electrons, and (b) electrons (say at the conduction-band minimum in a semiconductor) with an effective mass $m_e^* = 0.5m_e$?

6.12 Obtain expressions for the Hall coefficient and transverse magnetoresistivity for a two-band conductor.

(a) In general, the current $\boldsymbol{j}$ caused by an electric field $\boldsymbol{E}$ normal to a magnetic field can be written as $\boldsymbol{E} = \rho\boldsymbol{j}$, where the resistivity tensor has the form $\rho = \begin{pmatrix} \rho & -R_H B \\ R_H B & \rho \end{pmatrix}$, and where R_H is the Hall coefficient and ρ is the magnetoresistivity. For a metal with several part-filled bands, each band i produces a relation $\boldsymbol{E}_i = \rho_i \boldsymbol{j}_i$, where $\rho_i = \begin{pmatrix} \rho_i & -R_i B \\ R_i B & \rho_i \end{pmatrix}$.

Show that the effective resistivity tensor is given by $\rho = [\sum \rho_i^{-1}]^{-1}$.

(b) Show that, for the case where there are only *two* bands, 1 and 2, the Hall coefficient and magnetoresistivity are given by:

$$R_H = \frac{R_1\rho_2^2 + R_2\rho_1^2 + R_1R_2(R_1+R_2)B^2}{(\rho_1+\rho_2)^2 + (R_1+R_2)^2B^2}, \tag{1}$$

$$\rho = \frac{\rho_1\rho_2(\rho_1+\rho_2) + (\rho_1R_2^2 + \rho_2R_1^2)B^2}{(\rho_1+\rho_2)^2 + (R_1+R_2)^2B^2}. \tag{2}$$

(c) Hence obtain eqn. (6.133) for the Hall coefficient in the high-field limit for an electron and a hole band with closed orbits. Derive the corresponding equation in the *low-field* limit, viz.

$$R_H = (n_h\mu_h^2 - n_e\mu_e^2)/e(n_h\mu_h + n_e\mu_e)^2. \tag{3}$$

What is the behaviour in the compensated limit ($n_e = n_h$) in both cases?

6.13 Essay: Compare and contrast the behaviour of electrons in the superconducting and normal states of a metal.

6.14 Use the approximate relation for the temperature dependence of the critical field, $B_c(T)$ (eqn. (6.141)) to obtain expressions for the temperature dependences of the differences in the Gibbs free energy, entropy and heat capacity between superconducting and normal states. Show that the heat-capacity discontinuity per unit volume at T_c (in zero applied magnetic field) is $\Delta c = 4B_c^2(0)/\mu_0 T_c$.

6.15 Show that an attractive interaction, causing an electron (Cooper) pair to scatter from a state $(\boldsymbol{k}_\uparrow, -\boldsymbol{k}_\downarrow)$ to $(\boldsymbol{k}'_\uparrow, -\boldsymbol{k}'_\downarrow)$, leads to an energy for the pair *less* than that of the non-interacting ground state.

(a) The Schrödinger equation for an electron pair can be written as

$$-\frac{\hbar^2}{2m_e}(\nabla_1^2 + \nabla_2^2)\psi(\boldsymbol{r}_1,\boldsymbol{r}_2) + \mathcal{V}(\boldsymbol{r}_1,\boldsymbol{r}_2)\psi(\boldsymbol{r}_1,\boldsymbol{r}_2) = (\varepsilon + 2\mathscr{E}_F)\psi(\boldsymbol{r}_1,\boldsymbol{r}_2) \tag{1}$$

where ε is the energy of the pair relative to the non-interacting state ($\mathcal{V} = 0$) for which each electron at the Fermi level would have an energy $\mathscr{E}_F$. By inserting the pair wavefunction (eqn. (6.155)) into (1), show that

$$-\frac{\hbar^2k^2}{m_e}p(\boldsymbol{k}) + \frac{1}{V}\sum_{k'} p(\boldsymbol{k}')\mathcal{V}_{kk'} = (\varepsilon + 2\mathscr{E}_F)p(\boldsymbol{k}), \tag{2}$$

where the interaction matrix element is $\mathcal{V}_{kk'} = \int \mathcal{V}(\boldsymbol{r})e^{-i(\boldsymbol{k}-\boldsymbol{k}')\cdot\boldsymbol{r}}\,d\boldsymbol{r}$, and describes the scattering of the Cooper pair from $(\boldsymbol{k}_\uparrow, -\boldsymbol{k}_\downarrow)$ to $(\boldsymbol{k}'_\downarrow, -\boldsymbol{k}'_\uparrow)$. (Hint: multiply by exp(-i$\boldsymbol{k}'.\boldsymbol{r}$) and integrate over the normalization volume.)

(b) By assuming that $\mathcal{V}_{kk'}$ is independent of $\boldsymbol{k}$ and is also attractive, i.e. $\mathcal{V}_{kk'} = -\mathcal{V}_0$ for $0 < \varepsilon < \hbar\omega_D$ (see Fig. 6.31b), show that

$$1 = \frac{\mathcal{V}_0}{V}\sum_k (-\varepsilon + \hbar^2k^2/m_e - 2\mathscr{E}_F)^{-1}. \tag{3}$$

(Hint: write (2) with $A = (\mathcal{V}/V)\sum_{k'} p(k')$, and sum the former expression over k.)

(c) Replace the sum in (3) by an integral: $V^{-1}\sum_k \to (2\pi)^{-3}\int dk$ and obtain the expression

$$1 = \frac{\mathcal{V}_0}{(2\pi)^3}\int\int \frac{dS_{\mathscr{E}}}{|\nabla_k\mathscr{E}(k)|}\frac{d\mathscr{E}}{(2\mathscr{E} - \varepsilon - 2\mathscr{E}_F)}, \tag{4}$$

where $\mathscr{E} = \hbar^2k^2/2m_e$. Hence show that

$$1 = \frac{1}{2}\mathcal{V}_0 g(\mathscr{E}_F)\ln[(\varepsilon - 2\hbar\omega_D)/\varepsilon], \tag{5}$$

where $g(\mathscr{E}_F)$ is the density of states at the Fermi level for electrons of one spin orientation (cf. eqn. (5.131)) and, in the case of a *weak* interaction, $\mathcal{V}_0 g(\mathscr{E}_F) \ll 1$, that this reduces to

$$\varepsilon \simeq -2\hbar\omega_D \exp(-2/\mathcal{V}_0 g(\mathscr{E}_F)). \tag{6}$$

6.16 Derive eqn. (6.158) for the range of crystal momenta $\hbar\Delta k$ involved in the scattering of electrons in Cooper pairs in the BCS condensate (see Fig. 6.31b). What is a typical value for Δk?

6.17 Derive eqn. (6.166) for the Ginzberg–Landau order parameter of a superconductor condensate, $\psi(\boldsymbol{R}) = e^{i\boldsymbol{K}\cdot\boldsymbol{R}}\phi(\boldsymbol{r})$. By writing down an expression for the supercurrent density $\boldsymbol{j}_s$ carried by Cooper pairs, show that each Cooper pair experiences a total change in wavevector $\boldsymbol{K} = -2m_e\boldsymbol{j}_s/n_s e\hbar$, where n_s is the concentration of superconducting electrons (i.e. $n_s/2$ Cooper pairs). Hence obtain the expression for $\psi(\boldsymbol{R})$.

6.18 Show that the expression for the electrical current density

$$\boldsymbol{j} = \frac{i\hbar e}{2m_e}(\psi^*\nabla\psi - \psi\nabla\psi^*) - \frac{e^2}{m_e}\psi^*\psi\boldsymbol{A}$$

in the presence of a magnetic field $\boldsymbol{B} = \text{curl}\,\boldsymbol{A}$ is obtained as the expectation value of the quantity $-e\boldsymbol{v}$ where, for an electron, the momentum is given by $\boldsymbol{p} = m_e\boldsymbol{v} - e\boldsymbol{A}$, and where $\boldsymbol{A}$ is the magnetic vector potential, with the momentum being replaced by the operator $-i\hbar\nabla$. Confirm that for a *Cooper pair*, eqn. (6.167) is obtained.

6.19 Derive the London equation (eqn. (6.171)) by assuming that a superconductor can be described as being an ideal conductor of infinite conductivity. (Hint: make use of the Maxwell equation curl $\boldsymbol{E} = -\dot{\boldsymbol{B}}$, and invoke the Meissner effect exhibited by superconductors.)

6.20 Obtain an expression for the upper critical current density j_c for the existence of the superconducting state.

(a) Write down an expression for the energy of an electron in a Cooper pair in a current-carrying state in terms of the net wavevector $\boldsymbol{K}$ of the Cooper pair (see Problem 6.17), and hence obtain an estimate for the increase in energy $\delta\mathscr{E}$ of such an electron when current-carrying compared to the current-free case.

(b) By assuming that the condition for the destruction of the superconducting state is when the increase in energy $2\delta\mathscr{E}$ of a pair becomes comparable to 2Δ, obtain an expression for the corresponding critical current density. If $j_c = 2 \times 10^{11}\,\text{A m}^{-2}$ for Sn at $T \simeq 0$ K, calculate the concentration of superconducting electrons ($v_F = 1.88 \times 10^6$ m s^{-1}; $2\Delta(0) = 3.67 \times 10^{-4}$ eV).

6.21 Sketch the Giaever tunnelling current-voltage characteristic of a tunnel junction comprising *two* superconductors at temperatures (a) $T = 0$ K; (b) $0 < T < T_c$.

6.22 Essay: Review the latest theories for the mechanism of superconductivity in the high-T_c cuprate materials.

6.23 Explain why the position of the chemical potential (Fermi level) for an intrinsic semiconductor should depend on the curvature in $\boldsymbol{k}$-space of the valence and conduction bands in the vicinity of the gap (cf. eqn. (6.207)). For intrinsic crystalline InSb ($\mathscr{E}_g = 0.23$ eV, $m_e^* = 0.015m_e$, $m_h^* = 0.42m_e$ — see Table 6.1), calculate the position of the Fermi level and the carrier concentrations in the conduction and valence bands at (a) $T = 0$ K, and (b) $T = 300$ K.

6.24 Estimate the temperature at which the intrinsic carrier concentration in diamond ($\mathscr{E}_g = 5.4$ eV) would be equal to that in Ge ($\mathscr{E}_g = 0.74$ eV) at 100 K.

6.25 (a) Obtain an expression for the cyclotron-resonance frequency, ω_c, for the direction of a magnetic field $\boldsymbol{B}$ with direction cosines (l_1, l_2, l_3) with respect to the principal axes of an ellipsoidal constant-energy surface (having longitudinal and transverse values of the effective mass, m_l and m_t), for example near the conduction-band minima in crystalline Si (see eqn. (5.127)). (Hint: solve the equations of motion $m_i^*\frac{dv}{dt} = -e\boldsymbol{v} \times \boldsymbol{B}$ with oscillating solutions v_x, v_y, $v_z \propto \exp(i\omega t)$.)

(b) For the cyclotron-resonance data for c-Si shown in Fig. 6.48a, for which the cyclotron frequency is $\nu = 2.4 \times 10^{10}$ Hz, obtain values for m_l and m_t.

6.26 (a) Show that the concentration of un-ionized donors N_d^0 in an n-type extrinsic semiconductor containing a total concentration N_d of donors is given by $N_d^0 = N_d\left[1 + \frac{1}{2}\exp[(\mathscr{E}_c - \mathscr{E}_d - \mu)/k_BT]\right]^{-1}$. (Hint: assume that each donor level can be occupied by an electron with spin up *or* down and use the grand sum to obtain an expression for the electron occupancy.)

(b) Hence obtain an approximate general expression for the electron concentration in the conduction band if the hole concentration is neglected. Show that this expression reduces to eqns. (6.244) and (6.246) for the position of the chemical potential in the freeze-out and saturation régimes, respectively. How is the general expression altered if the holes in the valence band are taken into account?

(c) Obtain an expression for the concentration of *ionized* acceptors in a p-type doped tetrahedral semiconductor.

6.27 What concentration of donors N_d is necessary to make the high-temperature limit of the saturation region lie at 300 K for an n-type doped semiconductor with $\mathscr{E}_g = 1$ eV and $m_e^* = m_h^* = 0.5m_e$? Where does the chemical potential lie at this temperature? What is the corresponding hole concentration in the valence band?

6.28 (a) Obtain an expression for the energy of an electron in a dielectric-polaron state as the sum of three terms: (i) kinetic (associated with confinement of the electron to a volume of radius r_{sp}; (ii) electrostatic potential of the electron; (iii) electrostatic potential associated with polarization of the surrounding lattice. (Hint: for (iii) take the polarization energy to be the volume integral $-\frac{1}{2}\int_{r_{sp}}^{\infty} \boldsymbol{P}.\boldsymbol{E}\mathrm{d}V$, where $\boldsymbol{P}$ is the polarization (eqn. (4.140)).)

(b) By minimizing the expression for the energy, obtain equations for the equilibrium small-polaron radius $\bar{r}_{sp}$ and the overall binding energy (eqn. (6.251a)). (N.B., for comparison, an exact treatment gives $\bar{r}_{sp} = 5\hbar^2\varepsilon_p/m^*e^2$ if $r_{sp} \gg a$.)

6.29 Discuss the likelihood of self-trapping in small-polaron formation if there is a *minimum* value of configuration coordinate, $q = q_c$, below which polaron formation does not occur. (Hint: construct configuration-coordinate diagrams, with the electron-energy term given by $\mathscr{E}_{el} = -B(q - q_c)$, for various values of q_c.)

6.30 Show that an estimate for the maximum resistance for metallic behaviour exhibited by a thin wire is $R_c \simeq \hbar^2/e = 4.1$ kΩ. (Hint: use the uncertainty principle to estimate the shift in electron energy levels $\delta\mathscr{E}$ corresponding to the uncertainty in time of an electron diffusing the length of the wire. Use the Nernst–Einstein equation (3.84b), relating conductivity to diffusion coefficient to relate the resistance of the wire to $\delta\mathscr{E}$ and $\Delta\mathscr{E}$, the average spacing between energy levels. Couple two such wires together, and consider the condition necessary that the resistance scales normally with size.)

6.31 (a) Derive the Mott $T^{1/4}$ law (eqn. (6.265)) for the d.c. conductivity of an electron performing variable-range hopping between localized states. (Hint: Consider the electron hopping transition rate between two localized states separated in distance by r and energy by Δ to be given by $\gamma = \nu\exp(-2\alpha_L r - \Delta/k_B T)$, where α_L^{-1} is the electron localization length and ν is an attempt frequency of the order of a phonon frequency. Use the Nernst–Einstein relation (eqn. (3.84b)) to obtain an expression for the conductivity expressed in terms of the variables r and Δ. Optimize this expression, subject to the constraint that there be at least one state neighbouring another at a distance r and energy separation Δ.)

(b) Show that if, for a very thin film of thickness d, conduction is effectively constrained to *two* dimensions, the exponent in the temperature dependence of variable-range hopping conduction changes from 1/4 to 1/3. For an amorphous semiconductor with an electronic density of states at the Fermi energy of $D(\mathscr{E}_F) = 10^{25}$ m^{-3} eV^{-1} and a localization length of $\alpha_L^{-1} = 10$ Å, estimate the thickness at which the transition from 2D to 3D behaviour is expected.

6.32 Show that eqn. (6.280) must be satisfied for the maxima in the current passed by a SQUID. (Hint: find the conditions for both the individual Josephson tunnel currents (eqn. (6.272)) and the interference term involving the magnetic flux in eqn. (6.279) to be satisfied simultaneously.)

6.33 Essay: Discuss the application of Josephson junctions in electronic devices, e.g. SQUIDs, electrical switches and radiation detectors.

References

Alexandrov, A. S. and Mott, N. F., 1995, *Polarons and Bipolarons* (World Scientific: Singapore).
Anderson, P. W., 1958, *Phys. Rev.* **109**, 1492.

Ashcroft, N. W. and Mermin, N. D., 1976, *Solid State Physics* (Holt, Rinehart and Winston: New York).
Bardeen, J., Cooper, L. N. and Schrieffer, J. R., 1957, *Phys. Rev.* **108**, 1175.
Berman, R. and McDonald, D. K. C., 1951, *Proc. Roy. Soc. (Lond.)* A**209**, 368.
Bube, R. H., 1992, *Photoelectronic Properties of Semiconductors* (Cambridge University Press: Cambridge).
Burns, G., 1992, *High-Temperature Superconductivity* (Academic Press: San Diego).
Conwell, E. M., 1952, *Proc. IRE* **40**, 1327.
Cox, P. A., 1992, *Transition Metal Oxides* (Clarendon Press: Oxford).
Dresselhaus, G., Kip, A. F. and Kittel, C., 1955, *Phys. Rev.* **98**, 368.
Giaever, I. and Megerle, K., 1961, *Phys. Rev.* **122**, 1101.
Ibach, H. and Lüth, H., 1995, *Solid-State Physics*, 2nd edn (Springer: Berlin).
Jaklevic, R. C., Lambe, J., Mercereau, J. E. and Silver, A. H., 1965, *Phys. Rev.* **140**, A1628.
Jin, S., Tiefel, T. H., McCormack, M., Fastnacht, R. A., Ramesh, R. and Chen, L. H., 1994, *Science* **264**, 413.
Kamimura, H. and Aoki, H., 1989, *The Physics of Interacting Electrons in Disordered Systems* (Clarendon Press: Oxford).
Kittel, C., 1996, *Introduction to Solid State Physics*, 7th edn (Wiley: New York).
Knotek, M. L., 1974, Int. Conf. on Tetrahedrally Bonded Amorphous Semiconductors, eds M. H. Brodsky, S. Kirkpatrick and D. Weaire, AIP Conf. Proc. **20**, 297.
Knotek, M. L., Pollak, M. and Donovan, T. M., 1974, Proc. 5th Int. Conf. on Amorphous and Liquid Semiconductors, eds J. Stuke and W. Brenig (Taylor and Francis: London), p. 225.
Langenburg, D. N., Scalapino., D. J. and Taylor, B. N., 1966, *Proc. IEEE* **54**, 560.
Lück, R., 1966, *Phys. Stat. Sol.* **18**, 49.
McDonald, D. K. C. and Mendelssohn, K., 1950, *Proc. Roy. Soc. (Lond.)* A**202**, 103.
Madelung, O., 1978, *Introduction to Solid-State Theory* (Springer-Verlag: Berlin).
Moore, T. W. and Spong, F. W., 1962, *Phys. Rev.* **125**, 846.
Mott, N. F., 1993, *Conduction in Non-Crystalline Materials*, 2nd edn (Clarendon Press: Oxford).
Phillips, N. E., 1959, *Phys. Rev.* **114**, 676.
Reuszer, J. H. and Fischer, P., 1964, *Phys. Rev.* **135**, A1125.
Richards, P. L. and Tinkham, M., 1960, *Phys. Rev.* **119**, 575.
Ruggiero, S. T. and Rudman, D. A. (eds), 1990, *Superconducting Devices* (Academic Press: San Diego).
Sato, K., Hikata, T., Mukai, H., Ueyama, M., Shibata, N., Kato, T., Masuda, T., Nagata, M., Iwata, K. and Mitsui, T., 1991, *IEEE Trans. Mag.* **27**, 1231.
Satterthwaite, C. B., 1962, *Phys. Rev.* **125**, 873.
Smith, R. A., 1978, *Semiconductors*, 2nd edn (Cambridge University Press: Cambridge).
Tinkham, M., 1996, *Introduction to Superconductivity*, 2nd edn (McGraw Hill: New York).
Wiersma, D. S., Bartolini, P., Lagendijk, A. and Righini, R., 1997, *Nature* **390**, 671.
Yu, P. Y. and Cardona, M., 1996, *Fundamentals of Semiconductors* (Springer-Verlag: Berlin).
Zdetsis, A. D., Soukoulis, C. M., Economou, E. N. and Grest, G. S., 1985, *Phys. Rev.* B**32**, 7811.

CHAPTER 7 Dielectric and Magnetic Properties

Introduction

Dielectric properties of solids are determined by localized electrons (bound charges) forming *electrostatic dipole moments*; magnetic properties, on the other hand, involve intrinsic electron spins having an associated *magnetic dipole moment*. Although these two types of behaviour have different origins, there is some logic in discussing them together in a single chapter since both types of dipole moment can order structurally in very similar manners resulting, in these cases, in the same temperature dependence of a macroscopic dielectric or magnetic property.

Dielectric properties 7.1

7.1.1 Dielectric functions

Materials can be classified into one of two categories depending on their electrical response to a d.c. (or very low-frequency) electric field, $\boldsymbol{E}$. Solids in which a constant electric field produces an electric current, consisting of ions in the case of ionic conductors (§3.4.2.2) and/or electrons in the case of electronic conductors (§§6.3.2.1 and 6.5.1.2), are termed electrical conductors. In contrast, electrical insulators, in which no d.c. ionic or electronic current can flow, are dielectrics in which the only response of the bound charges (ions or electrons) to a d.c. electric field is a static spatial displacement causing a local change in a dipole moment associated, say, with the orientation of a permanent dipole, or an induced dipole moment in the case of electrons displaced with respect to the ion cores (Fig. 7.1). The overall polarization $\boldsymbol{P}$ (net dipole moment per volume) is related to the electric field (strictly, the macroscopic field inside the solid—see §7.1.2) by

$$\boldsymbol{P} = \varepsilon_0 \chi \boldsymbol{E}, \tag{7.1}$$

where ε_0 is the permittivity of free space, and χ is the first-order dielectric susceptibility, being generally a tensor in other than cubic or isotropic materials. Higher-order terms in the dielectric susceptibility (eqn. (5.226)) are responsible for non-linear optical behaviour (§5.8.4). The polarization can also be expressed as a function of the electric field in terms of the dielectric constant, ε, also a tensor quantity in general (cf. eqn. (4.140)):

$$\boldsymbol{P} = (\varepsilon - 1)\varepsilon_0 \boldsymbol{E}; \tag{7.2}$$

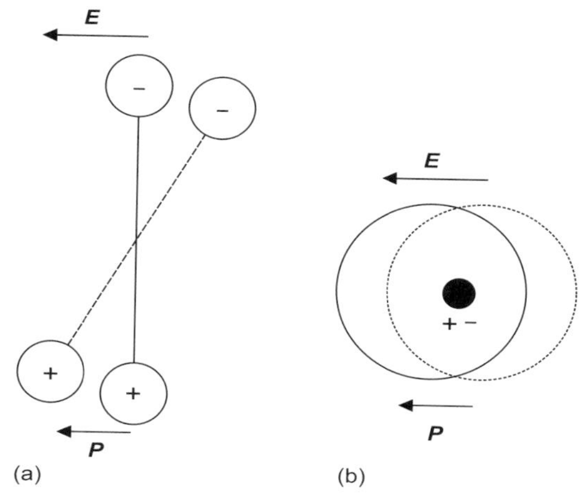

Fig. 7.1 Illustration of the origin of (a) a change in dipole moment for a bound ion-pair dipole; (b) an individual dipole moment for a bound electron distribution. The configuration before the application of a d.c. electric field is shown by the solid lines and that existing under the application of the field is represented by the dashed lines. The (change in) dipole moment is the vector $\boldsymbol{p}$.

hence the dielectric susceptibility and constant are inter-related via

$$\chi = \varepsilon - 1. \tag{7.3}$$

The dielectric displacement, $\boldsymbol{D}$, is defined (again in terms of macroscopic internal fields —see §7.1.2) as

$$\boldsymbol{D} = \varepsilon_0 \boldsymbol{E} + \boldsymbol{P} \equiv \varepsilon_0 \varepsilon \boldsymbol{E}. \tag{7.4}$$

In time-varying fields, a displacement current density, $\partial \boldsymbol{D}/\partial t$, associated with the bound charges can flow in addition to the electric current flux, $\boldsymbol{j}$, due to free charge carriers. These quantities are connected via one of the Maxwell equations:

$$\text{curl } \boldsymbol{H} = \boldsymbol{j} + \frac{\partial \boldsymbol{D}}{\partial t}. \tag{7.5}$$

The time-dependent fields are related, via Fourier transforms, to the corresponding frequency-dependent quantities which determine the *spectral* behaviour of the dielectric response:

$$\boldsymbol{E}(t) = \int_{-\infty}^{\infty} \boldsymbol{E}(\omega) \mathrm{e}^{-\mathrm{i}\omega t} \, \mathrm{d}\omega, \tag{7.6a}$$

$$\boldsymbol{D}(t) = \int_{-\infty}^{\infty} \boldsymbol{D}(\omega) \mathrm{e}^{-\mathrm{i}\omega t} \, \mathrm{d}\omega \tag{7.6b}$$

and, because the fields are real quantities, $\boldsymbol{E}(\omega) = \boldsymbol{E}^*(-\omega)$ and likewise for $\boldsymbol{D}(\omega)$; the star denotes the complex conjugate. Since the electrical current density $\boldsymbol{j}$ is related to the electric field via the conductivity σ (eqn. (6.1)), the Maxwell equation (7.5) can be rewritten in frequency-dependent form as

$$\text{curl } \boldsymbol{H}(\omega) = \sigma^\dagger(\omega) \boldsymbol{E}(\omega) - \mathrm{i}\omega\varepsilon_0\varepsilon^\dagger(\omega)\boldsymbol{E}(\omega) \tag{7.7a}$$

$$\equiv \tilde{\sigma}(\omega)\boldsymbol{E}(\omega), \tag{7.7b}$$

where the second term in eqn. (7.7a) arises from eqn. (7.6b), and $\sigma^\dagger(\omega)$ and $\epsilon^\dagger(\omega)$ are complex quantities.

A generalized expression for the conductivity that includes dielectric contributions can be defined as in eqn. (7.7), treating the response entirely in conductivity terms, with

$$\tilde{\sigma}(\omega) = \sigma^\dagger(\omega) - \mathrm{i}\omega\varepsilon_0\varepsilon^\dagger(\omega). \tag{7.8}$$

Alternatively, by regarding the contribution to curl $\boldsymbol{H}$ as being due entirely to the displacement-current density $\dot{\boldsymbol{D}}$, an equivalent generalized dielectric constant can be defined that includes the a.c. conductivity response:

$$\tilde{\varepsilon}(\omega) = \varepsilon^\dagger(\omega) + \frac{\mathrm{i}\sigma^\dagger}{\varepsilon_0 \omega}. \tag{7.9}$$

Equations (7.8) and (7.9) arise because it is only for *static* fields that a true distinction between bound and free charge carriers can be made: for alternating fields, alternating motion of bound charges contributes to the a.c. conductivity, and the oscillating motion of free charges contributes to the frequency-dependent dielectric constant.

The real and imaginary parts of the complex dielectric constant

$$\varepsilon^{\dagger}(\omega) = \varepsilon_1(\omega) + \mathrm{i}\varepsilon_2(\omega) \tag{7.10}$$

are inter-related via the Kramers–Kronig relations (eqns. (4.166)). The imaginary component, $\varepsilon_2(\omega)$, is associated with energy dissipation or loss in the dielectric material. Substitution of eqn. (7.10) into the time-derivative form of eqn. (7.4) (assuming for simplicity that $\varepsilon^{\dagger}(\omega)$ is a scalar) gives the displacement-current density:

$$\dot{\boldsymbol{D}} = \varepsilon^{\dagger}\varepsilon_0\dot{\boldsymbol{E}} = \varepsilon_0(\varepsilon_1 + \mathrm{i}\varepsilon_2)\dot{\boldsymbol{E}} \tag{7.11a}$$

$$= -\mathrm{i}\omega\varepsilon_0\varepsilon_1\boldsymbol{E} + \omega\varepsilon_0\varepsilon_2\boldsymbol{E}, \tag{7.11b}$$

where a periodic variation of electric field, $\boldsymbol{E} \propto \boldsymbol{E}_0\exp(-\mathrm{i}\omega t)$, has been assumed for eqn. (7.11b). The average dissipation per volume in one cycle resulting from the resistive (in-phase) component (the second term of eqn. (7.11b)) is

$$\langle \boldsymbol{E}\,.\,\dot{\boldsymbol{D}}\rangle = \varepsilon_0\varepsilon_2\omega E_0^2\langle\cos^2\omega t\rangle = \varepsilon_0\varepsilon_2\omega E_0^2/2. \tag{7.12}$$

This depends only on the imaginary component of the dielectric constant, ε_2, and not the real component, ε_1, that appears in the reactive (in-quadrature) component (the first term of eqn. (7.11b)). Equation (7.11b) can be represented in vector form in a complex-plane diagram (Fig. 7.2). Since a perfect dielectric is characterized by being completely reactive, with no resistive lossy component, the quality of a dielectric can be assessed in terms of the loss tangent, i.e. the tangent of the loss angle δ between resistive and reactive vectors in Fig. 7.2, given by

$$\tan\delta = \varepsilon_2/\varepsilon_1. \tag{7.13}$$

A perfect dielectric has $\delta = \tan\delta = 0$.

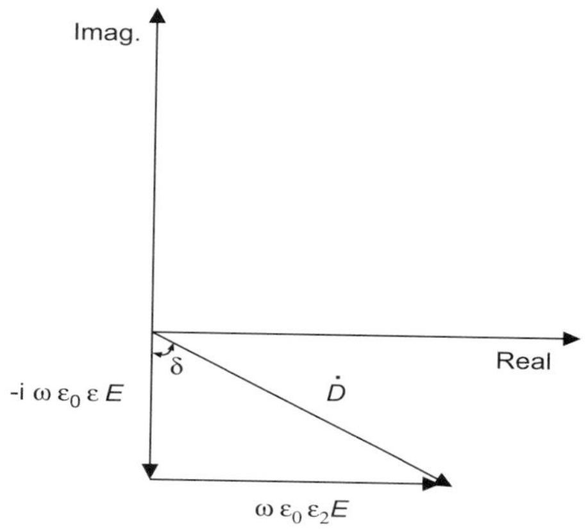

Fig. 7.2 Complex-plane representation of resistive $(\omega\varepsilon_0\varepsilon_2 E)$ and reactive $(-\mathrm{i}\omega\varepsilon_0\varepsilon_1 E)$ components of the displacement current density $\dot{D}$. The loss angle δ is indicated.

The complex dielectric constant $\varepsilon^{\dagger}(\omega)$ also controls optical properties since it is related to the complex refractive index $n^{\dagger}(\omega)$ via the relation

$$n^{\dagger}(\omega) \equiv n_{\rm r} + {\rm i}\kappa_{\rm i} = \sqrt{\varepsilon^{\dagger}(\omega)}. \tag{4.159}$$

The imaginary part of the dielectric constant, ε_2, also determines the energy loss of an electromagnetic wave, e.g. light, propagating through a material. The absorption coefficient $K(\omega)$ is given by

$$K(\omega) = \omega\varepsilon_2(\omega)/n_{\rm r}c. \tag{4.163}$$

7.1.2 Internal electric fields

In general, the electric field inside a polarizable medium is *not* the same as the applied external field, $\boldsymbol{E}_{\rm ext}$, because the electric dipoles in the material also produce an electric field. Moreover, the macroscopic internal field, $\boldsymbol{E}_{\rm mac}$, is not the same as the local field, $\boldsymbol{E}_{\rm loc}$, at a particular atom, and both these quantities depend on the particular sample geometry. Similar considerations apply also to *magnetic fields* (§7.2.2).

At the atomic level, the electric charge density, $\rho(\boldsymbol{r})$, of, say, an ionic insulator is a very rapidly varying function of $\boldsymbol{r}$ on the length scale of atomic spacings, a. Consequently, the microscopic electric field, $\boldsymbol{E}_{\rm mic}(\boldsymbol{r})$, related to the charge density by the Maxwell equation

$$\operatorname{div} \boldsymbol{E}_{\rm mic}(\boldsymbol{r}) = \rho_{\rm mic}(\boldsymbol{r})/\varepsilon_0, \tag{7.14}$$

is also a very rapidly varying quantity. The macroscopic electric field, $\boldsymbol{E}_{\rm mac}(\boldsymbol{r})$, is then an average of $\boldsymbol{E}_{\rm mic}(\boldsymbol{r})$ taken over a volume of spatial dimension $r_0 \gg a$ (but appreciably smaller than the sample volume) so that the very rapid spatial fluctuations in field are averaged out, but any slow spatial variations of the field on a length scale somewhat greater than r_0 are retained in $\boldsymbol{E}_{\rm mac}(\boldsymbol{r})$.

Application of a uniform external electric field $\boldsymbol{E}_{\rm ext}$ to a polarizable body produces a polarization $\boldsymbol{P}$ within it that henceforth will be assumed to be *uniform* throughout the sample. (This is true, for example, for ellipsoidal geometries, including spheres, discs and cylinders as limiting forms.) A uniform polarization $\boldsymbol{P}$ produces an electric field $\boldsymbol{E}_1$ that is exactly equivalent (see Problem 7.1) to the field *in vacuo* resulting from a fictitious surface charge density $\sigma_c = \hat{\boldsymbol{n}}\,.\,\boldsymbol{P}$ on the surface of the sample, where $\hat{\boldsymbol{n}}$ is the outward unit vector normal to the surface. Note that it is the component of $\boldsymbol{P}$ *normal* to the surface that determines $\sigma_{\rm c}$. The field $\boldsymbol{E}_1$ is called the depolarization field, for it opposes the external field (see Fig. 7.3). The depolarization field for a particular symmetry axis of an ellipsoidal geometry sample is given by (see Problem 7.1):

$$\boldsymbol{E}_1 = -\frac{N_i}{\varepsilon_0}\boldsymbol{P}. \tag{7.15}$$

Values of the depolarization factors $N_i (i = x, y, z)$ for various axes of simple-geometry ellipsoidal samples are given in Table 7.1 (see also Problem 7.1); the sum rule $N_x + N_y + N_z = 1$ holds. The internal macroscopic field is given by the sum of the external and depolarization fields:

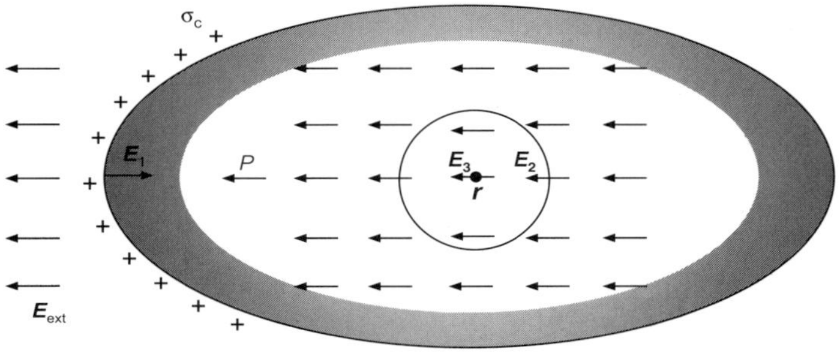

Fig. 7.3 Internal electric fields in a dielectric subject to a uniform external electric field, $\boldsymbol{E}_{\text{ext}}$. For a uniform polarization $\boldsymbol{P}$ induced within the material, the associated depolarization field $\boldsymbol{E}_1$ can be regarded as arising from fictitious charges induced on the surface, with a surface charge density given by the component of the polarization normal to the surface, i.e. $\sigma_c = \boldsymbol{P} \cdot \hat{\boldsymbol{n}}$. The local field at a particular point $\boldsymbol{r}$ within the sample can be calculated as $\boldsymbol{E}_{\text{loc}} = \boldsymbol{E}_{\text{ext}} + \boldsymbol{E}_1 + \boldsymbol{E}_2 + \boldsymbol{E}_3$, where the Lorentz cavity field $\boldsymbol{E}_2$ is the field from polarization charges on the surface of an imaginary sphere inscribed within the sample, and centred on $\boldsymbol{r}$, and $\boldsymbol{E}_3$ is the field from dipoles within the imaginary sphere.

$$\boldsymbol{E}_{\text{mac}} = \boldsymbol{E}_{\text{ext}} + \boldsymbol{E}_1. \tag{7.16}$$

Hence, for a thin disc perpendicular to the applied field:

$$\boldsymbol{E}_1 = -\boldsymbol{P}/\varepsilon_0; \qquad \boldsymbol{E}_{\text{mac}} = \boldsymbol{E}_{\text{ext}} - \boldsymbol{P}/\varepsilon_0; \qquad \boldsymbol{D}_{\text{mac}} = \varepsilon_0 \boldsymbol{E}_{\text{ext}}, \tag{7.17}$$

while for a long cylinder parallel to the field:

$$\boldsymbol{E}_1 = 0; \qquad \boldsymbol{E}_{\text{mac}} = \boldsymbol{E}_{\text{ext}}; \qquad \boldsymbol{D}_{\text{mac}} = \varepsilon_0 \boldsymbol{E}_{\text{ext}} + \boldsymbol{P}, \tag{7.18}$$

and for a sphere:

$$\boldsymbol{E}_1 = -\boldsymbol{P}/3\varepsilon_0; \qquad \boldsymbol{E}_{\text{mac}} = \boldsymbol{E}_{\text{ext}} - \boldsymbol{P}/3\varepsilon_0; \qquad \boldsymbol{D}_{\text{mac}} = \varepsilon_0 \boldsymbol{E}_{\text{ext}} + 2\boldsymbol{P}/3. \tag{7.19}$$

In general, the local field, $\boldsymbol{E}_{\text{loc}}$, at an atomic site is *not* the same as $\boldsymbol{E}_{\text{mac}}$. The local, or effective, field can be calculated by a trick. An imaginary surface is inscribed within the sample about the point $\boldsymbol{r}$ at which $\boldsymbol{E}_{\text{loc}}$ is required; for simplicity, this surface is taken to be spherical and centred on $\boldsymbol{r}$, with radius $\boldsymbol{R}$ much greater than the distance r_0 used to evaluate $\boldsymbol{E}_{\text{mac}}$ (Fig. 7.3). The local field can then be expressed as the sum of five terms:

$$\boldsymbol{E}_{\text{loc}}(\boldsymbol{r}) = \boldsymbol{E}_{\text{ext}} + \boldsymbol{E}_1 + \boldsymbol{E}_2 + \boldsymbol{E}_3(\boldsymbol{r}) + \boldsymbol{E}_4(\boldsymbol{r}), \tag{7.20}$$

Table 7.1 Values of depolarization factors for ellipsoidal-geometry samples

Shape	*Axis*	N_i
Sphere	any	1/3
Thin disc	longitudinal	1
Thin disc	transverse	0
Long circular cylinder	longitudinal	0
Long circular cylinder	transverse	1/2

where the Lorentz cavity field, $\boldsymbol{E}_2$, is the field due to dipoles in the 'far' region outside the fictitious spherical shell, effectively due to polarization charges on the surface of this shell, $\boldsymbol{E}_3(\boldsymbol{r})$ is the field due to atoms in the 'near' region, i.e. within the imaginary sphere, and $\boldsymbol{E}_4(\boldsymbol{r})$ is the contribution from the atom at the origin (which is henceforth neglected). The contribution $\boldsymbol{E}_3(\boldsymbol{r})$ can be taken for simplicity to be that at the *origin* of the atom, $\boldsymbol{E}_3(0)$: it is the only term that is dependent on the particular atomic structure of the material.

The Lorentz cavity field can be calculated in the same way as for the depolarization field of a sphere (Problem 7.1b). If θ is the angle with respect to $\boldsymbol{P}$, the surface charge density is given by $-P\cos\theta$ (see Fig. 7.4). The element of charge on a circular ring of radius $R\sin\theta$ on the surface of the cavity is $\mathrm{d}Q = -\boldsymbol{P}\cos\theta.2\pi R^2\sin\theta\ \mathrm{d}\theta$, and hence the field resolved along the direction of $\boldsymbol{P}$ is

$$\boldsymbol{E}_2 = \int_0^{\pi} \frac{2\pi R^2 \boldsymbol{P}\cos^2\theta\sin\theta\ \mathrm{d}\theta}{4\pi\varepsilon_0 R^2} = \frac{\boldsymbol{P}}{3\varepsilon_0}, \tag{7.21}$$

ie the negative of the depolarization field.

The contribution to the local field can only be evaluated knowing the atomic structure of the material. If the atoms are replaced by dipoles giving a moment p_i per cell, all assumed to be parallel to the z-axis (and $\boldsymbol{P}$), the z-component of the field at the centre of the imaginary sphere is

$$\boldsymbol{E}_3 = \sum_i \frac{3(\boldsymbol{p}_i\,.\,\boldsymbol{r}_i)\boldsymbol{r}_i - r_i^2\boldsymbol{p}_i}{r_i^5}. \tag{7.22}$$

For a *cubic* lattice (or a *random* distribution as in a liquid or gas and possibly an amorphous solid), the x-, y-and z-directions are all equivalent, and if there is only one type of dipole ($p_i = p$) eqn. (7.22) simplifies to

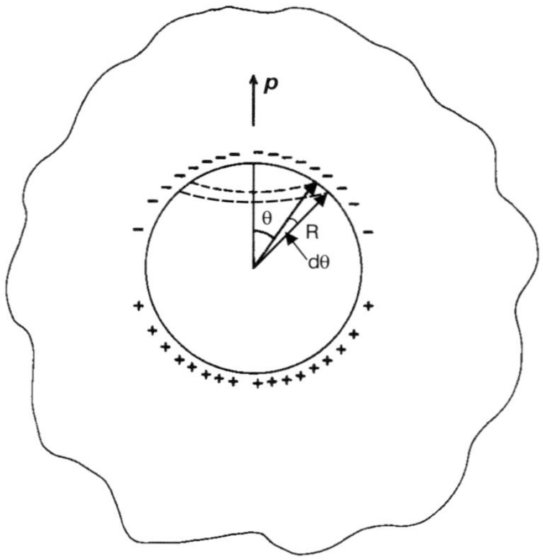

Fig. 7.4 Illustration of the geometry used to calculate the Lorentz cavity field $\boldsymbol{E}_2$.

$$\boldsymbol{E}_3 = \boldsymbol{p}\sum_i \frac{3z_i^2 - r_i^2}{r_i^5} = \boldsymbol{p}\sum_i \frac{2z_i^2 - x_i^2 - y_i^2}{r_i^5} \tag{7.23a}$$

$$= 0, \tag{7.23b}$$

since $\sum z_i^2/r_i^5 = \sum y_i^2/r_i^5 = \sum x_i^2/r_i^5$. Hence it can be seen from eqns. (7.20), (7.21), and (7.23b) that the local field for a cubic or isotropic material is given by the Lorentz relation:

$$\boldsymbol{E}_{\text{loc}} = \boldsymbol{E}_{\text{mac}} + \frac{\boldsymbol{P}}{3\varepsilon_0}. \tag{7.24}$$

For the particular case of a cubic (or isotropic) material in the shape of a sphere, $\boldsymbol{E}_{\text{loc}} = \boldsymbol{E}_{\text{max}}$ (cf. eqn. (7.19)), but this relation is not generally true.

7.1.3 Polarization and the dielectric constant

For an *isolated* atom, the electric dipole moment $\boldsymbol{p}$ induced by an external electric field is proportional to the field:

$$\boldsymbol{p} = \alpha_{\text{p}}\boldsymbol{E}_{\text{ext}}, \tag{7.25}$$

where α_p is the polarizability; it is a tensor for a non-spherical atom. For an assembly of n_i atoms of type i in unit volume, the polarization (dipole moment per unit volume) can be obtained by assuming that it is equal to a sum over the individual dipole moments:

$$\boldsymbol{P} = \sum_i n_i\boldsymbol{p}_i = \sum_i n_i\alpha_{\text{p,i}}\boldsymbol{E}_{\text{loc},i}, \tag{7.26}$$

where the *local* electric field (eqn. (7.24)) at each atom must now be used. The Lorentz relation for $\boldsymbol{E}_{\text{loc}}$, valid for *cubic* lattices or isotropic media, is the same for all atoms in such an assembly because the structure-dependent term $\boldsymbol{E}_3$ of the local field is identically zero for a cubic or random array of dipoles. In such a case, the polarization becomes

$$\boldsymbol{P} = \sum_i n_i\alpha_{\text{p},i}\left\{\boldsymbol{E}_{\text{mac}} + \frac{\boldsymbol{P}}{3\varepsilon_0}\right\} \tag{7.27}$$

and since the dielectric susceptibility (assumed to be a scalar) is given by $\chi = P/\varepsilon_0 E_{\text{mac}}$ (eqn. (7.1)), then

$$\chi = \frac{\sum_i n_i\alpha_{\text{p},i}}{\varepsilon_0 - \frac{1}{3}\sum_i n_i\alpha_{\text{p},i}}. \tag{7.28}$$

Making use of the relation $\chi = \varepsilon - 1$ (eqn. (7.3)) gives the Clausius–Mossotti relations

$$\frac{\varepsilon - 1}{\varepsilon + 2} \equiv \frac{\chi}{\chi + 3} = \frac{1}{3\varepsilon_0}\sum_i n_i\alpha_{\text{p},i}, \tag{7.29}$$

or equivalently after rearrangement:

$$\varepsilon = \frac{2\sum_i n_i\alpha_{\text{p},i} + 3\varepsilon_0}{3\varepsilon_0 - \sum_i n_i\alpha_{\text{p},i}}. \tag{7.30}$$

It should be stressed that eqns. (7.28–7.30) are only valid for *cubic* lattices or isotropic media.

The polarizability α_p that appears in eqns. (7.29) and (7.30) determining the dielectric constant can have several physical origins. One mechanism is the atomic polarizability associated with the distortion of the electronic charge distribution in an atom relative to the ion core. Another electronic mechanism operative in covalent solids is the bond polarizability resulting from the distortion of the electronic charge density in covalent bonds. Another type of mechanism is the ionic-displacement polarizability that is associated with the relative displacement of oppositely charged ions in a polar solid (see §4.4 for a discussion of this mechanism). Finally, a material containing permanent dipoles (e.g. OH groups in hydrous silicas or polar side-groups in polymers) exhibits a dipole-orientational polarizability. Not all of these mechanisms may be operative simultaneously: it will be seen later that particular polarizability mechanisms may be dominant in certain frequency ranges of the spectral dielectric response of a solid.

The atomic polarizability can be estimated by making use of the model for an atom sketched in Fig. 7.5: the Z electrons in the atom are assumed to be distributed homogeneously in a sphere of radius R. If the nucleus is displaced by distance x from the centre O of the spherical charge distribution, itself assumed to behave rigidly, it experiences a restoring electric field due to the electrons in the sphere of radius x centred on O; electrons outside this inner sphere do not contribute because the electric field inside a uniformly charged spherical shell is zero (Gauss's theorem). The electron charge contained within the inner sphere is $-Ze(x^3/R^3)$, and the associated field at the nucleus is

$$\boldsymbol{E} = \frac{-Ze(x/R)^3\hat{\boldsymbol{x}}}{4\pi\varepsilon_0 x^2} = \frac{-Zex}{4\pi\varepsilon_0 R^3}. \tag{7.31}$$

A local applied field $\boldsymbol{E}_{\text{loc}}$ acting on an atom will cause a displacement of the electron cloud and nucleus until the restoring field becomes equal and opposite, i.e. $\boldsymbol{E} = \boldsymbol{E}_{\text{loc}}$. The corresponding dipole moment induced in the atom is, from eqn. (7.31),

$$\boldsymbol{p} = Ze\boldsymbol{x} = 4\pi\varepsilon_0 R^3 \boldsymbol{E}_{\text{loc}}. \tag{7.32}$$

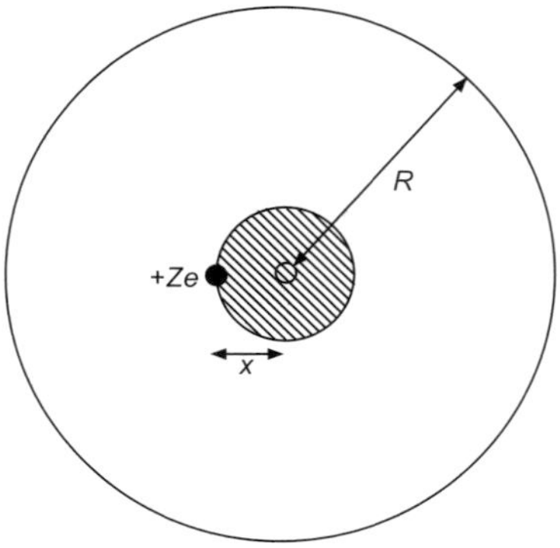

Fig. 7.5 Model for the calculation of the atomic polarizability. The Z electrons are assumed to be distributed homogeneously and are uniformly displaced by x relative to the nucleus by an applied static electric field. The restoring field (equal and opposite to $\boldsymbol{E}_{\text{loc}}$ at equilibrium) is caused by the electrons inside the sphere of radius x centred at O, the centre of the electron distribution.

Table 7.2 Atomic polarizabilities for alkali cations and halide anions

	Cations					*Anions*			
	Li^+	Na^+	K^+	Rb^+	Cs^+	F^-	Cl^-	Br^-	I^-
$\alpha_p^{at}/4\pi\varepsilon_0$ (Å^3)	0.029	0.408	1.334	1.979	3.335	0.0644	2.960	4.158	6.431

Values of α_p^{at} have been chosen to give the best fit between experimental values of the dielectric constant of alkali halide crystals (measured at the D-line frequency of the Na spectrum) and values calculated using the Clausius–Mossotti relation. (After Tessmann *et al.* (1953))

Table 7.3 Dielectric constants of alkali halide crystals

$\varepsilon(0)(\varepsilon(\infty))$	Li	Na	K	Rb	Cs
F	9.04 (1.92)	5.07 (1.74)	6.05 (1.85)	5.91 (1.93)	8.08 (2.2)
Cl	11.86 (2.79)	5.89 (2.35)	4.81 (2.20)	4.92 (2.91)	6.95 (2.67)
Br	13.33 (3.22)	6.40 (2.64)	4.90 (2.39)	5.0 (2.33)	6.66 (2.83)
I	11.0 (3.80)	7.28 (3.08)	5.09 (2.68)	4.94 (2.61)	6.59 (3.09)

The quantity not in parentheses is the static dielectric constant, $\varepsilon(0)$, and that in parentheses is the high-frequency value, $\varepsilon(\infty)$. (Data after Burns (1985). Reproduced by permission of Academic Press, Inc.)

Hence, the atomic polarizability is given by

$$\alpha_p^{at} = 4\pi\varepsilon_0 R^3. \tag{7.33}$$

Typical values of $\alpha_p^{at}/4\pi\varepsilon_0$ are $\simeq 10^{-30}$ m^3 for an atomic radius of $R \simeq 1$ Å (see Table 7.2 and Problem 7.2). The electronic polarizability associated with atoms (or bonds) will provide the dominant contribution to the dielectric constant, $\varepsilon(\infty)$, at high frequencies where the contributions of ions or permanent dipoles are frozen out (see Table 7.3).

The behaviour of the electronic dielectric constant in metals and semiconductors has been discussed previously in §5.8.1 in connection with intraband plasma oscillations (see eqn. (5.196)). The electronic contribution to the dielectric constant of semiconductors or insulators at higher frequencies is associated with *interband* transitions (see §5.8.2): the imaginary part of the dielectric constant, ε_2, is related to energy loss or, in this case, optical absorption involving electron excitation across the bandgap. A general expression for $\varepsilon_2(\omega)$ in terms of the momentum matrix element is given by eqn. (5.211). Use of the Kramers–Kronig relation (eqns. (4.166)) transforms eqn. (5.211) into the real part (Yu and Cardona (1996)):

$$\varepsilon_1(\omega) = 1 + \frac{e^2}{m_e^2\varepsilon_0}\sum_k \left(\frac{2}{m_e\hbar\omega_{cv}}\frac{|P_{cv}|^2}{(\omega_{cv}^2 - \omega^2)}\right). \tag{7.34}$$

The oscillator strength (so called because it represents the equivalent number of oscillators (see eqn. (7.43)) of the transition between valence (v) and conduction (c) bands is given by the quantity

$$f_{cv} = \frac{2}{m_e\hbar\omega_{cv}}|P_{cv}|^2. \tag{7.35}$$

The real part of the *static* dielectric constant, $\varepsilon_1(0)$, can be rewritten from eqn. (7.34) in terms of the Penn model, in which the electronic structure of a semiconductor or insulator can be represented as a free-electron gas with a single, average energy gap, the Penn gap, $\mathscr{E}_p$. This energy gap can be thought of as representing, say, the energy separation between band centres. Experimentally, an estimate for $\mathscr{E}_p$ can be obtained from $\varepsilon_2(\omega)$ spectra obtained from reflectance measurements. For example $\mathscr{E}_p$ can be taken as the energy of the high-energy (so-called E_2-transition) peak in ε_2 (see Fig. 5.79b) resulting from the van Hove singularity in the joint density of states at the X-point for tetrahedral semiconductors (§5.8.2). The Penn gap is related to covalent and ionic energies, $\mathscr{E}_{cov}$ and $\mathscr{E}_i$ respectively, for heteropolar materials by eqn. (2.7b):

$$\mathscr{E}_p = (\mathscr{E}_{cov}^2 + \mathscr{E}_i^2)^{1/2}.$$

For the case of covalent Ge, for example, $\mathscr{E}_p = \mathscr{E}_{cov} = 4.3$ eV($\mathscr{E}_i = 0$), whereas for the II–VI compound ZnSe with an ionicity (eqn. (2.8)) of $f_i = 0.64, \mathscr{E}_p = 7.0$ eV, with $\mathscr{E}_{cov} = 4.3$ eV and $\mathscr{E}_i = 5.6$ eV. The static dielectric constant can then be written, following eqn. (7.34), as

$$\varepsilon_1(0) = 1 + (\hbar\omega_p/\mathscr{E}_p)^2, \tag{7.36}$$

where ω_p is the plasma frequency given by $\omega_p^2 = ne^2/m_e$ (eqn. (5.166)). 'Static' in this context means frequencies much less than those corresponding to interband transitions, but higher than phonon frequencies. Some values of the static dielectric constant, and the corresponding Penn gaps, are given in Table 7.4

Permanent dipoles within a material can be oriented by an external electric field and so can also contribute to the net polarization. The interaction energy of such a dipole of moment $\boldsymbol{p}$ with a local field $\boldsymbol{E}_{loc}$ is

$$\mathscr{E} = -\boldsymbol{p} \cdot \boldsymbol{E}_{loc}. \tag{7.37}$$

The probability of a dipole being at an angle θ to $\boldsymbol{E}_{loc}$ is given by the Boltzmann factor

$$p(\theta) \propto \exp(\boldsymbol{p} \cdot \boldsymbol{E}_{loc}/k_B T) = \exp(pE_{loc}\cos\theta/k_B T) \tag{7.38}$$

if the dipoles behave independently and if they are in thermal equilibrium with the surroundings. This is a reasonable assumption for liquids and gases, but less so for solids where the interactions between dipoles may be so strong that they order structurally (see §7.1.5). Neglecting for the moment this effect and assuming that all orientations of the dipoles are equally probable (dielectric gas model), the average component of the dipole moment parallel to the field is

$$p_{par} = \langle p\cos\theta \rangle, \tag{7.39}$$

where the angular brackets denote an averaging using the Boltzmann probability as the weighting factor. The result is (Problem 7.3):

Table 7.4 Static dielectric constants and the Penn gap for tetrahedral semiconductors

	C (dia)	Si	Ge	GaAs	InP	GaP
$\varepsilon_1(0)$	5.66	12.0	16.0	10.9	9.6	9.1
$\mathscr{E}_p$ (eV)	13.5	4.8	4.3	5.2	5.2	5.75

(Data after Yu and Cardona (1996), *Fundamentals of Semiconductors*, p. 326, Table 6.9, © Springer-Verlag GmbH & Co. KG)

$$p_{\text{par}} = p(\coth x - 1/x) \equiv p\mathfrak{L}(x), \tag{7.40}$$

where $x = pE_{\text{loc}}/k_B T$ and $\mathfrak{L}(x)$ is known as the Langevin function. The small-x limit of eqn. (7.40) is normally valid in solids (see Problem 7.3), in which case $\coth x \simeq 1/x + x/3 + ..$, giving

$$p_{\text{par}} \simeq \frac{p^2 E_{\text{loc}}}{3k_B T}. \tag{7.41}$$

Hence, the dipolar polarizability is approximately given by

$$\alpha_{\text{p}}^{\text{dip}} \simeq \frac{p^2}{3k_B T}. \tag{7.42}$$

This $1/T$ temperature dependence is the dielectric equivalent of the Curie law for paramagnetism (§7.2.4.1). If the restriction is made that, in a solid (e.g. $NaNO_2$), the dipole (NO_2^-) has only *two* allowed states, parallel and antiparallel to the electric field, the same form for the polarizability is found as in eqn. (7.42), but *without* the factor of three in the denominator (see Problem 7.3).

7.1.4 Dielectric spectroscopy

The different sources of dielectric polarization described in the previous section contribute to the dielectric constant in different frequency régimes, depending on whether electrons or ions are responsible: the small inertial mass of electrons means that electronic mechanisms for the polarizability (i.e. atomic or bond polarizabilities) dominate at high frequencies, whereas polarizability mechanisms involving the motion of ions (i.e. ionic-displacement or dipole-orientational polarizabilities) contribute only at lower frequencies (see Fig. 7.6).

The frequency dependence of the polarizability (or dielectric constant) for any particular mechanism in which there is an *equilibrium configuration* of the polarizable species, applicable to all mechanisms except that of dipole orientation (a freely rotating dipole having no equilibrium configuration has the dynamics of orientation governed by a relaxational equation—see eqn. (7.46)), can be described in a semi-classical sense by a driven damped-oscillator equation (the Lorentz equation):

$$m\ddot{\boldsymbol{r}} + m\gamma\dot{\boldsymbol{r}} + k\boldsymbol{r} = q\boldsymbol{E}, \tag{7.43}$$

where m is the (reduced) mass of the oscillating species of charge q, $\boldsymbol{r}$ is the spatial displacement associated with the polarization, γ is a damping constant (e.g. arising from anharmonic coupling to other excitations, or alternatively representing optical absorption processes) and k is a spring constant characterizing the restoring force of the system in response to an oscillating driving field, $\boldsymbol{E} = \boldsymbol{E}_0\exp(-i\omega t)$. The frequency-dependent polarizability, $\alpha_p(\omega)$, can be obtained from the solution of eqn. (7.43), viz. $\boldsymbol{r}(\omega)$. A proper quantum-mechanical analysis treats the polarizability in terms of transitions between allowed states of the electronic or ionic system involved.

This driven, damped-oscillator equation has been used previously in a discussion of ionic-displacement polarizability and its relation to IR absorption (§4.5.1, eqn. (4.164)), and for plasma oscillations and intraband absorption (§5.8.1, eqn. (5.184)); see Problem

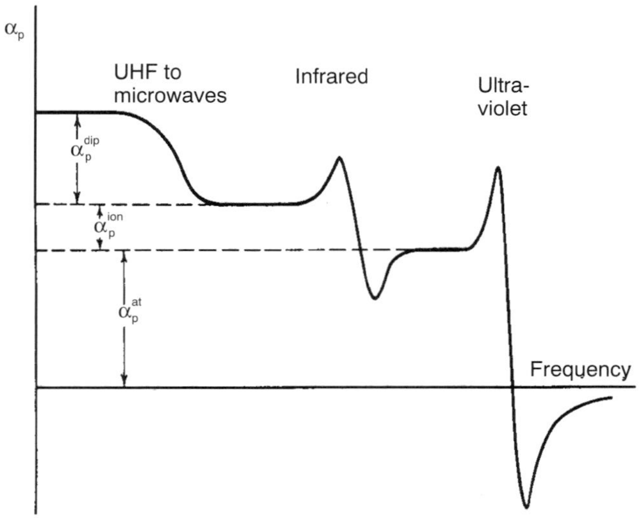

Fig. 7.6 Illustration of the frequency dependence of the real part of the dielectric polarizability (or dielectric constant), showing the relative contributions of different mechanisms. The imaginary part, corresponding to dielectric loss, exhibits Lorentzian peaks at frequencies corresponding to each discontinuity in the real part, and is zero otherwise.

7.4 for the case of atomic polarizability. The imaginary part of the polarizability, corresponding to the dielectric loss, ε_2, has a Lorentzian peak at a frequency close to the resonant frequency ω_0, corresponding to the resonant absorption of energy, the

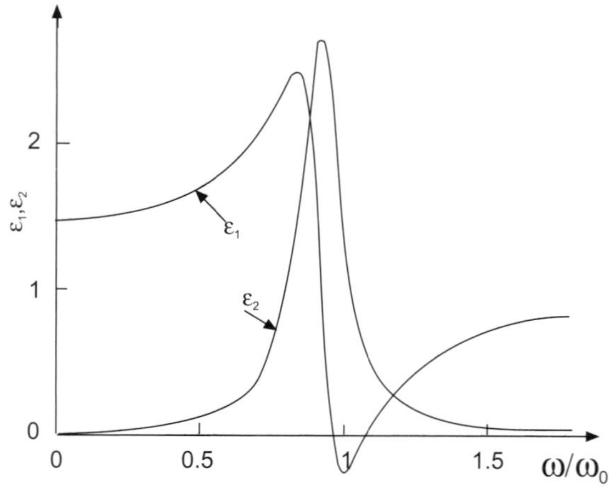

Fig. 7.7 Frequency dependence of the real (ε_1) and imaginary (ε_2) parts of the complex dielectric constant for atomic polarizability calculated using the driven damped oscillator model and the Clausius–Mossotti relation.

width of the peak increasing with the damping coefficient, γ. The real part, $\varepsilon_1(\omega)$, exhibits a characteristic discontinuity in the vicinity of ω_0 (Fig. 7.7). Neglecting local-field effects, the real and imaginary parts of the dielectric constant are simply proportional to the corresponding polarizabilities (cf. eqn. (7.2)):

$$\varepsilon_1 = \varepsilon(\infty) + \sum_i \frac{n_i q^2}{\varepsilon_0 m_e} \frac{(\omega_{0,i}^2 - \omega^2)}{\left((\omega_{0,i}^2 - \omega^2)^2 + \gamma_i^2 \omega^2\right)}, \tag{7.44}$$

$$\varepsilon_2 = \sum_i \frac{n_i q^2}{\varepsilon_0 m_e} \frac{\gamma_i \omega}{\left((\omega_{0,i}^2 - \omega^2)^2 + \gamma_i^2 \omega^2\right)}, \tag{7.45}$$

where the summations are over different mechanisms, i. This behaviour of the real part of the polarization is clearly evident in Fig. 7.6.

In experimental data of $\varepsilon_1(\omega)$ obtained, say, from capacitance experiments, there is often an extraneous contribution at very low frequencies due to space-charge polarization, arising for example from interfacial polarization at interfaces between sample and electrical contact, or between microcrystallite grains in the bulk of the sample.

The dielectric susceptibility and dielectric constant for the dipole-orientational mechanism can be calculated in terms of the ansatz of the Debye dipolar-relaxation model (Fig. 7.8). The sudden application of an electric field causes a polarization $\boldsymbol{P}(\infty) = \varepsilon_0(\varepsilon(\infty) - 1)\boldsymbol{E}$ that is instantaneous on the time scale of the dipolar rotation. Dipolar relaxation then causes the polarization to increase further, with a time-dependent component $\boldsymbol{P}'(t)$ until the static value $\boldsymbol{P}_s = \varepsilon_0(\varepsilon(0) - 1)\boldsymbol{E}$ is attained, where $\boldsymbol{P}_s = \boldsymbol{P}_\infty + \boldsymbol{P}'(t = \infty)$. The Debye relaxation equation is then

$$\frac{\mathrm{d}\boldsymbol{P}'}{\mathrm{d}t} = \frac{\boldsymbol{P}_s - \boldsymbol{P}}{\tau_D}, \tag{7.46}$$

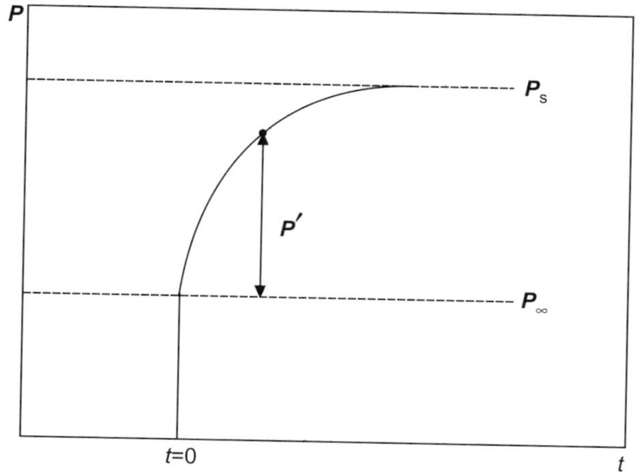

Fig. 7.8 Time dependence of the polarization $\boldsymbol{P}(t)$ after the sudden application of an electric field to a dielectric at time $t = 0$. The instantaneous increase to the level $\boldsymbol{P}_\infty$ is associated with electronic and ionic polarizations. Dipolar orientation then causes a slow increase in polarization to the static value, $\boldsymbol{P}_s$.

where the total polarization at time t is given by

$$\mathbf{P}(t) = \mathbf{P}_\infty + \mathbf{P}'(t), \tag{7.47}$$

and τ_D is the Debye dipolar relaxation time, assumed to be time-independent, which is determined by the particular mechanism by which the dipole interacts with its surroundings during the relaxational motion. The time dependence of the polarization is found by integrating eqn. (7.46), making use of eqn. (7.47), viz.:

$$\mathbf{P}'(t) = (\mathbf{P}_s - \mathbf{P}_\infty)(1 - e^{-t/\tau}). \tag{7.48}$$

Alternatively, the polarization decays exponentially with time after the electric field is removed.

For an applied (macroscopic) field $\mathbf{E}, \mathbf{P}_s = \chi(0)\varepsilon_0\mathbf{E} = \varepsilon_0(\varepsilon(0) - 1)\mathbf{E}$, $\mathbf{P}_\infty = \varepsilon_0(\varepsilon(\infty) - 1)\mathbf{E}$ and $P'(t) = \chi(t)\varepsilon_0\mathbf{E} = \chi(t)\varepsilon_0\mathbf{E}_0\exp(-i\omega t)$ for an assumed sinusoidal variation of the field. Using these relations, eqn. (7.46) can be transformed into a frequency-dependent form:

$$\chi(\omega) = \frac{(\varepsilon(0) - \varepsilon(\infty))}{(1 - i\omega\tau_D)}. \tag{7.49}$$

Taking into account the instantaneous response at $t = 0$ in producing P_∞, the complex dielectric constant is therefore given by (see also Problem 7.6):

$$\varepsilon^\dagger(\omega) = \varepsilon(\infty) + \frac{(\varepsilon(0) - \varepsilon(\infty))}{(1 - i\omega\tau_D)}, \tag{7.50}$$

with real and imaginary parts given separately by

$$\varepsilon_1(\omega) = \varepsilon(\infty) + \frac{(\varepsilon(0) - \varepsilon(\infty))}{(1 + \omega^2\tau_D^2)}, \tag{7.51a}$$

$$\varepsilon_2(\omega) = \frac{(\varepsilon(0) - \varepsilon(\infty))\omega\tau_D}{(1 + \omega^2\tau_D^2)}. \tag{7.51b}$$

The frequency behaviour of the terms ε_1 and ε_2 in the Debye model is shown in Fig. 7.9: note the difference in behaviour of $\varepsilon_1(\omega)$ from the result for the driven damped-oscillator model shown in Fig. 7.7.

★ Often, dielectric data are plotted as ε_2 versus ε_1 in the complex plane as an implicit function of frequency (a so-called Cole–Cole plot). For the case of the Debye model, the resulting Cole–Cole plot is simply a semicircle (Fig. 7.10). (Complex-plane (Nyquist) plots are employed also for displaying complex-impedance data for electrically conducting materials, particularly ionic conductors.) Very often, particularly for disordered dielectric materials, the Debye model is *not* obeyed: either the width of the peak in $\varepsilon_2(\omega)$ is wider than predicted theoretically, or the complex-plane plot of $\varepsilon_2(\omega)$ versus $\varepsilon_1(\omega)$ is not semicircular. Such non-Debye-like behaviour in the complex plane is often represented, for fitting purposes, by the purely *empirical* Havriliak–Negami expression

$$\varepsilon^\dagger(\omega) = \varepsilon(\infty) + \frac{(\varepsilon(0) - \varepsilon(\infty))}{(1 - (i\omega\tau_D)^{1-\alpha})^\beta}. \tag{7.52}$$

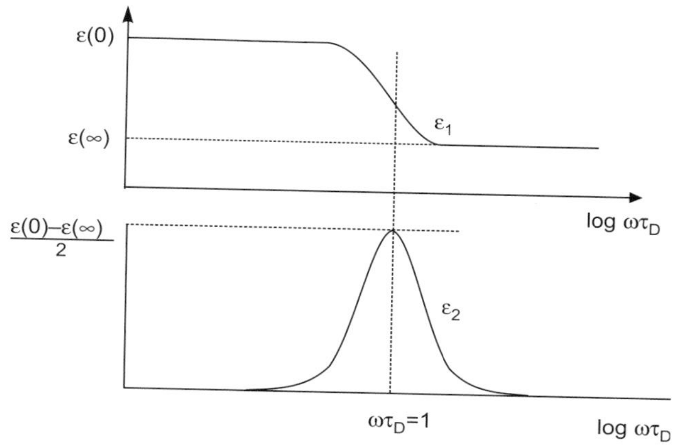

Fig. 7.9 Real and imaginary parts of the complex dielectric constant resulting from the Debye model for dipolar orientational polarization.

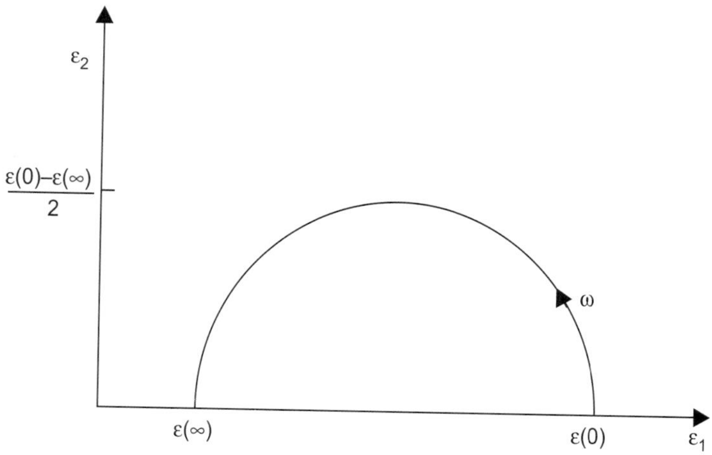

Fig. 7.10 Complex-plane, Cole–Cole plot of ε_2 versus ε_1 as an implicit function of ω for the Debye model of dipolar relaxation.

This has, as special cases, the Cole–Cole equation ($\beta = 1$) which gives a semicircular arc in the complex plane with the centre depressed below the real axis (Fig. 7.11a), and the Cole–Davidson expression ($\alpha = 0$) which gives a skewed arc (Fig. 7.11b); the Debye equation (7.50) is recovered for $\alpha = 0, \beta = 1$.

Non-Debye-like behaviour can also be simulated in terms of a *distribution* of relaxation times, $G(\tau_D)$ (cf. eqn. (7.50)):

$$\varepsilon^{\dagger}(\omega) = \varepsilon(\infty) + \int_{\tau_{\min}}^{\tau_{\max}} \frac{(\varepsilon(0) - \varepsilon(\infty))G(\tau_D)\mathrm{d}\tau_D}{1 - \mathrm{i}\omega\tau_D}, \tag{7.53}$$

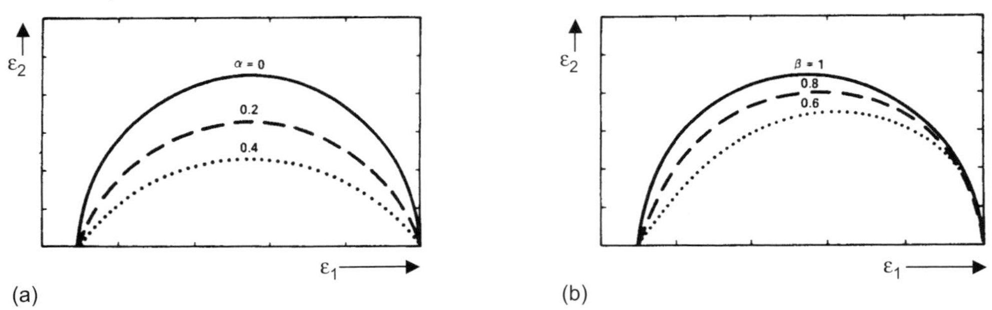

Fig. 7.11 Complex-plane plots of $\varepsilon_2(\omega)$ versus $\varepsilon_1(\omega)$ for (a) the Cole–Cole expression for different values of the parameter α (Debye-like behaviour is recovered for $\alpha = 0$); (b) the Cole–Davidson expression for different values of the parameter β (Debye-like behaviour is recovered for $\beta = 1$).

with $\int_{\tau_{\min}}^{\tau_{\max}} G(\tau_D)\mathrm{d}\tau_D = 1$. Thus, the overall dielectric response is regarded as a sum of Debye-like processes with different relaxation times, each contributing in parallel (see Problem 7.7).

Finally, yet another approach to fitting, if not interpreting, non-Debye-like dielectric relaxation spectra is in terms of the *temporal* response. A stretched exponential (or Kohlrausch–Williams–Watts) form for the decay of the polarization after the removal of a constant field has been suggested:

$$\phi(t) \equiv \frac{\boldsymbol{P}(t)}{\boldsymbol{P}_s - \boldsymbol{P}_\infty} = \exp(-t/\tau_{\mathrm{kww}})^{\gamma}, \quad 0 < \gamma \leqslant 1; \tag{7.54}$$

if $\gamma = 1$, the Debye-like form (eqn. (7.48)) is recovered. This empirical relationship has been found to fit reasonably satisfactorily many sets of dielectric data for amorphous polymers or ionic glasses, for example. Further details of dielectric (and impedance) spectroscopy can be found in MacDonald (1987). ★

7.1.5 Spontaneous polarization

Certain dielectrics generate an electrical polarization when subjected to an external mechanical stress that causes an internal strain; conversely, such materials develop a mechanical stress in an applied electric field. These materials are termed piezoelectric. A subset of piezoelectric solids are those in which a spontaneous electrical polarization in a crystal is caused by an intrinsic internal strain accompanying a change of crystal structure to one of lower symmetry. Such materials are termed pyroelectric because the natural spontaneous electrical polarization is usually masked by neutralizing counter-ions adsorbed onto the free surface; it can thus only be revealed by *heating* the sample, which removes some of the neutralizing counter-ions. A further subset of pyroelectric materials comprises those in which the electrical polarization can be reversed by the application of an external electric field. These materials are termed ferroelectric (by analogy with the ferromagnetic state — see §7.2.5); the net dipole moment in a unit cell is *parallel* to that in adjacent cells. Ferroelectric materials, together with ferromagnetic and ferroelastic solids, form a general class of materials known as ferroics, in which orientational states or domains can be switched from one configuration to another by the application of an appropriate driving 'force'. For ferroelectric, ferromagnetic and ferroelastic materials, the corresponding orientational states are, correspondingly, the spontaneous electrical polarization, magnetization and elastic strain; the driving forces are, respectively, electric and magnetic fields and mechanical stress.

7.1.5.1 *Piezoelectricity*

Crystals that are *piezoelectric*, i.e. where electrical polarization is generated by a mechanical stress (or vice versa), are, in general, *non-centrosymmetric* (without a centre of symmetry). From Table 2.5, it can be seen that of the 32 crystallographic point groups, 21 do not possess inversion ($i(\bar{1})$) symmetry elements; in addition the non-centrosymmetric cubic point group $O(432)$ has a combination of symmetry elements that precludes piezoelectric behaviour. Thus, only crystal structures with space groups that contain as a point group one of these 20 groups can be piezoelectric. Many crystals containing tetrahedral structural units (e.g. quartz (SiO_2), ZnO and ZnS) are piezoelectric since a shearing stress causes a distortional strain of the tetrahedra. Perhaps the most important piezoelectric material is PZT, an equimolar solid solution of lead zirconate and titanate: $PbZrO_3$–$PbTiO_3$.

The induced polarization $\boldsymbol{P}$ is related to the applied stress $\boldsymbol{\sigma}$ by the piezoelectric constant $\boldsymbol{d}$, which is a third-rank tensor, since polarization is a vector and stress is a second-rank tensor (§3.4.3):

$$P_i = \sum_{j,k} d_{ijk}\sigma_{jk}. \tag{7.55}$$

Alternatively, $\boldsymbol{d}$ relates the induced strain $\boldsymbol{e}$ (also a second-rank tensor) and an applied electric field $\boldsymbol{E}$:

$$e_{jk} = \sum_{i} d_{ijk}E_i. \tag{7.56}$$

The units of $\boldsymbol{d}$ are C/N or m/V. Of the possible 27 components of $\boldsymbol{d}$, only 18 are in principle independent because $d_{ijk} = d_{ikj}$ as a result, for example, of a strain being a symmetric tensor, $e_{jk} = e_{kj}$. However, many of these 18 components are, in fact, zero. For the cubic point group T_d ($\bar{4}3m$), for example, there is only *one* independent non-zero component, $d_{123}(= d_{213} = d_{312})$—see Table 7.5; for the D_2 (222) orthorhombic point group, these three components are unequal (Problem 7.8).

Table 7.5 Values of the piezoelectric constant d_{123} for crystals having the zinc-blende structure with point-group symmetry $T_d(\bar{4}3m)$

Crystal	CuCl	CuBr	CuI	GaAs	GaSb	InAs	InSb
$d_{123}(CN^{-1})$	27.2	16.0	7.0	−2.7	−2.9	−1.1	2.4

(Data after Burns (1985). Reproduced by permission of Academic Press, Inc.)

7.1.5.2 Pyroelectricity

Pyroelectric crystals are those having a permanent net electric dipole in each primitive unit cell. For example, ZnO is pyroelectric because the ZnO_4 tetrahedra (each possessing a net dipole moment) all point in the same direction. Although the term pyroelectric is general, it tends to be reserved for those non-ferroelectric materials whose net spontaneous polarization cannot be reversed by an external electric field. In this category fall, for example, lithium niobate ($LiNbO_3$) and lithium tantalate ($LiTaO_3$) for which the energy barrier between the two opposite spontaneous polarizations $\pm P_s$ corresponds to an electric field greater than the breakdown field of the material. An operational pyroelectric coefficient Π_p can be defined as the constant of proportionality between the observed *change* in polarization ΔP_s and the temperature change ΔT responsible for it:

$$\Delta P_s = \Pi_p \Delta T. \tag{7.57}$$

The coefficient Π_p is determined by the thermal- expansion characteristics of the crystal (thereby affecting the magnitude of P_s) and also by changes in the adsorbed neutralizing charge on the surface of the polarized crystal.

All pyroelectric crystals are piezoelectric (but the converse is not true). Of the 20 point groups associated with piezoelectricity (§7.1.5.1), only 10 can give rise to pyroelectricity, namely the C_n, C_{nv} and C_{1h} point groups. The reason for this is that the existence of a net polarization vector along a particular direction must be consistent with the symmetry elements operating along that axis; this is a consequence of Neumann's principle that a macroscopic physical property must have, at least, the point-group symmetry of the space group. Thus, for example, a two-fold rotation symmetry operation, $C_2(2)$, *perpendicular* to any component of $\boldsymbol{P}$ will transform it into the negative of itself; hence there can be *no* component along that axis. Similarly, there can be no component of a polarization vector perpendicular to a symmetry plane. A centre of inversion, $i(\bar{1})$, or the four $C_3(3)$ axes characteristic of cubic symmetry also cause $\boldsymbol{P}$ to be zero. For most of the C_n, C_{nv} and C_{1h} point groups, only vectors with components along the z-direction (the many symmetry axis) transform into themselves under *all* symmetry operations, and hence can be non-zero. For the triclinic point group, $C_2(1)$, $\boldsymbol{P}$ can point in any direction, and for the monoclinic point groups $C_{1h}(m)$, x-and y-components separately transform into themselves under the symmetry operations, and $\boldsymbol{P}$ can therefore point anywhere in the x–y plane.

An order-of-magnitude estimate of the electrical polarization per unit cell can be obtained by assuming that the maximum static dipole moment is equivalent to two charges $q^{+/-} = \pm e$ separated by say 1 Å; i.e. $p = 1.6 \times 10^{-29}$ C m. For a unit cell assumed to be 4 Å on a side (as in $BaTiO_3$), the polarization (dipole moment per unit volume) is

Table 7.6 Values of the electrical polarization P along the unique crystal axis

	Pyroelectric			*Ferroelectric*			
Crystal	$ZnS(C_{6v})$	$ZnO(C_{6v})$	$CdS(C_{6v})$	$BaTiO_3(C_{4v})$	$PbTiO_3(C_{4v})$	$LiNbO_3(C_{3v})$	$LiTaO_3(C_{3v})$
$P(Cm^{-2})$	0.02	0.06	0.03	0.26	0.81	0.71	0.50

(Data after Burns (1985). Reproduced by permission of Academic Press, Inc.)

of the order of $P \simeq 0.25$ C m^{-2}. Some representative values are given for pyro- and ferroelectric crystals in Table 7.6.

The material barium titanate, $BaTiO_3$, is a representative pyroelectric (and ferroelectric) compound. It exhibits a series of phase transitions between different crystal structures as the temperature is varied: at temperatures above about 134°C the material has a cubic perovskite structure and hence this phase is non-pyroelectric and is termed paraelectric. At lower temperatures, sequential transitions to three other crystal structures occur, all of which exhibit pyroelectricity (see Fig. 7.12). The structural distortion in $BaTiO_3$ associated with the displacive phase transition (in which the atomic displacements are small compared with interatomic separations) from the cubic to the tetragonal structure at 134°C is shown in Fig. 7.13. The atoms displace by about 0.1 Å as shown, to give a tetragonal structure that has $\boldsymbol{P}_s$ in the [001] (or equivalently [100] or [010]) direction. In a real crystal, domains, each of which has unit cells with $\boldsymbol{P}_s$ ordered in a particular direction, will tend to form in the pyroelectric phase. The orthorhombic phase stable below −5°C has $\boldsymbol{P}_s$ in the $\langle 110 \rangle$ directions of the original cube, and the low-temperature rhombohedral phase has $\boldsymbol{P}_s$ parallel to the original cubic $\langle 111 \rangle$ directions.

7.1.5.3 Ferroelectricity

A ferroelectric material is a pyroelectric solid in which the spontaneous electrical polarization in a unit cell can be *reversibly* changed between $\pm \boldsymbol{P}_s$ by the application of an electrical field of suitable polarity. In this way, the polarization can be *aligned* in neighbouring domains (Fig. 7.14a); it is this similarity to the corresponding behaviour of the magnetic moments in ferromagnets (§7.2.5) that gives the phenomenon of ferroelectricity its name. An intermediate type of polarization ordering is in ferrielectric materials, where the polarization vectors are, for example, canted in different directions

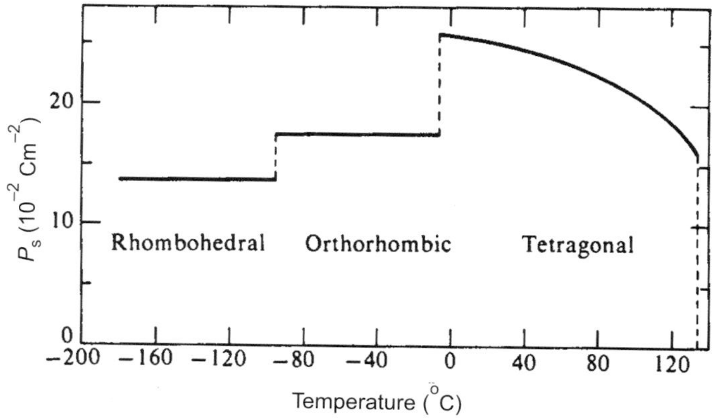

Fig. 7.12 The temperature dependence of the spontaneous electrical polarization P_s for the pyroelectric (and ferroelectric) compound $BaTiO_3$, corresponding to the three crystal phases stable below 134 °C. Above this temperature, the material has a cubic perovskite structure and hence is paraelectric. (Burns (1985). Reproduced by permission of Academic Press, Inc.)

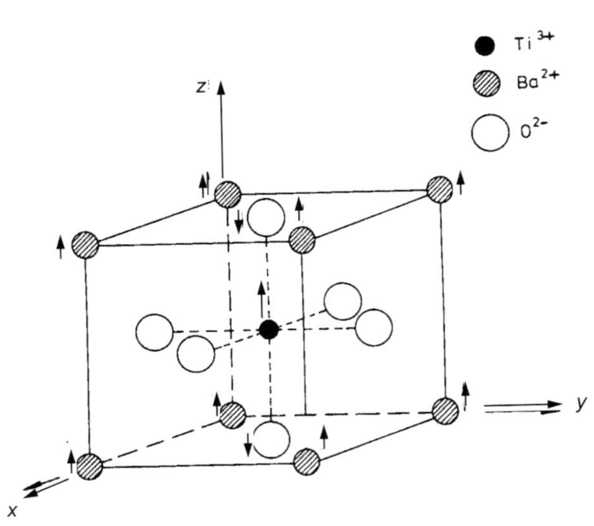

Fig. 7.13 The cubic perovskite (non-pyroelectric) structure of $BaTiO_3$ stable above 134 °C. The arrows show the relative magnitude and direction of the atomic displacements in the transition to the pyroelectric tetragonal phase stable between -5 and 134 °C. The oxygen atoms have been taken to be fixed (see also Fig. 7.19a).

in neighbouring domains (Fig. 7.14b), with the result that there is no net polarization in one direction but a net polarization in an orthogonal direction. Finally, antiferroelectric ordering (Fig. 7.14c) is that in which the polarization vectors are oriented in an anti-parallel fashion in neighbouring domains or unit cells. These latter two types of ordering also have magnetic equivalents (§7.2.5.6).

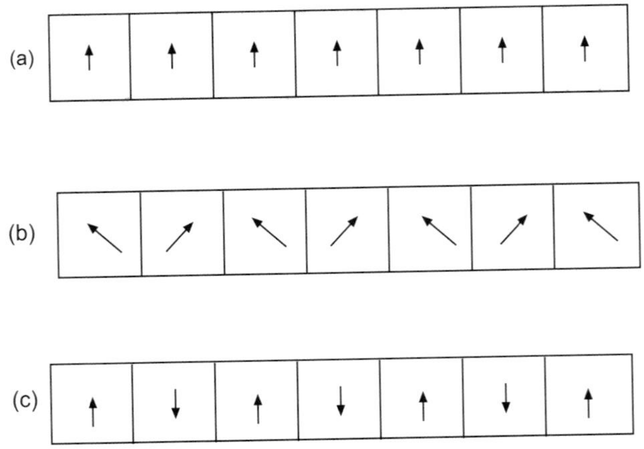

Fig. 7.14 Various types of ordering of polarization vectors between neighbouring domains: (a) ferroelectric; (b) ferrielectric; (c) antiferroelectric.

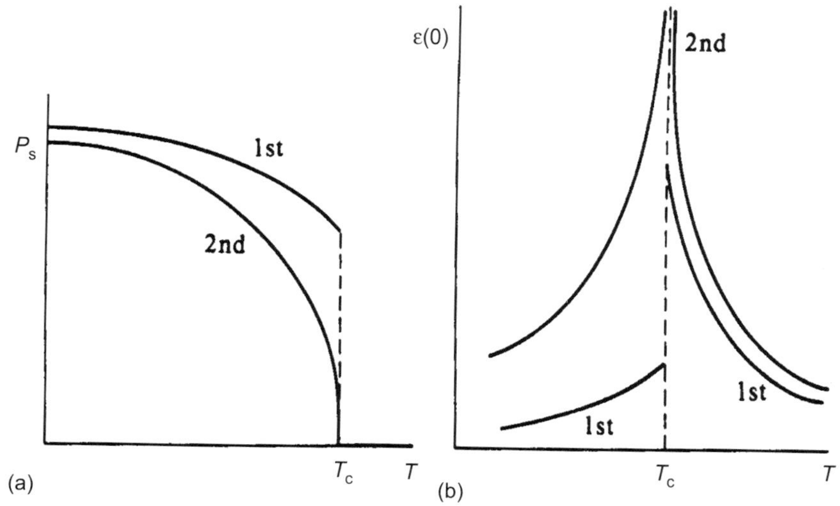

Fig. 7.15 First- and second-order ferroelectric transitions in the vicinity of the Curie temperature T_{cf}: (a) temperature dependence of P_s; (b) temperature dependence of $\varepsilon(0)$.

The temperature at which the transition takes place between the randomized paraelectric and the ordered ferroelectric crystal phases is termed the ferroelectric Curie temperature, T_{Cf}, by analogy with the corresponding temperature for ferromagnetic ordering (§7.2.5.2). In common with other types of phase transition, ferroelectric transitions can be divided into two categories: first-order and second-order transitions. First-order transitions are those in which there is a discontinuity in the *first* derivative of the free energy with respect to T or P, e.g. in the volume, $V = (\partial G/\partial p)_T$, and there is an associated latent heat (the melting transition of a crystal or the boiling transition of a liquid are examples). A first-order ferroelectric transition is characterized also by a discontinuity in $\mathbf{P}_s$ at T_{Cf} (see Fig. 7.15a): an example is the cubic–tetragonal transition in $BaTiO_3$ (cf. Fig. 7.12). The associated behaviour of the dielectric constant, taken to be the *clamped*† static value, $\varepsilon_1(0)$, is shown in Fig. 7.15b. Second-order transitions, in contrast, are those in which there is a discontinuity in *second* derivatives of the free energy: there is no change in volume (or $\mathbf{P}_s$) at T_{Cf}, but $\mathrm{d}P_s/\mathrm{d}T$ is discontinuous (as is the specific heat): the dielectric behaviour shows singular behaviour at T_{Cf} (Fig. 7.15). $LiTaO_3$ is an example of a material that exhibits a second-order ferroelectric transition.

The temperature dependence of $\varepsilon_1(0)$ in the paraelectric phase just above T_{Cf} has the Curie–Weiss form in which $\varepsilon_1(0)$ exhibits a singularity at a temperature θ_{CW}:

$$\varepsilon_1(T) = \frac{C}{(T - \theta_{CW})}, \tag{7.58}$$

† 'Clamped' indicates the value of dielectric constant excluding the contribution from mechanical resonances at $\simeq 10^5$ Hz due to the piezoelectric effect.

where C is the Curie constant and the Curie–Weiss temperature $\theta_{CW} = T_{Cf}$ for a second-order transition, but $\theta_{CW} \leqslant T_{Cf}$ for a first-order transition (see Fig. 7.15b). Thus, very large values indeed of dielectric constant, e.g. $\varepsilon_1(0) \simeq 10^4$, are reached at the ferroelectric transition. Ferroelectric materials can be classified on the basis of the magnitude of the Curie constant, C. Very large values of $C(\sim 10^5$ K) are found in oxide materials containing e.g. Ti^{4+}, Nb^{5+} and W^{6+} ions, or the 'lone-pair' ions Pb^{2+}, Bi^{3+} and Sb^{3+}; a large-scale displacive phase transition to the ferroelectric state is exhibited by such materials. Smaller values of $C(\sim 10^3$K) are characteristic of order–disorder transitions associated with the (small-scale) rotational ordering of dipolar ions (e.g. NO_2^- in $NaNO_2$) or the (small-scale) positional ordering of H atoms in hydrogen-bonded materials (e.g. KH_2PO_4).

Although for normal dielectrics the electric polarization $\boldsymbol{P}$ is proportional to the electric field $\boldsymbol{E}$ (eqn. (7.1)), this behaviour does *not* hold for ferroelectric materials even though reversal of an electric field can (eventually) reverse the saturation polarization $\boldsymbol{P}_s$. Instead, hysteresis is observed in the behaviour: the curve for the polarization for increasing E is different from that for decreasing E, and a hysteresis loop is obtained between the two saturation values, $\pm P_s$ (Fig. 7.16). The induced polarization is not destroyed at zero field, but instead at a negative field $-E_c$, the coercive field. As a corollary, the polarization remaining at zero field is termed the remanent polarization,

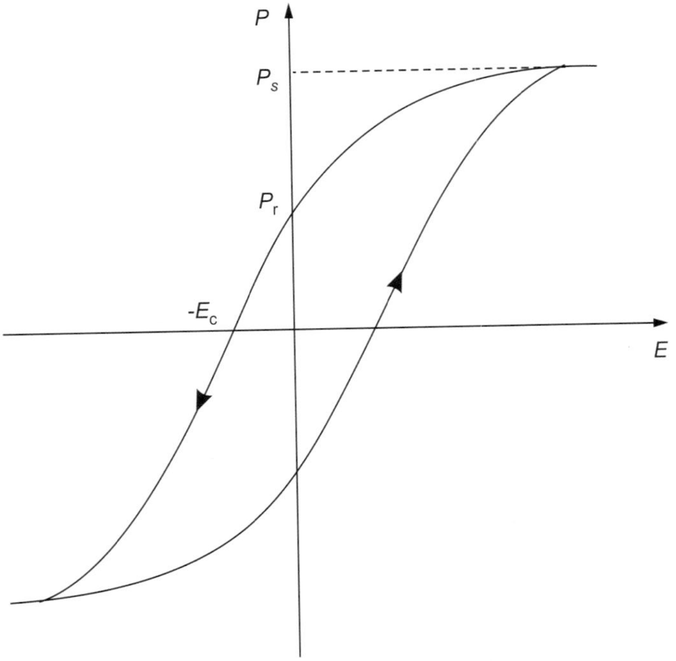

Fig. 7.16 Hysteresis loop for the P–E behaviour of a ferroelectric. The polarizations P_s and P_r are the saturation and remanent (zero-field) values, respectively; the coercive field $-E_c$ is necessary to reduce the polarization to zero.

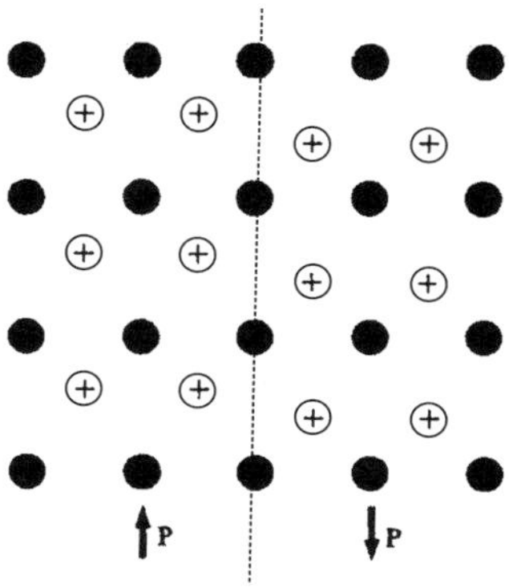

Fig. 7.17 Schematic illustration of ferroelectric domains having a 180° wall.

P_r. (Similar hysteretic behaviour is also exhibited by ferromagnetic materials—§7.2.5.5.) The area inside the hysteresis loop gives the loss in energy density per cycle, $\oint E\mathrm{d}P$.

The origin of hysteresis lies in the energy needed to move domain walls between domains having electric polarizations in different directions so that the net polarization of a sample can be changed, e.g. reversed, by altering the relative size of various domains. In the case of, say, tetragonal $BaTiO_3$, where the polarization is constrained by the crystallographic anisotropy to lie along one of the $\langle 100 \rangle$ directions, the domain boundaries are termed 180° or 90° walls, depending on whether the polarization directions in adjacent domains are antiparallel or orthogonal (Fig. 7.17). In the case of the low-temperature rhombohedral ferroelectric phase of $BaTiO_3$, in which P_s lies in the $\langle 111 \rangle$ directions, the domain walls can be 180°, 71° or 109°. Although it is relatively easy to move a 180° domain wall in the case of tetragonal $BaTiO_3$ since the structure is contiguous across the interface, this is not the case for a 90° wall; movement of a 90° wall therefore involves substantial deformation of the structure, and such strains can even be large enough to cause plastic deformation.

Although antiferroelectric materials exhibit spontaneous polarization in unit cells or domains below the antiferroelectric Curie temperature, T_{Caf} (above which they are paraelectric), there is no *net* polarization in this state because dipole moments are antiparallel in adjacent domains (Fig. 7.14c). However, at sufficiently high electric fields, polarization reversal is possible through domain-wall motion, and an antiferroelectric crystalline material can be driven into the ferroelectric state under the action of the field (Fig. 7.18a). In such a case, the P–E curve is linear and non-hysteretic through the origin when the material is in the antiferroelectric state, but hysteresis curves appear at field strengths high enough to cause ferroelectricity (Fig. 7.18b). Further information on ferroelectric materials is given in Rao and Rao (1992).

In the following two sections, ideas that can be used to explore the behaviour and origin of ferroelectric phase transitions will be discussed. Because these have, in fact, a more general applicability, they are afforded separate sections.

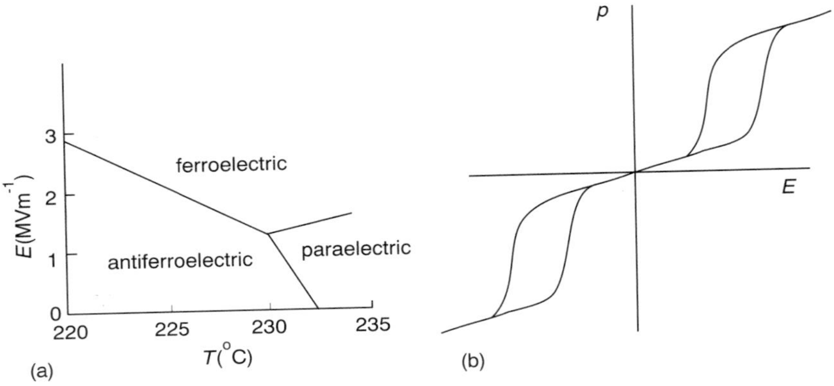

Fig. 7.18 (a) Phase diagram for $PbZrO_3$, showing the possibility of an antiferroelectric–ferroelectric transition at high electric fields. (b) Polarization–electric field curves for $PbZrO_3$. The linear portion corresponds to the antiferroelectric state and the hysteresis curves appear when the material is driven into the ferroelectric state at high fields. (After West (1987). Reproduced by permission of John Wiley & Sons Inc.)

*7.1.5.4 *Soft modes*

One way of regarding the transition from the paraelectric to the ferroelectric state (or indeed other types of displacive phase transitions) is in terms of a *softening* of a particular vibrational mode of the solid with decreasing temperature (Fig. 7.19); the restoring force associated with this motion becomes very small and hence the crystal can become mechanically unstable with respect to such atomic displacements. It can therefore transform to a lower symmetry structure in which the atomic displacements from the high-temperature phase can be regarded as those of the soft mode, but now they are static or 'frozen-in'. If the space group of the low-temperature phase is a sub-group of that of the high-temperature phase (certain symmetry elements being lost when atoms are displaced statically according to the soft mode), then the transition has the possibility of being second-order, but not otherwise.

The reason why a soft vibrational mode might be involved in displacive ferroelectric phase transitions can be gleaned from inspection of the Lyddane–Sachs–Teller (LST) equation (4.149) relating the (clamped) static and high-frequency dielectric constants to the longitudinal-optic (LO) and transverse-optic (TO) mode frequencies (at zero phonon wavevector, $\boldsymbol{k} = 0$):

$$\frac{\varepsilon_1(0)}{\varepsilon_1(\infty)} = \frac{\omega_{\mathrm{L}}^2}{\omega_{\mathrm{T}}^2}. \tag{7.59}$$

For ionic crystals with more than two atoms per unit cell, there can be more than one set of LO and TO modes. If there are p modes, the generalization of the LST relation is

$$\frac{\varepsilon_1(0)}{\varepsilon_1(\infty)} = \prod_{i=1}^{p} \frac{\omega_{\mathrm{L},i}^2}{\omega_{\mathrm{T},i}^2}. \tag{7.60}$$

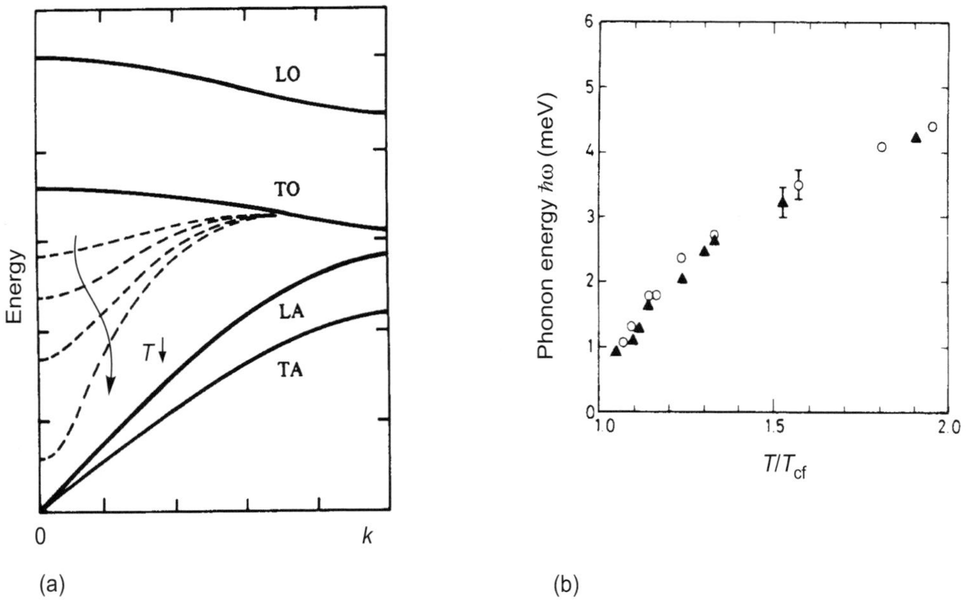

Fig. 7.19 (a) Schematic illustration of the softening of a TO phonon branch at $\boldsymbol{k} = 0$ with decreasing temperature. This soft-mode mechanism can be responsible for the transition from a paraelectric to a ferroelectric phase at T_{Cf}. (b) Soft-mode behaviour associated with a displacive phase transition in $SrTiO_3$. The phonon energies were obtained from inelastic neutron-scattering data. (Reprinted with permission from Feder and Pytte (1970), *Phys. Rev.* **B1**, 4803. © 1970. The American Physical Society)

We have previously seen (Fig. 7.15b) that the transition to the ferroelectric state is accompanied by a large enhancement of the static dielectric constant, $\varepsilon_1(0)$. Inspection of eqn. (7.59) or (7.60) shows that this can occur if there is a softening of one of the TO modes of a crystal.

The four optic modes of the paraelectric cubic perovskite phase of $BaTiO_3$ are shown in Fig. 7.20; of the three IR-active modes with dynamically induced dipoles (a, b, c in Fig. 7.20), the one with the lowest frequency is the mode shown in Fig. 7.20a, and this is the one that undergoes softening at $\boldsymbol{k} \simeq 0$ and is involved in the transition to the ferroelectric tetragonal phase at $T_{Cf} \simeq 134°C$. The fact that the softening in this case occurs at $\boldsymbol{k} \simeq 0$ means that the static 'frozen-in' atomic displacements (and hence dipole moment) in the tetragonal phase, characteristic of this TO mode, are repeated in every unit cell (in a given domain), i.e. the material is *ferroelectric*. Softening of optic modes can also occur at non-zero values of phonon wavevector. For example, a mode softening at the zone boundary at the X-point $k_x = \pm\pi/a$ (the face centre of the cubic Brillouin zone for a simple cubic crystal), means that the atomic displacements, and hence dipole moments, must be oriented in an antiparallel fashion in neighbouring cells; i.e. such a state of the material would be *antiferroelectric*.

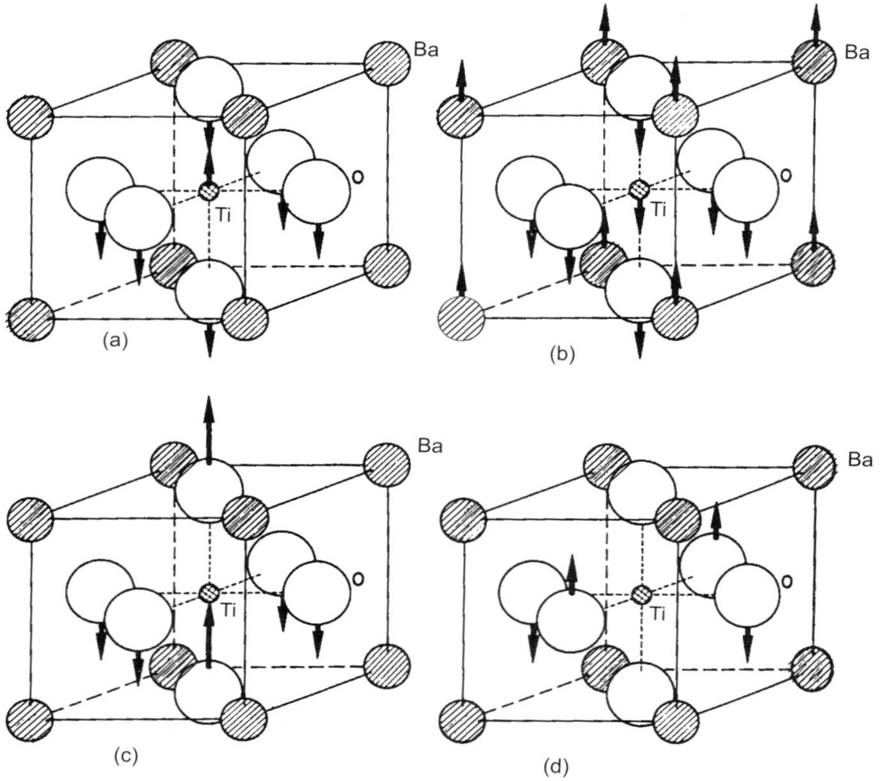

Fig. 7.20 The four vibrational optic modes for $BaTiO_3$ in the cubic perovskite (paraelectric) phase. All except (d) are IR-active with dynamically induced dipole moments. The lowest frequency mode (a) is the TO mode that undergoes a softening at $\boldsymbol{k} = 0$ in the transition to the tetragonal ferroelectric state. Note that all four modes are triply degenerate: the displacements can also occur along the two other orthogonal axes.

If the displacements of the soft TO mode in Fig. 7.20a are taken to be in the z-direction, giving rise to a transition to a tetragonal crystal with an unequal c-axis, in the conventional description, the TO modes with displacement directions in the two orthogonal directions can still be soft at $\boldsymbol{k} \simeq 0$. In this way, the transition at $-5°C$ in $BaTiO_3$ between the ferroelectric tetragonal and orthorhombic crystal phases (Fig. 7.12) can be understood. If the mode softening for this transition is now for displacements along say the y-direction, the net static polarization $\boldsymbol{P}_s$ in the orthorhombic phase will be along the $\langle 011 \rangle$ directions. At yet lower temperatures (-90°C), softening of the TO mode for displacements in the x-direction can cause a further transition (to a rhombohedral structure—Fig. 7.12), for which now $\boldsymbol{P}_s$ will be along the $\langle 111 \rangle$ directions, as observed.

The Curie–Weiss behaviour (eqn. (7.58)) found for the static dielectric constant just above T_{Cf} in the paraelectric phase implies, via the LST relation (eqn. (7.59)) and the idea of mode softening, that the temperature dependence of the TO-mode frequency should follow the relation

$$\omega_{\mathrm{T}}^2 = a(T - \theta_{\mathrm{CW}}), \tag{7.61}$$

where a is a constant. This behaviour can be obtained from a modification of the harmonic analysis used in §4.4 to obtain vibrational-mode frequencies for a polarizable ionic crystal. The equation of motion, for the case of a crystal containing just a cation and anion, is (eqn. (4.127)):

$$\mu_{\mathrm{M}}\ddot{\boldsymbol{u}} = -k\boldsymbol{u} + q\boldsymbol{E}_{\mathrm{loc}},$$

where μ_{M} is the reduced mass, k the force constant, q the charge on each ion, $\boldsymbol{u} = \boldsymbol{u}_+ - \boldsymbol{u}_-$ is the difference between the displacements of cation (+) and anion (−) and $\boldsymbol{E}_{\mathrm{loc}}$ is given by eqn. (7.24). The polarization for N cells in a volume V is (eqn. (4.125)):

$$\boldsymbol{P} = \frac{N}{V}\frac{(q\boldsymbol{u} + \alpha_{\mathrm{p}}\boldsymbol{E})}{(1 - N\alpha_{\mathrm{p}}/3\varepsilon_0 V)},$$

where $\alpha_{\mathrm{p}} = \alpha_{\mathrm{p}}^+ + \alpha_{\mathrm{p}}^-$ is the sum of the polarizabilities of the ions and $\boldsymbol{E}$ is the macroscopic field. The equation of motion can be rewritten correspondingly as

$$\mu_{\mathrm{M}}\ddot{\boldsymbol{u}} = \left[-k + \frac{Nq^2}{3\varepsilon_0 V(1 - N\alpha_{\mathrm{p}}/3\varepsilon_0 V)}\right]\boldsymbol{u} + \left[\frac{q}{(1 - N\alpha_{\mathrm{p}}/3\varepsilon_0 V)}\right]\boldsymbol{E}. \tag{7.62}$$

In this *harmonic* approximation, the TO frequency is given by eqn. (4.143):

$$\omega_{\mathrm{T}}^2 = \frac{1}{\mu_{\mathrm{M}}}\left[k - \frac{Nq^2}{3\varepsilon_0 V(1 - N\alpha_{\mathrm{p}}/3\varepsilon_0 V)}\right] = b_{\mathrm{s}} - b_{\mathrm{L}}, \tag{7.63}$$

where b_{S} and b_{L} are short-and long-range (Coulombic) contributions. The frequency is temperature-independent.

A temperature dependence can be produced by considering *anharmonicity* (§4.6.2): this will introduce a damping term $\gamma\dot{\boldsymbol{u}}$ into the equation of motion, but will also introduce a leading term linear in T into eqn. (7.63). (For example, anharmonicity causes thermal expansion, with the volume being approximately linearly temperature-dependent, $V = V_0(1 + \beta_T T)$ in an intermediate temperature régime—see §4.6.2.1) Hence, the TO-mode frequency becomes:

$$\omega_{\mathrm{T}}^2 = b_{\mathrm{s}} - b_{\mathrm{L}} + cT, \tag{7.64}$$

where c is a constant. This is of the form expected from the Curie–Weiss law (eqn. (7.61)), and hence $\theta_{\mathrm{CW}} = (b_{\mathrm{L}} - b_{\mathrm{S}})/c$ in this picture.

*7.1.5.5 *The Landau free-energy model*

The Landau model of phase transitions relates macroscopic property changes at a transition in terms of thermodynamics, specifically using the free energy: it says nothing, however, about the microscopic origin underlying the transition, as discussed, for example, in the previous section. The theory is based on the idea of an order parameter, which is a measure of the departure of, say, the atomic configuration of the low-temperature, lower-symmetry phase from that of the high-temperature symmetric phase. In the case of the ferroelectric phase transition, the spontaneous polarization, $\boldsymbol{P}_s$, can be used as the order parameter.

The (generalized Helmholtz) free-energy density can be written as the Landau power-law expansion as powers of the order parameter, in this case the polarization, P:

$$\hat{F}(T, P) = -EP + \hat{F}_0 + \alpha P^2 + \beta P^4 + \gamma P^6 + \ldots, \tag{7.65}$$

where the additional term $-EP$ represents the energy density of an already polarized system in an electric field, E, and the coefficients $\hat{F}_0 \equiv \hat{F}(P = 0), \alpha, \beta$ and γ are (in principle) temperature-dependent. Only even powers of P are included if the high-temperature, high-symmetry structure is centrosymmetric since the change $P \rightarrow -P$ leaves $\hat{F}$ unchanged. The equilibrium state at a particular temperature is determined by the minima of the free- energy density $\hat{F}(T, P)$. In the paraelectric phase above T_{Cf}, in the absence of an electric field, $P = 0$ and so $\hat{F}$ must have a minimum at this value. The ferroelectric state is associated with a minimum in $\hat{F}$ at a finite polarization value $P = P_s$ for $T < T_{\mathrm{Cf}}$. Two types of behaviour of the free-energy minimum can be distinguished (Fig. 7.21): in the first (Fig. 7.21a), there is a single (broad) minimum in $\hat{F}$ at $T = T_{\mathrm{Cf}}$ and so P increases *continuously* from zero in the ferroelectric state, as characteristic of a *second-order transition*; in contrast, in the second (Fig. 7.21b), at $T = T_{\mathrm{Cf}}$ there are *two* minima of equal depth, and correspondingly the polarization increases *discontinuously* at the transition, as characteristic of a *first-order transition*.

The behaviour shown in Fig. 7.21a can be reproduced by the abbreviated form of eqn. (7.65), $\hat{F} = \hat{F}_0 + \alpha P^2$: for $\alpha > 0$, the minimum of $\hat{F}$ is at $P = 0$, but for $\alpha < 0, \hat{F}$ has a minimum at a non-zero value of P. Thus, α must go through zero at the transition, and so, making a Taylor series expansion about $T = T_{\mathrm{Cf}}$, the temperature dependence of α can be represented as

$$\alpha = a(T - T_{\mathrm{Cf}}) \tag{7.66a}$$

with

$$\beta = b, \tag{7.66b}$$

where $a > 0$ and b are temperature-independent parameters.

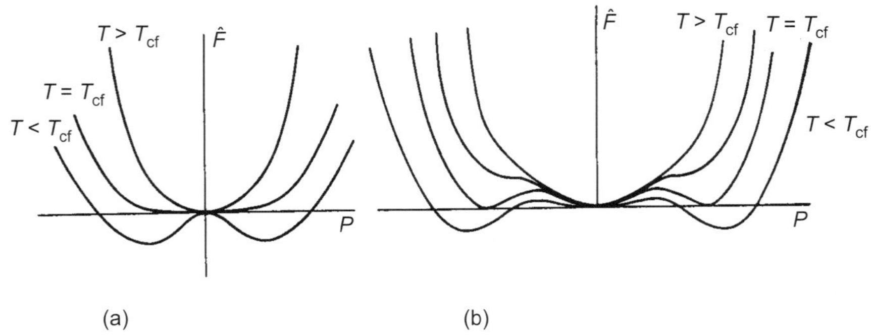

Fig. 7.21 Schematic illustration of the dependence of the zero-field Landau free-energy density $\hat{F}$ on polarization $\boldsymbol{P}$ at temperatures above, equal to and below the ferroelectric Curie temperature, T_{Cf}: (a) second-order phase transition; (b) first-order phase transition.

For the case of a second-order phase transition (Fig. 7.21a), only terms up to and including the quartic term in the zero-field free-energy expansion need be retained:

$$\hat{F} = \hat{F}_0 + a(T - T_{\mathrm{Cf}})P^2 + bP^4. \tag{7.67}$$

Minimization of eqn. (7.67) gives

$$\frac{\mathrm{d}\hat{F}}{\mathrm{d}P} = 0 = 2a(T - T_{\mathrm{Cf}})P + 4bP^3$$

or, since the minimum-energy state corresponds to the ferroelectric phase with spontaneous polarization P_s

$$P_s = (a/2b)^{1/2}(T_{\mathrm{Cf}} - T)^{1/2} \tag{7.68}$$

where b is a *positive* quantity. Thus, P_s is predicted to increase continuously below the second-order ferroelectric transition temperature (cf. Fig. 7.15a). The dielectric constant at temperatures just above T_{Cf} in the paraelectric phase can be evaluated from eqn. (7.65), neglecting the terms in P^4 and P^6 since P is so small there. The new equilibrium condition for a non-zero electric field is given by

$$\frac{\mathrm{d}\hat{F}}{\mathrm{d}P} = 0 = -E + 2a(T - T_{\mathrm{Cf}})P$$

or

$$\frac{P}{E} = \frac{1}{2a(T - T_{\mathrm{Cf}})} = \varepsilon_0\chi_1 \tag{7.69}$$

using eqn. (7.1), where χ_1 is the real part of the dielectric susceptibility. From eqn. (7.3), $\chi_1 = \varepsilon_1 - 1$, but since very close to T_{Cf} the dielectric constant diverges, $\varepsilon_1 \simeq \chi_1$ and hence

$$\varepsilon_1(T > T_{\mathrm{Cf}}) \simeq \frac{1}{2\varepsilon_0 a(T - T_{\mathrm{Cf}})}, \tag{7.70}$$

i.e. Curie–Weiss behaviour (eqn. (7.58)) is predicted.

For a first-order transition (Fig. 7.21b), all terms in the field-free Landau expansion of eqn. (7.65) need to be retained. However, in contrast to the second-order analysis, the coefficient β (eqn. (7.66b)) must be *negative*, $\beta = -b$, with a and γ both positive. Minimization of the free-energy density as before gives

$$\frac{\mathrm{d}\hat{F}}{\mathrm{d}P} = 0 = 2\alpha P + 4\beta P^3 + 6\gamma P^5 \tag{7.71}$$

or, since this equilibrium state corresponds to $P = P_s$, solving the resulting quadratic equation in P_s^2 gives

$$P_s^2 = \frac{|\beta| + \left(|\beta|^2 - 3\alpha\gamma\right)^{1/2}}{3\gamma}. \tag{7.72}$$

However, a relationship between the three coefficients α, β and γ, viz. $4\alpha\gamma = |\beta|^2$, can be obtained (Problem 7.10) by noting that, at the transition temperature $T = T_{\mathrm{Cf}}$, the two minima in the free energy have the same value of $\hat{F}$ (see Fig. 7.21b). As a result,

$$P_s^2 = (\alpha/\gamma)^{1/2} = 2\alpha/|\beta| \tag{7.73a}$$

or

$$P_s = \left(\frac{2a}{b}\right)^{1/2} (T_{Cf} - T)^{1/2}. \tag{7.73b}$$

Thus, the same behaviour of $P_s(T)$ is found as for second-order transitions (eqn. (7.68)).

Magnetic properties 7.2

Magnetic properties of solids can be divided into two categories: the first is the response of a material to an external applied magnetic field, due either to intrinsic magnetic dipole moments existing in the material or to the creation of an electrical current associated with free electronic charge carriers in response to the field; the second is the generation of a spontaneous magnetization as a result of the presence of intrinsic magnetic moments in a solid. For the most part (and exclusively in this chapter) magnetic effects in solids are associated with *electrons*. Although certain atomic nuclei also have intrinsic magnetic dipole moments (a feature exploited in nuclear magnetic resonance, NMR, the nuclear analogue of the technique of electron spin resonance, ESR — §3.3.2), nevertheless for most purposes nuclear-magnetic effects can be neglected since their magnitude is smaller than electron-magnetic effects by a factor $m_e/m_p \simeq 1/1830$.

Magnetism is inherently a *quantum* phenomenon: the Bohr–van Leeuwen theorem states that there can be no net magnetization for a collection of electrons, treated *classically*, in a state of thermal equilibrium and in the presence of applied magnetic or electric fields. Thus, magnetism is perhaps the oldest known quantum effect, the ancient Greeks having discovered the magnetic properties of 'lodestone' (magnetite, Fe_3O_4). The Bohr–van Leeuwen theorem can be demonstrated as follows. The magnetization M, the net magnetic dipole moment per unit volume, is given by the thermodynamic relation relating the change in Helmholtz free energy per volume, f_H, with a change in magnetic flux density B (cf. eqn. (6.143)):

$$M = -\frac{\partial f_H}{\partial B}. \tag{7.74}$$

In classical statistical mechanics, the free energy is related to the partition function which, for a continuous distribution of energies, can be written in the integral form:

$$f_H = -nk_B T \int\int [\exp(-\mathscr{H}(\boldsymbol{p},\boldsymbol{r})/k_B T]\mathrm{d}^3 p\mathrm{d}^3 r, \tag{7.75}$$

where $\mathscr{H}(\boldsymbol{p},\boldsymbol{r})$ is the Hamiltonian describing the energy of an electron as a function of momentum $\boldsymbol{p}$ and spatial coordinates $\boldsymbol{r}$. In a magnetic field, the quantity $m_e v$ for the momentum is replaced by the quantity involving the magnetic vector potential $\boldsymbol{A}(\boldsymbol{B} = \text{curl}\,\boldsymbol{A})$:

$$\boldsymbol{p} = m_e\boldsymbol{v} - e\boldsymbol{A}, \tag{7.76}$$

where the electronic charge is taken to be -e, and hence the Hamiltonian becomes

$$\mathscr{H}(\boldsymbol{p},\boldsymbol{r}) = \frac{1}{2m_e}(\boldsymbol{p} + e\boldsymbol{A})^2 + V(\boldsymbol{r}). \tag{7.77}$$

If the integration variable $\Pi = (\boldsymbol{p} + e\boldsymbol{A})$ is used instead of $\boldsymbol{p}$, integration of eqn. (7.75) shows that f_H is independent of $\boldsymbol{A}$ and hence $\boldsymbol{B}$. Thus, from eqn. (7.74), there can be no magnetization.

7.2.1 Magnetic quantities

The microscopic magnetic behaviour of a solid can be represented in terms of the magnetic dipole moment, $\boldsymbol{\mu}_m$, defined in terms of a current loop (Fig. 7.22) written in the Sommerfeld convention as:

$$\boldsymbol{\mu}_{\mathrm{m}} = Ia\hat{\boldsymbol{n}}, \tag{7.78}$$

where I is the current circulating around the loop (either resulting from an intrinsic angular motion of the electron or in response to an external magnetic field), a is the area of the loop and $\hat{\boldsymbol{n}}$ is a unit vector normal to the plane of the loop, directed according to the right-hand corkscrew rule depending on the sense of the current flow. Thus, μ_{m} is the magnetic analogue of the electrostatic dipole moment, $\boldsymbol{p}$; the units of μ_{m} are A m^2.

The magnetization, $\boldsymbol{M}$, is the net magnetic dipole moment per unit volume (the analogue of the electrical polarization $\boldsymbol{P}$ (eqn. (7.26)):

$$\boldsymbol{M} = \sum_i n_i \boldsymbol{\mu}_{\mathrm{m},i}, \tag{7.79}$$

where n_i is the concentration of magnetic moments of type i. The unit of $\boldsymbol{M}$ is A/m.

The magnetic flux density (or induction) $\boldsymbol{B}$ is defined in terms of the torque $\boldsymbol{T}$ exerted on a magnetic dipole by a magnetic field:

$$\boldsymbol{T} = \boldsymbol{\mu}_{\mathrm{m}} \times \boldsymbol{B}; \tag{7.80}$$

the unit for $\boldsymbol{B}$ is therefore N/(Am) or, equivalently, V s/m^2. The volt-second is also known as the *weber* (Wb), and the unit of induction is therefore the Wb/m^2, also called the *tesla* (T). The magnetic flux density and magnetization are related to the magnetic field intensity, $\boldsymbol{H}$, (unit, A/m) by the defining equation (cf. the dielectric analogue eqn. (7.4)):

$$\boldsymbol{B} = \mu_0(\boldsymbol{H} + \boldsymbol{M}), \tag{7.81}$$

where μ_0 is the vacuum permeability. Note that, as for the dielectric case, the field quantities in eqn. (7.81) relate to *macroscopic* quantities inside the material (§7.2.2). The magnetic flux density and field intensity are also related in general via an expression involving the relative permeability μ_{r} of a magnetic medium (which is not to be confused with μ_{m}):

$$\boldsymbol{B} = \mu_{\mathrm{r}}\mu_0\boldsymbol{H}. \tag{7.82}$$

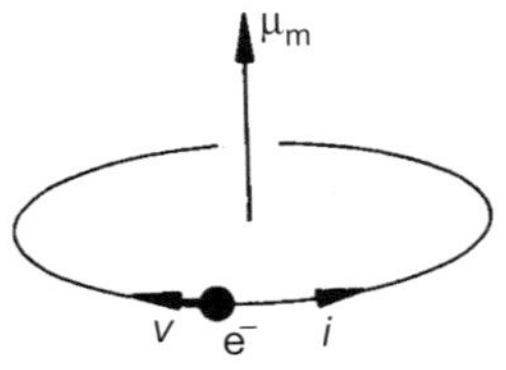

Fig. 7.22 Illustration of the magnetic dipole moment μ_{m} associated with a loop of area a carrying an electrical current i. (Note the direction of i is conventionally *opposite* to that of the electron flow.) Note the right-hand convention used to relate the direction of μ_{m} to the sense of the current.

Thus, μ_r plays a similar rôle in magnetic media to that the dielectric constant ε does in dielectric media (cf. eqn. (7.4)). In free space (vacuum), $\mu_r = 1$.

The magnetization is often linearly related to the field strength, the constant of proportionality being the magnetic susceptibility, χ_m:

$$\boldsymbol{M} = \chi_m \boldsymbol{H}, \tag{7.83}$$

where again the field strength is a macroscopic quantity. As in the analogous expression for the dielectric susceptibility, χ (eqn. (7.1)), the magnetic susceptibility is, in general, a *tensor* quantity. Moreover, if it represents temporal and spatial variations of $\boldsymbol{M}$, then χ_m is both frequency-and wavevector-dependent, $\chi = \chi(\omega, \boldsymbol{q})$. From eqns. (7.81), (7.82) and (7.83), the magnetic susceptibility and relative permeability are related (cf. eqn. (7.3) for the corresponding dielectric expression), via

$$\mu_r = (1 + \chi_m). \tag{7.84}$$

Hence, χ_m is zero *in vacuo*.

The quantity defined in eqn. (7.83) is known as the volume susceptibility, even though it is actually dimensionless, because the magnetization has been defined (eqn. (7.79)) as being the net magnetic dipole moment per unit *volume*. In some instances, the magnetization is defined instead as per unit mass or per mole: in such cases, the susceptibility per unit mass ($= \chi_m/\rho$) or molar susceptibility ($= \chi_m V_M$), respectively, are involved, where ρ is the mass density and V_M the molar volume.

The magnetic susceptibility can be either positive or negative. Magnetic materials with a *positive* value of χ_m are termed paramagnetic: the magnetization increases with increasing applied magnetic field. Materials with a *negative* value of χ_m are termed diamagnetic. As for the corresponding dielectric quantity, the magnetic susceptibility is, in general, a *complex* quantity: the imaginary part of the susceptibility is associated with energy-dissipative processes when the magnetization is varied by a change in the external magnetic field.

7.2.2 Internal magnetic fields

As for the corresponding case of electric fields inside dielectric media (§7.1.2), it is important to distinguish between various forms of the magnetic field inside magnetic media. An external magnetic field, of flux density $\boldsymbol{B}_{ext}$, applied to a magnetic material will produce a *macroscopic* internal field, of flux density $\boldsymbol{B}_{mac}$, that is an average over many atomic sites of a rapidly spatially fluctuating *microscopic* flux density $\boldsymbol{B}_{mic}$. The microscopic and macroscopic flux densities are related to the corresponding electric current densities via the Maxwell equation

$$\text{curl}\ \boldsymbol{B}_{mic/mac} = \mu_0\, \boldsymbol{j}_{mic/mac}, \tag{7.85}$$

together with

$$\text{div}\ B_{mic/mac} = 0. \tag{7.86}$$

The *local* flux density, $\boldsymbol{B}_{loc}$, different from all the above quantities, determines the energy of a magnetic dipole at the atomic level:

$$\mathscr{E} = -\boldsymbol{\mu}_m \cdot \boldsymbol{B}_{loc}. \tag{7.87}$$

The local magnetic field can be obtained in an analogous manner to that used to calculate the local electric field (§7.1.2 and Problem 7.1) by assuming that a demagnetizing field is set up in the material as a result of fictitious magnetic poles (the analogue of the surface charge density — see Problem 7.1) residing on the surface of the sample and reproducing the discontinuous change in the component of the magnetization normal to the surface, $M\tilde{\boldsymbol{n}}$. Hence, the relevant formulae for the internal magnetic fields can be obtained from the expressions for the corresponding electric fields by making the substitutions: $\boldsymbol{P} \to \mu_0 \boldsymbol{M}, \boldsymbol{E} \to \boldsymbol{H}, \boldsymbol{D} \to \boldsymbol{B}, \varepsilon_0 \to \mu_0$.

Alternatively, the internal magnetic fields can be evaluated by considering the associated surface electrical currents. If it is assumed that the magnetization $\boldsymbol{M}$ is *uniform* throughout the volume of a sample, the averaged macroscopic current density $\boldsymbol{j}_{\text{mac}}$ in the sample must everywhere vanish except at the surface, where there is a current per unit length equal to the discontinuity in $\boldsymbol{M}$ *parallel* to the surface, $i = M_{\parallel}$. Thus, the macroscopic flux density inside a sample is given by the sum of the external flux density and that, $\boldsymbol{B}_1$, generated by this surface current:

$$\boldsymbol{B}_{\text{mac}} = \boldsymbol{B}_{\text{ext}} + \boldsymbol{B}_1. \tag{7.88}$$

Moreover, rewriting eqn. (7.81) explicitly in terms of macroscopic quantities gives another expression for $\boldsymbol{B}_{\text{mac}}$:

$$\boldsymbol{B}_{\text{mac}} = \mu_0(\boldsymbol{H}_{\text{mac}} + \boldsymbol{M}). \tag{7.89}$$

In order to calculate the local flux density at an atom, the same trick is employed as for dielectrics (§7.1.2): an imaginary spherical surface is inscribed within the sample, centred at the point where $\boldsymbol{B}_{\text{loc}}$ is required. The contribution, $\boldsymbol{B}_2$, to the local flux density from the magnetization in the 'far' region, outside the volume of the fictitious sphere, is represented by a magnetization current flowing on the internal surface of this sphere in the *opposite* sense to that on the surface of the sample giving rise to $\boldsymbol{B}_1$ (Fig. 7.23). In addition, there is a contribution, $\boldsymbol{B}_3$, due to magnetic dipoles within the fictitious sphere, i.e.

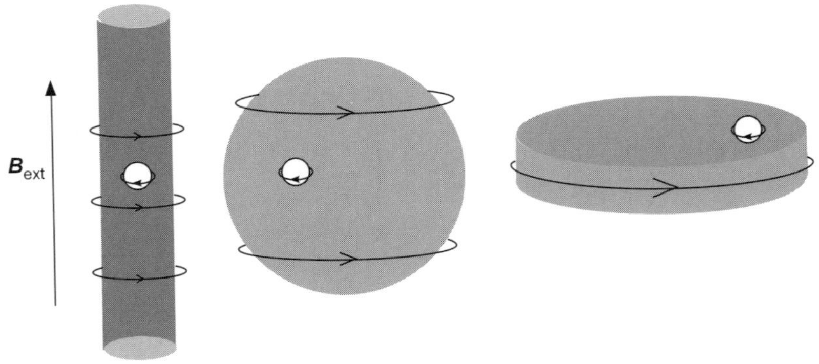

Fig. 7.23 Surface currents equivalent to a uniform magnetization $\boldsymbol{M}$ in samples of various shapes for an external magnetic field $\boldsymbol{B}_{\text{ext}}$ in the direction shown. The small inscribed spherical surfaces are used to calculate the local field $\boldsymbol{B}_{\text{loc}}$: the current flowing on the interior of such surfaces, equivalent to the magnetization in the region far from the spherical volume, is also shown.

$$\boldsymbol{B}_{\text{loc}} = \boldsymbol{B}_{\text{ext}} + \boldsymbol{B}_1 + \boldsymbol{B}_2 + \boldsymbol{B}_3$$
$$= \boldsymbol{B}_{\text{mac}} + \boldsymbol{B}_2 + \boldsymbol{B}_3. \tag{7.90}$$

As for the case of electrostatic dipoles (§7.1.2), in the special cases of either a cubic array or a random arrangement of magnetic dipoles, $\boldsymbol{B}_3 = 0$. It remains, therefore, to calculate $\boldsymbol{B}_1$ and $\boldsymbol{B}_2$ for various sample geometries, for example a sphere, a long cylinder and a thin disc.

Consider first the case of a spherical geometry, with $\boldsymbol{M}$ along the z-direction as in Fig. 7.24. The magnetic flux intensity at the centre of the sphere of radius R is calculated from the Biot–Savart law, considering the current flowing in a surface element subtending an angle $\mathrm{d}\theta$ at the centre of the sphere at a polar angle θ from the z-direction. The discontinuity in $M_{||}$, and hence the current element per length, is therefore $\mathrm{d}i = M\sin\theta$, and hence the current element is $\mathrm{d}I = MR\,\sin\theta\,\mathrm{d}\theta$. Therefore, for a length element $\mathrm{d}l$ along the circular strip, the contribution to the induction is

$$\mathrm{d}B = \frac{\mu_0}{4\pi}\frac{MR\sin\theta\,\mathrm{d}\theta\mathrm{d}l}{R^2} \tag{7.91}$$

in the direction indicated in Fig. 7.24. Summing the contributions from all the elements, with $\sum \mathrm{d}l = 2\pi R\sin\theta$, leads to a cancellation of components normal to $\hat{z}$, leaving only the component

$$\mathrm{d}\boldsymbol{B}_z = \frac{\mu_0}{2}\boldsymbol{M}\sin^3\theta\,\mathrm{d}\theta. \tag{7.92}$$

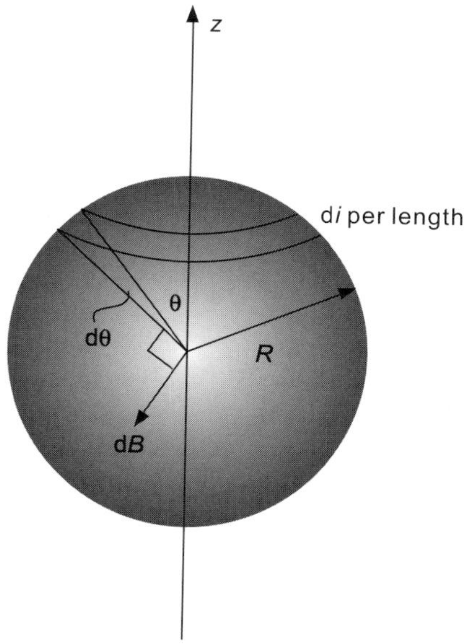

Fig. 7.24 Geometry used in the calculation, using the Biot–Savart law, of the magnetic field at the centre of a spherical current distribution.

Integrating this over the sphere gives the total field

$$\boldsymbol{B}_z = \frac{\mu_0 M}{2}\int_0^{\pi} \sin^3\theta \, \mathrm{d}\theta = \frac{2}{3}\mu_0 \boldsymbol{M}, \tag{7.93}$$

on making the substitution $x = \cos\theta$. If the sphere in Fig. 7.24 is regarded as the sample itself, the contribution to the macroscopic or local induction is

$$\boldsymbol{B}_1 = \frac{2}{3}\mu_0 \boldsymbol{M}. \tag{7.94}$$

On the other hand, if the spherical surface represents the fictitious sphere inserted within the sample, the contribution to the induction magnetization in the far region is

$$\boldsymbol{B}_2 = -\frac{2}{3}\mu_0 \boldsymbol{M}, \tag{7.95}$$

since the current flows are in the *opposite* directions in the two cases (Fig. 7.23b). Hence, for a *spherical geometry* of a sample

$$\boldsymbol{B}_{\text{mac}} = \boldsymbol{B}_{\text{ext}} + \frac{2}{3}\mu_0 \boldsymbol{M} \tag{7.96}$$

or (from eqn. (7.89)):

$$\boldsymbol{H}_{\text{mac}} = (\boldsymbol{B}_{\text{ext}}/\mu_0) - \boldsymbol{M}/3. \tag{7.97}$$

For the case of cubic (or isotropic) symmetry, the local flux density is therefore from eqn. (7.90):

$$\boldsymbol{B}_{\text{loc}} = \boldsymbol{B}_{\text{ext}}. \tag{7.98}$$

For the case of a *long cylinder* (Fig. 7.23a), the flux density $\boldsymbol{B}_1$ is that for a solenoid carrying a current per unit length $i = M$, and hence

$$\boldsymbol{B}_1 = \mu_0 \boldsymbol{M}. \tag{7.99}$$

Consequently,

$$\boldsymbol{B}_{\text{mac}} = \boldsymbol{B}_{\text{ext}} + \mu_0 \boldsymbol{M} \tag{7.100}$$

with

$$\boldsymbol{H}_{\text{mac}} = \boldsymbol{B}_{\text{ext}}/\mu_0. \tag{7.101}$$

For cubic (or isotropic) symmetry, the local flux density is

$$\boldsymbol{B}_{\text{loc}} = \boldsymbol{B}_{\text{ext}} + \frac{1}{3}\mu_0 \boldsymbol{M}. \tag{7.102}$$

Finally, for the case of a *very thin disc*, perpendicular to the field, the flux-density contribution to the current $id = Md$ flowing around the edge is approximately zero, i.e.

$$\boldsymbol{B}_1 \simeq 0, \tag{7.103}$$

since the thickness d of the disc is very small. Hence

$$\boldsymbol{B}_{\text{mac}} = \boldsymbol{B}_{\text{ext}} \tag{7.104}$$

with

$$\boldsymbol{H}_{\mathrm{mac}} = (\boldsymbol{B}_{\mathrm{ext}}/\mu_0) - \boldsymbol{M}. \tag{7.105}$$

The local flux density for this geometry is therefore

$$\boldsymbol{B}_{\mathrm{loc}} = \boldsymbol{B}_{\mathrm{ext}} - \frac{2}{3}\mu_0\boldsymbol{M}. \tag{7.106}$$

7.2.3 Diamagnetism

Diamagnetism is the most general of all magnetic properties, being exhibited by all atoms and even by the conduction electrons in metals. It can be regarded as being a variant of Lenz's law: a changing magnetic field produces an induced EMF (Faraday's law), thereby causing a motion of the electrons in the material that produces a magnetic field opposing—or 'screening'—the external field. Thus, the diamagnetic magnetic susceptibility, $\chi_{\mathrm{m}}^{\mathrm{dia}}$, is *negative* and is small in magnitude. In contrast, however, to the case of Lenz's law in electromagnetism, where the induced EMF is only produced by a *changing* magnetic flux, the diamagnetic screening currents in atoms or metals persist as long as the (steady) external magnetic field is applied. This is a quantum effect; in a classical system in thermal equilibrium, the screening currents would be reduced to zero, e.g. by collisions in a free-electron gas of conduction electrons, and therefore there would be no diamagnetism (the Born–van Leeuwen theorem).

For the case of atoms (§7.2.3.1), or of metals in the superconducting state (§7.2.3.3), the occurrence of diamagnetism can be associated with a 'rigidity' of the electron wavefunction with respect to magnetic fields; the wavefunction is perturbed only a little by a weak magnetic field. In the case of atoms, the wavefunction rigidity arises because different, orthogonal wavefunctions for the electrons correspond to states separated typically by large energies, of the order of electron volts, and consequently the wavefunctions cannot readily be perturbed.

In quantum systems, the momentum $\boldsymbol{p}$ is replaced by the quantum-mechanical operator $-\mathrm{i}\hbar\nabla$, and so eqn. (7.76) becomes

$$-\mathrm{i}\hbar\nabla = m_{\mathrm{e}}\boldsymbol{v} - e\boldsymbol{A} \tag{7.107}$$

for the case of an electron. From the argument above concerning wavefunction rigidity, it is expected, therefore, that the momentum $\boldsymbol{p} \equiv -\mathrm{i}\hbar\nabla$ would remain invariant in a magnetic field and, as a consequence, the electron velocity $\boldsymbol{v}$ must change when a magnetic field is applied. From eqn. (7.107), the induced velocity is therefore $\boldsymbol{v} = e\boldsymbol{A}/m_{\mathrm{e}}$, and hence the corresponding induced screening-current density is

$$\boldsymbol{j} = -ne\boldsymbol{v} = -\frac{ne^2}{m_{\mathrm{e}}}\boldsymbol{A} \tag{7.108}$$

for a concentration n of electrons. For a constant magnetic field, and hence vector potential, if n does not vary with time then $\mathrm{div}\boldsymbol{j} = 0$ (no sources of current). Hence, eqn. (7.108) is only satisfied for the particular gauge of the field

$$\mathrm{div}\ \boldsymbol{A} = 0. \tag{7.109}$$

Taking the curl of both sides of the Maxwell equation

$$\text{curl}\,\boldsymbol{B} \simeq \mu_0\,\text{curl}\,\boldsymbol{H} = \mu_0\boldsymbol{j} \tag{7.110}$$

(valid for small values of magnetization and for time-invariant fields), and using the vector identity

$$\text{curl}\,\text{curl}\,\boldsymbol{B} = \nabla(\text{div}\,\boldsymbol{B}) - \nabla^2\boldsymbol{B}, \tag{7.111}$$

together with the Maxwell relation

$$\text{div}\,\boldsymbol{B} = 0, \tag{7.112}$$

gives

$$\nabla^2\boldsymbol{B} = \frac{\mu_0 ne^2}{m_e}\boldsymbol{B} = \frac{1}{\lambda^2}\boldsymbol{B}. \tag{7.113}$$

The solution of eqn. (7.113) is a magnetic field B that exponentially decreases into the material from the surface, say in the x-direction:

$$B(x) = B_0\exp(-x/\lambda), \tag{7.114}$$

where the screening length $\lambda = (m_e/\mu_0 ne^2)^{1/2}$ is of the order of a few hundred angstroms; $\lambda = 170$ Å for $n = 10^{29}\text{m}^{-3}$. Since λ is much greater than atomic dimensions, the screening effect for atoms is very weak and hence the associated diamagnetism is very weak.

7.2.3.1 *Atomic diamagnetism*

For atoms or ions with closed shells of electrons (e.g. rare-gas atoms, positively charged alkali ions M^+ or negatively charged halogen ions X^-), there is no net unpaired spin or angular momentum associated with the electrons, and as a result such species cannot exhibit paramagnetism (see §7.2.4); the only magnetic response is therefore *diamagnetism*. The applied magnetic field with a flux density $\boldsymbol{B}_{ext}$ causes a screening current to flow within the electron distribution of the atom, thereby generating an opposing magnetic field at the centre of the atom.

A uniform magnetic field can be represented by the choice of vector potential

$$\boldsymbol{A} = -\tfrac{1}{2}(\boldsymbol{r} \times \boldsymbol{B}_{ext}) \tag{7.115}$$

which satisfies the gauge div$\boldsymbol{A} = 0$, eqn. (7.109) (see Problem 7.11). For the cylindrically symmetric geometry shown in Fig. 7.25, where a ring of charge of radius R normal to the direction of $\boldsymbol{B}_{ext}$ is considered, $\boldsymbol{A} = R\,B_{ext}\hat{\boldsymbol{u}}/2$, where $\hat{\boldsymbol{u}}$ is a unit vector tangential to the ring and normal to $\boldsymbol{B}_{ext}$. From eqn. (7.108), the induced current density is

$$\boldsymbol{j} = \frac{-ne^2}{m_e}\boldsymbol{A} = \frac{-ne^2R}{2m_e}B_{ext}\hat{\boldsymbol{u}}, \tag{7.116}$$

and this forms a current loop as in Fig. 7.22 with current $d\boldsymbol{I} = \boldsymbol{j}\text{d}R\text{d}z = -ne^2RB_{ext}\hat{\boldsymbol{u}}$ $\text{d}R\text{d}z/2m_e$ and area of loop $a = \pi R^2$, if n is assumed to be uniform throughout the ring. Thus, from eqn. (7.78), the contribution of this current loop to the induced magnetic dipole moment is

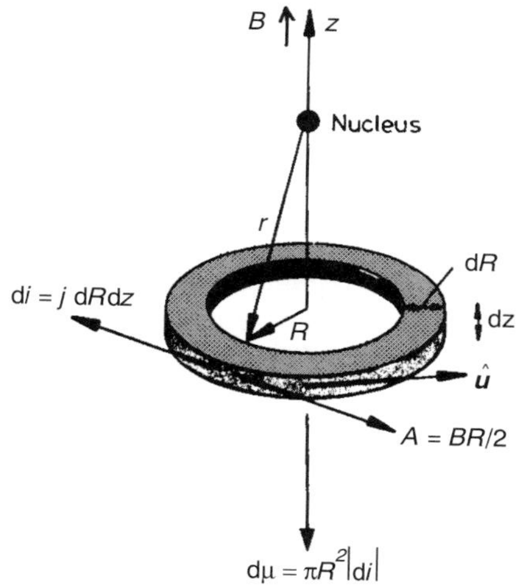

Fig. 7.25 Current loop within the electron distribution of an atom used for calculating the atomic diamagnetic susceptibility.

$$\begin{aligned} \mathrm{d}\mu_{\mathrm{m}} &= -\,|\mathrm{d}\boldsymbol{I}|\pi R^2\hat{\boldsymbol{z}} \\ &= \frac{-\pi n e^2}{2m_{\mathrm{e}}}\boldsymbol{B}_{\mathrm{ext}}R^3\mathrm{d}R\mathrm{d}z. \end{aligned} \tag{7.117}$$

Hence the total moment induced in the atom is

$$\begin{aligned} \mu_{\mathrm{m}} &= -\frac{e^2}{4m_{\mathrm{e}}}\boldsymbol{B}_{\mathrm{ext}}\iint nR^2 2\pi R\,\mathrm{d}R\,\mathrm{d}z \\ &= -\frac{e^2}{4m_{\mathrm{e}}}\boldsymbol{B}_{\mathrm{ext}}\int nR^2\mathrm{d}V, \end{aligned} \tag{7.118}$$

where $\mathrm{d}V = 2\pi R\mathrm{d}R\mathrm{d}z$ is the volume of the ring. If the number of electrons in the atom is $Z = \int n\mathrm{d}V$ (equal to the atomic number if the atom is un-ionized), the quantity $\int nR^2\mathrm{d}V = Z < R^2 >$ is the mean-square distance of the electron distribution from the z-axis. For a spherically symmetric charge distribution, $< R^2 >= \frac{2}{3} < r^2 >$, and hence from eqn. (7.118) the atomic dipole moment becomes

$$\mu_{\mathrm{m}} = -\frac{Ze^2}{6m_{\mathrm{e}}} < r^2 > \boldsymbol{B}_{\mathrm{ext}}. \tag{7.119}$$

The above calculation for an isolated atom can be extended to a solid consisting of n_{at} identical atoms per unit volume by assuming that the field at a given atom is given by $\boldsymbol{B}_{\mathrm{ext}}$, i.e. by ignoring the effects of the magnetization of the neighbouring atoms (approximately valid because diamagnetic effects are so weak). Hence, the overall induced magnetization is

$$\boldsymbol{M} = n_{\mathrm{at}}\mu_{\mathrm{m}} = -\frac{n_{\mathrm{at}}Ze^2}{6m_{\mathrm{e}}} < r^2 > \boldsymbol{B}_{\mathrm{ext}}. \tag{7.120}$$

In this approximation, where $\boldsymbol{M}$ is small, $\boldsymbol{B}_{\text{ext}} \simeq \boldsymbol{B}_{\text{mac}}$, and $\boldsymbol{H}_{\text{mac}} \simeq \boldsymbol{B}_{\text{ext}}/\mu_0$ (true, though, for long cylindrical geometry—see eqn. (7.101)), and so from eqn. (7.83), the Larmor atomic diamagnetic susceptibility is

$$\chi_{\text{m,d}}^{\text{at}} \simeq -n_{\text{at}} \frac{Ze^2 \mu_0}{6m_{\text{e}}} \langle r^2 \rangle. \tag{7.121}$$

Typical values of the volume atomic diamagnetic susceptibility are $\chi_{\text{m,d}}^{\text{at}} \simeq 10^{-5}$, or $\simeq 10^{-10}\ \text{m}^3\ \text{mol}^{-1}$ for the corresponding molar quantity.

★ The atomic diamagnetism can also be obtained as one of the terms resulting when the kinetic part of the Hamiltonian in the presence of a magnetic field (eqn. (7.77)) is considered with the particular choice of vector potential of eqn. (7.115). Thus:

$$\begin{aligned}\mathscr{H}_{\text{kin}} &= \frac{1}{2m_{\text{e}}} \sum_i (\boldsymbol{p}_i + e\boldsymbol{A})^2 = \frac{1}{2m_{\text{e}}} \sum_i \left(\boldsymbol{p}_i - \frac{e}{2}(\boldsymbol{r}_i \times \boldsymbol{B}_{\text{ext}})\right)^2 \\ &= \frac{1}{2m_{\text{e}}} \sum_i p_i^2 + \frac{e}{2m_{\text{e}}} \sum_i (\boldsymbol{r}_i \times \boldsymbol{p}_i)_z \boldsymbol{B}_{\text{ext},z} + \frac{e^2}{8m_{\text{e}}} B_{\text{ext},z}^2 \sum_i (x_i^2 + y_i^2),\end{aligned} \tag{7.122}$$

where it is assumed, as before, that the external magnetic field is along the z-direction, and the summation is over all electrons i in the atom. By analogy with eqn. (7.74), the expectation value of the atomic magnetic moment in a state $|\phi\rangle$ is therefore

$$\mu_{\text{m}} = -\frac{\partial}{\partial \boldsymbol{B}_{\text{ext},z}} (\langle\phi|\mathscr{H}|\phi\rangle) = -\frac{e\hbar}{2m_{\text{e}}} \langle\phi|\boldsymbol{L}_z|\phi\rangle - \frac{e^2}{4m_{\text{e}}} \boldsymbol{B}_{\text{ext},z} \langle\phi| \sum_i (x_i^2 + y_i^2)|\phi\rangle, \tag{7.123}$$

where the electronic orbital angular momentum is given by

$$\hbar \boldsymbol{L} = \sum_i \boldsymbol{r}_i \times \boldsymbol{p}_i. \tag{7.124}$$

The first term in eqn. (7.123) yields a contribution to the *paramagnetism* by the orbital motion of the electrons (see §7.2.4.4). The second term is the diamagnetic response. For a spherically symmetric electron distribution,

$$\langle\phi|x_i^2|\phi\rangle = \langle\phi|y_i^2|\phi\rangle = \tfrac{1}{3}\langle\phi|r_i^2|\phi\rangle, \tag{7.125}$$

and hence for a solid containing n_{at} atoms per volume, the susceptibility is (cf. eqn. (7.121)):

$$\chi_{\text{m,d}}^{\text{at}} \simeq -\frac{n_{\text{at}}}{6m_{\text{e}}} e^2 \mu_0 \sum_i \langle\phi|r_i^2|\phi\rangle. \tag{7.126}$$

In the summation, electrons in the outer shells make the dominant contribution since they are farthest from the nucleus. ★

7.2.3.2 *Diamagnetism of normal metals*

Conduction electrons in a normal (i.e. non-superconducting) metal (or a highly doped, degenerate semiconductor) are generally strongly perturbed by an applied magnetic field; the change in momentum $\boldsymbol{p}$ is of order $-e\boldsymbol{A}$ (eqn. (7.76)) and hence the induced electron velocity, and the magnitude of the screening current itself, is very small. Nevertheless, there is a non-zero contribution to the diamagnetic susceptibility from the conduction electrons, first calculated, for the case of the free-electron gas, by Landau.

The effect of an applied magnetic field on the conduction- electron states is to cause a collapse of the allowed states in $\boldsymbol{k}$-space onto *Landau tubes* lying parallel to the field direction (say $\hat{z}$)—see §6.3.3.1. The nth allowed energy levels for, say, a 3D free-electron gas are then given by (eqn. (6.119)):

$$\mathscr{E}_n = \frac{\hbar^2 k_z^2}{2m_e} + (n + \frac{1}{2})\hbar\omega_c, \tag{7.127}$$

where the cyclotron frequency, ω_c is given by (eqn. (6.117)):

$$\omega_c = eB/m_e. \tag{7.128}$$

The modified density of states $g(\mathscr{E})$ for a 3D quantum free-electron gas is then as in Fig. 7.26: there are discontinuities in $g(\mathscr{E})$ at odd multiples of the energy $\hbar\omega_c/2$ resulting from the 1D free motion remaining in the z-direction (see Problem 5.2), that become rounded off in practice because of electron scattering. The average energy $\langle\mathscr{E}\rangle$ of the electron gas in the presence of an applied magnetic field is *higher* than that ($= 3N\mathscr{E}_F/5$—eqn. (5.35b)) for the field-free case because there are no allowed states for electron energies below $\mathscr{E} = \hbar\omega_c/2$ for the magnetic case (see Fig. 7.26). This increase in energy in the presence of a magnetic field means that the material will be subject to a force, so that in an inhomogeneous magnetic field it moves from a region of higher field intensity to one of lower intensity, a characteristic of *diamagnetism*.

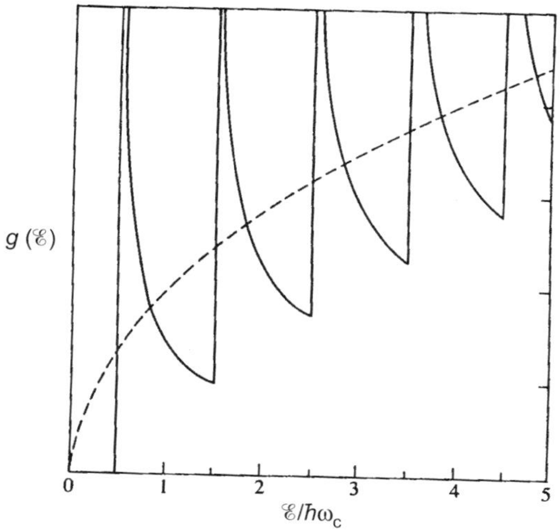

Fig. 7.26 The density of electron states of a free-electron gas in the presence of a magnetic field (solid line) compared with the parabolic dependence on energy in the field-free case (dashed line).

The expression for the diamagnetic susceptibility of a quantum free-electron gas (at zero wavevector, i.e. in the long-wavelength limit) was first given by Landau (see e.g. Peierls (1955) for a derivation), and is

$$\chi_{\mathrm{m,d}}^{\mathrm{fe}} = -\frac{n\mu_0\mu_{\mathrm{B}}^2}{2\mathscr{E}_{\mathrm{F}}} \equiv -\frac{n\mu_0 e^2}{4m_{\mathrm{e}}k_{\mathrm{F}}^2}, \tag{7.129}$$

where n is the conduction-electron concentration, $\mathscr{E}_{\mathrm{F}}$ and k_{F} are the Fermi energy and wavevector respectively, and μ_{B} is the Bohr magneton of the electron, the natural unit of magnetic moment, defined as

$$\mu_{\mathrm{B}} = \frac{e\hbar}{2m_{\mathrm{e}}}. \tag{7.130}$$

The value of μ_{B} is 9.27×10^{-24} J T^{-1}. A typical value of $\chi_{\mathrm{m,d}}^{\mathrm{fe}}$ is $\simeq -10^{-6}$, the same order of magnitude as the atomic diamagnetic susceptibility. For a real *crystal* in which the conduction electrons are subject to a periodic potential, eqn. (7.129) becomes scaled by the inverse of the electron effective mass m_{e}^* (§6.2.1), i.e.

$$\chi_{\mathrm{m,d}}^{\mathrm{cr}} = -\frac{n\mu_0\mu_{\mathrm{B}}^2}{2\mathscr{E}_{\mathrm{F}}}\left(\frac{m_{\mathrm{e}}}{m_{\mathrm{e}}^*}\right). \tag{7.131}$$

Thus, metals with very small values of m_{e}^*, like Bi (with $m_{\mathrm{e}}^* \simeq 0.01m_{\mathrm{e}}$), have much enhanced diamagnetic susceptibilities.

The above results are for the zero-wavevector limit, $\boldsymbol{q} = 0$. The full $\boldsymbol{q}$-dependent expression (for a quantum free-electron gas) is (see e.g. White (1983)):

$$\chi_{\mathrm{m,d}}^{\mathrm{fe}}(q) = \chi_{\mathrm{m,d}}^{\mathrm{fe}}(0)\frac{3k_{\mathrm{F}}^2}{2q^2}\left[1 + \frac{q^2}{4k_{\mathrm{F}}^2} - \frac{k_{\mathrm{F}}}{q}\left(1 - \frac{q^2}{4k_{\mathrm{F}}^2}\right)^2 \ln\left|\frac{2k_{\mathrm{F}}+q}{2k_{\mathrm{F}}-q}\right|\right]. \tag{7.132}$$

The $\boldsymbol{q}$-dependence represented by eqn. (7.132) is shown in Fig. 7.27.

The discussion of the effects of a magnetic field on the magnetic response of conduction electrons so far has been for a *constant* field. Very interesting non-linear, oscillatory

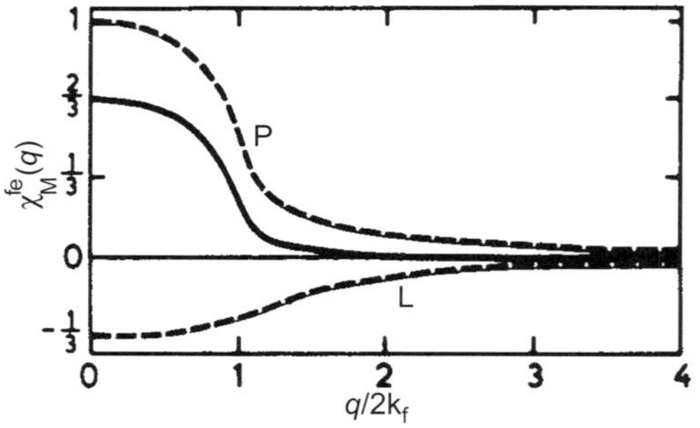

Fig. 7.27 The wavevector dependence of the diamagnetic Landau (L) and paramagnetic Pauli (P) susceptibilities of a free-electron gas. The solid line is the sum of the two contributions. (White (1983), *Quantum Theory of Magnetism*, p. 81, Fig. 3.3, © Springer-Verlag GmbH & Co. KG)

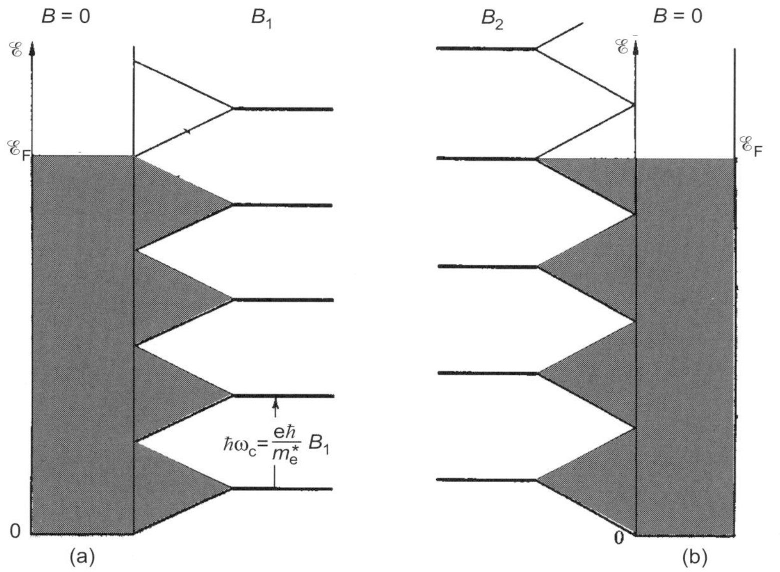

Fig. 7.28 Illustration of the origin of the de Haas–van Alphen oscillations for a free-electron gas: (a) the minimum-energy condition, when the Fermi level lies midway between two Landau levels for a magnetic flux density B_1; (b) the maximum-energy condition when the Fermi level coincides with a Landau level for a flux density $B_2 > B_1$.

behaviour occurs in the properties of the conduction electrons as the magnetic field is varied. For example, non-linear fluctuations occur in the magnetization as a function of $\boldsymbol{B}$, known as the de Haas–van Alphen (dHvA) effect. These oscillations occur because, as the magnetic flux density $\boldsymbol{B}$ increases, the radius or area (see eqn. (6.120)) of the Landau tubes also increases, and as a result the Landau tubes pass through the Fermi surface (see Fig. 6.20b). When the flux density is such that the Fermi energy lies exactly *midway* between two Landau tubes (see Fig. 7.28a), the total energy $\mathscr{E}_T$ of the free-electron gas is less than that ($\mathscr{E}_T^0 = 3N\mathscr{E}_F/5$) in the absence of a magnetic field and has a *minimum* value: the Landau level below $\mathscr{E}_F$ is filled and that above $\mathscr{E}_F$ is unoccupied, and half the electrons in the lower level have been shifted down in energy (on average by $\hbar\omega_c/4$, where ω_c is the cyclotron frequency - eqn. (6.117)). As $\boldsymbol{B}$ increases, the separation between the Landau levels increases (since $\omega_c \propto B$) and eventually the highest occupied level coincides with $\mathscr{E}_F$, i.e. when the Landau tube is just about to leave the Fermi surface (see Fig. 7.28b). Near this point, the total electron energy reaches a *maximum*. A further increase in $\boldsymbol{B}$ raises this level above $\mathscr{E}_F$ and the electrons in it therefore empty into the Landau levels lower in energy (below $\mathscr{E}_F$), a circumstance made possible by the fact that the degeneracy of each level also increases with $\boldsymbol{B}$ (eqn. (6.121)).

Hence, oscillations in the electron energy with increasing magnetic flux density are expected and consequently, from eqn. (7.74), oscillations occur also in the magnetization $\boldsymbol{M}$ (as well as in other properties of the electron gas, e.g. the electrical conductivity which is known as the Shubnikov–de Haas effect). As can be seen from Fig. 7.28b, the oscillations in $\boldsymbol{M}$ or energy occur whenever there are an integral number of Landau levels at and below $\mathscr{E}_F$, i.e.:

$$\mathscr{E}_F/\hbar\omega_c = n \tag{7.133}$$

where n is an integer, or from eqn. (6.117)

$$1/B = n\hbar e/m\mathscr{E}_F \tag{7.134a}$$

or equivalently

$$1/B = n2\pi e/\hbar S_F \tag{7.134b}$$

where $S_F = \pi k^2 k_F^2$ is the cross-sectional area of the Fermi surface in the plane of the cyclotron orbit (normal to B). Equation (7.134b) shows that the de Haas–van Alphen effect is periodic in the quantity $1/B$; the period of the oscillations is therefore

$$\delta(1/B) = 2\pi e/\hbar S_F. \tag{7.135}$$

Equation (7.135) can also be obtained by equating the quantized area between Landau tubes (eqn. (6.120)) and S_F.

The de Haas–van Alphen effect for electrons in *real metals*, as opposed to the ideal free-electron gas, is used to probe extremal closed cyclotron orbits associated with the Fermi surface (see e.g. Fig. 6.18 for the case of Cu); open circuits do not contribute to the dHvA effect. The effect can therefore be used to interpret complexities of the Fermi surface in favourable cases where the electrons can complete many cyclotron orbits between collisions, a situation achievable at low temperatures in very pure samples (see Problem 7.12). The dHvA oscillations for gold with $\boldsymbol{B}$ in the [110] direction are shown in Fig. 7.29 (see Problem 7.13).

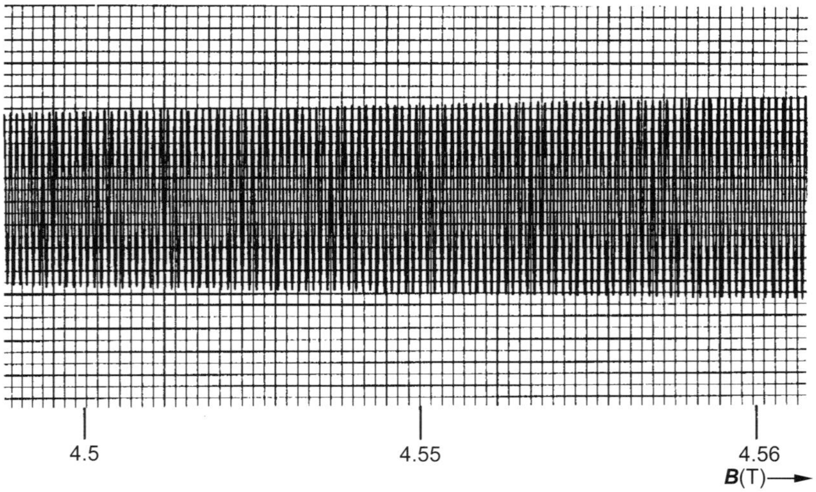

Fig. 7.29 De Haas–van Alphen oscillations for gold with $\boldsymbol{B}$ in the [110] direction. (After Kittel (1996). Reproduced by permission of John Wiley & Sons Inc.)

7.2.3.3 *Diamagnetism of superconductors*

One of the most remarkable effects exhibited by superconductors is the *Meissner effect* (sometimes also called the Meissner–Ochsenfeld effect) in which, for certain sample geometries, materials and strengths of magnetic field (see later), *all* magnetic flux is expelled from the interior of the superconductor (Fig. 7.30a). (See §6.4 for a discussion of the electronic properties of superconductors.) The simplest sample geometry to consider is that of a *long cylinder*, with the principal axis parallel to an external applied field, $\boldsymbol{B}_{\text{ext}}$. The complete Meissner effect can be written as

$$\boldsymbol{B}_{\text{mac}} = 0, \qquad (7.136)$$

and since $\boldsymbol{B}_{\text{mac}} = \mu_0(\boldsymbol{H}_{\text{mac}} + \boldsymbol{M})$, eqn. (7.136) implies (together with eqn. (7.101)) that

$$\boldsymbol{M} = -\boldsymbol{H}_{\text{mac}} = -\boldsymbol{B}_{\text{ext}}/\mu_0. \qquad (7.137)$$

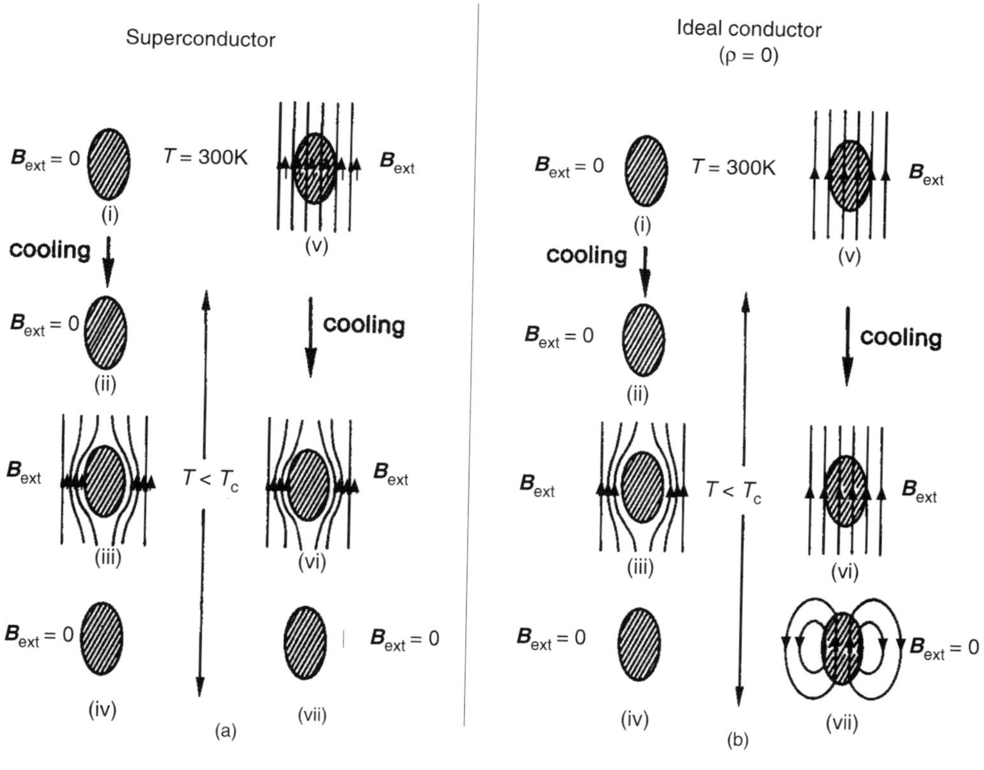

Fig. 7.30 Magnetic behaviour of: (a) a superconductor exhibiting the Meissner effect; (b) a perfect conductor with electrical resistivity $\rho = 0$. A series of steps are shown, beginning with the materials in the normal state ($T > T_c$), either without or with an applied external magnetic field, $\boldsymbol{B}_{\text{ext}}$. On cooling below T_c, and then applying a magnetic field, $\boldsymbol{B}_{\text{ext}}$, in the case of the originally field-free state, removal of $\boldsymbol{B}_{\text{ext}}$ causes the *same* state to be recovered for the superconductor, irrespective of path, whereas *different* states are found for the perfect conductor not exhibiting the Meissner effect.

Equation (7.137) further implies, from eqn. (7.83), that

$$\chi_m = -1; \tag{7.138}$$

i.e. a superconducting material exhibiting the complete Meissner effect (exclusion of flux) is a *perfect diamagnet*, as well as being an ideal electrical conductor.

The magnetization that opposes the external field, making the internal macroscopic flux density equal to zero ($\boldsymbol{B}_{mac} = \boldsymbol{B}_{ext} + \mu_0 M$ (eqn. (7.100)) with $\boldsymbol{H}_{mac} = -\boldsymbol{M} = \boldsymbol{B}_{ext}/\mu_0$ (eqn. (7.101))) is induced by *screening currents* throughout the material that, because they cancel within the bulk of the superconductor, effectively flow just at the surface of the superconductor within the London penetration depth, λ_L (eqn. (6.176)).

The superconducting state is distinct from that of an *ideal conductor* (with $\rho = 0$) because of the Meissner effect. For a superconductor, the final state of the material in zero magnetic field is the same whether or not a magnetic field initially permeated the material in the normal state at a temperature above T_c (Fig. 7.30a): on cooling below T_c, any flux in the material is expelled in the superconducting state, because of the Meissner effect, so that when the external field is removed once more, the material is in the *same* state, irrespective of the path taken. In this sense, the superconducting phase can be treated as a thermodynamic state (see §6.4.1.1).

An ideal conductor, however, behaves in a very different way (Fig. 7.30b). Starting from the magnetic field-free state for the normal material at a temperature above the transition to the zero-resistance state, application of an external magnetic field whilst in the zero-resistance state will lead to exclusion of the flux because of the screening currents induced. Subsequent removal of the field will leave the ideal conductor in a field-free state. However, if a magnetic field is first applied to the conductor in its normal state, so that the flux threads the material, on cooling down through the transition to the perfect-conductor state the magnetic flux remains trapped within the material: it is *not* expelled. The reason for this can be understood from the Maxwell equation

$$\text{curl}\,\boldsymbol{E} = -\frac{\partial \boldsymbol{B}}{\partial t}. \tag{7.139}$$

Inside a conductor with zero resistance, $\boldsymbol{E} = 0$, otherwise electrons will be accelerated without limit; thus, from eqn. (7.139), this implies that $\boldsymbol{B}_{mac}$ must remain constant with time in the interior. Hence, even after removal of the external field, magnetic flux will remain within the material, and hence the final state of the perfect conductor depends on the path taken.

In fact, the Meissner effect is rarely complete. First, only type-I superconductors in fields below the critical flux density B_c, or type-II superconductors below the lower critical flux density B_{c1} (see §6.4), can potentially exhibit complete exclusion of magnetic flux. Even for these, the Meissner effect may be incomplete. Materials that are not properly annealed can be inhomogeneous, and magnetic flux can get trapped in metastable regions that remain in the normal, non-superconducting state when the magnetic field is decreased below B_c.

The geometry of the sample can also influence the extent to which the Meissner effect is complete for certain magnetic fields below B_c. For example, a spherical superconducting sample exhibiting flux expulsion will have a flux-line pattern in its vicinity as shown in Fig. 7.31a; the flux density at the equatorial plane will be 50% larger than the uniform

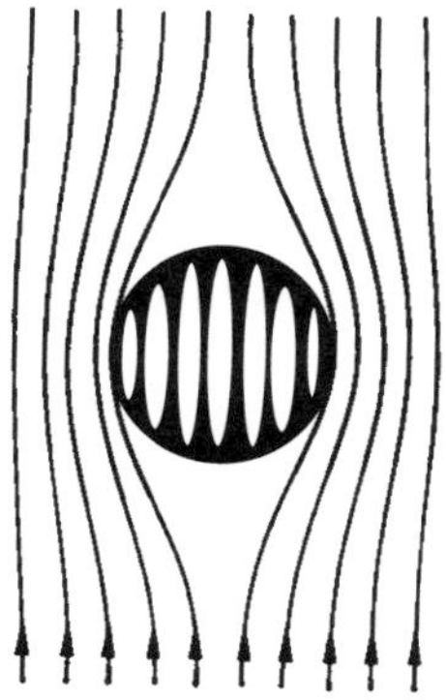

Fig. 7.31 Illustration of the origin of the intermediate Meissner state for a spherical sample of a superconductor. The flux density at the equatorial plane is $\boldsymbol{B}_{\text{eq}} = \frac{3}{2}\boldsymbol{B}_{\text{ext}}$ due to the excluded flux. For applied fields corresponding to flux densities $\frac{2}{3}\boldsymbol{B}_{\text{c}} < \boldsymbol{B}_{\text{ext}} < \boldsymbol{B}_{\text{c}}$, the sample is in the intermediate state consisting of normal (shaded) regions, within which the flux permeates, coexisting with (unshaded) superconducting regions in which flux expulsion is complete.

flux density of the applied field, $\boldsymbol{B}_{\text{eq}} = \frac{3}{2}\boldsymbol{B}_{\text{ext}}$. As long as $\boldsymbol{B}_{\text{eq}} < \boldsymbol{B}_{\text{c}}$, or $\boldsymbol{B}_{\text{ext}} < \frac{2}{3}\boldsymbol{B}_{\text{c}}$, for a type-I superconductor (or $\frac{2}{3}\boldsymbol{B}_{\text{c1}}$ for a type-II superconductor) the Meissner effect will be complete. However, in the range of flux densities $\frac{2}{3}\boldsymbol{B}_{\text{c}} < \boldsymbol{B}_{\text{ext}} < \boldsymbol{B}_{\text{c}}$ for a type-I material, the spherical sample is in an intermediate state, in which alternating regions of normal and superconducting material coexist (Fig. 7.31b); the flux permeates the normal regions but is completely excluded from the superconducting regions. This is because on the equatorial plane the critical flux density is reached, but the *whole* sample cannot revert to the normal state because otherwise the flux density inside the material would everywhere be equal to $\frac{2}{3}\boldsymbol{B}_{\text{c}}$, at which the normal state cannot exist. The system then adopts the intermediate configuration. (Note that this intermediate state should not be confused with the 'mixed' Shubnikov vortex phase of type-II superconductors for flux densities greater than $\boldsymbol{B}_{\text{c1}}$—see later).

Even in the cases where complete flux expulsion is expected, nevertheless the magnetic field penetrates some distance, the London penetration depth λ_{L} (eqn. (6.176)), into the interior of the superconductor. This is because the screening supercurrents that are responsible for the flux expulsion and the resulting diamagnetism themselves decay into the bulk of the material away from the surface (eqn. (6.175)). The decay of the flux density into a superconductor in the Meissner state can be calculated using the same method, involving the rigidity of a wavefunction in the presence of a magnetic field, as used in §7.2.3, but now the wavefunction is the macroscopic wavefunction representing the superconducting condensate, in other words the Ginzberg–Landau order parameter $\Psi(\boldsymbol{R})$ (eqn. (6.166)). Thus, eqn. (7.113) for the spatial variation of $\boldsymbol{B}$ becomes for the superconducting case

$$\nabla^2 \boldsymbol{B} = \frac{\mu_0 n_{\text{s}} e^2}{m_{\text{e}}} \boldsymbol{B} = \frac{1}{\lambda_{\text{L}}^2} \boldsymbol{B}, \tag{7.140}$$

with the solution for the decay of $\boldsymbol{B}$ in say the x-direction normal to a free surface:

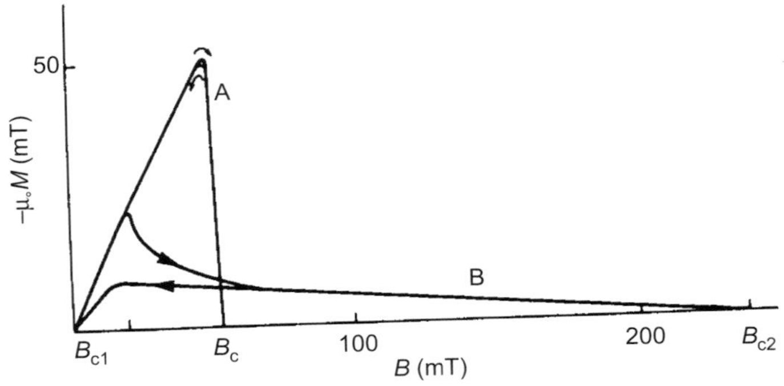

Fig. 7.32 Magnetization curves versus flux density of an applied external field, $\boldsymbol{B}_{ext}$: A is the curve for pure Pb (a type-I superconductor); B is the curve for Pb alloyed with 8.23 wt. % In (a type-II superconductor). (After Livingston (1963). Reprinted with permission from *Phys. Rev.* **129**, 1943. © 1963. The American Physical Society)

$$B(x) = B_0 \exp(-x/\lambda_L), \tag{7.141}$$

where $n_s/2$ is the concentration of Cooper pairs and λ_L is the London penetration depth. (Note the exact equivalence of eqns. (6.175) and (7.141) for the decay of the screening currents and of the magnetic flux density, respectively, into the bulk of a superconductor.) Thus, the Meissner effect is only complete for distances d away from the surface into the bulk of a superconductor greater than a few times λ_L, say $d \simeq 1000$ Å. Hence, *thin-film* superconductors, with thicknesses less than this order of magnitude, can never be completely diamagnetic.

For a type-I superconductor exhibiting complete flux exclusion, eqn. (7.137) shows that the magnetization $\boldsymbol{M}$ should vary linearly with the external flux density $\boldsymbol{B}_{ext}$ until the critical flux density $\boldsymbol{B}_c$ is reached, at which point the material reverts to the normal state and the magnetic susceptibility changes from the very large negative value ($\chi_m = -1$) characteristic of perfect diamagnetism to the very small value, $\chi_m \simeq 10^{-5}$, being the sum of paramagnetic (§7.2.4.5) and diamagnetic (§7.2.3.2) susceptibilities characteristic of normal metals. Curve A in Fig. 7.32 shows the magnetization behaviour of pure lead (a type-I superconductor - see Tables 6.3 and 6.4); the (reversible) linear behaviour of $\boldsymbol{M}$ versus $\boldsymbol{B}_{ext}$ for flux densities below $\boldsymbol{B}_c$ is clearly evident.

In §6.4.2.1, it was stated that whether a superconductor is type-I or type-II depends on the numerical value of the Ginzberg–Landau parameter $\kappa_{GL} = \lambda_m/\zeta_{eff}$ (eqn. (6.181)), where λ_m is the magnetic penetration depth (which can be much larger than λ_L, e.g. in the 'dirty' limit), and ζ_{eff} is the effective superconducting coherence length given by eqn. (6.180): $\zeta_{eff}^{-1} = \zeta_{BCS}^{-1} + \Lambda^{-1}$, where ζ_{BCS} is the BCS coherence length (eqn. (6.179)) and Λ is the (normal) electron-scattering mean-free path. Pure type-I materials are characterized by values of $\kappa_{GL} \ll 1$ since $\zeta_{eff} \gg \lambda_m$. Type-II materials, on the other hand, are characterized by the opposite limit, $\kappa_{GL} \gg 1$, and this arises when the electron mean-free path Λ is reduced (e.g. by alloying of another metal with the superconductor) so that $\zeta_{eff} \simeq \Lambda \ll \zeta_{BCS}, \lambda_m$. In fact, the demarcation value can be regarded as $\kappa_{GL} = 1/\sqrt{2}$; the cross-over in the surface energy, $\gamma_{n/s}$, between normal and super-

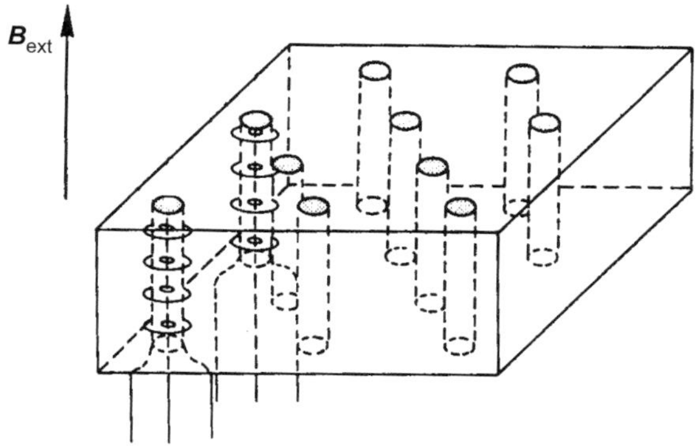

Fig. 7.33 Schematic illustration of the mixed, Shubnikov phase of a type-II superconductor that exists in the magnetic-flux range $B_{c1} < B < B_{c2}$. Tubes of quantized magnetic flux, or vortices, pass through the normal-state material surrounded by a superconducting phase.

conducting domains changes from positive to negative values for $\kappa \geqslant 1/\sqrt{2}$. Superconducting and normal regions are 'immiscible' when $\gamma_{n/s} < 0$ and so the mixed vortex (or Shubnikov) phase can form in type-II superconductors for values of magnetic flux densities in the range $B_{c1} < B < B_{c2}$: regions of superconducting material (exhibiting the Meissner effect and flux expulsion) coexist with regions of normal, non-superconducting material called vortices, parallel to the magnetic-field direction, within which the flux penetrates (see Fig. 7.33).

★ The surface energy associated with the normal-state–superconductor interface can be understood as follows. Consider such an interface in the x–y plane, with a magnetic field applied in this plane that decays into the superconductor with a characteristic magnetic penetration length $\lambda_m \geqslant \lambda_L$ (Fig. 7.34a). The expulsion of magnetic flux associated with the Meissner effect is associated with the magnetic-energy density

$$\tilde{\mathscr{E}}_m = \frac{B_c^2}{2\mu_0} \tag{7.142}$$

for a critical external flux density $\boldsymbol{B}_c$. The other relevant contribution to the overall energy density is that associated with the Cooper-pair condensation, $-\tilde{\mathscr{E}}_{cond}$: deep within the superconducting region, these two contributions are equal, $-\tilde{\mathscr{E}}_{cond} = B_c^2/2\mu_0$. However, in the vicinity of the interface, both contributions to the energy density decay to zero, but with *different* characteristic decay lengths: that for the magnetic-energy density is λ_m, and that for the condensate-energy density is the effective coherence length, ζ_{eff} (see Fig. 7.34b). Thus, the spatial variation of the energy-density difference is

$$\Delta\tilde{\mathscr{E}} = \tilde{\mathscr{E}}_{cond}(z) - \tilde{\mathscr{E}}_m(z) = \frac{B_c^2}{2\mu_0}\left[\left(1 - e^{-z/\zeta_{eff}}\right) - \left(1 - e^{-z/\lambda_m}\right)\right]. \tag{7.143}$$

Integrating this expression across the interface gives the surface energy $\gamma_{n/s}$ needed to create the boundary between normal and superconducting states:

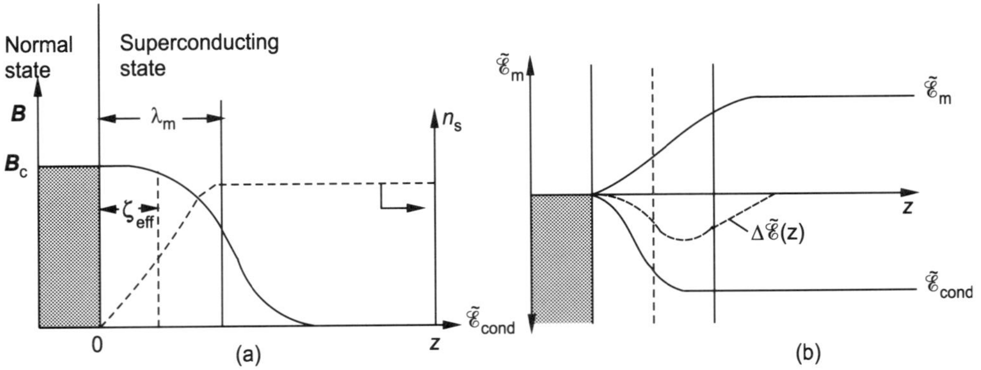

Fig. 7.34 Spatial variation, near an interface between a normal-state region (shaded) and a superconducting region (unshaded), of (a) the magnetic flux density $B(z)$ (solid line) and the Cooper-pair density $n_s(z)$ (dashed line); (b) the corresponding magnetic and condensate energy densities, $\tilde{\mathscr{E}}_m(z)$ and $\tilde{\mathscr{E}}_{cond}(z)$, respectively, with their difference $\Delta\tilde{\mathscr{E}}(z)$ shown by the dashed line. Integration of $\Delta\tilde{\mathscr{E}}(z)$ gives the surface energy $\gamma_{n/s}$ for the creation of a normal-state/superconducting interface.

$$\gamma_{n/s} = \int_0^\infty \Delta\tilde{\mathscr{E}}(z)\mathrm{d}z = (\zeta_{eff} - \lambda_m)\frac{B_c^2}{2\mu_0}. \tag{7.144}$$

Thus, if $\zeta_{eff} < \lambda_m$ (or $\kappa_{GL} > 1$), as in Fig. 7.34, $\gamma_{n/s} < 0$, and the mixed vortex state is stabilized. ★

From the above discussion it is clear that the magnetization behaviour of a type-II superconductor must be rather different from that of a type-I superconductor (curve A in Fig. 7.32). Above the upper critical flux density, B_{c2} (which may be very high, of order 50 T), a type-II material is in the normal state and hence the magnetization is effectively zero. Between this value and the lower critical flux density, B_{c1}, a type-II superconductor is in the Shubnikov vortex state, i.e. partially superconducting, and hence the magnetization must increase continuously (in a negative sense—see eqn. (7.137)) with decreasing flux density as the proportion of superconducting material increases. At the lower critical value of the flux density, B_{c1}, the material is *entirely* in the superconducting state, and hence the Meissner effect is complete. For yet lower flux densities, the linear behaviour of $\boldsymbol{M}$ characteristic of type-I superconductors is recovered (see Fig. 7.35 for a schematic illustration and curve B in Fig. 7.32 for an actual example, viz. Pb alloyed with In to transform it into a type-II superconductor).

Finally, we will discuss the characteristics and behaviour of the vortex 'lattice' in type-II superconductors in the mixed Shubnikov state. There is a magnetic *repulsive* interaction between the flux tubes (Fig. 7.33), at least for small separations, so the minimum-energy configuration will be that in which the average distance between vortices is maximized: this requirement is achieved by an ordered hexagonal array (the Abrikosov lattice) in two dimensions. Figure 7.36 shows such a hexagonal flux lattice for the case of Nb in the vortex state.

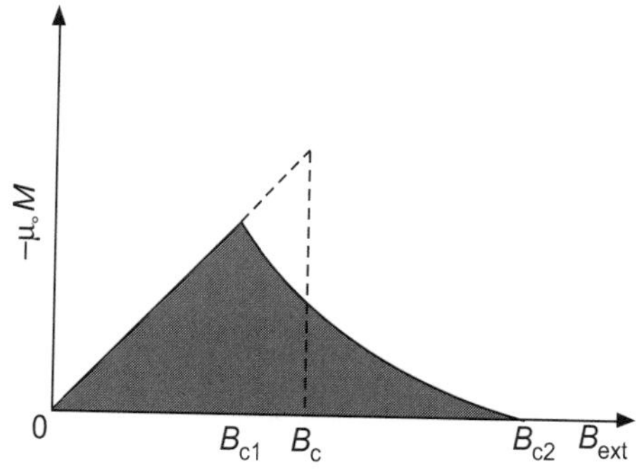

Fig. 7.35 Schematic illustration of the magnetization curves for a type-II superconductor (solid line) and a type-I superconductor (dashed line).

The presence of flux vortices in the mixed phase means that type-II superconductors can be electrically dissipative, i.e. they can exhibit a *finite* electrical resistance under certain circumstances. For a (super)current $\boldsymbol{j}$ flowing normal to $\boldsymbol{B}$ in a type-II superconductor in the vortex state, the Lorentz force (eqn. (6.109)) causes the flux vortices to move in the direction perpendicular to both $\boldsymbol{j}$ and $\boldsymbol{B}$. This flux motion induces an electric field parallel to $\boldsymbol{j}$, which is equivalent to an electric power loss, or resistance, in the

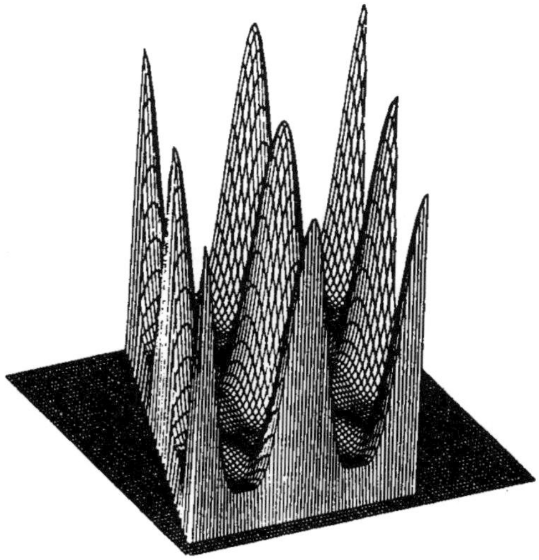

Fig. 7.36 The hexagonal flux lattice associated with the vortex (Shubnikov) phase present in Nb at 4.2 K in a magnetic field of flux density $B = 0.079$ T ($B_{c2} = 0.314$ T). The average nearest-neighbour separation of flux tubes is 177 nm. The results of a neutron-scattering study were used to ascertain the magnetic-field variation and the vortices shown here. (After Schelten *et al.* (1972). Reproduced by permission of Plenum Publishing Corp.)

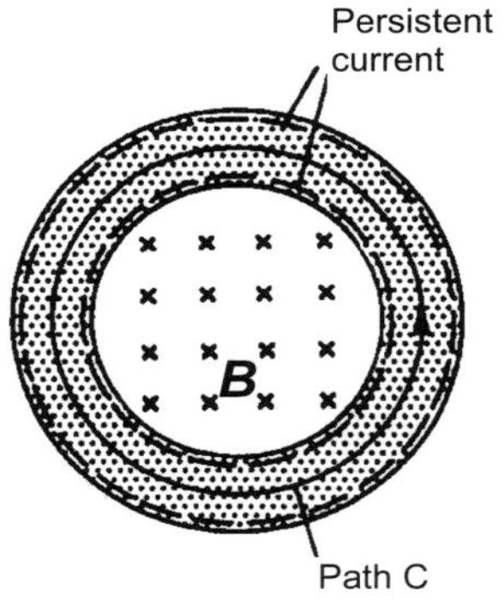

Fig. 7.37 Illustration of a superconducting ring (shaded) threaded by magnetic flux. The persistent supercurrents are associated with the flux flow on the inner and outer surfaces of the ring, but deep inside the superconductor (at distances greater than the magnetic penetration depth, λ_m) the current density is zero. Integration of the superconducting-condensate wavefunction along the contour C shows that only an integral number of flux quanta can exist within the ring. This is a model for a flux vortex in the mixed state of a type-II superconductor, in which a normal-state core, threaded by magnetic flux, is surrounded by a superconducting region.

sample. This dissipative loss can be prevented by flux pinning, i.e. the spatial trapping of the flux vortices by structural defects; this is particularly prevalent in oxide high-T_c materials, where the pinning defects can be oxygen vacancies or similar atomic defects associated with non-stoichiometry. Flux pinning can also be induced artificially by means of radiation damage: electron or ion bombardment of a superconducting material results in tracks of radiation-damaged material which efficiently trap the flux vortices. If flux pinning is dominant, then the flux-vortex arrangement will not be the ordered Abrikosov lattice, but instead a more-or-less random spatial arrangement of flux tubes, a vortex glass.

Lastly we consider the magnitude of the magnetic flux entrained within the normal regions containing the vortices. This flux can be evaluated by considering a *ring* of superconducting material threaded by magnetic flux, in the case of a type-II vortex, passing through normal-state material lying within the ring (Fig. 7.37). The persistent supercurrent associated with the flux flows on the inner surface of the ring, but along a circuit C deeper within the superconducting region of the ring (at distances greater than the magnetic or London penetration depth, λ_m or λ_L), the current is zero (eqn. (6.175)). In terms of the Ginzberg–Landau order-parameter wavefunction, $\psi(\boldsymbol{R}) = (n_s/2)^{1/2} \exp(i\theta)$ (eqn. (6.168)), the current density is (eqn. (6.169)):

$$\boldsymbol{j} = -\left[\frac{\hbar e}{2m_e}\nabla\theta + \frac{e^2}{m_e}\boldsymbol{A}\right]n_s.$$

Thus, the condition $\boldsymbol{j} = 0$ corresponds to

$$\hbar\nabla\theta = -2e\boldsymbol{A}. \tag{7.145}$$

Integration of both sides of this equation around the closed circuit C yields

$$\hbar \oint_{\mathrm{c}} \nabla \, . \, \mathrm{d}\boldsymbol{l} = \hbar \Delta\theta = -2e \oint_{\mathrm{c}} \boldsymbol{A} \, . \, \mathrm{d}\boldsymbol{l}. \tag{7.146}$$

The right-hand side of this equation can be rewritten using Stokes's theorem:

$$\oint_{\mathrm{c}} \boldsymbol{A} \, . \, \mathrm{d}\boldsymbol{l} = \oint_{\mathrm{c}} (\mathrm{curl}\, \boldsymbol{A}) \, . \, \mathrm{d}\boldsymbol{S} = \int_{\mathrm{c}} \boldsymbol{B} \, . \, \mathrm{d}\boldsymbol{S} = \Phi, \tag{7.147}$$

where Φ is the magnetic flux threading a surface bounded by the contour C, on which $\mathrm{d}\boldsymbol{S}$ is an element of area. Since the condensate wavefunction $\psi(\boldsymbol{R})$ must be single-valued, the change in phase going round the circuit C must be an integral number p of 2π radians, i.e. $\Delta\theta = \pm 2p\pi$. Hence, eqn. (7.146) becomes

$$\Phi = \pm \frac{2p\pi\hbar}{2e} = \pm \frac{ph}{2e} = \pm p\Phi_0, \tag{7.148}$$

where $\Phi_0 = 2.07 \times 10^{-15}\mathrm{Tm}^2$ is the flux quantum or fluxoid. A vortex containing p flux quanta has an energy proportional to $(p\Phi_0)^2$: hence, it is energetically favourable to have p individual vortices, each containing just a single flux quantum, rather than one larger vortex containing p fluxoids.

The critical flux density, B_{c1}, necessary to nucleate a single fluxoid in a vortex can be estimated as follows. The magnetic field will extend a distance λ_{m} into the superconducting region around the normal core of a vortex. The flux associated with the core will be of the order of $\pi\lambda_{\mathrm{m}}^2 B_{\mathrm{c1}}$, but this must equal the flux quantum Φ_0 from eqn. (7.148), and so

$$B_{\mathrm{c1}} \simeq \Phi_0/\pi\lambda_{\mathrm{m}}^2. \tag{7.149}$$

At the upper limit, B_{c2}, of the range of flux densities corresponding to the Shubnikov vortex state, the flux vortices will be packed as densely as possible, consistent with the superconducting phase remaining. In practice, this means that the closest distance of approach of two fluxoids will be of the order of the effective superconducting coherence length, ζ_{eff} (eqn. (6.180)). Thus, the area of a vortex in this case will be of the order of $\pi\zeta_{\mathrm{eff}}^2$, containing flux equal to $\pi\zeta_{\mathrm{eff}}^2 B_{\mathrm{c2}}$; this too must equal the flux quantum, and so

$$B_{\mathrm{c2}} \simeq \Phi_0/\pi\zeta_{\mathrm{eff}}^2. \tag{7.150}$$

7.2.4 Paramagnetism

Unpaired electrons possess an intrinsic spin angular momentum and, consequently, an intrinsic permanent magnetic dipole moment. In addition, electrons in part-filled (not half-filled or filled) shells in atoms have a net orbital angular momentum, and a magnetic dipole moment is associated with this motion too. As a result, atoms with incomplete shells, and even the conduction electrons in a metal, can exhibit permanent magnetic dipole moments. Hence, the application of an external magnetic field can cause alignment of these dipoles (in competition, in general, with the randomizing influence of temperature).

Therefore, the magnetization of a material containing such permanent dipoles will *increase* with increasing $\boldsymbol{H}$, and the paramagnetic susceptibility $\chi_{\mathrm{m,p}}$ is *positive*; the material is said to be paramagnetic. For the case of atomic paramagnetism, as we will see, the magnitude of the susceptibility is much larger than the corresponding value of diamagnetic susceptibility (eqn. (7.121)), $\chi^{at}_{m,p} \gg |\chi^{at}_{m,d}|$. On the other hand, the paramagnetic susceptibility and the diamagnetic susceptibility (eqn. (7.129)) for conduction electrons (e.g. in a free-electron gas) are of the same order of magnitude (in fact, $\chi^{fe}_{m,p} = 3|\chi^{fe}_{m,d}|$). It was shown in §7.2.3.1 (eqns. (7.122) and (7.123)) that the magnetic moment associated with the orbital motion of electrons in atoms and the resulting orbital angular momentum of electron i, $\hbar \boldsymbol{l}_i$, is given by

$$\begin{aligned} \mu^L_m &= \frac{-e}{2m_e}\sum_i \boldsymbol{r}_i \times \boldsymbol{p}_i = \frac{-e\hbar}{2m_e}\sum_i \boldsymbol{l}_i \\ &= -\mu_B \boldsymbol{L}, \end{aligned} \tag{7.151}$$

where the orbital angular-momentum eigenstate operator $\boldsymbol{L}$ is given by eqn. (7.124) and μ_B is the Bohr magneton (eqn. (7.130)).

The intrinsic spin $\boldsymbol{s}$ of an electron has an associated magnetic moment given by

$$\mu^s_m = -g_e \mu_B \boldsymbol{s}, \tag{7.152}$$

where the electronic g-factor (or spectroscopic splitting factor) has the value

$$g_e \simeq 2(1 + \alpha/\pi + ..) = 2.0023, \tag{7.153}$$

where α is the fine structure constant ($= \mu_0 c e^2/2h \simeq 1/137$). This value is different from that, $g_e = 2$, predicted by Dirac's theory of the electron because of quantum electrodynamic corrections. However, for simplicity in the following, we will assume henceforth that $g_e = 2$. The net electronic spin of an atom, and hence its net spin magnetic moment, is obtained by summing over all electrons:

$$\mu^S_m = -g_e \mu_B \sum_i \boldsymbol{s}_i = -g_e \mu_B \boldsymbol{S}. \tag{7.154}$$

Hence, it can be seen from eqns. (7.151) and (7.154) that permanent magnetic dipoles only exist where there is a net orbital or spin angular momentum, $\hbar \boldsymbol{L}$ or $\hbar \boldsymbol{S}$, respectively. For the case of atoms, this is only true for electronic shells that are part-filled (but not half-filled for the case of $\boldsymbol{L}$).

Comparison of eqns. (7.151) and (7.152) shows that electronic spin angular momentum is *twice* as effective in giving rise to a magnetic moment as is orbital angular momentum. Thus, the *total* magnetic moment of an atom can be represented as

$$\mu^{tot}_m = -\mu_B(\boldsymbol{L} + 2\boldsymbol{S}). \tag{7.155}$$

Note, as will be seen later, that this is *not*, in general, the value of the dipole moment that appears in the expression for the magnetic energy when an atom is placed in an external magnetic field.

7.2.4.1 Paramagnetism of isolated atoms

For an isolated atom containing a single part-filled shell, with orbital angular momentum quantum number l, there are $(2l + 1)$ associated energy levels which, together with the spin degeneracy $g_s = 2$, implies that the overall electronic degeneracy in principle should be $2(2l + 1)$. However, this neglects the intraionic interactions, such as electron–

electron Coulomb repulsion and the coupling between the spin of an electron and its orbital motion, i.e. the spin–orbit interaction. These interactions lift the level degeneracy and, except for the very heaviest atoms where the spin–orbit coupling is strongest and hence no longer a small perturbation to the Coulomb interactions, the Russell–Saunders (or LS) coupling scheme is operative, wherein the orbital and spin angular momenta *separately* couple (as assumed already for eqns. (7.151) and (7.154)); $\boldsymbol{L} = \sum_i \boldsymbol{l}_i$ and $\boldsymbol{S} = \sum_i \boldsymbol{s}_i$, rather than the individual $\boldsymbol{l}_i$ and $\boldsymbol{s}_i$ angular momenta coupling preferentially (so-called jj coupling). The net orbital and spin angular momenta, $\boldsymbol{L}$ and $\boldsymbol{S}$, then couple via the spin–orbit interaction

$$V_{\text{so}} = \lambda_{\text{so}} \boldsymbol{L}.\boldsymbol{S}, \tag{7.156}$$

where λ_{so} is the spin–orbit coupling constant. For a one-electron atom, the spin–orbit contribution to the Hamiltonian is

$$V_{\text{so}} = \frac{1}{2m_{\text{e}}^2 c^2} \frac{1}{r} \frac{\text{d}V}{\text{d}r} \boldsymbol{l}.\boldsymbol{s} \tag{7.157a}$$

and for a many-electron atom containing Z electrons it is (see e.g. Mandl (1992)):

$$V_{\text{so}} = \frac{1}{2m_{\text{e}}^2 c^2} \sum_{i=1}^{Z} \frac{1}{r_i} \frac{\text{d}V_i(r_i)}{\text{d}r_i} \boldsymbol{l}_i \,.\, \boldsymbol{s}_i, \tag{7.157b}$$

where $V_i(r_i)$ is the potential experienced by electron i due to all other electrons.

For a many-electron atom, the matrix element can be written as

$$\langle \phi | \mathscr{H}_{\text{so}} | \phi \rangle = \lambda_{\text{so}} \boldsymbol{L}.\boldsymbol{S}, \tag{7.158}$$

where $\lambda_{\text{so}} \propto \langle r^{-3} \rangle$, the expectation value of r^{-3} over the particular wavefunction.

As a result of spin-orbit coupling in the Russell–Saunders approximation, $\boldsymbol{L}$ and $\boldsymbol{S}$ couple to give the total angular momentum quantum number (see Fig. 7.38a):

$$\boldsymbol{J} = \boldsymbol{L} + \boldsymbol{S}, \tag{7.159}$$

where $\boldsymbol{J}$ can take all values between $(L + S)$ and $|L - S|$. This group of levels is called a multiplet, and the multiplicity of the system is *defined* as $(2S + 1)$. (In fact, if $L \geqslant S$ there are $(2S + 1)$ multiplets, but there are only $(2L + 1)$ if $L \leqslant S$.) The energy of the Jth state is given by (see Problem 7.18):

$$\mathscr{E}_{\text{J}} = \frac{\lambda_{\text{so}}}{2} [J(J + 1) - L(L + 1) - S(S + 1)]. \tag{7.160}$$

The magnetic moment of an atom with total angular momentum $\hbar \boldsymbol{J}$ is given by

$$\mu_{\text{m}}^J = -g_J \mu_{\text{B}} \boldsymbol{J} \equiv \gamma_{\text{m}} \hbar \boldsymbol{J}, \tag{7.161}$$

where γ_{m} is called the gyromagnetic (or magnetogyric) ratio and g_J is the Landé g-factor. Unlike J^2, $(\mu_{\text{m}}^{\text{tot}})^2$ (cf. eqn. (7.155)) is not a good quantum number, so that only the component of $\mu_{\text{m}}^{\text{tot}}$ along the $\boldsymbol{J}$-direction, μ_{m}^J, contributes to the magnetic properties (e.g. in the presence of an applied magnetic field, whose direction will fix that of $\boldsymbol{J}$). The total moment $\mu_{\text{m}}^{\text{tot}}$ can be thought of classically as precessing rapidly around the $\boldsymbol{J}$-axis (Fig. 7.38b), and hence the time-averaged component of $\mu_{\text{m}}^{\text{tot}}$ perpendicular to $\boldsymbol{J}$ will be

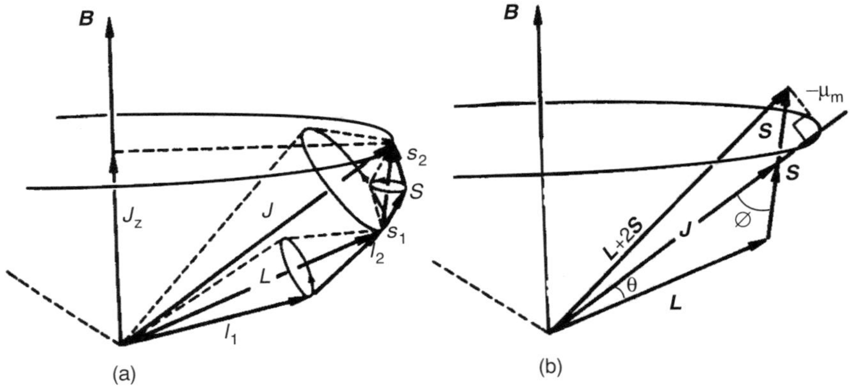

Fig. 7.38 (a) Russell–Saunders coupling of two electrons; (b) Vector diagram to aid in the calculation of the Landé g-factor, g_J.

zero. From Fig. 7.38b, the effective moment acting along the $\boldsymbol{J}$-direction (note in antiparallel with $\boldsymbol{J}$) can be written as

$$\mu_{\mathrm{m}}^{J} = -\frac{\boldsymbol{J}}{|\boldsymbol{J}|}\mu_{\mathrm{B}}(|\boldsymbol{L}|\cos\theta + 2|\boldsymbol{S}|\cos\phi). \tag{7.162}$$

Application of the cosine rule to the triangle formed by the vectors $\boldsymbol{J}$, $\boldsymbol{L}$ and $\boldsymbol{S}$ gives

$$\cos\theta = (|\boldsymbol{J}|^2 + |\boldsymbol{L}|^2 - |\boldsymbol{S}|^2)/2|\boldsymbol{J}||\boldsymbol{L}|,$$
$$\cos\phi = (|\boldsymbol{J}|^2 + |\boldsymbol{S}|^2 - |\boldsymbol{L}|^2)/2|\boldsymbol{J}||\boldsymbol{S}|.$$

Substitution of these expressions into eqn. (7.162), and replacing $|\boldsymbol{J}|^2$ by the eigenvalue $J(J+1)$ etc., gives for the Landé g-factor (assuming that $g_e = 2$):

$$g_J = 1 + \frac{J(J+1) + S(S+1) - L(L+1)}{2J(J+1)}. \tag{7.163}$$

The *ground-state* configuration [*JLS*] of an isolated atom can be found by the application of Hund's rules. The three rules, in the order that they must be obeyed, are:

1) S has the *maximum* value consistent with the Pauli exclusion principle (i.e. as many of the electrons as possible should have *parallel* spins).

2) L takes the *maximum* value consistent both with the value of S found from rule 1 and the exclusion principle.

3) The value of J in the ground state depends on the extent of the filling of the electron shell: $J = |L - S|$ when the shell is less than half-full, $J = (L + S)$ when more than half-full, and $J = S$ for a half-filled shell (since $L = 0$).

Rules 1 and 2 arise because of Coulomb interactions between electrons, together with the exclusion principle that prevents two electrons with the same spin being simultaneously at the same place (see §7.2.5.1). Concerning rule 1, the Coulombic repulsion energy is lowered if electrons are not in the same orbital, but are in different orbitals, where they may adopt parallel or antiparallel spin configurations. However, the exchange interaction for two orthogonal orbitals is positive (§7.2.5.1) and hence the parallel-spin configuration is favoured. The second rule is more difficult to rationalize, but has been confirmed by detailed calculations. The third rule is less robust than the other two: since it is determined by spin–orbit coupling, i.e. associated with *internal* magnetic fields ($\simeq 10$ T) within the atom caused by the orbital motion of the electrons, application of external fields of this order of magnitude can vitiate the rule. Other effects, too (e.g. crystal fields), can interfere with the operation of this rule, as will be seen later (§7.2.4.3).

It would be logical to denote the electronic configuration of the Hund-rule ground state of an atom simply by the set of quantum numbers $[J\ L\ S]$. However, it is conventional, instead, to denote it by the spectroscopic term symbol, $^{2S+1}L_J$, where, even more confusingly, the orbital angular momentum quantum number is denoted by a letter according to the prescription

$$\begin{array}{cccccc} L=0, & 1, & 2, & 3, & 4, & 5\ldots \\ \downarrow & \downarrow & \downarrow & \downarrow & \downarrow & \downarrow \\ \mathrm{S}, & \mathrm{P}, & \mathrm{D}, & \mathrm{F}, & \mathrm{G}, & \mathrm{H}\ldots \end{array} \tag{7.164}$$

The operation of Hund's rules can be understood by reference to two examples, the two-f-electron rare-earth ion, Pr^{3+} and the six-d-electron transition-metal ion, Fe^{2+}. For Pr^{3+}, the two electrons will occupy in a spin–parallel fashion the two highest-lying orbital angular momentum levels corresponding to $l = 3$ and $l = 2$ (Fig. 7.39a). Thus, $S = 1$ and $L = 5$, and since the shell is less than half-full, $J = |L - S| = 4$. Hence, the ground-state configuration of Pr^{3+} is $[J\ L\ S] = [4\ 5\ 1]$ with term symbol $^3\mathrm{H}_4$. For the Fe^{2+} ion, five of the six electrons will singly occupy in a spin–parallel fashion each of the five d-levels, and the sixth electron will therefore occupy the highest l-state ($l = 2$) in a spin-paired relationship to the other electron in the level (Fig. 7.39b). Thus, $S = 2$ and $L = 2$, and since the shell is more than half-filled, $J = (L + S) = 4$. Therefore, the ground-state configuration of Fe^{2+} is [4 2 2] with term symbol $^5\mathrm{D}_4$. See Problem 7.19 and Tables 7.7 and 7.8 for the ground-state configurations of other transition-metal and rare-earth ions.

7.2.4.2 Paramagnetism of a magnetic gas

The magnetization of a collection of paramagnetic atoms or ions in a solid can be calculated readily assuming that there are *no* interactions between the magnetic moments, i.e. the material behaves as a magnetic gas. In fact, the *magnetic dipole interaction* between magnetic moments can generally be neglected in comparison with the much stronger *exchange interactions* resulting from Coulomb effects (§7.2.5.1), but for very dilute concentrations of magnetic species, the atomic magnetic moments can be assumed to act independently of each other.

An external magnetic field, applied to a paramagnetic solid in the z-direction, lifts the $(2J + 1)$-degeneracy of the J-states and introduces an equal splitting of levels, in place of the energy level $\mathscr{E}_J$ (eqn. (7.160)), given by (cf. the corresponding equation for electric dipoles, eqn. (7.37)):

$$\mathscr{E}_\mathrm{B} = g_J \mu_\mathrm{B} J_z B_\mathrm{loc}, \tag{7.165}$$

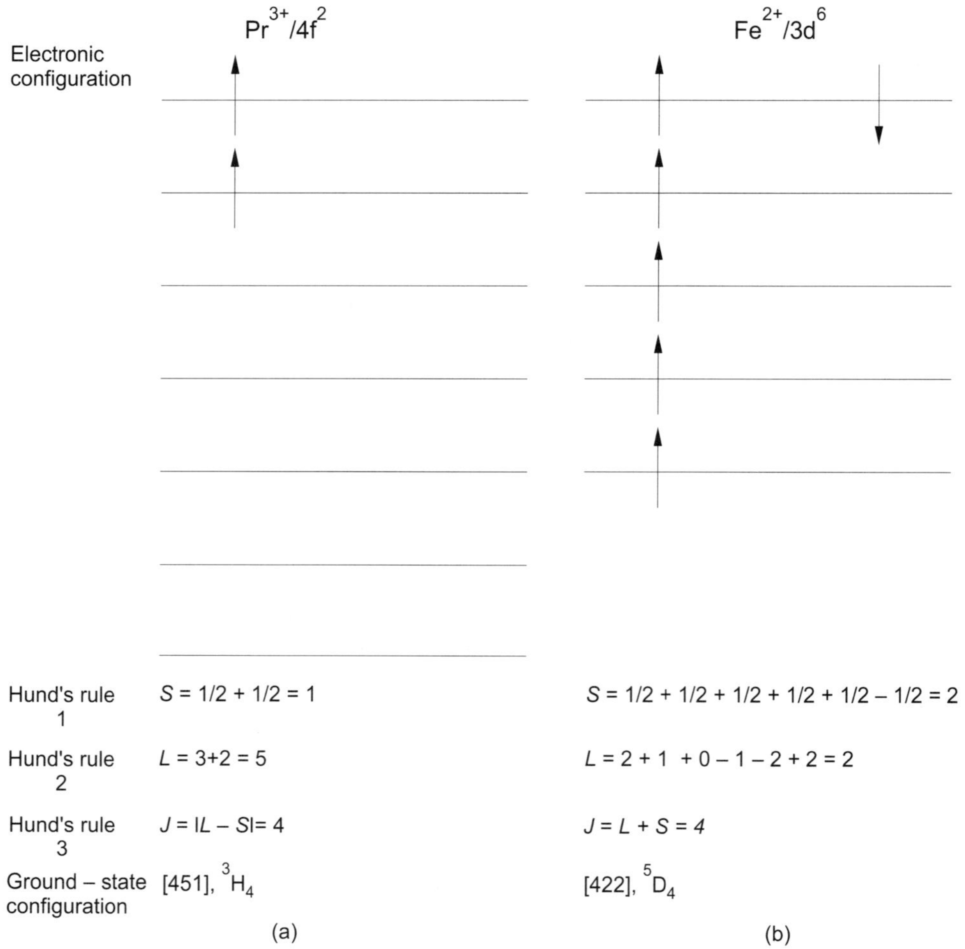

Fig. 7.39 Illustration of the operation of Hund's rules for: (a) Pr^{3+}; (b) Fe^{2+}.

with $J_z = -J, \ldots, 0, \ldots J$ (see Fig. 7.40), and B_{loc} is the *local* flux density at the magnetic moment (§7.2.2). If the atomic magnetic moments are in thermal equilibrium with the host lattice, the relative occupation of the energy levels given by eqn. (7.165) (cf. Fig. 7.40) is determined by the Boltzmann factor $\exp(-g_J \mu_B J_z B_{\text{loc}}/k_B T)$. Therefore, the net magnetization for a concentration n of independent atomic moments is given by

$$M = n \frac{\sum_{J_z=-J}^{J} -g_J \mu_B J_z \exp(-g_J \mu_B J_z B_{\text{loc}}/k_B T)}{\sum_{J_z=-J}^{J} \exp(-g_J \mu_B J_z B_{\text{loc}}/k_B T)}. \tag{7.166}$$

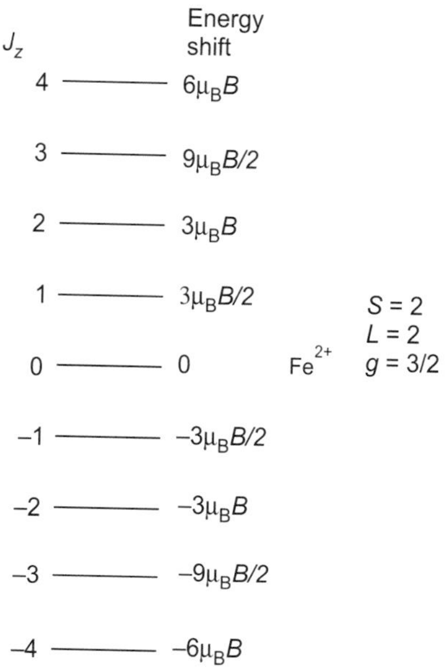

Fig. 7.40 Splitting of the ground-state degeneracy of Fe^{2+} ions by a magnetic field of flux density B.

This expression can be rewritten in terms of the partition function Z for an atomic moment,

$$Z = \sum_{J_z=-J}^{J} \exp(-g_J \mu_B J_z B_{loc}/k_B T), \tag{7.167}$$

as

$$M = -\frac{nk_B T^2}{B_{loc}} \left(\frac{\partial \ln Z}{\partial T}\right)_B. \tag{7.168}$$

The partition function can be evaluated, since it is a geometric progression, as

$$Z = \frac{e^x(1 - e^{-(2J+1)x/J})}{(1 - e^{-x/J})} = \frac{\sinh[(2J + 1)x/2J]}{\sinh(x/2J)}, \tag{7.169}$$

where

$$x = g_J \mu_B J B_{loc}/k_B T. \tag{7.170}$$

Hence, eqn. (7.168) can be rewritten as

$$M = -\frac{nk_B T^2}{B_{loc}} \left(\frac{\partial \ln Z}{\partial x}\right)\left(\frac{\partial x}{\partial T}\right)_B = n g_J \mu_B J B_J(x), \tag{7.171}$$

where the Brillouin function $B_J(x)$ is given by

$$B_J(x) = \frac{(2J + 1)}{2J} \coth[(2J + 1)x/2J] - \frac{1}{2J} \coth(x/2J). \tag{7.172}$$

The Brillouin function increases linearly with x for small x, $B_J(x) \simeq (J+1)x/3J$, and hence the magnetization increases linearly with magnetic flux density for small B:

$$M = \frac{ng_J^2\mu_B^2 J(J+1)B_{loc}}{3k_B T}. \tag{7.173}$$

For small values of magnetic susceptibility, χ_m, $\mu_0 \boldsymbol{M} \ll \boldsymbol{B}$ and hence the difference between local and external flux densities, $\boldsymbol{B}_{loc}$ and $\boldsymbol{B}_{ext}$, is negligible, as is the difference between $\boldsymbol{B} = \boldsymbol{B}_{ext} \simeq \boldsymbol{B}_{loc}$ and $\mu_0 \boldsymbol{H}$ (cf. eqn. (7.81)). In any case, for the case of a *spherical* sample geometry and a cubic, or random, array of magnetic dipoles, $\boldsymbol{B}_{loc} = \boldsymbol{B}_{ext}$ (eqn. (7.98)). Hence, from eqns. (7.83) and (7.173), the weak-field paramagnetic susceptibility of an assembly of independent atomic magnetic moments (i.e. a magnetic gas) is

$$\chi_{m,p}^{mg} = \frac{na_m^2 \mu_0 \mu_B^2}{3k_B T}, \tag{7.174}$$

where a_m is the magnetic moment expressed as the number of Bohr magnetons:

$$a_m = \mu_m^{eff}/\mu_B = g_J[J(J+1)]^{1/2}. \tag{7.175}$$

This is the Curie law of paramagnetism, $\chi_{m,p}^{mg} = C/T$, with Curie constant

$$C = \frac{na_m^2 \mu_0 \mu_B^2}{3k_B}. \tag{7.176}$$

Values of the parameter a_m for rare-earth ions, deduced from measured values of the Curie constant and compared with the value expected from Hund's rules (eqn. (7.175)) are given in Table 7.7, where good agreement is seen in general. The marked

Table 7.7 Values of $a_m = \mu_m^{eff}/\mu_B$ for rare-earth ions deduced from measured values of the Curie constant C (eqn. (7.176)) and compared with the theoretical value (eqn. (7.175)) expected for the Hund's rule ground state

Ion	*Number of 4f electrons*	*Ground-state configuration* [*JLS*]	$^{2S+1}L_J$	a_m^{meas}	a_m^{calc}
La^{3+}	0	[0 0 0]	$^1S_0$	0	0
Ce^{3+}	1	[5/2 3 1/2]	$^2F_{5/2}$	2.4	2.54
Pr^{3+}	2	[4 5 1]	$^3H_4$	3.5	3.58
Nd^{3+}	3	[9/2 6 3/2]	$^4I_{9/2}$	3.5	3.62
Pm^{3+}	4	[4 6 2]	$^5I_4$	–	2.68
Sm^{3+}	5	[5/2 5 5/2]	$^6H_{5/2}$	1.5	0.84
Eu^{3+}	6	[0 3 3]	$^7F_0$	3.4	0
Gd^{3+}	7	[7/2 0 7/2]	$^8S_{7/2}$	8.0	7.94
Tb^{3+}	8	[6 3 3]	$^7F_6$	9.5	9.72
Dy^{3+}	9	[15/2 5 5/2]	$^6H_{15/2}$	10.6	10.63
Ho^{3+}	10	[8 6 2]	$^5I_8$	10.4	10.6
Er^{3+}	11	[15/2 6 3/2]	$^4I_{15/2}$	9.5	9.59
Tm^{3+}	12	[6 5 1]	$^3H_6$	7.3	7.57
Yb^{3+}	13	[7/2 3 1/2]	$^2F_{7/2}$	4.5	4.54
La^{3+}	14	[0 0 0]	$^1S_0$	0	0

(After Kubo and Nagamiya (1968), *Solid State Physics*. Reproduced by permission of The McGraw-Hill Companies)

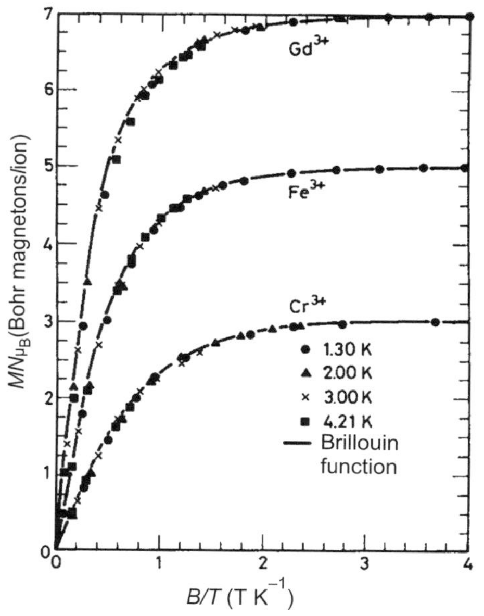

Fig. 7.41 Magnetization data for Gd^{3+} ions in gadolinium sulphate octahydrate, Fe^{3+} ions in iron ammonium alum and Cr^{3+} ions in potassium chromium alum as a function of the quantity B/T. The solid curves are plots of the magnetization as a function of the Brillouin function $B_J(x)$ (eqn. (7.172)) for $g_e = 2$ and J values of $\frac{7}{2}(Gd^{3+})$, $\frac{5}{2}(Fe^{3+})$, and $\frac{3}{2}(Cr^{3+})$. The orbital angular momentum for Cr^{3+} is quenched by crystal-field effects so that $J \simeq S$ with $L \simeq 0$. (After Henry (1952). Reprinted with permission from *Phys. Rev.* **88**, 559. © 1952. The American Physical Society)

discrepancies between the experimental and theoretical values for Sm^{3+} and Eu^{3+} ($J = 0$ for the latter ion) can be understood in terms of low-lying levels characterized by different J-values within an energy $\simeq k_B T$ of the ground state.

At *large* values of x or, equivalently, of the ratio B/T (eqn. (7.170)), the Brillouin function saturates at the value of unity. Thus, the paramagnetic magnetization also saturates, at the value

$$M_{sat} = ng_J \mu_B J, \qquad (7.177)$$

corresponding to all the ions being in the same $J_z = -J$ state. Figure 7.41 shows magnetization data for one rare-earth and two transition-metal ions. It can be seen that the Brillouin function gives a good fit to the data. Note, however, that the value of a_m needed to fit the data for Cr^{3+} is *not* given by eqn. (7.175) in terms of J, but instead S must be used. Thus, L appears to be zero, or 'quenched'; this is a consequence of crystal-field effects (see the next section).

7.2.4.3 Crystal-field effects

It might be thought that the paramagnetic behaviour of transition-metal ions in solids would also be well described by the Brillouin function (eqns. (7.171), (7.172)), as

Table 7.8 Values of $a_m = \mu_m^{eff}/\mu_B$ for transition-metal ions deduced from measured values of the Curie constant C (eqn. (7.176)) and compared with the theoretical value (eqn. (7.175)) expected for the Hund's rule ground state or if angular momentum is quenched

Ion	*Number of 3d electrons*	*Ground-state configuration* $[JLS]$	$^{2S+1}L_J$	a_m^{meas}	$a_m^{calc} = g_J[J(J+1)]^{1/2}$	$a_m^{calc} = 2[S(S+1)]^{1/2}$
V^{4+}	1	[3/2 2 1/2]	$^2D_{3/2}$	1.8	1.55	1.73
V^{3+}	2	[2 3 1]	$^3F_2$	2.8	1.63	2.83
V^{2+}	3	[3/2 3 3/2]	$^4F_{3/2}$	3.8	0.77	3.87
Cr^{3+}	3	[3/2 3 3/2]	$^4F_{3/2}$	3.7	0.77	3.87
Cr^{2+}	4	[0 2 2]	$^5D_0$	4.8	0	4.90
Mn^{4+}	3	[3/2 3 3/2]	$^4F_{3/2}$	4.0	0.77	3.87
Mn^{3+}	4	[0 2 2]	$^5D_0$	5.0	0	4.90
Mn^{2+}	5	[5/2 0 5/2]	$^6S_{5/2}$	5.9	5.92	5.92
Fe^{3+}	5	[5/2 0 5/2]	$^6S_{5/2}$	5.9	5.92	5.92
Fe^{2+}	6	[4 2 2]	$^5D_4$	5.4	6.70	4.90
Co^{2+}	7	[9/2 3 3/2]	$^4F_{9/2}$	4.8	6.54	3.87
Ni^{2+}	8	[4 3 1]	$^3F_4$	3.2	5.59	2.83
Cu^{2+}	9	[5/2 2 1/2]	$^2D_{5/2}$	1.9	3.55	1.73

discussed in the previous section. However, this is found not to be the case: values of the parameter $a_m = \mu_m^{eff}/\mu_B$ found from experimentally measured values of the Curie constant (eqn. (7.176)) for transition-metal ion-containing materials disagree markedly with the values expected on the basis expected from the third of Hund's rules (Table 7.8). Instead of a_m being determined by J as in eqn. (7.175), much better agreement is achieved if alternatively the net spin $\boldsymbol{S}$ is used (Table 7.8); it appears that $\boldsymbol{L} \simeq 0$, i.e. the angular momentum seems to be quenched. This is true for all the transition-metal ions; for the case of the $3d^5$ ions, Fe^{3+} and Mn^{2+}, the net angular momentum is already zero, of course, because of Hund's first rule.

This behaviour is the result of perturbations of the electronic wavefunction on a given ion by the electric fields (strictly, the electric-field *gradients*) due to the charges on surrounding ions. These inhomogeneous electric fields are known as crystal fields; they are not centrosymmetric, but have only the symmetry of the site at which the ion is based. The crystal fields do not affect Hund's first two rules, but crystal-field effects can be comparable to the spin–orbit interaction, and so Hund's third rule is affected. If the crystal-field interaction is much larger than the spin–orbit coupling (as in transition-metal ions), the latter interaction may be neglected, and so crystal-field effects give rise to what becomes a new third rule. Of the $(2S+1)(2L+1)$-degenerate set of levels resulting from the application of the first two Hund's rules, the crystal field will not affect the spin degeneracy, since this depends only on spatial variables and hence commutes with $\boldsymbol{S}$. However, in a non-central field, the orbital angular momentum precesses, in a classical picture, and its component L_z can average to zero even though the magnitude $(L(L+1))$ remains unchanged. Thus, when L_z averages to zero, the angular momentum, and hence also the magnetic moment μ_m^L (eqn. (7.151)), is *quenched*.

Alternatively, the influence of crystal fields can be understood in terms of a Stark (electric-field-induced) splitting of the otherwise $(2L+1)$-degenerate levels (see

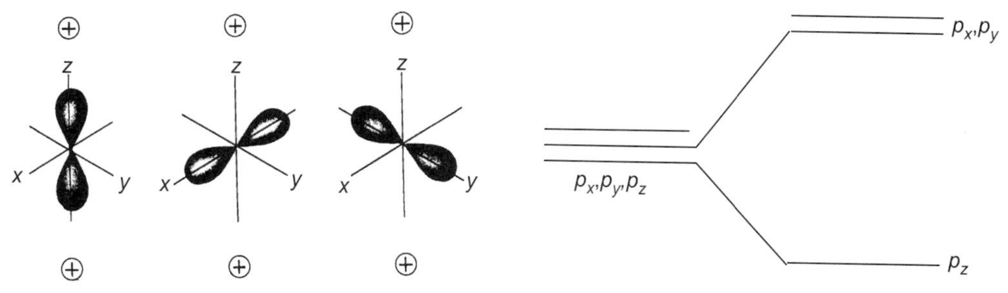

Fig. 7.42 Illustration of the quenching of orbital angular momentum by crystal fields. In an isolated atom, the three p-orbitals ($l = 1$) characterized by $m_l = 0, \pm 1$, are degenerate in energy. In a crystal field, here represented by a uniaxial electric field caused by two neighbouring ions in the z-direction, the p_z orbital has a lower energy than the p_x, p_y orbitals (which remain degenerate).

Fig. 7.42). For strong crystal fields, the energy difference between such Stark-split states will be larger than $g_J \mu_B B$, the Zeeman magnetic-field-induced splitting; hence the states will be unaffected by the external magnetic field, and cannot contribute to the magnetic susceptibility.

Why transition-metal, rather than rare-earth, ions should be particularly susceptible to crystal-field effects can be seen from the following argument. Consider a cubic crystal (e.g. with the NaCl structure), in which six charges surround a central ion, at a distance d, with $O_h(m3m)$ point symmetry. For one of the charges, say at $x = -d$, $y = z = 0$, the electrostatic potential near the origin is $V = q/4\pi\varepsilon_0[(d+x)^2 + y^2 + z^2]^{1/2}$, which can be expanded as $V \simeq q[1 - (x/d) + \ldots]/4\pi\varepsilon_0 d$. The leading term is thus $V = 6q/4\pi\varepsilon_0 d$, but this is centrosymmetric and hence will not contribute to the crystal field. The first non-centrosymmetric, crystal-field contribution is

$$V_{cf} = \frac{35q}{16\pi\varepsilon_0 d^5}[x^4 + y^4 + z^4 - \frac{3}{5}r^4]. \tag{7.178}$$

Hence, the contribution of this to the crystal-field splitting will be

$$\Delta =< \phi| - eV_{cf}|\phi > \propto < r^4 > /d^5. \tag{7.179}$$

The essential differences of relevance here between rare-earth and transition-metal ions are in the overall ion size (d) and in the spatial extent of the 4f-and 3d-electron shells: thus, $d_{4f} > d_{3d}$, and the expectation value $\langle r^4 \rangle$ over the respective wavefunctions gives $\langle r^4_{4f} \rangle < \langle r^4_{3d} \rangle$, since the 4f-electrons lie deep within the rare-earth ion (beneath filled 5s and 5p shells) but the 3d-electrons form the outermost electron shell of transition-metal ions. Consequently, the effect of crystal fields, i.e. angular-momentum quenching, is expected to be much greater for transition-metal ions than for rare-earth ions. Furthermore, spin–orbit coupling is expected to be weaker for 3d-electrons than for 4f-electrons because $\lambda_{so} \propto \langle r^{-3} \rangle$ (§7.2.4.1), making the relative influence of crystal-field effects over spin–orbit coupling even larger for 3d-electrons.

*7.2.4.4 Van Vleck paramagnetism

Ions that have one electron less than is necessary to half-fill a shell have $J = 0$ (e.g. Cr^{2+}, Mn^{3+}—see Table 7.8, or Eu^{3+}—see Table 7.7) and hence it would be expected that consequently they would have no magnetic moment (cf. eqn. (7.161)). Although this is true for the ground state, $|0>$, mixing with a low-lying excited state $|1>$, at an energy $\Delta = \mathscr{E}_1 - \mathscr{E}_0$ above the ground state, can produce a moment. This is known as Van Vleck paramagnetism.

The susceptibility can be calculated by considering the kinetic part of the electron Hamiltonian in the presence of a magnetic field (eqn. (7.122)), but including also the contribution of the total intrinsic electron spin of the ion, $\boldsymbol{S}$. Thus, retaining only the field-dependent terms:

$$\mathscr{H}_{\text{kin}} = \mu_{\text{B}} \boldsymbol{B}_{\text{ext}} \cdot (\boldsymbol{L} + g_{\text{e}} \boldsymbol{S}) + \frac{e^2}{8m_{\text{e}}} B_{\text{ext}}^2 \sum_i x_i^2 + y_i^2. \tag{7.180}$$

In order to calculate the magnetic susceptibility, the energy $\mathscr{E}$ in the magnetic field must be evaluated keeping terms *quadratic* in the field (since $\chi_{\text{m}} = \partial M/\partial H$ (eqn. (7.83)) and $M = -\partial(\mathscr{E}/V)/\partial B$ (eqn. (7.74))). Thus, the effect of the field on the energy must be calculated by second-order perturbation theory, giving a shift $\Delta\mathscr{E}_0$ in the position of the energy level from $\mathscr{E}_0$ to $\mathscr{E}_0 + \Delta\mathscr{E}_0$, where

$$\Delta\mathscr{E}_0 = \langle 0|\mathscr{H}_{\text{kin}}|0\rangle + \frac{|\langle 0|\mathscr{H}_{\text{kin}}|1\rangle|^2}{\mathscr{E}_0 - \mathscr{E}_1}. \tag{7.181}$$

The second term should really be a summation over all excited states, but for simplicity we consider here only a single level above the ground state. Thus, the perturbation in energy can be written, on substituting eqn. (7.180) into (7.181), as

$$\begin{aligned}\Delta\mathscr{E}_0 &= \mu_{\text{B}} \boldsymbol{B}_{\text{ext}} \cdot \langle 0|\boldsymbol{L} + g_{\text{e}}\boldsymbol{S}|0\rangle + \frac{|\langle 0|\mu_{\text{B}}\boldsymbol{B}_{\text{ext}} \cdot (\boldsymbol{L} + g_{\text{e}}\boldsymbol{S})|1\rangle|^2}{\mathscr{E}_0 - \mathscr{E}_1} \\ &\quad + \frac{e^2}{8m_{\text{e}}} B_{\text{ext}}^2 \langle 0| \sum_i (x_i^2 + y_i^2)|0\rangle.\end{aligned} \tag{7.182}$$

The last term in eqn. (7.182) is just the *diamagnetic* contribution (§7.2.3.1) and will be neglected henceforth. The paramagnetic susceptibility for a collection of n ions per unit volume is

$$\chi_{\text{m,p}} = -n\mu_0 \frac{\partial^2 \mathscr{E}}{\partial B^2}. \tag{7.183}$$

The first term in eqn. (7.182) (cf. the first term in eqn. (7.123)) is only a *linear* function of B, and hence does not contribute to the susceptibility; only the second term in eqn. (7.182) gives the Van Vleck paramagnetic susceptibility:

$$\chi_{\text{m,p}}^{\text{VV}} = 2\mu_0\mu_{\text{B}}^2 \frac{|\langle 0|\boldsymbol{L}_z + g_{\text{e}}\boldsymbol{S}_z|1\rangle|^2}{\mathscr{E}_1 - \mathscr{E}_0}. \tag{7.184}$$

Note that, in contrast to the Curie paramagnetic susceptibility for a set of ions (eqn. (7.174)), the Van Vleck susceptibility is temperature-independent.

7.2.4.5 Paramagnetism of normal metals

Thus far, we have discussed atomic paramagnetic behaviour arising mostly from unpaired electrons in ions. However, because of their intrinsic spin, unpaired conduction electrons in a (normal) metal can also contribute to the paramagnetic response of a material. If the electrons were *distinguishable* particles, governed by (classical) Boltzmann statistics, use of the Curie expression (eqn. (7.174)) for the paramagnetic susceptibility, with $J = S = 1/2$ and $g_J = g_e \simeq 2$, would predict that for a concentration, n, of free electrons:

$$\chi_{m,p} = \frac{n\mu_0\mu_B^2}{k_B T}. \tag{7.185}$$

However, this Curie-law behaviour of conduction electrons is *not* observed; the observed paramagnetic susceptibility of conduction electrons is temperature-*independent*. The reason, of course, is that electrons are *indistinguishable* particles and obey Fermi–Dirac statistics (eqn. (5.23)). In particular, the Pauli exclusion principle is obeyed, which greatly reduces the number of electrons able to reverse their spin in response to an external magnetic field, since most orbitals will already be occupied by electrons having the *same* spin orientation. Thus, as for the case of the electronic heat capacity (§5.1.3.1), for example, it is only those electrons within an energy range of the order of $k_B T$ of the Fermi level that are able to be excited and can reverse their spin in an external magnetic field, and hence can contribute to the paramagnetic susceptibility. Therefore, it is expected that the Curie susceptibility, eqn. (7.185), will be reduced by a factor of (T/T_F), making it much smaller than the classical prediction, and also temperature-independent, as observed.

An expression for the paramagnetism of conduction electrons can be calculated by reference to electron occupations relating to the density of states shown in Fig. 7.43 (assumed there to be parabolic, i.e. free-electron-like, for simplicity). The density of states for electrons in an external magnetic field $\boldsymbol{B}$ can be viewed as being split into two parts: one part is for electrons that have their spin *parallel* to $\boldsymbol{B}$, and consequently that have their energies *lowered* by an amount $\mathscr{E}_B = -g_J\mu_B J_z B_{loc}$ (eqn. (7.165)), or $\mathscr{E}_B = -\mu_B B$, taking $g_J = g_e = 2$ and $J = S = 1/2$ for electrons, and assuming that $\boldsymbol{B}_{ext} = \boldsymbol{B}_{loc} = \boldsymbol{B}$. Likewise, electrons with their spins *antiparallel* to $\boldsymbol{B}$ have their energies *raised* by $\mathscr{E}_B = \mu_B B$. Figure 7.43 shows the two density-of-states distributions 'back-to-back', with their zeros shifted relatively in energy by $2\mu_B B$. The Fermi level in the two distributions must be the same, and is unaffected by the magnetic field; it is only the relative number of electron spins with orientations parallel and antiparallel to $\boldsymbol{B}$ that is different in the two cases.

Neglecting thermal-excitation effects, i.e. at $T = 0$ K where the Fermi–Dirac function $f(\mathscr{E})$ is unity for $\mathscr{E} \leq \mathscr{E}_F$, the concentration n_+ of electrons with their spins parallel to $\boldsymbol{B}$ is given by an integral over the appropriate energy-shifted density of states:

$$n_+ = \frac{1}{2}\int_{-\mu_B B}^{\mathscr{E}_F} D(\mathscr{E} + \mu_B B)\mathrm{d}\mathscr{E} \simeq \frac{1}{2}\int_0^{\mathscr{E}_F} D(\mathscr{E})\mathrm{d}\mathscr{E} + \frac{1}{2}\mu_B B D(\mathscr{E}_F), \tag{7.186}$$

where it has been assumed that $\mu_B B < \mathscr{E}_F$. Similarly, the concentration n_- of electrons with spins antiparallel to $\boldsymbol{B}$ is given by

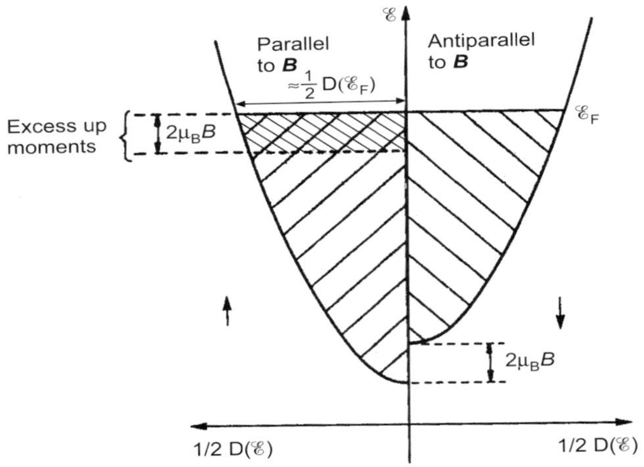

Fig. 7.43 Illustration of the origin of Pauli paramagnetism of conduction electrons regarded, for simplicity, as a free-electron gas. The parabolic density of electron states in the presence of a magnetic field, $\boldsymbol{B}$, splits into two, and is shifted in energy by $\pm\mu_B B$ with respect to the field-free curve. Here the split distributions are placed back-to-back for purposes of illustration. The excess moments parallel to $\boldsymbol{B}$ correspond to the shaded area, corresponding to an average density-of-states value $\simeq \frac{1}{2}D(\mathscr{E}_F)$ and energy interval $2\mu_B B$.

$$n_- = \frac{1}{2}\int_{\mu_B B}^{\mathscr{E}_F} D(\mathscr{E} - \mu_B B)\mathrm{d}\mathscr{E} \simeq \frac{1}{2}\int_0^{\mathscr{E}_F} D(\mathscr{E})\mathrm{d}\mathscr{E} - \frac{1}{2}\mu_B B D(\mathscr{E}_F). \tag{7.187}$$

The net magnetization of the electron gas is given by $M = \mu_B(n_+ - n_-)$ and so

$$M = \mu_B^2 D(\mathscr{E}_F)B. \tag{7.188}$$

For a free-electron gas, $D(\mathscr{E}_F) = 3n/2\mathscr{E}_F = 3n/2k_B T_F$ (cf. eqn. (5.16)) and hence the Pauli paramagnetic susceptibility for free electrons is given by

$$\chi_{m,p}^{fe} = \frac{3n\mu_0\mu_B^2}{2k_B T_F}, \tag{7.189}$$

where it has been assumed that $\boldsymbol{B}_{ext} \simeq \boldsymbol{B}_{mac} = \mu_0 \boldsymbol{H}_{mac}$ since the susceptibility is so small. Comparison of eqn. (7.189) with the expression (eqn. (7.185)) resulting from the application of classical statistics shows that indeed the latter is multiplied by a factor of order (T/T_F). Note also that the magnitude of the Pauli paramagnetic susceptibility is exactly three times that of the Landau diamagnetic susceptibility (eqn. (7.129)).

For a *real* crystal containing unpaired conduction electrons, e.g. a normal metal or a highly doped extrinsic semiconductor, the electron mass m_e that appears in the numerator of the expression for the free-electron density of states (eqn. (5.15)) must be replaced by the effective mass m_e^* arising from band-structure effects (§6.2.1). Thus, the Pauli susceptibility for a crystal becomes

$$\chi_{\mathrm{m,p}}^{\mathrm{cr}} = \frac{3n\mu_0\mu_{\mathrm{B}}^2}{2k_{\mathrm{B}}T_{\mathrm{F}}}\left(\frac{m_{\mathrm{e}}^*}{m_{\mathrm{e}}}\right). \tag{7.190}$$

Hence, the total (paramagnetic plus diamagnetic) susceptibility of conduction electrons in a crystal is given by the sum of eqns. (7.131) and (7.190) (making use of the relation $D(\mathscr{E}_{\mathrm{F}}) = 3n/2\mathscr{E}_{\mathrm{F}}$ (eqn. (5.16))):

$$\chi_{\mathrm{m}}^{\mathrm{cr}} = \mu_0\mu_{\mathrm{B}}^2 D(\mathscr{E}_{\mathrm{F}})\left[1 - \frac{1}{3}\left(\frac{m_{\mathrm{e}}}{m_{\mathrm{e}}^*}\right)^2\right]. \tag{7.191}$$

For those semiconductors and metals (notably Bi) for which $m_{\mathrm{e}}^* \ll m_{\mathrm{e}}$, the diamagnetic contribution can dominate the paramagnetic term.

7.2.5 Spontaneous magnetization

For diamagnetic and paramagnetic materials, a net magnetization in a sample only occurs on the application of an external magnetic field. The permanent magnetic dipole moments that exist in paramagnetic materials were assumed in §7.2.4.2 *not* to interact with one another, e.g. as a result of dilution (ideal magnetic gas). In any case, if the temperature is such that the thermal energy $k_{\mathrm{B}}T$ is greater than any interaction energy between magnetic moments, the paramagnetic state will be favoured as a result of the randomizing influence of thermal fluctuations.

At low temperatures (compared to that corresponding to the interaction energy), however, a cooperative ordering of magnetic moments in a solid is expected, as for electrical dipoles (§7.1.5). Thus, ferromagnetic ordering occurs when the inter-moment interaction is such as to ensure that all magnetic moments are parallel to one another; hence, they all contribute equally to the overall spontaneous magnetization present even in zero field (cf. ferroelectric ordering—§7.1.5.3). Also, by analogy with the corresponding electrical case, antiferromagnetic order occurs when the interaction produces an antiparallel arrangement of the moments and gives zero net magnetization. Ferrimagnetic ordering falls in between the above two categories: antiparallel moments do not cancel entirely, resulting in a (small) spontaneous magnetization.

It might be thought at first that *magnetic dipolar interactions* between magnetic moments might be responsible for magnetic ordering (ferromagnetism, antiferromagnetism, etc.). However, this interaction is far too weak at any but very low temperatures to overcome thermal-disordering effects and hence is unable to cause an ordering of magnetic moments (see Problem 7.23). The interaction that is responsible instead is the *exchange interaction* which is effectively a spin-dependent Coulombic interaction.

7.2.5.1 The exchange interaction

The Pauli exclusion principle exerts a powerful influence on the energetics of a collection of electrons through their relative intrinsic spin orientations. Two electrons with parallel spins are not permitted to occupy the same region of space, and thus tend to keep apart. As a result, the Coulombic repulsion energy between pairs of electrons with a parallel spin configuration is lower than for pairs of electrons with antiparallel spins: in this case, ferromagnetic (parallel) spin ordering is favoured. This difference in Coulombic energies for different spin configurations is the exchange-interaction energy.

Fermions, such as electrons, are indistinguishable particles characterized by Fermi–Dirac statistics, underlying which is the assertion that the wavefunction of two electrons must be *antisymmetric* under interchange of the particles, i.e. interchange of both spatial ($\boldsymbol{r}$) and spin (σ) coordinates. Thus, for two electrons with coordinates $\boldsymbol{r}_1, \sigma_1; \boldsymbol{r}_2, \sigma_2$:

$$\Psi(\boldsymbol{r}_1, \sigma_1; \boldsymbol{r}_2, \sigma_2) = -\Psi(\boldsymbol{r}_2, \sigma_2; \boldsymbol{r}_1, \sigma_1). \tag{7.192}$$

This is simply a mathematical restatement of the Pauli exclusion principle since the wavefunction vanishes, and hence there is zero probability of finding two electrons at the same place with the same spins, when $\boldsymbol{r}_1 = \boldsymbol{r}_2$ and $\sigma_1 = \sigma_2$.

The two-electron wavefunctions (Ψ) for a system of two electrons (e.g. in a hydrogen molecule) can be written as the product of orbital wavefunctions $\phi(\boldsymbol{r})$ and spin functions (spinors) $\eta(\sigma)$. Assuming that electron 1 can occupy a (non-degenerate) orbital a with wavefunction ϕ_a and electron 2 can occupy an orbital b with wavefunction ϕ_b, then a particular *symmetric* combination of the spatial variables alone is

$$\Phi^{\mathrm{s}} = \phi_a(1)\phi_b(2) + \phi_a(2)\phi_b(1). \tag{7.193}$$

This combination represents the Heitler–London approximation for the H_2 molecule (see Problem 7.24(a)). A corresponding antisymmetric spatial combination is

$$\Phi^{\mathrm{as}} = \phi_a(1)\phi_b(2) - \phi_a(2)\phi_b(1). \tag{7.194}$$

However, these two functions do not include the spin variables. If there is an up-spin at a site, $\eta(\sigma) = \alpha$, and if the spin is down, $\eta(\sigma) = \beta$. Hence there are also antisymmetric and symmetric combinations of the spin variables:

$$\Xi^{\mathrm{as}} = \alpha(1)\beta(2) - \alpha(2)\beta(1) \tag{7.195}$$

and

$$\Xi^{\mathrm{s}} = \alpha(1)\alpha(2) \tag{7.196a}$$

or

$$= \beta(1)\beta(2) \tag{7.196b}$$

or

$$= \alpha(1)\beta(2) + \alpha(2)\beta(1). \tag{7.196c}$$

The antisymmetric combination (eqn. (7.195)) corresponds to a spin-paired state, i.e. a singlet state; the symmetric combinations (eqns. (7.196)) refer to parallel-spin states, i.e. a triplet state (triply degenerate in the absence of spin–orbit coupling).

For the overall wavefunction Ψ to be antisymmetric under interchange of particles (or $\{\boldsymbol{r}, \sigma\}$), a symmetric spatial combination (eqn. (7.193)) must be coupled with an antisymmetric (singlet) spin combination (eqn. (7.195)), or vice versa. For example, for an up-spin at both sites (a triplet configuration), the total two-electron wavefunction is (neglecting normalization constants):

$$\begin{aligned}\Psi_1 &= \Phi^{\mathrm{as}}\Xi^{\mathrm{s}} \\ &= [\phi_a(1)\phi_b(2) - \phi_a(2)\phi_b(1)]\alpha(1)\alpha(2).\end{aligned} \tag{7.197}$$

This can be written more succinctly as a Slater determinant:

$$\Psi_1 = \begin{vmatrix} \phi_a(1)\alpha(1) & \phi_a(2)\alpha(2) \\ \phi_b(1)\alpha(1) & \phi_b(2)\alpha(2) \end{vmatrix}. \tag{7.198}$$

In general, for N electrons at N positions, $r_1, \ldots r_N$, the Slater determinant is

$$\Psi = \frac{1}{\sqrt{N!}} \begin{vmatrix} \phi_1(\boldsymbol{q}_1) & \ldots & \phi_\mathrm{N}(\boldsymbol{q}_1) \\ \vdots & & \vdots \\ \phi_1(\boldsymbol{q}_\mathrm{N}) & \ldots & \phi_\mathrm{N}(\boldsymbol{q}_\mathrm{N}) \end{vmatrix} \tag{7.199}$$

where $\phi_i(q_i)$ is the (orthogonal) wavefunction for the ith electron with spatial and spin variables $\boldsymbol{q}_i = \{\boldsymbol{r}_i; \sigma_i\}$. The Slater determinant satisfies the conditions of the Pauli exclusion principle: exchange of any two electrons is equivalent to the exchange of the corresponding columns of the determinant and hence gives, by the property of determinants, a change in sign of Ψ; if any two electrons i and j have the same spatial and spin coordinates, $\boldsymbol{q}_i = \boldsymbol{q}_j$, two columns of the determinant are identical and Ψ is zero.

Equations (7.197) and (7.198) represent one wavefunction of the triplet state. Another representation, Ψ_2, is for there instead to be a down-spin at both sites, in which case Ξ^s is given by eqn. (7.196b) and $\alpha(i)$ is replaced by $\beta(i)$ in the Slater determinant (eqn. (7.198)). The other two wavefunctions, one for the singlet state and the remaining one for the triplet state, can be constructed from product wavefunctions in which there is an up-spin on site 1 and a down-spin on site 2, and vice versa. Thus, in the former case:

$$\Phi_1 = \begin{vmatrix} \phi_a(1)\alpha(1) & \phi_\mathrm{a}(2)\alpha(2) \\ \phi_b(1)\beta(1) & \phi_\mathrm{b}(2)\beta(2) \end{vmatrix}, \tag{7.200}$$

and Φ_2 for the latter case is obtained by exchanging α and β in eqn. (7.200). Then, orthogonal (unnormalized) wavefunctions can be obtained from Φ_1 and Φ_2 as the triplet wavefunction:

$$\Psi_3 = \Phi_1 + \Phi_2, \tag{7.201}$$

and the singlet wavefunction:

$$\Psi_4 = \Phi_1 - \Phi_2. \tag{7.202}$$

It can be verified that eqns. (7.201) and (7.202) correspond to the spin combinations Ξ given by eqns. (7.196c) and (7.195), respectively.

Returning now to the problem of the H_2 molecule with protons a and b at $\boldsymbol{R}_a$ and $\boldsymbol{R}_b$, the two-electron Hamiltonian for electrons 1 and 2 can be written as

$$\mathscr{H} = \mathscr{H}_1 + \mathscr{H}_2 + \mathscr{H}_3 \tag{7.203a}$$

$$= \left(-\frac{\hbar^2}{2m_\mathrm{e}}\nabla_1^2 - \frac{e^2}{4\pi\varepsilon_0 r_{1a}}\right) + \left(-\frac{\hbar^2}{2m_\mathrm{e}}\nabla_2^2 - \frac{e^2}{4\pi\varepsilon_0 r_{1b}}\right) + \frac{e^2}{4\pi\varepsilon_0}\left(\frac{1}{R_\mathrm{ab}} + \frac{1}{r_{12}} - \frac{1}{r_{1b}} - \frac{1}{r_{2a}}\right), \tag{7.203b}$$

where the first two terms are those for isolated hydrogen atoms with electron 1 associated with proton a and electron 2 with proton b, and the quantities in the third term respectively refer to inter-proton repulsion ($\boldsymbol{R}_{ab} = |\boldsymbol{R}_a - \boldsymbol{R}_b|$), inter-electron repulsion ($\boldsymbol{r}_{12} = |\boldsymbol{r}_1 - \boldsymbol{r}_2|$) and the other two terms are attractive interactions between an electron and the other proton from that considered in the first two terms. (It is assumed that the proton mass is infinite so that proton motion is non-existent, and hence there is no corresponding kinetic-energy term.) The one-particle contributions, $\mathscr{H}_1$ and $\mathscr{H}_2$, to the Hamiltonian have eigenfunctions $\phi_a(1)$ and $\phi_b(2)$ respectively, with eigenenergy $\mathscr{E}_0$:

$$\mathscr{H}_{1,2}\,\phi_{a,\,b}(1,2) = \mathscr{E}_0\phi_{a,b}(1,2). \tag{7.204}$$

The overall Hamiltonian can be diagonalized by using the symmetric and antisymmetric real-space functions given by eqns. (7.193) and (7.194). A variational calculation (see Problem 7.24(b)) gives for the two eigenvalues

$$\mathscr{E}_{\pm} = 2\mathscr{E}_0 + \frac{C \pm A}{1 \pm B^2} \tag{7.205}$$

where the positive sign refers to the spatially symmetric (i.e. spin–singlet) solution and the negative sign gives the spatially antisymmetric (spin–triplet) solution. The quantities C, B and A are given respectively by the Coulomb integral

$$C = \frac{e^2}{4\pi\varepsilon_0}\int\left(\frac{1}{R_{ab}} + \frac{1}{r_{12}} - \frac{1}{r_{a2}} - \frac{1}{r_{b1}}\right)|\phi_a(1)|^2|\phi_b(2)|^2 \mathrm{d}\boldsymbol{r}_1\mathrm{d}\boldsymbol{r}_2, \tag{7.206}$$

which is the net normal Coulomb energy between charge densities $-e|\phi_a(1)|^2$ and $-e|\phi_b(2)|^2$ separated by various distances, the overlap integrals between orbitals on different sites are

$$B = \int \phi_a^*(1)\phi_b(1)\mathrm{d}\boldsymbol{r}_1 = \int \phi_b^*(2)\phi_a(2)\mathrm{d}\boldsymbol{r}_2, \tag{7.207}$$

and finally the exchange integral is

$$A = \frac{e^2}{4\pi\varepsilon_0}\int\left(\frac{1}{R_{ab}} + \frac{1}{r_{12}} - \frac{1}{r_{a1}} - \frac{1}{r_{b2}}\right)\phi_a^*(1)\phi_a(2)\phi_b(1)\phi_b^*(2)\mathrm{d}\boldsymbol{r}_1\mathrm{d}\boldsymbol{r}_2. \tag{7.208}$$

The exchange term is a purely quantum-mechanical effect and has no classical counterpart. The reason for its name can be understood by comparing eqns. (7.206) and (7.208). The triplet–singlet energy splitting is therefore from eqn. (7.205)

$$\mathscr{E}_{\mathrm{t}} - \mathscr{E}_{\mathrm{s}} = \frac{2(CB^2 - A)}{1 - B^4} \tag{7.209}$$

where $\mathscr{E}_{\mathrm{t}} \equiv \mathscr{E}_-$ and $\mathscr{E}_{\mathrm{s}} \equiv \mathscr{E}_+$.

Considering for the moment the case of different orbitals in the *same* atom, eqn. (7.209) can be used, but with the overlap term B taken to be *zero* (the atomic orbitals are orthogonal to each other), in which case

$$\mathscr{E}_{\mathrm{t}} - \mathscr{E}_{\mathrm{s}} \equiv -\mathfrak{J} = -2A \tag{7.210}$$

where $\mathfrak{J}$ is the exchange constant (or parameter). Since, in this case, A is simply the self-energy of the charge distribution $-e\phi_a^*\phi_b$, the exchange constant is a positive quantity, and the triplet configuration has the *lower* energy. This is the origin of Hund's first rule (§7.2.4.1).

Returning to the case of the H_2 molecule, it can be seen from eqn. (7.209) that whether the triplet or the singlet spin configuration has the lower energy depends on whether A is bigger or smaller than the term CB^2, since now the overlap is non-zero (eqn. (7.207)). In fact, it is found that the *singlet* configuration has the lower energy. The effective exchange constant $\mathfrak{J} = 2(A - CB^2)/(1 - B^4)$ is negative in this case; i.e. antiferromagnetic spin ordering is favoured.

★ Thus far, we have considered the exchange interaction between electrons localized on atoms (as in ions) or in bonds (as in molecules). The exchange interaction between *free electrons*, as a model for a metal, can also be evaluated; it is *positive*, and therefore there is a correlation between like electron spins. Consider two electrons, i and j, with the *same* spin, for which the antisymmetrized real-space two-electron wavefunction can be written as

$$\Psi_{ij} = \frac{1}{\sqrt{2V}}(e^{i\boldsymbol{k}_i\,.\,\boldsymbol{r}_i}e^{i\boldsymbol{k}_j\,.\,\boldsymbol{r}_j} - e^{i\boldsymbol{k}_i\,.\,\boldsymbol{r}_j}e^{i\boldsymbol{k}_j\,.\,\boldsymbol{r}_i}) \tag{7.211a}$$

$$= \frac{1}{\sqrt{2V}}e^{i(\boldsymbol{k}_i\,.\,\boldsymbol{r}_i+\boldsymbol{k}_j\,.\,\boldsymbol{r}_j)}(1 - e^{-i(\boldsymbol{k}_i-\boldsymbol{k}_j)\,.\,(\boldsymbol{r}_i-\boldsymbol{r}_j)}). \tag{7.211b}$$

The probability of finding electron i in volume element $d\boldsymbol{r}_i$ and electron j in $d\boldsymbol{r}_j$ is

$$|\Psi_{ij}|^2 d\boldsymbol{r}_i d\boldsymbol{r}_j = \frac{1}{V^2}[1 - \cos(\boldsymbol{k}_i - \boldsymbol{k}_j)\,.\,(\boldsymbol{r}_i - \boldsymbol{r}_j)]d\boldsymbol{r}_i d\boldsymbol{r}_j. \tag{7.212}$$

Note that this probability is zero for two electrons with the same spin when $\boldsymbol{r}_i = \boldsymbol{r}_j$ for all $\boldsymbol{k}_i$ and $\boldsymbol{k}_j$. As a consequence, electrons with the same spin as a particular electron cannot effectively screen it locally from the positive charges of the ion cores; as a result, the energy of electrons with parallel spins is lowered. The probability of finding a second spin-up electron at a distance $\boldsymbol{r} = \boldsymbol{r}_i - \boldsymbol{r}_j$ from a given spin-up electron in a volume element $d\boldsymbol{r}$ is

$$\rho^{ex}(\boldsymbol{r})d\boldsymbol{r} = n_\uparrow d\boldsymbol{r} < 1 - \cos(\boldsymbol{k}_i - \boldsymbol{k}_j)\,.\,\boldsymbol{r}\rangle, \tag{7.213}$$

where $n_\uparrow = n/2$, i.e. half the total electron concentration n, and the angular brackets denote an averaging process to be made over the Fermi surface. The quantity $\rho^{ex}(\boldsymbol{r})$, the exchange (or Hartree–Fock) electron density, can be written as

$$\begin{aligned}\rho^{ex}(\boldsymbol{r}) &= \frac{en}{2}\left\{1 - \frac{1}{(4\pi k_F^3/3)^2}\int d\boldsymbol{k}_i \int d\boldsymbol{k}_j\left(e^{i(\boldsymbol{k}_i-\boldsymbol{k}_j)\,.\,\boldsymbol{r}} + e^{-i(\boldsymbol{k}_i-\boldsymbol{k}_j)\,.\,\boldsymbol{r}}\right)/2\right\}\\ &= \frac{en}{2}\left\{1 - \frac{1}{(4\pi k_F^3/3)^2}\int e^{i\boldsymbol{k}_i\,.\,\boldsymbol{r}}d\boldsymbol{k}_i \int e^{i\boldsymbol{k}_j\,.\,\boldsymbol{r}}d\boldsymbol{k}_j\right\},\end{aligned} \tag{7.214}$$

and evaluation of the integral over the Fermi sphere gives (see e.g. Madelung (1978)):

$$\rho^{ex}(r) = \frac{en}{2}\left\{1 - \frac{9(\sin k_F r - k_F r \cos k_F r)^2}{(k_F r)^6}\right\}. \tag{7.215}$$

The total charge density experienced by a given electron is given by the sum of the exchange charge density given by eqn. (2.115) for electrons of the same spin and the (homogeneous) charge density $en/2$ for electrons of the *opposite* spin (for which the Pauli exclusion principle is not applicable, and hence exchange is not appropriate), i.e.

$$\rho^{\text{tot}}(r) = en\left\{1 - \frac{9(\sin k_{\text{F}}r - k_{\text{F}}r\cos k_{\text{F}}r)^2}{2(k_{\text{F}}r)^6}\right\}. \tag{7.216}$$

This distribution has been plotted in Fig. 2.45 and represents the *exchange hole* (see §§2.5.3.4 and 5.1.3.2). The corresponding exchange (Hartree–Fock) energy is given by eqns. (5.37) and (5.38) (see e.g. Madelung (1978) for a derivation). ★

Although exchange is a well-defined concept in the case of two electrons, as discussed above, its extension to many-electron systems can be somewhat problematic. The use of non-orthogonal orbitals leads to divergences in the solution of the Schrödinger equation, a circumstance that can be avoided by using orthogonalized tight-binding orbitals, e.g. Wannier wavefunctions (eqn. (5.93)).

Normally, the exchange interaction is very short-ranged, essentially confined to electrons in orbitals either on the same atom or on nearest-neighbour atoms where there is appreciable overlap (in the case of a non-orthogonal wavefunction representation). This is known as direct exchange. In this case, the exchange interaction can be expressed in terms of the *spins* $\boldsymbol{S}_1$ and $\boldsymbol{S}_2$ of two electrons on neighbouring atoms as the Heisenberg exchange Hamiltonian:

$$\mathscr{H}_{\text{Heis}} = -\mathfrak{J}_{12}\boldsymbol{S}_1 \,.\, \boldsymbol{S}_2, \tag{7.217a}$$

$$\equiv -\mathfrak{J}_{12}\left[\frac{(\boldsymbol{S}_1 + \boldsymbol{S}_2)^2}{2} - \frac{3}{4}\right], \tag{7.217b}$$

where in the latter equation use has been made of the fact that the eigenvalue of the operator S^2 is $S(S+1)$ and it has been assumed that $S_1 = S_2 = \frac{1}{2}$. (Note, however, that the accepted convention in ferromagnetism is to use $\boldsymbol{S}$ in general to represent the *total* angular momentum of an electron in an atom, even though this quantity is usually denoted elsewhere by $\boldsymbol{J}$ (see e.g. §7.2.4.1).) Note also that in many texts the Heisenberg exchange energy is written with a factor of two multiplying the exchange constant in eqn. (7.217a). For the Heitler–London model of the H_2 molecule, the singlet state ($S_1 = \frac{1}{2}, S_2 = -\frac{1}{2}$) has an energy $+\frac{3}{4}\mathfrak{J}_{12}$ from eqn. (7.217b), while the triplet state, e.g. $S_1 = \frac{1}{2}, S_2 = \frac{1}{2}$) has the energy $-\frac{1}{4}\mathfrak{J}_{12}$; the singlet–triplet splitting is therefore $\mathfrak{J}_{12}$ and hence, from eqn. (7.209), $\mathfrak{J}_{12} = -2(CB^2 - A)/(1 - B^4)$ in this case.

Exchange interactions can be effective over larger distances than those for which direct exchange is operative, however. One mechanism that produces such an effect is superexchange, the effective exchange coupling of magnetic cations in an insulating solid via an intervening non-magnetic anion when there is extensive orbital overlap (mixing) between cations and anions. An excited state can be produced wherein an electron from an anion orbital 'hops' to a non-orthogonal cation orbital (satisfying Hund's rules in so doing) and the other anion electron can couple ferromagnetically to another cation if the two orbitals involved are orthogonal, and antiferromagnetically otherwise. A perturbation calculation of the total energy, involving such excited states and the ground state, allows the effective exchange interaction to be obtained. An example is the case of MnF_2, in which Mn^{2+} ions effectively couple antiferromagnet-

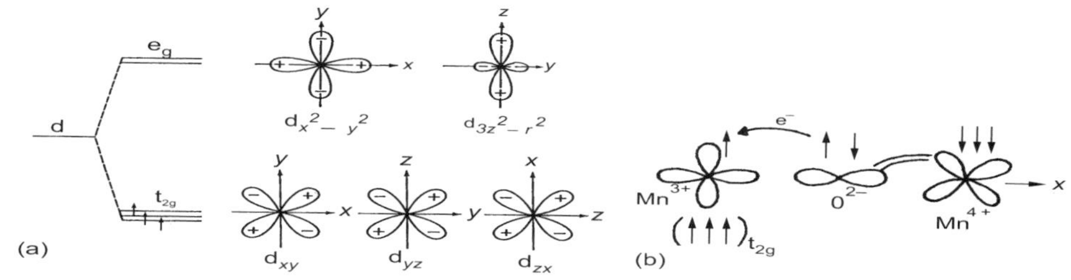

Fig. 7.44 (a) Energy levels and wavefunctions of d-electrons in a cubic crystal field, e.g. for Mn^{4+} in $CaMnO_3$. The three d-electrons in Mn^{4+} occupy the three t_{2g} states in a spin-parallel fashion because of Hund's first rule. (b) Illustration of the superexchange interaction between Mn^{4+} d-orbitals and an O^{2-} p_σ orbital. Hopping of an up-spin electron from O^{2-} can only occur to a non-orthogonal $d_{x^2-y^2}$ orbital in the e_g states on an Mn^{4+} ion having up-spin electrons in the t_{2g} states; direct ferromagnetic exchange coupling then occurs between the remaining down-spin O^{2-} p_σ electron and the three t_{2g} electrons in orthogonal d_{xy}, d_{yz}, d_{zx} orbitals on another Mn^{4+} ion.

ically through intervening collinear F^- ions; other transition-metal fluorides, such as FeF_2 and CoF_2, behave in the same way.

Oxide materials containing magnetic ions can also exhibit superexchange (see Cox 1992). An example is $CaMnO_3$ which is antiferromagnetic because of superexchange between Mn^{4+} ions via collinear oxygen ions. The large cubic crystal field experienced by Mn^{4+} ions causes the five-fold degenerate d-states to split into a three-fold degenerate state (t_{2g}) and a two-fold degenerate state (e_g); see Fig. 7.44a. The three d-electrons in an Mn^{4+} ion then occupy each of the three t_{2g} states in a spin-parallel configuration as a result of residual intraionic Coulomb interactions causing Hund's first rule to be obeyed. Superexchange involves the $p_\sigma(p_x)$ orbitals of intervening O^{2-} ions which are orthogonal to all but the $d_{x^2-y^2}$ orbitals of the e_g set on the Mn^{4+} ions. Thus, an up-spin electron in a $p_\sigma O^{2-}$ orbital can transfer to a neighbouring Mn^{4+} ion also having up-spin electrons in the t_{2g} states by hopping to a $d_{x^2-y^2}$ orbital in the e_g states, thereby satisfying Hund's first rule. The remaining down-spin electron on the O^{2-} p_z-orbital must couple *ferromagnetically* to the three electrons in t_{2g} states on an Mn^{4+} ion on the other side of the O^{2-} ion because such orbitals are mutually orthogonal. The result is an overall antiferromagnetic coupling between Mn^{4+} ions (Fig. 7.44b).

Although superexchange interactions result in an overall effective *antiferromagnetic* coupling in the case of *collinear* arrangements of magnetic and non-magnetic ions (Fig. 7.44b), *ferromagnetic* coupling can result for linkages in which the intervening non-magnetic anion subtends an angle of 90°, and where an electron on one magnetic cation interacts with a *different* orbital on the anion than does an electron on the other cation. Such structural configurations are more commonly found in layer-like halide crystals than in oxides. The origin of this ferromagnetic coupling is illustrated pictorially in Fig. 7.45. The ground-state configuration (Fig. 7.45a) mixes with the excited-state configuration shown in Fig. 7.45b. This particular configuration, with *parallel* electron spins in different orbitals on the anion, has a lower energy than that with antiparallel spins on the anion (Fig. 7.45c), which would result from an antiferromagnetic ground-state arrangement of spins on the magnetic ions, because the exchange interaction is ferromagnetic for orbitals that are orthogonal (as are the two orbitals on the same anion).

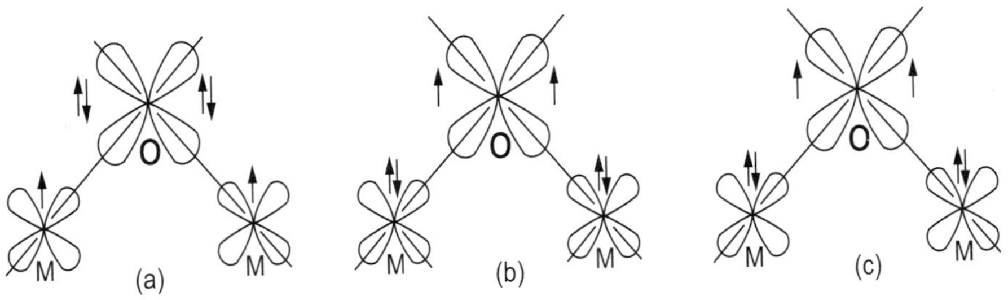

Fig. 7.45 Illustration of the origin of ferromagnetic coupling between two magnetic ions, M, via superexchange involving an intervening non-magnetic ion (e.g. the oxide anion) in a structural configuration where the angle subtended at the anion is 90°. (a) Ground-state configuration, with ferromagnetic ordering of the electron spins on the magnetic ions. (b) Excited-state configuration resulting from (a). (c) Excited-state configuration resulting from an antiferromagnetic ground state (not shown); this excited state has a higher energy than (b).

Exchange interactions can also effectively exist between magnetic atoms at distances greater than that at which direct exchange becomes inoperative by the mediation of *conduction electrons*; this, then, is the metallic equivalent of the superexchange mechanism operative in magnetic insulators, and is responsible for the magnetic behaviour of rare-earth metals where the 4f ion cores are too spatially localized to participate in direct-exchange interactions. This mechanism is termed indirect exchange, or the RKKY (Ruderman–Kittel–Kasuya–Yosida) interaction.

An electron spin $\boldsymbol{S}_i$ on a magnetic atom polarizes the conduction electrons in its vicinity. The response of the electron gas is determined by the susceptibility $\chi^{\mathrm{fe}}_{\mathrm{m,d}}(\boldsymbol{q})$ (7.132), and the spatial variation of the conduction-electron spin density, $\boldsymbol{s}(\boldsymbol{r})$ (that appears in the Heisenberg exchange Hamiltonian, eqn. (7.217a)) is given by the Fourier transform of the susceptibility. This has the same functional form as the Friedel oscillations (eqn. (5.159) and Fig. 5.61) resulting from the Lindhard screening of a test charge by an electron gas. A second, distant magnetic atom, j, will then interact with the spin $\boldsymbol{S}_i$ indirectly via the polarized electron gas, and the *sign* of this exchange interaction will depend on whether the position of atom j coincides with a peak or a trough in $\boldsymbol{s}(\boldsymbol{r})$.

7.2.5.2 *Ferromagnetism due to localized moments*

Ferromagnetism in materials where spatially localized, rather than delocalized, electrons are responsible for the magnetic moments to be coupled can be understood via the Heisenberg Hamiltonian (eqn. (7.217a)) if it is assumed that the couplings between an assembly of spins in a solid can be treated in a pairwise manner:

$$\mathscr{H}_{\mathrm{Heis}} = -\frac{1}{2}\sum_{i}\sum_{j\neq i}\mathfrak{J}_{ij}\boldsymbol{S}_i\,.\,\boldsymbol{S}_j. \tag{7.218}$$

In general, this expression is very difficult to analyse because of the product of the spin operators.

Progress can be made, however, in a mean-field approximation where the exchange interaction between one spin $\boldsymbol{S}_i$ and another spin $\boldsymbol{S}_j$ can be approximated by replacing $\boldsymbol{S}_j$ by the *average* value $\langle \boldsymbol{S}_j \rangle$. Thus, substituting

$$\boldsymbol{S}_j \equiv \langle \boldsymbol{S}_j \rangle = \langle \boldsymbol{S} \rangle, \tag{7.219}$$

into eqn. (7.218) gives

$$\begin{aligned} \mathscr{H}_{\text{Heis}} &= -\frac{1}{2}\sum_i \sum_j \mathfrak{J}_{ij} \langle \boldsymbol{S}_j \rangle \cdot \boldsymbol{S}_i \\ &\simeq -\sum_i \left(\sum_j \mathfrak{J}_{ij} \right) \boldsymbol{S}_i \,.\, \langle \boldsymbol{S} \rangle, \end{aligned} \tag{7.220}$$

where the factor of $\frac{1}{2}$ introduced to correct for double counting is omitted if j runs over nearest neighbours of i. For an assembly of spins, with a concentration n, each of which has a magnetic moment (cf. eqn. (7.161))

$$\mu_{\text{m},i} = -g_J \mu_{\text{B}} \boldsymbol{S}_i \tag{7.221}$$

(recall that, in the present context, $\boldsymbol{S}$ is being used for the *total* angular momentum rather than $\boldsymbol{J}$), the magnetization is given by

$$\boldsymbol{M} = -n g_J \mu_{\text{B}} \langle \boldsymbol{S} \rangle, \tag{7.222}$$

where g_J is the Landé g-factor (eqn. (7.163)) and μ_B is the Bohr magneton (eqn. (7.130)). Defining the dimensionless Weiss constant $\lambda_w (\gg 1)$ as

$$\lambda_{\text{w}} = \frac{\sum_{j \neq i} \mathfrak{J}_{ij}}{n \mu_0 g_J^2 \mu_{\text{B}}^2}, \tag{7.223}$$

means that eqn. (7.220) for the Heisenberg Hamiltonian can be rewritten as:

$$\mathscr{H}_{\text{Heis}} \simeq -\sum_i \lambda_{\text{w}} \mu_0 \mu_{\text{m},i} \,.\, \boldsymbol{M}. \tag{7.224}$$

This mean-field approximation is the origin of the fictitious internal or Weiss molecular field $\boldsymbol{B}_w = \lambda_w \mu_0 \boldsymbol{M}$ originally proposed by Weiss to explain ferromagnetism, in which the effective magnetic field acting on a magnetic moment was taken to have the flux density

$$\boldsymbol{B}_{\text{eff}} = \boldsymbol{B}_{\text{loc}} + \lambda_{\text{w}} \mu_0 \boldsymbol{M}. \tag{7.225}$$

Since the Hamiltonian for a magnetic moment μ_m experiencing a magnetic field of flux density $\boldsymbol{B}$ is $\mathscr{H} = -\mu_m \,.\, \boldsymbol{B}$ (cf. eqn. (7.165)), use of eqn. (7.225) for the flux density gives rise to the molecular-field term of the same form as in eqn. (7.224).

The mean-field approximation, leading to the effective field given by eqn. (7.225), greatly simplifies the solution of the problem. The eigenvalues are given by (cf. eqn. (7.165))

$$\mathscr{E} = \pm \frac{1}{2} g_J \mu_{\text{B}} \boldsymbol{B}_{\text{eff}} \tag{7.226}$$

for spins taken to be $S = J = \frac{1}{2}$ (i.e. $L = 0$) and with $g_J = g_e \simeq 2$. The net magnetization is proportional to the difference in the spin densities parallel $(n_\uparrow)$ and antiparallel $(n_\downarrow)$ to the effective field, i.e.

$$\begin{aligned} M &= \frac{1}{2} g_e \mu_B (n_\uparrow - n_\downarrow) \\ &\simeq n\mu_B \tanh(g_e \mu_B B_{eff}/2k_B T), \end{aligned} \tag{7.227}$$

where, in thermal equilibrium,

$$n_\downarrow / n^\uparrow = \exp(-g_e \mu_B B_{eff}/k_B T). \tag{7.228}$$

Equation (7.227) can also be obtained from the Brillouin function (eqn. (7.172)) for $J = \frac{1}{2}$.

Equation (7.227) has non-zero solutions for the magnetization even in the absence of an external magnetic field, i.e. it predicts a spontaneous magnetization at temperatures lower than a critical temperature, the Curie–Weiss temperature, θ_{CW}, given by

$$\theta_{CW} = \left(\sum_{j \neq i} \mathfrak{J}_{ij} \right) / 4k_B. \tag{7.229}$$

Unfortunately, except near $T = 0$ K, eqn. (7.227) for the magnetization, with the effective flux density given by eqn. (7.225), cannot be solved analytically, although a graphical solution is possible. These expressions can be rewritten respectively in dimensionless form as

$$y = \tanh x \tag{7.230a}$$

and

$$x = \frac{\theta_{CW}}{T} y \tag{7.230b}$$

using the dimensionless quantities $y = M/n\mu_B$ and $x = \mu_B B_{eff}/k_B T$. It has been assumed in obtaining eqn. (7.230b) that the external magnetic flux density, $\boldsymbol{B}_{ext}$, is zero and that the remaining contribution to $\boldsymbol{B}_{loc}$ of order $\mu_0 \boldsymbol{M}$ (eqns. (7.102), (7.106)) is negligible compared with the quantity $\lambda_w \mu_0 \boldsymbol{M}$ (λ is typically of order 10^3), so that $\boldsymbol{B}_{eff} \simeq \lambda_w \mu_0 \boldsymbol{M}$. A simultaneous solution of eqns. (7.230) can be obtained by plotting both functions on the same graph and looking for intersections between the two curves. As seen from Fig. 7.46, a non-zero spontaneous magnetization, characteristic of ferromagnetic ordering, appears in zero applied magnetic field only for $T < \theta_{CW}$ (see Problem 7.25). Near $T = 0$ K, x is large and $y = \tanh x \simeq 1 - 2\exp(-2x) = 1 - 2\exp(-2y\theta_{CW}/T)$. Thus:

$$M \simeq n\mu_B [1 - 2\exp(-2\theta_{CW}/T)], \tag{7.231}$$

where the approximation $y = 1$ has been used in the exponent. However, this exponential temperature dependence is not observed experimentally (see §7.2.5.4). For temperatures just below θ_{CW}, x is small and $y = \tanh x \simeq x - x^3/3$. Hence, solving for $y = Tx/\theta_{CW}$ gives

$$M \simeq \sqrt{3} n\mu_B \frac{T}{\theta_{CW}} \left(1 - \frac{T}{\theta_{CW}}\right)^{1/2}. \tag{7.232}$$

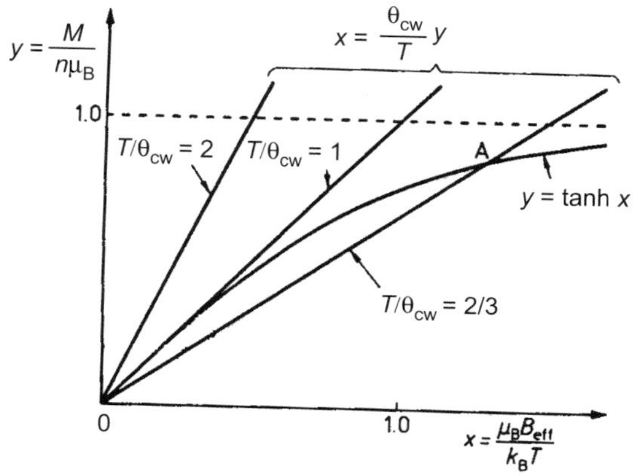

Fig. 7.46 Graphical solution of eqns. (7.230) for the existence of spontaneous ferromagnetism in the mean-field (Weiss) model.

Note that this parabolic temperature dependence is the same as that predicted by the Landau free-energy model for the second-order transition in *ferroelectrics* (eqn. (7.68)). The temperature dependence of the spontaneous magnetization of a ferromagnet predicted by mean-field theory is shown in Fig. 7.47, compared with experimental data for Ni.

As seen from Fig. 7.46, there is no spontaneous magnetization above the Curie–Weiss temperature, θ_{CW}; thermal fluctuations destroy the cooperative ferromagnetic ordering, and the material reverts to the paramagnetic state (§7.2.4.2). In the high-temperature limit, $\mu_B B_{eff}/k_B T \ll 1$, tanh $x \simeq x$, and the expression for the magnetization (eqn. (7.227)) becomes

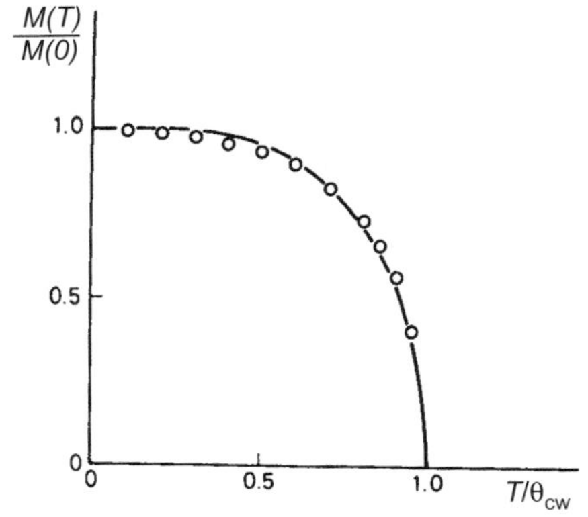

Fig. 7.47 Temperature dependence of the spontaneous magnetization of a ferromagnet in the mean-field approximation (solid line) compared with experimental data for Ni. (Hook and Hall (1991). Reproduced by permission of John Wiley & Sons Inc.)

$$M \simeq \frac{n\mu_B^2}{k_B T} B_{eff} = \frac{n\mu_B^2}{k_B T}(B_{loc} + \lambda_w \mu_0 M). \tag{7.233}$$

Neglecting, as before, the contribution of order $\mu_0 M$ to the local magnetic field (cf. eqns. (7.102), (7.106)) relative to the Weiss term $\lambda_w \mu_0 M$, i.e. $B_{loc} \simeq \mu_0 H_{mac}$, the magnetic susceptibility (eqn. (7.83)) becomes

$$\chi_m = \frac{M}{H_{mac}} \simeq \frac{n\mu_0\mu_B^2/k_B}{T - \theta_{CW}}, \tag{7.234}$$

where θ_{CW} is given by eqn. (7.229). This, the Curie–Weiss law, is found also for ferroelectrics (§7.1.5.3).

However, in the vicinity of the critical temperature, θ_{CW}, mean-field theory breaks down because of the presence of large thermal fluctuations of the magnetic moments around the mean value. As a result, the experimental critical behaviour of the magnetic susceptibility

$$\chi_m \propto (T - \theta_{CW})^{-\gamma} \tag{7.235}$$

and of the magnetization

$$M \propto (T - \theta_{CW})^{\beta} \tag{7.236}$$

in the ferromagnetic régime for $T \le \theta_{CW}$ do not have the critical exponents $\gamma = 1$ (eqn. (7.234)) and $\beta = 0.5$ (eqn. (7.232)) predicted by mean-field theory. Instead, values of $\gamma \simeq 1.35$ and $\beta \simeq 0.35$ are generally found. Examples of simple ferromagnetic insulators are EuO and EuS.

7.2.5.3 *Ferromagnetism due to itinerant electrons*

The treatment of ferromagnetism in terms of exchange interactions between localized magnetic moments given in the previous section is not suitable for ferromagnetic metals, such as the 3d transition metals Fe, Co and Ni, where the electrons responsible for the magnetism are itinerant and form a band, albeit narrow, of delocalized states (§5.4.3). Ferromagnetism in this case can be understood in terms of the following picture.

The exchange energy (eqn. (7.208)) can be assumed simply to add to the energy of Bloch states (§5.2.1), thereby causing a displacement in energy of electron states having a given spin direction from those with the opposite spin direction: the exchange interaction acts as an *internal* magnetic field (cf. eqn. (7.225)), causing a displacement of the densities of states of electrons (in this case, d-electrons) with different spin directions, in an analogous fashion to the displacement caused by an *external* field in the case of paramagnetism of conduction electrons (see Fig. 7.43). As a result, the numbers of spin-up and spin-down 3d-electrons are different, and hence a spontaneous magnetic moment appears in the absence of an applied magnetic field.

This exchange splitting is particularly pronounced for 3d-electrons, for which the density of states is high (due to narrow bands). The transition metals Fe, Co and Ni, with free-atom electronic configurations $3d^6 4s^2$, $3d^7 4s^2$ and $3d^8 4s^2$ respectively, have the Fermi level lying in the d-band (Fig. 5.55), in contrast to the case of non-magnetic Cu ($3d^{10} 4s^1$) where $\mathscr{E}_F$ lies midway in the s-band above the, now-filled, d-band (Fig. 5.56a). For the case of Ni, for example (see Fig. 7.48), of the ten conduction electrons per atom, five completely fill the lower $3d_\uparrow$ sub-band, but only 4.46 electrons occupy the upper $3d_\downarrow$

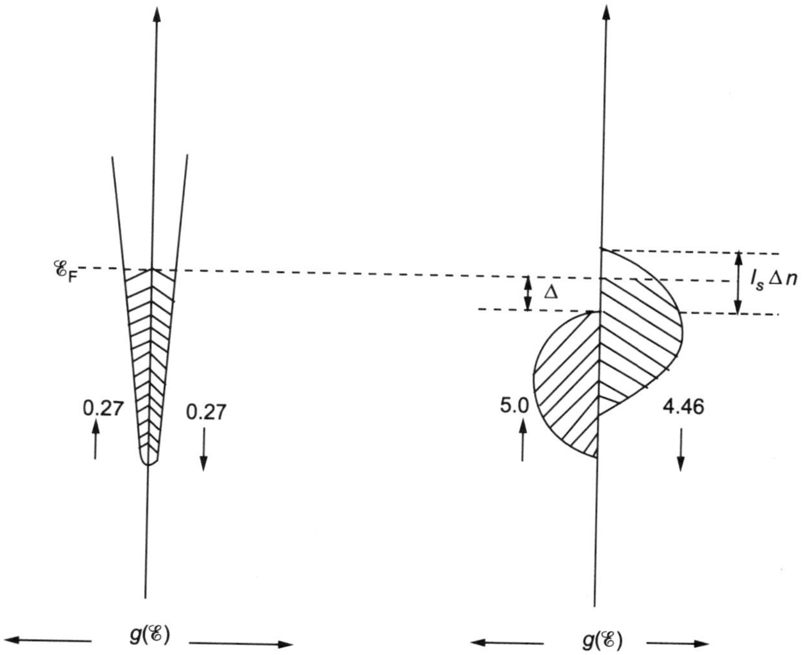

Fig. 7.48 Schematic illustration of the Stoner model for ferromagnetism in metals due to itinerant 3d electrons, e.g. in Ni. The exchange interaction causes a displacement in energy of the 3d density of states according to the electron-spin direction (↑ or ↓), but a negligible displacement of the 4s states spin states. The Stoner gap, Δ, is the energy separation between the Fermi level and the upper edge of the majority-spin band. Of the 10 electrons per atom, 5 completely fill the lower $3d_\uparrow$ band and 4.46 occupy the upper $3d_\downarrow$ band; the remaining 0.54 of an electron per atom is in the 4s band. Thus, there is a net magnetic moment of $0.54\mu_B$ per atom due to the imbalance in the 3d spin populations.

band; the remaining 0.54 of an electron is distributed (approximately equally) between the $4s_\downarrow$ and $4s_\uparrow$ bands (for which the exchange splitting is negligible). Thus, there is a net magnetic moment of 0.54 μ_B per atom (pointing in the [111] direction), resulting from the difference in the populations of spin-up and spin-down 3d-electrons (or equivalently the presence of the 0.54 of a hole in the $3d_\downarrow$ band). This value is close to the absolute-zero value of atomic magnetic moment for Ni of 0.6 μ_B, the small difference being due to orbital angular momentum contributions. For the case of Fe, the eight conduction electrons per atom are partitioned such that 4.8 occupy the $3d_\uparrow$ band and 2.6 occupy the $3d_\downarrow$ band; the remaining 0.6 of an electron is partitioned approximately equally between the $4s_\uparrow$ and $4s_\downarrow$ sub-bands. Hence, the magnetic moment is 2.2 μ_B per atom in this case and points along the [100] direction. Note that this, the Stoner model, naturally permits *non-integral* values of magnetic moment per atom, a circumstance not understandable from the localized-electron picture of ferromagnetism (§7.2.5.2).

The criterion for the existence of itinerant ferromagnetism in the Stoner model can be evaluated as follows. Assume that the effect of the exchange interaction is simply to shift the energy of the one-electron Bloch states $\mathscr{E}_0(\boldsymbol{k})$ for spin-up (↑) or spin-down (↓) electrons:

$$\mathscr{E}_{\uparrow,\downarrow}(\boldsymbol{k}) = \mathscr{E}_0(\boldsymbol{k}) \mp I_S n_{\uparrow,\downarrow}, \tag{7.237}$$

where I_S is the Stoner parameter reflecting the strength of the exchange interaction (assumed to be $\boldsymbol{k}$-independent) and $n_\uparrow (n_\downarrow)$ is the number of spin-up (spin-down) electrons per atom. If the excess population of spin-up over spin-down electrons is

$$\Delta n = n_\uparrow - n_\downarrow, \tag{7.238}$$

a symmetric pair of equations for the energies of the two spin sub-bands can be obtained from eqn. (7.237) by subtracting the quantity $I_S(n_\uparrow + n_\downarrow)/2$ from the one-electron energies $\mathscr{E}(\boldsymbol{k})$ to give

$$\mathscr{E}_\uparrow(\boldsymbol{k}) = \bar{\mathscr{E}}(\boldsymbol{k}) - I_S \Delta n/2, \tag{7.239a}$$

$$\mathscr{E}_\downarrow(\boldsymbol{k}) = \bar{\mathscr{E}}(\boldsymbol{k}) + I_S \Delta n/2, \tag{7.239b}$$

with

$$\bar{\mathscr{E}}(\boldsymbol{k}) = \mathscr{E}(\boldsymbol{k}) - I_S(n_\uparrow + n_\downarrow)/2. \tag{7.240}$$

The value of this splitting (illustrated schematically for the 3d-band in Ni in Fig. 7.48) depends on Δn, which in turn is determined by Fermi statistics:

$$\begin{aligned}\Delta n &= \sum_k (f_\uparrow(\boldsymbol{k}) - f_\downarrow(\boldsymbol{k})) \\ &= \sum_k \{\exp[\bar{\mathscr{E}}(\boldsymbol{k}) - I_S\Delta n/2 - \mu] + 1\}^{-1} - \{\exp[\bar{\mathscr{E}}(\boldsymbol{k}) + I_S\Delta n/2 - \mu] + 1\}^{-1}.\end{aligned} \tag{7.241}$$

A Taylor series expansion of the function, i.e.

$$f(x - \Delta x/2) - f(x + \Delta x/2) \simeq -\frac{\partial f}{\partial x}\Delta x - \frac{2}{3!}\frac{\partial^3 f}{\partial x^3}\left(\frac{\Delta x}{2}\right)^3, \tag{7.242}$$

gives

$$\Delta n = -\sum_k \frac{\partial f(\boldsymbol{k})}{\partial \bar{\mathscr{E}}(\boldsymbol{k})} I_S \Delta n - \frac{1}{24}\sum_k \frac{\partial^3 f(\boldsymbol{k})}{\partial \bar{\mathscr{E}}(\boldsymbol{k})^3}(I_S\Delta n)^3. \tag{7.243}$$

The first derivative of the Fermi function is negative but the third derivative is positive, so that the condition for the existence of a spontaneous magnetic moment, $\Delta n > 0$, is

$$I_S \sum_k \frac{\partial f(\boldsymbol{k})}{\partial \bar{\mathscr{E}}(\boldsymbol{k})} > 1. \tag{7.244}$$

The first derivative of the Fermi function is greatest at $T = 0\text{K}$ when $f(\bar{\mathscr{E}})$ is a step function and $\partial f/\partial \bar{\mathscr{E}} = -\delta(\bar{\mathscr{E}} - \mathscr{E}_F)$ (eqn. (6.48)). Hence

$$\begin{aligned}-\sum_k \frac{\partial f}{\partial \bar{\mathscr{E}}} &= \frac{V}{(2\pi)^3}\int\left(-\frac{\partial f}{\partial \xi}\right)\mathrm{d}\boldsymbol{k} = \frac{V}{(2\pi)^3}\int \delta(\bar{\mathscr{E}} - \mathscr{E}_F)\mathrm{d}\boldsymbol{k} \\ &= \bar{g}(\mathscr{E}_F),\end{aligned} \tag{7.245}$$

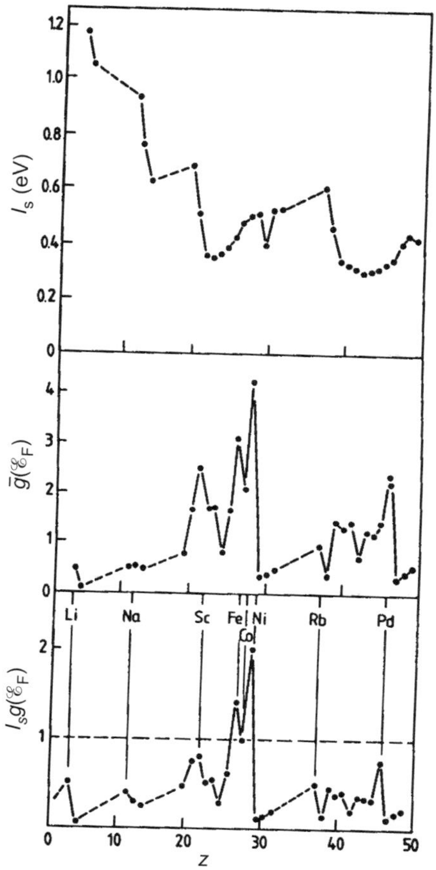

Fig. 7.49 Calculated parameters for the Stoner model for ferromagnetism for elements in the first half of the periodic table: (a) the Stoner exchange parameter I_S; (b) the density of states per atom at the Fermi level, $\bar{g}(\mathscr{E}_F)$; (c) the Stoner criterion, $I_S\bar{g}(\mathscr{E}_F) > 1$. ((a), (b) after Janak (1977) Reprinted with permission from *Phys. Rev.* B**16**, 255. © 1977. The American Physical Society; (c) after Ibach and Lüth (1995), *Solid State Physics*, p. 168, Fig. 8.5(c), © Springer-Verlag GmbH & Co. KG)

where $\bar{g}(\mathscr{E}_F)$ is the density of states (per energy per atom) at the Fermi level for electrons of a *particular* spin type (i.e. $\bar{g}(\mathscr{E})$ is half the normal density of states, eqn. (5.132)). Thus, from eqn. (7.244), the Stoner criterion for ferromagnetism in a band model is simply

$$I_S\bar{g}(\mathscr{E}_F) > 1. \tag{7.246}$$

Figure 7.49 shows calculated values of the Stoner parameter I_S, the density of states at the Fermi level $\bar{g}(\mathscr{E}_F)$, and their product, for the elements composing the first half of the periodic table. It can be seen that only Fe, Co and Ni are predicted (correctly) to be ferromagnetic, principally because of the high values of $\bar{g}(\mathscr{E}_F)$ associated with the

3d-states characteristic of these metals. However, Ca, Sc and Pd come close to satisfying the Stoner criterion and, although these elements are not ferromagnetic, the exchange interactions lead to a strong enhancement of the Pauli paramagnetic susceptibility given by (see Problem 7.26):

$$\chi_{m,p} = \chi^0_{m,p}/(1 - I_S\bar{g}(\mathscr{E}_F)), \tag{7.247}$$

where $\chi^0_{m,p}$ is given, for example, by eqn. (7.190).

Although vanadium ($Z = 23$) is not ferromagnetic, nor is predicted to be by the simple Stoner criterion as a result of a low value of density of states (Fig. 7.49), alloying it with gold increases the V–V distance from 2.49 Å in b.c.c. V to 3.78 Å in Au_4V and decreases the width of the d-band, and as a consequence $\bar{g}(\mathscr{E}_F)$ increases sufficiently to make the alloy ferromagnetic with a magnetic moment of $\simeq 1\mu_B$ per vanadium atom. The effect of alloying two different transition metals, with part-filled 3d-bands, is shown in Fig. 7.50: the effective magnetic moment (a_m Bohr magnetons) per atom depends on the average number of conduction electrons per atom, which can be understood in terms of the Stoner model if the 3d-bands are assumed to be unaffected on alloying. Ferromagnetism is generally favoured for almost full (or empty) d-bands: half-filled bands tend to favour *antiferromagnetism* (§7.2.5.6).

Note also that, although the discussion so far has implicitly concerned *crystalline* materials, the existence of spatial periodicity is not a prerequisite for the occurrence of ferromagnetism. Since the exchange interaction between spins is essentially a short-range effect, the existence of ferromagnetism in amorphous metals is not precluded. Glassy metal alloys, such as $Fe_{80}B_{20}$, are indeed ferromagnetic.

The temperature dependence of the magnetization, $M = \mu_B \Delta n/V$, predicted by the Stoner model can be analysed simply by treating the d-electron density of states as a delta function lying at the upper edge of the actual distribution:

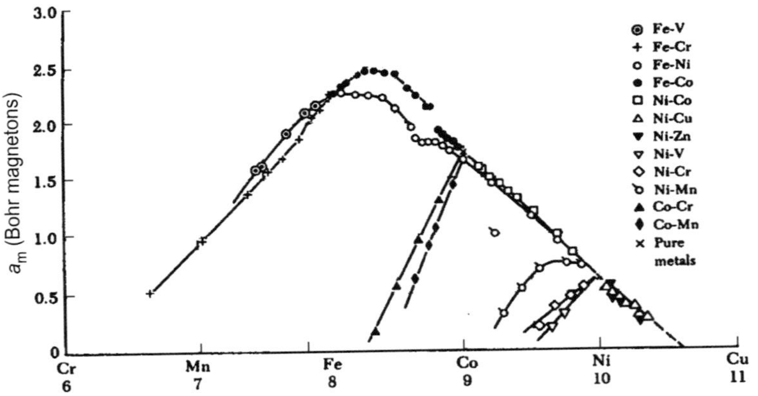

Fig. 7.50 Effective magnetic moment per atom, given as a_m Bohr magnetons, for binary transition-metal alloys as a function of the average number of conduction electrons per atom. (Crangle and Hallam (1963). Reproduced by permission of The Royal Society)

$$\bar{g}(\mathscr{E}) \simeq \frac{\mu_B^{\text{eff}}}{\mu_B}[\delta(\mathscr{E} - \mathscr{E}_F - I_S\Delta n/2) + \delta(\mathscr{E} - \mathscr{E}_F + I_S\Delta n/2)]. \tag{7.248}$$

In this case, eqn. (7.241) for the difference in spin populations, in the absence of an external field, becomes

$$\Delta n = \frac{\mu_B^{\text{eff}}}{\mu_B}\{[\exp(-I_S\Delta n/2k_BT)] + 1]^{-1} - [\exp(+I_S\Delta n/2k_BT) + 1]^{-1}]. \tag{7.249}$$

Defining the dimensionless quantities $\bar{y} = \Delta n\mu_B/\mu_B^{\text{eff}}$ and $\bar{x} = \bar{y}\theta_{CW}/T$, where the Curie–Weiss temperature, θ_{CW}, in this case is given by (cf. eqn. (7.229)):

$$\theta_{CW} = \mu_B^{\text{eff}} I_S/4\mu_B k_B, \tag{7.250}$$

the spin-population difference, and hence the magnetization, from eqn. (7.249) obeys the relationship $\bar{y} = \tanh\bar{x}$, exactly as for the Heisenberg model (eqn. (7.230a)). Thus, all the discussion given in the previous section for the behaviour of $\boldsymbol{M}$ is valid also, in this approximation, for the Stoner model. The temperature dependence of the spontaneous magnetization of Ni is shown in Fig. 7.47 compared with the experimental prediction. The temperature dependence of the magnetic susceptibility of, e.g., Fe, above θ_{CW} does not exhibit a critical exponent of $\gamma = 1$ as predicted theoretically (cf. eqn. (7.234)) but instead a higher value of 1.33 (Fig. 7.51).

Of course, the Stoner model is highly simplified, and as a result does not give very good agreement with experimental values of magnetic moments, e.g. of the 3d

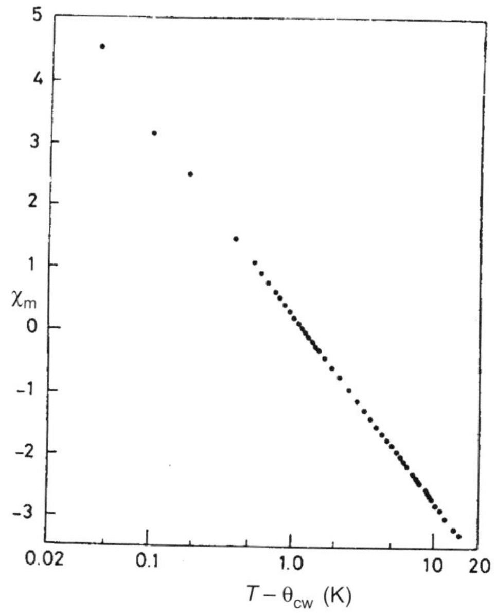

Fig. 7.51 Experimental data for the temperature dependence of the magnetic susceptibility of Fe (with 0.16 at. % W impurity), showing a critical exponent $\gamma = 1.33$. (Noakes *et al.* (1966))

transition-metal ferromagnets. For materials with narrow bands, intraionic (Hubbard) correlation effects (§5.6.3) are also important (see e.g. White (1983) and Mattis (1988)).

*7.2.5.4 *Spin waves: magnons*

The mean-field theory of ferromagnetism predicts an exponential temperature dependence of the magnetization at very low temperatures (eqn. (7.231)) in disagreement with the $T^{3/2}$ power-law behaviour observed experimentally. This discrepancy arises because mean-field theory does not properly take account of low-energy spin excitations. For example, in the simple band model of ferromagnetism described in §7.2.5.3, a spin flip can only be accomplished by means of an interband transition of an electron from one exchange-shifted band to another (Stoner excitation); the minimum energy required is the energy separation, the Stoner gap, between the Fermi level and the upper edge of the majority-spin band (Fig. 7.48). It is not surprising, therefore, that the temperature dependence of $\boldsymbol{M}$ should be thermally activated in this model, the exponential behaviour reflecting this energy gap. The above process describes a spin flip in the one-particle (band) approximation: one spin is inverted independently of all the others, which remain in a ferromagnetically ordered state (Fig. 7.52b). However, a lower-energy spin excitation can be achieved by means of a *collective* excitation of all the spins accompanying the inversion of a particular spin. A chain of spins, for example, can adopt a helical arrangement of orientations, with neighbouring spins being canted by a small angle (Fig. 7.52c); in this way, the cost in exchange energy is minimized. However, the energy of a spin configuration as in Fig. 7.52c is very large in the mean-field model since the *mean* magnetization, and thus the internal field, vanishes in such a case.

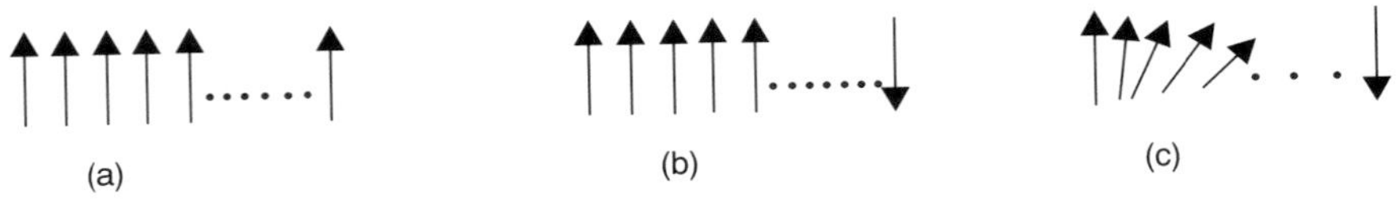

Fig. 7.52 Spin configurations in a ferromagnetic 1D chain: (a) the ground state; (b) an excited state, with one spin inverted (Stoner excitation); (c) a low-energy collective spin excitation.

The collective dynamical behaviour of spins can be analysed analytically for the case of a 1D ferromagnet, i.e. a chain of spins as in Fig. 7.52, in an analogous fashion to the treatment given for atomic vibrations of a 1D chain (§4.2.2). If it is assumed that only exchange interactions between nearest neighbours are important, with constant exchange parameter $\mathfrak{J}$, the exchange energy of the jth spin in an insulating ferromagnetic chain is, from eqn. (7.217a)

$$\mathscr{E}_j = -\mathfrak{J}\boldsymbol{S}_j \cdot (\boldsymbol{S}_{j-1} + \boldsymbol{S}_{j+1}). \tag{7.251}$$

This energy can be re-expressed as $\mathscr{E}_j = -\boldsymbol{\mu}_{\mathrm{m}}^j \cdot \boldsymbol{B}_j$, where the magnetic moment of spin j is $\boldsymbol{\mu}_{\mathrm{m}}^j = -g_{\mathrm{J}}\mu_{\mathrm{B}}\boldsymbol{S}_j$ (eqn. (7.161)) and the effective magnetic flux density $\boldsymbol{B}_j$ acting on this spin due to the exchange interaction with its neighbouring spins is given by

$$\boldsymbol{B}_j = -\frac{\mathfrak{J}}{g_J\mu_B}(\boldsymbol{S}_{j-1} - \boldsymbol{S}_{j+1}). \tag{7.252}$$

The resulting torque on the spin, $\boldsymbol{T}_j = \boldsymbol{\mu}_m^j \times \boldsymbol{B}_j$ (eqn. (7.80)), causes a change in angular momentum of

$$\hbar\frac{d\boldsymbol{S}_j}{dt} = \mathfrak{J}\boldsymbol{S}_j \times (\boldsymbol{S}_{j-1} + \boldsymbol{S}_{j+1}). \tag{7.253}$$

This non-linear equation can be linearized by assuming that displacements $\boldsymbol{s}_j$ of the jth spin only take place in the x–y plane if the overall magnetization is oriented in the z-direction (and hence the spin in the ground-state configuration is aligned in the $-z$-direction—see eqn. (7.161)), i.e.

$$\boldsymbol{S}_j = -S\hat{\boldsymbol{z}} + \boldsymbol{s}_j. \tag{7.254}$$

Note that $\hat{\boldsymbol{z}}$ need have no special orientational relationship with the chain direction. Substitution of eqn. (7.254) into eqn. (7.253), and retaining only those terms that are first-order in $\boldsymbol{s}_j$, gives

$$\hbar\frac{d\boldsymbol{s}_j}{dt} = -\mathfrak{J}S\hat{\boldsymbol{z}} \times (\boldsymbol{s}_{j-1} - 2\boldsymbol{s}_j + \boldsymbol{s}_{j+1}). \tag{7.255}$$

The x-and y-components of eqn. (7.255) are

$$\left(\hbar\frac{ds_j}{dt}\right)_{x,y} = \pm\mathfrak{J}S(s_{j-1} - 2s_j + s_{j+1})_{y,x}, \tag{7.256}$$

where the positive sign corresponds to the y-component of the right-hand side of the equation. In terms of the complex spin variable

$$s_j^\dagger = s_{jx} - \mathrm{i}s_{jy}, \tag{7.257}$$

eqns. (7.256) can be written in compact form as

$$\mathrm{i}\hbar\frac{d\sigma_j^\dagger}{dt} = -\mathfrak{J}S(s_{j-1}^\dagger - 2s_j^\dagger + s_{j+1}^\dagger). \tag{7.258}$$

Note the formal similarity between this equation and eqn. (4.33) describing atomic lattice vibrations if use is made of the wave-like solution

$$s_j^\dagger = \sum_k C_k \exp[\mathrm{i}(kja - \omega_k t)], \tag{7.259}$$

where a is the lattice spacing of the chain. Substitution of eqn. (7.259) for one particular k-value into eqn. (7.258) yields, after cancellation of the common terms C_k and $\exp[i(kja - \omega_k t)]$:

$$\begin{aligned}\hbar\omega_k &= -\mathfrak{J}S(\mathrm{e}^{-\mathrm{i}ka} - 2 + \mathrm{e}^{\mathrm{i}ka}) \\ &= 2\mathfrak{J}S[1 - \cos ka].\end{aligned} \tag{7.260}$$

From eqns. (7.257) and (7.259), the components of the spin displacement for a particular k-vector behave as $s_{jx} \propto \cos(kja - \omega_k t)$ and $s_{jy} \propto -\sin(kja - \omega_k t)$, and thus the spins (regarded as classical entities) precess about the $-\hat{\boldsymbol{z}}$-direction (see Fig. 7.53);

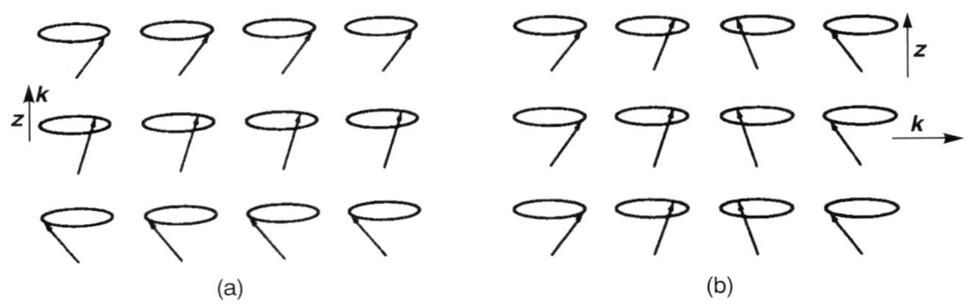

Fig. 7.53 Schematic illustration of a spin wave (or magnon) in an insulating crystalline ferromagnet. In a classical sense, the spins localized at each lattice site precess about the net magnetization direction $\hat{z}$, with the relative phases of neighbouring spins being determined by the wavevector $\boldsymbol{k}$ in the direction of the spin wave: (a) spin wave propagating parallel to $\hat{z}$; (b) spin wave propagating perpendicular to $\hat{z}$.

such a collective excitation is termed a spin wave. Its dispersion relation is given by eqn. (7.260); this is a periodic function of wavevector k, with period $2\pi/a$, as for lattice vibrations (§4.2.2), but for $k \simeq 0$, the spin-wave dispersion relation (eqn. (7.260)) depends *quadratically* on wavevector:

$$\hbar\omega_k \simeq \mathfrak{J} S a^2 k^2, \tag{7.261}$$

in contrast to the linear dispersion relation exhibited by acoustic phonons (eqns. (4.39), (4.50a)). Note that eqn. (7.261) implies that the energy to displace a spin in a chain tends to zero in the infinite-wavelength limit, in marked contrast to one-electron Stoner excitations involving a single spin flip (Fig. 7.52b). In fact, the spin-wave frequency is small but finite at $\boldsymbol{k} = 0$ because of *magnetocrystalline anisotropy* (see §7.2.5.5): a finite energy is required to rotate a spatially uniform magnetization, corresponding to an infinite-wavelength spin wave, if there exists an easy direction for the magnetization. Spin-wave dispersion curves can be measured by magnetic neutron scattering (see Fig. 7.54).

A proper quantum-mechanical treatment of the spin-wave problem leads to the same dispersion relation (eqn. (7.260)) as obtained from the above semi-classical analysis, but with the difference that the energies of the spin waves are quantized and obey the harmonic-oscillator relation, as for other bosons such as photons and phonons (§4.2.5):

$$\mathscr{E}_k = \left(n + \frac{1}{2}\right)\hbar\omega_k; \tag{7.262}$$

such quantized spin-wave excitations are called magnons. Excitation of one magnon corresponds to the inversion of one spin, i.e. a reduction in the magnetization by an amount of $g_J\mu_B$ per unit volume. The magnetization at temperature T is therefore given by

$$M(T) = M_s(0) - g_J\mu_B n_{mag}, \tag{7.263}$$

where $M_S(0) = n_0 g_J \mu_B S$ is the saturation magnetization at absolute zero. The density of thermally excited magnons is

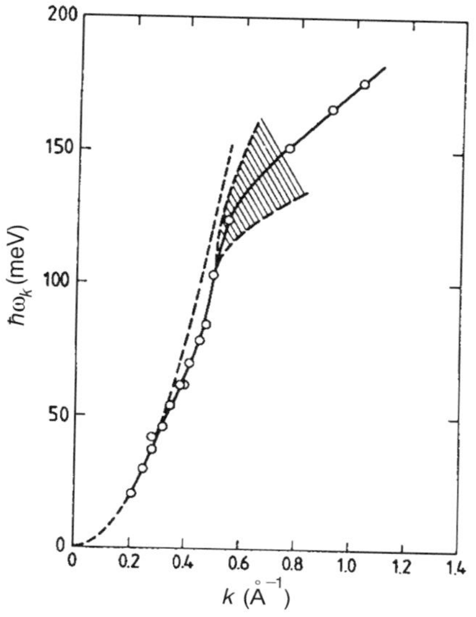

Fig. 7.54 Experimental dispersion curve for the spin waves (magnons) propagating in Ni in the [111] direction at 295 K, measured by inelastic magnetic neutron scattering (Reprinted with permission from Mook and Paul (1985), *Phys. Rev. Lett.* **54**, 227. © 1985. The American Physical Society). The dashed line shows the theoretical quadratic limiting behaviour of the dispersion relation. At high energies, Stoner excitations are also produced, which cause a reduction in the spin-wave lifetimes and a concomitant lifetime broadening of the spectra (hatched region)

$$n_{\text{mag}} = \frac{1}{V}\int_0^{\infty} n(\omega)g(\omega)\,\mathrm{d}\omega, \tag{7.264}$$

where the average number of magnons in thermal equilibrium, $n(\omega, T)$, is given by the Planck distribution law (eqn. (4.68)), $n(\omega, T) = [\exp(\hbar\omega_k/k_B T) - 1]^{-1}$. The density of magnon modes, $g(\omega)$, is given by

$$\begin{aligned} g(\omega)\,\mathrm{d}\omega &= g(k)\mathrm{d}k \\ &= \frac{V}{4\pi^2}\left(\frac{\hbar}{\mathfrak{J}Sa^2}\right)^{3/2}\omega^{1/2}\,\mathrm{d}\omega, \end{aligned} \tag{7.265}$$

where use has been made of eqn. (4.21) for $g(k)$, and of the approximate form of the magnon dispersion relation (eqn. (7.261)) appropriate at very low temperatures where only low-energy, small-k magnons are excited. Thus, the magnetization in the vicinity of absolute zero can be written as

$$M(T) = M_s(0)\left\{1 - \frac{1}{4\pi^2 n_0 S}\left(\frac{k_B T}{\mathfrak{J}Sa^2}\right)^{3/2}\int_0^{\infty}\frac{x^{1/2}\mathrm{d}x}{(\mathrm{e}^x - 1)}\right\}, \tag{7.266}$$

where $x = \hbar\omega_k/k_B T$ This is the limiting Bloch $T^{3/2}$ law for the low-temperature behaviour of the magnetization of ferromagnets that is observed experimentally (Fig. 7.55), in marked contrast to the exponential temperature dependence predicted by mean-field

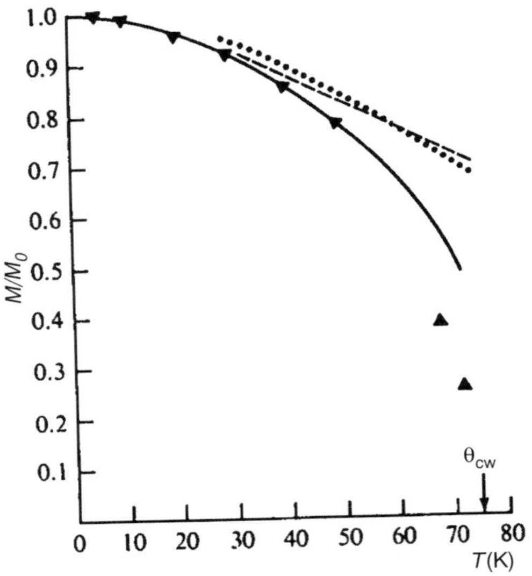

Fig. 7.55 Temperature dependence of the magnetization of EuO (data points after Low (1963)). The solid curve includes dynamical interactions between pairs of magnons; the dashed curve corresponds to non-interacting spin waves; and the dotted curve corresponds to a series expansion with $T^{3/2}$ and $T^{5/2}$ terms. (After White (1983), *Quantum Theory of Magnetism*, p. 198, Fig. 6.7, © Springer-Verlag GmbH & Co. KG)

theory (eqn. (7.231)). If higher-order terms proportional to k^4, k^6 etc. are retained in the Taylor expansion of the dispersion relation (eqn. (7.260)), corresponding terms in the magnetization varying as $T^{5/2}$, $T^{7/2}$ etc. are included. See Problem 7.27 for an analysis of the heat capacity associated with magnon excitations.

7.2.5.5 *Ferromagnetic domains*

Although ferromagnetic materials exhibit a spontaneous magnetization on the *microscopic* length scale below the Curie–Weiss temperature θ_{CW}, this does not mean that a macroscopic sample, in the absence of an external magnetic field, also exhibits a large net magnetic moment. In fact, ferromagnetically ordered samples generally exhibit very small moments in zero field. This behaviour is due to the formation of ferromagnetic domains, i.e. small regions of material within each of which the individual spins are ferromagnetically ordered; the different domains have the magnetization pointing in different directions, so that the net magnetization is zero in the absence of an external field. Similar behaviour is also exhibited by *ferroelectric* materials (§7.1.5.3).

The reason that domain formation is favoured is that the magnetic-energy density, $U_M = B^2/2\mu_0$, associated with the magnetic-flux density B exerted outside a sample, assumed for example to be a single domain (Fig. 7.56a), is decreased if domains with opposing magnetizations are created that reduce B. The additional creation of small

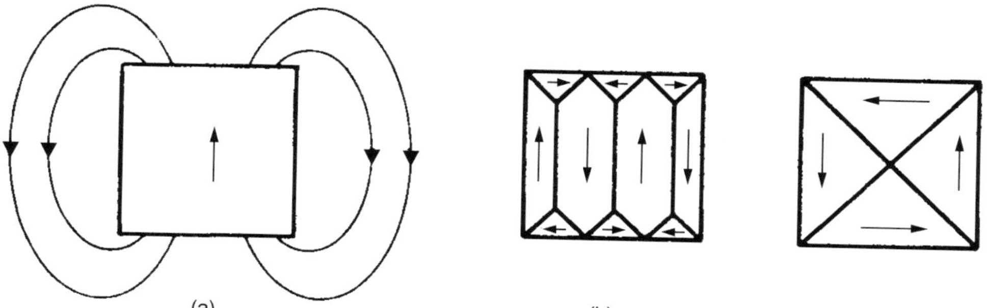

Fig. 7.56 Domain formation in a ferromagnet below the Curie–Weiss temperature. (a) A sample consisting of a single domain producing flux density B outside the material with associated energy density $U_M = B^2/2\mu_0$. (b) The formation of closure domains results in a negligible external flux density B, and concomitant reduction in U_M. (c) A domain configuration in which U_M is also zero but which has a higher magneto-elastic energy.

closure domains, with magnetization directions normal to that of the principal domains (Fig. 7.56b), can reduce B almost to zero. Although U_M is decreased by domain formation, there is a cost in energy involved in forming the transition region between domains, the Bloch wall, where the magnetization changes continuously from the value characteristic of one domain to the different value characteristic of another domain. However, overall, the net energy is lowered in going from a single-domain to a closed-domain configuration. Since it costs energy to create a Bloch wall, it might be thought that a domain configuration where the Bloch wall area is minimized, such as that shown in Fig. 7.56c, would have a lower net energy than that involving closure domains (Fig. 7.56b). However, such a consideration neglects magnetostriction effects, whereby a magnetically aligned crystal expands or contracts along the magnetization direction. The incompatibility of such strains for a domain configuration such as that shown in Fig. 7.56c results in a large elastic stress energy; this positive magneto-elastic energy is reduced by forming smaller closure domains, as in Fig. 7.56b.

The spin configuration of a 180° Bloch wall separating two domains, in which the directions of the spontaneous magnetization are antiparallel, is shown schematically in Fig. 7.57. The gradual canting of spins in the wall region minimizes the cost in exchange energy in inverting the spins; indeed, if exchange interactions were the only energetic consideration, the domain-wall width would increase without limit to reduce the increase in exchange energy to zero. However, there is another effect operative which acts to *minimize* the wall width; this is magnetocrystalline anisotropy.

In crystals, there are directions of easy magnetization along which it is energetically favourable for the magnetization to point, e.g. the ⟨100⟩ directions in b.c.c. Fe, the ⟨0001⟩ directions along the hexagonal axis of h.c.p. Co, and the ⟨111⟩ directions in f.c.c. Ni; other directions in each case are termed *hard* magnetization directions. The energy of a spin configuration polarized along a hard direction is greater than that directed along an easy direction by the magnetocrystalline or anisotropy energy. Asymmetry of overlap of non-spherical electron distributions (e.g. resulting from spin–orbit interactions) on adjacent ions is one cause of magnetocrystalline anisotropy.

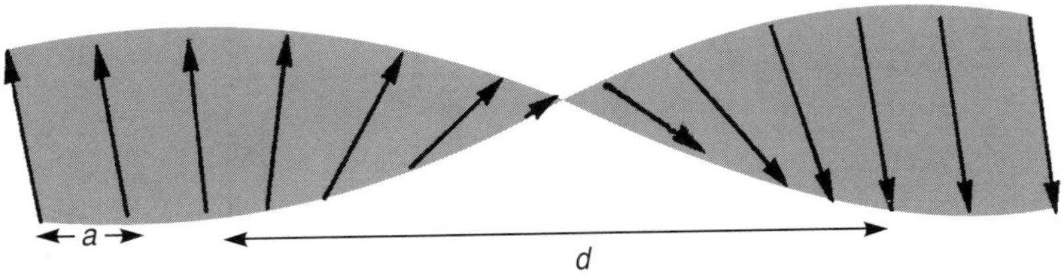

Fig. 7.57 Schematic illustration of a 180° Bloch wall, of thickness d, comprising N spins at a spacing a, between two domains having spontaneous magnetizations pointing in opposite directions.

The expression for the anisotropy energy density U_K is particularly simple for the case of uniaxial anisotropy, e.g. for Co where there is just a single easy axis, the c-axis of the unit cell:

$$U_K = K_1 \sin^2\theta + K_2 \sin^4\theta, \tag{7.267}$$

where θ is the angle between the magnetization direction and the c-axis; $K_1 = 4.1 \times 10^5 \text{Jm}^{-3}$ and $K_2 = 1 \times 10^5 \text{Jm}^{-3}$ for Co. Thus, the anisotropy energy-density difference for the magnetization direction lying in the basal plane ($\theta = 90°$) and along the c-axis ($\theta = 0°$) is $\Delta U_k = K_1 + K_2 \equiv K \simeq 5 \times 10^5 \text{Jm}^{-3}$.

Magnetocrystalline anisotropy restricts the width of Bloch walls because the anisotropy energy increases with increasing wall width. Assuming for simplicity that, for a wall of width $d = Na$, comprising N spins each separated by a distance a (e.g. the lattice constant in a crystal)—see Fig. 7.57—approximately half the spins are pointing in a hard direction, the anisotropy energy per unit area of the wall, given by half the energy density multiplied by the wall width, is then approximately proportional to the wall width:

$$\sigma_{\text{aniso}} \simeq \tfrac{1}{2} KNa. \tag{7.268}$$

In the direction normal to the interface (180° Bloch wall) between the two domains, i.e. along the canted chain of spins, a given spin has two neighbours at an average relative angle of π/N, and hence the associated exchange energy experienced by the spin is, from eqn. (7.217a),

$$\mathscr{E}_{\text{exch}} = -2\mathfrak{J}S^2 \cos(\pi/N) \simeq \mathfrak{J}S^2\pi^2/N^2 - 2\mathfrak{J}S^2, \tag{7.269}$$

on expanding the cosine term for small angles. Hence, the contribution to the exchange energy per unit area of wall that depends on the thickness of the wall is the first term on the extreme right-hand side of eqn. (7.269) multiplied by N/a^2, the number of spins per unit area, i.e.

$$\sigma_{\text{exch}} = \pi^2\mathfrak{J}S^2/Na^2. \tag{7.270}$$

An estimate for the equilibrium wall thickness can then be obtained by minimizing with respect to N the expression for the total areal energy density of the Bloch wall:

$$\sigma_{\rm B} \simeq \frac{1}{2}KNa + \pi^2\mathfrak{J}S^2/Na^2, \tag{7.271}$$

giving the optimum number of spins making up the wall thickness as

$$N = (2\pi^2\mathfrak{J}S^2/Ka^3)^{1/2}, \tag{7.272}$$

with a total energy per unit area of the Bloch wall given by

$$\sigma_{\rm B} = (2\pi^2 K\mathfrak{J}S^2/a)^{1/2}. \tag{7.273}$$

Taking typical values of $\mathfrak{J} \simeq 0.06$eV, $K \simeq 5 \times 10^5$J m^{-3} and $a \simeq 2.5$ Å means that the width of a Bloch wall involves of the order of 100 atoms.

The presence of magnetic domains also explains the behaviour of the magnetization curves of ferromagnetic materials below the Curie–Weiss temperature. A representative *initial* magnetization curve is shown in Fig. 7.58: the net magnetization of a ferromagnetic sample increases from zero, rapidly increases, and then saturates at the saturation value of magnetization, M_s, with increasing magnetic field intensity, H. The zero-magnetization initial state results from the overall cancellation of the magnetic moments of individual domains arranged in a closure configuration (Fig. 7.59a). The appearance of a net magnetization at finite applied magnetic fields is due to the movement of Bloch walls through the material under the action of an applied field, thereby causing the growth of those domains with dominant components of magnetization parallel to $\boldsymbol{H}$ at the expense of those for which the dominant component of $\boldsymbol{M}$ is antiparallel to $\boldsymbol{H}$. At very low applied fields (region I), this domain-wall movement is reversible (figs. 7.59b, c), but at higher fields (region II) the process is irreversible due to the complete

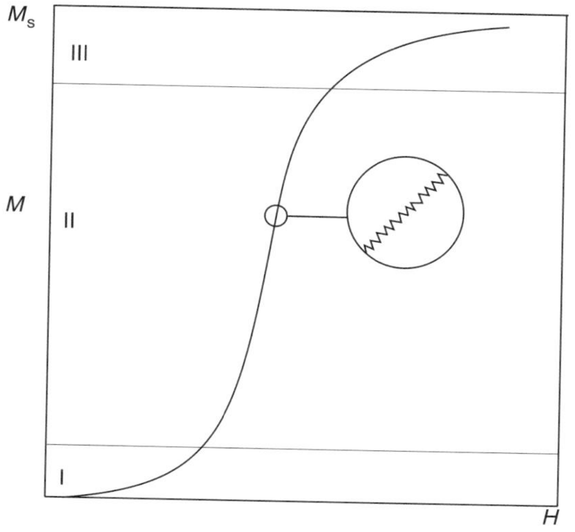

Fig. 7.58 Schematic illustration of the initial magnetization curve of a ferromagnet in a magnetic field of intensity H. The behaviour associated with the magnetic domains at each stage of the curve is indicated. Magnification of the curve in region II shows a step-like profile, i.e. Barkhausen jumps.

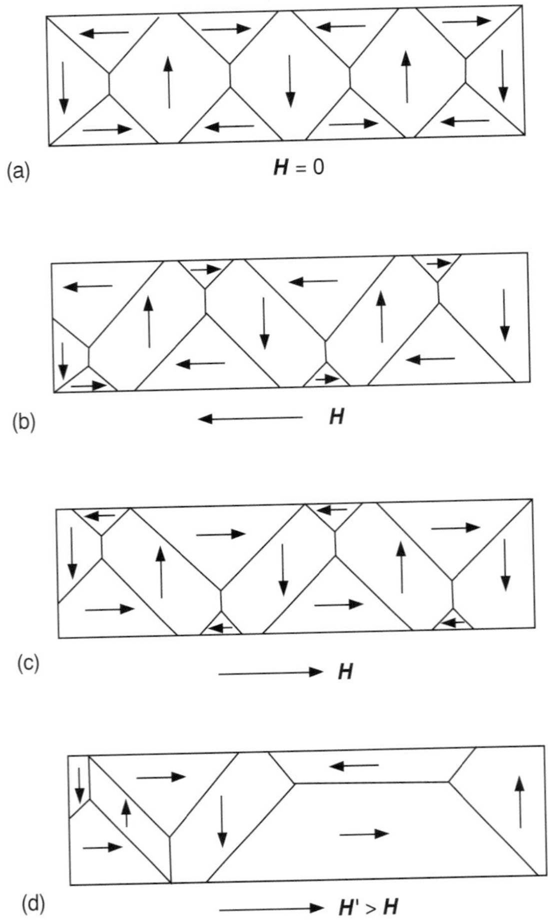

Fig. 7.59 Representation of the movement of domain walls in a ferromagnet in causing a net magnetization in an applied magnetic field, H. (a) Zero net magnetization due to closure domains. (b, c) Reversible domain-wall movement causing enlargement of domains with magnetization parallel to $\boldsymbol{H}$ for small fields. (d) Irreversible domain-wall motion leading to the disappearance of domains at higher fields. Domain walls can be made visible by the Bitter magnetic-powder pattern technique, whereby small magnetic particles are attracted to the regions where Bloch walls intersect an external surface because the magnetic-field gradient is greatest there.

disappearance of domains (Fig. 7.59d) and effects due to magnetic inhomogeneities. At yet higher fields (region III), the saturation magnetization of the sample is achieved by gradual rotation of the magnetization direction of the domains away from the easy direction towards the field direction; this last process is resisted by the magnetocrystalline anisotropy.

Although the magnetization curve shown in Fig. 7.58 appears to increase smoothly with $\boldsymbol{H}$, in fact in the irreversible region II the curve actually consists of a disjointed series of discontinuous jumps in $\boldsymbol{M}$ with increasing values of $\boldsymbol{H}$. These Barkhausen jumps are due to sudden movements of domain walls and arise because sample inhomogeneities such as defects, compositional variations etc., mean that the Bloch wall surface energy (eqn. (7.273)) depends on its position in a sample. For a uniaxial crystal of width D, containing a single 180° Bloch wall of unit area at distance x from one edge separating two domains with the saturation magnetization M_s lying respectively parallel and antiparallel to the easy axis, the total energy density can be written as:

$$U = \sigma_B(x) - \mu_0 M_s H x + \mu_0 M_s H(D - x) \qquad (7.274)$$

in the presence of an external field H directed along the easy axis. A stable, equilibrium position of the wall is determined by the conditions $dU/dx = 0$ and $d^2U/dx^2 > 0$, corresponding to a field intensity

$$H = \frac{1}{2\mu_0 M_s} \frac{d\sigma_B(x)}{dx}. \qquad (7.275)$$

Consider the representative spatial profile of the derivative $d\sigma_B(x)/dx$ shown in Fig. 7.60a, where point A corresponds to a saturation of the magnetization for a magnetic field applied along the easy direction. On decreasing H, the wall moves such that the curve in Fig. 7.60a is followed to point C where $H = 0$ (i.e. $d\sigma_B/dx = 0$). Reversal of the field (i.e. giving $d\sigma_B/dx < 0$) causes a further shift of the wall such that the curve in Fig. 7.60a is followed to point D where $d\sigma_B/dx$ is a minimum, i.e. where $d^2\sigma_B/dx^2 = 0$. However, at this position, the wall is in an unstable position and hence makes a discontinuous (Barkhausen) jump, corresponding in Fig. 7.60a to the transition from D to E where $d^2\sigma_B/dx^2$ is positive and the wall is energetically stable (minimum wall energy, σ_B).

The irreversibility of the magnetization curve, $M(H)$, in region II, shown in Fig. 7.58 implies that, on reducing the magnetic field from the value required to induce the

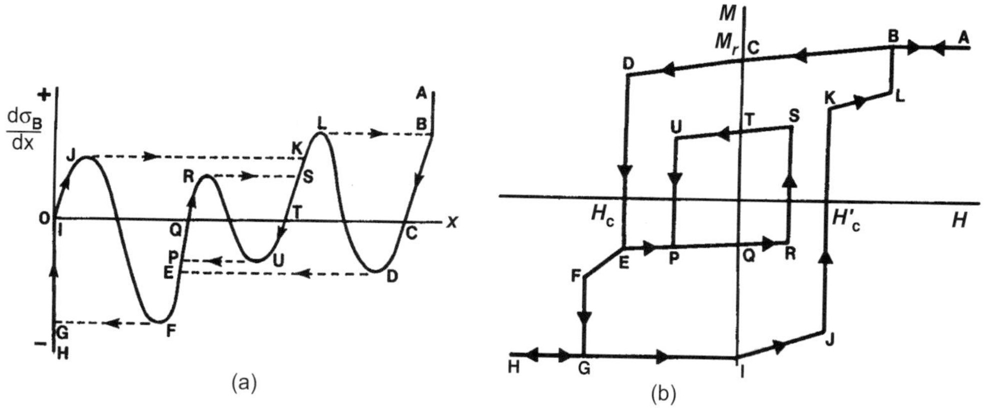

Fig. 7.60 (a) Dependence of the spatial rate of change of Bloch-wall energy density, $d\sigma_B/dx$, with the position x of a single 180° wall separating two magnetic domains in an inhomogeneous ferromagnet. (b) Major and minor hysteresis loops in the magnetization curve corresponding to the behaviour of $d\sigma_B/dx$ shown in (a). (After McCurrie (1994) *Ferromagnetic Materials: Structure and Properties*, © 1994, by permission of the publisher Academic Press Ltd., London)

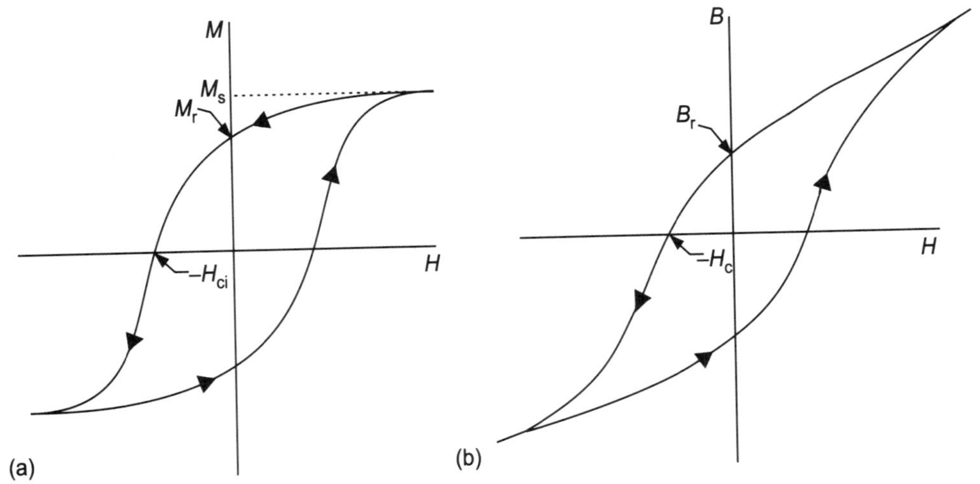

Fig. 7.61 Hysteresis curves for a ferromagnetic material, showing the intrinsic coercivity, H_{ci}, the flux coercivity H_c, the flux remanence B_r, the remanent magnetization M_r and the saturation magnetization M_s in: (a) magnetization, $M(H)$; (b) magnetic flux density, $B(H)$.

saturation magnetization, the initial magnetization curve is *not* retraced, but instead another curve is followed (Fig. 7.61a): $M(H)$ curves of ferromagnets exhibit *hysteresis* in exactly the same way, and for the same reason, as do electrical polarization curves, $P(E)$, of ferroelectrics (see §7.1.5.3 and Fig. 7.16). The hysteresis curve of $B(H)$ (Fig. 7.61b) is very similar to that of $M(H)$, except that saturation of $B(H)$ is not reached at very high field intensities, corresponding to the saturation magnetization M_s since, from eqn. (7.81), $\boldsymbol{B} = \mu_0(\boldsymbol{H} + \boldsymbol{M}_s)$.

Hysteresis curves can be characterized by a number of parameters. For $M(H)$ curves, the remanent magnetization, M_r, is the value remaining at zero field, $H = 0$, and the intrinsic coercivity, H_{ci}, is the (reverse) magnetic field intensity required to reduce the magnetization to zero. For $B(H)$ curves, the remanence, B_r, is the flux intensity at $H = 0$, and the flux coercivity H_c is the field intensity required to reduce the magnetic induction to zero. These quantities are defined for the situation where the applied magnetic field is reduced from the value at which the saturation magnetization is achieved.

The spatial profile of the variation of the spatial derivative of the energy density of a single Bloch wall, $d\sigma_B/dx$, shown in Fig. 7.60a can be used to obtain the corresponding magnetization hysteresis curve (Fig. 7.60b). Note the similarity between this curve and that shown in Fig. 7.61a: however, the $M(H)$ curve for real polycrystalline samples is the superposition of many slightly different hysteresis curves for different domains, resulting in a smoothing of the abrupt variations shown in Fig. 7.60b. In this picture, the coercivity is related to the rate of variation of σ_B with position, i.e. $d^2\sigma_B/dx^2$, since $d\sigma_B/dx$ is a measure of the energy barrier that must be overcome in the motion of a domain wall. Thus, the coercivity is the magnetic field intensity corresponding to the point D in Fig. 7.60a at which the quantity $d\sigma_B/dx$ is a (local) minimum, i.e.

$$H_{ci} = \frac{1}{2\pi M_s}\left[\frac{d\sigma_B(x)}{dx}\right]_{min}. \tag{7.276}$$

Soft ferromagnetic materials are those with *low* coercivities (say less than 10^3 A/m), and hard magnetic materials are those with *high* values of $H_c > 10^4$ A/m. Thus, soft materials must be spatially homogeneous, since $d\sigma_B/dx$ must be as small as possible and spatially invariant; this is achieved by having very small anisotropy energies, K (cf. eqn. (7.273)). In contrast, hard materials must have very large values of $d\sigma_B/dx$ and large spatial variations of this quantity; this can be achieved by making the material inhomogeneous (e.g. two-phase).

Permanent magnet materials must be magnetically hard, having a high coercivity, associated with a very high uniaxial magnetocrystalline anisotropy energy. The coercivity can be enhanced by making the material very inhomogeneous; in addition to the influence of the microstructure on the coercivity described above, structural defects can act as pinning centres for domain-wall motion, thereby increasing H_c. In addition, the materials must have a high saturation magnetization, and a high remanent magnetization. As a result, the $M(H)$ hysteresis loop of a permanent magnet should enclose the largest possible area. An *ideal* permanent magnet would have a *rectangular* $M(H)$ hysteresis loop (Fig. 7.62a), with $M_r = M_s$ and a vertical demagnetization line at H_{ci}. The corresponding $B(H)$ hysteresis loop is a parallelogram (Fig. 7.62b), for which the energy product $(BH)_{max}$, given by the shaded area in Fig. 7.62b, is maximal. It can be shown that $H = M_s/2$ at the theoretical maximum value of BH, given by (Problem 7.28):

$$(BH)_{max}^{theor} = \frac{\mu_0 M_s^2}{4} = \frac{B_r^2}{4\mu_0}. \tag{7.277}$$

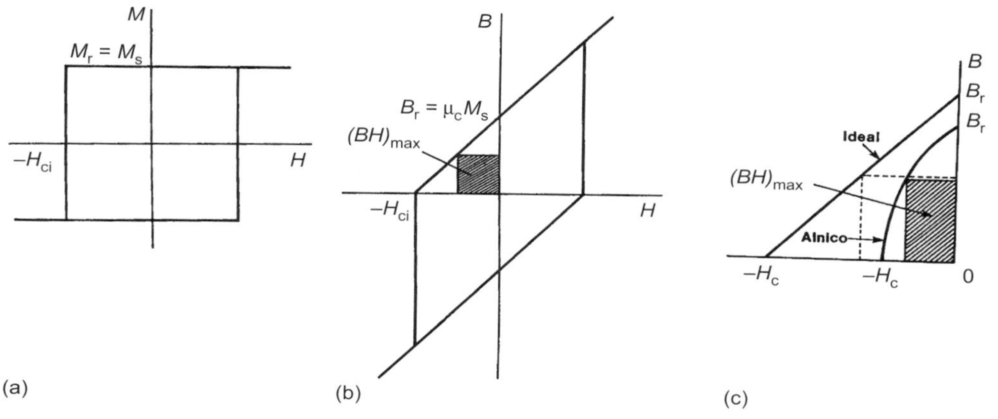

Fig. 7.62 (a) Rectangular $M(H)$ hysteresis loop characteristic of an ideal permanent-magnet material. (b) Ideal $B(H)$ hysteresis loop corresponding to (a), in which the flux and intrinsic coercivities are identical. The shaded area gives the energy product $(BH)_{max}$. (c) Comparison of the demagnetizing quadrants of the $B(H)$ hysteresis curve for an Alnico magnet and an ideal permanent-magnet material with the same value of M_s. (After McCurrie (1994) *Ferromagnetic Materials: Structure and Properties*, © 1994, by permission of the publishers Academic Press Ltd., London)

This value can only be attained if $H_c \geq M_s/2$.

Most permanent-magnet materials exhibit a non-linear demagnetizing $B(H)$ curve, with $B_r < \mu_0 M_s$ and $H_c < M_s/2$, and as a result the maximum energy product is considerably less than the theoretical value (see Fig. 7.62c). However, the sintered intermetallic material $SmCo_5$ has a $(BH)_{max}$ value (180 kJ m^{-3}) that is about 80% of the theoretical maximum. This material is also characterized by a very high value of flux coercivity ($H_c = 700$ kA m^{-1}), due to an exceptionally large value of uniaxial magnetocrystalline energy coefficient, $K_1 = 1.3 \times 10^7$ J m^{-3} (the crystal structure of $SmCo_5$ is hexagonal) and a remanence of 0.95 T. The recently developed tetragonal material $Nd_2Fe_{14}B$ has even higher values of coercivity and $(BH)_{max}$ product, namely 1.1 MA m^{-1} and 350 kJ m^{-3}, respectively, although the Curie–Weiss temperature is rather low, viz. $\theta_{CW} \simeq 400$ °C (cf. 720 °C for $SmCO_5$). By contrast, the widely used Alnico alloys (of Fe, Al, Ni, Co and perhaps Cu and Ti) have typical values of $(BH)_{max} \simeq 10$ kJ m^{-3}, $H_c \simeq 50$ kA m^{-1} and $B_r \simeq 0.7$ T, and the electrically insulating magnetic materials, sintered barium or strontium ferrite ($(Ba/Sr)Fe_{12}O_{19}$), have at best values of $(BH)_{max} \simeq 30$ kJ m^{-3}, $H_c \simeq 300$ kA m^{-1} and $B_r \simeq 0.4$ T. Further details of permanent-magnet materials can be found in McCurrie (1994).

Another major use of ferromagnetic materials is as cores in transformers. However, here the need is for materials that are magnetically soft (ideally with zero remanence and coercivity), and with high magnetic permeabilities (to increase the flux density in the core). A high permeability is associated with a high saturation magnetization. Since the direction of magnetization in a mains transformer changes at a frequency of say 50 Hz, it is necessary that domain-wall motion is unimpeded (i.e. a very low coercivity is required). This requirement can be achieved by ensuring that the material is homogeneous, with no defects that can act as pinning centres for domain walls, and that it has a very small magnetocrystalline anisotropy energy K_1, ideally with many easy directions of magnetization. As a result of the small value of H_c, the hysteresis curves of soft magnetic materials are very narrow, and hence hysteretic energy-density losses (equal to the area of the hysteresis loop, $\oint H \mathrm{d}B = \mu_0 \oint H \mathrm{d}M$) are minimized. Eddy-current losses can be reduced by using a material with a high electrical resistivity, ρ. The material most commonly used in transformer cores is Fe alloyed with about 3 at. % Si (which reduces K_1 and increases ρ) with a (110) [001] grain-oriented texture; the crystallite grains (optimally a few millimetres in diameter to reduce losses to a minimum) have the (110) plane parallel to the [001] easy direction, itself parallel to the rolling direction used in producing the laminar foils that are stacked together to form the cores. This material has a coercive field of $H_c = 12$ A m^{-1} and a relative permability $\mu_r = 4 \times 10^4$. Glassy metallic alloys, e.g. $Fe_{80}B_{20}$ or more complex alloys involving Ni or P, are very soft magnetically. Evidently, there is no magnetocrystalline anisotropy energy in such materials, but nevertheless there is a very small residual uniaxial magnetic anisotropy associated with mechanical strains introduced in producing the material in thin-ribbon form by 'melt spinning' (the very rapid cooling of a jet of molten material directed onto the edge of a rapidly spinning Cu disc—see §1.2.2). Glassy $Fe_{80}B_{20}$ has a flux coercivity of 3 A m^{-1}.

7.2.5.6 *Ferrimagnetism and antiferromagnetism*

Thus far, in this section on spontaneous magnetization, it has generally been assumed that the exchange parameter $\mathfrak{J}_{ij}$ between nearest-neighbour spins is positive, favouring a

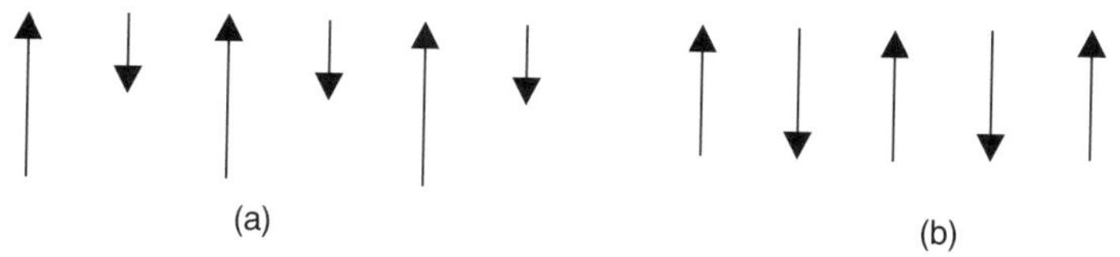

Fig. 7.63 Possible ordered spin arrangements resulting from a negative value of exchange parameter, $\mathfrak{J}_{ij} < 0$: (a) ferrimagnetic configuration (unequal spins on neighbouring sites); (b) antiferromagnetic configuration (equal spins on neighbouring sites).

parallel, i.e. ferromagnetic, ordering of the spins. However, for example, many compounds of Fe, Co or Ni are characterized by *negative* values of $\mathfrak{J}_{ij}$ (e.g. as a result of superexchange—see §7.2.5.1) which favours antiparallel ordering of the spins (associated with the transition-metal 3d-electrons). A ferrimagnetic spin configuration is one in which unequal magnetic moments on neighbouring sites are arranged in an antiparallel arrangement (Fig. 7.63a) resulting in a net magnetization. An antiferromagnetic configuration is a special case of a ferrimagnetic arrangement, in which the magnetic moments on neighbouring sites are equal in magnitude (Fig. 7.63b), resulting in *zero* net magnetization even in this spin-ordered state.

Ferrimagnetism is exhibited by a class of magnetic metal oxides known as ferrites with the general chemical formula $MO . Fe_2O_3$, where M is a divalent cation (e.g. Cd, Co, Cu, Fe, Mg, Ni or Zn), and which have the cubic spinel ($MgAl_2O_4$) structure in which there are eight occupied tetrahedral (A) sites and twice as many octahedral (B) sites in a unit cube (see also Table 2.2): in the normal spinel arrangement, the divalent ions occupy all the A sites and the trivalent ions the B sites, whereas in the inverse spinel structure, the A sites are occupied instead by trivalent ions, with half the B sites occupied by divalent ions and the other half by the remaining trivalent ions.

The most familiar example of a ferrite is magnetite (or lodestone) with $M = Fe^{2+}$ and which has the inverse spinel structure. From Table 7.8, it can be seen that Fe^{3+} ions have the spin state $S = \frac{5}{2}$ with quenched angular momentum $L = 0$ and Fe^{2+} ions have a spin of $S = 2$; thus each Fe^{3+} ion should contribute a magnetic moment of $5\mu_B$ and each Fe^{2+} ion a moment of $4\mu_B$. If all the spins were aligned in a *ferromagnetic* arrangement, it would be expected therefore that the magnetic moment per Fe_3O_4 formula unit would be $14\mu_B$; the experimental value of $4.1\mu_B$ can be explained if the Fe^{3+} spin sub-lattices order antiferromagnetically with respect to each other, leaving a net moment due to the Fe^{2+} spins (Fig. 7.64a). The Fe^{3+} ions on neighbouring A and B sites are strongly coupled antiferromagnetically due to superexchange (§7.2.5.1) via the intervening oxygen anions; this interaction is stronger than the weaker A–A and B–B antiferromagnetic interactions. The effective ferromagnetic coupling between the spins on Fe^{2+} ions and Fe^{3+} on neighbouring sites in the B sub-lattice can also be understood pictorially in terms of a double-exchange mechanism (Fig. 7.64b): an inter-ion electron transfer, responsible for the electrical conductivity in this material, simultaneously maintaining the high-spin state required by Hund's first rule (§7.2.4.1) on the two ions (hence 'double exchange') can only involve the minority spin on an Fe^{2+} ion moving to an Fe^{3+} ion that has a spin configuration parallel to that of the Fe^{2+} ion; otherwise, the transfer would be

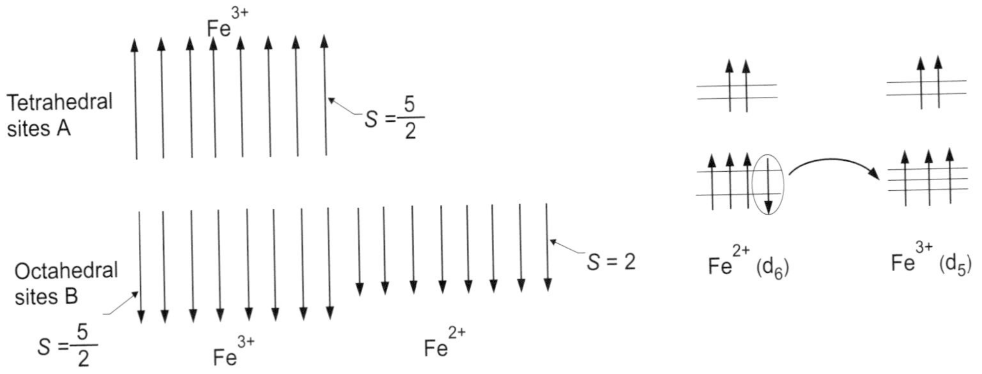

Fig. 7.64 (a) Spin arrangements on Fe ions in magnetite, $Fe_3O_4(FeO \cdot Fe_2O_3)$ having the inverse spinel structure. The Fe^{2+} and Fe^{3+} ions on the octahedral B sites are coupled ferromagnetically via a double-exchange mechanism associated with inter-ion electron transfer (see (b)). The Fe^{3+} ions on tetrahedral A and B sites are coupled antiferromagnetically via superexchange through the oxygen anions. (b) Illustration of the mechanism of double exchange associated with inter-ion electron transfer in mixed-valency compounds, e.g. Fe_3O_4. A ferromagnetic alignment of the spins on neighbouring sites is necessary to maintain the high-spin configuration on both the electron-donating and -accepting ions.

inhibited by the Pauli exclusion principle. This double-exchange mechanism is particularly appropriate in describing the behaviour of manganites exhibiting colossal magnetoresistance (§6.3.3.2).

Another common class of ferrimagnetic insulators comprises the cubic iron garnets with the general formula $M_3Fe_5O_{12}$ (where M is a trivalent ion, e.g. Y^{3+} or rare-earth ions). The garnet ($Ca_3Al_2Si_3O_{12}$) structure of e.g. yttrium iron garnet (YIG) has, per formula unit, two Fe^{3+} ions in tetrahedral (a) sites and three Fe^{3+} ions in octahedral (d) sites. The spins on the a and d sites are each aligned ferromagnetically, but in an antiparallel configuration between sites: as a result, there is a net moment of $(3 \times 5 - 2 \times 5)\mu_B = 5\mu_B$ per formula unit.

Antiferromagnetic spin ordering is a special case of ferrimagnetic alignment, in which the net moments of different sub-lattices are equal and opposite. A particularly simple example is given by the transition-metal monoxides, MnO, FeO, CoO and NiO, which have the rocksalt structure, comprising two interpenetrating f.c.c. lattices, one for the transition-metal ions and the other for the oxygen ions. At low temperatures, in the antiferromagnetically ordered state, the spins, e.g. of the Mn^{2+} ions, within a particular (111) plane are all aligned ferromagnetically, but they are aligned antiparallel to the spins lying on adjacent planes (Fig. 7.65). This spin arrangement arises because of superexchange interactions in $\langle 100 \rangle$ directions between neighbouring Mn^{2+} ions on adjacent (111) planes via the intervening, collinear O anion (cf. Fig. 7.44). As a consequence of this type of spin ordering, the magnetic unit cell in the ordered state (taking into account the spin orientations of the cations) is *twice* the size of the chemical unit cell.

This structural distinction can be detected using diffraction (§2.6.1), either of neutrons (making use of the intrinsic spin of the neutron as a magnetic probe) or of circularly polarized X-rays from a synchrotron radiation source (which are sensitive

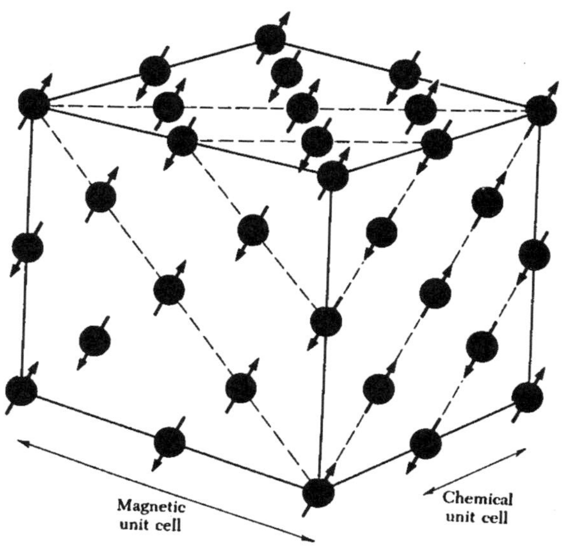

Fig. 7.65 Antiferromagnetic ordering of spins of Mn^{2+} ions in MnO. The O^{2-} ions are not shown, but lie midway between the Mn^{2+} ions in $\langle 100 \rangle$ directions. Spins are ferromagnetically aligned within $\{111\}$ planes, and antiferromagnetically aligned between adjacent planes. Note that the magnetic unit cell is twice the size of the chemical unit cell.

to the spin and orbital angular momentum of a scattering atom). In the case of magnetic neutron scattering, the magnetic neutron-scattering length is given by

$$b_m = b_0 \pm p \sin\alpha, \tag{7.278}$$

where b_0 is the normal scattering length (§2.6.1.3), p is a coefficient proportional to the magnetic moments of the neutron and of the atom (decreasing with scattering vector $\boldsymbol{K}$ in a similar manner to the X-ray atomic scattering factor—see Problem 7.29), α is the angle between $\boldsymbol{K}$ and the net atomic magnetic moment μ_m, and the +/− signs depend on the orientation of the neutron spin to the direction perpendicular to both $\boldsymbol{K}$ and μ_m. Thus, up-and down-spin atoms, as in the magnetic structure of Fig. 7.65, scatter spin-polarized neutrons *differently*. Figure 7.66 shows neutron powder-diffraction patterns for MnO at 80 K in the antiferromagnetic state, below the Néel ordering temperature (see below) of $T_N = 120$ K, and in the magnetically disordered, paramagnetic state above T_N. Note the appearance of the extra peak at small scattering angle in the low-temperature pattern, resulting from diffraction from $\{111\}$ planes in the larger magnetic unit cell.

A simple model for antiferromagnetism can be developed in which the exchange interaction between spins on neighbouring atoms is given by the Heisenberg Hamiltonian (eqn. (7.217a)), but with a *negative* value for the exchange constant $\mathfrak{J}_{ij}$. In the ideal case, such an interaction ensures that the spin on an atom is antiparallel to the spins on the nearest-neighbour atoms. Such a configuration can be achieved for *all* atoms if the crystal structure of the magnetic ions can be separated into two interpenetrating

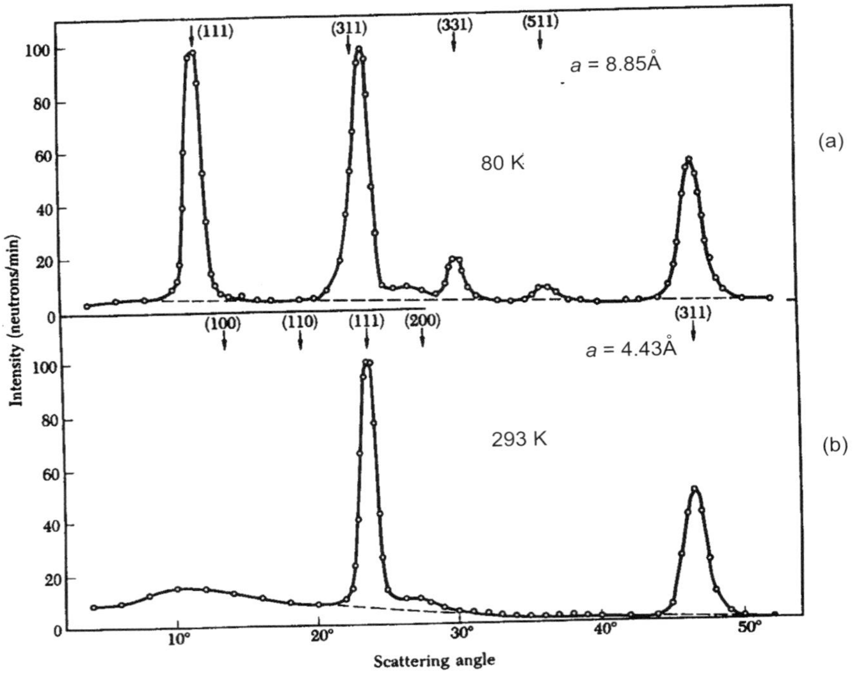

Fig. 7.66 Neutron powder-diffraction patterns of MnO at a temperature (a) below the Néel temperature, θ_N; (b) above θ_N. The unit-cell parameters, *a*, differ by a factor of two. (After Shull and Smart (1949). Reprinted with permission from *Phys. Rev.* **76**, 1256. © 1949. The American Physical Society)

sub-lattices, so that one sub-lattice is decorated by spins of one orientation and the other is decorated by spins with the opposite orientation. Examples of such structural decompositions are a monatomic simple cubic structure, which can be separated into two interpenetrating f.c.c. lattices, and the monatomic b.c.c. structure (§2.2.3.2) in which body-centred sites form one sub-lattice and the corner sites form the other. (Note that the simple cubic structure of MnO shown in Fig. 7.65 does not satisfy this criterion since the two interpenetrating f.c.c. lattices are decorated respectively in this case by Mn and O atoms; the f.c.c. lattice itself cannot be further divided into two sub-lattices and so *complete* antiferromagnetic ordering is not possible in this structure (see Problem 7.30), as seen in Fig. 7.65.)

The Néel model of antiferromagnetism generalizes the Weiss mean-field theory for ferromagnetism (§7.2.5.2) by assuming that magnetic atoms on one sub-lattice (A) experience a molecular field proportional, and opposite in direction, to the magnetization $\boldsymbol{M}_B$ of the *other* sub-lattice (B), and vice versa. Thus, neglecting local-field corrections:

$$\boldsymbol{B}_{\text{eff}}^{A,B} = \mu_0(\boldsymbol{H} - \lambda_N \boldsymbol{M}_{B,A}). \tag{7.279}$$

By analogy with the Weiss constant, λ_W, for ferromagnetism (eqn. (7.223)), a corresponding expression for the Néel constant in eqn. (7.279) is

$$\lambda_N = \frac{-2\sum_{j\neq i}\mathfrak{J}_{ij}}{n\mu_0 g_J^2 \mu_B^2}, \tag{7.280}$$

with $\sum_{j\neq i}\mathfrak{J}_{ij} = z\mathfrak{J}$, where z is the nearest-neighbour coordination number and the exchange interaction has been taken to be constant for all pairs of interacting spins. Equation (7.280) differs by a factor of 2 from eqn. (7.223) because there are only $n/2$ atoms per volume in each sub-lattice in the antiferromagnetic case. Assuming for simplicity, as for the ferromagnetic case (§7.2.5.2), that $J = S = \frac{1}{2}$, with $g_J = 2$, the Brillouin function (eqn. (7.172)) reduces to a tanh function as in eqn. (7.227); thus, the sub-lattice magnetizations are given by

$$M_{A,B} = \frac{n}{2}\mu_B \tanh[\mu_B B_{eff}^{A,B}/k_B T]. \tag{7.281}$$

In the *high-temperature* limit, $\tanh x \simeq x$, and eqn. (7.281) can be rewritten as

$$M_{A,B} + \frac{\lambda_N C}{2T} M_{B,A} = \frac{CH}{2T}, \tag{7.282}$$

where $C = n\mu_0\mu_B^2/k_B$ is the Curie constant (cf. eqn. (7.234)). The net magnetization is therefore

$$M = M_A + M_B = \frac{CH}{T + \frac{1}{2}\lambda_N C} \tag{7.283}$$

and hence the magnetic susceptibility is

$$\chi_m = \frac{C}{T + T_N}, \tag{7.284}$$

where

$$T_N = \frac{1}{2}\lambda_N C. \tag{7.285}$$

The expression (7.284) for the high-temperature behaviour of the susceptibility of an antiferromagnet should be compared with that for a ferromagnet (eqn. (7.234)).

In the absence of an external magnetic field ($H = 0$), antiferromagnetic ordering sets in as the temperature becomes less than the Néel temperature, θ_N, determined by the determinant of the coefficients of the pair of equations (7.282) being zero:

$$\begin{vmatrix} 1 & \frac{\lambda C}{2T} \\ \frac{\lambda C}{2T} & 1 \end{vmatrix} = 1 - \frac{\lambda^2 C^2}{4T^2} = 0$$

or

$$T = \theta_N = \frac{1}{2}\lambda C \equiv T_N. \tag{7.286}$$

In general, however, it is found that the ratio T_N/θ_N is greater than the value of unity predicted by the simple mean-field model; e.g. for MnO, $T_N/\theta_N = 5.3$. At the Néel transition, and with $H = 0$, eqn. (7.283) shows that $M_A = -M_B$, as expected for an antiferromagnetic state. At lower temperatures, $T < \theta_N$, again with $H = 0$, the magnetization of one sub-lattice is found by substituting eqn. (7.279) into eqn. (7.281) to give

$$M_A = -M_B = \frac{n\mu_B}{2}\tanh\left(\frac{\theta_N}{T}\frac{2M_A}{n\mu_B}\right). \tag{7.287}$$

For temperatures *below* the Néel temperature, the magnetic susceptibility depends on the relative direction of $\boldsymbol{H}$ and the spin axis. For a field direction *normal* to the spin axis, spins are relatively easily canted away from the field-free axis so as to create a component of magnetization parallel to $\boldsymbol{H}$. The susceptibility is expected to be approximately independent of temperature; a mean-field analysis (Problem 7.31) gives for all temperatures:

$$\chi_m^{\perp} = \frac{C}{2T_N}, \tag{7.288}$$

i.e. the value at the Néel temperature (cf. eqn. (7.285)). For the case of $\boldsymbol{H}$ *parallel* to the spins, at $T = 0$K where both spin sub-lattices are perfectly aligned, the applied field cannot change the magnetization and so $\chi_m^{\parallel}(0) = 0$. With increasing temperature, the effect of the molecular field diminishes progressively and hence $\chi_m^{\parallel}(T)$ increases, becoming equal to $\chi_m(T_N) = C/2T_N$ at the Néel transition. A mean-field treatment (Problem 7.31) gives

$$\chi_m^{\parallel}(T) = \frac{\alpha C}{(1+\alpha)T_N}, \tag{7.289}$$

with

$$\alpha = \frac{\lambda_N C}{2T}\operatorname{sech}^2\left(\frac{\mu_0 \lambda_N M_0}{k_B T}\right). \tag{7.290}$$

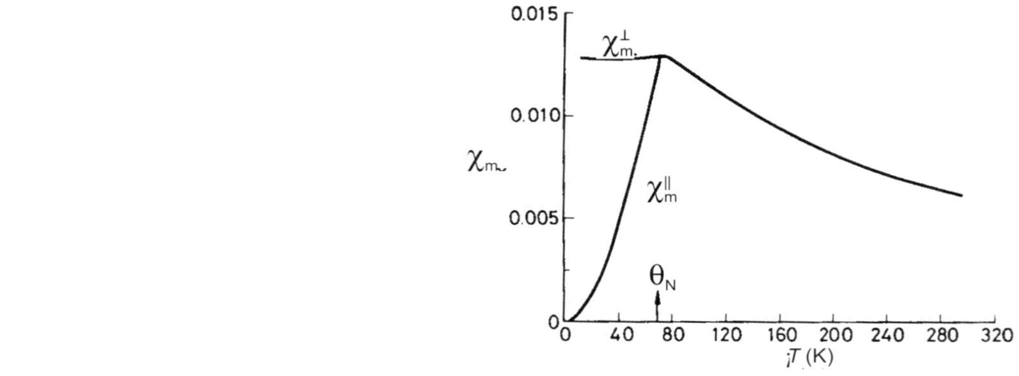

Fig. 7.67 Experimental magnetic susceptibility of the antiferromagnetic material MnF_2. The susceptibility is different below the Néel temperature, $\theta_N = 67$K, for fields parallel or perpendicular to the spin directions of the sub-lattices, but independent of spin direction above this temperature. (Hook and Hall (1991). Reproduced by permission of John Wiley & Sons Inc.)

The experimental temperature dependence of the susceptibility of the antiferromagnetic material MnF_2 (Fig. 7.67) approximately follows these predictions.

★ Finally, it should be mentioned that antiferromagnets below the Néel temperature can support spin waves (magnons), as for ferromagnets (§7.2.5.4). However, instead of the dispersion law having the limiting (low-k) quadratic behaviour characteristic of ferromagnets (eqn. (7.261)), that for antiferromagnets has a linear dependence, $\omega_k \propto k$ (see, e.g. Kittel (1996)). ★

7.2.6 Applications

The magnetic properties of materials are widely exploited in very many different areas of application. Some magnetic applications have already been touched upon in §7.2.5.5: for example, permanent magnets are fabricated from hard magnetic materials, and soft magnetic materials are used in electrical-power applications, e.g. as cores in transformers. Another very important application of magnetic materials is in information technology, namely in magnetic data recording, whereby information is stored magnetically as a preferred spin orientation in the material.

7.2.6.1 Magnetic data recording

Information can be stored in materials in a form suitable for electronic (or optical) processing in several ways. For example, it may be stored as electrical charge, e.g. in semiconductor-based memories, as variations in optical reflectivity or refractive index for optical processing, or as variations in electron-spin alignment in magnetic memories. This latter format is the most widespread method of digital and analogue data recording used at present.

Magnetic recording makes use of the hysteresis in the variation of magnetization with applied magnetic field intensity that is characteristic of ferro- and ferrimagnetic materials (§7.2.5.5). That is, the remanent magnetization is a record of the last maximum field intensity experienced by the magnetic solid, in terms of both magnitude and direction of the field. The electrical signal to be recorded is converted, by passing an electrical current through a recording head, into a time-varying, spatially confined magnetic field. The magnetic recording medium, for example in the form of ferro/ferrimagnetic particles bonded to the surface of a flexible tape or of a disc or a thin film of magnetic material deposited on a disc substrate, is then moved mechanically relative to the recording head so that electrical signals varying in time are transformed into spatially varying magnetization regions in the recording medium. Alternatively, data may be stored in a ferrimagnetic film by thermomagnetic means. A highly focused laser beam heats a small region of a magnetized material to above its Curie temperature, and the material is allowed to cool from the paramagnetic state in a reverse magnetic field which induces a reverse magnetization in the previously irradiated spot. Thus, digital data can be stored, e.g. the bit '0' corresponding to the original magnetization direction and '1' to the reversed direction. Once the data are written magnetically, for long-term storage the magnetization should not be easily altered by stray magnetic fields: this requires the magnetic material to have a high coercivity. However, the coercivity should not be *too* large if the medium is to be reusable, since the magnetic fields produced by the recording head should be capable of re-magnetizing the material in order to record new data; typical values of coercivity for this purpose are in the range 20–100 kA/m.

For the recording medium to be at all useful, the data stored as spatial variations of magnetization must be able to be 'read', but in such a manner that the reading operation does not perturb the stored information. In order that the output signal during the reading process is as large as possible, the magnetic material should have a high value of saturation magnetization. Two methods of reading magnetically stored data can be envisaged. The simplest method is to move the magnetic storage medium (tape or disc) past a coil of metallic wire (having a soft magnetic core): the motion of the spatially varying magnetization in the medium produces a time-varying magnetic flux density passing through the coil which, due to Lenz's law, produces a time-varying output voltage signal.

Alternatively, signals recorded magnetically can be read by making use of a magneto-resistive sensor (§6.3.3.2), made either of a metallic material like permalloy ($Ni_{81}Fe_{19}$) with $\Delta R/R \simeq 2.5\%$, or one using the larger effect, termed 'giant magnetoresistance' (GMR), where $\Delta R/R \lesssim 200\%$, which is observed in multilayers of thin ferromagnetic metal films (e.g. Fe/Cr) with antiferromagnetic coupling between the layers (Baibich *et al.* (1988)). (Note that this effect should not be confused with the much larger (negative) 'colossal' magnetoresistance observed in manganites (§6.3.3.2). The GMR effect occurs because of spin-dependent electron scattering at the interfaces in the multilayer stack.

A more sophisticated method of reading magnetically stored data employs magneto-optical effects, such as the Faraday and the Kerr effects, in which the polarization direction of a linearly polarized light beam is rotated by the interaction of the light with a magnetic field: the Kerr effect is when this phenomenon occurs for light *reflected* from the surface of a magnetic material; the Faraday effect occurs for light *transmitted* through a (semi-)transparent magnetic material (e.g. a magnetic insulator or a very thin magnetic metal) and is correspondingly less useful in this regard. The process based on the Kerr effect is particularly suited to the reading of *digital* data stored magnetically, since the presence or absence of reversely magnetized domains created thermomagnetically can readily be detected by measuring the associated change in polarization direction of the reflected light; analogue signals, where the variation of magnetization can be much less pronounced, are more difficult to read because the corresponding angular changes θ of the polarization direction given by

$$\theta = K_r M, \tag{7.291}$$

where K_r is a constant, are so small (typical rotations are 9 arc minutes for saturated Ni and 20 arc minutes for saturated Fe or Co).

Oxide materials suitable for magnetic recording are the ferrimagnetic iron oxides magnetite, Fe_3O_4, and maghemite, $\gamma - Fe_2O_3$, and the ferromagnetic material CrO_2. Perhaps the most widely used material for magnetic-tape applications is maghemite, which has a defect inverse-spinel structure, with some of the octahedral B sites vacant, so that, in contrast to the magnetite structure (Fig. 7.64a), the Fe^{3+} ions are divided unequally between the A and B sites, this imbalance giving rise to a net magnetic moment. The chemical formula for maghemite can be written as $\frac{4}{3}[Fe_2O_3] \equiv FeFe_{5/3}O_4$ to make the discussion in terms of the spinel structure more transparent.

Since there are twice as many B sites (16) as A sites (8) in the spinel structure, this implies that there is one vacancy for every six sites on the B sub-lattice; i.e. the spin and chemical composition in the spinel representation is

$$\left[Fe^{3+}\right]^{A}_{\downarrow}\left[\square_{1/3}Fe^{3+}_{5/3}\right]^{B}_{\uparrow}O_4,$$

where the symbol $\square$ denotes a vacancy. The magnetic moment per spinel formula unit is thus $\mu_m = \left(\frac{5}{3} - 1\right) \times 5\mu_B = 3.33\mu_B$ (since each Fe^{3+} ion contributes $5\mu_B$—see §7.2.5.6), or per chemical formula unit, $\mu_m = \frac{3}{4} \times 3.33\mu_B = 2.5\mu_B$. This material has a coercivity of $H_c \simeq 3 \times 10^4$ A m^{-1}. An improved material is Co-modified $\gamma - Fe_2O_3$ which has a greater value of $H_c = 5 \times 10^4$ A m^{-1} (associated with an increase in the magnetocrystalline anisotropy) the Co is incorporated at the surface of the maghemite grains. CrO_2 also has a reasonably large value of magnetocrystalline anisotropy that contributes to the coercivity, but in all these materials the dominant cause of the coercivity is *not* this intrinsic magnetic characteristic but rather an extraneous effect depending on the *microstructure* of the magnetic grains, namely shape anisotropy.

Shape anisotropy energy is a measure of the difference in magnetic energies for magnetization parallel or perpendicular to the long axis of an anisotropically shaped magnetic sample: the magnetic energy for $\boldsymbol{M}$ parallel to the axis is lower than for $\boldsymbol{M}$ normal to the axis. This effect can be understood in terms of an internal demagnetizing field, $\boldsymbol{H}_d$, given by

$$\boldsymbol{H}_d = -N_i\boldsymbol{M}, \tag{7.292}$$

where N_i is a dimensionless (positive) demagnetizing factor. Values for N_i for various axes of some representative ellipsoidal samples are given in Table 7.1: the demagnetizing field is exactly analogous to the depolarization field for dielectrics (eqn. (7.15)). The magnetic energy of a sample can therefore be written as the integral over the sample volume:

$$\mathscr{E}_m = -\frac{1}{2}\int \mu_0\boldsymbol{H}_d\,.\,\boldsymbol{M}\mathrm{d}V. \tag{7.293}$$

For a sample in the shape of an ellipsoid of revolution (with the long axis along the z-axis) and with $\boldsymbol{M}$ uniform and directed at an angle θ to the z-axis (Fig. 7.68), $\boldsymbol{M} = \hat{x}M\sin\theta + \hat{z}M\cos\theta$, and from eqn. (7.292) the demagnetizing field is

$$\boldsymbol{H}_d = -\hat{x}N_xM\sin\theta - \hat{z}N_zM\cos\theta, \tag{7.294}$$

which is also uniform. Thus, the magnetic energy density is

$$\begin{aligned} U_m &= -\frac{1}{2}\mu_0\boldsymbol{H}_d\,.\,\boldsymbol{M} = \frac{1}{2}\mu_0M^2(N_x\cos^2\theta + N_y\sin^2\theta) \\ &= \frac{1}{2}\mu_0M^2(N_x - N_z)\sin^2\theta + \text{const.} \end{aligned} \tag{7.295}$$

Equating this expression to the leading term in $\sin^2\theta$ for the anisotropy energy density (eqn. (7.267)) gives for the anisotropy constant

$$K_1 = \frac{1}{2}\mu_0M^2(N_x - N_z). \tag{7.296}$$

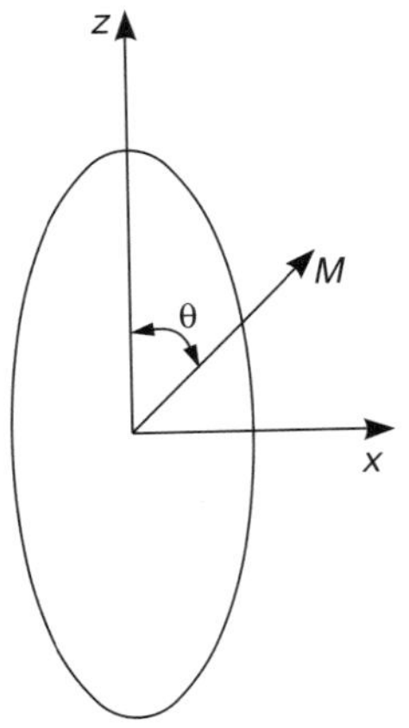

Fig. 7.68 An ellipsoidal sample with a uniform magnetization lying in the x–z plane at an angle θ to the z-axis as a model for calculating shape anisotropy.

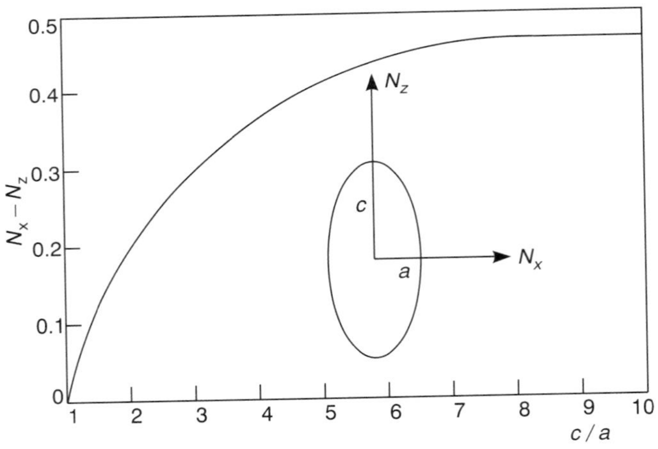

Fig. 7.69 Shape anisotropy for a prolate ellipsoid of revolution, showing the variation with axial ratio c/a of the difference in principal demagnetizing factors, $(N_x - N_z)$.

Inserting this equation into the expression for the anisotropy field (see Problem 7.32):

$$H_A = \frac{2K_1}{\mu_0 M_s} \tag{7.297}$$

gives the *shape* anisotropy field:

$$H_s = (N_x - N_z)M_s, \tag{7.298}$$

which is simply the intrinsic coercivity H_{ci} when shape-anisotropy effects are dominant. Figure 7.69 shows the variation of $(N_x - N_z)$ for a prolate ellipsoid of revolution as a function of the axial ratio c/a. Single-domain, acicular particles of γ-Fe_2O_3 and CrO_2 can be grown relatively readily with axial ratios of 5:1 or higher, for which shape anisotropy is therefore the dominant contribution to the coercivity.

For the case of magnetic thin films used as magnetic data-storage media, two modes of operation are possible. In-plane magnetic recording utilizes variations in magnetization that lies in the plane of the film: electroplated Co–Ni alloy films can be used in this regard, and these have a high in-plane magnetocrystalline anisotropy. Alternatively, perpendicular magnetic recording has the magnetization normal to the film surface: sputtered Co–Cr films having a columnar microstructure normal to the surface are used for this purpose. For further details on magnetic-recording materials, see e.g. McCurrie (1994).

Problems

7.1 (a) Show that, if the electrostatic potential of a dipole $\boldsymbol{p}$ is given by $\phi(\boldsymbol{r}) = (1/4\pi\varepsilon_0)\boldsymbol{p}\cdot\nabla(1/r)$, the electric field due to a uniform polarization $\boldsymbol{P}$ in a sample is equal to the field *in vacuo* of a fictitious surface charge density, $\sigma_c = \hat{\boldsymbol{n}}\cdot\boldsymbol{P}$, where $\hat{\boldsymbol{n}}$ is the outward unit vector normal to the surface. (Hint: use a vector identity for $\nabla\cdot(\boldsymbol{P}/r)$.)
(b) Hence obtain expressions for the depolarizing field (and the depolarization factors N_i) for samples with geometries of
(i) a thin disc;
(ii) a long cylinder;
(iii) a sphere.

7.2 Calculate the atomic polarizability, and hence the dielectric constant, of liquid Ar, for which the atomic number density is $2.128 \times 10^{28}\text{m}^{-3}$ and the atomic radius is $R = 1.18$ Å. (The experimental value is $\varepsilon = 1.538$.)

7.3 (a) Show that the average component of a dipole moment $\boldsymbol{p}$ parallel to a local electric field $\boldsymbol{E}_{\text{loc}}$, when independent, and able to adopt any orientation, in thermal equilibrium is given by $p_{\text{par}} = p\mathfrak{L}(x)$, where $\mathfrak{L}(x)$ is the Langevin function (eqn. (7.40)) and $x = pE_{\text{loc}}/k_BT$. Show that, under normal circumstances, $x \ll 1$, and hence that $p_{\text{par}} \simeq p^2E_{\text{loc}}/3k_BT$.
(b) Prove that for an orientable dipole with only two allowed orientations, i.e. parallel and antiparallel to the electric field, $p_{\text{par}} \simeq p^2E_{\text{loc}}/k_BT$.

7.4 Obtain an expression for the frequency dependence of the atomic polarizability based on the driven damped-oscillator model (eqn. (7.43)). Show that for the model of atomic polarizability illustrated in Fig. 7.5, the corresponding resonant frequency is given by $\omega_0 = (Ze^2/4\pi\varepsilon_0R^3m_e)^{1/2}$.

7.5 Show that if the Debye relaxation time is thermally activated, $\tau_D = \tau_0\exp(W/k_BT)$, with activation energy W, an Arrhenius plot of $\varepsilon_2(T)$ at a fixed frequency exhibits a peak at a temperature $T_0 = W/k_B\ln(1/\omega\tau_0)$.

7.6 Show that an equivalent parallel electrical circuit, consisting of a capacitance C_1 in one arm and a series combination of a resistor R and capacitance C_2 in the other, gives the Debye expression for the dielectric constant (eqn. (7.50)) if the following identities hold: $C_1 \equiv -\varepsilon_0\varepsilon(\infty)$, $C_2 \equiv \varepsilon_0(\varepsilon(0) - \varepsilon(\infty))$, $R = \tau_D/C_2$. (Hint: obtain an expression for the complex admittance $Y^\dagger = i\omega\varepsilon_0\varepsilon^\dagger \equiv (Z^\dagger)^{-1}$, where $Z^\dagger$ is the complex impedance.)

7.7 Assume that the microscopic mechanism for dielectric relaxation in an amorphous solid (e.g. an ionic conductor) can be described by the Debye dipolar-orientational model, with a distribution $G(\tau_D)$ of relaxation times. If the relaxation time is taken to be thermally activated, with a *random* distribution of activation energies, show that the overall real part of the a.c. conductivity has a linear frequency dependence, $\sigma_1(\omega) \propto \omega$.

7.8 Show that for the orthorhombic point group $D_2(222)$, with symmetry operations $\{E, C_2, C_2^x, C_2^y\}$, the only non-zero components of the piezoelectric constant tensor $\boldsymbol{d}$ are d_{123}, d_{213} and d_{312}.

7.9 Many hydrogen-containing crystals are ferroelectric, e.g. Rochelle salt, NaK $(C_4H_4O_6) \cdot 4H_2O$, and potassium dihydrogen phosphate, (KDP), KH_2PO_4, and its isomorphic compounds (K replaced by Rb, Cs or NH_4; PO_4 replaced by AsO_4). The orthorhombic structure of KDP consists of PO_4 tetrahedra arranged in a staggered fashion such that the top edge of one tetrahedron is approximately at the same height along the c-axis as the lower edges of two adjacent tetrahedra. Each PO_4 tetrahedron forms four hydrogen bonds with adjacent tetrahedra. In terms of motion of H atoms account for
(a) the paraelectric phase above the Curie temperature, T_{Cf};
(b) the ferroelectric phase below T_{Cf}.
(c) How can the dipole moment associated with a given PO_4 tetrahedron be easily reversed?
(d) What do you expect to be the effect on T_{Cf} of substitution of D for H?

7.10 By analysing the field-free expression for the Landau free-energy expression (eqn. (7.65)), show that for a first-order transition the saturation polarization is given by eqns. (7.73).

7.11 Show that the particular choice of vector potential given by eqn. (7.115) is equivalent to a uniform magnetic field and satisfies the gauge div $\boldsymbol{A} = 0$.

7.12 De Haas–van Alphen (dHvA) oscillations of the magnetization in Cu, corresponding to the belly orbit, have a period in $(1/B)$ of $1.83 \times 10^{-5}\mathrm{T}^{-1}$. Obtain a value for the cross-sectional area of Fermi surface responsible, and compare your answer with the value for the free-electron Fermi sphere (the atomic density of Cu is $8.45 \times 10^{28}\mathrm{m}^{-3}$). Estimate the maximum temperature, and the maximum impurity density, at which the dHvA effect will be observable in a field of 10 T, if the impurity concentration n_i and scattering time τ are related via $n_i\tau = 10^{14}\mathrm{m}^{-3}\mathrm{s}$.

7.13 In gold, two de Haas–van Alphen periods are found for $\boldsymbol{B}$ in the [111] direction, $\delta(1/B) = 2.05 \times 10^{-5}\mathrm{T}^{-1}$ and $6 \times 10^{-4}\mathrm{T}^{-1}$. Calculate values for the extremal areas of the Fermi surface probed in this experiment, and account for these results in terms of the shape of the Fermi surface of Au (like that of Cu — see Fig. 6.18). Analyse the dHvA oscillations shown in Fig. 7.29 for $\boldsymbol{B}$ in the [110] direction. To which orbit do these oscillations correspond?

7.14 Obtain an expression for the magnetic flux density in a thin film of a superconductor of thickness $d < \lambda_L$, subject to an applied field of flux density $\boldsymbol{B}_0$ on both sides of the film. (Hint: use the solutions of the expression for the spatial variation of $\boldsymbol{B}$ given by eqn. (7.140).)

7.15 Flux quantization in a superconducting ring was measured for a sample consisting of a hollow cylinder of the superconductor Sn (electroplated on a non-superconducting Cu core of radius 6.6 μm) by applying a magnetic field of flux density B to the sample in the normal state, and cooling through T_c in the presence of the field which was then removed, thereby trapping magnetic flux Φ in the superconducting cylinder. What would you expect the plot of Φ versus B to look like for $B \leqslant 50$ mT?

7.16 Estimate the duration of persistent currents in a superconducting ring, of thickness $R \simeq 1\mu$m, for a type-I superconductor with a critical field $B_c = 0.1$T and superconducting coherence length $\zeta_{\mathrm{eff}} = 1\mu$m, by finding the time taken for one flux quantum to leak out of the ring.

7.17 Essay: Discuss the behaviour of flux vortices in both 'conventional' and high-T_c type-II superconductors.

7.18 Prove eqn. (7.160) for the energy of the Jth level in a Russell–Saunders-coupled atom. (Hint: the eigenvalue for an angular-momentum operator X is $\boldsymbol{X} \cdot \boldsymbol{X} = X(X+1)$).)

7.19 Obtain the ground-state electronic configuration, and hence give the corresponding term symbol, for the transition-metal ions, Ti^{3+}, V^{3+}, Cr^{2+} and Cr^{3+}, Mn^{2+}, Mn^{3+} and Mn^{4+}, Fe^{3+}, Co^{2+}, Ni^{2+} and Cu^{2+} and for the rare-earth ions La^{3+}, Ce^{3+}, Nd^{3+}, Pm^{3+}, Sm^{3+}, Eu^{3+}, Gd^{3+}, Tb^{3+}, Dy^{3+}, Ho^{3+}, Er^{3+}, Tm^{3+}, Yb^{3+} and Lu^{3+}.

7.20 Show that the Brillouin function $B(x)$ (eqn. (7.172)) reverts to the Langevin function (eqn. (7.40)) in the classical limit.

7.21 The dominant component of the paramagnetism of copper sulphate is associated with the Cu^{2+} ions with quenched configuration $[JLS] = [\frac{1}{2}0\frac{1}{2}], g = 2)$. Show that the magnetization for a concentration of n ions in a magnetic field of flux density B is given by

$M = n\mu_B \tanh(\mu_B B/k_B T)$. Calculate an expression for the *magnetic* contribution to the heat capacity per unit volume, c_B, at constant flux density B. (Hint: calculate first the internal energy associated with the ions.) See also eqn. (4.204) and Problem 4.21 for another instance of a two-level system.

7.22 Essay: Discuss adiabatic (isentropic) demagnetization of a paramagnetic sample as a method for producing ultra-low temperatures.

7.23 Show that the magnetic dipolar interaction between two magnetic moments, of magnitude μ_B, at a separation of 3 Å, is only important at temperatures below 0.02 K.

7.24 (a) In the limit of zero electron–electron interactions, the two-electron wavefunction for say the H_2 molecule with nuclei a and b, can be written as the product of one-electron wavefunctions for electrons 1 and 2: $\Phi = [\phi_a(1) + \phi_b(1)][\phi_a(2) + \phi_b(2)]$. Give a physical interpretation of all the product terms in this expression, and show why the Heitler–London approximation (eqn. (7.193)) is appropriate when electron–electron interactions are not negligible.

(b) Obtain the eigenenergies given by eqn. (7.205) for the case of the H_2 molecule by a variational method, taking the wavefunction to be $\Psi = c_1\psi_1 + c_2\psi_2$, where $\psi_1 = \phi_a(1)\phi_b(2)$ and $\psi_2 = \phi_a(2)\phi_b(1)$, where $\phi_i(n)$ are one-electron wavefunctions as in (a).

7.25 By considering a fluctuation in magnetization, demonstrate that the intersection at the origin of the curves in Fig. 7.46 corresponding to eqns. (7.230) represents an unstable solution to the magnetization of a ferromagnet.

7.26 By generalizing eqn. (7.241) for the spin-population difference in the Stoner model to include the effects of an external magnetic field of flux density $\boldsymbol{B}_0$, obtain an expression for the net magnetization, and hence show that the exchange enhancement of the Pauli paramagnetic susceptibility is given by eqn. (7.247).

7.27 Obtain an expression for the heat capacity, valid at low temperatures, associated with the excitation of spin waves in a ferromagnet, and show that this contribution behaves as $C_v \propto T^{3/2}$. Compare this expression with that predicted by mean-field theory for low temperatures. A plot of $C_v/T^{3/2}$ versus $T^{3/2}$ for a ferromagnet gives a straight line. What physical properties of the material can be deduced from the slope and intercept?

7.28 Show that the theoretical maximum energy product $(BH)_{\max}$, for an ideal permanent magnet material is given by eqn. (7.277). (Hint: consider only the demagnetizing quadrant of the hysteresis curve (Fig. 7.62b).)

7.29 (a) Calculate the angular dependence of magnetic scattering of polarized neutrons from an atomic magnetic moment μ_m. (Hint: assume that the contribution to the scattering from a point anywhere within the atom is proportional to the local magnetic-moment density, and that this is uniformly distributed over a spherical surface of radius R and is zero elsewhere.)

(b) Describe qualitatively the difference in magnetic neutron scattering from an antiferromagnetic crystal above and below the Néel temperature.

7.30 Extend the Néel theory of antiferromagnetism to the case of a ferrimagnet.

(a) Show that, even if the exchange parameters $\mathfrak{J}_{AA'}\mathfrak{J}_{AB}$ and $\mathfrak{J}_{BB}$ for interactions between nearest-neighbour spins on the A and B sub-lattices are all *antiferromagnetic*, i.e. $\mathfrak{J}_{ij} < 0$, nevertheless separate ferromagnetic ordering on the A and B sub-lattices occurs, with ferrimagnetic alignment overall, as long as the AB interaction is the strongest. (Hint: consider the interaction energy density, $U = -\frac{1}{2}(\boldsymbol{B}_A \cdot \boldsymbol{M}_A + \boldsymbol{B}_B \cdot \boldsymbol{M}_B)$, and take $\boldsymbol{B}_{\text{eff}}^{A,B} = \boldsymbol{B}_{\text{loc}} - \lambda_1 \boldsymbol{M}_{B,A} - \lambda_2 \boldsymbol{M}_{A,B}$.)

(b) If a spin has z_a antiparallel and z_p parallel nearest-neighbour spins, show that $\lambda_1/\lambda_2 = z_a/z_p$. Show also that the high-temperature magnetic susceptibility has the Néel form $\chi_m = C/(T+\theta)$, where θ is related to the Néel temperature θ_N by $\theta/\theta_N = (\lambda_1+\lambda_2)/(\lambda_1-\lambda_2) = (z_a+z_p)/z_a - z_p)$.

(c) For spins arranged on an f.c.c. lattice, it is impossible to find a truly antiferromagnetic ordering of nearest-neighbour spins. Consider the case where the spins in alternate (200) planes are arranged ferromagnetically either up or down: evaluate the ratio θ/θ_N.

7.31 Obtain the expressions (7.288) and (7.289) for the perpendicular and parallel components of the magnetic susceptibility of an antiferromagnet below the Néel temperature in the Néel model.

7.32 By considering a spherical single-domain sample of a magnetically uniaxial material, and its magnetic energy density, show that the magnetic susceptibility is $\chi_m = \mu_0 M_s^2/2K$ and, hence, that the anisotropy field is $H_A = 2K_1/\mu_0 M_s$ (eqn. (7.297)).

References

Baibich, M. N., Broto, J. M., Fert, A., Nguyen Van Dau, F., Petroff, F., Etienne, P., Greuzet, B., Friederich, A. and Chazelas, J., 1988, *Phys. Rev. Lett.* **61**, 2472.
Burns, G., 1985, *Solid State Physics* (Academic Press: New York).
Cox, P. A., 1992, *Transition Metal Oxides* (Clarendon Press: Oxford).
Crangle, J. and Hallam, G. C., 1963, *Proc. Roy. Soc.* A**272**, 119.
Feder, J. and Pytte, E., 1970, *Phys. Rev.* **B1**, 4803.
Henry, W. E., 1952, *Phys. Rev.* **88**, 559.
Hook, J. R. and Hall, H. E., 1991, *Solid State Physics*, 2nd edn (Wiley: Chichester).
Ibach, H. and Lüth, H, 1995, *Solid-State Physics*, 2nd edn (Springer-Verlag: Berlin)
Janak, J. F., 1977, *Phys. Rev.* **B16**, 255.
Kittel, C., 1996, *Introduction to Solid State Physics*, 7th edn (Wiley: New York).
Kubo, R. and Nagamiya, T., 1968, *Solid State Physics*, 2nd edn (McGraw-Hill: New York).
Livingston, J. P., 1963, *Phys. Rev.* **129**, 1943.
Low, G. 1963, *Proc. Phys. Soc. Lond.* **82**, 992.
McCurrie, R. A., 1994, *Ferromagnetic Materials: Structure and Properties* (Academic Press: London).
MacDonald, J. R., 1987, *Impedance Spectroscopy* (Wiley: New York).
Madelung, O., 1978, *Introduction to Solid-State Theory* (Springer-Verlag: Berlin).
Mandl, F., 1992, *Quantum Mechanics* (Wiley: Chichester).
Mattis, D. C., 1988, *The Theory of Magnetism I*, 2nd edn (Springer-Verlag: Berlin).
Mook, H. A. and Paul, D. McK., 1985, *Phys. Rev. Lett.* **54**, 227.
Noakes, J. E., Tornberg, N. E. and Arrott, A., 1966, *J. Appl. Phys.* **37**, 1264.
Peierls, R. E., 1955, *Quantum Theory of Solids* (Clarendon Press: Oxford).
Rao, C. N. R. and Rao, K. J., 1992, in *Solid State Chemistry: Compounds*, eds A. K. Cheetham and P. Day (Clarendon Press: Oxford).
Schelten, J., Ullmeier, H., Lippmann, G. and Schmatz, W., 1972, in *Low Temperature Physics*, eds K. D. Timmerhaus *et al.*, LT13, vol. 3 (Plenum: New York), p. 54.
Shull, C. G. and Smart, J. S., 1949, *Phys. Rev.* **76**, 1256.
Tessmann, J. R., Kaha, A. H. and Shockley, W., 1953, *Phys. Rev.* **92**, 890.
West, A. R., 1987, *Solid State Chemistry and Its Applications* (Wiley: Chichester).
White, R. M., 1983, *Quantum Theory of Magnetism* (Springer-Verlag: Berlin).
Yu, P. Y. and Cardona, M., 1996, *Fundamentals of Semiconductors* (Springer-Verlag: Berlin).

CHAPTER

8 Reduced Dimensionality

Introduction

Dimensionality often has a profound effect on the behaviour of materials, such as changing the functional form of the dependence of one quantity on another (e.g. the energy dependence of the electronic density of states of one-dimensional (1D), two-dimensional (2D) and three-dimensional (3D) systems is very different in each case (Problem 5.2)). In other cases, phenomena *only* occur in systems of reduced dimensionality (e.g. the integer Hall effect observed in 2D — see §8.4.5). Since much of the discussion so far in this book deals with bulk properties of materials, i.e. involving 3D properties, it is of interest to finish with a discussion of the effect of reduced dimensionality. Note, however, that although we will discuss 2D systems in §8.4, for reasons of space and balance one major instance of 2D behaviour, namely 'surface science', i.e. the interaction of surfaces of materials with molecules in the gas (or liquid) phase and the behaviour of the surfaces themselves (e.g. structural reconstructions), will be omitted. Such a broad subject merits a book to itself—see e.g. Zangwill (1988) and Lüth (1995).

Definitions 8.1

Several ways of defining the dimensionality of a system can be envisaged, depending on the behaviour of interest. Here, we will concentrate on two approaches, one being based on the structure and bonding of a material and the other based on its behaviour (e.g. electronic).

The first, structure-based, microscopic classification scheme (I) defines dimensionality in terms of percolation along, say, the covalent bonds of a structure: the space so traced out corresponds to the dimensionality of the system. Thus, if the process of bond percolation inevitably ends up with the starting point being revisited, the structure can be said to be zero-dimensional (0D): examples are discrete rings or molecular clusters of atoms. Alternatively, if bond percolation traces out a linear topology, which may be kinked or linear in shape, the structure is 1D-like. Finally if bond percolation traces out the topology of a surface, which may be buckled as well as planar, the structure is 2D-like.

Figure 8.1 shows four examples of materials in which the (iono-)covalent bonding is restricted to 0D (molecular P_4Se_3), 1D (crystalline SiS_2) and 2D (crystalline $GeSe_2$), compared with the 3D (giant molecular) structure of amorphous SiO_2. Note that in the last three cases, all the structures are based on AX_4 tetrahedra (A = Si, Ge; X = O, S, Se) in which the relative degree of corner-sharing and edge-sharing of neighbouring tetrahedra varies: a 1D structure results from complete edge-sharing of tetrahedra, while the 3D structure involves a completely corner-sharing mode of connection; mixed edge-and corner-sharing of tetrahedra produces a 2D structure. Of course, very many different other types of structural unit, and different modes of connection between them, can produce structures of reduced dimensionality: however, in all cases, from this structural viewpoint, a distinction is made between the strong (covalent) bonds *within* regions of the structure that define the dimensionality and the weak (e.g. van der Waals) bonds *between* such structural regions that bind the regions together to form a structure that is three-dimensional overall. Where such a distinction between intra- and inter-region bonding cannot be sustained, this microscopic scheme for determining reduced dimensionality breaks down and the structure is inherently three-dimensional from this viewpoint.

The other classification scheme (II) for defining the dimensionality of a system is a more macroscopic approach, based on the size dependence of some physical behaviour of the system. This could involve, for example, transport, usually of electrons (but in principle also of other excitations, such as phonons), in which case an important length scale is the mean-free path, Λ (§§4.6.2.2 and 6.3.1.1). In this scheme, a system is said to be of reduced dimensionality if the size, L_i, of a macroscopic sample of material is reduced sufficiently in one or more orthogonal directions, $i = \{x, y, z\}$, so that in those directions the mean-free path (m.f.p.) is determined by boundary scattering and not by some other intrinsic mechanism (e.g. see §6.3.1 for scattering mechanisms for electrons, or §4.6.2.2 for phonons). Thus, reduced dimensionality occurs in this picture if $\Lambda_{\text{int}} > L_i$ so that $\Lambda_{\text{tot}} \simeq L_i$ for transport in the ith direction, since the total m.f.p. is given by a reciprocal sum of intrinsic and extrinsic (boundary-scattering) mechanisms, $\Lambda_{\text{tot}}^{-1} = \Lambda_{\text{int}}^{-1} + \Lambda_{\text{ext}}^{-1}$ (cf. eqn. (4.233)). Alternatively, the important length scale can be related to the 'size', L_0, of the electron wavefunction, such as the Fermi wavelength,

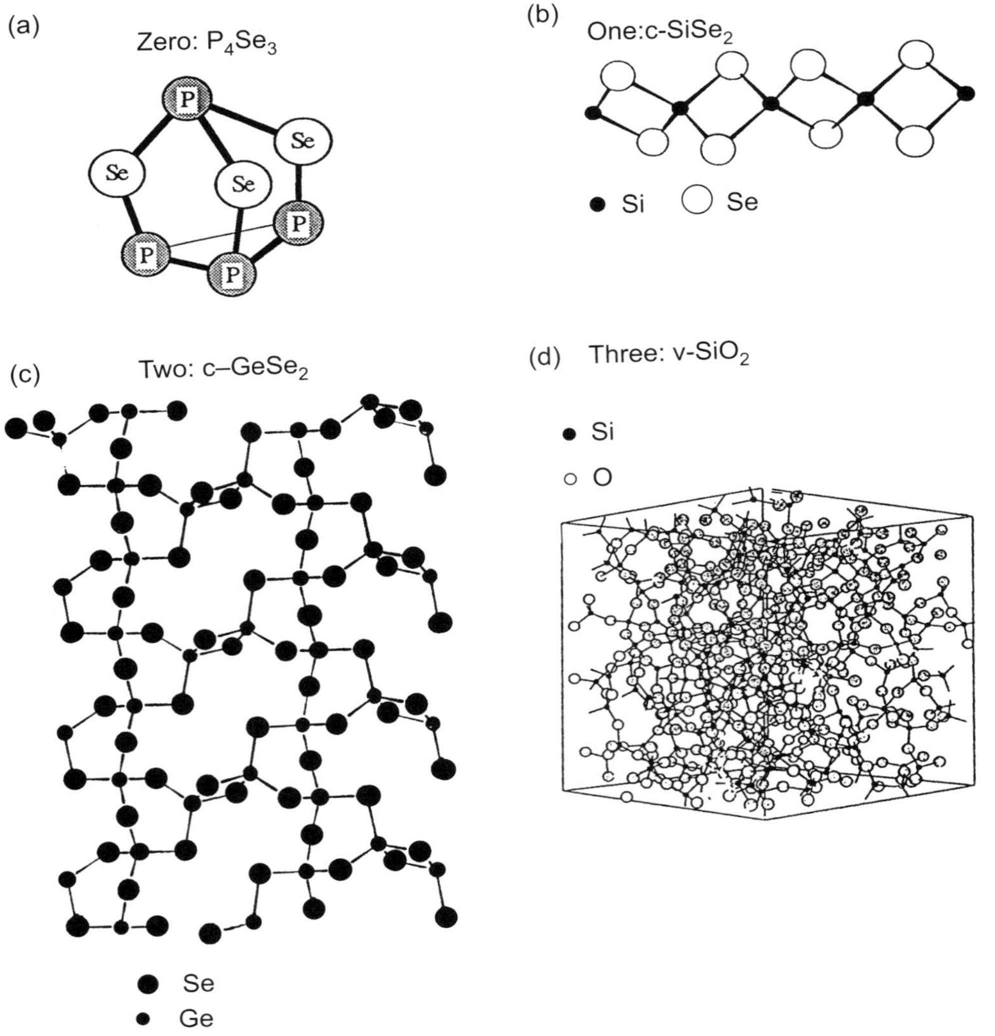

Fig. 8.1 Examples of reduced-dimensional materials defined by a bond-percolation categorization: (a) zero-dimensional (molecular P_4Se_3); (b) one-dimensional (crystalline SiS_2); (c) two-dimensional (crystalline $GeSe_2$); (d) a three-dimensional (giant-molecular) structure of amorphous SiO_2, for comparison.

$\lambda_F = 2\pi/k_F$, or the effective Bohr radius, a_0^*, of an exciton (§5.8.3). In this case, size-quantization sets in if $L_i < L_0$. Hence, in this scheme, a zero-dimensional system is one in which all *three* orthogonal lengths of a sample are less than Λ_{int} and a quantum dot results when the lengths are such that $L_{x,y,z} < L_0$. A one-dimensional system is one in which *two* spatial dimensions are smaller than Λ_{int}; transport (e.g. of electrons) is then allowed along the remaining one dimension unencumbered by boundary scattering (Fig. 8.2b). A quantum wire is a 1D conducting sample with size quantization in two dimensions ($L_{y,z} < L_0$). Finally, a two-dimensional system is that in which only *one* spatial dimension is less than Λ_{int} and hence transport is allowed in two dimensions limited only by intrinsic scattering mechanisms (Fig. 8.2c). A quantum well is a 2D conducting sample with size quantization in the third dimension ($L_z < L_0$).

It should be noted that these two classification schemes are not necessarily mutually inclusive: a 3D material (from a bond-percolation point of view), e.g. GaAs, can be readily fabricated in a low-dimensional sample form (quantum dot, wire or well) by means of modern lithographic and etching techniques; conversely, for example, a 2D (layer) crystal (e.g. graphite—see §8.4.1) can, if in polycrystalline form with no preferential orientational ordering of the crystallites, exhibit 3D (isotropic) transport behaviour in a macroscopic sample with a size in all directions of $L_i > \Lambda_{\text{int}}$.

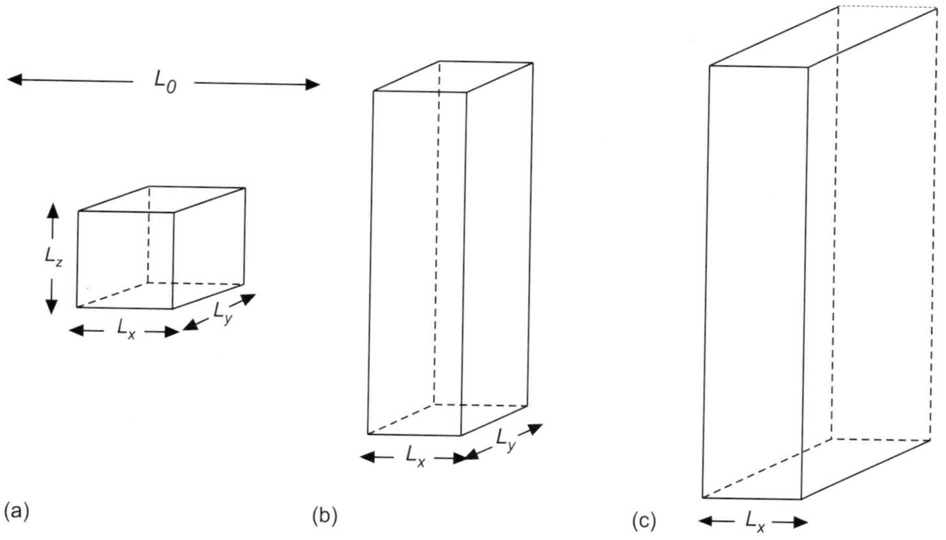

Fig. 8.2 Examples of reduced-dimensional samples defined with respect to a characteristic length, $L_0 = \lambda_F$, the electron Fermi wavelength: (a) quantum dot ($\{L_x, L_y, L_z\} < L_0$); (b) quantum wire ($\{L_x, L_y\} < L_0$); (c) quantum well ($L_x < L_0$).

8.2 Zero-dimensional systems

8.2.1 Fullerenes

Perhaps the most interesting of 0D materials as classified by scheme I are the fullerenes, the newly discovered cage-like polymorphs of carbon with general formula C_{20+2n}, consisting of hollow closed nets with 12 pentagonal faces (rings) and n hexagonal faces (rings), and formed from three-coordinated carbon atoms. The first such cluster found was C_{60} which has been named 'buckminsterfullerene', after Buckminster Fuller who designed the geodesic dome that has an identical geometry (as has a soccer ball). Since the initial discovery of C_{60}, other related structures of carbon have been found, called generically fullerenes or, more colloquially, 'buckyballs'; some examples are illustrated in Fig. 8.3. The structures with lowest energies have *isolated* pentagonal rings: the smallest cluster satisfying such a rule is C_{60} itself (Fig. 8.3a), followed by C_{70} (Fig. 8.3b), the two most stable and prevalent forms found in carbon-arc soot. In addition species consisting of concentric shells of fullerene structures, known colloquially as onions, have also been found.

Although fullerenes are zero-dimensional according to classification scheme I when embedded in normal, Euclidean space, one could alternatively regard them as planar, 2D structures, but in a *curved* space. The carbon allotrope that *is* planar in Euclidean space is, of course, graphite (§8.4.1), each layer of which consists *only* of three-coordinated sp^2–hybridized carbon atoms forming hexagonal rings. Curvature can then be induced in such a planar layer by the introduction of *disclination lines* (§3.1.2), in this case associated with the centres of pentagonal rings replacing the hexagonal rings. In this sense, fullerenes (and also carbon tubes—see §8.3.4) and graphite can be regarded as being structurally related, rather than being separate allotropes.

The C_{60} cluster has *icosahedral* (I_h) point-group symmetry, associated with its inherent five-fold rotational axes (see Fig. 8.3). However, sublimed crystalline films of C_{60}

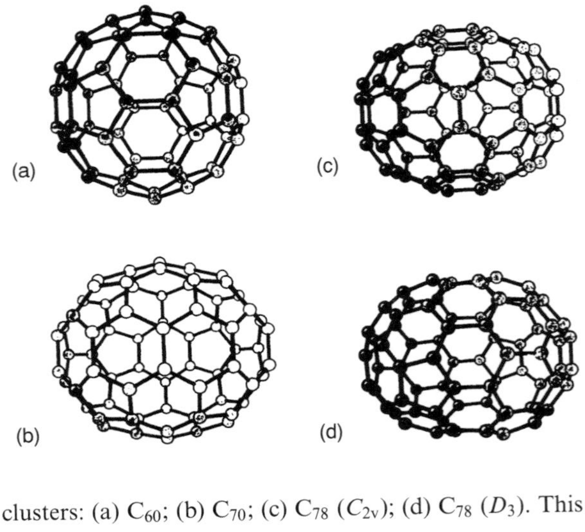

Fig. 8.3 Fullerene clusters: (a) C_{60}; (b) C_{70}; (c) C_{78} (C_{2v}); (d) C_{78} (D_3). This cluster is chiral.

consist of an f.c.c. packing of such almost spherical units, the clusters being held together by weak van der Waals interactions, with an inter-centre spacing of $\simeq 10$ Å and a minimum inter-cage separation of about 3 Å. The intramolecular bonding, on the other hand, is covalent with some contribution from π-bonding.

Part of the electronic band structure of f.c.c. C_{60}, in the vicinity of the top of the valence band and the bottom of the conduction band, is shown in Fig. 8.4, together with the corresponding density of states. The uppermost states in the valence band derive from the 30 π-orbitals for an isolated C_{60} cluster, grouped into seven variously degenerate molecular-orbital energy levels, and are occupied by 60 electrons (one from each carbon atom). The HOMO is five-fold degenerate for the isolated cluster: it can be seen that the highest groups of bands of states in the valence band of the solid derive from these molecular levels. The widths of all bands in the valence and conduction bands are very narrow, $\simeq 0.5$ eV, because of weak π–π interactions between C_{60} clusters in the solid state. Pure crystalline C_{60} is expected from the band structure, therefore, to be a semiconductor, with a gap of $\mathscr{E}_g \simeq 1.5$ eV.

Since the spacing between C_{60} molecules in condensed films is so large, the three interstitial sites (two tetrahedral, one octahedral) per cluster in the f.c.c. packing (cf. §2.2.4.2) are large enough to accept extraneous atoms, e.g. alkali atoms. The insertion (e.g. by in-diffusion) of such foreign atoms, a process termed intercalation (§1.2.5), is accompanied by charge transfer between the alkali atoms M and the C_{60} host, forming a fulleride: an electron is donated by each intercalated alkali atom and occupies the lowest group of (three) bands forming the bottom of the conduction band. Thus, at the composition M_3C_{60}, all interstitial sites of the f.c.c. lattice are filled with alkali atoms and if ionization of the alkali atoms is complete, the lowest band of conduction states will be half-filled with electrons; since the chemical potential therefore lies in the middle of a band of delocalized states, this intercalated compound will behave electrically as a *metal* (Fig. 8.4(b)).

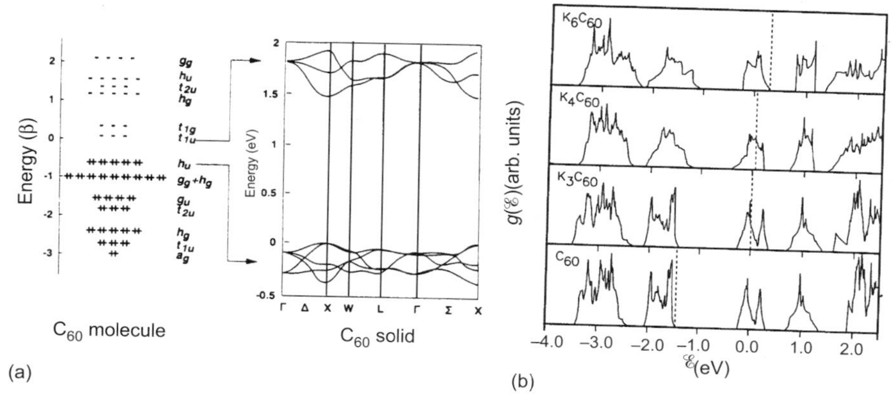

Fig. 8.4 Electronic structure of C_{60} and alkali fullerides. (a) Band structure of solid C_{60}, showing the top of the valence band and the bottom of the conduction bands, together with the molecular-orbital energy levels (and occupancies) obtained from a Hückel calculation for molecular C_{60}. (After Weaver and Poirier (1994). Reproduced by permission of Academic Press, Inc.) (b) Densities of states for solid C_{60} and some potassium fullerides. The position of the Fermi level is shown in each case by a dashed line. (After Erwin (1993). Reproduced by permission of John Wiley & Sonc Inc.)

Although the f.c.c. structure can only accept a maximum of three intercalant atoms per C_{60} cluster, expansion of the structure of the C_{60} packing to b.c.c. allows a further three foreign atoms to be intercalated, to give a limiting composition of M_6C_{60}. For this composition, the group of bands at the bottom of the conduction band have a different detailed shape from those for the f.c.c. structure (see Fig. 8.4b) but nonetheless are still narrow and well separated from the other conduction bands); for M_6C_{60}, they are completely filled and hence the material reverts to being a semiconductor (Fig. 8.4b).

Perhaps the most interesting behaviour exhibited by the solid form of C_{60} is that metallic intercalated compounds, M_3C_{60}, become *superconducting* at quite elevated transition temperatures, higher than 'conventional' intermetallic alloy superconducting materials (e.g. Nb_3Ge) but lower than the cuprate high-T_c materials (§6.4.3); M_3C_{60} materials are type-II superconductors. The very interesting feature is that the superconducting-transition temperature, T_c, increases with the type (i.e. size) of the intercalant cation or, alternatively, with the size of the expanded lattice constant, a, of the intercalated compound (see Fig. 8.5). The highest transition temperature is found for the compound $(Tl_2Rb)\,C_{60}$, with $T_c = 45$ K. This correlation between T_c and a can be understood, for example, in terms of the BCS expression for the superconducting transition temperature (eqn. (6.159)), namely $T_c = (1.14\hbar\omega_D/k_B)\exp(-1/\mathcal{V}_0 g(\mathscr{E}_F))$. If the electron–lattice interaction $\mathcal{V}_0$ is assumed to be independent of a, that is if *intra*molecular vibrational modes are responsible for the Cooper pairing, then as a increases, the electronic overlap between orbitals on adjacent C_{60} clusters will decrease, resulting in a reduced bandwidth and a corresponding increase in $g(\mathscr{E}_F)$ and hence of T_c. A review of the superconducting properties of fullerides has been given by Gunnarsson (1997), and a general survey of fullerenes and their compounds is given in the volume edited by Ehrenreich and Spaepen (1994).

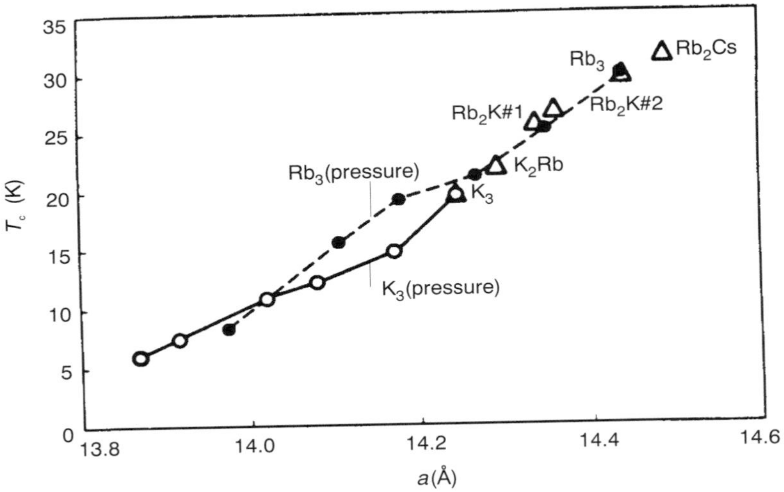

Fig. 8.5 Variation of the superconducting-transition temperature, T_c, with the lattice constant, a, for intercalated C_{60} films with general composition M_3C_{60}. (After Weaver and Poirier (1994). Reproduced by permission of Academic Press, Inc.)

8.2.2 Quantum dots

Material systems exhibiting size-quantization in three orthogonal directions, i.e. quantum dots, can be fabricated in a number of ways. Nanocrystals of a material (crystallites having spatial dimensions of the order of nanometres) can be grown with a well-defined size distribution in one of two ways: either as colloids, resulting from liquid-phase homogeneous precipitation or by the controlled growth of crystallites, e.g. of II–VI materials (e.g. CdS or CdSe) or IB–VII materials (e.g. CuCl) in either a glassy matrix (e.g. a borosilicate glass) or, in the latter case, in NaCl crystals as well. Alternatively, clusters of atoms can be grown (e.g. by vapour-phase condensation) to a limited size on the nanoscale. Finally, instead of the size of crystallites being restricted during their growth, the reverse procedure is possible, namely *removal* of material (e.g. in a film) by lithographic etching, resulting in a protuberance of nanoscaled dimensions (called a mesa) standing proud of the substrate. The optical and electronic behaviour characteristic of nanoscale samples will be discussed in the next two sections.

8.2.2.1 *Quantum dots: optical properties*

For nanocrystals (assumed to be spherical) of radius $\boldsymbol{R}$, the restricted size can have a marked influence on their optical properties. This is because the electronic wavefunction for a nanocrystal is strongly confined within its volume: there is a very high potential barrier at the interface between the semiconductor nanocrystal and the insulating glass matrix. Two spatial régimes can be distinguished, depending on the relative values of $\boldsymbol{R}$ and the Bohr radius, a_0^*, of the Mott–Wannier exciton (§5.8.3), with a_0^* for the Coulombically–coupled electron–hole pair being given by (cf. eqn. (6.235)):

$$a_0^* = 4\pi\varepsilon\varepsilon_0\hbar^2/\mu e^2, \tag{8.1}$$

where μ is the reduced mass ($\mu^{-1} = m_e^{*-1} + m_h^{*-1}$) for the relative (rotational) motion of electron and hole of effective masses m_e^* and m_h^*, respectively, and ε is the dielectric constant of the material containing the exciton.

The *weak-confinement* régime (Kayanuma (1988)) is for $R/a_0^* \gtrsim 4$: in this case, the kinetic energy associated with the centre-of-mass motion of the exciton (§5.8.3) is increased because of quantum-confinement effects (essentially because of the Heisenberg uncertainty principle) so that the exciton binding energy is decreased from the value it has for an infinite-sized system in terms of the Rydberg energy R^* (eqn. (5.223)), i.e.

$$\mathscr{E} = -\frac{R^*}{n^2} + \frac{\hbar^2}{2M}\left(\frac{\pi}{R}\right)^2 \tag{8.2}$$

where M is the exciton mass ($M = m_e^* + m_h^*$). The exciton can still be regarded as a quasiparticle, but its translational motion is quantized (see Fig. 8.6a).

The other limit is the *strong-confinement* régime (Kayanuma (1988), where $R/a_0^* \lesssim 2$ and the confinement kinetic energy is greater than the Coulomb-attraction interaction: for such small sizes of confinement box, the *relative* motion of the exciton can be regarded as being so strongly hindered that the spatial correlation between electron and hole is lost to a considerable extent, and the electron and hole occupy quantized energy levels corresponding to the particle-in-a-box problem. The wavefunction $\psi_n(r)$ with s-like symmetry for a particle of mass m confined in a spherical box of radius R is

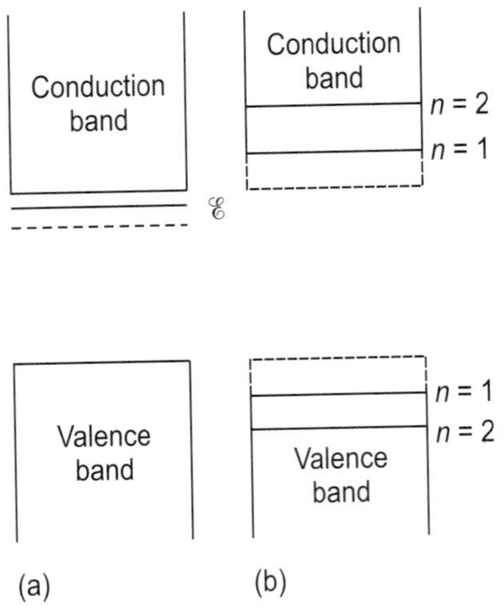

Fig. 8.6 Effects of quantum confinement on the electronic energy levels associated with optical absorption for semiconductor nanocrystals in: (a) the weak-confinement régime. The exciton binding energy $\mathscr{E}$ is reduced from the value it has for an infinite crystal (dashed line). (b) the strong-confinement régime. The valence and conduction bands split into a series of sub-bands corresponding to the energy levels of a particle-in-a-box.

$$\Psi_n(r) = \frac{C_n}{r}\sin(n\pi r/R), \tag{8.3}$$

where C_n is a constant and n are integers ($n = 1, 2, \ldots$), and the corresponding energy levels are given by

$$\mathscr{E}_n = -\frac{\hbar^2\pi^2 n^2}{2mR^2}, \tag{8.4}$$

so that the energy of the electron–hole pair in the $n = 1$ (ls) state is given by (Brus (1986), Katayuma (1988)):

$$\mathscr{E} \simeq \frac{\hbar^2}{2\mu}\left(\frac{\pi}{R}\right)^2 - \frac{1.8e^2}{4\pi\varepsilon\varepsilon_0 R}, \tag{8.5}$$

where the second term is the remnant of the exciton interaction. Thus, in both cases, a *blue*-shift in optical absorption is expected because of quantum-confinement effects—see Fig. 8.7. Since the (ls) exciton Bohr radius a_0^* in CuCl is 7 Å and in CdS is 29 Å, for likely sizes of nanocrystals ($R \simeq 30$ Å), CuCl will therefore be in the weak-confinement régime and CdS in the strong-confinement régime. The effect of quantum confinement on the optical-absorption spectrum of CdS nanocrystals in a borosilicate glass matrix is illustrated in Fig. 8.8: as the radius of the nanocrystals decreases from $\simeq 50$ Å to 17 Å, the optical-absorption edge shifts to higher energies. See also plate VI. ★

Quantum confinement also has an important effect on the oscillator strength f of the optical absorption involving excitons in nanocrystallites. The oscillator strength, f_1, of the optical transition to the first excited ls exciton state can be written as (Kayanuma (1988)):

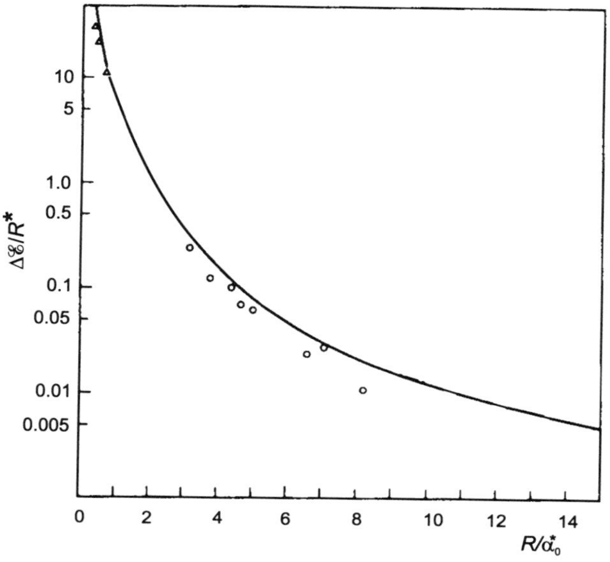

Fig. 8.7 Calculated blue-shift (as a ratio of the infinite–crystal exciton Rydberg energy) of the ground state of an electron–hole system in a spherical quantum dot for $m_h^*/m_e^* = 10$ as a function of the radius of the confinement box (as a ratio of the exciton Bohr radius). Experimental data for CuCl nanocrystals in NaCl (O) and CdS nanocrystals in a silicate glass (Δ) are shown for comparison (Reprinted with permission from Kayanuma (1988), *Phys. Rev.* **B38**, 9797. © 1988. The American Physical Society)

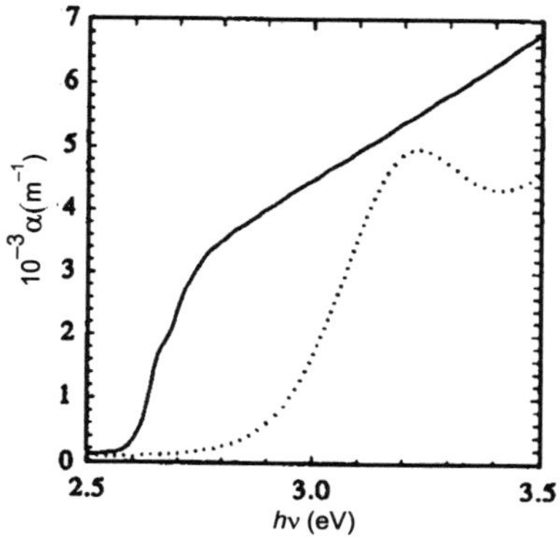

Fig. 8.8 Room-temperature optical-absorption spectra for CdS nanocrystals in a borosilicate glass matrix with different crystallite radii: $R = 1.7$ nm (dotted line) and $\simeq 5$ nm (solid line). (Yükselici and Persans, reprinted from *J. Non-Cryst. Sol.* **203**, 206, © 1996 with kind permission from Elsevier Science - NL, Sara Burgerhartstraat 25, 1055 KV Amsterdam, The Netherlands)

$$\frac{f_1}{f} = \frac{\pi a_0^{*3}}{V} \left| \int \Psi_1(r_e, r_h, r_{eh} = 0) \mathrm{d}r \right|^2, \tag{8.6}$$

where f is the ls-exciton oscillator strength per unit volume in a *bulk* crystal, Ψ is the exciton wavefunction ($r_{eh} = r_e - r_h$ being the difference between electron and hole spatial coordinates), and V is the volume of the nanocrystal. Thus, from eqn. (8.6), the overall oscillator strength of the nanocrystal is Vf_1, and so scales as R^3. In the strong-confinement régime, moreover, the oscillator strength is concentrated into the lowest-energy transitions; as a result, excitonic effects (i.e. peaks near the gap energy) in the optical-absorption spectra are accentuated for nanocrystals (see Fig. 8.8).

This size-enhancement of the oscillator strength of nanocrystals also produces an enhancement of the *non-linear* optical response of the material (see §5.8.4), specifically of the third-order susceptibility $\chi^{(3)}$, the real part of which is related to the non-linear refractive index, n_2 (eqn. (5.231)) and the imaginary part of which is related to the change in the optical-absorption coefficient, ΔK. The value of n_2 for bulk CuCl is 7.5×10^{-13} m^2 W^{-1}, but a glass matrix containing just 0.5 vol. % of nanocrystallites of CuCl of radius 34 Å has an effective $n_2 = 3 \times 10^{-11}$ m^2 W^{-1} (Justus *et al.* (1990)): taking into account the small overall volume of the nanocrystallites, the actual value of n_2 for CuCl in nanocrystalline form must be about three orders of magnitude higher still. ⋆

Finally, quantum confinement has an effect on the rate of radiative interband transitions in semiconductors that, as bulk materials, have *indirect* gaps between states at the top of the valence band and those at the bottom of the conduction band. Normally, the rate of indirect optical transitions is very low because of the necessity of phonon involvement in the electronic transition in order to conserve momentum (§5.8.2). However, the lack of translational symmetry resulting from the restricted size of nanocrystals means that the electron wavevector $\boldsymbol{k}$ is no longer a good quantum number, and hence optical transitions are allowed between any states and phonons need not be involved. As a result, the radiative transition rate increases. This behaviour is found for *porous* silicon; bulk c-Si is, of course, an indirect-gap semiconductor (§5.4.2). Porous Si is made by electrolysing doped c-Si wafers in an aqueous HF solution; most of the material is etched away, leaving isolated columns of large aspect ratio (with diameters of the order of a few nanometres) standing normal to the substrate. The cores of such columns comprise nanocrystals of c-Si embedded in a disordered matrix of oxidized silicon. As a result of the loss of translational invariance, the Si nanocrystals in porous silicon can exhibit efficient photoluminescence (Canham (1990)), or even electroluminescence, which is the emission of light following the absorption of other light or the application of an external electric field (§8.5.2.3). Hence the optical behaviour of porous Si, in contrast to its bulk counterpart, is effectively that of a 'direct-gap' material.

Thus far, it has been assumed that the separation between quantum dots is sufficiently great that they do not interact electronically. However, it is possible to envisage controlling the distance separating quantum dots during the fabrication process, so that inter-dot interactions can be changed in a controllable way. Such a procedure can be achieved by a colloidal aggregation of nanocrystallites (whose surfaces are passivated with organic ligands) from a dispersion in a liquid, following evaporation of the liquid. Colloidal crystals, i.e. 3D quantum-dot superlattices, several tens of microns in size can be made in this way by self-organization of CdSe nanocrystals each with diameters of a few tens of nanometres (Murray *et al.* (1995))—see Fig. 8.9. (Opal is a naturally

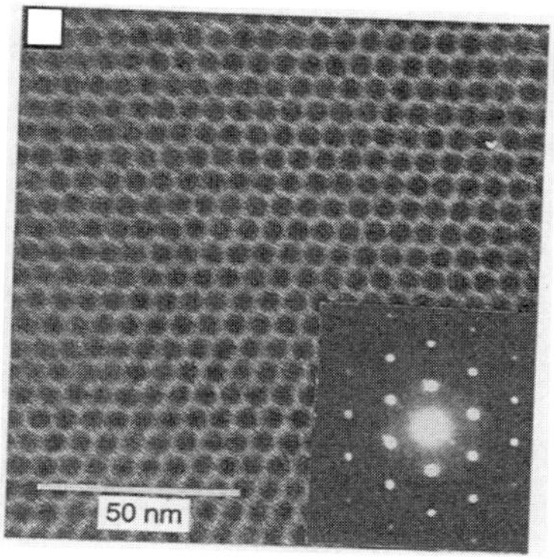

Fig. 8.9 High-resolution transmission electron micrograph and, inset, the corresponding electron diffraction pattern of a $(10\bar{1})$ projection of an f.c.c. superlattice array of quantum dots of nanocrystals of CdSe, each 48 Å in diameter. (Reprinted with permission from Murray *et al.*, *Science* **270**, 1335. © 1995 American Association for the Advancement of Science)

occurring colloidal crystal, consisting of ordered 3D arrays of monodisperse, micron-sized silica particles; its iridescence results from Bragg diffraction (§2.6.1.1) of light from planes of particles.) The electronic interaction between CdSe quantum dots in such a superlattice causes a small red-shift of the optical absorption spectrum with respect to that of an isolated nanocrystal.

8.2.2.2 *Quantum dots: electronic properties*

Size can have dramatic effects on the electronic properties of samples in the nanoscale range. Electrons confined within a small-radius spherical potential well will occupy discrete energy levels given by the particle-in-a-box solution (see Fig. 8.10) rather than, for example, the Bloch functions (§5.2.1) characteristic of systems with (infinite) translational periodicity. It can be seen from Fig. 8.10 that a closed-shell electronic structure, as in atoms, is evident: clusters of metal atoms containing a total number of valence electrons sufficient completely to fill a number of shells will be particularly stable energetically. This picture explains the so-called 'magic numbers' of the sizes of clusters of, say, Na atoms preferentially found in supersonic beams, produced by mixing the metal vapour with an inert carrier gas and ejecting the mixture through a nozzle (see e.g. de Heer (1993)). Figure 8.11 shows that magic clusters, with numbers of Na atoms $N = 8, 20, 40, 58, 92, 138$, etc., are found experimentally: examination of the energy-level diagrams shown in Fig. 8.10 shows that such numbers correspond to the occupancy of closed shells of electrons in a spherical square-well (or similar) potential. In fact, preferred sizes are found even for clusters of Na atoms containing up to $N = 25000$ atoms. The spherical shell-closure series of numbers continues only up to $N = 1430$. After that, preferred cluster sizes seem to be determined more by structural-packing than by electronic shell-filling considerations: the resulting set of other magic numbers 1980, 2820 . . . 21300 correspond to complete Mackay icosahedra (§2.2.2.4) of different sizes. Thus, it is only for clusters of metal atoms containing several tens of thousands of atoms that behaviour characteristic of the bulk material is recovered.

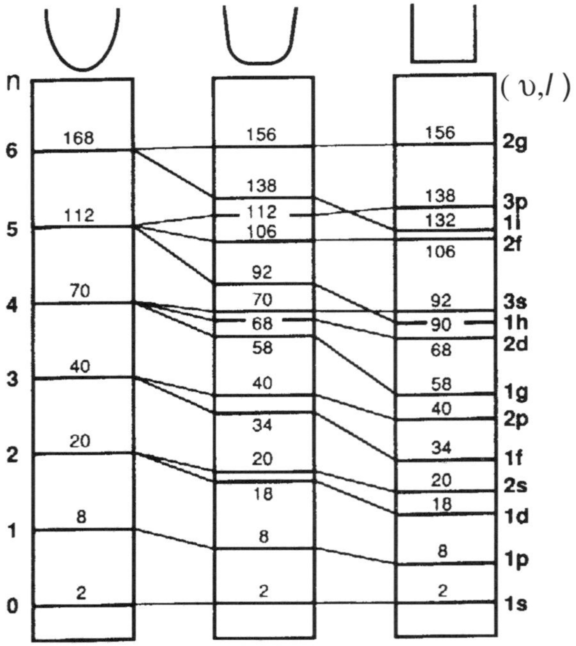

Fig. 8.10 Energy levels for an electron in spherical 3D harmonic, intermediate and square-well potentials (Reprinted with permission from de Heer (1993), *Rev. Mod. Phys.* **65**, 611. © 1993. The American Physical Society). The levels for the square-well and intermediate potentials are labelled according to the nuclear (not atomic) convention whereby, for a given angular momentum quantum number l, the lowest radial quantum number is $\upsilon = 1$. The cumulative maximum occupation numbers are shown for each level, from which it can be seen that 'magic' numbers (8, 20, 40, 58, 92, etc.) correspond to filled shells of electrons.

Perhaps the most remarkable effect exhibited by quantum dots is that the capacitance of, or the electrical current passing through, a dot can be changed drastically simply by the addition of a *single* extra electron as a result of one-electron tunnelling. In order to observe this phenomenon, it is necessary to confine a few electrons in a region of mesoscopic (i.e. sub-micron) dimensions, insulated from the surroundings, and to measure the electrical current passing through this region via source and drain electrodes; the charge carriers pass between the quantum dot and the electrodes by quantum-mechanical tunnelling through the intervening (thin) insulating barrier. A number of configurations can be envisaged by which electron confinement can be achieved. Perhaps the simplest configuration has a quantum dot formed from a metallic film of mesoscopic dimensions, separated by narrow insulating regions from source and drain electrodes in a *coplanar* arrangement, with a 'gate' electrode underneath the device, also separated by an insulating layer (Fig. 8.12a). A related configuration is that of an etched mesa, in which a quantum dot (e.g. of the semiconductor GaAs) is sandwiched between thin semi-insulating layers of $Al_xGa_{1-x}As$ alloy (with a bandgap larger by $\simeq 0.5$ eV than that of pure GaAs for $x = 0.4$); source and drain electrodes contact the top and

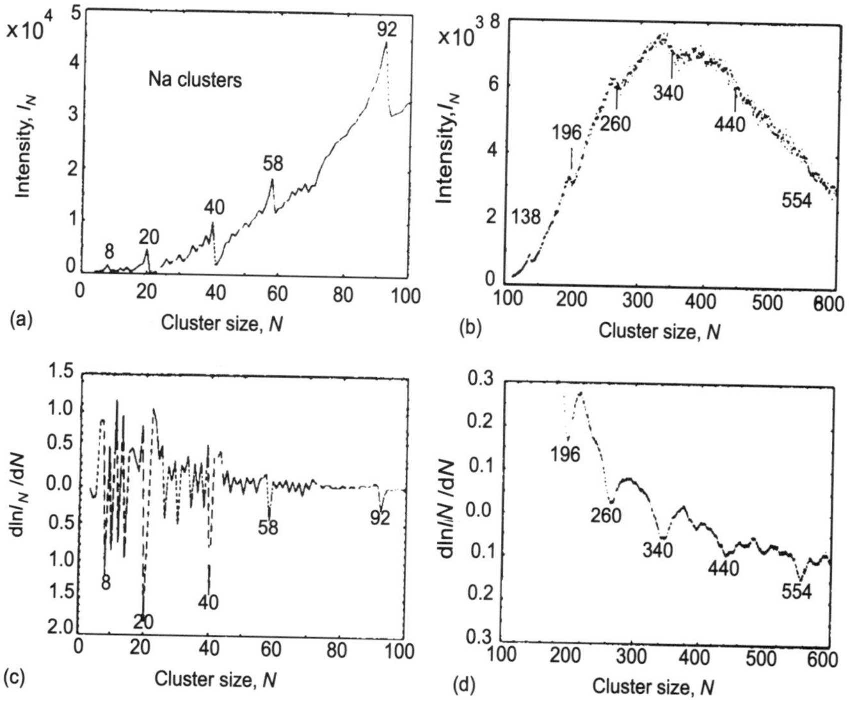

Fig. 8.11 Abundances of sizes of sodium clusters produced in a supersonic beam, showing the 'magic' numbers marked (8, 20, 40 . . . 554) (after Bjørnholm *et al.* (1991)). The data are plotted as intensity (a, b), and also as logarithmic differences (c, d) to accentuate the peaks in (b).

bottom surfaces of the insulator–semiconductor–insulator sandwich (Fig. 8.12b). A more sophisticated way to confine electrons is to use electric-field gradients. A device geometry that can be used to achieve this is shown in Fig. 8.12c. A semi-insulating $Al_xGa_{1-x}As$ layer insulates a gate electrode from a semiconducting layer of GaAs; a positive voltage V_g applied to the gate causes a 2D electron gas to accumulate at the AlGaAs/GaAs interface (see §8.4.2.2). However, additional shaped electrodes placed on the top surface of the GaAs, biased negatively with respect to the source and drain electrodes, repel electrons in the region under the shaped electrodes, with the result that the electrons are confined by electrostatic potential barriers (Fig 8.12d) to the mesoscopic region between constricting protuberances of the shaped electrodes. In this case, electrons can only join, or leave, the pool of confined electrons by tunnelling from the source, or to the drain, electrode through the potential barriers, whose height can be varied by changing the voltage applied to the confining electrodes.

In such configurations of quantum dots, the conductance G_0 of the device (equal to the source–drain current divided by the voltage dropped between source and drain) exhibits a series of sharp peaks that are almost periodic as a function of the gate voltage at very low temperatures (Fig. 8.13). Each peak in G_0 corresponds to the addition of a *single* electron to the confined electron pool: hence, such devices have been termed single-electron transistors (Kastner 1992, 1993).

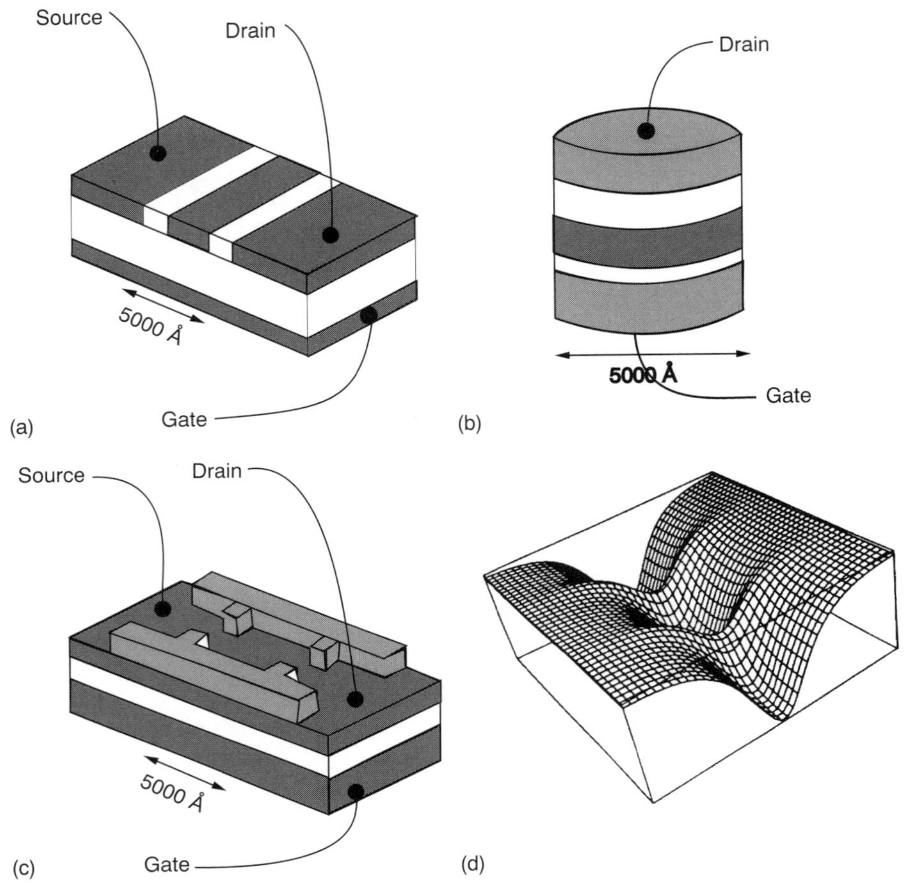

Fig. 8.12 Configurations of quantum dots that act as one-electron transistors. (a) Metal quantum dot in a coplanar configuration. (b) Semiconductor (GaAs) quantum-dot mesa in a sandwich configuration. (c) Semiconductor (GaAs) quantum dot defined by the electric-field gradients associated with the constrictions in shaped electrodes on top of the semiconductor. (d) The electrostatic potential surface associated with the shaped-electrode configuration in (c). The pool of confined electrons occupies the central potential well.

The extraordinary pseudo-periodic behaviour evident in Fig. 8.13 can be explained by the Coulomb-blockade model. Suppose that a quantum dot is initially electrically neutral; addition of an electron to it requires an energy $e^2/2C$, where C is the total capacitance between the dot and its surroundings. Thus, in order for a current to flow, an energy barrier (the Coulomb blockade) of $e^2/2C$ must be surmounted: for an electron to tunnel onto the dot (or Coulomb island), its energy must be greater than the Fermi energy of the contact by $e^2/2C$, and for a hole to tunnel, likewise its energy must be below the contact Fermi energy also by $e^2/2C$. Thus, there is a total gap of width e^2/C in the tunnelling density of states of the quantum dot. If the temperature is sufficiently low that $k_{\mathrm{B}}T < e^2/2C$, neither electrons nor holes can flow from source to

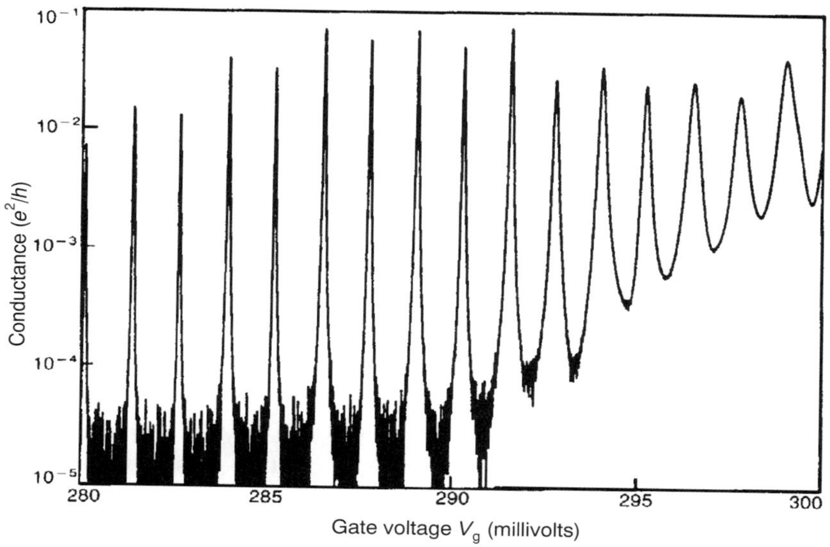

Fig. 8.13 Conductance, in terms of the quantity e^2/h, of a one-electron transistor with a configuration as in Fig. 8.12c, measured at 60 mK as a function of gate voltage (Kastner (1993). Reprinted with permission from *Phys. Today* **46**, 1, 24. © 1993 American Institute of Physics)

drain via the quantum dot. Since the capacitance of such quantum dots is typically $C \simeq 10^{-16}$F, this means that Coulomb-blockade effects will only be evident at very low temperatures, $T \lesssim 1$ K.

However, a tunnelling current can be made to flow by varying the gate voltage for the configurations (a) and (c) in Fig. 8.12, since this changes the energy needed to add charge to the quantum dot. The electrostatic energy of a quantum dot having a charge Q and subject to a gate voltage, V_g, is

$$\mathscr{E} = QV_g + Q^2/2C. \tag{8.7}$$

For negative Q, and with a positive V_g, eqn. (8.7) shows that the energy is a parabolic function of Q, with its minimum at $Q_0 = -CV_g$ (Fig. 8.14). However, since electrical charge is quantized in units of the electronic charge e, the energy given by eqn. (8.7) also can only take discrete values. When V_g is such that the minimum energy is determined by $Q_0 = -Ne$, an integral number of electrons, there is a Coulomb blockade of $\pm e^2/2C$ as before for changing N by ± 1. For other values of Q_0, that are not integral numbers of electrons (or holes), and except also for the case $Q_0 = -(N + 1/2)e$, there is a small but finite energy gap between $\mathscr{E}_F$ and the nearest available quantum-dot state and so no electrical current will flow at low T (Fig. 8.14). However, for the special case $Q_0 = -(N + 1/2)e$, the two charge states of the Coulomb island with $Q = -Ne$ and $-(N + 1)e$ are energetically degenerate and, as a consequence, there is a fluctuation in charge between these two values even at zero kelvin. At this charge-degeneracy point, the gap in tunnelling energies disappears and current can flow via the quantum dot (Fig. 8.14c). Thus, as V_g is increased continuously, the tunnelling states for the Coulomb island are pulled down with respect to the Fermi level in the contacts until, at periodic values of V_g with period e/C, and corresponding to the transfer of a single electron (hole), the charge-degeneracy condition is achieved and the source–drain conductance is high: otherwise, there is a gap in tunnelling states at the Fermi energy and the device switches off.

This Coulomb-blockade picture is valid for metallic quantum dots (Fig. 8.12a) where, due to screening effects (§5.6.1) associated with the very large number $N \simeq 10^7$) of free electrons, extra charges reside on the surface of the metallic region and the Coulomb-blockade energy is given accurately by $\pm e^2/2C$. For the semiconducting quantum-dot structure shown in Fig. 8.12c, the number of trapped electrons is very much smaller ($N \simeq 50$) and so the above expression for the Coulomb-blockade energy is only approximate. Moreover, energy quantization due to the restricted size of the Coulomb island is also important for the semiconductor case, as indicated in Fig. 8.14. Since size-quantization leads to energy levels whose spacing decreases in general with energy (see Fig. 8.10), this discreteness will be more evident for lower electron-occupation numbers than for very high occupation numbers.

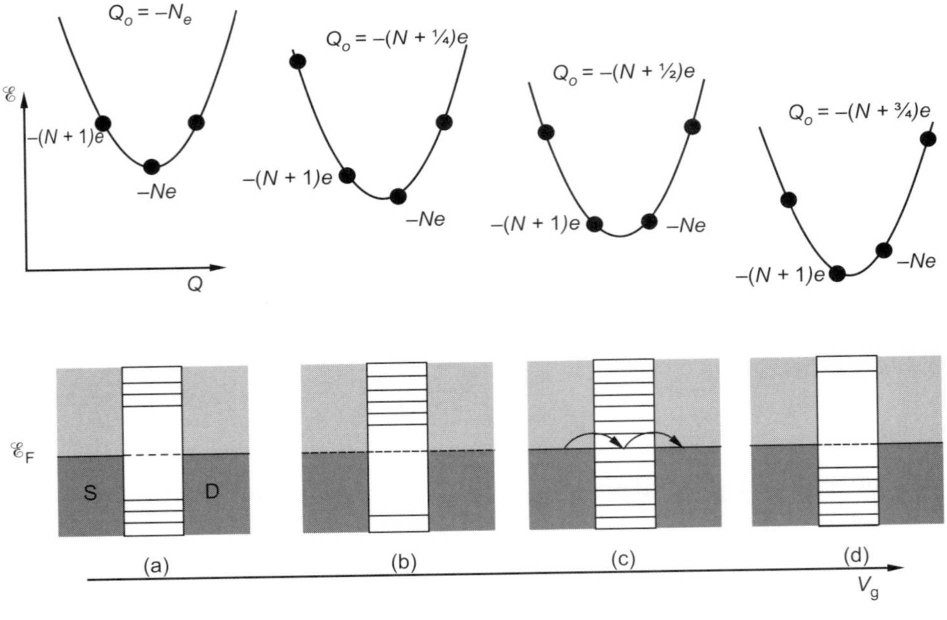

Fig. 8.14 Illustration of the origin of the pseudo-periodic variation in source–drain conductance of one-electron transistors as a function of the gate voltage, V_g. It is only at the charge-degeneracy point (c), where two charge states (-Ne and $-(N+1)e$) of a quantum dot have the same energy, that an electron state for the dot coincides with the Fermi energy of the contacts, thereby allowing a tunnelling source-drain current to flow via the dot. (After Kastner (1993))

One-dimensional systems 8.3

8.3.1 One-dimensional metals

How can one produce a one-dimensional metal? What is required from a microscopic structural point of view (i.e. in terms of classification scheme I (§8.1)) is a very strong overlap between orbitals on neighbouring atoms along a chain of atoms or within a stack of molecules, with negligible electronic interactions between chains or stacks; the resulting 1D band must also be part-filled. One type of structure that satisfies these criteria is the platinocyanate chain compound $K_2Pt(CN)_4Br_{0.3}3H_2O$, known as KCP. In this, square-planar $Pt(CN)_4^{2-}$ molecular units form 1D stacks as in Fig. 8.15, with direct metal–metal bonding involving overlap of the Pt $5d_{z^2}$ orbitals; the average Pt–Pt separation within the chain is 2.88 Å, not much larger than in elemental Pt metal (2.78 Å). The cyanide ligands, together with the other ions and molecules (K^+, Br^-, H_2O) situated between the chains ensure that the interchain Pt–Pt separation is large (9.9 Å) and hence that the electronic coupling *between* chains is very small. The presence of the non-stoichiometric amount of Br^- ions ensures that 0.3 holes per Pt atom are introduced into the otherwise filled 1D $5d_{z^2}$ band, ensuring metallic behaviour (at least at not-too-low temperatures — see §8.3.2).

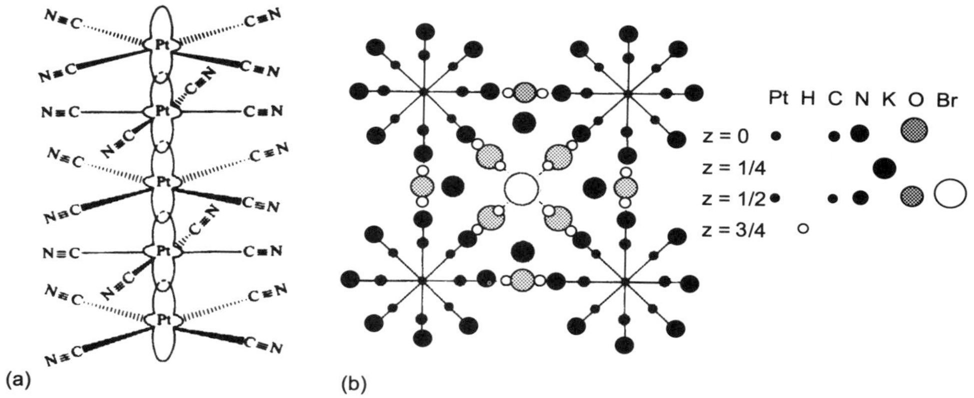

Fig. 8.15 Structure of the 1D metal, KCP ($K_2Pt(CN)_4Br_{0.3}.3H_2O$), consisting of a stack of square-planar $Pt(CN)_4^{2-}$ clusters with Pt–Pt bonding, via overlap of $5d_{z^2}$ orbitals, along the chain. (a) side view; (b) projection.

As expected, the electrical and optical behaviour of KCP are very anisotropic. The electrical conductivity along the chains is $10^4 - 10^5$ times larger than that perpendicular to the chains; above 200 K, the conductivity parallel to the chains, $\sigma_{||}$ is only weakly temperature-dependent and has a value characteristic of a metal (Fig. 8.16a). (The strongly thermally activated temperature dependence of $\sigma_{||}$ below 200 K evident in Fig. 8.16a is evidence of a breakdown in metallic behaviour characteristic of 1D metals, associated with a (Peierls) structural transition of the conducting chains of atoms — see

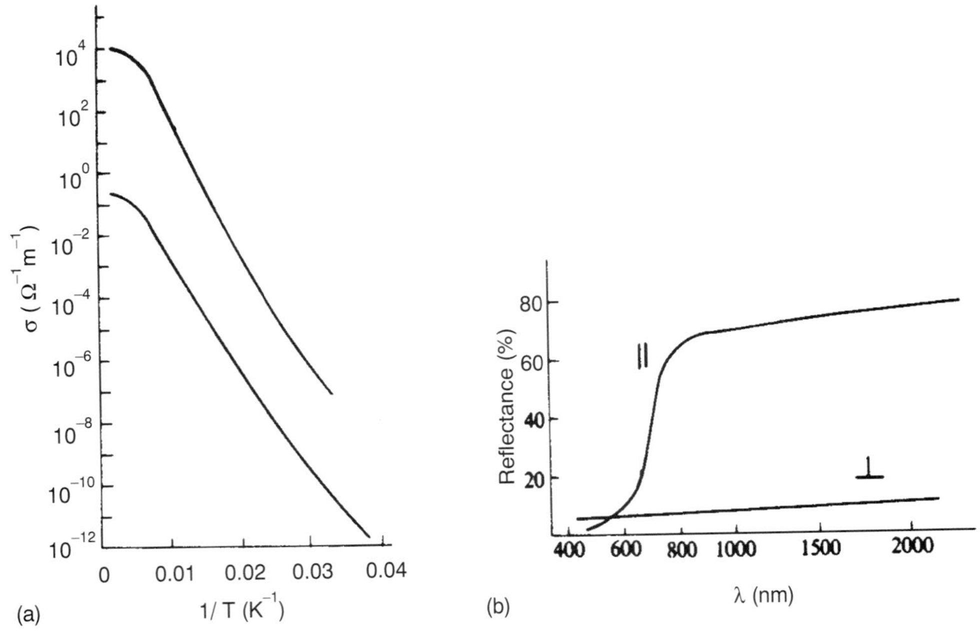

Fig. 8.16 Anisotropic electronic behaviour of KCP (a 1D metal above 200 K): (a) electrical conductivity parallel and perpendicular to the chains; (b) optical reflectivity for light polarized parallel and perpendicular to the chains.

§8.3.2.) The 1D metallic character of KCP (above 200 K) is evident also in the optical reflectivity (Fig. 8.16b): light with the ***E***-vector polarized *parallel* to the chains is strongly reflected for wavelengths longer than that ($\simeq$ 600 nm) corresponding to the plasma edge (§5.8.1), whereas the reflectivity of KCP for light polarized *orthogonal* to the chain direction is very small, characteristic of insulating materials (see Plate VII). A general discussion of the behaviour of KCP is given in Keller (1975).

Another class of materials that are 1D metals (at least at not-too-low temperatures) are the so-called molecular metals which consist of stacks of organic molecules where the electronic (π-) interactions between molecules within a stack are much stronger than those between stacks. The canonical example of such a material is the charge-transfer organic compound TTF:TCNQ (TTF = tetrathiafulvalene; TCNQ = tetracyanoquinodimethane). The crystal structure consists of alternate 1D stacks consisting entirely of one or the other type of molecule (Fig. 8.17). The TTF molecule is a donor, losing charge relatively readily; in contrast, TCNQ is an acceptor molecule. In TTF:TCNQ, there is a partial charge transfer of 0.69 electron per molecule between donor and acceptor molecules, and so the 1D bands for *both* the TTF and TCNQ stacks are part-filled and hence both contribute to the metallic behaviour. Indeed, TTF:TCNQ is a 1D metal above 60 K, with a maximum metallic conductivity some 10^2 times as large as that of KCP, and a room-temperature value of $\sigma_{\|} \simeq 10^4\ \Omega^{-1}\ \mathrm{m}^{-1}$. A discussion of TTF:TCNQ and similar materials is given in Keller (1977).

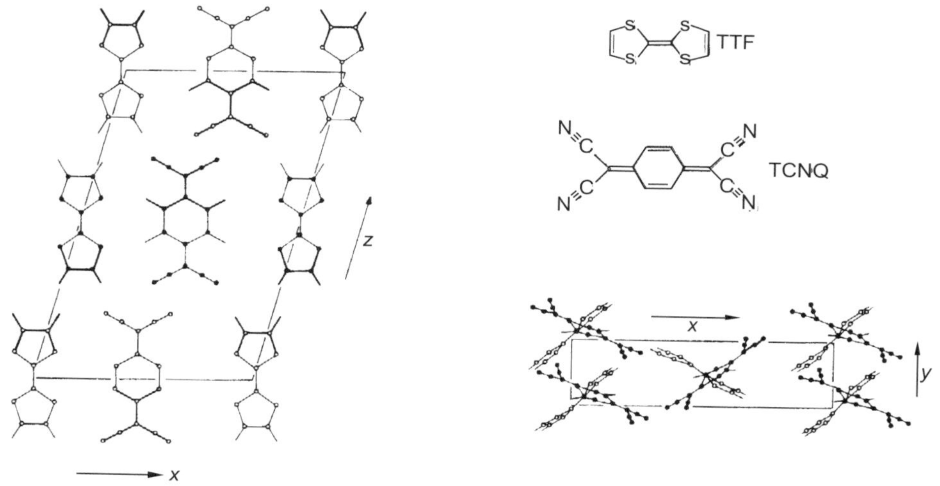

Fig. 8.17 Crystal structure of the 1D metal, charge-transfer organic compound, TTF: TCNQ (the component molecules are shown in the inset) for two projections of the structure.

8.3.2 Peierls distortion

We have seen in the previous section that certain materials behave as 1D metals at elevated temperatures; however, as the temperature is lowered, metallic behaviour is not maintained as in normal 3D metals (§6.3.2.1), but instead semiconducting behaviour occurs (see Fig. 8.16a). A gap must therefore open up in the electron states at the Fermi energy at low temperatures. This opening up of a gap at $\mathscr{E}_F$ is due to a static structural distortion termed the Peierls distortion, and it occurs because of a strong coupling between electrons and phonons of particular wavevectors.

The Peierls distortion can most easily be understood for the simple case of a half-filled 1D band, corresponding to a linear chain of atoms with (undistorted) real-space periodicity a, where the Fermi level is at the Fermi points $k_F = \pm\pi/2a$. If, in a gedanken experiment, a 1D chain of atoms is assumed to distort so that the new real-space periodicity is *twice* the original value, then new bandgaps will open up at values of electron wavevector equal to *half* the original zone-boundary values, i.e. $k = \pm\pi/2a$, in other words at the Fermi 'points' where the Fermi level intersects the 1D bands (Fig. 8.18). Thus, the electronic energy is lowered, since filled electron states ($|k| \leqslant |k_F|$) are pulled down in energy while the states that are pushed up ($|k| > |k_F|$) are empty; as long as this decrease in electronic energy is greater than the increase in elastic strain energy associated with the atomic distortion of the chain the reconstruction will be favoured energetically.

By analogy with the results for the NFE model (§5.2.3), the new reciprocal-lattice vector $\boldsymbol{G}'$ for the Peierls-distorted state is simply given in general by the minimum spanning vector (i.e. $2k_F$) linking the parts of the band structure where the new bandgaps occur (Fig. 8.18), and it is also related to the new spatial period, D, via

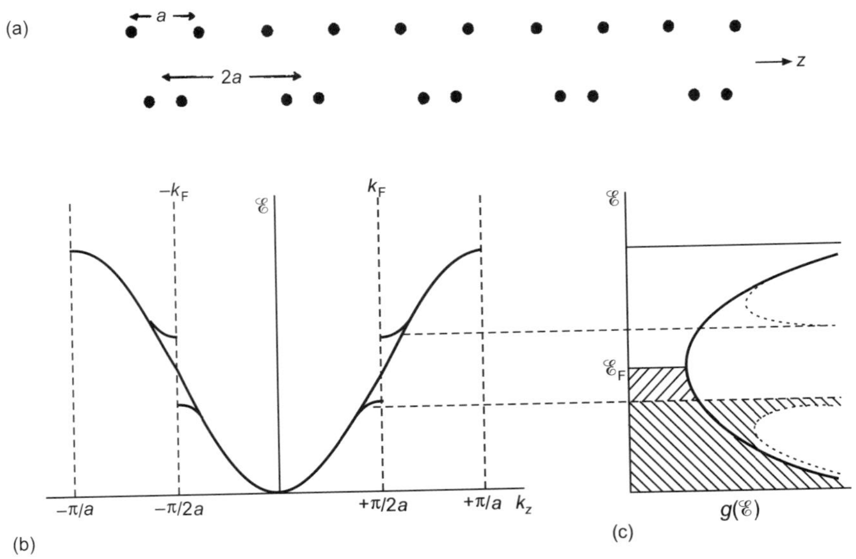

Fig. 8.18 Illustration of the origin of the Peierls distortion in a 1D chain of atoms resulting in the opening of a bandgap at $\mathscr{E}_F$ in a partially filled 1D band. For the case of half-filling, as shown here, (a) a distortion giving a doubling of the spatial periodicity of the chain leads to (b) gaps opening up in the band structure at $k = \pm\pi/2a = \pm k_F$. (c) The corresponding redistribution of the density of states.

$$G' = 2k_F \tag{8.8a}$$

$$= \frac{2\pi}{D}. \tag{8.8b}$$

Hence, from eqn. (8.8), the value of the spatial period of the Peierls distortion is set by the value of k_F (i.e. simply by the electron occupancy of the 1D band) according to

$$D = \pi/k_F. \tag{8.9}$$

(Note that eqns. (8.8b) and (8.9) are valid only for less than, or exactly, half-filling of the band, $k_F \leqslant \pi/2a$—see Problem 8.2.) For the case of the half-filled band, $k_F = \pi/2a$ and, from eqn. (8.9), $D = 2a$, as postulated. When k_F is a rational fraction of the zone-boundary wavevector $k_F = \frac{1}{n}(\pi/a)$, the period D of the distortion in the Peierls-distorted state is an integral number of fundamental lattice constants, $D = na$, from eqn. (8.9): such a distortion is said to be commensurate with the underlying lattice. However, k_F need *not* be a rational fraction of π/a, in which case the corresponding Peierls distortion is incommensurate. For example, for the case of KCP, the Peierls-distortion periodicity is $D = 6.67a$ (Problem 8.2).

The Peierls distortion is a solid-state version of the Jahn–Teller distortion that is found for molecules: a coupling of degenerate electron states to a vibrational normal mode of the chain of atoms causes a structural distortion to a structural configuration of lower symmetry. The involvement of the phonon mode in the Peierls distortion can be demonstrated pictorially for the simple case of a chain of atoms, each atom

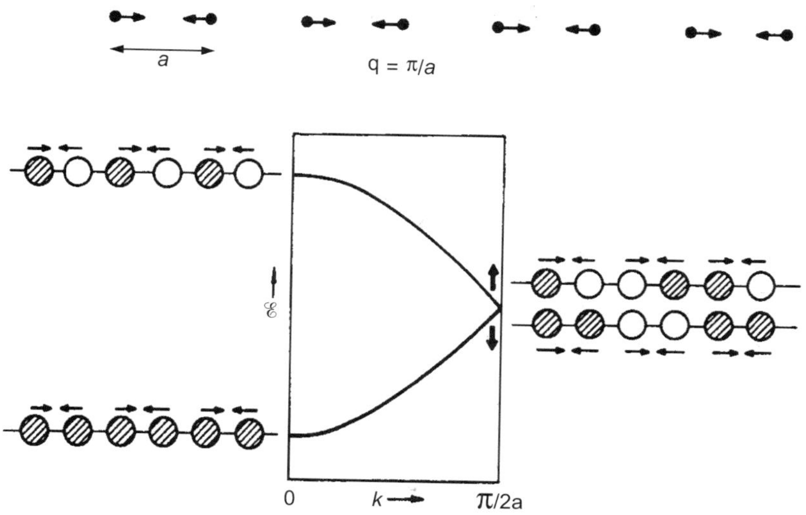

Fig. 8.19 Pictorial illustration of the electron–phonon interaction between electrons at $k = 0$ and k_F ($= \pi/2a$ for a half-filled 1D band) and a phonon mode with $q = \pi/a$. In the Peierls-distorted state, the periodicity in this case is twice that of the undistorted chain; hence the Brillouin zone for electron states is halved in extent, and the band is folded back. The two configurations of the s-orbital coefficients of the Bloch electronic wavefunctions (open circles = positive value; hatched circle = negative value of same magnitude) that are degenerate for the undistorted chain become non-degenerate when subject to a static distortion with wavevector $q = 2k_F = \pi/a$.

contributing one s-electron to the conduction band, which is therefore half-filled. In the Peierls-distorted case, with $D = 2a$, the new Brillouin zone is half that of the undistorted case, resulting in a folding of the original band for $k > \pi/2a$ in the reduced-zone scheme (§5.2.2). The phonon mode that drives the reconstruction of the regular chain into the Peierls state in this case is shown in Fig. 8.19: this symmetric pairing mode is, in fact, simply a zone-boundary LA phonon (for the undistorted chain), with wavevector $q = \pi/a$ (cf. Fig. 4.12). If the phases of the s-orbital coefficients a_{kn} of the Bloch electronic wavefunctions are represented as open circles (positive values) and hatched circles (negative values of the same magnitude), the bonding and antibonding combinations at $\boldsymbol{k} = 0$ (infinite electron wavelength) are as shown in Fig. 8.19. The $q = \pi/a$ LA phonon mode causes the two orbital combinations having an electron wavevector $\boldsymbol{k} = \pi/2a$, which are degenerate in the case of the original undistorted chain, to have *different* energies; the lower state has more bonding character and the upper has more antibonding character after the static displacement occurs.

At elevated temperatures, the vibrational amplitude of the atoms increases (§4.2.6) and the static displacement characteristic of the Peierls distortion is washed out; the material then reverts to the metallic state.

The electron–phonon coupling is apparent also in the dynamical behaviour of the material *above* the Peierls transition temperature in the phonon spectra, as revealed by inelastic neutron scattering. This occurs because the Lindhard expression (eqn. (5.157)) for the electronic screening of the ion–ion Coulombic interaction exhibits a logarithmic singularity when $q = 2k_F$, translating into singularities (Kohn anomalies) in the phonon

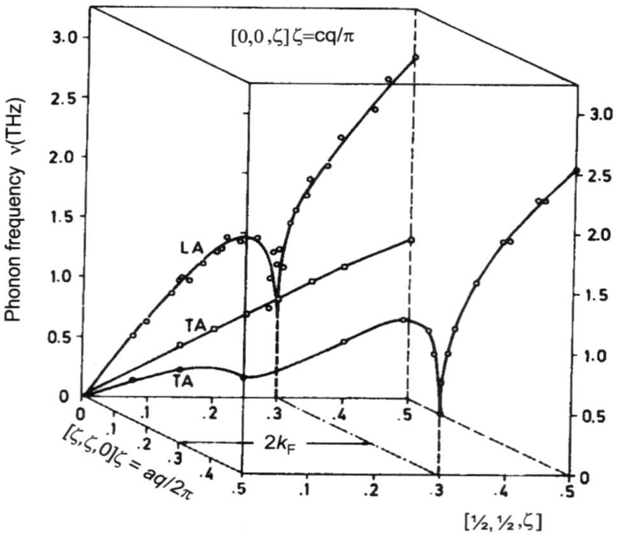

Fig. 8.20 Acoustic-phonon branches in deuterated KCP at room temperature measured by inelastic neutron scattering (Renker and Comès (1975)). A Kohn anomaly is evident along the line $q = 2k_F = 0.3\pi/c$. Reproduced by permission of Plenum Publishing Corp.

dispersion curves, $\omega(\boldsymbol{q})$. Such Kohn anomalies are evident in the acoustic-phonon branches of (deuterated) KCP in the metallic state above the Peierls transition temperature (Fig. 8.20). It appears, therefore, that the Peierls transition could be viewed as arising from a soft-mode mechanism (§7.1.5.4) for a phonon with $q = 2k_F$. However, fluctuations also play an important rôle in the actual transition. Further discussion of the Peierls transition can be found, for example, in Keller (1975, 1977).

8.3.3 Conjugated polymers

Another important class of one-dimensional conductors comprises organic polymers in which the carbon atoms exhibit $sp^2 + p_z$ hybridization and there is the possibility of *intra*molecular π-bond formation along the chains involving the p_z orbitals (in contrast to the *inter*molecular π-interactions present in TTF:TCNQ, for example). Extended bonding of the π-electrons along the polymer chain is possible when the polymer is conjugated, i.e. when single (σ) and double ($\sigma + \pi$) bonds alternate. The simplest conjugated polymer is the polymerized form of acetylene, $(CH)_x$, known as polyacetylene (PA); two conformational isomers of this polymer are shown in Fig. 8.21. In general, the polymers do not consist only of aliphatic chains but also include aromatic groups: Table 8.1 shows some representative monomer units that form conjugated polymers.

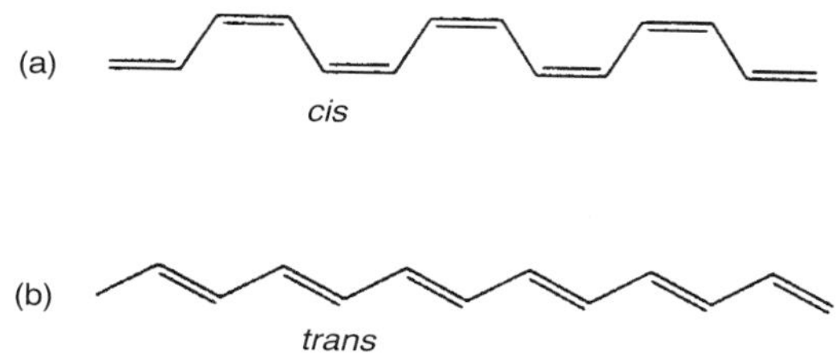

Fig. 8.21 Conformational isomers of polyacetylene (PA): (a) *cis*-PA; (b) *trans*-PA (thermodynamically stable at room temperature). (N.B. for simplicity neither the single hydrogen atom bonded to each carbon atom, nor the carbon atoms themselves positioned at the intersections of the bonds, are shown.)

Table 8.1 Some representative conjugated polymers

Polymer	*Chemical name*	*Monomer*	π–π* *energy gap* (eV)
t-PA	*trans*-polyacetylene		1.5
PPP	poly (*p*-phenylene)	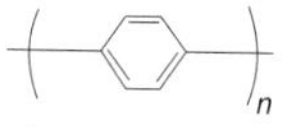	3.0
PPV	poly (*p*-phenylenevinylene)	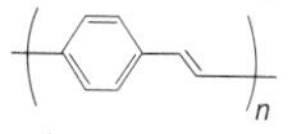	2.5
PPy	polypyrrole	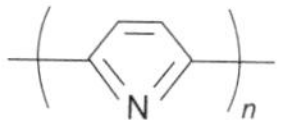	3.1

(After Greenham and Friend (1995). Reproduced by permission of Academic Press, Inc.)

The electronic structure of conjugated polymers is dominated by the π electrons. In the 'classical' chemist's representation of the bonding in a conjugated system, e.g. *trans*-PA, *single* bonds resulting from overlap between sp^2–hybrids alternate with *double* bonds that comprise an sp^2 σ-bond and a π-bond, resulting from the overlap of p_z electrons on neighbouring atoms pointing in the direction normal to the x–y plane in which the zigzag polymer chain lies (Fig. 8.22a). Alternatively, the p_z electrons could be regarded as being delocalized along the chain, in which case all intrachain bonds would be equivalent (Fig. 8.22b). The electrical behaviour of these two electronic configurations is very different: the conjugated state is electrically insulating, since the p_z electrons completely fill a π-bonding band separated by an energy gap from an empty π*-antibonding band (Fig. 8.22a). In contrast, the delocalized configuration is *metallic* (along the chain) since now the p_z electrons half-fill a π-band (Fig. 8.22b).

At first sight, it appears that the two bonding configurations of the polyene chain shown in Fig. 8.22 are connected by a Peierls distortion, as found for 1D conductors (§8.3.2). However, the situation for *trans*-PA is a little more complicated than for, say, KCP, since the zigzag polyene chain is not truly one-dimensional, but only quasi-1D. Note also that the period *a* of the chain (the next-nearest C–C distance) is *not* different

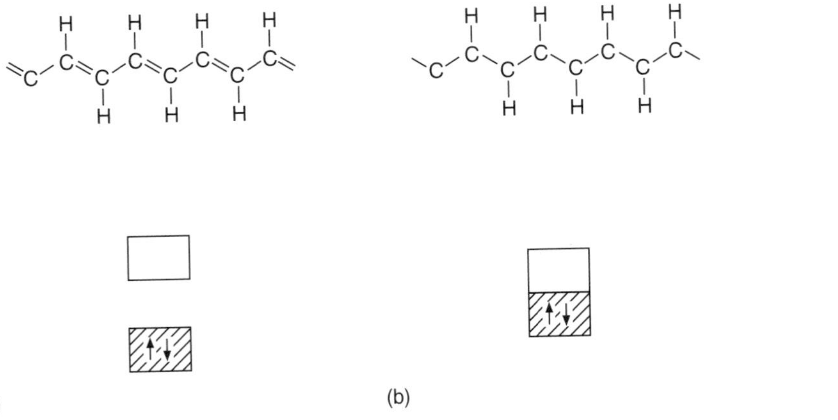

Fig. 8.22 Two configurations of the bonding in *trans*-polyacetylene: (a) conjugated system in which the p_z-electrons are localized in π-bonds and σ- and π-bonds alternate; (b) delocalized configuration in which the p_z-electrons are delocalized along the polymer chain and all bonds are equivalent. A block representation of the bands, and their occupancy by p_z- electrons, is also shown for each case.

for the unreconstructed and reconstructed configurations if the average length of the single and double bonds of the conjugated configuration (Fig. 8.22a) is the same as the bond length in the unreconstructed, delocalized configuration (Fig. 8.22b). If the simple Peierls mechanism outlined in §8.3.2 were operative, it would be expected that the conjugated configuration would have *twice* the period of the unreconstructed configuration since, in this case, the π-band is half-filled (one p_z electron per C atom)—see eqn. (8.9). The symmetry-lowering involved in the transition from the unreconstructed to the reconstructed configuration of *trans*-PA therefore does not involve the periodicity of the chain but, instead, involves the loss of a *glide-plane* symmetry operation (§2.3.3).

★ The particular glide-plane operation can be written as

$$\gamma = \left\{ \sigma \middle| \frac{1}{2}\boldsymbol{a} \right\}; \tag{8.10}$$

the unreconstructed structure of an infinite chain with equal C–C bond lengths can be generated by reflecting half of the chain through the mirror plane σ and translating it by a vector $\frac{1}{2}\boldsymbol{a}$ along the chain direction (Fig. 8.23). Both structural configurations in Figs 8.22 and 8.23 have a basis of *two* C atoms per primitive cell and, since both have the same real-space period $\boldsymbol{a}$, both also have first Brillouin zones in reciprocal space extending between $-\pi/a$ and π/a that are completely filled.

This circumstance might be expected to produce an electrically insulating state (§5.2.5) in both cases, a prediction at variance with the expectation that the unreconstructed structure with delocalized π-electrons should be metallic. However, the glide-plane symmetry characteristic of the unreconstructed configuration means that, in fact, there is *no* gap at the first Brillouin-zone boundary in this case (cf. Problem 5.13); the two Bloch states ψ_k and ψ'_k, related by

$$\psi'_k = \gamma \mathfrak{C} \psi_k = \left\{ \sigma \middle| \frac{1}{2}\boldsymbol{a} \right\} \mathfrak{C}\, \psi_k, \tag{8.11}$$

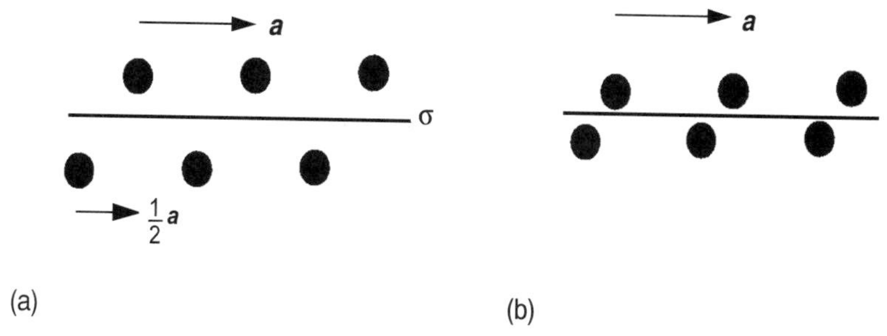

Fig. 8.23 A polyene chain, e.g. *trans*-PA, with circles representing C atoms, in (a) the unreconstructed state, with equal C–C bond lengths; (b) the conjugated state, with shorter π-bonds and longer σ-bonds. Both configurations have the same period $\boldsymbol{a}$ along the chain, but only (a) contains a glide-plane symmetry operation.

are degenerate at $k = \pm\pi/a$, and so there is no gap there (Fig. 8.24a); $\mathfrak{C}$ is the conjugator operator that causes the complex conjugation of the operand:

$$\mathfrak{C}f(x) = f^{\dagger}(x^{\dagger}). \tag{8.12}$$

Altmann (1991) gives a proof of this statement. There *is* a gap at the second boundary at $k = \pm 2\pi/a$, but since this Jones zone can contain a total of four electrons per primitive cell, it is half-filled and hence the unreconstructed structure is metallic, as expected intuitively. Reconstruction of the polymer structure to give the conjugated configuration lifts the glide-plane symmetry and, with it, the wavefunction degeneracy at $k = \pm\pi/a$. Thus, a gap opens up at the first Brillouin-zone boundary, and hence the pure form of *trans*-PA is a semiconductor (the $\pi - \pi^*$ energy gap is $\simeq 1.5$ eV—see Table 8.1). This structural configuration is energetically favoured at room temperature.

★ Although pure conjugated *trans*-PA is non-metallic, its electrical conductivity can be greatly increased by chemical doping with electron donors (e.g. Na) or acceptors (e.g. AsF_5 or I_2): such dopant species (with molar fractions of up to 20% per C atom) reside between the polymer chains, and charge transfer occurs between dopants and π or π^* bands of the polymer, resulting in high, almost metallic-like, values of d.c. conductivity (e.g. $\sigma_0 \simeq 10^5$ S m^{-1} at 300 K).

Although it might be thought that the extra electrons or holes introduced into conjugated polymer chains by chemical doping, or by optical excitation or electrical

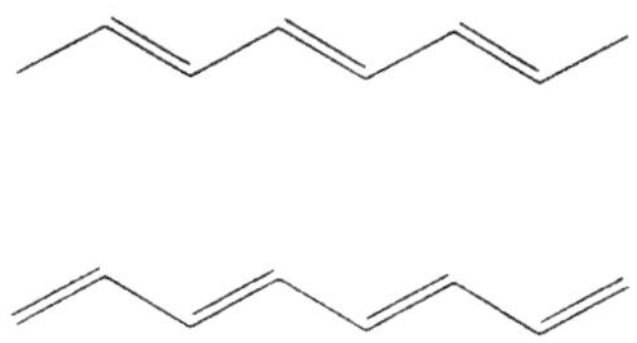

Fig. 8.24 The two different senses of bond alternation in *trans*-PA. For an infinite chain, this means that the electronic ground state is doubly degenerate.

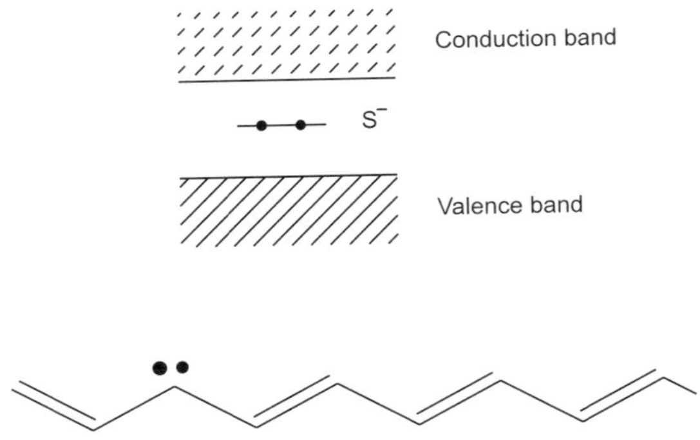

Fig. 8.25 Schematic illustration of the formation of a negatively charged soliton state, S^-, on a chain of *trans*-PA following the introduction of an excess electron into the π^* conduction band. The soliton is associated with the change in phase of bond conjugation (and actually extends over 10–15 C atoms, rather than being localized at one as shown here). The soliton electronic energy level lies at midgap between the π- and π^*- bands and is doubly occupied; the state is negatively charged. Midgap optical absorption involving occupied soliton states is possible in addition to the normal π–π^* transitions.

injection, would enter the (π^*) conduction band or the (π) valence band respectively, as in conventional 3D inorganic semiconductors (§6.5.2), in fact this is not the most stable electronic state for conjugated-polymer semiconductors. For the simple case of an infinite chain of *trans*-PA, because the electronic ground state is doubly degenerate (there are two different senses of bond alternation—see Fig. 8.24), the lowest-energy electronically excited states consist instead of phase kinks in the conjugation sequence (see Fig. 8.25). Such bond-alternation defects are known as solitons because of the similarity in their behaviour to the solitary waves that are the solution of the non-linear Schrödinger wave equation; they propagate without change of shape. A soliton on a *trans*-PA chain actually extends over 10–15 carbon atoms, rather than being localized at a single carbon site as shown schematically in Fig. 8.25, but can occur anywhere along the chain and is therefore mobile along the chain direction. The electronic energy level of a soliton lies midway between π-and π^*-band edges since the state has p_z non-bonding character. An excess electron on a *trans*-PA chain results in a doubly occupied, negatively charged soliton state, S^-; likewise, an excess hole results in an empty, positively charged state, S^+. Note that the presence of S^- states leads to the possibility of additional optical transitions between their levels and the π^* band, giving midgap optical absorption in addition to normal $\pi - \pi^*$ interband transitions.

For conjugated polymers that do not have the special characteristic of a doubly degenerate electronic ground state, solitons do not occur. Excess electronic charge also causes a change in the bond-alternation sequence, but now it is strongly localized at a particular point on the polymer chain in order to minimize the cost in energy of introducing the higher energy excited state (see Fig. 8.26). Such excited states are called *polarons* (§6.6), since they are associated with a local distortion of the chain (Fig. 8.27). A polaron on a conjugated chain can be regarded as being equivalent to a localized

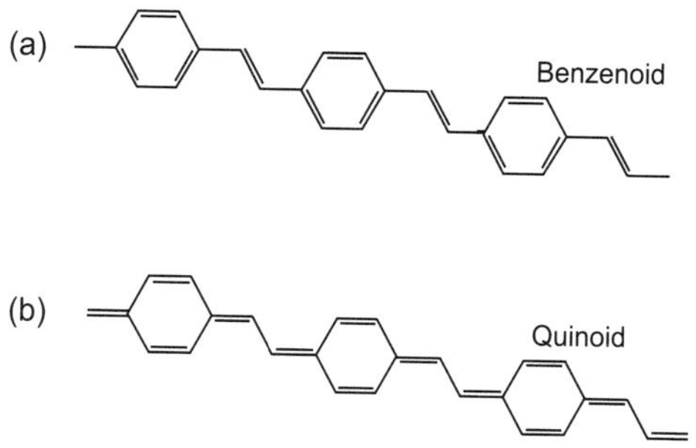

Fig. 8.26 Two different, non-degenerate senses of bond alternation in PPV: (a) benzenoid; (b) quinoid (higher energy).

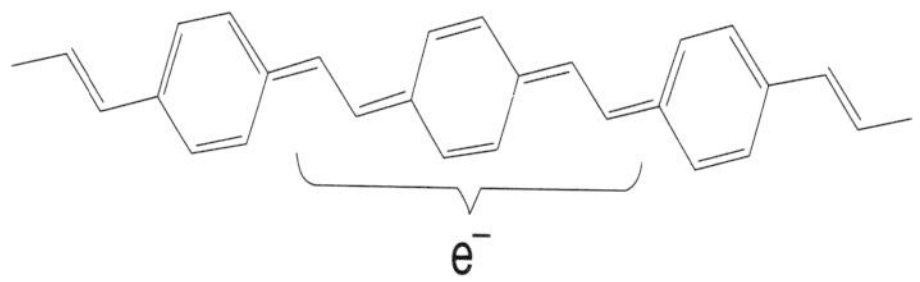

Fig. 8.27 Excess negative charge on a PPV chain, represented as a region of local quinoid character.

pair of solitons, the interaction of which causes two midgap soliton states to form bonding and antibonding combinations with energy levels symmetrically situated about midgap (Fig. 8.28). Such polaron states can accommodate up to four electrons, giving a positive bipolaron, bp^{2+} (0), positive polaron, p^+ (1), negative polaron, p^-(3), and negative bipolaron, bp^-(4), where the numbers in parentheses denote the numbers of electrons involved. A full discussion of conjugated polymers and their applications has been given by Greenham and Friend (1995).

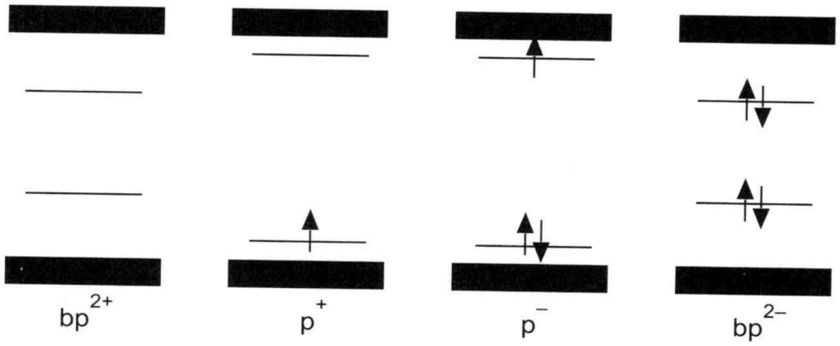

Fig. 8.28 Polaron and bipolaron energy levels in a non-degenerate ground-state conjugated polymer (e.g. PPV).

8.3.4 Nanotubules

Another member of the carbon fullerene family of materials, intermediate in dimensionality between the zero-dimensional buckyballs and onions (§8.2.1) and the two-dimensional hexagonal-net layer structure of graphite (§8.4.1) comprises nanotubules. These are one-dimensional tubes with diameters of a few nanometres, formed from rolled graphitic sheets, and capped with hemispherical sections of fullerene structures. An example of such a structure, a chiral fibre consisting of a single sheet, with end-caps based on an icosahedral C_{140} fullerene, is shown in Fig. 8.29. Actual carbon nanotubules produced in a d.c. arc struck in a low-pressure inert-gas atmosphere between graphite electrodes (Iijima (1991)) consist of a series of concentric tubes like that shown in Fig. 8.29. Since BN is isostructural with graphite (each type of atom occupying the two crystallographically distinct sites in the honeycomb lattice of graphite—§8.4.1), mixed $B_xC_yN_z$ nanotubules can also be made in the same way.

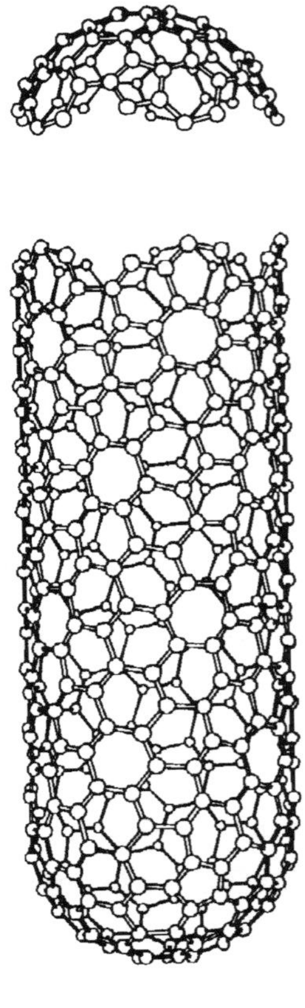

Fig. 8.29 Illustration of a chiral, single-shell carbon nanotubule, of diameter $d = 10.36$ Å, based on an icosahedral C_{140} fullerene, sections of which form the two end-caps. The chiral vector is $\mathbf{c}_h = (15, 5)$, with helicity angle $\theta = -19.11°$. (After Saito *et al.* (1992). Reprinted with permission from *Appl. Phys. Lett.* **60**, 2204. © 1992 American Institute of Physics)

Since these nanotubules can be thought of as being made from rolled-up sections of graphitic-like layers, the characteristics of such tubes, e.g. diameter and degree of helicity, can be defined uniquely in terms of a helical, or chiral, vector

$$c_h = n_1\hat{a}_1 + n_2\hat{a}_2 \tag{8.13}$$

linking two lattice sites in the planar graphitic layer from which the tubule is constructed. The vectors $\hat{a}_1$ and $\hat{a}_2$ are the unit vectors of the honeycomb lattice, defined as in Fig. 8.30a, and n_1 and n_2 are positive or negative integers (or zero). The coordinates (n_1, n_2) of some lattice points are shown in Fig. 8.30b. A tubule corresponding to the vector c_h is constructed by rolling up the section cut out of the graphitic layer along lines perpendicular to the ends of the vector c_h (Fig. 8.30a), and joining up the cylinder along these lines. If either of the coordinates n_1 or $n_2 = 0$, the resulting tubule is non-chiral; the structure is the same whether the sheet is rolled up in one sense or the other. However, for non-zero values of n_1 and n_2, a *chiral* tubular structure results; left-hand and right-hand helicities (chiralities) are produced by rolling up the graphitic sheet in one sense or the reverse. The helical, or chiral, angle, θ_h, defined as the angle between c_h and the 'zigzag' direction $\hat{a}_1$ (Fig. 8.30a), determines the degree of helicity. The diameter of the tubule is given by $d = |c_h|/\pi$.

The most extraordinary feature of carbon nanotubules is that their electrical characteristics, whether they are metallic or semiconducting along the tubular axis, are predicted to depend on the helicity parameter, c_h (Hamada *et al.* (1992), Mintmire *et al.* (1992), Saito *et al.* (1992)). For example, if the zigzag axis ($\hat{a}_1$) of the honeycomb net

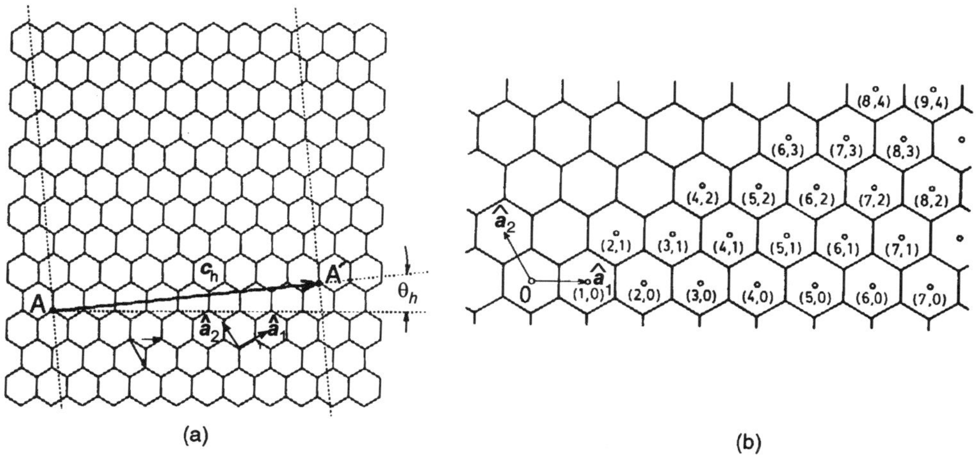

Fig. 8.30 (a) Definition of the characteristic parameters describing graphitic-like nanotubules in terms of the planar honeycomb graphite lattice from which the tubule is constructed. The helicity, or chiral, vector $c_h = n_1\hat{a}_1 + n_2\hat{a}_2$ connects two lattice sites A and A′ in the net ($\hat{a}_1$ and $\hat{a}_2$ are the unit vectors of the primitive cell); the helicity angle θ_h is the angle between the vectors c_h and $\hat{a}_1$. The dotted lines, drawn perpendicular to c_h at A and A′ show the extent of the graphitic layer which is used to construct the tubule, by rolling along an axis perpendicular to c_h and joining to make a cylinder along the lines. (b) Labelling of lattice sites (n_1 , n_2) connected to the origin (0) by the helicity vector c_h.

is *normal* to the tubular axis, $\boldsymbol{c}_{\mathrm{h}} = (n, 0)$, semiconducting behaviour is predicted, with a narrow gap (almost semi-metallic) if n is a multiple of three, and a wider gap otherwise. Figure 8.31a shows band structures calculated by a tight-binding method for two values of n (12 and 13) for (n, 0) tubules that exhibit this behaviour. The reason why tubules with $n = 3p$, where p is an integer, have very small gaps is that for them, the periodic boundary conditions around the circumference of the tubule mean that only n wavevectors are allowed in the corresponding reciprocal-space direction. In the tubular-axis direction, the allowed wavevectors are essentially continuous if the tubule is very long. For the Brillouin zone shown in Fig. 8.31a, $n = 6$, and in this case, and for $n = 3p$ in general, the allowed wavevectors intersect (in zeroth order) the K-point where, for a graphite layer, the bonding and antibonding combinations of p-states are degenerate (§8.4.1) and semi-metallic behaviour would be expected, as for graphite. However, the degeneracy point for a tubule is not exactly at the equivalent of the K-point because the p_z orbitals involved on neighbouring C atoms are not parallel to each other, but make a small non-zero angle to each other since they point radially away from the tubular axis (parallel to $\boldsymbol{c}_{\mathrm{h}}$). As a result, electron overlap is enhanced in the circumferential direction because of the consequent involvement also of σ-orbital contributions.

However, one carbon tubule configuration *is* predicted to be metallic (Mintmire *et al.* (1992), Hamada *et al.* (1992)) and this is for $\boldsymbol{c}_{\mathrm{h}} = (n, m) = (2, 1)m$, where m is a common factor (i.e. $n = 2m$). For this structure, the zigzag axis of the honeycomb lattice is *parallel* to the tubular axis (Fig. 8.31b). In this case, no matter what the value of m, an allowed wavevector will *always* pass through the K-point of the graphitic layer (from Γ), and hence the tubule should be metallic. The tight-binding band structure shown in Fig. 8.31b for a (2, 1)6 tubule shows two bands crossing at the Fermi level very near the

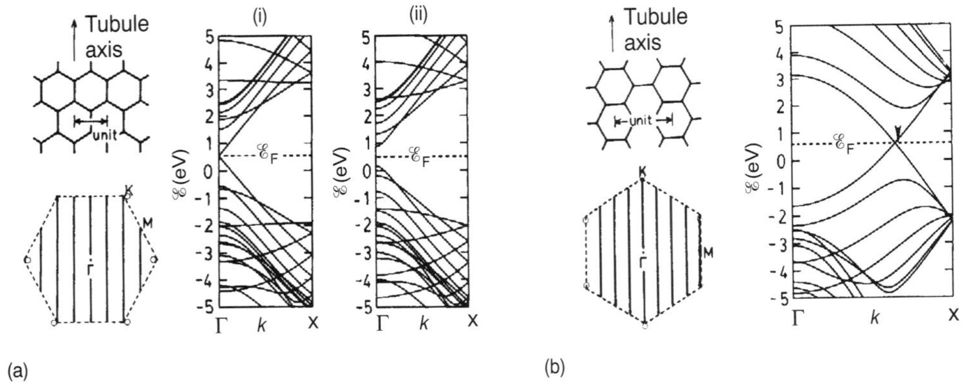

Fig. 8.31 Electronic band structures calculated by tight-binding methods for carbon nanotubules (Reprinted with permission from Hamada *et al.* (1992), *Phys. Rev. Lett.* **68**, 1579. © 1992. The American Physical Society): (a) (n, 0) tubules for (i) $n = 12$ and (ii) $n = 13$. The X-point is near π/a. Also shown is the relationship between this tubular configuration and the honeycomb lattice from which it is derived, and the hexagonal first Brillouin zone of the honeycomb lattice with the allowed ranges of $\boldsymbol{k}$-vectors (shown as solid vertical lines) for $n = 6$. (b) The (2, 1)6 tubule. The X-point is near $\sqrt{3}\pi/a$, and the arrow marks the position of the wavevector corresponding to the K-point in the Brillouin zone of graphite. Also shown is the relationship between this tubule and the graphite lattice. The allowed ranges of $\boldsymbol{k}$-vectors for $n = 4$ are shown superposed on the Brillouin zone of graphite.

K-point, confirming the metallic character of this class of tubule. In general, a carbon tubule with helicity vector (n, m) $(n \geqslant 2m \geqslant 0)$ is metallic if $n - 2m = 0$, is a narrow-gap semiconductor if $n - 2m = 3p (p = 1, 2, \ldots)$, and a wider-gap semiconductor otherwise.

In practice, however, real carbon nanotubules tend to be semiconducting only for the smallest diameters: otherwise, the semi-metallic behaviour of graphite is recovered at larger diameters. However, BN-containing nanotubules are semiconducting irrespective of tube diameter and helicity: tubules made only from BN have a bandgap of about 5.5 eV and those with overall composition C_2BN have a gap around 2 eV. These large values are due to the fact that, because of the large ionicity difference between B and N (and C), ionic interactions determine the magnitude of the bandgap; the degeneracy of electron states at the K-point, characteristic of graphite, is lifted as a consequence (see §8.4.1).

8.3.5 Quantum wires

The final 1D systems that we will discuss fall into classification scheme II (§8.1), i.e. they are manufactured so that they exhibit 1D behaviour, rather than having an intrinsically 1D structure at the atomic level as in §§8.3.1–4. The behaviour of electrons in a solid can be made to be one-dimensional in one of two ways: either by restricting the dimensions of a sample by selective etching to the form of a very thin wire of material (so that the lateral dimensions are mesoscopic in size), or alternatively by means of electrostatic confinement of electrons in an otherwise two-dimensional gas at the interface between two semiconductors (§8.4.2.2). The latter method is achieved by means of electrodes, separated by a very narrow (sub-micron) gap and placed over the 2D electron gas,

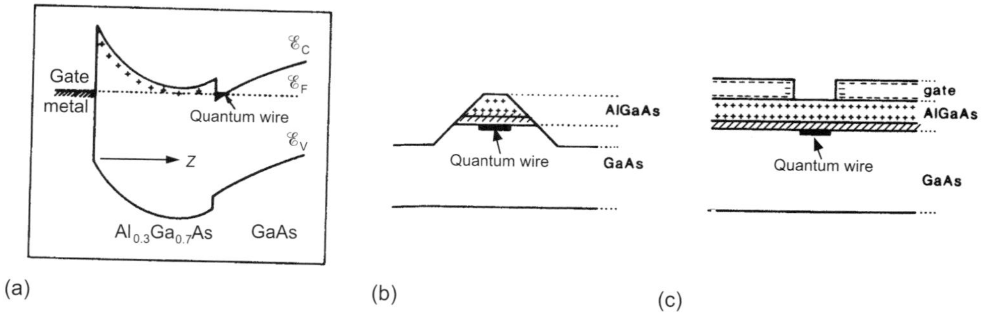

Fig. 8.32 Methods of confining a 2D electron gas to form a 1D quantum wire. (a) Formation of a 2D electron gas at the interface between an undoped GaAs layer and an n-type doped AlGaAs layer, itself in contact with a metal gate electrode where a Schottky junction is formed. The bending of the electron bands (at one representative $\boldsymbol{k}$-point) as a function of distance, z, through the modulation-doped heterostructure is shown; this results from the electrostatic potential due to the ionized donors in the AlGaAs layer (shown as +). The 2D degenerate electron gas forms in a well in the conduction band resulting from the band bending of the GaAs conduction band and the (repulsive) 0.3 eV conduction-band discontinuity at the interface with the AlGaAs. (b) Restriction of the 2D gas to a 1D channel by means of the etching of the n-type AlGaAs layer to form a linear mesa (extending into the page). The confined 1D electron gas is shown by the thick black line. (c) Use of a split gate, biased negatively, to confine the 1D gas by electrostatic means. In both (b) and (c), the + signs refer to positively charged ionized donors in the AlGaAs layer, and the hatched layer is an *undoped* AlGaAs spacer layer to reduce scattering of the 1D gas by the ionized donors.

which serve to confine the electrons, in the layer underneath, in the gap region between the 'split-gate' electrodes as a result of the shape of the associated electrostatic potential (as for quantum dots—see §8.2.2.2). Figures 8.32b–c show schematic representations of these two approaches of confining a 2D electron gas lying at the interface between a (doped) AlGaAs and an undoped GaAs layer in a modulation-doped heterostructure (Fig. 8.32a). A heterostructure is the name given to a sample consisting of layers of at least two different materials, often semiconductors, grown one on top of the other in a sandwich configuration (see §8.4.2); modulation doping refers to the selective electrical doping (§6.5.2), for example, of only *one* of the materials making up the layers in a heterostructure (see §8.5.1.3).

Two limiting types of behaviour for electron transport through narrow wires can be envisaged. The first is the diffusive-transport régime (Fig. 8.33a), for which $\Lambda \ll w, L$, with the (elastic) electron mean-free path (resulting from elastic scattering from impurities — §6.3.1.2) given by $\Lambda = v_F \tau$ (eqn. (6.12)), where v_F is the Fermi velocity and τ is the scattering relaxation time, and w and L are the width and the length of the wire, respectively. The opposite limiting behaviour is the ballistic-transport régime, in which $\Lambda \gg w, L$ (Fig. 8.33b): in this case elastic (specular) boundary scattering is dominant (as in the Casimir limit for *phonon* conduction—§4.6.2.2). Although it might be thought that ballistic transport would be resistance-free, in fact a non-zero electrical resistance can result from backscattering of electrons at the connection between the narrow wire and the wide 2D-electron-gas reservoirs, as shown schematically in Fig. 8.33b. An intermediate régime can also be distinguished, the quasi-ballistic régime, in which $w < \Lambda < L$ and impurity scattering and boundary scattering in the narrow wire are of comparable importance.

Elastic collisions (with defects and impurities or boundary walls) do not destroy the phase information encoded in the electron wavefunction but merely shift the phase by an additive amount. Inelastic scattering events (e.g. involving phonons—§6.3.1.3), on the other hand, do completely destroy the phase memory of the electrons. If electron phase coherence is taken into account, then the minimal length scale determining electron transport is not the elastic mean-free path Λ, but instead the phase-coherence length, l_ϕ, the distance over which electron phase coherence is maintained, given by the length an electron diffuses between inelastic-scattering events:

$$l_\phi = (D\tau_\phi)^{1/2}, \tag{8.14}$$

where D is the electron diffusion coefficient and τ_ϕ is the time between inelastic collisions. Since the number of phonons decreases with decreasing temperature (eqn. (4.68)), l_ϕ can be correspondingly rather large at cryogenic temperatures ($T < 1K$), e.g. several microns for metals, or several tenths of a micron in the case of semiconductor 2D electron gases. Modern lithographic etching technique allow sub-micron features to be fabricated, and samples having such very small dimensions can therefore exhibit electron phase coherence extending over rather large parts of the sample. In such cases, even if in the diffusive-transport régime (for elastic impurity scattering), the electrical conductance G_0 of the sample does not depend on the conductivity, scaling with sample size according to Ohm's law which, for a 2D electron gas, is written as (cf. eqn. (6.3) for the corresponding 3D case):

$$G_0 = (w/L)\sigma_0, \tag{8.15}$$

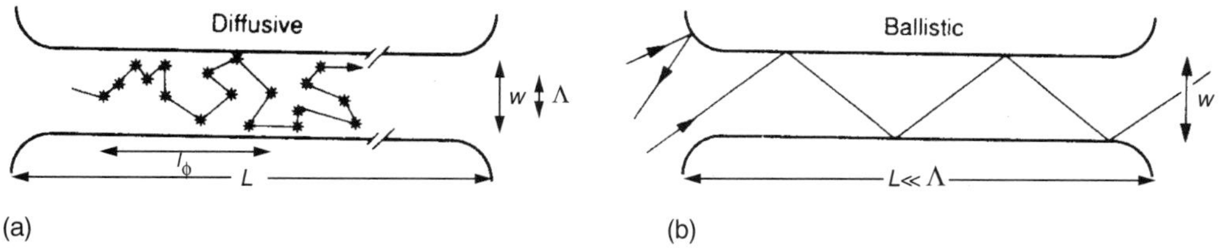

Fig. 8.33 Schematic illustration of electron trajectories in a narrow conducting region ('wire'), of width w and length L, formed, say, from a 2D electron gas in two limiting cases: (a) diffusive-transport régime (elastic mean-free path $\Lambda \ll w, L$). The asterisks represent elastic-scattering events of electrons with impurities; (b) ballistic-transport régime ($\Lambda \gg w, L$). Specular boundary scattering is dominant within the conducting channel. A finite electrical resistance arises from backscattering of electrons at the junction between the wire and the electrical reservoir connected to it.

where σ_0 is the 2D conductivity. (In 2D, σ_0 and G_0 have the *same* units, viz. Ω^{-1} or S.) Instead, it is the *conductance* that is the fundamental quantity.

The same is true also for ballistic transport along a narrow channel (Fig. 8.33b), which can be regarded as equivalent to the propagation of guided modes along a waveguide (as in electromagnetism). In this picture, the conductance of the wire is simply proportional to the transmission probability T for current injected from an electron reservoir in contact with one end of the channel, given by the Landauer formula (see Problem 8.4):

$$G_0 = (e^2/h)T, \tag{8.16}$$

(neglecting electron-spin degeneracy).

Size-quantization sets in, i.e. the electronic density of states becomes characteristic of a 1D system, $g(\mathscr{E}) \propto \mathscr{E}^{-1/2}$ (Problem 5.2), when the width of the conducting channel, w, becomes less than the Fermi wavelength of the electrons, $w < \lambda_F = 2\pi/k_F$. For metals, $\lambda_F \simeq 5$ Å (see Table 5.1), and hence this régime cannot readily be reached. However, for channels made from 2D electron gases at semiconductor heterojunctions, λ_F can be much larger (typically $\lesssim 0.1\mu$m), essentially because of the much smaller electron densities therein (which are, moreover, controllable via changes of bias on the gate electrode) compared with the very high (and invariant) electron densities characteristic of metals. Thus, patterned semiconductor devices of mesoscopic dimensions exhibiting size-quantization effects are feasible.

The study of electron transport in metallic and semiconductor nanostructures is one of the most exciting and active fields in solid-state physics today, and in this short section we cannot hope to do full justice to the many and varied phenomena being discovered. Instead, we will concentrate on just *three* aspects of the behaviour of electrons in confined 1D geometries, namely universal conductance fluctuations and weak localization in the diffusive-transport régime, and quantized point-contact conductance in the ballistic-transport régime. Beenakker and van Houten (1991) and Kelly (1995) give an extensive review of these and other related phenomena, especially those observed in the presence of magnetic fields.

**8.3.5.1 *Universal conductance fluctuations*

Normally, for large samples at not-too-low temperatures, that is, in the 'classical' régime where the sample size L is much bigger than the elastic mean-free path $L \gg \Lambda$, fluctuations in the electrical conductance G_0, say from sample to sample, are completely negligible. Assuming that a wire of length L is divided into L/Λ *independently* fluctuating segments connected in series, the r.m.s. deviation of conductance is $\delta G_0 = (L/\Lambda)^{1/2}\langle G_0 \rangle$, where $\langle G_0 \rangle$ is the average conductance, and hence δG_0 is negligible. This is not the case, however, in the phase-coherent transport régime where the phase-coherence length, l_ϕ, is comparable to the sample size, $l_\phi \simeq L$. In this case, the Al'tshuler–Lee–Stone theory (Al'tshuler (1985), Lee and Stone (1985)) predicts that, at $T = 0$ K, the conductance fluctuations have the universal value

$$\delta G_0 \simeq e^2/h, \tag{8.17}$$

where the conductance quantity $e^2/h \simeq 4 \times 10^{-5}\Omega^{-1}$ appears also in many other related phenomena such as quantized point-contact conductance (§8.3.5.3), weak-localization corrections to the Drude conductivity (§8.3.5.2), and the quantized Hall effect in 2D (§8.4.5).

The origin of these universal conductance fluctuations can be understood in terms of the Landauer expression (eqn. (8.16)) relating the conductance to the transmission probability of propagation of a mode through a waveguide. This equation can be generalized to the case where N modes, termed quantum channels, propagate simultaneously, to give

$$G_0 = \frac{e^2}{h}\sum_{\alpha,\beta=1}^{N} \left|\boldsymbol{t}_{\alpha\beta}\right|^2, \tag{8.18}$$

where $t_{\alpha\beta}$ is the quantum-mechanical transmission probability amplitude in going from an incident channel α from a source reservoir to an outgoing channel β. This process can be envisaged as in Fig. 8.34. A conducting wire, in the phase-coherent régime ($l_\phi > L, w$), is connected to source and drain electron reservoirs in thermal equilibrium, in which it is assumed that inelastic scattering takes place, leading to a randomization of electron-wavefunction phases and lack of phase coherence of the incoming modes, α. These modes interact with an intervening disordered region in the wire (containing, for example, impurities), where only elastic scattering takes place, and transform into the outgoing (transmitted and backscattered) channels, β.

The Drude conductance for a 2D electron gas can be written, from eqn. (6.9), as

$$G_0 = \frac{w}{L} n_s \frac{e^2\tau}{m_e} \tag{8.19a}$$

$$= \frac{w}{L}\frac{e^2}{h}\frac{k_F\Lambda}{2} \tag{8.19b}$$

$$= \frac{e^2}{h}\frac{\pi\Lambda}{2L} N, \tag{8.19c}$$

where n_s is the areal electron density, and N is the number of transverse waveguide modes (or 1D sub-bands) that are occupied at energy $\mathscr{E}_F$ in a wire of width w, given by

$$N = k_F w/\pi; \tag{8.20}$$

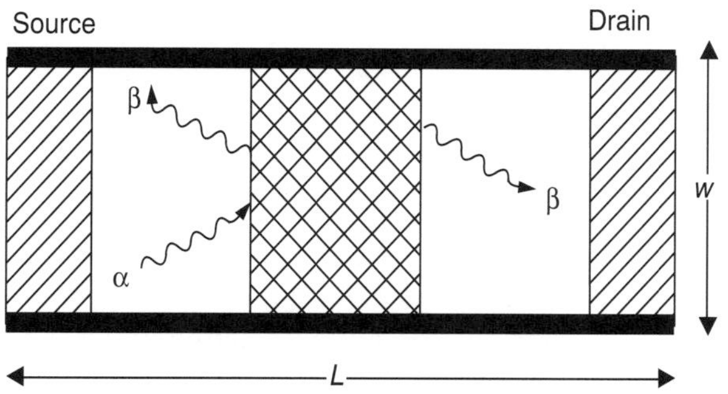

Fig. 8.34 Schematic representation of a conducting wire in the phase-coherent régime ($l_\phi > L$, w), connected to source and drain electron reservoirs, containing a disordered region (hatched) in which only elastic collisions take place. Incoming quantum channels (corresponding to transverse waveguide modes) are denoted by α, and outgoing (transmitted and backscattered) channels by β.

eqn. (8.19c) is written in this way to correspond with the Landauer relation. Equation (8.19b) is obtained from eqn. (8.19a) by substituting $\tau = \Lambda/v_F$ (eqn. (6.12)) and $n_s = D(\mathscr{E})\mathscr{E}_F = m_e\mathscr{E}_F/\pi\hbar^2$ (where $D(\xi)$ is the 2D areal density of states—Problem 5.2), and using $\mathscr{E}_F = m_e v_F^2/2 = \hbar^2 k_F^2/2m_e$.

The ensemble-averaged transmission probability is independent of α and β and is given by

$$\langle |t_{\alpha\beta}|^2 \rangle = \pi\Lambda/2NL \tag{8.21}$$

on comparing eqns. (8.18) and (8.19c). Current conservation means that the probabilities of transmission $|t_{\alpha\beta}|^2$ and reflection $|r_{\alpha\beta}|^2$ are related via

$$\sum_{\alpha,\beta=1}^{N} |t_{\alpha\beta}|^2 + \sum_{\alpha,\beta=1}^{N} |r_{\alpha\beta}|^2 = N \tag{8.22}$$

and hence the average reflection probability, also independent of α and β, is given by

$$\langle |r_{\alpha\beta}|^2 \rangle = \frac{1}{N}(1 - \langle |t_{\alpha\beta}|^2 \rangle) \tag{8.23a}$$

$$= \frac{1}{N}\left(1 - \frac{\pi\Lambda}{2L}\right). \tag{8.23b}$$

It can be shown that the reflection probabilities $|r_{\alpha\beta}|^2$ for different pairs α, β and α', β' of incident and reflected channels are uncorrelated, since reflection back into the source reservoir is controlled by only a few scattering events. On the other hand, it cannot be assumed that the transmission probabilities $|t_{\alpha\beta}|^2$ are similarly uncorrelated: transmission between source and drain involves multiple scattering events, and hence correlations between different channels are unavoidable. Thus, using the Landauer multiple-channel expression with the transmission probabilities (eqn. (8.18)) cannot be used directly to calculate the conductance fluctuation, δG_0; instead, the averaging must be done using the reflection probability, $|r_{\alpha\beta}|^2$ (Lee (1986)).

We require the variance of the conductance, defined as

$$\mathrm{Var}(G_0) \equiv (\Delta G_0)^2 = \langle G_0^2 \rangle - \langle G_0 \rangle^2. \tag{8.24}$$

From the current-conservation relation (eqn. (8.22)) and the Landauer formula (eqn. (8.18)), the conductance variance can be related to the variance in the reflection probabilities,

$$\mathrm{Var}(G) = \left(\frac{e^2}{h}\right)^2 \mathrm{Var}\left(\sum_{\alpha,\beta}^{N} |r_{\alpha\beta}|^2\right) = \left(\frac{e^2}{h}\right)^2 N^2 \mathrm{Var}(|r_{\alpha\beta}|^2), \tag{8.25}$$

where $\mathrm{Var}(|r_{\alpha\beta}|^2) = \langle |r_{\alpha\beta}|^4 \rangle - \langle |r_{\alpha\beta}|^2 \rangle^2$. A given reflection probability amplitude, $r_{\alpha\beta}$, is determined by a number, M, of scattering events with amplitude $A_i (i = 1 - M)$, where the A_i can be taken to be uncorrelated. Thus,

$$\begin{aligned}\langle |r_{\alpha\beta}|^4 \rangle &= \sum_{i,j,k,l=1}^{M} \langle A_i^\dagger A_j A_k^\dagger A_l \rangle \\ &= \sum_{i,j,k,l=1}^{M} \left[\langle |A_i|^2 \rangle \langle |A_k|^2 \rangle \delta_{ij} \delta_{kl} + \langle |A_i|^2 \rangle \langle |A_j|^2 \rangle \delta_{il} \delta_{jk} \right] \\ &= 2 \langle |r_{\alpha\beta}|^2 \rangle^2. \end{aligned} \tag{8.26}$$

Hence

$$\mathrm{Var}(|r_{\alpha\beta}|^2) = \langle |r_{\alpha\beta}|^2 \rangle^2; \tag{8.27}$$

the r.m.s. variation of $|r_{\alpha\beta}|^2$ is simply equal to the ensemble average of $|r_{\alpha\beta}|^2$. Substituting eqn. (8.27) into eqn. (8.25), and making use of eqn. (8.23b), gives

$$\delta G_0 = (\mathrm{Var}(G_0))^{1/2} = \frac{e^2}{h}\left(1 - \frac{\pi}{2}\frac{\Lambda}{L}\right). \tag{8.28}$$

Since we are considering the diffusive-transport limit ($L \gg \Lambda$), eqn. (8.28) shows that universal fluctuations in conductance of $\delta G_0 \simeq e^2/h$ should be observed as, say, the chemical potential (or the magnetic field) is varied.

An example of these fluctuations in conductance is shown in Fig. 8.35 for the case of a very small wire of Sb ($L = 0.6\mu$m, $w = 0.1\mu$m), measured at $T = 0.01$K (and with an

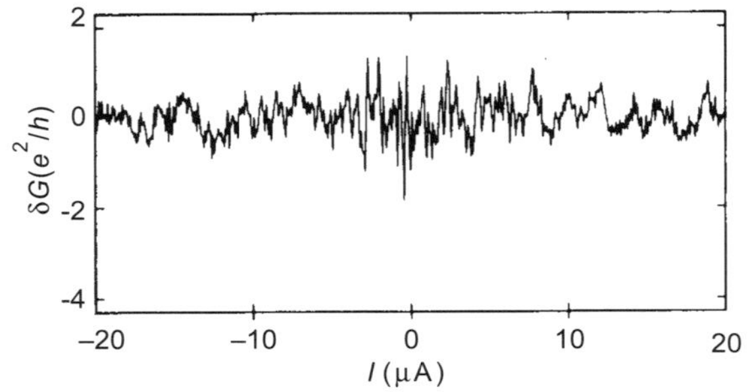

Fig. 8.35 Universal conductance fluctuations in a small Sb wire, of length 0.6 μm and width 0.1 μm, measured at 0.01 K (and with an applied magnetic field of 3 T). Note that the magnitude of the fluctuations is of the order of e^2/h, and that Ohm's law is not obeyed ($G(I) \neq G(-I)$) in the phase-coherent transport régime. (After Webb *et al.* (1988). Reprinted with permission from *Phys. Rev.* B**37**, 8455. © 1988. The American Physical Society)

applied magnetic field of 3 T). It can be seen that the conductance fluctuations are indeed of order e^2/h in magnitude; changes in voltage dropped along the wire lead to random fluctuations in the quantum interference of the electrons. Note also that the conductance is *not* symmetric upon reversal of the current direction; Ohm's law breaks down. This breakdown is also expected in the coherent-phase transport régime; for the sample relating to Fig. 8.35, the phase-coherence length l_ϕ at $T < 0.2$ K is 1.2μm, i.e. $l_\phi > L, w$.

**8.3.5.2 *Weak localization*

The quantum interference that electrons experience in the phase-coherent régime (where the phase-coherence length is comparable to or greater than the sample size, $l_\phi > L, w$) has other effects on the electrical-transport properties that are broadly referred to as being due to weak localization. The quantum interference causes a marked enhancement of the probability for an electron being *backscattered* by potential fluctuations in a disordered system in the metallic régime: this behaviour thus reduces the electron diffusion coefficient, and hence the electrical conductivity. It is termed *weak* localization because it can be thought of as a precursor to the régime of proper localization found in strongly disordered systems (§6.7). An instance of this weak-localization effect has already been touched on in Problem 6.30 dealing with the *minimum* conductance $G_{\min}$ for which metallic behaviour is exhibited by a thin wire in the phase-coherent régime. This turns out to have a value of the order of magnitude of the ubiquitous quantity e^2/h. However, this startling prediction that a wire of sufficient length that its conductance is less than $G_{\min}$ would be insulating is only strictly valid at $T = 0$ K where there is no inelastic scattering due to phonons; at higher temperatures, the inelastic scattering will cause a loss in phase coherence of the electrons and cause l_ϕ to decrease. However, a vestige of this behaviour can be seen in the upturn in electrical resistivity observed at the very lowest temperature for very thin wires, as shown in Fig. 8.36 for the case of narrow 1D channels in a 2D electron gas in a GaAs–AlGaAs heterostructure (cf. Fig. 8.32).

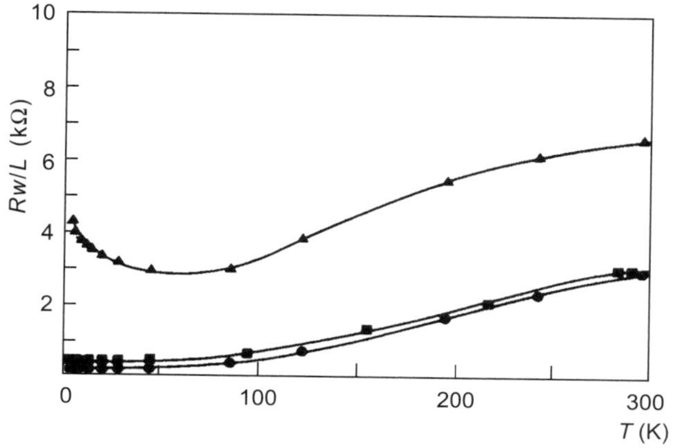

Fig. 8.36 Temperature dependence of the electrical resistance of narrow 1D channels in a 2D electron gas in a GaAs–AlGaAs heterostructure, of length $L = 10\,\mu$m and widths $w = 1.5\mu$m(■) and 0.5 μm (▲), compared with that of a wide 2D electron gas (●). (After Beenakker and van Houten (1991). Reproduced by permission of Academic Press, Inc.)

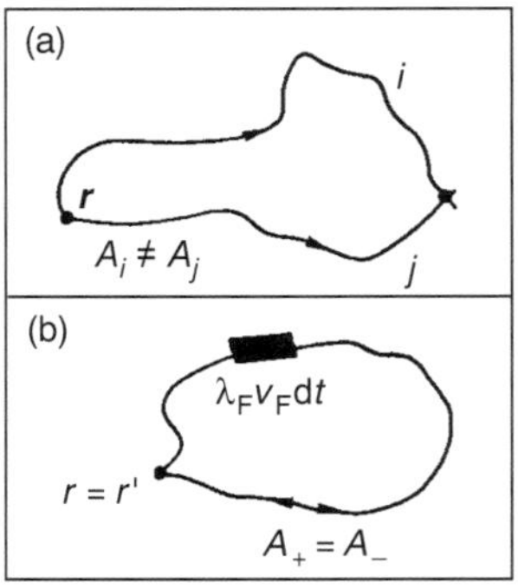

Fig. 8.37 Illustration of two types of electron propagation paths in solids in the phase-coherent régime: (a) two paths i and j between points r and r' for which the phases of the corresponding probability amplitudes A_i, A_j are uncorrelated; (b) time-reversed backscattering paths for which the two probability amplitudes are equal, $A_+ = A_-$, causing an enhancement of the probability of return of an electron to the origin. The region indicated in black is given by $\lambda_F V_F dt$ swept out in time dt.

The process of coherent backscattering of electrons at the heart of weak localization can be understood by reference to Fig. 8.37. The propagation of an electron between two sites $\boldsymbol{r}$, $\boldsymbol{r}'$ in a solid along two different trajectories i, j have probability amplitudes A_i, A_j that have uncorrelated phases in general (Fig. 8.37a). However, for a backscattering trajectory that involves a return of the electron to the origin (Fig. 8.37b), the amplitudes A_+ for propagation in one sense along the trajectory and A_- for the *time-reversed* trajectory along the same path are identical. In a Feynman path description, the probability $\boldsymbol{P}(\boldsymbol{r}, \boldsymbol{r}', t)$ for diffusion between $\boldsymbol{r}$ and $\boldsymbol{r}'$ in time is given by

$$\boldsymbol{P}(\boldsymbol{r}, \boldsymbol{r}'t) = \left| \sum_i A_i \right|^2 = \sum_i |A_i|^2 + \sum_{i \neq j} A_i A_j^\dagger. \tag{8.29}$$

Classical diffusion corresponds to the first term on the right-hand side of this expression; quantum interference is accounted for by the second term, which averages out to zero for trajectories in which the propagation amplitudes have uncorrelated phases (as in Fig. 8.37a). However, the special time-reversed backscattering paths shown in Fig. 8.37b produce a finite quantum-interference contribution, so that the coherent backscattering probability is *twice* the classical (incoherent) result. This two-fold augmentation of backscattering probability has been demonstrated in an *optical* experiment for laser light backscattered from a colloidal suspension (Fig. 8.38).

The weak-localization correction to the Drude free-electron-gas expression (eqn. (6.9)) for the d.c. electrical conductivity is proportional to the probability that an electron returns to the origin of a trajectory (Chakravarty and Schmid (1986)). If $\mathfrak{W}(t)$ is the probability that an electron returns to within a distance $d\boldsymbol{r}$ of the original starting point in time t, the correction to the conductivity is given by (Chakravarty and Schmid (1986)):

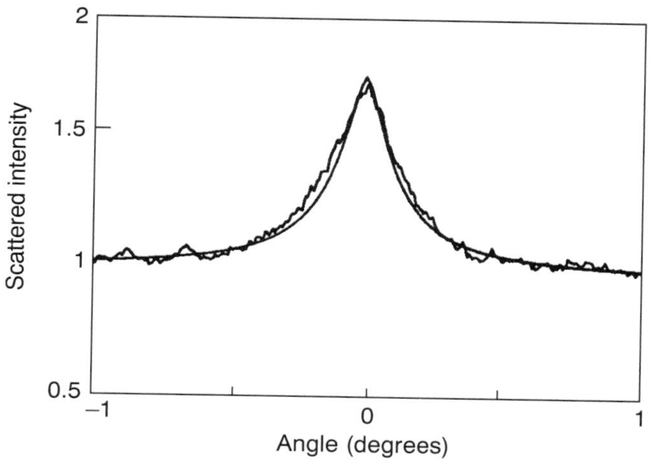

Fig. 8.38 Backscattered intensity of laser light incident on a colloidal dispersion of latex microspheres. Note the almost two-fold enhancement in the 180° backscattering direction due to coherent interference, as in the quantum interference of electrons responsible for weak-localization behaviour. (After Akkermans *et al.* (1986). Reprinted with permission from *Phys. Rev. Lett.* **56**, 1471. © 1986. The American Physical Society)

$$\frac{\delta\sigma_0}{\sigma_0} = -\frac{2\hbar}{m_e}\int_0^\infty \mathfrak{W}(t)\mathrm{e}^{-t/\tau_\phi}\mathrm{d}t, \tag{8.30}$$

where the factor $\exp(-t/\tau_\phi)$ accounts for the time-dependent loss of phase coherence, with a characteristic rate τ_ϕ^{-1}, due to inelastic scattering. The quantity $\hbar/m_e \propto \lambda_F v_F$ arises because of the flux tube of width λ_F and length $v_F\mathrm{d}t$ swept out along a propagation path (Fig. 8.37b). For 1D diffusion in a channel of width w (in, say, a 2D electron gas), $\mathfrak{W}(t)$ is given by (cf. eqn. (3.40)):

$$\mathfrak{W}(t) = w^{-1}(4\pi Dt)^{-1/2}. \tag{8.31}$$

In this case, the weak-localization correction (neglecting spin degeneracy) to the conductivity can be evaluated from eqn. (8.30) to be

$$\delta\sigma_0 \simeq -\frac{e^2}{\pi\hbar}\frac{l_\phi}{w}, \tag{8.32}$$

where $\sigma_0 = e^2 g(\mathscr{E}_F)D$, and the 2D density of electron states has been used. Since in the phase-coherent régime, $l_\phi \gg w$, this correction is appreciable in the case of 1D wires (see Fig. 8.36). In the case of 2D conduction, the effect is smaller (Problem 8.6) with only logarithmic corrections being involved. Nevertheless, the associated $\ln T$ temperature dependence of the surface conductance predicted by the weak-localization approach has been observed in thin metal films.

Weak-localization effects are suppressed by the application of a magnetic field, since the presence of a magnetic field breaks time-reversal invariance; a negative magnetoresistance is found in thin metal films and 2D electron gases in semiconductor heterostructures. Note that this behaviour is *opposite* to that found in bulk metals in the Boltzmann transport régime (§6.3.3.2).

8.3.5.3 *Quantized point-contact conductance*

As the size of a conducting sample is reduced, ultimately boundary scattering dominates and the ballistic-transport régime is reached (Fig. 8.33b). An interesting effect associated with this régime is the observation of steps in the electrical conductance of a narrow channel of a 2D electron gas, defined by two point contacts formed in a split-gate electrode (see inset to Fig. 8.39), as the gate voltage is varied (Fig. 8.39); the value of the conductance is quantized in units of $2e^2/h \simeq (13k\Omega)^{-1}$. The sample configuration shown in Fig. 8.39 can be regarded as a quasi-1D ideal conducting channel in a two-terminal contact arrangement, as in Fig. 8.34, and as such, the Landauer expression (eqn. (8.18)) can be used, multiplied by a factor $g_s = 2$ to take account of the spin degeneracy in GaAs. If scattering between the quantum channels (i.e. the sub-bands of the 1D system) is neglected, then the transmission probability in eqn. (8.18) is given by $|t_{\alpha\beta}|^2 = \delta_{\alpha,\beta}$, and hence

$$G_0 = \frac{2e^2}{h}\sum_{\alpha,\beta}^{N}|t_{\alpha\beta}|^2 = \frac{2e^2}{h}N, \tag{8.33}$$

where $N = \mathrm{Int}(k_F w/\pi)$ (eqn. (8.20)) is the number of occupied sub-bands in a wire of width w. Thus, a contact conductance, quantized in units of $2e^2/h$ is predicted by eqn. (8.33), each step occurring when ΔN changes by unity as a result of a change in the gate voltage, as observed in Fig. 8.39.

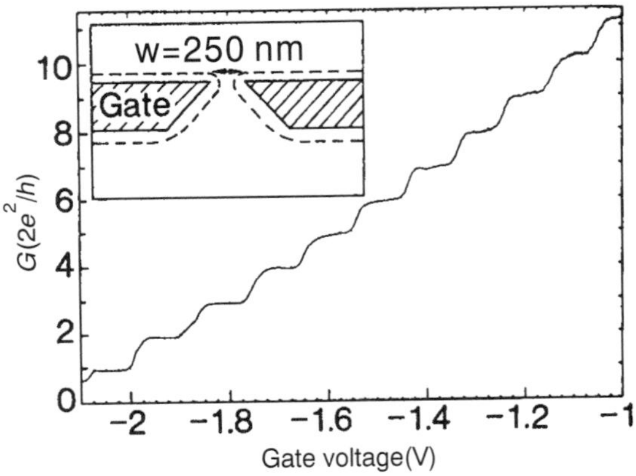

Fig. 8.39 Quantized point-contact conductance (in units of $2e^2/h$) measured at 0.6 K as a function of gate voltage, for a narrow channel formed in a 2D electron gas in a GaAs–AlGaAs heterostructure underneath a split gate whose shape is shown in the inset. The width of the constriction decreases as the negative gate voltage is increased. (After van Wees *et al.* (1988). Reprinted with permission from *Phys. Rev. Lett.* **60**, 848. © 1988. The American Physical Society)

Two-dimensional systems 8.4

Two-dimensional systems are of much interest from a scientific point of view, and are of great importance technologically: the entire microelectronics industry is based on the behaviour of electrons moving at interfaces between a semiconductor and another semiconductor, insulator or metal. As for the other instances of reduced dimensionality discussed previously, 2D systems can occur naturally as layered crystals (category I) or as artificial structures involving one or more interfaces (category II). Examples in both categories will be discussed in the following sections.

8.4.1 Layered crystals

Perhaps the canonical 2D layered crystal is the graphite allotrope of carbon, which is simple because it only contains one type of atom. The crystal structure consists of parallel sheets of atoms, with an ABAB... stacking sequence, the interplanar spacing (along the c-axis) being 3.35 Å. Each sheet has a honeycomb structure, as shown in Fig. 8.40: the rhombus-shaped primitive unit cell (of cell parameter a) contains a basis of two atoms. The intralayer bonding is due partly to overlap between sp^2-hybridized orbitals forming σ-bonds between nearest-neighbour C atoms: the planar trigonal arrangement of sp^2-hybrids (Problem 5.18) is responsible for the formation of the three-coordinated honeycomb lattice. In addition, π-bonding takes place between the remaining p_z-orbitals on each atom that point perpendicularly to the layer; such π-electrons are delocalized throughout the layer so that individual single and double carbon bonds cannot be recognized. The intralayer nearest-neighbour C–C distance in graphite is 1.415 Å (cf. $r_{C-C} = 1.39$ Å in the molecule benzene, C_6H_6), intermediate in length between that of a single carbon bond (as in sp^3-hybridized diamond, $r_{C-C} = 1.54$ Å) and of a double carbon bond ($r_{C-C} = 1.33$ Å). Individual graphitic layers are held together by van der Waals interactions.

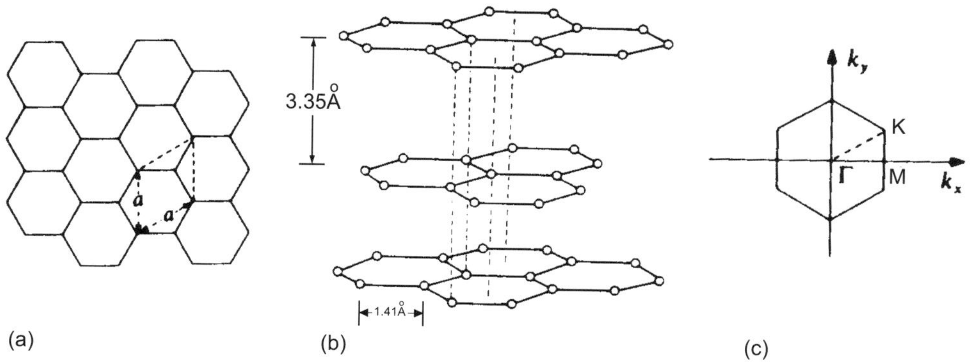

Fig. 8.40 Structure of graphite in real and reciprocal space: (a) real-space honeycomb lattice of a single layer, showing the primitive unit cell containing a basis of two atoms; (b) ABAB stacking sequence of layers; (c) first Brillouin zone with high-symmetry positions as marked.

Aspects of the electronic structure of 'ideal' graphite, i.e. of a *single* graphitic layer, have been touched on previously in §8.3.4 in a discussion of graphitic nanotubules (i.e. cylinders formed from rolled-up graphitic sheets). The electronic band structure consists of a deep-lying sp^2-σ bonding band followed by a p_z-π bonding band at the top of the valence band, then a π^*-band at the bottom of the conduction band and a σ^*-band at the highest electron energy: the σ- and π-bands are completely filled, and the π^*- and σ^*-bands are completely empty, as we shall see. The states that are of most interest are the adjacent π-and π^*-bands, for which a tight-binding treatment (§5.3.1) gives as the dispersion relation:

$$\mathscr{E}(\boldsymbol{k}) = \pm\beta_{p_z}\left[1 + 4\cos\left\{\frac{\sqrt{3}k_x a}{2}\right\}\cos\left\{\frac{k_y a}{2}\right\} + 4\cos^2\left\{\frac{k_y a}{2}\right\}\right]^{1/2}, \tag{8.34}$$

where β_{p_z} is the overlap term for p_z-states.

The first Brillouin zone corresponding to the real-space hexagonal honeycomb lattice is also a hexagon, rotated by 60° with respect to the real-space lattice (Problem 2.18), with high-symmetry points at the corner between two faces (K-point) and at the mid-face position (M) (Fig. 8.40c). The corresponding 3D surface representing the electron energy dispersion, $\mathscr{E}(\boldsymbol{k})$, is shown in Fig. 8.41a for a segment of $\boldsymbol{k}$-space defined by the vectors Γ, K and M (one-twelfth of the area of the Brillouin zone), together with 1D cuts through the energy surface for specific directions in $\boldsymbol{k}$-space (Fig. 8.41b). It can be seen that there is a large gap (of width $6\beta_{p_z}$) between π-bonding and π^*-antibonding states at the Γ-point ($\boldsymbol{k} = (0,\ 0)$). There is a smaller gap also at the M-point $(2\pi/\sqrt{3}a, 0)$ of magnitude $2\beta_{p_z}$. However, the interesting feature is that at the K-point $(2\pi/\sqrt{3}a, 2\pi/3a)$ the two states are *degenerate*: the gap vanishes there. (This feature plays a crucial rôle in determining whether a particular carbon nanotubule is metallic or not—see §8.3.4.) The corresponding 2D electronic density of states (cf. eqn. (4.59)) is as shown in Fig. 8.41c. The lower π-bonding band is completely filled and the upper π^*-antibonding band is completely empty, with the Fermi level $\mathscr{E}_F$ lying midway between the bands: 'ideal' graphite (i.e. a single layer) is a 2D semi-metal (§5.2.5). In practice, residual interlayer interactions cause the density of states at the Fermi level to be small but finite. Note also

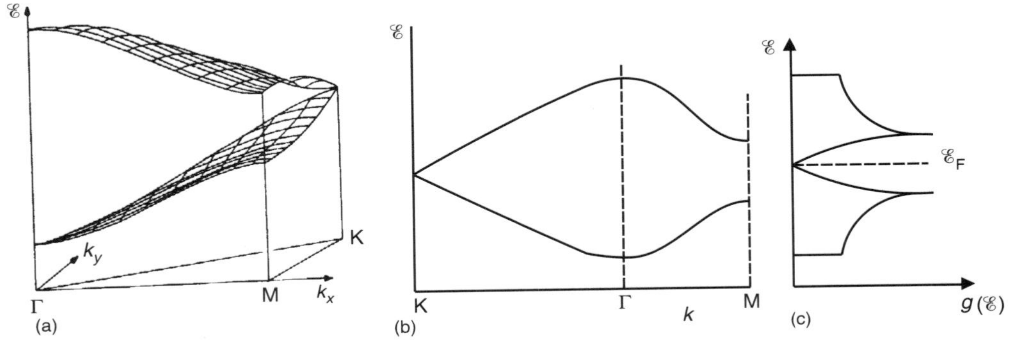

Fig. 8.41 The electronic structure of a single graphitic layer: (a) energy surfaces for the triangular section of the Brillouin zone, ΓMK; (b) energy bands for particular directions in k-space (K–Γ–M); (c) density of states. The lower π-bonding band is full and the upper π^*-antibonding band is empty.

the van Hove singularities in the density of states at the band edges and at the energies $\pm\beta_{p_z}$ where $\nabla_k \mathscr{E} = 0$, resulting from the flat dispersion of the bands in the vicinity of the K-and M-points.

The p_z-orbital contributions to the Bloch electron π-states at the high-symmetry points Γ, M and K in the Brillouin zone are shown in Fig. 8.42. For the Γ-point, a given atom is seen to have either completely bonding or completely antibonding interactions with its nearest neighbours (Fig. 8.42a), and this accounts for the fact that the largest bandgap occurs at Γ. At the M-point (Fig. 8.42b), a given atom is seen to have either one bonding and two antibonding interactions, or vice versa, with nearest neighbours, thereby producing a gap one-third the width of that at Γ. However, at

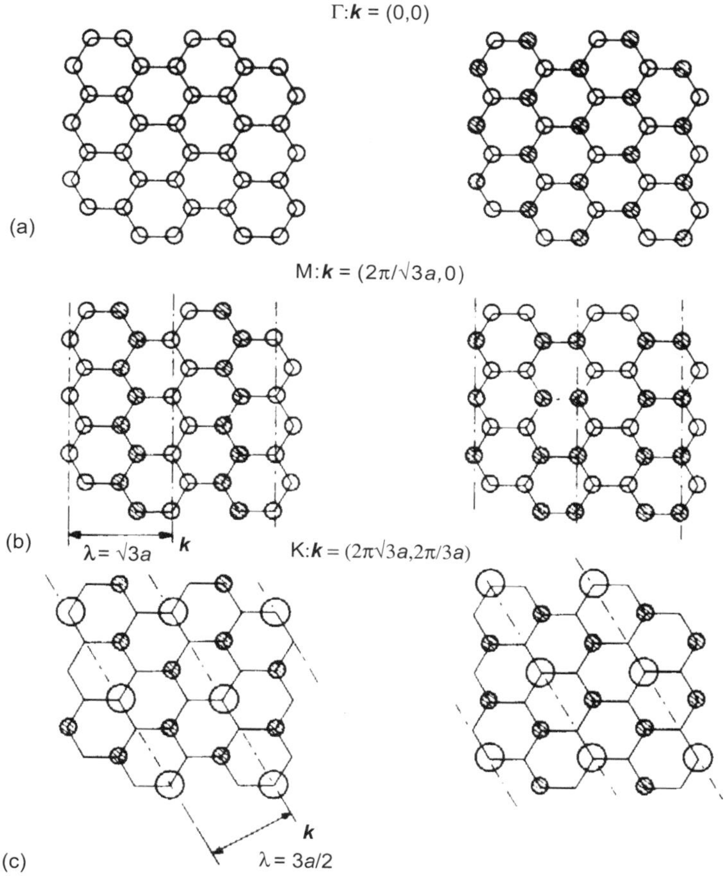

Fig. 8.42 Pictorial illustration of the origin of electron states for a graphitic layer at high-symmetry points in the Brillouin zone, showing Bloch sums of p_z-orbitals contributing to the π-bands for the $\boldsymbol{k}$-values: (a) Γ ($\boldsymbol{k} = (0, 0)$); (b) M ($\boldsymbol{k} = (2\pi/\sqrt{3}a, 0)$; (c) K ($\boldsymbol{k} = (2\pi/\sqrt{3}a, 2\pi/3a)$. In (a) and (b), the open and hatched circles denote orbital coefficients with values of plus and minus one, respectively, whereas in (c) the non-zero coefficients have values of +1 (large open circles) and $-\frac{1}{2}$. (small hatched circles). The wavelengths and directions of the associated Bloch waves are indicated in each case.

the K-point, the three nearest neighbours of a non-zero orbital atom have coefficients with values +1 and $-\frac{1}{2}$ (twice), leading to a doubly degenerate non-bonding state.

Boron nitride (BN) is isostructural with graphite, with B and N atoms occupying, respectively, one or other of the two crystallographically distinct sites in the unit cell. However, the electronic character of BN is very different from that of graphite: instead of being a semi-metal, BN is an insulator with a (minimum) gap of $\simeq 5.5$ eV. The reason for this is that, because the orbital combinations for the K-point shown in Fig. 8.42c are all either on the B atoms or on the N atoms (having different energies because of the large electronegativity difference between the two types of atoms), the degeneracy characteristic of graphite is lifted.

One interesting property exhibited by graphite (and some other layered crystals, such as the transition-metal dichalcogenides), but not BN (Problem 8.7), is that foreign atoms or molecules can easily be inserted or intercalated (§1.2.5) between the layers to form compounds, often with a well-defined stoichiometry. The intercalation reaction is one of charge transfer: electropositive intercalants (e.g. alkali atoms) donate electron charge to the graphite π^*-bands, thereby causing the Fermi level to be raised into the π^*-band, whereas electronegative intercalants (e.g. Br, AsF_5, PtF_6) accept charge taken from the graphite π-band, thereby causing $\mathscr{E}_F$ to fall into the π-band (Fig. 8.43). In both cases, the Fermi level for the intercalated compounds lies in a band of delocalized states and hence the materials become *metallic*.

Intercalation invariably causes an expansion of the interlayer spacing: for example, for the intercalation compound C_8K, the interlayer spacing where the intercalants lie is 5.41 Å, compared with the value of 3.35 Å for pure graphite. Moreover, the intercalants

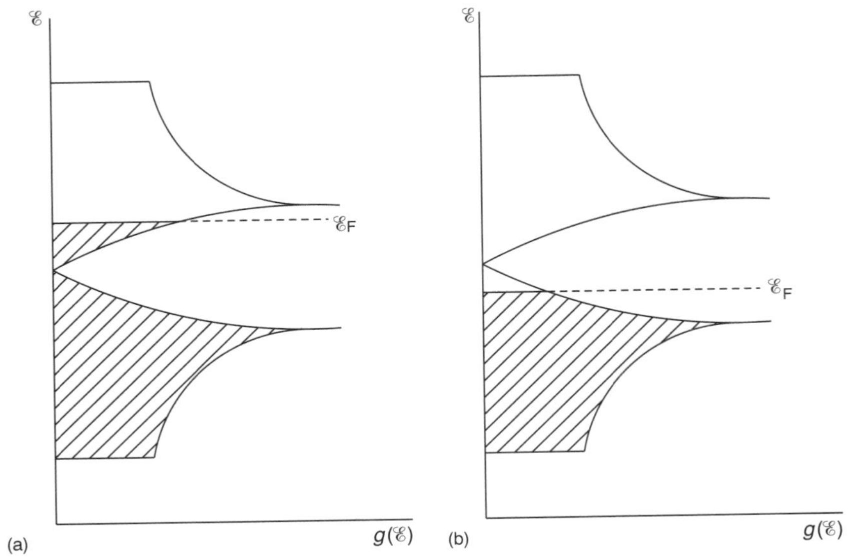

Fig. 8.43 Illustration of the effect of intercalation of electron-donating or -accepting species between the layers of graphite on the filling of the π–π^* band system: (a) donors (e.g. Li); (b) acceptors (e.g. Br).

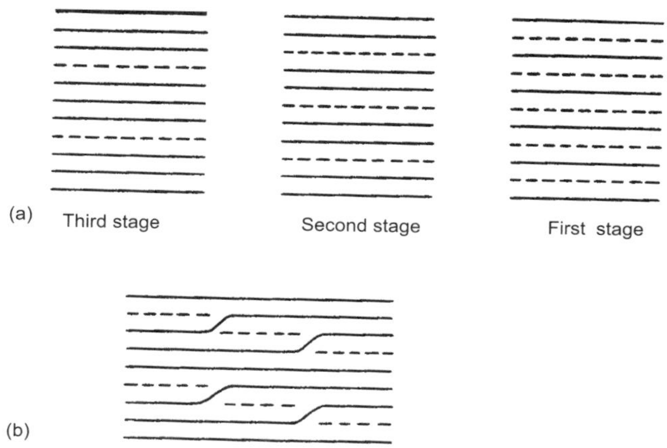

Fig. 8.44 (a) Schematic illustration of staging in graphite (or other) intercalation compounds. The solid lines denote the graphite layers; the dashed lines the intercalant layers. The stage number is the number of intervening graphite layers between intercalant layers. (b) A model for domain formation in a third-stage intercalation compound.

are *not* distributed homogeneously among the graphite layers at low concentrations: the strain energy is minimized if the intercalant atoms instead are situated only in equally spaced layers, the intervening layers being devoid of foreign atoms. The stage of the extent of intercalation is the number of graphite layers between layers of intercalants (Fig. 8.44a). Staging may, in fact, occur inhomogeneously, in the form of domains, each domain having the same stage but with a different relative stacking sequence of intercalant layers with respect to other domains (Fig. 8.44b). The relative positions of the graphite layers may also be different in the intercalated compound from pure graphite. In the case of C_8K, for example, the graphitic-layer stacking sequence is . . . A A A A . . .; K^+ ions are sandwiched between pairs of stacked hexagonal carbon rings, and therefore are 12-fold coordinated by C atoms (Fig. 8.45). If all such sites were occupied in a stage-1 compound, the composition would be C_2K; the stoichiometry C_8K is achieved if only one-quarter of the sites are occupied in an ordered way (Fig. 8.45).

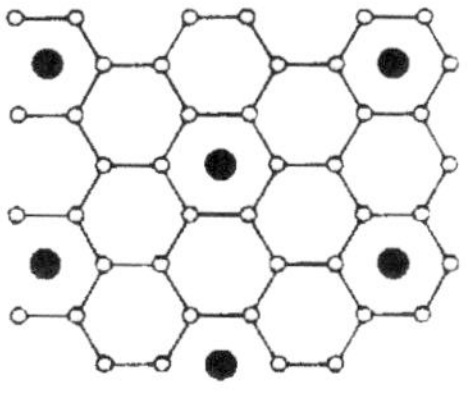

Fig. 8.45 Illustration of the structural arrangement of K^+ ions relative to the graphite lattice in the intercalation compound K_8C. Note that the graphite layers are stacked in a direct . . . A A A A . . . sequence.

As mentioned previously (§1.2.5), many other types of layered crystals, in addition to graphite, can form intercalation compounds (Jacobson (1992)). Perhaps the most well-known hosts are the transition-metal dichalcogenides, MX_2 (M = Group IV_A, V_A, VI_A transition metals, e.g. Ti, Nb, Mo, etc.; X = S, Se). In these, the transition metal occupies either octahedral (trigonal antiprismatic) or trigonal prismatic sites between close-packed layers of chalcogen atoms: the X–M–X sandwich layers are then stacked together with a large van der Waals gap between chalcogen layers of adjacent sandwiches. One of the simplest of such intercalation systems is Li_xTiS_2 $(0 < x < 1)$ which forms a *single* phase over the entire composition range: as a result, the *c*-axis lattice parameter increases smoothly from 5.7 to 6.2 Å between $x = 0$ and 1 as Li is intercalated between the layers (see Fig. 3.50). In the charge-transfer process accompanying the intercalation of lithium, reduction of Ti^{4+} to Ti^{3+} occurs: in other words, electrons from the Li atoms enter the Ti 3d band. This system is of interest as a cathode material for battery applications (see §3.5.2). Other alkalis intercalated into TiS_2 form staged compounds ($n = 2$ for all alkalis except Li; $n = 4$ for K, Rb and Cs).

8.4.2 Heterojunctions

In the remainder of this section, we will consider the behaviour of 'artificial' or 'engineered' materials with 2D character, i.e. those consisting of one or more heterojunctions between layers of different materials, or homojunctions between differently doped regions of the same semiconductor, grown one on top of the other. We will consider heterojunctions between metal and metal, metal and semiconductor, semiconductor and semiconductor, and semiconductor and insulator. Size-quantization, resulting in 2D transport behaviour, occurs when the spacing d between *two* heterojunctions is very small ($d < \lambda_F$); this is known as a quantum well. Furthermore, new properties can arise when an *ordered* array of many heterojunctions, i.e. a superlattice, is fabricated. Much of the exciting new science, and technological application, of solid-state physics is emerging from this area.

8.4.2.1 Metal–semiconductor heterojunctions

Metal–semiconductor heterojunctions are unavoidable when electrical contacts are made to a semiconducting material in order to measure its electrical properties. In order to probe the behaviour of the sample alone, it is essential that the electrical contacts themselves do not perturb the measurement, for example by altering the current extracted from the sample when it is subjected to a voltage difference. One contact must replenish the electrical charge that is being extracted from the material by the other contact. Hence, for this purpose, the metal–semiconductor contact must be *ohmic*, i.e. the current drawn is linearly proportional to the applied voltage (Fig. 8.46a). However, ohmic behaviour of contacts is the exception rather the rule. Often, blocking behaviour is observed instead (Fig. 8.46b), in which the contact is unable fully to replenish the charge being extracted from the sample by the electric field, or injecting behaviour is found (Fig. 8.46c), in which *extra* charge is injected into the material from a contact (generally at high applied fields).

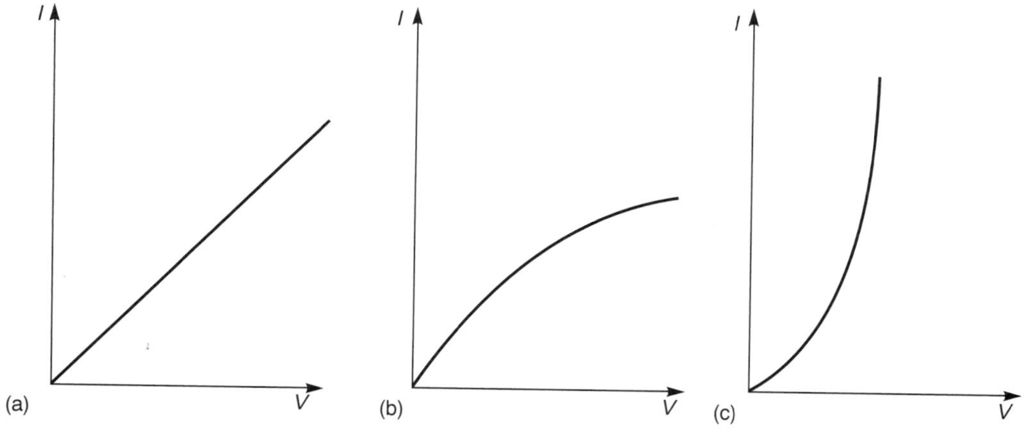

Fig. 8.46 Behaviour of different types of electrical contact in terms of current–voltage (I–V) characteristics: (a) ohmic; (b) blocking; (c) injecting.

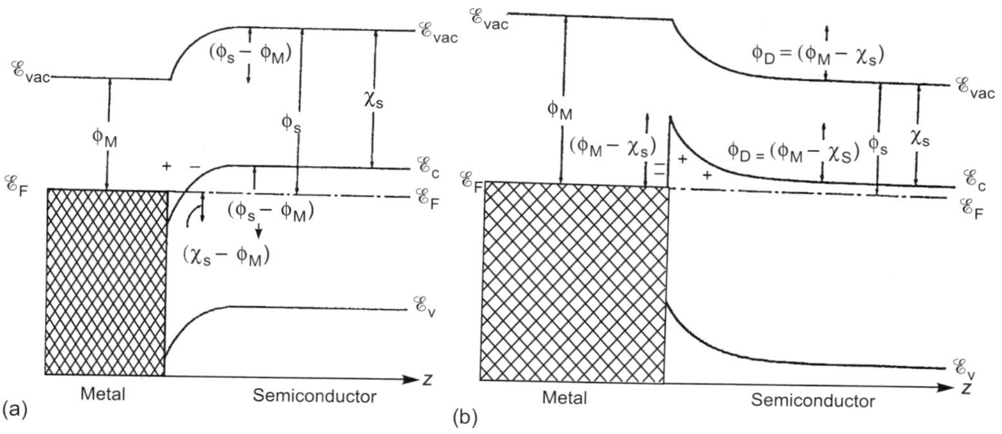

Fig. 8.47 Illustration of band bending in an n-type semiconductor at a heterojunction with a metal (the energy of the bands at a representative $\boldsymbol{k}$-point is plotted as a function of distance z in a direction normal to the interface): (a) ohmic contact ($\phi_S > \phi_M$); (b) blocking contact (Schottky barrier: $\phi_S < \phi_M$).

The criterion that determines whether a particular metal–semiconductor heterojunction forms an ohmic or blocking contact concerns the relative magnitudes of the work functions of the metal and of the semiconductor. The work function ϕ of a material is the difference in energy between the Fermi level and the vacuum level (see §5.7.2). If the work function of an n-type semiconductor ϕ_S is *greater* than that of the metal, ϕ_M, i.e. $\phi_S > \phi_M$, then the heterojunction forms an *ohmic* contact (Fig. 8.47a); if the relative magnitudes are reversed, i.e. $\phi_S < \phi_M$, then the heterojunction between a metal and an n-type semiconductor forms a *blocking contact* or Schottky barrier (Fig. 8.47b). For the case of heterojunctions between a metal and a p-type semiconductor, the above criteria are simply reversed.

The bands bend in the semiconductor in the vicinity of the heterojunction because, in order to equalize the chemical potentials in both metal and semiconductor when in contact (in the absence of an external electric field), electrical charge must flow from one material to the other. In the case, for example, of the metal–n-type semiconductor Schottky barrier shown in Fig. 8.48a, electrons flow from the semiconductor to the metal, leaving the semiconductor positively charged within a depletion region of width d (the positive charge residing on the ionized donor atoms) and the metal becomes negatively charged; the band bending is a response to the internal electrical field that results from the charge separation at the junction. Note that in the region deep within the semiconductor, far away from the heterojunction, the band bending is zero and the position of the chemical potential relative to the band edges of the semiconductor is the same as in the bulk material. For the case of an ohmic contact with an n-type semiconductor (Fig. 8.47a), the charge flow at the junction needed to equalize the chemical potentials is in the opposite direction to that of the Schottky barrier, and an electron accumulation layer forms in the semiconductor near the junction; the semiconductor bands bend in the opposite sense in the two cases. Note that in the accumulation region, the semiconductor is *degenerate*: the chemical potential lies in the conduction band for an n-type semiconductor (valence band for a p-type semiconductor). This is the reason why charge flow between metal and semiconductor is so facile in this case, i.e. why the contact is ohmic.

For the ideal case of a Schottky barrier to an n-type semiconductor having no interface states (Fig. 8.47b), however, there is an energy barrier of height $\mathscr{E}_{B,n} = \phi_M - \chi_S$ (where χ_S is the electron affinity of the semiconductor—see §5.7.2) for transfer of electrons from the level of the chemical potential of the metal into the conduction band of the semiconductor: the barrier height can also be expressed as the sum of the diffusion potential energy $\phi_D = \phi_M - \phi_S$, the difference of the work functions of metal and semiconductor, and the difference in energy between the conduction-band

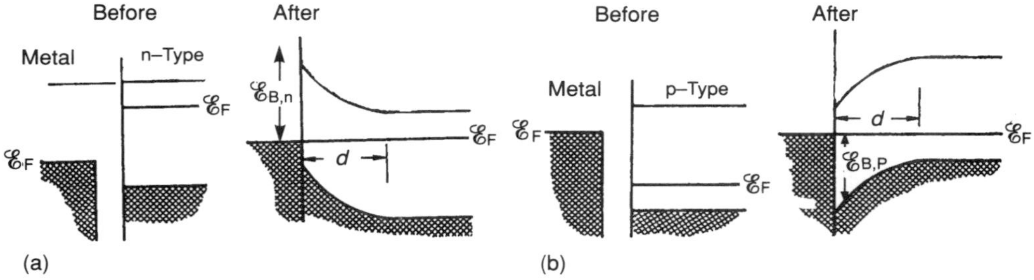

Fig. 8.48 Illustration of the band bending that occurs at a metal–semiconductor heterojunction (with zero applied bias) in the case of Schottky-barrier formation involving: (a) an n-type semiconductor; (b) a p-type semiconductor. In both cases the left-hand diagram illustrates the relative chemical-potential and band positions when metal and semiconductor are separated, and the right-hand diagram illustrates the situation after contact. The region of band bending (the depletion region for a Schottky barrier) has a width, d.

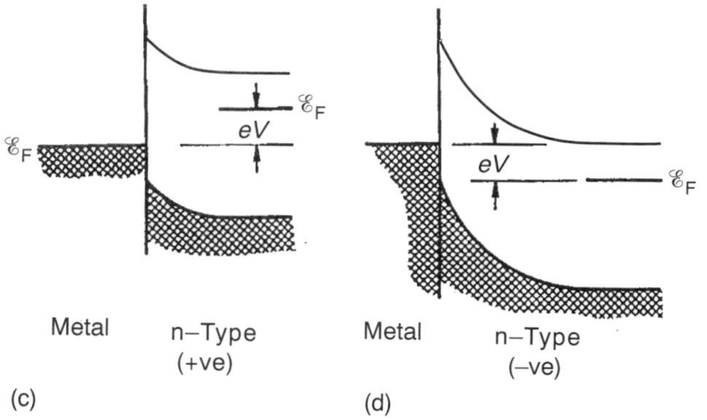

Fig. 8.48 (*contd.*) The effect of an external voltage V on the band bending of an n-type semiconductor-metal Schottky barrier: (c) forward bias; (d) reverse bias.

edge and the chemical potential, $\mathscr{E}_c - \mu \equiv \mathscr{E}_c - \mathscr{E}_F$, in a region of the semiconductor far from the junction, i.e.

$$\mathscr{E}_{B,n} = (\phi_M - \phi_S) + (\mathscr{E}_c - \mathscr{E}_F) = \phi_M - \chi_S. \tag{8.35}$$

Equation (8.35) implies that the barrier height for a given (n-type) semiconductor (with fixed electron affinity, χ_S) should be different for different contact metals and should scale linearly with the magnitude of the metal work function, ϕ_M. However, invariably the experimental variation of $\mathscr{E}_{B,n}$ with ϕ_M is weaker than this prediction, being strongest for wide-gap compound semiconductors (e.g. ZnS), but very weak or almost non-existent for small-gap, more covalent semiconductors (e.g. Si or GaAs). Thus, instead of the vacuum levels of metal and semiconductor being coincident at the interface (as assumed in Fig. 8.47 and the text leading to eqn. (8.35)), the observed near-independence of $\mathscr{E}_{B,n}$ on ϕ_M can be understood if, alternatively, the Fermi level is 'pinned' at a particular energy level in a band of *interface states*; these states derive from the bulk states of the semiconductor and lie in the otherwise forbidden gap between valence and conduction bands (Fig. 8.49a). If the density of interface states in much larger than the areal density (in a layer) of bulk donor states for, say, an n-type semiconductor (as in Fig. 8.49a), electrons flowing from the bulk of the semiconductor to the interface, prior to their equilibration and establishment of a constant Fermi level, will enter the band of interface states but, because its density of states is so relatively high, the demarcation level in the band between filled and empty states will hardly change. Thus, the Fermi level is pinned at this position. Note that the filling of some interface states will create electrical dipoles at the interface, resulting in the vacuum levels of metal and semiconductor being displaced by an energy δ_M (Fig. 8.49a).

The interface states must be induced by the presence of the metal overlayer itself, since clean semiconductor surfaces, containing surface dangling bonds, invariably reconstruct so as to remove the dangling bonds by forming bonds between them, the midgap dangling-bond states being removed and converted to bonding states lying in the valence band (Zangwill (1988)). One picture of the origin of *metal-induced gap states*

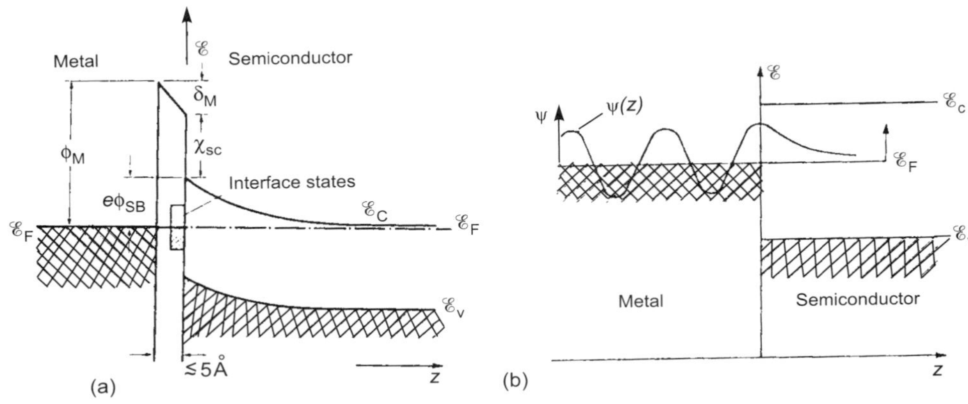

Fig. 8.49 Schottky-barrier formation involving interface states. (a) Band diagram showing a Schottky barrier on an n-type semiconductor, the Fermi level being pinned in the middle of a band of midgap interface states. (b) Schematic illustration of the origin of metal-induced gap states, namely the decay of a Bloch-like wavefunction in the metal into the forbidden gap of the semiconductor.

(MIGS) is given in Fig. 8.49b: electron density, associated with Bloch-like wavefunctions deep inside the metal, tails exponentially into the forbidden-gap region of the semiconductor in the energy range where the conduction band of the metal overlaps the bandgap of the semiconductor. The interface states, into which this charge density 'leaks', are distributed more-or-less uniformly throughout the gap region and are derived from valence- and conduction-band states of the bulk semiconductor. These states are donor- (acceptor-) like in the lower (upper) part of the gap: if $\mathscr{E}_F$ is near the valence (conduction) band, some of the donor (acceptor) MIGS are ionized and empty (full), resulting in a positive (negative) interfacial charge. There is, therefore, a charge-neutrality level, $\mathscr{E}_n$, separating donor-like from acceptor-like MIGS; the Fermi level is pinned near to this energy.

Tersoff (1985) has proposed that the neutrality level lies approximately in the middle of the *indirect* gap of the semiconductor (whether or not it has a direct or indirect minimum gap), and that the Fermi level is shifted from this position by δ_M, i.e.

$$\mathscr{E}_F \simeq \frac{1}{2}(\bar{\mathscr{E}}_v + \mathscr{E}_c^i) + \delta_M, \tag{8.36a}$$

where $\mathscr{E}_c^i$ is the indirect conduction-band minimum energy and $\bar{\mathscr{E}}_v$ is the valence-band maximum energy (at the Γ-point for tetrahedral semiconductors—see Fig. 5.52) in the *absence* of spin–orbit splitting (§5.4.2); i.e. $\bar{\mathscr{E}}_v = \mathscr{E}_v - \Delta_{SO}/3$, where $\mathscr{E}_v$ is the actual valence-band energy maximum and $\Delta_{SO}/3$ is the energy by which the $p_{3/2}$ states are pushed up in energy relative to the valence band as a whole (the $p_{1/2}$ states are pushed down by $2\Delta_{SO}/3$). The *indirect*-gap conduction-band minima are considered because states in their vicinity are more representative of the conduction band as a whole than those in the vicinity of the direct-gap minimum at Γ, and also because states that are spatially localized (e.g. the interface states) can only be constructed from components

Table 8.2 Indirect bandgap energies and, for direct-gap materials, direct bandgap energies ($\mathscr{E}_g^i$ and $\mathscr{E}_g$, respectively), spin–orbit splittings (Δ_{SO}) and n-type Schottky-barrier heights, $\mathscr{E}_{B,n}$, for gold contacts to various semiconductors, obtained both experimentally and theoretically from eqn. (8.36c) (with $\delta_M = -0.2$ eV)

Semiconductor	$\mathscr{E}_g^i$(eV)	$\mathscr{E}_g$(eV)	Δ_{SO}(eV)	$\mathscr{E}_{B,n}^{exp}$(eV)	$\mathscr{E}_{B,n}^{theo}$(eV)
Si	1.11	—	0.04	0.80	0.76
Ge	0.66	—	0.29	0.59	0.58
Gap	2.64	2.27	0.08	1.30	1.16
GaAs	1.81	1.43	0.34	0.90	0.78
GaSb	0.82	0.75	0.75	0.60	0.67
AlAs	2.15	—	0.28	1.20	1.32
AlSb	1.62	—	0.70	1.08	1.13
InP	1.84	1.34	0.11	0.52	0.64

(Experimental data of $\mathscr{E}_g^i$, $\mathscr{E}_g$ and Δ_{SO} from Madelung (1996) and of $\mathscr{E}_{B,n}^{exp}$ from Sze (1981))

having a wide range of $\boldsymbol{k}$-values. Thus, the pinned Fermi-level position is, from eqn. (8.36a), given approximately by (Tersoff (1985)):

$$\mathscr{E}_F \simeq \frac{1}{2}(\mathscr{E}_g^i + 2\mathscr{E}_v - \Delta_{SO}/3) + \delta_M, \tag{8.36b}$$

where the indirect gap is given by $\mathscr{E}_g^i = \mathscr{E}_c^i - \mathscr{E}_v$. The p-type Schottky-barrier height is given by $\mathscr{E}_{B,p} = \mathscr{E}_F - \mathscr{E}_v$ (see Fig. 8.48b), i.e. $\mathscr{E}_{B,p} \simeq \frac{1}{2}(\mathscr{E}_g^i - \Delta_{SO}/3) + \delta_M$. The n-type barrier height is given by $\mathscr{E}_{B,n} = \mathscr{E}_c - \mathscr{E}_F$, where $\mathscr{E}_c$ is the *actual* conduction-band minimum energy, i.e.

$$\begin{aligned}\mathscr{E}_{B,n} &\simeq \mathscr{E}_c - \mathscr{E}_v - \frac{1}{2}(\mathscr{E}_g^i - \Delta_{SO}/3) - \delta_M \\ &= \mathscr{E}_g - \frac{1}{2}(\mathscr{E}_g^i - \Delta_{SO}/3) - \delta_M,\end{aligned} \tag{8.36c}$$

where $\mathscr{E}_g$ is the actual minimum gap energy (direct or indirect). Values of $\mathscr{E}_{B,n}$ calculated using eqn. (8.36c) are in very good agreement with experimental values (see Table 8.2).

The application of an external electric field to a heterojunction changes the extent of the band bending that occurs in the absence of the field. In the presence of the external field, the quantity that is constant through the junction region is not the chemical potential but the *electrochemical potential* η (eqn. 6.86):

$$\eta = \mu - eV,$$

where V is the applied electrostatic potential. Figures 8.48(c, d) show the effect of an external voltage V on the band bending of an n-type Schottky barrier: forward bias (the semiconductor biased positively with respect to the metal) *reduces* the barrier $\mathscr{E}_{B,n}$ by an amount eV (Fig. 8.48c), and reverse bias *increases* it by eV (Fig. 8.48d).

It is apparent from Figs 8.48(c, d) that a Schottky barrier is a rectifying contact: it is markedly non-ohmic in allowing a large current to flow (electrons from semiconductor to metal) in forward bias when the barrier to electron transfer is *reduced* by an amount eV (see Figs. 8.48a and 8.48c), but only a very small current ($-j_0$) flows in reverse bias (due to thermionic emission over the barrier $\mathscr{E}_B$, the height of which is unaffected by the bias field). As a result, the net current density flowing across the Schottky-barrier heterojunction in forward bias can be written as

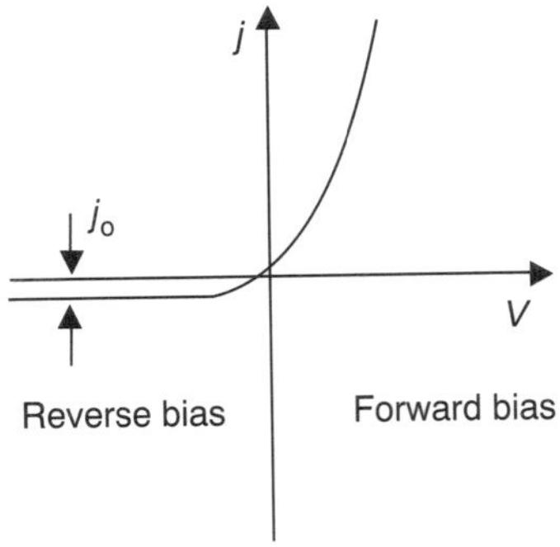

Fig. 8.50 Variation of current density j with applied voltage V for a rectifying junction, e.g. a Schottky barrier, or a p–n junction.

$$j = j_0[\exp(eV/k_B T) - 1], \tag{8.37a}$$

with the sign of V reversed in the case of reverse bias. The thermionic-emission current density j_0 is given by the Richardson–Dushman equation:

$$j_0 = AT^2 \exp(-\mathscr{E}_{B,n}/k_B T), \tag{8.37b}$$

where A is a constant. Neglecting diffusion processes in the depletion layer of the semiconductor, $A = 4\pi e m_e^* k_B^2/h^3$ (Lüth (1995)). Figure 8.50 shows the current–voltage characteristics of a rectifying junction, such as a Schottky barrier.

Low-resistance, quasi-ohmic contacts to a semiconductor that would otherwise form a rectifying Schottky barrier can be made by heavily doping the surface of the semiconductor in contact with the metal to form a so-called n^+ (or p^+) layer. The constraint of charge neutrality in the junction region causes the n^+ layer to have a depletion region of reduced width; this allows electron tunnelling through the barrier, rather than thermal activation over it, resulting in approximately the same magnitude of current for both polarities of applied field. Further details on Schottky barriers are given in Sze (1981).)

8.4.2.2 *Semiconductor–semiconductor heterojunctions and homojunctions*

Two different types of semiconductor–semiconductor junctions can be envisaged: either *heterojunctions* between *different* semiconducting materials (with different bandgaps), or *homojunctions* between differently doped regions of the *same* semiconducting material. In both cases, if the work functions of the two semiconductors forming the heterojunction are different, band bending in the vicinity of the interface will take place, as for metal–semiconductor heterojunctions (§8.4.2.1).

Heterojunctions between different semiconducting materials have already been mentioned in §8.3.5, namely the junction in the modulation-doped heterostructure between n-type AlGaAs and intrinsic GaAs, in connection with the 2D electron gas formed in the inversion layer at the surface of the GaAs layer at the interface; the electrons are confined in the potential well in the GaAs conduction band formed by the band bending on one side due to the junction field and the conduction-band-edge discontinuity between AlGaAs and GaAs on the other (Fig. 8.32a). AlGaAs–GaAs heterojunctions feature also in quantum wells (tri-layer heterostructures, with two heterojunctions and an intervening layer of the lower-gap material, GaAs—see §8.4.3) and in semiconductor superlattices (§8.4.4).

Semiconductor homojunctions, involving differently doped regions of the *same* semiconductor material, are known as p–n junctions. These structures, almost invariably made using crystalline Si as the host semiconductor, are very important technologically and are used in many electronic applications. The origin of the band bending at a p–n junction can be seen by reference to Fig. 8.51. The majority carriers in the n-type material are electrons in the conduction band and are holes in the valence band of the p-type material (Fig. 8.51a): when placed in contact, the concentration gradients of electrons and holes cause a net diffusion of electrons into the p-type material, and of holes into the n-type material, until the chemical potentials are equalized (Fig. 8.51b). (It is assumed that the temperature is sufficiently low that the dopant atoms themselves cannot diffuse.) As a result of this electron and hole diffusion, the n-type layer within a distance d_n of the junction becomes depleted of electrons and positively charged (with the charge located on ionized donors) and the p-type material within a distance d_p of the junction becomes negatively charged (with the charge located on ionized acceptors); the total width of the depletion region is thus $d = d_n + d_p$.

The electrostatic potential associated with the charged double layer is responsible for the band bending in the depletion region. This contact potential (or diffusion potential), ϕ_c, can be calculated by equating expressions (e.g. eqn. (6.246)) for the position of the chemical potential (relative to the respective band edges) for n-type and p-type materials (valid far from the junction), to give (see Problem 8.10):

$$\phi_c = \frac{k_B T}{e} \ln\left(\frac{N_d N_a}{n_i^2}\right) \tag{8.38a}$$

$$= \frac{\mathscr{E}_g}{e} - \frac{k_B T}{e} \ln\left[\frac{N_c N_v}{N_d N_a}\right] \tag{8.38b}$$

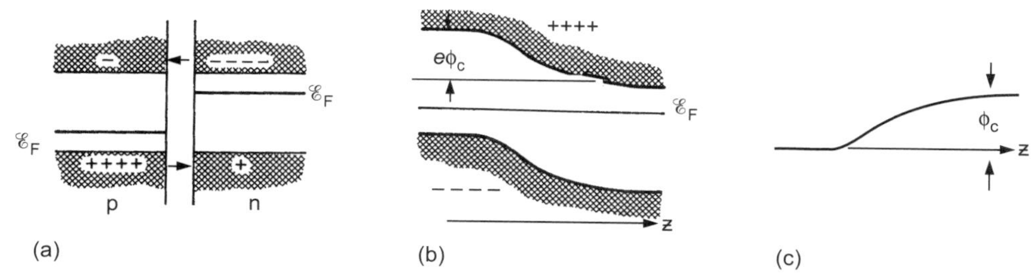

Fig. 8.51 Illustration of the origin of band bending at a p–n junction. (a) Two separated p-type and n-type samples of the same semiconducting material. The directions of the diffusion of the majority carriers, which occurs to equalize the chemical potentials when the heterojunction is formed, are shown. (b) Band bending at the p–n junction, showing the contact potential energy $e\phi_c$ and the junction double layer of space charge responsible for the potential and band bending. (c) Spatial profile of the contact potential, $\phi_c(z)$. Note that it is the mirror image of the profile of the electron energy bands in (b).

making use of eqn. (6.203), and where N_d and N_a are the donor and acceptor concentrations in the n-type and p-type materials, respectively, N_c and N_v are the effective concentration of states at the conduction- and valence-band edges, respectively, and n_i is the intrinsic carrier (electron or hole) concentration of the semiconductor. A typical room-temperature value of $\phi_c \simeq 0.7\,\text{V}$ is found for c-Si, doped such that $N_d = N_a = 10^{22}\,\text{m}^{-3}$ ($n_i \simeq 10^{16}\,\text{m}^{-3}$ at this temperature—see §6.5.1.1); this value corresponds to a sizeable fraction of the bandgap ($\simeq 1.1\,\text{eV}$). By solving Poisson's equation for the homojunction, another expression for the contact potential can be found in terms of the depletion lengths d_n and d_p (Problem 8.10):

$$\phi_c = \frac{e}{2\varepsilon_r\varepsilon_0}\left(N_d d_n^2 + N_a d_p^2\right), \tag{8.39}$$

where ε_r is the dielectric constant of the semiconductor. Charge neutrality of the homojunction in dynamic equilibrium requires that

$$N_d e d_n = N_a e d_p, \tag{8.40}$$

whence expressions for the individual depletion widths can be found:

$$d_{n,p} = \left[\frac{2\varepsilon_r\varepsilon_0 N_{a,d}\phi_c}{eN_{d,a}(N_d + N_a)}\right]^{1/2}. \tag{8.41}$$

Note that the spatial profile for the contact potential, $\phi_c(z)$, in the direction (z) normal to the junction (see Problem 8.10) is simply the mirror image of the profile of the electron energy bands (Fig. 8.51c), since electrostatic potential is conventionally defined with respect to a *positive* test charge whereas energy bands are plots of the energy of electrons (*negatively* charged particles).

The p–n junction exhibits rectifying behaviour in the current–voltage characteristic depending on the polarity of the applied bias, similar to that exhibited by the Schottky barrier (§8.4.2.1) and for the same reason: forward bias (a positive potential applied to the p-type layer of the junction) reduces the contact-potential barrier and hence produces a large increase in the current; negative bias (a positive potential applied to the n-type layer) increases the contact-potential barrier and hence greatly decreases the current. In the presence of an applied electrostatic potential $\pm V$, it is the electrochemical potential $\eta = \mu \pm eV$ that is constant through the junction, rather than the chemical potential μ being constant as in the equilibrium case.

In order to calculate the current passed by a p–n junction as a function of bias, consider first the equilibrium case for zero bias (Fig. 8.52a). In dynamic equilibrium, the electron flow from the p-layer to the n-layer through the conduction bands is equal to the reverse flow (and the same is true for the hole currents through the valence bands). Consider the case of the electron current I_e^0 in the p $\rightarrow$ n direction. This originates from the very small concentration n_p of electrons thermally generated in the conduction band in the p-type material. If the extrinsic semiconductor is in the saturation régime (§6.5.2.2), i.e. all acceptors in the p-layer are ionized, then the concentration of holes in the valence band of this layer is $p_p \simeq N_a$, and hence from the law of mass action (eqn. (6.202)):

$$n_h = n_i^2/N_a, \tag{8.42}$$

where n_i is the intrinsic carrier concentration of electrons or holes. If the lifetime of the electrons in the p-layer before they recombine with holes is τ_p, the recombination rate is

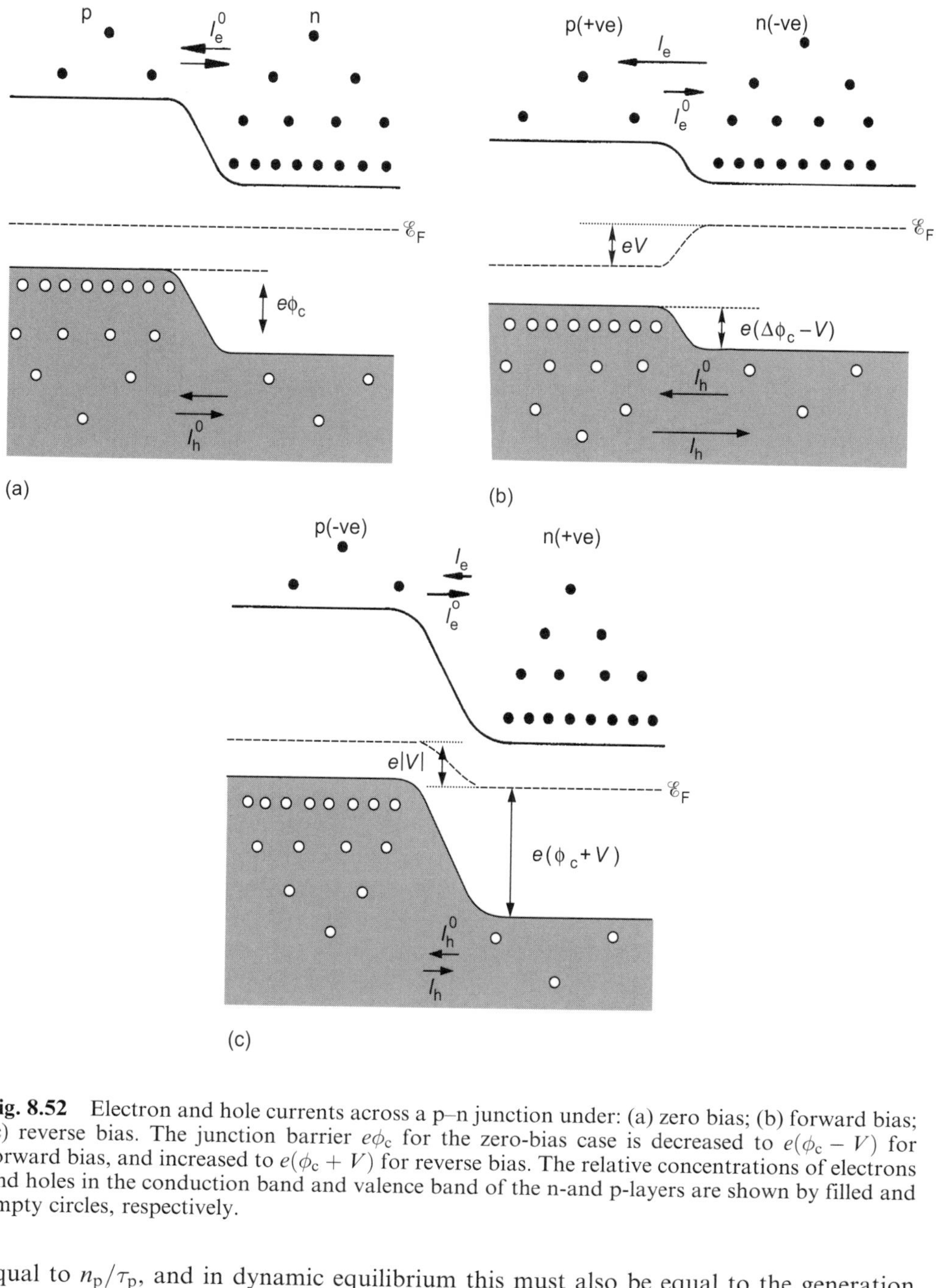

Fig. 8.52 Electron and hole currents across a p–n junction under: (a) zero bias; (b) forward bias; (c) reverse bias. The junction barrier $e\phi_c$ for the zero-bias case is decreased to $e(\phi_c - V)$ for forward bias, and increased to $e(\phi_c + V)$ for reverse bias. The relative concentrations of electrons and holes in the conduction band and valence band of the n-and p-layers are shown by filled and empty circles, respectively.

equal to n_p/τ_p, and in dynamic equilibrium this must also be equal to the generation rate. The distance that a newly generated electron moves before it recombines is the electron diffusion length $L_e = (D_e\tau_p)^{1/2}$, where D_e is the electron diffusion coefficient.

Electrons generated in the conduction band of the p-layer within a distance L_e of the interface with the n-layer are likely to be able to diffuse across the boundary before recombination, and hence contribute to the current. Thus, the p → n current can be expressed approximately as the product of the electron generation rate per unit volume and the volume L_eA within the p-type depletion layer (where A is the area of the heterojunction); there is no potential barrier to electron motion in this direction unlike the reverse case, n → p (see Fig. 8.52a). Hence the current is

$$I_e^0 \simeq e\left(\frac{n_p}{\tau_p}\right)L_eA \tag{8.43a}$$

$$= \frac{eD_en_i^2A}{L_eN_a}. \tag{8.43b}$$

Electrons moving in the n → p direction have to surmount the contact-potential barrier ϕ_c, and so the current in this direction can be written as

$$I_e^0 = c\exp(-e\phi_c/k_BT), \tag{8.44}$$

where c is a constant.

With the application of an external voltage V, the p → n current is unchanged: $I^e_{p\to n} = I_e^0$. However, the reverse (n → p) current is greatly altered since the bias either decreases the band offset (contact potential) to $e(\phi_c - V)$ (for forward bias—see Fig. 8.52b), or increases it to $e(\phi_c + V)$ (for reverse bias—see Fig. 8.52c). Thus, the n → p electron current is given by

$$I^e_{n\to p} = c\exp[-e(\phi_c - V)/k_BT]$$
$$= I_e^0\exp(eV/k_BT) \tag{8.45}$$

from eqn. (8.44). In eqn. (8.45), V is positive for forward bias and negative for negative bias. Therefore, the net electron current is

$$I^e = I^e_{n\to p} - I^e_{p\to n} = I_e^0[\exp(eV/k_BT) - 1]. \tag{8.46}$$

Similar considerations can be applied to the case of the hole currents, giving by analogy with eqn. (8.43):

$$I_h^0 \simeq e\left(\frac{p_n}{\tau_n}\right)L_hA \tag{8.47a}$$

$$= \frac{eD_hn_i^2A}{L_hN_d}. \tag{8.47b}$$

Thus, the total current is the sum of electron and hole currents:

$$I = I_0[\exp(eV/k_BT) - 1], \tag{8.48}$$

where

$$I_0 = en_i^2A\left[\frac{D_e}{L_eN_a} + \frac{D_h}{L_hN_d}\right]. \tag{8.49}$$

Note that the rectifying characteristic of the p–n junction has the same functional form as that of the Schottky barrier (eqn. (8.37a)), and so the shape of the I–V curve is as shown in Fig. 8.50.

Departures in the rectifying behaviour of a p–n junction from that described above occur if the bias voltage is large and if the level of doping in the p- and n-layers is high. In this case, *tunnelling* of electrons and holes through the contact-potential barrier at the homojunction can take place, since a high dopant concentration leads to a reduction in the width of the depletion layer (eqn. (8.41)). The tunnelling results in a marked enhancement of the junction current at particular values of bias.

An instance of this behaviour is the reverse breakdown of a p–n junction. A large reverse bias (say 2–3 V) causes the conduction-band minimum of the n-type material to become *lower* than the valence-band maximum of the p-layer (Fig. 8.53a). For a heavily doped structure, in which the depletion region is very narrow, quantum-mechanical tunnelling of electrons in the valence band of the p-layer can occur into vacant conduction-band states in the n-layer causing Zener breakdown—see Problem 8.12(b). Hence, at a certain critical negative breakdown voltage $-V_c$, a large (negative) tunnelling current begins to flow (Fig. 8.53b). Reverse breakdown can also occur in lightly doped p–n junctions, but at higher values of negative voltage than is characteristic of Zener breakdown. In this case, impact ionization of electrons from valence-band states to conduction-band states occurs, due to collisions with high-energy mobile electrons accelerated by the large electric field. This process can lead to avalanche breakdown, since electron hole pairs created by impact ionization are in turn accelerated by the field and can thus lead to subsequent impact-ionization events (see Problem 8.12(c)).

In very heavily doped p–n junctions, the semiconductor becomes *degenerate*: the chemical potential lies in one or other of the bands outside the depletion region (Fig. 8.54a). As a result, even with *zero* applied bias, the conduction-band minimum of the n-layer can lie below the valence-band maximum of the p-layer. Since the width of the depletion layer in the case of high dopant concentrations is very small (see eqn. (8.41)),

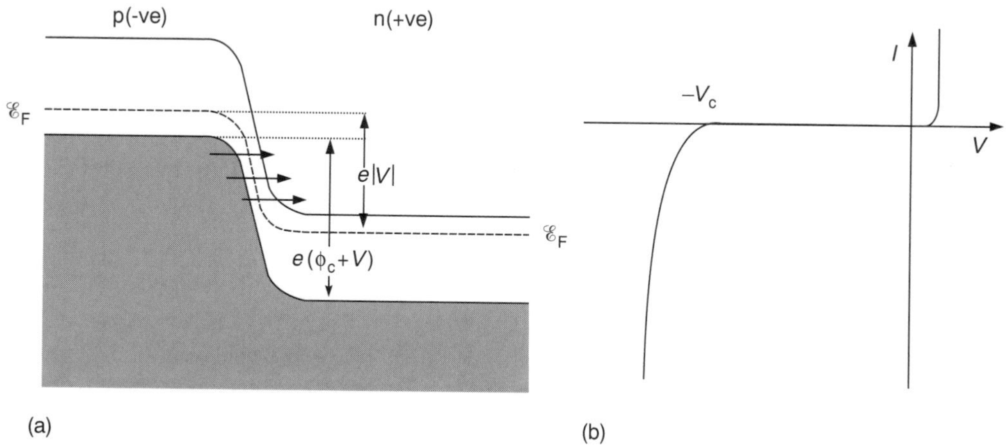

Fig. 8.53 Zener breakdown in a heavily doped p–n junction in reverse bias. (a) A large reverse bias causes the conduction-band minimum of the n-layer to drop below the valence-band maximum of the p-layer. Electron tunnelling can take place from the p-layer valence band into the n-layer conduction band. (b) The I–V characteristic showing the Zener tunnelling current at a reverse bias of $-V_c$.

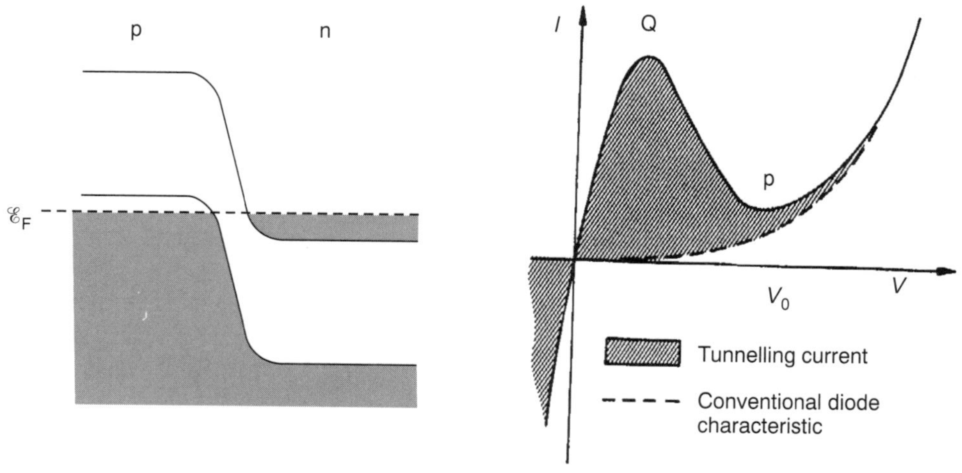

Fig. 8.54 Esaki tunnelling in a heavily doped (degenerate) p–n junction in forward bias. (a) In zero bias, the conduction-band minimum of the n-layer lies below the valence-band maximum of the p-layer. For small values of forward bias, electrons can tunnel from the n-layer conduction band into ezmpty states in the p-layer valence band. The tunnelling current ceases at a value of forward bias, V_o, when the band overlap vanishes. (b) The forward-bias I–V characteristic of an Esaki tunnel diode. The peak in the current is due to tunnelling: this contribution vanishes at a point (P) where the band overlap in (a) vanishes. The region Q–P is one of negative differential resistance.

tunnelling of electrons from the conduction band of the n-layer into empty states in the valence band of the p-layer can occur at small values of forward bias, leading to a large current. However, as the forward bias is increased, the overlap of the bands decreases until it ceases at a particular voltage, V_0; at this point (P in Fig. 8.54b) the excess tunnelling current also vanishes. At higher values of forward bias, the normal forward-bias behaviour of the simple p–n junction is recovered. Devices exhibiting the I–V characteristics shown in Fig. 8.54b are known as Esaki tunnel diodes; they are of interest because of the negative differential resistance ($\mathrm{d}V/\mathrm{d}I < 0$) exhibited in the region Q–P of the I–V characteristic.

8.4.2.3 *Insulator–semiconductor heterojunctions*

A technologically important heterojunction is that between a semiconductor and an insulator, specifically between crystalline Si and its (amorphous) oxide, SiO_2, which forms the basis of so-called 'metal–oxide–semiconductor' (MOS) devices. A metal gate electrode is deposited on top of a thin ($\simeq$ 50 nm) insulating oxide layer formed on the (100) face of p-type Si (Fig. 8.55a). Application of a positive bias to the gate electrode causes the semiconductor bands to bend down at the insulator–semiconductor interface. The majority hole carriers are repelled away from this interface into the bulk of the semiconductor, leaving a hole-depletion layer in the semiconductor, starting at the interface, of width $\simeq$ 1000 Å. As the gate voltage is progressively increased, eventually it

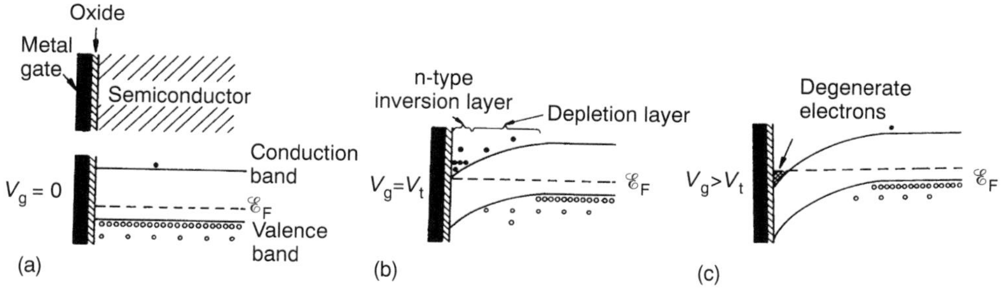

Fig. 8.55 Band bending in an MOS device consisting of: (a) a metallic gate electrode on a thin insulating layer (SiO_2) on a semiconductor (c-Si); (b) formation of an inversion layer in the bent-down conduction band of the semiconductor at the interface with the insulator for a threshold gate voltage $V_g = V_t$; (c) formation of a degenerate 2D gas in the inversion layer for $V_g > V_t$.

reaches a threshold value V_t at which the semiconductor conduction band is bent down below the level of the chemical potential (Fig. 8.55b). At this point, an inversion layer is formed: electrons (normally the minority carriers) congregate in the approximately triangular-shaped potential well (of width $\simeq$ 10–100 Å) in the semiconductor near the interface with the insulator to form a 2D electron gas. The higher the gate voltage V_g, the greater is the areal (sheet) density n_a of electrons in the inversion layer:

$$n_a = \frac{C_a}{e}(V_g - V_t) \tag{8.50a}$$

$$= \frac{\varepsilon_{ox}}{ed_{ox}}(V_g - V_t), \tag{8.50b}$$

where C_a is the capacitance per unit area of the gate electrode relative to the electron gas, and ε_{ox} (= 3.9) is the dielectric constant of the a-SiO_2 layer of thickness d_{ox}. The sheet electron density can be varied by a couple of orders of magnitude simply by varying V_g; the upper limit is set by the dielectric-breakdown field of the insulator ($\simeq 10^9$ V/m for a-SiO_2—see Problem 8.13).

The degree of perfection required of the insulator– semiconductor heterojunction is high in order to be able to produce and make use of the electron inversion layer. The surface charge density ρ_s in the oxide at the interface, due to imperfections (impurities, dangling bonds) must be kept very low ($\rho_s < 10^{17}\,m^{-2}$) in order that band bending in the Si can be controlled by the gate voltage, and also to reduce scattering of electrons in the inversion layer, which otherwise would limit the mobility and conductivity. Present oxide-growth technology can produce oxide films with $\rho_s \lesssim 10^{14}\,m^{-2}$ (cf. the surface density of atoms on a (100) face of c-Si is $\simeq 6.7 \times 10^{18}\,m^{-2}$!). Furthermore, the interface must be atomically smooth (with a rugosity much less than the width of the inversion layer) in order to reduce surface (boundary) scattering to a minimum.

8.4.3 Quantum wells

A quantum well is a 2D conducting system (in the x–y plane), with size-quantization occurring in the third (z) dimension due to the width, d, of the well being less than the

Fermi wavelength, $d < \lambda_F$. This condition is almost impossible to achieve for metals, for which $\lambda_F \simeq 5$ Å (see Table 5.1). However, with modern thin-film-deposition technology (e.g. molecular-beam epitaxy—see §1.1.1) it is possible to produce quantum wells, with widths of a few tens of angstroms, using semiconductors such as c-Si or GaAs, since for them the Fermi wavelength is much larger than in metals. The Fermi wavelength of a 2D electron gas, e.g. in an inversion layer at an Si–SiO_2 interface in an MOS device (§8.4.2.3) or at the GaAs–n-type AlGaAs interface in a modulation-doped heterostructure (§8.4.2.2), can be calculated by analysing the behaviour of electrons assumed to be moving freely in 2D in the x–y plane. In this case, the kinetic energy for this 2D motion is

$$\mathscr{E}(x, y) = \frac{\hbar^2}{2m_x^*}k_x^2 + \frac{\hbar^2}{2m_y^*}k_y^2; \tag{8.51}$$

the energy associated with motion in the third confined, dimension (z) will be considered later. In eqn. (8.51), $m_{x,y}^*$ are the effective masses for electrons moving in the x, y directions. The corresponding 2D electron density of states (per unit area) is (see Problem 5.2):

$$D_{2D}(\mathscr{E}) = g_s g_v \left(m_x^* m_y^*\right)^{1/2} / 2\pi\hbar^2, \tag{8.52}$$

which is *independent* of energy. The factor g_s relates to electron-spin degeneracy; for both Si and GaAs, $g_s = 2$. The factor g_v is the valley degeneracy, reflecting the degeneracy of the states at the bottom of the conduction band. The case of the direct-gap semiconductor GaAs is particularly simple since there is only a single conduction-band minimum at $\Gamma = 0$ in the Brillouin zone, and hence $g_v = 1$. In addition, $m_x^* = m_y^* = m_e^*$ ($= 0.067 m_e$ for GaAs—Table 6.1).

The case of the Si{100} interface generally used in MOS devices is a little more complicated since Si is an indirect-gap semiconductor with conduction-band minima, or conduction-band valleys, lying at non-zero $\boldsymbol{k}$-values in the Brillouin zone. These minima for Si occur in the $\langle 100 \rangle$ directions in reciprocal space; the constant-energy surfaces near the bottom of the conduction band are ellipsoidal (eqn. (5.127)), with the major axes of the ellipsoids pointing along the $\langle 100 \rangle$ directions (Fig. 5.51a). These ellipsoidal energy surfaces are characterized by so-called *longitudinal* and *transverse* effective masses (eqn. (5.127) and §6.5.1.4), with $m_l^* \simeq 0.92 m_e$ and $m_t^* \simeq 0.19 m_e$ (Problem 6.25). For a quantum well (inversion layer) oriented such that the 2D electron gas is in the x–y plane, the valley degeneracy, and hence the density of states, depends on which of the six conduction-band states (valleys) shown in Fig. 5.51a are involved in the motion; there are two choices for the Si{100} surface. One possibility is for the two valleys in the $\boldsymbol{k}_z$-direction to be involved, in which case the valley degeneracy is $g_v = 2$ and the effective masses appearing in the expression for the density of states (eqn. (8.52)) are $m_x^* = m_y^* = m_t^*$. The other possibility involves the other four valleys in the $\boldsymbol{k}_x - \boldsymbol{k}_y$ plane, in which case $g_v = 4$ and the product $m_x^* m_y^* = m_t^* m_l^*$.

The Fermi energy $\mathscr{E}_F$ and the areal (sheet) electron density n_a are linearly related because the 2D electron-gas density of states (eqn. (8.52)) is energy-independent:

$$\begin{aligned} n_a &= D_{2D}(\mathscr{E}_F)\mathscr{E}_F \\ &= g_s g_v (m_x^* m_y^*)^{1/2} \mathscr{E}_F / 2\pi\hbar^2. \end{aligned} \tag{8.53}$$

The Fermi wavevector k_{F}, and hence the wavelength $\lambda_{\mathrm{F}} = 2\pi/k_{\mathrm{F}}$, can be obtained from eqn. (8.53) by substituting into it eqn. (8.51) evaluated at the Fermi level. In the case of a single conduction-band valley (GaAs), or the simpler case for Si{100} of the two valleys in the $\boldsymbol{k}_z$-direction, a simple expression for k_{F} is obtained since the *single* effective mass involved in eqns. (8.51) and (8.53) (m_{e}^* for GaAs and m_{t}^* for Si) cancels out, leaving:

$$k_{\mathrm{F}} = 2\pi/\lambda_{\mathrm{F}} = (4\pi n_{\mathrm{a}}/g_{\mathrm{s}}g_{\mathrm{v}})^{1/2}. \tag{8.54}$$

For the case of GaAs{100} $\lambda_{\mathrm{F}} = 400$ Å, and for Si{100} $\lambda_{\mathrm{F}} = 350$–$1100$ Å, depending on n_{a}.

The simplest case of a quantum well is when the electrons are confined in the z-direction by a *square-well potential* with infinitely high potential barriers, $V = \infty$, at the two interfaces at $z = 0$ and d, and $V = 0$ for $0 < z < d$ (Fig. 8.56); this is just the 1D particle-in-a-box problem. The solutions are sinusoidal standing waves that satisfy the boundary conditions ($\psi_z(0) = \psi_z(d) = 0$)—see Problem 3.6, i.e.

$$\psi_z = A \sin k_z z, \tag{8.55}$$

where A is a normalization constant, with

$$k_z = n\pi/d \tag{8.56}$$

and n is a non-zero integer. The associated energy levels are quantized according to:

$$\mathscr{E}_n(z) = \frac{\hbar^2\pi^2}{2m_z^* d^2} n^2 \qquad (n = 1, 2, 3, \ldots), \tag{8.57}$$

where m_z^* is the effective mass for motion in the z-direction. For a 10 nm thick layer of GaAs, $\mathscr{E}_1(z) \simeq 50$ meV. Note that the energies of successive levels increase quadratically with quantum number, n.

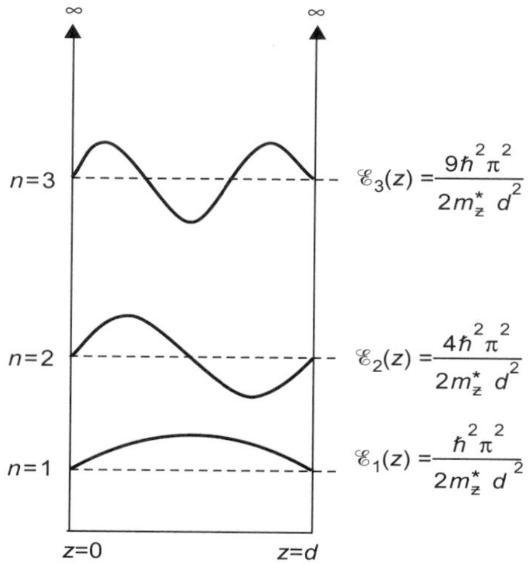

Fig. 8.56 Wavefunctions and energy levels of the three lowest-bound states ($n = 1, 2, 3$) for an electron confined in a 1D square potential well, of width d, with infinitely high potential barriers.

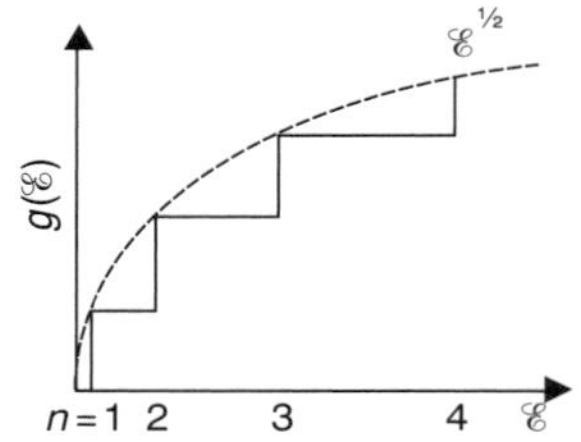

Fig. 8.57 Density of states for electron motion in a quantum well. The steps correspond to the population of successively higher sub-bands labelled by the quantum number, n. The dashed line is the free-electron density of states for 3D motion.

The total energy of the 2D electron gas in an infinitely high square quantum well is thus given by the sum of eqns. (8.51) and (8.57):

$$\mathscr{E} = \mathscr{E}_n(z) + \mathscr{E}(x, y) = \frac{\hbar^2 \pi^2}{2m_z^* d^2} n^2 + \frac{\hbar^2}{2} \left(\frac{k_x^2}{m_x^*} + \frac{k_y^2}{m_y^*} \right). \tag{8.58}$$

Each energy $\mathscr{E}_n(z)$ for a particular value of n marks the bottom of a continuous sub-band of states for 2D motion. The overall density of states for a quantum well is a stepwise function of energy (Fig. 8.57). Starting from zero energy, there are no allowed states until $\mathscr{E} = \mathscr{E}_1(z)$: the zero-point energy of the particle-in-a-box is non-zero. The energy $\mathscr{E} = \mathscr{E}_1(z)$ corresponds to $\mathscr{E}(x, y) = 0$; an increase in total energy of an electron in the quantum well is achieved by progressively increasing the kinetic energy in the x, y directions by increasing k_x and k_y. Since the 2D free-electron density of states is independent of energy (eqn. (8.52)), the total density of states above the step corresponding to $n = 1$ remains constant until the total energy reaches $\mathscr{E} = \mathscr{E}_2(z)$, at which point the next highest sub-band corresponding to $n = 2$ can start to be populated, and another step occurs in the density of states (of magnitude given by eqn. (8.52)). Note from Fig. 8.57 that the envelope of the 2D stepwise density of states is given by the parabolic $\mathscr{E}^{1/2}$-dependence of the free-electron density of states for 3D motion (eqn. (5.15)): the reason for this is that each step in the 2D density of states is of equal height but the energies at which the steps occur increase quadratically.

The above model of a quantum well involving infinitely high potential barriers at the interfaces cannot, however, be realized in practice. One way of fabricating a square quantum well is by sandwiching a thin layer ($d \simeq 50$ Å) of a low-bandgap semiconductor (e.g. GaAs, $\mathscr{E}_g \simeq 1.5$ eV) between thick layers of a high-bandgap material (e.g. $Al_xGa_{1-x}As$, $\mathscr{E}_g \simeq 1.42 + 1.26x$ eV $\lesssim 2$ eV for $x \lesssim 0.45$). The GaAs–AlGaAs heterojunction is of particular interest because the unit-cell parameters of the two materials are so similar that epitaxial growth of one on the other is possible, without interfacial strain or defects. The energy-band line-up for this heterojunction is a straddled configuration, in which the bandgap of the large-gap material (AlGaAs) straddles that of the other (GaAs)—see Fig. 8.58. Note that the conduction-band and valence-band discontinuities are generally *not* equal: in the case of GaAs–AlGaAs, $\Delta\mathscr{E}_c > \Delta\mathscr{E}_v$. Thus, electrons in the conduction band of the GaAs layer (and holes in the valence band) are confined in

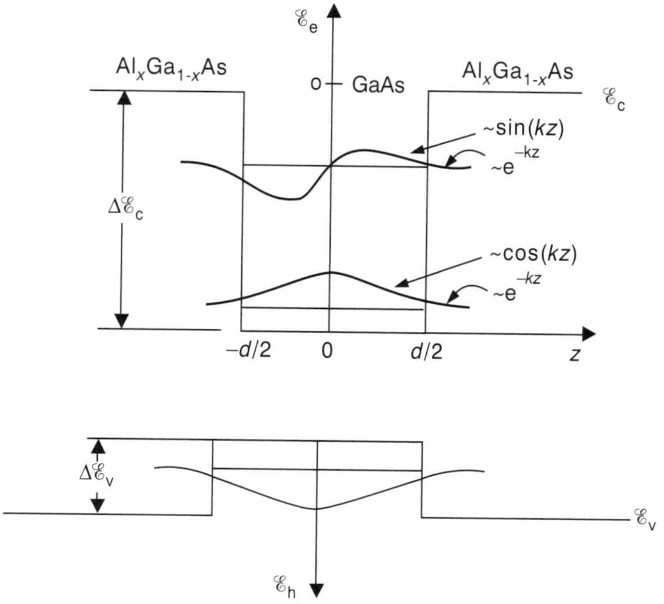

Fig. 8.58 Schematic illustration of a quantum well formed by a thin layer of GaAs, of thickness d, sandwiched between two semi-infinitely thick layers of $Al_xGa_{1-x}As$. The conduction-band and valence-band discontinuities, $\Delta\mathscr{E}_c$ and $\Delta\mathscr{E}_v$, are indicated; because $\Delta\mathscr{E}_c > \Delta\mathscr{E}_v$ for this heterostructure system, more electron states can be bound in the conduction band of GaAs than hole states in the valence bands. The wavefunctions of the quantized states are shown superimposed on their respective energy levels: note that the wavefunction amplitude is not zero in the gap region of the AlGaAs.

the z-direction (normal to the x–y plane of the heterojunctions) by the existence of the forbidden energy gap of the AlGaAs on either side of the GaAs layer for energies below the onset of the AlGaAs conduction band for electrons, or above the onset of the AlGaAs valence band for holes.

Since the confining potential in real quantum wells is not infinitely large, only a *finite* number of electron or hole states can be bound. The maximum number of bound states is related to the well depth V_0 and width d via

$$n_{\max} = 1 + \mathrm{Int}\left[(2m_z^*|V_0|d^2/\hbar^2\pi^2)^{1/2}\right]; \tag{8.59}$$

at least one level is bound for a non-zero-depth potential well. This relation can be confirmed by reference to Fig. 8.59, showing the variation of (normalized) energy of the levels versus the (normalized) well depth.

For the *triangular* potential well characterizing the quantum well (or inversion layer) formed at the Si–SiO_2 junction in MOS devices, or at the GaAs–n-AlGaAs heterojunction, and due to the linear variation in the confining electrostatic potential Ez (see Figs. 8.32a and 8.55), quantized energy levels are found as for the square-well potential (eqn. (8.57)), but the functional form is slightly different (see Problem 8.14):

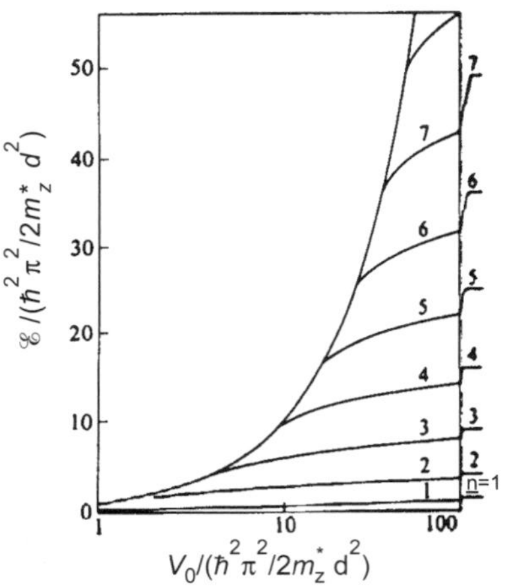

Fig. 8.59 Normalized energies of levels of a particle in a finite square well as a function of normalized well depth V_0. The levels for an infinitely deep square-well potential are also indicated for comparison.

$$\mathscr{E}_v(z) \simeq \left(\frac{3}{2}\pi\hbar e\right)^{2/3} \left(\frac{E^2}{2m_z^*}\right)^{1/3} \left(v + \frac{3}{4}\right)^{2/3} \quad (v = 0, 1, 2, \ldots). \tag{8.60}$$

(The exact eigenvalues are 0.7587, 1.7540, 2.7575, ... instead of $(v + 3/4)$ for $v = 0, 1, 2$, etc.)

Perhaps the most striking physical property of quantum wells concerns their optical-absorption behaviour: simply by varying the well thickness, d, the optical-absorption edge of the semiconductor involved can be moved in energy by up to several hundred meV. The reason for this behaviour is that, in the quantum-confined state, the threshold energy for optical absorption does *not* involve electron states at the conduction-band minimum and valence-band maximum of the *bulk* semiconductor composing the well, since these states do not exist for the quantum well. The lowest allowed state in the conduction band, and highest allowed state in the valence band, is instead the first quantized level ($n = 1$ for a square well, $v = 0$ for a triangular well) for the quantum well. Since the energy of this level scales inversely with the width, d, of the well (as d^{-2} for the square-well potential (eqn. (8.57)), and as $E^{2/3}$ for the triangular well, the width of an inversion layer decreasing with increasing field), large shifts in the energies of the levels of the bound states relative to the band edges are possible. Hence, the threshold energy for optical absorption is increased with respect to the bulk-semiconductor behaviour for very thin quantum wells.

This 'engineering' of the bandgap by tailoring the width of a quantum well is illustrated in Fig. 8.60 for the case of a GaAs-in-$(Al_{0.2}Ga_{0.8})As$ well. It can be seen that the absorption edge shifts to higher energies with decreasing well width for the two thinnest films, as expected. For GaAs thicknesses significantly greater than $\lambda_F = 400$ Å, size-quantization effects are absent, as seen for the optical-absorption profile of the $d = 4000$ Å film in Fig. 8.60. This spectrum is the same as that for bulk GaAs,

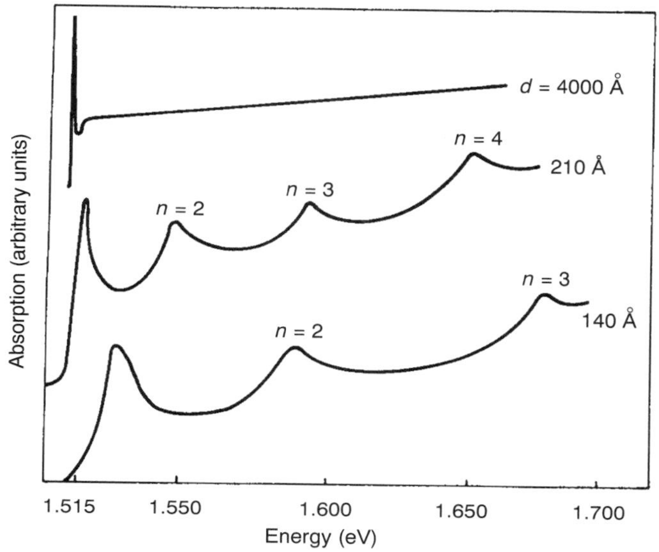

Fig. 8.60 Experimental optical-absorption spectra taken at 2 K for films of GaAs of various thicknesses d, sandwiched between thick layers of $(Al_{0.2}Ga_{0.8})As$. The two thinner GaAs films, with $d < \lambda_F = 400$ Å, exhibit quantum-well behaviour: the threshold energy for absorption increases with decreasing well thickness, and evidence is seen for a finite number of bound states in the well at higher energies. The thickest film behaves like bulk GaAs. (After Dingle *et al.* (1974). Reprinted with permission from *Phys. Rev. Lett.* **33**, 827. © 1974. The American Physical Society)

exhibiting a sharp exciton peak (§5.8.3) at the absorption threshold, followed by a parabolic increase in absorption mirroring the 3D free-electron density of states (eqn. (5.15)). The quantum wells with thinner layers of GaAs can bind a finite number of levels, i.e. $n_{\max} = 4$ for $d = 210$ Å and $n_{\max} = 3$ for $d = 140$ Å. The optical-absorption profile for these samples is similar to the stepped density of states characteristic of a quantum well (Fig. 8.57), with exciton peaks occurring at each discontinuity in $D(\mathscr{E})$ (see also Problem 8.15).

The threshold exciton peak in the spectrum for the thinnest quantum well shown in Fig. 8.60 shows some evidence of structure. This splitting is fully resolved in the spectrum for a very thin GaAs well ($d = 50$ Å), capable of binding only a single level (Fig. 8.61). The reason for this behaviour is that the uniaxial potential associated with the quantum well lifts the degeneracy of the otherwise quadrupally degenerate $j = 3/2$ p-states ($j_z = \pm 3/2, \pm 1/2$) forming the top of the valence band at $\boldsymbol{k} = \mathbf{0}$ in covalent tetrahedral semiconductors such as GaAs (§5.4.2). (It is assumed that the spin orbit splitting is sufficiently large that the split-off $j = 1/2$ band need not be considered.) The well potential causes the $j_z = \pm 1/2$ (normally light-hole) states to be pushed down in energy more than the $j_z = \pm 3/2$ (normally heavy-hole) states because the confinement energy scales inversely with mass (eqn. (8.58)). Hence the $n = 1$ quantum-well state in the valence band, corresponding to electron motion in the confined z-direction, is split into two (Fig. 8.62a). Electronic transitions between these two quantized valence-band

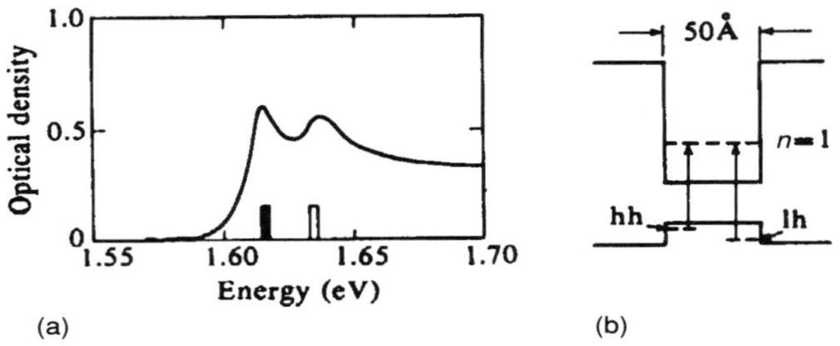

Fig. 8.61 Optical absorption spectrum (a) of a very thin ($d = 50$ Å) quantum well of GaAs, (b) sandwiched between thick layers of ($Al_{0.25}Ga_{0.75}$) As, which can support only a single bound state in valence and conduction bands. The splitting in the exciton peak is caused by the quantum-well-induced splitting of the $n = 1$ level in the valence band: transitions from the upper heavy-hole band, consisting of $p_{3/2}(j_z = 3/2)$ states, are responsible for the lower-energy peak; transitions from the lower light-hole band, consisting of $p_{3/2}(j_z = 1/2)$ states, contribute to the higher-energy peak. (After Dingle *et al.* (1975). Reprinted with permission from *Phys. Rev. Lett.* **34**, 1327. © 1975, The American Physical Society)

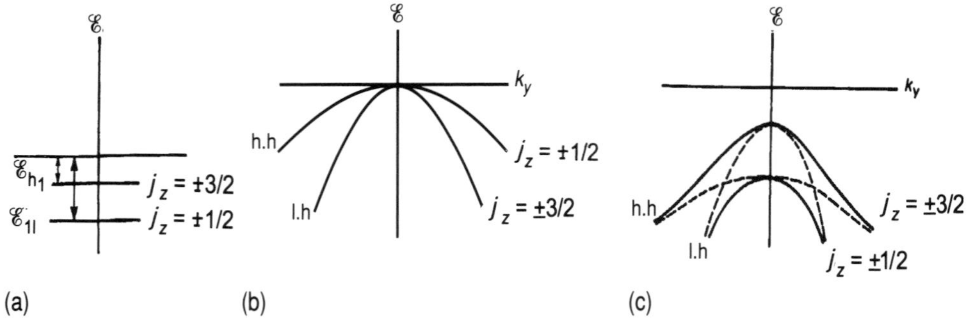

Fig. 8.62 Evolution of hole dispersion in the valence band of a quantum well of GaAs associated with the $p_{3/2}(j_z = \pm 3/2, \pm 1/2)$ states. (a) Energy levels for electron motion in the confined z-direction: the $n = 1$ level is split into two levels corresponding to $j_z = \pm 3/2$ (heavy holes, h.h.) and $\pm\, 1/2$ (light holes, l.h.). (b) Band dispersion in the k_y-direction for the $p_{3/2}$ states at the top of the valence band for bulk GaAs. (c) Band dispersion in the k_y-direction for the quantum well. The band dispersions for the bulk case are shown by the dashed lines: mixing of the states and the no-crossing rule produces the final band dispersions shown by the solid lines.

levels and the single $n = 1$ level in the conduction band give rise to the split exciton peak evident in Fig. 8.61.

The upper ($j_z = \pm 3/2$) states in Fig. 8.61a are referred to as *heavy-hole* states because the energy band corresponding to electron motion in the x–y plane, e.g. $\mathscr{E}(k_y)$, is flatter than that corresponding to the lower ($j_z = \pm 1/2$) states, which are therefore denoted as *light-hole* states. The evolution of the heavy-and light-hole bands in going from the bulk to the quantum well is illustrated in Fig. 8.62b, c: note that in the quantum-well case, the upper band ($j_z = \pm 3/2$) that is the *light*-hole band in the bulk material is transformed into the *heavy*-hole band for the quantum well by mixing with the states of the other band in order to satisfy the no-crossing rule.

8.4.4 Artificial structures

Using molecular-beam-epitaxy deposition techniques (§1.1.1), it is entirely feasible to produce a series of heterojunctions (between different semiconductors or metals) or homojunctions (between differently doped regions of a semiconductor) in a stacked sequence in the growth (z) direction. It is possible, therefore, to fabricate ordered arrays of hetero- or homojunctions in which layers of one material (1) with thickness d_1 alternate with layers of another (2) with thickness d_2 (Fig. 8.63a); thus, an artificial structure can be produced with a new period

$$d = d_1 + d_2. \tag{8.61}$$

(Note, also, that periodic stacks of different amorphous materials can be fabricated so that for them an artificial periodicity can be introduced where none existed previously.)

In the case of a periodic array of semiconductor quantum wells (§8.4.3), where the thickness of the well is d_1 and of the barrier layer is d_2, two scenarios may be envisaged. In one, the spacing of the barrier layer is much greater than the well spacing ($d_2 \gg d_1 \simeq \lambda_F$), in which case the quantum wells are isolated (no tunnelling of electrons through the intervening barrier layers): such structures are called multiple quantum wells. However, if the inter-well spacing is *comparable* to the individual well width ($d_1 \lesssim d_2$), then there is a very considerable electronic interaction between neighbouring quantum wells (see Problem 8.14), and this superlattice has very different electronic properties in the growth direction from those of the bulk constituent materials. (In the

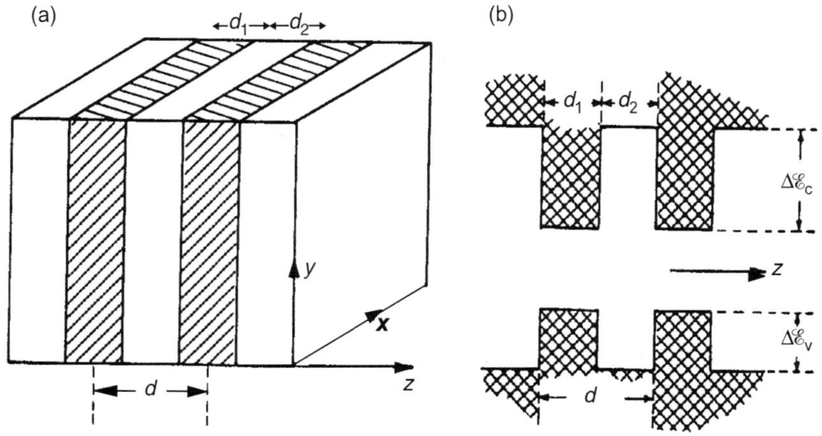

Fig. 8.63 (a) An artificial structure consisting of a stack of layers of two materials 1 and 2, of thicknesses d_1 and d_2, with a periodicity $d = d_1 + d_2$. (b) Spatial profile of the conduction-band minimum and valence-band maximum, i.e. the energy gap, through a semiconductor superlattice with $d_1 \simeq d_2$.

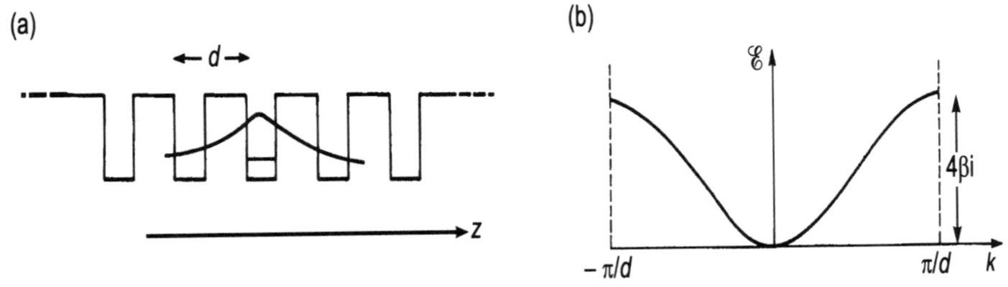

Fig. 8.64 Electron states in a semiconductor superlattice. (a) Representation of the wavefunction, $\psi(z)$, centred on a particular conduction-band well, for the ground-state level of the quantum well. (b) Dispersion of the corresponding superlattice miniband.

x–y plane, normal to the growth direction of the superlattice, the behaviour of the layers is controlled by the microscopic structure of the materials involved, modified, if appropriate, by the 2D nature of the layers—see §8.4.3.)

The effect of the artificial periodicity d in a superlattice on the electronic structure can be understood from a tight-binding model (§5.3.1) for electrons moving in periodic wells in the conduction band (Fig. 8.63b). If the envelope function for the nth well, centred at $z = nd$ and denoted $\psi_n(z - nd)$, overlaps sufficiently with the two neighbouring wells in the growth (z) direction (Fig. 8.64a), then the overall Bloch function for an array of N quantum wells subject to periodic boundary conditions can be written as (cf. eqn. (5.93)):

$$\Psi_k(z) = N^{-1/2} \sum_n e^{iknd} \psi_n(z - nd). \tag{8.62}$$

The envelope function for a well in eqn. (8.62) is the equivalent of the atomic orbital or Wannier function used in the tight-binding treatment of normal crystals made from arrays of atoms. By analogy with eqn. (5.95), the tight-binding expression for the electron energy associated with electron motion through the superlattice (in the z-direction) is

$$\mathscr{E}(k) = \mathscr{E}_i - \alpha_i - 2\beta_i \cos kd, \tag{8.63}$$

where $\mathscr{E}_i$ is the quantized energy of the ith level of a quantum well ($i = 1, 2, 3, \ldots$), α_i is the self-energy integral (cf. eqn. (5.96a)) and is a positive quantity, and β_i is the overlap integral (cf. eqn. (5.96b)) and is also positive; the potential appearing in such integrals is, in this case, the well potential (e.g. a square well—Fig. 8.63b). Thus, the energy dispersion in the z-direction has a cosinusoidal dependence, as shown in Fig. 8.64b; such a band of 'superlattice states' is known as a miniband, and it has an energy width of $4|\beta_i|$.

The corresponding density of states for a superlattice is shown in Fig. 8.65. It can be seen that the step-like profile of a single quantum well has the discontinuities in the density of states, corresponding to different energy levels i in the well, smeared out: the energy width of this smearing in the density of states corresponds to the bandwidth shown in Fig. 8.64b, and the density of states in this region is given by

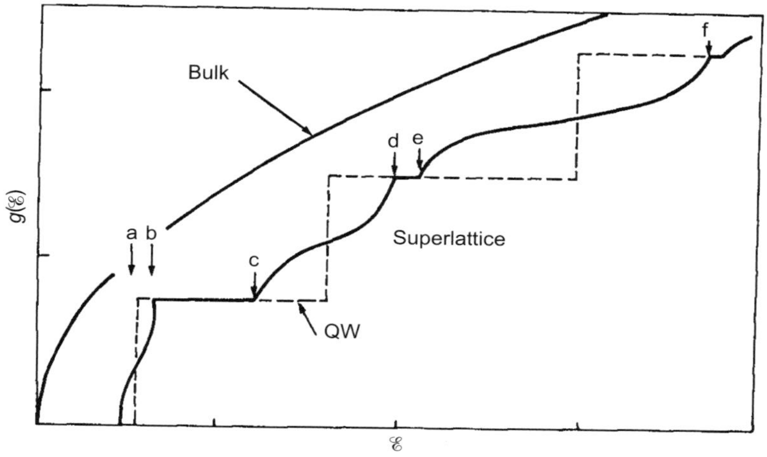

Fig. 8.65 Density of states of a semiconductor superlattice. The regions a–b, c–d and e–f, etc. mark the energy ranges of the minibands, such as shown in Fig. 8.64b. The densities of free-electron states for a single quantum well (QW) and of the bulk material are also shown.

$$g(\mathscr{E}) = \frac{Nm^*}{\hbar^2\pi^2}\cos^{-1}[-(\mathscr{E} - \mathscr{E}_i + \alpha)/2\beta_i] \quad \text{for } |\mathscr{E} - \mathscr{E}_i + \alpha_i| < 2|\beta_i|. \tag{8.64}$$

There are forbidden gaps for 'superlattice' states between minibands, each band being associated with a different quantum-well state; the gaps occur at wavevector values that are multiples of π/d (Fig. 8.64b) and hence considerably smaller than the Brillouin-zone boundaries of the underlying crystal (multiples of π/a) since $d \gg a$, where a is the unit-cell parameter. For electron energies within such gaps, the electronic behaviour of the heterostructure array reverts to that of a 2D quantum well, for which the density of states is independent of energy (eqn. (8.52)), since no periodic extended state through the superlattice in the z-direction can then exist. This behaviour is evident in Fig. 8.65.

Semiconductor superlattices were originally fabricated in order to see whether Bloch oscillations could be observed. This is the prediction that a *d.c.* electric field applied to a material having a periodic potential for electrons should induce an *oscillatory* electron velocity, i.e. an a.c. current, as a result of the motion of the electrons within the concomitant periodically varying band of states in $\boldsymbol{k}$-space. Consider the cosinusoidally varying band structure shown in Fig. 8.66, $\mathscr{E}(\boldsymbol{k}) = \mathscr{E}_0 - \mathscr{E}_1 \cos kd$. Since the real-space group velocity (of a wave packet) of an electron is given by the reciprocal-space gradient of the electron energy, $\boldsymbol{v}_g(\boldsymbol{k}) = (1/\hbar)\partial\mathscr{E}(k)/\partial k = (d\mathscr{E}_1/\hbar)\sin kd$ (eqn. (6.5a)), the velocity for the band shown in Fig. 8.66 increases with increasing k, reaching a maximum at the point of inflection of the band, and thereafter decreasing to zero at the zone boundary, $|k| = \pi/d$. This behaviour can be ascribed to Bragg scattering at $k = \pi/d$ (§6.2). Thus, the electron velocity is a periodic function of k and hence also of time t since, on integrating the equation of motion $\hbar\dot{\boldsymbol{k}} = -e\boldsymbol{E}$ (eqn. 6.22)

$$k = k_0 - eEt/\hbar. \tag{8.65}$$

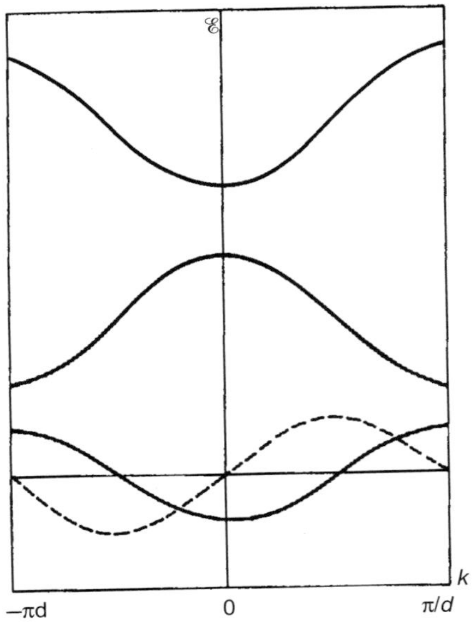

Fig. 8.66 Minibands for a semiconductor superlattice corresponding to three quantum-well levels, together with the k-variation of the electron group velocity (dashed line).

Thus, the period for the motion in reciprocal space of an electron between $k = \pm\pi/d$, the Bloch oscillation, is given by

$$T = \frac{2\pi}{d}\frac{\hbar}{eE}. \tag{8.66}$$

Bloch oscillations can in principle only be observed if the period is shorter than the scattering relaxation time $\tau, T < \tau$. This condition is not satisfied for *bulk* crystals, where the spatial period in eqn. (8.66) is the unit-cell parameter, a (Problem 6.2(b)). Nevertheless, for a superlattice with $d \simeq 10$–$100a$, there is a prospect of observing the Bloch oscillations (see also Problem 8.16).

However, the above picture is rather too simple in its neglect of certain aspects of the effect of an applied electric field on the superlattice minibands. In particular, if an applied electric field E acts uniformly on all wells in the superlattice, overlap of the well wavefunctions between neighbouring wells will cease, i.e. the concept of miniband formation will become meaningless, once the field-induced energy displacement of the quantum-well levels is greater than the miniband width, viz.

$$eEd \gtrsim 4|\beta_i|. \tag{8.67}$$

The arrangement of energy levels in a superlattice when the applied electric field is sufficient to prevent miniband formation is called a Stark ladder (Fig. 8.67). As a result, unambiguous experimental evidence for Bloch oscillations, even in superlattices, is scanty.

The periodicity of a superlattice can also have profound effects on *phonons* propagating in such structures in a somewhat analogous manner to those found for electron

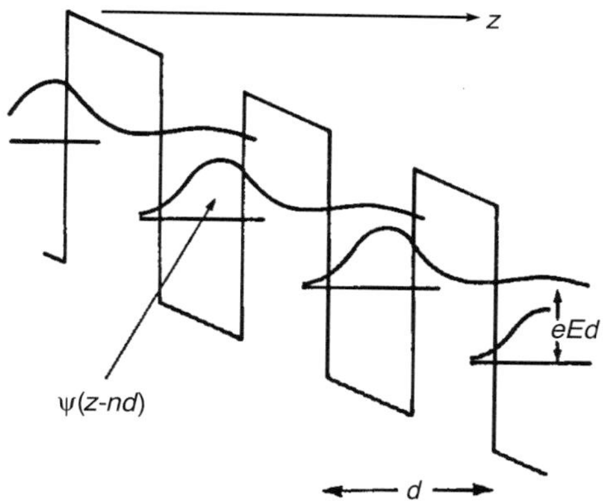

Fig. 8.67 Formation of a Stark ladder in a superlattice subject to an applied electric field E strong enough to prevent overlap of well wavefunctions and the formation of minibands.

states (see Problem 8.17). Thus, *acoustic* phonons (§4.2.3) are *folded back* in the new Brillouin zone with boundaries at $k = \pm\pi/d$, where d is the superlattice periodicity: this is analogous to the formation of electron minibands. In contrast, the atomic displacements associated with *optic* phonons (§4.2.3) in a superlattice tend to be confined to one or other of the different layers of materials, and are known as confined modes: this is somewhat analogous to the 2D motion of electrons within the narrow-gap semiconductor layers (but not the large-gap layers) for electron energies in the gaps between the minibands. Further details on phonon behaviour in superlattices are given in Yu and Cardona (1996).

Thus far, it has been assumed that a semiconductor superlattice is fabricated as a periodic array of compositional *heterojunctions* (Fig. 8.63). However, an artificial structure can also be produced as a periodic array of *homojunctions* or p–n junctions (§8.4.2.2), in which case it is called a doping superlattice. An alternative name is a nipi structure, since there is an effectively intrinsic layer (or even a deliberate extra intrinsic layer) between p- and n-type doped regions of the same semiconductor. Modulation of the conduction-and valence-band edges in real space, along the growth direction, occurs in the case of doping superlattices because of the presence of a spatially periodic space charge, resulting from ionized donors or acceptors in the n-type or p-type layers, respectively (Fig. 8.68), rather than being due to band offsets as for compositional superlattices. For nipi structures, the magnitude of the band-edge modulations is determined by the doping level: the higher the dopant concentration, the larger the modulation energy $\mathscr{E}_0$. The symmetric real-space modulation of the band edges in the growth (z) direction characteristic of a doping superlattice, where at any point z in the superlattice the energy gap between valence and conduction bands is $\mathscr{E}_g$, means that an 'indirect' minimum gap exists in *real space* (rather than reciprocal space) of size

$$\mathscr{E}_g^{\text{eff}} = \mathscr{E}_g - \mathscr{E}_0. \tag{8.68}$$

An estimate for $\mathscr{E}_0$ can be obtained by assuming that a given homojunction can be treated as a capacitor, with a capacitance per unit area of

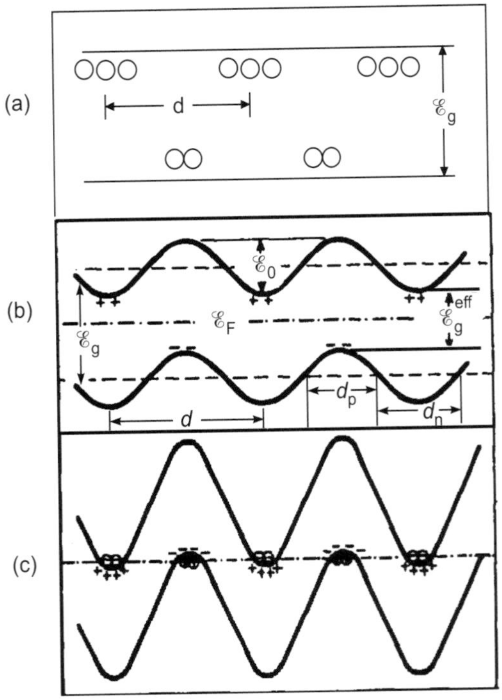

Fig. 8.68 Illustration of doping superlattice (nipi) artificial structures. (a) Spatial variation of n-type and p-type doping of a semiconductor with an intrinsic bandgap $\mathscr{E}_g$. It has been assumed, unphysically, that the dopants are un-ionized and so no band bending occurs. (b) Spatial variation of the conduction- and valence-band edges in the growth direction, z. The modulation $\mathscr{E}_0$ of the bands is due to the spatially varying space charge associated with the ionized donors and acceptors in the n- and p-layers, respectively. Note that the 'direct' gap in real space is everywhere equal to $\mathscr{E}_g$, but the minimum 'indirect' gap is equal to $\mathscr{E}_g - \mathscr{E}_0$. (c) A semi-metallic nipi structure with a very high doping level.

$$C/A \simeq \frac{\varepsilon\varepsilon_0}{d/2}, \tag{8.69}$$

where ε is the dielectric constant of the semiconductor. If it is assumed that the space charge per unit area in the n-type and p-type regions of widths d_n and d_p, respectively, are the same, i.e.

$$Q/A = N_d d_n = N_a d_p, \tag{8.70}$$

then, since $C = Q/V$, where V is the electrostatic potential difference ($= \mathscr{E}_0/e$):

$$\mathscr{E}_0 \simeq \frac{e^2 N_d d_n d}{2\varepsilon\varepsilon_0} \tag{8.71a}$$

$$= \frac{e^2 N_d d^2}{4\varepsilon\varepsilon_0}. \tag{8.71b}$$

for $d_n = d_p = d/2$. Note that $\mathscr{E}_0$ scales with the dopant concentration. For very high dopant levels, the modulation of the band edges can become so pronounced that the

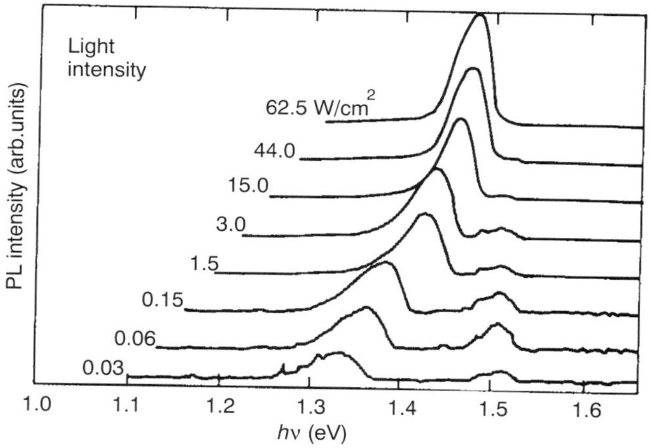

Fig. 8.69 Photoluminescence (PL) spectra at 2 K of a GaAs nipi superlattice with period $d \simeq 800$ Å (C_{As} and Si_{Ga} dopants) for different light excitation powers as shown ($h\nu = 1.92$ eV). (Reprinted from *J. Cryst. Growth* **81**, 270, Heinecke *et al.*, © 1987 with kind permission from Elsevier Science - NL, Sara Burgerhartstraat 25, 1055 KV Amsterdam, The Netherlands)

conduction-band minimum in the n-layers is lowered in energy *below* that of the valence-band maximum in the p-layers: the artificial material is then a real-space nipi semi-metal (Fig. 8.68c).

Doping superlattices are interesting because the band modulation, and hence the effective bandgap, $\mathscr{E}_g^{eff}$, can be controlled simply by varying the space-charge densities in the n- and p-layers. This can be achieved by injecting *excess* electrons and holes (with respect to the thermal-equilibrium concentrations) into the structure, and these then neutralize some of the charged donors and acceptors, respectively, thereby reducing $\mathscr{E}_0$ (eqn. (8.7)) and *increasing* $\mathscr{E}_g^{eff}$. Excess carriers can be injected *electrically* from injecting contacts, or *optically* as electron–hole pairs following the absorption of photons: the higher the light intensity in the latter case, the higher the excess electron–hole density and consequently the larger the effective gap energy, $\mathscr{E}_g^{eff}$. An experimental demonstration of this effect is shown in Fig. 8.69, where photoluminescence spectra of a GaAs nipi structure are shown as a function of light-excitation power: a progressive shift of $\mathscr{E}_g^{eff}$ towards the bandgap value for bulk GaAs ($\mathscr{E}_g \simeq 1.5$ eV) with increasing power is seen. Photoluminescence is the light emitted due to radiative recombination of excess, optically generated electron–hole pairs: the luminescence energy will correspond to the value of $\mathscr{E}_g^{eff}$ pertaining to the particular excitation power.

Extensive discussion of the behaviour of semiconductor superlattices is given in Capasso (1990) and Kelly (1995).

**8.4.5 Quantum Hall effect

It might be thought naïvely that the behaviour of the Hall effect and related galvanomagnetic properties (§6.3.3.2) for a 2D electron gas should be qualitatively the same as in 3D, namely a Hall (transverse) resistivity ρ_{xy} that is proportional to the magnetic field $\boldsymbol{B} = B\hat{\boldsymbol{z}}$ (cf. eqn. (6.126))

$$\rho_{xy} = -\frac{B}{n_{\mathrm{a}} e} \tag{8.72}$$

and a field-independent magnetoresistivity (cf. eqn. (6.9))

$$\rho_{xx} = \frac{m_{\mathrm{e}}}{n_{\mathrm{a}} e^2 \tau}, \tag{8.73}$$

where n_{a} is the areal electron density of the 2D electron gas. The 2D resistivities for current flow in an arbitrary direction in the x–y plane are defined by the set of equations:

$$E_x = \rho_{xx} j_x + \rho_{xy} j_y, \tag{8.74a}$$

$$E_x = \rho_{yx} j_x + \rho_{yy} j_y, \tag{8.74b}$$

where $\rho_{xx} = \rho_{yy}$ and $\rho_{xy} = -\rho_{yx}$. (Note that the 2D electrical resistivity has the unit of ohms (cf. Ω m in 3D) and the 2D current density, e.g. $j_x = I_x/w$, the current per unit width of a film of width w, has the units of A m^{-1}.) Equations (8.74) can also be inverted to give

$$j_x = \sigma_{xx} E_x + \sigma_{xy} E_y \tag{8.75a}$$

$$j_y = \sigma_{yx} E_x + \sigma_{yy} E_y, \tag{8.75b}$$

where $\sigma_{xx} = \sigma_{yy}$ and $\sigma_{xy} = -\sigma_{yx}$, and where the components of the conductivities and resistivities are related via

$$\sigma_{xx} = \frac{\rho_{xx}}{\rho_{xx}^2 + \rho_{xy}^2}, \tag{8.76a}$$

$$\sigma_{xy} = -\frac{\rho_{xy}}{\rho_{xx}^2 + \rho_{xy}^2}, \tag{8.76b}$$

with similar expressions relating resistivities to conductivities, obtained by substituting ρ_{ij} for σ_{ij}, and vice versa.

However, the actual galvanomagnetic behaviour of a 2D electron gas, e.g. that confined at the interface of a modulation-doped AlGaAs/GaAs heterostructure (Fig. 8.32a) is startlingly different at high magnetic field ($\omega_c \tau \gg 1, \hbar\omega_{\mathrm{c}} \gg k_{\mathrm{B}} T$) from the simple predictions of eqns. (8.72) and (8.73) (see Fig. 8.70a). The Hall resistivity exhibits a series of plateaux: values of the Hall resistivity are *quantized* in units of $h/e^2 (= 25\,812.807\,\Omega)$ divided by consecutive integers

$$\rho_{xy} = -\frac{h}{pe^2}, \quad p = 1, 2, 3, \ldots. \tag{8.77}$$

This is the integral quantum Hall effect first discovered by von Klitzing *et al.* (1980) for an inversion layer in an Si MOSFET. Furthermore, at the values of magnetic fields corresponding to the sharp rises in transverse resistivity between quantized Hall plateaux, there are sharp peaks in the longitudinal magnetoresistivity $\rho_{\mathrm{xx}}(B)$, an instance of the Shubnikov–de Haas effect (§7.2.3.2).

For the ranges of magnetic fields corresponding to a particular Hall plateau, the longitudinal magnetoresistivity is *zero* (see Fig. 8.70b). However, this does not mean that the system becomes a perfect conductor, with an infinite conductivity, because the transverse resistivity (and conductivity) remains finite (Fig. 8.70a). Instead, the set of equations (8.76) have the curious property that, if $\rho_{xy} \neq 0, \rho_{xx} \to 0$ simultaneously as

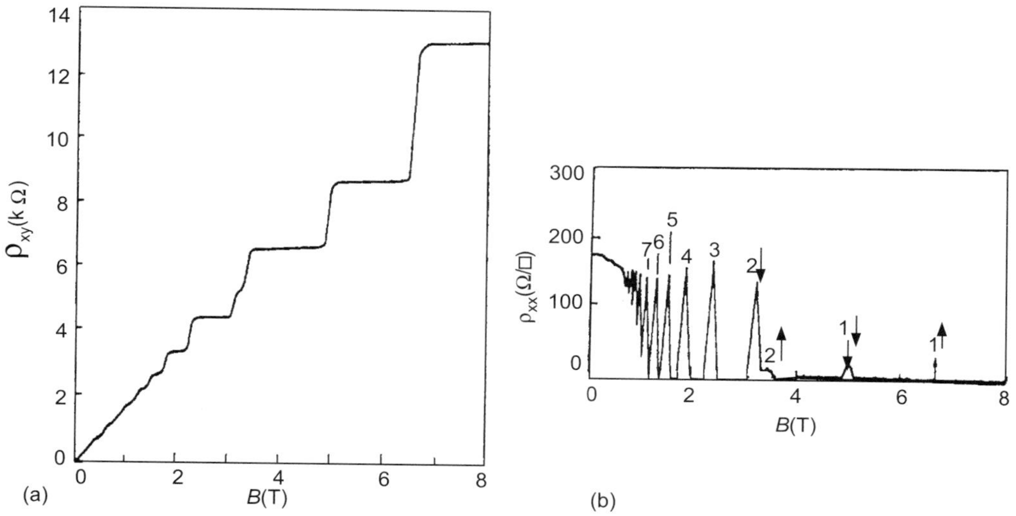

Fig. 8.70 Quantum Hall effect for the 2D electron gas confined in a modulation-doped AlGaAs/GaAs heterostructure, measured at 4 K. (After Paalonen *et al.* (1982). Reprinted with permission from *Phys. Rev.* **B25**, 5566. © 1982. The American Physical Society) (a) Hall (transverse) resistivity as a function of magnetic flux density. (b) Shubnikov–de Haas oscillations in the longitudinal resistivity.

$\sigma_{xx} \to 0$, and vice versa. In other words, electrons move longitudinally along a bar of material (in the x-direction) under the action of a *transverse* electric field, with zero longitudinal resistance, as seen from eqns. (8.74) and (8.75). The trajectories of electrons in this situation are as shown in Fig. 8.71a: electron cyclotron orbits are confined to the *edges* of the sample, and are known as skipping orbits associated with edge states. Such skipping orbits do not permit back scattering, and hence edge-channel transport is resistanceless. The net electron motions along the long edges of the sample are in opposite directions. The four-terminal sample configuration used to measure the Hall and longitudinal resistivities is shown in Fig. 8.71b. From this, it is evident that the relationship between the edge states and the contacts is analogous to the situation encountered in quantized point-contact conductance (§8.3.5.3); the origin of the quantum Hall effect can therefore be understood in terms of the Landauer formalism (eqn. (8.18)).

Quantization of the Hall resistivity in a 2D electron gas can also be understood very simply in terms of the 2D behaviour of the Landau levels (§6.3.3.1). For the case of a three-dimensional system, the density of states for the Landau levels is a continuous (albeit spiky) function of electron energy (Fig. 6.20d) in that there are no energy gaps: the $\mathscr{E}^{-1/2}$ behaviour of $g(\mathscr{E})$ for the residual motion parallel to the magnetic field is superimposed on the discrete Landau levels. However, if the motion of the electron gas is confined to two dimensions by a potential well (as in a quantum well), the electron energies in the presence of an external magnetic field of flux density B are given by

$$\mathscr{E} = \mathscr{E}_1 + \left(n + \tfrac{1}{2}\right)\hbar\omega_c \pm \mu_B B, \tag{8.78}$$

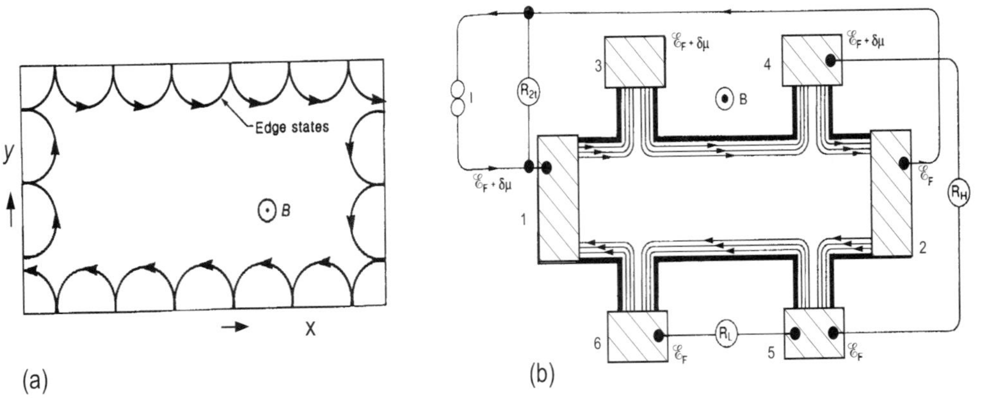

Fig. 8.71 (a) Skipping cyclotron orbits (edge states) in a bar of material containing a 2D electron gas exhibiting the quantum Hall effect. (b) Measurement configuration for a 'Hall bar'. A current is passed down the length of the bar between source and drain current electrodes 1 and 2, and electrodes 3–6 are voltage contacts; the chemical potentials of all contacts are shown. The edge states and localized cyclotron orbits are shown by the continuous thin lines, with the directions of electron motion marked by the arrows. The four-terminal Hall resistance, R_{xy} (for current and voltage contacts alternating at the sample boundary), and the two-terminal resistance, R_{2t}, are the same ($= h/e^2p$): the four-terminal longitudinal resistance, R_{xx} (for non-alternating current and voltage contacts) is zero. (After Beenakker and van Houten (1991). Reproduced by permission of Academic Press, Inc.)

where $\mathscr{E}_1$ is the ground-state level for the well. The second term is the quantized energy of the nth Landau level (cf. eqn. (6.119)), where $\omega_c = eB/m_c^*$ is the cyclotron frequency (eqn. (6.117)); this replaces the free-electron term (eqn. (8.58)) in the presence of a magnetic field. The last term is the Zeeman energy (eqn. (7.222)) for the interaction of an electron spin with a magnetic field and $\mu_B = e\hbar/2m_e$ is the Bohr magneton (eqn. (7.130)); the positive/negative signs refer to the electron-spin magnetic moment pointing parallel/antiparallel to $\boldsymbol{B}$, or spin-down/up, respectively. Thus, the corresponding density of states of a 2D system (without disorder) in a magnetic field consists of a series of delta functions (Fig. 8.72), with a spin doublet for every Landau level, in contrast to the constant density of states for a 2D system without a magnetic field. For the case when the cyclotron effective mass, m_c^* (eqn. (6.118)) equals the free-electron mass m_e, the Zeeman (spin-) and cyclotron splittings are equal, i.e. $\hbar\omega_c = 2\mu_B B$ (cf. eqns. (6.117) and (7.130)). As the magnetic field increases, the degeneracy per unit area of each Landau level (for a particular spin direction) increases (cf. eqn. (6.121)):

$$g_n = \frac{eB}{h}, \tag{8.79}$$

and the levels also move upwards in energy since the Landau-level spacing, $\hbar\omega_c$, is proportional to B.

A quantum Hall effect can then be understood from the magnetic-field dependence of the chemical potential, $\mu(B)$, for such a density of states. Consider the case when, for a particular areal density of electrons, n_a, of a 2D electron gas in a given magnetic field B,

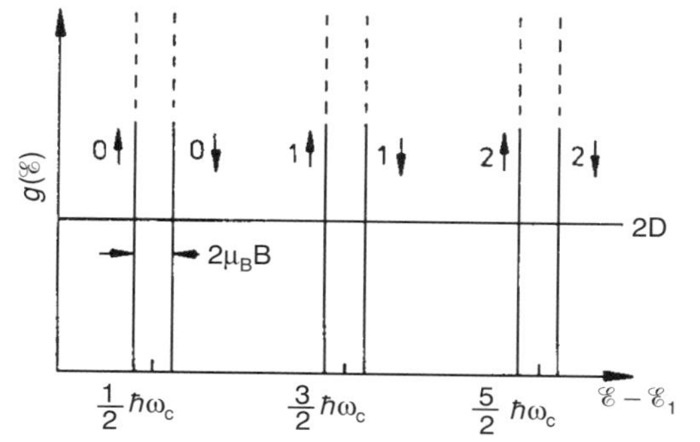

Fig. 8.72 Density of Landau levels in a 2D electron gas, drawn for the case when $m_c^* < m_e$. The density of states consists of a series of delta functions (in the absence of disorder), each doublet for the nth Landau level corresponding to spin-up and spin-down configurations of the electron spin with respect to the magnetic-field direction being labelled as such. The constant 2D density of states in the absence of a magnetic field is also shown.

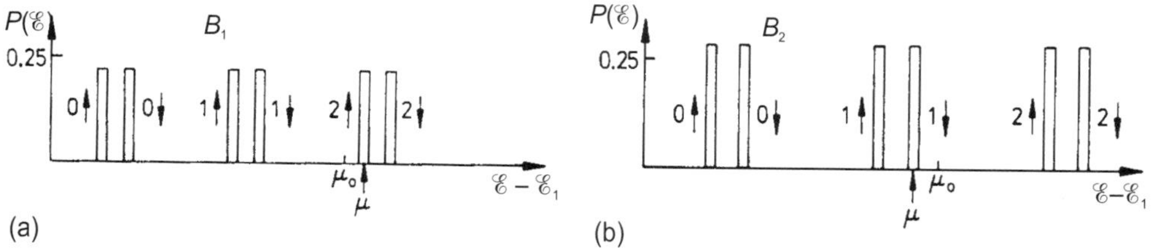

Fig. 8.73 Landau-level occupancy $P(\mathscr{E})$ in a 2D electron gas at two magnetic fields, (a) B_1; (b) $B_2 > B_1$. The delta functions characteristic of the density of states for an ideal system have been broadened for ease of illustration to show the occupancy. The chemical potential is pinned to the position of the part-filled Landau level, and oscillates about the value μ_0 for the field-free system.

one particular Landau level ($n = 2 \uparrow$ in Fig. 8.73a) is partly occupied: all Landau levels below or above this level are fully occupied or completely empty, respectively. In this case, the chemical potential is pinned to the Landau-level position. As B increases, the level will move up in energy and with it, μ. However, the degeneracy of the lower-lying Landau levels also increases with B (eqn. (8.79)), and therefore for these levels to remain completely filled the part-filled level must be increasingly depleted of electrons. The point will come when all electrons are drained from this Landau level and hence the chemical potential will decrease *discontinuously* to the position of the next lowest Landau level ($1 \downarrow$ in Fig. 8.73b) which will then start to be depleted of electrons; the position of the chemical potential will now be pinned to this level as long as it is part-filled. The discontinuous jumps in $\mu(B)$ occur, therefore, whenever an integral number of Landau levels are completely occupied; the condition for this is that $n_a/g_n = p$, where p is an integer, or from eqn. (8.79)

$$B_p = \frac{hg_n}{e} = \frac{hn_a}{pe}. \tag{8.80}$$

Substitution of this value of magnetic field, at which discontinuous jumps in $\mu(B)$ occur, into the semi-classical expression for the transverse resistivity (eqn. (8.72)) shows that the Hall resistivity is expected to be quantized (cf. eqn. (8.77)):

$$\rho_{xy} = -\frac{B_p}{n_a e} = -\frac{h}{pe^2}.$$

The vanishing of the longitudinal conductivity, σ_{xx} (and hence, from eqn. (8.76a), ρ_{xx}) is also understandable from this picture since, at the point when all Landau levels are either completely full or empty, the density of states at the Fermi level is zero.

However, although this simple picture predicts a quantization of ρ_{xy}, it cannot account for the *ranges* of magnetic fields, corresponding to the Hall plateaux, for which $\sigma_{xx} = \rho_{xx} = 0$; the model predicts this behaviour only at the discrete fields, B_p (eqn. (8.80)). A modification of this model that goes some way towards explaining the experimental data (Fig. 8.70) invokes the presence of *disorder* in the 2D system, due perhaps to structural defects at the heterojunction. Disorder can have *two* effects on electron states: bands of states in the density of states can become *broadened* (§5.5), and, if the disorder is sufficiently great, electron states can become spatially *localized* (§6.7). The effect of disorder on the Landau-level density of states of 2D electron gas is shown schematically in Fig. 8.74: the delta functions of the ideal system (Fig. 8.72) are broadened, with delocalized (extended) states existing only for energies close to the peak maxima and intervening states being localized. The variation of $\mu(B)$ is still oscillatory but is now a smoothly varying function, increasing when near one of the peaks in the density of states, and decreasing when near a minimum between two peaks. When $\mu(B)$ lies in such a band of localized states, the system is a *Fermi glass* (§7.7) and

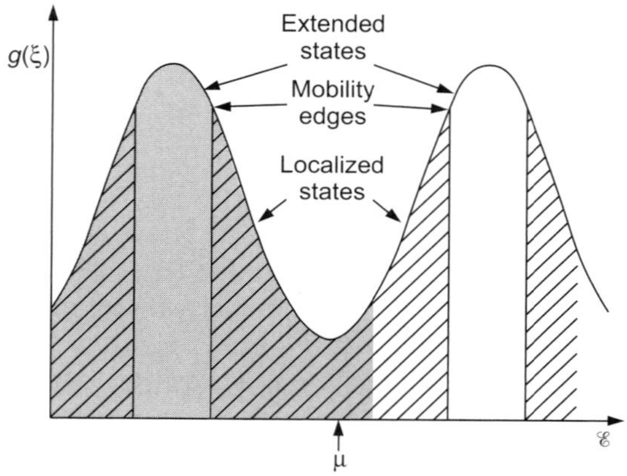

Fig. 8.74 Illustration of the density of states for a disordered 2D electron-gas system in a magnetic field. The delta-function peaks are broadened by disorder, and there are mobility edges between extended and localized states. If the chemical potential lies in a region of localized states, $\sigma_{xx} = 0$ at $T = 0\,\mathrm{K}$.

σ_{xx} will be zero (at $T = 0\,\text{K}$, where hopping conduction cannot occur). Hence, the reason for σ_{xx} being zero for a *range* of magnetic fields can be understood in this way.

It might be thought that at sufficiently high magnetic fields, where only the lowest Landau level is occupied, the quantum Hall effect should disappear. Indeed, the *integral* quantum Hall effect, as discussed above, does vanish in this régime. However, the Hall resistivity is found still to be quantized (Fig. 8.75), but now the quantization factor is not integral but non-integral, $p = n/m$, where $n(< m)$ and m are integers. This is the fractional (or non-integral) quantum Hall effect. Although the behaviour is superficially similar to that of the integral quantum Hall effect, the underlying mechanism is very different and is believed to result from electron–electron interactions. However, the theory for this is beyond the level of this text (see, for example, Prange and Girvin (1990)).

Laughlin (1982) has formulated a theory for the fractional quantum Hall effect wherein, because of electron–electron interactions, the 2D electron gas becomes an incompressible quantum fluid in which, for a fractional Landau-level filling factor, say $p = 1/m$, the quasiparticle excitations have a charge Q that is a *fraction* of the electronic charge, i.e. $Q = e/m$. This extraordinary prediction of fractional charge has

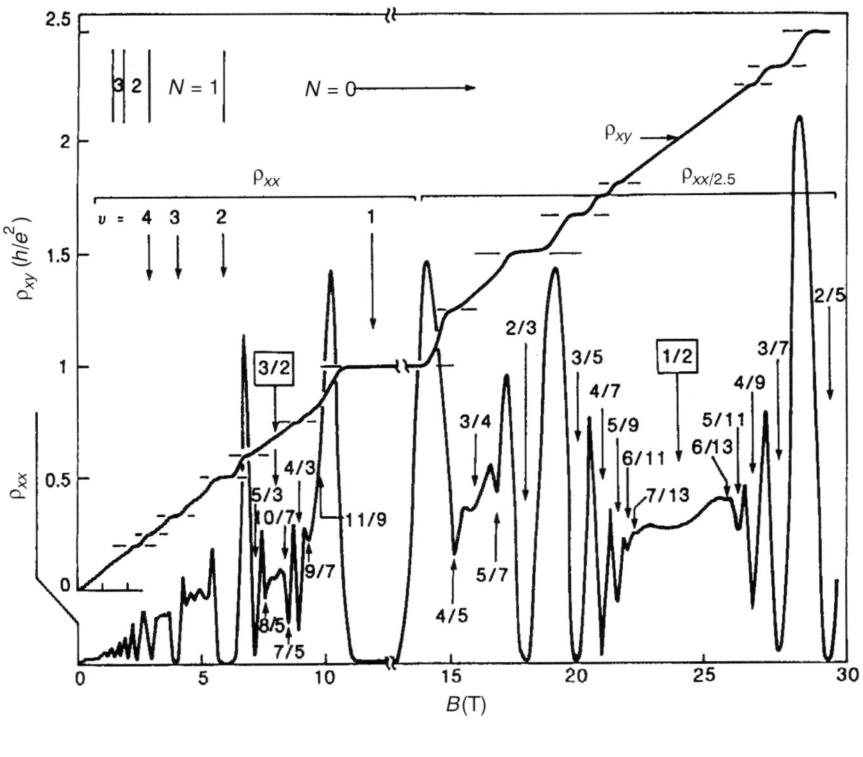

Fig. 8.75 The fractional quantum Hall effect in a 2D electron gas at a GaAs/AlGaAs interface, measured at 150 mK (except for the highest-field Hall data, measured at 85 mK). (After Willett *et al.* (1987). Reprinted with permission from *Phys. Rev. Lett.* **59**, 1776. © 1987. The American Physical Society)

now been confirmed experimentally (de-Picciotto *et al.* 1997, Saminadayar *et al.* 1997) by measurements of the noise in the current passing through a constriction in the 2D electron gas subject to a very large magnetic field (in the fractional quantum-Hall-effect régime), i.e. between quantum point contacts (cf. §8.3.5.3) formed by the electrostatic repulsion of electrons underneath two shaped, negatively charged split-gate electrodes placed above the electron gas (see inset to Fig. 8.39).

Any electrical current is not, in fact, a *uniform* flow of charge carriers but, if the particles flow independently (characterized by an uncorrelated Poisson distribution), there will be fluctuations in the number of carriers passing a particular point in a given time interval; these current fluctuations are known as shot noise (by analogy with the random patter of lead shot on a target). This is a classical phenomenon, and it allows the determination of the charge Q of the carriers since, in the simplest case, Q is proportional to the mean-square fluctuation in the current, $\Delta(\overline{I^2})$ and the average current, I_0 (see e.g. Bleaney and Bleaney (1976)):

$$\Delta(\overline{I^2}) = 2QI_0\Delta f, \tag{8.81a}$$

where Δf is a frequency interval. The current is a 2D electron gas in a strong magnetic field is carried by edge states (Fig. 8.71a); this is true in both the integral and fractional quantum-Hall-effect régimes (although in the latter case the electrons do not form a Fermi liquid). A more general expression for the noise associated with the current passing through a quantum point contact connecting such 1D conducting channels, valid at finite temperatures, is given by (de-Picciotto *et al.* (1997)):

$$\Delta(\overline{I^2}) = 2G_0t(1-t)\Delta f[QV\coth(QV/2k_BT) - 2k_BT] + 4k_BTG_0t\Delta f, \tag{8.81b}$$

where G_0 is the quantized conductance ($G_0 = Qe/h$), t is the transmission coefficient for transport between the point contacts and V is the applied voltage. The second term in eqn. (8.81b) is the thermal (Johnson) noise generated in any resistor by spontaneous thermal fluctuations in voltage (Nyquist theorem)—see e.g. Kittel (1969). At zero kelvin and for very weak transmission, eqn. (8.81b) reduces to eqn. (8.81a), since $VG_0t(1-t) \simeq VG_0t = I_0t \simeq I_0$.

The split-gate electrode geometry leading to 1D confinement of the 2D electron gas (quantum point contacts) allows two régime to be explored, depending on the magnitude of the applied (negative) gate voltage, V_g: for relatively low values of V_g ('weak pinch-off'), the 2D electron gas is restricted in the region between the gate electrodes but is still continuous (Fig. 8.76a); at higher values of V_g ('strong pinch-off'), the 2D electron gas is completely separated into two parts (Fig. 8.76b). In the strong pinch-off limit, electrons can quantum-mechanically tunnel from one branch of the edge states (say in the left-hand region of the electron gas) to the other branch (in the right-hand region) through the gap separating the two regions where the electron gas has been repelled electrostatically (Fig. 8.76b); even in the fractional quantum-Hall-effect régime, charge can be added to, or subtracted from, the incompressible electron fluid only in multiples of whole electrons. However, in the weak pinch-off limit, where the 2D electron fluid is continuous in the region of the split gate, quasiparticles can tunnel through the fluid from one edge channel (say the top) flowing in one direction to the other (the bottom) flowing in the opposite direction; as a result, the quasiparticle is effectively backscattered at the point contacts (Fig. 8.76a). Measurements of the shot

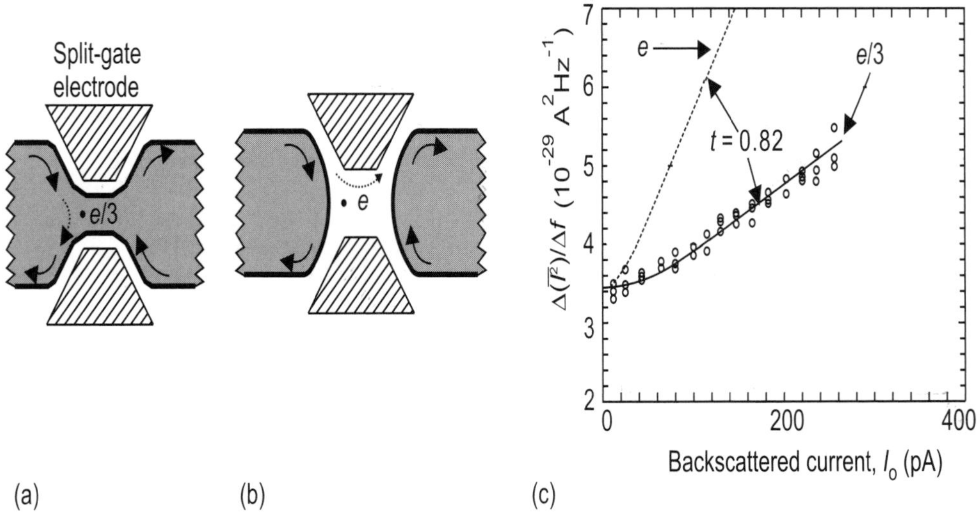

Fig. 8.76 Quantum point contacts in a two-dimensional electron liquid, showing the conducting edge channels (bold lines) and the direction of current flow: (a) The weak pinch-off limit. Fractionally charged quasiparticles can tunnel between edge channels (i.e. backscatter) through the electron liquid. (b) The strong pinch-off limit. Electrons can tunnel between edge channels in separated regions of electron fluid. (c) Measured shot noise in the backscattered current (circles) at quantum point contacts in a 2D electron fluid at an AlGaAs/GaAs interface for a transmission coefficient $t = 0.82$, together with the behaviour predicted by eqn. (8.81b) for the current carried by fractionally charged quasiparticles with charge $Q = e/3$ (solid line) and electrons with charge $Q = e$ (dashed line). (de-Picciotto et al. (1997). Reprinted with permission from *Nature* **389**, 162. © 1997 Macmillan Magazines Ltd.)

noise in a 2D electron fluid at an AlGaAs/GaAs interface in the weak pinch-off limit, at a very high magnetic field where the Landau-level filling factor was $p = 1/3$, could be fitted to eqn. (8.81b) only by assuming that the charge of the tunnelling quasiparticle was $Q = e/3$ (de-Picciotto *et al.* 1997, Saminadayar *et al.* 1997)—see Fig. 8.76c.

Applications 8.5

There are very many applications of some of the topics that have been discussed in this chapter. However, the majority of these applications relate to semiconductor homo- and heterojunctions, and so here we will concentrate on some electronic and opto-electronic applications of devices that incorporate such heterostructures. However, even this field is so vast that entire books have been devoted to its description, and hence in this short section we can do no more than give a flavour of the subject, stressing the physical operation of the devices involved. The interested reader is referred, for example, to Sze (1981) for further details on semiconductor devices.

8.5.1 Semiconductor transistors

The writing of this book has coincided with the fiftieth anniversary of the development of the semiconductor transistor (*tran*sfer re*sistor*) in 1947 by Bardeen, Brattain and Shockley at Bell Laboratories (and simultaneously also with the centenary of the discovery by J. J. Thomson of the electron!). The invention of the transistor has been responsible for the unprecedented information-technology revolution that the world has recently been experiencing. Although the functional operation of a transistor is the same as that of a vacuum triode, the extreme miniaturization made possible with the use of solid-state devices has meant that new applications (e.g. high-density computer memories) or more powerful variants (e.g. ultra-fast electronic computers) have been developed that could never have been feasible using vacuum tubes.

Essentially, two types of transistor action can be distinguished on the basis of the direction of current flow with respect to the homo/heterojunctions: the bipolar transistor involves current flowing *normal* to the homo/heterojunctions, and a field-effect transistor (FET) has the current flow *parallel* to the plane of the heterojunction. We will briefly describe both types of device in the following two sections.

8.5.1.1 Bipolar transistors

Bipolar transistors comprise a trilayer of doped semiconductor (usually c-Si), with a thin ($< 1\mu$m) layer of one doping type sandwiched between more heavily doped layers of the other type, e.g. p–n–p or n–p–n. The intermediate layer is called the base, and the other two layers are termed the emitter and collector (Fig. 8.77a). In the case of the p–n–p bipolar transistor, holes (the minority carriers in the base region) are injected into the base by the emitter (which is forward-biased) and collected by the reverse-biased collector (Fig. 8.77b). For the n–p–n device, since the rôles of holes and electrons are interchanged, all polarities are reversed. It can be seen from Fig. 8.77a that the bipolar transistor can be regarded as simply two p–n junctions placed back-to-back. Thus, the operation of this device can be understood in terms of the behaviour of the p–n junction (§8.4.2.2).

The bias arrangement of the bipolar transistor indicated in Fig. 8.77b leads to the band-bending spatial profile through the device shown in Fig. 8.77c; in this configuration, the device operates as a current-controlled amplifier. The emitter current, I_e, is that of a forward-biased p–n junction (cf. eqn. (8.45)):

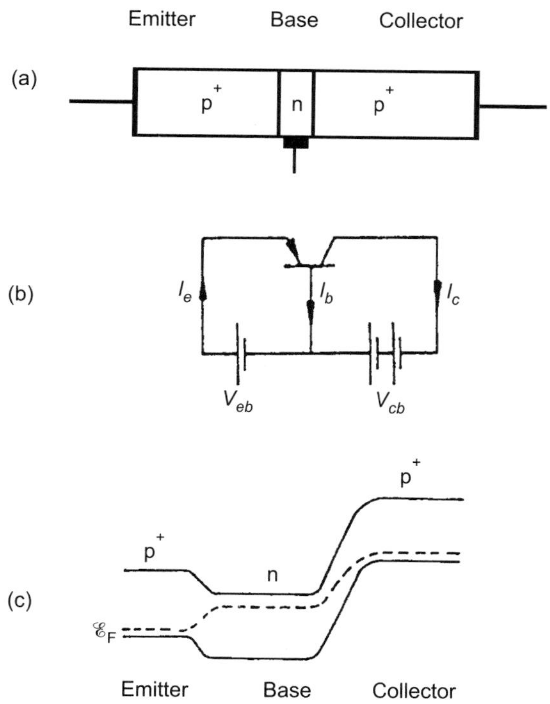

Fig. 8.77 The p–n–p bipolar transistor. (a) Schematic representation of the structure of the device. (b) Bias arrangement of the bipolar transistor: forward bias of the emitter–base junction and reverse bias of the collector–base junction. The emitter, base and collector currents are labelled I_e, I_b, I_c, respectively. (c) Spatial profile of the conduction-band minimum and valence-band maximum through a p–n–p bipolar transistor when biased as in (b).

$$I_{\rm e} \simeq I_0 \exp(eV_{\rm eb}/k_{\rm B}T), \tag{8.82}$$

where $V_{\rm eb}$ is the emitter–base bias voltage. The constant I_0 is given by eqn. (8.47b), except with the hole diffusion length, $L_{\rm h}$, replaced by the thickness $d_{\rm b}(< L_{\rm h})$ of the base layer; injected holes therefore reach the base collector interface in general before recombination takes place, and they are thereupon accelerated into the collector region by the junction electric field. Hence, $I_{\rm c} \simeq I_{\rm e}$, and the difference of the two is the base current:

$$I_{\rm b} = I_{\rm e} - I_{\rm c}. \tag{8.83}$$

The current $I_{\rm b}$ provides the electrons for the electron component of the emitter current (neglected above) and to replace those lost through recombination with the very few holes that fail to diffuse across the base region. Thus, the base current is some (very small) fraction f of the emitter current and hence the current gain of the amplifier is

$$\gamma_I \equiv \frac{I_e}{I_b} = \frac{1}{f}. \tag{8.84}$$

Typically, γ_I has a value of order 100, and bipolar transistors are commonly used as power amplifiers. Further details can be found in Sze (1981).

8.5.1.2 *Field-effect transistors*

As the name implies, a field-effect transistor (FET) is a voltage-controlled resistor, in contrast to the current-controlled bipolar transistor discussed in the previous section. The most common FET configuration is that involving a metal–insulator–semiconductor capacitor structure. Usually, the semiconductor is crystalline Si and the insulator is its oxide, SiO_2, in which case the device is known as a metal–oxide– semiconductor FET (MOSFET). A section through such a device is shown in Fig. 8.78. The general behaviour of the semiconductor bands in the vicinity of the interface with the insulator has already been discussed in §8.4.2.3. For a sufficiently large positive (negative) bias on the gate electrode, an *inversion layer* is formed in the p-type (n-type) semiconductor at this interface (Fig. 8.55) in which the charge carriers in the resulting 2D electron gas are of the *opposite* type to the majority carriers of the semiconductor substrate. Thus, an n-channel of enhanced electrical conductivity is created in the inversion layer of a p-type semiconductor (as in Fig. 8.55) and a p-channel in an n-type semiconductor. Thus, it is a truly 2D device.

The voltage-controlled MOSFET can work in two ways, either as a switch or as an amplifier. In digital circuits, a MOSFET functions as an on–off switch (corresponding to the bits 1, 0) by operating with two gate voltages, V_g, greater than, or less than, the threshold voltage V_t required to form the inversion layer, corresponding to the creation, or not, of a conducting n-channel (for a p-type semiconductor). For $V_g < V_t$, no current flows between source and drain electrodes (consisting of n^+ regions) since one of the associated p–n junctions must be reverse-biased.

In analogue circuits, the FET can work as a voltage amplifier as a result of the source–drain resistance in the conducting (n-channel) state being larger than the gate

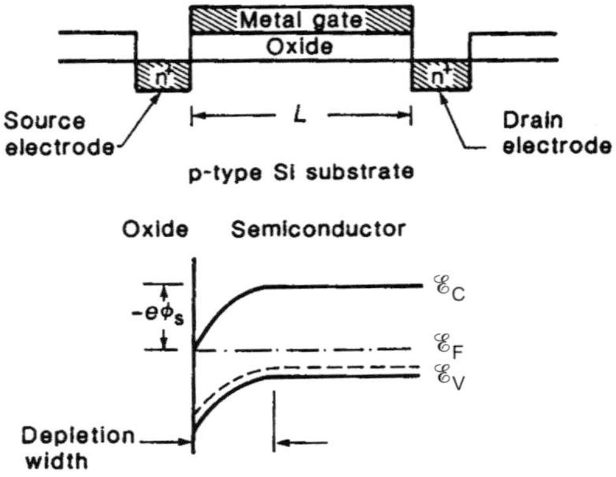

Fig. 8.78 Metal–insulator–semiconductor FET, e.g. involving p-type Si (MOSFET).

resistance, so that small changes in gate voltage (current) produce larger changes in source–drain voltage (current). The gate voltage controls the areal density of carriers in the inversion layer (cf. eqn. (8.50a)):

$$n_a = \frac{C}{eA}(V_g - V_t), \tag{8.85}$$

where A is the surface area of the FET. The source–drain current depends on n_a and the drift velocity v_d:

$$I_{SD} = n_a e w v_d, \tag{8.86}$$

where w is the width of the device (normal to the section shown in Fig. 8.78). A figure of merit is the transconductance, G_t, which is the derivative of the output current I_{SD} with respect to the input gate voltage, V_g, i.e.

$$G_t = \partial I_{SD}/\partial V_g = C v_d/L, \tag{8.87}$$

where L is the distance between source and drain electrodes. The time $t_{SD} = L/v_d$ taken for electrons (or holes) to transit between the electrodes determines the maximum rate at which the device can respond to a time-varying input gate signal.

The ability to place very many MOSFETs (and bipolar transistors) on a single slice of c-Si has led to the development of integrated circuits in which such devices intercommunicate, and miniaturization leading to sub-micron feature sizes (e.g. the length of a gate in an FET or the base width in a bipolar transistor) has led to an extraordinary increase in device density and speed. Over the last 30 years, every three years the minimum feature size has decreased by 30%, the number of transistors per chip has increased four-fold and the chip speed has increased by a factor of 1.5: now chips each contain four million transistors with a clock frequency of 200 MHz. For further details, see, for example, Sze (1981), and Weisbuch and Vinter (1991). In many respects, GaAs is a better material than Si for this purpose since it has a higher electron mobility, resulting in even faster devices. However, diamond appears potentially to be the best material of all, based on the Johnson criterion linking the maximum gate voltage, V_g^{max}, and the minimum transit time, t_{SD}, viz.

$$V_g^{max}/t_{SD} = E_b v_d, \tag{8.88}$$

where the maximum gate voltage is determined by the breakdown electric-field strength, E_b, of the semiconductor. The figure of merit, $E_b v_d = 6 \times 10^{12} \mathrm{V\,s^{-1}}$ for GaAs, is three times that for Si, but the figure for diamond is 16 times larger than for GaAs. Diamond also has the distinct advantage that it has an extremely high (phonon) thermal conductivity (§4.6.2.2), meaning that diamond-based devices could in principle operate at higher powers since the extra heat generated thereby can be readily dissipated. However, it is proving difficult to introduce shallow dopants into diamond films in order to create the doped regions required in device manufacture, and so the general use of this material remains a distant prospect.

8.5.1.3 *Modulation-doped devices*

A modulation-doped heterostructure is one in which the doping level of one type of layer is different from that of another type of layer. An example is the n-AlGaAs/i-GaAs heterostructure mentioned in §8.3.5 which can support a 2D electron gas in the intrinsic layer near the interface (Fig. 8.32a). An interesting feature is that the extra carriers originating from the ionized donors in the high-bandgap material (AlGaAs) transfer to, and reside in, the smaller-gap intrinsic material. As a result, the low-temperature mobility of these charge carriers is increased by several orders of magnitude with respect to the mobility that would be found if the heterostructure were *uniformly* doped.

The reason for this enhancement is that, for a uniformly doped bulk material or heterostructure, the dominant cause of scattering of electrons that limits the mobility at low temperatures is ionized-impurity scattering (§6.5.2.3)—see Fig. 8.79a. However, in a modulation-doped heterostructure (or artificial structure), the conduction electrons in the intrinsic layer are physically separated from the ionized donors whence they originated and, as a result, ionized-impurity scattering is obviated and the mobility is increased. Thus, at low temperatures, the only scattering mechanisms remaining are those involving extrinsic effects (*neutral* defects) and intrinsic effects (piezoelectric electron–acoustic-phonon scattering). Fig. 8.79b shows the steady increase over the years in the low-temperature mobility in modulation-doped AlGaAs/GaAs devices, as improvements in fabrication techniques have led to the gradual elimination of neutral

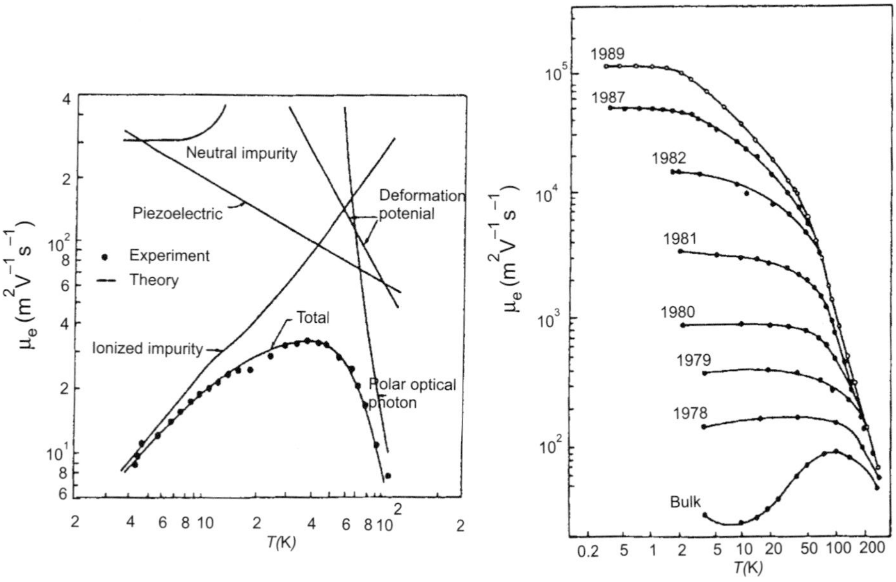

Fig. 8.79 (a) Temperature dependence of the electron mobility in bulk GaAs. The points are experimental data and the solid curves are the theoretical temperature dependences of various electron-scattering mechanisms as marked. (After Stillman and Wolfe (1976). Reprinted from *Thin Solid Films* **31**, 69, Stillman and Wolfe, © 1976 with kind permission from Elsevier Science Ltd, The Boulevard, Langford Lane, Kidlington OX 1GB5, UK) (b) Evolution with time of the electron mobility (as a function of temperature) in GaAs in modulation-doped n-AlGaAs/GaAs HEMTs. The residual scattering mechanism at low temperatures is neutral-defect scattering (After Pfeiffer *et al.* (1989). Reprinted with permission from *Appl. Phys. Lett.* **55**, 1888. © 1989 American Institute of Physics)

impurities in the intrinsic GaAs material. Mobilities in excess of $10^3\ m^2\ V^{-1}\ s^{-1}$ are now achievable. Field-effect transistors that incorporate modulation doping are referred to as high electron mobility transistors (HEMTs) or modulation-doped field-effect transistors (MODFETs). These devices are used to make the lowest-noise solid-state amplifiers available.

8.5.2 Opto-electronic devices

Opto-electronic devices can essentially be divided into two categories, the functions of which may be summarized as either photons in, electrons out, or electrons in, photons out (where 'electrons' should be read as electrons *and* holes). Examples of the former category of devices incorporating heterojunctions are solar cells and photon detectors. Examples of the latter category include semiconductor light-emitting diodes and lasers. These applications will be discussed in the following sections.

8.5.2.1 Solar cells

The conversion of solar light to electrical power using solar cells is one of the few available renewable sources of energy. Solar cells with efficiencies up to 30% can be made using semiconductor hetero- or homo-(p–n) junctions. The basic physical principle involved is the absorption of photons in the semiconductor, resulting in the creation of electron–hole pairs; the excess electrons in the conduction band, and holes in the valence band, generated in the depletion region of a hetero- or homojunction, are then separated by the internal junction field before recombination can take place and hence produce an open-circuit voltage, V_{oc}, or a short-circuit current density, j_{sc}.

The intensity of solar radiation in space, just above the earth's atmosphere, is termed the solar constant and has a value of 1353 W m^{-2}. This situation is referred to as AM 0 (air mass zero) since there is zero attenuation of the light by the atmosphere. The solar spectral irradiance under these conditions can be approximated by a black body of temperature 5800 K (Fig. 8.80). At the earth's surface, with the sun at its zenith (AM 1 condition), the solar intensity is reduced to 925 W/m^2 as a result of absorption of the light by water vapour (responsible for dips in the solar spectrum) and scattering by dust and aerosols. For the sun at 45° to the horizon (AM 1.5), the solar irradiance spectrum is as shown in Fig. 8.80. It can be seen from this figure that the maximum in the solar spectrum corresponds to a photon energy of the order of 1 eV. Hence, it is expected that semiconductors with bandgaps of this magnitude (e.g., Si, InP, GaAs) will be most efficient at solar conversion.

The equivalent circuit of a p–n junction solar cell connected to a load resistor R_L can be represented as in Fig. 8.81, i.e. a constant-current source I_s (resulting from the excitation of excess electrons and holes by solar radiation) in parallel with the diode current, $I_d = I_0[\exp(eV/k_BT) - 1]$ (eqn. (8.48)). Thus, the I–V characteristics of the cell can be written as

$$I = I_0[\exp(eV/k_BT) - 1] - I_s. \qquad (8.89)$$

The open-circuit voltage of the solar cell (corresponding to $I = 0$) is therefore, from the above equation

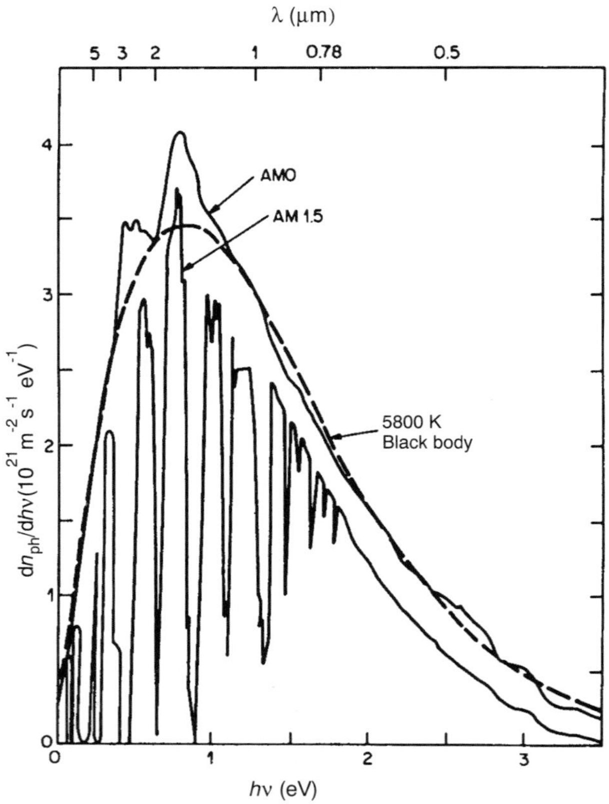

Fig. 8.80 Solar power spectrum as a function of photon energy for AM 0 and AM 1.5 conditions. The emission spectrum of a black body at 5800 K is shown by the dashed line for comparison. (After Henry (1980). Reprinted with permission from *J. Appl. Phys.* **51**, 4494. © 1980 American Institute of Physics)

$$V_{\rm oc} = \frac{k_{\rm B}T}{e}\ln\left(\frac{I_{\rm s}}{I_0}+1\right) \simeq \frac{k_{\rm B}T}{e}\ln\left(\frac{I_{\rm s}}{I_0}\right). \tag{8.90}$$

Since $I_s \ll I_0 \ll 1\text{A}$, $V_{\rm oc}$ is dominated by the behaviour of I_0: from eqn. 8.49, $I_0 \propto n_{\rm i}^2$, the square of the intrinsic carrier concentration and, since from eqn. (6.202), $I_0 \propto \exp(-\mathscr{E}_{\rm g}/k_{\rm B}T)$, $V_{\rm oc}$ is essentially determined by the bandgap of the semiconductor used in making the solar cell. The short-circuit current, $I_{\rm sc}$, is simply the solar-generated current, $I_{\rm s}(V = 0$ in eqn. (8.89)).

The theoretical maximum power that can be extracted from the device is $P^{\rm th}_{\rm max} = V_{\rm oc}I_{\rm sc}$ but, because the diode I–V characteristic is never perfectly L-shaped, the actual maximum power achievable is $P_{\rm max} = V_{\rm max}I_{\rm max}$ which is the area of the largest-area rectangle that can be inscribed within the I–V characteristic (see Fig. 8.82). The solar-cell efficiency is thus defined as

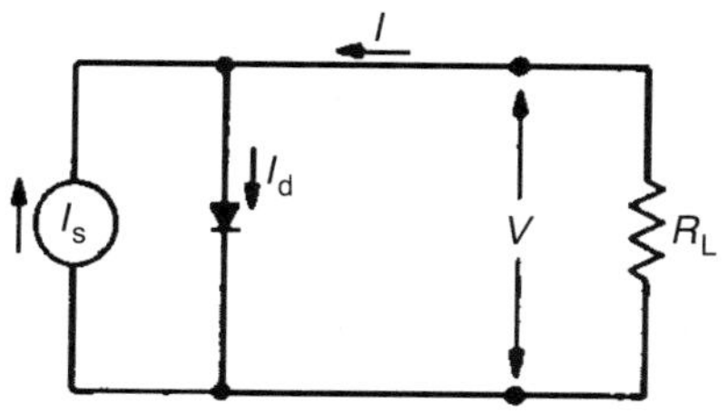

Fig. 8.81 Equivalent electrical circuit of a p–n junction solar cell.

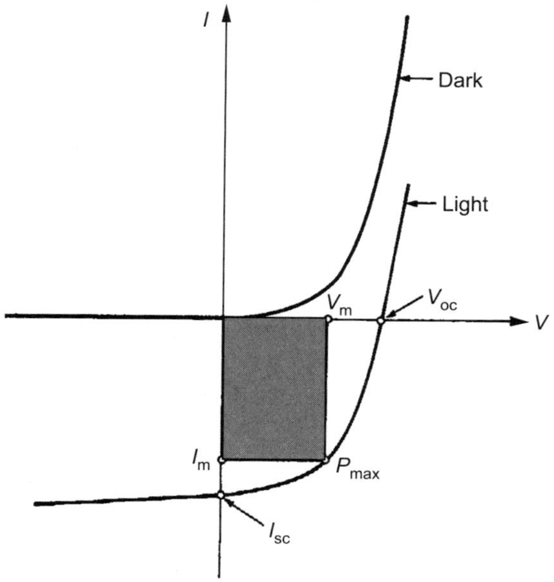

Fig. 8.82 Current–voltage characteristic of an Si solar cell under illumination. The diode characteristic has been displaced down in current by I_{sc}, the short-circuit cell current (eqn. (8.89)). The maximum-power rectangle is indicated.

$$\eta = \frac{P_{max}}{P_{inc}}, \tag{8.91}$$

where P_{inc} is the solar power incident on the device. The fill factor, FF ≤ 1, defined as

$$\text{FF} = \frac{P_{max}}{P_{max}^{th}} = \frac{V_{max} I_{max}}{V_{oc} I_{sc}} \tag{8.92}$$

is a measure of how square is the diode I–V characteristic. Typical values for solar cells made from crystalline Si or GaAs have FF $\simeq 0.8$ and efficiencies in the range 20–30%. The efficiency is optimized by appropriate device engineering, for example, utilizing anti-reflection coatings on the top surface, stacked multilayers of various materials with differing bandgaps to absorb solar radiation with a range of wavelengths, etc. For further details, see e.g. Sze (1981), Kelly (1995).

Single-crystal solar cells, although having the highest efficiencies, are nonetheless very expensive. Although this economic drawback is of little account for space applications where efficiency, size and weight are at a premium and cost is generally not a problem, nevertheless it has precluded the widespread terrestrial use of such solar cells. In this connection, amorphous hydrogenated silicon, a-Si:H, is a material of considerable promise. Although the efficiency of solar cells fabricated using a-Si:H is no more than about half that achievable using single-crystal Si (the defect states in the bandgap of the amorphous material act as recombination centres that reduce the numbers of light-induced electrons and holes, and so reduce I_{sc}) nevertheless, because the material is made by vapour deposition (e.g. 'glow-discharge decomposition' or plasma-enhanced chemical vapour deposition — §1.1.2), it can be made relatively cheaply in large areas. This capability and economic benefit makes it attractive for use in large-area solar arrays for electrical-power generation as well as in inexpensive consumer products (e.g. solar cells in pocket calculators).

8.5.2.2 Photodetectors

Photodetectors work by measuring the photocurrent resulting from the excess charge carriers induced in a semiconductor by the absorption of photons and swept out by an applied electric field (§6.5.1.2). In the simplest case, a photodetector merely consists of a bulk semiconductor to which are bonded ohmic contacts in a two-terminal configuration. Three distinct mechanisms of photocarrier creation can be envisaged in general: *intrinsic absorption* of photons with $\hbar\omega \geq \mathscr{E}_g$ in which an electron is excited into the conduction band from the valence band (see §5.8.2); *extrinsic absorption* of photons with $\hbar\omega \geq \mathscr{E}_d$ (or $\mathscr{E}_a$) in which an electron (or hole) is excited into the conduction band (or valence band) from a donor level at $\mathscr{E}_d$ below the conduction-band edge (or an acceptor level at $\mathscr{E}_a$ above the valence-band edge); or *free-carrier* (*intraband*) *absorption* in which an electron in a conduction band is excited into a higher-lying empty conduction band (§5.8.1).

Infrared detectors, used for thermal-imaging and night-vision applications, are one example of the use of photodetectors. Such IR detectors have to work in the wavelength band 8–14 μm in order to avoid the absorption bands due to atmospheric water vapour. Of the intrinsic photodetectors, the alloy $Hg_xCd_{1-x}Te$ is the most widely used; it has a bandgap $\mathscr{E}_g \simeq 0.1$eV (corresponding to a photon wavelength of $\simeq 10\mu$m) for $x = 0.18$. Although this alloy material has the advantages that it can be grown epitaxially on CdTe or GaAs substrates and it has long photocarrier lifetimes (several microseconds at 77 K), it is mechanically very soft and prone to defect formation during high-temperature processing.

An alternative method of detecting IR photons is by creating photocarriers using *inter-sub-band absorption* in semiconductor quantum wells (§8.4.3). The sample configuration for quantum-well photoelectron detection is shown in Fig. 8.83. A modulation-doped, multiple-quantum-well (MQW) array, of n-type GaAs wells sandwiched between intrinsic $Al_xGa_{1-x}As$ barrier layers, is connected to n^+-type contacts in a two-terminal configuration. The n-doping of the GaAs layers ensures that the $n = 1$

ground state of the conduction-band well is occupied, and the well width, d, is chosen so that the energy separation between the $n = 1$ and $n = 2$ sub-bands of the well (cf. eqn. (8.57)) matches the energy of the photons to be detected ($d \simeq 7$ nm for $\lambda \simeq 10\,\mu$m). The conduction-band offset between well and barrier layer, determined by the composition x of the $Al_xGa_{1-x}As$ layer, is chosen so that the $n = 2$ well level is close to the top of the well when a small bias electric field is applied to the electrodes (Fig. 8.83). In this case, the excited electrons can tunnel through the small remnant barrier in the up-field direction into continuum states of the conduction band, and hence be swept out by the applied field and collected at the electrode. There is an optical problem, however, associated with the use of MQW arrays as photon detectors: the envelope function associated with each quantum well (Fig. 8.64a) means that only the component of the polarization vector of the light that is *normal* to the quantum-well direction (parallel to the barrier–well–barrier axis) can cause inter-sub-band transitions. Thus, light that is incident normal to the MQW stack (i.e. through one of the (semi-transparent) electrodes) is not absorbed. In order to circumvent this difficulty, the back face of the substrate supporting the MQW stack can be bevelled at 45° in order to allow some light with the correct polarization to reach the stack (see inset to Fig. 8.83).

Diodes, such as p–n and p–i–n junctions (§8.4.2.2) and metal–semiconductor Schottky barriers may be used as intrinsic-absorption photodetectors, the spectral sensitivity depending on the bandgap of the semiconductor. The principle of operation

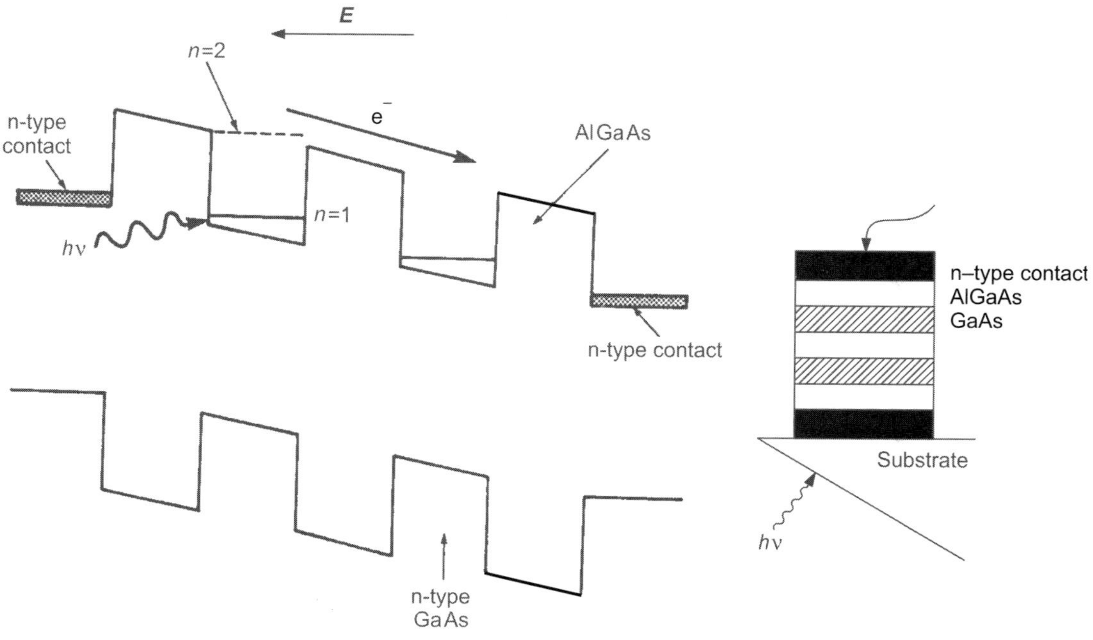

Fig. 8.83 Schematic illustration of the use of a modulation-doped, multiple-quantum-well (MQW) array as an IR detector, utilizing inter-sub-band ($n = 1 \rightarrow 2$) optical transitions and subsequent charge collection in a bias electric field. Shown in the inset is the bevelled shape of the substrate necessary to allow incident light to be absorbed in the MQW stack.

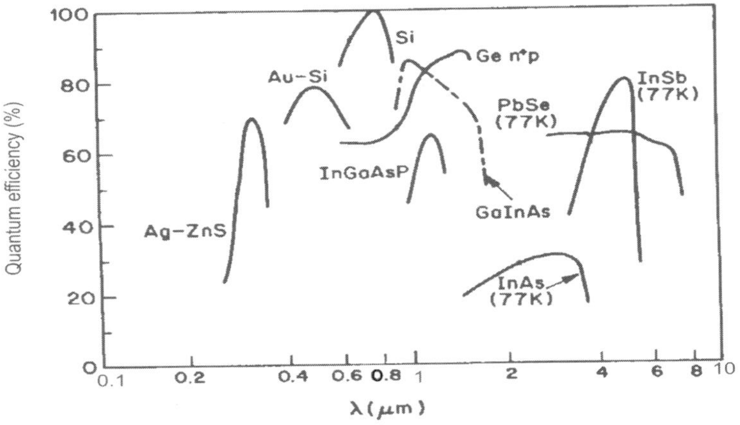

Fig. 8.84 Spectral dependence of the quantum efficiency of some semiconductors used in diode photodetectors. (After Sze (1981). Reproduced by permission of John Wiley & Sons Inc.)

is somewhat similar to that of a diode solar cell: photons absorbed in the depletion region of the junction generate electron–hole pairs that can be separated and swept out by the junction electric field. The measured photocurrent or photovoltage is thus a measure of the photon flux. For photodiodes, two characteristics are important for device performance. The quantum efficiency (the number of electron–hole pairs generated per incident photon) determines the sensitivity of the detector; the greater the thickness of the depletion layer (or equivalently the intrinsic-layer thickness in a p–i–n diode), the larger the fraction of the incident light that is usefully absorbed and hence the greater is the quantum efficiency. On the other hand, for high-speed operation, the carrier transit time through the device must be kept to a minimum and this can be achieved by reducing the depletion-layer thickness, D. Hence, for optimized overall performance, there must be a compromise in the value of D. The spectral dependence of the quantum efficiency, $\eta_q(\lambda)$, of a photodiode is a peaked function (Fig. 8.84). The long-wavelength cut-off is associated with the bandgap of the semiconductor; for photon wavelengths longer than $\lambda_c = hc/\mathscr{E}_g$, corresponding to the gap energy, few photons are absorbed. The short-wavelength limit is determined by surface recombination. At very small photon wavelengths, $\lambda < \lambda_c$, the absorption coefficient $K(\lambda)$ of the semiconductor is very large and so the absorption region where the light is preferentially absorbed is very close to the surface, where the recombination time is very short because of defects in the surface region. Further details about photodetectors can be found, for example, in Sze (1981).

8.5.2.3 Light-emitting diodes

In the opto-electronic applications considered in the previous two sections (involving photon → electron–hole pair processes), strenuous efforts were made to reduce electron–hole recombination to a minimum in order to extract the maximum photocurrent. In the applications discussed in this section on light-emitting diodes (LEDs) and the next section on semiconductor lasers (both involving electron–hole pair → photon processes), the converse is true: radiative recombination of excess electron–hole pairs is encouraged. All such light-emitting devices based on semiconductors utilize homo- or heterojunctions, into the depletion region of which excess electrons and holes can be electrically injected under forward bias (see §8.4.2.2). If such excess carriers are not quickly separated by the junction field, they can recombine radiatively, thereby emitting photons with energy comparable to the bandgap of the semiconductor. This process is termed electroluminescence. Radiative recombination in crystalline semiconductors is a first-order process (i.e. no phonon involvement) in the case of *direct-gap* materials (§5.8.2); as a result, the quantum efficiency of the process is much higher than for an indirect-gap material. Thus, in general, direct-gap semiconductors are used in electroluminescent applications.

Injection electroluminescent devices rely on the injection of minority carriers (electrons into the p-layer, holes into the n-layer) in a p–n junction subject to forward bias: these injected minority carriers then recombine radiatively with the majority carriers already present in the layers (Fig. 8.85). Although the *internal* (microscopic) quantum efficiency for electroluminescent radiative recombination in a direct-gap p–n diode may be very high, nevertheless the light output can be severely attenuated by internal reflection of the emitted light at surfaces and subsequent absorption of the light by the semiconductor making up the junction or the substrate. The light output is enhanced by making the light-emitting top surface of the semiconductor curved (e.g. hemispherical), rather than planar, in order to decrease the level of total internal reflection.

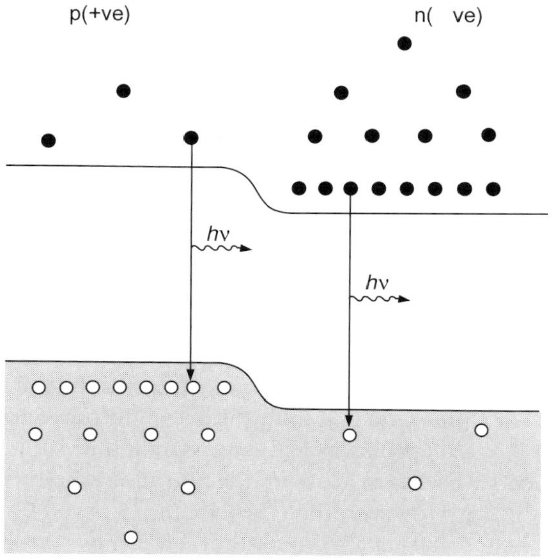

Fig. 8.85 Schematic representation of injection electroluminescence in a forward-biased p–n junction light-emitting diode (LED). The injected minority carriers recombine radiatively with the majority carriers in each layer making up the junction.

Recombination of excess carriers is a competition between processes that are radiative and those that are non-radiative, in which the excess (electronic) energy is carried away by emission of photons or phonons, respectively. A particular recombination rate R_i can be written as a first-order kinetic equation involving the excess concentration of electrons, Δn, injected into the p-layer with an equivalent equation for excess holes in the n-layer), i.e.

$$R_i = -\frac{\mathrm{d}}{\mathrm{d}t}(\Delta n) = \frac{\Delta n}{\tau_i} = \frac{n - n_0}{\tau_i} \tag{8.93}$$

where n is the total (injected plus thermal-equilibrium) concentration of electrons and n_0 is the thermal-equilibrium concentration. Radiative and non-radiative recombination processes will have their own particular lifetimes $\tau_i (i = \mathrm{r}, \mathrm{nr}$ respectively), and the total recombination rate (i = tot) is simply

$$R_{\mathrm{tot}} = R_{\mathrm{r}} + R_{\mathrm{nr}} = \frac{n - n_0}{\tau_{\mathrm{tot}}}, \tag{8.94}$$

with

$$\tau_{\mathrm{tot}}^{-1} = \left(\tau_{\mathrm{r}}^{-1} + \tau_{\mathrm{nr}}^{-1}\right). \tag{8.95}$$

The quantum efficiency η_{q} for radiative recombination is the fraction of excited carriers that recombine radiatively to the total recombination, i.e.

$$\eta_{\mathrm{q}} = R_{\mathrm{r}}/R_{\mathrm{tot}} = \tau_{\mathrm{tot}}/\tau_{\mathrm{r}} \equiv \tau_{\mathrm{nr}}/(\tau_{\mathrm{r}} + \tau_{\mathrm{nr}}) \equiv 1/(1 + \tau_{\mathrm{r}}/\tau_{\mathrm{nr}}). \tag{8.96}$$

Thus, a high quantum efficiency is associated with a small radiative lifetime, τ_{r}; for a direct-gap LED material, $\eta_{\mathrm{q}} \simeq 0.5$.

The radiative recombination rate can also be written as (see Problem 8.18):

$$R_{\mathrm{r}} = A_{\mathrm{em}} np, \tag{8.97}$$

where A_{em} is the Einstein coefficient of spontaneous emission. The optical power generated by an LED per unit volume can therefore be written as

$$P = A_{\mathrm{em}} np \eta_{\mathrm{q}} h\bar{\upsilon}, \tag{8.98}$$

where $h\bar{\upsilon}$ is the average photon energy emitted.

The wavelength of the emitted light is governed by the bandgap of the semiconductor. The visible region of the electromagnetic spectrum (violet $\rightarrow$ red, $\lambda = 0.45$–$0.7\ \mu$m) can be mostly covered by inorganic crystalline semiconductors with bandgaps in the range 1.8–2.8 eV, such as $GaAs_{1-x}P_x$, SiC, ZnSe and GaN (Fig. 8.86). The $GaAs_{1-x}P_x$ system is of particular importance since light emission in the IR by pure GaAs LEDs can be used as light sources for fibre-optic communication applications (§5.9.1), and the alloy system $GaAs_{1-x}P_x (0 < x < 1)$ produces light emission in the visible range (red to green, with increasing x). Red LEDs are made from the alloy $GaAs_{0.6}P_{0.4}$ which, like GaAs, is a direct-gap semiconductor. However, for $x > 0.45$, the $GaAs_{1-x}P_x$ system is an *indirect* semiconductor and hence has a low quantum efficiency for electroluminescence. This drawback is circumvented by doping the material with the isovalent impurity, nitrogen, substituting for phosphorus (§6.5.2); the resulting N_P defects act as efficient recombination centres since for them the $\boldsymbol{k}$-selection rule for optical transitions, operative in translationally periodic materials, does not apply. Crystalline Si is another

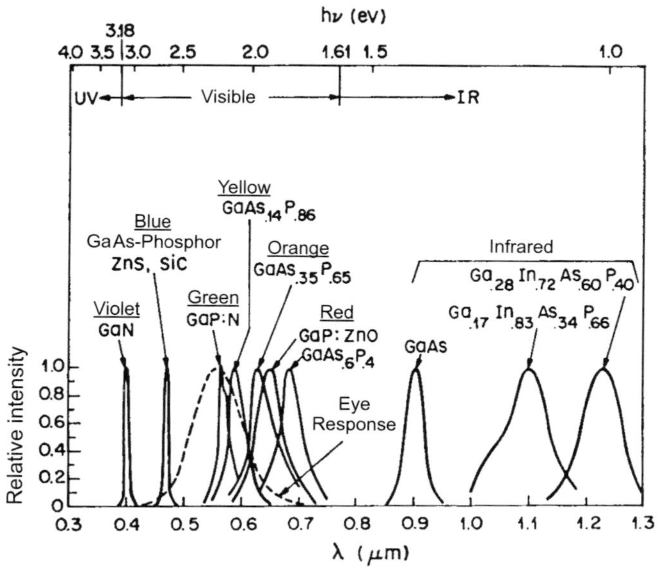

Fig. 8.86 Relative spectral emission of some representative LED materials. The dashed curve shows the spectral response of the human eye. (After Sze (1981). Reproduced by permission of John Wiley & Sons Inc.)

indirect-gap material which, because of its widespread use in microelectronic applications, would be very valuable if it could be made to electroluminesce efficiently, since then combined opto-electronic processing on the same chip would be possible. *Porous* Si can be made to electroluminesce quite efficiently, probably as a result of quantum-confinement effects (§8.2.2.1), but device stability and lifetime are problematic.

For a long time, it was difficult to make LEDs that produced blue light. Recent research on II–VI semiconductors has shown some promise, e.g. the production of (Zn, Cd)Se/Zn(S, Se) multiple-quantum-well LEDs, in which the ZnSe-based layers are the active regions. However, there are many materials-processing problems associated with the use of such materials, notably the high concentration of structural defects that act as non-radiative centres, thereby limiting the quantum efficiency. See Nurmikko and Gunshor (1995) for further details on II–VI LEDs and lasers. However, the most promising technology seems to be based on nitride semiconductors, e.g. GaN and InGaN; c.w. LEDs based on these materials emitting in the spectral range yellow-to-purple are now commercially available–see Plate VIII (see Nakamura and Fasol (1997) for a review).

Although LEDs based on inorganic crystalline semiconductors are very widely used, they have the disadvantage that they cannot easily be made into large-area light-emitting displays. An emerging technology that might answer this need is based on the use of conjugated polymers (§8.3.3), rather than inorganic crystalline semiconductors, as the light-emitting materials (see Plate VIII and Greenham and Friend (1995) for a review). The conjugated polymer poly (*p*-phenylenevinylene), or PPV, emits electroluminescence (EL) when electrons are injected into the material from a negatively biased metal electrode (e.g. Ca or Al) and holes are injected from a semi-transparent electrode (such as indium tin oxide, ITO) in a sandwich configuration. The electrons and holes form singlet excitons which then recombine radiatively. This system has the processing advantage that a thin PPV film can be deposited onto a large-area substrate by spin-coating: a solution of a precursor polymer in a suitable solvent is dropped onto the rapidly spinning substrate and forms a thin ($\simeq$ 100 nm) layer, which is subsequently converted to the polymer PPV by heat treatment. The peak EL emission for PPV is at a wavelength of 565 nm, but this can be shifted towards the blue region by introducing non-conjugated units into the PPV to reduce the average conjugation length, or by using poly (*p*-phenylene), PPP (λ = 465 nm). However, lifetimes of these polymer LEDs are currently problematic.

8.5.2.4 Semiconductor lasers

The acronym LASER stands for 'light amplification by the stimulated emission of radiation', thereby highlighting two of the special features that characterize lasers as light sources. Two other particular features of the light emitted by lasers are that it is phase-coherent and is generally monochromatic (see, e.g. Davis (1996) for more details). The aspect which sets a laser apart from other (incoherent) light sources is that the light emission resulting from a radiative transition is triggered or *stimulated* by the presence of a pre-existing photon having the same energy as that for the transition. Thus, this optical process is to be distinguished from the related process, namely *spontaneous* emission (luminescence), in which the relevant optical transition occurs spontaneously without the triggering effect of another photon (Fig. 8.87). We have already briefly mentioned other stimulated optical processes, namely *stimulated optical absorption* (§5.8.2) and *stimulated Raman emission* which is an example of a third-order non-linear optical effect (§5.8.4).

Semiconductor lasers, which comprise the majority of all lasers in use today (e.g. in CD players) are essentially light-emitting diodes (§8.5.2.3) having a configuration which

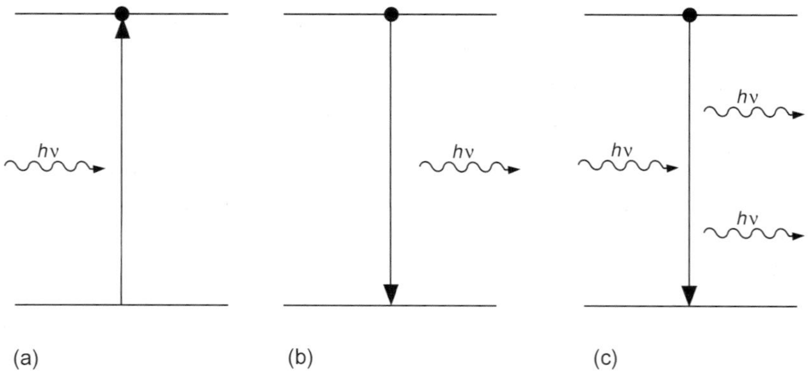

Fig. 8.87 Spontaneous and stimulated optical processes: (a) stimulated absorption; (b) spontaneous emission (luminescence); (c) stimulated (laser) emission.

favours stimulated emission. The optical process responsible for the laser emission in the case of semiconductor lasers can be termed stimulated electroluminescence (since population of the excited, lasing state occurs by electrical injection of electrons), to distinguish it from the process in other types of solid-state lasers, namely stimulated photoluminescence, involving *optical* pumping of the lasing state, e.g. an electronic level of an ion, e.g. of Cr^{3+} in the ruby laser (ruby is sapphire, i.e. alumina, doped with $\simeq 0.05$ wt % Cr^{3+}). Lasing action, i.e. light amplification or gain, occurs when the rate of stimulated emission of many photons caused by a single photon is greater than the rate of stimulated absorption of the photon. This can be achieved by 'trapping' the light emitted by normal spontaneous emission, as in normal LED operation, so that the radiation density eventually builds up to the point at which stimulated emission becomes favourable. Such light trapping can be achieved by polishing and mirroring two opposing faces, of the four faces of a semiconductor-junction laser normal to the light-emitting plane (i.e. the junction), leaving the other two opposing faces rough (so that they scatter radiation); one of the mirrored faces is made semi-transparent (Fig. 8.88). In this manner, spontaneous light radiation emitted from the junction under forward bias is reflected back and forth between the two mirrored surfaces, making up the optical cavity, until stimulated laser emission occurs, whereupon the output light emerges from the device through the semi-transparent mirrored face.

The condition for optical gain, or amplification, can be obtained from the expressions for the photon-induced transition rates for absorption (valence $\rightarrow$ conduction band) and emission (conduction $\rightarrow$ valence band), involving states at $\mathscr{E}$ in the conduction band and $(\mathscr{E} - h\upsilon)$ in the valence band, with densities of states $g_c(\mathscr{E})$ and $g_v(\mathscr{E})$, respectively, i.e.

$$R_{\rm abs}^{\rm spont}(h\upsilon) = B_{\rm abs}\int g_{\rm v}(\mathscr{E} - h\upsilon)g_{\rm c}(\mathscr{E})f_{\rm v}(\mathscr{E} - h\upsilon)[1 - f_{\rm c}(\mathscr{E})]|M|^2 {\rm d}\mathscr{E}, \qquad (8.99a)$$

$$R_{\rm em}^{\rm spont}(h\upsilon) = B_{\rm em}\int g_{\rm c}(\mathscr{E})g_{\rm v}(\mathscr{E} - h\upsilon)f_{\rm c}(\mathscr{E})[1 - f_{\rm v}(\mathscr{E} - h\upsilon)]|M|^2 {\rm d}\mathscr{E}, \qquad (8.99b)$$

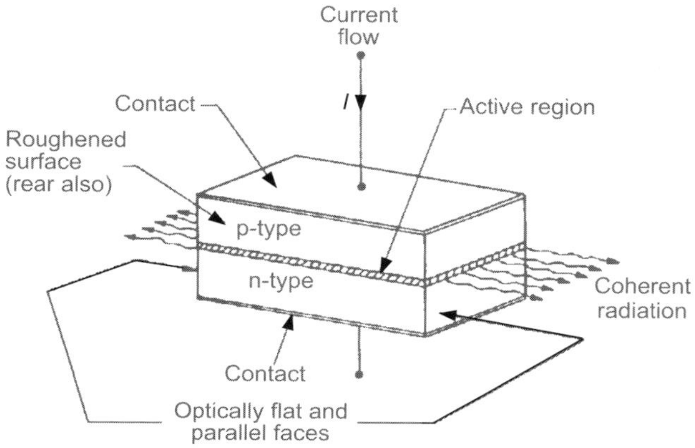

Fig. 8.88 Schematic illustration of a semiconductor laser based on a single p–n junction. The optical cavity is formed by two plane-parallel mirrored surfaces on opposite sides of the device, one of which is semi-transparent to allow the coherent laser light to emerge.

where the Einstein coefficients for stimulated emission, $B_{\rm em}$, and absorption, $B_{\rm abs}$, are equal from the principle of detailed balance (Problem 8.19), i.e. $B_{\rm em} = B_{\rm abs} = B$. There is optical gain if

$$R_{\rm em}^{\rm spont}(h\nu) - R_{\rm abs}^{\rm spont}(h\nu) > 0, \tag{8.100}$$

or, from eqns. (8.99), if

$$f_{\rm c}(\mathscr{E}) - f_{\rm v}(\mathscr{E} - h\nu) > 0. \tag{8.101}$$

In the case when the concentration of electrons injected (e.g. electrically in the case of semiconductor lasers) into the conduction band (and of holes into the valence band) is in *excess* of the thermally generated concentration n_0 (or p_0), obviously thermal equilibrium cannot exist overall. Nevertheless, there is a vestige of thermal equilibrium remaining, in that quasi-Fermi levels can be defined *separately* for electron and hole distributions in conduction and valence bands, i.e. $\mathscr{E}_{\rm F,n}$ and $\mathscr{E}_{\rm F,p}$, respectively: for energies $\mathscr{E} - \mathscr{E}_{\rm F,n} > 0$ (or $\mathscr{E} - \mathscr{E}_{\rm F,p} < 0$), the electrons (or holes) are in thermal equilibrium with each other, but this is *not* the case for energies $\mathscr{E}_{\rm F,p} < \mathscr{E} < \mathscr{E}_{\rm F,n}$. The quasi-Fermi level position is defined in terms of the *total* carrier concentration in a band, e.g. $n = n_0 + \Delta n$, where n_0 is the thermal-equilibrium (intrinsic or extrinsic) electron concentration and Δn is the excess steady-state density, via the relation (cf. eqns. (6.194) and (6.195):

$$\mathscr{E}_{\rm c} - \mathscr{E}_{\rm F,n} = k_{\rm B}T\ln(N_{\rm c}/n) \tag{8.102a}$$

and the corresponding relation for holes (cf. eqn. (6.200)):

$$\mathscr{E}_{\rm F,p} - \mathscr{E}_{\rm v} = k_{\rm B}T\ln(N_{\rm v}/p). \tag{8.102b}$$

For the case of a direct-gap semiconductor used as a lasing material, it is assumed that there are no (defect) states in the gap between the conduction- and valence-band edges ($\mathscr{E}_{\rm c} > \mathscr{E} > \mathscr{E}_{\rm v}$), in which case the quasi-Fermi levels reside in the respective bands (see Fig. 8.89). (However, for the case of, say, photoconductivity of a semiconductor with trap states in the gap (see §6.5.1.2), the quasi-Fermi levels can lie in the gap, near the respective band edges.)

Since thermal equilibrium can be defined for energies above the electron quasi-Fermi level $\mathscr{E}_{\rm F,n}$ (or below $\mathscr{E}_{\rm F,p}$), the occupation probability of such states is given by the Fermi–Dirac distribution function (eqn. (5.23)), modified so that the thermal-equilibrium chemical potential (Fermi energy) is replaced by the steady-state quasi-Fermi levels, i.e.

$$f_{\rm c}(\mathscr{E}) = \frac{1}{\exp[(\mathscr{E} - \mathscr{E}_{\rm F,n})/k_{\rm B}T] + 1}, \tag{8.103a}$$

$$f_{\rm v}(\mathscr{E}) = \frac{1}{\exp[(\mathscr{E} - \mathscr{E}_{\rm F,p})/k_{\rm B}T] + 1}. \tag{8.103b}$$

Thus, from the inequality (8.101), the condition for optical gain can also be written as

$$\mathscr{E}_{\rm F,n} - \mathscr{E}_{\rm F,p} > h\nu. \tag{8.104}$$

This condition is illustrated graphically in Fig. 8.89. The condition that $f_{\rm c}(\mathscr{E}) - f_{\rm v}(\mathscr{E} - h\nu) > 0$ implies a population inversion with respect to the thermal-equilibrium occupations of the bands.

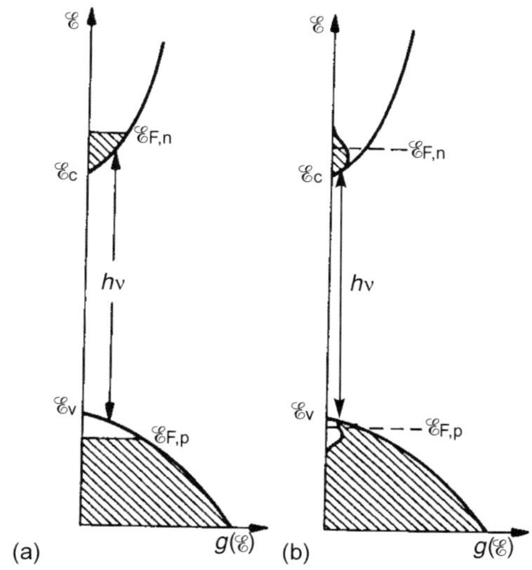

Fig. 8.89 Occupancy of electron states in the valence and conduction bands, showing the positions of the respective quasi-Fermi levels, $\mathscr{E}_{F,p}$ and $\mathscr{E}_{F,n}$, for (a) $T = 0$ K; (b) $T > 0$ K. The condition for a population inversion, leading to optical gain, is that $\mathscr{E}_{F,n} - \mathscr{E}_{F,p,} > h\upsilon$, as shown.

The simple p–n junction device shown in Fig. 8.88, e.g. based on GaAs, does indeed lase, but population inversion can only be achieved by the use of very high injection currents, necessitating pulsed operation. The threshold injection current density can be reduced by two orders of magnitude, making continuous (c.w.) operation feasible, by fabricating double-heterojunction devices, in which a doped lasing layer, e.g. heavily doped p-GaAs, is sandwiched between two doped layers of a material with a *larger* bandgap, e.g. p-and n-$Al_xGa_{1-x}As$, i.e. a P–p–N structure (Fig. 8.90). When the P-layer is biased positively and the N-layer biased negatively, electrons are injected from the N-layer into the conduction band of the p-layer (and similarly holes are injected into the valence band of the p-layer from the P-layer), and there they are confined by the large conduction-band barrier at the p–P heterojunction and the valence-band barrier at the N–p heterojunction (Fig. 8.90). Thus, a population inversion can be created with a lower injection current compared with the simple p–n-junction laser diode. There is also an additional advantage conferred by the double-heterojunction configuration concerning *optical* confinement of the emitted light. Since the larger-gap AlGaAs layers have a *lower* refractive index than that of the lasing GaAs layer, the double-heterojunction structure acts as a *planar waveguide*: the emitted light travelling parallel to the heterojunctions is confined to the GaAs layer by total internal reflection, thereby increasing the stimulated emission in the active layer.

The width d of the active GaAs layer in double-heterojunction lasers is typically $d \simeq 0.5\,\mu$m, i.e. much larger than the Fermi wavelength of electrons in GaAs ($\lambda_F \simeq$ 40 nm). However, single (and multiple) quantum-well semiconductor lasers can also be fabricated, for which $d < \lambda_F$; such lasers have the advantage that the lasing transition

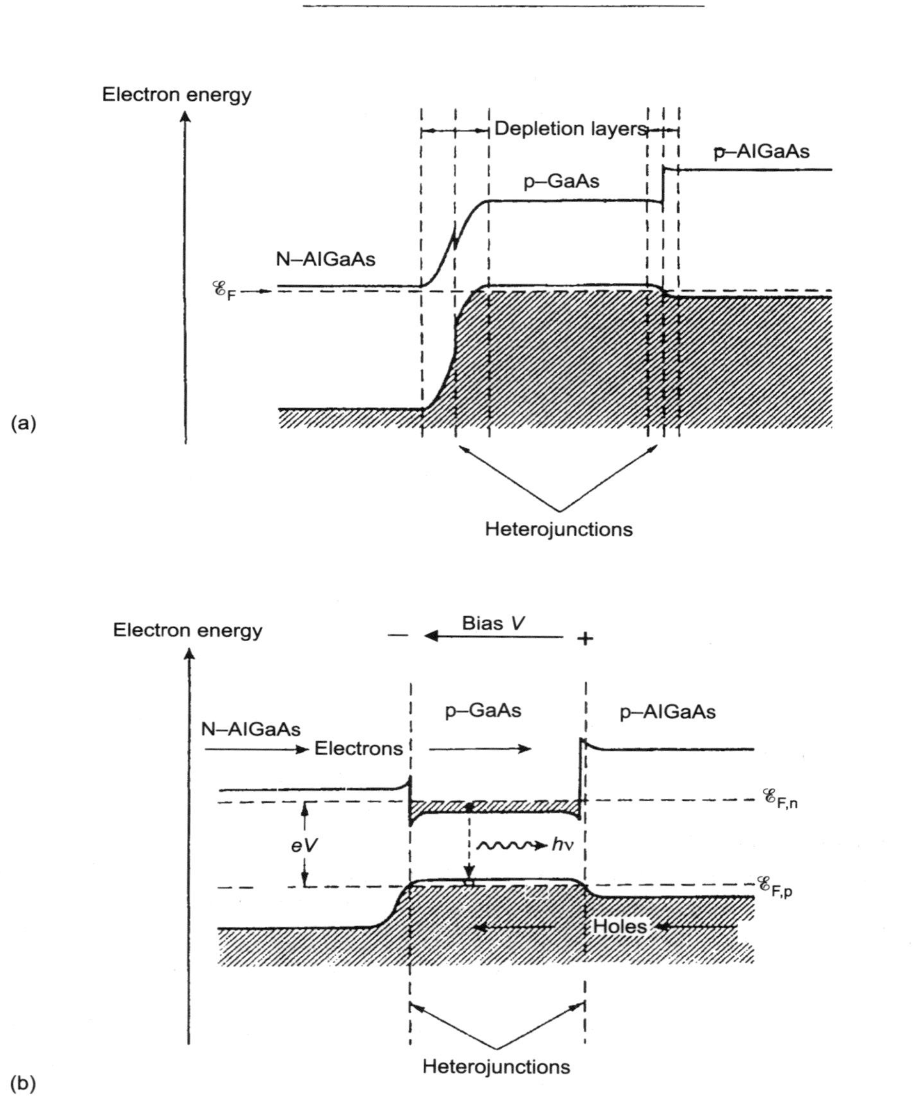

Fig. 8.90 Schematic illustration of the spatial profiles of the bands for a P–p–N double-heterojunction semiconductor laser, consisting of an active heavily doped p-type layer of GaAs sandwiched between barrier layers of p- and n-type $Al_xGa_{1-x}As$ having a larger bandgap and smaller refractive index: (a) zero applied bias; (b) positive bias applied to the P-layer.

(between sub-bands) can be tuned by varying the well width. Optical confinement within the lasing-layer waveguide is ineffective for thicknesses of the layer $d \lesssim 100$ nm; thus, quantum-well lasers suffer because the emitted light is no longer confined to the

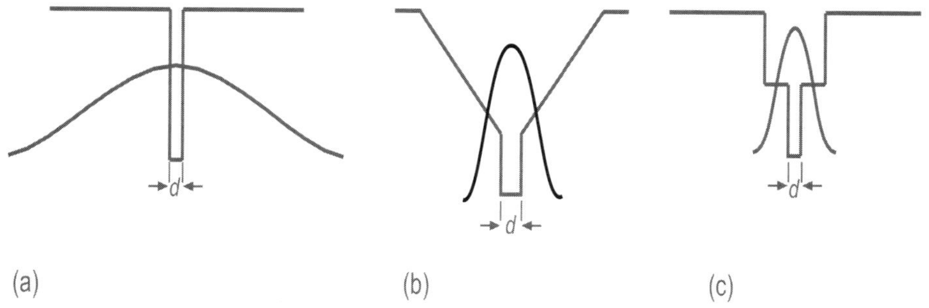

Fig. 8.91 Schematic illustration of the confinement of an optical guided wave for: (a) a quantum well; (b) a graded-index (GRIN) profile; (c) a separate confinement by heterojunctions (SCH) configuration.

active layer (Fig. 8.91a). In order to circumvent this problem, confinement of the optical field is achieved by having a graded-index (GRIN) configuration, in which the refractive index of the $Al_xGa_{1-x}As$ barrier layers is varied continuously outside the well region by varying x (see Fig. 8.91b) and/or separate confinement by heterojunctions (SCH), in which additional heterojunctions (having refractive-index discontinuities) are grown on

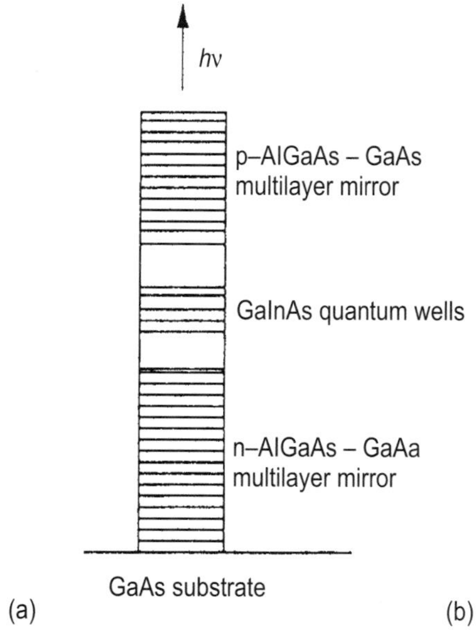

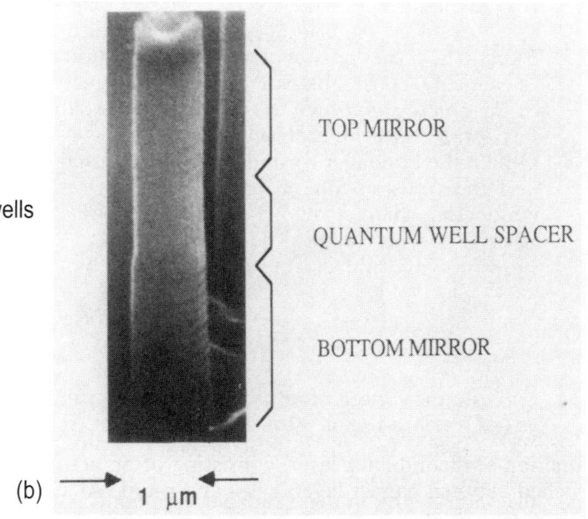

Fig. 8.92 Device structure with multilayer (Bragg-reflector) mirrors used for a surface-emitting semiconductor laser: (a) schematic illustration showing the various regions; (b) actual device. (After Scherer *et al.* (1989). Reprinted with permission from *Appl. Phys. Lett.* **55**, 2724. © 1989 American Institute of Physics)

either side of the quantum well at a separation for which optical confinement is efficient (Fig. 8.91c).

All the above planar device configurations produce laser radiation from one of the edges, *parallel* to the plane of the active layer (as for LEDs — §8.5.2.3). This arrangement is rather limiting, and a *surface*-emitting semiconductor-laser configuration, in which laser radiation emerges *normal* to the active layer(s), would be much more useful in many applications. One way of realizing this objective is to fabricate a multilayer device in which the active region consists of a stack of quantum wells (say of GaInAs); this is sandwiched between additional multilayers of AlGaAs/GaAs, each layer of which has a thickness of $\simeq \lambda/4$, where λ is the wavelength of the laser light, and which therefore acts as a multilayer mirror ('Bragg reflector') with nearly 100% reflectivity (Fig. 8.92). This high reflectivity allows the laser cavity length to be reduced to just a few microns in length; the lateral dimensions of the device can be restricted by etching through the layers to the substrate leaving a mesa-like structure (Fig. 8.92). Two-dimensional arrays of such devices can thus be fabricated.

And on this luminous note that, for the time being, is that.

Problems

8.1 Can long-range ferromagnetic order exist at a finite temperature in a 1D material having only nearest-neighbour magnetic interactions? (Hint: consider spin fluctuations and the influence of entropy.)

8.2 (a) Show that, for the case of KCP ($K_2Pt(CN)_4Br_{0.3} \cdot 3H_2O$), the spatial period of the Peierls-distorted state of the Pt chain is $D = 6.67c$, where $c = 2.88$ Å is the lattice parameter (Pt–Pt distance) in the undistorted state.

(b) Explain why, whereas TTF:TCNQ exhibits metallic behaviour only at elevated temperatures, Na is a metal at *all* temperatures.

8.3 Explain how the concept of a Peierls distortion can be used to relate the crystalline structures of elemental As and black P (which are both layer-like structures, with trigonal coordination of the pnictogen atoms) to that of a simple cubic lattice. What influence does the distortion have on the electronic structure?

8.4 Obtain the Landauer formula (eqns. (8.16) and (8.18)) relating the electrical conductance G_0 and the transmission probability T for electrical current to propagate along a narrow conducting channel in a 2D electron gas with spin degeneracy g_s, between two wide electron reservoirs, with chemical potentials, μ and $\mu + \delta\mu$, respectively.

(a) Show that the current density can be written as:

$$j = e \int_{\mathscr{E}_F}^{\mathscr{E}+\delta\mu} g_s \left(2\pi \frac{d\mathscr{E}(k)}{dk} \right)^{-1} \frac{d\mathscr{E}(k)}{\hbar dk} d\mu = \frac{g_s \delta\mu}{h}$$

(Note the cancellation of terms for the group velocity and density of states.)

(b) Obtain an expression for the Einstein relation valid in the *degenerate* limit (cf. eqn. (3.84a) for the non-degenerate case), and hence show that for 2D conduction $G_0 = e^2 g_{2D}(\mathscr{E}_F) D$, where D is the diffusion coefficient.

(c) Hence obtain the Landauer formula $G_0 = (2e^2/h)T$ for $g_s = 2$.

8.5 The Aharonov–Bohm effect in quantum mechanics states that the phase ϕ of an electron wavefunction is changed by an amount $\Delta\phi = -(e/\hbar) \int \boldsymbol{A}.d\boldsymbol{s}$ where $\boldsymbol{A}$ is the magnetic vector potential and $d\boldsymbol{s}$ is an element of the path traversed by an electron, even when the electron trajectory is restricted to regions where the magnetic-field intensity is *zero*. Consider a sample geometry (in cross-section) consisting of a conducting ring, enclosing an area S, with source and drain electrodes attached to opposite sides and a magnetic field normal to and threading the ring. By considering trajectories that involve phase-coherent interference of electrons after either a half or one revolution, show that Aharonov–Bohm oscillations as a function of magnetic field in the magneto-conductance have periods of $(h/e)\ S^{-1}$ and $(h/2e)\ S^{-1}$ for a sample in the form of a 2D ring, but only $(h/2e)\ S^{-1}$ for a long hollow metal cylinder.

8.6 Show that weak localization leads to a correction to the Drude expression for the electrical conductivity in two dimensions that depends logarithmically on temperature. (Hint: take $\mathfrak{W}(t) = (4\pi Dt)^{-1}$ in eqn. (8.30).)

8.7 Explain why intercalation of guest species into crystalline BN is extremely difficult (the acceptor molecule SO_3F is the only known intercalant).

8.8 By solving Poisson's equation for the depletion region (width d) of a Schottky barrier involving an n-type semiconductor, show that a plot involving the capacitance of the junction versus applied voltage gives the diffusion potential ϕ_D and the ionized donor concentration N_{d^+}.

8.9 Show that the space-charge-limited (SCL) current density, j_{SCL}, arising from single injection of carriers into an n-type semiconductor in a metal–semiconductor–metal heterostructure from the negatively–biased ohmic contact ('virtual cathode'), separated by a thickness d of semiconductor from a positively biased ohmic contact, is proportional to V^2/d^3, where V is the applied voltage. (Hint: consider the transit time, t_{tr}, through the semiconductor.) Show that the critical voltage onset for SCL currents is $V_c = \sigma d^2/\varepsilon_r\varepsilon_0\mu_e$, where σ, μ_e and ε_r are the conductivity, electron mobility and dielectric constant, respectively, of the semiconductor.

8.10 (a) Obtain eqn. (8.38a) for the contact potential of a p–n junction formed from an n-type region of a semiconductor with donor concentration N_d and from a p-type region of the same semiconductor material with acceptor concentration N_a (where the intrinsic electron (or hole) concentration is n_i), starting from expressions for the chemical-potential position in an n-type semiconductor (eqn. (6.246)) and a similar expression for a p-type material.

(b) By solving Poisson's equation for a p–n junction at $z = 0$, show that the spatial profile of the contact potential has the quadratic behaviour: $\phi_c(z) = eN_a(z + d_p)^2/2\varepsilon_r\varepsilon_0$ for $-d_p < z < 0$, and $\phi_c(z) = \phi_c - eN_d(z - d_n)^2/2\varepsilon_r\varepsilon_o$ for $0 < z < d_n$. Hence prove eqn. (8.39) for the contact potential in terms of the depletion widths d_n and d_p.

(c) Find the spatial profile $\phi_c(z)$ for the contact potential, and the depletion-layer width, for a *graded* p–n junction in which the doping level varies linearly with position throughout the depletion layer: $N_d - N_a = kz$.

8.11 (a) Calculate the capacitance of a p–n junction, subject to an applied voltage V, for a junction with contact potential ϕ_c formed from p-type and n-type semiconducting layers with a concentration N_a of acceptors and N_d of donors, respectively.

(b) A p–n junction is used with a 50 μH inductance to produce a resonant electrical circuit. Calculate the change in resonant frequency when the bias applied to the room-temperature junction is changed from − 1 to − 5 V. (Assume that $N_d = N_a = 10^{22}\text{m}^{-3}$, the semiconductor is c-Si and that the area of the junction is 10^{-6}m^2.)

8.12 (a) Obtain an expression for the maximum electric-field strength, E_{max}, in the depletion region of a p–n junction subject to a non-zero bias voltage $| V |$.

(b) Hence estimate the width D of the potential barrier through which electrons must tunnel in Zener breakdown, and show that the associated current-voltage characteristic is $I \propto \exp(-a | V |^{-1/2})$ for $| V | >> \phi_c$, where a is a constant.

(c) Obtain an expression for the critical electric field for avalanche breakdown of a p–n junction. (Hint: consider the maximum velocity of an electron achieved in a time τ between collisions.) Hence estimate the breakdown voltage for a p–n junction formed from c-Si (with $\mathscr{E}_g = 1.1$ eV, $\varepsilon = 11.7$, $\tau = 10^{-12}$s and $m_e^* = 0.1m_e$) with a doping level $N_d = N_a = 10^{19}\text{m}^{-3}$.

8.13 Estimate the maximum electric-field strength that can be exerted in the Si layer of an MOS device if the dielectric-breakdown field of a-SiO_2 is $E_b = 10^9$ V m^{-1}($\varepsilon(SiO_2) = 3.9$;

$\varepsilon(\mathrm{Si}) = 11.7$). Hence obtain an estimate for the ground-state energy of the quantized 2D electron gas confined to the inversion layer at the Si/SiO_2 interface.

8.14 Obtain an expression of the form of eqn. (8.60) for the ground-state energy level of an electron in a triangular quantum well in an MOS device, formed by an infinitely high barrier at the insulator–semiconductor interface ($z \neq 0$) with a potential increasing linearly with depth z into the semiconductor as Ez, where E is the surface electric field. (Hint: use the Heisenberg uncertainty relation.)

8.15 Compare the optical-absorption spectrum of a single GaAs–AlGaAs quantum well (Fig. 8.61) with that expected for *two* such quantum wells separated by a very thin layer of AlGaAs (d(AlGaAs) $\leqslant$ d(GaAs)). What would be the effect on the optical properties of forming a superlattice of a very large number of such coupled quantum wells?

8.16 Show that when the effect of scattering on electron motion in a semiconductor superlattice is considered, the electron drift velocity (and hence mobility or conductivity) exhibits a negative differential region at high electric fields. (Hint: the drift velocity can be written as $v_d = \int \exp(-t/\tau)\mathrm{d}v_g$, where τ is the scattering relaxation time and v_g is the group velocity for motion in the growth (z) direction of the superlattice.) Obtain an expression for the superlattice effective mass, m^*_{SL}.

8.17 The Raman spectrum of a superlattice consisting of alternating layers of GaAs (of thickness $d_1 = 13.6$ Å) and of AlAs ($d_2 = 11.4$ Å) exhibits peaks at $\simeq 65$, $\simeq 280$ and $\simeq 380\ \mathrm{cm}^{-1}$, whereas the Raman spectra of bulk GaAs and AlAs exhibit peaks at $\simeq 280$ and $300\ \mathrm{cm}^{-1}$, and $\simeq 360$ and $400\ \mathrm{cm}^{-1}$. Account for these observations. The peak at $\simeq 65\ \mathrm{cm}^{-1}$ for the superlattice is actually a doublet. Why should this be? ($c_{11}(\mathrm{GaAs}) = 11.9 \times 10^{10}\ \mathrm{N\,m^{-2}}$, $c_{11}(\mathrm{AlAs}) = 12.02 \times 10^{10}\ \mathrm{N\,m^{-2}}$, $\rho(\mathrm{GaAs}) = 5.32 \times 10^3\ \mathrm{kg\,m^{-3}}$, $\rho(\mathrm{AlAs}) = 3.76 \times 10^3\ \mathrm{kg\,m^{-3}}$)

8.18 Using eqn. (8.97) for the spontaneous radiative-recombination rate, $R_r = A_{em}np$, where A_{em} is the Einstein coefficient for spontaneous emission, and n and p are electron and hole densities in the conduction and valence band of an LED, show that the light-emission spectrum has the lineshape $I(v) \propto (hv - \mathscr{E}_g)\exp[-(hv - \mathscr{E}_g)/k_B T]$, if it is assumed that n and p decay exponentially away from the band edges.

8.19 (a) Show that, for an empty cavity containing black-body radiation, the radiation energy density is $\rho(v) = (8\pi h v^3/c^3)(\exp(hv/k_B T) - 1)^{-1}$.

(b) By considering radiative processes between two energy levels 1 and 2, with energies $\mathscr{E}_1$ and $\mathscr{E}_2$ ($\mathscr{E}_2 - \mathscr{E}_1 = hv$), show that the two Einstein B coefficients for stimulated transitions are equal, $B_{12} = B_{21}$, and that $A_{em}/B = 8\pi h v^3/c^3$, where A_{em} is the Einstein coefficient for spontaneous emission.

(c) In a dispersive medium, where the refractive index varies with wavelength, $n_r = n_r(\lambda)$, the group refractive index can be defined as $n_g = n_r - \lambda \frac{\mathrm{d}n_r}{\mathrm{d}\lambda}$. Show that $n_g = n_r + v\frac{\mathrm{d}n_r}{\mathrm{d}v}$.

(d) Hence show that, for a cavity filled with a dispersive medium, the radiation mode density is $p(v) = 8\pi v^2 n_r^2 n_g / c^3$.

8.20 Estimate the laser gain at $hv = 1.48$ eV of an n-doped GaAs-based double-heterojunction laser, for the case when there are $N_d = 10^{24}\mathrm{m}^{-3}$ donors in the active GaAs layer, $A_{em} = 10^{-16}\ \mathrm{m^3\,s^{-1}}, n_r = 3.6, n_g = 4, m_e^* = 0.067m_e, m_h^* = 0.48m_e$, and the laser device, with an internal quantum efficiency of $\eta_q = 0.6$, is operated with a current of 10 mA entering an active region $w = 100\ \mu$m wide and $L = 10\ \mu$m high with a thickness between heterojunctions $d = 0.5\ \mu$m.

References

Akkermans, E., Wolf, P. E. and Maynard R., 1986, *Phys. Rev. Lett.* **56**, 1471.

Altmann, S. L., 1991, *Band Theory of Solids: An Introduction from the Point of View of Symmetry* (Clarendon Press: Oxford).

Al'tshuler, B. L., 1985, *JETP Lett.*, **41**, 648.

Beenakker, C. W. J. and van Houten, H., 1991, *Solid State Physics*, vol. **44**, eds H. Ehrenreich and F. Spaepen, (Academic Press: New York), p. 1.

Bjørnholm, S., Borggren, J., Echt, O., Hansen, K., Pederson, J. and Rasmussen, H. D., 1991, *Z. Phys.* D**19**, 47.
Bleaney, B.I. and Bleaney, B., 1976, *Electricity and Magnetism*, 3rd edn (Oxford University Press: Oxford), p. 686.
Brus, L. E., 1986, *J. Chem. Phys.* **80**, 4403.
Canham, L. T., 1990, *Appl. Phys. Lett.* **57**, 1064.
Capasso, F., ed., 1990, *Physics of Quantum Electron Devices*, Springer Series in Electronics and Photonics (Springer-Verlag: Berlin).
Chakravarty, S. and Schmid, A., 1986, *Phys. Rep.* **140**, 193.
Davis, C. C., 1996, *Lasers and Electro-optics* (Cambridge University Press: Cambridge).
de Heer, W., 1993, *Rev. Mod. Phys.* **65**, 611.
de-Picciotto, R., Reznikov, M., Heiblum, M., Umansky, V., Bunin, G. and Mahalu, D., 1997, *Nature* **389**, 162.
Dingle, R., Gossard, A. C. and Wiegmann, W., 1975, *Phys. Rev. Lett.* **34**, 1327.
Dingle, R., Wiegmann, W. and Henry, C. H., 1974, *Phys. Rev. Lett.* **33**, 827.
Ehrenreich, H. and Spaepen, F., eds, 1994, *Solid State Physics*, vol. **48** (Academic Press: New York).
Erwin, S. C., 1993, in *Buckminsterfullerenes*, eds W. E. Billups and M. A. Ciufolini (VCH: New York), p. 217.
Greenham, N. C. and Friend, R. H., 1995, *Solid State Physics*, vol. **49**, eds H. Ehrenreich and F. Spaepen, (Academic Press: New York), p. 1.
Gunnarsson, O., 1997, *Rev. Mod. Phys.* **69**, 575.
Hamada, N., Sawada, S. and Oshiyama, A., 1992, *Phys. Rev. Lett.* **68**, 1579.
Heinecke, H., Werner K., Weyers, M., Lüth, H. and Balk, P., 1987, *J. Cryst. Growth* **81**, 270.
Henry, C. H., 1980, *J. Appl. Phys.* **51**, 4494.
Iijima, S., 1991, *Nature* **354**, 56.
Jacobson, A. J., 1992, in *Solid State Chemistry: Compounds*, eds A. K. Cheetham and P. Day (Clarendon Press: Oxford).
Justus, B. L., Seaver, M. E., Ruller, J. A. and Campillo, A. J., 1990, *Appl. Phys. Lett.* **57**, 1381.
Kastner, M. A., 1992, *Rep. Progr. Phys.* **64**, 849.
Kastner, M.A., 1993, *Phys. Today* **46**.1, 24.
Kayanuma, Y., 1988, *Phys. Rev.* B**38**, 9797.
Keller, H. J., ed., 1975, *Low-Dimensional Cooperative Phenomena* (Plenum: New York).
Keller, H. J., ed., 1977, *Chemistry and Physics of One-Dimensional Metals* (Plenum: New York).
Kelly, M. J., 1995, *Low-Dimensional Semiconductors* (Clarendon Press: Oxford).
Kittel, C., 1969, *Thermal Physics* (Wiley: New York), p. 403.
Laughlin, R.B., 1982, *Phys. Rev. Lett.* **50**, 1395.
Lee, P. A., 1986, *Physica* **140A**, 169.
Lee, P. A. and Stone, A. D., 1985, *Phys. Rev. Lett.* **55**, 1622.
Lüth, H., 1995, *Surfaces and Interfaces of Solid Materials*, 3rd edn (Springer-Verlag: Berlin).
Madelung O. (ed.), 1996, *Semiconductors–Basic Data*, 2nd edn (Springer-Verlag: Berlin).
Mintmire, J. W., Dunlap, B. I. and White, C. T., 1992, *Phys. Rev. Lett.* **68**, 631.
Murray, C. B., Kagan, C. R. and Bawendi, M. G., 1995, *Science* **270**, 1335.
Nakamura, S. and Fasol, G., 1997, *The Blue Laser Diode: GaN-based Light Emitters and Lasers* (Springer-Verlag: Berlin).
Nurmikko, A. V. and Gunshor, R. L., 1995, *Solid State Physics* vol. **49**, eds H. Ehrenreich and F. Spaepen (Academic Press: New York), p. 205.
Paalonen, M. A., Tsui, D. C. and Gossard, A. C., 1982, *Phys. Rev.* B**25**, 5566.
Pfeiffer, L., West, K. W., Stormer, H. L. and Baldwin, K. W., 1989, *Appl Phys. Lett.* **55**, 1888.
Prange, R. E. and Girvin, S. M., 1990, *The Quantum Hall Effect*, 2nd edn (Springer-Verlag: New York).
Renker, B. and Comès, R., 1975, in *Low-Dimensional Cooperative Phenomena*, ed. H. J. Keller (Plenum: New York), p. 235.
Saito, R., Fujita, M., Dresselhaus, G. and Dresselhaus, M. S., 1992, *Appl. Phys. Lett.* **60**, 2204.
Saminadayar, L., Glattli, D. C., Jin, Y. and Etienne, B., 1997, *Phys. Rev. Lett.* **79**, 2526.

Scherer, A., Jewell, J. L., Lee, Y. H., Harbison, J. P. and Florez, L. T., 1989, *Appl. Phys. Lett.* **55**, 2724.
Stillman, G. E. and Wolfe, C. M., 1976, *Thin Solid Films* **31**, 69.
Sze, S. M., 1981, *Physics of Semiconductor Devices*, 2nd edn (Wiley: New York).
Tersoff, J., 1985, *Phys. Rev.* **B32**, 6968.
von Klitzing, K., Dorda, G. and Pepper, M., 1980, *Phys. Rev. Lett.* **45,** 494.
van Wees, B. J., van Houten, H., Beenakker, C. W. J., Williamson, J. G., Kouwenhoven, L. P., van der Marel, D. and Foxon, C. T., 1988, *Phys. Rev. Lett.* **60**, 848.
Weaver, J. H. and Poirier, D. M., 1994, in *Solid State Physics*, vol. **48**, eds H. Ehrenreich and F. Spaepen, (Academic Press: New York), p. 1.
Webb, R. A., Washburn, S. and Umbach, C. P., 1988, *Phys. Rev.* B**37**, 8455.
Weisbuch, C. and Vinter, B., 1991, *Quantum Semiconductor Structures* (Academic Press: London).
Willett, R., Eisenstein, J. P., Stormer, H. L., Tsui, D. C., Hwang, J. C. M. and Gossard, A. C., 1987, *Phys. Rev. Lett.* **59**, 1776.
Yu, P. Y. and Cardona, M., 1996, *Fundamentals of Semiconductors* (Springer-Verlag: Berlin).
Yükselici, H. and Persans, P. D., 1996, *J. Non-Cryst. Sol.* **203**, 206.
Zangwill, A., 1988, *Physics at Surfaces* (Cambridge University Press: Cambridge).

Chemical Formula Index

Subject Index